Lecture Notes in Computer Science 10976

Commenced Publication in 1973
Founding and Former Series Editors:
Gerhard Goos, Juris Hartmanis, and Jan van Leeuwen

More information about this series at http://www.springer.com/series/7407

Lusheng Wang · Daming Zhu (Eds.)

Computing
and Combinatorics

24th International Conference, COCOON 2018
Qing Dao, China, July 2–4, 2018
Proceedings

 Springer

Editors
Lusheng Wang
City University of Hong Kong
Hong Kong
China

Daming Zhu
Shandong University
Jinan
China

ISSN 0302-9743   ISSN 1611-3349   (electronic)
Lecture Notes in Computer Science
ISBN 978-3-319-94775-4   ISBN 978-3-319-94776-1   (eBook)
https://doi.org/10.1007/978-3-319-94776-1

Library of Congress Control Number: 2018947438

LNCS Sublibrary: SL1 – Theoretical Computer Science and General Issues

Printed on acid-free paper

This Springer imprint is published by the registered company Springer International Publishing AG part of Springer Nature
The registered company address is: Gewerbestrasse 11, 6330 Cham, Switzerland

Preface

This volume contains the papers presented at the 24th International Computing and Combinatorics Conference (COCOON 2018), held during July 2–4, 2018, in Qing Dao, China. COCOON 2018 provided a forum for researchers working in the areas of algorithms, theory of computation, computational complexity, and combinatorics related to computing.

The technical program of the conference included 62 contributed papers selected by the Program Committee from 120 full submissions received in response to the call for papers. All the papers were peer reviewed by at least two (2.83 in average) Program Committee members or external reviewers. The papers cover various topics, including algorithms and data structures, complexity theory and computability, algorithmic game theory, computational learning theory, cryptography, computational biology, computational geometry and number theory, graph theory, and parallel and distributed computing. Some of the papers were selected for publication in special issues of *Algorithmica, Theoretical Computer Science* (TCS), and *Journal of Combinatorial Optimization* (JOCO), with the journal version of the papers being in a more complete form.

The conference also included three invited presentations, delivered by Michael Segal (Ben-Gurion University of the Negev), Ming Li (University of Waterloo), and Russell Schwartz (Carnegie Mellon University). Abstracts of their talks are included in this volume. We would like to thank all the authors for contributing high-quality research papers to the conference. We express our sincere thanks to the Program Committee members and the external reviewers for reviewing the papers. We thank Springer for publishing the proceedings in the *Lecture Notes in Computer Science* series. We thank the Shandong University for hosting COCOON 2018. We are also grateful to all members of the Organizing Committee and to their supporting staff. electronic Program Committee meetings, and to assist with the assembly of the proceedings.

May 2018   Daming Zhu

Organization

Program Co-chairs

Lusheng Wang   City University of Hong Kong, SAR China
Daming Zhu   Shandong University, China

Program Committee

Lenwood Heath	Virginia Tech University, USA
Valentine Kabanets	Simon Fraser University, Canada
Ming-Yang Kao	Northwestern University, USA
Donghyun Kim	Kennesaw State University, USA
Nam Nguyen	Towson University, USA
Desh Ranjan	Old Dominion University, USA
Marc Uetz	University of Twente, The Netherlands
Bhaskar Dasgupta	University of Illinois at Chicago, USA
Raffaele Giancarlo	University of Palermo, Italy
Mohammadtaghi Hajiaghayi	University of Maryland at College Park, USA
Kazuo Iwama	Kyoto University, Japan
Iyad Kanj	DePaul University, Illinois, USA
Monaldo Mastrolilli	Dalle Molle Institute for Artificial Intelligence, Switzerland
Youming Qiao	University of Technology Sydney, Australia
Ryuhei Uehara	Japan Advanced Institute of Science and Technology, Japan
Jarek Byrka	University of Wroclaw, Poland
Rajesh Chitnis	Weizmann Institute of Science, Israel
Funda Ergun	Indiana University, Indiana, USA
Pinar Heggernes	University of Bergen, Norway
Jianxin Wang	Central South University, China
Xiuzhen Huang	Arkansas State University, USA
Hans-Joachim Boeckenhauer	ETH Zurich, Switzerland
Dennis Komm	ETH Zurich, Switzerland
Chee Yap	New York University, USA
Dorothea Wagner	Karlsruhe Institute of Technology, Germany
Vinod Variyam	University of Nebraska-Lincoln, USA
Zhizhong Chen	Tokyo Denki University, Japan
Xiaowen Liu	Indiana University, USA
Bin Ma	University of Waterloo, Canada
Shuaicheng Li	City University of Hong Kong, SAR China
Xiaoming Sun	Institute of Computing Technology, China

Boting Yang   University of Regina, Canada
Zhenhua Duan   Xidian University, China
Peng Zhang   Shandong University, China

Abstracts of Invited Talks

Privacy Aspects in Data Querying

Michael Segal

Communication Systems Engineering Department,
Ben-Gurion University of the Negev, Beer-Sheva, Israel
segal@bgu.ac.il

Abstract. Vast amounts of information of all types is collected daily about people by governments, corporations and individuals. The information is collected, for example, when users register to or use online applications, receive health related services, use their mobile phones, utilize search engines, or perform common daily activities. As a result, there is an enormous quantity of privately-owned records that describe individuals finances, interests, activities, and demographics. These records often include sensitive data and may violate the privacy of the users if published. The common approach to safeguarding user information, or data in general, is to limit access to the storage (usually a database) by using and authentication and authorization protocol. This way, only users with legitimate permissions can access the user data. However, even in these cases some of the data is required to stay hidden or accessible only to a specific subset of authorized users. Our talk focuses on possible malicious behavior by users with both partial and full access to queries over data. We look at privacy attacks that meant to gather hidden information and show methods that rely mainly on the underlying data structure, query types and behavior, and data format of the database. The underlying data structure may vary between graphs, trees, lists, queues, and so on. Each of these behaves differently with regard to data storage and querying, allow for different types of attacks, and require different methods of defense. The data stored in databases can be just about anything, and may be a combination of many different data types such as text, discrete numeric values, coordinates, continuous numeric values, timestamps, and others. We will show how to identify the potential weaknesses and attack vectors for each of these combinations of data structures and data types, and offer defenses against them. This is a joint work with Eyal Nussbaum.

Challenges from Cancer Immunotherapy

Ming Li

School of Computer Science, University of Waterloo
mli@uwaterloo.ca

There are currently two revolutions happening in the scientific world: deep learning and cancer immunotherapy. The former we have all heard, but I believe it is the latter [1–4] that is more closely related to the CPM/COCOON community and personally to each of us.

In principle, cancer immunotherapy is to activate our own defense system to kill cancer cells. When a cell in our bodies (for all vertebrates) becomes sick beyond repair, the MHC complex brings fragments of 8-15 amino acids, or (neo)antigens, from the foreign invader or cancerous proteins, to the surface of the cell inviting the white blood cells to kill that cell.

Short peptide immunotherapy uses these short sequences (of 8-15 amino acids) as the vaccine. One key obstacle for this treatment to become a clinical reality is how to identify and validate these somatic mutation loaded neoantigens (peptides of 8-15 amino acids) that are capable of eliciting effective anti-tumor T-cell responses for each individual. Currently, to treat a patient, we take a biopsy, do exome sequencing, perform somatic mutation analysis and MHC binding prediction. This process is a long, unreliable, and very expensive detour to predicting the neoantigens that are brought to the cancer cell surface [3, 4]. This process potentially can be validated by mass spectrometry (MS) [3–5] or even replaced by MS altogether if MS has sufficient sensitivity to capture the low abundant neoantigens on the cancer cell surface.

There is a promising MS technology called Data-Independent Acquisition (DIA) [6, 7] that has unbiased fragmentation of all precursor ions within a certain range of m/z. In this talk we will present our preliminary work [8] on how to find these mutated peptide sequences (de novo sequencing) from the cancer cell surface using deep learning and DIA data. We will discuss major open problems.

This is joint work with NH. Tran, R. Qiao, L. Xin, X. Chen, C. Liu, X. Zhang, and B. Shan. This work is partially supported by China's National Key R&D Program under grants 2018YFB1003202 and 2016YFB1000902, Canada's NSERC OGP0046506, Canada Research Chair Program, MITACS, and BSI.

References

1. Ott, P.A., et al.: Nature **547**, 217–221 (2017)
2. Sahin, U., et al.: Nature **547**, 222–226 (2017)
3. Editorial: The problem with neoantigen. Nat. Biotech. **35**(2) (2017)

4. Vitiello, A., Zanetti, M.: Nat. Biotech. **9**, 35 (2017)
5. Bassani-Sternberg, M., et al.: Nat. Commun. **7**, 13404 (2016)
6. Venable, J.D., et al.: Nat. Methods **1**, 39–45 (2004)
7. Röst, H.L., et al.: Nat. Biotechnol. **32**, 219–223 (2014)
8. Tran, N.H., et al.: De novo peptide sequencing by deep learning. PNAS **114**(31) (2017)

Reconstructing Tumor Evolution and Progression in Structurally Variant Cancer Cells

Russell Schwartz

Biological Sciences and Computational Biology, Carnegie Mellon University, Pittsburg, PA 15213, USA
russells@andrew.cmu.edu

Abstract. Cancer is disease governed by the process of evolution, in which a process of accelerated genomic diversification and selection leads to the formation of tumors and a process of generally increasing aggressiveness over time. As a result, computational algorithms for reconstructing evolution have become a crucial tool for making sense of the immense complexity of tumor genomic data and the molecular mechanisms that produce them. While cancers are evolutionary systems, though, they follow very different rules than standard species evolution. A large body of research known as cancer phylogenetics has arisen to develop evolutionary tree reconstructions adapted to the peculiar mechanisms of tumor evolution and the limitations of the data sources available for studying it. Here, we will explore computational challenges in developing phylogenetic methods for reconstructing evolution of tumors by copy number variations (CNVs) and structural variations (SVs). CNVs and SVs are the primary mechanisms by which tumors functionally adapt during their evolution, but require very different models and algorithms than are used in traditional species phylogenetics. We will examine variants of this problem for handling several forms of tumor genomic data, including particular challenges of working with various bulk genomic and single-cell technologies for profiling tumor genetic variation. We will further see how the resulting models can help us develop new insight into how tumors develop and progress and how we can predict their future behavior.

Contents

Constructing Independent Spanning Trees on Bubble-Sort Networks   1
   Shih-Shun Kao, Jou-Ming Chang, Kung-Jui Pai, and Ro-Yu Wu

Exact Algorithms for Finding Partial Edge-Disjoint Paths.   14
   Yunyun Deng, Longkun Guo, and Peihuang Huang

A Randomized FPT Approximation Algorithm for Maximum
Alternating-Cycle Decomposition with Applications   26
   Haitao Jiang, Lianrong Pu, Letu Qingge, David Sankoff, and Binhai Zhu

Contextual Dependent Click Bandit Algorithm for Web Recommendation . . .   39
   Weiwen Liu, Shuai Li, and Shengyu Zhang

LP-Based Pivoting Algorithm for Higher-Order Correlation Clustering.   51
   Takuro Fukunaga

Approximation Algorithms for a Two-Phase Knapsack Problem   63
   Kameng Nip and Zhenbo Wang

More Routes for Evacuation.   76
   Katsuhisa Yamanaka, Yasuko Matsui, and Shin-ichi Nakano

Fine-Grained Parameterized Complexity Analysis of Knot-Free
Vertex Deletion – A Deadlock Resolution Graph Problem   84
   Alan Diêgo Aurélio Carneiro, Fábio Protti, and Uéverton S. Souza

Approximating Global Optimum for Probabilistic Truth Discovery   96
   Shi Li, Jinhui Xu, and Minwei Ye

Online Interval Scheduling to Maximize Total Satisfaction.   108
   Koji M. Kobayashi

Properties of Minimal-Perimeter Polyominoes.   120
   Gill Barequet and Gil Ben-Shachar

Computing Convex-Straight-Skeleton Voronoi Diagrams for Segments
and Convex Polygons   130
   Gill Barequet, Minati De, and Michael T. Goodrich

Polygon Queries for Convex Hulls of Points   143
   Eunjin Oh and Hee-Kap Ahn

Synergistic Solutions for Merging and Computing Planar Convex Hulls.   156
  Jérémy Barbay and Carlos Ochoa

Cophenetic Distances: A Near-Linear Time Algorithmic Framework   168
  Paweł Górecki, Alexey Markin, and Oliver Eulenstein

Computing Coverage Kernels Under Restricted Settings.   180
  Jérémy Barbay, Pablo Pérez-Lantero, and Javiel Rojas-Ledesma

Weak Mitoticity of Bounded Disjunctive and Conjunctive Truth-
Table Autoreducible Sets   192
  Liyu Zhang, Mahmoud Quweider, Hansheng Lei, and Fitra Khan

Approximation Algorithms for Two-Machine Flow-Shop Scheduling
with a Conflict Graph   205
  Yinhui Cai, Guangting Chen, Yong Chen, Randy Goebel, Guohui Lin,
  Longcheng Liu, and An Zhang

On Contact Representations of Directed Planar Graphs   218
  Chun-Hsiang Chan and Hsu-Chun Yen

Computation and Growth of Road Network Dimensions   230
  Johannes Blum and Sabine Storandt

Car-Sharing Between Two Locations: Online Scheduling with Flexible
Advance Bookings   242
  Kelin Luo, Thomas Erlebach, and Yinfeng Xu

Directed Path-Width and Directed Tree-Width of Directed Co-graphs   255
  Frank Gurski and Carolin Rehs

Generalized Graph k-Coloring Games   268
  Raffaello Carosi and Gianpiero Monaco

On Colorful Bin Packing Games.   280
  Vittorio Bilò, Francesco Cellinese, Giovanna Melideo,
  and Gianpiero Monaco

Nonbipartite Dulmage-Mendelsohn Decomposition for Berge Duality   293
  Nanao Kita

The Path Set Packing Problem   305
  Chenyang Xu and Guochuan Zhang

Manipulation Strategies for the Rank-Maximal Matching Problem.   316
  Pratik Ghosal and Katarzyna Paluch

Finding Maximal Common Subgraphs via Time-Space Efficient
Reverse Search   328
    Alessio Conte, Roberto Grossi, Andrea Marino, and Luca Versari

An FPT Algorithm for Contraction to Cactus   341
    R. Krithika, Pranabendu Misra, and Prafullkumar Tale

An Approximation Framework for Bounded Facility Location Problems   353
    Wenchang Luo, Bing Su, Yao Xu, and Guohui Lin

Reconfiguration of Satisfying Assignments and Subset Sums: Easy to Find,
Hard to Connect   365
    *Jean Cardinal, Erik D. Demaine, David Eppstein, Robert A. Hearn,
    and Andrew Winslow*

Solving the Gene Duplication Feasibility Problem in Linear Time   378
    Alexey Markin, Venkata Sai Krishna Teja Vadali, and Oliver Eulenstein

An Efficiently Recognisable Subset of Hypergraphic Sequences   391
    Syed M. Meesum

Partial Homology Relations - Satisfiability in Terms of Di-Cographs   403
    *Nikolai Nøjgaard, Nadia El-Mabrouk, Daniel Merkle, Nicolas Wieseke,
    and Marc Hellmuth*

Improved Algorithm for Finding the Minimum Cost of Storing
and Regenerating Datasets in Multiple Clouds   416
    Yingying Wang, Kun Cheng, and Zimao Li

Reconfiguring Spanning and Induced Subgraphs   428
    *Tesshu Hanaka, Takehiro Ito, Haruka Mizuta, Benjamin Moore,
    Naomi Nishimura, Vijay Subramanya, Akira Suzuki,
    and Krishna Vaidyanathan*

Generalizing the Hypergraph Laplacian via a Diffusion Process
with Mediators   441
    T.-H. Hubert Chan and Zhibin Liang

Efficient Enumeration of Bipartite Subgraphs in Graphs   454
    Kunihiro Wasa and Takeaki Uno

Bipartite Graphs of Small Readability   467
    *Rayan Chikhi, Vladan Jovičić, Stefan Kratsch, Paul Medvedev,
    Martin Milanič, Sofya Raskhodnikova, and Nithin Varma*

Maximum Colorful Cliques in Vertex-Colored Graphs   480
    *Giuseppe F. Italiano, Yannis Manoussakis, Nguyen Kim Thang,
    and Hong Phong Pham*

Partial Sublinear Time Approximation and Inapproximation for
Maximum Coverage 492
   Bin Fu

Characterizing Star-PCGs. 504
   Mingyu Xiao and Hiroshi Nagamochi

Liar's Dominating Set in Unit Disk Graphs 516
   Ramesh K. Jallu, Sangram K. Jena, and Gautam K. Das

Minimum Spanning Tree of Line Segments 529
   Sanjana Dey, Ramesh K. Jallu, and Subhas C. Nandy

Improved Learning of k-Parities 542
   Arnab Bhattacharyya, Ameet Gadekar, and Ninad Rajgopal

On a Fixed Haplotype Variant of the Minimum Error Correction Problem . . . 554
   Axel Goblet, Steven Kelk, Matúš Mihalák, and Georgios Stamoulis

Non-monochromatic and Conflict-Free Coloring on Tree Spaces
and Planar Network Spaces 567
   Boris Aronov, Mark de Berg, Aleksandar Markovic,
   and Gerhard Woeginger

Amplitude Amplification for Operator Identification
and Randomized Classes 579
   Debajyoti Bera

Reconstruction of Boolean Formulas in Conjunctive Normal Form 592
   Evgeny Dantsin and Alexander Wolpert

A Faster FPTAS for the Subset-Sums Ratio Problem. 602
   Nikolaos Melissinos and Aris Pagourtzis

A Linear-Space Data Structure for Range-LCP Queries
in Poly-Logarithmic Time 615
   Paniz Abedin, Arnab Ganguly, Wing-Kai Hon, Yakov Nekrich,
   Kunihiko Sadakane, Rahul Shah, and Sharma V. Thankachan

Non-determinism Reduces Construction Time in Active Self-assembly
Using an Insertion Primitive. 626
   Benjamin Hescott, Caleb Malchik, and Andrew Winslow

Minimum Membership Hitting Sets of Axis Parallel Segments 638
   N. S. Narayanaswamy, S. M. Dhannya, and C. Ramya

Minimum Transactions Problem 650
   Niranka Banerjee, Varunkumar Jayapaul, and Srinivasa Rao Satti

Heuristic Algorithms for the Min-Max Edge 2-Coloring Problem   662
  Radu Stefan Mincu and Alexandru Popa

Geometric Spanners in the MapReduce Model   675
  Sepideh Aghamolaei, Fatemeh Baharifard, and Mohammad Ghodsi

SDP Primal-Dual Approximation Algorithms for Directed Hypergraph
Expansion and Sparsest Cut with Product Demands.   688
  T.-H. Hubert Chan and Bintao Sun

Lower Bounds for Special Cases of Syntactic Multilinear ABPs   701
  C. Ramya and B. V. Raghavendra Rao

Approximation Algorithms on Multiple Two-Stage Flowshops   713
  Guangwei Wu and Jianer Chen

Constant Factor Approximation Algorithm for *l*-Pseudoforest Deletion
Problem.   726
  Mugang Lin, Bin Fu, and Qilong Feng

New Bounds for Energy Complexity of Boolean Functions   738
  Krishnamoorthy Dinesh, Samir Otiv, and Jayalal Sarma

Hitting and Covering Partially   751
  *Akanksha Agrawal, Pratibha Choudhary, Pallavi Jain,
  Lawqueen Kanesh, Vibha Sahlot, and Saket Saurabh*

Author Index   765

Constructing Independent Spanning Trees on Bubble-Sort Networks

Shih-Shun Kao[1], Jou-Ming Chang[1]($\boxtimes$), Kung-Jui Pai[2], and Ro-Yu Wu[3]

[1] Institute of Information and Decision Sciences,
National Taipei University of Business, Taipei, Taiwan
{10566011,spade}@ntub.edu.tw
[2] Department of Industrial Engineering and Management,
Ming Chi University of Technology, New Taipei City, Taiwan
poter@mail.mcut.edu.tw
[3] Department of Industrial Management,
Lunghwa University of Science and Technology, Taoyuan, Taiwan
eric@mail.lhu.edu.tw

Abstract. A set of spanning trees in a graph G is called independent spanning trees (ISTs for short) if they are rooted at the same vertex, say r, and for each vertex $v(\neq r)$ in G, the two paths from v to r in any two trees share no common vertex except for v and r. Constructing ISTs has applications on fault-tolerant broadcasting and secure message distribution in reliable communication networks. Since Cayley graphs have been used extensively to design interconnection networks, the study of constructing ISTs on Cayley graphs is very significant. It is well-known that star networks S_n and bubble-sort network B_n are two of the most attractive subclasses of Cayley graphs. Although it has been dealt with about two decades for the construction of ISTs on S_n (which has been pointed out that there is a flaw and has been corrected recently), so far the problem of constructing ISTs on B_n has not been dealt with. In this paper, we present an efficient algorithm to construct $n - 1$ ISTs of B_n. It seems that our work is the latest breakthrough on the problem of ISTs for all subclasses of Cayley graphs except star networks.

Keywords: Independent spanning trees · Bubble-sort networks
Interconnection networks · Cayley graphs

1 Introduction

Let G be a graph with the vertex set $V(G)$ and edge set $E(G)$. A set of spanning trees in G is called *independent spanning trees* (ISTs for short) if all the trees are rooted at the same vertex, say r, and for each vertex $v \in V(G) \setminus \{r\}$, the two paths from v to r in any two trees are *internally vertex-disjoint* (i.e., there exists no common vertex in the two paths except the two end vertices v and r). Constructing multiple ISTs in networks has been studied not only from a

© Springer International Publishing AG, part of Springer Nature 2018
L. Wang and D. Zhu (Eds.): COCOON 2018, LNCS 10976, pp. 1–13, 2018.
https://doi.org/10.1007/978-3-319-94776-1_1

theoretical point of view but also for some practical applications, such as fault-tolerant broadcasting [3,9] and secure message distribution [3,15,22] in reliable communication networks.

A long-standing conjecture proposed by Zehavi and Itai [23] says that a k-connected graph G admits k ISTs rooted at an arbitrary vertex of G. This conjecture has been affirmed for k-connected graphs with $k \leqslant 4$ (see [6,7,9,23], for $k = 2,3,4$, respectively), but it remains open for $k \geqslant 5$. Afterward, subsequent studies tend to favor the construction of ISTs on some restricted classes of graphs. Especially, those graphs related to interconnection networks (e.g. see recent papers [4,5] and references quoted therein). Although a lot of research of ISTs focused on variations of hypercubes, to the best of our knowledge, the construction of ISTs in the family of Cayley graphs was known only for star networks, which was proposed by Rescigno [15]. Unfortunately, Ko et al. [11] recently pointed out that there is a flaw in Rescigno's algorithm and provided an amendatory scheme to correct it. In fact, due to more subgraphs being produced in the recursive decomposition, constructing ISTs on star networks is harder than that on variations of hypercubes. In this paper, we make a further investigation of constructing ISTs on another famous subclass of Cayley graphs called bubble-sort networks.

Let B_n denote the n-dimensional bubble-sort network (defined later in Sect. 2). The following are known results of B_n. For $n \geqslant 4$, B_n is vertex transitive, but is not edge transitive (see [13]). B_n has connectivity $n - 1$ and diameter $n(n-1)/2$ (see [1,17]). Algorithms for hamiltonian laceability, pancyclicity, and node-to-node disjoint paths in B_n are obtained in [2,10], and [16,17], respectively. In particular, finding a shortest path between two vertices in B_n can be accomplished by using the familiar bubble-sort algorithm [1]. Also, research results related to fault tolerance, diagnosability, and reliability on bubble-sort networks can be found in [8,14,18–21,24].

The rest of this paper is organized as follows. Section 2 formally gives the definition of bubble-sort networks and introduces some necessary notations. Section 3 presents our constructing scheme of ISTs for B_n and provides some auxiliary example for illustration. Section 4 shows the correctness of our algorithm. The final section contains our concluding remarks.

2 Preliminaries

Let Σ_n be the set of all permutations on $\{1, 2, \ldots, n\}$. For a permutation $p \in \Sigma_n$ and an integer $i \in \{1, 2, \ldots, n\}$, we use the following notations. The symbol at the ith position of p is denoted by $p(i)$, and the position where the symbol i appears in p is denoted by $p^{-1}(i)$. For notational convenience, we also write p_i instead of $p(i)$, so $p = p_1 \cdots p_n$. A symbol i is said to be at the *right position* of p if $p_i = i$. For $i \in \{1, \ldots, n-1\}$, let $p\langle i \rangle = p_1 p_2 \cdots p_{i-1} p_{i+1} p_i p_{i+2} \cdots p_n$ be the permutation of Σ_n obtained from p by swapping two consecutive symbols at positions i and $i + 1$. Hence, $p\langle p^{-1}(i) \rangle$ is a permutation obtained from p by swapping symbol i and its immediately succeeding symbol. Also, if $p \in \Sigma_n$ with

$p_n = n$, we denote by $p \ominus \{n\}$ the permutation of Σ_{n-1} that removes the last symbol of p. By contrast, if $p \in \Sigma_{n-1}$, we denote by $p \oplus \{n\}$ the permutation of Σ_n that is obtained from p by adding n as its last symbol.

The *n-bubble-sort network*, denoted by B_n, is an undirected graph consisting of the vertex set $V(B_n) = \Sigma_n$ and edge set $E(B_n) = \{(v, v\langle i \rangle) : v \in \Sigma_n, 1 \leqslant i \leqslant n-1\}$, where the edge $(v, v\langle i \rangle)$ is called an *i-edge* of B_n. Thus, B_n is a Cayley graph generated by the transposition set $\{(i, i+1) : 1 \leqslant i \leqslant n-1\}$, which is specified by an *n*-path $P_n = (1, 2, \ldots, n)$ as its transposition graph [1,13]. For example, Fig. 1(a) depicts B_3 and B_4, where each edge is labeled by a number i to indicate that it is an *i*-edge, and (b) shows the transposition graph P_n. Clearly, for B_n, the transposition graph P_n contains only two subgraphs isomorphic to an $(n-1)$-path: one is $(1, 2, \ldots, n-2)$ and the other is $(2, 3, \ldots, n-1)$. Thus, for $n \geqslant 3$, there are exactly two ways to decompose B_n into n disjoint subgraphs that are isomorphic to B_{n-1}. Let B_n^i denote the graph obtained from B_n by removing the set of all *i*-edges. Then, both B_n^1 and B_n^{n-1} consist of n disjoint subgraphs isomorphic to B_{n-1}.

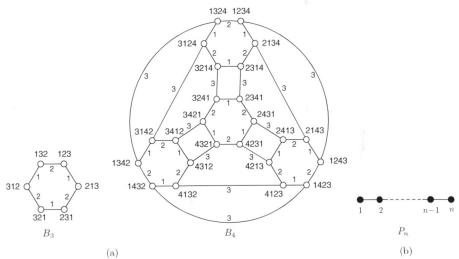

Fig. 1. (a) Bubble-sort networks B_3 and B_4; (b) the transposition graph P_n.

3   Constructing ISTs on B_n

In this section, we present an algorithm for constructing $n-1$ ISTs of B_n. Since B_n is vertex transitive, without loss of generality, we may choose the identity $\mathbf{1}_n = 12 \cdots n$ as the common root of all ISTs. Also, since B_n has connectivity $n-1$, the root in every spanning tree has a unique child. For $1 \leqslant i \leqslant n-1$, if the root of a spanning tree takes $\mathbf{1}_n\langle i \rangle = 12 \cdots (i-1)(i+1)i(i+2) \cdots n$ as its unique child, then the spanning tree of B_n is denoted by T_i^n. To describe such a spanning tree, for each vertex $v = v_1 \cdots v_n \in V(B_n)$ except the root $\mathbf{1}_n$, we denote by Parent(v, i, n) as the parent of v in T_i^n. Since B_3 is isomorphic to a 6-cycle, we have

$$\text{Parent}(v,1,3) = \begin{cases} 123 & \text{if } v = 213; \\ 213 & \text{if } v = 231; \\ 231 & \text{if } v = 321; \\ 321 & \text{if } v = 312; \\ 312 & \text{if } v = 132; \end{cases} \quad \text{and} \quad \text{Parent}(v,2,3) = \begin{cases} 231 & \text{if } v = 213; \\ 321 & \text{if } v = 231; \\ 312 & \text{if } v = 321; \\ 132 & \text{if } v = 312; \\ 123 & \text{if } v = 132. \end{cases}$$

That is, the two paths

$$T_1^3 = (132, 312, 321, 231, 213, 123) \text{ and } T_2^3 = (213, 231, 321, 312, 132, 123)$$

are ISTs of B_3 that take $\mathbf{1}_3 = 123$ as the common root. In general, for B_n with $n \geqslant 4$, we define the function $\alpha(v, i, n)$ for each $v \in V(B_n) \setminus \{\mathbf{1}_4\}$ and $i \in \{1, 2, \ldots, n-1\}$ as follows:

$$\alpha(v, i, n) = \begin{cases} n & \text{if } i = 1; \\ n-3 & \text{if } i = n-1 \text{ and } v_{n-3} \neq n-3; \\ n-2 & \text{if } i = n-2 \text{ and } v_{n-3} \neq n-3; \\ \alpha(v, i-1, n-1) & \text{if } v_{n-3} = n-3; \\ i-1 & \text{otherwise.} \end{cases}$$

Then, the construction of ISTs of B_n can be accomplished by using the function Parent(v, i, n) (see Fig. 2) to determine the parent of each vertex (except the root) in every spanning tree.

Function Parent(v, i, n)

    if $v_n = n$ **then**
(1) | **if** $i \neq n-1$ **then** $p = \text{Parent}(v \ominus \{n\}, i, n-1) \oplus \{n\}$;
(2) | **else** $p = v\langle n-1 \rangle$;

    else
      | **if** $v_n = n-1$, $v_{n-1} = n$, and $v\langle n-1 \rangle \neq \mathbf{1}_n$ **then**
(3) | | **if** $i = 1$ or $v_{n-2} = n-2$ **then** $p = v\langle v^{-1}(\alpha(v, i, n)) \rangle$;
(4) | | **else** $p = v\langle v^{-1}(i-1) \rangle$;

      | **else**
(5) | | **if** $v_n = i$ **then** $p = v\langle v^{-1}(n) \rangle$;
(6) | | **else** $p = v\langle v^{-1}(i) \rangle$;

    return p;

Fig. 2. The function Parent(v, i, n).

For example, in Table 1, we calculate the parent of every vertex $v \in V(B_4) \setminus \{\mathbf{1}_4\}$ in T_i^4 for $i \in \{1, 2, 3\}$. In this table, the column 'Rule' indicates which rule is used for computing the parent p. For example, we consider $v = 2143$ and $i = 2$. Since $v_4 = 3$, $v_3 = 4$ and $v\langle 3 \rangle = 2134 \neq \mathbf{1}_4$, it follows from Rule (4) that $p = v\langle v^{-1}(1) \rangle = v\langle 2 \rangle = 2413$. Also, we consider $v = 3214$ and $i = 1$. Since $v_4 = 4$ and $i \neq 3$, it follows from Rule (2) that $p = \text{Parent}(321, 1, 3) \oplus \{4\} = 2314$. As a consequence, three ISTs rooted at vertex $\mathbf{1}_4$ for B_4 are shown in Fig. 3.

Table 1. Computing the parent of every vertex $v \in V(B_4) \setminus \{1_4\}$ in T_i^4 for $i \in \{1, 2, 3\}$.

v	i	v_n	$v\langle n-1\rangle$	Rule	p	v	i	v_n	$v\langle n-1\rangle$	Rule	p
1234	-	-	-	-	-	3124	1	4	3142	(1)	3214
							2			(1)	1324
							3			(2)	3142
1243	1	3	1234	(6)	2143	3142	1	2	3124	(6)	3412
	2			(6)	1423		2			(5)	3124
	3			(5)	1234		3			(6)	1342
1324	1	4	1342	(1)	3124	3214	1	4	3241	(1)	2314
	2			(1)	1234		2			(1)	3124
	3			(2)	1342		3			(2)	3241
1342	1	2	1324	(6)	3142	3241	1	1	3214	(5)	3214
	2			(5)	1324		2			(6)	3421
	3			(6)	1432		3			(6)	2341
1423	1	3	1432	(6)	4123	3412	1	2	3421	(6)	3421
	2			(6)	1432		2			(5)	3142
	3			(5)	1243		3			(6)	4312
1432	1	2	1423	(6)	4132	3421	1	1	3412	(5)	3241
	2			(5)	1342		2			(6)	3412
	3			(6)	1423		3			(6)	4321
2134	1	4	2143	(1)	1234	4123	1	3	4132	(6)	4213
	2			(1)	2314		2			(6)	4132
	3			(2)	2143		3			(5)	1423
2143	1	3	2134	(3)	2134	4132	1	2	4123	(6)	4312
	2			(4)	2413		2			(5)	1432
	3			(4)	1243		3			(6)	4123
2314	1	4	2341	(1)	2134	4213	1	3	4231	(6)	4231
	2			(1)	3214		2			(6)	4123
	3			(2)	2341		3			(5)	2413
2341	1	1	2314	(5)	2314	4231	1	1	4213	(5)	2431
	2			(6)	3241		2			(6)	4321
	3			(6)	2431		3			(6)	4213
2413	1	3	2431	(6)	2431	4312	1	2	4321	(6)	4321
	2			(6)	4213		2			(5)	3412
	3			(5)	2143		3			(6)	4132
2431	1	1	2413	(5)	2341	4321	1	1	4312	(5)	3421
	2			(6)	4231		2			(6)	4312
	3			(6)	2413		3			(6)	4231

4   Correctness

In this section, we show the correctness of the algorithm. If P and Q are two paths
with a common end vertex in a graph, we denote by $P \cup Q$ the concatenation
of P and Q. If T is a tree and $u, v \in V(T)$, we use $T[u, v]$ to denote the unique
path joining u and v in T. For two spanning trees T and T' with the common
root r in a graph G, if $v \in V(G) \setminus \{r\}$, we use $T[v, r] \mathbin{/\!/} T'[v, r]$ to mean that
$T[v, r]$ and $T'[v, r]$ are internally vertex-disjoint.

Lemma 1. *T_1^4, T_2^4 and T_3^4 are three ISTs of B_4.*

Proof. From Table 1 and Fig. 3, we assure that the function $\mathrm{Parent}(v, i, n)$ can
construct three spanning trees rooted at $\mathbf{1}_4 (= 1234)$ for B_4. So, the independency
of the three spanning trees can be verified by a brute-force checking.   □

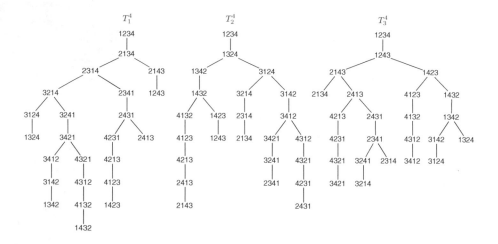

Fig. 3. A set of three ISTs of B_4.

Theorem 1. *For $n \geqslant 4$, $T_1^n, T_2^n, \ldots, T_{n-1}^n$ are $n-1$ ISTs of B_n.*

Proof. The proof is by induction on n. Lemma 1 establishes the validity of the
base case with $n = 4$. Suppose that $n \geqslant 5$ and the result is true for all $n < 5$.
Let $r = \mathbf{1}_n (= 12 \cdots n)$. The proof is by showing the existence of a unique path
from any vertex $v(= v_1 v_2 \cdots v_n) \in V(B_n) \setminus \{r\}$ to r, and thereby proving the
independence. Consider the following three cases:

CASE 1: $v_n = n$. If $i \in \{1, 2, \ldots, n-2\}$, by Rule (1) in the function $\mathrm{Parent}()$,
we have $v = u \oplus \{n\}$ where u is a vertex (except the root) obtained from
T_i^{n-1}. By induction hypotheses, we know that $T_i^{n-1}[u, \mathbf{1}_{n-1}] \mathbin{/\!/} T_j^{n-1}[u, \mathbf{1}_{n-1}]$
for $i, j \in \{1, 2, \ldots, n-2\}$ with $i \neq j$, and thus it immediately follows that
$T_i^n[v, r] \mathbin{/\!/} T_j^n[v, r]$. On the other hand, if $i = n-1$, by Rule (2), v is adjacent
to its parent, say $p(= p_1 p_2 \cdots p_n)$, by using an $(n-1)$-edge in T_{n-1}^n (see Fig. 4,

where v is a leaf represented by a circle with label n in T_{n-1}^n). If $v_{n-1} = p_n = k$ for $1 \leqslant k \leqslant n-2$, we can show later in CASE 2 that there exists a path $T_{n-1}^n[p, r]$ such that $(v, p) \cup T_{n-1}^n[p, r]$ contains internal vertices with the last symbol either k or $n-1$ (see the bold line of T_{n-1}^n in Fig. 4). If $v_{n-1} = p_n = n-1$, we can show later in CASE 3 that there exists a path $T_{n-1}^n[p, r]$ such that $(v, p) \cup T_{n-1}^n[p, r]$ contains internal vertices with the last symbol $n-1$ (see the dashed line of T_{n-1}^n in Fig. 4). Since every vertex in the path $T_i^n[v, r]$ for $i \in \{1, 2, \dots, n-2\}$ has the last symbol $n(\neq k)$, this shows that $T_i^n[v, r] // (v, p) \cup T_{n-1}^n[p, r]$.

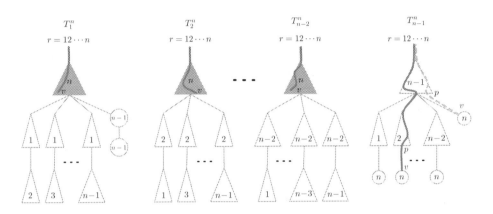

Fig. 4. The structure of $n-1$ ISTs of B_n. For $i = 1, 2, \dots, n-1$, a subtree with a label $k \in \{1, 2, \dots, n-1\}$ in T_i^n means that every vertex in the subtree has the last symbol k. In particular, a subtree with label n covered by shadows as well as the root $r(= 12 \cdots n)$ in T_i^n for $1 \leqslant i \leqslant n-2$ forms a tree isomorphic to T_i^{n-1}. A circle labeled by $k \in \{n-1, n\}$ means a vertex with the last symbol k. Bold lines and an dashed line indicate the paths described in the proof of Case 1 Theorem 1.

CASE 2: $v_n = j$ and $1 \leqslant j \leqslant n-2$. For $i \in \{1, 2 \dots, n-1\}$, if $i = j$ (resp., $i \neq j$), by Rule (5) (resp., Rule (6)) the parent of v in T_i^n is a vertex obtained from v by swapping symbol n (resp., symbol i) and its immediately succeeding symbol. Let $p(= p_1 p_2 \cdots p_n)$ be the parent of v in T_i^n. If $p_n \neq n$ (resp., $p_n \neq i$), we can obtain the parent of a vertex by using Rule (5) (resp., Rule (6)) repeatedly until a vertex $u^{(i)}(= u_1^{(i)} u_2^{(i)} \cdots u_n^{(i)})$ with $u_{n-1}^{(i)} = j$ and $u_n^{(i)} = n$ (resp., with $u_{n-1}^{(i)} = j$ and $u_n^{(i)} = i$) is reached. Thus, we obtain the path $T_i^n[v, u^{(i)}]$. Note that every internal vertex in the path $T_i^n[v, u^{(i)}]$ for $i \in \{1, 2 \dots, n-1\}$ takes j as the last symbol. For distinct paths, since exchanges of symbols start at different positions and are operated in a sequence, they can only share a common vertex v. That is, $T_i^n[v, u^{(i)}] \cap T_{i'}^n[v, u^{(i')}] = \{v\}$ for $i, i' \in \{1, 2 \dots, n-1\}$ with $i \neq i'$. In addition, if $i \neq j$, since $u_n^{(i)} = i$, a similar argument shows that, using Rule (5) repeatedly until a vertex $w^{(i)}(= w_1^{(i)} w_2^{(i)} \cdots w_n^{(i)})$ with $w_{n-1}^{(i)} = i$ and $w_n^{(i)} = n$ being reached, we can derive a path $T_i^n[u^{(i)}, w^{(i)}]$ such that every internal vertex has the last symbol i. In particular, if $i = n-1$, we can show later in CASE 3 that $w^{(n-1)} = r$. Also, since $w_n^{(i)} = n$ for $i \in \{1, 2, \ldots, n-2\} \setminus \{j\}$ (resp. $u_n^{(i)} = n$ for $i = j$), from CASE 1 we know that there exists a path $T_i^n[w^{(i)}, r]$ (resp. a path $T_j^n[u^{(j)}, r]$) such that every vertex in these paths takes n as the last symbol. Therefore, we obtain the following paths in all spanning trees (see bold lines in Fig. 5 for an illustration of these paths described in Eq. (1)):

$$
T_i^n[v, r]
= \begin{cases}
T_i^n[v, u^{(i)}] \cup T_i^n[u^{(i)}, w^{(i)}] \cup T_i^n[w^{(i)}, r] & \text{if } i \in \{1, 2, \ldots, n-2\} \setminus \{j\}; \\
T_i^n[v, u^{(i)}] \cup T_i^n[u^{(i)}, r] & \text{if } i = j; \\
T_i^n[v, u^{(i)}] \cup T_i^n[u^{(i)}, r(= w^{(n-1)})] & \text{if } i = n-1.
\end{cases}
\tag{1}
$$

We now verify the independence of the above paths. For each $i \in \{1, 2, \ldots, n-2\} \setminus \{j\}$, we observe that the change of labels for vertices in the path $T_i^n[w^{(i)}, r]$ have the following recursive structure. Firstly, $w^{(i)}$ has the symbol i at position $n-1$ and the symbol n at the right position. Then, for each k from $n-1$ down to 2, vertices along the path keep the symbol i at position k until a vertex with the symbol i at position $k-1$ and the symbol k at the right position is reached (e.g., consider the path

$$
T_1^5[43215, 1_5] = (432\underline{15}, 342\underline{15}, 32\underline{41}5, 32\underline{14}5, 23\underline{14}5, 21\underline{34}5, \underline{12}345),
$$

where the symbols with underscore are at positions $k-1$ and k. In fact, the path $T_1^5[43215, 1_5]$ is constructed from the path

$$
T_1^4[4321, 1_4] = (4321, 3421, 3241, 3214, 2314, 2134, 1234)
$$

in Fig. 3). Accordingly, it guarantees that $T_i^n[w^{(i)}, r] \cap T_{i'}^n[w^{(i')}, r] = \{r\}$ for $i, i' \in \{1, 2, \ldots, n-2\} \setminus \{j\}$ with $i \neq i'$. Since the path $T_j^n[u^{(j)}, r]$ has the similar structure, it follows that $T_j^n[u^{(j)}, r] \cap T_i^n[w^{(i)}, r] = \{r\}$ for $i \in \{1, 2, \ldots, n-2\} \setminus \{j\}$. Consequently, for all paths described in Eq. (1), the independence not mentioned above can be verified by checking the last symbol of every vertex in each subpath.

CASE 3: $v_n = n-1$. Let $p(= p_1 p_2 \cdots p_n)$ be the parent of v in T_i^n. Consider the following subcases.

CASE 3.1: $i = 1$. There are three situations as follows:

CASE 3.1.1: $v_{n-1} = n$ and $v\langle n-1 \rangle \neq r$. By Rule (3), $p = v\langle v^{-1}(\alpha(v, 1, n)) \rangle = v\langle v^{-1}(n) \rangle$. Thus, $p_n = v_{n-1} = n$. From CASE 1, there is a path $T_1^n[p, r]$ such that every internal vertex has the last symbol n. Thus, $T_1^n[v, r] = (v, p) \cup T_1^n[p, r]$ (see Fig. 6, where the path is drawn by a dashed line in T_1^n).

CASE 3.1.2: $v_{n-1} = n$ and $v\langle n-1 \rangle = r$. By Rule (6), we have $p = v\langle v^{-1}(1) \rangle = 213 \cdots (n-2)n(n-1)$. Since $p_n = n-1$, $p_{n-1} = n$ and $p\langle n-1 \rangle \neq r$, p is at a status that matches the condition of Rule (3). Thus, v is a child of a vertex described in CASE 3.1.1 (see Fig. 6, where v is a leaf represented by a circle with label $n-1$ in T_1^n).

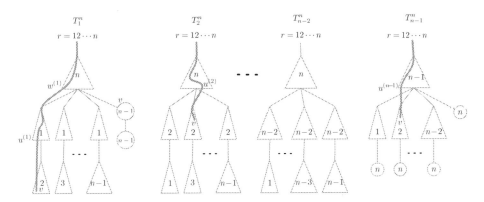

Fig. 5. The bold lines indicate the paths defined by Eq. (1) in the proof of Case 2 of Theorem 1 when $j = 2$.

CASE 3.1.3: $v_{n-1} \neq n$. By Rule (6), we have $p = v\langle v^{-1}(1)\rangle$. In this case, an argument similar to CASE 2 shows that $T_1^n[v, r]$ can be constructed like the first description of Eq. (1) such that it contains vertices with the last symbol $n - 1$, 1 or n (see Fig. 6, where the path is drawn by a bold line in T_1^n).

CASE 3.2: $i = n - 1$. If $v_{n-1} \neq n$ or $v\langle n - 1\rangle = r$, since $v_n = n - 1 = i$, by Rule (5) we have $p = v\langle v^{-1}(n)\rangle$. Moreover, if $v_{n-1} \neq n$, we can obtain the parent of a vertex by using Rule (5) repeatedly until the path reaches a vertex $u(= u_1 u_2 \cdots u_n)$ with $u_{n-1} = n$. In particular, if $u\langle n-1\rangle = r$ (resp., $v\langle n-1\rangle = r$), then u (resp., v) is the child of the root in T_{n-1}^n. Otherwise (i.e., $v_{n-1} = n$ and $v\langle n - 1\rangle \neq r$), there are three situations as follows: (i) if $v_{n-2} \neq n - 2$, by Rule (4) we have $p = v\langle v^{-1}(n - 2)\rangle$; (ii) if $v_{n-2} = n - 2$ and $v_{n-3} \neq n - 3$, by Rule (3) we have $p = v\langle v^{-1}(\alpha(v, n - 1, n))\rangle = v\langle v^{-1}(n - 3)\rangle$; and (iii) if $v_{n-2} = n - 2$ and $v_{n-3} = n - 3$, by Rule (3) we have $p = v\langle v^{-1}(\alpha(v, n - 1, n))\rangle = v\langle v^{-1}(\alpha(v, n - 2, n - 1))\rangle$. That is, the rules first make the symbols $n - 2$ and $n - 3$ to be at the right positions, and then recursively make the symbol k to be at the right position for all k from $n - 4$ down to 1. Accordingly, we can construct a path that starts from v and reaches a vertex u with $u\langle n - 1\rangle = r$, i.e., the child of the root in T_{n-1}^n. Note that all internal vertices in the above constructed path take $n - 1$ as their last symbol (see Fig. 6, where the path is drawn by a bold line in T_{n-1}^n).

CASE 3.3: $i = n - 2$. If $v_{n-1} \neq n$ or $v\langle n - 1\rangle = r$, since $v_n \neq i$, by Rule (6) we have $p = v\langle v^{-1}(n - 2)\rangle$. Moreover, if $v_{n-1} \neq n$, we can obtain the parent of a vertex by using Rule (6) repeatedly until the path reaches a vertex $u(= u_1 u_2 \cdots u_n)$ with $u_n = n - 2$. Also, if $v\langle n - 1\rangle = r$, then $p = 12 \cdots (n - 3)n(n - 2)(n - 1)$. Since $p_{n-1} \neq n$, it follows that there exists a path $(v, p) \cup T_{n-2}^n[p, u]$ where u is a vertex with $u_n = n - 2$ (e.g., consider the path $(v = 12354, p = 12534, u = 12543)$ in T_3^5). On the other hand (i.e., $v_{n-1} = n$ and $v\langle n - 1\rangle \neq r$), there are three situations as follows: (i) if $v_{n-2} \neq n - 2$, by Rule (4) we have

$p = v\langle v^{-1}(n-3)\rangle$; (ii) if $v_{n-2} = n-2$ and $v_{n-3} \neq n-3$, by Rule (3) we have $p = v\langle v^{-1}(\alpha(v, n-2, n))\rangle = v\langle v^{-1}(n-2)\rangle$; and (iii) if $v_{n-2} = n-2$ and $v_{n-3} = n-3$, by Rule (3) we have $p = v\langle v^{-1}(\alpha(v, n-2, n))\rangle = v\langle v^{-1}(\alpha(v, n-3, n-1))\rangle$. For all situations, using Rule (3) or Rule (4) repeatedly, we can construct a path that starts from v and reaches a vertex $w(= w_1 w_2 \cdots w_n)$ with $w_{n-1} \neq n$. Note that the path constructed by situation (iii) is similar to that of the Case 3.2. Since $w_{n-1} \neq n$, it follows that there exists a path $T_{n-2}^n[v, w] \cup T_{n-2}^n[w, u]$ where u is a vertex with $u_n = n-2$ (e.g., consider the path $(v = 23154, 32154, 31254, w = 31524, 13524, 15324, 15234, u = 15243)$ in T_3^5 for situation (i), and the path $(v = 21354, w = 21534, u = 21543)$ in T_3^5 for situation (ii)). From CASE 2, we have known that there is a path $T_{n-2}^n[u, r]$. Thus, $T_{n-2}^n[v, r] = T_{n-2}^n[v, u] \cup T_{n-2}^n[u, r]$. Note that all internal vertices in the above constructed path take $n-1$ or $n-2$ as their last symbol (see Fig. 6, where the path is drawn by a bold line in T_{n-2}^n).

CASE 3.4: $i \notin \{1, n-1, n-2\}$. Clearly, $v_n = n-1 \neq i$. If $v_{n-1} \neq n$, by Rule (6) we have $p = v\langle v^{-1}(i)\rangle$. Thus, we can obtain the parent of a vertex by using Rule (6) repeatedly until the path reaches a vertex $u(= u_1 u_2 \cdots u_n)$ with $u_n = i$. If $v_{n-1} = n$ and $v\langle n-1\rangle \neq r$, there are three situations as follows: (i) if $v_{n-2} \neq n-2$, by Rule (4) we have $p = v\langle v^{-1}(i-1)\rangle$; (ii) if $v_{n-2} = n-2$ and $v_{n-3} \neq n-3$, by Rule (3) we have $p = v\langle v^{-1}(\alpha(v, i, n))\rangle = v\langle v^{-1}(i-1)\rangle$; and (iii) if $v_{n-2} = n-2$ and $v_{n-3} = n-3$, by Rule (3) we have $p = v\langle v^{-1}(\alpha(v, i, n))\rangle = v\langle v^{-1}(\alpha(v, i-1, n-1))\rangle$. For all situations, using Rule (3) or Rule (4) repeatedly, we can construct a path that starts from v and reaches a vertex $w(= w_1 w_2 \cdots w_n)$ with $w_{n-1} \neq n$. Note that the path constructed by situation (iii) is similar to that of the Case 3.2. Since $w_{n-1} \neq n$, it follows that there exists a path $T_{n-2}^n[v, w] \cup T_{n-2}^n[w, u]$ where u is a vertex with $u_n = i$ (e.g., consider the path $(v = 13254, 31254, 32154, w = 32514, 35214, 35124, u = 35142)$ in T_3^5 for situation (i), and the path $(v = 21354, 23154, w = 23514, 32514, 35214, 35124, u = 35142)$ in T_3^5 for situation (ii)). Finally, if $v\langle n-1\rangle = r$, by Rule (6) we have $p = v\langle v^{-1}(i)\rangle$. Thus, we have $p_n = v_n = n-1$, $p_{n-1} = v_{n-1} = n$ and $p\langle n-1\rangle \neq r$. Since p is at a position described as before, it follows that there exists a path $T_i^n[v, u] = (v, p) \cup T_{n-2}^n[p, u]$ where u is a vertex with $u_n = i$. From CASE 2, we have known that there is a path $T_i^n[u, r]$. Thus, $T_i^n[v, r] = T_i^n[v, u] \cup T_i^n[u, r]$. Note that all internal vertices in the above constructed path take $n-1$ or i as their last symbol (see Fig. 6, where the path drawn by a bold line in T_2^n is an example).

From the paths described above, we only need to inspect vertices with last symbol n or $n-1$ in the trees for verifying the independence. For the vertices with the last symbol n, the proof is exactly the same as Case 2. For the vertices with the last symbol $n-1$, the proof is similar to Case 2 except that the last symbol of vertices we concerned is $n-1$ instead of j. This completes the proof. □

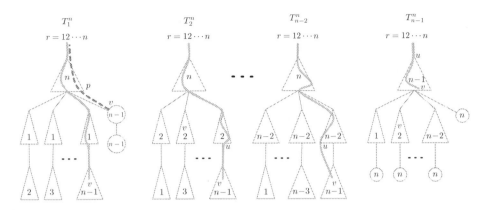

Fig. 6. An illustration of the paths described in the proof of Case 3 of Theorem 1.

The *height* of a rooted tree T, denoted by $h(T)$, is number of edges from the root to a farthest leaf. To analyze the height of our constructed ISTs for B_n, we define $H_n = \max_{i \leqslant i \leqslant n-1} h(T_i^n)$. Clearly, $H_3 = 5$ and $H_4 = 9$. From the recursive function $\mathrm{Parent}(v, i, n)$, it is easy to prove that $H_n = n(n+1)/2 - 1$ by induction. Since the construction includes $n-1$ ISTS, each IST contains $n!$ vertices, and the algorithm requires $n-1$ recursions, the total complexity is obviously in $\mathcal{O}(n^2 n!)$ time. According to Theorem 1 and the above discussion, we have the following corollary.

Corollary 1. *For bubble-sort network B_n, using the function $\mathrm{Parent}(v, i, n)$, the algorithm can correctly construct $n - 1$ ISTs of B_n with height at most $n(n + 1)/2 - 1$ in $\mathcal{O}(n^2 n!)$ time.*

5 Concluding Remarks

In this paper, we study the problem of constructing ISTs on bubble-sort networks. To the best of our knowledge, it seems that our work is the latest breakthrough on the problem of ISTs for all subclasses of Cayley graphs except star networks. Since B_n is vertex transitive, with a slight modification, we can easily derive a set of $n - 1$ ISTs with a common root at arbitrary vertex of B_n. Also, since B_n is a regular graph with connectivity $n - 1$, the number of constructed ISTs reaches the upper limit. In addition, for providing a full instance of ISTs of B_5 and the detail of our implementation, a website is now available on [12].

One interesting open problem remaining from our work is whether our algorithm can be improved to determine the parent of every vertex in each spanning tree directly (i.e., the computation requires no recursion and can be done only by referring the label of a vertex and the index of a tree). The advantage of such an improved algorithm is that it can easily be parallelized.

Acknowledgments. This research was partially supported by MOST grants MOST 104-2221-E-141-002-MY3 (Jou-Ming Chang), 105-2221-E-131-027 (Kung-Jui Pai) and 104-2221-E-262-005 (Ro-Yu Wu) from the Ministry of Science and Technology, Taiwan.

References

1. Akers, S.B., Krishnamurty, B.: A group theoretic model for symmetric interconnection networks. IEEE Trans. Comput. **28**, 555–566 (1989)
2. Araki, T., Kikuchi, Y.: Hamiltonian laceability of bubble-sort graphs with edge faults. Inform. Sci. **177**, 2679–2691 (2007)
3. Bao, F., Funyu, Y., Hamada, Y., Igarashi, Y.: Reliable broadcasting and secure distributing in channel networks. In: Proceedings of 3rd International Symposium on Parallel Architectures, Algorithms and Networks (ISPAN 1997), Taipei, pp. 472–478 (1997)
4. Chang, Y.-H., Yang, J.-S., Hsieh, S.-Y., Chang, J.-M., Wang, Y.-L.: Construction independent spanning trees on locally twisted cubes in parallel. J. Comb. Optim. **33**, 956–967 (2017)
5. Chang, J.-M., Yang, T.-J., Yang, J.-S.: A parallel algorithm for constructing independent spanning trees in twisted cubes. Discret. Appl. Math. **219**, 74–82 (2017)
6. Cheriyan, J., Maheshwari, S.N.: Finding nonseparating induced cycles and independent spanning trees in 3-connected graphs. J. Algorithms **9**, 507–537 (1988)
7. Curran, S., Lee, O., Yu, X.: Finding four independent trees. SIAM J. Comput. **35**, 1023–1058 (2006)
8. Hao, R.-X., Tian, Z.-X., Xu, J.-M.: Relationship between conditional diagnosability and 2-extra connectivity of symmetric graphs. Theor. Comput. Sci. **627**, 36–53 (2012)
9. Itai, A., Rodeh, M.: The multi-tree approach to reliability in distributed networks. Inform. Comput. **79**, 43–59 (1988)
10. Kikuchi, Y., Araki, T.: Edge-bipancyclicity and edge-fault-tolerant bipancyclicity of bubble-sort graphs. Inform. Process. Lett. **100**, 52–59 (2006)
11. Kao, S.-S., Chang, J.-M., Pai, K.-J., Yang, J.-S., Tang, S.-M., Wu, R.-Y.: A parallel construction of vertex-disjoint spanning trees with optimal heights in star networks. In: Gao, X., Du, H., Han, M. (eds.) COCOA 2017. LNCS, vol. 10627, pp. 41–55. Springer, Cham (2017). https://doi.org/10.1007/978-3-319-71150-8_4
12. Kao, S.-S., Chang, J.-M., Pai, K.-J., Wu, R.-Y.: Open source for "Constructing independent spanning trees on bubble-sort networks". https://sites.google.com/ntub.edu.tw/ist-bs/. Accessed 8 Jan 2018
13. Lakshmivarahan, S., Jwo, J., Dhall, S.K.: Symmetry in interconnection networks based on Cayley graphs of permutation groups: a survey. Parallel Comput. **19**, 361–407 (1993)
14. Kung, T.-L., Hung, C.-N.: Estimating the subsystem reliability of bubblesort networks. Theor. Comput. Sci. **670**, 45–55 (2017)
15. Rescigno, A.A.: Vertex-disjoint spanning trees of the star network with applications to fault-tolerance and security. Inform. Sci. **137**, 259–276 (2001)
16. Suzuki, Y., Kaneko, K.: An algorithm for disjoint paths in bubble-sort graphs. Syst. Comput. Jpn. **37**, 27–32 (2006)
17. Suzuki, Y., Kaneko, K.: The container problem in bubble-sort graphs. IEICE Trans. Inf. Syst. **E91–D**, 1003–1009 (2008)

18. Wang, M., Guo, Y., Wang, S.: The 1-good-neighbour diagnosability of Cayley graphs generated by transposition trees under the PMC model and MM* model. Int. J. Comput. Math. **94**, 620–631 (2017)
19. Wang, M., Lin, Y., Wang, S.: The 2-good-neighbor diagnosability of Cayley graphs generated by transposition trees under the PMC model and MM* model. Theor. Comput. Sci. **628**, 92–100 (2016)
20. Wang, S., Yang, Y.: Fault tolerance in bubble-sort graph networks. Theor. Comput. Sci. **421**, 62–69 (2012)
21. Yang, Y., Wang, S., Li, J.: Subnetwork preclusion for bubble-sort graph networks. Inform. Process. Lett. **115**, 817–821 (2015)
22. Yang, J.-S., Chan, H.-C., Chang, J.-M.: Broadcasting secure messages via optimal independent spanning trees in folded hypercubes. Discret. Appl. Math. **159**, 1254–1263 (2011)
23. Zehavi, A., Itai, A.: Three tree-paths. J. Graph Theory **13**, 175–188 (1989)
24. Zhou, S., Wang, J., Xu, X., Xu, J.-M.: Conditional fault diagnosis of bubble sort graphs under the PMC model. Intel. Comput. Evol. Comput. **180**, 53–59 (2013). AISC

Exact Algorithms for Finding Partial Edge-Disjoint Paths

Yunyun Deng[1], Longkun Guo[1(✉)], and Peihuang Huang[2]

[1] College of Mathematics and Computer Science, Fuzhou University, Fuzhou, China
deng.yunyun@foxmail.com, lkguo@fzu.edu.cn
[2] College of Physics and Information Engineering, Fuzhou University, Fuzhou, China
peihuang.huang@foxmail.com

Abstract. For a given graph G with non-negative integral edge length, a pair of distinct vertices s and t, and a given positive integer δ, the k partial edge-disjoint shortest path ($kPESP$) problem aims to compute k shortest st-paths among which there are at most δ edges shared by at least two paths. In this paper, we first present an exact algorithm with a runtime $O(mn \log_{(1+m/n)} n + \delta n^2)$ for $kPESP$ with $k = 2$. Then observing the algorithm can not be extended for general k, we propose another algorithm with a runtime $O(\delta 2^k n^{k+1})$ in $DAGs$ based on graph transformation. In addition, we show the algorithm can be extended to $kPESP$ with an extra *edge congestion constraint* that each edge can be shared by at most C paths for a given integer $C \leq k$.

Keywords: Partial edge-disjoint path · Exact algorithm
Directed acyclic graph · Restricted shortest path

1 Introduction

Network congestion is an important issue arising in networks when some nodes or links are overloaded, resulting in possibly data packet loss and failure of new connections. Although many techniques (like exponential backoff in CSMA/CA in 802.11 and CSMA/CD in the original Ethernet, window reduction in TCP, etc.) were developed to ease the consequences, it remains a long standing open challenge to eradicate network congestion. The fundamental reason causing congestion is that, data transmission in networks is mainly based on single path approaches which use a single optimal path (i.e. a path with minimum cost or delay) for transferring data. In the context, the transferred data is likely to concentrate on those links with less cost or delay, causing congestion when the quantity of data over the links exceeds its capacity. So network congestion is likely to exist as long as the single path approaches are employed for data transmission.

The research is supported by Natural Science Foundation of China (Nos. 61772005, 61300025) and Natural Science Foundation of Fujian Province (No. 2017J01753).

Disjoint path routing, the routing technique simultaneously using multiple disjoint paths instead of a single optimal path, is considered as an ultimate approach to settle the challenge against network congestion. In the scenario, the data transmission would be carried out over a set of disjoint paths, such that no nodes or links will be overloaded. However, the disjoint path routing has not yet been widely deployed in practice, mainly because the requirement of completely disjointness is not necessary for many practical applications while being resource-consuming. Therefore, we propose the following generalization of the edge-disjoint path problem, with a tunable disjointness degree:

Definition 1. *(The k Partial Edge-disjoint Shortest Path problem, kPESP). For a digraph $G = (V, E)$ with a non-negative integral length on each edge, a pair of distinct vertices s, $t \in V$ as the source and the destination, and an integer $\delta > 0$ as the disjointness factor, the problem aims to compute k paths connecting s and t, such that among the k paths there are at most δ edges shared by at least two paths and the length sum is minimized.*

Moreover, for the sake of further reducing congestion, many network applications further require each edge to be shared by at most $C \leq k$ paths. This brings the *kPESP* problem with congestion *(kPESPwC)*. Note that the δ-Vertex shared k Edge-Disjoint Shortest Path (δV-kEDSP) problem [21], aiming to compute 2 edge-disjoint paths sharing at most δ common vertices, is similar to the vertex version of *kPESP* but further requires the computed paths to be edge-disjoint.

1.1 Related Work

To the best of our knowledge, this paper is the first formally addressing *kPESP* and *kPESPwC*. Although the two problems are new, when $\delta = 0$ both of them reduce to the *min-sum* edge-disjoint path problem of computing completely edge-disjoint shortest *st*-paths, which was well studied by scientists from the community of computer science and networking. For any fixed integer $k > 0$, the reduced problem is known solvable in $O(n^2 \log n)$ time [18], which was improved to $O\left(m \log_{(1+m/n)} n\right)$ later in [17]. These algorithms have attracted many research interests and were used in parallel and distributed systems. In [13], an distributed algorithm was developed to construct a pair of disjoint paths of minimum total cost from every vertex to a destination with communication complexity $O\left(mn + n^2 D\right)$ and time complexity $O\left(nD\right)$, where D is the depth of the shortest-path spanning tree. Then, Sidhu et al. [16] devised a distributed distance vector algorithm which found more paths than the algorithm in [13]. Other than the *min-sum* problem, some other network applications require to minimize the length of the shorter path among the disjoint paths, which brings the *min-min* problem. The vertex-disjoint version and the directed edge-disjoint version of the min-min problem were shown $\mathcal{NP}$-complete by Xu et al. in [20]. Then, Guo and Shen showed the $\mathcal{NP}$-completeness for the edge-disjoint min-min problem in undirected graphs in [8] as a complementary to the work of Xu et al. in [20], and proved that the edge-disjoint min-min problem remains

$\mathcal{NP}$-complete in planar digraph in [7]. Opposite to the min-min problem, the min-max problem is to minimize the length of the longer path among the disjoint paths. The problem was shown strongly $\mathcal{NP}$-complete in both directed and undirected graphs in [12], while admits pseudo polynomial-time algorithms in directed acyclic graphs (DAGs). For $kPESPwC$, there exists literature on close related problems. The Edge-Disjoint Path problem with Congestion ($EDPwC$) is similar to $kPESPwC$, except that it is with a different aim of maximizing the number of disjoint paths that are respectively connecting k given pairs of vertices, say $s_1t_1, \ldots, s_it_i, \ldots, s_kt_k$, and without the bound of shared edges. The most recent results on $EDPwC$ can be found in [1,2].

There also exists literature on problems similar to but not exactly the vertex-disjoint version of $kPESP$. The δ-Vertex shared k Edge-Disjoint Shortest Path problem ($\delta V\text{-}kEDSP$), aiming to compute a pair of edge-disjoint st-paths with bounded shared vertices, was shown solvable in time $O(mn^2+n^3 \log n)$ when $k = 2$ by Yallouz et al. in [21]. Recently, the runtime was improved to $O(\delta m+n \log n)$ by Guo et al. in [5], where δ is the bound on the number of shared vertices. However, for general k it remains open whether $\delta V\text{-}kEDSP$ is in $\mathcal{P}$.

Our algorithm will solve $kPESP$ via computing restricted shortest path (RSP). We note that general RSP was shown $\mathcal{NP}$-complete in [4], and Joksch [10] was the first formally solving RSP in pseudo-polynomial time via dynamic programming technique. Later, Hassin [9] gave another pseudo-polynomial algorithm that consequently results in an $FPTAS$ with a run-time $O\left(m(\frac{n^2}{\epsilon}) + \log(\frac{n}{\epsilon})\right)$, which was then improved by Lorena and Raz to $O\left(mn\left(\log \log n + 1/\epsilon\right)\right)$, where $\epsilon > 0$ is any fixed real number. Recently, there were increasing research interest on k disjoint RSP instead of a single RSP, namely the k-edge (vertex) disjoint restricted shortest paths ($kRSP$) problem which combines RSP and disjoint paths, because using $kRSP$ in many nowaday applications would compare favorably to using a single RSP subject to QoS constraint. Orda and Sprintson [14] gave an interesting approximation algorithm of a ratio $(1+r, 1+\frac{1}{r})$, $r > 0$, for $2RSP$ to find two disjoint paths satisfying the given QoS constraints at minimum cost, which was then improved to a polynomial-time approximation algorithm with an improved bifactor approximation ratio $(1 + \epsilon, 2 + \epsilon)$ for general k in [6].

1.2 Our Results

In this paper, we first propose an exact algorithm for $kPESP$ with a runtime $O(nm \log_{(1+m/n)} n + \delta n^2)$ when $k = 2$, via constructing an auxiliary graph. The main observation is that, $2PESP$ is feasible subjected to δ if and only if there exists an st-path with a *new cost* bounded by δ in the constructed auxiliary graph, where the *new cost* is to capture the number of shared edges. Consequently, our algorithm needs only to solve a special case of RSP in the auxiliary graph, which can be done in polynomial time.

Observing the algorithm for $2PESP$ is not applicable to $kPESP$, we propose an algorithm for $kPESP$ in DAGs by constructing another auxiliary graph which reduces *kPESP* to a special case of *RSP* in a different way. We show the algorithm runs in $O(\delta 2^k n^{k+1})$ time and solves *kPESP* optimally. Note that our algorithm grants the new ability of computing partial edge disjoint paths in *DAGs* to the elegant algorithm by Perl and Shilaoch in [15], which computes a pair of disjoint paths in *DAGs* by finding a single path in the proposed multi-dimensional graph[1]. Moreover, in *DAGs*, our algorithm for *kPESP* can be extended to solve *kPESPwC*, a generalization of *kPESP* in which the number of paths sharing an identical common edge is bounded by a given parameter, say $C \leq k$. It remains open whether *kPESP* admits polynomial algorithms for general k in either general graphs or *DAGs*.

2 An Exact Algorithm to the *2PESP* problem

In this section, we will present an exact polynomial-time algorithm for the *2PESP* problem. The algorithm is inspired by the famous 2-approximation for the minimum Steiner tree problem, known as the Kou-Markowsky-Berman algorithm [11], which constructs a metric closure induced by the terminals and then computes a spanning tree therein, and also the algorithm for solving the δ-Vertex shared k Edge-Disjoint Shortest Path (δV-*kEDSP*) problem [21]. Similarly, our algorithm is composed by two phases: Firstly, construct the promising auxiliary graph G' which is an enhanced version of the metric closure; Secondly, find a shortest st-path with cost bounded by δ in G', which equivalently solves *2PESP*. For briefness, we define the Binary Restricted Shortest Path (*BRSP*) problem as follows:

Definition 2 *(Binary Restricted Shortest Path, BRSP). For a digraph* $G = (V, E)$ *with a pair of distinct vertices* $s, t \in V$, *a length function* $l: E \rightarrow Z_0^+$ *and a Boolean function* $c: E \rightarrow \{0, 1\}$ *over the edges, the BRSP problem is to compute an st-path in* G, *such that* $\sum_{e \in E} l(e)$ *is minimized subject to* $\sum_{e \in E} c(e) \leq \delta$, *where* $\delta \in Z$ *is the given cost constraint.*

Apparently, the *BRSP* problem is a special case of the restricted shortest path (*RSP*) problem since each edge therein is with a cost of 0 or 1. Following *Joksch's algorithm* [10], we can immediately obtain a polynomial algorithm for *BRSP*:

Proposition 3 [10]. *The BRSP problem can be solved in time* $O(\delta|E|)$.

Hence, the second task of computing a shortest path with at most δ common edges in the auxiliary graph can be done within time $O(\delta|E|)$. So it remains only to construct the auxiliary graph G'.

The main observation for the construction is that, an optimal solution of *2PESP* is composed by a set of edge-disjoint path pairs alongside a set of shared edges (of G), with the path pairs connecting the common edges.

---

[1] Note that the algorithm was previously extended to solve many other related problems including [3, 19].

Lemma 4. *Let $\{P_1, P_2\}$ be an optimal solution to 2PESP. Let $E(P_1 \cap P_2) = \{e_1, e_2, \cdots, e_\delta\}$, $e_i = (u_i, v_i)$, be the set of common edges. Then we have:*

1. *Each edge of $E(P_1 \cap P_2)$ separates s and t;*
2. *Let $\mathcal{P} = \{\{P_{i,j} \mid j = 1, 2\} \mid i = 1, \ldots, \delta + 1\}$ be the set of components resulting from $(P_1 \cup P_2) \setminus E(P_1 \cap P_2)$, where $P_{i,1}$ and $P_{i,2}$ connects v_{i-1} to u_i. Then $P_{i,1}$ and $P_{i,2}$ are edge disjoint for any $i \in [\delta + 1]^+$ [2]. Moreover, for any $i \neq j$, $P_{i,1} \cup P_{i,2}$ and $P_{j,1} \cup P_{j,2}$ shares no common edges.*

Proof. Omitted due to space limit.   □

Following the main observation as in Lemma 4, the auxiliary graph G' is mainly composed by two parts of edges. The first part of edges compose a complete graph $\overline{G}$ (the metric closure), in which each edge (u, v) represents a pair of shortest edge-disjoint paths from u to v; the second part contains the edges of the original graph G, which are actually possibly shared edges.

By employing *Suurballe and Tarjan's algorithm* [17], we can find a pair of shortest edge-disjoint paths from u to v in $O(m \log_{(1+m/n)} n)$ time. So the first part $\overline{G}$ can be constructed in polynomial time since it has $O(n(n-1))$ edges. Then, the length of each $e \in \overline{G}$ is set to the length of the shortest edge-disjoint paths pair it represents, while the length of the edges of the second part, say $e \in G' \setminus \overline{G}$, remains the same as in G. Moreover, we set the cost of each edge of $\overline{G}$ to 0 and each $e \in G' \setminus \overline{G}$ to 1 because each edge in $e \in G' \setminus \overline{G}$ is a possibly shared edge. Note that G' can be a multigraph. The detailed construction is formally as in Algorithm 1.

Then we need only to solve *BRSP* wrt G' to find an optimal solution to *2PESP* by the following Lemma:

Lemma 5. *There exists a solution $\{P_1, P_2\}$ against an instance of 2PESP in G iff there exists a shortest st-path Q with $c(Q) \leq \delta$ and $l(Q) = l(P_1 \cup P_2)$ in the corresponding auxiliary graph $G' = (V', E')$ output by Algorithm 1, where $\delta \in \mathbb{Z}_0^+$ is the bound on the number of shared edges.*

Proof. Omitted due to space constraint.   □

Lemma 6. *The above algorithm runs in time $O(mn \log_{(1+m/n)} n + \delta n^2)$.*

Proof. The algorithm takes $O(m \log_{(1+m/n)} n)$ time to run *Suurballe and Tarjan's Algorithm* to compute a pair of edge-disjoint paths from a vertex to every other vertex in graph G [17]. Since there are $O(n)$ vertices, the algorithm takes $O(mn \log_{(1+m/n)} n)$ time to construct the auxiliary graph G', which has $O(n^2)$ edges. Then, the algorithm calls *Joksch's algorithm* to solve *BRSP* with a runtime $O(\delta n^2)$ [10]. Therefore, the total runtime is $O(mn \log_{(1+m/n)} n + \delta n^2)$.   □

Combining Lemmas 5 and 6, we immediately have the following theorem:

---

[2] Following the tradition in mathematics, the notation $[n]^+$ denotes the set $\{1, \ldots, n\}$.

Algorithm 1. The construction of auxiliary graph G'

Input: A digraph $G = (V, E)$, a specified vertices s and t, a length function $l(e)$ and
      an upper bound $\delta \in Z^+$ for the number of common edges;

Output: An auxiliary graph $G' = (V', E')$.

 1: Initially set $G' := \emptyset$;

 2: **For** each pair of $u, v \in V$ $(u \not= v)$ **do** /*Construct $\overline{G}$. */

 3:    Find a pair of edge-disjoint shortest paths (P_1^*, P_2^*) from u to v using *Suurballe*
        and Tarjan's Algorithm [17];

 4:    **If** (P_1^*, P_2^*) exists **then**

 5:       Set $E' := E' \cup \{e'(u, v)\}$; /*$e'(u, v)$ represents (P_1^*, P_2^*). */

 6:       $l(e'(u, v)) := l(P_1^*(u, v)) + l(P_2^*(u, v))$ and $c(e'(u, v)) := 0$;

 7:    **Endif**

 8: **Endfor**

 9: **For** each $e(u, v) \in G$ **do** /*Add the possibly shared edges to G'. */

10:    Set $G' := G' \cup \{e(u, v)\}$, $c(e(u, v)) := 1$ and $l(e(u, v))$ equal to $e(u, v) \in G$;

11: **Endfor**

Theorem 7. *2PESP problem can be optimally solved within time complexity*
$O(mn \log_{(1+m/n)} n + \delta n^2)$.

Although the above algorithm can efficiently solve the *2PESP*, it can not be easily extended to solve *kPESP* for general k. The reason is that, when $k \geq 3$ we can not construct the auxiliary graph G' by Algorithm 1 via using the cost function to capture the number of shared edges. So we will propose another method to solve *kPESP* for general k but only in *DAGs*.

3 Exact Algorithms for *kPESP* in *DAGs*

In this section, we will present an algorithm for the *kPESP* problem in *DAGs* mainly based on graph transformation, in which we construct a k-dimensional auxiliary graph $\mathcal{G}_k$ wrt G, such that *kPESP* in G is transformed to *BRSP* in $\mathcal{G}_k$. That is, the key idea is to reduce the aimed k partial disjoint paths to a single path in the auxiliary graph, extending the technique in [15, 19].

3.1 An Exact Algorithm for *2PESP*

To accomplish the reduction, each node in $\mathcal{G}_k$ is assigned with k dimensions which actually represent k vertices in the original graph, and consequently an edge between two k-dimensional nodes is added if it can possibly correspond to part of a feasible solution to *kPESP*. For briefness, we will first present our algorithm for *2PESP*, and then show how to extend to both *kPESP* and *kPESPwC* for general k.

    The construction of the auxiliary graph $\mathcal{G}_2$ is composed by two phases: Firstly, add all the 2-dimensional nodes to $\mathcal{G}_2$, n^2 nodes in total; Secondly, repeatedly add edges each of which with a length and a cost, where the cost is to capture the number of shared edges and hence its sum must be bounded by δ.

Algorithm 2. Construction of $\mathcal{G}_2$

Input: A simple $DAG\ G = (V,\ E)$, distinct source s and destination t, the common
      edges upper bound $\delta \in Z^+$, a length function $l\,(e)$;
Output: A 2-dimensional auxiliary graph $\mathcal{G}_2 = (\mathcal{N}_2,\ \mathcal{E}_2)$.

1: Label vertices in $V\,(G)$ with numbers $\{1,2,\cdots,|V|\}$ in a way that $(u,\ v) \in E\,(G)$
    only if $u < v$;   /*W.l.o.g. assume that $s = 1$ and $t = |V|$.*/
2: Set $\mathcal{N}_2 := \{\,\langle u,\ v \rangle\,|\,u,\ v \in V\,(G)\}$, $\mathcal{E}_2 := \emptyset$, and $\mathcal{S} := \{\langle s,\ s \rangle\}$ being a queue;
3: **While** $\mathcal{S} \neq \emptyset$ **do**
4:   Select a node $N = \langle u,\ v \rangle$ from the top of the queue $\mathcal{S}$, and set $\mathcal{S} := \mathcal{S} \setminus \{N\}$;
5:   **If** $u < v$ **then**
6:      **For** each $(u,u') \in E\,(G)$ **do**
7:         Set $\mathcal{S} := \mathcal{S} \cup \langle u',v \rangle$;
8:         Set edge $e := (N,\ \langle u',v \rangle)$, $\mathcal{E}_2 := \mathcal{E}_2 \cup \{e\}$, $l(e) := l(u,u')$ and $c(e) := 0$;
9:      **Endfor**
10: **Endif**
11: **If** $u > v$ **then**
12:     **For** each $(v,v') \in E\,(G)$ **do**
13:        Set $\mathcal{S} := \mathcal{S} \cup \langle u,v' \rangle$;
14:        Set edge $e := (N,\ \langle u,v' \rangle)$, $\mathcal{E}_2 := \mathcal{E}_2 \cup \{e\}$, $l(e) := l(v,v')$ and $c(e) := 0$;
15:     **Endfor**
16: **Endif**
17: **If** $u = v$ **then**
18:     **For** each pair of u' and v' that (u,u') and (v,v') both in $E\,(G)$ **do**
19:        Set $\mathcal{S} := \mathcal{S} \cup \langle u',v' \rangle$;
20:        Set edge $e := (N,\ \langle u',v' \rangle)$, $\mathcal{E}_2 := \mathcal{E}_2 \cup \{e\}$;
21:        **If** $u' \neq v'$ **then**
22:           Set $c(e) := 0$ and $l(e) := l(u,u') + l(v,v')$;
23:        **Else** $c(e) := 1$ and $l(e) := l(u,u')$;
24:     **Endfor**
25: **Endif**
26: **Endwhile**
27: Return $\mathcal{G}_2 := (\mathcal{N}_2,\ \mathcal{E}_2)$.

For any given $DAG\ G = (V,E)$ with a source s and a destination t, our algorithm relabels the vertices with number 1 to $|V|$, such that $s = 1$, $t = |V|$, and u has a path to v iff $u < v$. Then the auxiliary graph $\mathcal{G}_2 = (\mathcal{N}_2,\mathcal{E}_2)$ can be obtained from G as follows. First, we set $\mathcal{N}_2 = \{\langle u,v \rangle\,|u,v \in V\}$, where $u = v$ is allowed. Then the set of edges of $\mathcal{G}_2$ is as below:

$$\mathcal{E}_2 = \{\langle u,v \rangle \rightarrow \langle u,z \rangle\,|(v,z) \in E\ and\ v < u\}$$
$$\cup\,\{\langle v,u \rangle \rightarrow \langle z,u \rangle\,|(v,z) \in E\ and\ v < u\}$$
$$\cup\,\{\langle u,u \rangle \rightarrow \langle v,z \rangle\,|(u,v),(u,z) \in E\ and\ v \neq z\}$$
$$\cup\,\{\langle u,u \rangle \rightarrow \langle v,v \rangle\,|(u,v) \in E\}$$

Note that edges in form of $\langle u,u \rangle \rightarrow \langle v,v \rangle$ of $\mathcal{E}_2$ correspond to common edges in a solution against *2PESP*, while the other edges correspond to non-common edges. Accordingly, the lengths and costs to edges in $\mathcal{E}_2$ are as follows:

$$l(\langle u,v \rangle \rightarrow \langle u,z \rangle) = l(v,z), \quad c(\langle u,v \rangle \rightarrow \langle u,z \rangle) = 0$$
$$l(\langle v,u \rangle \rightarrow \langle z,u \rangle) = l(v,z), \quad c(\langle v,u \rangle \rightarrow \langle z,u \rangle) = 0$$
$$l(\langle u,u \rangle \rightarrow \langle v,z \rangle) = l(u,v) + l(u,z), \quad c(\langle u,u \rangle \rightarrow \langle v,z \rangle) = 0$$
$$l(\langle u,u \rangle \rightarrow \langle v,v \rangle) = l(u,v), \quad c(\langle u,u \rangle \rightarrow \langle v,v \rangle) = 1$$

Although $\mathcal{N}_2$ contains $O(n^2)$ nodes, many of them are unreachable from $\langle s, s \rangle$ in $\mathcal{G}_2$, i.e. have no path from $\langle s, s \rangle$. Hence, when constructing $\mathcal{E}_2$, our algorithm will actually add only the edges reachable from $\langle s, s \rangle$. Similar to *Dijkstra's algorithm* for the shortest path problem [11], our construction conducts a broad first search (*BFS*) traversal over all the nodes reachable from $\langle s, s \rangle$, via maintaining a queue $\mathcal{S}$ which is initially $\mathcal{S} := \langle s, s \rangle$. In the traversal, we repeatly add all the edges leaving the current visiting node to $\mathcal{E}_2$ until $\mathcal{S}$ is empty. The detailed layout of the algorithm is as in Algorithm 2 (An example executing the construction is as depicted in Fig. 1).

For clearance, we say $\mathcal{G}_2$ is *full* if it contains both edges reachable and unreachable from $\langle s, s \rangle$, and is *reduced* if it contains only edges reachable from $\langle s, s \rangle$ (like the graph resulting from Algorithm 2). Clearly, *full* $\mathcal{G}_2$ and *reduced* $\mathcal{G}_2$ contain exactly the same set of ST-paths, because *reduced* $\mathcal{G}_2$ already contains all the edges reachable from $\langle s, s \rangle$. So it suffices to show the correctness for *full* $\mathcal{G}_2$, which can be obtained from the lemma below:

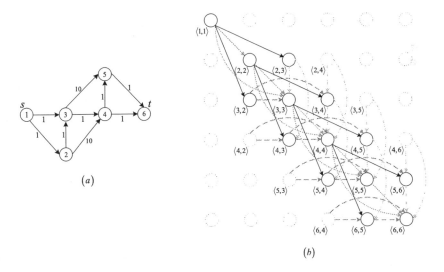

(a)

(b)

Fig. 1. An example executing Algorithm 2: (a) The original *DAG* G; (b) The 2-dimensional auxiliary graph $\mathcal{G}_2$, in which the edges of different colors correspond to the four different types of edges in $\mathcal{E}_2$, the blue dotted nodes are the reducible nodes, and the path $(1,1) \rightarrow (2,3) \rightarrow (3,3) \rightarrow (4,4) \rightarrow (5,6) \rightarrow (6,6)$ corresponds to an optimal solution to *2PESP* subject to $\delta = 1$ in G. (Color figure online)

Algorithm 3. An exact algorithm for *2PESP*

Input: A *DAG* $G = (V, E)$, a pair of specified vertices s and t, a length function $l(e)$
    and a common edge upper bound $\delta \in Z^+$;

Output: Two shortest st-paths with at most δ common edges.

1: Construct a 2-dimensional auxiliary graph $\mathcal{G}_2$ wrt G by Algorithm 2;
2: Find a shortest path $\mathcal{P}$ in $\mathcal{G}_2$ such that $c(P) \leq \delta$ by employing *Joksch's algorithm*
    for *BRSP* [10];
3 Decompose $\mathcal{P}$ into 2 paths $\{P_1, P_2\}$ where P_i goes through the vertex sequence in
    the ith dimension of $\mathcal{P}$;
4: Return P_1 and P_2.

Lemma 8. *There exists in full $\mathcal{G}_2$ a path $\mathcal{P}$ from $\langle u, v \rangle$ to $\langle u', v' \rangle$, $u \neq u'$ and $v \neq v'$, with a cost bounded by σ iff there exist in G a pair of paths P_1 and P_2, respectively connecting u, u' and v, v' and sharing at most σ edges.*

Proof. Omitted due to the length limit.   □

From Lemma 8, by setting $u = v = s$ and $u' = v' = t$ (note that $s \neq t$), we immediately have the correctness of Algorithm 2:

Theorem 9. *There exists an ST-path with cost bounded by δ in $\mathcal{G}_2$ iff there exists a pair of st-paths that share at most δ edges.*

Lemma 10. *Algorithm 2 runs in time $O(n^3)$, and constructs a graph $\mathcal{G}_2$ with at most $O(n^2)$ nodes and $O(n^3)$ edges.*

Proof. Let $m = |E(G)|$ and $n = |V(G)|$. Step 2 in Algorithm 2 brings at most $O(n^2)$ nodes to $\mathcal{G}_2$. Then let $\mathcal{N}_{dg} =| u \in V(G)$, we will count the edges by summing up firstly the number of edges leaving nodes in $\mathcal{N}_{dg}$ and secondly those that are not. For the first, from each node in $\mathcal{N}_{dg}$ there can leave $O(n^2)$ edges, because as in Steps 18–24, there are $O(n^2)$ pairs of u' and v' that $(u,u') \in E(G)$ and $(v,v') \in E(G)$ both hold for $u = v$. Then, since $|\mathcal{N}_{dg}| = O(n)$, there are $O(n^3)$ edges leaves the nodes of $\mathcal{N}_{dg}$. For the second, there are at most $O(n^2)$ nodes in $\mathcal{N}_2 \setminus \mathcal{N}_{dg}$, from each of which there leave at most $O(n)$ edges. So there are at most $O(n^3)$ edges leaving the nodes of $\mathcal{N}_2 \setminus \mathcal{N}_{dg}$. Summing up both parts, we have $|E(\mathcal{G})| = O(n^3)$. Note that the time complexity is actually equal to the size of the graph, so the runtime of Algorithm 2 is $O(n^3)$.   □

By constructing $\mathcal{G}_2(\mathcal{N}_2, \mathcal{E}_2)$, *2PESP* is reduced to *BRSP* of finding a shortest *ST*-path in $\mathcal{G}_2$ with cost at most δ to capture the number of shared edges. The formal layout of the whole algorithm is as in Algorithm 3.

Lemma 11. *Algorithm 3 runs in time $O(\delta n^3)$ and correctly produces an optimal solution to 2PESP.*

Proof. For the correctness, following Theorem 9, a shortest *ST*-path $\mathcal{P}$ with cost bounded by δ in $\mathcal{G}_2$ is clearly corresponding to a minimum length solution to

2PESP; since otherwise, there exists a solution to *2PESP* with less length, indicating the existence of an *ST*-path with cost bounded by δ but of less length than $\mathcal{P}$, which contradicts with the minimality of $\mathcal{P}$.

For the runtime, from Lemma 10, Step 1 takes $O(n^3)$ time to run Algorithm 2, which constructs the 2-dimensional auxiliary graph $\mathcal{G}_2$ with $|E(\mathcal{G}_2)| = O(n^3)$. In step 2, the algorithm for *BRSP* runs in $O(\delta|E(\mathcal{G}_2)|) = O(\delta n^3)$ following Proposition 3. Other steps obviously take relatively trivial time. Therefore, the total runtime of Algorithm 3 is $O(\delta n^3)$.   □

3.2   Extension to *kPESP* for $k \geq 3$ in *DAGs*

The key idea of solving *kPESP* with general k in *DAGs* is similar to *2PESP*, except that the algorithm will construct a k-dimensional auxiliary graph $\mathcal{G}_k$ instead of $\mathcal{G}_2$. However, there are still some tricky differences in details between constructing $\mathcal{G}_k$ and $\mathcal{G}_2$. Since constructing a reduced $\mathcal{G}_k$ is similar to Algorithm 2, we will only give the construction of *full* $\mathcal{G}_k$ for the sake of briefness.

1. Construction of $\mathcal{N}_k$:
   Similar to $\mathcal{G}_2$, the k-dimensional auxiliary graphs $\mathcal{G}_k$ has $|V|^k$ nodes that $\mathcal{N}_k = \{ \langle v_1, \ldots, v_i, \ldots, v_k \rangle \,|\, v_1, \ldots, v_k \in V(G) \}$.
2. Construction of $\mathcal{E}_k$:
   For each node $N = \langle v_1, \ldots, v_{i_1}, \ldots, v_{i_2}, \ldots, v_{i_l}, \ldots, v_k \rangle$ in $\mathcal{N}_k$, where $v_{i_1} = v_{i_2} = \cdots = v_{i_l}$ and $v_{i_1} < u, \forall u \in \{v_1, \ldots v_k\} \setminus \{v_{i_1}, \ldots, v_{i_l}\}$, if there exists an edge $(v_{i_1}, v_{i'_j}) \in E(G)$ for $\forall j \in [l]^+$, then add an edge e to $\mathcal{E}_k$:

$$ e = \langle v_1, \ldots, v_{i_1}, \ldots, v_{i_l}, \ldots, v_k \rangle \rightarrow \langle v_1, \ldots, v_{i'_1}, \ldots, v_{i'_l}, \ldots, v_k \rangle. $$

It remains to set the cost and length for each edge in $\mathcal{E}_k$. Assume that the vertices $v_{i'_1}, \ldots, v_{i'_l}$ after $v_{i_1} = v_{i_2} = \cdots = v_{i_l}$ (according to e) contain q different vertices, among which there are $q' \leq q$ vertices appearing at least two times. That is, edge e contains exactly q' different shared edges, so we set the cost of each edge as $c(e) := q'$. Let U be the set of different vertices in $\{v_{i'_1}, \ldots, v_{i'_l}\}$. Then $\{(v_{i_1}, u) | u \in U\}$ is the set of different edges leaving v_{i_1}, so $l(e)$ equals the length sum of the edges in $\{(v_{i_1}, u) \mid u \in U\}$, i.e. $l(e) = \sum_{u \in U} l(v_{i_1}, u)$. Eventually, similar to Lemma 10, we have the size of *full* $\mathcal{G}_k$:

Lemma 12. *The graph full $\mathcal{G}_k$ resulted from the above construction has $O(n^k)$ nodes and $O(2^k n^{k+1})$ edges.*

Proof. The case for the nodes is obvious. It remains to count the number of the edges. For a node $N = \langle v_1, \ldots, v_{i_1}, \ldots, v_{i_2}, \ldots, v_{i_l}, \ldots, v_k \rangle$ with $v_{i_1} = v_{i_2} = \cdots = v_{i_l}$, there are $\binom{k}{l}$ combinations for selecting l dimensions from the k dimensions to place the vertices $v_{i_1}, \ldots, v_{i_l}$. For the other $k - l$ dimensions, each dimension is apparently free to choose a vertex from the other $n - 1$ vertices except v_{i_1}, and hence $O((n-1)^{k-l})$ combinations. Because $v_{i_1} = v_{i_2} = \cdots = v_{i_l}$

can be any vertex of $V(G)$, there are $O(n)$ choices for v_{i_1}. So there are at most $O\left(\binom{k}{l} n^{k-l+1}\right)$ different combinations for a node N with a fixed l. Moreover, there are at most n^l edges leaving N according to the construction. Therefore, the total number of different edges in *full* $\mathcal{G}_k$ for a fixed l is $\binom{k}{l} \cdot O(n^{k-l+1}) \cdot O(n^l)$. So by summing up all ls, we have $\sum_{l=1}^{k} \binom{k}{l} \cdot O(n^{k-l+1}) \cdot O(n^l) = O(2^k n^{k+1})$.

$\square$

Similar to Theorem 9, we have the following theorem for the correctness of the construction:

Theorem 13. *There exists an ST-path with cost bounded by δ in $\mathcal{G}_k$ iff there exist k paths connecting s to t and sharing at most δ edges.*

Because there are in total $O(2^k n^{k+1})$ edges in $\mathcal{G}_k$, following Proposition 3 we have:

Corollary 14. *The kPESP problem admits an exact algorithm with a runtime $O(\delta 2^k n^{k+1})$.*

It is worth noting that our algorithm can be easily extended to solve *kPESPwC*, observing that the only difference between *kPESPwC* and *kPESP* is in *kPE-SPwC* an edge can be shared by at most C paths among $P_1, P_2, \cdots, P_k$. The extension needs only to remove from $\mathcal{E}_k$ every edge e that corresponds to an edge shared for more than C times. That is, it needs only to prune out $e = \langle v_1, \ldots, v_{i_1}, \ldots, v_{i_2}, \ldots, v_{i_l}, \ldots, v_k \rangle \to \langle v_1, \ldots, v_{i'_1}, \ldots, v_{i'_2}, \ldots, v_{i'_l}, \ldots, v_k \rangle$ with $v_{i_1} = v_{i_2} = \cdots = v_{i_l}$ in which there exists a vertex appearing in $\{v_{i'_1}, \ldots, v_{i'_l}\}$ more than C times.

4   Conclusion

In this paper, we studied the problem of finding k shortest st-paths with a bounded number of common edges, namely *kPESP*. We first proposed an exact algorithm for *2PESP* with a runtime $O(mn \log_{(1+m/n)} n + \delta n^2)$, where m, n and δ are respectively the number of edges, vertices, and shared edges. Because the algorithm can not be easily extended for general k, we proposed a different exact algorithm with a runtime $O(\delta 2^k n^{k+1})$ by graph transformation, which can solve *kPESP* with general k but only in *DAGs*. Our technique can be easily extended to solve *kPESPwC* for any fixed integer $k \geq 2$ in *DAGs*, in which an edge is allowed to be shared by at most $C \leq k$ paths. We are currently investigating how to efficiently solve *kPESP* with $k > 2$ in general graphs.

References

1. Chuzhoy, J.: Routing in undirected graphs with constant congestion. SIAM J. Comput. **45**(4), 1490–1532 (2016)
2. Chuzhoy, J., Li, S.: A polylogarithmic approximation algorithm for edge-disjoint paths with congestion 2. J. ACM (JACM) **63**(5), 45 (2016)
3. Fleischer, R., Ge, Q., Li, J., Zhu, H.: Efficient algorithms for k-disjoint paths problems on DAGs. In: Kao, M.-Y., Li, X.-Y. (eds.) AAIM 2007. LNCS, vol. 4508, pp. 134–143. Springer, Heidelberg (2007). https://doi.org/10.1007/978-3-540-72870-2_13
4. Garey, M.R., Johnson, D.S.: Computers and Intractability, vol. 29. WH Freeman, New York (2002)
5. Guo, L., Deng, Y., Liao, K., He, Q., Sellis, T., Hu, Z.: A fast algorithm for optimally finding partially disjoint shortest paths. In: Accepted by the 27th International Joint Conference on Artificial Intelligence and the 23rd European Conference on Artificial Intelligence, IJCAI (2018)
6. Guo, L., Liao, K., Shen, H., Li, P.: Brief announcement: efficient approximation algorithms for computing k disjoint restricted shortest paths. In: Proceedings of the 27th ACM Symposium on Parallelism in Algorithms and Architectures, pp. 62–64. ACM (2015)
7. Guo, L., Shen, H.: On the complexity of the edge-disjoint min-min problem in planar digraphs. Theor. Comput. Sci. **432**, 58–63 (2012)
8. Guo, L., Shen, H.: On finding min-min disjoint paths. Algorithmica **66**(3), 641–653 (2013)
9. Hassin, R.: Approximation schemes for the restricted shortest path problem. Math. Oper. Res. **17**(1), 36–42 (1992)
10. Joksch, H.C.: The shortest route problem with constraints. J. Math. Anal. App. **14**(2), 191–197 (1966)
11. Korte, B., Vygen, J.: Combinatorial Optimization, vol. 2. Springer, Heidelberg (2012). https://doi.org/10.1007/978-3-642-24488-9
12. Li, C.-L., McCormick, S.T., Simchi-Levi, D.: The complexity of finding two disjoint paths with min-max objective function. Discret. Appl. Math. **26**(1), 105–115 (1990)
13. Ogier, R.G., Rutenburg, V., Shacham, N.: Distributed algorithms for computing shortest pairs of disjoint paths. IEEE Trans. Inf. Theory **39**(2), 443–455 (1993)
14. Orda, A., Sprintson, A.: Efficient algorithms for computing disjoint QOS paths. In: INFOCOM 2004. Twenty-Third AnnualJoint Conference of the IEEE Computer and Communications Societies, vol. 1. IEEE (2004)
15. Shiloach, Y., Perl, Y.: Finding two disjoint paths between two pairs of vertices in a graph. J. ACM (JACM) **25**(1), 1–9 (1978)
16. Sidhu, D., Nair, R., Abdallah, S.: Finding disjoint paths in networks. ACM SIGCOMM Comput. Commun. Rev. **21**, 43–51 (1991)
17. Suurballe, J.W., Tarjan, R.E.: A quick method for finding shortest pairs of disjoint paths. Networks **14**(2), 325–336 (1984)
18. Suurballe, J.W.: Disjoint paths in a network. Networks **4**(2), 125–145 (1974)
19. Wu, B.Y.: A note on approximating the min-max vertex disjoint paths on directed acyclic graphs. J. Comput. Syst. Sci. **77**(6), 1054–1057 (2011)
20. Xu, D., Chen, Y., Xiong, Y., Qiao, C., He, X.: On the complexity of and algorithms for finding the shortest path with a disjoint counterpart. IEEE/ACM Trans. Netw. (TON) **14**(1), 147–158 (2006)
21. Yallouz, J., Rottenstreich, O., Babarczi, P., Mendelson, A., Orda, A.: Optimal link-disjoint node-"somewhat disjoint" paths. In: 2016 IEEE 24th International Conference on Network Protocols (ICNP), pp. 1–10. IEEE (2016)

A Randomized FPT Approximation Algorithm for Maximum Alternating-Cycle Decomposition with Applications

Haitao Jiang[1], Lianrong Pu[1], Letu Qingge[2], David Sankoff[3], and Binhai Zhu[2(✉)]

[1] School of Computer Science and Technology, Shandong University, Jinan, China
htjiang@sdu.edu.cn
[2] Gianforte School of Computing, Montana State University, Bozeman, MT 59717-3880, USA
letu.qingge@msu.montana.edu, bhz@montana.edu
[3] Department of Mathematics and Statistics,
University of Ottawa, Ottawa, Ont K1N 6N5, Canada
sankoff@uottawa.ca

Abstract. Comparing genomes in terms of gene order is a classical combinatorial optimization problem in computational biology. Some of the popular distances include translocation, reversal, and double-cut-and-join (abbreviated as DCJ), which have been extensively used while comparing two genomes. Let d_x, $x \in \{$translocation, reversal, DCJ$\}$, be the distance between two genomes such that one can be sorted/converted into the other using the minimum number of x-operations. All these problems are NP-hard when the genomes are unsigned. Computing d_x, $x \in \{$translocation, reversal, DCJ$\}$, between two unsigned genomes involves computing a proper alternating cycle decomposition of its breakpoint graph, which becomes the bottleneck for computing the genomic distance under almost all types of genome rearrangement operations and prohibits to obtain approximation factors better than 1.375 in polynomial time. In this paper, we devise an FPT (fixed-parameter tractable) approximation algorithm for computing the DCJ and translocation distances with an approximation factor $4/3+\varepsilon$, and the running time is $O^*(2^{d^*})$, where d^* represents the optimal DCJ or translocation distance. The algorithm is randomized and it succeeds with a high probability. This technique is based on a new randomized method to generate approximate maximum alternating cycle decomposition.

1 Introduction

Computing genomic distance on gene order is a fundamental problem in computational biology. In the last two decades, a variety of biological operations, such as reversals, translocations, fusions, fissions, transpositions and block-interchanges,

L. Wang and D. Zhu (Eds.): COCOON 2018, LNCS 10976, pp. 26–38, 2018.
https://doi.org/10.1007/978-3-319-94776-1_3

have been proposed to handle gene order. The *double-cut-and-join* operation, introduced by Yancopoulos *et al.* [22], unifies all the classical operations. In the past, the rearrangement distance for signed genomes is well studied by single operations (like reversals) [15], combinations of operations (reversals, translocations, fusions and fissions) [16] and universal operations (double-cut-and-join) [2, 22].

Unfortunately, as for unsigned genomes, almost all these problems are NP-hard. Then people resort to approximation algorithms. Christie devised a factor-1.5 approximation algorithm for sorting unsigned genomes by reversals [8], and the approximation factor was improved to 1.375 by Berman *et al.* in 2002 [4]. Cui *et al.* investigated the problem of sorting by unsigned translocations and proposed an algorithm with an approximation factor $1.5 + \varepsilon$ [9]. This bound was improved to $1.408 + \epsilon$ [17] and recently further to 1.375 [21]. The problem of Sorting by Transpositions was first studied by Bafna and Pevzner [1], who devised an 1.5-approximation algorithm which runs in quadratic time. The bound was improved to 1.375 by Elias and Hartman in 2006 [11]. As far as we know, the best polynomial-time approximation algorithms for the unsigned DCJ distance problem has a factor $1.408 + \epsilon$ [7]. Among almost all these problems, a bottleneck to break the 1.375 barrier seems to be on decomposing the breakpoint graph (to be defined formally) into maximum (edge-disjoint) alternating-cycles. We make fundamental contributions in this paper on using FPT approximation algorithms. The design of FPT algorithms for genome rearrangement problems was started very recently. With the help of weak kernels, sorting unsigned genomes by either reversals, translocations or DCJs admits small weak kernels, hence are in FPT with a running time $O^*(4^k)$, where k is the solution value [18, 19]. However, this algorithm is only practical for k bounded from above by around 20 to 25.

In this paper, we devise a new randomized algorithm for maximum alternating-cycle decomposition. Consequently, we design an FPT approximation algorithm for sorting unsigned genomes by DCJ operations (resp. by translocations), the approximation factor reaches $4/3 + \varepsilon$, and the running time is $O^*(2^{d^*})$, where d^* represents the optimal DCJ distance (resp. translocation distance). The algorithm is randomized and it succeeds with a high probability.

2   Preliminaries

We first define the basics regarding gene, chromosome and genome. An unsigned gene is a sequence of DNA, which is denoted by a positive integer. A chromosome can be viewed as a sequence of genes and denoted by a permutation, while a genome is a set of chromosomes. A gene that lies at the end of some linear chromosome is called an *ending-gene*. Gene g_i and g_j form an *adjacency* if they are consecutive in some chromosome. An adjacency (g_i, g_{i+1}) is *trivial* if it satisfies $|g_{i+1} - g_i| = 1$. A chromosome is *trivial* if every adjacency is trivial. A genome is *trivial* if all its chromosomes are trivial.

In the context of sorting genomes, the comparative order of the genes in the same chromosome does matter, but not the order of chromosomes and the direction of a whole chromosome, which implies that each chromosome can be viewed in both directions. In the case of signed genomes, a chromosome $\langle g_i, g_{i+1}, \cdots, g_j \rangle$ is equivalent to $\langle -g_j, \cdots, -g_{i+1}, -g_i \rangle$; and in the case of unsigned genomes, a chromosome $\langle g_i, g_{i+1}, \cdots, g_j \rangle$ is equivalent to $\langle g_j, \cdots, g_{i+1}, g_i \rangle$.

2.1 Breakpoint Graph of Signed and Unsigned Genomes

Now, we recall the well-known tool for computing the genomic rearrangement distance, the *Breakpoint Graph*. Given signed genomes X and Y over the same set of genes (WLOG, assume that Y is trivial), the breakpoint graph $G_s(X, Y)$ can be obtained as follows: for each chromosome $S = [x_1, x_2, \ldots, x_{n_i}]$ of X, replace each x_i with an ordered pair $(l(x_i), r(x_i))$ of vertices. If x_i is positive, then $(l(x_i), r(x_i)) = (x_i^t, x_i^h)$; and if x_i is negative, then $(l(x_i), r(x_i)) = (x_i^h, x_i^t)$. If the genes x_i and x_{i+1} are adjacent in X, then we connect $r(x_i)$ and $l(x_{i+1})$ by a black edge in $G_s(X, Y)$. If the genes x_i and x_{i+1} are adjacent in Y, then we connect $r(x_i)$ and $l(x_{i+1})$ by a gray edge in $G_s(X, Y)$. Every vertex (except the ones at the two ends of a chromosome) in $G_s(X, Y)$ is incident to one black and one gray edge. Therefore, $G_s(X, Y)$ can be uniquely decomposed into cycles, on which the black edges and gray edges appear consecutively. A cycle containing exactly i black (gray) edges is called an *i-cycle*.

As for unsigned genomes, the breakpoint graph is a bit different. Given two unsigned genomes X and Y on the same set of n genes, the *Breakpoint Graph* $G_u(X, Y) = (V, E_b \cup E_g)$, where $|V| = n$ and each vertex in V corresponds to a gene, every adjacency in X forms a black edge belonging to E_b and every adjacency in Y forms a gray edge belonging to E_g. The breakpoint graph $G_u(X, Y)$ can be decomposed into a set of edge-disjoint cycles, denoted as **D**, and on each cycle, the black edges and gray edges appear alternatively.

2.2 The Signed DCJ Distance Formula

Let b (resp. c) be the number of black edges (resp. cycles) in $G_s(X, Y)$. Yancopoulos *et al.* proved the following theorem [22].

Theorem 1. *Let $d_s(X, Y)$ be the (optimal) signed DCJ distance between X and Y. Then $d_s(X, Y) = b - c$.*

2.3 The UDCJ Problem

The Double-Cut-and-Join Operations. The Double-Cut-and-Join operation (abbreviated as DCJ) unifies all the traditional genome rearrangement operations such as reversal, translocation, fusion, fission, transposition and block-interchange, as well as excision, integration, circularization and linearization. The formal definition of the DCJ operation on the breakpoint graph (for both signed and unsigned genomes) is as follows.

Definition 1. *The Double-Cut-and-Join operation acts on the Breakpoint Graph in the following four ways (Fig. 1):*

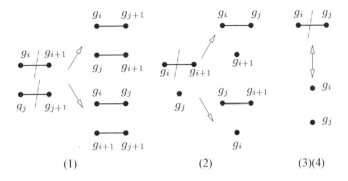

Fig. 1. The DCJ Operation.

1. *For two black edges $b_1 = (g_i, g_{i+1})$ and $b_2 = (g_j, g_{j+1})$, cut them, and either form two new black edges $b'_1 = (g_i, g_{j+1})$ and $b'_2 = (g_j, g_{i+1})$ or form two new black edges $b'_1 = (g_i, g_j)$ and $b'_2 = (g_{i+1}, g_{j+1})$.*
2. *For a black edge $b = (g_i, g_{i+1})$ and an end-gene g_j, cut the black edge, and either form a new black edge $b' = (g_i, g_j)$ and a new end-gene g_{i+1} or form a new black edge $b' = (g_j, g_{i+1})$ and a new end-gene g_i.*
3. *For two end-genes g_i and g_j, join them with a black edge (g_i, g_j).*
4. *For a black edge $b = (g_i, g_{i+1})$, cut it into two end-genes g_i and g_{i+1}.*

We now formally formulate the problem to be investigated in this paper.

Sorting Unsigned Genomes by the DCJs (UDCJ):
Input: Two unsigned linear genomes X and Y, Y is trivial, and an integer k.
Question: Can X be converted into Y by a series of k DCJs $\rho_1, \rho_2, \cdots, \rho_k$.

The minimum k is the unsigned *DCJ distance* between X and Y.

Throughout this paper, we assume that the ending-gene sets of X and Y are the same, since the details to handle genomes with different ending-gene sets is not the main purpose of this paper.

Coming back to the technical details, since each ending-gene of X and Y is incident to only one black edge and one gray edge; and each of the rest genes is incident to exactly two black edges and two gray edges, the ways to decompose $G_u(X, Y)$ into cycles might not be unique. Caprara showed that computing a maximum alternating-cycle decomposition (MAX-ACD) of the breakpoint graph is NP-hard [5], which implies that UDCJ is also NP-hard. We comment that the best polynomial-time approximation for MAX-ACD only has a factor $1.4193 + \epsilon$ [20].

2.4 Converting Unsigned Genomes into Signed Ones

A natural way to solve UDCJ is to convert the unsigned genome into a signed one, then resort to the algorithm for computing the signed DCJ distance. But, how to convert an unsigned genome into a 'good' signed genome, which would result in a smaller DCJ distance? Once we have a cycle-decomposition **D** of

$G_u(X,Y)$, we can obtain two signed genomes $\bar{X}$ and $\bar{Y}$ by assigning a sign to each gene in X and Y such that $G_s(\bar{X},\bar{Y}) = \mathbf{D}$.

As Y is trivial, we arrange all its chromosomes monotonously increasing, then assign all its genes positive. Therefore, all gray edges in $G_s(\bar{X},\bar{Y})$ have the form $((x_i)^h, (x_i+1)^t)$.

Next, we show how to assign a proper sign to each gene in X (to obtain $\bar{X}$). An ending-gene is positive if it lies at the same (i.e., both left or both right) ends of some chromosome in X and some chromosome in Y; otherwise, it is negative in X. For a non-ending gene x_i, according to the two gray edges, $((x_i)^h, (x_i+1)^t)$ and $((x_i-1)^h, (x_i)^t)$ in the cycle decomposition, we assign x_i positive if $((x_i-1)^h, l(x_i))$ is a gray edge in the given cycle decomposition; if $((x_i-1)^h, r(x_i))$ is a gray edge in the given cycle decomposition, then x_i is assigned a negative sign. See Fig. 2 for an example.

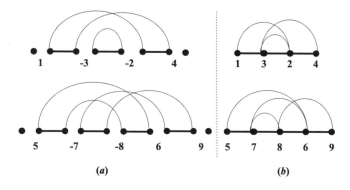

Fig. 2. (a) The breakpoint graph $G_s(\bar{X},\bar{Y})$ for the signed case, where $\bar{X}$={[1, −3, −2, 4], [5, −7, −8, 6, 9]} and $\bar{Y}$={[1,2,3,4], [5,6,7,8,9]}. (b) The breakpoint graph $G(X,Y)$ for the unsigned case, where X={[1,3,2,4], [5,7,8,6,9]} and Y={[1,2,3,4], [5,6,7,8,9]}. (a) is a cycle decomposition of (b).

Finally, a fixed-parameter tractable (FPT) algorithm for a decision problem Π with solution value k is an algorithm which solves the problem in $O(f(k)n^{O(1)}) = O^*(f(k))$ time, where f is any function only on k, n is the input size. FPT also stands for the set of problems which admit such an algorithm [10,12].

In summary, to solve UDCJ efficiently, we need to find a proper cycle decomposition of the breakpoint graph. We handle this NP-hard problem by designing an FPT approximation algorithm. This is the main content in the next section.

3   An FPT-time Approximation Algorithm

In this section, we present a factor-$(4/3 + \epsilon)$ FPT-approximation algorithm for UDCJ, which runs in $O^*(2^{d^*})$ time, where d^* is the optimal DCJ distance. We try to decompose the breakpoint graph into enough number of small cycles (i.e., cycles containing 1, 2, and 3 black edges, formally called *1-cycles, 2-cycles* and *3-cycles* henceforth), with a new randomized method. The algorithm is randomized and it succeeds with a high probability.

3.1   General Sketch

Jiang *et al.* [19] showed that all the possible 1-cycles could be kept in the cycle decomposition by the following lemma.

Lemma 1. *There exists some optimal cycle decomposition containing all the existing 1-cycles.*

Hence, our cycle-decomposition will keep all the 1-cycles, and is always compared with an optimal cycle-decomposition which also keeps all the 1-cycles.

Let c_i^* be the number of i-cycles in some optimal cycle-decomposition, from the signed DCJ distance formula $d_s(X, Y) = b - \sum_{i \geq 1} c_i^*$. Since $b = \sum_{i \geq 1} i * c_i^*$, we have $\sum_{i \geq 3} c_i^* \leq \frac{1}{3}(b - c_1^* - 2c_2^*)$, that results in, $d_s(X, Y) \geq b - c_1^* - c_2^* - \frac{1}{3}(b - c_1^* - 2c_2^*) = \frac{2}{3}(b - c_1^*) - \frac{1}{3}c_2^* = \frac{2}{3}(b - c_1^* - \frac{1}{2}c_2^*)$, which implies that, to achieve an approximation factor of 1.5, it is sufficient to find a half number of 2-cycles of the optimal cycle-decomposition.

Moreover, if we can find an α portion of 2-cycles and a β portion of 2-cycles and 3-cycles, then use the bound, $d_s(X, Y) = b - \sum_{i \geq 1} c_i^* \geq b - c_1^* - c_2^* - c_3^* - \frac{1}{4}(b - c_1^* - 2c_2^* - 3c_3^*)$, together with the conditions $c_2 \geq \alpha * c_2^*$, $c_2 + c_3 \geq \beta * (c_2^* + c_3^*)$, and $d_{alg}(X, Y) \leq b - c_1^* - c_2 - c_3$, the approximation factor could become $max\{\frac{4}{3}, 2 - \alpha, \frac{3-\beta}{2}, \frac{3\alpha - \beta - \alpha\beta}{2\alpha - \beta}\}$ [6].

Our idea is to find an α portion of 2-cycles, and a β portion of 2-cycles as well as a γ portion of 3-cycles. We will show that, ignoring some small constant ϵ, $\alpha \geq \frac{5}{6}$, $\beta \geq \frac{3}{5}$, and $\gamma \geq \frac{3}{5} * \frac{57}{64}$, which leads to an approximation 4/3.

Now we give the details of our algorithm. By cycle decomposition, we aim at finding small cycles, but searching cycles directly from the breakpoint graph will not guarantee enough number, that is because cycles in the breakpoint graph could possibly share some edges, and it is necessary to compute an independent set from them. Our main idea is to fix the sign of some genes, then find cycles from the partly decomposed breakpoint graph, so that we can obtain more cycles.

3.2   Finding 2-Cycles

Let V_2 be the set of vertices, each of which is involved in at least two 2-cycles in the breakpoint graph. (Following [6], the intersection graph of the 2-cycles has a maximum degree of 6. By the following random selection procedure, together with the enumeration of the signs of the selected vertices, we show that the intersection graph of these 'partial' 2-cycles has a maximum degree of 3.) We randomly choose $|V_2|/2$ vertices and enumerate all possible combinations of signs for them. Under each combination of sign assignment, the breakpoint graph could be partly decomposed. A *candidate* 2-cycle is a 2-cycle which could exist subject to the current sign assignment, and at least one vertex has a sign fixed. Nonetheless, in the partly decomposed breakpoint graph, some of the candidate 2-cycles could share edges and could not co-exist in any cycle decomposition. Let C_{c2} be the set of candidate 2-cycles. Construct a conflict graph $G_{c2} = (C_{c2}, E_{c2})$, where each 2-cycle of C_{c2} corresponds to a vertex, and there is an edge between two vertices if and only if their corresponding 2-cycles share edges in the partly decomposed breakpoint graph. Thus, an independent set of G_{c2} represents the conflict-free 2-cycles we find.

Algorithm 1. Finding 2-cycles

1: Identify vertices of V_2.
2: Choose $|V_2|/2$ vertices randomly.
3: **for** each combination of sign assignment for these chosen vertices **do**
4:     Construct the conflict graph G_{c2}.
5:     Compute an approximate independent set I_{c2} of G_{c2}.
6: **end for**
7: Keep the 2-cycles corresponding to the maximum I_{c2}.

Lemma 2. *There exists an approximation algorithm with ratio $\frac{5}{r+3} - \varepsilon$, for any $\varepsilon > 0$, for the maximum independent set problem on a graph with maximum degree r (see reference [3]).*

3.3   Finding 2,3-Cycles

Let a 2,3-cycle be either a 2-cycle or a 3-cycle in the breakpoint graph. Let V_{23} be the set of vertices, each of which is involved in at least two 2,3-cycles in the breakpoint graph. We randomly choose $|V_{23}|/2$ vertices, enumerate all possible combinations of signs for them. Under each combination of sign assignment, the breakpoint graph has been partly decomposed. A *candidate* 2,3-cycle is a 2-cycle or a 3-cycle which could exist under the current sign assignment, and at least two vertices have their signs fixed. In the partly decomposed breakpoint graph, some of the 2-cycles and candidate 3-cycles could share edges and could not co-exist in any cycle decomposition. Each 2-cycle and candidate 3-cycles is composed of paths, where each path is composed of black edges, or gray edges, or composed of black edges and gray edges appearing alternatively. A candidate 2,3-cycles is composed of at most four such paths. We view paths as elements, and candidate 2,3-cycles as sets of elements. We can construct a set packing system S_{c23}, whose basic elements are the paths and each candidate 2,3-cycles is a subset of at most four elements. Thus, a set packing could be the 2,3-cycles we find.

Lemma 3. *There is a ratio $\frac{3}{p+1} - \varepsilon$ approximation algorithm, for any $\varepsilon > 0$, for the maximum set packing problem with set size at most p and set degree bounded (see reference [13]).*

Algorithm 2. Finding 2,3-cycles

1: Identify vertices of V_{23}.
2: Choose $|V_{23}|/2$ vertices randomly.
3: **for** each combination of sign assignment for these chosen vertices **do**
4:    Construct the set packing system S_{c23}.
5:    Compute an approximate set packing P_{c23}.
6: **end for**
7: Keep the 2,3-cycles corresponding to the maximum P_{c23}.

Algorithm 3. UDCJ-Final

1: Run Algorithm 1 n times, among the computed I_{c2}'s, pick the largest one $I_{c2}^{\max}$.
2: Run Algorithm 2 n times, among the computed P_{c23}'s, pick the largest one $P_{c23}^{\max}$.
3: If $|I_{c2}^{\max}| \geq |P_{c23}^{\max}|$, then keep the 2-cycles corresponding to $I_{c2}^{\max}$,
4: Otherwise, keep the 2,3-cycles corresponding to $P_{c23}^{\max}$.
5: Arbitrarily assign signs to the rest of genes so that every gene has a sign.
6: Compute the DCJ distance between the signed genomes.
7: Simulate the signed DCJ sorting process to the original unsigned genomes.

---

4   Performance Analysis

4.1   The Approximation Factor

As aforementioned, the approximation factor performance is determined by the number of 2-cycles and 3-cycles we have found.

Lemma 4. G_{c2} *is a graph with maximum degree 3.*

Proof. Omitted due to space constraint.   □

Corollary 1. I_{c2} *is an* $\frac{5}{6} - \varepsilon$ *approximation for the maximum independent set of* G_{c2}.

Lemma 5. S_{c23} *is a set packing system with set size at most 4 and set degree bounded.*

Proof. Each 3-cycle has exactly 6 vertices, since at least two of them have a fixed sign, each fixed-sign vertex brings a fixed connection of a black edge and a gray edge. Then each 3-cycle has at most four undetermined connections, i.e., it is composed of at most four paths.

Now we show that each path can be shared by at most 4 such 2,3-cycles. Note that if three paths of such a 3-cycle are fixed, then the 2,3-cycle is obtained. Standing on an ending vertex of a path, we have at most two choices. After any choice, the path is extended, i.e., two paths are connected together. To form 2,3-cycles, we have twice opportunities to make choices, which would result in at most four 2,3-cycles. Thus, each subset could share elements with at most $4 \times (4 - 1) = 12$ other subsets. The set degree is bounded.   □

Corollary 2. P_{c23} *is an* $\frac{3}{5} - \varepsilon$ *approximation for the maximum set packing of* S_{c23}.

Next, we bound the number of vertices in G_{c2} and the number of subsets in S_{c23}. Let c_i^* be the number of i-cycles in the optimal cycle decomposition.

Lemma 6. *With probability* $1 - \frac{1}{e^{O(n)}}$, *the maximum independent set of one* G_{c2} *at Step 1 in Algorithm 3 has a size greater than* $(1 - \delta)\frac{15}{16}c_2^*$ *for any* $0 < \delta < 1$.

Proof. We show that each 2-cycle of the optimal cycle decomposition has a probability of $\frac{15}{16}$ to fall into G_{c2}. If the 2-cycle has a vertex with a fixed sign, then it will surely become a candidate 2-cycle. If all its vertices do not have a fixed sign, then they are all in V_2. The probability that none of them is chosen is $\frac{C_{|v_2|-4}^{|v_2|/2}}{C_{|v_2|}^{|v_2|/2}} \leq \frac{1}{16}$. Hence, we can conclude that, with probability $\frac{15}{16}$, we have $\frac{15}{16}$ portion of 2-cycles of the optimal cycle decomposition in G_{c2}; moreover, they form an independent set of G_{c2}. If we view X_i as a random variable to put a 2-cycle of the optimal cycle decomposition into G_{c2} and define $X = \sum_i X_i$, then $E[X] = \mu = \frac{15}{16}c_2^*$.

By Chernoff bounds, for any $0 < \delta < 1$,

$$P[X \leq (1 - \delta)\mu] \leq e^{-\frac{\delta^2 \mu}{2}},$$

which means $P[X \leq (1-\delta)\mu] \leq \frac{1}{e^{O(1)}}$ when μ is only a constant. For Algorithm 3, as we repeat Algorithm 1 n times, the probablity that the MIS of all the n G_{c2}'s has a size at most $(1 - \delta)\frac{15}{16}c_2^*$ is at most $(\frac{1}{e^{O(1)}})^n = \frac{1}{e^{O(n)}}$.

Therefore, with probability $1 - \frac{1}{e^{O(n)}}$, the MIS of one of the G_{c2} at Step 1 in Algorithm 3, has a size greater than $(1 - \delta)\frac{15}{16}c_2^*$ for any $0 < \delta < 1$. □

As δ could be arbitrarily small, we would use this size as $\frac{15}{16}c_2^*$ in the proof of Theorem 2.

Lemma 7. *With probability* $1 - \frac{1}{e^{O(n)}}$, *the maximum set packing of one of the* S_{c23} *at Step 2 in Algorithm 3 has a size greater than* $(1 - \delta)(c_2^* + \frac{57}{64}c_3^*)$ *for any* $0 < \delta < 1$.

Proof. We show that each 3-cycle of the optimal cycle decomposition has a probability of $\frac{57}{64}$ to be a candidate 3-cycle. If the 3-cycle has two vertices whose signs are fixed, then it will surely become a candidate 3-cycle. If the 3-cycle has exactly one vertex whose sign is fixed, then the other five vertices do not have a fixed sign, so they are all in V_{23}. The probability that none of them is chosen is $\frac{C_{|v_{23}|-5}^{|v_{23}|/2}}{C_{|v_{23}|}^{|v_{23}|/2}} \leq \frac{1}{32}$. If all its vertices do not have a fixed sign, then they are all in V_{23}. The probability that none of them is chosen is $\frac{C_{|v_{23}|-6}^{|v_{23}|/2}}{C_{|v_{23}|}^{|v_{23}|/2}} \leq \frac{1}{64}$. The probability that exact one of them is chosen is $6 \times \frac{C_{|v_{23}|-6}^{|v_{23}|/2}}{C_{|v_{23}|}^{|v_{23}|/2}} \leq \frac{6}{64}$. Hence, we can conclude that an expected $1 - \frac{1}{64} - \frac{6}{64} = \frac{57}{64}$ portion of 3-cycles of the optimal cycle decomposition are in S_{c23}; moreover, together with the c_2^* 2-cycles of the optimal cycle decomposition, they form a set packing of S_{c23}. Note that a 2-cycle has only 4 edges, by Lemma 5, all of the c_2^* 2-cycles will be in the set packing solution.

By an argument similar to Lemma 6, we conclude that, with probability $1 - \frac{1}{e^{O(n)}}$, the maximum set packing one of the S_{c23}'s at Step 2 of Algorithm 3, has size greater than $(1 - \delta)(c_2^* + \frac{57}{64}c_3^*)$ for any $0 < \delta < 1$. $\square$

Again, as δ could be arbitrarily small, we would use this size as $c_2^* + \frac{57}{64}c_3^*$ in the proof of Theorem 2.

Theorem 2. *With probability* $1 - \frac{1}{e^{O(n)}}$, *Algorithm 3 approximates the DCJ distance within a factor* $4/3+\varepsilon$.

Proof. Let c_i be the number of i-cycles computed by Algorithm 3, c_i^* be the number of i-cycles in the optimal cycle decomposition. Let d^* and d be the optimal DCJ distance and approximated DCJ distance respectively. Let the approximation factor be ρ, where $\frac{4}{3} \leq \rho \leq \frac{3}{2}$.

We have, $d^* = b - c_1^* - c_2^* - c_3^* - \sum_{i \geq 4} c_i^*$ Since, $\sum_{i \geq 4} c_i^* \leq (b - c_1^* - 2c_2^* - 3c_3^*)/4$, and $b - c_1^* \geq 2c_2^* + 3c_3^*$, then,

$$
\begin{aligned}
d^* &\geq b - c_1^* - c_2^* - c_3^* - (b - c_1^* - 2c_2^* - 3c_3^*)/4 \\
&= \frac{3}{4}(b - c_1^*) - \frac{1}{2}c_2^* - \frac{1}{4}c_3^* \\
&= \frac{1}{\rho}((b - c_1^*) + (\frac{3}{4}\rho - 1)(b - c_1^*) - \frac{1}{2}\rho c_2^* - \frac{1}{4}\rho c_3^*) \\
&\geq \frac{1}{\rho}((b - c_1^*) + (\frac{3}{4}\rho - 1)(2c_2^* + 3c_3^*) - \frac{1}{2}\rho c_2^* - \frac{1}{4}\rho c_3^*) \\
&= \frac{1}{\rho}(b - c_1^* - (2 - \rho)c_2^* - (3 - 2\rho)c_3^*)
\end{aligned}
\tag{1}
$$

This implies that, if the number of cycles we found satisfy $c_2 \geq (2 - \rho)c_2^*$ and $c_3 \geq (3 - 2\rho)c_3^*$, then the approximated DCJ distance computed by our algorithm satisfies $d \leq b - c_1^* - c_2 - c_3$; consequently, the approximation factor reaches to ρ.

From Corollary 1 and Lemma 6, with high probability, we have

$$
|I_{c2}^{\max}| \geq (\frac{5}{6} - \varepsilon) \times \frac{15}{16}c_2^*.
\tag{2}
$$

From Corollary 2 and Lemma 7, with high probability, we have

$$
|P_{c23}^{\max}| \geq (\frac{3}{5} - \varepsilon') \times (c_2^* + \frac{57}{64}c_3^*).
\tag{3}
$$

Now we show that, by a balanced analysis, at least one of $|I_{c2}^{\max}|$ and $|P_{c23}^{\max}|$ are greater than $(2 - \rho)c_2^* + (3 - 2\rho)c_3^*$.

We have two cases: either (I) $(\frac{5}{6} - \varepsilon) \times \frac{15}{16}c_2^* \geq (\frac{3}{5} - \varepsilon') \times (c_2^* + \frac{57}{64}c_3^*)$ or (II) not.

In case (I): $(\frac{5}{6} - \varepsilon) \times \frac{15}{16}c_2^* \geq (\frac{3}{5} - \varepsilon') \times (c_2^* + \frac{57}{64}c_3^*)$. Since ε and ε' are very small comparing with the other constants, for the sake of computation simplification, we ignore them here. Then, $(\frac{25}{32} - \frac{3}{5})c_2^* \geq \frac{3}{5} \times \frac{57}{64}c_3^*$; that is, $c_2^* \geq \frac{171}{58}c_3^*$. Therefore, if $\frac{25}{32}c_2^* \geq (2 - \rho)c_2^* + (3 - 2\rho)c_3^*$, we are done. Solving this inequality,

$$(2 - \rho)c_2^* + (3 - 2\rho)c_3^* \leq (2 - \rho)c_2^* + (3 - 2\rho) \times \frac{58}{171}c_2^* \leq \frac{25}{32}c_2^*,$$

it holds provided that $\rho \geq 1.332$. Together with the constraint that $\rho \geq \frac{4}{3}$, we choose $\rho = \frac{4}{3}$.

In case (II): $(\frac{5}{6} - \varepsilon) \times \frac{15}{16}c_2^* < (\frac{3}{5} - \varepsilon') \times (c_2^* + \frac{57}{64}c_3^*)$. Similarly, we have $c_2^* < \frac{171}{58}c_3^*$. If $\frac{3}{5} \times (c_2^* + \frac{57}{64}c_3^*) \geq (2 - \rho)c_2^* + (3 - 2\rho)c_3^*$, then we are done. Hence we need to prove,

$$(\frac{3}{5} \times \frac{57}{64} - 3 + 2\rho)c_3^* \geq (2 - \frac{3}{5} - \rho)c_2^*.$$

Again, this inequality holds when $\rho \geq 1.332$. Together with the constraint that $\rho \geq \frac{4}{3}$, we have $\rho = \frac{4}{3}$. $\square$

4.2   The Time Complexity

Theorem 3. *The time complexity of Algorithm 3 is $O^*(2^{d^*})$, where d^* is the optimal DCJ distance.*

Proof. The most time-consuming parts are enumerating all possible sign combination of V_2 and V_{23}. $d^* \geq b - c_1^* - c_2^* - c_3^*$ and $c_2^* + c_3^* \leq (b - c_1^*)/2$, then, $d^* \geq (b - c_1^*)/2$. Each vertex of V_2 or V_{23} is connected to two black edges, while each black edge has at most two unsigned genes as its endpoints, which means that the number of unsigned genes of V_2 or V_{23} is smaller than that of black edges, e.g., $|V_2| \leq b - c_1^* \leq 2d^*$ and $|V_{23}| \leq b - c_1^* \leq 2d^*$.

Hence while we choose a half of $|V_2|$ vertices or a half of $|V_{23}|$ vertices, and enumerating all their combination of signs, the time complexity is at most $O^*(2^{d^*})$. $\square$

In fact, we could extend our method to Sorting by Translocations [9,14,17,21]. We summarize the result as follows. The details will be given in the full version.

Theorem 4. *With probability $1 - \frac{1}{e^{O(n)}}$, there is an FPT algorithm which approximates the translocation distance within a factor $\alpha = 4/3 + \varepsilon$.*

5   Concluding Remarks

We design a factor $4/3 + \epsilon$ FPT-approximation algorithm for the DCJ distance, improving the previous (polynomial-time approximation) factor of $1.408 + \epsilon$. The algorithm is randomized and it succeeds with a high probability. The running time is bounded by $O^*(2^k)$; in fact, by $O^*(2^x)$ where $x \leq k$, which makes it practical for k at least as large as 40–50. The exact FPT algorithm for the same problem takes $O^*(4^k)$ time, which is only practical for k bounded by 20–25 from above. Our algorithm involves a new randomized method to decompose the breakpoint graph into the maximum number of alternating-cycles and can be used to improve the approximation factor for Sorting by Translocations — again in a similar FPT time, which admits a factor-1.375 (polynomial-time) approximation and uses maximum alternating-cycle decomposition as a subroutine.

For Sorting by Reversals, note that special care must be taken as in the optimal solution the 1-cycles might not be all kept. For instance, for the sequence $S = \langle 3, 4, 1, 2 \rangle$, if we keep the 1-cycles (3,4) and (1,2) then three reversals are needed to sort S into $\langle 1, 2, 3, 4 \rangle$. On the other hand we could sort $S = \langle 3, 4, 1, 2 \rangle$ into $\langle 1, 2, 3, 4 \rangle$ using two reversals: $\langle 3, \underline{4, 1, 2} \rangle \rightarrow \langle \underline{3, 2, 1}, 4 \rangle \rightarrow \langle 1, 2, 3, 4 \rangle$.

A related open question is whether one can design an FPT-approximation algorithm with a factor better than 1.375 for the problem of Sorting by Transpositions. Note that the technique of giving signs to some genes does not seem to work for this problem.

Acknowledgments. This research is partially supported by NSF of China under project 61472222 and 61628207.

References

1. Bafna, V., Pevzner, P.A.: Sorting by transpositions. SIAM J. Discret. Math. **11**(2), 224–240 (1998)
2. Bergeron, A., Mixtacki, J., Stoye, J.: A unifying view of genome rearrangements. In: Bücher, P., Moret, B.M.E. (eds.) WABI 2006. LNCS, vol. 4175, pp. 163–173. Springer, Heidelberg (2006). https://doi.org/10.1007/11851561_16
3. Berman, P., Fürer, M.: Approximating maximum independent set in bounded degree graphs. In: Proceedings of the 5th Annual ACM-SIAM Symposium on Discrete Algorithms (SODA 1994), pp. 365–371 (1994)
4. Berman, P., Hannenhalli, S., Karpinski, M.: 1.375-approximation algorithm for sorting by reversals. In: Möhring, R., Raman, R. (eds.) ESA 2002. LNCS, vol. 2461, pp. 200–210. Springer, Heidelberg (2002). https://doi.org/10.1007/3-540-45749-6_21
5. Caprara, A.: Sorting permutations by reversals and Eulerian cycle decompositions. SIAM J. Discret. Math. **12**(1), 91–110 (1999)
6. Caprara, A., Rizzi, R.: Improved approximation for breakpoint graph decomposition and sorting by reversals. J. Comb. Optim. **6**(2), 157–182 (2002)
7. Chen, X., Sun, R., Yu, J.: Approximating the double-cut-and-join distance between unsigned genomes. BMC Bioinform. **12**(S-9), S17 (2011)
8. Christie, D.A.: A 3/2-approximation algorithm for sorting by reversals. In: Proceedings of the 9th Annual ACM-SIAM Symposium on Discrete Algorithms (SODA 1998), pp. 244–252 (1998)

9. Cui, Y., Wang, L., Zhu, D., Liu, X.: A $(1.5 + \epsilon)$-approximation algorithm for unsigned translocation distance. IEEE/ACM Trans. Comput. Biol. Bioinform. **5**(1), 56–66 (2008)
10. Downey, R., Fellows, M.: Parameterized Complexity. Springer, Heidelberg (1999). https://doi.org/10.1007/978-1-4612-0515-9
11. Elias, I., Hartman, T.: A 1.375-approximation algorithm for sorting by transpositions. IEEE/ACM Trans. Comput. Biol. Bioinform. **3**(4), 369–379 (2006)
12. Flum, J., Grohe, M.: Parameterized Complexity Theory. Springer, Heidelberg (2006). https://doi.org/10.1007/3-540-29953-X
13. Halldórsson, M.M.: Approximating discrete collections via local improvements. In: Proceedings of the 6th Annual ACM-SIAM Symposium on Discrete Algorithms (SODA 1995), pp. 160–169 (1995)
14. Hannenhalli, S.: Polynomial-time algorithm for computing translocation distance between genomes. Discret. Appl. Math. **71**(1–3), 137–151 (1996)
15. Hannenhalli, S., Pevzner, P.A.: Transforming cabbage into turnip: polynomial algorithm for sorting signed permutations by reversals. J. ACM **46**(1), 1–27 (1999)
16. Hannenhalli, S., Pevzner, P.A.: Transforming men into mice (polynomial algorithm for genomic distance problem). In: Proceedings of the 36th Annual IEEE Symposium on Foundations of Computer Science (FOCS 1995), pp. 581–589 (1995)
17. Jiang, H., Wang, L., Zhu, B., Zhu, D.: A factor-$(1.408 + \epsilon)$ approximation for sorting unsigned genomes by reciprocal translocations. Theor. Comput. Sci. **607**, 166–180 (2015)
18. Jiang, H., Zhang, C., Zhu, B.: Weak Kernels. ECCC Report, TR10-005, September 2010
19. Jiang, H., Zhu, B., Zhu, D.: Algorithms for sorting unsigned linear genomes by the DCJ operations. Bioinformatics **27**(3), 311–316 (2011)
20. Lin, G., Jiang, T.: A further improved approximation algorithm for breakpoint graph decomposition. J. Comb. Optim. **8**(2), 183–194 (2004)
21. Pu, L., Zhu, D., Jiang, H.: A new approximation algorithm for unsigned translocation sorting. In: Frith, M., Storm Pedersen, C.N. (eds.) WABI 2016. LNCS, vol. 9838, pp. 269–280. Springer, Cham (2016). https://doi.org/10.1007/978-3-319-43681-4_22
22. Yancopoulos, S., Attie, O., Friedberg, R.: Efficient sorting of genomic permutations by translocation, inversion and block interchange. Bioinformatics **21**, 3340–3346 (2005)

Contextual Dependent Click Bandit Algorithm for Web Recommendation

Weiwen Liu[1], Shuai Li[1], and Shengyu Zhang[1,2]([✉])

[1] The Chinese University of Hong Kong, Sha Tin, Hong Kong
[2] Tencent, Shenzhen, China
{wwliu,shuaili,syzhang}@cse.cuhk.edu.hk

Abstract. In recommendation systems, it has been an increasing emphasis on recommending potentially novel and interesting items in addition to currently confirmed attractive ones. In this paper, we propose a contextual bandit algorithm for web page recommendation in the dependent click model (DCM), which takes user and web page features into consideration and automatically balances between exploration and exploitation. In addition, unlike many previous contextual bandit algorithms which assume that the click through rate is a linear function of features, we enhance the representability by adopting the generalized linear models, which include both linear and logistic regressions and have exhibited stronger performance in many binary-reward applications. We prove an upper bound of $\tilde{O}(d\sqrt{n})$ on the regret of the proposed algorithm. Experiments are conducted on both synthetic and real-world data, and the results demonstrate significant advantages of our algorithm.

1 Introduction

Given a search query, a web page recommendation algorithm recommends a list of related web pages based on a certain model of past user behavior and page information [1]. An online learning algorithm for personalized recommender systems aims at learning user preferences and incorporating the user feedback at each time step, while maintaining a high Click-Through Rate (CTR) over a long period of time. Earlier recommendation algorithms mostly focus on recommending the currently confirmed attractive items, and put less emphasis on the potentially valuable items in the future, e.g., the Logistic Regression (LR) [2] and the Factorizations Machines (FM) [3]. It was observed that such algorithms usually lead to suboptimal recommendations in a long term [4]. Besides, though accuracy is a typical target for recommendation, the diversity and the long-term user satisfactory of a recommender system have shown more and more importance [1]. Therefore, special attention should be paid to a balance between exploiting immediate yet suboptimal rewards (exploitation) and exploring uncertain but potentially interesting items which may produce large benefits later (exploration).

Multi-armed bandit (MAB) is a general framework of sequential decision problems, in which a balance between exploration and exploitation is needed [5].

© Springer International Publishing AG, part of Springer Nature 2018
L. Wang and D. Zhu (Eds.): COCOON 2018, LNCS 10976, pp. 39–50, 2018.
https://doi.org/10.1007/978-3-319-94776-1_4

In the basic stochastic setting, we have a number of arms each with an unknown reward distribution. At each time step, we need to select one of them, receiving the reward randomly drawn from the corresponding distribution. The goal is to maximize the total reward over the time, or equivalently, to minimize the regret, which is the difference between our cumulative reward and the reward of always pulling the best arm. Numerous algorithms have been proposed for MAB and they have been successfully applied in many scenarios, such as personalized recommendation [6], clinical trials [7], etc.

The Cascade Model (CM) is a widely used click model in which the recommended web pages are listed in a sequence and the user examines the list from top to bottom until she finds a satisfactory one [8]. This model is particularly suitable for characterizing the user browsing behavior on mobile devices. A number of bandit algorithms were developed and have exhibited prominent effectiveness in cascade model [9–11]. One limit of the model is its assumption that the user clicks at most one of the recommended items, and a natural extension to allowing multiple clicks is the *dependent click model* (DCM), where the user may click more than one items before finding a satisfactory one [12].

In the DCM bandit setting, at each time step t, the learning agent displays an ordered list of K items out of L ground items to the user. The user examines the items in the displayed order and clicks on the *attracted* items. After an item is clicked, the user may either be satisfied and leave, or unsatisfied and proceed to the next item. The user leaves if all K items have been examined, regardless of whether the user has found any satisfactory item or not. If the user leaves with satisfaction, then the learning agent receives a reward of 1; otherwise the reward is 0. However, this reward is not observed by the learning agent, as the agent cannot distinguish between the user leaving with satisfaction or leaving because she has exhausted all items. All the feedback the learning agent receives is the clicking pattern such as 0100110000, in which case the learning agent knows that the user is attracted by the 2nd, 5th and 6th items, but not by the 1st, 3rd and 4th items. However, whether the user is attracted by the rest (the 7th and beyond) remains unknown to the learning agent.

In many modern personalized news/apps/ads recommendation systems, certain features of users are available through registration or historical behaviors, which can be exploited to provide more accurate recommendations [1]. In the bandit setting, these features are usually called the *context*, modeled as a d-dimensional vector that contains information of users or items. In previous studies on contextual bandit in the cascade model, the attraction weight is assumed to be the inner product of the vector of the contextual vector and a fixed but unknown vector θ [6,11,13], i.e. a linear function of the contextual vector (thereby the name *linear bandit*). However, the reward function in real-world applications can be complicated and hardly confined to being linear. With an increasing amount of historical data, stronger models may be preferred for better representability. Besides, logistic regression (LR) has exhibited empirical improvements over the linear model in news recommendation [14]. In this paper, we go beyond the linear reward model and consider the more general *exponential family distributions*, which include LR as a special case.

Our work has four main contributions. First, we incorporate contextual information into DCM bandit model, and strengthen the linear model by including exponential family distributions. Second, we present a computationally efficient version of our algorithm which may be valuable for practical use. Third, we prove an upper bound of $\tilde{O}(d\sqrt{n})$ on the regret. Fourth, experiments are conducted on both synthetic and real world data, which demonstrate the substantial advantage of our algorithm compared to the typical LR algorithm and the one without utilizing contextual information.

2   Problem Formulation

In this paper, we consider the contextual DCM bandit problem with the generalized linear payoff for list recommendation. Let n be the total number of time steps. Suppose that we have a set $E = \{1, \ldots, L\} = [L]$ of ground items. At each time step t, the learning agent receives a user query. Combining the user query and each arm i gives a contextual vector $x_{i,t} \in \mathbb{R}$ known to the learning agent, whose action is to recommend an ordered list $\mathbf{A}_t = (\mathbf{a}_1^t, \ldots, \mathbf{a}_K^t)$ of K distinct items from E to the user.[1] We say that such an action has length K, and denote by $\Pi_K(E)$ the feasible action set of all ordered lists of K distinct items from E. The user checks the list of items one by one from top to bottom. For each item a, the user is *attracted* with probability $\bar{w}_t(a) \in [0,1]$, and we will use $\mathbf{w}_t(a) \in \{0,1\}$ to denote the *attraction weight*, a Bernoulli random variable with mean $\bar{w}_t(a)$ indicating whether the user is attracted by a or not. Denote by $\mathbf{w}_t \in \{0,1\}^E$ the random vector of these indicators, and by P_w the distribution of $\mathbf{w}_t$. We assume that the attraction vectors are independent across time steps and items, namely $\{\mathbf{w}_t\}_{t=1}^n$ are i.i.d. drawn from a probability distribution P_w.

If the user is attracted by the k-th item a_k in the recommended list, i.e. $\mathbf{w}_t(a_k) = 1$, then she clicks it and examines the item. The user may be satisfied and leave, which happens with probability $\bar{v}_t(k)$ and then the learning agent receives a reward of 1. The user may also find the item unsatisfactory (which happens with probability $1 - \bar{v}_t(k)$, and then continues to check the next item. If all items have been checked and the user has not found any satisfactory item, then the user leaves and the learning agent receives reward 0. The *termination weight* $\mathbf{v}_t(k) \in \{0,1\}$ is the Bernoulli random variable with mean $\bar{v}_t(k)$. We denote by $\mathbf{v}_t \in \{0,1\}^K$ the random vector of the termination weights, by P_v its distribution, and assume $\{\mathbf{v}_t\}_{t=1}^n$ to be i.i.d. drawn from P_v.

The above process defines a random $\{0,1\}$ reward, but note that this reward is not revealed to the learning agent, as the user just leaves after checking some items and does not report whether she finds the item she wants. Indeed, the search engine does not even know when the user leaves. All the feedback that the search engine receives is a sequence of k click indicators $(\mathbf{w}_1', \ldots, \mathbf{w}_K')$. Note that $\mathbf{w}_i'$ may not be the same as $\mathbf{w}_i$ as. For example, if the sequence is 0100110000, it

---

[1] Here and throughout the paper, we use bold letters for random variables.

may be the case that the user leaves at the sixth item with satisfaction. Another case is that the user checks all items without finding anyone satisfactory, but has to leave at the end. This feedback is too limited to admit any good learning algorithm. Therefore, we adopt the same assumption as in [15] that the order $\pi(\bar{v})$ of $\bar{v} = (\bar{v}(1), \ldots, \bar{v}(K))$ is known to the agent, where the order π is a permutation satisfying that $\bar{v}(\pi(1)) \geq \ldots \geq \bar{v}(\pi(K))$. This assumption is practically reasonable as in many cases, though we may not have a precise estimation of each value $\bar{v}(k)$, we do know their relative comparison. (For instance, for typical search engines it may well be the case that π is identity, namely $\bar{v}_t(1) \geq \ldots \geq \bar{v}_t(K)$.) Under this assumption, it can be easily shown that the expected reward is maximized when the items are listed in the decreasing order of their attractiveness.

To give a more formal treatment of the award, consider the reward function $f : \Pi_k(E) \times [0,1]^E \times [0,1]^K \to [0,1]$ defined by

$$f(A, v, w) = 1 - \prod_{k=1}^{K} (1 - v(k)w(a_k)), \tag{1}$$

where $A = (a_1, \ldots, a_K)$. In this notation, the reward in time step t is $\mathbf{r}_t = f(\mathbf{A}_t, \mathbf{v}_t, \mathbf{w}_t)$. Due to the assumed independence of all $\{\mathbf{v}_t\}$ and $\{\mathbf{w}_t\}$, it is easily seen that for any fixed action A, the expected reward is $f(A, \bar{v}_t, \bar{w}_t)$.

The performance of the learning agent is evaluated by the pseudo-regret, the difference of cumulative reward of the optimal actions and that of the actions of the agent:

$$\mathcal{R}(n) = \mathbb{E}\Big[\sum_{t=1}^{n} \big(f(A_t^*, \bar{v}_t, \bar{w}_t) - f(\mathbf{A}_t, \bar{v}_t, \bar{w}_t) \big) \Big], \tag{2}$$

where

$$A_t^* = \operatorname{argmax}_{A \in \Pi_K(E)} f(A, \bar{v}_t, \bar{w}_t)$$

is the optimal list that maximizes the expected reward in step t.

We adopt the standard assumption that in contextual bandits that all contextual vectors $x_{t,a} \in \mathbb{R}^d$ are assumed to have bounded norm $\|x_{t,a}\|_2 \leq 1$. Besides, we assume that the attraction weight $\mathbf{w}_t(a)$ satisfies the *generalized linear model* (GLM), a flexible extension of the ordinary linear model that previous cascading bandit studies assumed. More precisely, assume that

$$\bar{w}_t(a) = \mathbb{E}[\mathbf{w}_t(a)|\mathcal{H}_t] = \mu(\theta_*^\top x_{t,a}), \tag{3}$$

where $\{\mathcal{H}_t\}_{t=1}^n$ represents the history containing clicks and features up to time t, and θ_* is a fixed but unknown vector $\theta_* \in \mathbb{R}^d$. The *inverse link function* μ is chosen such that $0 \leq \mu(\theta_*^\top x_{t,a}) \leq 1$ for any a and t. This GLM admits a wider range of nonlinear distributions such as Gaussian, binomial, Poisson, gamma distributions, etc. In particular, when the feedback is binary or count variables, the logistic or Poisson regression can be used. Especially in the present DCM setting, the logistic regression fits the web page recommendation better than the linear model [14].

3   Algorithm and Results

3.1   Algorithm

To maximize user satisfaction, two sets of parameters, $\bar{w}_t$ and $\bar{v}_t$ need to be estimated. We assume that the order of the expected termination weight is known to the agent, which in practice can be easily estimated using historical click data. The problem then reduces to the estimation of the mean and variance of the expected attraction weight. Due to the limited feedback, it is unclear whether the user is attracted by the item of the last click position, which is denoted by $\mathcal{C}_t \in \{0, 1, \ldots, K\}$, where $\mathcal{C}_t = 0$ means no item has been clicked. The algorithm therefore simply uses the feedback before $\mathcal{C}_t$ for updates. As introduced before, the random variable $\mathbf{w}_t(a)$ satisfies Eq. (3) with the inverse link function μ assumed to be twice continuously differentiable and strictly increasing. We further assume that μ is a k_μ-Lipschitz function (namely, the first order derivative of μ is upper bounded by k_μ), and that $c_\mu := \inf_{\{\|x\|_2 \le 1, \|\theta - \theta^*\|_2 \le 1\}} \mu'(\theta^\top x) > 0$. For logistic regression, $\mu(x) = 1/(1 + e^{-x})$ and it is easily verified that $c_\mu = 0.1$, $k_\mu = 0.25$ suffice for the requirements. Given the historical information $\{(x_{s,a}, \mathbf{w}_s(\mathbf{a}_k^s)) : s \in [t], a \in E, k \in [\mathcal{C}_s]\}$, where $(x_s, \mathbf{w}_s) \in \mathcal{H}_s$, the estimator $\hat{\theta}_t$ can be efficiently obtained by solving the following equation:

$$\sum_{s=1}^{t} \sum_{k=1}^{\mathcal{C}_s} \left(\mathbf{w}_s(\mathbf{a}_k^s) - \mu(\theta^\top x_{s,\mathbf{a}_k^s}) \right) x_{s,\mathbf{a}_k^s} = 0. \tag{4}$$

For logistic regression, this step can be computed by Newton method. Next, we design an upper confidence bound of the expected attraction weight. Define $\mathbf{V}_t = \lambda I + \sum_{s=1}^{t} \sum_{k=1}^{\mathcal{C}_s} x_{s,\mathbf{a}_k^s} x_{s,\mathbf{a}_k^s}^\top$, we have the following fact by Lemma 3 in [16].

Lemma 1. *For any $\delta \in [1/n, 1)$, with probability at least $1 - \delta$, for all $1 \le t \le n$, we have*

$$\|\hat{\theta}_t - \theta_*\|_{\mathbf{V}_t} \le \frac{\sigma}{c_\mu} \sqrt{\frac{d}{2} \log(1 + t/(\lambda d)) + \log(1/\delta)}. \tag{5}$$

Here the l_2-norm of x based on a positive definite matrix A is defined by $\|x\|_A = \sqrt{x^\top A x}$. Building on this, we can bound $|\mu(\hat{\theta}_t^\top x_{t,a}) - \mu(\theta_*^\top x_{t,a})|$ by first applying the definition of k_μ-Lipschitz of function μ and then using the Cauchy-Schwartz inequality.

$$|\mu(\hat{\theta}_t^\top x_{t,a}) - \mu(\theta_*^\top x_{t,a})| \le k_\mu |\hat{\theta}_t^\top x_{t,a} - \theta_*^\top x_{t,a}| \le k_\mu \|\hat{\theta}_t - \theta_*\|_{\mathbf{V}_t} \|x_{t,a}\|_{\mathbf{V}_t^{-1}}$$

$$\le \frac{k_\mu \sigma}{c_\mu} \sqrt{\frac{d}{2} \log(1 + t/(\lambda d)) + \log(1/\delta)} \|x_{t,a}\|_{\mathbf{V}_t^{-1}}$$

Let $\rho(t) = \frac{k_\mu \sigma}{c_\mu} \sqrt{\frac{d}{2} \log(1 + t/(\lambda d)) + \log(1/\delta)}$, and define the upper confidence bound of the expected attraction weight for item a at time t by

$$\mathbf{U}_t(a) = \min\{\mu(\hat{\theta}_{t-1}^\top x_{t,a}) + \rho(t-1)\|x_{t,a}\|_{\mathbf{V}_{t-1}^{-1}}, 1\}, \tag{6}$$

where the first term of $\mathbf{U}_t(a)$ is for exploitation and the second term for exploration. Choosing an item with the maximum $\mathbf{U}_t(a)$ balances the exploration and exploitation. Based on the above discussion, we propose an algorithm given in box Algorithm 1. Firstly, for each item in the ground item set, an upper confidence bound $\mathbf{U}_t \in [0,1]^E$ for the expected attraction weight is calculated. Then the agent uses any $\tilde{v}_t$ that has the same order as $\bar{v}_t$, gets a maximizer $\mathbf{A}_t = \operatorname{argmax}_{A \in \Pi_K(E)} f(A, \tilde{v}_t, \mathbf{U}_t)$, and recommends the list. After user examines the list, the agent observes the last click position $\mathcal{C}_t$, and $\mathbf{w}_t(\mathbf{a}_k^t)$, $k \in [\mathcal{C}_t]$ (Here we adopt the notation that $[0] = \emptyset$). The estimator $\hat{\theta}_t$ of θ_* is then updated based on new feedback. Finally, the related statistics are updated for the next time step.

Algorithm 1. Contextual DCM Bandits with Generalized Linear Payoff (GL-CDCM)

1: *Parameters* : $\delta = \frac{1}{\sqrt{n}}$; $\lambda \geq K$
2: *Initialization* : $\hat{\theta}_0 = 0$, $\rho(0) = 1$, $\mathbf{V}_0 = \lambda I$
3: **for** $t = 1$ to n **do**
4:     Obtain context $x_{t,a}$ for all $a \in E$
5:     $\forall a \in E$, compute
        $\mathbf{U}_t(a) = \min\{\mu(\hat{\theta}_{t-1}^\top x_{t,a}) + \rho(t-1)\|x_{t,a}\|_{\mathbf{V}_{t-1}^{-1}}, 1\}$
6:     $\mathbf{A}_t \leftarrow \operatorname{argmax}_{A \in \Pi_K(E)} f(A, \tilde{v}_t, \mathbf{U}_t)$
7:     Play $\mathbf{A}_t$ and observe $\mathcal{C}_t$, $\mathbf{w}_t(\mathbf{a}_k^t)$, $k \in [\mathcal{C}_t]$
8:     Solve $\hat{\theta}_t$ from
        $\sum_{s=1}^{t} \sum_{k=1}^{\mathcal{C}_s} (\mathbf{w}_s(\mathbf{a}_k^s) - \mu(\theta_t^\top x_{s,\mathbf{a}_k^s})) x_{s,\mathbf{a}_k^s} = 0$
9:     $\mathbf{V}_t \leftarrow \mathbf{V}_{t-1} + \sum_{k=1}^{\mathcal{C}_t} x_{t,\mathbf{a}_k^t} x_{t,\mathbf{a}_k^t}^\top$
10: **end for**

3.2   Results

The result on the upper bound on the regret for the proposed contextual DCM bandits is presented in this section. Denote $p_v = \max_{1 \leq t \leq n} \max_{i=1,\dots,K}(\bar{v}_t(i) - \bar{v}_t(i+1))$ by the maximal difference of expected termination weights between two consecutive positions over all time. The main theorem on the regret is stated as follows.

Theorem 1. *For $n \geq 1$, and the reward function $f(A, v, w) = 1 - \prod_{k=1}^{K}(1 - v(k)w(a_k))$, the pseudo-regret $\mathcal{R}(n)$ of Algorithm 1 has the following bound*

$$\mathcal{R}(n) \leq \frac{4dK p_v k_\mu \sigma}{c_\mu} \sqrt{nK \log\left(\frac{1 + n/(\lambda d)}{\delta}\right) \log(1 + Kn/(\lambda d))}. \qquad (7)$$

The theorem shows a $\tilde{O}(d\sqrt{n})$ pseudo-regret bound, which is independent of L, and improves the previous regret bound of [17] by a $\sqrt{\log(n)}$ term, though our result is under the combinatorial setting. With an additional assumption on item generating process, the result may be further improved by a $\sqrt{d}$-order while sacrificing an increase on order of $\log(n)$ by using Theorem 1 of [16].

Proof. To begin with, we bound the one-step regret at time t, denoted by $\mathcal{R}_t = f(A_t^*, \mathbf{v}_t, \mathbf{w}_t) - f(\mathbf{A}_t, \mathbf{v}_t, \mathbf{w}_t)$, then

$$\mathbb{E}[\mathcal{R}_t | \mathcal{H}_t] = f(A_t^*, \bar{v}_t, \bar{w}_t) - f(\mathbf{A}_t, \bar{v}_t, \bar{w}_t)$$

$$\leq \sum_{k=1}^{K} \bar{v}_t(k) \bar{w}_t(a_k^*) - \sum_{k=1}^{K} \bar{v}_t(k) \bar{w}_t(\mathbf{a}_k^t) \tag{8}$$

$$= \sum_{i=1}^{K} (\bar{v}_t(i) - \bar{v}_t(i+1)) \sum_{k=1}^{i} (\bar{w}_t(a_k^*) - \bar{w}_t(\mathbf{a}_k^t))$$

$$\leq p_v \sum_{i=1}^{K} \sum_{k=1}^{i} (\bar{w}_t(a_k^*) - \bar{w}_t(\mathbf{a}_k^t)), \tag{9}$$

where $\bar{v}_t(i+1) = 0$. The inequality (8) is because of the definition of A_t^* and f, while (9) is by definition of the p_v. We can observe that the problem has reduced to the cascading problem of bounding $\sum_{k=1}^{i}(\bar{w}_t(a_k^*) - \bar{w}_t(\mathbf{a}_k^t))$, which is equal to $\sum_{k=1}^{i} \mu(\theta_*^\top x_{t,a_k^*}) - \mu(\theta_*^\top x_{t,\mathbf{a}_k^t})$. We need the following Lemma 2 to bound this cascade difference.

Lemma 2. *Let $t \geq 1$ and $\mathbf{A}_t = (\mathbf{a}_1^t, ..., \mathbf{a}_i^t)$, $i \in [K]$, we have:*

$$\sum_{k=1}^{i}(\mu(\theta_*^\top x_{t,a_k^*}) - \mu(\theta_*^\top x_{t,\mathbf{a}_k^t})) \leq 2 \sum_{k=1}^{i} \rho(t-1) \|x_{t,\mathbf{a}_k^t}\|_{\mathbf{V}_{t-1}^{-1}}.$$

Proof. Let $A_t^* = (a_1^*, \ldots, a_K^*)$. By the definition of $\mathbf{A}_t$, which is set of items with the largest UCBs placed to the most terminating position, we have $\sum_{k=1}^{i} \mathbf{U}_t(\mathbf{a}_k^*) \leq \sum_{k=1}^{i} \mathbf{U}_t(\mathbf{a}_k^t)$, $i = [K]$, that is,

$$\sum_{k=1}^{i} \mu(\hat{\theta}^\top x_{t,a_k^*}) + \rho(t-1) \|x_{t,a_k^*}\|_{\mathbf{V}_{t-1}^{-1}}$$

$$\leq \sum_{k=1}^{i} \mu(\hat{\theta}^\top x_{t,\mathbf{a}_k^t}) + \rho(t-1) \|x_{t,\mathbf{a}_k^t}\|_{\mathbf{V}_{t-1}^{-1}}. \tag{10}$$

Then

$$\sum_{k=1}^{i} \mu(\theta_*^\top x_{t,u_k^*}) - \mu(\theta_*^\top x_{t,\mathbf{a}_k^t})$$

$$= \sum_{k=1}^{i} \mu(\theta_*^\top x_{t,a_k^*}) - \mu(\hat{\theta}^\top x_{t,a_k^*}) + \mu(\hat{\theta}^\top x_{t,a_k^*}) - \mu(\hat{\theta}^\top x_{t,\mathbf{a}_k^t}) +$$

$$\mu(\hat{\theta}^\top x_{t,\mathbf{a}_k^t}) - \mu(\theta_*^\top x_{t,\mathbf{a}_k^t})$$

$$\leq \rho(t-1) \sum_{k=1}^{i} \|x_{t,a_k^*}\|_{\mathbf{V}_{t-1}^{-1}} + \left(\|x_{t,\mathbf{a}_k^t}\|_{\mathbf{V}_{t-1}^{-1}} - \|x_{t,a_k^*}\|_{\mathbf{V}_{t-1}^{-1}} \right) + \|x_{t,a_k^t}\|_{\mathbf{V}_{t-1}^{-1}} \tag{11}$$

$$= 2\rho(t-1) \sum_{k=1}^{i} \|x_{t,a_k^t}\|_{V_{t-1}^{-1}},$$

where Eq. (11) is obtained by applying (10). Our next step is to bound $\sum_{s=1}^{t} \sum_{k=1}^{i} \|x_{s,\mathbf{a}_k^s}\|_{V_t^{-1}}^2$ by Lemma 4.4 in [11], when $\lambda \geq K$,

Lemma 3. *If $\lambda \geq K$, then*

$$\sum_{s=1}^{t} \sum_{k=1}^{i} \|x_{s,\mathbf{a}_k^s}\|_{V_t^{-1}}^2 \leq 2d \log(1 + \frac{Kt}{\lambda d}).$$

Building upon the previous discussion, we have:

$$\mathcal{R}(n) = \sum_{t=1}^{n} \mathbb{E}[\mathbb{E}[\mathcal{R}_t|\mathcal{H}_t]]$$

$$\leq p_v \sum_{t=1}^{n} \mathbb{E}\left[\sum_{i=1}^{K} \sum_{k=1}^{i} (\bar{w}_t(a_k^*) - \bar{w}_t(\mathbf{a}_k^t)) \right] \tag{12}$$

$$\leq p_v \sum_{t=1}^{n} \mathbb{E}\left[\sum_{i=1}^{K} \sum_{k=1}^{i} 2\rho(t-1)\|x_{t,\mathbf{a}_k^t}\|_{V_t^{-1}} \right]$$

$$\leq 2\rho(n)p_v \sum_{t=1}^{n} \mathbb{E}\left[\sum_{i=1}^{K} \sum_{k=1}^{i} \|x_{t,\mathbf{a}_k^t}\|_{V_t^{-1}} \right]. \tag{13}$$

where Eq. (12) is due to the tower rule and the inequality (13) holds since $\rho(t)$ increases with t. Applying the Cauchy-Schwarz inequality on the current result, we can derive that:

$$\mathcal{R}(n) \leq 2\rho(n)p_v \mathbb{E}\left[\sqrt{\left(n \sum_{i=1}^{K} \sum_{k=1}^{i} 1^2\right) \left(\sum_{t=1}^{n} \sum_{i=1}^{K} \sum_{k=1}^{i} \|x_{t,\mathbf{a}_k^t}\|_{V_t^{-1}}^2\right)} \right].$$

Substituting $\rho(n)$ back and applying Lemma 3 back yields our claimed result.

3.3 Computationally Efficient Updates

Though our proposed GL-CDCM enjoys good theoretical properties, the computational cost may be high in some applications. The inverse of a $d \times d$ matrix is computed at each time step while the MLE is calculated using samples up to the current time step, which is increased linearly over time. We provide an iterative optimization solution for GL-CDCM for the logistic regression where $\mu(x) = 1/(1 + \exp(-x))$, denoted by GL-CDCM (SGD).

Instead of solving Eq. (4), we use the stochastic logistic gradient at time t

$$\mathbf{g}_t = \sum_{k=1}^{\mathcal{C}_t} \left(\mu(\hat{\theta}_t^{\top} x_{t,\mathbf{a}_k^t}) - \mathbf{w}_t(\mathbf{a}_k^t) \right) x_{t,\mathbf{a}_k^t}, \tag{14}$$

and we can update on $\hat{\theta}_t$ by

$$\hat{\theta}_t = \hat{\theta}_{t-1} - \eta \mathbf{g}_t, \tag{15}$$

where η is the learning rate.

Let $\mathbf{C}_t \in \mathbb{R}^{\mathcal{C}_t \times d}$ be the matrix whose rows are the feature vectors of the observed items at time t. Then $\mathbf{V}_t = \mathbf{V}_{t-1} + \mathbf{C}_t^\top \mathbf{C}_t$. Let $\mathbf{G}_t = I + \mathbf{C}_t \mathbf{V}_t^{-1} \mathbf{C}_t^\top$, based on the Woodbury matrix identity [18], $\mathbf{V}_t^{-1}$ can be calculated efficiently using

$$\mathbf{V}_t^{-1} = \mathbf{V}_{t-1}^{-1} - \mathbf{V}_{t-1}^{-1} \mathbf{C}_t^\top \mathbf{G}_t^{-1} \mathbf{C}_t \mathbf{V}_{t-1}^{-1}, \tag{16}$$

in time $O(Kd^2)$.

Therefore, the burden of computing the inverse of a $d \times d$ matrix of is reduced to computing the inverse of a square matrix of dimension at most K, which is always smaller than d and can be much smaller in practice.

4   Experiments

4.1   Synthetic Data

In this section, we compare our algorithms (GL-CDCM) with the dcmKL-UCB algorithm proposed in [15] (denoted as *KL-DCM* in our comparisons) and the logistic regression (LR) on the synthetic data. Here LR means for each time step t, it conducts logistic regression on all historical data and uses the obtained parameters to choose the current items, which corresponds to selecting arms by values of $\mu(\hat{\theta}_{t-1}^\top x_{t,a})$, instead of $\mathbf{U}_t(a)$ (which has an additional exploration term) in Line 5-6 for our Algorithm 1.

We simulate a scenario of web search as follows. First, we randomly select the model parameter θ_*. Then at each time step t, randomly select contextual vectors $x_{t,a}$ for each item a and expected termination weights $\bar{v}_t$. Then according to Eq. (3), the expected attraction weight $\bar{w}_t$ is computed by the given θ_*. Both attraction weights $\mathbf{w}_t$ and termination weights $\mathbf{v}_t$ are then drawn from Bernoulli distribution with the respective mean. The sigmoid function $\mu(x) = 1/(1 + \exp(-x))$ serves as the inverse link function. The evaluation criterion is the cumulative pseudo-regret defined in Eq. (2).

The curves of the cumulative regrets for these algorithms, i.e. GL-CDCM, GL-CDCM (SGD), LR, LR (SGD) and KL-DCM, under $n = 10^4$ are shown in Fig. 1(a). To further demonstrate the estimation ability of GL-CDCM and LR, the cosine distances between $\hat{\theta}_t$ and θ_*, i.e., $1 - \frac{\hat{\theta}_t^\top \theta_*}{\|\hat{\theta}_t\|_2 \|\theta_*\|_2}$, are calculated and shown in Fig. 1(b), where the value 0 indicates that the learning agent correctly estimates θ^*. We do not show KL-DCM in Fig. 1(b) since it does not estimate the parameter θ_*. As depicted in Fig. 1(a), KL-DCM has the largest regret since it ignores the contextual information. For both GL-CDCM and LR, the SGD version generally has higher regret, which is a price to pay for efficiency. Compared to the LR algorithm, the bandit algorithm balances the exploitation and exploration and therefore has a better performance. Furthermore, the error curve shows that the GL-CDCM converges more quickly than LR.

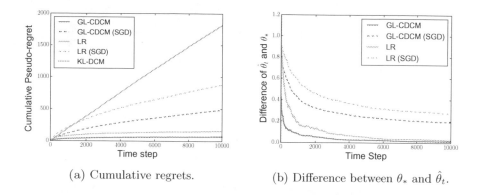

(a) Cumulative regrets.   (b) Difference between θ_* and $\hat{\theta}_t$.

Fig. 1. Experimental results of different recommendation algorithms on synthetic data.

4.2   Web Page Recommendation

In this section, we test our algorithms on the Yandex Personalized Web Search dataset [19], which contains 35 million search sessions. Let M be the number of users and L be the number of web pages. We use top 3 most frequent queries for evaluation. Each query corresponds to one DCM which is estimated using PyClick library [8]. In all the algorithms, we assume that the higher positions have higher expected termination weight. In order to derive the feature vectors for web pages, we first construct a sparse matrix $A \in \mathbb{Z}^{M \times L}$ where $A(i, j) \in \mathbb{Z}$ denotes the number that user i clicked on web page j. Then the feature vector is obtained through the SVD decomposition of A, i.e. $A = USV^\top$. We use $V = [v_1; \dots; v_L] \in \mathbb{R}^{L \times d}$ as the contextual information for the L web pages. We set $d = 200$, $K = 10$, and $L = 100$. The cumulative pseudo-regret over 5000 rounds for our proposed GL-CDCM, GL-CDCM (SGD), LR, LR (SGD) and KL-DCM are shown in Fig. 2. To incorporate the user features, we concatenate user and item features as the contextual information. Let $U = [u_1; \dots; u_M] \in \mathbb{R}^{M \times d}$, then $x_{i,j} = [u_i, v_j] \in \mathbb{R}^{2d}$ for user i and web page j. The features derived from outer product where $x_{i,j} = u_i \otimes v_j$ are also tested, but the performance is not as good as $x_{i,j} = [u_i, v_j]$. At each time step, a user is randomly selected. Follow the previous setting of the parameters, the results are displayed in Fig. 2(b).

For the setting that only the item features are used, after 5000 rounds, the proposed GL-CDCM obtains a regret of 32.28, which is much lower than 59.08 for LR and 99.09 for KL-DCM. Furthermore, the curve for KL-CDCM forms a stair-step pattern since the ground item set is changing and the algorithm needs to learn from the cold start from time to time. In contrast, GL-CDCM and LR make use of the contextual information, and therefore achieve a better estimation. Compared with LR, which is always exploiting, GL-CDCM explores more and achieves a lower cumulative pseudo-regret. The SGD versions generally have a higher regret for both GL-CDCM and LR, 81.71 for GL-CDCM (SGD) and 114.96 for LR (SGD), but the time complexity reduces significantly. In addition, the proposed GL-CDCM (SGD) still outperforms LR (SGD) because

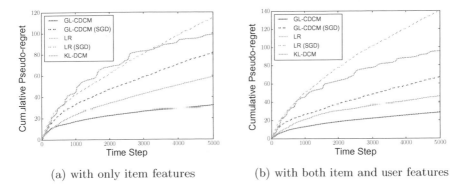

(a) with only item features   (b) with both item and user features

Fig. 2. Experimental results of different recommendation algorithms on Yandex dataset.

of exploration. A similar pattern is also observed in the setting of involving both user and item features, where more useful information are provided and the regrets of GL-CDCM and LR decrease to 28.51 and 46.00, respectively. The experimental results are consistent with our previous discussions and show that our proposed algorithm has better performance even for practical problems, where the assumptions might be violated.

5 Conclusion

In this paper, we present a bandit algorithm (and SGD variant) for web page recommendation that automatically balances the exploration and exploitation. We formulate the problem of DCM bandits with contextual information. The dependent click model (DCM) covers the scenario of multiple clicks and is a popular click model in web search. The contextual information is incorporated in our work to better estimate the expected attraction weight. Under a reasonable assumption on knowing the order of the expected termination weight, we prove a regret bound of $\tilde{O}(d\sqrt{n})$ for the algorithm. A computationally efficient version is also given by removing the expensive step of computing the MLE on a linearly increasing sample set, and reducing the cost of inverting a $d \times d$ matrix. Experimental results confirm the value of exploring, utilizing the contextual information and adopting a generalized linear model.

Acknowledgment. This work is sponsored by Huawei Innovation Research Program.

References

1. Aggarwal, C.C.: Recommender Systems. Springer, Heidelberg (2016). https://doi.org/10.1007/978-3-319-29659-3
2. Richardson, M., Dominowska, E., Ragno, R.: Predicting clicks: estimating the click-through rate for new ads. In: Proceedings of the 16th International Conference on World Wide Web, pp. 521–530. ACM (2007)

3. Rendle, S.: Factorization machines. In: 2010 IEEE 10th International Conference on Data Mining (ICDM), pp. 995–1000. IEEE (2010)
4. Wang, X., Wang, Y., Hsu, D., Wang, Y.: Exploration in interactive personalized music recommendation: a reinforcement learning approach. ACM Trans. Multimed. Comput. Commun. Appl. (TOMM) **11**(1), 7 (2014)
5. Gittins, J., Glazebrook, K., Weber, R.: Multi-armed Bandit Allocation Indices. Wiley, Hoboken (2011)
6. Li, L., Chu, W., Langford, J., Schapire, R.E.: A contextual-bandit approach to personalized news article recommendation. In: Proceedings of the 19th International Conference on World Wide Web, pp. 661–670. ACM (2010)
7. Villar, S.S., Bowden, J., Wason, J.: Multi-armed bandit models for the optimal design of clinical trials: benefits and challenges. Stat. Sci.: Rev. J. Inst. Math. Stat. **30**(2), 199 (2015)
8. Chuklin, A., Markov, I., de Rijke, M.: Click models for web search. Synth. Lect. Inf. Concepts, Retrieval, Serv. **7**(3), 1–115 (2015)
9. Kveton, B., Wen, Z., Ashkan, A., Szepesvari, C.: Cascading bandits: learning to rank in the cascade model. In: Proceedings of the 32th International Conference on Machine Learning (2015)
10. Kveton, B., Wen, Z., Ashkan, A., Szepesvari, C.: Combinatorial cascading bandits. In: Advances in Neural Information Processing Systems, pp. 1450–1458 (2015)
11. Li, S., Wang, B., Zhang, S., Chen, W.: Contextual combinatorial cascading bandits. In: Proceedings of The 33rd International Conference on Machine Learning, pp. 1245–1253 (2016)
12. Guo, F., Liu, C., Wang, Y.M.: Efficient multiple-click models in web search. In: Proceedings of the Second ACM International Conference on Web Search and Data Mining, pp. 124–131. ACM (2009)
13. Abbasi-Yadkori, Y., Pál, D., Szepesvári, C.: Improved algorithms for linear stochastic bandits. In: Advances in Neural Information Processing Systems, pp. 2312–2320 (2011)
14. Li, L., Chu, W., Langford, J., Moon, T., Wang, X.: An unbiased offline evaluation of contextual bandit algorithms with generalized linear models. In: Proceedings of the Workshop on On-line Trading of Exploration and Exploitation vol. 2, pp. 19–36 (2012)
15. Katariya, S., Kveton, B., Szepesvári, C., Wen, Z.: DCM bandits: learning to rank with multiple clicks. In: Proceedings of The 33rd International Conference on Machine Learning (2016)
16. Li, L., Lu, Y., Zhou, D.: Provable optimal algorithms for generalized linear contextual bandits. In: Proceedings of The 34rd International Conference on Machine Learning (2017)
17. Filippi, S., Cappe, O., Garivier, A., Szepesvári, C.: Parametric bandits: the generalized linear case. In: Advances in Neural Information Processing Systems, pp. 586–594 (2010)
18. Hager, W.W.: Updating the inverse of a matrix. SIAM Rev. **31**(2), 221–239 (1989)
19. Yandex: Yandex personalized web search challenge (2013)

LP-Based Pivoting Algorithm for Higher-Order Correlation Clustering

Takuro Fukunaga[✉]

RIKEN Center for Advanced Intelligence Project, Tokyo, Japan
takuro.fukunaga@riken.jp

Abstract. Correlation clustering is an approach for clustering a set of objects from given pairwise information. In this approach, the given pairwise information is usually represented by an undirected graph with nodes corresponding to the objects, where each edge in the graph is assigned a nonnegative weight, and either the positive or negative label. Then, a clustering is obtained by solving an optimization problem of finding a partition of the node set that minimizes the disagreement or maximizes the agreement with the pairwise information. In this paper, we extend correlation clustering with disagreement minimization to deal with higher-order relationships represented by hypergraphs. We give two pivoting algorithms based on a linear programming relaxation of the problem. One achieves an $O(k \log n)$-approximation, where n is the number of nodes and k is the maximum size of hyperedges with the negative labels. This algorithm can be applied to any hyperedges with arbitrary weights. The other is an $O(r)$-approximation for complete r-partite hypergraphs with uniform weights. This type of hypergraphs arise from the coclustering setting of correlation clustering.

1 Introduction

Problem Formulation. In this paper, we consider approximation algorithms for the hypergraph correlation clustering. In the hypergraph correlation clustering, a problem instance consists of an undirected hypergraph $G = (V, E)$ with the node set V and the hyperedge set E, and the label and the weight of each hyperedge in E. The label on a hyperedge is either positive or negative. We call a hyperedge positive if it is assigned the positive label, and negative otherwise. The sets of positive and negative hyperedges are denoted by E_+ and E_-, respectively (i.e., E is the disjoint union of E_+ and E_-). The weight of each hyperedge e is a nonnegative real number, denoted by $w(e)$.

The hypergraph correlation clustering is an optimization problem of finding a clustering of the given hypergraph $G = (V, E)$. A *clustering* $\mathcal{C}$ of G is defined as a partition of V into nonempty subsets. Each node set in $\mathcal{C}$ is called a *cluster*. A hyperedge e in G is defined to *disagree* with a clustering $\mathcal{C}$ if either of the following statements is true:

- e is a positive hyperedge, and some two end nodes of it belong to different clusters of $\mathcal{C}$;

L. Wang and D. Zhu (Eds.): COCOON 2018, LNCS 10976, pp. 51–62, 2018.
https://doi.org/10.1007/978-3-319-94776-1_5

– e is a negative hyperedge, and all of its end nodes belong to the same cluster of $\mathcal{C}$.

Then, the objective of the problem is to find a clustering minimizing the total weight of hyperedges that disagree with the clustering.

In a part of this paper, we focus on a special type of hypergraphs called *complete r-partite* hypergraphs. A hypergraph is called r-partite if its node set can be divided into disjoint r subsets $V_1, \ldots, V_r$ so that each hyperedge includes exactly one node from V_j for each $j = 1, \ldots, r$. An r-partite hypergraph is complete if each tuple $\{v_1, \ldots, v_r\} \in V_1 \times \cdots \times V_r$ is included as a hyperedge. We refer to the set of instances with complete r-partite hypergraphs and uniform hyperedge weights as *coclustering setting*; the reason for this name will be explained below.

Motivation. Correlation clustering is originally an approach for computing a clustering from given pairwise information. It was introduced by Bansal et al. [3]. They proposed representing the pairwise information as a graph with nodes corresponding to the objects to be clustered. As in the hypergraph correlation clustering, each edge in the graph is associated with a nonnegative weight and the positive or negative label. A positive edge indicates that its two end nodes should belong to the same cluster, while a negative edge indicates that the end nodes should belong to different clusters. Informations represented by the edge labels are possibly inconsistent due to existence of noise or observation errors. The weight of each edge represents the reliability of the information represented by it. The purpose of correlation clustering is to find a clustering matching the pairwise information to the greatest degree possible. This purpose presents two optimization problems defined on the graph naturally; one seeks a clustering that minimizes the disagreement, and the other seeks a clustering that maximizes the agreement. Since these problems are NP-hard, several approximation algorithms have been proposed for them, and have been successfully applied to numerous applications in machine learning and computational biology [4,11,15,23]. We will review these previous studies briefly in Sect. 2.

In several applications, pairwise information does not give enough information for the extraction of precise clusterings, and hence it is motivated to study clustering from higher-order information, which is modeled as the hypergraph correlation clustering. Even in the hypergraph correlation clustering, we can consider both the disagreement minimization and the agreement maximization. However, since this paper discusses only the disagreement minimization, we gave the disagreement minimization formulation above. A straightforward idea for the hypergraph correlation clustering is to reduce the problem to the graph correlation clustering by expanding each hyperedge to some graphs like cliques. However, this idea does not give efficient algorithms as we will see in Sect. 3.2.

Study on the hypergraph correlation clustering was initiated by Kim et al. [20] for an application to the image segmentation. Subsequently Kappes et al. [19] and Kim et al. [21] also considered the same problem. All of these studies are similar in that:

- they proposed linear programming (LP) relaxations for the hypergraph correlation clustering (with the disagreement minimization objective), and presented algorithms using LP solvers or tailored cutting-plane algorithms;
- when the solution output by the algorithms is not integer, they round it by a simple procedure of rounding up an arbitrary non-zero variable into an integer;
- they empirically showed that introducing higher-order relationships into correlation clustering improves the quality of image segmentation.

These observations indicate that efficient algorithms for the hypergraph correlation are useful in practice. On the other hand, to the best of our knowledge, no approximation algorithm with a provable performance guarantee is known for the problem. Motivated by this fact, our aim is to present performance guarantee of approximation algorithms for the hypergraph correlation clustering.

In addition to the general case of the hypergraph correlation clustering, we will study the coclustering setting of the problem. Coclustering denotes the task of clustering the objects which are categorized into two or more classes, and relationships of objects from different classes are considered. For example, this setting arises when we find a clustering of documents and words from word occurrences in documents. It is also known to be useful for clustering of gene expression data. To distinguish clustering from pairwise relationships and higher-order relationships, in this paper, we call the former by *biclustering*, and the latter by coclustering.

In correlation clustering, the biclustering setting implies that the given information is represented by bipartite graphs. This setting has been studied extensively [1,2,8]. These previous studies show that the disagreement minimization problem with complete bipartite graphs and uniform edge weights admits constant-factor approximation algorithms. In contrast, the coclustering setting has not been studied in the context of correlation clustering although it seems useful; for example, consider clustering users, search key words, and goods from purchase records in an E-commerce website; if a user i purchased a good j after searching with a key word s, then the category of i, j, and s are likely same, and hence solving the hypergraph correlation clustering defined from these order-3 relationships gives a more precise clustering rather than computing from pairwise relationships. We note that the hypergraph defined in this situation is 3-partite.

Contributions. We present two approximation algorithms with approximation guarantees for the hypergraph correlation clustering. One of the algorithms is for general hypergraphs. It has an $O(k \log n)$-approximation guarantee (Theorem 1), where n is the number of nodes and k is the maximum size of hyperedges assigned the negative label. In other words, for any instance of the hypergraph correlation clustering, our algorithm outputs a clustering the objective function value of which is within a factor of $O(k \log n)$ from the optimal. The other algorithm is for the coclustering setting (the given hypergraph is complete r-partite and hyperedge weights are uniform). It achieves an $O(r)$-approximation guarantee (Theorem 2). Note that this approximation factor is a constant when r is a constant, and hence it extends the constant-approximation guarantees of [1,2,8] for the disagreement minimization problem with complete bipartite graphs and uniform edge weights.

Our algorithms are so-called *pivoting algorithms*. The pivoting algorithms are those computing a clustering by repeating the following operations: (i) choose a pivoting node v; (ii) let the cluster including v be the set of nodes near to v in a metric; (iii) remove the nodes in the cluster from the graph (or hypergraph). Most of the known approximation algorithms for correlation clustering are this type of algorithms. Selection of the pivoting node v in Step (i) and the precise definition of the cluster including v in Step (ii) are tailored for each variation.

In our algorithms, Steps (i) and (ii) are based on an LP relaxation of the problem. Our LP relaxation has decision variables that represent metrics on the nodes of the given hypergraph. The metrics are then optimized such that two nodes are located closer to each other if they share more hyperedges assigned the positive label, and they are located further from each other if they share more hyperedges assigned the negative label. This LP is a straightforward extension of the one considered in [6,8,12] for the disagreement minimization problem with graphs. Moreover, it is almost same as or simpler than those used in the previous studies [19–21] on the hypergraph correlation clustering. Indeed, our algorithm works even with the LP relaxations considered in [19–21], and hence it can replace the rounding algorithms therein.

In our $O(k \log n)$-approximation algorithm, we use the *region-growing* idea to define the cluster including a chosen pivoting node. Indeed, our algorithm generalizes the $O(\log n)$-approximation algorithms in [6,12] for graphs. On the other hand, our $O(r)$-approximation algorithm for the coclustering setting is based on a new idea. When $r = 2$, the coclustering setting is equivalent to the disagreement minimization on complete bipartite graphs with uniform weights. Although several constant-factor approximation algorithms are known for this case [1,2,8], it seems difficult to extend them to $r \geq 3$ because they crucially relies on a structure of graphs representing inconsistent informations. Hence we design a new algorithm from scratch. It achieved a slightly worse approximation factor for $r = 2$ compared with the previous studies on the complete bipartite graphs.

Organization. The rest of this paper is organized as follows. Section 2 surveys related previous studies. Section 3 introduces notations, the LP relaxation used in our algorithms, and an outline of our algorithms. Section 3 also explains that reducing the hypergraph correlation problem to the graph correlation clustering is not efficient. Section 4 presents our $O(k \log n)$-approximation algorithm, and Sect. 5 gives our $O(r)$-approximation algorithm for the correlation clustering setting. Section 6 concludes the paper.

2   Related Work

Correlation Clustering. Both of the agreement maximization and the disagreement minimization formulations of correlation clustering were introduced by

Bansal et al. [3]. Charikar et al. [6] gave a factor 0.7664 approximation algorithm for the agreement maximization. For the disagreement minimization, Charikar et al. [6] and Demaine et al. [12] gave factor $O(\log n)$ approximations. Demaine et al. also proved that the disagreement minimization is equivalent to the minimum multicut problem. This equivalence indicates that obtaining a constant-factor approximation for the disagreement minimization is unique-games hard because of the hardness result on the minimum multicut problem given in [7].

Several special cases of the disagreement minimization also have been studied well. For example, Amit [2], Ailon et al. [1], and Chawla et al. [8] considered the case where the graph is complete bipartite and weights are uniform. They gave constant-factor approximation algorithms for this case, and the current best approximate factor among them is 3 due to Chawla et al. [8]. Chawla et al. also considered complete graphs, and presented a 2.06-approximation algorithm for uniform weights, and 1.5-approximation algorithm for weights satisfying the triangle inequality.

Note that the above studies on correlation clustering all consider graphs. To the best of our knowledge, the correlation clustering over hypergraphs have been studied only in [19–21], and no algorithm with a performance guarantee is known.

Coclustering. Biclustering of data represented by a matrix has been studied since the 1970s [16]. There has been a huge number of algorithms proposed so far, and we name a few of them [14,26,27]. These algorithms have been successfully applied to numerous unsupervised learning tasks [9,29]. In particular, clustering on gene expression data [10,22] and document classification [5,13,18] are studied actively. Compared with biclustering, coclustering of higher-order relational data has not been extensively studied so far. Zhao and Zaki [28] proposed a graph-based algorithm for coclustering. Hatano et al. [17] proposed a coclustering algorithm based on sampling hypergraph multicuts. Other previous studies [24,25] depend on an algebraic approach known as tensor rank decomposition.

3   Preliminaries

3.1   Notations

Let $G = (V, E)$ be a hypergraph with the node set V and the hyperedge set E. Throughout this paper, we let n denote the cardinality of V (i.e., $n = |V|$). The cardinality of a hyperedge e is called the *rank* of e, and the rank of a hypergraph G is defined as the maximum rank of hyperedges in G. Note that a hyperedge of rank 2 and a hypergraph of rank 2 are an edge and a graph, respectively. For $U \subseteq V$ and $H \subseteq E$, let $\delta(U; H)$ denote the set of hyperedges in H that include nodes both in U and $V \setminus U$, and $H[U]$ denote the set of hyperedges in H that include no nodes from $V \setminus U$. $G[U]$ denotes the sub-hypergraph of G induced by U (i.e., hypergraph with the node set U and the hyperedge set $E[U]$).

3.2 Reduction to the Graph Correlation Clustering

A natural approach for solving the hypergraph correlation clustering is to apply an existing algorithm for the graph correlation clustering to the graph obtained by transforming the given hypergraph. However, the transformation of a hypergraph into a graph may change the structure of the problem drastically. To see this, suppose that a positive hyperedge of rank k' is replaced by a clique that consists of $\binom{k'}{2}$ positive edges. In the original hypergraph, the weight of a positive hyperedge is counted in the disagreement only once if the nodes in the hyperedge belong to more than one cluster. In contrast, in the corresponding graph, the contribution of the edges in the clique to the measured disagreement depends on how the clique is divided. For example, if all but one of the nodes in the clique belong to the same cluster, the weights of $k' - 1$ edges are counted, whereas if the nodes are all divided into different clusters, the weights of $\binom{k'}{2}$ edges are counted. Thus, the contributions are very different when the clique is divided into two clusters and when it is divided into k' clusters. Because of this fact, applying the best-known graph correlation clustering algorithm to the obtained graph only gives an $O(k' \log n)$-approximation even if all negative hyperedges in the given hypergraph are order-2, while our algorithm given in Sect. 4 attains an $O(\log n)$-approximation in this case. It seems hard to avoid this phenomenon even if we consider other ways of transformation. When the given hypergraph includes a negative hyperedge of order larger than 3, it seems difficult to bound the approximation factor given by the above approach; even if a clustering partitions a negative hyperedge into at least two clusters (and hence it incurs no cost from the hyperedge), an edge generated by transforming the negative hyperedge belongs to the same cluster in the clustering (and it incurs a positive cost).

3.3 Overview of Our Algorithms

In this subsection, we introduce our algorithms for the hypergraph correlation clustering. Our algorithms are based on an LP obtained by relaxing an integer programming (IP) formulation of the problem. We first introduce this IP formulation.

This formulation optimizes the following variables, which take numbers in $\{0, 1\}$:

– A variable x_{uv} for each pair of nodes $u, v \in V$; it is 0 if u and v belong to the same cluster, and it is 1 otherwise;
– A variable x_e for each hyperedge e; it is 0 if all nodes in e are included in a cluster, and it is 1 otherwise.

If a positive hyperedge $e \in E_+$ is included in a cluster, this implies that any two nodes u and v included in e belong to the same cluster. The following constraint formulates this condition:

$$x_{uv} \le x_e, \quad \forall e \in E_+, \forall \{u, v\} \subseteq e. \tag{1}$$

If a negative hyperedge $e = \{v_1, \ldots, v_r\} \in E_-$ intersects more than one cluster, then a node v_1 and some of the other nodes $v_2, \ldots, v_r$ belong to different clusters. This is represented by

$$x_e \leq \sum_{i=1}^{r-1} x_{v_i v_{i+1}}, \quad \forall e = \{v_1, \ldots, v_r\} \in E_-. \tag{2}$$

Here, the ordering of nodes included in e is fixed arbitrarily.

If two nodes u and v belong to the same cluster, and if v and z also do, then all of these three nodes belong to the same cluster. This means that if $x_{uv} = x_{vz} = 0$, then $x_{uz} = 0$ must hold. Thus, the variables satisfy the following triangle inequalities:

$$x_{uz} \leq x_{uv} + x_{vz}, \quad \forall u, v, z \in V. \tag{3}$$

Our IP formulation optimizes over these constraints. The disagreement objective function is $\sum_{e \in E_+} w(e) x_e + \sum_{e \in E_-} w(e)(1 - x_e)$. The LP relaxation is obtained from the IP formulation by relaxing the range of each variable to $[0, 1]$. Specifically, it is described as follows:

$$\begin{aligned}
\text{minimize} \quad & \sum_{e \in E_+} w(e) x_e + \sum_{e \in E_-} w(e)(1 - x_e) \\
\text{subject to} \quad & (1), (2), (3), \\
& x_{uv} \in [0, 1], \quad \forall u, v \in V, \\
& x_e \in [0, 1], \quad \forall e \in E.
\end{aligned} \tag{4}$$

For convenience, we let $x_{vv} = 0$ for all $v \in V$ in the rest of this paper although these variables do not appear in LP (4).

Our algorithms first compute an optimal solution x for the LP relaxation (4). Then they construct a clustering from x by repeating the three steps described in Sect. 1. Let U denote the set of nodes that belong to no cluster yet at the beginning of a certain iteration. In this iteration, the pivoting node v is chosen from U. The cluster containing v is defined as $B_{x,v,U}(\xi) := \{u \in U : x_{uv} < \xi\}$ from some radius $\xi \in [0, 1]$. Selection of v and the definition of ξ are customized in two variations of our algorithms. Roughly speaking, we optimize them so that the ratio of weights of disagreed hyperedges incident to $B_{x,v,U}(\xi)$ to the fractional weights of hyperedges incident to $B_{x,v,U}(\xi)$ is minimized. Refer to Sects. 4 and 5 for the details. The algorithms are described in Algorithm 1.

4 $O(k \log n)$-Approximation for General Hypergraphs

In this section, we discuss general hypergraphs with arbitrary hyperedge weights. First, let us introduce several notations. We let x and L refer to an optimal solution for (4) and its objective value. For a hyperedge e and a node v, let $d(e, v)$ and $d'(e, v)$ denote $\min_{u \in e} x_{vu}$ and $\max_{u \in e} x_{vu}$, respectively. For $\xi \in [0, 1]$, we define $F_{v,x,E}(\xi)$ and $C_{v,x,E}(\xi)$ by

---

Algorithm 1. LP-based pivoting algorithm for hypergraph correlation clustering

Input: a hypergraph $G = (V, E)$ with $E = E_+ \cup E_-$ and a nonnegative weight $w(e)$
for each $e \in E$

Output: a clustering $\mathcal{C}$ of V,

1: $\mathcal{C} \longleftarrow \emptyset$, $U \longleftarrow V$
2: compute an optimal solution x of (4)
3: **while** $U \neq \emptyset$ **do**
4:     compute $v \in U$ and $\xi \in [1, 0]$ (details are described in Sects. 4 and 5)
5:     $\mathcal{C} \longleftarrow \mathcal{C} \cup \{B_{x,v,U}(\xi)\}$, $U \longleftarrow U \setminus B_{x,v,U}(\xi)$
6:     remove all hyperedges intersecting $B_{x,v,U}(\xi)$ from E
7: **end while**
8: output $\mathcal{C}$

$$F_{v,x,E}(\xi) := \frac{L}{n} + \sum_{e \in E_+[B_{x,v,U}(\xi)]} w(e)x_e + \sum_{e' \in \delta(B_{x,v,U}(\xi);E_+)} w(e')x_{e'} \frac{\xi - d(e', v)}{d'(e', v) - d(e', v)}$$

and

$$C_{v,x,E}(\xi) := \sum_{e \in \delta(B_{x,v,U}(\xi);E_+)} w(e),$$

where $C_{v,x,E}(\xi)$ is defined to be $+\infty$ if $\delta(B_{x,v,U}(\xi); E_+) = \emptyset$. We note that if $e' \in \delta(B_{x,v,U}(\xi); E_+)$, then $d(e', v) \leq \xi \leq d'(e', v)$ holds, and hence the third term in $F_{v,x,E}(\xi)$ is at most $\sum_{e' \in \delta(B_{v,x,U}(\xi);E)} w(e')x_{e'}$. Below, we omit the subscripts of $B_{x,v,U}(\xi)$, $F_{v,x,E}(\xi)$, and $C_{v,x,E}(\xi)$ when they are clear from the context. Roughly speaking, the second and the third terms of $F(\xi)$ represent how much the objective value of x in (4) is reduced when the plus hyperedges incident to $B(\xi)$ are removed from the hypergraph, and $C(\xi)$ represents how much the disagreement of the positive hyperedges is increased when $B(\xi)$ is added as a cluster to a clustering.

Now, we are ready to describe details of our algorithm for general hypergraphs. In this variation, the pivoting node v is chosen arbitrarily from U. The radius ξ defining the cluster $B(\xi)$ including v is chosen from $[0, 1/(2k)]$ so that $C(\xi)/F(\xi)$ is minimized. Although this is a continuous optimization problem, it can be done in $O(n)$ evaluations of the objective because of the following reason. Call the nodes in U by $u_1, \ldots, u_{|U|}$ so that $x_{vu_1} \leq x_{vu_2} \cdots \leq x_{vu_{|U|}}$ holds. Let $i \in \{1, \ldots, |U| - 1\}$. For any $\xi', \xi'' \in (x_{vu_i}, x_{vu_{i+1}}]$ with $\xi' \leq \xi''$, we have $B(\xi') = B(\xi'')$, from which $F(\xi') \leq F(\xi'')$ and $C(\xi') = C(\xi'')$ follow. These two relationships indicate $C(\xi')/F(\xi') \geq C(\xi'')/F(\xi'')$. Therefore, the radius ξ minimizing $C(\xi)/F(\xi)$ can be found from $\{0, 1/(2k)\} \cup \{x_{vu} : u \in U, x_{vu} \leq 1/(2k)\}$, the size of which is $O(|U|) = O(n)$.

The approximation performance of our algorithm depends on $C(\xi)/F(\xi)$; if $C(\xi)/F(\xi) \leq \alpha$ for any iterations, it achieves $2 \max\{k, \alpha\}$-approximation. Lemma 1 guarantees that there always exists a radius $\xi \in [0, 1/(2k)]$ such that $C(\xi)/F(\xi) \leq 2k \log(n + 1)$.

Lemma 1. *For any $v \in U$, there exists $\xi \in [0, 1/(2k)]$ such that $C_{v,x,E}(\xi) \leq 2k \log(n + 1)F_{v,x,E}(\xi)$.*

Theorem 1. *If the radius ξ is defined as in Lemma 1, the approximation factor of Algorithm 1 is $4k \log(n + 1)$.*

As mentioned in Sect. 2, for the disagreement minimization on graphs, the best known approximation factor is $O(\log n)$ [6, 12], and the approximation factor of our algorithm matches it up to a constant when the problem is restricted to graphs. It is an obvious open problem to improve this factor, even for graphs. It is unique-games hard to obtain a constant-factor approximation algorithm for graphs [12]. Since hypergraphs include graphs, the same hardness result applied to the hypergraph correlation clustering.

5 $O(r)$-Approximation for the Coclustering Setting

In this section, we consider the coclustering setting. In other words, the node set of the input hypergraph is the disjoint union of $V_1, \ldots, V_r$, the set of hyperedges coincides with $V_1 \times V_2 \times \cdots \times V_r$, and each hyperedge is associated with a unit weight. The task is to find a partition $\mathcal{C}$ of $\bigcup_{i=1}^{k} V_i$ that minimizes $|\{e \in E_+ : e \in \delta(\mathcal{C})\}| + |\{e \in E_- : e \in E(\mathcal{C})\}|$.

In our pivoting algorithm for this case, the pivoting node v is chosen from $U \cap V_1$ in a certain way whenever $U \cap V_1 \neq \emptyset$, and the radius ξ is set to $1/\sqrt{2(r-1)(2r-1)}$. When $U \cap V_1 = \emptyset$, the clustering of the remaining nodes in U makes no effect on the objective value of the solution because no hyperedge remains in the hypergraph. Hence the algorithm stops the iterations and terminates after adding an arbitrary clustering of the remaining nodes to the solution.

To describe the choice of the pivoting node, let us introduce notations. Let $\mathbf{1}$ be the indicator function for events; if an even $\mathcal{E}$ happens, then $\mathbf{1}(\mathcal{E}) = 1$, and $\mathbf{1}(\mathcal{E}) = 0$ otherwise. In what follows, we rewrite the radius ξ as $1/\theta$ for notational convenience, and assume that θ is a fixed parameter; later, we show that $\theta = \sqrt{2(r-1)(2r-1)}$ minimizes the approximation factor. In each iteration, for a node $v \in U$ and a remaining hyperedge e, we define two costs $L_v(e)$ and $A_v(e)$ as follows:

$$L_v(e) = \begin{cases} x_e \cdot \mathbf{1}(B(v, 1/\theta) \cap e \neq \emptyset) & \text{if } e \in E_+, \\ (1 - x_e) \cdot \mathbf{1}(B(v, 1/\theta) \cap e \neq \emptyset) & \text{if } e \in E_-, \end{cases}$$

$$A_v(e) = \begin{cases} \mathbf{1}(B(v, 1/\theta) \cap e \neq \emptyset \neq e \setminus B(v, 1/\theta)) & \text{if } e \in E_+, \\ \mathbf{1}(e \subseteq B(v, 1/\theta)) & \text{if } e \in E_-. \end{cases}$$

If v is chosen as a pivoting node in this iteration, the cost of the LP solution is decreased by $\sum_{e \in E_+ \cup E_-} L_v(e)$, and the cost of the solution is increased by $\sum_{e \in E_+ \cup E_-} A_v(e)$. In our algorithm, in each iteration, we choose a node $v \in V_1 \cap U$ minimizing $\sum_{e \in E_+ \cup E_-} A_v(e) / \sum_{e \in E_+ \cup E_-} L_v(e)$ as the pivoting node.

If the pivoting node v satisfies $\sum_{e \in E_+ \cup E_-} A_v(e) / \sum_{e \in E_+ \cup E_-} L_v(e) \leq \alpha$ for some $\alpha \geq 1$ in each iteration, then it can be proven that the algorithm achieves α-approximation. Below, we prove that the condition is satisfied with

$$\alpha = \frac{2r - 2}{\sqrt{2(r-1)(2r-1)} - 2r + 2} - \frac{2r-1}{\sqrt{2(r-1)(2r-1)} - 2r + 1}$$
$$= 2\sqrt{2(r-1)(2r-1)} + 4r - 3 \tag{5}$$

when $\theta = \sqrt{2(r-1)(2r-1)}$. Indeed, we prove that

$$\sum_{v \in V_1 \cap U} \sum_{e \in E_+ \cup E_-} A_v(e) \leq \alpha \sum_{v \in V_1 \cap U} \sum_{e \in E_+ \cup E_-} L_v(e) \tag{6}$$

holds with α satisfying (5). Notice that this implies that the node chosen as the pivoting node satisfies the required condition.

In the rest of this section, we prove (6) under an assumption that $U = V$; if $U \subset V$, (6) is proven by applying the following discussion to the sub-hypergraph induced by U. First, we bound $\sum_{v \in V_1} \sum_{e \in E_-} A_v(e)$ in the following lemma.

Lemma 2. $\sum_{v \in V_1} \sum_{e \in E_-} A_v(e) \leq \frac{\theta}{\theta - 2r + 2} \sum_{v \in V_1} \sum_{e \in E_-} L_v(e)$.

Next, we bound $\sum_{v \in V_1} \sum_{e \in E_+} A_v(e)$. We introduce a parameter β that satisfies $0 \leq \beta \leq 1/\theta$. Let us remark that $1 - 1/\theta - (r-1)\beta \geq 0$ holds because $1/\theta \leq (1 - 1/\theta)/(r-1)$ follows from $\theta \geq 2r - 2 \geq r$. We first bound $\sum_{v \in V_1} \sum_{e \in E_+ : x(e) \geq \beta} A_v(e)$.

Lemma 3. $\sum_{v \in V_1} \sum_{e \in E_+ : x(e) \geq \beta} A_v(e) \leq \frac{1}{\beta} \sum_{v \in V_1} \sum_{e \in E_+} L_v(e)$.

Lemma 4. $\sum_{v \in V_1} \sum_{e \in E_+ : x(e) < \beta} A_v(e) \leq \frac{\theta}{1 - \theta\beta} \sum_{v \in V_1} \sum_{e \in E_+ \cup E_-} L_v(e)$.

From Lemmas 2, 3, and 4, we obtain the following inequality:

$$\sum_{v \in V_1} \sum_{e \in E_+ \cup E_-} A_v(e) \leq \left(\max \left\{ \frac{\theta}{\theta - 2r + 2}, \frac{1}{\beta} \right\} + \frac{\theta}{1 - \theta\beta} \right) \sum_{v \in V_1} \sum_{e \in E_+ \cup E_-} L_v(e).$$

Hence, (6) is satisfied when α is the minimum value of $\max\{\theta/(\theta - 2r + 2), 1/\beta\} + \theta/(1 - \theta\beta)$ subject to $\theta \geq 2r - 2$ and $0 \leq \beta \leq 1/\theta$. This minimum value is equal to the right-hand side of (5), which is attained by $\theta = \sqrt{2(r-1)(2r-1)}$ and $\beta = 1 - \sqrt{2(r-1)/(2r-1)}$.

Theorem 2. *If $\xi = 1/\sqrt{2(r-1)(2r-1)}$ and the pivoting node v is the one in $U \cap V_1$ minimizing $\sum_{e \in E_+ \cup E_-} A_v(e) / \sum_{e \in E_+ \cup E_-} L_v(e)$, then the approximation factor of Algorithm 1 is $2\sqrt{2(r-1)(2r-1)} + 4r - 3$ for the coclustering setting.*

6   Conclusion

We considered the hypergraph correlation clustering, and gave two approximation guarantees for the LP-based pivoting algorithms. One is an $O(k \log n)$-approximation guarantee, and the other is an $O(r)$-approximation guarantee. In practice, the former guarantee is more useful because it deals with arbitrary weights while the latter is restricted to coclustering setting. Nevertheless, the latter guarantee is interesting in relationship with previous studies on the disagreement minimization with bipartite graphs [1,2,8].

Acknowledgments. The author is grateful to anonymous referees for their comments. This study was supported by JSPS KAKENHI Grant Number JP17K00040.

References

1. Ailon, N., Avigdor-Elgrabli, N., Liberty, E., van Zuylen, A.: Improved approximation algorithms for bipartite correlation clustering. SIAM J. Comput. **41**(5), 1110–1121 (2012)
2. Amit, N.: The bicluster graph editing problem. Master thesis, Tel Aviv University (2004)
3. Bansal, N., Blum, A., Chawla, S.: Correlation clustering. Mach. Learn. **56**(1–3), 89–113 (2004)
4. Ben-Dor, A., Shamir, R., Yakhini, Z.: Clustering gene expression patterns. J. Comput. Biol. **6**(3/4), 281–297 (1999)
5. Bisson, G., Hussain, S.F.: Chi-sim: a new similarity measure for the co-clustering task. In: Proceedings of the Seventh ICMLA, pp. 211–217 (2008)
6. Charikar, M., Guruswami, V., Wirth, A.: Clustering with qualitative information. J. Comput. Sys. Sci. **71**(3), 360–383 (2005)
7. Chawla, S., Krauthgamer, R., Kumar, R., Rabani, Y., Sivakumar, D.: On the hardness of approximating multicut and sparsest-cut. Comput. Complex. **15**(2), 94–114 (2006)
8. Chawla, S., Makarychev, K., Schramm, T., Yaroslavtsev, G.: Near optimal LP rounding algorithm for correlation clustering on complete and complete k-partite graphs. In: Proceedings of the ACM STOC, pp. 219–228 (2015)
9. Chen, X., Ritter, A., Gupta, A., Mitchell, T.M.: Sense discovery via co-clustering on images and text. In: Proceedings IEEE CVPR, pp. 5298–5306 (2015)
10. Cheng, Y., Church, G.M.: Biclustering of expression data. In: Proceedings of the Eighth International Conference on Intelligent Systems for Molecular Biology, pp. 93–103 (2000)
11. Cohen, W.W., Richman, J.: Learning to match and cluster large high-dimensional data sets for data integration. In: Proceedings of the Eighth ACM SIGKDD, pp. 475–480 (2002)
12. Demaine, E.D., Emanuel, D., Fiat, A., Immorlica, N.: Correlation clustering in general weighted graphs. Theor. Comput. Sci. **361**(2–3), 172–187 (2006)
13. Dhillon, I.S.: Co-clustering documents and words using bipartite spectral graph partitioning. In: Proceedings of the Seventh ACM SIGKDD, pp. 269–274 (2001)
14. Dhillon, I.S., Mallela, S., Modha, D.S.: Information-theoretic co-clustering. In: Proceedings of the Ninth ACM SIGKDD, pp. 89–98 (2003)

15. Filkov, V., Skiena, S.: Integrating microarray data by consensus clustering. In: Proceedings of the 15th IEEE International Conference on Tools with Artificial Intelligence, pp. 418–425 (2003)
16. Hartigan, J.A.: Direct clustering of a data matrix. J. Am. Stat. Assoc. **67**(337), 123–129 (1972)
17. Hatano, D., Fukunaga, T., Kawarabayashi, K.: Scalable algorithm for higher-order co-clustering via random sampling. In: Proceedings of the Thirty-First AAAI, pp. 1992–1999 (2017)
18. Hussain, S.F., Bisson, G., Grimal, C.: An improved co-similarity measure for document clustering. In: Proceedings of the Ninth ICMLA, pp. 190–197 (2010)
19. Kappes, J.H., Speth, M., Reinelt, G., Schnörr, C.: Higher-order segmentation via multicuts. Comput. Vis. Image Underst. **143**, 104–119 (2016)
20. Kim, S., Nowozin, S., Kohli, P., Yoo, C.D.: Higher-order correlation clustering for image segmentation. In: Proceedings of the 25th NIPS, pp. 1530–1538 (2011)
21. Kim, S., Yoo, C.D., Nowozin, S., Kohli, P.: Image segmentation using higher-order correlation clustering. IEEE Trans. Pattern Anal. Mach. Intell. **36**(9), 1761–1774 (2014)
22. Madeira, S.C., Teixeira, M.C., Sá-Correia, I., Oliveira, A.L.: Identification of regulatory modules in time series gene expression data using a linear time biclustering algorithm. IEEE/ACM Trans. Comput. Biol. Bioinform. **7**(1), 153–165 (2010)
23. McCallum, A., Wellner, B.: Toward conditional models of identity uncertainty with application to proper noun coreference. In: Proceedings of the IIWeb, pp. 79–84 (2003)
24. Papalexakis, E.E., Sidiropoulos, N.D., Bro, R.: From K-means to higher-way co-clustering: multilinear decomposition with sparse latent factors. IEEE Trans. Sig. Process. **61**(2), 493–506 (2013)
25. Peng, W., Li, T.: Temporal relation co-clustering on directional social network and author-topic evolution. Knowl. Inf. Syst. **26**(3), 467–486 (2011)
26. Shan, H., Banerjee, A.: Bayesian co-clustering. In: Proceedings of the 8th IEEE ICDM, pp. 530–539 (2008)
27. Zha, H., He, X., Ding, C.H.Q., Gu, M., Simon, H.D.: Bipartite graph partitioning and data clustering. In: Proceedings of the ACM CIKM, pp. 25–32 (2001)
28. Zhao, L., Zaki, M.J.: TriCluster: an effective algorithm for mining coherent clusters in 3D microarray data. In: Proceedings of the ACM SIGMOD, pp. 694–705 (2005)
29. Zhu, Y., Yang, H., He, J.: Co-clustering based dual prediction for cargo pricing optimization. In: Proceedings of the 21th ACM SIGKDD, pp. 1583–1592 (2015)

Approximation Algorithms
for a Two-Phase Knapsack Problem

Kameng Nip[1(✉)] and Zhenbo Wang[2]

[1] School of Mathematics (Zhuhai), Sun Yat-sen University, Zhuhai, China
niejm3@mail.sysu.edu.cn
[2] Department of Mathematical Sciences, Tsinghua University, Beijing, China
zwang@math.tsinghua.edu.cn

Abstract. We consider a natural generalization of the knapsack problem and the multiple knapsack problem, which has two phases of packing decisions. In this problem, we have a set of items, several small knapsacks called boxes, and a large knapsack called container. Each item has a size and profit, each box has a size and the container has a capacity. The first phase is to select some items to pack into the boxes, and the second phase is to select the boxes (each includes some packed items) to pack into the container. The total profit of the problem is determined by the items that are selected and packed into the container within some packed box, and the objective is to maximize the total profit. This problem is motivated by various practical applications, e.g., in logistics. It is a generalization of the multiple knapsack problem, and hence is strongly NP-hard. We mainly propose three approximation algorithms for it. Particularly, the first one is a $\frac{1}{4}$-approximation algorithm based on its linear programming relaxation; the second one is based on applying the algorithms for the multiple knapsack problem and the knapsack problem, and has an approximation ratio $\frac{1}{3} - \epsilon$ for any small enough $\epsilon > 0$. We finally provide a polynomial time approximation scheme for this problem.

Keywords: Knapsack · Multiple knapsack
Approximation algorithms · Polynomial time approximation scheme

1 Introduction

The knapsack problem and its generalizations are fundamental and well-studied problems in combinatorial optimization. Given a knapsack with capacity W and a set of n items, each item j has a profit p_j and a size w_j, the classic 0–1 knapsack problem (KP for short) is to find a subset of items that can be packed into the knapsack and has a maximum total profit. The multiple knapsack problem (MKP for short) is a generalization of KP, in which there are m knapsacks, and each knapsack has a size s_i. Each item can be packed into at most one knapsack, and the objective is also to pack the items into the knapsacks with maximum total profit. In this research, we consider a variant of KP and MKP which is naturally

© Springer International Publishing AG, part of Springer Nature 2018
L. Wang and D. Zhu (Eds.): COCOON 2018, LNCS 10976, pp. 63–75, 2018.
https://doi.org/10.1007/978-3-319-94776-1_6

motivated in some practical fields such as logistics and storage. In our problem, there are m small knapsacks and a large knapsack, as well as a set of n items. Each item has also a profit and a size, and can only be packed into the small knapsack. Each small knapsack has a size, and can be packed into the large knapsack which has a certain capacity. Throughout this paper, we will call a small knapsack "box" and a large knapsack "container" for clarity. The decision maker is facing a two-phase packing decisions, the first phase is to pack the items into the boxes, and the second phase is to pack the boxes (each includes some packed items) into the container, such that the size of each packed box and the capacity of the container are not violated. The total profit is gained from the items that are successfully packed into some box and then the container, and the goal is to find the packing of items and boxes with maximum total profit. We call this problem a two-phase knapsack problem (2-PKP for short).

The 2-PKP problem naturally arises in various practical areas of knapsack problem and multiple knapsack problem. Consider such a scenario in logistic, the goods to be transported are usually packed into some boxes or cartons, and then these boxes are packed into a large container, such as a truck or a cargo ship, and the total profit depends on how many (or how much) goods that are packed. In many applications, it is common that the goods are indivisible and can be packed into at most one box. Therefore, each different box probably has distinct unused space while packing items, depending on its own size. The unused spaces and the selection of boxes determine which items can be selected and packed, and thus has a serious impact on the total profit (we will see some concrete examples in the following sections). Similar instances can be found in the fields such as storage management, transportation and production, where knapsack problem and multiple knapsack problem have a lot of applications [1].

The 2-PKP problem can be viewed as a generalization of KP and MKP, since it reduces to MKP when the capacity of the container is at least the total size of all the boxes, and further reduces to KP when there is only one box. In addition, the 2-PKP problem can be viewed as a combination of MKP and KP, in the sense that each knapsack of a multiple knapsack problem (packing the items into the boxes) is also an item of another knapsack problem (packing the boxes into the container). For more details about the combination of two combinatorial optimization problems, readers can refer to [2–6].

It is well-known that KP is NP-hard [7], and MKP is strongly NP-hard if m is an input of the instance and does not admit a fully polynomial time approximation schemes (FPTAS) even for the case with two identical knapsacks [8]. Therefore, the 2-PKP problem is also strongly NP-hard and is unlikely to admit an FPTAS unless $P = NP$. As a result, we mainly focus on designing approximation algorithms (see, e.g., [9]) for the problem. For KP, a natural strategy is to greedily pack the items in an non-increasing order of their profit-to-size ratio p_j/w_j. It can be shown that the greedy algorithm is a $\frac{1}{2}$-approximation algorithm, if it returns the maximum among the above result and the largest profit item. Moreover, it is known that FPTAS exists for KP [10,11]. For MKP, Chekuri and Khanna [8] proposed the first PTAS with running time $n^{O(\log(1/\epsilon)/\epsilon^8)}$ for any

$\epsilon > 0$, and later Jansen [12] proposed an efficient polynomial time approximation scheme (EPTAS) with running time $2^{O(\log^4(1/\epsilon)/\epsilon)} + n^{O(1)}$. Furthermore, KP and MKP are the special cases of the generalized assignment problem (GAP for short), where both the profit p_{ij} and the size of an item w_{ij} are a function of the knapsack. Shmoys and Tardos [13] proposed a linear programming (LP) rounding algorithm for the minimum version of GAP, and Chekuri and Khanna [8] applied it to obtain a $\frac{1}{2}$-approximation algorithm for MKP and GAP. The approximation ratio for GAP is subsequently improved to $1 - \frac{1}{e} - \epsilon$ based on different LP techniques [14,15] for any small enough $\epsilon > 0$. For more results about the knapsack problem and its generalizations, readers can refer to [1].

There is a wide literature on the variants of knapsack problem. Perhaps the most similar problems with the 2-PKP problem are the nested knapsack problem [16,17] and the knapsack problems with shelf divisions [18]. In the nested knapsack problem in [17] and decomposed knapsack problem in [16] (which can be seen as a special case with two stages of nested knapsack problem), there are multiple stage of knapsacks. Each lower stage i contains several small knapsacks, and each small knapsack is associated with one (and one only) large knapsack in the upper stage $i + 1$. The objective is to select the most profitable subset of items, such that for each stage i, these items can be packed into some knapsacks, and the items packed into each small knapsack, can also be packed into its associated large knapsack in the next stage $i + 1$. The existed studies are mainly focused on the integer programming and branch-and-bound approaches [16,17]. In the knapsack problems with shelf divisions, besides the items, there are some shelf divisors with identical sizes that are required to be packed into a knapsack, and divide the packed items into groups each has total size within a certain range $[\delta, \Delta]$. The problem does not admit any approximation algorithm in general, and has a PTAS when $\delta = 0$ [18]. The key difference is that, in these problems, the objects packed into the larger knapsack are still the items, whereas in our 2-PKP problem, the objects packed into the larger knapsack (the container) are the smaller knapsacks (the boxes), instead of the items themselves (see Fig. 1 for an illustration). Other related work on the variants of KP and MKP include the multiperiod knapsack problem [16,19], in which there are m periods, and the cumulative capacity of knapsack in each period i cannot be exceeded by items chosen from period 1 to i. The two stage knapsack problem [20,21] is a stochastic knapsack problem with random sizes, where in the first stage the items are selected without knowing their sizes, then in the second stage the sizes are revealed and some items can be added or removed with some payments. The bi-level knapsack problem [22–24] usually has two decision makers called a leader and a follower. One classic problem is, the leader decides the capacity of the knapsack, and where the follower decides which items are packed into the knapsack [22]. The problem of packing groups of items into multiple knapsacks [25] have multiple knapsack and a set of items which are partitioned into groups in advance, and the profit of a group can be gained if all items are packed. To the best of our knowledge, we are not aware of the study about the approximation algorithms for the 2-PKP problem in the literature.

2-PKP:

2-NKP:

SKP:

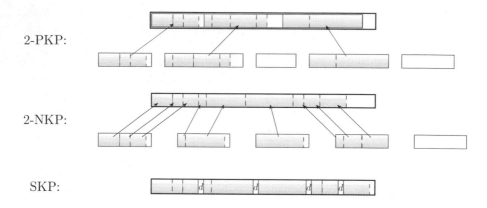

Fig. 1. An illustration of 2-PKP problem, two-stage nested knapsack/decomposed knapsack (2-NKP) problem and knapsack problem with shelf divisions (SKP)

In this paper, we introduce the 2-PKP problem and discuss its approximability. First, we observe that when the number of distinct size values of boxes is a fixed constant, then 2-PKP problem can be solved by applying any MKP algorithms, and thus admits a PTAS. Then we propose three approximation algorithms for the general case of the 2-PKP problem. The first two approximation algorithms are based on packing the largest size boxes. The first one is a $\frac{1}{4}$-approximation algorithm with $O(n^2 \log n)$ time, which is based on the LP rounding algorithm for GAP [13]. The second one is a $1/\left(\frac{2}{r_{MKP}} + \frac{1}{r_{KP}}\right)$-approximation algorithm, where r_{MKP} and r_{KP} are the performance ratios of the MKP algorithm and the KP algorithm, respectively. This algorithm has best possible approximation ratio $\frac{1}{3} - \epsilon$, when the PTAS of MKP and the FPTAS of KP are applied. Finally, we propose a PTAS, by extending several ideas from the PTAS for the MKP [8] to guess the boxes that are packed.

The remainder is organized as follows: In Sect. 2, we formally state the 2-PKP problem studied in this paper. We then present an approximation algorithm based on LP rounding in Sect. 3.1, and an approximation algorithm based on applying the MKP and KP algorithms in Sect. 3.2. In Sect. 4, we propose a PTAS. Finally, we provide some concluding remarks in Sect. 5.

2   Problem Description

First, we introduce the formal definition of the two-phase knapsack problem.

Definition 1 (2-PKP problem). *There are n items $J = \{1, \ldots, n\}$, m boxes $B = \{1, \ldots, m\}$ and a container. Each item j has profit p_j and size w_j, each box i has size s_i, and the container has capacity W. The goal of the two-phase knapsack (2-PKP) problem is to find a subset of items $J' \subset J$ and boxes $B' \subset B$, and a packing in which the total size of items packed into each box i is at most* *its size s_i, and the total size of boxes packed into the container is at most its capacity W, such that the total profit of items (included in some packed box) packed into the container is maximized.*

Without loss of generality, we assume that $w_j \leq \max_{i=1,\ldots,m} s_i$ for all $j = 1,\ldots,n$ and $s_i \leq W$ for all $i = 1,\ldots,m$. Moreover, we first sort the items and the boxes such that $p_1/w_1 \geq p_2/w_2 \geq \cdots \geq p_n/w_n$, and $s_1 \geq \cdots \geq s_m$. If the total size of all the boxes is no more than W, then it is trivial to select all the boxes and the problem is reduced to MKP. Therefore, we assume that $\sum_{i=1}^{m} s_i > W$ in the rest of the discussion.

A natural idea to solve the 2-PKP problem, is to first pack the items into the boxes by some MKP algorithm, then take the profit of each box as the total profit of items inside it, and finally pack the boxes into the container by a KP algorithm. However, it is not hard to find such an example that, even though we can optimally solve the two problems, the obtained solution could be far away from the optimal solution, see e.g., the example shown in Fig. 2. Let n be a sufficiently large number, and ϵ be a sufficiently small value in $(0, \frac{1}{2})$. Obviously, the unique optimal solution to MKP between the items and the boxes, is to pack all the items. However, it gains total profit at most $1 + \epsilon$ to the 2-PKP problem. Indeed, we can pack each profit 1 item (which has size $\frac{1}{2} + \epsilon$) solely into a size 1 box (although it is suboptimal to MKP in the first phase), and pack all these boxes into the container, which has total profit n. The approximation ratio can be arbitrary closed to 0 as n grows up.

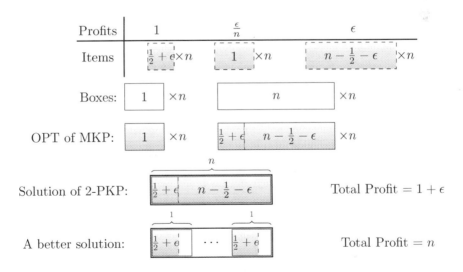

Fig. 2. A bad example for the 2-PKP problem

Suppose we know which boxes are used in an optimal solution, then the 2-PKP problem is simply reduced to MKP, and hence can be solved by any MKP algorithm. As a corollary, when the number of distinct size of boxes is a fixed constant K, we can first guess all the boxes packed in an optimal solution in $O(m^K)$ time (there are at most m boxes for each size of boxes), which is a polynomial in the input size. Therefore, we can conclude that 2-PKP problem has a PTAS in this case, by applying the PTAS of MKP [8].

Theorem 1. *If the number of distinct size values of boxes is fixed, then 2-PKP problem can be reduced to the MKP in polynomial time, and thus has a PTAS.*

3 Approximation Algorithms Using the Largest Size Boxes

From the previous section, the main difficulty to deal with the 2-PKP problem is to consider which boxes should be packed into the container. In this section, we present two approximation algorithms. Both of them pack the boxes into the container one by one in a non-increasing order of their sizes, and compare the solution with the one using the right next box that cannot be packed.

3.1 A $\frac{1}{4}$-Approximation LP Rounding Algorithm

The 2-PKP problem can be formulated as an integer programming problem:

$$\max \quad \sum_{i=1}^{m}\sum_{j=1}^{n} p_j x_{ij} \tag{IP}$$

$$\text{s.t.} \sum_{j=1}^{n} w_j x_{ij} \le s_i y_i \quad \forall i = 1, \ldots, m$$

$$\sum_{i=1}^{m} x_{ij} \le 1 \quad \forall j = 1, \ldots, n$$

$$\sum_{i=1}^{m} s_i y_i \le W$$

$$x_{ij}, y_i \in \{0, 1\} \quad \forall i = 1, \ldots, m, j = 1, \ldots, n.$$

Consider the LP relaxation which is analogous to that for the generalized assignment problem (GAP) in [13]:

$$\max \quad \sum_{i=1}^{m}\sum_{j=1}^{n} p_j x_{ij} \tag{LPR}$$

$$\text{s.t.} \sum_{j=1}^{n} w_j x_{ij} \le s_i y_i, \quad \forall i = 1, \ldots, m \tag{LPR1}$$

$$\sum_{i=1}^{m} x_{ij} \leq 1 \qquad\qquad \forall j = 1, \ldots, n \qquad\qquad\qquad \text{(LPR2)}$$

$$\sum_{i=1}^{m} s_i y_i \leq W \qquad\qquad\qquad\qquad\qquad\qquad \text{(LPR3)}$$

$$x_{ij} \geq 0, \quad 0 \leq y_i \leq 1 \quad \forall i = 1, \ldots, m, j = 1, \ldots, n \qquad \text{(LPR4)}$$

$$x_{ij} = 0, \text{ if } w_j > s_i \quad \forall i = 1, \ldots, m, j = 1, \ldots, n. \qquad \text{(LPR5)}$$

First, we show that there exists an optimal solution to (LPR), which packs the largest size boxes $1, 2, \ldots$ (the last box is possibly packed fractionally).

Lemma 1. *There exists an optimal solution to (LPR) with $y_1 = \cdots = y_{b-1} = 1$, and $y_b = \frac{W - \sum_{l=1}^{b-1} s_l}{s_b}$, where $b = \min \left\{ k \in B \,\middle|\, \sum_{l=1}^{k} s_l > W \right\}$, $y_i = 0$ otherwise.*

The proofs of this and all subsequent lemmas/theorems are omitted, and will be provided in the full version. Given Lemma 1, we can find the optimal solution to (LPR) by considering only the b-largest size boxes, which leads to a LP relaxation of GAP:

$$\max \quad \sum_{i=1}^{b} \sum_{j=1}^{n} p_j x_{ij} \qquad\qquad\qquad\qquad\qquad \text{(LPR}_b)$$

$$\text{s.t.} \sum_{j=1}^{n} w_j x_{ij} \leq s_i, \qquad \forall i = 1, \ldots, b-1 \qquad \text{(LPR}_b 1)$$

$$\sum_{j=1}^{n} w_j x_{bj} \leq s_b y_b, \qquad\qquad\qquad\qquad \text{(LPR}_b 2)$$

$$\sum_{i=1}^{b} x_{ij} \leq 1 \qquad\qquad \forall j = 1, \ldots, n \qquad\qquad \text{(LPR}_b 3)$$

$$x_{ij} \geq 0 \qquad\qquad \forall i = 1, \ldots, b, j = 1, \ldots, n \qquad \text{(LPR}_b 4)$$

$$x_{ij} = 0, \text{ if } w_j > s_i \quad \forall i = 1, \ldots, b, j = 1, \ldots, n. \qquad \text{(LPR}_b 5)$$

In other words, let OPT_{LPR} and OPT_{LPR_b} be the optimal values to (LPR) and (LPR$_b$) respectively, Lemma 1 implies that $OPT_{\text{LPR}} = OPT_{\text{LPR}_b}$. Furthermore, the optimal solution to (LPR$_b$) can be efficiently obtained in $O(n \log n)$ time instead of solving a linear program, as shown in Lemma 2.

Lemma 2. *The optimal solution to (LPR$_b$) can be found in $O(n \log n)$ time.*

Given an optimal solution to (LPR$_b$), we next apply Shmoys and Tardos [8, 13]'s LP rounding procedure for GAP to obtain an integral solution. Altogether, we can obtain an algorithm for the 2-PKP problem, which is summarized in Algorithm 1.

Theorem 2. *Algorithm 1 returns a $\frac{1}{4}$-approximate solution for the 2-PKP problem in $O(n^2 \log n)$ time, and the bound is tight.*

Algorithm 1. A LP rounding based $\frac{1}{4}$-approximation algorithm

---

1: Let $b = \min\{k \in B | \sum_{l=1}^{k} s_l > W\}$, $y_1 = y_2 = \cdots = y_{b-1} = 1$, and $y_b = \frac{W - \sum_{l=1}^{b-1} s_l}{s_b}$, $y_i = 0$ otherwise

2: Find the optimal solution to (LPR$_b$), and apply Shmoys and Tardos's rounding procedure [13] to obtain a packing of items into boxes 1, ..., b

3: Pack the items and boxes 1, ..., $b - 1$ obtained in Step 2 into the container. Denote P_1 as the total profit of this solution

4: Pack the items of J which have size $w_j \le s_b$ into box b, according to the non-increasing order of their profit-to-size ratios. Pack these items and box b into the container. Denote P_2 as the total profit of this solution

5: Let i_{bmax} be the single item which has highest profit among those can be packed into box b, select i_{bmax} and box b. Denote P_3 as its total profit

6: **Return** the solution with highest total profit among P_1, P_2, P_3

3.2 Approximation Algorithm Based on MKP and KP Algorithms

Recall that in Algorithm 1, we first select the largest b boxes (where b is the box that adding it would violate the capacity of the container) that fits in the container, and then pack the items by the 2-approximation LP rounding algorithm for GAP and the greedy algorithm for KP. A natural question is, can we obtain a better performance if we use algorithms with better ratio? We summarize the idea in Algorithm 2. Let r_{MKP}, r_{KP} be the performance ratios of the used MKP and KP algorithms, and T_{MKP}, T_{KP} be their running times, respectively.

Algorithm 2. An approximation algorithm based on MKP and KP algorithms

---

1: Let $b = \min\{k \in B | \sum_{l=1}^{k} s_l > W\}$, $y_1 = y_2 = \cdots = y_{b-1} = 1$, and $y_b = \frac{W - \sum_{l=1}^{b-1} s_l}{s_b}$, $y_i = 0$ otherwise

2: Pack the items of J into the boxes 1, ..., $b - 1$ by any MKP algorithm, and these boxes into the container. Denote P_{MKP} as its total profit.

3: Pack the items of J into the box b by any KP algorithm, and the box b into the container. Denote P_{KP} as its total profit.

4: **Return** the solution with highest total profit among P_{MKP}, P_{KP}.

Theorem 3. *Algorithm 2 returns a* $1 / \left(\frac{2}{r_{MKP}} + \frac{1}{r_{KP}} \right)$-*approximate solution for the 2-PKP problem in* $O(T_{MKP} + T_{KP})$ *time.*

It is known that MKP has a PTAS [8] and KP has a FPTAS [10], thus the best possible performance ratio that Algorithm 2 can attain is $\frac{1}{3} - \epsilon$.

Finally, it is worthwhile to notice that, the performance ratio obtained by Theorem 3 is $\frac{1}{6}$ if we simply apply Shmoys and Tardos's algorithm for MKP and the greedy algorithm for KP to Algorithm 2 (both are $\frac{1}{2}$-approximation algorithms). However, from the previous discussion in Sect. 3.1, we have already seen that by more careful implementation and analysis, this approach actually admits a $\frac{1}{4}$-approximation algorithm and the bound is tight.

4   A Polynomial Time Approximation Scheme

In this section, we present a PTAS for the 2-PKP problem. It adapts several ideas and the framework from the PTAS proposed by Chekuri and Khanna [8] for MKP, except for quite a few details.

Let ϵ be a constant in $(0, 3/4)$. During the process, we might use an $O(\epsilon)$ times of additional boxes to pack the items, but we will eliminate these boxes at the end and maintain a solution with total size of boxes at most W and losing at most $O(\epsilon)$ fractional of total profit. Given a set of items $J' \subset J$ and boxes $B' \subset B$, we denote $P(J')$ as the total profit of J', $S(J')$ and $S(B')$ as the total size of J' and B' respectively. As in [8], we start with an instance in which items have nice structures and can be packed into the container within some boxes.

Lemma 3. *Given an instance for the 2-PKP problem, we can obtain in $n^{O(1/\epsilon^3)}$ time an instance consists of the items $J' \subset J$ which have at most $O(\epsilon^{-1} \ln n)$ distinct profit values and $O(\epsilon^{-2} \ln n)$ distinct size values, and, J' can be packed into some boxes $B' \subset B$ with $S(B') \leq W$, and $P(J') \geq (1 - O(\epsilon))OPT$.*

Next we describe how to partition the boxes into groups, which is a bit more involved than the PTAS of MKP. In [8], the groups are simply divided as small and large according to the number of boxes. But in our problem, we do not know in advance the number of boxes can be used in an optimal solution for each group. Therefore, we need to do some guess for the boxes that are packed in an optimal solution. Let $\widetilde{W}$ be a guess for S^* (which will be precisely specified in Lemma 4), where S^* is the total size of boxes packed in an optimal solution. Then we order the boxes in non-decreasing order of their sizes and group them into groups $B_0, B_1, \ldots, B_l$. Note that we temporarily ignore B_0, which consists of all boxes with size less than $\frac{\epsilon^6 \widetilde{W}}{m}$, and guess the total size of boxes packed in the optimal solution for each other group.

Lemma 4. *Let B_i be the set of boxes each has size in $\left[\frac{\epsilon^6 \widetilde{W}}{m}(1 + \epsilon)^i,\right.$ $\left.\frac{\epsilon^6 \widetilde{W}}{m}(1 + \epsilon)^{i+1}\right)$, and $S^*(B_i)$ be the total size of boxes from group B_i that are packed in the optimal solution, $\forall i = 1, \ldots, l$, where $l = 2\left\lceil \epsilon^{-1} \ln \frac{m}{\epsilon^6} \right\rceil$. We can guess the values $\widetilde{S}(B_1), \ldots, \widetilde{S}(B_l)$ in polynomial time $m^{O(1/\epsilon^3)}$, such that $\widetilde{S}(B_i) \geq S^*(B_i)$ for each i, and if $\widetilde{S}(B_i) \geq \frac{\epsilon W}{l}$, then $\widetilde{S}(B_i) \geq S^*(B_i) \geq (1 - O(\epsilon))\widetilde{S}(B_i)$.*

Given a guess of the total size of boxes packed in an optimal solution for $B_1, \ldots, B_l$, we are now able to classify the groups of boxes. Suppose now we are in a certain iteration of the guesses. We call the group B_i small group if $\widetilde{S}(B_i) \leq \frac{\epsilon^3 \widetilde{W}}{m}(1 + \epsilon)^{i+1}$, and a group B_i large group if $\widetilde{S}(B_i) > \frac{\epsilon^3 \widetilde{W}}{m}(1 + \epsilon)^{i+1}$. Denote by SMALL, LARGE the sets of small and large groups respectively. Note that by definition the number of boxes in each small group is still a fixed number, since each item in B_i has size at least $\frac{\epsilon^6 \widetilde{W}}{m}(1+\epsilon)^i$, and $|B_i| \leq S^*(B_i)/\left(\frac{\epsilon^6 \widetilde{W}}{m}(1 + \epsilon)^i\right) \leq$

$\widetilde{S}(B_i)/\left(\frac{\epsilon^6\widetilde{W}}{m}(1+\epsilon)^i\right) \leq \frac{1+\epsilon}{\epsilon^3}$ if B_i is in SMALL. If LARGE is empty, then by Lemma 6 stated later, we only need to focus on the small groups. The packing of the boxes hence can be enumerated in polynomial time, and then a PTAS is obtained by simply solving the MKP. Therefore, we assume that LARGE is nonempty in the rest of this section. Similar to [8], it is sufficient to remain a fixed number of small groups.

Lemma 5 (Lemma 2.8, [8]). *Let J be the set of items that can be packed in the boxes from SMALL. There exists a set $J' \subseteq J$ such that J' can be packed into the boxes from the largest $1/\epsilon + \lceil 4\epsilon^{-1}\ln 1/\epsilon \rceil + 1$ groups in SMALL, and $P(J') \geq (1-\epsilon)P(J)$.*

Furthermore, we show that it is insignificant to ignore the boxes in B_0, since we can pack all the items in J' (which we guessed in Lemma 4) by a few additional number of boxes in the SMALL and LARGE.

Lemma 6. *Suppose that we have a feasible packing for the item set J', then there exists a packing that can pack all the items in J', and uses boxes only from SMALL and LARGE. Moreover, if the total size of boxes exceed W, then the exceeded part consists of at most $O(\epsilon)$ times of additional boxes from LARGE.*

For the rest of the discussion, when we say guessing the optimal solution, we implicitly refer to the packing that only uses boxes from small and large groups, and a fixed number of small groups (by Lemmas 5 and 6). Now we guess for each box in SMALL, whether it is selected in the knapsack, as well as the $1/\epsilon$ most profitable items that are packed in the box in the optimal solution. In total, the total number of guess is $O\left(2^{\sum_{B_k \in \text{SMALL}}|B_k|}n^{O\left(\frac{1}{\epsilon}\sum_{B_k \in \text{SMALL}}|B_k|\right)}\right) = n^{O\left(\frac{\ln(1/\epsilon)}{\epsilon^5}\right)}$.

Note that by the previous discussion, there must be an enumeration that all the boxes from SMALL in the optimal solution (with $S^*(B_i) \leq \frac{\epsilon^3\widetilde{W}}{m}(1+\epsilon)^{i+1}$) are correctly selected, since the total number of such boxes is a fixed number.

Next we deal with the groups with $S^*(B_i) > \frac{\epsilon^3\widetilde{W}}{m}(1+\epsilon)^{i+1}$. Similar to [8], if the size of an item is at least ϵ times the capacity of the box to which it is packed, then we say that it is packed as large, otherwise it is packed as small.

Lemma 7 ([8]). *In polynomial time, we can guess all the items that are packed as large and also to which groups they are packed.*

By Lemma 7, at least one enumeration can correctly pack the items to the groups that they are packed as large, and now our goal is to finding a feasible packing for these items within each group. Let J_i be the items that are guessed to be packed in group B_i as large. The group B_i consists of several boxes with sizes $\left[\frac{\epsilon^6\widetilde{W}}{m}(1+\epsilon)^i, \frac{\epsilon^6\widetilde{W}}{m}(1+\epsilon)^{i+1}\right)$, and each item in J_i has size in $\left[\frac{\epsilon^7\widetilde{W}}{m}(1+\epsilon)^i, \frac{\epsilon^6\widetilde{W}}{m}(1+\epsilon)^{i+1}\right)$. The main idea is to adapt the shifting technique and the configuration integer program for the variable size bin packing problem [26], which is different from the PTAS of MKP in [8], as it uses the APTAS for the identical size bin packing problem [27].

Lemma 8. *Let J_i and B_i defined as above. There is a polynomial time algorithm that either decides that there is no feasible packing for the items or returns a feasible packing using at most $(1 + O(\epsilon))S^*(B_i)$ total size of boxes.*

Observe that at this stage we have correctly guessed all the boxes that are from the small groups in the optimal solution (i.e., with $S^*(B_i) \leq \frac{c^3\widetilde{W}}{m}(1+\epsilon)^{i+1}$), and have packed some boxes from some large groups in the optimal solution with total size at most $(1+O(\epsilon))S^*(B_i)$. Now we consider to fill the remaining capacity of the container by some boxes, and then pack the remaining items. The idea is to use the LP rounding algorithm in Sect. 3.1.

Lemma 9. *The remaining items can be packed by applying a LP rounding procedure, which uses at most $O(\epsilon)$ times of additional boxes from* LARGE, *and a fixed number of additional boxes from* SMALL.

We finally discard $O(\epsilon)$ fractional of boxes with small total profit and obtain a feasible packing that does not violate the capacity of the knapsack. Altogether, we obtain a PTAS for the 2-PKP problem.

Theorem 4. *There is a PTAS for the 2-PKP problem.*

5   Conclusions

In this paper, we considered a two-phase knapsack problem, which packs the items into the boxes, and packs the boxes into a container. We proposed several approximation algorithms for this problem. There are several future directions for research. For the 2-PKP problem itself, theoretically, it would be interesting to see if there is EPTAS for the problem, as the MKP problem admits an EPTAS [12]. Moreover, it would also be interesting to find some efficient approximation algorithms with performance ratio better than $\frac{1}{4}$ and $\frac{1}{3} - \epsilon$, or some practical algorithms for the 2-PKP problem.

There are several interesting extensions of the 2-PKP problem, such as when each box has a cost, or the weight of the box in the second phase (packing into container) is independent to the first phase (the size of the box), or the knapsack problem with multiple phases. One can also consider other packing problems under this scenario.

Acknowledgments. This work has been supported by NSFC No.11771245 and No. 11371216.

References

1. Kellerer, H., Pferschy, U., Pisinger, D.: Knapsack Problems. Springer, Heidelberg (2004). https://doi.org/10.1007/978-3-540-24777-7
2. Wang, Z., Cui, Z.: Combination of parallel machine scheduling and vertex cover. Theoret. Comput. Sci. **460**, 10–15 (2012)

3. Nip, K., Wang, Z.: Combination of two-machine flow shop scheduling and short-est path problems. In: Du, D.-Z., Zhang, G. (eds.) COCOON 2013. LNCS, vol. 7936, pp. 680–687. Springer, Heidelberg (2013). https://doi.org/10.1007/978-3-642-38768-5_60

4. Nip, K., Wang, Z., Xing, W.: Combinations of some shop scheduling problems and the shortest path problem: complexity and approximation algorithms. In: Xu, D., Du, D., Du, D. (eds.) COCOON 2015. LNCS, vol. 9198, pp. 97–108. Springer, Cham (2015). https://doi.org/10.1007/978-3-319-21398-9_8

5. Nip, K., Wang, Z., Talla Nobibon, F., Leus, R.: A combination of flow shop schedul-ing and the shortest path problem. J. Comb. Optim. **29**(1), 36–52 (2015)

6. Nip, K., Wang, Z., Xing, W.: A study on several combination problems of classic shop scheduling and shortest path. Theoret. Comput. Sci. **654**, 175–187 (2016)

7. Garey, M.R., Johnson, D.S.: Computers and Intractability: A Guide to the Theory of NP-Completeness. Freeman, New York (1979)

8. Chekuri, C., Khanna, S.: A polynomial time approximation scheme for the multiple knapsack problem. SIAM J. Comput. **35**(3), 713–728 (2005)

9. Williamson, D.P., Shmoys, D.B.: The Design of Approximation Algorithms. Cam-bridge University Press, New York (2011)

10. Ibarra, O.H., Kim, C.E.: Fast approximation algorithms for the knapsack and sum of subset problems. J. ACM **22**(4), 463–468 (1975)

11. Lawler, E.L.: Fast approximation algorithms for knapsack problems. Math. Oper. Res. **4**(4), 339–356 (1979)

12. Jansen, K.: Parameterized approximation scheme for the multiple knapsack prob-lem. SIAM J. Comput. **39**(4), 1392–1412 (2010)

13. Shmoys, D.B., Tardos, E.: An approximation algorithm for the generalized assign-ment problem. Math. Program. **62**(3), 461–474 (1993)

14. Feige, U., Vondrak, J.: Approximation algorithms for allocation problems: improv-ing the factor of $1 - 1/e$. In: FOCS 2006, pp. 667–676 (2006)

15. Fleischer, L., Goemans, M.X., Mirrokni, V.S., Sviridenko, M.: Tight approximation algorithms for maximum separable assignment problems. Math. Oper. Res. **36**(3), 416–431 (2011)

16. Dudziński, K., Walukiewicz, S.: Exact methods for the knapsack problem and its generalizations. Eur. J. Oper. Res. **28**(1), 3–21 (1987)

17. Johnston, R.E., Khan, L.R.: Bounds for nested knapsack problems. Eur. J. Oper. Res. **81**(1), 154–165 (1995)

18. Xavier, E., Miyazawa, F.: Approximation schemes for knapsack problems with shelf divisions. Theoret. Comput. Sci. **352**(1), 71–84 (2006)

19. Faaland, B.H.: The multiperiod knapsack problem. Oper. Res. **29**(3), 612–616 (1981)

20. Kosuch, S., Lisser, A.: On two-stage stochastic knapsack problems. Discrete Appl. Math. **159**(16), 1827–1841 (2011)

21. Kosuch, S.: Approximability of the two-stage stochastic knapsack problem with discretely distributed weights. Discrete Appl. Math. **165**, 192–204 (2014)

22. Dempe, S., Richter, K.: Bilevel programming with knapsack constraint. CEJOR **8**, 93–107 (2000)

23. Chen, L., Zhang, G.: Approximation algorithms for a bi-level knapsack problem. Theoret. Comput. Sci. **497**, 1–12 (2013)

24. Caprara, A., Carvalho, M., Lodi, A., Woeginger, G.J.: A study on the computa-tional complexity of the bilevel knapsack problem. SIAM J. Optim. **24**(2), 823–838 (2014)

25. Chen, L., Zhang, G.: Packing groups of items into multiple knapsacks. In: Ollinger, N., Vollmer, H. (eds.) STACS 2016. LIPIcs, vol. 47, pp. 28:1–28:13 (2016)
26. Murgolo, F.D.: An efficient approximation scheme for variable-sized bin packing. SIAM J. Comput. **16**(1), 149–161 (1987)
27. Fernandez de la Vega, W., Lueker, G.: Bin packing can be solved within $1 + \epsilon$ in linear time. Combinatorica **1**(4), 349–355 (1981)

More Routes for Evacuation

Katsuhisa Yamanaka[1($\boxtimes$)], Yasuko Matsui[2], and Shin-ichi Nakano[3]

[1] Iwate University, Morioka, Japan
yamanaka@cis.iwate-u.ac.jp
[2] Tokai University, Hiratsuka, Japan
yasuko@tokai-u.jp
[3] Gunma University, Kiryu, Japan
nakano@cs.gunma-u.ac.jp

Abstract. In this paper, we consider the problem of enumerating spanning subgraphs with high edge-connectivity of an input graph. Such subgraphs ensure multiple routes between two vertices. We first present an algorithm that enumerates all the 2-edge-connected spanning subgraphs of a given plane graph with n vertices. The algorithm generates each 2-edge-connected spanning subgraph of the input graph in O(n) time. We next present an algorithm that enumerates all the k-edge-connected spanning subgraphs of a given general graph with m edges. The algorithm generates each k-edge-connected spanning subgraph of the input graph in O(mT) time, where T is the running time to check the k-edge-connectivity of a graph.

1 Introduction

Evacuation route planning in a road network requires at least one route from any point to a shelter. For example, a spanning tree of a network gives one evacuation route for each point. However, in time of disaster, it is easy to imagine that a lot of roads are broken. In the situation that we know only one route between the current place to a shelter, nobody can ensure that the route can be passed through in safety. Hence, we are required to ensure "multiple" evacuation routes to a shelter from every place. Moreover, to avoid traffic congestion at the time of evacuation, we need some evacuation routes. From these points of view, finding spanning subgraphs with high edge-connectivity is important, since such graphs ensure multiple routes between two points. In this paper, we consider the problem of enumerating spanning subgraphs with high edge-connectivity. From the enumerated subgraphs, we can choose good spanning subgraphs for various criteria. These can be used as candidates of suitable evacuation route planning.

Enumerating designated subgraphs is a fundamental and important problem. The subgraph enumeration is one of the strong and appealing strategies to discover valuable knowledge from enormous graph data in various research areas such as data mining, bioinformatics, and artificial intelligence. To discover valuable knowledge from practical graphs, enumeration algorithms for subgraphs with some properties are studied, such as paths [2,11], cycles [2,11], subtrees [15],

© Springer International Publishing AG, part of Springer Nature 2018
L. Wang and D. Zhu (Eds.): COCOON 2018, LNCS 10976, pp. 76–83, 2018.
https://doi.org/10.1007/978-3-319-94776-1_7

spanning trees [11,12,14], cliques [5,8], pseudo cliques [13], k-degenerate sub-graphs [4], matchings [7,14], connected induced subgraphs [1,9], and so on.

In this paper, we focus on the problem of enumerating spanning subgraphs in a graph. Khachiyan et al. [6] studied the problem of enumerating all the minimal 2-vertex-connected spanning subgraphs. This enumeration problem is a natural extension of the well-known spanning tree enumeration problem. Boros et al. [3] proposed an algorithm that enumerates all the k-vertex-connected spanning sub-graphs. Both papers focused on the vertex-connectivity of spanning subgraphs. It is well known that the vertex-connectivity is one of the most fundamental concepts in network reliability. On the other hands, edge-connectivity is also the well-known fundamental measure of network reliability. To the best of our knowledge, there is no result for the problem of enumerating k-edge-connected spanning subgraphs. In this paper, we present algorithms for the problem.

In this paper, we present the following two enumeration algorithms for span-ning subgraphs using reverse search method by Avis and Fukuda [1]. We first present an algorithm that enumerates all the 2-edge-connected spanning sub-graphs of a given plane graph with n vertices. The algorithm generates each 2-edge-connected spanning subgraph of the input graph in $O(n)$ time. We next present an algorithm that enumerates all the k-edge-connected spanning subgraphs of a given general graph with n vertices and m edges. The algo-rithm generates each k-edge-connected spanning subgraph of the input graph in $O(mT)$ time, where T is the running time to check the k-edge-connectivity of a graph. From the result by Nagamochi and Ibaraki [10], it can be observed that $T = O(m + \min\{kn^2, nm + n^2 \log n\})$ holds.

2 Preliminary

Let $G = (V(G), E(G))$ be an undirected unweighted graph with vertex set $V(G)$ and edge set $E(G)$. We always denote $|V(G)|$ and $|E(G)|$ by n and m, respec-tively. A graph G is *simple* if G has no multi-edge and no self-loop. Throughout this paper, we suppose that graphs are simple unless otherwise noted. A graph G is *k-edge-connected* if the removal of any $k-1$ edges in $E(G)$ does not discon-nect G. Let us remark that, for $k = 1$, a 1-edge-connected graph is a just con-nected graph. A graph $H = (V(H), E(H))$ is a *subgraph* of G if $V(H) \subseteq V(G)$ and $E(H) \subseteq E(G)$ hold. A subgraph $H = (V(H), E(H))$ of G is *spanning* if $V(H) = V(G)$ and $E(H) \subseteq E(G)$ hold. Throughout this paper, we assume that the edges in $E(G)$ are labeled such as $E = \{e_1, e_2, \ldots, e_m\}$. Let e_i and e_j, $i < j$, be two edges in G. We say that e_i is *smaller* than e_j, denoted by $e_i \prec e_j$. Let e be an edge of G. For a subgraph H of G, we denote by $H - e$ the graph obtained from H by removing e. We denote by $H + e$ the graph obtained from H by inserting e.

A graph is *planar* if it can be embedded in the plane so that no two edges intersect geometrically except at a vertex to which they are both incident. A *plane* graph is a planar graph with a fixed planar embedding. A plane graph divides the plane into connected regions called *faces*. The *contour* of a face is the list of edges on the boundary of the face. Let e be an edge in a plane graph G. From the definition of plane graphs, e is shared by the two faces. We denote the two faces by $f_1(e)$ and $f_2(e)$. The *dual graph* $D = (V(D), E(D))$ of a plane graph G is a multi-graph where $V(D)$ is the set of the faces of G and $E(D) = \{\{f, g\} \mid \text{The contours of } f \text{ and } g \text{ share an edge}\}$. If two faces f and g share two or more edges, then the dual graph has multi-edges between f and g.

3 Enumerating All Spanning Subgraphs

In this section, we first present an algorithm that enumerates all the 2-edge-connected spanning subgraphs of a given plane graph. We then present an algorithm that enumerates all the k-edge-connected spanning subgraphs of a given general graph.

3.1 2-Edge-Connected Spanning Subgraphs in Plane Graphs

Let $G = (V(G), E(G))$ be a plane graph. In this section, we give an algorithm that enumerates all the 2-edge-connected spanning subgraphs of G. If G is not 2-edge-connected, then there is no 2-edge-connected spanning subgraph of G. Hence, without loss of generality, we suppose that G is 2-edge-connected. We first define a tree structure among the set of all the 2-edge-connected spanning subgraphs of G. Our algorithm based on the reverse search [1] traverses the tree structure in depth-first manner and enumerates all the 2-edge-connected spanning subgraphs.

We denote by $S_2(G)$ the set of 2-edge-connected spanning subgraphs of G. Note that G itself is in $S_2(G)$. To define a tree structure, we define the *parent* for each 2-edge-connected spanning subgraph of G except G. Let H be a 2-edge-connected spanning subgraph in $S_2(G) \setminus \{G\}$. Let $\mathsf{sm}(H)$ be the smallest edge in $E(G) \setminus E(H)$. Then, we define $\mathsf{par}(H) := H + \mathsf{sm}(H)$. From the definition, it is easy to observe that $\mathsf{par}(H)$ is also 2-edge-connected spanning subgraph of G. By repeatedly finding the parents starting from H, we obtain a sequence $H, \mathsf{par}(H), \mathsf{par}(\mathsf{par}(H)), \ldots$ of 2-edge-connected spanning subgraphs. We call such a sequence the *appending sequence* of H. The sequence starts with H and ends with G. We have the following lemma.

Lemma 1. *Let $H \neq G$ be a 2-edge-connected spanning subgraph of a plane graph G. Then, the appending sequence of H always ends up with G.*

Proof. Let us define a potential function ϕ for a subgraph H as $\phi(H) := |E(H)|$. Then, from the definition of the parent, $\mathsf{par}(H)$ is 2-edge-connected spanning subgraph and $\phi(\mathsf{par}(H)) = \phi(H) + 1$. Since the parent is defined for every 2-edge-connected spanning subgraph except G, we finally have G in the appending sequence. □

Now, we are ready to define a tree structure. By merging all the appending sequences for all the subgraphs in $S_2(G)$, we have a rooted tree structure, called the *family tree*, such that (1) its root corresponds to G, (2) each vertex in the tree corresponds to a 2-edge-connected spanning subgraph of G, and (3) each edge in the tree corresponds to a parent-child relation. Figure 1 shows an example of the family tree.

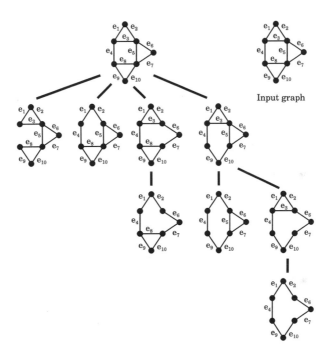

Fig. 1. The family tree of the input graph.

It is easy to see that, if we can traverse the family tree, we can enumerate all the 2-edge-connected spanning subgraphs of G, since the family tree contains all the 2-edge-connected spanning subgraphs of G. To traverse it, we design an algorithm that enumerates all the children of a given subgraph in $S_2(G)$. By recursively applying the child-enumeration algorithm from the root, we can traverse the family tree with depth-first manner.

Now, let us give a condition to be a child of a given 2-edge-connected spanning subgraph. Let H be a 2-edge-connected spanning subgraph in $S_2(G)$. If an edge e is removed from H, then we may obtain a child. However, if $H - e$ is not 2-edge-connected, then such $H - e$ is not a child. Similarly, if e is not the smallest edge in $E(G) \setminus E(H - e)$, that is $e \neq \mathsf{sm}(H - e)$, then such $H - e$ is not a child either. From the observations above, we have the following lemma.

Lemma 2. *Let H be a 2-edge-connected spanning subgraph in a plane graph G, and let e be an edge of H. Then, $H - e$ is a child of H if and only if $e \prec \mathsf{sm}(H)$ holds and $H - e$ is 2-edge-connected.*

From the lemma above, we have the algorithm shown in Algorithm 1. In the for-loop of the algorithm, we choose only the edges with $e = \mathsf{sm}(H - e)$. For each such edge, we check 2-edge-connectivity of $H - e$. The algorithm has no explicit condition of the property for spanning, since $H - e$ is always a spanning subgraph if H is 2-edge-connected.

Algorithm 1. FIND-CHILDREN-2-CONNECTED(H)

---

1 /* G is an input plane graph and is stored in a global memory. H is
    a 2-edge-connected spanning subgraph of G.   */
2 Output H.
3 **foreach** $e \in E(H)$ *with* $e \prec \mathsf{sm}(H)$ **do**
4   **if** $H - e$ *is 2-edge-connected spanning subgraph of* G **then**
5   FIND-CHILDREN-2-CONNECTED($H - e$)

Now, let us estimate the running time of Algorithm 1. We estimate the running time to be required for a vertex in a family tree. In the worst case, We check 2-edge-connectivity for each edge. For each edge e, 2-edge-connectivity of the graph $H - e$ can be checked in $O(m)$ time using depth first search. Hence, we need $O(m^2)$ time in total. Since for a plane graph $m \leq 3n - 6$ holds, the following theorem is obtained.

Theorem 1. *Let G be a plane graph. One can generate each 2-edge-connected spanning subgraph of G in $O(n^2)$ time for each.*

Now, let us improve the running time of our algorithm. The bottleneck of Algorithm 1 is the running time to check 2-edge-connectivity when an edge is removed. To check the connectivity more efficiently, we introduce an observation and a data structure.

Lemma 3. *Let H be a 2-edge-connected plane graph, and let e be an edge of H. Then, $H - e$ is 2-edge-connected if and only if $f_1(e)$ and $f_2(e)$ share only the edge e in H (recall that $f_1(e)$ and $f_2(e)$ are the two faces sharing e).*

Proof. ($\rightarrow$) We assume for a contradiction that $f_1(e)$ and $f_2(e)$ share two or more edges in H. Let e' be an edge shared by $f_1(e)$ and $f_2(e)$ except e. Then e' is a bridge in $H - e$, which is a contradiction.

($\leftarrow$) The removal of e combines $f_1(e)$ and $f_2(e)$ into a face. Let f be the face in $H - e$ obtained by removing e. Then, any edge in $H - e$ is included in at least one cycle. Hence, $H - e$ is still 2-edge-connected.   □

We can use the lemma above to check the 2-edge-connectivity of $H - e$. Our algorithm maintains the dual graph of G as an adjacency matrix representation, where each element for two faces f and g in the matrix stores the number of edges shared by f and g. Using the matrix of the dual graph, we can know the number of edges shared by any two faces. Hence, we can decide whether or not $H - e$ is 2-edge-connected. This check can be done in constant time. To generate a child, an edge e is removed. When e is removed, $f_1(e)$ and $f_2(e)$ are merged into a face, say f'. The face f' is adjacent to the faces adjacent to $f_1(e)$ or $f_2(e)$ except $f_1(e)$ and $f_2(e)$. This update is reflected to the matrix in $O(n)$ time. Therefore, we have the following theorem.

Algorithm 2. FIND-CHILDREN-k-CONNECTED(H)

---

1 /* G is an input graph and stored in a global variable. H is a
     k-edge-connected spanning subgraph of G.   */
2 Output H.
3 **foreach** $e \in E(H)$ *with* $e \prec \mathsf{sm}(H)$ **do**
4   **if** $H - e$ *is k-edge-connected spanning subgraph of G* **then**
5   FIND-CHILDREN-k-CONNECTED($H - e$)

Theorem 2. *Let G be a plane graph with n vertices. One can enumerate every 2-edge-connected spanning subgraph of G in $O(n)$ time for each.*

Proof. Let us estimate the running time required for a subgraph $H \in S_2(G)$ in the family tree. When H is generated from its parent $\mathsf{par}(H)$, the dual graph is updated. This takes $O(n)$ time. To generate a child of H, for each edge e in H with $e \prec \mathsf{sm}(H)$, we check 2-edge-connectivity of $H - e$. Recall that, from Lemma 3, it is sufficient to check the number of edges shared by the corresponding two faces in the dual graph of H. Hence, this check can be done in $O(1)$ time for each and $O(m)$ time in total. Since $m < 3n$ holds, the total running time for H is $O(n)$ time. □

3.2 k-Edge-Connected Spanning Subgraphs in General Graphs

The discussion in the previous subsection can be applied to a general cases: we are given a general graph G and are required to enumerate all the k-edge-connected spanning subgraphs of G. This section shows that Algorithm 1 can be applied to the problem with a slight modification.

Let G be an input graph. We denote by $S_k(G)$ the set of all the k-edge-connected spanning subgraphs of G. If G is not k-edge-connected, then G has no k-edge-connected spanning subgraph. Thus, we assume that G is k-edge-connected.

In a similar way, we define a family tree among the set of k-edge-connected spanning subgraphs of G, as follows. Let H be a k-edge-connected spanning subgraph in $S_k(G)$. Then, we define $\mathsf{par}(H) := H + \mathsf{sm}(H)$. By repeatedly finding the parents starting from H, we obtain a sequence of subgraphs in $S_k(G)$. By merging all the sequences for all the subgraphs in $S_k(G)$, we have a rooted tree structure, called the *family tree*, among $S_k(G)$ such that (1) its root corresponds to G, (2) each vertex in the tree corresponds to a spanning subgraph in $S_k(G)$, and (3) each edge in the tree corresponds to a parent-child relation.

Now, let us consider how to traverse the family tree. The algorithm is almost same to our first algorithm. For the self-containment of this section, we show the pseudo-code in Algorithm 2 which enumerates k-edge-connected spanning subgraph of a given graph G. The difference is how to check edge-connectivity when a child is generated. Let H be a k-edge-connected spanning subgraph of G. For each edge e of H, $H - e$ is a child of H if and only if $H = \mathsf{par}(H - e)$ holds. More precisely, the condition is equivalent to the following two conditions: (1) $H - e$ is k-edge-connected and (2) e is the smallest edge in $E(G) \setminus E(H - e)$, that is $e = \mathsf{sm}(H - e)$. The condition (1) can be checked using the Nagamochi and Ibaraki's algorithm [10] in $O(T)$ time, where $T = O(m + \min \{kn^2, nm + n^2 \log n\})$.

The condition (2) can be checked in $O(1)$ time by maintaining a sorted list of edges in $E(G) \setminus E(H)$. (It is sufficient to compare an edge e with the smallest element in the list.) Besides, the list can be updated in $O(1)$ time. (If the conditions are satisfied, we update the sorted list by inserting an edge e as the first element.) Hence, we can check whether or not $H - e$ is a child of H in $O(T)$ time. To enumerate all the children of H, we check the edge-connectivity of $H - e$ for each edge e in $E(G) \setminus E(H)$. Therefore, we have the following theorem.

Theorem 3. *Let G be a graph. One can generate each k-edge-connected spanning subgraph of G in $O(mT)$ time for each, where $O(T)$ is the running time to check whether a graph is k-edge-connected.*

If $k = 1$, we check whether or not $H - e$ is connected for each edge e in $E(G) \setminus E(H)$. This connectivity can be checked in $O(m)$ time using a depth-first search on $H - e$.

Corollary 1. *Let G be a graph. One can generate each connected spanning subgraph of G in $O(m^2)$ time for each.*

4   Conclusions

We have designed two algorithms for enumerating spanning subgraphs with edge-connectivity at least k. Our first algorithm enumerates all the 2-edge-connected spanning subgraphs of a given plane graph with n vertices in $O(n)$ time for each. The second algorithm enumerates all the k-edge-connected spanning subgraphs of a given general graph with m edges in $O(mT)$ time for each, where T is the running time to check the k-edge-connectivity of a graph.

Future works include improving the running time of our algorithms. Can we enumerate all the 2-edge-connected spanning subgraphs of a given plane graph in constant time for each? Our algorithm enumerates "all" the k-edge-connected spanning subgraphs of a given graph. Can we enumerate only all the "minimal" k-edge-connected spanning subgraphs efficiently?

Acknowledgement. This work was supported by JSPS KAKENHI Grant Numbers JP16K00002, JP17K00003, and JP18H04091.

References

1. Avis, D., Fukuda, K.: Reverse search for enumeration. Discrete Appl. Math. **65**(1–3), 21–46 (1996)
2. Birmelé, E., Ferreira, R., Grossi, R., Marino, A., Pisanti, N., Rizzi, R., Sacomoto, G.: Optimal listing of cycles and st-paths in undirected graphs. In: Proceedings of the 24th Annual ACM-SIAM Symposium on Discrete Algorithms, pp. 1884–1896, January 2012
3. Boros, E., Borys, K., Elbassioni, K., Gurvich, V., Makino, K., Rudolf, G.: Generating minimal k-vertex connected spanning subgraphs. In: Lin, G. (ed.) COCOON 2007. LNCS, vol. 4598, pp. 222–231. Springer, Heidelberg (2007). https://doi.org/10.1007/978-3-540-73545-8_23
4. Conte, A., Kanté, M.M., Otachi, Y., Uno, T., Wasa, K.: Efficient enumeration of maximal k-degenerate subgraphs in a chordal graph. In: Cao, Y., Chen, J. (eds.) COCOON 2017. LNCS, vol. 10392, pp. 150–161. Springer, Cham (2017). https://doi.org/10.1007/978-3-319-62389-4_13
5. Conte, A., Virgilio, R.D., Maccioni, A., Patrignani, M., Torlone, R.: Finding all maximal cliques in very large social networks. In: Proceedings of the 19th International Conference on Extending Database Technology, pp. 173–184 (2016)
6. Khachiyan, L., Boros, E., Borys, K., Elbassioni, K., Gurvich, V., Makino, K.: Enumerating spanning and connected subsets in graphs and matroids. In: Azar, Y., Erlebach, T. (eds.) ESA 2006. LNCS, vol. 4168, pp. 444–455. Springer, Heidelberg (2006). https://doi.org/10.1007/11841036_41
7. Kurita, K., Wasa, K., Uno, T., Arimura, H.: Efficient enumeration of induced matchings in a graph without cycles with length four. CoRR, abs/1707.02740 (2017)
8. Makino, K., Uno, T.: New algorithms for enumerating all maximal cliques. In: Hagerup, T., Katajainen, J. (eds.) SWAT 2004. LNCS, vol. 3111, pp. 260–272. Springer, Heidelberg (2004). https://doi.org/10.1007/978-3-540-27810-8_23
9. Maxwell, S., Chance, M.R., Koyutürk, M.: Efficiently enumerating all connected induced subgraphs of a large molecular network. In: Dediu, A.-H., Martín-Vide, C., Truthe, B. (eds.) AlCoB 2014. LNCS, vol. 8542, pp. 171–182. Springer, Cham (2014). https://doi.org/10.1007/978-3-319-07953-0_14
10. Nagamochi, H., Ibaraki, T.: Computing edge-connectivity in multigraphs and capacitated graphs. SIAM J. Discrete Math. **5**(1), 54–66 (1992)
11. Read, R.C., Tarjan, R.E.: Bounds on backtrack algorithms for listing cycles, paths, and spanning trees. Networks **5**(3), 237–252 (1975)
12. Shioura, A., Tamura, A., Uno, T.: An optimal algorithm for scanning all spanning trees of undirected graphs. SIAM J. Comput. **26**(3), 678–692 (1997)
13. Uno, T.: An efficient algorithm for solving pseudo clique enumeration problem. Algorithmica **56**(1), 3–16 (2010)
14. Uno, T.: Constant time enumeration by amortization. In: Dehne, F., Sack, J.-R., Stege, U. (eds.) WADS 2015. LNCS, vol. 9214, pp. 593–605. Springer, Cham (2015). https://doi.org/10.1007/978-3-319-21840-3_49
15. Wasa, K., Kaneta, Y., Uno, T., Arimura, H.: Constant time enumeration of subtrees with exactly k nodes in a tree. IEICE Trans. Inf. Syst. **97–D**(3), 421–430 (2014)

Fine-Grained Parameterized Complexity Analysis of Knot-Free Vertex Deletion – A Deadlock Resolution Graph Problem

Alan Diêgo Aurélio Carneiro, Fábio Protti, and Uéverton S. Souza[✉]

Fluminense Federal University, Niterói, RJ, Brazil
{aaurelio,fabio,ueverton}@ic.uff.br

Abstract. A knot in a directed graph G is a strongly connected subgraph Q of G with size at least two, such that no vertex in $V(Q)$ is an in-neighbor of a vertex in $V(G) \setminus V(Q)$. Knots are a very important graph structure in the networked computation field, because they characterize deadlock occurrences into a classical distributed computation model, the so-called OR-model. Given a directed graph G and a positive integer k, in this paper we present a parameterized complexity analysis of the KNOT-FREE VERTEX DELETION (KFVD) problem, which consists of determining whether G has a subset $S \subseteq V(G)$ of size at most k such that $G[V \setminus S]$ contains no knot. KFVD is a graph problem with natural applications in deadlock resolution, and it is closely related to DIRECTED FEEDBACK VERTEX SET. It is known that KFVD is NP-complete on planar graphs with bounded degree, but it is polynomial time solvable on subcubic graphs. In this paper we prove that: KFVD is W[1]-hard when parameterized by the size of the solution; it can be solved in $2^{k \log \varphi} n^{O(1)}$ time, but assuming SETH it cannot be solved in $(2 - \epsilon)^{k \log \varphi} n^{O(1)}$ time, where φ is the size of the largest strongly connected subgraph of G; it can be solved in $2^{\phi} n^{O(1)}$ time, but assuming ETH it cannot be solved in $2^{o(\phi)} n^{O(1)}$ time, where ϕ is the number of vertices with out-degree at most k; unless $PH = \Sigma_p^3$, KFVD does not admit polynomial kernel even when $\varphi = 2$ and k is the parameter.

Keywords: Knot · Deadlock resolution · FPT · W[1]-hard · ETH

1 Introduction

Distributed computations are usually represented by directed graphs called *wait-for graphs*. In a wait-for graph $G = (V, E)$, the vertex set V represents processes, and the set E of directed arcs represents wait conditions [2]. An arc exists in E directed away from $v_i \in V$ towards $v_j \in V$ if v_i is blocked waiting a signal from v_j. The graph G changes dynamically according to a set of prescribed rules (the *deadlock model* or *dependency model*), as the computation progresses. In essence, the deadlock model governs how processes should behave throughout computation, i.e., the deadlock model specifies rules for vertices that are not

© Springer International Publishing AG, part of Springer Nature 2018
L. Wang and D. Zhu (Eds.): COCOON 2018, LNCS 10976, pp. 84–95, 2018.
https://doi.org/10.1007/978-3-319-94776-1_8

sinks in G to become sinks [1]. The classical deadlock models related to our work are presented below.

OR-model – In this model, for a process v_i to become a sink, it suffices to receive a signal from *at least one* of the processes from which it is waiting a signal.

AND-model – In this model, a process v_i can only become a sink when it receives a signal from *all* the processes from which it is waiting a signal.

The study of deadlocks is fundamental in computer science and it can be divided into four fields: prevention, avoidance, detection, and resolution (or recovery). Whenever the prevention and avoidance techniques are not applied, and deadlocks are detected, they must be broken through some intervention such as aborting one or more processes to break the circular wait condition causing the deadlock.

In this paper, we consider a scenario where deadlock was detected in a system and some minimum cost deadlock-breaking set must be found and removed from the system. Distributed computations are dynamic, however, as deadlock is a stable property, whenever we refer to G we mean the wait-for graph that corresponds to a *snapshot* of the distributed computation in the usual sense of a consistent global state [7].

The deadlock resolution as an optimization problem differs according to the considered deadlock model, i.e. according to the graph structure that characterizes the deadlock situation. Although prevention, avoidance, and detection of deadlocks have been widely studied in the literature, only few studies have been dedicated to deadlock resolution [6,12,21,24], most of them considering only the AND model. The characterization of a deadlock occurrence in each model and the corresponding decision problem is defined below.

Deadlock in the OR-model – the occurrence of deadlocks in wait-for graphs G working according to the OR-model are characterized by the existence of knots in G [3,17]. A knot in a directed graph G is a strongly connected subgraph Q of G, such that $|V(Q)| \geq 2$ and no vertex in $V(Q)$ is an in-neighbour of a vertex in $V(G) \setminus V(Q)$. Given a graph G and a positive integer k, the KNOT-FREE VERTEX DELETION (KFVD) problem consists of determining whether there exists a subset $S \subseteq V(G)$ of size at most k such that $G[V \setminus S]$ is knot-free. This problem was proved to be NP-hard in [5].

Deadlock in the AND-model – the occurrence of deadlocks in wait-for graphs G working according to the AND-model is characterized by the existence of cycles in G [1,3]. Thus, given a graph G and a positive integer k, the problem of determining whether there exists a subset $S \subseteq V(G)$ of size at most k such that $G[V \setminus S]$ is cycle-free is the well-known DIRECTED FEEDBACK VERTEX SET (DFVS) problem, proved to be NP-hard in the seminal paper of Karp [20], and proved to be fixed-parameter tractable in [8].

In this work we mainly consider the KNOT-FREE VERTEX DELETION problem. The KFVD problem is closed related to DFVS problem not only because of their relation with deadlocks, but some structural similarities between them: the goal of DFVS is to obtain a direct acyclic graph (DAG) via vertex deletion (in such graphs _all_ maximal directed paths end into a sink); the goal of KFVD is to obtain a knot-free graph, and in such graphs for every vertex v there _exists_ at least one maximal path containing v that ends into a sink. Finally, every directed feedback vertex set is a knot-free vertex deletion set; thus, the size of a minimum directed feedback vertex set is an upper bound for KFVD.

In [5], Carneiro et al. present a polynomial-time algorithm for KFVD on graphs with maximum degree three. They also show that the problem is NP-complete even restricted to planar bipartite graphs G with maximum degree equal to four ($\Delta(G) = 4$) and maximum out-degree equal to two ($\Delta^+(G) = 2$).

The remainder of this work is organized as follows. In Sect. 2 we give the main definitions and concepts used during the work. In Sect. 3 we show that KFVD is W[1]-hard when parametered by k, the size of the solution. In Sect. 4 we present two FPT-algorithms for KFVD considering different parameters as well as tight lower bounds based on SETH and ETH, and some proofs of infeasibility of polynomial kernelization.

2   Preliminaries

Graphs. We use standard graph-theoretic notation and concepts, and any undefined notation can be found in [4]. A directed graph $G = (V, E)$ consists of a set of vertices V with $n = |V|$ and a set of direct edges E with $m = |E|$. Let $G[X]$ denote the subgraph of G induced by the vertices in $X \subseteq V$. For $v_i \in V(G)$, let D_i denote the set of descendants of v_i in G (nodes that are reachable from v_i, including itself). Let A_i denote the set of ancestors of v_i in G (nodes that reach v_i, including itself). Let $O_i \subseteq D_i$ be the set of immediate descendants of $v_i \in G$ (descendants that are one arc away from v_i). The out-degree (resp., in-degree) of a vertex v is denoted by $deg^+(v)$ (resp., $deg^-(v)$). In addition, $\delta^+(G)$ (resp., $\delta^-(G)$) denotes the minimum out-degree (resp., in-degree) of a vertex in G.

Parameterized Complexity. Basic concepts, notation, and definitions on parameterized complexity can be found in [11,14,15,23].

A parameterized problem Π is called Fixed Parameter Treatable (FPT) if there is an algorithm A (called FPT-algorithm) that computes every instance $I = (\chi, k)$ correctly and decides if I is a _yes-_ or _no-_ instance in time $f(k).|\chi|^{O(1)}$ for some computable function f. Thus, if k is set to a small value, the growth of the function in relation to χ is relatively small. Kernelization is a powerful technique commonly used to give FPT-algorithms for parameterized problems that mainly consist of, in polynomial time, transforming an instance input I into a new instance I' in such way that the size of I is somehow bounded by the parameter k. In order to show lower bounds on the kernel size we use a parameterized polynomial transformation (called PPT-reduction). Such reduction is defined next.

Definition 1 PPT-reduction: *Let $\Pi(k)$ and $\Pi'(k')$ be parameterized problems where $k' \leq g(k)$ for some polynomial function $g : \mathbb{N} \to \mathbb{N}$. An FPT-reduction from $\Pi(k)$ to $\Pi'(k')$ is a reduction R such that: (i) for all χ, we have $x \in \Pi(\chi)$ if and only if $R(\chi) \in \Pi'(k')$; (ii) R is computable in polynomial time (in relation to k).*

Exponential Time Hypothesis. The Exponential Time Hypothesis (ETH) and its strong variant, the Strong Exponential Time Hypothesis (SETH), are well-known and accepted conjectures that first appeared in [18] and are commonly used to proof lower bounds in parameterized computation. In the literature, several lower bounds have been found to many well-known problems, under such conjectures [11].

Conjecture 1. [19, 22]
Exponential Time Hypothesis (ETH): There is a positive real c such that 3-CNF-Sat cannot be solved in time $2^{cn}(n + m)^{O(1)}$, where n is the number of variables, and m is the number of clauses. In particular, 3-CNF-Sat cannot be solved in $2^{o(n)}(n + m)^{O(1)}$ time.

Conjecture 1 is commonly used together with the Sparsification Lemma [19], meaning that 3-CNF-Sat cannot be solved in $2^{o(n+m)}(n + m)^{O(1)}$ time. In this work, without loss of generality, whenever we refer to ETH we mean to the latter version of the hypothesis.

Conjecture 2. [19, 22]
A consequence of the Strong Exponential Time Hypothesis (SETH): CNF-Sat cannot be solved in time $(2 - \epsilon)^n(n + m)^{O(1)}$, where n is the number of variables, and m is the number of clauses.

Conjecture 2 is an immediate consequence of the Strong Exponential Time Hypothesis (SETH), whose formal definition is omitted due to space constraints.

Additional Concepts and Notation. Sinks are vertices with out-degree zero and sources are vertices with in-degree zero. We use PH to denote the polynomial hierarchy, and Σ_p^3 to denote its third level.

3 W[1]-Hardness

Let k-KFVD stand for the parameterized version of KFVD, where it is asked whether there is a set S with at most k vertices such that $G[V(G) \setminus S]$ is knot-free. We show next that unless $FPT = W[1]$, there is no FPT-algorithm for k-KFVD.

Theorem 1. *The k-KFVD problem is W[1]-hard.*

Proof. The proof is based on an FPT-reduction from MULTICOLORED INDEPENDENT SET, a well-known W[1]-complete problem [9]. Let (G', k') be an instance of MULTICOLORED INDEPENDENT SET, and let $V^1, V^2, \ldots, V^{k'}$ be the color classes of G'. We construct an instance (G, k) of KNOT-FREE VERTEX DELETION as follows (see Fig. 1):

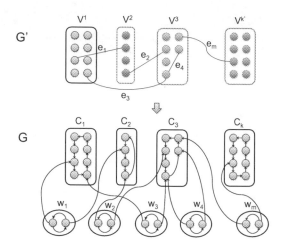

Fig. 1. Instance (G', k') of MULTICOLORED INDEPENDENT SET and instance (G, k) of k-KFVD. (Color figure online)

1. for each vertex v' in G', create a vertex v in G;
2. for a color class V^i in G', create a direct cycle C_i with its corresponding vertices in G;
3. for each edge $e_j = (u', v')$ in G' create a strongly connected component (scc) W_j with two artificial vertices, u_j^w and v_j^w;
4. for each artificial vertex v_j^w, create an edge from v_j^w towards v in G.
5. finally, set $k = k'$.

Suppose that S' is a k-independent set with exactly one vertex of each set V^i of G'. By construction, G has k knots, one for each color class V^i in G'. Thus, at least k vertex removals are necessary to make G free of knots. We set $S = \{v \mid v' \in S'\}$. Next, we show that $G[V \setminus S]$ is knot-free. Each knot C_i is an induced cycle of G, and it is associated to a color class V^i of G'. As S' has one vertex of each color class V^i, all induced cycles C_i will become a directed path after the removal of S. Now, it only remains to show that no new knots appear after the removal of S. Notice that S' is a k-independent set of G'; thus, each scc W_j in G is adjacent to at least one vertex that is not in S. Hence, each scc W_j will have at least one of its exits preserved, i.e., no new knots are created.

Conversely, suppose that G has a set of vertices S of size k such that $G[V \setminus S]$ is knot-free. Note that G have k knots. Then, exactly one vertex of each cycle C_i is in S. By deleting S, each cycle C_i related to V^i will be turned into a path, and no new knots are created after the deletion of S; thus, every scc W_j will have at least one of its exits preserved. We set $S' = \{v' \mid v \in S\}$. Since each scc W_j corresponds to an edge of G', and at least one vertex of each edge of G' is not in S' (otherwise $G[V \setminus S]$ is not knot-free), S' has no pair of adjacent vertices; moreover, S' is composed by one vertex of each C_i. Therefore S' is a multicolored independent set of G'. □

Corollary 2. *Assuming ETH, there is no $f(k)n^{o(k)}$ time algorithm for KFVD for any computable function f.*

Proof. It is known that MULTICOLORED INDEPENDENT SET does not admit a $f(k)n^{o(k)}$ time algorithm, unless ETH fails (see [11]). As the parameterized reduction present in Theorem 1 have linear parameter dependence, we obtain the tight lower bound for KFVD. □

4 Additional Parameters

In this section we present two FPT-algorithms for the KFVD problem. The first algorithm takes into account the size of a largest scc and the size k of the solution as aggregated parameters. The second algorithm uses the number of vertices with maximum out-degree at most k as parameters.

4.1 The Size of the Largest Strongly Connected Component as a Parameter

In this section we consider the size of the largest scc of the input as an additional parameter. The choice of the size of a largest scc as a parameter is mainly inspired by the reductions presented in [5] that prove the NP-hardness of KFVD (even for restricted graph classes). Such reductions result in graphs with scc's of size at most three, and planar graphs with scc's of size at most six.

From the W[1]-hardness of k-KFVD, and the NP-completeness of KFVD on graphs having only scc's of small size, the following question arises: "What is the complexity of k-KFVD restricted to graphs G having only scc's of bounded size?". That question motivates the following parameterized problem:

$[k, \varphi]$–KFVD
Instance: A directed graph $G = (V, E)$, and a positive integer k;
Parameter: k and φ (the size of a largest scc of G);
Goal: Determine if G has a set $S \subset V(G)$ such that $|S| \leq k$ and $G[V \setminus S]$ are knot-free.

We first describe a $2^{k \log \varphi} n^{O(1)}$ time algorithm for $[k, \varphi]$–KFVD.

Lemma 3. *$[k, \varphi]$–KFVD can be solved in $2^{k \log \varphi} n^{O(1)}$ time.*

Proof. Algorithm 1 produces a bounded search tree to solve $[k, \varphi]$–KFVD in $2^{k \log \varphi} n^{O(1)}$ time. In each node of the search tree all possible vertices to be removed of the smallest knot of the current graph are analyzed (their number is bounded by φ). Next, for each possibility, one selected vertex is removed generating a new branch, where the previous steps will be recursively applied until obtaining a directed graph free of knots, or removing exactly k vertices. Since any knot has at most φ vertices and the branching is bounded by k, Algorithm 1 is performed in $2^{k \log \varphi} n^{O(1)}$ time.

It is easy to see that Algorithm 1 is correct due to the following observations:

Algorithm 1. KFVD(G, k)

Result: *true* if G has a knot-free vertex deletion set of size k, and *false* otherwise.

1 **if** G *is knot-free* **then**
2 | **return** *true*;
3 **else**
4 | **if** $k = 0$ **then**
5 | | **return** *false*;
6 | **end if**
7 **end if**
8 answer := *false*;
9 $Q \leftarrow$ set of vertices of the smallest knot in G;
10 **foreach** $v_i \in Q$ **do**
11 | **if** $G - v_i$ *is knot-free* **then**
12 | | **return** "true";
13 | **else**
14 | | answer := answer $\wedge$ KFVD$(G - v_i, k - 1, \varphi)$;
15 | **end if**
16 | **return** answer;
17 **end foreach**

(i) Algorithm 1 effectively checks all possible sets of size at most k that produce a solution;

(ii) all knots in a direct graph can be found and enumerated in linear time with a depth-first search [10];

(iii) the deletion of a vertex cannot increase the size of a largest scc;

(iv) any knot of a directed graph must have at least one vertex removed.   □

Lower Bounds Based on SETH

Now, we show that $[k, \varphi]$–KFVD cannot be solved in $(2 - \epsilon)^{k \log \varphi} n^{O(1)}$ time, unless SETH fails. To show this lower bound we present a SERF-reduction from CNF-Sat to KFVD.

Theorem 4. *Assuming SETH, there is no $(2 - \epsilon)^{k \log \varphi} n^{O(1)}$ time algorithm for KFVD for any $\epsilon > 0$, where φ is the size of a largest strongly connected subgraph of the input.*

Proof. Let F be an instance of CNF-Sat [16] with n variables and m clauses. From F we build a graph $G_F = (V, E)$ which will contain a set $S \subseteq V(G)$ of size $k = n$ such that $G[V \setminus S]$ is knot-free if and only if F is satisfiable. The construction of G_F is described below:

1. For each variable x_i in F, create a directed cycle with two vertices ("variable cycle"), t_{x_i} and f_{x_i}, in G_F.

2. For each clause C_j in F create a directed cycle with two vertices ("clause cycle"), $\ell^1_{c_j}$ and $\ell^2_{c_j}$, in G_F.

3. for each literal x_i (resp. $\bar{x}_i$) in a clause C_j, create an arc from $\ell^1_{c_j}$ to t_{x_i} (resp. f_{x_i}) (Fig. 2).

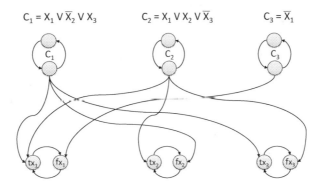

Fig. 2. The resulting graph $G = (V, E)$ from a formula $F = (x_1 \lor x_2 \lor x_3) \land (x_1 \lor x_2 \lor x_3) \land (\overline{x_1})$ where $|V| = O(n + m)$, $|E| = O(n + m)$ and $\varphi = 2$.

At this point it is easy to see that F has a truth assignment if and only if G_F has a set S of vertices containing exactly one vertex of each knot of G_F, such that the removal of S from G_F creates n sinks, for which any clause cycle reaches at least one of them.

Notice that the construction of G_F can be done in polynomial time, $\varphi = 2$ and $k = n$. Therefore, if KFVD can be solved in $(2 - \epsilon)^{k \log \varphi} |V(G_F)|^{O(1)}$ time for $\epsilon > 0$, then we can solve CNF-Sat in $(2 - \epsilon)^n (n + m)^{O(1)}$ time, i.e., SETH fails. □

Lower Bound on the Kernelization

Now, we present some lower bounds on the size of a kernel to $[k, \varphi]$–KFVD and k-KFVD.

Theorem 5. *Unless $PH = \Sigma_p^3$, k-KFVD does not admit a polynomial kernel, even when a largest scc of the input graph G has size 2.*

Proof. In RED-BLUE DOMINATING SET (RBDS) we are given a bipartite graph $G = (B \cup R, E)$ and an integer k and asked whether there exists a vertex set $R' \subset R$ of size at most k such that every vertex in B has at least one neighbor in R'. RBDS parameterized by $(|B|, k)$ is equivalent to SMALL UNIVERSE SET COVER, and RBDS parameterized by $(|R|, k)$ is equivalent to SMALL UNIVERSE HITTING SET. Both problems were shown to have no polynomial kernel (see [13]), unless $PH = \Sigma_p^3$.

The proof is a PPT-reduction from RBDS parameterized by $(|R|, k)$. Let (G, k) be an instance of RBDS parameterized by $(|R|, k)$. We build an instance (G', k') of KNOT-FREE VERTEX DELETION as follows (see Fig. 3):

1. for each vertex v_i in R, create in G' a weakly connected component C_i as follows:
   (a) create two directed cycles of size two, (c_i^1, c_i^2) and (c_i^3, c_i^4);

(b) create an edge from c_i^3 towards c_i^2.

2. for each vertex u_j in B create a set $W_j = \{C_j^1, C_j^2, \ldots, C_j^{k'+1}\}$, were each C_j^z is a directed cycle of size two;

3. for each edge (u_i, u_j) in G, create one directed edge from a vertex of each $C_j^z \in W_j$ to the vertex c_i^2.

4. finally, set $k' = |R| + k$.

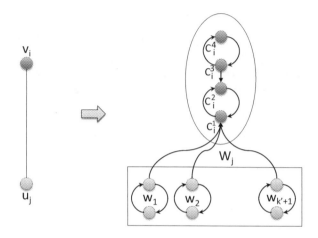

Fig. 3. PPT-reduction from RBDS parameterized by $(|R|, k)$ to k-KFVD with $\varphi = 2$. (Color figure online)

Suppose that S is a red/blue dominating set of G with size k. We build from S a knot-free vertex deletion set S' of G' with size $|R| + k$ as follows: for each vertex $v_i \in R$ we add c_i^1 to S' if $v_i \notin S$, and add c_i^2 and c_i^3 to S' if $v_i \in S$. Since S is a red/blue dominating set of G, every cycle in each W_j will have an arc pointing to one sink c_i^1 in $G'[V \setminus S']$. In addition, all other vertices have either turned into sinks or reach a sink in $G'[V \setminus S']$. Therefore $G'[V \setminus S']$ is knot-free, and $|S'| = |R| + k$.

Conversely, suppose that G' has a set S' of size $k' = |R| + k$ such that $G'[V \setminus S']$ is knot-free. We build from S' a red/blue dominating set S of G with size at most k as follows: add vertex v_i in S if $c_i^1 \notin S'$. Now, we show that S is a red/blue dominating set of G. First observe that G' has $|R|$ knots, and for each $v_i \in R$, $\{c_i^1, c_i^2\}$ induces a knot of G'; then either $c_i^1 \in S'$ or $c_i^2 \in S'$. In addition, for any $v_i \in R$ the removal of c_i^2 creates another knot induced by $\{c_i^3, c_i^4\}$; thus $c_i^1 \notin S' \Rightarrow c_i^2 \in S' \Rightarrow \{c_i^3, c_i^4\} \cap S' \neq \emptyset$. Therefore, as G' has $|R|$ knots and $|S'| = |R| + k$, it follows that $|S| \leq k$. Since each W_j has $|R| + k + 1$ cycles, without loss of generality we can assume that no vertex in W_j is in S', and as $G'[V \setminus S']$ is knot-free, any vertex in W_j (representing a blue vertex) reaches a sink in $G'[V \setminus S']$, which by construction is a vertex c_i^1 (representing a red vertex). Then S is a set of red vertices with size at most k that dominates all blue vertices of G. □

Corollary 6. *Unless $PH = \Sigma_p^3$, $[k, \varphi]$–KFVD does not admit a kernel of size $k^{f(\varphi)}$.*

Proof. This follows from Theorem 5 and the fact that a kernel of size $k^{f(\varphi)}$ for $[k, \varphi]$–KFVD, would be a polynomial kernel for k–KFVD when a largest scc of the input graph G has size 2.

4.2   Number of Vertices with Few Out-Edges as a Parameter

An interesting property related to the degree of the vertices is that if we are interested in removing a set S with k vertices to obtain a knot-free graph, then the out-neighbors of the vertices that will be turned into sinks are contained in S. Thus, if we look for only k removals to obtain a knot-free graph then the candidate vertices to become sinks are the vertices with out-degree at most k. At this point, we consider the number of vertices with out-degree at most k as a parameter.

ϕ-KFVD
Instance: A directed graph $G = (V, E)$, and a positive integer k;
Parameter: ϕ (the number of vertices $v \in G$ with $deg^+(v) \leq k$);
Goal: Determine if G has a set $S \subset V(G)$ such that $|S| \leq k$ and $G[V \setminus S]$ are knot-free.

Lemma 7. *ϕ-KFVD can be solved in $2^\phi n^{O(1)}$ time. In addition, it cannot be solved in $2^{o(\phi)} n^{O(1)}$ time, unless ETH fails.*

Proof. Let L be a set of vertices with $deg^+(v_i) \leq k$ of an input graph G. To solve ϕ-KFVD in $2^\phi n^{O(1)}$ time, it is only needed to try the deletion of all out-neighbors of the subsets of L, checking if the deletion does not exceed k vertices and if the resulting graph is knot-free.

In order to show a lower bound based on ETH to ϕ-KFVD, we can transform an instance F of 3-CNF-Sat to an instance G_F of KFVD using the SERF-reduction presented in Theorem 4, obtaining in polynomial time a graph with $\phi = O(n + m)$. □

Corollary 8. *Unless $PH = \Sigma_p^3$, KFVD does not admit polynomial kernel when parameterized by k and ϕ.*

Proof. We start by making the same transformation as in Theorem 5, obtaining a graph G. Now, for each scc associated with a blue vertex, we add k auxiliary vertices, and add edges in order to transform the component into a complete directed subgraph with $k + 2$ vertices and $(k + 2)(k + 1)$ arcs. Now, the resulting graph G has $\phi = 4|R|$, and the rest of the proof follows as in Theorem 5. □

5   Conclusions

In this work we study the KNOT-FREE VERTEX DELETION problem from a parameterized complexity point of view. We proved that KFVD with the natural parameter k is W[1]-hard through a FPT-reduction from KNOT-FREE VERTEX DELETION, a well-known W[1]-complete problem [9]. Next we proposed two FPT-algorithms, each exploring a different additional parameter. The first parameter, φ, is the maximum size of a scc of the input graph. We show that KFVD can be solved in $2^{k \log \varphi} n^{O(1)}$ time and unless SETH fails it cannot be solved in $(2 - \epsilon)^{(k \log \varphi)} n^{O(1)}$ time. Using a PPT-reduction from RBDS parameterized by $(|R|, k)$ we show that k-KFVD has no polynomial kernel even if the input graph has only scc's with size bounded by 2. The second algorithm runs in $2^{\phi} n^{O(1)}$ time and it is appropriate for graphs where there are few vertices, ϕ, with small out-degree. In addition, assuming ETH, we show that it cannot be done in $2^{o(\phi)} n^{O(1)}$ time. We also show that KFVD has no polynomial kernel when the number of vertices with out-degree at most k is a parameter.

Table 1 summarizes the results presented in this work.

Table 1. Fine-grained parameterized complexity of KNOT-FREE VERTEX DELETION.

Parameter	Complexity	Running time	Lower bounds assuming (S)ETH
k	W[1]-hard	n^k	no $f(k) \times n^{o(k)}$ alg.
k, φ	FPT	$2^{k \log \varphi} \times n^{O(1)}$	no $(2 - \epsilon)^{k \log \varphi} \times n^{O(1)}$ alg.
ϕ	FPT	$2^{\phi} \times n^{O(1)}$	no $2^{o(\phi)} \times n^{O(1)}$ alg.

Recall that knots characterize the presence of deadlock. The algorithms presented in this work have also practical value. The most common approach to deal with deadlock is to forbid the formation of cycles in the direct graph as the computation proceeds. This approach, although simple and easy to implement, is very restrictive. Having an algorithm that breaks the knots of a graph (therefore removing deadlocks) in exponential time, but over a controlled characteristic, allows the construction of a more permissive deadlock prevention. For example, as Algorithm 1 is FPT with respect to k and the size of a largest scc, it is possible to forbid only the formation of large knots, rather than cycles.

References

1. Barbosa, V.C.: The combinatorics of resource sharing. In: Corrêa, R., Dutra, I., Fiallos, M., Gomes, F. (eds.) Models for Parallel and Distributed Computation. APOP, vol. 67, pp. 27–52. Springer, Boston (2002). https://doi.org/10.1007/978-1-4757-3609-0_2
2. Barbosa, V.C., Benevides, M.R.: A graph-theoretic characterization of AND-OR deadlocks. Technical report COPPE-ES-472/98, Federal University of Rio de Janeiro, Rio de Janeiro, Brazil (1998)
3. Barbosa, V.C., Carneiro, A.D.A., Protti, F., Souza, U.S.: Deadlock models in distributed computation: foundations, design, and computational complexity. In: Proceedings of the 31st ACM/SIGAPP Symposium on Applied Computing, pp. 538–541 (2016)

4. Bondy, J.A., Murty, U.S.R.: Graph Theory with Applications, vol. 290. Macmilan, London (1976)
5. Carneiro, A.D.A., Protti, F., Souza, U.S.: Deletion graph problems based on deadlock resolution. In: Cao, Y., Chen, J. (eds.) COCOON 2017. LNCS, vol. 10392, pp. 75–86. Springer, Cham (2017). https://doi.org/10.1007/978-3-319-62389-4_7
6. Chahar, P., Dalal, S.: Deadlock resolution techniques; an overview. Int. J. Sci. Res. Publ. 3(7), 1–5 (2013)
7. Chandy, K.M., Lamport, L.: Distributed snapshots: determining global states of distributed systems. ACM Trans. Comput. Syst. 3, 63–75 (1985)
8. Chen, J., Liu, Y., Lu, S., O'sullivan, B., Razgon, I.: A fixed-parameter algorithm for the directed feedback vertex set problem. J. ACM (JACM) 55(5), 21 (2008)
9. Chen, J., Meng, J.: On parameterized intractability: hardness and completeness. Comput. J. 51(1), 39–59 (2007)
10. Cormen, T.H., Leiserson, C.E., Rivest, R.L., Stein, C.: Introduction to Algorithms. MIT Press, Cambridge (2009)
11. Cygan, M., Fomin, F.V., Kowalik, Ł., Lokshtanov, D., Marx, D., Pilipczuk, M., Pilipczuk, M., Saurabh, S.: Parameterized Algorithms. Springer, Cham (2015). https://doi.org/10.1007/978-3-319-21275-3
12. de Mendívil, J.G., Fariña, F., Garitagotia, J.R., Alastruey, C.F., Bernabeu-Auban, J.M.: A distributed deadlock resolution algorithm for the and model. IEEE Trans. Parallel Distrib. Syst. 10(5), 433–447 (1999)
13. Dom, M., Lokshtanov, D., Saurabh, S.: Kernelization lower bounds through colors and IDs. ACM Trans. Algorithms (TALG) 11(2), 13 (2014)
14. Downey, R.G., Fellows, M.R.: Parameterized Complexity. Monographs in Computer Science, p. 87. Springer, New York (1999). https://doi.org/10.1007/978-1-4612-0515-9
15. Flum, J., Grohe, M.: Parameterized Complexity Theory. Texts in Theoretical Computer Science. An EATCS Series. Springer, Heidelberg (2006). https://doi.org/10.1007/3-540-29953-X
16. Gary, M.R., Johnson, D.S.: Computers and intractability: a guide to the theory of NP-completeness (1979)
17. Holt, R.C.: Some deadlock properties of computer systems. ACM Comput. Surv. (CSUR) 4(3), 179–196 (1972)
18. Impagliazzo, R., Paturi, R.: Complexity of k-SAT. In: 1999 Proceedings of the Fourteenth Annual IEEE Conference on Computational Complexity, pp. 237–240. IEEE (1999)
19. Impagliazzo, R., Paturi, R., Zane, F.: Which problems have strongly exponential complexity? In: 1998 Proceedings 39th Annual Symposium on Foundations of Computer Science, pp. 653–662. IEEE (1998)
20. Karp, R.M.: Reducibility among combinatorial problems. In: Miller, R.E., Thatcher, J.W., Bohlinger, J.D. (eds.) Complexity of Computer Computations. IRSS, pp. 85–103. Springer, Boston (1972). https://doi.org/10.1007/978-1-4684-2001-2_9
21. Leung, E.K., Lai, J.Y.-T.: On minimum cost recovery from system deadlock. IEEE Trans. Comput. 9(C–28), 671–677 (1979)
22. Lokshtanov, D., Marx, D., Saurabh, S., et al.: Lower bounds based on the exponential time hypothesis. Bull. EATCS 3(105), 41–72 (2013)
23. Niedermeier, R.: Invitation to fixed-parameter algorithms (2006)
24. Terekhov, I., Camp, T.: Time efficient deadlock resolution algorithms. Inf. Process. Lett. 69(3), 149–154 (1999)

Approximating Global Optimum
for Probabilistic Truth Discovery

Shi Li, Jinhui Xu, and Minwei Ye[✉]

State University of New York at Buffalo, Buffalo, USA
{shil,jinhui,minweiye}@buffalo.edu

Abstract. The problem of truth discovery arises in many areas such
as database, data mining, data crowdsourcing and machine learning. It
seeks trustworthy information from possibly conflicting data provided by
multiple sources. Due to its practical importance, the problem has been
studied extensively in recent years. Two competing models were proposed for truth discovery, weight-based model and probabilistic model.
While $(1+\epsilon)$-approximations have already been obtained for the weight-
based model, no quality guaranteed solution has been discovered yet
for the probabilistic model. In this paper, we focus on the probabilistic
model and formulate it as a geometric optimization problem. Based on
a sampling technique and a few other ideas, we achieve the first $(1+\epsilon)$-
approximation solution. The general technique we developed has the
potential to be used to solve other geometric optimization problems.

Keywords: Geometric optimization · Truth discovery
High-dimension · Data mining

1 Introduction

Truth discovery has received a great deal of attention in recent years in databases,
data crowdsourcing, machine learning and data mining [9,10,13,14,16]. It
emerges from various practical scenarios such as copying detection [5], data
fusion [3] and conflicting information resolving on the web [16]. In a typical
scenario, the unknown truth for one or multiple objects can be viewed as a
vector in a high-dimension space. The information about the truth vector may
come from multiple sources. Those sources may be inaccurate, conflicting or even
biased from the beginning if they come from subjective evaluation. Our goal is
to infer the truth vector from these noisy information.

A naive method for this problem is to take the average of all the vectors
from sources as the ground truth (for coordinates correspondent to categorical
data, take the majority vote). However, this approach, which inherently treats all

The research of the first author was supported in part by NSF grants CCF-1566356
and CCF-1717134. The research of the last two authors was supported in part by
NSF through grants CCF-1422324, IIS-1422591, and CCF-1716400.

L. Wang and D. Zhu (Eds.): COCOON 2018, LNCS 10976, pp. 96–107, 2018.
https://doi.org/10.1007/978-3-319-94776-1_9

sources as equally important, is vulnerable to unreliable and malicious sources. Such sources can provide information that pulls the average away from the truth. A more robust type of approaches is to give weights to sources to indicate their reliability and use the weighted average or weighted majority as the ground truth. However, since the weights are often unknown, the goal of finding the ground truth is coupled with the task of reliability estimation. This type of approaches is referred as a *truth discovery* approach. Among all, there are two competing and sometimes complementary frameworks that are widely accepted and used for different data types.

Weight-Based Truth Discovery. In this framework, both the truth and the weights are treated as variables. An objective function is defined on these variables [10]. Then an alternating minimization algorithm can be used to solve the problem. In each iteration, the algorithm fixes one set of variables (either the truth variables, or the weight variables) and optimizes the other. This procedure continues until a stable solution is reached. Many existing methods [4,7,11,16] follow this framework and justify themselves by experimenting with different types of real-world datasets. However, none of these methods provides any theoretical guarantee regarding the quality of solution. Recently, Ding et al. [2] gave the first algorithm that achieves a theoretical guarantee (*i.e.*, a $(1 + \epsilon)$-approximation) for a well-known weight-based model of truth discovery introduced in [10]. Later, Huang et al. [19] further improved the running time to near quadratic.

Probabilistic Truth Discovery. Probabilistic models lie in a different category of models for truth discovery. They were also studied extensively in the literature [12,15,17,18]. Instead of giving weights to indicate the reliability of all sources, these models assume that the information for each source is generated independently from some distribution that depends on the truth and the reliability of the source. Then the goal under these models is to find the truth that maximizes the likelihood of the generated information from all sources. The probabilistic models have been shown to outperform the weight-based methods on numerical data [17]. They also prevail other models in the case where sources come from subjective evaluation [13]. For the qualify of the optimization, [15] gave an iterative algorithm with guaranteed fast convergence to a local optimum.

1.1 Our Results

We propose a probabilistic truth discovery model, reformulate it as an optimization problem and give a PTAS (Polynomial-Time Approximation Scheme) to solve it. We assume that each observation of a source is generated around the truth vector with variance corresponding to the reliability of the source. Then, the goal of finding the truth vector with the maximum likelihood can be formulated as an optimization problem. Instead of directly solving the optimization problem, we convert it to the following more general geometric optimization problem:

$$\text{Given } \{p_1, p_2, \cdots, p_n\} \subset \mathbb{R}^d, \text{ find } x \in \mathbb{R}^d \text{ to minimize } \sum_{i=1}^{n} f(\|x - p_i\|),$$

where f is a function satisfying some reasonable properties.

This general problem encloses as special cases the classic 1-median and 1-mean problems, and the more general problem of minimizing p-th power of distances. Moreover, by considering the corresponding functions with an upper-threshold, i.e, $f(\ell) = \min\{\ell, B\}$, $f(\ell) = \min\{\ell^2, B\}$ and $f(\ell) = \min\{\ell^p, B\}$, one can capture the outlier versions of all these problems.

We give a sampling-based method that solves the above optimization problem up to a factor of $1 + \epsilon$ for any $\epsilon > 0$ in quadratic running time. Thus, it not only solves our truth discovery problem but also gives a unified approach to solve all the above problems under this framework.

1.2   Our Techniques

One property that we *do not* impose on the function f is convexity. Requiring f to be convex will make our problem too restrictive. For example, the cost function f_{truth}(defined later) is non-convex in our truth discovery problem. The threshold functions that are used to model the outlier versions of the 1-center problems are also non-convex. Without the convexity property, iterative approaches such as gradient descent and EM do not guarantee the global optimality. General coreset technique (such as the one in [6]) which reduces the size of the problem will not work, either. The dimensionality is not reduced by those techniques so that the problem is still hard even for the coreset.

Instead of using methods in continuous optimization or general sampling technique, our algorithm is based on the elegant method Badoiu, Har-Peled and Indyk developed to give fast algorithms for many clustering problems [1,8]. Roughly speaking, [1] showed that a small set of sample points X can guarantee that the affine subspace span(X) contains a $(1 + \epsilon)$ approximate solution for these clustering problems. Therefore both the size and the dimensionality can be reduced.

Directly applying [1] does not work for non-convex cost function. In this paper, we extend [1] to a more general family of cost functions, including the non-convex cost function for our truth discovery problem. We will elaborate the challenges in Sect. 3.2.

2   Problem Formulation and Main Results

2.1   Probabilistic Truth Discovery

We first set the stage for the problem. The unknown truth can be represented as a d dimensional vector p^*, as justified in [10]. There are n sources, and the observation/evaluation made by the i-th source is denoted as p_i which also lies in the d dimensional space $\mathbb{R}^d$. In our model, we assume that each observation/evaluation is a random variable following a multi-variate Gaussian distribution centered at the truth p^* with covariance $\sigma_i^2 I_d$.[1] Each unknown parameter $\sigma_i \geq 0$ represents the reliability of the source; the smaller the variance, the more reliable the source is.

We formulate the problem as finding the $(p^*, \sigma = (\sigma_i)_{i \in [n]})$ that maximizes the likelihood of the random procedure generating p^*. We impose a hyper-parameter $\sigma_0 > 0$ and require $\sigma_i \geq \sigma_0$ for every $i \in [n]$. It is naturally interpreted as an upper bound of the reliability of all sources, but there is another interpretation that we will discuss later.

Given the set of observation $P = \{p_i\}_{i=1}^n \subset \mathbb{R}^d$ under this probabilistic model and a hyper-parameter σ_0, we need to find a point x that maximizes the following likelihood function:

$$\prod_{i=1}^n \mathcal{N}(p_i \mid x, \sigma_i^2 I_d) = \prod_{i=1}^n \left(\frac{1}{\sqrt{2\pi}\sigma_i}\right)^d \exp\left[-\frac{\|p_i - x\|^2}{2\sigma_i^2}\right].$$

Taking negative logarithm and optimizing the quantity over all valid vectors $\sigma = (\sigma_i)_{i \in [n]}$, we obtain the following optimization problem:

$$\min_{x \in \mathbb{R}^d, \sigma} \left\{\frac{nd}{2}\ln(2\pi) + \sum_{i=1}^n \left(d\ln\sigma_i + \frac{\|p_i - x\|^2}{2\sigma_i^2}\right)\right\}, \quad \text{s.t.} \quad \sigma_i \geq \sigma_0, \forall i \in [n]. \tag{1}$$

Lemma 1. *For a fixed $x \in \mathbb{R}^d$, the following vector σ minimizes the objective function in (1):*

$$\sigma_i = \max\left\{\sigma_0, \|p_i - x\|/\sqrt{d}\right\}, \quad \forall i \in [n].$$

Applying Lemma 1, the optimization problem now only depends on the point $x \in \mathbb{R}^d$:

$$\min_{x \in \mathbb{R}^d} \left\{\frac{nd}{2}\ln(2\pi) + \sum_{\|p_i - x\| < \sigma_0 \sqrt{d}} \left(\frac{\|p_i - x\|^2}{2\sigma_0^2} + d\ln\sigma_0\right)\right.$$
$$\left. + \sum_{\|p_i - x\| \geq \sigma_0 \sqrt{d}} \left(\frac{d}{2} + d\ln\frac{\|p_i - x\|}{\sqrt{d}}\right)\right\}.$$

Notice that scaling x, σ_0 and all points p_i by a fact of c only changes the value of the function by a constant additive term $(nd \ln c)$. For simplicity, we will apply

---

[1] For categorical data, the Gaussian distribution may cause fractional answers, which can be viewed as a probability distribution over possible truths. In practice, variance for different coordinates of the truth vector may be different and there might be some non-zero covariance between different coordinates; however, up to a linear transformation, we may assume the covariance matrix is $\sigma_i^2 I_d$.

a scaling to the triple $(x, \sigma_0, \{p_i\}_{i=1}^n) \mapsto (x', \sigma_0'.\{p_i'\}_{i=1}^n)$ so that $\sigma_0' = 1/\sqrt{d}$ and drop the prime symbol if there is not ambiguity. The objective function becomes:

$$\min_{x \in \mathbb{R}^d} \left\{ \frac{nd}{2} \ln(2\pi) + \sum_{\|p_i - x\| < 1} \left(\frac{d\|p_i - x\|^2}{2} - \frac{d\ln d}{2} \right) \right.$$

$$\left. + \sum_{\|p_i - x\| \geq 1} \left(d \left(\frac{1}{2} + \ln\|p_i - x\| \right) - \frac{d\ln d}{2} \right) \right\}.$$

Moreover, we can drop the constant term $\frac{nd}{2} \ln(2\pi) - \frac{nd}{2} \ln d$, and then divide the whole function by $d/2$, the final optimization problem becomes:

$$\min_{x \in \mathbb{R}^d} \sum_{i=1}^n f_{truth}(\|x - p_i\|) \quad \text{where} \quad f_{truth}(\ell) = \begin{cases} \ell^2 & 0 \leq \ell < 1 \\ 1 + \ln \ell^2 & \ell \geq 1 \end{cases}. \quad (2)$$

This objective function can be seen as the summation of costs from each individual point. The cost function f for each p_i is quadratic when its distance to the variable p is close, and it grows logarithmically when p_i is far away.

The function $\sum_{i=1}^n f_{truth}(\|x - p_i\|)$ can be served as an alternative way of evaluating the solution's quality other than the negative log-likelihood since:

(1) It has non-negative objective function value so that multiplicative approximation factor can be properly defined, which serves as a criterion of the solution's quality.

(2) The $(1 + \epsilon)$ approximation of $\sum f_{truth}$ gives the following guarantee. Let $Q_0 = \left(\frac{1}{\sqrt{2\pi}\sigma_0} \right)^d$ be the maximum possible likelihood for the optimum solution of *any instance* with n points and d dimensions. Let Q^* be the likelihood for the optimum solution to the given instance. If $Q^* = Q_0 e^{-t}$, then we shall give a solution with likelihood at least $Q_0 e^{-(1+\epsilon)t}$.

Interpretation of the Parameter σ_0. σ_0 in our model is introduced to reflect the overall reliability of the dataset. If each σ_i is unconstrained, or in other words $\sigma_0 = 0$, then quantity (1) can tend to $-\infty$ by letting $p_i = x$ and $\sigma_i \to 0$ for some $i \in [n]$. At this point, it may seem that the introduction of the parameter σ_0 is a little bit unnatural. However, we argue that this issue caused by the singular solutions does not only exist in our model; it comes with the truth discovery problem itself. If one does not impose any assumption on the reliability of the sources, then a solution (in any model) can be: one source is 100% reliable, all the other sources are not reliable at all and the truth is the data given by the reliable source. Such a model will not be general enough. Any meaningful model needs to be able to capture more than this type of solutions.

With the understanding that σ_0 gives an upper bound on the reliability of the sources, we can discuss how σ_0 affects the optimum solution of our problem. In one extreme, σ_0 is very small, meaning that any source can be very reliable. Then in our final optimization problem (2), the points p_i's are far away from each other. (Recall that to obtain (2), we scaled the original $p_i \mapsto p_i' = \frac{p_i}{\sqrt{d}\sigma_0}$.)

Then for a typical center point x, most p_i's will have large $\|p_i - x\|$. For these points, the f values are logarithmic in their distances to the center and thus are very insensitive to the location of the center. In this case, the optimum solution x will be very close to some input point p_i.

Consider the other extreme where σ_0 is very large. Then, the points p_i are close to each other. In this case, the cost function will be distance square when x is close to all points. The problem then becomes the classic 1-mean problem. This coincides with our intention of setting the "overall confidence" σ_0: σ_0 being very large indicates that all sources are unreliable when considered alone, and it is wiser to take the average than to favor a particular source.

It might seem unreasonable to set a hyper parameter in "truth discovery" problem because "truth" is usually assumed to be invariant to some hyper-parameter we select in our model. Indeed, the truth should be invariant if it is a numerical fact such as the height of a mountain or today's weather forecast at some location. But if we are talking about the rating of a movie or evaluation of an instructor, it is presumptuous to suggest that there exists some "truth discovery" model which can somehow "calculate" such truth exactly or approximately. In such setting, the best we can guarantee is providing a model that can rule out some outliers for the users. The hyper-parameter is provided for the users to decide how much portion of the sources are outliers to him/her.

Here we present our main result for probabilistic truth-discovery problem. It is directly implied by our main theorem, Theorem 3.

Theorem 1. *Let $0 < \epsilon \leq 1$. Let P be a set of n points in $\mathbb{R}^d$ and $G(x) = \sum_{p \in P} f_{truth}(\|x - p\|)$. A $(1 + \epsilon)$-approximate solution can be obtained in time $O(2^{(1/\epsilon)^{O(1)}} d + n^2 d)$.*

3   Solution for General 1-Center Optimization Problem

3.1   General Description of the Algorithm

The following notations are used throughout this section. Given the point set $P \subset \mathbb{R}^d$, a cost function $f : \mathbb{R}_{\geq 0} \to \mathbb{R}$, let $G(x) = \sum_{p \in P} f(\|x - p\|)$ denote the objective function. We reuse the variable p_{opt} as the optimizer of $G(x)$.

We show in advance the following three properties that a general cost function f need to satisfy in order to apply our extended sampling method.

Property 1. (Regularity) f is a continuous, non-negative, monotonically increasing function.

Property 2. (Sub-proportionality)[2] $\exists \alpha \geq 1 : f(kx) \leq k^\alpha f(x)$ for any $k \geq 1$, $x \geq 0$. We say α is the *proportional degree* of f if it is the smallest α satisfies such property.

---

[2] Also referred as polynomial growing function or Log-Log Lipschitz function in literature.

Property 3. The function f can be computed in polynomial time with respect to the size of the input. The inverse of f, defined as $f^{-1}(y) = \sup_x\{x : f(x) = y\}$, should also be able to calculate in polynomial time w.r.t to the size of x when $y \leq 2f(x)$.

Remark 1. The only place Property 3 is used is in Theorem 3. It is imposed to ensure a polynomial running time in arithmetic calculation.

Remark 2. Continuity can be implied by Property 2 by taking $k \to 1$. Also, by taking $x = 0$ in Property 2, one can infer that $f(0) \geq 0$. With the fact that f is non-decreasing, one can also infer that f is non-negative.

Thus essentially the first two properties are (i) monotonically increasing, which is a common assumption when a function is referred as a "cost" function; (ii) sub-proportionality, which can be roughly thought of as requiring the function not growing exponentially. Intuitively speaking, an equivalent statement is that for every $a > 0$, the graph of the unique function $g(x) = Cx^\alpha$ going through $(0,0)$ and $(a, f(a))$ is completely above (can overlap) the graph of $f(x)$ when $x \geq a$.

From now on, these three properties are always assumed for a cost function f unless stated otherwise.

To approximate the optimizer p_{opt} of $G(x)$, we generalize an existing result from Bādoiu, et al. [1] (for convex functions) to our problem where the function can be non-convex. The key idea is to sample a core-set X from the input points P such that the affine subspace $\text{span}(X)$ contains a $(1+\epsilon)$-approximate solution. We summarize the method in a general way in the following procedures:

1. The value L is chosen so that the following two things can both happen:
    (a) It's possible to sample a few points and guarantee that with constant probability, the Euclidean distance from one of the sample is close enough to the optimizer p_{opt}, i.e. $\|s_i - p_{opt}\| \leq L$ for some sample s_i.
    (b) If the distance from p' to p_{opt} is $O(\epsilon L)$ for sufficiently small constant in this big O notation, p' is guaranteed to be $(1 + \epsilon)$ approximate solution.
2. Continue the sampling in batches so that for each batch of samples, either the $(1 + \epsilon)$-approximate solution is already in the affine subspace spanned by the sampled points, or the subspace becomes closer to p_{opt} by a factor about $1 + \epsilon$. It is also required that the size of each batch is $\text{poly}(1/\epsilon)$.
3. Repeat step 2 until the distance from $\text{span}(X)$ to p_{opt} is smaller than $O(\epsilon L)$, where X is the set of sampled points.
4. Inside $\text{span}(X)$, draw a grid around each point in X. The radius of the grid is $2L$ and the side length is ϵL. Then there is an $(1 + \epsilon)$ approximate solution in these grid points.

Remark 3. To be able to shorten from the initial gap L to the desired gap $O(\epsilon L)$, the number of batches required on average is bounded by $\text{poly}(1/\epsilon)$, which means it only depends on the approximation factor. Since each batch contains $\text{poly}(1/\epsilon)$ many samples, in total the sample set X is of size $\text{poly}(1/\epsilon)$.

Remark 4. Notice that in Step 4 we need to approximately know the value L to perform the actual algorithm. This is guaranteed in our algorithm for general cost function f, as we showed in the next section.

3.2   The Choice of L

Let us first focus on the choice of L for a general cost function f. Denote $\texttt{AVG} = G(p_{opt})/n$. If $f(x) = x$, L can be chosen to be $2G(p_{opt})/n$ in Step 1, as shown in [1]. We can think of this L as the average cost contributed from points in P. So for condition (b), it is trivial that $\epsilon L/2$ is the necessary distance from p' to p_{opt} to make p' a $(1 + \epsilon)$-approximated solution. At the same time, L is also roughly the "average" of Euclidean distance from each point in P to p_{opt} since the cost function f is an identity function. So for condition (a), a point $s \in P$ such that $\|s - p_{opt}\| \leq L$ can be regarded as an "average" case. An average case is easy to approximate using sampling.

However, such coincidence will not happen for general f. If f is a slowly growing function (e.g. $log(x)$, $1 - 1/x$) and L is chosen like above, condition (b) still holds but L is far from the "average" of Euclidean distances to p_{opt} in some of worse cases. To compromise, we do not require L to be "average". We only require roughly ϵn points in P satisfying that the distance from them to p_{opt} is less than L. Then on average, we can obtain such point after $O(1/\epsilon)$ samples. Consequently, condition (a) and (b) can both be satisfied again. The following lemma shows the exact choice of the value L, the unknown variables A and B will be removed later:

Lemma 2. *Let $0 < \epsilon \leq 1$. Let $P \subset \mathbb{R}^d$ and $|P| = n$, $G(x) = \sum_{p \in P} f(\|x - p\|)$ with α as the proportion degree of f. Suppose $\tilde{p}$ is the $\lceil \epsilon n \rceil$-th closest point to the optimal solution p_{opt} among the points in P. Choose L accordingly if the following two cases apply:*
(i) If we know a value A such that $f(\|\tilde{p} - p_{opt}\|) \in [A, (1 + \epsilon/3)A)$, choose $L = f^{-1}((1 + \epsilon)A)$.
(ii) If $f(\|\tilde{p} - p_{opt}\|) \leq \epsilon AVG/B$ for some constant $B \geq 3$, choose a value $L \in [f^{-1}(\epsilon AVG/B), f^{-1}(\epsilon AVG/3)]$.
Then p' is a $(1 + \epsilon)$−approximate solution of G if $\|p' - p_{opt}\| \leq \epsilon L/(4\alpha)$.

Proof. We prove case (i) first. Let $P = \{p_1, p_2, \cdots, p_n\}$ so that $\|p_1 - p_{opt}\| \leq \|p_2 - p_{opt}\| \leq \cdots \leq \|p_n - p_{opt}\|$. Then $i \geq \lceil \epsilon n/4 \rceil$ implies $f(\|p_i - p_{opt}\|) \geq A$ since f is non-decreasing. By Markov's inequality the value A can not be greater than $\texttt{AVG}/(1 - \epsilon/4)$. This is a fact we are going to use in the following argument and later in Lemma 5. Now assume p' is a point satisfies $\|p' - p_{opt}\| \leq \epsilon L/(4\alpha)$. For p_i with $i < \lceil \epsilon n/4 \rceil$, the total increase of cost by moving p_{opt} to p' is at most

$$\sum_{i < \lceil \epsilon n/4 \rceil} f(\|p_i - p'\|) \leq \epsilon \frac{n}{4} f(L + \frac{L\epsilon}{4\alpha})$$

$$\leq \epsilon \frac{n}{4} (1 + \frac{\epsilon}{4\alpha})^\alpha f(L) \leq c \frac{n}{4} (1 + \frac{\epsilon}{3})(1 + \epsilon/3)A$$

$$\leq \epsilon \frac{n}{4} (1 + \frac{\epsilon}{3})(1 + \epsilon/3) \frac{AVG}{(1 - \epsilon/4)} < \frac{16}{27} \epsilon G(p_{opt})$$

The second inequality comes from the sub-proportionality of f. For the remaining points, If $\|p_i - p_{opt}\| < L - \epsilon L/(4\alpha)$ but $\|p_i - p_{opt}\| \geq \|\tilde{p} - p_{opt}\|$, then

$$f(\|p_i - p'\|) \leq f(\|p_i - p_{opt}\| + \|p_{opt} - p'\|) \leq f(L) = (1 + \epsilon/3)A$$

With the fact that $f(\|p_i - p_{opt}\|) \geq A$, we have

$$f(\|p_i - p'\|) - f(\|p_i - p_{opt}\|) \leq \frac{\epsilon}{3}A \leq \frac{\epsilon}{3}f(\|p_i - p_{opt}\|)$$

If $\|p_i - p_{opt}\| \geq L - \epsilon L/(4\alpha)$, the cost from moving p_{opt} to p' is increased by a factor of at most $(1 + 11\epsilon/27)$:

$$f(\|p_i - p'\|) \leq f(\|p_i - p_{opt}\| + \|p_{opt} - p'\|) \leq f(\|p_i - p_{opt}\| + \frac{\epsilon L}{4\alpha})$$

$$\leq (1 + \frac{\epsilon}{4\alpha - \epsilon})^\alpha f(\|p_i - p_{opt}\|) \leq e^{\epsilon/3} f(\|p_i - p_{opt}\|)$$

$$\leq (1 + \frac{11}{27}\epsilon)f(\|p_i - p_{opt}\|)$$

In sum, the total difference between $G(p') = \sum_i f(\|p_i - p'\|)$ and $G(p_{opt})$ is at most $\epsilon G(p_{opt})$, therefore p' is a $(1 + \epsilon)$-approximate solution of G.

For case (ii), for $i < \lfloor \epsilon n/4 \rfloor$, in other words, $\|p_i - p'\| \leq \|\tilde{p} - p'\|$, the total increase of cost by moving p_{opt} to p' is at most:

$$\sum_{i < \lceil \epsilon n/4 \rceil} f(\|p_i - p'\|) \leq \epsilon \frac{n}{4}(1 + \frac{\epsilon}{4\alpha})^\alpha f(L) \leq \epsilon \frac{n}{4}(1 + \frac{\epsilon}{3})\frac{\epsilon \mathsf{AVG}}{3} < \frac{1}{9}\epsilon G(p_{opt})$$

When $\|\tilde{p} - p_{opt}\| \leq \|p_i - p_{opt}\| < L - \epsilon L/(4\alpha)$ we have:

$$f(\|p_i - p'\|) \leq f(\|p_i - p_{opt}\| + \|p_{opt} - p'\|) \leq f(L) = \epsilon \mathsf{AVG}/3$$

Lastly, if $\|p_i = p_{opt}\| \geq L - \epsilon L/(4\alpha)$, the argument is the same as in case(i):

$$f(\|p_i - p'\|) \leq (1 + \frac{11}{27}\epsilon)f(\|p_i - p_{opt}\|)$$

In sum, the total difference between $G(p')$ and $G(p_{opt})$ is $< \epsilon G(p_{opt})$. So p' is a $(1 + \epsilon)$-approximate solution of G.   □

The above lemma shows that if we choose L in this way, condition (b) of Step 1 is satisfied. Furthermore, the following lemma indicates that condition (a) can also be achieved.

Lemma 3. *Let* ϵ, P, G, f, $\tilde{p}$, L *be defined as in Lemma 2. By uniformly sampling* $|X| = O(1/\epsilon)$ *points in* P, *there will be a point* $s \in X$ *satisfying inequality* $\|s - p_{opt}\| \leq L$ *with constant probability.*

Proof. Since for both case(i) and case(ii) there are at least $\lfloor \epsilon n \rfloor$ points in P having $\|p_i - p_{opt}\| \leq \|\tilde{p} - p_{opt}\| \leq L$, after $2/\epsilon$ samples there will be at least one point falling in this set of points with probability $\geq 1/2$ by Markov's inequality.   □

3.3   Main Result

In this subsection, we omit the details of most of the proofs due to the space limit. First we present the theorem which guarantees the correctness of Step 2. For a set of points $X \subset \mathbb{R}^d$, we denote by span(X) the affine subspace spanned by the set of points in X.

Theorem 2 (Core-set). *Let $0 < \epsilon < 1$. Let P be a point set in $\mathbb{R}^d$. $G(x) = \sum_{p \in P} f(\|x - p\|)$ with α as the proportion degree of f. L is chosen as in Lemma 2. If X is a set of points obtained from sampling $O(\log(1/\epsilon)/\epsilon^{3+\alpha})$ points in P, then with constant probability, the following two events happen: (i) The distance from the affine subspace span(X) to the optimizer p_{opt} is at most $\epsilon L/(8\alpha)$, and (ii) X contains a point in distance $\leq L$ from p_{opt}.*

The above theorem gives the existence of a $(1 + \epsilon)$-approximate solution in the affine subspace of a small sample. To actually find the solution is the final issue. We provide one of the possible approaches in the following.

The lemma below shows that we know a value $t = \Theta(\mathtt{AVG})$. It also shows that trust the best source alone gives a constant approximate factor solution.

Lemma 4 (a $2^\alpha-$ approximated solution). *Let P be a set of n points in $\mathbb{R}^d$ and $G(x) = \sum_{p \in P} f\|x - p\|$ with α as the proportion degree of f. We can try every point in P to achieve a 2^α-approximate solution for the function G, and the total running time is $O(n^2 d)$.*

Proof. Let $p' \in P$ be the one closest to the optimal point p_{opt}. Then

$$G(p') = \sum_{p \in P} f(\|p - p'\|) \leq \sum_{p \in P} f(\|p - p_{opt}\| + \|p' - p_{opt}\|)$$

$$\leq \sum_{p \in P} f(2\|p - p_{opt}\|) \leq 2^\alpha \cdot G(p_{opt}).$$

The last inequality comes from the sub-proportionality of f. The minimum among $G(p_1), G(p_2), \cdots, G(p_n)$ must be less than $G(p')$. The function G can be evaluated in $O(nd)$ time. Therefore, the $2^\alpha-$approximate solution can be found in $O(n^2 d)$ time.   □

There are more efficient ways to bound the value of $\mathtt{AVG}$ for special f. For example, when $f(x) = x$, it is shown [8] that $\mathtt{AVG}$ can be approximated in linear time.

Now we settle the unknown variables A and B in Lemma 2. We will show that if choosing B properly, A is approximately bounded in the way that $A = \Theta_\epsilon(\mathtt{AVG})$. Thus the search of the value A takes at most poly$(1/\epsilon)$ time. The effect on the whole algorithm is a multiplicative factor of poly$(1/\epsilon)$, which is small comparing to the time for drawing grid points.

Lemma 5. *Let ϵ, P, G, f be defined as in Lemma 2. Let $\tilde{p}$ be the $\lceil \epsilon n \rceil$-th closest point to the optimal solution p_{opt} among the points in P. There exists a set $\mathcal{L}$ of size $O(\log(1/\epsilon)/\epsilon)$ such that for every possible values of $f(\|\tilde{p} - p_{opt}\|)$, there is a member $L \in \mathcal{L}$ such that it satisfies condition (a) and (b) in Step 1.*

The next theorem summarizes the complete algorithm.

Theorem 3. *Let $0 < \epsilon \leq 1$. Let P be a set of n points in $\mathbb{R}^d$ and $G(x) = \sum_{p \in P} f(\|x - p\|)$ with α as the proportion degree of f. Let X be a set of random samples from P of size $O(\log(1/\epsilon)/\epsilon^{3+\alpha})$. We can construct a set of grid points Y of size $O(2^{(1/\epsilon)^{O(1)}})$ such that with constant probability there is at least one point p' in Y being a $(1 + \epsilon)$-approximate solution of G. The time complexity is $O(2^{(1/\epsilon)^{O(1)}} d + n^2 d)$ for the construction of Y.*

Synopsis of the Proof: For each $L \in \mathcal{L}$, denote Y_L as the union of the grid points around each $x \in X$, where the diameter of the grid is $4L$ and the side length is roughly $O(\epsilon L)$. Let $Y = \cup_{L \in \mathcal{L}} Y_L$. Then Theorem 2 guarantees the desired result.

References

1. Bădoiu, M., Har-Peled, S., Indyk, P.: Approximate clustering via core-sets. In: Proceedings of the Thiry-Fourth Annual ACM Symposium on Theory of Computing, pp. 250–257. ACM (2002)
2. Ding, H., Gao, J., Xu, J.: Finding global optimum for truth discovery: entropy based geometric variance. In: LIPIcs-Leibniz International Proceedings in Informatics, vol. 51. Schloss Dagstuhl-Leibniz-Zentrum fuer Informatik (2016)
3. Dong, X., Gabrilovich, E., Heitz, G., Horn, W., Lao, N., Murphy, K., Strohmann, T., Sun, S., Zhang, W.: Knowledge vault: a web-scale approach to probabilistic knowledge fusion. In: Proceedings of the 20th ACM SIGKDD International Conference on Knowledge Discovery and Data Mining, pp. 601–610. ACM (2014)
4. Dong, X.L., Berti-Equille, L., Srivastava, D.: Integrating conflicting data: the role of source dependence. Proc. VLDB Endow. **2**(1), 550–561 (2009)
5. Dong, X.L., Berti-Equille, L., Srivastava, D.: Truth discovery and copying detection in a dynamic world. Proc. VLDB Endow. **2**(1), 562–573 (2009)
6. Feldman, D., Langberg, M.: A unified framework for approximating and clustering data. In: Proceedings of the Forty-Third Annual ACM Symposium on Theory of Computing, pp. 569–578. ACM (2011)
7. Galland, A., Abiteboul, S., Marian, A., Senellart, P.: Corroborating information from disagreeing views. In: Proceedings of the Third ACM International Conference on Web Search and Data Mining, pp. 131–140. ACM (2010)
8. Kumar, A., Sabharwal, Y., Sen, S.: Linear time algorithms for clustering problems in any dimensions. In: Caires, L., Italiano, G.F., Monteiro, L., Palamidessi, C., Yung, M. (eds.) ICALP 2005. LNCS, vol. 3580, pp. 1374–1385. Springer, Heidelberg (2005). https://doi.org/10.1007/11523468_111
9. Li, F., Lee, M.L., Hsu, W.: Entity profiling with varying source reliabilities. In: Proceedings of the 20th ACM SIGKDD International Conference on Knowledge Discovery and Data Mining, pp. 1146–1155. ACM (2014)
10. Li, Q., Li, Y., Gao, J., Zhao, B., Fan, W., Han, J.: Resolving conflicts in heterogeneous data by truth discovery and source reliability estimation. In: Proceedings of the 2014 ACM SIGMOD International Conference on Management of Data, pp. 1187–1198. ACM (2014)

11. Pasternack, J., Roth, D.: Knowing what to believe (when you already know something). In: Proceedings of the 23rd International Conference on Computational Linguistics, pp. 877–885. Association for Computational Linguistics (2010)
12. Welinder, P., Branson, S., Belongie, S.J., Perona, P.: The multidimensional wisdom of crowds. NIPS **23**, 2424–2432 (2010)
13. Whitehill, J., Wu, T.-F., Bergsma, J., Movellan, J.R., Ruvolo, P.L.: Whose vote should count more: optimal integration of labels from labelers of unknown expertise. In: Advances in Neural Information Processing Systems, pp. 2035–2043 (2009)
14. Xiao, H., Gao, J., Li, Q., Ma, F., Su, L., Feng, Y., Zhang, A.: Towards confidence in the truth: a bootstrapping based truth discovery approach. In: Proceedings of the 22nd ACM SIGKDD International Conference on Knowledge Discovery and Data Mining, pp. 1935–1944. ACM (2016)
15. Xiao, H., Gao, J., Wang, Z., Wang, S., Su, L., Liu, H.: A truth discovery approach with theoretical guarantee. In: Proceedings of the 22nd ACM SIGKDD International Conference on Knowledge Discovery and Data Mining, pp. 1925–1934. ACM (2016)
16. Yin, X., Han, J., Philip, S.Y.: Truth discovery with multiple conflicting information providers on the web. IEEE Trans. Knowl. Data Eng. **20**(6), 796–808 (2008)
17. Zhao, B., Han, J.: A probabilistic model for estimating real-valued truth from conflicting sources. In: Proceedings of QDB (2012)
18. Zhao, B., Rubinstein, B.I., Gemmell, J., Han, J.: A bayesian approach to discovering truth from conflicting sources for data integration. Proc. VLDB Endow. **5**(6), 550–561 (2012)
19. Huang, Z., Ding, H., Xu, J.: Faster algorithm for truth discovery via range cover. Algorithms and Data Structures. LNCS, vol. 10389, pp. 461–472. Springer, Cham (2017). https://doi.org/10.1007/978-3-319-62127-2_39

Online Interval Scheduling to Maximize Total Satisfaction

Koji M. Kobayashi$^{(\boxtimes)}$

The University of Tokyo, Tokyo, Japan
kojikoba@mi.u-tokyo.ac.jp

Abstract. The interval scheduling problem is one variant of the scheduling problem. In this paper, we propose a novel variant of the interval scheduling problem, whose definition is as follows: given jobs are specified by their *release times, deadlines* and *profits*. An algorithm must start a job at its release time on one of m identical machines, and continue processing until its deadline on the machine to *complete* the job. All the jobs must be completed and the algorithm can obtain the profit of a completed job as a user's satisfaction. It is possible to process more than one job at a time on one machine. The profit of a job is distributed uniformly between its release time and deadline, that is its interval, and the profit gained from a subinterval of a job decreases in reverse proportion to the number of jobs whose intervals intersect with the subinterval on the same machine. The objective of our variant is to maximize the total profit of completed jobs.

This formulation is naturally motivated by best-effort requests and responses to them, which appear in many situations. In best-effort requests and responses, the total amount of available resources for users is always invariant and the resources are equally shared with every user. We study online algorithms for this problem. Specifically, we show that for the case where the profits of jobs are arbitrary, there does not exist an algorithm whose competitive ratio is bounded. Then, we consider the case in which the profit of each job is equal to its length, that is, the time interval between its release time and deadline. For this case, we prove that for $m = 2$ and $m \geq 3$, the competitive ratios of a greedy algorithm are at most 4/3 and at most 3, respectively. Also, for each $m \geq 2$, we show a lower bound on the competitive ratio of any deterministic algorithm.

1 Introduction

The interval scheduling problem is one of the variants of the scheduling problem, which has been widely studied. One of the most basic definitions is as follows: We have $m \geq 1$ identical machines and jobs are given. A job is characterized by the *release time, deadline* and *weight* (or *value*). To *complete* a job, we must start to process it at its release time on a machine of the m machines, and continue processing it until its deadline on that machine. That is, the *processing time* (or *length*) of the job is the time interval between its release time and deadline. The

© Springer International Publishing AG, part of Springer Nature 2018
L. Wang and D. Zhu (Eds.): COCOON 2018, LNCS 10976, pp. 108–119, 2018.
https://doi.org/10.1007/978-3-319-94776-1_10

number of jobs which can be processed on one machine at a time is at most one. The objective of an algorithm is to maximize the total weight of completed jobs. There are many applications of the interval scheduling problem, such as bandwidth allocation and vehicle assignment (see e.g., [13,14]). Many variants of this problem have been proposed and extensively studied. Furthermore, research on online settings has also been considered. In an online variant of the interval scheduling problem, a job arrives at its release time and an online algorithm must decide whether it processes the job before the next job arrives. The performance of online algorithms is evaluated using *competitive analysis* [3,19]. For any input, if the total weight gained by an optimal offline algorithm is at most c times that gained by an online algorithm, the online algorithm is *c-competitive*.

In this paper, we introduce a novel variant of the interval scheduling problem. In many existing variants of the interval scheduling problem, jobs (or users) require resources for an algorithm, and the algorithm assigns the required resources of a machine to the job. Thus, the number of jobs assigned to one machine at a time is subject to the maximum amount of resources of the machine. The amount is generally one; that is, at most one job can be processed at a time on one machine in most variants. Therefore, we can regard such existing variants as formulating *resource reservation* requests by users, who designate the amount of resources they want to use in advance and the responses to them. However, it is not always possible for users to designate the amount of resources they want when they issue requests. Additionally, there are not necessarily sufficient resources of a machine to meet users' requests. Thus, we focus on a *best-effort* method to manage situations, which is often considered paired with resource reservation methods. In this method, the amount of resources of a machine is always invariant and the resources are equally shared by users who want to use the resources at the same time. Then, we formulate best-effort requests and responses to them as a variant of the interval scheduling problem. Specifically, we remove the capacity constraints from machines in our variant, which makes it possible to assign jobs unlimitedly on one machine at a time. To the best of our knowledge, this is the first such formulation of the interval scheduling problem. Consider a given job as a user's request. If a machine processes the request using sufficient resources, the user is sufficiently satisfied with the result obtained from the process. Conversely, if there are not sufficient resources to process the request, the user is less satisfied with the result than usual. Then, the objective of our variant is to maximize the total satisfaction gained by users. Bandwidth allocation in networks is one of the most suitable examples for best-effort requests and responses. In this example, the total bandwidth which may be supplied to users on the same communication link is fixed in advance, and all users share the bandwidth. Hence, the fewer users which use the communication link at a time, the greater the bandwidth which each one can use, which means that the effective speed of the communication link is higher for the users. Conversely, the more people there are using link simultaneously, the lower the effective speed for each user. As a result, if the bandwidth for a user is high, then the user's satisfaction is high. Otherwise, it is low. Best-effort requests and responses such as bandwidth allocation could happen in many cases, for example, the use of facilities, such as swimming pools and gyms, passenger trains without reservations, and buffet style meals. Therefore, we have sufficient incentives to study our variant.

Our Results. In this paper, we propose and analyze a novel variant of the interval scheduling problem. We study online algorithms for this problem. Specifically, in the case where the profits of jobs are arbitrary; that is, the profits are not relevant to the lengths of jobs, we show that the competitive ratio of any deterministic algorithm is unbounded. Then, we introduce the profits of jobs are equal to their lengths, which is a more natural case, called the *uniform profit case*. In this case, the total amount of time during which at least one job is scheduled on a machine is equal to the total amount of the satisfaction gained on the machine. That is, the objective of this case can be regarded as maximizing the working hours of all the machines. We analyze the performance of a greedy algorithm GR in this case. Since GR is a significant algorithm from a practical point of view, it is worthwhile to evaluate its performance. When $m = 2$ and $m \geq 3$, we show that the competitive ratios of GR are at most 4/3 and at most 3, respectively. When $m = 2$, we prove that a lower bound on the competitive ratio of GR is 4/3. That is, for $m = 2$, our analysis of GR is tight. Also, we show lower bounds of any deterministic online algorithms for each $m \geq 2$, which are summarized in Tables 1 and 2 in Sect. 5.

Table 1. Our results

m	Upper bound	Lower bound
2	$4/3 \leq 1.334$	$(10 - \sqrt{2})/7 \geq 1.226$
3	3	$7/6 \geq 1.166$
4		$(22 - 2\sqrt{2})/17 \geq 1.127$
5		$(420 - 15\sqrt{7})/333 \geq 1.142$
6		$(51 - 6\sqrt{2})/41 \geq 1.140$
∞		$(48 - 2\sqrt{2})/41 \geq 1.101$

Related Results. Much research on the interval scheduling problem has been conducted. Arkin and Silverberg [1] and Bouzina and Emmons [4] provided polynomial time algorithms to solve the interval scheduling problem.

There is also much research on online interval scheduling problems. If an online algorithm aborts a job J which was placed on a machine, then we say that the algorithm *preempts* J. In the case in which preemption is allowed, Faigle and Nawijn [9] designed a 1-competitive algorithm to maximize the number of completed jobs. This algorithm was independently discovered by Carlisle and Lloyd [6] but used only for the offline setting. Moreover, for the variant in which the objective is to maximize the total weight of completed jobs, Woeginger [20] showed that no any competitive deterministic algorithm exists (even) for $m = 1$. Canetti and Irani [5] provided a randomized online algorithm whose competitive ratio is $O(\log \Delta)$ and proved that a lower bound on the competitive ratio of any randomized algorithm is $\Omega(\sqrt{\log \Delta / \log \log \Delta})$, where Δ is the ratio of the longest length to the shortest length. This result indicates that the competitive ratio of an online algorithm may become worse depending on a given input even if it is supported by randomization. Additionally, the setting in which the jobs are unit length has been extensively studied. For the one machine setting, Woeginger [20] designed a deterministic algorithm whose competitive ratio is at most 4 and showed that this is the best possible ratio. There has also been much work regarding randomized algorithms (e.g. [8, 10–12, 16, 18]). When $m = 1$, the current best upper and lower bounds on the competitive ratios of randomized algorithms are 2 by Fung et al. [12] and $1 + \ln 2 \geq 1.693$ by Epstein and Levin [8], respectively. For $m \geq 2$, Fung et al. [11] proved that, if m is even, an upper bound is 2, and otherwise $2 + 2/(2m - 1)$. However, for $m = 2$, the current best lower bound is 2 by Fung et al. [10]. When each $m \geq 3$, Fung et al. [11] indicated that we can obtain a lower bound of $1 + \ln 2 \geq 1.693$ in a similar manner to the lower bound of Epstein and Levin [8]. If preemption is not allowed, Lipton and Tomkins [15] proposed a randomized algorithm whose competitive ratio is $O((\log \Delta)^{1+\epsilon})$ and proved that a lower bound of any randomized algorithms is $\Omega(\log \Delta)$.

For a job given in the interval scheduling problem, its length is equal to the length of the time between its release time and deadline. On the other hand, a variant in which the job length is generalized has also been studied. Specifically, a parameter *slack* $\varepsilon > 0$ is introduced, whose value is known to an algorithm in advance, and the length of a job is at most x times as long as the length of the time between its release time and deadline, in which $x = 1/(1 + \varepsilon)$. In this variant, preemption is allowed and to complete a job, an algorithm must process it during its length by its deadline after its release time. For several m, optimal online algorithms were designed [2, 7, 17], whose competitive ratios are $1 + 1/\varepsilon$.

2   Model Description

We have $m(\geq 2)$ identical machines. A list consisting of $n(\geq 1)$ jobs is provided as an *input*. A job J is specified by a triplet (r, d, v), where $r(J)$ is the *release time* of J, $d(J)$ is the *deadline* of J, and $v(J)$ is the *profit* of J. An algorithm ALG must place each job onto one of the m machines. It is possible to place more than one job at a time on one machine. The profit of a job is distributed uniformly between its release time and deadline, that is its interval, and the profit gained from a subinterval of a job decreases in reverse proportion to the number of jobs whose intervals intersect with the subinterval on the same machine. Specifically, the profit from the subinterval is defined as follows: For an algorithm ALG, if the numbers of jobs placed at any two points in an interval (x, y) $(x < y)$ are equal on ALG's $a(\in [1, m])$th machine and (x, y) does not contain any endpoint of the interval of a job placed on the machine after processing of the input, then we call the interval a *P-interval* on *ALG*'s *a*th machine. Also, let $k_{ALG}(a, x, y)$ denote the number of the jobs. If an algorithm *ALG* places a job J onto the *a*th machine, then we define $m_{ALG}(J) = a$. For an algorithm *ALG* and a job J, suppose that the interval $(r(J), d(J))$ consists of $b(\geq 1)$ *P*-intervals (x_i, x_{i+1}) $(i = 1, \ldots, b-1)$ on *ALG*'s $m_{ALG}(J)$th machine such that $r(J) = x_1 < x_2 < \cdots < x_b = d(J)$. Then, we define the satisfaction (*profit*) which is yielded from $[x_i, x_{i+1}]$ of J and *ALG* gains as

$$V_{ALG}(J, i) = \frac{x_{i+1} - x_i}{d(J) - r(J)} \frac{v(J)}{k_{ALG}(m_{ALG}(J), x_i, x_{i+1})}.$$

We define the satisfaction (*profit*) of J gained by *ALG* as

$$V_{ALG}(J) = \sum_{i=1}^{b-1} V_{ALG}(J, i).$$

The *profit* of *ALG* for an input σ is defined as

$$V_{ALG}(\sigma) = \sum_{J \in \mathcal{L}} V_{ALG}(J),$$

where $\mathcal{L}$ is a list consisting of the n given jobs. The objective is to maximize the total satisfaction of the n jobs.

In this paper, we consider an online variant of this problem. Specifically, n jobs are given one by one. The jobs are not necessarily given in order of release time. An online algorithm must place a given job to a machine before the next job is given. Once a job is placed on a machine, it cannot be removed later. That is, preemption is not allowed. The total number n of given jobs is not known to the online algorithm, and it does not require this information until after all the jobs arrive. We say that the competitive ratio of an online algorithm A is at most c or A is c-competitive if, for any input, the profit gained by an offline optimal algorithm OPT is at most c times the profit gained by A.

3   General Profit Case

Due to page limitations, we omit almost all of the proofs in this paper. The full version of this paper is available at https://arxiv.org/abs/1805.05436.

In this section, we consider the case in which the profits of jobs are arbitrary. First, we consider the case $m = 2$ for better understanding of any $m \geq 3$.

Theorem 1. *When $m = 2$, there does not exist any deterministic online algorithm whose competitive ratio is bounded.*

Theorem 2. *For any m, there does not exist a competitive deterministic algorithm.*

4   Upper Bounds for Uniform Profit Case

In this section, we consider the uniform profit case, that is, the case in which the profit of a job is equal to its length. In this case, the total amount of time during which at least one job is scheduled on a machine is equal to the total amount of the satisfaction gained on the machine. That is, the objective of this case can be regarded as maximizing the working hours of all the machines.

4.1   Preliminaries

After the end of the input, we need to evaluate the profit from each job by OPT using the profits yielded from intervals of jobs scheduled by GR to analyze the performance of GR. Then, we classify intervals (or points) in a job J by GR or OPT into the following four categories depending on the behaviors of GR and OPT for J.

For any two intervals $I = [t_1, t_2]$ and $I' = [t'_1, t'_2]$, we say that I intersects with I' if $t'_1 < t_2$ and $t_1 < t'_2$. For any job J, we call the interval $[r(J), d(J)]$ the interval of J. If an algorithm ALG places two jobs onto the same machine and they intersect, then we say that they overlap. For any interval $I = [t, t']$, we call the value of $t' - t$ the length of I, written as $|I|$.

We give the definition of a greedy algorithm GR and analyze its performance in this section. GR places a given job J onto the machine on which GR gains the largest profit from J. The tie-breaking rule selects the minimum indexed machine.

For ease of analyzing, we introduce the following idea. Suppose that two jobs J_1 and J_2 are placed onto the same machine, and they overlap in an interval I. Also, suppose that J_1 is the first job placed in I on the machine. Then, pretend that the profits from I of J_1 and J_2 are $|I|$ and zero, respectively. That is, we pretend that a job which is placed chronologically first in an interval on a machine monopolizes the machine power in the interval. Note that in the uniform profit case, the total profit gained from an interval of jobs placed on a machine depends not on how large the number of the jobs in the interval is but on whether there exists at least one job placed in the interval. That is why this assumption does not affect the profit of any algorithm.

4.2   Overview of Analysis

To evaluate the performance of GR, that is, its competitive ratio, we bound the profit of OPT at the end of the input using that of GR. Then, we classify intervals of jobs placed by either GR or OPT into four categories.

For any job J and any interval $I \subseteq [r(J), d(J)]$, if the profit gained from I of J by GR is zero and that by OPT is $|I|$, then we call I of J an OPT extra interval of J (denoted as an oe-interval, for short). Also, if the profit gained from I of J by OPT is zero and that by GR is $|I|$, then we call I of J a GR extra interval of J (a ge-interval, for short). If the profits gained from I of J by GR and OPT are both $|I|$, we call I of J a common interval of J (a c-interval, for short). For ease of presentation, we call an interval which is a c-interval or a ge-interval a *profit interval* (a *p-interval*, for short). If the profits gained from I of J by GR and OPT are both zero, we call I of J a *non-profit interval of J* (an *n-interval*, for short). Further, we call a point in an oe-interval (a ge-interval, a c-interval, and a p-interval, respectively) of J an *oe-fraction* (a *ge-fraction*, a *c-fraction*, and a *p-fraction*, respectively) of J.

We evaluate the competitive ratio of GR by "assigning" p-fractions (i.e., p-intervals) to all oe-fractions (i.e., oe-intervals) according to a routine, which is defined later. This "assignment" is realized by some functions. Let $V_{oe}(\sigma)$ be the total length of oe-intervals to which c-intervals are assigned. Let $V_{oe'}(\sigma)$ be the total length of oe-intervals to which ge-intervals are assigned. Also, let $V_c(\sigma)$ be the total length of c-intervals and $V_{ge}(\sigma)$ be the total length of ge-intervals. Then, we have by definition,

$$V_{GR}(\sigma) = V_c(\sigma) + V_{ge}(\sigma) \tag{1}$$

and

$$V_{OPT}(\sigma) = V_c(\sigma) + V_{oe}(\sigma) + V_{oe'}(\sigma). \tag{2}$$

We will show the following three properties of the assignments by the routine:

1. Each oe-fraction is assigned a p-fraction,
2. a c-fraction of a job given to GR is assigned at most twice, and
3. a ge-fraction is assigned at most three times.

To show these, we will construct sequentially three functions M_1, M_2 and M_3 from oe-intervals to p-intervals satisfying the following properties: Initially, for any oe-fraction f and any $i \in \{1,2,3\}$, $M_i(f) = \varnothing$. At the end of the input, for any oe-fraction f, $M_1(f) \cup M_2(f) \cup M_3(f) \neq \varnothing$. There exists a p-fraction f' such that $M_1(f) = f'$ if $M_1(f) \neq \varnothing$. There exists a ge-fraction f' such that $M_2(f) = f'$ if $M_2(f) \neq \varnothing$. There exists a p-fraction f' such that $M_3(f) = f'$ if $M_3(f) \neq \varnothing$. For any oe-fractions f and $f'(\neq f)$ and any $i \in \{1,2,3\}$, $M_i(f) \cap M_i(f') = \varnothing$. Then, we have by these functions,

$$V_{oe}(\sigma) \leq 2V_c(\sigma) \tag{3}$$

and

$$V_{oe'}(\sigma) \leq 3V_{ge}(\sigma). \tag{4}$$

By Eq. (2), we have

$$\begin{aligned}
V_{OPT}(\sigma) &= V_c(\sigma) + V_{oe}(\sigma) + V_{oe'}(\sigma) \\
&\leq V_c(\sigma) + 2V_c(\sigma) + 3V_{ge}(\sigma) \quad \text{(by Eqs. (3) and (4))} \\
&= 3(V_c(\sigma) + V_{ge}(\sigma)) = 3V_{GR}(\sigma), \quad \text{(by Eq. (1))}
\end{aligned}$$

which leads to the following theorem:

Theorem 3. *For any $m \geq 2$, the competitive ratio of GR is at most three.*

4.3   Analysis of GR

For any job J and any point $t \in [r(J), d(J)]$, let $E(J, t)$ denote the total length of oe-intervals of J in the interval $[r(J), t]$. For any job J, any job J' given before J, any interval $[t_1, t_2]$ and any $a(\in [1, m])$, let $P_a(J, J', t_1, t_2)$ denote the total length of p-intervals of GR's jobs placed on the ath machine which are in $[t_1, t_2]$ immediately after J is placed and are not intersecting with any n-interval of J'. For any $a(\in [1, m])$, any job J, any job J' given before J, and any point $t \in [r(J'), d(J')]$, define $h_a(J, J', t) = t'$ in which t' is the point such that $P_a(J, J', r(J'), t') = E(J', t)$ and $t' \in [r(J'), d(J')]$ immediately after J is placed onto the machine. (t' exists by Lemma 1, which is shown later.) For any $i \in \{1, 2, 3\}$ and any p-fraction f', define $M_i^{-1}(f') = \{f \mid M_i(f) = f'\}$. We say that a c-fraction f' such that $M_1^{-1}(f') = \varnothing$ is 1-assignable. We say that a ge-fraction f' such that $M_2^{-1}(f') = \varnothing$ is 2-assignable. We say that a ge-fraction f' such that $M_2^{-1}(f') \neq \varnothing$ and $M_1^{-1}(f') = \varnothing$ is 1-assignable. If a p-fraction is 1-assignable or 2-assignable, we say that it is $assignable$. Now we give the definition of the routine mentioned in the previous section.

ASSIGNMENTROUTINE

Consider a moment immediately after the jth job J_j is placed. $\mathcal{J} :=$ (the set of J_j plus each job $J_{j'}$ ($j' \leq j - 1$) whose interval intersects with the interval of J_j). For any oe-fraction f of each $J \in \mathcal{J}$, execute the following.

Step 1: For each $i \in \{1, 2, 3\}$, $M_i(f) := \varnothing$. $t_1 := h_1(J_j, J, t)$, in which f exists at a point t.

Step 2: Execute one of the following two cases.

   Case 2.1 (An assignable p-fraction f_1 exists at t_1): If f_1 is 1-assignable, $M_1(f) := f_1$. Otherwise, if f_1 is 2-assignable, $M_2(f) := f_1$.

   Case 2.2 (No assignable p-fraction exists at t_1): By Lemma 2, there exists a p-fraction f_a at the point t_a on some $a(\in \{1, m\})$th machine such that $M_3^{-1}(f_a) = \varnothing$, in which $t_a = h_a(J_j, J, t)$. (For any $a' \in \{1, m\}$, there exists $t_{a'}$ by Lemma 1.)

In the following, we first show the existence of t_a in Case 2.2. Next, we show that there exists p_a in Case 2.2. That is, we prove that the routine can assign a p-fraction to each oe-fraction.

Lemma 1. *For any $a(\in [1, m])$, any job J, any job J' which is given before J, and any point $t \in [r(J'), d(J')]$, there exists the point t' such that $h_a(J, J', t) = t'$ and $t' \in [r(J'), d(J')]$ immediately after J is placed.*

Lemma 2. *Case 2.2 is executable. That is, when Case 2.2 is executed for an oe-fraction f, f can be assigned a p-fraction f_a such that $M_3^{-1}(f_a) = \varnothing$ immediately before executing Case 2.2.*

4.4   Upper Bound for $m = 2$

When $m = 2$, we also evaluate the competitive ratio of GR by assigning p-fractions to all oe-fractions. In this case, we obtain a better upper bound on the competitive ratio of GR than one for general m by implementing more detailed assignments. If the routine assigns one ge-fraction to one oe-fraction, we say that the routine ge-$assigns$ the ge-fraction to the oe-fraction. Also, if the routine assigns three p-fractions to one oe-fraction, we say that the routine $3p$-$assigns$ each of the p-fractions to the oe-fraction. We will show the following three properties by the assignments according to the routine defined later:

1. Each oe-fraction is ge-assigned or $3p$-assigned,
2. a c-fraction of a job given to GR is $3p$-assigned at most once, and
3. a ge-fraction is ge-assigned at most once and is $3p$-assigned at most once.

We will show them by sequentially constructing two functions N_1 and N_2 from oe-intervals to p-intervals satisfying the following properties: Initially, for any oe-fraction f and any $i \in \{1,2\}$, $N_i(f) = \varnothing$. At the end of the input, for any oe-fraction f, $N_1(f) \cup N_2(f) \neq \varnothing$. There exist three distinct p-fractions f_1, f_2 and f_3 such that $N_1(f) = \{f_1, f_2, f_3\}$ if $N_1(f) \neq \varnothing$. There exists a ge-fraction f' such that $N_2(f) = f'$ if $N_2(f) \neq \varnothing$. For any oe-fractions f and $f'(\neq f)$ and any $i \in \{1,2\}$, $N_i(f) \cap N_i(f') = \varnothing$. Let $V_{\overline{oe}}(\sigma)$ denote the total length of oe-intervals to which the routine $3p$-assigns, and let $V_{\overline{oe'}}(\sigma)$ denote the total length of oe-intervals to which the routine ge-assigns. Thus,

$$V_{\overline{oe}}(\sigma) \leq V_{GR}(\sigma)/3$$

and

$$V_{\overline{oe'}}(\sigma) \leq V_{ge}(\sigma).$$

Then, using these inequalities, we have

$$V_{OPT}(\sigma) = V_c(\sigma) + V_{\overline{oe}}(\sigma) + V_{\overline{oe'}}(\sigma)$$

$$\leq V_c(\sigma) + V_{GR}(\sigma)/3 + V_{ge}(\sigma) = \frac{4}{3} V_{GR}(\sigma).$$

Therefore, we have the following theorem:

Theorem 4. *When $m = 2$, the competitive ratio of GR is at most $4/3$.*

For any $i \in \{1,2\}$ and any p-fraction f', define $N_i^{-1}(f') = \{f \mid N_i(f) = f'\}$. We say that a p-fraction f' is 1-assignable if $N_1^{-1}(f') = \varnothing$. Also, we say that a ge-fraction f' is 2-assignable if $N_2^{-1}(f') = \varnothing$. Now we give the definition of the routine to construct the above two functions.

ASSIGNMENTROUTINE2

Consider a moment immediately after a job J is placed. For any oe-fraction f of J, execute the following.

Step 1: $m_2 := m_{GR}(J)$ and $m_1 := \{1,2\} \setminus \{m_2\}$. $t_1 := h_{m_1}(J, J, t)$, in which f

exists at a point t.

Step 2: Let f' be the p-fraction at t on the m_2th machine (f' exists by the definition of oe-fractions). Execute one of the following two cases.

   Case 2.1 (f' is 2-assignable): $N_2(f) := f'$.

   Case 2.2 (Otherwise): $N_1(f) := \{f', f_1, f_2\}$, in which f_1 is the p-fraction at t_1 on GR's m_1th machine (f_1 exists by Lemma 1), and f_2 is the p-fraction at t_1 on GR's m_2th machine (f_2 exists because the interval of J contains t_1 by the definition of h_{m_1}). (By Lemma 3, f', f_1 and f_2 are 1-assignable.)

Lemma 3. *Case 2.2 is executable. That is, when Case 2.2 is executed for an oe-fraction f, f can be assigned 3 p-fractions (i.e., 3p-assigned) each of which is 1-assignable immediately before executing Case 2.2.*

We show that our analysis of GR for $m = 2$ is tight in the following theorem.

Theorem 5. *When $m = 2$, for any $\varepsilon > 0$, the competitive ratio of GR is at least $4/3 - \varepsilon$.*

5   Lower Bounds for Uniform Profit Case

In this section, we show lower bounds on the competitive ratios of online algorithms for the uniform profit case. For better understanding, we first consider the case of $m = 2$.

Theorem 6. *When $m = 2$, the competitive ratio of any deterministic online algorithm is at least $(10 - \sqrt{2})/7 \geq 1.226$.*

Proof. Consider an online algorithm ON. The first given job is J_1 such that $r(J_1) = 0$ and $d(J_1) = 1$. The second job is J_2 such that $r(J_2) = 1 + x$ and $d(J_2) = 2 + x$. Note that x is set later. Without loss of generality, we may assume that both ON and OPT place J_1 onto the first machine.

In the following, we use two inputs. First, we consider the case where ON places J_1 and J_2 on two different machines. That is, suppose that ON places J_2 on the second machine. Then, the third job J_3 such that $r(J_3) = 0$ and $d(J_3) = 2 + x$ is given, and no further job arrives. We call this input σ_1. If ON places J_3 onto the first machine, we have $V_{ON}(\sigma_1) = 2 + x + 1 = 3 + x$. ON also gains the same profit if ON places J_3 onto the second machine. On the other hand, the machine onto which OPT places both J_1 and J_2 is different from that onto which J_3 is placed. Thus, $V_{OPT}(\sigma_1) = 2 + 2 + x = 4 + x$. By the above argument,

$$\frac{V_{OPT}(\sigma_1)}{V_{ON}(\sigma_1)} = \frac{4 + x}{3 + x}. \tag{5}$$

Second, we consider the case where ON places J_1 and J_2 onto the first machine. The third job J_1' such that $r(J_1') = 1 - y$ and $d(J_1') = 1 + x$ and the fourth job J_2' such that $r(J_2') = 1$ and $d(J_2') = 1 + x + y$ are given, where y is fixed later. No further job is given; we call this input σ_2. We first consider the case where ON places J_1' and J_2' on different machines. If J_1' is placed onto the first machine, on which J_1 and J_2 are placed,

$$V_{ON}(\sigma_2) = 1 + x + 1 + x + y = 2 + 2x + y. \tag{6}$$

ON gains the same profit if J_2' is placed onto the first machine. Next, we consider the case in which ON places J_1' and J_2' onto the machine. If the machine is the second one, then it is clear that ON gains larger profits than it does in the other case. Hence,

$$V_{ON}(\sigma_2) = 2 + x + 2y. \tag{7}$$

Now, set $y = x$ and we have $V_{ON}(\sigma_2) = 2 + 3x$ by Eqs. (6) and (7). On the other hand, OPT places both J_1 and J_2' onto the first machine and both J_2 and J_1' onto the second machine. Thus, $V_{OPT}(\sigma_2) = 2(1 + x + y) = 2 + 4x$. By the above argument,

$$\frac{V_{OPT}(\sigma_2)}{V_{ON}(\sigma_2)} = \frac{2 + 4x}{2 + 3x}. \tag{8}$$

Therefore, by Eqs. (5) and (8),

$$\frac{V_{OPT}(\sigma)}{V_{ON}(\sigma)} \geq \min\left\{\frac{4 + x}{3 + x}, \frac{2 + 4x}{2 + 3x}\right\} = \frac{4 + \sqrt{2}}{3 + \sqrt{2}} = \frac{10 - \sqrt{2}}{7},$$

where we choose $x = \sqrt{2}$.

The following theorem provides lower bounds for $m \geq 3$ by generalizing the input used to prove Theorem 6.

Theorem 7. *The competitive ratio of any deterministic algorithm is at least 1.101. It is better for fixed m and then refer to Table 2 for details.*

Table 2. Lower bounds for each $m(\geq 3)$.

m	Lower bound	m	Lower bound	m	Lower bound
3	$7/6 \geq 1.166$	6	$\frac{51 - 6\sqrt{2}}{41} \geq 1.140$	9	$9/8 \geq 1.125$
4	$\frac{22 - 2\sqrt{2}}{17} \geq 1.127$	7	$\frac{280 - 70\sqrt{11}}{227} \geq 1.158$	10	$\frac{290 - 15\sqrt{2}}{239} \geq 1.124$
5	$\frac{420 - 15\sqrt{7}}{333} \geq 1.142$	8	$28/25 \geq 1.12$	∞	$\frac{48 - 2\sqrt{2}}{41} \geq 1.101$

Acknowledgments. This work was supported by JSPS KAKENHI Grant Number 26730008.

References

1. Arkin, E.M., Silverberg, E.B.: Scheduling jobs with fixed start and end times. Discrete Appl. Math. **18**(1), 1–8 (1987)
2. Baruah, S.K., Haritsa, J.R.: Scheduling for overload in real-time systems. IEEE Trans. Comput. **46**(9), 1034–1039 (1997)
3. Borodin, A., El-Yaniv, R.: Online Computation and Competitive Analysis. Cambridge University Press, Cambridge (1998)
4. Bouzina, K.I., Emmons, H.: Interval scheduling on identical machines. J. Global Optim. **9**(3–4), 379–393 (1996)
5. Canetti, R., Irani, S.: Bounding the power of preemption in randomized scheduling. SIAM J. Comput. **27**(4), 993–1015 (1998)
6. Carlisle, M.C., Lloyd, E.L.: On the k-coloring of intervals. Discrete Appl. Math. **59**(3), 225–235 (1995)
7. DasGupta, B., Palis, M.A.: Online real-time preemptive scheduling of jobs with deadlines on multiple machines. J. Sched. **4**(6), 297–312 (2001)
8. Epstein, L., Levin, A.: Improved randomized results for the interval selection problem. Theoret. Comput. Sci. **411**(34–36), 3129–3135 (2010)
9. Faigle, U., Nawijn, W.M.: Note on scheduling intervals on-line. Discrete Appl. Math. **58**(1), 13–17 (1995)
10. Fung, S.P.Y., Poon, C.K., Zheng, F.: Online interval scheduling: randomized and multiprocessor cases. J. Comb. Optim. **16**(3), 248–262 (2008)
11. Fung, S.P.Y., Poon, C.K., Yung, D.K.W.: On-line scheduling of equal-length intervals on parallel machines. Inf. Process. Lett. **112**(10), 376–379 (2012)
12. Fung, S.P.Y., Poon, C.K., Zheng, F.: Improved randomized online scheduling of intervals and jobs. Theor. Comput. Syst. **55**(1), 202–228 (2014)
13. Kolen, A.W.J., Lenstra, J.K., Papadimitriou, C.H., Spieksma, F.C.R.: Interval scheduling: a survey. Naval Res. Logist. **54**(5), 530–543 (2007)
14. Kovalyov, M.Y., Ng, C.T., Cheng, T.C.E.: Fixed interval scheduling: models, applications, computational complexity and algorithms. Eur. J. Oper. Res. **178**(2), 331–342 (2007)
15. Lipton, R.J., Tomkins, A.: Online interval scheduling. In: Proceedings of the Fifth Annual ACM-SIAM Symposium on Discrete Algorithms, pp. 302–311 (1994)
16. Miyazawa, H., Erlebach, T.: An improved randomized on-line algorithm for a weighted interval selection problem. J. Sched. **7**(4), 293–311 (2004)
17. Sankowski, P., Zaroliagis, C.: The power of migration for online slack scheduling. In: Proceedings of the 24th Annual European Symposium on Algorithms, pp. 75:1–75:17 (2016)
18. Seiden, S.S.: Randomized online interval scheduling. Oper. Res. Lett. **22**(4–5), 171–177 (1998)
19. Sleator, D.D., Tarjan, R.E.: Amortized efficiency of list update and paging rules. Commun. ACM **28**(2), 202–208 (1985)
20. Woeginger, G.J.: On-line scheduling of jobs with fixed start and end times. Theoret. Comput. Sci. **130**(1), 5–16 (1994)

Properties of Minimal-Perimeter Polyominoes

Gill Barequet$^{(\boxtimes)}$ and Gil Ben-Shachar

Center for Graphics and Geometric Computing, Department of Computer Science,
Technion—Israel Institute of Technology, 3200003 Haifa, Israel
{barequet,gilbe}@cs.technion.ac.il

Abstract. A polyomino is a set of connected squares on a grid. In this
paper we address the class of polyominoes with minimal perimeter for
their area, and we show a bijection between minimal-perimeter polyominoes of certain areas.

Keywords: Polyominoes · Lattice animals · Perimeter

1 Introduction

A polyomino is an edge-connected set of cells on the square lattice. The area of a
polyomino is the number of cells it contains. The problem of counting polyominoes dates back to the 1950s when it was studied in parallel in the fields of combinatorics [8] and statistical physics [6]. Let $A(n)$ denote the number of polyominoes of area n. A general formula for $A(n)$ is still unknown. Klarner [10] showed
the existence of the *growth rate* of $A(n)$, denoting it by $\lambda := \lim_{n\to\infty} \sqrt[n]{A(n)}$.
The exact value of λ is also unknown yet, and its best estimate, 4.06, is by
Jensen [9]. The current best lower and upper bounds on λ are 4.0025 [3] and
4.6496 [11], respectively. Several works provide enumeration by area of special
classes of polyominoes, such as column-convex [7], convex [5], and directed [4]
polyominoes.

The perimeter of a polyomino is the set of empty cells that are adjacent
to at least one polyomino cell, where, as above, two cells are adjacent if thy
share a common edge of the lattice. Although less explored than the area, some
works studied the perimeter of polyominoes. Asinowski et al. [2] showed that
$2n + 2$ is the maximum possible perimeter size for a polyomino of area n, and
provided a few formulae for the numbers of polyominoes with area n and perimeter $2n+2-k$, for some small values of k. In this paper, we shed some light on the
opposite aspect of this type of polyominoes, namely, polyominoes with the minimal perimeter for their area. Closely related works are by Altshuler et al. [1] and
by Sieben [13], both providing a formula for the maximum area of a polyomino
with a certain perimeter size. Sieben [13] also gave a formula for the minimum

Work on this paper by both authors has been supported in part by ISF Grant 575/15.

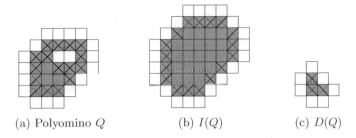

(a) Polyomino Q (b) $I(Q)$ (c) $D(Q)$

Fig. 1. A polyomino Q, its inflated polyomino, and its deflated polyomino. The gray cells are the polyomino cells, while the white cells are the perimeter. Border cells are marked with crosses.

perimeter of a polyomino of area n. Both works also characterized all the polyominoes that have the maximum area for a given perimeter size. Ranjan and Zubair [12] showed a similar result about the minimum perimeter of a directed graph on an infinite grid. In this paper, we study the number of polyominoes which have the minimum perimeter for their area. We define the operation of inflating a polyomino as the extension of a polyomino by all its perimeter cells, and show the following: (1) All minimal-perimeter polyominoes of some area n are inflated into polyominoes of the same area n'. (Polyominoes of the same area, which are not minimal perimeter, may be inflated into polyominoes of different areas.) (2) The inflation operation induces a bijection between the sets of minimal-perimeter polyominoes of area n and the set of minimal-perimeter polyominoes of area n'.

In Sect. 2 we define the notions used throughout this paper. In Sect. 3 we discuss some properties of polyominoes through an analysis of the patterns that may appear in the perimeter. Section 4 is where we reach our main result, proving that the inflating operation induces a bijection between sets of minimal-perimeter polyominoes. We end in Sect. 5 with some concluding remarks.

2 Preliminaries

Let Q be a polyomino, and let $\mathcal{P}(Q)$ be the perimeter of Q. Define $\mathcal{B}(Q)$, the *border* of Q, to be the set of cells of Q that have at least one empty neighboring cell. Given a polyomino Q, its *inflated* polyomino, $I(Q)$, is defined as $I(Q) = Q \cup \mathcal{P}(Q)$. Notice that the border of $I(Q)$ is a subset of the perimeter of Q. Analogously, the *deflated* polyomino, $D(Q)$, is defined as $D(Q) = Q \setminus \mathcal{B}(Q)$, which is obtained by "shaving" the outer layer, i.e., the border cells from the polyomino. Notice that the perimeter of $D(Q)$ is a subset of the border of Q. Also note that $D(Q)$ is not necessarily a valid polyomino since the removal of the border of Q may break it into disconnected pieces. Figure 1 demonstrates all the above definitions.

Following the notation of Sieben [13], we denote by $\epsilon(n)$ the minimum size of the perimeter of all polyominoes of area n. Sieben showed that $\epsilon(n) =$

$\lceil 2 + \sqrt{8n-4} \rceil$. A polyomino Q of area n will be called a minimal-perimeter polyomino if $|\mathcal{P}(Q)| = \epsilon(n)$.

3   Border, Perimeter, and Excess

In this section, we express the size of the perimeter of a polyomino, $|\mathcal{P}(Q)|$, as a function of the border size, $|\mathcal{B}(Q)|$, and the number of excess cells as defined below. The excess of a perimeter cell [2] is defined as the number of polyomino cells that are adjacent to it minus one, and the total excess of a polyomino Q, e_P, is defined as the sum of excess over all the cells of the perimeter of Q. Similarly, the excess of a border cell is defined as the number of perimeter cells adjacent to it minus one, and the border excess, denoted by e_B, is defined as the sum of excess over all the border cells. Let $\pi = |\mathcal{P}(Q)|$ and $\beta = |\mathcal{B}(Q)|$.

Observation 1. *The following holds for any polyomino:* $\pi + e_P = \beta + e_B$. *Equivalently,*

$$\pi = \beta + e_B - e_P. \tag{1}$$

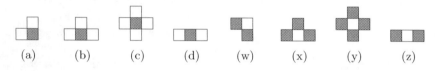

Fig. 2. All possible patterns of excess cells. The gray cells are polyomino cells, while the white cells are perimeter cells. Patterns (a–d) exhibit excess border cells and their surrounding perimeter cells, while Patterns (w–z) exhibit excess perimeter cells and their surrounding polyomino cells.

Equation (1) holds since both $\pi + e_P$ and $\beta + e_B$ are equal to the total length of the polygons forming the boundary of the polyomino. This quantity can be calculated either by summing up over the perimeter cells, where each cell contributes 1 plus its excess for a total of $\pi + e_P$, or by summing up over the border cells for a total of $\beta + e_B$. Figure 2 shows all possible patterns of border and perimeter excess cells, while Fig. 3 shows a sample polyomino with some cells tagged with the corresponding patterns.

Fig. 3. A sample polyomino with marked patterns.

Let $\#\square$ be the number of excess cells of a certain type in a polyomino as classified in the figure, where '$\square$' is one of the symbols a–d or w–z, as in Fig. 2. Counting e_P and e_B as functions of the different patterns of excess cells, we see that $e_B = \#a + 2\#b + 3\#c + \#d$ and $e_P = \#w + 2\#x + 3\#y + \#z$. Substituting e_B and e_P in Eq. (1), we obtain

$$\pi = \beta + \#a + 2\#b + 3\#c + \#d - \#w - 2\#x - 3\#y - \#z.$$

Since Pattern (c) is a singleton cell, we can ignore it in the general formula. Thus, we have

$$\pi = \beta + \#a + 2\#b + \#d - \#w - 2\#x - 3\#y - \#z.$$

3.1 Properties of Minimal-Perimeter Polyominoes

We now discuss the relation between the perimeter and the border of minimal-perimeter polyominoes.

Lemma 1. *Any minimal-perimeter polyomino is simply connected (that is, it does not contain holes).*

Proof. The sequence $\epsilon(n)$ is monotone increasing in the wide sense[1] [13]. Assume that there exists a minimal-perimeter polyomino Q with a hole. Consider the polyomino Q' that is obtained by filling this hole. The area of Q' is clearly larger than the area of Q, and its perimeter size is smaller since we eliminated the perimeter cells inside the hole and did not introduce new perimeter cells. This is a contradiction to $\epsilon(n)$ being monotone increasing.   □

Lemma 2. *For a simply connected polyomino, we have* $\#a + 2\#b - \#w - 2\#x = 4$.

Proof. The boundary of a polyomino without holes is a simple polygon, thus, the sum of its internal angles is $(180(v - 2))°$, where v is the complexity of the polygon. Notice that Pattern (a) (resp., (b)) adds one (resp., two) 90°-vertex to the polygon. Similarly, Pattern (w) (resp. (x)) adds one (resp., two) 270°-vertex. All other patterns do not involve vertices. Let $L = \#a + 2\#b$ and $R = \#w + 2\#x$. Then, the sum of angles of the boundary polygon implies that $L \cdot 90° + R \cdot 270° = (L + R - 2) \cdot 180°$, that is, $L - R = 4$. The claim follows.   □

Theorem 2. *(Stepping Theorem) For a minimal-perimeter polyomino (except the singleton cell), we have that* $\pi = \beta + 4$.

Proof. Lemma 2 tells us that $\pi = \beta + 4 + \#d - \#z$. We will show that any minimal-perimeter polyomino contains neither Pattern (d) nor Pattern (z).

Let Q be a minimal-perimeter polyomino. For the sake of contradiction, assume first that there is a cell $f \in \mathcal{P}(Q)$ as part of Pattern (z). Assume w.l.o.g. that the two adjacent polyomino cells are to the left and to the right of f. These two cells must be connected, thus, the area below (or above) f must be bounded by polyomino cells. Let, then, Q' be the polyomino with the area below f, and the cell f itself, filled with polyomino cells. The cell directly above f becomes a perimeter cell, the cell f ceases to be a perimeter cell, and at least one perimeter cell in the area filled below f is eliminated, thus, $|\mathcal{P}(Q')| < |\mathcal{P}(Q)|$ and $|Q'| > |Q|$, which is a contradiction to the sequence $\epsilon(n)$ being monotone

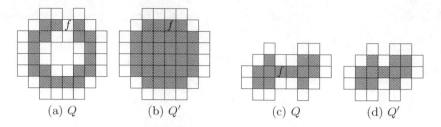

(a) Q (b) Q' (c) Q (d) Q'

Fig. 4. Examples for the first and second parts of the proof of Theorem 2.

increasing. Therefore, the polyomino Q does not contain perimeter cells that fit Pattern (z). Figures 4(a, b) demonstrate this argument.

Now assume for contradiction that Q contains a cell f, forming Pattern (d). Let Q' be the polyomino obtained from Q by removing f and then "pushing" together the two cells adjacent to f. This is always possible since Q is of minimal perimeter, hence, by Lemma 1, it is simply connected, and thus, removing f breaks Q into two separate polyominoes. Any two separated polyominoes can be shifted by one cell without colliding, thus, the transformation described above is valid. The area of Q' is one less than the area of Q, and the perimeter of Q' is smaller by at least two than the perimeter of Q, since the perimeter cells below and above f cease to be part of the perimeter, and connecting the two parts does not create new perimeter cells. From the formula of $\epsilon(n)$ we know that $\epsilon(n) - \epsilon(n-1) \leq 1$ for $n \geq 3$, but $|Q| - |Q'| = 1$ and $|\mathcal{P}(Q)| - |\mathcal{P}(Q')| = 2$, hence, Q is not a minimal-perimeter polyomino, which contradicts our assumption. Thus, there are no cells in Q that fit Pattern (d). Figures 4(c, d) demonstrate this argument. This completes the proof.   □

4   Inflating a Minimal-Perimeter Polyomino

In this section we reach our main results. First, we show that inflating a minimal-perimeter polyomino results in a minimal-perimeter polyomino as well. Second, we show that if any minimal-perimeter polyomino of a certain area n' is created by inflating a minimal-perimeter polyomino of area n, then *all* minimal-perimeter polyominoes of area n' are created by inflating polyominoes of area n. Furthermore, inflating different minimal-perimeter polyominoes of area n results in different minimal-perimeter polyominoes of area n', and so, this operation induces a bijection between the two sets.

Lemma 3. *If Q is a minimal-perimeter polyomino, then $I(Q)$ is simply connected.*

Proof. Let Q be a minimal-perimeter polyomino, and assume that $I(Q)$ is not simply connected, i.e., it contains a hole. Let Q' be the polyomino obtained

---
[1] In the sequel we simply say "monotone increasing.".

by filling the holes of Q by polyomino cells. Notice that this operation only eliminates border cells of $I(Q)$, and does not create any new border cells, thus, $\mathcal{B}(Q') \subset \mathcal{B}(I(Q))$. Let us now compare the polyomino $D(Q')$ to the original polyomino Q. On the one hand, by the way Q' is constructed, $|Q'| > |I(Q)|$, and since $\mathcal{B}(Q') \subset \mathcal{B}(I(Q))$, we have that $|D(Q')| > |D(I(Q))|$. Using the fact that the border of an inflated polyomino is a subset of the perimeter of the original polyomino, we have that $\mathcal{B}(I(Q)) \subset \mathcal{P}(Q)$, thus, $|D(I(Q))| \geq |Q|$, and, hence, $|D(Q')| > |Q|$. On the other hand, we have that $|\mathcal{P}(D(Q'))| \leq |\mathcal{B}(Q')|$ since the perimeter of a deflated polyomino is a subset of the border of the original polyomino, and we have already established that $|\mathcal{B}(Q')| < |\mathcal{B}(I(Q))| \leq |\mathcal{P}(Q)|$, thus, $|\mathcal{P}(D(Q'))| < |\mathcal{P}(Q)|$. But we have shown above that $|D(Q')| > |Q|$, which is in contradiction to $\epsilon(n)$ being an increasing sequence. $\square$

Lemma 4. *If Q is a minimal-perimeter polyomino, then $|\mathcal{P}(I(Q))| \leq |\mathcal{P}(Q)| + 4$.*

Proof. By Lemma 3 we have that $I(Q)$ is simply connected. Thus, by Lemma 2, we have that $|\mathcal{P}(I(Q))| = |\mathcal{B}(I(Q))| + 4 + \#d - \#z$. Since $|\mathcal{B}(I(Q))| \leq |\mathcal{P}(Q)|$, all that remains to show is that Pattern (d) does not occur in $I(Q)$. Assume to the contrary that there is a cell f forming Pattern (d) in $I(Q)$. This cell is a "bridge" in the polyomino. Since $I(Q)$ is simply connected, removing f will break it into exactly two pieces, denoted by Q_1 and Q_2. Both Q_1 and Q_2 must contain cells of the original Q since any cell in $I(Q)$ either belongs to Q or is adjacent to a cell of Q. However, this implies that Q is not connected, which is a contradiction. Hence, Q cannot contain a pattern of type (d), as required. $\square$

Theorem 3. *(Inheritance Theorem) If Q is a minimal-perimeter polyomino, then $I(Q)$ is a minimal-perimeter polyomino as well.*

Proof. Let Q be a minimal-perimeter polyomino. Assume to the contrary that $I(Q)$ is not a minimal-perimeter polyomino, i.e., there exists a polyomino Q' with the same area as $I(Q)$, such that $|\mathcal{P}(Q')| < |\mathcal{P}(I(Q))|$. From Lemma 4 we know that $|\mathcal{P}(I(Q))| \leq |\mathcal{P}(Q)| + 4$, thus, the perimeter of Q' is at most $|\mathcal{P}(Q)| + 3$, and since Q' is a minimal-perimeter polyomino, we know by Theorem 2 that the size of its border is at most $|\mathcal{P}(Q)| - 1$. Consider now the polyomino $D(Q')$. The area of Q' is $|Q| + |\mathcal{P}(Q)|$, thus, the size of $D(Q')$ is at least $|Q| + 1$, and its perimeter size is at most $\epsilon(n) - 1$ (since the perimeter of $D(Q')$ is a subset of the border of Q'). This is a contradiction to the fact that the sequence $\epsilon(n)$ is monotone increasing. Hence, the polyomino Q' cannot exist, and $I(Q)$ is a minimal-perimeter polyomino. Figure 5 demonstrates this theorem. It shows a minimal-perimeter polyomino Q of area 6 and the two minimal-perimeter polyominoes of areas 15 and 28 obtained by inflating Q twice. $\square$

Corollary 1. *The minimum perimeter size of a polyomino of area $n + k\epsilon(n) + 2k(k-1)$ (for $n \neq 1$ and any $k \in \mathbb{N}$) is $\epsilon(n) + 4k$.*

Proof. The claim follows from repeatedly applying Theorem 3 to a minimal-perimeter polyomino of area n. Indeed, inflating once a minimal-perimeter polyomino Q of area n increases the area by $\epsilon(n)$, and the new border size is $\epsilon(n)$.

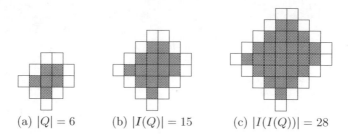

(a) $|Q| = 6$ (b) $|I(Q)| = 15$ (c) $|I(I(Q))| = 28$

Fig. 5. A demonstration of Theorem 3.

Thus, by Theorem 2, the perimeter size increases by 4, becoming $\epsilon(n)+4$. Inflating the new polyomino again increases the area by $\epsilon(n) + 4$, forming yet a new polyomino with perimeter $\epsilon(n) + 8$. Thus, by induction, the kth inflation of the polyomino increases the area by $\epsilon(n) + 4(k-1)$, forming a polyomino with perimeter size $\epsilon(n) + 4k$. As to the area, summing up the contributions in all steps yields $\sum_{i=1}^{k}(\epsilon(n) + 4(i-1)) = k\epsilon(n) + 2k(k-1)$, implying the claim. □

Lemma 5. *Let Q be a minimal-perimeter polyomino of area $n + \epsilon(n)$ (for $n \geq 3$). Then, $D(Q)$ is a valid (connected) polyomino.*

Proof. Assume to the contrary that $D(Q)$ is not connected and that it is composed of at least two parts. Assume first that $D(Q)$ is composed of exactly two parts, Q_1 and Q_2. Define the *joint perimeter* of the two parts, $\mathcal{P}(Q_1, Q_2)$, to be $\mathcal{P}(Q_1) \cup \mathcal{P}(Q_2)$. Since Q is a minimal-perimeter polyomino of area $n+\epsilon(n)$, we know that its perimeter size is $\epsilon(n)+4$ and its border size is $\epsilon(n)$, by Corollary 1 and Theorem 2, respectively. Thus, the size of $D(Q)$ is exactly n regardless of whether or not $D(Q)$ is connected. Since Q_1 and Q_2 are the result of deflating Q, the polyomino Q must have an (either horizontal, vertical, or diagonal) "bridge" of border cells which disappeared in the deflation. The width of the bridge is at most 2, thus, $|\mathcal{P}(Q_1) \cap \mathcal{P}(Q_2)| \leq 2$. Hence, $|\mathcal{P}(Q_1)|+|\mathcal{P}(Q_2)|-2 \leq |\mathcal{P}(Q_1,Q_2)|$. Since $\mathcal{P}(Q_1, Q_2)$ is a subset of $\mathcal{B}(Q)$, we have that $|\mathcal{P}(Q_1,Q_2)| \leq \epsilon(n)$. Therefore,

$$\epsilon(|Q_1|) + \epsilon(|Q_2|) - 2 \leq \epsilon(n). \tag{2}$$

Recall that $|Q_1| + |Q_2| = n$. It is easy to observe that $\epsilon(|Q_1|) + \epsilon(|Q_2|)$ is minimized when $|Q_1| = 1$ and $|Q_2| = n-1$ (or vice versa). Had the function $\epsilon(n)$ (shown in Fig. 6) been $2+\sqrt{8n-4}$ (without rounding up), this would be obvious. But since $\epsilon(n) = \lceil 2 + \sqrt{8n-4} \rceil$, it is a step function (with an infinite number of intervals), where the gap between all successive steps is exactly 1, except the gap between the two leftmost steps which is 2. This guarantees that despite the rounding, the minimum of $\epsilon(|Q_1|) + \epsilon(|Q_2|)$ occurs as claimed. Substituting this into Eq. (2), and using the fact that $\epsilon(1) = 4$, we see that $\epsilon(n-1) + 2 \leq \epsilon(n)$. However, we know [13] that $\epsilon(n)-\epsilon(n-1) \leq 1$ for $n \geq 3$, which is a contradiction. Thus, $D(Q)$ cannot split into two parts unless it splits into two singleton cells, which is indeed the case for a minimal-perimeter polyomino of size 8.

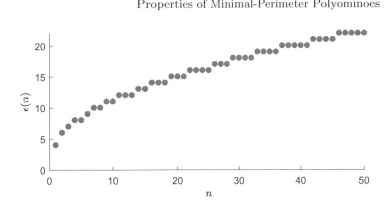

Fig. 6. Values of $\epsilon(n)$.

The same method can be used to show that $D(Q)$ cannot be composed of more then two parts. Note that this proof does not hold for polyominoes of area which is not of the form $n + \epsilon(n)$, but it suffices for the proof of Theorem 4 below. □

Lemma 6. *Let Q_1, Q_2 be two different minimal-perimeter polyominoes. Then, regardless of whether or not Q_1, Q_2 have the same area, the polyominoes $I(Q_1)$ and $I(Q_2)$ are different as well.*

Proof. Assume to the contrary that $Q = I(Q_1) = I(Q_2)$. By definition, this means that $Q = Q_1 \cup \mathcal{P}(Q_1) = Q_2 \cup \mathcal{P}(Q_2)$. Furthermore, since $Q_1 \neq Q_2$, and since a cell can belong to either a polyomino or to its perimeter, but not to both, it must be that $\mathcal{P}(Q_1) \neq \mathcal{P}(Q_2)$. The border of Q is a subset of both $\mathcal{P}(Q_1)$ and $\mathcal{P}(Q_2)$, that is, $\mathcal{B}(Q) \subset \mathcal{P}(Q_1) \cap \mathcal{P}(Q_2)$. Since $\mathcal{P}(Q_1) \neq \mathcal{P}(Q_2)$, we have that either $|\mathcal{B}(Q)| < |\mathcal{P}(Q_1)|$ or $|\mathcal{B}(Q)| < |\mathcal{P}(Q_2)|$; assume w.l.o.g. the former case. Now consider the polyomino $D(Q)$. Its area is $|Q| - |\mathcal{B}(Q)|$. The area of Q is $|Q_1| + |\mathcal{P}(Q_1)|$, thus, $|D(Q)| > |Q_1|$, and since the perimeter of $D(Q)$ is a subset of the border of Q, we conclude that $|\mathcal{P}(D(Q))| < |\mathcal{P}(Q_1)|$. However, Q_1 is a minimal-perimeter polyomino, which is a contradiction to $\epsilon(n)$ being monotone increasing. □

Theorem 4. *(Chain Theorem) Let M_n be the set of minimal-perimeter polyominoes of area n. Then, for $n \geq 3$, we have that $|M_n| = |M_{n+\epsilon(n)}|$.*

Proof. By Theorem 3, if $Q \in M_n$, then $I(Q) \in M_{n+\epsilon(n)}$, and hence, by Lemma 6, we have that $|M_n| \leq |M_{n+\epsilon(n)}|$. Let us now show the opposite relation, namely, that $|M_n| \geq |M_{n+\epsilon(n)}|$. The combination of the two relations will imply the claim.

Let $I(M_n) = \{I(Q) \mid Q \in M_n\}$. For a polyomino $Q \in M_{n+\epsilon(n)}$, our goal is to show that $Q \in I(M_n)$. Since $Q \in M_{n+\epsilon(n)}$, we have by Corollary 1 that $|\mathcal{P}(Q)| = \epsilon(n) + 4$. Moreover, by Theorem 2, we have that $|\mathcal{B}(Q)| = \epsilon(n)$, thus, $|D(Q)| = n$ and $|\mathcal{P}(D(Q))| \geq \epsilon(n)$. Since the perimeter of $D(Q)$ is a subset of the border of Q, and $|\mathcal{B}(Q)| = \epsilon(n)$, we conclude that the perimeter of $D(Q)$ and the border of Q are the same set of cells. Thus, $I(D(Q)) = Q$. Since $|\mathcal{P}(D(Q))| = \epsilon(n)$, we have that $D(Q)$ is a minimal-perimeter polyomino, thus, $Q \in I(M_n)$ as required. Hence, $M_{n+\epsilon(n)} \subseteq I(M_n)$, implying that $\left|M_{n+\epsilon(n)}\right| \leq |I(M_n)| = |M_n|$.

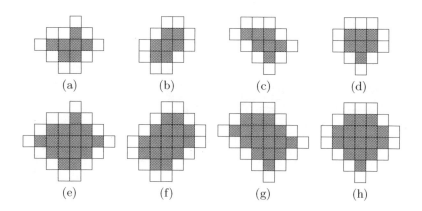

Fig. 7. A demonstration of Theorem 4.

Figure 7 shows, for example, all minimal-perimeter polyominoes of area 7. When they are inflated, they become the entire set of minimal-perimeter polyominoes of area 17. □

Corollary 2. *For* $n \geq 3$ *and any* $k \in \mathbb{N}$, *we have that* $|M_n| = \left|M_{n+k\epsilon(n)+2k(k-1)}\right|$.

Proof. The claim follows from applying Theorem 4 repeatedly on M_n. □

5   Conclusion

We have shown that inflating a set of minimal-perimeter polyominoes of a certain area creates a new set, of the same cardinality, of minimal-perimeter polyominoes of some other area. This creates chains of sets of minimal-perimeter polyominoes of the same area.

In the future, we would like to characterize the roots of these chains and to determine how many minimal-perimeter polyominoes the sets of each chain contains. One may take an algorithmic approach in order to calculate the number of minimal-perimeter polyominoes of a certain area. It seems to be feasible to calculate efficiently the number of minimal-perimeter polyominoes of some area using dynamic programing, utilizing the constraints induced by the special geometric structures of such polyominoes.

References

1. Altshuler, Y., Yanovsky, V., Vainsencher, D., Wagner, I.A., Bruckstein, A.M.: On minimal perimeter polyminoes. In: Kuba, A., Nyúl, L.G., Palágyi, K. (eds.) DGCI 2006. LNCS, vol. 4245, pp. 17–28. Springer, Heidelberg (2006). https://doi.org/10.1007/11907350_2

2. Asinowski, A., Barequet, G., Zheng, Y.: Enumerating polyominoes with fixed perimeter defect. In: Proceedings of 9th European Conference on Combinatorics, Graph Theory, and Applications, vol. 61, pp. 61–67. Elsevier, Vienna, August 2017

3. Barequet, G., Rote, G., Shalah, M.: $\lambda > 4$: an improved lower bound on the growth constant of polyominoes. Commun. ACM **59**(7), 88–95 (2016)

4. Bousquet-Mélou, M.: New enumerative results on two-dimensional directed animals. Discrete Math. **180**(1–3), 73–106 (1998)

5. Bousquet-Mélou, M., Fédou, J.M.: The generating function of convex polyominoes: the resolution of a q-differential system. Discrete Math. **137**(1–3), 53–75 (1995)

6. Broadbent, S., Hammersley, J.: Percolation processes: I. Crystals and mazes. In: Mathematical Proceedings of the Cambridge Philosophical Society, vol. 53, pp. 629–641. Cambridge University Press (1957)

7. Delest, M.P.: Generating functions for column-convex polyominoes. J. Comb. Theory Ser. A **48**(1), 12–31 (1988)

8. Golomb, S.: Checker boards and polyominoes. Am. Math. Mon. **61**(10), 675–682 (1954)

9. Jensen, I., Guttmann, A.: Statistics of lattice animals (polyominoes) and polygons. J. Phys. A: Math. Gen. **33**(29), L257 (2000)

10. Klarner, D.: Cell growth problems. Can. J. Math. **19**, 851–863 (1967)

11. Klarner, D., Rivest, R.: A procedure for improving the upper bound for the number of n-ominoes. Can. J. Math. **25**(3), 585–602 (1973)

12. Ranjan, D., Zubair, M.: Vertex isoperimetric parameter of a computation graph. Int. J. Found. Comput. Sci. **23**(04), 941–964 (2012)

13. Sieben, N.: Polyominoes with minimum site-perimeter and full set achievement games. Eur. J. Comb. **29**(1), 108–117 (2008)

Computing Convex-Straight-Skeleton Voronoi Diagrams for Segments and Convex Polygons

Gill Barequet[1], Minati De[2(✉)], and Michael T. Goodrich[3]

[1] Department of Computer Science, The Technion—Israel Institute of Technology, Haifa, Israel
barequet@cs.technion.ac.il
[2] Department of Mathematics, Indian Institute of Technology Delhi, New Delhi, India
minati@maths.iitd.ac.in
[3] Department of Computer Science, University of California, Irvine, Irvine, CA, USA
goodrich@uci.edu

Abstract. We provide efficient algorithms for computing compact representations of Voronoi diagrams using a convex-straight-skeleton (i.e., convex polygon offset) distance function when sites are line segments or convex polygons.

Keywords: Polygon-offset distance · Voronoi diagrams
Straight skeletons

1 Introduction

Voronoi diagrams (VD) are well-studied in a variety of fields, including, of course, computational geometry, but also ecology, biology, astro-physics, robot motion planning, and medical diagnosis (e.g., see [3,5]). Given a collection of disjoint geometric objects, such as points, segments, or polygons, which are called *sites*, a *Voronoi diagram* is a subdivision of the plane into *cells* such that all the points in a given cell have the same nearest site according to some distance metric.

There are many different types of distance functions that can be used to determine such nearest sites (e.g., see [5,15,19,20]), depending on the application, with one of particular interest for this paper being based on offsets of a convex polygon. Conceptually, this distance function is measured by locally translating the edges of an underlying convex polygon by some amount, either inwardly or outwardly. Such offset distance functions are motivated by applications in three-dimensional modeling and folding (e.g., see [1,7,8]) and are related to a structure known as the *straight skeleton* [2,10,13]. For this reason, we refer to such functions as *convex-straight-skeleton* distance functions.

M. De—Partially supported by DST-INSPIRE Faculty Grant (DST-IFA14-ENG-75).

L. Wang and D. Zhu (Eds.): COCOON 2018, LNCS 10976, pp. 130–142, 2018.
https://doi.org/10.1007/978-3-319-94776-1_12

Assuming that sites are line segments or convex polygons, the combinatorial complexities of convex-straight-skeleton Voronoi diagrams are given in a recent paper by Barequet and De [6], but they do not give efficient algorithms for computing such structures. Our interest in the present paper is the study of such efficient algorithms, for computing a compact representation of a convex-straight-skeleton Voronoi diagram for segments or convex polygons (where Voronoi edges comprising polygonal chains are represented implicitly).

1.1   Related Work

The Voronoi diagram of point sites is extensively studied in the literature (e.g., see [3,5]). Using the Euclidean metric, the combinatorial complexity for the Voronoi diagram is $O(n)$, where the sites are n points [12], or n disjoint line segments [4,18]. These diagrams can be constructed in $O(n \log n)$ time, which is worst-case optimal. For a set of n disjoint convex polygonal sites, each with complexity k, the Voronoi Diagram in the Euclidean metric for this set of sites has combinatorial complexity $O(kn)$ [14,16,20].

McAllister *et al.* [17] introduced the concept of a compact representation of a Voronoi diagram of convex polygonal sites, with distance defined either by the standard Euclidean metric or by scaling a convex polygon (which is related to but nevertheless different from the offset-polygon distance functions we study in this paper). They represent chains of piecewise-algebraic curves as single segments and they show that their compact representation can be used to quickly answer nearest-site queries and that, given a compact Voronoi diagram representation, one can compute the original Voronoi diagram in time proportionate to its combinatorial complexity. They show that such a compact Voronoi diagram can be constructed to have total size $O(n)$, where n is the total number of sites. They provide an algorithm running in time $O(n(\log n + \log k) \log m + m)$ for constructing such a compact Voronoi diagram of n convex polygons, each of size k, using a scaled distance function based on a convex m-gon.

Recently, Cheong *et al.* [11] showed that in the Euclidean metric, the farthest-site counterpart to the compact Voronoi diagram also has combinatorial complexity $O(n)$, and it can be computed in $O(n \log^3 n)$ time. On a related note, Bohler *et al.* [9] recently introduced the related notion of an abstract higher-order Voronoi diagram and studied its combinatorial complexity.

With respect to convex polygon-offset distance functions, the combinatorial complexity of the convex-straight-skeleton Voronoi diagram of a set of n point sites is shown by Barequet *et al.* [7] to be $O(nm)$, where m is the combinatorial complexity of the underlying convex polygon defining distance. Furthermore, they show that compact representations of such diagrams can be computed in $O(n(\log n + \log^2 m) + m)$ time. Recently, Barequet and De [6] show that the combinatorial complexity of a convex-straight-skeleton Voronoi diagram is $O(nm)$ for n line-segment sites and $O(n(m+k))$ for n convex polygons having at most k sides each. We are not aware of any previous results for efficient algorithms for computing a compact representation for a convex-straight-skeleton Voronoi diagram for line-segment or convex-polygon sites, however.

1.2   Our Contributions

In this paper, we show that it is possible to compute a compact representation of a convex-straight-skeleton Voronoi diagram of n line segments in $O(n(\log n + \log^2 m) + m^2)$ time and of n convex polygon sites, each of complexity at most k, in $O(n(\log n + \log k \log^2 m) + m^2)$ time.

Our algorithms are based on new insights into the geometry of convex-straight-skeleton distance functions with line segment and convex polygon sites, which allow us to show how to compute a number of geometric primitives efficiently for segments and convex polygons when the distance is defined by a convex offset-polygon distance function. For instance, we present an $O(\log m)$-time algorithm for computing the distance, $D_\mathcal{P}(z, s)$, between a point, z, and a line segment, s, using an offset distance defined by the polygon, $\mathcal{P}$. We also present an $O(\log^2 m)$-time algorithm for computing another elementary query, $vertex(s_1, s_2, s_3)$: Given three line segments s_1, s_2, and s_3, find the point which is equidistant from them. Our data structures for answering both of these types of queries require $O(m^2)$ preprocessing time. For convex polygon sites, we show that the elementary query operation, $D_\mathcal{P}(z, q)$, can be answered in $O(\log k \log m)$ time, where z is a point and q is a convex polygon with at most k sides. We also show that the primitive, $vertex(q_1, q_2, q_3)$, can be answered in $O(\log k \log^2 m)$ time, where z is a point and the q_i's are convex polygons with at most k sides. Both of these results use data structures having $O(m^2)$ preprocessing time.

2   Preliminaries

Let us borrow a few definitions from earlier papers [6,7]. Given a convex polygon, $\mathcal{P}$, described by the intersection of m closed half-planes, $\{H_i\}$, an *offset copy* of $\mathcal{P}$, denoted as $O_{\mathcal{P},\varepsilon}$, is defined as the intersection of the closed half-planes $\{H_i(\varepsilon)\}$, where $H_i(\varepsilon)$ is the half-plane parallel to H_i with bounding line translated by ε. Depending on whether the value of ε is positive or negative, the translation is respectively done outward or inward of P. See Fig. 1(a). Let $\varepsilon_0 < 0$ be the value for which $O_{\mathcal{P},\varepsilon_0}$ degenerates into a single point c (or a line segment s). We call the value, ε_0, the *negative radius* of $\mathcal{P}$, and the point c (or any point on s) the *center* of $\mathcal{P}$.

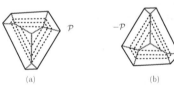

Fig. 1. (a) Offsets of a convex polygon, $\mathcal{P}$, along its straight skeleton (which is the same as its medial axis inside $\mathcal{P}$). (b) Offsets of the convex polygon $(-\mathcal{P})$, including its straight skeleton.

Using the above concept, the polygon-offset distance function $D_\mathcal{P}$ from one point to another point [7] and to an object [6] are defined as follows.

Definition 1 (Point to point distance [7]). *Let z_1 and z_2 be two points in $\mathbb{R}^2$ and $O_{\mathcal{P},\varepsilon}$ be an offset of $\mathcal{P}$ such that a translated copy of $O_{\mathcal{P},\varepsilon}$, centered at z_1, contains z_2 on its boundary. The offset distance is defined as*

$$D_\mathcal{P}(z_1, z_2) = \frac{\varepsilon + |\varepsilon_0|}{|\varepsilon_0|} = \frac{\varepsilon}{|\varepsilon_0|} + 1.$$

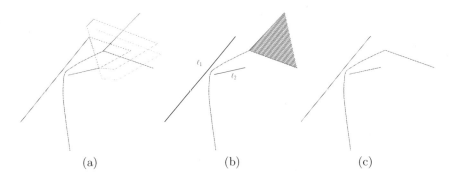

| (a) | (b) | (c) |

Fig. 2. (a) Three different positions of the offset polygons from where both line segments are equidistant; (b) The bisector (colored with blue) of two line segments according to the definition used in [7]; and (c) The bisector (colored with blue) according to our definition. (Color figure online)

Note that this distance function is not a metric since it is not symmetric. On the other hand, observe that $D_\mathcal{P}(z_1, z_2) = D_{(-\mathcal{P})}(z_2, z_1)$, where $(-\mathcal{P}) = \{-z | z \in \mathcal{P}\}$ is a "centrally-mirrored" copy of $\mathcal{P}$. See Fig. 1(b). This fact is widely used to compute the Voronoi diagram for point sites in [7].

Definition 2 (Point to object distance [6]). *Let z be any point, and let o be any object in $\mathbb{R}^2$. The offset distance $D_\mathcal{P}(z, o)$ is defined as $D_\mathcal{P}(z, o) = \min\limits_{z' \in o} D_\mathcal{P}(z, z')$.*

2.1 Convex-Straight-Skeleton Voronoi Diagrams

Under the convex polygon offset distance function, the bisector of two points (as defined originally [7]) can be 2-dimensional instead of 1-dimensional (see Fig. 2 for an illustration). This makes the Voronoi diagram of points unnecessarily complicated. To make it simple, as is also done by Klein and Woods [15], the bisector and Voronoi diagram with respect to the offset distance function $D_\mathcal{P}$ is defined as follows [6].

Let z be a point, and $\Sigma = \{\sigma_i\}$ a set of objects in the plane. In order to avoid 2-dimensional bisectors between two objects in Σ, we define the index of the objects as the "tie breaker" for the relation '$\prec$' between distances from z to the sites. That is, $D_\mathcal{P}(z, \sigma_i) \prec D_\mathcal{P}(z, \sigma_j)$, if $D_\mathcal{P}(z, \sigma_i) < D_\mathcal{P}(z, \sigma_j)$ or, in case $D_\mathcal{P}(z, \sigma_i) = D_\mathcal{P}(z, \sigma_j)$, if $i < j$. Note that the relation '$\prec$' does not allow equality if $i \neq j$. Therefore, the definition below uses the *closure* of portions of the plane in order to have proper boundaries between the regions of the diagram.

Definition 3 (Convex-Straight-Skeleton Voronoi diagram). *Let $\Sigma = \{\sigma_1, \sigma_2, \ldots, \sigma_n\}$ be a set of n sites in $\mathbb{R}^2$. For any $\sigma_i, \sigma_j \in \Sigma$, we define the region of σ_i with respect to σ_j as $NV_\mathcal{P}^{\sigma_j}(\sigma_i) = \{z \in \mathbb{R}^2 | D_\mathcal{P}(z, \sigma_i) \prec D_\mathcal{P}(z, \sigma_j)\}$. The bisecting curve $B_\mathcal{P}(\sigma_i, \sigma_j)$ is defined as $\overline{NV_\mathcal{P}^{\sigma_j}(\sigma_i)} \cap \overline{NV_\mathcal{P}^{\sigma_i}(\sigma_j)}$, where $\overline{X}$ is* *the closure of X. The region of a site σ_i in the convex straight-skeleton Voronoi diagram of Σ is defined as*

$$NV_{\mathcal{P}}(\sigma_i) = \{z \in \mathbb{R}^2 \mid D_{\mathcal{P}}(z, \sigma_i) \prec D_{\mathcal{P}}(z, \sigma_j) \ \ \forall j \neq i\}.$$

The nearest-site convex straight-skeleton Voronoi diagram is the union of the regions

$$NVD_{\mathcal{P}}(\Sigma) = \bigcup_i NV_{\mathcal{P}}(\sigma_i).$$

In other words, the diagram $NVD_{\mathcal{P}}(\Sigma)$ is a partition of the plane, such that if a point $p \in \mathbb{R}^2$ has more than one closest site, then it belongs to the region of the site with the smallest index. The bisectors between regions are defined by taking the closures of the open regions.

3   Tools for Constructing Convex Straight-Skeleton Voronoi Diagrams

Let us generalize our distance function to object-to-point distance as follows.

Definition 4 (Object-to-point distance). *Let z be any point, and let o be any object in $\mathbb{R}^2$. The offset distance $D_{\mathcal{P}}(o, z)$ is defined as $D_{\mathcal{P}}(o, z) = \min\limits_{z' \in o} D_{\mathcal{P}}(z', z)$.*

The following lemma is crucial for the correctness of our algorithms.

Lemma 5. $D_{\mathcal{P}}(z, o) = D_{(-\mathcal{P})}(o, z)$.

3.1   Tools for Line Segments

In [7], the strong relationship between the continuous change of $O_{(-\mathcal{P}),\varepsilon}$ (as a function of ε) and the medial axis of $(-\mathcal{P})$ was observed. The *medial axis*, which is also the *straight skeleton* of the convex polygon $(-\mathcal{P})$ is defined as the set of points inside $(-\mathcal{P})$ that have more than one closest point among the points of $\partial(-\mathcal{P})$. It was noticed that if we change the value of ε continuously by fixing the center, then the vertices of $O_{(-\mathcal{P}),\varepsilon}$ slide along the edges of the medial axis. Outside the polygon, the medial axis and straight skeleton differ in that the straight skeleton extends outward as bisectors of edge offsets, whereas the medial axis extends "rounded"

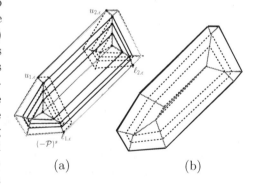

(a)   (b)

Fig. 3. (a) A convex polygon $(-\mathcal{P})^s$ (marked with red), and $O_{(-\mathcal{P})^s,\varepsilon}$ for different values of ε; the extruded medial axis of $(-\mathcal{P})^s$ is marked with blue; and (b) The medial axis of $(-\mathcal{P})^s$ (marked with blue). (Color figure online)

edges from vertices. This information was widely used to efficiently compute $D_{\mathcal{P}}(z_1, z_2)$ for any two points z_1 and z_2 in $\mathbb{R}^2$. As a result, the preprocessing of the medial axis of $(-\mathcal{P})$ in a tree-like data structure was sufficient to answer $D_{\mathcal{P}}(z_1, z_2)$ queries in $O(\log m)$ time, for any two points z_1 and z_2 in $\mathbb{R}^2$.

Lemma 6 *[7, Theorem 10]. Allowing $O(m)$ time preprocessing of the polygon $(-\mathcal{P})$, the distance function $D_{\mathcal{P}}(z_1, z_2)$ can be computed in $O(\log m)$ time, where z_1 and z_2 are two points.*

For our purposes, we would like to preprocess the underlying polygon and compute a similar data structure which will enable us to answer $D_{\mathcal{P}}(z, s)$ queries efficiently, for any point z and line segment s in $\mathbb{R}^2$.

Let $(-\mathcal{P})^s$ be the convex polygon obtained by taking the union of all the translated copies of $(-\mathcal{P})$ centered at all the points in the line segment s. Similarly, we define $O_{(-\mathcal{P})^s, \varepsilon}$ as the convex polygon obtained by taking the union of all the translated copies of $O_{(-\mathcal{P}), \varepsilon}$ centered at all the points in the line segment s. Note that $(-\mathcal{P})^s$ (resp., $O_{(-\mathcal{P})^s, \varepsilon}$) is the convex-hull of two translated copies of $(-\mathcal{P})$ (resp., $O_{(-\mathcal{P}), \varepsilon}$) centered at the two endpoints of s. Note that when ε takes the value ε_0, which is the negative radius of $(-\mathcal{P})$, then $O_{(-\mathcal{P})^s, \varepsilon_0}$ degenerates into the line segment s. We refer to the line segment s as the *center* of $O_{(-\mathcal{P})^s, \varepsilon}$, for $\varepsilon \geq \varepsilon_0$. Note that even in the worst case, the complexity of $(-\mathcal{P})^s$ is not twice the complexity of $\mathcal{P}$, but simply $|(-\mathcal{P})^s| = |\mathcal{P}| + 2$. We define the *extruded medial axis* of $(-\mathcal{P})^s$ as the set of points inside $(-\mathcal{P})^s$ such that if we change the value of ε continuously by fixing the center at s, then the vertices of $O_{(-\mathcal{P})^s, \varepsilon}$ slide along the edges of the extruded medial axis (see Fig. 3(a)). Also note that the extruded medial axis of $(-\mathcal{P})^s$ may not be the same as the medial axis of $(-\mathcal{P})^s$ (see Figs. 3(a–b) for a comparison).

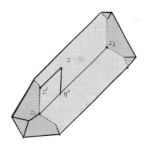

Fig. 4. The *extruded medial axis* of $(-\mathcal{P})^s$: The medial region and parallel region are colored with light green and light blue, respectively; two common boundaries are marked with red. (Color figure online)

We define the following distance function.

Definition 7 $(D_{(-\mathcal{P})^s}(s, z))$. *Let z be any point and s be any line segment in $\mathbb{R}^2$, and $O_{(-\mathcal{P})^s, \varepsilon}$ be an offset of $(-\mathcal{P})^s$ (centered at s) such that $O_{(-\mathcal{P})^s, \varepsilon}$ contains z on its boundary. The offset distance $D_{(-\mathcal{P})^s}(s, z)$ is defined as $D_{(-\mathcal{P})^s}(s, z) =$*
$$\frac{\varepsilon + |\varepsilon_0|}{|\varepsilon_0|} = \frac{\varepsilon}{|\varepsilon_0|} + 1.$$

Lemma 8. $D_{(-\mathcal{P})}(s,z) = D_{(-\mathcal{P})^s}(s,z)$.

Combining Lemmata 5 and 8, we have the following.

Lemma 9. $D_{\mathcal{P}}(z,s) = D_{(-\mathcal{P})^s}(s,z)$.

Let us now show how to compute efficiently the distance $D_{\mathcal{P}}(z,s)$ for any point z and line segment s in $\mathbb{R}^2$. Following Lemma 9, we know that it is sufficient to compute $D_{(-\mathcal{P})^s}(s,z)$. Provided that the extruded medial axis of $(-\mathcal{P})^s$ is computed in a preprocessing step, $D_{(-\mathcal{P})^s}(s,z)$ can be computed as in Lemma 6 [7]. Note that $D_{\mathcal{P}}(z,s)$ is a primitive operation for the computation of the compact Voronoi diagram. Thus, simply preprocessing the extruded medial axis of $(-\mathcal{P})^s$ for every segment $s \in \mathscr{S}$ would result in increased preprocessing space (i.e., $O(nm)$ space). However, we prove here that even with preprocessing only the medial axis of $(-\mathcal{P})$, we can compute $D_{\mathcal{P}}(z,s)$ with the same query time as that of computing $D_{\mathcal{P}}(z_j, z_\ell)$, $z_j, z_\ell \in \mathbb{R}^2$.

Let us now illustrate the properties of the extruded medial axis of $(-\mathcal{P})^s$. Place two translated copies $T_{1,\varepsilon}$ and $T_{2,\varepsilon}$ of the offset polygon $O_{(-\mathcal{P}),\varepsilon}$, centered at two endpoints z_1 and z_2 of the line segment s (see Fig. 3(a)). Let $u_{1,\varepsilon}$ and $u_{2,\varepsilon}$ be the two vertices of the upper tangent[1] of the convex-hull joining $T_{1,\varepsilon}$ and $T_{2,\varepsilon}$. Observe that $u_{1,\varepsilon}$ and $u_{2,\varepsilon}$ are the same vertex of the offset polygon $O_{(-\mathcal{P}),\varepsilon}$. Similarly, let $\ell_{1,\varepsilon}$ and $\ell_{2,\varepsilon}$ be the two vertices of the lower tangent of the convex-hull joining $T_{1,\varepsilon}$ and $T_{2,\varepsilon}$. Note that if we change the value of ε, then both $u_{i,\varepsilon}$ and $\ell_{i,\varepsilon}$ change along the medial axis of $T_{i,\varepsilon}$, $i \in \{1,2\}$, and the upper (resp., lower) tangent moves and is always parallel to s. Thus, the extruded medial axis of $(-\mathcal{P})^s$ is a subset of the union of the medial axes of $T_{1,\varepsilon}$ and $T_{2,\varepsilon}$. Specifically, the portion of the medial axis of $T_{i,\varepsilon}$, that lies between $u_{i,\varepsilon}$ and $\ell_{i,\varepsilon}$ and whose end vertices are in the convex-hull, are in the extruded medial axis of $(-\mathcal{P})^s$. We refer to this part of the medial axis of $T_{i,\varepsilon}$ as the *medial region* with respect to the line segment s. We define the *parallel region* as the part of the polygon $(-\mathcal{P})^s$ that does not have dominating edges from the medial region (see Fig. 4 for an illustration). For a point z which belongs to the parallel region of $(-\mathcal{P})^s$, let z' be the projection of z on the *common boundary* of the medial region of T_i, $i \in \{1,2\}$ and the parallel region (see Fig. 4). Then, $D_{\mathcal{P}}(z,s)$ can be calculated by simply computing $D_{\mathcal{P}}(z', z_i)$.

Lemma 10. *Allowing $O(m^2)$ time preprocessing of the polygon $(-\mathcal{P})$, $D_{\mathcal{P}}(z,s)$ can be computed in $O(\log m)$ time, where z is a point and s is a line segment. Along with that, a point $q^* \in s$ satisfying $D_{\mathcal{P}}(z, q^*) = D_{\mathcal{P}}(z,s)$ can be reported in the same amount of time.*

---

[1] If s is vertical, then we arbitrarily choose the left tangent as the upper.

Proof. We keep two copies, T_1 and T_2, of the processed medial axis of $(-\mathcal{P})$ as required by Lemma 6. In addition, we preprocess $(-\mathcal{P})$ such that both traversing and binary searching is possible along any vertex-to-center path of $(-\mathcal{P})$. Since there are m vertices, there are m such paths. By simply storing each path as a list, we need a total of $O(m^2)$ space and time to preprocess this data structure[2].

To answer the query $D_\mathcal{P}(z, s)$, we do the following:

Step 1. Compute the upper and lower tangents of the two translated copies T_1 and T_2, centered at z_1 and z_2, respectively, where z_1 and z_2 are the endpoints of the line segment s. Let u_i and ℓ_i be the points in which the upper and lower tangents T_i, $i \in \{1, 2\}$, touch the two polygons.

Step 2.1. If z is in the relevant region of T_i with respect to s, then answer $D_\mathcal{P}(z, s)$, return the point q^* by evoking $D_\mathcal{P}(z, z_i)$, and stop.

Step 2.2. Otherwise (if z is in the parallel region), we answer a ray-shooting query on the common boundary B_i between the medial region of T_i and the parallel region of $(-\mathcal{P})^s$ to find z', the projection of z on B_i (see Fig. 4). We obtain $D_\mathcal{P}(z, s)$ by computing $D_\mathcal{P}(z', z_i)$. We can then find the point q^* by translating the point z_i by $d(z, z')$ along the line segment s, where $d(z, z')$ is the Euclidean distance between z and z'.

Step 1 takes $O(\log m)$ time, assuming that $(-\mathcal{P})$ is available as a cyclic list of vertices. Step 2.1 can be done in $O(\log m)$ time as in Lemma 6. Since B_i is a path from a vertex of $(-\mathcal{P})$ to its center along the medial axis where we can perform a binary search, we can find z' in $O(\log m)$ time. Hence, Step 2.2 takes $O(\log m)$ time. In total, the query time complexity is $O(\log m)$.   □

Another primitive operation is finding the Voronoi vertex v^* where the Voronoi cells with respect to the polygon-offset distance $D_\mathcal{P}$ for sites s_1, s_2, s_3 occur in counterclockwise order around v^*. Applying the tentative prune-and-search paradigm, we can find v^* similarly to the method in [7]. The only difference is that here, three different polygons $(-\mathcal{P})^{s_1}$, $(-\mathcal{P})^{s_2}$, and $(-\mathcal{P})^{s_3}$ come into the picture instead of three identical copies of $(-\mathcal{P})$. Thus, we have the following:

Lemma 11. *Given three line segments s_1, s_2, s_3 in the plane, and a convex polygon $\mathcal{P}$ of m sides, a Voronoi vertex v^*, where the Voronoi cells with respect to the polygon-offset distance $D_\mathcal{P}$ for sites s_1, s_2, s_3 occur in counterclockwise order around v^*, can be computed in $O(\log^2 m)$ time (allowing $O(m^2)$ time for preprocessing).*

---

[2] This preprocessing step can probably be implemented in a more efficient way, but since it's not the bottleneck of the algorithm, such an improvement will not affect the total running time of the algorithm for computing $D_\mathcal{P}(z, s)$.

3.2   Tools for Convex Polygonal Sites

Here, we generalize the formerly described tool of line segments for convex polygons. Let q be a convex polygon with k sides, and ∂q be the boundary of q. Let $(-\mathcal{P})^q$ be the convex polygon obtained by uniting all the translated copies of $(-\mathcal{P})$ centered at all the points $z \in q$. Similarly, we define an offset copy $O_{(-\mathcal{P})^q,\varepsilon}$ as the convex polygon obtained by uniting all the translated copies of $O_{(-\mathcal{P}),\varepsilon}$ centered at all the points $z \in q$. Note that

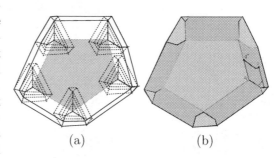

(a)   (b)

Fig. 5. (a) $(-\mathcal{P})^q$: The polygon q is colored gray. Translated copies of $(-\mathcal{P})$, centered at the vertices of q, are shown with dotted lines. The extruded medial axis of $(-\mathcal{P})^q$ is marked with blue. (b) The medial region and parallel region are colored light green and light blue, respectively. (Color figure online)

$(-\mathcal{P})^q$ (resp., $O_{(-\mathcal{P})^q,\varepsilon}$) is the convex-hull of k translated copies of $(-\mathcal{P})$ (resp., $O_{(-\mathcal{P}),\varepsilon}$) centered at the k vertices of the polygon q. Note that when ε takes the value ε_0, which is the negative radius of $(-\mathcal{P})$, then $O_{(-\mathcal{P})^q,\varepsilon_0}$ degenerates into the polygon q. We refer to the polygon q as the *center* of $O_{(-\mathcal{P})^q,\varepsilon}$, for $\varepsilon \geq \varepsilon_0$. The complexity of $(-\mathcal{P})^q$ is $|(-\mathcal{P})^q| = |\mathcal{P}| + k$ because each side of $(-\mathcal{P})$ can appear at most once along the boundary of $(-\mathcal{P})^q$, and there are exactly k tangents. We define the *extruded medial axis* of $(-\mathcal{P})^q$ as the set of points inside $(-\mathcal{P})^q$ such that if we change the value of $\varepsilon \geq \varepsilon_0$ continuously by fixing the center at q, then the vertices of $O_{(-\mathcal{P})^q,\varepsilon}$ slide along the edges of the extruded medial axis (see Fig. 5). Similarly to the previous section, the extruded medial axis of $(-\mathcal{P})^q$ may differ from the medial axis of $(-\mathcal{P})^q$.

We define the following distance function.

Definition 12 $\left(D_{(-\mathcal{P})^q}(q,z)\right)$**.** *Let z be any point and q be any convex polygon with k sides in $\mathbb{R}^2$, and $O_{(-\mathcal{P})^q,\varepsilon}$ be an offset copy of $(-\mathcal{P})^q$ (centered at q) such that $O_{(-\mathcal{P})^q,\varepsilon}$ contains z on its boundary. The offset distance $D_{(-\mathcal{P})^q}(q,z)$ is defined as* $D_{(-\mathcal{P})^q}(q,z) = \dfrac{\varepsilon + |\varepsilon_0|}{|\varepsilon_0|} = \dfrac{\varepsilon}{|\varepsilon_0|} + 1.$

Lemma 13. $D_{(-\mathcal{P})}(q,z) = D_{(-\mathcal{P})^q}(q,z).$

Proof. The proof is similar to the proof of Lemma 8.   □

Combining Lemmata 5 and 13, we have the following.

Lemma 14. $D_{\mathcal{P}}(z,q) = D_{(-\mathcal{P})^q}(q,z).$

Let us illustrate the properties of the extruded medial axis of $(-\mathcal{P})^q$. Place k translated copies $T_{i,\varepsilon}$, $i \in \{1,\ldots,k\}$, of the offset polygon $O_{(-\mathcal{P}),\varepsilon}$ centered at the k vertices $\{z_i\}$, $i \in \{1, \ldots, k\}$, of the polygon q (see Fig. 5(a)). Here $T_{i,\varepsilon}$ and $T_{i+1,\varepsilon}$ are two clockwise consecutive copies, and let t_i be the outer common tangent of $T_{i,\varepsilon}$ and $T_{i+1,\varepsilon}$, where the addition of subtraction of the index i are modulo k. Let $u_{i_1,\varepsilon}$ and $u_{i_2,\varepsilon}$ be the two vertices of the outer tangent of the convex-hull joining $T_{i,\varepsilon}$ and $T_{i+1,\varepsilon}$. Observe that $u_{i_1,\varepsilon}$ and $u_{i_2,\varepsilon}$ are the same vertex of the offset polygon $O_{(-\mathcal{P}),\varepsilon}$. Note that if we change the value of ε, then all vertices $u_{i_j,\varepsilon}$ move along the medial axis of $T_{i,\varepsilon}$, $i \in \{1, 2, \ldots, k\}$, $j \in \{1, 2\}$, and all k outer tangents move parallel to themselves. Thus, the extruded medial axis of $(-\mathcal{P})^q$ is a subset of the union of medial axes of $T_{i,\varepsilon}$, $i \in \{1, 2, \ldots, k\}$ (see Fig. 5(b)). Specifically, the portion of the medial axis of $T_{i,\varepsilon}$ that lies between $u_{(i-1)_2,\varepsilon}$ and $u_{i_1,\varepsilon}$ and whose end vertices appear in $O_{(-\mathcal{P})^q,\varepsilon}$ are in the extruded medial axis of $(-\mathcal{P})^q$. We refer to this part of the medial axis of $T_{i,\varepsilon}$ as the *medial region* of $T_{i,\varepsilon}$ with respect to the polygon q. We define the *ith parallel region* as the part of the polygon $O_{(-\mathcal{P})^q,\varepsilon}$ that does not have dominating edges from the medial region and lies between $T_{i,\varepsilon}$ and $T_{i+1,\varepsilon}$ (see Fig. 5(b)). For a point z which belongs to the ith parallel region of $O_{(-\mathcal{P})^q,\varepsilon}$, let z' be the projection of z on the *common boundary* of the medial region of T_i and the ith parallel region. Then, $D_{\mathcal{P}}(z, q)$ can be found by simply computing $D_{\mathcal{P}}(z', z_i)$.

Lemma 15. *Allowing $O(m^2)$ time and space for preprocessing of the polygon $\mathcal{P}$, the function $D_{\mathcal{P}}(z, q)$ can be computed in $O(\log k \log m)$ time, where z is a point, and q is a convex polygon with at most k sides.*

Proof. We keep three copies T_j, $j \in \{-1, 0, 1\}$ of the processed medial axis of $(-\mathcal{P})$ as required by Lemma 6. In addition, we preprocess $(-\mathcal{P})$ such that both traversing and binary searching are possible along each vertex-to-center path of $(-\mathcal{P})$. Since there are m vertices, there are m such paths. By simply storing each path as a list, we need a total of $O(m^2)$ space and time to preprocess this data structure (See footnote 2).

To answer the query $D_{\mathcal{P}}(z, s)$, we perform a binary search along the cyclic list of vertices of the polygon q. At each step of the binary search, we select, say, z_i, the ith vertex of q, and decide whether either (i) z is in the medial region; (ii) z is in the parallel region of the corresponding translated copy of $(-\mathcal{P})$; or (iii) z is in the left or right side of z_i in the cyclic order list of vertices of q.

To decide whether z is in the medial region or in the parallel region of the corresponding translated copy of $(-\mathcal{P})$, centered at the ith vertex, we do the following:

Step 1. Place three copies T_j, $j \in \{-1, 0, 1\}$ at the $(i-1)$st, ith, and $(i+1)$st vertices of q. Let z_i be the ith vertex of q.

Step 2. Find the outer common tangents of T_{-1}, T_0 and of T_0, T_1. This will allow us to detect the medial region and parallel region of T_0 with respect to q.

Step 3.1. If z is in the medial region of T_0, then we can answer $D_{\mathcal{P}}(z, q)$ and report the point q^* by invoking $D_{\mathcal{P}}(z, z_i)$. We stop after reporting.

Step 3.2. Else, if z is in the parallel region (see Fig. 5(b)), we perform ray-shooting on the common boundary B_i between the medial region and the parallel region of T_i to find z', the projection of z on B_i. We obtain $D_{\mathcal{P}}(z, s)$ by computing $D_{\mathcal{P}}(z', z_i)$. We can find the point q^* by translating the point z_i by $d(z, z')$ along the line segment s, where $d(z, z')$ is the Euclidean distance between z and z'.

Step 3.3. Otherwise, determine the side where z lies with respect to z_i in the cyclic list of vertices of q.

Step 1 takes constant time, Step 2 takes $O(\log m)$ time assuming that the polygon $(-\mathcal{P})$ is available as a cyclic list of vertices. Step 3.1 can be done in $O(\log m)$ time as in Lemma 6. Since B_i is a path from a vertex of $(-\mathcal{P})$ to its center along the medial axis where we can do binary search, we can find z' in $O(\log m)$ time. Therefore, Step 3.2 takes $O(\log m)$ time. Step 3.3 needs constant time. Thus, at each step of the binary search, we need $O(\log m)$ time. In total, the query time complexity is $O(\log k \log m)$.   □

The other primitive operation is to find the Voronoi vertex v^*, where the Voronoi cells, with respect to the polygon-offset distance $D_{\mathcal{P}}$ for three polygonal sites q_1, q_2, q_3, occur in counterclockwise order around v^*. Applying the tentative prune-and-search paradigm, we can find v^* similarly to the method used in [7]. The main difference is that here, three different polygons $(-\mathcal{P})^{q_1}$, $(-\mathcal{P})^{q_2}$ and $(-\mathcal{P})^{q_3}$ come into the picture instead of three identical copies of $(-\mathcal{P})$. On the other hand, each $O(\log m)$-time distance evaluation function is replaced by an $O(\log k \log m)$-time operation for evaluating $D_{\mathcal{P}}(z, q)$ (by Lemma 15). Thus, we have the following:

Lemma 16. *Given three polygons p_1, p_2, p_3 in the plane, each having at most k sides, and a convex polygon $\mathcal{P}$ with m sides, the point v^*, equidistant from p_1, p_2, p_3 with respect to the polygon-offset distance $D_{\mathcal{P}}$, can be computed in $O(\log k \log^2 m)$ time (allowing $O(m^2)$ time for preprocessing).*

4   Computing a Compact Convex Straight-Skeleton Voronoi Diagram

As mentioned above, McAllister *et al.* [17] presented an algorithm for computing a *compact* representation of the nearest-site Voronoi diagram of a set of convex polygonal sites with respect to a convex (scaled) distance function. Here, we show how to adapt their method to obtain a compact representation of $NVD_{\mathcal{P}}(\mathcal{Q})$, where $\mathcal{Q} = \{q_1, q_2, \ldots, q_n\}$ is a set of n convex polygonal sites, each having at most k sides, and $\mathrm{NVD}_{\mathcal{P}}(\mathcal{Q})$ is the nearest-site convex-straight-skeleton Voronoi diagram of these sites, with respect to the convex polygon-offset distance function $D_{\mathcal{P}}$, where $\mathcal{P}$ is an m-sided convex polygon.

For any point z and a polygonal site q, $spoke(z, q)$ is defined as a line segment $\overline{z, z^*}$, such that $D_{\mathcal{P}}(z, z^*) = \min_{z' \in q} D_{\mathcal{P}}(z, z')$. Here, z^* is referred to as the *attachment point* of the spoke. Note that $spoke(z, q)$ can be computed in $O(\log k \log m)$ time (with $O(m^2)$ preprocessing time), where the polygon q has k vertices/edges (Lemma 15).

For three sites q_1, q_2, q_3, $vertex(q_1, q_2, q_3)$ is defined as the point v equidistant from q_1, q_2, q_3 with respect to the polygon-offset distance function $D_\mathcal{P}$. From Lemma 16 we know that $vertex(q_1, q_2, q_3)$ can be computed in $O(\log k \log^2 m)$ time (with $O(m^2)$ preprocessing time).

The compact Voronoi diagram is a simplified version of the full Voronoi diagram. Here, we maintain a set of spokes from the Voronoi vertices around the cell, and each polygonal site q is replaced by its *core*, where a core is the convex hull of the attachment points that lie on the boundary of q (see [17, Fig. 4]). As a result, we obtain a compact representation whose combinatorial complexity is $O(n)$. Note that the combinatorial complexity of $NVD_\mathcal{P}(\mathcal{Q})$ is $O(n(m+k))$ [6], which is higher than the combinatorial complexity of the compact representation.

Note that each cell of this compact diagram is actually composed of portions of two cells of the full Voronoi diagram. Thus, we can do the point location as follows. Given a point z, we can obtain the compact cell and the two corresponding candidate sites q_i and q_j in $O(\log n)$ time. Then, spending additional $O(\log k \log m)$ time to compare $D_\mathcal{P}(z, q_i)$ and $D_\mathcal{P}(z, q_j)$, we can determine the identity of the cell of the full Voronoi diagram in which the point is located.

As observed in [7, Sect. 4], the geometric properties of the compact Voronoi diagram are preserved when we use a convex polygon-offset distance function instead of a convex distance function. Hence, we can apply Theorem 3.10 of [17], which states that the compact representation of the Voronoi diagram can be computed in $O(n(\log n + T_v))$ time, where T_v is the time needed for performing primitive operations like $spoke(z, q)$ and $vertex(q_1, q_2, q_3)$. For convex polygonal sites, T_v is $O(\log k \log^2 m)$ (this follows from Lemmata 15 and 16). Thus, we can compute a compact Voronoi diagram for $NVD_\mathcal{P}(\mathcal{Q})$ in $O(n(\log n + \log k \log^2 m) + m^2)$ time, where $\mathcal{Q}$ is a set of n disjoint convex polygonal sites, each having at most k sides.

Theorem 17. *For a set $\mathcal{Q}$ of n convex polygonal sites, each having at most k sides, the compact representation of the Voronoi diagram $NVD_\mathcal{P}(\mathcal{Q})$ can be computed in expected $O(n(\log n + \log k \log^2 m) + m^2)$ time, where m is the number of sides of the underlying convex polygon $\mathcal{P}$.*

Following the same arguments as in the proof of Theorem 17, we have the following result.

Theorem 18. *For a set $\mathcal{S}$ of n line segments, the compact representation of the nearest-site Voronoi diagram $NVD_\mathcal{P}(\mathcal{S})$ can be computed in $O(n(\log n + \log^2 m) + m^2)$ time, where m is the number of sides of the underlying convex polygon $\mathcal{P}$.*

142   G. Barequet et al.

References

1. Abel, Z., Demaine, E.D., Demaine, M.L., Itoh, J.-I., Lubiw, A., Nara, C., O'Rourke, J.: Continuously flattening polyhedra using straight skeletons. In: 13th Symposium on Computational Geometry (SoCG), pp. 396:396–396:405 (2014)
2. Aichholzer, O., Aurenhammer, F.: Straight skeletons for general polygonal figures in the plane. In: COCOON, pp. 117–126 (1996)
3. Aurenhammer, F.: Voronoi diagrams - a survey of a fundamental geometric data structure. ACM Comput. Surv. **23**(3), 345–405 (1991)
4. Aurenhammer, F., Drysdale, R.L.S., Krasser, H.: Farthest line segment Voronoi diagrams. Inf. Process. Lett. **100**(6), 220–225 (2006)
5. Aurenhammer, F., Klein, R., Lee, D.: Voronoi Diagrams and Delaunay Triangulations. World Scientific, Singapore (2013)
6. Barequet, G., De, M.: Voronoi diagram for convex polygonal sites with convex polygon-offset distance function. In: Gaur, D., Narayanaswamy, N.S. (eds.) CALDAM 2017. LNCS, vol. 10156, pp. 24–36. Springer, Cham (2017). https://doi.org/10.1007/978-3-319-53007-9_3
7. Barequet, G., Dickerson, M.T., Goodrich, M.T.: Voronoi diagrams for convex polygon-offset distance functions. Discret. Comput. Geom. **25**(2), 271–291 (2001)
8. Barequet, G., Goodrich, M.T., Levi-Steiner, A., Steiner, D.: Contour interpolation by straight skeletons. Graph. Models **66**(4), 245–260 (2004)
9. Bohler, C., Cheilaris, P., Klein, R., Liu, C., Papadopoulou, E., Zavershynskyi, M.: On the complexity of higher order abstract Voronoi diagrams. Comput. Geom. **48**(8), 539–551 (2015)
10. Cheng, S.-W., Mencel, L., Vigneron, A.: A faster algorithm for computing straight skeletons. In: Schulz, A.S., Wagner, D. (eds.) ESA 2014. LNCS, vol. 8737, pp. 272–283. Springer, Heidelberg (2014). https://doi.org/10.1007/978-3-662-44777-2_23
11. Cheong, O., Everett, H., Glisse, M., Gudmundsson, J., Hornus, S., Lazard, S., Lee, M., Na, H.: Farthest-polygon Voronoi diagrams. Comput. Geom. **44**(4), 234–247 (2011)
12. Fortune, S.: A sweepline algorithm for Voronoi diagrams. Algorithmica **2**, 153–174 (1987)
13. Huber, S., Held, M.: A fast straight-skeleton algorithm based on generalized motorcycle graphs. Int. J. Comput. Geom. Appl. **22**(05), 471–498 (2012)
14. Kirkpatrick, D.G.: Efficient computation of continuous skeletons. In: 20th IEEE Symposium on Foundations of Computer Science (FOCS), pp. 18–27 (1979)
15. Klein, R., Wood, D.: Voronoi diagrams based on general metrics in the plane. In: Cori, R., Wirsing, M. (eds.) STACS 1988. LNCS, vol. 294, pp. 281–291. Springer, Heidelberg (1988). https://doi.org/10.1007/BFb0035852
16. Leven, D., Sharir, M.: Planning a purely translational motion for a convex object in two-dimensional space using generalized Voronoi diagrams. Discret. Comput. Geom. **2**, 9–31 (1987)
17. McAllister, M., Kirkpatrick, D.G., Snoeyink, J.: A compact piecewise-linear Voronoi diagram for convex sites in the plane. Discret. Comput. Geom. **15**(1), 73–105 (1996)
18. Papadopoulou, E., Dey, S.K.: On the farthest line-segment Voronoi diagram. Int. J. Comput. Geom. Appl. **23**(6), 443–460 (2013)
19. Papadopoulou, E., Zavershynskyi, M.: The higher-order Voronoi diagram of line segments. Algorithmica **74**(1), 415–439 (2016)
20. Yap, C.: An $O(n \log n)$ algorithm for the Voronoi diagram of a set of simple curve segments. Discret. Comput. Geom. **2**, 365–393 (1987)

Polygon Queries for Convex Hulls
of Points

Eunjin Oh[1] and Hee-Kap Ahn[2](✉) (iD)

[1] Max Planck Institute for Informatics, Saarbrücken, Germany
eoh@mpi-inf.mpg.de
[2] Pohang University of Science and Technology, Pohang, Korea
heekap@postech.ac.kr

Abstract. We study the following range searching problem: Preprocess a set P of n points in the plane with respect to a set $\mathcal{O}$ of k orientations in the plane so that given an $\mathcal{O}$-oriented convex polygon Q as a query, the convex hull of $P \cap Q$, and its perimeter and area, can be reported efficiently, where an $\mathcal{O}$-oriented polygon is a polygon whose edges have orientations in $\mathcal{O}$. We present a data structure with $O(nk^3 \log^2 n)$ space and $O(nk^3 \log^2 n)$ construction time, and a query algorithm to compute the perimeter or area of the convex hull of $P \cap Q$ in $O(s \log^2 n)$ time for any query $\mathcal{O}$-oriented convex s-gon Q. For reporting the convex hull, $O(h)$ is added to the running times of query algorithms, where h is the complexity of the convex hull.

1 Introduction

Range searching is one of the most thoroughly studied problems in computational geometry for decades from 1970s. Range trees and kd-trees were proposed as data structures for orthogonal range searching, and their sizes and query times had been improved over the years. The most efficient data structures for orthogonal range searching for points in the plane [6] and in higher dimensions [7] are due to Chazelle.

There are variants of the range searching problem that allow other types of query ranges, such as circles or triangles. Many of them can be solved using partition trees or a combination of partition trees and cutting trees. The simplex range searching problem, which is a higher dimensional analogue of the triangular range searching, has gained much attention in computational geometry as many other problems with more general ranges can be reduced to them. As an application, it can be used to solve the hidden surface removal in computer graphics [4, 11].

This research was supported by the NRF grant 2011-0030044 (SRC-GAIA) funded by the Korea government and the MSIT (Ministry of Science and ICT), Korea, under the SW Starlab support program (IITP-2017-0-00905) supervised by the IITP (Institute for Information & communications Technology Promotion).

L. Wang and D. Zhu (Eds.): COCOON 2018, LNCS 10976, pp. 143–155, 2018.
https://doi.org/10.1007/978-3-319-94776-1_13

The polygon range searching is a generalization of the simplex range searching in which the search domain is a convex polygon. Willard [18] gave a data structure, called the polygon tree, with $O(n)$ space and an $O(n^{0.77})$-time algorithm for counting the number of points lying inside an arbitrary query polygon of constant complexity. The query time was improved later by Edelsbrunner and Welzl [13] to $O(n^{0.695})$. By using the stabbing numbers of spanning trees, Chazelle and Welzl [8] gave a data structure of size $O(n \log n)$ that counts the number of points lying inside a query convex k-gon in $O(\sqrt{kn} \log n)$ time for arbitrary values of k with $k \leq n$. When k is fixed for all queries, the size of the data structure drops to $O(n)$. Quite a few heuristic techniques and frameworks have been proposed to process polygon range queries on large-scale spatial data in a parallel and distributed manner on top of MapReduce [12]. For overviews of results on range searching, see the survey by Agarwal and Erickson [2].

In this paper, we consider the following polygon range searching problem: Preprocess a set P of n points with respect to a set $\mathcal{O}$ of k orientations in the plane so that given an $\mathcal{O}$-oriented convex polygon Q as a query, the convex hull of $P \cap Q$, and its perimeter and area, can be reported efficiently, where an $\mathcal{O}$-oriented polygon is a polygon whose edges have orientations in $\mathcal{O}$.

Whereas orthogonal and simplex range queries can be carried out efficiently, it is quite expensive for queries of arbitrary polygons in general. This is a phenomenon that occurs in many other geometry problems. In an effort to overcome such inefficiency and provide robust computation, there have been quite a few works on "finite orientation geometry", for instance, computing distances [17] in fixed orientations, finding the contour of the union of a collection of polygons with edges of restricted orientations [16], and constructing Voronoi diagrams [3,9] using a distance metric induced by a convex k-gon. In the line of this research, we suggests the polygon queries whose edges have orientations from a fixed set of orientations. Such a polygon query, as an approximation of an arbitrary polygon, can be used in appropriate areas of application, for instance, in VLSI-design, and possibly takes advantages of the restricted number of orientations and robustness in computation.

Previous Works. Brass et al. [5] gave a data structure on P for a query range Q and a few geometric extent measures, including the convex hull of $P \cap Q$ and its perimeter or area. For any axis-parallel query rectangle Q, they gave a data structure with $O(n \log^2 n)$ space and $O(n \log^3 n)$ construction time that reports the convex hull of $P \cap Q$ in $O(\log^5 n + h)$ time and its perimeter or area in $O(\log^5 n)$ time, where h is the complexity of the convex hull.

Both the data structure and query algorithm for reporting the convex hull of $P \cap Q$ were improved by Modiu et al. [14]. They gave a data structure with $O(n \log n)$ space and $O(n \log n)$ construction time that given a query axis-parallel rectangle Q, reports the convex hull of $P \cap Q$ in $O(\log^2 n + h)$ time.

For computing the perimeter of the convex hull of $P \cap Q$, the running time of the query algorithm by Brass et al. was improved by Abrahamsen et al. [1]. For a query axis-parallel rectangle Q, their data structure supports $O(\log^3 n)$ query time. Also, they presented a data structure of size $O(n \log^3 n)$ for supporting $O(\log^4 n)$ query time for a 5-gon whose edges have three predetermined orientations.

Our Result. Let $\mathcal{O}$ be a set of k orientations in the plane, and P be a set of n points in the plane.

- We present a data structure on P that allows us to compute the perimeter or area of the convex hull of points of P contained in any query $\mathcal{O}$-oriented convex s-gon in $O(s \log^2 n)$ time. We can construct the data structure with $O(nk^3 \log^2 n)$ space in $O(nk^3 \log^2 n)$ time. Note that s is at most $2k$ because Q is convex. When the query polygon has a constant complexity, as for the case of $\mathcal{O}$-oriented triangle queries, the query time is only $O(\log^2 n)$.
- For queries of reporting the convex hull of the points contained in a query $\mathcal{O}$-oriented convex s-gon, the query algorithm takes $O(h)$ time in addition to the query times for the perimeter or area cases, without increasing the size and construction time for the data structure, where h is the complexity of the convex hull.
- For $k = 2$, we can construct the data structure with $O(n \log n)$ space in $O(n \log n)$ time whose query time is $O(\log^2 n)$ for computing the perimeter or area of the convex hull of $P \cap Q$ and $O(\log^2 n + h)$ for reporting the convex hull.
- Our data structure can be used to speed up the $O(n \log^4 n)$-time algorithm by Abrahamsen et al. [1] for computing the minimum perimeter-sum bipartition of P. Their data structure requires $O(n \log^3 n)$ space and allows to compute the perimeter of the convex hull of points of P contained in a 5-gon whose edges have three predetermined orientations. If we replace their data structure with ours, we can obtain an $O(n \log^2 n)$-time algorithm for their problem using $O(n \log^2 n)$ space.

In the following sections, we give descriptions on the case of computing the perimeter of the convex hull of points of P contained in a query polygon. The description for the cases of computing the area and reporting the convex hull of points in a query polygon and all missing proofs can be found in the full version of this paper.

2   Axis-Parallel Rectangle Queries for Convex Hulls

We first consider axis-parallel rectangle queries. Given a set P of n points in the plane, Modiu et al. [14] gave a data structure on P with $O(n \log n)$ space that reports the convex hull of $P \cap Q$ in $O(\log^2 n + h)$ time for any query axis-parallel rectangle Q, where h is the complexity of the convex hull. We show that their data structure with a modification allows us to compute the perimeter of the convex hull of $P \cap Q$ in $O(\log^2 n)$ time.

2.1   Two-Layer Grid-Like Range Tree

The data structure given by Modiu et al. is a *two-layer grid-like range tree* on P, a variant of the two-layer standard range tree on P. The two-layer standard range tree on P is a two-level balanced binary search tree [10]. The level-1 tree is a balanced binary search tree T_x on the points of P with respect to their x-coordinates. Each node α in T_x corresponds to a vertical slab $I(\alpha)$. The node α has a balanced binary search tree on the points of $P \cap I(\alpha)$ with respect to their y-coordinates as its level-2 tree. In this way, each node v in a level-2 tree corresponds to an axis-parallel rectangle $B(v)$.

For any axis-parallel rectangle Q, there is a set $\mathcal{V}$ of $O(\log^2 n)$ nodes of the level-2 trees such that the rectangles $B(v)$ of $v \in \mathcal{V}$ are pairwise interior disjoint, $Q \cap B(v) \neq \emptyset$ for every $v \in \mathcal{V}$, and $\bigcup_{v \in \mathcal{V}}(P \cap B(v)) = P \cap Q$. For $v \in \mathcal{V}$, we call $B(v)$ a *canonical cell* for Q. One drawback of this structure is that the canonical cells for Q are not aligned with respect to their horizontal sides in general.

To overcome this drawback, Modiu et al. [14] gave the *two-layer grid-like range tree* so that the canonical cells for any query axis-parallel rectangle Q are aligned across all nodes α in the level-1 tree with $I(\alpha) \cap Q \neq \emptyset$. The two-layer grid-like range tree is also a two-level tree whose level-1 tree is a balanced binary search tree T_x on the points of P with respect to their x-coordinates. Each node α of T_x is associated with the level-2 tree $T_y(\alpha)$ which is a binary search tree on the points of $P \cap I(\alpha)$. But, unlike the standard range tree, $T_y(\alpha)$ is obtained from T_y by removing the subtrees rooted at all nodes whose corresponding rectangles do not have any point in $P \cap I(\alpha)$ and by contracting all nodes which have only one child, where T_y is a balanced binary search tree on the points of P with respect to their y-coordinates. Therefore, $T_y(\alpha)$ is not balanced but a full binary tree of height $O(\log n)$, and it is called a *contracted tree* on $P \cap I(\alpha)$. By construction, the canonical cells for any axis-parallel rectangle Q are aligned.

Lemma 1 ([14]). *The two-layer grid-like range tree on a set of n points in the plane can be computed in $O(n \log n)$ time. Moreover, its size is $O(n \log n)$.*

Information Stored on Each Node. To compute the perimeter of the convex hull of $P \cap Q$ for a query axis-parallel rectangle Q efficiently, we store additional information on each node v of the level-2 trees as follows. It has two children in the level-2 tree that v belongs to. Let u_1 and u_2 be the two children of v such that $B(u_1)$ lies above $B(u_2)$. By construction, $B(v)$ is partitioned into $B(u_1)$ and $B(u_2)$.

Consider the convex hull $\text{CH}(v)$ of $B(v) \cap P$ and the convex hull $\text{CH}(u_i)$ of $B(u_i) \cap P$ for $i = 1, 2$. There are at most two edges of $\text{CH}(v)$ that appear on neither $\text{CH}(u_1)$ nor $\text{CH}(u_2)$. We call such an edge a *bridge* of $\text{CH}(v)$ with respect to $\text{CH}(u_1)$ and $\text{CH}(u_2)$, or simply a *bridge* of $\text{CH}(v)$. Note that a bridge of $\text{CH}(v)$ has one endpoint on $\text{CH}(u_1)$ and the other endpoint on $\text{CH}(u_2)$.

For each node v of the level-2 trees, we store the two bridges of $\text{CH}(v)$ and the length of each polygonal chain of $\text{CH}(v)$ lying between the two bridges. In addition, we store the length of each polygonal chain connecting an endpoint e of a bridge of CH(v) and an endpoint e' of a bridge of CH($p(v)$) for the parent node $p(v)$ of v with $e, e' \in B(v)$ along the boundary of CH(v). We do this for every pair of the endpoints of the bridges of CH(v) and CH($p(v)$) that are contained in $B(v)$. Since only a constant number of bridges are involved, the information stored at v is also of constant size. Each bridge can be computed in time linear to the number of vertices of CH(u) which do not appear on CH(v) for a child u of v. The lengths of each polygonal chain we store for v can also be computed in this time. Notice that a vertex of CH(u) which do not appear on CH(v) does not appear again on CH(v') for any ancestor v' of v. Therefore, the total running time for computing the bridges is linear in the total number of points corresponding to the leaf nodes of the level-2 trees, which is $O(n \log^2 n)$.

Lemma 2. *Given a node v of a level-2 tree and two vertices x, y of CH(v), we can compute the length of the part of the boundary of CH(v) from x to y in clockwise order along the boundary of CH(v) in $O(\log n)$ time.*

2.2 Query Algorithm

Let Q be an axis-parallel rectangle. We present an algorithm for computing the perimeter of the convex hull of $P \cap Q$ in $O(\log^2 n)$ time. We call the part of the convex hull from its topmost vertex to its rightmost vertex in clockwise order along its boundary the *urc-hull* of $P \cap Q$. In the following, we compute the length of the urc-hull γ of $P \cap Q$ in $O(\log^2 n)$ time. The lengths of the other parts of the convex hull of $P \cap Q$ can be computed analogously.

We use the algorithm by Overmars and van Leeuwen [15] for computing the outer tangents between any two convex polygons.

Lemma 3 ([15]). *Given any two convex polygons stored in two binary search trees of height $O(\log n)$, we can compute the outer tangents between them in $O(\log n)$ time, where n is the total complexity of the convex hulls.*

We compute the set $\mathcal{V}$ of the canonical cells for Q in $O(\log^2 n)$ time. Recall that the size of $\mathcal{V}$ is $O(\log^2 n)$. We consider the cells of $\mathcal{V}$ as grid cells of a grid with $O(\log n)$ rows and $O(\log n)$ columns. We use $C(i, j)$ to denote the grid cell of the ith row and jth column such that the leftmost cell in the topmost row is $C(1, 1)$. Notice that a grid cell $C(i, j)$ might not be contained in $\mathcal{V}$.

Recall that we want to compute the urc-hull of $P \cap Q$. To do this, we compute the point p_x with largest x-coordinate and the point p_y with largest y-coordinate from $P \cap Q$ in $O(\log n)$ time using the range tree [10]. Then we find the cells of $\mathcal{V}$ containing each of them in the same time. Let $C(i_1, j_1)$ and $C(i_2, j_2)$ be the cells of $\mathcal{V}$ containing p_y and p_x, respectively.

We traverse the cells of $\mathcal{V}$ starting from $C(i_1, j_1)$ until we reach $C(i_2, j_2)$ as follows. We find every cell $C(i, j) \in \mathcal{V}$ with $i_1 \leq i \leq i_2$ and $j_1 \leq j \leq j_2$ such that no cell $C(i', j')$ with $i < i'$ and $j > j'$ is in $\mathcal{V}$. There are $O(\log n)$ such cells, and we call them *extreme cells*. We can compute all extreme cells in $O(\log^2 n)$ time. Note that the urc-hull of $P \cap Q$ is the urc-hull of points contained in the extreme cells. To compute the urc-hull of $P \cap Q$, we traverse the extreme cells in the lexicographical order with respect to the first index and then the second index.

During the traversal, we maintain the urc-hull of the points contained in the cells we visited so far using a binary search tree of height $O(\log n)$. Imagine that we have just visited a cell $C \in \mathcal{V}$ in the traversal. Let δ_1 and δ_2 denote the urc-hulls of the points contained in the cells we visit before and after the visit to C in the traversal, respectively. Due to the data structure we maintained, we have a binary search tree of height $O(\log n)$ for the convex hull CH of the points contained in C. Moreover, we have a binary search tree of height $O(\log n)$ for δ_1 from the traversal to the cells we visited so far. Therefore, we compute the outer tangents (bridges) between them in $O(\log n)$ time by Lemma 3. The urc-hull δ_2 is the concatenation of three polygonal curves: a part of CH, the bridge, and a part of δ_1. Thus we can represent δ_2 using a binary search tree of height one plus the maximum of the heights of the binary search trees for CH and δ_1.

Since we traverse $O(\log n)$ cells in total, we obtain a binary search tree of height $O(\log n)$ representing the urc-hull of $P \cap Q$ after the traversal. The traversal takes $O(\log^2 n)$ time. Notice that the urc-hull consists of $O(\log n)$ polygonal curves that are parts from the convex hulls stored in cells of $\mathcal{V}$ and $O(\log n)$ bridges connecting them. We can compute the length of the polygonal curve in $O(\log^2 n)$ time in total by Lemma 2.

Theorem 1. *Given a set P of n points in the plane, we can construct a data structure with $O(n \log n)$ space in $O(n \log n)$-time preprocessing that allows us to compute the perimeter of the convex hull of $P \cap Q$ in $O(\log^2 n)$ time for any query axis-parallel rectangle Q.*

Since the data structure with its construction and the query algorithm can be used for any pair orientations which are not necessarily orthogonal through an affine transformation, they work for any pair of orientations with the same space and time complexities.

Corollary 1. *Given a set P of n points and a set $\mathcal{O}$ of two orientations in the plane, we can construct a data structure with $O(n \log n)$ space in $O(n \log n)$-time preprocessing that allows us to compute the perimeter of the convex hull of $P \cap Q$ in $O(\log^2 n)$ time for any query $\mathcal{O}$-oriented rectangle Q.*

3 $\mathcal{O}$-oriented Triangle Queries for Convex Hulls

In this section, we are given a set P of n points and a set $\mathcal{O}$ of k distinct orientations in the plane. We preprocess the two sets so that we can compute the perimeter of $P \cap Q$ for any $\mathcal{O}$-oriented triangle Q in the plane efficiently. We construct a *three-layer grid-like range tree* on P with respect to every 3-tuple (o_1, o_2, o_3) of the orientations in $\mathcal{O}$, which is a generalization of the two-layer grid-like range tree described in Sect. 2.1. A straightforward query algorithm takes

$O(\log^3 n)$ time since there are $O(\log^2 n)$ canonical cells for a query $\{o_1, o_2, o_3\}$-oriented triangle Q. However, it is unclear how to obtain a faster query algorithm as the query algorithm described in Sect. 2 does not generalize to this problem directly. A main reason is that a canonical cell for any query $\{o_1, o_2, o_3\}$-oriented triangle is a $\{o_1, o_2, o_3\}$-polygon, not a parallelogram. This makes it unclear how to apply the approach in Sect. 2 to this case.

In this section, we present an $O(\log^2 n)$-time query algorithm for this problem. Our algorithm improves this straightforward algorithm by a factor of $\log n$. To do this, we classify canonical cells for Q into two types. We can handle the cells of the first type as we do in Sect. 2 and compute the convex hull of the points of P contained in them. Then we handle the cells of the second type by defining a specific ordering to these cells so that we can compute the convex hull of the points of P contained in them efficiently. Then we merge the two convex hulls to obtain the convex hull of $P \cap Q$.

3.1 Three-Layer Grid-Like Range Tree

We construct a *three-layer grid-like range tree* on P with respect to every 3-tuple of the orientations in $\mathcal{O}$. Let (o_1, o_2, o_3) be a 3-tuple of the orientations in $\mathcal{O}$. For an index $i = 1, 2, 3$, we call the projection of a point in the plane onto a line orthogonal to o_i the o_i-*projection* of the point. Let T_i be a balanced binary search tree on the o_i-projections of the points of P for $i = 1, 2, 3$.

The level-1 tree of the grid-like range tree is T_1. Each node of T_1 corresponds to a slab of orientation o_1. For each node of the level-1 tree, we construct the contracted tree of the o_2-projections of the points contained in the slab corresponding to the node with respect to T_2. A node of a level-2 tree corresponds to an $\{o_1, o_2\}$-oriented parallelogram. For each node of a level-2 trees, we construct the contracted tree of the o_3-projections of the points contained in the $\{o_1, o_2\}$-oriented parallelogram corresponding to the node with respect to T_3. A node v of a level-3 tree corresponds to an $\{o_1, o_2, o_3\}$-oriented polygon $B(v)$ with at most six vertices.

Information Stored on Each Node. Without loss of generality, we assume o_3 is parallel to the x-axis. To compute the perimeter of $P \cap Q$ for a query $\mathcal{O}$-oriented triangle Q, we store additional information on each node v of the level-3 trees as follows. The node v has two children u_1 and u_2 in the level-3 tree that v belongs to such that $B(u_1)$ lies above $B(u_2)$. By construction, $B(v)$ is partitioned into $B(u_1)$ and $B(u_2)$.

Consider the convex hull CH(v) of $P \cap B(v)$ and the convex hull CH(u_i) of $P \cap B(u_i)$ for $i = 1, 2$. There are at most two edges of CH(v) that appear on neither CH(u_1) nor CH(u_2). We call such an edge a *bridge* of CH(v) with respect to CH(u_1) and CH(u_2), or simply a bridge of CH(v). Note that a bridge of CH(v) has one endpoint on CH(u_1) and the other endpoint on CH(u_2). As we do in Sect. 2, for each node v of the level-3 trees, we store two bridges of CH(v) and the length of each polygonal chain of CH(v) lying between the two bridges. Also, we store the length of each polygonal chain connecting an endpoint of a bridge of $\text{CH}(v)$ and an endpoint of a bridge of $\text{CH}(p(v))$ for the parent $p(v)$ of v along the boundary of $\text{CH}(v)$ if the two endpoints appear on $\text{CH}(v)$. We can prove the following lemma in a similar way to Lemma 2.

Lemma 4. *Given a node v of a level-3 tree and two vertices x, y of $\text{CH}(v)$, we can compute the length of the part of the boundary of $\text{CH}(v)$ from x to y in clockwise order along the boundary of $\text{CH}(v)$ in $O(\log n)$ time.*

3.2 Query Algorithm

In this subsection, we present an $O(\log^2 n)$-time query algorithm for computing the perimeter of the convex hull of $P \cap Q$ for a query $\{o_1, o_2, o_3\}$-oriented triangle Q. Let T be the three-layer grid-like range tree constructed with respect to (o_1, o_2, o_3).

Canonical Cells. We obtain $O(\log^2 n)$ cells of T, called *canonical cells* of Q, such that the union of $P \cap C$ coincides with $P \cap Q$ for all the canonical cells C as follows. We first search the level-1 tree of T along the endpoints of the o_1-projection of Q. Then we obtain $O(\log n)$ nodes such that the union of the slabs corresponding to the nodes contains Q. Then we search the level-2 tree associated with each such node along the endpoints of the o_2-projection of Q. Then we obtain $O(\log^2 n)$ nodes in total such that the union of the $\{o_1, o_2\}$-parallelograms corresponding to the nodes contains Q. We discard all $\{o_1, o_2\}$-parallelograms not intersecting Q. Some of the remaining $\{o_1, o_2\}$-parallelograms are contained in Q, but the others intersect the boundary of Q in their interiors. For the nodes corresponding to the $\{o_1, o_2\}$-parallelograms intersecting the boundary of Q, we search their level-3 trees along the o_3-projection of Q.

As a result, we obtain $\{o_1, o_2\}$-parallelograms from level-2 trees and $\{o_1, o_2, o_3\}$-polygons from level-3 trees of size $O(\log^2 n)$ in total. See Fig. 1(a). We call them the *canonical cells* of Q and denote the set of them by $\mathcal{V}$. Also, we use $\mathcal{V}_p$ and $\mathcal{V}_h$ to denote the subsets of $\mathcal{V}$ consisting of $\{o_1, o_2\}$-parallelograms from level-2 trees and $\{o_1, o_2, o_3\}$-polygons from level-3 trees, respectively. We can compute them in $O(\log^2 n)$ time.

Computing Convex Hulls for Each Subset. We first compute the convex hull CH_p for $\mathcal{V}_p$ and the convex hull CH_h for $\mathcal{V}_h$. Then we merge them into the convex hull of $P \cap Q$ in Sect. 3.2. We can compute the convex hull CH_p of the points contained in the cells of $\mathcal{V}_p$ in $O(\log^2 n)$ time due to Corollary 1. This is because the cells are aligned with respect to two axes which are parallel to o_1 and o_2 each. Then we obtain a binary search tree of height $O(\log n)$ representing the convex hull CH_p. Thus in the following, we focus on compute CH_h.

Without loss of generality, assume that Q lies above the x-axis. Let ℓ be the side of Q of orientation o_3. We assign a pair of indices to each cell of $\mathcal{V}_h$, which consists of a *row index* and a *column index* as follows. The cells of $\mathcal{V}_h$ come

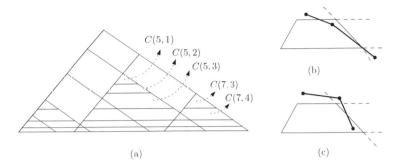

Fig. 1. (a) Canonical cells for a triangle. Four $\{o_1, o_2\}$-oriented parallelogram cells from level-2 trees and 26 $\{o_1, o_2, o_3\}$-oriented polygon cells from level-3 trees. (b) It is not sufficient to choose the only cells $C(i,j)$ such that there is no cell $C(i',j') \in \mathcal{V}$ with $i < i'$ and $j > j'$ since the urc-hull might have its vertices in such a cell. (c) Similarly, it is not sufficient to choose the only cells $C(i,j)$ such that there is no cell $C(i',j') \in \mathcal{V}$ with $i < i'$ and $j < j'$.

from $O(\log n)$ level-3 trees of the range tree. This means that each cell of $\mathcal{V}_h$ is contained in the cell corresponding to the root of one of such level-3 trees. These root cells are pairwise interior disjoint and intersect ℓ. For each cell v of $\mathcal{V}_h$ contained in the ith leftmost root cell along ℓ, we assign i to it as the row index of v. The bottom side of a cell of $\mathcal{V}_h$ is parallel to the x-axis. Consider the y-coordinates of all bottom sides of the cells of $\mathcal{V}_h$. By construction, there are $O(\log n)$ distinct y-coordinates although the size of $\mathcal{V}_h$ is $O(\log^2 n)$. We assign an index j to the cells of $\mathcal{V}_h$ whose bottom side has the jth largest y-coordinates as their column indices. Thus, each cell of $\mathcal{V}_h$ has an index (i,j), where i is its row index and j is its column index. Any two distinct cells of $\mathcal{V}_h$ have distinct indices. We let $C(i,j)$ be the cell of $\mathcal{V}_h$ with index (i,j).

Due to the indices we assigned, we can apply a procedure similar to Graham's scan algorithm for computing CH_h. We show how to compute the urc-hull of CH_h only. The other parts of the boundary of CH_h can be computed analogously. To do this, we choose $O(\log n)$ cells as follows. Note that a cell of $\mathcal{V}_h$ is a polygon with at most 6 vertices. A trapezoid cell $C(i,j)$ of $\mathcal{V}_h$ is called an *extreme cell* if there is no cell $C(i',j') \in \mathcal{V}_h$ such that $i < i'$ and $j > j'$, or if there is no cell $C(i',j') \in \mathcal{V}_h$ such that $i < i'$ and $j < j'$. Here, we need the disjunction. Otherwise, we cannot find some trapezoidal cell containing a vertex of the urc-hull. See Fig. 1(b, c). There are $O(\log n)$ extreme cells of $\mathcal{V}_h$. In addition to these extreme cells, we choose every cell of $\mathcal{V}_h$ which are not trapezoids, that is, convex t-gons with $t = 3, 5, 6$. Note that there are $O(\log n)$ such cells because such cells are incident to the corners of the cells of $\mathcal{V}_p$. In this way, we choose $O(\log n)$ cells of $\mathcal{V}_h$ in total.

Lemma 5. *A cell of $\mathcal{V}_h$ containing a vertex of the urc-hull of CH_h is an extreme cell of $\mathcal{V}_h$ if it is a trapezoid.*

By Lemma 5, the convex hull CH_h coincides with the convex hull of the convex hulls of points in the cells chosen by the previous procedure. For each column j, we consider the cells with column index j chosen by the previous procedure one by one in increasing order with respect to their row indices, and compute the convex hull of points contained in those cells. Then we consider the column indices one by one in increasing order, and compute the convex hull of the convex hulls for column indices. This takes $O(\log^2 n)$ time in total as we do in Sect. 2.2.

In this way, we can obtain a binary search tree of height $O(\log n)$ representing the urc-hull of CH_h. The urc-hull consists of $O(\log n)$ polygonal curves that are parts of the boundaries of the convex hulls stored in cells of $\mathcal{V}_h$ and $O(\log n)$ bridges connecting them. Therefore, we can compute the lengths of the polygonal curves in $O(\log^2 n)$ time in total.

Merging the Two Convex Hulls. The convex hull CH of $P \cap Q$ coincides with the convex hull of CH_p and CH_h. To compute it, we need the following lemma.

Lemma 6. *The boundary of CH_p intersects the boundary of CH_h at most $O(\log n)$ times. We can compute the intersection points in $O(\log^2 n)$ time in total.*

We first compute the intersection points of the boundaries of CH_p and CH_h in $O(\log^2 n)$ time by Lemma 6, and then sort them along the boundary of their convex hull in clockwise order in $O(\log n \log \log n)$ time. Note that this order is the same as the clockwise order along the boundary of CH_p (and CH_h). Then we locate each intersection point on the boundary of each convex hull with respect to the bridges in $O(\log n)$ time in total.

There are $O(\log n)$ edges of the convex hull CH of CH_p and CH_h that do not appear on the boundaries of CH_p and CH_h. To distinguish them with the bridges on the boundaries of CH_p and CH_h, we call the edges on the boundary of CH appearing neither CH_p nor CH_h the *hull-bridges*. Also we call the bridges on CH_p and CH_h with endpoints in two distinct cells of $\mathcal{V}$ the *node-bridges*.

The boundary of the convex hull of CH_p and CH_h consists of $O(\log n)$ hull-bridges and $O(\log n)$ polygonal curves each of which connects two hull-bridges along CH_p or CH_h. We compute all hull-bridges in $O(\log^2 n)$ time.

Lemma 7. *All hull-bridges can be computed in $O(\log^2 n)$ time in total.*

As a result, we obtain a binary search tree of height $O(\log n)$ representing the convex hull CH of $P \cap Q$. We can compute the length of each polygonal curve connecting two hull-bridge in $O(\log n)$ time by Lemma 2 and the fact that there are $O(\log n)$ node-bridges lying on CH. Therefore, we have the following theorem.

Theorem 2. *Given a set P of n points and a set $\mathcal{O}$ of k orientations in the plane, we can construct a data structure with $O(nk^3 \log^2 n)$ space in $O(nk^3 \log^2 n)$ time that allows us to compute the perimeter of $P \cap Q$ in $O(\log^2 n)$ time for any query $\mathcal{O}$-oriented triangle Q.*

4 $\mathcal{O}$-oriented Polygon Queries for Convex Hulls

The data structure in Sect. 3 can be used for more general queries. We are given a set P of n points in the plane and a set $\mathcal{O}$ of k orientations. Let Q be a query $\mathcal{O}$-oriented convex s-gon. Since Q is convex, s is at most $2k$. Assume that we are given the three-layer grid-like range tree on P with respect to the set $\mathcal{O}$ including the axis-parallel orientations. We want to compute the perimeter of the convex hull of $P \cap Q$ in $O(s \log^2 n)$ time.

We draw vertical line segments through the vertices of Q to subdivide Q into at most $2k$ trapezoids. We subdivide each trapezoid further using the horizontal lines passing through its vertices into at most three triangles. The edges of each triangle $\triangle$ has orientations in the set $\mathcal{O}$ including the axis-parallel orientations. Thus, we can compute the convex hull of $\triangle \cap P$ in $O(\log^2 n)$ time and represent it using a binary search tree of height $O(\log n)$. By Lemma 3, we can compute the convex hull of the points contained in each trapezoid in $O(s \log^2 n)$ time in total and represent them using balanced binary search trees of height $O(\log n)$.

Let $A_1, \ldots, A_t$ be the trapezoids from the leftmost one to the rightmost one for $t \leq k$. We consider the trapezoids one by one from A_1 to A_t. Assume that we have just handled the trapezoid A_i and we want to handle A_{i+1}. Assume further that we already have the convex hull CH_i of the points contained in A_j for all $j \leq i$. Since the convex hull of the points in A_{i+1} is disjoint from CH_i, we can compute CH_{i+1} in $O(\log n)$ time using Lemma 3. In this way, we can compute the convex hull of $P \cap Q$ in $O(s \log^2 n)$ time in total. Moreover, we can compute its perimeter in the same time as we did before. If s is a constant as for the case of $\mathcal{O}$-oriented triangle queries, it takes only $O(\log^2 n)$ time.

Theorem 3. *Given a set P of n points in the plane and a set $\mathcal{O}$ of k orientations, we can construct a data structure with $O(nk^3 \log^2 n)$ space in $O(nk^3 \log^2 n)$ time that allows us to compute the perimeter of the convex hull of $P \cap Q$ in $O(s \log^2 n)$ time for any $\mathcal{O}$-oriented convex s-gon.*

As mentioned in Introduction, our data structure can be used to improve the algorithm and space requirement by Abrahamsen et al. [1]. They considered the following problem: Given a set P of n points in the plane, partition P into two subsets P_1 and P_2 that minimizes the sum of the perimeters of $\text{CH}(P_1)$ and $\text{CH}(P_2)$, where $\text{CH}(A)$ is the convex hull of a point set A. They gave an $O(n \log^4 n)$-time algorithm for this problem using $O(n \log^3 n)$ space. Using our data structure, we can improve their running time to $O(n \log^2 n)$ and their space complexity to $O(n \log^2 n)$.

Corollary 2. *Given a set P of n points in the plane, we can compute a minimum perimeter-sum bipartition of P in $O(n \log^2 n)$ time using $O(n \log^2 n)$ space.*

We also have the following results for the cases of computing the area and reporting the convex hull of points in a query polygon.

Theorem 4. *Given a set P of n points and a set $\mathcal{O}$ of k orientations in the plane, we can construct a data structure with $O(nk^3 \log^2 n)$ space in*

$O(nk^3 \log^2 n)$ *time that allows us to compute the area of* $P \cap Q$ *in* $O(s \log^2 n)$ *time for any* $\mathcal{O}$-*oriented convex s-gon.*

Theorem 5. *Given a set* P *of* n *points and a set* $\mathcal{O}$ *of* k *orientations in the plane, we can construct a data structure with* $O(nk^3 \log^2 n)$ *space in* $O(nk^3 \log^2 n)$ *time that allows us to report all edges of the convex hull of* $P \cap Q$ *in* $O(s \log^2 n + h)$ *time for any* $\mathcal{O}$-*oriented convex s-gon, where* h *is the number of edges of the convex hull.*

References

1. Abrahamsen, M., de Berg, M., Buchin, K., Mehr, M., Mehrabi, A.D.: Minimum perimeter-sum partitions in the plane. In: Proceedings of the 33rd International Symposium on Computational Geometry (SoCG 2017), pp. 4:1–4:15 (2017)
2. Agarwal, P.K., Erickson, J.: Geometric range searching and its relatives. In: Chazelle, B., Goodman, J.E., Pollack, R. (eds.) Advances in Discrete and Computational Geometry. Contemporary Mathematics, vol. 223, pp. 1–56. American Mathematical Society Press (1999)
3. Agarwal, P.K., Kaplan, H., Rubin, N., Sharir, M.: Kinetic Voronoi diagrams and Delaunay triangulations under polygonal distance functions. Discrete Comput. Geom. **54**(4), 871–904 (2015)
4. Agarwal, P.K., Matoušek, J.: Ray shooting and parametric search. SIAM J. Comput. **22**(4), 794–806 (1993)
5. Brass, P., Knauer, C., Shin, C.S., Schmid, M., Vigan, I.: Range-aggregate queries for geometric extent problems. In: Proceedings of the 19th Computing: Australasian Theory Symposium (CATS 2013), vol. 141, pp. 3–10 (2013)
6. Chazelle, B.: Filtering search: a new approach to query answering. SIAM J. Comput. **15**(3), 703–724 (1986)
7. Chazelle, B.: Lower bounds for orthogonal range searching: I. the reporting case. J. ACM **37**(2), 200–212 (1990)
8. Chazelle, B., Welzl, E.: Quasi-optimal range searching in spaces of finite VC-dimension. Discrete Comput. Geom. **4**(5), 467–489 (1989)
9. Chen, Z., Papadopoulou, E., Jinhui, X.: Robustness of k-gon Voronoi diagram construction. Inf. Process. Lett. **97**(4), 138–145 (2006)
10. de Berg, M., Cheong, O., van Kreveld, M., Overmars, M.: Computational Geometry: Algorithms and Applications. Springer, Heidelberg (2008). https://doi.org/10.1007/978-3-540-77974-2
11. de Berg, M., Halperin, D., Overmars, M., Snoeyink, J., van Kreveld, M.: Efficient ray shooting and hidden surface removal. Algorithmica **12**(1), 30–53 (1994)
12. Dean, J., Ghemawat, S.: MapReduce: simplified data processing on large clusters. Commun. ACM **51**, 107–113 (2008)
13. Edelsbrunner, H., Welzl, E.: Halfplanar range search in linear space and $O(n^{0.695})$ query time. Inf. Process. Lett. **23**, 289–293 (1986)
14. Moidu, N., Agarwal, J., Kothapalli, K.: Planar convex hull range query and related problems. In: Proceedings of the 25th Canadian Conference on Computational Geometry (CCCG 2013), pp. 307–310 (2013)
15. Overmars, M.H., van Leeuwen, J.: Maintenance of configurations in the plane. J. Comput. Syst. Sci. **23**(2), 166–204 (1981)

16. Souvaine, D.L., Bjorling-Sachs, I.: The contour problem for restricted-orientation polygons. Proc. IEEE **80**(9), 1449–1470 (1992)
17. Widmayer, P., Ying-Fung, W., Wong, C.-K.: On some distance problems in fixed orientations. SIAM J. Comput. **16**(4), 728–746 (1987)
18. Willard, D.E.: Polygon retrieval. SIAM J. Comput. **11**(1), 149–165 (1982)

Synergistic Solutions for Merging and Computing Planar Convex Hulls

Jérémy Barbay$^{(\boxtimes)}$ and Carlos Ochoa

Departamento de Ciencias de la Computación, Universidad de Chile, Santiago, Chile
jeremy@barbay.cl, cochoa@dcc.uchile.cl

Abstract. We describe and analyze the first adaptive algorithm for merging k convex hulls in the plane. This merging algorithm in turn yields a synergistic algorithm to compute the convex hull of a set of planar points, taking advantage both of the positions of the points and their order in the input. This synergistic algorithm asymptotically outperforms all previous solutions for computing the convex hull in the plane.

Keywords: Convex hull · Merging · Multivariate analysis · Synergistic

1 Introduction

One way to close the gap between practical performance and the worst case complexity over instances of fixed input size is to refine the latter, considering smaller classes of instances defined via difficulty measures. The computation of the CONVEX HULL of a set of n points in the plane is a good example of the variety of such techniques. The Gift Wrapping algorithm proposed by Chand and Kapur [6] in 1970 is adaptive to the size h of the CONVEX HULL output, with a running time within $O(nh) \subseteq O(n^2)$. In 1973, Graham [8] described an algorithm known as Graham's scan, running in time within $O(n \log n)$. On instances where the output is small ($h \in o(\log n)$), Gift Wrapping asymptotically outperforms Graham's scan, while the reverse is true on other instances.

In 1986 (13 years later!), Kirkpatrick and Seidel [9] described an algorithm computing the CONVEX HULL of size h in time within $O(n \log h)$, which asymptotically outperforms both Gift Wrapping and Graham's scan. This was further improved when Afshani et al. [1] observed that a minor variant of Kirkpatrick and Seidel's algorithm [9] takes optimal advantage of the positions of the points, and proved its instance optimality among algorithms ignoring the order of the input, in a decision tree model where the tests involve only multilinear functions with a constant number of arguments. They showed that the time complexity of this variant is within $O(n(1 + \mathcal{H}(n_1, \ldots, n_h))) \subseteq O(n(1 + \log h))$, where $n_1, \ldots, n_h$ are the sizes of a partition of the points by enclosing triangles,

C. Ochoa is supported by CONICYT-PCHA/Doctorado Nacional/2013-63130161 (Chile).

L. Wang and D. Zhu (Eds.): COCOON 2018, LNCS 10976, pp. 156–167, 2018.
https://doi.org/10.1007/978-3-319-94776-1_14

such that every triangle is completely below the upper hull of the points, with the minimum possible value for $\mathcal{H}(n_1, \ldots, n_h) = \sum_{i=1}^{h} \frac{n_i}{n} \log \frac{n}{n_i} \leq \log h$.

Levcopoulos et al. [10] described in 2002 an algorithm to compute the convex hull of a set of planar points that do consider the order of the input, and thus could break Afshani et al.'s lower bound [1]. The algorithm uses a decomposition of the points into simple polygonal chains. A *polygonal chain* is specified by a sequence of points, and consists of the line segments connecting the pairs of consecutive points. A polygonal chain is *simple* if it does not have a self-intersection. They prove that the time complexity of this algorithm is within $O(n(1 + \log \kappa))$, where κ is the minimum number of simple subchains into which the sequence of n points can be partitioned. Note that κ depends only of the input order: by reordering the points, one can always reduce it to one, or increase it to within $\Theta(n)$.

A dovetailing combination[1] of the algorithms described by Kirkpatrick and Seidel [9] and Levcopoulos et al. [10] takes advantage of the order in which the points are given while maintaining (input order oblivious) instance optimality. But this solution is inefficient: many operations will be repeated, and some opportunities to quickly solve the instance are lost due to this lack of communication between the two parallel branches of the algorithm. To address this problem, we describe an algorithm for computing the CONVEX HULL of a set of planar points that takes advantage both of the positions of the points and their order in the input, synergistically, in the sense that it never performs asymptotically worse than the algorithms described by Kirkpatrick and Seidel [9] and Levcopoulos et al. [10], and on large classes of instances asymptotically outperforms both by more than a constant factor (see Example 1 in Sect. 3.2).

In order to yield a synergistic algorithm, we obtain several new results. (1) We generalize Demaine et al.'s algorithm and corresponding analysis [7] from the MERGING OF MULTISETS to the MERGING OF CONVEX HULLS, in Sect. 2 (this is the most technical part of this work). (2) We present an algorithm to partition a sequence of points into simple subchains, which is faster than the one described by Levcopoulos et al. [10], and (3) we refine Levcopoulos et al.'s measure of difficulty and analysis, in Sect. 3.1. (4) We combine those results into a synergistic algorithm to compute the CONVEX HULL in the plane, and (5) we prove that in large classes of instances this algorithm asymptotically outperforms the best previous solutions [1,10], and never asymptotically performs worse than them, in Sect. 3.2.

2   Computing the Union of Upper Hulls

The computation of convex hulls in the plane reduces to the computation of upper hulls [9]. Given ρ upper hulls in the plane, where the points in each upper hull are given sorted by their x-coordinates, the UNION OF UPPER HULL problem consists in computing the upper hull of the union of the ρ upper hulls.

---

[1] A dovetailing combination of k algorithms executes the k algorithms in parallel and stops as soon as one of the algorithms finishes.

Algorithm 1. Quick Union Hull

Input: A set $\mathcal{U}_1, \ldots, \mathcal{U}_\rho$ of ρ upper hulls
Output: The upper hull of the union of $\mathcal{U}_1, \ldots, \mathcal{U}_\rho$
 1: Compute the median μ of the slopes of the middle edges of the ρ upper hulls;
 2: Identify the "pivot" point p in the input that has a supporting line of slope μ;
 3: Partition the ρ upper hulls by the vertical line through p;
 4: For each upper hull $\mathcal{V}$, compute the (at most) two tangents of $\mathcal{V}$ through p: the ones to the left and right of p, and discard the blocks of consecutive points below the line segments determined by the points of tangency;
 5: Output a block of points in the upper hull $\mathcal{U}$ containing p that forms part of the union, by computing common tangents between $\mathcal{U}$ and the other upper hulls;
 6: Discard all points that lie below the line segments determined by the points in the common tangents between $\mathcal{U}$ and the other upper hulls;
 7: Recurse on the resulting upper hulls to the left and to the right of p.

We describe the algorithm Quick Union Hull, that solves the UNION OF UPPER HULL problem, in Sect. 2.1, and we analyze its time complexity in Sect. 2.2. This algorithm is inspired by the algorithms Simplified Ultimate Planar Convex Hull described by Chan et al. [5], and Quick Synergy Sort described by Barbay et al. [3]. The Quick Union Hull algorithm is an essential building block towards the synergistic algorithm for computing the convex hull of a set of planar points, described and analyzed in Sect. 3.

2.1   Description of the Algorithm Quick Union Hull

In the context of the UNION OF UPPER HULL problem, in each upper hull the points are given sorted by their x-coordinates, and the slopes of the edges monotonically decrease from left to right. The algorithm Quick Union Hull takes advantage of these facts: its pseudocode is described in Algorithm 1. For each upper hull, the algorithm identifies blocks of consecutive points that form part of the output and blocks of consecutive points that lie underneath the upper hull of the union. The algorithm uses a divide-and-conquer approach to take advantage of the positions of the points.

We next define some key concepts that are used in the description of the algorithm. Let S be a finite set of planar points. A *supporting line* of S is a straight line that contains a point p of S and that leaves all the points of S in the same half-plane (i.e., p is a vertex of the convex hull of S). Let μ be the slope of a supporting line passing through a point q in the upper hull $\mathcal{U}$ of S. If there is a pair of points of S to the left of q such that the line through the pair has slope less than μ, then the rightmost point in the pair cannot be part of $\mathcal{U}$. A symmetric situation arises if the pair of points is to the right of q and the slope of the line through the pair is greater than μ: the leftmost point in the pair cannot be part of $\mathcal{U}$. If the points in S are paired, a good candidate to discard points that cannot be part of $\mathcal{U}$ is the point p that has a supporting line whose slope is the median of the slopes of the lines passing through the pairs.

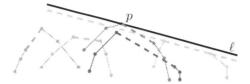

Fig. 1. An instance of the UNION OF UPPER HULLS problem. The middle edges of the upper hulls are marked by thick dashed segments, and the one whose slope is the median μ of the slopes of the middle edges has been extended into a line. The straight line ℓ is the supporting line of slope μ. The line ℓ passes through the "pivot" vertex p.

Once the points have been discarded, the choice of p guarantees that at most a constant fraction of the points in S remains on each side of p [5].

In the UNION OF UPPER HULLS problem, in each upper hull, the slopes of the edges monotonically decrease from left to right. So, in each upper hull $\mathcal{V}$, the edge at the middle position is the one whose slope is the median among the slopes of the edges of $\mathcal{V}$. We show that the point that has a supporting line whose slope is the median of the slopes of the middle edges of the upper hulls is also a good candidate to discard points that cannot be part of the upper hull of the union. Note that the time complexity of computing the median of the slopes of the middle edges of the upper hulls is linear in the number ρ of upper hulls, but that the time complexity of pairing the points and computing the median of the slopes of the lines through the pairs is linear in the number n of points [5].

The algorithm `Quick Union Hull` identifies a "pivot" vertex p of the upper hull of the union, and uses p to discard blocks of consecutive points that cannot be part of the output. It computes the median μ of the slopes of the middle edges of the upper hulls, and identifies p as the point that has a supporting line ℓ of slope μ. Note that p is the extreme point in the direction orthogonal to ℓ. Taking advantage that in each upper hull $\mathcal{V}$ the slopes of the edges are sorted, the algorithm identifies the extreme point in the direction orthogonal to ℓ by performing a doubling search[2] for the value μ in the list of slopes of the edges of $\mathcal{V}$. (See Fig. 1 for a graphical representation of these steps.)

To know which points are to the left and which ones are to the right of p, the algorithm partitions the points in the upper hulls by the vertical line $x = p_x$, where p_x is the x-coordinate of the point p, by performing doubling searches for the value p_x in the x-coordinates of the points in the upper hulls.

For each upper hull $\mathcal{V}$, the algorithm then computes the (at most) two tangents of $\mathcal{V}$ through p: the one passing through a point to the left of p in $\mathcal{V}$, and the one passing through a point to the right of p in $\mathcal{V}$. In $\mathcal{V}$, the algorithm discards the blocks of consecutive points below the line segments determined by the points of tangency. It computes all the tangents via doubling searches [2].

Before the recursive step, in the upper hull $\mathcal{U}$ containing p, the algorithm identifies a block $\mathcal{B}$ of consecutive points that forms part of the output (p is

---

[2] Doubling search is a technique for searching sorted unbounded arrays in which an element of rank k is found by performing $2 \log k$ comparisons [4].

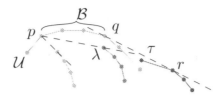

Fig. 2. The state of the algorithm `Quick Union Hull` during an execution of the step that computes the block $\mathcal{B}$ that forms part of the upper hull of the union. The upper hull $\mathcal{U}$ contains the point p. λ marks the tangent of maximum slope between p and the upper hulls to the right of p. τ marks the common tangent between the portion of $\mathcal{U}$ above λ and one of the upper hulls below λ passing through the point nearest to p in $\mathcal{U}$. The points q and r lie in τ.

included in $\mathcal{B}$). The algorithm certifies that $\mathcal{B}$ forms part of the output by computing common tangents between a portion of $\mathcal{U}$ and the other upper hulls. Computing a common tangent between two upper hulls could be costly, but if there is a line separating them, then the time complexity is logarithmic [2]. The algorithm takes advantage of this fact by using as separating lines two tangents through p computed in the previous step (i.e., ignoring the portion of $\mathcal{U}$ in the same half plane as the other upper hulls). The block $\mathcal{B}$ is determined by the common tangents passing through the points nearest to p in $\mathcal{U}$ (one point to the left of p and the other one to the right). To avoid the computation of all common tangents, the algorithm interweaves the different tangent computations (similarly to how Demaine et al.'s algorithm [7] interweaves doubling searches to compute the intersection of sorted sets). We devote the rest of the section to describe this step in more details.

We describe how to identify the part of $\mathcal{B}$ to the right of p (the left counterpart is symmetric). Let λ be the tangent of maximum slope between p and the upper hulls to the right of p (i.e., the tangent of maximum slope among those computed in the previous step of the algorithm). Let $\mathcal{U}'$ be the portion of the upper hull $\mathcal{U}$ containing p above λ. The tangent λ separates $\mathcal{U}'$ from the upper hulls below λ. Among the common tangents between $\mathcal{U}'$ and the upper hulls below λ, let τ be the one passing through the nearest point to p in $\mathcal{U}'$. Let q and r be the points that lie in τ, such that q belongs to $\mathcal{U}'$ and r belongs to one of the upper hulls below of λ. The point q determines the end of the right portion of $\mathcal{B}$ (see Fig. 2 for a graphical representation of these definitions).

Given two upper hulls $\mathcal{A}$ and $\mathcal{B}$ separated by a vertical line, Barbay and Chen [2] described an algorithm that computes the common tangent τ between them, in time within $O(\log a + \log b)$, where a and b are the ranks of the points that lie in τ in the sequences of points representing $\mathcal{A}$ and $\mathcal{B}$, respectively. At each step this algorithm considers two points: one from $\mathcal{A}$ and the other one from $\mathcal{B}$, and in at least one upper hull, it can certify, in constant time, if the point that lies in τ is to the right or to the left of the point considered. A minor variant manages the case where the separating line is not vertical: as the first

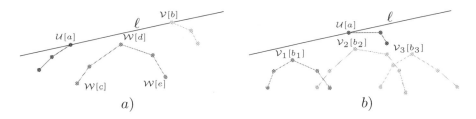

Fig. 3. Example of arguments: (a) an eliminator argument formed by 3 blocks and (b) a convex argument formed by 4 blocks.

step, in each upper hull, the algorithm computes the supporting line of slope equal to the slope of the separating line, by performing doubling searches.

To compute the point q that determines the right portion of $\mathcal{B}$, the algorithm Quick Union Hull executes several instances of the algorithm described by Barbay and Chen [2] for computing the common tangents between $\mathcal{U}'$ and the upper hulls below λ, always considering the same point u in $\mathcal{U}'$. Once all decisions about the point u are reached, the upper hulls below λ can be divided into two sets: (i) those whose common tangents pass through a point to the left of u in $\mathcal{U}'$, and (ii) those whose common tangents pass through a point to the right of u in $\mathcal{U}'$. If the set (i) is not empty, then the algorithm stops the computation in the set (ii). For each upper hull $\mathcal{V}$ in the set (ii), the algorithm discards the block of points in $\mathcal{V}$ to the left of the penultimate point considered. This step continues until there is just one instance running, and computes the tangent τ in this instance. The algorithm discards all points to the left of r (i.e., all points that lie below the arc of the output that leaves $\mathcal{U}$ clockwise and follows τ).

After identifying the block $\mathcal{B}$ of the output, the algorithm recurses on the resulting upper hulls to the left and right of p.

2.2   Complexity Analysis of the Quick Union Hull Algorithm

Each algorithm that solves the Union of Upper Hulls problem needs to certify that some blocks of points in the upper hulls cannot participate in the upper hull of the union, and that some other blocks are indeed in the upper hull of the union. In the following, we formalize the notion of partition certificate, which can be used to check the correctness of the output in less time than to recompute the output itself. A *partition certificate* of an instance is a partition of the points of the upper hulls into regions so that, in each region, it is "easy" to certify whether the points form part of the output or not. This notion of partition certificate yields a measure of the difficulty of an instance ("short" partition certificates characterize "easy" instances, while "long" partition certificates suggest "difficult" instances). We define a language of basic arguments for such partition certificates: *eliminator* arguments discard points from the input and *convex* arguments justify the presence of points in the output. A partition certificate is formed by eliminator and convex arguments and will be verified by checking each of its arguments. See Fig. 3 for a graphical representation of such arguments.

Fig. 4. A partition certificate of size 7 of an instance of the UNION OF UPPER HULL problem. The thick black lines mark the division between the 7 regions.

Definition 1. *Consider the upper hulls $\mathcal{U}$, $\mathcal{V}$, and $\mathcal{W}$. Let ℓ be the straight line through the points $\mathcal{U}[a]$ and $\mathcal{V}[b]$. $\langle \mathcal{U}[a], \mathcal{V}[b] \supset \mathcal{W}[c..d..e] \rangle$ is an Eliminator Argument if the points of the block $\mathcal{W}[c..e]$ are between the vertical lines through $\mathcal{U}[a]$ and $\mathcal{V}[b]$, the slope of ℓ is between the slopes of the two edges in $\mathcal{W}$ that precede and follow the point $\mathcal{W}[d]$, and the point $\mathcal{W}[d]$ lies below ℓ.*

If $\langle \mathcal{U}[a], \mathcal{V}[b] \supset \mathcal{W}[c..d..e] \rangle$ is an eliminator argument, then the points of the block $\mathcal{W}[c..e]$ cannot contribute to the upper hull of the union. Several blocks that are "eliminated" by the same pair of points can be combined into a single argument. These eliminator arguments are the ones used in the Steps 4 and 6 of the algorithm `Quick Union Hull`.

It is not enough to discard some points that do not contribute to the output. Certifying still requires additional work: a correct algorithm must justify the exactness of its output. To this end we define convex arguments.

Definition 2. *Consider the upper hulls $\mathcal{U}, \mathcal{V}_1, \ldots, \mathcal{V}_t$. $\langle \mathcal{U}[a] \dashv \mathcal{V}_1[b_1], \ldots, \mathcal{V}_t[b_t] \rangle$ is a Convex Argument if there is a straight line ℓ through $\mathcal{U}[a]$ such that the slope of ℓ is between the slopes of the edges that precede and follow the points $\mathcal{V}_1[b_1], \ldots, \mathcal{V}_t[b_t]$, respectively, and the points $\mathcal{V}_1[b_1], \ldots, \mathcal{V}_t[b_t]$ lie below ℓ.*

If $\langle \mathcal{U}[a] \dashv \mathcal{V}_1[b_1], \ldots, \mathcal{V}_t[b_t] \rangle$ is a convex argument, then the point $\mathcal{U}[a]$ is a vertex of the upper hull of the union of $\mathcal{U}, \mathcal{V}_1, \ldots, \mathcal{V}_t$. Blocks of points can also be "easily" certified as part of the output using similar arguments: when the first and last points p and q, respectively, in such blocks are vertices of the output, and all other points in the instance lie below the line through p and q. These convex arguments are the ones used in Step 5 of `Quick Union Hull`.

Those arguments are a two-dimensional generalization of the arguments from Demaine et al. [7] for computing the UNION OF SORTED SETS, and are inspired by the ones introduced by Barbay and Chen [2] for the binary UNION OF UPPER HULLS. Those atomic arguments combine into a general definition of *partition certificate* that any correct algorithm for solving the UNION OF UPPER HULLS problem in the algebraic decision tree model can be modified to output (see Fig. 4 for an example of such a partition certificate).

Definition 3. *Given an instance $\mathcal{I}$ of the UNION OF UPPER HULL problem, a Partition Certificate of $\mathcal{I}$ is a partition of the points into regions, so that in each region, the points of $\mathcal{I}$ that belong to the output can be decided using a constant number of eliminator and convex arguments. The Size of $\mathcal{I}$ is the number of regions which compose it.*

The algorithm `Quick Union Hull` partitions the upper hulls into blocks of consecutive points, where each block is either discarded or output. A block is discarded if it is underneath the upper hull of the union, or is output if it forms part of the upper hull of the union. Each of such blocks forms part of an argument of the partition certificate computed by the algorithm. We separate the analysis of the steps that discard or output blocks of points (i.e., Steps 2, 3, 4, 5, and 6)[3] from the steps that compute the medians of the slopes of the middle edges (i.e., Step 1). The following lemma states that the asymptotic time complexity for discarding a block s is logarithmic in the number of points in s.

Lemma 1. *Given an upper hull $\mathcal{U}$, the cumulated time complexity of the steps that discard blocks of points of the algorithm* `Quick Union Hull` *considering only points of $\mathcal{U}$ is within $O(\sum_{j=1}^{\beta} \log s_j)$, where $s_1, \ldots, s_\beta$ are the sizes of the β blocks into which the whole algorithm partitions $\mathcal{U}$.*

We state the following lemmas in function of the partition certificate computed by the algorithm. The blocks that are discarded in each execution of the Steps 4 and 6 are certified using a single eliminator argument. In the same way, the block that is output in Step 5 is certified using a single convex argument.

Lemma 2. *Given a block $\mathcal{B}$ that forms part of the output, the time complexity of the step that outputs $\mathcal{B}$ of the algorithm* `Quick Union Hull` *is within $O(w \log s)$, where s is the size of $\mathcal{B}$ and w is the number of arguments in the convex argument used by the algorithm to certify that $\mathcal{B}$ forms part of the output.*

This is a consequence of the w searches for the common tangent in Step 5. The amount of arguments in the partition certificate and the number of blocks in each of the arguments are related to the time complexity of Step 1.

Lemma 3. *Given ρ upper hulls, the cumulated time complexity of the steps that compute the medians of the slopes of the middle edges of the algorithm* `Quick Union Hull` *is within $O(\sum_{i=1}^{\delta} \log \binom{\rho}{m_i})$, where δ is the size of the partition certificate $\mathcal{C}$ computed by the algorithm, and $m_1, \ldots, m_\delta$ is a sequence where m_i is the number of blocks in the i-th argument of $\mathcal{C}$.*

We describe an analysis of the algorithm `Quick Union Hull` in function of the smallest possible size δ of a partition certificate for a particular instance.

Theorem 1. *Given ρ upper hulls of sizes $r_1, \ldots, r_\rho$ such that the upper hull of their union admits a partition certificate of size δ, there is an algorithm that computes the upper hull of their union in time within $O(\rho \delta \log \frac{h}{\delta} + \delta \sum_{i=1}^{\rho} \log \frac{r_i}{\delta})$, where h is the number of points in the upper hull of their union.*

Proof. The size of the partition certificate $\mathcal{C}$ computed by the algorithm `Quick Union Hull` in an instance $\mathcal{I}$ is a constant factor of the size δ of a partition

---

[3] Even though the Steps 2 and 3 do not discard or output blocks of points by themselves we include them in the same analysis as the Steps 4 and 6.

certificate $\mathcal{P}$ of minimal size for $\mathcal{I}$, such that in each region there is just one block that forms part of the output, and this block can be certified using a single convex argument. Indeed, if a region $\mathcal{R}$ of $\mathcal{P}$ contains a block $\mathcal{B}$ that forms part of the upper hull $\mathcal{U}$ of the union, then the algorithm Quick Union Hull can certify that $\mathcal{B}$ forms part of $\mathcal{U}$ using a constant number of arguments. This is a consequence of the step that computes the blocks that form part of the output. This step computes the block of maximum size (p included) that can be certified that forms part of the output using a single convex argument. In addition, the algorithm partitions each upper hull in at most a constant factor of δ blocks. Combining the results from Lemmas 1, 2, and 3 with the concavity of the logarithm function, we obtain that the time complexity is within $O(\sum_{i=1}^{\rho} \sum_{j=1}^{\delta} s_{ij} + \sum_{k=1}^{\delta} w_k \log n_k) \subseteq O(\delta \sum_{i=1}^{\rho} \log \frac{r_i}{\delta} + \rho\delta \log \frac{h}{\delta})$, where s_{ij} is the size of the j-th block of the i-th upper hull, w_k is the number of arguments in the k-th convex argument, and n_k is the size of the k-th block of $\mathcal{U}$.   □

In the following section, we combine the union algorithm with an algorithm that partitions the sequence of points into "easy" instances, to obtain a synergistic algorithm that computes the convex hull of a set of planar points.

3   Synergistic Computation of Convex Hulls

We describe a synergistic algorithm for computing the convex hull of a set of planar points. It is synergistic in the sense that it takes advantage of both the order of the points and their positions at once, whereas all previous solutions take advantage only of one of those. As a consequence, this algorithm outperforms the best previous solutions [1,10], as well as any dovetailing combination of them. This algorithm decomposes first the input sequence of points into simple subchains (Sect. 3.1), computes their convex hulls [10], and then merges their convex hulls (Sect. 2). There are two noteworthy advantages to this approach: (1) the algorithm decomposes the points into "easy" instances (these "easy" instances are determined by the order in which the points are given), and computes their convex hulls, both steps in time linear in the number of points; and (2) when merging the resulting convex hulls it takes advantage of the number of convex hulls, that the points in the convex hulls are given in sorted order, and the positions of the points (analyzed in Sect. 3.2).

3.1   Linear Time Partitioning Algorithm

A *polygonal chain* is a curve specified by a sequence of points $p_1, \ldots, p_n$. The curve itself consists of the line segments connecting the pairs of consecutive points. A polygonal chain is *simple* if it does not have a self-intersection. Levcopoulos et al. [10] described an algorithm to compute the convex hull of n points in the plane in time within $O(n(1 + \log \kappa))$, where κ is the minimum number of simple subchains into which the input sequence of points can be partitioned. The algorithm tests if the polygonal chain $\mathcal{P}$ given as input is simple:

Algorithm 2. Doubling Search Partition

Input: A sequence of n planar points $p_1, \ldots, p_n$
Output: A sequence of simple polygonal chains
1: Initialize i to 1;
2: **for** $t = 1, 2, \ldots$ **do**
3:     **if** $i + 2^t - 1 > n$ or the chain $p_i, \ldots, p_{i+2^t-1}$ is not simple **then**
4:         Output the chain $p_i, \ldots, p_{i+2^{t-1}-1}$
5:         Update $i \leftarrow i + 2^{t-1}$ and $t \leftarrow 1$

---

if $\mathcal{P}$ is simple, it computes the convex hull of $\mathcal{P}$ in time linear in the size of $\mathcal{P}$. Otherwise, if $\mathcal{P}$ is not simple, it partitions $\mathcal{P}$ into two subchains, whose sizes differ at most by one; recurses on each of them; and merges the resulting convex hulls. The time complexity of the partitioning and merging steps are both within $\Theta(n(1 + \log \kappa))$.

We describe an improved partitioning algorithm running in time linear in the size of the input, which is key to the synergistic result. The Doubling Search Partition algorithm searches one by one for the largest integer t such that the subchain formed by the first 2^t points is simple. It identifies this subchain as simple and restarts the computation in the rest of the sequence. Its pseudocode is described in Algorithm 2. This algorithm identifies a simple subchain of size k in time within $O(k)$, because the sizes of the tested subchains form a geometric progression of ratio 2. The time complexity of this partitioning algorithm is linear in the number n of points in the sequence, but we prove that the entropy $\mathcal{H}(r_1, \ldots, r_k) = \sum_{i=1}^{k} \frac{r_i}{n} \log \frac{n}{r_i}$ of the sizes $r_1, \ldots, r_k$ of the resulting k simple subchains is a constant factor of the entropy of the sizes of any partition of the sequence of n points into the minimum possible number κ of simple subchains:

Theorem 2. *Given a sequence $\mathcal{S}$ of n planar points, the algorithm Doubling Search Partition computes in linear time a partition of $\mathcal{S}$ into k simple subchains of sizes $r_1, \ldots, r_k$, such that $n(1 + \mathcal{H}(r_1, \ldots, r_k)) \in O(n(1 + \alpha))$, where α is the minimum value for the entropy function $\mathcal{H}(s_1, \ldots, s_\kappa)$ of any partition of $\mathcal{S}$ into κ simple subchains, of respective sizes $s_1, \ldots, s_\kappa$.*

Proof. Consider a partition π of $\mathcal{S}$ into κ simple subchains of sizes $s_1, \ldots, s_\kappa$. Fix the subchain c_i of size s_i. The subchain c_i contributes $\frac{s_i}{n} \log \frac{n}{s_i}$ to the value of $\mathcal{H}(s_1, \ldots, s_\kappa)$. The algorithm Doubling Search Partition partitions c_i into simple subchains. One of such subchains is at least of size $\frac{s_i}{2}$, and in the worst case, the sizes of the rest of them form a decreasing geometric progression of ratio $\frac{1}{2}$. Hence, the subchains into which the algorithm partitions c_i contribute $O(\sum_{i=1}^{\infty} \frac{s_i}{2^i} \log \frac{2^i n}{s_i}) = O(s_i + \frac{s_i}{n} \log \frac{n}{s_i})$ to the entropy of the partition obtained by the algorithm. The result follows.   □

Given the convex hulls of the subchains obtained by the algorithm Doubling Search Partition, an algorithm that merges two by two the shortest ones takes advantage of the potential disequilibrium in the distribution of their sizes, a result that improves upon the algorithm described by Levcopoulos et al. [10]:

Corollary 1. *Given a sequence $\mathcal{S}$ of n planar points that can be partitioned into κ simple subchains of respective sizes $r_1, \ldots, r_\kappa$, there is an algorithm that computes the convex hull of $\mathcal{S}$ in time within $O(n(1 + \mathcal{H}(r_1, \ldots, r_\kappa))) \subseteq O(n(1 + \log \kappa))$.*

3.2 Synergistic Algorithm to Compute the Convex Hull

Given a set $\mathcal{S}$ of planar points, the algorithm `Quick Synergy Hull` computes the upper hull of $\mathcal{S}$. It proceeds in two phases. It first partitions $\mathcal{S}$ into simple subchains using the algorithm `Doubling Search Partition` (described in Sect. 3.1), and computes the upper hulls of the simple subchains [10], both steps in time linear in the number of points in $\mathcal{S}$. Then it merges those upper hulls using the algorithm `Quick Union Hull` (described in Sect. 2).

The algorithm `Quick Synergy Hull` outperforms both the algorithm described by Levcopoulos et al. [10] and the one described by Kirkpatrick and Seidel [9] (even when analyzed by Afshani et al. [1]), as well as any dovetailing combination of them. We prove this more formally in the following theorem:

Theorem 3. *Consider a sequence $\mathcal{S}$ of n planar points that can be partitioned into κ simple subchains of sizes $r_1, \ldots, r_\kappa$ (such that $\sum_{i=1}^{\kappa} r_i = n$); and also can be partitioned into h sets of sizes $n_1, \ldots, n_h$ (such that $\sum_{i=1}^{h} n_i = n$), where each set can be enclosed by a triangle completely below the upper hull of $\mathcal{S}$. There is an algorithm that computes the upper hull of $\mathcal{S}$ in time within $O(n + \sum_{j=1}^{\delta} w_j \log s_j + \sum_{i=1}^{\delta} \log \binom{\kappa}{m_i}) \subseteq O(n(1 + \min(\mathcal{H}(r_1, \ldots, r_\kappa), \mathcal{H}(n_1, \ldots, n_h)))) \subseteq O(n(1 + \min(\log \kappa, \log h))) \subseteq O(n \log n)$, where the union of the upper hulls of the simple subchains admits a partition certificate $\mathcal{C}$ of minimum size δ (such that $\delta \leq h$), $m_1, \ldots, m_\delta$ is a sequence where m_i is the number of blocks in the i-th argument of $\mathcal{C}$ (such that $m_i \leq \kappa$ for $i \in [1..\delta]$), w_j is the number of arguments in the j-th convex argument of $\mathcal{C}$, and s_j is the size of the j-th block of the output.*

Proof. This result is a consequence of Theorems 1 and 2. For example, if the simple subchains obtained by the partitioning algorithm are all of constant size (i.e., the algorithm cannot take advantage of the order of the points), then the time complexity of the algorithm `Quick Synergy Hull` and the one described by Kirkpatrick and Seidel [9] (as analyzed by Afshani et al. [1]) are asymptotically the same. This algorithm also takes advantage of the positions of the points to improve upon the algorithm described in Corollary 1. □

Example 1. Consider for example the family of instances depicted in Fig. 5: on such instances, the time complexity of the algorithm described by Kirkpatrick and Seidel [9], as refined by Afshani et al. [1], is within $O(n + h \log n)$ (all the points in the sequences 1, 2 and 3 can be enclosed by a triangle completely below the upper hull of the points, hence $n_1 = \cdots = n_{h-1} = 1$ and $n_h = n - h + 1$ in the formula $O(n(1 + \mathcal{H}(n_1, \ldots, n_h)))$, where $h - 1$ is the number of points in the sequence 4). The time complexity on such instances of the algorithm described by Levcopoulos et al. [10], as refined in Sect. 3.1, is within $O(n + \kappa \log n)$ (the

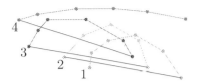

Fig. 5. A sequence of points and its decomposition into $\kappa = 4$ simple subchains. The numbers indicate the order in which the sequence of points are given: each from left to right internally, and mark the simple subchains.

whole sequence of points can be partitioned into κ simple subchains, suppose that the sizes of the simple subchains labeled 1 to $\kappa - 1$ are a constant c and that the size of the simple subchain labeled κ is $n - (\kappa - 1)c$, then $r_1 = \cdots = r_{\kappa-1} = c$ and $r_\kappa = n - (\kappa - 1)c$ in the formula $O(n(1 + \mathcal{H}(r_1, \ldots, r_\kappa))))$. On the other hand, the time complexity on such instances of the algorithm Quick Synergy Hull is within $O(n)$: once it computes the first vertex of the output, it discards all the points except the points in the upper hull labeled 4. If $h \in \Theta(n)$ and $\kappa \in \Theta(n)$, then the algorithm Quick Synergy Hull is faster than the previous algorithms [1,10] by a factor logarithmic in the size of the input.

References

1. Afshani, P., Barbay, J., Chan, T.M.: Instance-optimal geometric algorithms. J. ACM **64**(1), 3:1–3:38 (2017)
2. Barbay, J., Chen, E.Y.: Convex hull of the union of convex objects in the plane: an adaptive analysis. In: Proceedings of the Annual Canadian Conference on Computational Geometry (CCCG) (2008)
3. Barbay, J., Ochoa, C., Satti, S.R.: Synergistic solutions on MultiSets. In: 28th Annual Symposium on Combinatorial Pattern Matching, CPM 2017, Warsaw, Poland, 4–6 July 2017, pp. 31:1–31:14 (2017)
4. Bentley, J.L., Yao, A.C.C.: An almost optimal algorithm for unbounded searching. Inf. Process. Lett. **5**(3), 82–87 (1976)
5. Chan, T.M., Snoeyink, J., Yap, C.K.: Primal dividing and dual pruning: output-sensitive construction of four-dimensional polytopes and three-dimensional voronoi diagrams. Discrete Comput. Geom. (DCG) **18**(4), 433–454 (1997)
6. Chand, D.R., Kapur, S.S.: An algorithm for convex polytopes. J. ACM **17**(1), 78–86 (1970)
7. Demaine, E.D., López-Ortiz, A., Munro, J.I.: Adaptive set intersections, unions, and differences. In: Proceedings of the 11th ACM-SIAM Symposium on Discrete Algorithms (SODA), pp. 743–752 (2000)
8. Graham, R.L.: An efficient algorithm for determining the convex hull of a finite planar set. Inf. Process. Lett. **1**, 132–133 (1972)
9. Kirkpatrick, D.G., Seidel, R.: The ultimate planar convex hull algorithm? SIAM J. Comput. (SICOMP) **15**(1), 287–299 (1986)
10. Levcopoulos, C., Lingas, A., Mitchell, J.S.B.: Adaptive algorithms for constructing convex hulls and triangulations of polygonal chains. In: Penttonen, M., Schmidt, E.M. (eds.) SWAT 2002. LNCS, vol. 2368, pp. 80–89. Springer, Heidelberg (2002). https://doi.org/10.1007/3-540-45471-3_9

Cophenetic Distances: A Near-Linear Time Algorithmic Framework

Paweł Górecki[1(✉)], Alexey Markin[2], and Oliver Eulenstein[2]

[1] Faculty of Mathematics, Informatics and Mechanics,
University of Warsaw, Warsaw, Poland
gorecki@mimuw.edu.pl
[2] Department of Computer Science, Iowa State University, Ames, USA
{amarkin,oeulenst}@iastate.edu

Abstract. Tree metrics that compare pairs of trees are an elementary tool for analyzing phylogenetic trees. The cophenetic distance is a classic vector-based tree metric introduced by Cardona et al. that originates from the pioneering work of Sokal and Rohlf more than 50 years ago. However, when faced with phylogenetic analyses where sets of large-scale trees are compared, the quadratic runtime of the current best-known (naïve) algorithm to compute the cophenetic distance becomes prohibitive. Here we describe an algorithmic framework that computes the cophenetic distance under the L_1-norm in $O(n \log^2 n)$ time, where n is the size of the compared pair of trees. Based on the work from Sokal and Rohlf, we introduce a natural class of cophenetic distances and show that our algorithmic framework can compute each member of this class in $O(n \log^2 n)$ time. In addition, we present a modification of this framework for computing these distances under the L_2-norm in $O(n \log n)$ time. Finally, we demonstrate the scalability of our algorithm.

Keywords: Phylogenetic tree · Distance · Metric · Cophenetic metric

1 Introduction

Phylogenetic trees depict the evolutionary relationships among a set of genes, genomes, or species, and thus provide a fundamental understanding of how entities have evolved over time the way they are today. Analyzing phylogenetic trees benefits a vast variety of fundamental research areas including biology, ecology, epidemiology, conservation biology, and linguistic [12,14,25].

Studying phylogenetic trees typically requires the comparative evaluation of their similarities and differences that has become an elementary task in computational phylogenetics [11,18]. A large variety of *tree metrics* has been proposed and analyzed that measure the distance in evolutionary terms between pairs of phylogenetic trees [4,19,21,27,29].

The cophenetic distance [6] is a popular tree metric that has been established based on the notions from the pioneering work of Sokal and Rohlf more than

© Springer International Publishing AG, part of Springer Nature 2018
L. Wang and D. Zhu (Eds.): COCOON 2018, LNCS 10976, pp. 168–179, 2018.
https://doi.org/10.1007/978-3-319-94776-1_15

50 years ago [28]. This distance is defined between a pair of rooted phylogenetic trees over the same taxon set and based on their representation as cophenetic vectors. The *cophenetic vector* for a rooted tree specifies a value for each pair of the tree's taxa; that is, the depth of the least common ancestor of the taxon pair in the given tree. An example is depicted in Fig. 1. Cophenetic vectors equivalently encode their corresponding trees, and consequently, the distance between a pair of trees can be measured in the cophenetic vector space [6]. Therefore, the cophenetic distance can be formulated under various vector norms. Most natural are the L_1-norm and the L_2-norm that have received considerable attention in the literature [5,6,20,23], and consequently, are the focus of this paper.

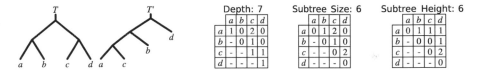

Fig. 1. Examples from the class of cophenetic distances under the L_1-norm. Given trees T and T', the matrices represent the differences between the cophenetic vectors defined by the depths, subtree heights, and subtree size with respect to the least common ancestors. The distance is the sum of the elements of the corresponding matrix. For example, in the matrix Subtree Height the entry for $\langle a, d \rangle$ is $1 = |2 - 3|$, since the height of the least common ancestor of a and d in T is 2, while in T' it equals 3.

Unlike most other popular tree metrics, such as the widely-used Robinson-Foulds metric [3,24], the triplet/quartet metric [26], and the nearest neighbor interchange metric [7,9,22], which only apply to phylogenetic trees without edge lengths, the cophenetic distance also applies to trees with edge lengths. Comparing trees with edge lengths often allows for a much more refined analysis, in particular when similar tree topologies are compared [21].

The cophenetic distance has been well-researched. The distribution of this distance under the uniform model is positively skewed and has a low-rank correlation with the Robinson-Foulds metric and the path-difference metric [6]. Further the diameter of the cophenetic distance under the L_p norm for trees with n taxa is $O(n^{(p+2)/p})$, where $p \in \{1, 2\}$ [6]. While the cophenetic distance is a vector-based metric, it does not share many similarities with other such metrics, like the family of path-difference metrics [29,30]. Perhaps, as experiments have demonstrated, most closely related to the cophenetic distance are the gene tree parsimony measures that are based on biological models. This may be in parts because all of these measures rely primarily on least common ancestors [6,10,13].

Given the reputation of the cophenetic distance, there has been an increased interest by practitioners to use this distance for large-scale phylogenetic analyses [20,23]. For example, recently, the cophenetic distance has been used to infer credible species trees from a collection of input trees by estimating their median tree under this distance [23]. Despite these promising results, the best-known (naïve) approaches for computing the cophenetic distance under the L_1-norm and the L_2-norm requires quadratic time. In the face of large-scale phylogenetic studies, where trees with tens of thousands of taxa are compared or synthesized, this quadratic runtime becomes prohibitive.

Here, we first describe an algorithmic framework that computes the cophenetic distance for a pair of trees with n taxa under the L_1-norm in $O(n \log^2 n)$ time. Then we present a modification of this framework that computes the cophenetic distance under the L_2-norm in $O(n \log n)$ time. In addition, similar to Cardona et al. [6], we introduce a novel class of cophenetic distances based on the groundwork laid by Sokal and Rohlf [28]. Like the original cophenetic distance, each metric in this class is encoding a phylogenetic tree as a cophenetic vector that is primarily based on the least common ancestors of taxa pairs. Unlike the original cophenetic vector that lists as values the depth of least common ancestors in the input tree, these values can be defined in various other ways suitable for comparing trees, such as the size of the subtrees of the encoded tree rooted at the least common ancestors, or the height of these subtrees (see Fig. 1 for an example). Note that if ultrametric trees are compared, then the cophenetic vector defined by the heights of the subtrees is equivalent to what has been originally proposed by Sokal and Rohlf [28]. Considering all types of distances that relate to least common ancestors by satisfying a particular monotonicity property, the metrics corresponding to these distances define our new class of cophenetic distances, which contains the original cophenetic distance. We show that our algorithmic frameworks can compute each member of the cophenetic distance class in the same time that is needed to compute the original cophenetic distance under the L_1-norm and the L_2-norm. Finally, using a scalability study, we demonstrate the performance of our algorithmic frameworks.

Related Work. Today's, phylogenetic tree databases host an unprecedented wealth of trees that have been estimated using various inference methods and different data sets. As a consequence, such trees often represent conflicting evolutionary relationships for the same set of taxa [11]. It is an important part of phylogenetic analyses to study such conflicts, and therefore, a large variety of metrics have been proposed to quantify the differences and similarities between a pair of trees [4,19,21,27]. For example, metrics originating from tree edit operations assess the difference between a pair of trees by counting the minimum number of such operations that transform the trees into each other, such as the nearest neighbor interchange distance [22], the subtree prune and regraft distance [1,2,15], and tree bisection and reconnection distance [1]. Computing these distances is an NP-hard problem [1,8,16,17]. Other metrics assess the distance between encodings of the compared pair of trees. When the encodings of the trees are sets, e.g., the set of clusters or triplets induced by the tree, then the distance between a pair of trees is the cardinality of the symmetric difference of their encodings. Metrics corresponding to cluster and triplet encodings are the widely-used Robinson-Foulds metric [3,24] and the triplet metric [7], respectively. Another classic encoding of a tree is a vector that contains a value related to the tree for every pair of taxa. Then the trees can be compared in the corresponding vector space using different norms, like the L_1-, L_2-, or L_∞-

norm. Such metrics include the path-difference distance and the cophenetic distance. Unlike the metrics originating from edit distances, the presented metrics using encodings are efficiently computable. However, large-scale phylogenetic analyses require near-linear time runtime solutions in practice. Addressing this challenge, recent algorithmic advances provided such solutions for the standard path-difference metric for unrooted trees under any L_p-norm. For a more detailed treatment of tree metrics, the reader is referred to [11,19].

Contribution. We introduce a natural class of cophenetic distances that includes the original cophenetic distance for rooted pairs of trees over n taxa and show that these distances can be computed under the L_1-norm in $O(n \log^2 n)$ time, and under the L_2-norm in $O(n \log n)$ time.

For computing the cophenetic distances we present a divide and conquer framework, which we briefly overview. First, we describe a procedure that identifies a *median node* in a given rooted tree that divides it into two subtrees of sizes between $\frac{n}{4}$ and $\frac{3n}{4} + 1$. Our framework divides the input trees by using median nodes and then performs four recursive calls to compute partial distances between all pairs of the subtrees. To incorporate the remaining pairs of leaves, we distinguish two cases whose difficulty depends on the location of the leaves. We show that the first case can be handled in $O(n)$ time for both the L_1- and L_2-norms, while the second case requires $O(n \log n)$ time for the L_1-norm and $O(n)$ time for the L_2-norm. Then, we solve the recurrences and prove the resulting $O(n \log^2 n)$ and $O(n \log n)$ time complexities of our divide and conquer frameworks for the L_1 and L_2 cophenetic metrics respectively.

The paper is organized the way that the results for the more complex L_1-norm are described first and later the related results for the L_2-norm are derived from that. In more detail, given two trees, we show how to compute the distance that is defined as the sum of $(|\xi(\mathsf{lca}_T(x,y)) - \xi'(\mathsf{lca}_{T'}(x,y))|)^p$ over all pairs of taxa $\langle x, y \rangle$ present in the input trees (note that x and y can be the same element), where lca refers to the least common ancestor function, and ξ (and ξ') define the contribution of a node in T (and T'). The only assumption is that the contribution mapping is *path-monotonic*, i.e., every path that starts in the root induces a monotonic sequence of ξ values. We define all metrics that can be computed this way as the class of *cophenetic distances*.

In conclusion, using the cophenetic distances under the L_1-norm as an example, we demonstrate that an implementation of our divide and conquer framework significantly outperforms the runtime of the (best-known) quadratic algorithm when pairs of trees with more than roughly 1800 taxa are compared.

2 Definitions

Let $T = \langle V_T, E_T \rangle$ be a rooted binary tree and v, w be nodes in T. By $\mathsf{root}(T)$ we denote the root of T. The least common ancestor of v and w in T is denoted by $\mathsf{lca}_T(v,w)$. By $v \preceq w$ we denote that w is on the path between v and the root of T. Note that $v \prec w$ is equivalent to $v \preceq w$ and $v \neq w$. A node v is a *branch* if it is strictly internal in T, i.e., v is neither a leaf nor the root. In algorithms, for a non-root node v, $v.\mathsf{parent}$ and $v.\mathsf{sibling}$ denotes the parent and the sibling of v, respectively. The set of leaves in T is denoted by L_T. By $|T|$ we denote the number of leaves in T. Similarly, by L_v we denote the set of leaves visible from v, and by $|v|$ the size of L_v. A *weighted* (time-annotated) tree T is an extension of a rooted binary tree with an edge weight function $e\colon E_T \to \mathbb{R}_+$.

We say that a function $\xi\colon V_T \to \mathbb{R}$ is *path-monotonic*, if for every v and w such that $v \preceq w$, $\xi(v) \geq \xi(w)$ (descending) or for every v and w such that $v \preceq w$, $\xi(w) \leq \xi(v)$ (ascending). In this article, we study three natural *contribution functions* ξ_depth, ξ_height and ξ_size such that $\xi_\mathsf{depth}(v)$ [6] is the depth of a node v (depth is defined as the number of edges on the path from v to $\mathsf{root}(T)$ for unweighted trees; and as the sum of edge weights on that path for weighted trees), $\xi_\mathsf{height}(v)$ is the height of the subtree rooted at v, and $\xi_\mathsf{size}(v) = |L_v|$. Note that ξ_depth is descending, while ξ_height and ξ_size are ascending functions.

Given two trees T, T' having the same set of leaves and two path-monotonic contribution functions $\xi\colon V_T \to \mathbb{R}$ and $\xi'\colon V_{T'} \to \mathbb{R}$, both ascending or both descending, let $(x_1, x_2, \ldots, x_n)$ be a fixed ordering of leaves of T and T'. A *cophenetic vector* of T is defined as $\phi(T, \xi) := [\xi(\mathsf{lca}_T(x_i, x_j))]_{i \leq j}$ (similarly for T'). Then *a (generalized) cophenetic distance* with respect to contribution functions ξ and ξ' is defined for $p > 0$ as follows:

$$d(T, T') := \|\phi(T, \xi) - \phi(T', \xi')\|_p.$$

That is, d is an L_p norm of the difference between two cophenetic vectors induced by ξ and ξ'. For this work we define the L_p *cophenetic distance* as the cophenetic distance under the L_p-norm, for $p \in \{1, 2\}$. It should be clear that d is a metric [6] as long as both ξ and ξ' are the same functions. Figure 1 depicts an example.

3   Results

In this section, we present details of our algorithms for computing L_p cophenetic distances. We begin with the existence of a median node in a rooted tree. Then, we classify pairs of taxa depending on their location in the parts of divided rooted trees. Further, focusing on the L_1 cophenetic distance, we show three types of computing partial distances. Next, we solve the recurrence proving $O(n \log^2 n)$ time complexity of the algorithm for the L_1 cophenetic distance. Finally, we demonstrate how the algorithm can be adapted to obtain the $O(n \log n)$ algorithm for computing the L_2 cophenetic distance. We omit some proofs for brevity.

3.1   Median Node in a Rooted Tree

A node t of a rooted tree T divides a tree into two parts: the subtree of T rooted at t, denoted T_t and called the *lower tree* with respect to t, and the tree T^t, called the *upper tree*, obtained from T by replacing T_t with a leaf. Node t from the next lemma is called a *median node* and is $O(n)$ time computable.

Lemma 1 (The existence of a median node). *For every rooted tree T of size $n \geq 2$ there is a node t such that $\frac{n}{2} \leq |T^t| \leq \frac{3n}{4} + 1$ and $\frac{n}{4} \leq |T_t| \leq \frac{n}{2}$.*

3.2   Classification of Taxa Pairs

From now on we assume that T and T' are two trees with the same set of leaves, where t and t' are fixed median nodes of T and T', respectively. We also assume that ξ and ξ' are the contribution functions of T and T', respectively. Without loss of generality, we assume that all contribution functions are descending, i.e., if for every v and w such that $v \preceq w$, $\xi(v) \geq \xi(w)$.

The path connecting the median node with the root will be called a *median path*. By A and B we denote the set of leaves (excluding the median node) in the upper tree and the lower tree of T respectively, and, similarly, we denote A' and B' for T'. Note that $|A| + |B| = |A'| + |B'| = n$. Then, we have four variants for a pair of leaves $\langle x, y \rangle$ (note that x can be equal to y) from T depending on their location: AA - if both leaves are located in the upper tree (i.e., $x, y \in A$), BB - if $x, y \in B$, AB - if $x \in A$ and $y \in B$, and BA - if $x \in B$ and $y \in A$. Note that types AB and BA are identical due to symmetry.

When considering $\langle x, y \rangle$ in both trees, we have 16 possible *types* of locations (disregarding the symmetry) denoted in the form $XY|X'Y'$ where $X, Y \in \{A, B\}$ and $X', Y' \in \{A', B'\}$. We say that $\langle x, y \rangle \in L_T \times L_{T'}$ has type $XY|X'Y'$ if $x \in X \cap X'$ and $y \in Y \cap Y'$. To compute $d(T, T')$ for the L_1-norm we show, for every type $XY|X'Y'$ such that $X, Y \in \{A, B\}$ and $X', Y' \in \{A', B'\}$, how to compute the *partial distances*

$$d_{XY|X'Y'} = \sum_{x \in X \cap X'} \sum_{y \in Y \cap Y'} |\xi(\mathsf{lca}_T(x, y)) - \xi'(\mathsf{lca}_{T'}(x, y))|.$$

3.3   The Partial Distance of Non-mixed Types

There are four non-mixed types: $AA|A'A'$, $AA|B'B'$, $BB|A'A'$, and $BB|B'B'$. To compute the partial distance for a non-mixed type, we first contract the trees T and T' to a set of leaves present in pairs determined by a given type. For example, for $BB|A'A'$, T and T' are contracted to $B \cap A'$. Next, we compute the distance d recursively for such trees. If $f(n)$ is the complexity of the algorithm for computing the distance between trees of the size n, then, the computation of these four partial distances requires $\sum_{i=1}^{4} f(c_i)$ time, where $c_1 = |A \cap A'|$, $c_2 = |A \cap B'|$, $c_3 = |B \cap A'|$ and $c_4 = |B \cap B'|$. Note, that $\sum_i c_i = n$.

3.4   The Partial Distance of Double Mixed Types

Accounting for symmetry, there are the two double-mixed types $AB|A'B'$ and $AB|B'A'$. The type $AB|A'B'$ denotes pairs $\langle x, y \rangle$ when x is in upper trees and y is in lower trees of T and T'. In this case, the lca of $\langle x, y \rangle$ is located on the median path in both trees. Moreover, $\mathsf{lca}_T(x, y)$ equals $\mathsf{lca}_T(x, t)$. Similarly, we have $\mathsf{lca}_{T'}(x, y) = \mathsf{lca}_{T'}(x, t')$. Thus, the partial distance $d_{AB|A'B'}$ equals

$$\sum_{x \in A \cap A'} |\xi(\mathsf{lca}_T(x, t)) - \xi'(\mathsf{lca}_{T'}(x, t'))| \cdot |B \cap B'|, \tag{1}$$

Algorithm 1. The contribution of type $AB|B'A'$

1: **Input:** T and T' with median nodes t and t', resp. **Output:** $d_{AB|B'A'}(T, T')$.
2: **Function** GetCntr(G, g, X) where g is a median node of G
3:     **For** every v on the median path of G: $v.c := 0$.
4:     **For** every leaf l in X: lca$_G(g, l).c$ += 1.
5:     $\gamma := []$. # the empty sequence (a list)
6:     **For** every node v on the median path:
7:         **For** $i = 1$ to $v.c$: append $\xi(v)$ to γ. # insert $\xi(v)$ in γ $v.c$ times
8:     **Return** γ.
9: **Function** SeqPrd$(\alpha_1, \alpha_2, \ldots, \alpha_k, \beta_1, \beta_2, \ldots, \beta_m)$:
10:     $i := j := 1$; **If** $\alpha_k > \beta_m$ **Then** swap α and β
11:     **While** $j \leq m$:
12:         **If** $i \leq k$ and $\alpha_i \leq \beta_j$ **Then** σ_j += α_i; λ_j += 1; i += 1
13:         **Else** j += 1; $\sigma_j := \sigma_{j-1}$; $\lambda_j = \lambda_{j-1}$
14:     **Return** $m(\Sigma_{i=1}^k \alpha_i) + \Sigma_{j=1}^m 2\lambda_j\beta_j - 2\sigma_j - k\beta_j$
15: **Return** SeqPrd(α, β) where $\alpha =$ GetCntr$(T, t, A \cap B')$, $\beta =$ GetCntr$(T', t', B \cap A')$.

---

which can be computed in $O(n)$ time.

For $AB|B'A'$ the naïve approach requires $\Theta(n^2)$ steps. Below, we show an $O(n)$ time solution. We begin with the following problem.

Problem 1 (SeqProduct). Given two sequences of numbers: $\alpha_1 \leq \alpha_2 \leq \cdots \leq \alpha_k$ and $\beta_1 \leq \beta_2 \leq \cdots \leq \beta_m$. Compute: $\sum_{i,j} |\alpha_i - \beta_j|$.

Lemma 2. SeqPrd *from Algorithm 1 computes* $\sum_{i,j} |\alpha_i - \beta_j|$ *in* $O(m+k)$ *time.*

Proof. We have that λ_j is the number of elements from α that are smaller or equal to β_j, and $\sigma_j = \sum_{i=1}^{\lambda_j} \alpha_i$, i.e., it is the sum of all elements from α that are smaller or equal to β_j. Then, for a fixed j, $\sum_{i=1}^k |\alpha_i - \beta_j| = (\lambda_j\beta_j - \sigma_j) + ((\sum_i \alpha_i) - \sigma_j) - (k - \lambda_j)\beta_j)$. Easy transformations are left to the reader.   □

Now, $d_{AB|B'A'}$ is computed by calling SeqPrd(α, β), where α is the sequences of contributions of lca$_T(x, t)$'s for x in $A \cap B$ and β is the sequence of contributions of lca$_{T'}(y, t')$'s for y in $B \cap A'$. Such sequences are inferred in $O(n)$ steps by the function GetCntr in Algorithm 1.

Lemma 3. *Algorithm 1 computes* $d_{AB|B'A'}$ *in* $O(n)$ *time.*

Proof. Note, that the call of function GetCntr(G, g, X) returns the ordered sequence of all $\xi(\text{lca}_G(x, g))$'s for $x \in X$. It is clear that the time complexity of such a call is $O(|G|)$. Now, by Lemma 2, the rest follows easily.   □

3.5 The Partial Distance of Single Mixed Types

Accounting for symmetry, there are four single mixed variants $AA|A'B'$, $AB|A'A'$, $BB|A'B'$ and $AB|B'B'$. These variants can be solved similarly. Thus, for brevity, we only show the solution for the first variant. Let us assume that $\langle x, y \rangle$ has type $AA|A'B'$, that is, $x \in A \cap A'$ and $y \in A \cap B'$. Then, lca$_T(x, y)$ is

---

Algorithm 2. The contribution of $AA|A'B'$

1: **Input:** T and T' with median nodes t and t', resp. **Output:** $d_{AA|A'B'}(T,T')$.
2: **For** every v in the upper tree of T: $v.\delta := v.\sigma^+ := v.\sigma^- := \langle 0,0 \rangle$; $v.\kappa := 0$.
3: **For** $x \in A \cap A'$: # *The preprocessing loop*
4:      $\beta - \xi'(\mathsf{lca}_{T'}(x,t'))$.
5:      **If** $\beta \leq \xi(x)$ **Then** $\omega_x := \arg\min_w\{\xi(w)|x \preceq w \text{ and } \beta \leq \xi(w)\}$; $\omega_x.\delta += \langle \beta,1 \rangle$
6:      **If** $\beta \leq \xi(x.\mathrm{parent})$ **Then** $x.\sigma^+ := \langle \beta,1 \rangle$ **Else** $x.\sigma^- := \langle \beta,1 \rangle$
7: **For** every non-root v in the upper tree of T in postfix order: # *The main loop*
8:      $v.\sigma^+ := v.\mathsf{lft}.\sigma^+ + v.\mathsf{rgh}.\sigma^+ - v.\delta$; $v.\sigma^- := v.\mathsf{lft}.\sigma^- + v.\mathsf{rgh}.\sigma^- + v.\delta$
9:      $v.\kappa := |A \cap B' \cap L(v.\mathrm{sibling})| \cdot (\xi(v.\mathrm{parent}) \cdot (v.\sigma_2^+ - v.\sigma_2^-) + v.\sigma_1^- - v.\sigma_1^+)$
10: **Return** $\sum_{v \in T} v.\kappa$.

---

a node from the upper tree of T, while $\mathsf{lca}_{T'}(x,y)$ is located on the median path of T'. If v is a non-leaf node, then its right child is denoted by $v.\mathrm{rgh}$ and its left child by $v.\mathrm{lft}$. While the proof of the next lemma is omitted for brevity, please see the example in Fig. 2.

Lemma 4. *If v is a branch in T, then after the main loop of Algorithm 2 we have*

$$v.\kappa = \sum_{x \in A \cap A' \cap L(v)} \sum_{y \in A \cap B' \cap L(v.\mathrm{sibling})} |\xi(\mathsf{lca}_T(x,y)) - \xi'(\mathsf{lca}_{T'}(x,y))|. \qquad (2)$$

Lemma 5. *Algorithm 2 computes $d_{AA|A'B'}$ in $O(n \log n)$ time.*

Proof. Correctness: Let I be the set of non-root nodes from the upper tree of T. Then, every pair of leaves $\langle x,y \rangle$ of type $AA|A'B'$ uniquely determines a node $v \in I$ such that $\mathsf{lca}_T(x,y) = v.\mathrm{parent}$ and $x \in L(v)$. It also follows that $y \in L(v.\mathrm{sibling})$. Let $x \oplus y$ denote such node v. For a given v, $A \cap A' \cap L(v) \times A \cap B' \cap L(v.\mathrm{sibling})$ is the set of all pairs $\langle x,y \rangle$ such that $x \oplus y = v$. Hence,

$$d_{AA'|B'A'} = \sum_{v \in I} \sum_{x \in A \cap A' \cap L(v)} \sum_{y \in A \cap B' \cap L(v.\mathrm{sibling})} |\xi(\mathsf{lca}_T(x,y)) - \xi'(\mathsf{lca}_{T'}(x,y))|.$$

By Lemma 4 the above sum equals the value returned in the last line of Algorithm 2.

The crucial line for the complexity is line 5. We show that ω can be found by a binary search in $O(\log n)$ time, that seeks for the value in an ordered array composed of nodes on the path connecting a given leaf x with the root of T. Such an array can be constructed by an infix traversal of T. Then, a node is inserted into the array when it is visited for the first time. When a node is visited for the last time, it is removed from the array. Thanks to the monotonic ordering of paths, the array is always sorted, and its size is limited by n. □

3.6   Time Complexity

Let $f(n)$ be the worst-case time complexity of the algorithm. By Sect. 3.3, the computation of partial distances of non-mixed types requires $f(c_1) + f(c_2) +$

$f(c_3) + f(c_4) + O(n)$ time where $\sum c_i = n$ and $0 \le c_i \le .75n + 1$ for each i (by Lemma 1), while the computation of mixed types requires $O(n \log n)$ time. Therefore, for some $k \ge 1$ we can write that, $f(n) = 1$ if $n \le 5$ and $f(n) = kn \log n + \max_{0 \le c_i \le .75n+1} f(c_1) + f(c_2) + f(c_3) + f(c_4)$, otherwise.

Theorem 1. *The time complexity of the D&C algorithm is* $O(n \log^2 n)$.

Proof. We show that there are constants $b \ge 1$ and $d > 0$ such that for every $n > 0$, $f(n) \le dn \log^2 n + b$. The proof is by induction on n. For $n \le 5$ we have $f(n) = 1$ and the inequality is satisfied. For $n > 5$, $f(n) = kn \log n + \max \sum_i f(c_i) \le kn \log n + 4b + d \max \sum_i c_i \log^2 c_i \le kn \log n + 4b + dn \log^2(.75n + 1)$.

Let $d = -bk/\log \frac{11}{12} \approx bk7.966$. Then, for $n > 5$, $\log(.75 + \frac{1}{n}) \le \log(.75 + \frac{1}{6}) = \log \frac{11}{12}$. Hence, $d(\log^2(.75n + 1) - \log^2 n) = d \log(.75 + \frac{1}{n}) \log((.75n + 1)n) \le -bk \log((.75n + 1)n) \le -bk \log n - bk$. Finally, for $n > 5$, $f(n) - dn \log^2 n - b \le kn \log n + 3b + dn(\log^2(.75n + 1) - \log^2 n) \le (1 - b)kn \log n + 3b - nbk < 0$. $\square$

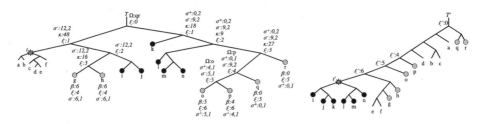

Fig. 2. An example of two trees T and T' with attributes from Algorithm 2 with $d_{AA|A'B'} = 118$. Here the contribution function is based on the depth. Stars denote the median nodes t and t'. Leaves marked by green circles are from $A \cap A'$ while black circles denote the elements of $A \cap B'$. For example, the contribution of the parent of t is 48 (κ), which is computed from the product of two green leaves (g, h) and four black leaves from its sibling (k, l, m, n), that is, $48 = ((6 - 0) + (6 - 0)) \cdot 4$. (Color figure online)

3.7 Partial Distances Under the L_2 Cophenetic Distance

The above algorithm results are derived for computation of the L_1 cophenetic metric. We demonstrate that they can be applied to compute the L_2 cophenetic distance as well. Using the same notation for locations of a pair of leaves $\langle x, y \rangle \in L(T) \times L(T)$, the (square) partial distance for an $XY|X'Y'$ type under the L_2-norm is as follows:

$$d_{XY|X'Y'} := \sum_{x \in X \cap X'} \sum_{y \in Y \cap Y'} (\xi(\text{lca}_T(x, y)) - \xi'(\text{lca}_{T'}(x, y))^2.$$

Observe, the non-mixed types can be computed using the recursive calls the same way as in Sect. 3.3. Next, we demonstrate how the key representatives from the double and single mixed types can be computed for the L_2 norm.

Double Mixed $AB|A'B'$. Adopting Eq. 1 we have a linear time computable

$$d_{AB|A'B'} = \sum_{x \in A \cap A'} (\xi(\mathsf{lca}_T(x,t)) - \xi'(\mathsf{lca}_{T'}(x,t')))^2 \cdot |B \cap B'|.$$

Double Mixed $AB|B'A'$. The difference with Sect. 3.4 lies within function SeqPrd, which has to be adopted for the L_2 norm. Given two non-decreasing sequences α and β of sizes m and k respectively, let $S^p(\beta) := \sum_{j=1}^{k} \beta_j^p$ for some $p > 0$. Then SeqPrd can be computed for the L_2-norm as follows:

$$\mathsf{SeqPrd}(\alpha, \beta) = \sum_{i=1}^{m} \sum_{j=1}^{k} (\alpha_i - \beta_j)^2 = \sum_{i=1}^{m} \left(k\alpha_i^2 - 2\alpha_i S^1(\beta) + S^2(\beta) \right).$$

This sum can be computed in $O(m+k)$ time by precomputing $S^1(\beta)$ and $S^2(\beta)$.

Single Mixed $AA|A'B'$. The key value computed on each step of the main loop of Algorithm 2 is $v.\kappa$. Adopting $v.\kappa$ definition for the L_2 case, we show how it can be efficiently computed.

$$v.\kappa := \sum_{x \in A \cap A' \cap L(v)} \sum_{y \in A \cap B' \cap L(v.\mathsf{sibling})} \left(\xi(\mathsf{lca}_T(x,y)) - \xi'(\mathsf{lca}_{T'}(x,y)) \right)^2.$$

Lemma 6. *Values $v.\kappa$ can be computed for all non-root nodes v in the upper tree of T in $O(n)$ time overall.*

The partial distance $d_{AA|A'B'}$ then can be computed in linear time as a sum of $v.\kappa$ over all nodes in the upper tree of T. Similarly, for other single mixed types.

Time Complexity. Besides four recursive calls, the algorithm spends $O(n)$ time for the partial distances of mixed types. It is not difficult to adapt the proof of Theorem 1 to show that this leads to an $O(n \log n)$ algorithm for the L_2 cophenetic distance.

4 Scalability Analysis

Here we study the scalability of our algorithm by comparing its runtime to the runtime of the previously best-known algorithm on different pairs of trees.

Experimental Setting. The near linear time algorithm for computing the L_1 cophenetic distance as well as the naïve (quadratic) algorithm were implemented using Java 1.8. The runtimes for both algorithms were evaluated using randomly generated pairs of binary trees varying in size from 200 taxa to 10000 taxa with the step of 200. Each algorithm was executed on each pair of trees 20 times. The experiment was carried out on an Intel Core i7 2.5 GHz CPU under Windows 7.

Results. We compared the median runtimes among 20 runs for each pair of trees. We observed that the naïve solution outperforms our algorithm for pairs of trees containing less than 1800 leaves; beyond that point the quadratic solution becomes drastically less efficient as compared to our algorithm. For example, at the extreme, on 10000 taxa the quadratic algorithm took around 15 s to complete on average, while our method took less than 2 s to compute the distance. These observations suggest a cutoff for the divide-and-conquer strategy, i.e., an efficient implementation will use the quadratic solution for the recursive calls with trees containing less than ≈ 1800 leaves.

5   Conclusion

In the age of high-throughput next-generation sequencing, practitioners are challenged to perform large-scale studies that compare phylogenetic trees having tens of thousands of taxa. Tree metrics that do not allow for near-linear time solutions thereof become infeasible for such studies, while they are of potential theoretical and practical importance. Prior to this work, the popular cophenetic distance remained one of such computationally prohibitive metrics.

Here, we presented a novel algorithmic framework for computing the L_1 cophenetic distance in $O(n \log^2 n)$ time, while the previously best-known (naïve) algorithm requires $\Theta(n^2)$ time. Moreover, our modification of this framework can compute the L_2 cophenetic distance in only $O(n \log n)$ time. These significantly improved runtimes make the cophenetic distance much more applicable for today's large-scale phylogenetic comparative studies and species tree inferences.

The framework has a largely broader application, as it is generalized to metrics satisfying the path-monotonic property. We refer to the set of such metrics as the class of cophenetic distances. While this class includes the original (depth) cophenetic distance, it also contains many other metrics that can be of interest to practitioners, perhaps most notable are the height and size cophenetic distances that in parts relate closely to tree measures introduced by Sokal and Rohlf [28].

Finally, we demonstrated the scalability of our algorithmic framework on an example of the more involved L_1-norm algorithm. This algorithm significantly outperforms the previously best-known quadratic solution for pairs of large trees.

References

1. Allen, B.L., Steel, M.: Subtree transfer operations and their induced metrics on evolutionary trees. Ann. Comb. **5**(1), 1–15 (2001)
2. Bordewich, M., Semple, C.: On the computational complexity of the rooted subtree prune and regraft distance. Ann. Comb. **8**(4), 409–423 (2005)
3. Bourque, M.: Arbres de Steiner et réseaux dont varie l'emplacement de certains sommets. Ph.D. thesis, University of Montréal Montréal, Canada (1978)
4. Bryant, D.: Hunting for trees, building trees and comparing trees: theory and method in phylogenetic analysis. Ph.D. thesis, University of Canterbury, New Zealand (1997)
5. Cardona, G., Mir, A., Rosselló, F., Rotger, L.: The expected value of the squared cophenetic metric under the yule and the uniform models. Math. Biosci. **295**, 73–85 (2018)

6. Cardona, G., Mir, A., Rosselló, F., Rotger, L., Sánchez, D.: Cophenetic metrics for phylogenetic trees, after Sokal and Rohlf. BMC Bioinform. **14**(1), 3 (2013)
7. Critchlow, D., Pearl, D., Qian, C.: The triples distance for rooted bifurcating phylogenetic trees. Syst. Biol. **45**, 323–334 (1996)
8. DasGupta, B., et al.: On distances between phylogenetic trees. In: SODA, vol. 97, pp. 427–436 (1997)
9. Estabrook, G., McMorris, F., Meacham, C.: Comparison of undirected phylogenetic trees based on subtrees of four evolutionary units. Syst. Zool. **34**, 193–200 (1985)
10. Eulenstein, O., Huzurbazar, S., Liberles, D.: Reconciling phylogenetic trees. In: Evolution After Gene Duplication. Wiley, Hoboken (2010)
11. Felsenstein, J.: Inferring Phylogenies. Sinauer Associates, Inc., Sunderland (2004)
12. Forster, P., Renfrew, C.: Phylogenetic Methods and the Prehistory of Languages. McDonald Inst of Archeological, Cambridge (2006)
13. Górecki, P., Eulenstein, O., Tiuryn, J.: Unrooted tree reconciliation: a unified approach. IEEE/ACM Trans. Comput. Biol. Bioinform. **10**(2), 522–536 (2013)
14. Harris, S., et al.: Whole-genome sequencing for analysis of an outbreak of meticillin-resistant staphylococcus aureus: a descriptive study. Lancet. Infect. Dis. **13**(2), 130–136 (2013)
15. Hein, J.: Reconstructing evolution of sequences subject to recombination using parsimony. Math. Biosci. **98**(2), 185–200 (1990)
16. Hein, J., et al.: On the complexity of comparing evolutionary trees. Discrete Appl. Math. **71**(1–3), 153–169 (1996)
17. Hickey, G., et al.: SPR distance computation for unrooted trees. Evol. Bioinform. online **4**, 17–27 (2008)
18. Hoef-Emden, K.: Molecular phylogenetic analyses and real-life data. Comput. Sci. Eng. **7**(3), 86–91 (2005)
19. Katherine, S.J.: Review paper: the shape of phylogenetic treespace. Syst. Biol. **66**(1), e83–e94 (2017)
20. Kendall, M., Colijn, C.: Mapping phylogenetic trees to reveal distinct patterns of evolution. Mol. Biol. Evol. **33**(10), 2735–2743 (2016)
21. Kuhner, M.K., Yamato, J.: Practical performance of tree comparison metrics. Syst. Biol. **64**(2), 205–214 (2015)
22. Li, M., Tromp, J., Zhang, L.: On the nearest neighbour interchange distance between evolutionary trees. J. Theor. Biol. **182**(4), 463–467 (1996)
23. Markin, A., Eulenstein, O.: Cophenetic median trees under the manhattan distance. In: ACM-BCB 2017, pp. 194–202. ACM, New York (2017)
24. Robinson, D.F., Foulds, L.R.: Comparison of phylogenetic trees. Math. Biosci. **53**(1–2), 131–147 (1981)
25. Roux, J., et al.: Resolving the native provenance of invasive fireweed (Senecio madagascariensis Poir.) in the Hawaiian Islands as inferred Poir.) in the Hawaiian Islands as inferred from phylogenetic analysis. Div. Distr. **12**, 694–702 (2006)
26. Sand, A., et al.: Algorithms for computing the triplet and quartet distances for binary and general trees. Biology **2**(4), 1189–1209 (2013)
27. Semple, C., Steel, M.A.: Phylogenetics. University Press, Oxford (2003)
28. Sokal, R.R., Rohlf, F.J.: The comparison of dendrograms by objective methods. Taxon **11**(2), 33–40 (1962)
29. Steel, M.A., Penny, D.: Distributions of tree comparison metrics. Syst. Biol. **42**(2), 126–141 (1993)
30. Williams, W., Clifford, H.: On the comparison of two classifications of the same set of elements. Taxon **20**(4), 519–522 (1971)

Computing Coverage Kernels Under Restricted Settings

Jérémy Barbay$^{1(\boxtimes)}$, Pablo Pérez-Lantero^2, and Javiel Rojas-Ledesma^1

^{1} Departamento de Ciencias de la Computación,
Universidad de Chile, Santiago, Chile
jeremy@barbay.cl, jrojas@dcc.uchile.cl
^{2} Departamento de Matemática y Ciencia de la Computación,
Universidad de Santiago, Santiago, Chile
pablo.perez.l@usach.cl

Abstract. We consider the MINIMUM COVERAGE KERNEL problem:
given a set $\mathcal{B}$ of d-dimensional boxes, find a subset of $\mathcal{B}$ of minimum
size covering the same region as $\mathcal{B}$. This problem is NP-hard, but as
for many NP-hard problems on graphs, the problem becomes solvable in
polynomial time under restrictions on the graph induced by $\mathcal{B}$. We consider various classes of graphs, show that MINIMUM COVERAGE KERNEL
remains NP-hard even for severely restricted instances, and provide two
polynomial time approximation algorithms for this problem.

1 Introduction

Given a set P of n points, and a set $\mathcal{B}$ of m boxes (i.e. axis-aligned closed
hyper-rectangles) in d-dimensional space, the BOX COVER problem consists in
finding a set $\mathcal{C} \subseteq \mathcal{B}$ of minimum size such that $\mathcal{C}$ covers P. A special case is the
ORTHOGONAL POLYGON COVERING problem: given an orthogonal polygon $\mathcal{P}$
with n edges, find a set of boxes $\mathcal{C}$ of minimum size whose union covers $\mathcal{P}$. Both
problems are NP-hard [8,11], but their known approximabilities in polynomial
time are different: while BOX COVER can be approximated up to a factor within
$\mathcal{O}(\log \texttt{OPT})$, where $\texttt{OPT}$ is the size of an optimal solution [5,7]; ORTHOGONAL
POLYGON COVERING can be approximated up to a factor within $\mathcal{O}(\sqrt{\log n})$ [14].
In an attempt to better understand what makes these problems hard, and why
there is such a gap in their approximabilities, we introduce the notion of coverage
kernels and study its computational complexity.

Given a set $\mathcal{B}$ of n d-dimensional boxes, a *coverage kernel* of $\mathcal{B}$ is a subset
$\mathcal{K} \subseteq \mathcal{B}$ covering the same region as $\mathcal{B}$, and a minimum coverage kernel of $\mathcal{B}$ is a
coverage kernel of minimum size. The computation of a minimum coverage kernel
(namely, the MINIMUM COVERAGE KERNEL problem) is intermediate between
the ORTHOGONAL POLYGON COVERING and the BOX COVER problems. This
problem has found applications (under distinct names, and slight variations) in

This work was supported by projects CONICYT Fondecyt/Regular nos 1170366 and
1160543, and CONICYT-PCHA/Doctorado Nacional/2013-63130209 (Chile).

© Springer International Publishing AG, part of Springer Nature 2018
L. Wang and D. Zhu (Eds.): COCOON 2018, LNCS 10976, pp. 180–191, 2018.
https://doi.org/10.1007/978-3-319-94776-1_16

the compression of access control lists in networks [9], and in obtaining concise descriptions of structured sets in databases [15,18]. Since ORTHOGONAL POLYGON COVERING is NP-hard, the same holds for the MINIMUM COVERAGE KERNEL problem. We are interested in the exact computation and approximability of MINIMUM COVERAGE KERNEL in various restricted settings:

1. **Under which restrictions is the exact computation of Minimum Coverage Kernel still NP-hard?**
2. **How precisely can one approximate a Minimum Coverage Kernel** in polynomial time?

When the interactions between the boxes in a set $\mathcal{B}$ are simple (e.g., when all the boxes are disjoint), a minimum coverage kernel of $\mathcal{B}$ can be computed efficiently. A natural way to capture the complexity of these interactions is through the intersection graph. The intersection graph of $\mathcal{B}$ is the un-directed graph with a vertex for each box, and in which two vertices are adjacent if and only the respective boxes intersect. When the intersection graph is a tree, for instance, each box of $\mathcal{B}$ is either completely covered by another, or present in any coverage kernel of $\mathcal{B}$, and thus a minimum coverage kernel can be computed efficiently. For problem on graphs, a common approach to understand when does an NP-hard problem become easy is to study distinct restricted classes of graphs, in the hope to define some form of "boundary classes" of inputs separating "easy" from "hard" instances [2]. Based on this, we study the hardness of the problem under restricted classes of the intersection graph of the input.

Our Results. We study the MINIMUM COVERAGE KERNEL problem under three restrictions of the intersection graph, commonly considered for other problems [2]: planarity of the graph, bounded clique-number, and bounded vertex-degree. We show that the problem remains NP-hard even when the intersection graph of the boxes has clique-number at most 4, and the maximum degree is at most 8. For the BOX COVER problem we show that it remains NP-hard even under the severely restricted setting where the intersection graph of the boxes is planar, its clique-number is at most 2 (i.e., the graph is triangle-free), the maximum degree is at most 3, and every point is contained in at most two boxes.

We complement these hardness results with two approximation algorithms for the MINIMUM COVERAGE KERNEL problem running in polynomial time. We describe a $\mathcal{O}(\log n)$-approximation algorithm which runs in time within $\mathcal{O}(\text{OPT} \cdot n^{\frac{d}{2}+1} \log^2 n)$; and a randomized algorithm computing a $\mathcal{O}(\log \text{OPT})$-approximation in expected time within $\mathcal{O}(\text{OPT} \cdot n^{\frac{d+1}{2}} \log^2 n)$, with high probability (at least $1 - \frac{1}{n^{\Omega(1)}}$). Our main contribution in this matter is not the existence of polynomial time approximation algorithms (which can be inferred from results on BOX COVER), but a new data structure which allows to significantly improve the running time of finding those approximations (when compared to the approximation algorithms for BOX COVER). This is relevant in applications where a minimum coverage kernel needs to be computed repeatedly [1,9,15,18].

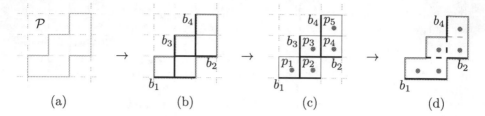

Fig. 1. (a) An orthogonal polygon $\mathcal{P}$. (b) A set of boxes $\mathcal{B} = \{b_1, b_2, b_3, b_4\}$ covering exactly $\mathcal{P}$, and such that in any cover of $\mathcal{P}$ with boxes, every box is either in $\mathcal{B}$, or fully covered by a box in $\mathcal{B}$. (c) A set of points $\mathcal{D}(\mathcal{B}) = \{p_1, p_2, p_3, p_4, p_5\}$ such that any subset of $\mathcal{B}$ covering $\mathcal{D}(\mathcal{B})$, covers also $\mathcal{P}$. (d) The subset $\{b_1, b_2, b_4\}$ is an optimal solution for the ORTHOGONAL POLYGON COVER problem on $\mathcal{P}$, the MINIMUM COVERAGE KERNEL problem on $\mathcal{B}$, and the BOX COVER problem on $\mathcal{D}(\mathcal{B}), \mathcal{B}$.

In the next section we review the reductions between the three problems we consider, and introduce some basic concepts. We then present the hardness results in Sect. 3, and describe in Sect. 4 the two approximation algorithms. We conclude in Sect. 5 with a discussion on the results and future work.

2   Preliminaries

To better understand the relation between the ORTHOGONAL POLYGON COV-ERING, the BOX COVER and the MINIMUM COVERAGE KERNEL problems, we briefly review the reductions between them. We describe them in the Cartesian plane, as the generalization to higher dimensions is straightforward.

Let $\mathcal{P}$ be an orthogonal polygon with n horizontal/vertical edges. Consider the grid formed by drawing infinitely long lines through each edge of $\mathcal{P}$ (see Fig. 1a for an illustration), and let G be the set of $\mathcal{O}(n^2)$ points of this grid lying on the intersection of two lines. Create a set $\mathcal{B}$ of boxes as follows: for each pair of points in G, if the box having those two points as opposed vertices is completely inside $\mathcal{P}$, then add it to $\mathcal{B}$ (see Fig. 1b) Let $\mathcal{C}$ be any set of boxes covering $\mathcal{P}$. Note that for any box $c \in \mathcal{C}$, either the vertices of c are in G, or c can be extended horizontally and/or vertically (keeping c inside $\mathcal{P}$) until this property is met. Hence, there is at least one box in $\mathcal{B}$ that covers each $c \in \mathcal{C}$, respectively, and thus there is a subset $\mathcal{B}' \subseteq \mathcal{B}$ covering $\mathcal{P}$ with $|\mathcal{B}'| \leq |\mathcal{C}|$. Therefore, any minimum coverage kernel of $\mathcal{B}$ is also an optimal covering of $\mathcal{P}$ (and thus, transferring the NP-hardness of the ORTHOGONAL POLYGON COVERING problem [8] to the MINIMUM COVERAGE KERNEL problem).

Now, let $\mathcal{B}$ be a set of n boxes, and consider the grid formed by drawing infinite lines through the edges of each box in $\mathcal{B}$. This grid has within $\mathcal{O}(n^2)$ cells ($\mathcal{O}(n^d)$ when generalized to d dimensions). Create a point-set $\mathcal{D}(\mathcal{B})$ as follows: for each cell c which is completely inside a box in $\mathcal{B}$ we add to $\mathcal{D}(\mathcal{B})$ the middle point of c (see Fig. 1c for an illustration). We call such a point-set a *coverage discretization* of $\mathcal{B}$, and denote it as $\mathcal{D}(\mathcal{B})$. Note that a set $\mathcal{C} \subseteq \mathcal{B}$ covers $\mathcal{D}(\mathcal{B})$ if and only if $\mathcal{C}$ covers the same region as $\mathcal{B}$ (namely, $\mathcal{C}$ is a coverage kernel of $\mathcal{B}$).

Therefore, the MINIMUM COVERAGE KERNEL problem is a special case of the BOX COVER problem.

The relation between the BOX COVER and the MINIMUM COVERAGE KERNEL problems has two main implications. Firstly, hardness results for the MINIMUM COVERAGE KERNEL problem can be transferred to the BOX COVER problem. In fact, we do this in Sect. 3, where we show that MINIMUM COVERAGE KERNEL remains NP-hard under severely restricted settings, and extend this result to the BOX COVER problem under even more restricted settings. The other main implication is that polynomial-time approximation algorithms for the BOX COVER problem can also be used for MINIMUM COVERAGE KERNEL. However, in scenarios where the boxes in $\mathcal{B}$ represent high dimensional data [9,15,18] and COVERAGE KERNELS need to be computed repeatedly [1], using approximation algorithms for BOX COVER can be unpractical. This is because constructing $\mathcal{D}(\mathcal{B})$ requires time and space within $\Theta(n^d)$. We deal with this in Sect. 4, where we introduce a data structure to index $\mathcal{D}(\mathcal{B})$ without constructing it explicitly. Then, we show how to improve two existing approximation algorithms [5,16] for the BOX COVER problem by using this index, making possible to use them for the MINIMUM COVERAGE KERNEL problem in the scenarios commented on.

3   Hardness Under Restricted Settings

We prove that MINIMUM COVERAGE KERNEL remains NP-hard for restricted classes of the intersection graph of the input set of boxes. We consider three main restrictions: when the graph is planar, when the size of its largest clique (namely the clique-number of the graph) is bounded by a constant, and when the degree of a vertex with maximum degree (namely the vertex-degree of the graph) is bounded by a constant.

Consider the k-COVERAGE KERNEL problem: given a set $\mathcal{B}$ of n boxes, find whether there are k boxes in $\mathcal{B}$ covering the same region as the entire set. Proving that k-COVERAGE KERNEL is NP-complete under restricted settings yields the NP-hardness of MINIMUM COVERAGE KERNEL under the same conditions. To prove that k-COVERAGE KERNEL is NP-hard under restricted settings we reduce instances of the PLANAR 3-SAT problem (a classical NP-complete problem [17]) to restricted instances of k-COVERAGE KERNEL. In the PLANAR 3-SAT problem, given a boolean formula in 3-CNF whose incidence graph[1] is planar, the goal is to find whether there is an assignment which satisfies the formula. The (planar) incidence graph of any planar 3-SAT formula φ can be represented in the plane as illustrated in Fig. 2 for an example, where all variables lie on a horizontal line, and all clauses are represented by *non-intersecting* three-legged combs [13]. We refer to such a representation of φ as the *planar embedding* of φ. Based on this planar embedding we proof the results in Theorem 1. Although our arguments are described in two dimensions, they extend trivially to higher dimensions.

---

[1] The *incidence graph* of a 3-SAT formula is a bipartite graph with a vertex for each variable and each clause, and an edge between a variable vertex and a clause vertex for each occurrence of a variable in a clause.

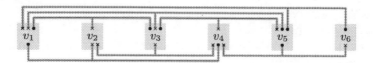

Fig. 2. Planar embedding of the formula $\varphi = (v_1 \vee \overline{v_2} \vee v_3) \wedge (v_3 \vee \overline{v_4} \vee \overline{v_5}) \wedge (\overline{v_1} \vee \overline{v_3} \vee v_5) \wedge (v_1 \vee \overline{v_2} \vee v_4) \wedge (\overline{v_2} \vee \overline{v_3} \vee \overline{v_4}) \wedge (\overline{v_4} \vee v_5 \vee \overline{v_6}) \wedge (\overline{v_1} \vee v_5 \vee v_6)$. The crosses and dots at the end of the clause legs indicate that the connected variable appears in the clause negated or not, respectively.

Theorem 1. *Let $\mathcal{B}$ be a set of n boxes in the plane and let G be the intersection graph of $\mathcal{B}$. Solving k-COVERAGE KERNEL over $\mathcal{B}$ is NP-complete even if G has clique-number at most 4, and vertex-degree at most 8.*

Due to space restrictions, we provide a brief intuition of the proof here, and defer the details to the extended version [4]. Given any set $\mathcal{B}$ of n boxes in $\mathbb{R}^d$, and any subset $\mathcal{K}$ of $\mathcal{B}$, certifying that $\mathcal{K}$ covers the same region as $\mathcal{B}$ can be done in time within $\mathcal{O}(n^{d/2})$ using Chan's algorithm [6] for computing the volume of the union of the boxes in $\mathcal{B}$. Therefore, k-COVERAGE KERNEL is in NP. To prove that it is NP-complete we construct, given a planar 3-SAT formula φ with n variables and m clauses, a set $\mathcal{B}$ of $\mathcal{O}(n + m)$ boxes which has a coverage kernel of size $31m + 3n$ if and only if there is an assignment of the variables satisfying φ. We use the planar embedding of φ as a starting point, and replace the components corresponding to variables and clauses, respectively, by gadgets composed of several boxes. Figure 3 illustrates the general layout of such construction for the formula $\varphi = (\overline{v_1} \vee v_2 \vee v_3) \wedge (v_1 \vee \overline{v_2} \vee v_4) \wedge (v_1 \vee \overline{v_3} \vee v_4)$. This construction can be obtained in polynomial time, and thus any polynomial time solution to k-COVERAGE KERNEL yields a polynomial time solution for PLANAR 3-SAT.

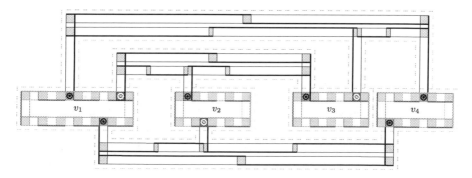

Fig. 3. Variable and clause gadgets for $\varphi = (\overline{v_1} \vee v_2 \vee v_3) \wedge (v_1 \vee \overline{v_2} \vee v_4) \wedge (v_1 \vee \overline{v_3} \vee v_4)$. The bold lines highlight one side of each rectangle, and the dashed lines delimit the regions of the variable and clause components in the planar embedding of φ. Finding the minimum subset of rectangles covering the non-white regions answers the satisfiability of φ.

To complete the proof we show that the instance of the construction meets all the restrictions of Theorem 1.

Since the MINIMUM COVERAGE KERNEL problem is a special case of the BOX COVER problem, the result of Theorem 1 also applies to the BOX COVER problem. However, in Theorem 2 we show that this problem remains hard under even more restricted settings (the proof is deferred to the the extended version [4]).

Theorem 2. *Let $P, \mathcal{B}$ be a set of m points and n boxes in the plane, respectively, and let G be the intersection graph of $\mathcal{B}$. Solving* BOX COVER *over $\mathcal{B}$ and P is* NP*-complete even if every point in P is covered by at most two boxes of $\mathcal{B}$, and G is planar, has clique-number at most 2, and vertex-degree at most 4.*

In the next section, we complement these hardness results with two approximation algorithms for the MINIMUM COVERAGE KERNEL problem.

4 Efficient Approximation of Minimum Coverage Kernels

Let $\mathcal{B}$ be a set of n boxes in $\mathbb{R}^d$, and let $\mathcal{D}(\mathcal{B})$ be a coverage discretization of $\mathcal{B}$ (as defined in Sect. 2). A *weight index* for $\mathcal{D}(\mathcal{B})$ is a data structure which can perform the following operations:

- *Initialization:* Assign an initial unitary weight to every point in $\mathcal{D}(\mathcal{B})$;
- *Query:* Given a box $b \in \mathcal{B}$, find the total weight of the points in b.
- *Update:* Given a box $b \in \mathcal{B}$, multiply the weights of all the points within b by a given value $\alpha \geq 0$;

We assume that the weights are small enough so that arithmetic operations over the weights can be performed in constant time. There is a trivial implementation of a weight index with initialization and update time within $\mathcal{O}(n^d)$, and with constant query time. In this section we describe a weight index for $\mathcal{D}(\mathcal{B})$ which can be initialized in time within $\mathcal{O}(n^{\frac{d+1}{2}})$, and with query and update time within $\mathcal{O}(n^{\frac{d-1}{2}} \log n)$. We combine this data structure with two existing approximation algorithms for the BOX COVER problem [5,16] and obtain improved approximation algorithms (in the running time sense) for the MINIMUM COVERAGE KERNEL problem.

A Weight Index for a Set of Intervals. Consider first the case of a set I of n intervals. A trivial weight index which explicitly saves the weights of each point in $\mathcal{D}(I)$ can be initialized in time within $\mathcal{O}(n \log n)$, has linear update time, and constant query time. We show that by sacrificing query time (by a factor within $\mathcal{O}(\log n)$) one can improve update time to within $\mathcal{O}(\log n)$. The main idea is to maintain the weights of each point of $\mathcal{D}(I)$ indirectly using a tree.

Consider a balanced binary tree whose leafs are in one-to-one correspondence with the values in $\mathcal{D}(I)$ (from left to right in a non-decreasing order). Let p_v denote the point corresponding to a leaf node v of the tree. In order to represent the weights of the points in $\mathcal{D}(I)$, we store a value $\mu(v)$ at each node v of the tree subject to the following invariant: for each leaf v, the weight of the point p_v equals the product of the values $\mu(u)$ of all the ancestors u of v (including v itself). The μ values allow to increase the weights of many points with only a few changes. For instance, if we want to double the weights of all the points we simply multiply by 2 the value $\mu(r)$ of the root r of the tree. Besides the μ values, to allow efficient query time we also store at each node v three values $min(v), max(v), \omega(v)$: the values $min(v)$ and $max(v)$ are the minimum and maximum p_u, respectively, such that u is a leaf of the tree rooted at v; the value $\omega(v)$ is the sum of the weights of all p_u such that u is a leaf of the tree rooted at v.

Initially, all the μ values are set to one. Besides, for every leaf l of the tree $\omega(l)$ is set to one, while $min(l)$ and $max(l)$ are set to p_l. The min, max and ω values of every internal node v with children l, r, are initialized in a bottom-up fashion as follows: $min(v) = min\{min(l), min(r)\}$; $max(v) = max\{max(l), max(r)\}$; $\omega(v) = \mu(v) \cdot (\omega(l) + \omega(r))$. It is simple to verify that after this initialization, the tree meets all the invariants mentioned above. We show in Theorem 3 that this tree can be used as a weight index for $\mathcal{D}(I)$.

Theorem 3. *Let I be a set of n intervals in $\mathbb{R}$. There exists a weight index for $\mathcal{D}(I)$ which can be initialized in time within $\mathcal{O}(n \log n)$, and with query and update time within $\mathcal{O}(\log n)$.*

Proof. Since intervals have linear union complexity, $\mathcal{D}(I)$ has within $\mathcal{O}(n)$ points, and it can be computed in linear time after sorting, for a total time within $\mathcal{O}(n \log n)$. We store the points in the tree described above. Its initialization can be done in linear time since the tree has within $\mathcal{O}(n)$ nodes, and when implemented in a bottom-up fashion, the initialization of the μ, ω, min, and max values, respectively, cost constant time per node.

To analyze the query time, let $\mathsf{totalWeight}(a, b, t)$ denote the procedure which finds the total weight of the points corresponding to leafs of the tree rooted at t that are in the interval $[a, b]$. This procedure can be implemented as follows:

1. if $[a, b]$ is disjoint to $[min(t), max(t)]$ return 0;
2. if $[a, b]$ completely contains $[min(t), max(t)]$ return $\omega(r)$;
3. if both conditions fail (leafs must meet either 1. or 2.), let l, r be the left and right child of t, respectively;
4. if $a > max(l)$ return $\mu(t) \cdot \mathsf{totalWeight}(a, b, r)$;
5. if $b < min(r)$ return $\mu(t) \cdot \mathsf{totalWeight}(a, b, l)$;
6. otherwise return $\mu(t)(\mathsf{totalWeight}(a, \infty, l) + \mathsf{totalWeight}(-\infty, b, r))$.

Due to the invariants to which the min and max values are subjected, every leaf l of t corresponding to a point in $[a, b]$ has an ancestor (including l itself) which is visited during the call to $\mathsf{totalWeight}$ and which meets the condition in step 2. For this, and because of the invariants to which the ω and μ values are subjected, the procedure $\mathsf{totalWeight}$ is correct. Note that the number of nodes visited is at most 4 times the height h of the tree: when both children need to be visited, one of the endpoints of the interval to query is replaced by $\pm\infty$, which ensures that in subsequent calls at least one of the children is completely covered by the query interval. Since $h \in \mathcal{O}(\log n)$, and the operations at each node consume constant time, the running time of totalWeight is within $\mathcal{O}(\log n)$.

Similarly, to analyze the update time, let updateWeights(a, b, t, α) denote the procedure which multiplies by a value α the weights of the points in the interval $[a, b]$ stored in leafs descending from t. This can be implemented as follows:

1. if $[a, b]$ is disjoint to $[min(t), max(t)]$, finish;
2. if $[a, b]$ completely contains $[min(t), max(t)]$ set $\mu(r) = \alpha \cdot \mu(r)$, set $\omega(r) = \alpha \cdot \omega(r)$, and finish;
3. if both conditions fail, let l, r be the left and right child of t, respectively;
4. if $a > max(l)$, call updateWeights(a, b, r, α);
5. else if $b < min(r)$, call updateWeights(a, b, l, α);
6. otherwise, call updateWeights(a, ∞, l, α), and updateWeights$(-\infty, b, r, \alpha)$;
7. finally, after the recursive calls set $\omega(t) = \mu(t) \cdot (\omega(l) + \omega(r))$, and finish.

Note that, for every point p_v in $[a, b]$ corresponding to a leaf v descending from t, the μ value of exactly one of the ancestors of u changes (by a factor of α): at least one changes because of the invariants to which the min and max values are subjected (as analyzed for totalWeight); and no more than one can change because once μ is assigned for the first time to some ancestor u of v, the procedure finishes leaving the descendants of v untouched. The analysis of the running time is analogous to that of totalWeight, and thus within $\mathcal{O}(\log n)$. □

A weight index for a set of intervals can be used to obtain an index for d-dimensional sets of boxes. The main idea is to split the space into cells such that, within each cell, the weights of the points in the cell can be represented by combining d one dimensional weight indexes. This space partition is stored in a binary tree where each node represents a cell of the space. We state this result in Theorem 4, and due to space restrictions we defer the proof to the extended version [4].

Theorem 4. *Let $\mathcal{B}$ be a set of n d-dimensional boxes. There is a weight index for $\mathcal{D}(\mathcal{B})$ which can be initialized in time within $\mathcal{O}(n^{\frac{d+1}{2}})$, and with query and update time within $\mathcal{O}(n^{\frac{d-1}{2}} \log n)$.*

Approximating the MINIMUM COVERAGE KERNEL of a set $\mathcal{B}$ of boxes via approximation algorithms for the BOX COVER problem requires that $\mathcal{D}(\mathcal{B})$ is explicitly constructed. However, the weight index described in the proof of Theorem 4 can be used to significantly improve the running time of these algorithms. We describe below two examples.

Practical Approximation Algorithms. The first algorithm we consider is the greedy $\mathcal{O}(\log n)$-approximation algorithm by Lovász [16]. The greedy strategy applies naturally to the MINIMUM COVERAGE KERNEL problem: iteratively pick the box which covers the most yet uncovered points of $\mathcal{D}(\mathcal{B})$, until there are no points of $\mathcal{D}(\mathcal{B})$ left to cover. To avoid the explicit construction of $\mathcal{D}(\mathcal{B})$ three operations most be simulated: (*i.*) find how many uncovered points are within a given a box $b \in \mathcal{B}$; (*ii.*) delete the points that are covered by a box $b \in \mathcal{B}$; and (*iii.*) find whether a subset $\mathcal{B}'$ of $\mathcal{B}$ covers all the points of $\mathcal{D}(\mathcal{B})$.

For the first two we use the weight index described in the proof of Theorem 4: to delete the points within a given box $b \in \mathcal{B}$ we simply multiply the weights of all the points of $\mathcal{D}(\mathcal{B})$ within b by $\alpha = 0$; and finding the number of uncovered points within a box b is equivalent to finding the total weight of the points of $\mathcal{D}(\mathcal{B})$ within b. For the last of the three operations we use the following observation:

Observation 1. *Let $\mathcal{B}$ be a set of d-dimensional boxes, and let $\mathcal{B}'$ be a subset of $\mathcal{B}$. The volume of the region covered by $\mathcal{B}'$ equals that of $\mathcal{B}$ if and only if $\mathcal{B}'$ and $\mathcal{B}$ cover the exact same region.*

Let OPT denote the size of a minimum coverage kernel of $\mathcal{B}$, and let N denote the size of $\mathcal{D}(\mathcal{B})$ ($N \in \mathcal{O}(n^d)$). The greedy algorithm of Lovász [16], when run over the sets $\mathcal{B}$ and $\mathcal{D}(\mathcal{B})$ works in $\mathcal{O}(\text{OPT} \log N)$ steps; and at each stage a box is added to the solution. The size of the output is within $\mathcal{O}(\text{OPT} \log N) \subseteq \mathcal{O}(\text{OPT} \log n)$. This algorithm can be modified to achieve the following running time, while achieving the same approximation ratio:

Theorem 5. *Let $\mathcal{B}$ be a set of n boxes in $\mathbb{R}^d$ with a minimum coverage kernel of size OPT. Then, a COVERAGE KERNEL of $\mathcal{B}$ of size within $\mathcal{O}(\text{OPT} \log n)$ can be computed in time within $\mathcal{O}(\text{OPT} \cdot n^{\frac{d}{2}+1} \log^2 n)$.*

Proof. We initialize a weight index as in Theorem 4, which can be done in time $\mathcal{O}(n^{\frac{d+1}{2}})$, and compute the volume of the region covered by $\mathcal{B}$, which can be done in time within $\mathcal{O}(n^{d/2})$ [6]. Let C be an empty set. At each stage of the algorithm, for every box $b \in \mathcal{B} \setminus C$ we compute the total weight of the points inside b (which can be done in time within $n^{\frac{d-1}{2}} \log n$ using the weight index). We add to C the box with the highest total weight, and update the weights of all the points within this box to zero (by multiplying their weights by $\alpha = 0$) in time within $n^{\frac{d-1}{2}} \log n$. If the volume of the region covered by C (which can be computed in $\mathcal{O}(n^{d/2})$-time [6]) is the same as that of $\mathcal{B}$, then we stop and return C as the approximated solution. The total running time of each stage is within $\mathcal{O}(n^{\frac{d+1}{2}} \log n)$. This, and the fact that the number of stages is within $\mathcal{O}(\text{OPT} \log n)$ yield the result of the theorem.   □

Now, we show how to improve Brönnimann and Goodrich's $\mathcal{O}(\log \text{OPT})$ approximation algorithm [5] via a weight index. First, we describe their main idea. Let $w : \mathcal{D}(\mathcal{B}) \to \mathbb{R}$ be a weight function for the points of $\mathcal{D}(\mathcal{B})$, and for a subset $\mathcal{P} \subseteq \mathcal{D}(\mathcal{B})$ let $w(\mathcal{P})$ denote the total weight of the points in $\mathcal{P}$. A point p is said to be ε-heavy, for a value $\varepsilon \in (0, 1]$, if $w(p) \geq \varepsilon w(\mathcal{D}(\mathcal{B}))$, and ε-light otherwise. A subset $\mathcal{B}' \subseteq \mathcal{B}$ is said to be an ε-net with respect to w if for every ε-heavy point $p \in \mathcal{D}(\mathcal{B})$ there is a box in $\mathcal{B}'$ which contains p. Let OPT denote the size of a minimum coverage kernel of $\mathcal{B}$, and let k be an integer such that $k/2 \leq \text{OPT} < k$. The algorithm initializes the weight of each point in $\mathcal{D}(\mathcal{B})$ to 1, and repeats the following *weight-doubling step* until every range is $\frac{1}{2k}$-heavy: find a $\frac{1}{2k}$-light point p and double the weights of all the points within every box

$b \in \mathcal{B}$. When this process stops, it returns a $\frac{1}{2k}$-net C with respect to the final weights as the approximated solution.

Since each point in $\mathcal{D}(\mathcal{B})$ is $\frac{1}{2k}$-heavy, C covers all the points of $\mathcal{D}(\mathcal{B})$. Hence, if a $\frac{1}{2k}$-net of size $\mathcal{O}(kg(k))$ can be computed efficiently, this algorithm computes a solution of size $\mathcal{O}(kg(k))$. Besides, Brönnimann and Goodrich [5] showed that for a given k, if more than $\mu_k = 4k\log(n/k)$ weight-doubling steps are performed, then $\mathsf{OPT} > 2k$. This allows to guess the correct k via exponential search, and to bound the maximum weight of any point by n^4/k^3 (which allows to represent the weights with $\mathcal{O}(\log n)$ bits). See Brönnimann and Goodrich's article [5] for the complete details of their approach.

We simulate the operations over the weights of $\mathcal{D}(\mathcal{B})$ again using a weight index, this time with a minor variation to that of Theorem 4: in every node of the space partition tree, besides the ω, μ values, we also store the minimum weight of the points within the cell corresponding to the node. During the initialization and update operations of the weight index this value can be maintained as follows: for a node v with children l, r, the minimum weight $min_\omega(v)$ of a point in the cell of v can be computed as $min_\omega(v) = \omega(v) \cdot \min\{min_\omega(l), min_\omega(r)\}$. This value allows to efficiently detect whether there are $\frac{1}{2k}$-light points, and to find one in the case of existence by tracing down, in the partition tree, the path from which that value comes.

To compute a $\frac{1}{2k}$-net, we choose a sample of $\mathcal{B}$ by performing at least $(16k\log 16k)$ random independent draws from $\mathcal{B}$. We then check whether it is effectively a $\frac{1}{2k}$-net, and if not, we repeat the process, up to a maximum of $\mathcal{O}(\log n)$ times. Haussler and Welzl [12] showed that such a sample is a $\frac{1}{2k}$-net with probability at least $1/2$. Thus, the expected number of samples needed to obtain a $\frac{1}{2k}$-net is constant, and since we repeat the process up to $\mathcal{O}(\log n)$ times, the probability of effectively finding one is at least $1 - \frac{1}{n^{\Omega(1)}}$. We analyze the running time of this approach in the following theorem.

Theorem 6. *Let $\mathcal{B}$ be a set of n boxes in $\mathbb{R}^d$ with a minimum coverage kernel of size OPT. A coverage kernel of $\mathcal{B}$ of size within $\mathcal{O}(\mathsf{OPT}\log\mathsf{OPT})$ can be computed in $\mathcal{O}(\mathsf{OPT}n^{\frac{d+1}{2}}\log^2 n)$-expected time, with probability at least $1 - \frac{1}{n^{\Omega(1)}}$.*

Proof. The algorithm performs several stages guessing the value of k. Within each stage we initialize a weight index in time within $\mathcal{O}(n^{\frac{d+1}{2}})$. Finding whether there is a $\frac{1}{2k}$-light point can be done in constant time: the root of the partition tree stores both $w(\mathcal{D}(\mathcal{B}))$ and the minimum weight of any point in the ω and min_ω values, respectively. For every light point, the weight-doubling steps consume time within $\mathcal{O}\left(n \times \left(n^{\frac{d-1}{2}}\log n\right)\right) \subseteq \mathcal{O}(n^{\frac{d+1}{2}}\log n)$ (by Theorem 4). Since at each stage at most $4k\log(n/k)$ weight-doubling steps are performed, the total running time of each stage is within $\mathcal{O}(kn^{\frac{d+1}{2}}\log n \log \frac{n}{k}) \subseteq \mathcal{O}(kn^{\frac{d+1}{2}}\log^2 n)$. Given that k increases geometrically while guessing its right value, and since the running time of each stage is a polynomial function, the sum of the running times of all the stages is asymptotically dominated by that of the last stage, for which we have that $k \leq \mathsf{OPT} \leq 2k$. Thus the result of the theorem follows. $\square$

Compared to the algorithm of Theorem 5, this last approach obtains a better approximation factor on instances with small COVERAGE KERNELS ($\mathcal{O}(\log n)$ vs. $\mathcal{O}(\log \mathsf{OPT})$), but the improvement comes with a sacrifice, not only in the running time, but in the probability of finding such a good approximation. In two and three dimensions, weight indexes might also help to obtain practical $\mathcal{O}(\log \log \mathsf{OPT})$ approximation algorithms for the MINIMUM COVERAGE PROBLEM. We discuss this, and other future directions of research in the next section.

5   Discussion

Whether it is possible to close the gap between the factors of approximation of BOX COVER and ORTHOGONAL POLYGON COVERING has been a long standing open question [14]. The MINIMUM COVERAGE KERNEL problem, intermediate between those two, has the potential of yielding answers in that direction, and has natural applications of its own [9,15,18]. Trying to understand the differences in hardness between these problems, we studied distinct restricted settings. We show that while MINIMUM COVERAGE KERNEL remains NP-hard under severely restricted settings, the same can be said for the BOX COVER problem under even more extreme settings; and show that while the BOX COVER and MINIMUM COVERAGE KERNEL can be approximated by at least the same factors, the running time of obtaining some of those approximations can be significantly improved for the MINIMUM COVERAGE KERNEL problem.

Another approach to understand what makes a problem hard is Parameterized Complexity [10], where the hardness of a problem is analyzed with respect to multiple parameters of the input, with the hope of finding measures gradually separating "easy" instances form the "hard" ones. The hardness results described in Sect. 3 show that for the MINIMUM COVERAGE KERNEL and BOX COVER problems, the vertex-degree and clique-number of the underlaying graph are not good candidates of such kind of measures, opposed to what happens for other related problems [2].

In two and three dimensions, the BOX COVER problem can be approximated up to $\mathcal{O}(\log \log \mathsf{OPT})$ [3]. We do not know whether the running time of this algorithm can be also improved for the case of MINIMUM COVERAGE KERNEL via a weight index. We omit this analysis since the approach described in Sect. 4 is relevant when the dimension of the boxes is high (while still constant), as in distinct applications [9,15,18] of the MINIMUM COVERAGE KERNEL problem.

Acknowledgments. We thank an anonymous reviewer for carefully reading our manuscript, and providing many insightful comments and suggestions.

References

1. Agarwal, P.K., Pan, J.: Near-linear algorithms for geometric hitting sets and set covers. In: Proceedings of the 30th Annual Symposium on Computational Geometry (SoCG), pp. 271:271–271:279. ACM, New York (2014)

2. Alekseev, V.E., Boliac, R., Korobitsyn, D.V., Lozin, V.V.: NP-hard graph problems and boundary classes of graphs. Theor. Comput. Sci. (TCS) **389**(1–2), 219–236 (2007)

3. Aronov, B., Ezra, E., Sharir, M.: Small-size ε-nets for axis-parallel rectangles and boxes. SIAM J. Comput. (SICOMP) **39**(7), 3248–3282 (2010)

4. Barbay, J., Pérez-Lantero, P., Rojas-Ledesma, J.: Computing coverage kernels under restricted settings. arXiv e-prints (2018)

5. Brönnimann, H., Goodrich, M.T.: Almost optimal set covers in finite VC-dimension. Discrete Comput. Geom. (DCG) **14**(4), 463–479 (1995)

6. Chan, T.M.: Klee's measure problem made easy. In: 54th Annual IEEE Symposium on Foundations of Computer Science (FOCS), Berkeley, CA, USA, 26–29 October 2013. IEEE Computer Society, pp. 410–419 (2013)

7. Clarkson, K.L., Varadarajan, K.R.: Improved approximation algorithms for geometric set cover. Discrete Comput. Geom. (DCG) **37**(1), 43–58 (2007)

8. Culberson, J.C., Reckhow, R.A.: Covering polygons is hard. J. Algorithms (JALG) **17**(1), 2–44 (1994)

9. Daly, J., Liu, A.X., Torng, E.: A difference resolution approach to compressing access control lists. IEEE/ACM Trans. Netw. (TON) **24**(1), 610–623 (2016)

10. Downey, R.G., Fellows, M.R.: Parameterized Complexity. Monographs in Computer Science. Springer, New York (1999). https://doi.org/10.1007/978-1-4612-0515-9

11. Fowler, R.J., Paterson, M., Tanimoto, S.L.: Optimal packing and covering in the plane are NP-complete. Inf. Process. Lett. (IPL) **12**(3), 133–137 (1981)

12. Haussler, D., Welzl, E.: ε-nets and simplex range queries. Discrete Comput. Geom. **2**, 127–151 (1987)

13. Knuth, D.E., Raghunathan, A.: The problem of compatible representatives. SIAM J. Discrete Math. **5**(3), 422–427 (1992)

14. Kumar, V.S.A., Ramesh, H.: Covering rectilinear polygons with axis-parallel rectangles. SIAM J. Comput. (SICOMP) **32**(6), 1509–1541 (2003)

15. Lakshmanan, L.V.S., Ng, R.T., Wang, C.X., Zhou, X., Johnson, T.: The generalized MDL approach for summarization. In: Proceedings of 28th International Conference on Very Large Data Bases (VLDB), Hong Kong, China, 20–23 August 2002, pp. 766–777. Morgan Kaufmann (2002)

16. Lovász, L.: On the ratio of optimal integral and fractional covers. Discrete Math. (DM) **13**(4), 383–390 (1975)

17. Mulzer, W., Rote, G.: Minimum-weight triangulation is NP-hard. J. ACM (JACM) **55**(2), 1–29 (2008)

18. Pu, K.Q., Mendelzon, A.O.: Concise descriptions of subsets of structured sets. ACM Trans. Database Syst. (TODS) **30**(1), 211–248 (2005)

Weak Mitoticity of Bounded Disjunctive and Conjunctive Truth-Table Autoreducible Sets

Liyu Zhang$^{(\boxtimes)}$, Mahmoud Quweider, Hansheng Lei, and Fitra Khan

Department of Computer Science, University of Texas Rio Grande Valley,
One West University Boulevard, Brownsville, TX 78520, USA
{liyu.zhang,mahmoud.quweider,hansheng.lei,fitra.khan}@utrgv.edu

Abstract. Glaßer et al. (SIAMJCOMP 2008 and TCS 2009 (The two papers have slightly different sets of authors)) proved existence of two sparse sets A and B in EXP, where A is 3-tt (truth-table) polynomial-time autoreducible but not weakly polynomial-time Turing mitotic and B is polynomial-time 2-tt autoreducible but not weakly polynomial-time 2-tt mitotic. We unify and strengthen both of those results by showing that there is a sparse set in EXP that is polynomial-time 2-tt autoreducible but not even weakly polynomial-time Turing mitotic. All these results indicate that polynomial-time autoreducibilities in general do not imply polynomial-time mitoticity at all with the only exceptions of the many-one and 1-tt reductions. On the other hand, however, we proved that every autoreducible set for the polynomial-time bounded disjunctive or conjunctive tt reductions is weakly mitotic for the polynomial-time tt reduction that makes logarithmically many queries only. This shows that autoreducible sets for reductions making more than one query could still be mitotic in some way if they possess certain special properties.

1 Introduction

Let r be a reduction between two languages as defined in computational complexity such as the common *many-one* and *Turing* reductions. We say that a language L is *r-autoreducible* if L is reducible to itself via the reduction r where the reduction does not query on the same string as the input. In case that r is the many-one reduction, we require that r outputs a string different from the input in order to be an autoreduction. Researchers started investigating on autoreducibility as early as 1970's [12] although much of the work done then was in the recursive setting. Ambos-Spies [1] translated the notion of autoreducibility to the polynomial-time setting, and Yao [13] considered autoreducibility in the probabilistic polynomial-time setting, which he called *coherence*.

More recently polynomial-time autoreducibilities, which correspond to polynomial-time reductions, gained attention due to its candidacy as a structural property that can be used in the *"Post's program for complexity theory"*

L. Zhang—Research supported in part by NSF CCF grant 1218093.

L. Wang and D. Zhu (Eds.): COCOON 2018, LNCS 10976, pp. 192–204, 2018.
https://doi.org/10.1007/978-3-319-94776-1_17

[3] that aims at finding a structural/computational property that complete sets of two complexity classes don't share, hereby separating the two complexity classes. Autoreducibility is believed to be possibly one of such properties that will lead to new separation results in the future [2]. We refer the reader to Glaßer et al. [7] and Glaßer et al. [6] for recent surveys along this line of research.

In this paper we continue to study the relation between the two seemingly different notions, autoreducibility and mitoticity. Glaßer et al. [9] proved that among polynomial-time reductions, autoreducibility coincides with polynomial-time mitoticity for the many-one and 1-tt reductions, but not for the 3-tt reduction or any reduction weaker than 3-tt. In a subsequent paper Glaßer et al. [4] further proved that 2-tt autoreducibility does not coincide with 2-tt mitoticity. However, the set they construct is weakly 5-tt mitotic. So the technical question remained open whether one can construct a language that is 2-tt autoreducible but not weakly Turing-mitotic. We solve this problem in the positive way. More precisely, we proved that there exists a sparse set in EXP that is 2-tt autoreducible but not even weakly Turing-mitotic. This result unifies and strengthens both of the previous results.

In attempting to strengthen our results further we asked the question whether one can even construct a language that is r-autoreducible but not weakly Turing-mitotic for any reduction r that is weaker than the 1-tt reduction but stronger than 2-tt reduction such as 2-dtt and 2-ctt reductions. We proved that any language that is k-dtt or k-ctt autoreducible is also weakly $k^{O(2^c \log(c-1)n)}$-tt mitotic for any integers $k, c \geq 2$. Glaßer et al. [9] and Glaßer et al. [5] showed that k-dtt and/or k-ctt complete set for many common complexity classes including NP, PH, PSPACE and NEXP are k-dtt and/or k-ctt autoreducible, respectively. In light of that we have the interesting corollary that k-dtt and/or k-ctt complete sets of those complexity classes are weakly dtt- and/or ctt-mitotic, respectively.

We give definitions and notations needed to present our results in Sect. 2 below. We then describe our main results in more details in Sect. 3. Due to space limit we have to omit the proofs for Lemma 1, Theorems 4, and 5 in this proceeding paper. Those proofs will be available upon request and in the journal version of the paper.

2 Definitions and Notations

We assume familiarity with basic notions in complexity theory and particularly, common complexity classes such as P, NP, PH, PSPACE and EXP, and polynomial-time reductions including many-one ($\leq^p_m$), truth-table ($\leq^p_{tt}$) and Turing reductions ($\leq^p_T$) [10,11]. Without loss of generality, we use the alphabet $\Sigma = \{0, 1\}$ and all sets we referred to in this paper are either languages over Σ or sets of *integers*. Let $\mathbb{N}$ denote the set of natural numbers and $\mathbb{N}^+$ denote $\mathbb{N}\backslash\{0\}$. We use a pairing function $\langle \cdot, \cdot \rangle$ that satisfies $\langle x, y \rangle > x + y$. For every string/integer x, we use $|x|/abs(x)$ to denote the length/absolute value of x. For every function f, we use $f^{(i)}(x)$ to denote $\underbrace{f(f(\cdots f(x)))}_{i}$ for every $i \in \mathbb{N}$, where

$f^{(0)}(x) = x$.

Throughout the paper, we use the two terms *Turing machines* and *algorithms* interchangeably. Following Glaßer et al. [9], we define a non-trivial set to be a set L where both $L|$ and $\overline{L}$ contain at least two distinct elements. This allows us present our results in a simple and concise way. All reductions used in this paper are *polynomial-time computable* unless otherwise specified. A language L is *complete* for a complexity class $\mathcal{C}$ for a reduction r if every language in $\mathcal{C}$ is reducible to L via r. For any algorithm or Turing machine $\mathcal{A}$, we use $\mathcal{A}(x)$ to denote both the execution and output of $\mathcal{A}$ on input x, i.e., "$\mathcal{A}(x)$ accepts" has the same meaning as "$\mathcal{A}(x) = accept$". We use $\mathcal{A}^B(x)$ or $\mathcal{A}^g(x)$ to denote the same for algorithm/Turing machine $\mathcal{A}$ that has oracle access to a set B or a function g. Also $L(\mathcal{A})$ ($L(\mathcal{A}^B)$ or $L(\mathcal{A}^g)$) denotes the language accepted by $\mathcal{A}$ ($\mathcal{A}^B$ or $\mathcal{A}^g$).

We provide detailed definitions for the most relevant reductions considered in this paper below.

Definition 1. *Define a language A to be* polynomial-time truth-table reducible *($\leq_{tt}^p$) to a language B, if there exists a polynomial-time algorithm $\mathcal{A}$ that accepts A with oracle access to B. In addition, there exists a polynomial-time computable function g that on input x outputs all queries $\mathcal{A}(x)$ makes to B.*

Truth-table reductions are also called *nonadaptive Turing reductions* in the sense that they are the same as the general Turing reductions except that all queries the reductions make can be computed from the input in polynomial time without knowing the answer to any query.

Definition 2. *For any positive integer k, define a language A to be* polynomial-time k-tt reducible *($\leq_{k\text{-}tt}^p$) to a language B, if there exists a polynomial-time truth-table reduction r from A to B that makes at most k queries on every input x. If in addition the k queries $q_0, q_1, \cdots, q_{k-1}$ that r makes are such that $x \in A$ if and only if some/every $q_i \in B$, then r is called a* disjunctive/conjunctive truth-table reduction *and A is said to be* disjunctive/conjunctive truth-table reducible *($\leq_{k\text{-}dtt}^p/\leq_{k\text{-}ctt}^p$) to B.*

Now we define autoreducible and mitotic languages formally.

Definition 3. *Given any reduction r, a language is* autoreducible for r or r-autoreducible, *if the language is reducible to itself via r that does not query on the input.*

Definition 4. *Given any reduction r, a language L is* weakly mitotic for r or weakly r-mitotic, *if there exists another language S, where L, $S \cap L$ and $\overline{S}$ are all equivalent under the reduction r, i.e., $L \equiv_r S \cap L \equiv_r \overline{S} \cap L$. If in addition $S \in \mathrm{P}$, then we say that L is* mitotic for r or r-mitotic.

The proof of our second result uses *log derivative sequences* based on *log-distance functions*. We define both concepts below.

Let sgn denote the common sign function defined on integers, i.e., for every $z \in \mathbb{Z}$, $sgn(z) = 1$ if $z \geq 0$ and $sgn(z) = -1$ otherwise.

Definition 5. *For every pair of integers or strings x and y, we define the following* log-distance *function, $logD$, as follows.*

$$logD(x, y) = \begin{cases} sgn(y - x)\lfloor \log |y - x| \rfloor & if\ x \neq y\ and\ \infty \notin \{x, y\} \\ \infty & otherwise. \end{cases}$$

In case where x and y are strings, $y - x$ is defined to be their lexicographical *difference.*

The above function is the same as the "distance function" defined by Glaßer et al. [9], except that we define $logD(x, y) = \infty$ instead of 0 when $x = y$, or either x or y is ∞.

Definition 6. *Let $X = \{x_j\}_{j \geq 0}$ be a sequence of strings or integers, where x_j denotes the j-th element in X. Define the i-th log derivative sequence of X, written $X^{(i)}$ as follows:*

– *$X^{(0)} = X$, and*
– *For $i \geq 1$, $X^{(i)} = \{x_j^{(i)}\}$, where $x_j^{(i)} = logD(x_j^{(i-1)}, x_{j+1}^{(i-1)})$.*

In case X is a finite sequence $\{x_j\}_{s \leq j \leq t}$, where $s, t \in \mathbb{N}$ and $s \leq t$, then $X^{(i)} = \{x_j^{(i)}\}_{s \leq j \leq t-i}$ for every $i \in [0, t - s]$. For every $i \geq 2$, we say that $X^{(i)}$ is a higher-order log derivative *of X.*

3   Results

Our first main result is that there exists a sparse set in EXP that is 2-tt autoreducible but not weakly mitotic even for the polynomial-time Turing reduction, the most general polynomial-time reduction.

Overall the proof of our first main result follows the approach of the proof by Glaßer et al. [4] that there exists a sparse set in EXP that is 2-tt autoreducible but not 2-tt mitotic. The proof is in general a *diagonalization* against all possible partitions of a constructed language L into L_1 and L_2, as well as all possible polynomial-time oracle Turing machines M_i and M_j, where $L \leq_{2\text{-}tt}^p L_1$ via M_i and $L \leq_{2\text{-}tt}^p L_2$ via M_j. The construction of L proceeds in stages, where in each stage, only polynomially many strings of a particular length are added to L. The gaps between lengths of strings added to L in different stages are made super-exponential so that strings added to L in later stages won't affect the computations of considered Turing machines on strings added to L in previous stages. In light of the fact that the set constructed as described above is actually weakly 5-tt mitotic [4], it was assumed that a straightforward adaption of the above construction won't be sufficient for proving a stronger result that there exists a sparse set in EXP that is 2-tt autoreduction but not weakly Turing mitotic.

However, something overlooked here is that the aforementioned proof actually proved a stronger statement than stated - The proof actually shows that there is a set L, where for every partition $\{L_1, L_2\}$ of L, either $L \not\leq^p_{2\text{-}tt} L_1$ or $L \not\leq^p_{2\text{-}tt} L_2$. In order to prove that L is not 2-tt mitotic we only need to show $L \not\leq^p_{2\text{-}tt} L_1$, $L_1 \not\leq^p_{2\text{-}tt} L_2$ or $L_2 \not\leq^p_{2\text{-}tt} L$. The latter is clearly a weaker statement. In light of this observation, we adapt the previous proof by considering *three* oracle Turing machines M_i, M_j and M_k instead for the purpose of diagonalization in each stage of constructing the language L. It turns out that this is critical for the proof to go through.

We now state our first main result in detail below.

Theorem 1. *There exists* $L \in \text{SPARSE} \cap \text{EXP}$ *such that*

- *L is 2-tt-autoreducible, but*
- *L is not weakly Turing-mitotic.*

Theorem 1 indicates that 2-tt autoreducibility does not imply weak mitoticity even for the Turing reduction, the most general polynomial-time reduction. This shows that in general autoreducibility does not even imply the weakest form of mitoticity in the polynomial-time setting among reductions making more than one query, despite that autoreducibility and mitoticity are equivalent for the many one and 1-tt reductions. A further question that is natural to ask is whether autoreducibility implies any form of mitoticity at all for reductions that lie between the 2-tt reduction and 1-tt reduction, or reductions with special properties that are incomparable to 2-tt and/or 1-tt reductions, such as honest and positive reductions.

Here we consider *bounded disjunctive* and *conjunctive* truth-table reductions. We prove that if a language is k-dtt or k-ctt autoreducible for some integer $k \geq 2$, then the language is weakly truth-table mitotic. In addition, the reduction can be made to query on at most $k^{O(2^c \log^{(c-1)} n)}$ strings for every integer $c \geq 2$. Our proof adapts and generalizes in a significant way the proof strategy used by Glaßer et al. [9], where they showed that every nontrivial language is many-one or 1-tt autoreducible if and only if the language is many-one or 1-tt mitotic, respectively. We review their proof strategy below at a higher level and then describe the changes needed in order to establish the weak mitoticity of any k-dtt or k-ctt autoreducible sets.

Let L be a nontrivial many-one ($\leq^p_m$) autoreducible set, where there exists a polynomial-time computable function f such that $f(x) \neq x$ and $f(x) \in L$ if and only if $x \in L$ for every $x \in \Sigma^*$. It is sufficient to prove that there exists a polynomial-time decidable set S and a function f', where for every $x \in \Sigma^*$,

(i) $f'(x) \in L$ if and only if $x \in L$, and
(ii) $f'(x) \in S$ if and only if $x \in \overline{S}$.

The idea of finding a function f' that satisfies conditions (i) and (ii) as stated above is to define $f'(x) = f^{(i)}(x)$ for an appropriate $i \leq p(|x|)$, where p is a polynomial. Since $f(x)$ is an autoreduction, it is obvious that f' defined in this way satisfies condition (i). Now we need to construct a set S where for each x we can find a correct i so that Condition (ii) also holds, i.e., $f'(x) = f^{(i)}(x) \in S$ if and only if $x \in \overline{S}$.

To construct such set S we consider the sequence $x, f(x), f(f(x)), \ldots$, called *trajectory* of x in Glaßer et al. [9]. Note that every string on the trajectory of x has the same membership in L as x and also that every two consecutive strings on the trajectory are unequal to each other. We first partition Σ^* into a sequence of *segments*, each of which contains all strings of some consecutive lengths. In addition, those segments are assigned to S and $\overline{S}$ in an alternate way, i.e., the i-th segment is assigned to S if and only if the $(i+1)$-st is assigned to $\overline{S}$. Then we look for changes of monotonicity in the trajectory of x and assign x to S or $\overline{S}$ accordingly. For instance, assign x to S if $x < f(x) > f(f(x))$ and to $\overline{S}$ if $x > f(x) < f(f(x))$. Clearly if we can find more than one change in monotonicity along the trajectory of x within polynomially many strings, then we will find a string y where $y \notin S$ if and only if $x \in S$. Otherwise, we can find a strictly monotonic sub-trajectory within the trajectory of x. If within that sub-trajectory, f increases or decreases fast enough, i.e., leading to a change in length, then we can find strings on the trajectory of x that are polynomially many strings from x but belong to different segments and hence have different memberships of S by x.

The most difficult case arises when the trajectory of x is a strictly monotonic but does not increase or decrease fast enough so that trajectory can reach a neighboring segment from the segment containing x within polynomially many strings away from x. Glaßer et al. [9] dealt with this case essentially by dividing each segment into smaller segments of increasing sizes based on a *log distance* function applied on strings on the trajectory of x. This way the trajectory will contain strings in neighboring segments of the same size depending on how fast the autoreduction function increases or decreases. Then we can find a y in a neighboring segment from x on the trajectory, where $y \in S$ if and only $x \in \overline{S}$.

The above strategy obviously does not apply to reductions making more than one queries as it is. We, however, found a way to adapt the strategy to apply it on bounded dtt or ctt reductions and prove the weak tt mitoticity of bounded dtt or ctt autoreducible sets.

Consider a k-dtt autoreducible set L for some integer $k \geq 2$. Then there exists a polynomial-time computable function f, where for every $x \in \Sigma^*$, $f(x) = \langle y_0, y_1, \cdots, y_{k-1} \rangle$ such that

(i) $y_j \neq x$ for every $j \in [0, k-1]$, and
(ii) $f(x) \in L$ if and only if $y_j \in L$ for some $j \in [0, k-1]$.

Now define a function g where $g(x)$ is the *lexicographically least* string in $f(x)$ that has the same membership of L as x. Then it is clear that $g(x) \neq x$ and $g(x) \in L$ if and only if $x \in L$. We can apply Glaßer et al.'s construction [9] as described above on g and establish the tt mitoticity of any k-dtt autoreducible set. The problem with this approach is, however, that the function g might not be polynomial-time computable. We circumvented this problem by considering all *possible* values of $g(x)$ for every x, i.e., any of the k values in $f(x)$, This means that we will look for a $y \neq x$ where $y \in S$ if and only if $x \notin S$, along all *possible* trajectories from x. Hence, we no longer can afford traversing along a trajectory from x for polynomially many strings before we can find the desired y since there could be exponentially many possible trajectories. Instead, we need another way to construct the set S so that there exists a y on the g-trajectory that is at most $O(\log n)$ strings away from x, where $x \in S$ if and only if $y \notin \overline{S}$.

We solve this problem by considering *higher-order log derivatives*, defined in Sect. 2, of the sequence consisting of strings on the trajectory of x, i.e., $X = \{x_j = g^{(j)}\}_{j \geq 1}$. The log-distance function used by Glaßer et al. [9] can be viewed as the *first-order log derivative* of the sequence X. We will attempt to find changes of monotonicity in X, the 1st-order log derivative of X, the 2nd-order log derivative of X and so on until the c-th order log derivative for some integer $c \geq 2$, in that order, and then assign x to either S or $\overline{S}$ accordingly. If we don't find enough changes of monotonicity among all those high-order log derivative sequences of X, then we will show that the function g increases fast enough already so that for some $j \in [1, O(log|x|)]$, $g^{(j)}(x)$ belongs to the next segment after the one containing x. Hence, $g^{(j)}(x)$ will be assigned to $\overline{S}$ if and only if x is assigned to S. We now provide the detailed proof for our main theorem.

We first need the following lemma that was essentially proved in Glaßer et al. [9].

Lemma 1 [9]. *Let $\{x_j\}_{0 \leq j \leq 2}$ be a strictly monotonic sequence of integers, where there exists some $d \in \mathbb{Z}$ such that $log D(x_j, x_{j+1}) = d$ for $0 \leq j \leq 1$. Then the set $X = \{\lfloor \frac{x_j}{2^{abs(d)+1}} \rfloor \mid 0 \leq j \leq 2\}$ contains at least one even number and one odd number.*

Theorem 2. *Let g be a polynomially-bounded function and $g(x) \neq x$ for every $x \in \Sigma^*$. Then for every positive integer $c \geq 2$, there is a polynomial-time algorithm $\mathcal{S}_c$ with oracle access to function g, and a polynomial r, where for every $x \in \Sigma^*$ an integer $j_x \in [1, \lceil 2^c \log^{(c-1)} |x| \rceil]$ exists such that*

(i) for each $j \in [0, j_x]$, $|g^j(x)| \leq r(|x|)$,
(ii) for each $j \in [0, j_x - 1]$, $\mathcal{S}_c^g$ accepts x if and only if $\mathcal{S}_c^g$ accepts $g^{(j)}(x)$, and
(iii) $\mathcal{S}_c^g$ accepts x if and only if $\mathcal{S}_c^g$ rejects $g^{(j_x)}(x)$.

Proof. Let g be an $(n^l + l)$-bounded function for some $l \in \mathbb{N}^+$ as given in the premise. Let t be a tower function defined by $t(0) = 0$ and $t(i + 1) = t(i)^l + l$ for $i \in \mathbb{N}$. Define the inverse tower function as $t^{-1}(n) = \min\{i \mid t(i) \geq n\}$. Note that t^{-1} is polynomial-time computable. Now consider the algorithm $\mathcal{S}_c^g$ given below (Algorithm 1).

Let $m = \lceil 2^c \log^{(c-1)} |x| \rceil$. We first observe that Algorithm $\mathcal{S}_c^g$ queries on strings $g^{(j)}(x)$ for $1 \leq j \leq c + 2$ only, each of which is polynomially bounded since g is a polynomially bounded function. Hence, $\mathcal{S}_c^g$ runs in polynomial time assuming the value of $g^{(j)}(u)$ for any u queried on can be obtained instantly.

We now turn to the proof for conditions (i)–(iii) of Theorem 2. Let x be an arbitrary input string and let X denote the sequence $\{x_j = g^{(j)}(x)\}_{0 \leq j \leq m}$. The

---

Input   : An arbitrary string $w \in \{0,1\}^*$, where $|w| = n$

Output : ACCEPT or REJECT

1 $m \leftarrow k\lfloor \log n \rfloor$;

2 $D[0,0] \leftarrow x$, $D[0,1] \leftarrow g(x)$;

3 **if** $t^{-1}(|D[0,0]|) < t^{-1}(|D[0,1]|)$ **then**

4 | ACCEPT iff $t^{-1}(|D[0,0]|)$ is odd

5 **end**

6 $D[1,0] \leftarrow logD(D[0,0], D[0,1])$;

7

8 // Compute the log derivatives of X at $x_0 = x$, $x_1 = g(x)$ and $x_2 = g(g(x))$;

9 **for** $i \leftarrow 0$ **to** k **do**

10

11 | // Compute the i-th order log derivatives;

12 | **for** $j \leftarrow i + 2$ **to** 0 **do**

13 | | **if** $j = i + 2$ **then**

14 | | | $D[0,j] \leftarrow g(D[0, j-1])$;

15 | | | **if** $t^{-1}(D[0, j-1])) < t^{-1}(|D[0,j]|)$ **then**

16 | | | | ACCEPT iff $t^{-1}(|D[0,j-1]|)$ is even

17 | | | **end**

18 | | **end**

19 | | **else**

20 | | | $D[i+2-j, j] \leftarrow logD(D[i+1-j, j+1], D[i+1-j, j])$

21 | | **end**

22 | **end**

23

24 | // Accept or reject x based on the computed log derivatives;

25 | // Here $D[i,0] = x_0^{(i)}$, $D[i,1] = x_1^{(i)}$, $D[i,2] = x_2^{(i)}$;

26 | $u \leftarrow D[i,0]$, $v \leftarrow D[i,1]$, $w \leftarrow D[i,2]$;

27 | **if** $u = v = w$ **then**

28 | | ACCEPT iff $\lfloor \frac{D[i-1,0]}{2^{abs(u)}} \rfloor$ is odd

29 | **end**

30 | **if** $u < v \geq w$ *or* $u = v > w$ **then** ACCEPT;

31 | **if** $u > v \leq w$ *or* $u = v < w$ **then** REJECT;

32 **end**

33

34 ACCEPT iff *isEvenStage* ;

Algorithm 1. The Splitting Algorithm $\mathcal{S}_c^g$ based on log-derivative sequence.

algorithm $\mathcal{S}_c^g$ uses an array D to compute and store the log derivatives of the sequence X. More precisely, every time the execution of the algorithm $\mathcal{S}_c^g$ reaches Line 26, $D[p,q]$ will contain the value of $x_q^{(p)}$, the p-th order log derivative of X at x_q, for each $p \in [0,i]$ and $q \in [0, i+2]$, where $p + q \leq i + 2$. In particular, $D[i,0]$, $D[i,1]$ and $D[i,2]$ will store the values of $x_0^{(i)}$, $x_1^{(i)}$ and $x_2^{(i)}$, in that order.

We consider the following cases in that order so that the proof for Case e assumes that none of the cases $1, 2, \cdots, e-1$ holds. We also assume that $t^{-1}(|x|)$ is even.

Case 1: There exists $j_1 \in [1, m-1]$, where $t^{-1}(|x_{j_1}|) < t^{-1}(|x_{j_1+1}|)$. Let j_1 be the smallest such number. Then S_c^g accepts $x_{j_1} = x_{j_1}^{(0)}$ at Line 4 if and only if $t^{-1}(|x_{j_1}|)$ is odd.

Subcase 1(a): $t^{-1}(|x_{j_1-1}|) \geq t^{-1}(|x_{j_1}|)$. In this subcase S_c^g accepts $x_{j_1-1} = x_{j_1-1}^{(0)}$ at Line 16 if and only if $t^{-1}(|x_{j_1}|)$ is even. It follows that S_c^g accepts x_{j_1-1} if and only if S_c^g rejects x_{j_1}.

Subcase 1(b): $t^{-1}(|x_{j_1-1}|) < t^{-1}(|x_{j_1}|)$. In this subcase, S_c^g accepts $x_{j_1-1} = x_{j_1-1}^{(0)}$ at Line 4 if and only if $t^{-1}(|x_{j_1-1}|)$ is odd. Note that for each $j \in [1, m]$, it holds that

$$|x_j^{(0)}| = |x_j| \leq |x_{j-1}|^l + l = |x_{j-1}^{(0)}|^l + l$$

since $x_j = g(x_{j-1})$. This implies that $t^{-1}(|x_j|) \leq t^{-1}(|x_{j-1}|) + 1$ for each $j \in [1, m]$. Hence, it follows from the hypothesis of this subcase that $t^{-1}(|x_{j_1-1}| = t^{-1}(|x_{j_1}|) - 1$. Then we derive again in this case that S_c^g accepts x_{j_1-1} if and only if S_c^g rejects x_{j_1}.

Let $j_2 = j_1 - 1$. Then we have shown in both subcases (a) and (b) of Case 1 that S_c^g accepts x_{j_2} if and only if S_c^g rejects x_{j_1}, where $\{j_1, j_2\} \subseteq [0, m]$.

If Case 1 does not hold, then

$$\forall j \in [1, m-1], t^{-1}(|x_j|) \geq t^{-1}(|x_{j+1}|) \tag{1}$$
$$\forall j \in [0, m], t^{-1}(|x_j|) \leq t^{-1}(|x_1|) \leq t^{-1}(|x|) + 1 \tag{2}$$

We assume that statements (1) and (2) is true for all the subsequent cases.

Case 2.i: For each $i \in [0, c]$, we consider Case 2.i in the increasing order of i, which consists of the following subcases 2.$i(a-d)$.

If $i = 0$, let Z_0 be $X^{(0)} \backslash \{x\} = X \backslash \{x\}$, which is a consecutive subsequence of $X^{(0)} = X$ with start index $s_0 = 1$ and ending index $t_0 = m$, respectively. Otherwise, Z_i is a consecutive subsequence of $X^{(i)}$ constructed in Case 2.$(i-1)(c)$ or 2.$(i-1)(d)$ if applicable, with start index s_i and ending index t_i. Note the following statement:

Statement 3. *If Cases 2.i needs to be considered, then S_c does not accept or reject any string x_j, where $s_i \leq j \leq t_i - 2$, before the i-th iteration of the outer loop.*

Statement 3 is true for $i = 0$ in light of Case 1: We will consider Case 2.0 only if S_c^g does not accept any $x_j = x_j^{(0)}$ for $1 \leq j \leq m-2$ at Line 4. We will see that Statement 3 holds true through all cases 2.i until the smallest i where S_c^g makes output during the i-th iteration of the outer loop on $x_j = x_j^{(0)}$ for some $j \in [s_i, t_i - 2]$. In addition, if the execution of S_c^g reaches Line 26 during the i-th iteration, then none of the elements $u = x_0^{(i)}$, $v = x_1^{(i)}$, and $w = x_2^{(i)}$ is ∞, for otherwise two elements among $\{x_j^{(i-1)} \mid 0 \le j \le 3\}$ must equal each other since $x_j^{(i)} = logD(x_{j+1}^{(i-1)}, x_j^{(i-1)})$ for every $j \in [0, m_i - i]$. That will make $\mathcal{S}_c^g$ halt in the $(i-1)$-st iteration of the outer loop already, at lines 29–31.

Subcase 2.i(a): Z_i contains 5 consecutive equal elements $\{x_j^{(i)}\}_{a \le j \le a+4}$, where $a \in [s_i, t_i - 4]$. Then for $a \le j \le a + 2$, $\mathcal{S}_c^g$ accepts x_j at Line 29 if and only if $\lfloor \frac{x_j^{(i-1)}}{2^{abs(d)+1}} \rfloor$ is odd.

Define

$$E_i = \left\{ \left\lfloor \frac{x_j^{(i-1)}}{2^{abs(d)+1}} \right\rfloor \right\}_{r \le j \le r+2} \quad , \text{ where } d = x_r^{(i)}.$$

Note that $x_j^{(i)} = logD(x_{j+1}^{(i-1)}, x_j^{(i-1)})$ for each $j \in [a, a+2]$. Then by Lemma 1, E_i contains at least one even number and one odd number. Therefore, $\mathcal{S}_c^g$ accepts at least one string x_{j_1} and rejects at least one string x_{j_2}, where $j_1, j_2 \in [a, a+2]$.

Now we assume that there don't exist 5 consecutive equal strings in Z_i.

Subcase 2.i(b): The sequence Z_i contains two elements $x_{j_1}^{(i)}$ and $x_{j_2}^{(i)}$, where $\{j_1, j_2\} \subseteq [s_i, t_i - 2]$ and

$- \quad x_{j_1}^{(i)} < x_{j_1+1}^{(i)} \ge x_{j_1+2}^{(i)}$ or $x_{j_1}^{(i)} = x_{j_1+1}^{(i)} < x_{j_1+2}^{(i)}$, and
$- \quad x_{j_2}^{(i)} > x_{j_2+1}^{(i)} \le x_{j_2+2}^{(i)}$ or $x_{j_2}^{(i)} = x_{j_2+1}^{(i)} > x_{j_2+2}^{(i)}$.

In this subcase Algorithm $\mathcal{S}_c^g$ accepts x_{j_1} at Line 30 and rejects x_{j_2} at Line 31.

Subcase 2.i(c): There does not exist $j_1 \in [s_i, t_i - 2]$ as required by Subcase 2.i(b), then both of the following hold for each string $j \in [s_i, t_i - 2]$:

$-$ If $x_j^{(i)} < x_{j+1}^{(i)}$, then $x_{j+2}^{(i)} < x_{j+1}^{(i)}$.
$-$ If $x_j^{(i)} = x_{j+1}^{(i)}$, then $x_{j+2}^{(i)} \ge x_{j+1}^{(i)}$.

This shows that Z_i is of the following forms, where $s_i \le s_i' \le t_i' \le t_i$:

$$x_{s_i}^{(i)} < x_{s_i+1}^{(i)} < \cdots < x_{s_i'}^{(i)} = x_{s_i'+1}^{(i)} \cdots = x_{t_i'}^{(i)} < x_{t_i'+1}^{(i)} < \cdots < x_{t_i}^{(i)} \tag{3}$$

Hence, both $Y_{i_1} = \{x_{s_i}^{(i)}, x_{s_i+1}^{(i)}, \cdots, x_{s_i'}^{(i)}\}$ and $Y_{i_2} = \{x_{t_i'}^{(i)}, x_{t_i'+1}^{(i)}, \cdots < x_{t_i}^{(i)}\}$ are strictly monotonic consecutive subsequences of Z_i. We set $Y_i = Y_{i_1}$ if $|Y_{i_1}| \ge |Y_{i_2}|$ and $Y_i = Y_{i_2}$ otherwise. Note that Y_i is a consecutive subsequence of $Y_{i-1}^{(1)}$, the log derivative sequence of Y_{i-1}.

If $i = 0$, Subcase 2.i(a) does not apply since $x_j \ne x_{j-1}$ for every $j \in [1, m]$. Consequently $|Y_0| \ge m_0 = \lceil m/2 \rceil$. Otherwise, $t_i' - s_i' \le 4$ due to Subcase 2.i(a) and Eq. (3). Assume $|Y_{i-1}| \ge m_{i-1}$. Then $|Z_i| = |Y_{i-1}| - 1$. Hence, $|Y_i| \ge m_i = \lceil ((m_{i-1} - 1) - 3)/2 \rceil = \lceil m_{i-1}/2 \rceil - 2$.

Subcase 2.i(d): There does not exist $j_2 \in [0, m_i - 2]$ as required by Subcase 2.i(b). This subcase is symmetric to Subcase 2.i(c). We again obtain a strictly monotonic consecutive subsequence Y_i of length at least m_i within Z_i. We let Y_i be that subsequence with starting index s_i and ending index t_i. Again, Y_i is a subsequence of the log derivative sequence of Y_{i-1}.

For both subcases 2.i(c) and 2.i(d) we define $Z_{i+1} = Y_i^{(1)}$ and proceed to Case 2.$(i+1)$. Clearly Z_{i+1} is a consecutive subsequence of $X^{(i+1)}$ since Y_i is a consecutive subsequence of $X^{(i)}$. Also, the start and ending indices of Z_{i+1} are $s_{i+1} = s_i$ and $t_{i+1} = s_i' - 1$, or $s_{i+1} = t_i'$ and $t_{i+1} = t_i - 1$, respectively, depending on how Y_i is formed in Case 2.i(c) or Case 2.i(d).

Summary of Case 2.i: In Case 2.i, we either find $\{j_1, j_2\} \subseteq [0, m-2]$, where $\mathcal{S}_c^g$ accepts x_{j_1} if and only if $\mathcal{S}_c^g$ rejects x_{j_2} (subcases 2.i(a) and 2.i(b)) or we obtain a strictly monotonic and consecutive subsequence Y_i of both $X^{(i)}$ and Y_{i-1}, where $|Y_i| \geq m_i$.

If none of the subcases 2.i(a) and 2.i(b) apply for all $0 \leq i \leq c$, then we arrive at a set of sequences Y_i, where

- Y_0 is a strictly monotonic and consecutive subsequence of $X^{(0)}$ with $|Y_0| \geq m_0 = \lceil m/2 \rceil$, and
- for each $i \in [1, c]$, Y_i is a strictly monotonic and consecutive subsequence of both $X^{(i)}$ and $Y_{i-1}^{(1)}$ with $|Y_i| \geq m_i = \lceil m_{i-1}/2 \rceil - 2$

Note that the length of each string in $X^{(0)} = X$ is at most $n^l + l$ due to Eq. 2. Hence, every element in Y_1 has an absolute value no more than $\log(2 \cdot 2^{(n^l + l + 1)}) = O(n^l)$ for sufficiently large n. This in turn implies that every element in Y_2 has an absolute value no more than $O(\log n)$ using the same argument. Continuing applying this argument on Y_3, Y_4,... through Y_c, we obtain that every element in Y_c for $c \geq 2$ should have an absolute value no more than $O(\log^{(c-1)} n)$.

However, a simple induction proof shows that $m_c = \Omega(\log^{(c)} n)$. The maximal absolute value of elements in Y_c is no less than $m_c/2 = \Omega(\log^{(c)} n)$, since Y_c is a strictly monotonic sequence of integers of length at least m_c. This is a contradiction to the argument above that every element in Y_c for $c \geq 2$ should have an absolute value no more than $O(\log^{(c-1)} n)$. Hence, either subcase 2.i(a) or 2.i(b) must hold for some $i \in [0, c]$ if Case 1 does not hold. This will ensure that $\mathcal{S}_c^g$ accepts $x_{j_1} = g^{(j_1)}(x)$ if and only if $\mathcal{S}_c^g$ rejects $x_{j_2} = g^{(j_2)}(x)$ for some $\{j_1, j_2\} \subseteq [0, m]$. This proves (ii) and (iii) of Theorem 2.

Regarding Condition (i) of the theorem, we observe that if Case 1 applies then $j_x \leq j_1$, where j_1 is the smallest number $j \in [1, m-1]$ such that $t^{-1}(|x_j|) < t^{-1}(|x_{j+1}|)$. This implies that $t^{-1}(|x_j|) \leq t^{-1}(|x_1|)$, for each $j \in [1, j_1]$. Hence, it follows that $|x_j| \leq |x_1|^l + l$ for each $j \in [1, j_x]$. Note that $|x_1| \leq |x|^l + l$ since $x_1 = g(x_0) = g(x)$. So for each $j \in [0, j_x]$, $|g^{(j)}(x)| \leq |x|^l + l$.

Now assume that Case 1 does not apply. Then by Eq. (2), for each $j \in [0, m]$, it holds that $t^{-1}(|x_j|) \leq t^{-1}(|x_1|) \leq t^{-1}(|x|) + 1$, or equivalently, $|g^{(j)}(x)| = |x_j| \leq (|x|^l + l)^l + l \leq 2|x|^{2l}$. Therefore, Condition (i) holds for $r(n) = 2n^{2l}$.

This finishes the proof of Theorem 2. $\square$

With Theorem 2 we can now establish the rest of our main results.

Theorem 4. *For every $k \in \mathbb{N}^+$ and positive integer $c \geq 2$, if a non-trivial language L is $\leq^p_{k\text{-}dtt}$-autoreducible, then L is weakly $\leq^p_{k^{O(2^c \log^{(c-1)} n)}\text{-}tt}$-mitotic.*

Proof. We assume that $k \geq 2$ since it is already known that a non-trivial language is $\leq^p_{1\text{-}tt}$ autoreducible if and only if it is $\leq^p_{1\text{-}tt}$ mitotic [9].

Let L be a non-trivial and $\leq^p_{k\text{-}dtt}$-autoreducible language. Then there exists a polynomial-time computable function f, where $f(w) = \langle u_0, u_1, \ldots, u_{k-1} \rangle$ for each $x \in \Sigma^*$ such that

- for each $i \in [0, k-1]$, $x \neq u_i$, and
- $x \in L$ if and only if $\exists i \in [0, k-1]$, $u_i \in L$.

When there is no confusion we also use $f(x)$ to denote the set $\{u_0, u_1, \ldots, u_{k-1}\}$. For a set of strings W, let $lex-min\ (W)$ denote the *lexicographically least* string in W.

Now define

$$g(w) = \begin{cases} lex\text{-}min\ (f(x)) & \text{if } x \notin L \\ lex\text{-}min\ (f(x) \cap L) & \text{if } x \in L \end{cases}$$

It's clear that function g is polynomial bounded and for every $x \in \Sigma^*$, $g(x) \neq x$. Hence we can apply Theorem 2 on function g and any positive integer $c \geq 2$ to obtain an algorithm $\mathcal{S}^g_c$ and a polynomial r that satisfies all the conditions as stated in Theorem 2. Let $S = L(\mathcal{S}^g_c)$. Then one can argue that S can be used to show that

$$S \cap L \equiv^p_{k^{O(c)}\text{-}tt} L \equiv^p_{k^{O(2^c \log^{(c-1)} n)}\text{-}tt} \overline{S} \cap L.$$

$\square$

Using a similar argument we prove the same result for k-ctt autoreducible sets:

Theorem 5. *For every $k \in \mathbb{N}^+$ and integer $c \geq 2$, if a non-trivial language L is $\leq^p_{k\text{-}ctt}$-autoreducible, then L is weakly $\leq^p_{k^{O(2^c \log^{(c-1)} n)}\text{-}tt}$-mitotic.*

In light that the k-dtt complete sets of many common complexity classes have been proven to be k-dtt autoreducible for $k \geq 2$ [5,8] we have the following corollary providing a better understanding of (weak) mitoticity of complete sets in complexity theory.

Corollary 1. *For every integer $k \geq 2$ and $c \geq 2$, every k-dtt complete set for the following classes is weakly $k^{O(2^c \log^{(c-1)} n)}$-dtt mitotic:*

- PSPACE,
- *the levels Σ^P_h, Π^P_h and Π^P_h of the polynomial time hierarchy for $h \geq 2$*
- 1NP,
- *the levels of the Boolean hierarchy over NP,*
- *the levels of the MODPH hierarchy, and*
- NEXP.

Proof. Glaßer et al. [8] showed that all k-dtt complete sets of the complexity classes listed above except NEXP are k-dtt autoreducible. In addition, Glaßer et al. [5] recently showed that any k-dtt complete set for NEXP is k-dtt autoreducible. The corollary follows immediately by applying Theorem 4.

$\square$

References

1. Ambos-Spies, K.: On the structure of the polynomial time degrees of recursive sets. Habilitationsschrift, Zur Erlangung der Venia Legendi Für das Fach Informatik an der Abteilung Informatik der Universität Dortmund, September 1984
2. Buhrman, H., Fortnow, L., van Melkebeek, D., Torenvliet, L.: Using autoreducibility to separate complexity classes. SIAM J. Comput. **29**(5), 1497–1520 (2000)
3. Buhrman, H., Torenvliet, L.: A Post's program for complexity theory. Bull. EATCS **85**, 41–51 (2005)
4. Glaßer, C., Selman, A., Travers, S., Zhang, L.: Non-mitotic sets. Theoret. Comput. Sci. **410**(21–23), 2011–2033 (2009)
5. Glaßer, C., Nguyen, D.T., Reitwießner, C., Selman, A.L., Witek, M.: Autoreducibility of complete sets for log-space and polynomial-time reductions. In: Fomin, F.V., Freivalds, R., Kwiatkowska, M., Peleg, D. (eds.) ICALP 2013. LNCS, vol. 7965, pp. 473–484. Springer, Heidelberg (2013). https://doi.org/10.1007/978-3-642-39206-1_40
6. Glaßer, C., Nguyen, D.T., Selman, A.L., Witek, M.: Introduction to autoreducibility and mitoticity. In: Day, A., Fellows, M., Greenberg, N., Khoussainov, B., Melnikov, A., Rosamond, F. (eds.) Computability and Complexity. LNCS, vol. 10010, pp. 56–78. Springer, Cham (2017). https://doi.org/10.1007/978-3-319-50062-1_5
7. Glaßer, C., Ogihara, M., Pavan, A., Selman, A., Zhang, L.: Autoreducibility and mitoticity. ACM SIGACT News **40**(3), 60–76 (2009)
8. Glaßer, C., Ogihara, M., Pavan, A., Selman, A.L., Zhang, L.: Autoreducibility, mitoticity, and immunity. J. Comput. Syst. Sci. **73**, 735–754 (2007)
9. Glaßer, C., Pavan, A., Selman, A., Zhang, L.: Splitting NP-complete sets. SIAM J. Comput. **37**(5), 1517–1535 (2008)
10. Hemaspaandra, L., Ogihara, M.: The Complexity Theory Companion. Springer, Heidelberg (2002). https://doi.org/10.1007/978-3-662-04880-1
11. Homer, S., Selman, A.: Computability and Complexity Theory. Texts in Computer Science, 2nd edn. Springer, New York (2011). https://doi.org/10.1007/978-1-4614-0682-2
12. Trakhtenbrot, B.: On autoreducibility. Dokl. Akad. Nauk SSSR **192**(6), 1224–1227 (1970). Transl. Soviet Math. Dokl. **11**(3), 814–817 (1790)
13. Yao, A.: Coherent functions and program checkers. In: Proceedings of the 22nd Annual Symposium on Theory of Computing, pp. 89–94 (1990)

Approximation Algorithms
for Two-Machine Flow-Shop Scheduling
with a Conflict Graph

Yinhui Cai[1], Guangting Chen[2], Yong Chen[3(✉)], Randy Goebel[4],
Guohui Lin[4(✉)], Longcheng Liu[4,5], and An Zhang[3]

[1] School of Sciences, Hangzhou Dianzi University, Hangzhou, China
[2] Taizhou University, Taizhou, Zhejiang, China
`gtchen@hdu.edu.cn`
[3] Department of Mathematics, Hangzhou Dianzi University, Hangzhou, China
`{chenyong,anzhang}@hdu.edu.cn`
[4] Department of Computing Science, University of Alberta, Edmonton, AB, Canada
`{rgoebel,guohui}@ualberta.ca`
[5] School of Mathematical Sciences, Xiamen University, Xiamen, China
`longchengliu@xmu.edu.cn`

Abstract. Path cover is a well-known intractable problem whose goal
is to find a minimum number of vertex disjoint paths in a given graph to
cover all the vertices. We show that a variant, where the objective function is not the number of paths but the number of length-0 paths (that is,
isolated vertices), turns out to be polynomial-time solvable. We further
show that another variant, where the objective function is the total number of length-0 and length-1 paths, is also polynomial-time solvable. Both
variants find applications in approximating the two-machine flow-shop
scheduling problem in which job processing constraints are formulated as
a conflict graph. For the unit jobs, we present a 4/3-approximation algorithm for the scheduling problem with an arbitrary conflict graph, based
on the exact algorithm for the variants of the path cover problem. For
arbitrary jobs where the conflict graph is the union of two disjoint cliques
(i.e., all the jobs can be partitioned into two groups such that the jobs in
a group are pairwise conflicting), we present a simple 3/2-approximation
algorithm.

Keywords: Flow-shop scheduling · Conflict graph · b-matching
Path cover · Approximation algorithm

1 Introduction

Scheduling is a well established research area that finds numerous applications
in modern manufacturing industries and in operations research at large. All

---

Y. Cai and L. Liu—Co-first authors.

© Springer International Publishing AG, part of Springer Nature 2018
L. Wang and D. Zhu (Eds.): COCOON 2018, LNCS 10976, pp. 205–217, 2018.
https://doi.org/10.1007/978-3-319-94776-1_18

scheduling problems modeling real-life applications have at least two components: the machines and the jobs. One intensively studied problem focuses on scheduling constraints are imposed between a machine and a job, such as a time interval during which the job is allowed to be processed *nonpreemptively* on the machine, while the machines are considered as independent from each other, so are the jobs. For example, the parallel machine scheduling (the *multiprocessor scheduling* in [15]) is one of the first studied problems, denoted as $Pm \parallel C_{\max}$ in the three-field notation [20], in which a set of jobs each needs to be processed by one of the m given identical machines, with the goal to minimize the maximum job completion time (called the *makespan*); the m-machine flow-shop scheduling (the *flow-shop scheduling* in [15]) is another early-studied problem, denoted as $Fm \parallel C_{\max}$, in which a set of jobs each needs to be processed by all the m given machines in the same sequential order, with the goal to minimize the makespan.

In another category of scheduling problems, additional but limited resources are required for the machines to process the jobs [13]. The resources are renewable but normally non-sharable in practice; the jobs competing for the same resource have to be processed at different time if their total demand for a certain resource exceeds the supply. Scheduling with resource constraints [13,14] or scheduling with conflicts (SwC) [11] also finds numerous applications [3,9,22] and has attracted as much attention as the non-constrained counterpart. In this paper, we use SwC to refer to nonpreemptive scheduling problems with additional constraints or conflicting relationships among the jobs to disallow them to be processed concurrently on different machines. We note that SwC is also presented as the scheduling with agreements (SwA), in which a subset of jobs can be processed concurrently on different machines if and only if they agree with each other [4,5]. While in the most general scenario a conflict could involve multiple jobs, in this paper we consider only those conflicts each involves two jobs and consequently all the conflicts under consideration can be presented as a *conflict graph* $G = (V, E)$, where V is the set of jobs and an edge $e = (J_{j_1}, J_{j_2}) \in E$ represents a conflicting pair such that the two jobs J_{j_1} and J_{j_2} cannot be processed concurrently on different machines in any feasible schedule.

Extending the three-field notation [20], the parallel machine SwC with a conflict graph $G = (V, E)$ (also abbreviated as SCI in the literature) [11] is denoted as $Pm \mid G = (V,E), p_j \mid C_{\max}$, where the first field Pm tells that there are m parallel identical machines, the second field describes the conflict graph $G = (V, E)$ over the set V of all the jobs, where the job J_j requires a non-preemptive processing time of p_j on any machine, and the last field specifies the objective function to minimize the makespan $C_{\max}$. One clearly sees when $E = \emptyset$, $Pm \mid G = (V,E), p_j \mid C_{\max}$ reduces to the classical multiprocessor scheduling $Pm \parallel C_{\max}$, which is already NP-hard for $m \geq 2$ [15]. Indeed, with m either a given constant or part of input, $Pm \mid G = (V,E), p_j \mid C_{\max}$ is more difficult to approximate than the classical multiprocessor scheduling, as $P3 \mid G = (V,E), p_j = 1 \mid C_{\max}$ and $P2 \mid G = (V,E), p_j \in \{1,2,3,4\} \mid C_{\max}$ are APX-hard [11,26]. There is a rich line of research to consider the unit jobs (that is, $p_j = 1$) and/or to consider certain special classes of conflict graphs. The interested reader might see [11] and the references therein.

In the general m-machine (also called m-stage) flow-shop [15] denoted as $Fm \parallel C_{\max}$, there are $m \geq 2$ machines $M_1, M_2, \ldots, M_m$, a set V of jobs each job J_j needs to be processed through $M_1, M_2, \ldots, M_m$ sequentially with processing times $p_{1j}, p_{2j}, \ldots, p_{mj}$ respectively. When $m = 2$, the two-machine flow-shop problem is polynomial time solvable, by Johnson's algorithm [23]; the m-machine flow-shop problem, when $m \geq 3$, is *strongly* NP-hard [16]. After several efforts [10,16,19,23], Hall presented a polynomial-time approximation scheme (PTAS) for the m-machine flow-shop problem, for any fixed integer $m \geq 3$ [21]. When m is part of input (*i.e.*, an arbitrary integer), there is no known constant ratio approximation algorithm, and the problem cannot be approximated within 1.25, unless P = NP [33].

The m-machine flow-shop SwC was first studied in 1980's. Blazewicz et al. [8] considered multiple resource characteristics including the number of resource types, resource availabilities and resource requirements; they expanded the middle field of the three-field notation to express these resource characteristics, for which the conflict relationships are modeled by complex structures such as hypergraphs. At the end, they proved complexity results for several variants in which either the conflict relationships are simple enough or only the unit jobs are considered. Further studies on more variants can be found in [6,7,28–30]. In this paper, we consider those conflicts each involves only two jobs such that all the conflicts under consideration can be presented as a conflict graph $G = (V, E)$. The m-machine flow-shop scheduling with a conflict graph $G = (V, E)$ is denoted as $Fm \mid G = (V, E), p_{ij} \mid C_{\max}$. We remark that our notation is slightly different from the one introduced by Blazewicz et al. [8], which uses a prefix "res" in the middle field for describing the resource characteristics.

Several applications of the m-machine flow-shop scheduling with a conflict graph have been mentioned in the literature. In a typical example of scheduling medical tests in an outpatient health care facility where each patient (the job) needs to do a sequence of m tests (the machines), a patient must be accompanied by their doctor during a test, so two patients under the care of the same doctor cannot go for tests simultaneously. That is, two patients of the same doctor are conflicting to each other, and all the conflicts can be effectively described as a graph $G = (V, E)$, where V is the set of all the patients and an edge represents a conflicting pair of patients.

In two recent papers [31,32], Tellache and Boudhar studied the problem $F2 \mid G = (V, E), p_{ij} \mid C_{\max}$, which they denote as FSC. In [32], the authors summarized and proved several complexity results; to name a few, $F2 \mid G = (V, E), p_{ij} \mid C_{\max}$ is strongly NP-hard when $G = (V, E)$ is the complement of a complete split graph [8,32] (that is, G is the union of a clique and an independent set), $F2 \mid G = (V, E), p_{ij} \mid C_{\max}$ is weakly NP-hard when $G = (V, E)$ is the complement of a complete bipartite graph [32] (that is, G is the union of two disjoint cliques), and for an arbitrary conflict graph $G = (V, E)$, $F2 \mid G = (V, E), p_{ij} = 1 \mid C_{\max}$ is strongly NP-hard [32]. In [31], the authors proposed three mixed-integer linear programming models and a branch and bound algorithm to solve the last variant $F2 \mid G = (V, E), p_{ij} = 1 \mid C_{\max}$ exactly; their empirical study shows that the branch and bound algorithm outperforms and can solve instances of up to $20,000$ jobs.

In this paper, we pursue approximation algorithms with provable performance for the NP-hard variants of the two-machine flow-shop scheduling with a conflict graph. In Sect. 2, we present a $4/3$-approximation algorithm for the strongly NP-hard problem $F2 \mid G = (V, E), p_{ij} = 1 \mid C_{\max}$ for the unit jobs with an arbitrary conflict graph. In Sect. 3, we present a simple $3/2$-approximation algorithm for the weakly NP-hard problem $F2 \mid G = K_\ell \cup K_{n-\ell}, p_{ij} \mid C_{\max}$ for arbitrary jobs with a conflict graph that is the union of two disjoint cliques. Some concluding remarks are provided in Sect. 4.

2   Approximating $F2 \mid G = (V, E), p_{ij} = 1 \mid C_{\max}$

Tellache and Boudhar proved that $F2 \mid G = (V, E), p_{ij} = 1 \mid C_{\max}$ is strongly NP-hard by a reduction from the well known Hamiltonian path problem, which is strongly NP-complete [15]. Furthermore, they remarked that $F2 \mid G = (V, E), p_{ij} = 1 \mid C_{\max}$ has a feasible schedule of makespan $C_{\max} = n + k$ if and only if the complement $\overline{G}$ of the conflict graph G, called the agreement graph, has a path cover of size k (that is, a collection of k vertex-disjoint paths that covers all the vertices of the graph $\overline{G}$), where n is the number of jobs (or vertices) in the instance. This way, $F2 \mid G = (V, E), p_{ij} = 1 \mid C_{\max}$ is polynomially equivalent to the PATH COVER problem, which is NP-hard even on some special classes of graphs including planar graphs [17], bipartite graphs [18], chordal graphs [18], chordal bipartite graphs [24] and strongly chordal graphs [24]. In terms of approximability, to the best of our knowledge there is no $o(n)$-approximation algorithm for the Path Cover problem.

We begin with some terminologies. The conflict graphs considered in this paper are all simple graphs. All paths and cycles in a graph are also simple. The number of edges on a path/cycle defines the length of the path/cycle. A length-k path/cycle is also called a k-path/cycle for short. Note that a single vertex is regarded as a 0-path, while a cycle has length at least 3. For an integer $b \geq 1$, a b-matching of a graph is a spanning subgraph in which every vertex has degree no greater than b; a maximum b-matching is a b-matching that contains the maximum number of edges. A maximum b-matching of a graph can be computed in $O(m^2 \log n \log b)$-time, where n and m are the number of vertices and the number of edges in the graph, respectively [12]. Clearly, a graph could have multiple distinct maximum b-matchings.

Given a graph, a path cover is a collection of vertex-disjoint paths in the graph that covers all the vertices, and the size of the path cover is the number of paths therein. The PATH COVER problem is to find a path cover of a given graph of the minimum size, and the well known Hamiltonian path problem is to decide whether a given graph has a path cover of size 1. Many variants of the PATH COVER problem have been studied in the literature [1,2,25,27]. We mentioned earlier that Tellache and Boudhar proved that $F2 \mid G = (V, E), p_{ij} = 1 \mid C_{\max}$ is polynomially equivalent to the PATH COVER problem, but to the best of our knowledge there is no approximation algorithm designed for $F2 \mid G = (V, E), p_{ij} = 1 \mid C_{\max}$. Nevertheless, one easily sees that, since $F2 \mid G = (V, E), p_{ij} = 1 \mid C_{\max}$ has a feasible schedule of makespan $C_{\max} - n + k$ if and only if the complement $\overline{G}$ of the conflict graph G has a path cover of size k, a trivial algorithm simply processing the jobs one by one (each on the first machine M_1 and then on the second machine M_2) produces a schedule of makespan $C_{\max} = 2n$, and thus is a 2-approximation algorithm.

In this section, we will design two approximation algorithms with improved performance ratios for $F2 \mid G = (V, E), p_{ij} = 1 \mid C_{\max}$. These two approximation algorithms are based on our polynomial time exact algorithms for two variants of the PATH COVER problem, respectively. We start with the first variant called the *Path Cover with the minimum number of 0-paths*, in which we are given a graph and we aim to find a path cover that contains the minimum number of 0-paths. In the second variant called the *Path Cover with the minimum number of $\{0, 1\}$-paths*, we aim to find a path cover that contains the minimum total number of 0-paths and 1-paths. We remark that in both variants, we do not care about the size of the path cover.

2.1 Path Cover with the Minimum Number of 0-Paths

Recall that in this variant of the PATH COVER problem, given a graph, we aim to find a path cover that contains the minimum number of 0-paths. The given graph is the complement $\overline{G} = (V, \overline{E})$ of the conflict graph $G = (V, E)$ in $F2 \mid G = (V, E), p_{ij} = 1 \mid C_{\max}$. We next present a polynomial time algorithm that finds for $\overline{G}$ a path cover that contains the minimum number of 0-paths.

In the first step, we apply any polynomial time algorithm to find a maximum 2-matching in $\overline{G}$, denoted as M; recall that this can be done in $O(m^2 \log n)$-time, where $n = |V|$ and $m = |\overline{E}|$. M is a collection of vertex-disjoint paths and cycles; let $\mathcal{P}_0$ ($\mathcal{P}_1$, $\mathcal{P}_2$, $\mathcal{P}_{\geq 3}$, $\mathcal{C}$, respectively) denote the sub-collection of 0-paths (1-paths, 2-paths, paths of length at least 3, cycles, respectively) in M. That is, $M = \mathcal{P}_0 \cup \mathcal{P}_1 \cup \mathcal{P}_2 \cup \mathcal{P}_{\geq 3} \cup \mathcal{C}$.

Clearly, if $\mathcal{P}_0 = \emptyset$, then we have a path cover containing no 0-paths after removing one edge per cycle in $\mathcal{C}$. In the following discussion we assume the existence of a 0-path, which is often called a *singleton*. We also call an ending vertex of a k-path with $k \geq 1$ as an endpoint for simplicity. The following lemma is trivial due to the edge maximality of M.

Lemma 1. *All the singletons and endpoints in the maximum 2-matching M are pairwise non-adjacent to each other in the underlying graph $\overline{G}$.*

Let v_0 be a singleton. If v_0 is adjacent to a vertex v_1 on a cycle of $\mathcal{C}$ in the underlying graph $\overline{G}$, then we may delete a cycle-edge incident at v_1 from M while add the edge (v_0, v_1) to M to achieve another maximum 2-matching with one less singleton. Similarly, if v_0 is adjacent to a vertex v_1 on a path of $\mathcal{P}_{\geq 3}$ (note that v_1 has to be an internal vertex on the path by Lemma 1) in the underlying graph $\overline{G}$, then we may delete a certain path-edge incident at v_1 from M while add the edge (v_0, v_1) to M to achieve another maximum 2-matching with one less singleton. In either of the above two cases, assume the edge deleted from M is (v_1, v_2); then we say the *alternating* path v_0-v_1-v_2 *saves the singleton* v_0.

In the general setting, in the underlying graph $\overline{G}$, v_0 is adjacent to the middle vertex v_1 of a 2-path P_1, one endpoint v_2 of P_1 is adjacent to the middle vertex v_3 of another 2-path P_2, one endpoint v_4 of P_2 is adjacent to the middle vertex v_5 of another 2-path P_3, and so on, one endpoint v_{2i-2} of P_{i-1} is adjacent to the middle vertex v_{2i-1} of another 2-path P_i, one endpoint v_{2i} of P_i is adjacent to a vertex v_{2i+1} of a cycle of $\mathcal{C}$ or a path of $\mathcal{P}_{\geq 3}$ (see an illustration in Fig. 1), on which the edge (v_{2i+1}, v_{2i+2}) is to be deleted. Then we may delete the edges $\{(v_{2j+1}, v_{2j+2}) \mid j = 0, 1, \ldots, i\}$ from M while add the edges $\{(v_{2j}, v_{2j+1}) \mid j = 0, 1, \ldots, i\}$ to M to achieve another maximum 2-matching with one less singleton; and we say the *alternating* path v_0-v_1-v_2-$\ldots$-v_{2i}-v_{2i+1}-v_{2i+2} *saves the singleton* v_0.

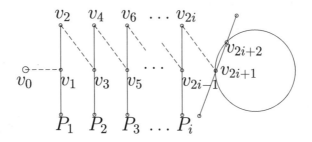

Fig. 1. An alternating path v_0-v_1-v_2-$\ldots$-v_{2i}-v_{2i+1}-v_{2i+2} that saves the singleton v_0, where the last two vertices are on a path of $\mathcal{P}_{\geq 3}$ or a cycle. In the figure, solid edges are in and dashed edges are outside of the maximum 2-matching M.

ALGORITHM A($\overline{G} = (V, \overline{E})$):

Step 1. Compute a maximum 2-matching M;

Step 2. repeatedly find an alternating path to save a singleton in M, till either no singleton exists or no alternating path is found;

Step 3. break cycles in M by removing one edge per cycle, and return the resulting path cover.

Fig. 2. A high-level description of ALGORITHM A for computing a path cover in the agreement graph $\overline{G} = (V, \overline{E})$.

The second step of the algorithm is to iteratively find a simple alternating path to save a singleton; it terminates when no alternating path is found. The resulting maximum 2-matching is still denoted as M.

In the last step, we break the cycles in M by deleting one edge per cycle to produce a path cover. Denote our algorithm as ALGORITHM A, of which a high-level description is provided in Fig. 2. We will prove in the next theorem that the path cover produced by ALGORITHM A contains the minimum number of 0-paths.

Theorem 1. ALGORITHM A *is an* $O(m^2 \log n)$-*time algorithm for computing a path cover with the minimum number of* 0-*paths in the agreement graph* $\overline{G}$.

2.2 Path Cover with the Minimum Number of $\{0, 1\}$-Paths

In this variant of the PATH COVER problem, given a graph, we aim to find a path cover that contains the minimum total number of 0-paths and 1-paths. Again, the given graph is the complement $\overline{G} = (V, \overline{E})$ of the conflict graph $G = (V, E)$ in $F2 \mid G = (V, E), p_{ij} = 1 \mid C_{\max}$. We next present a polynomial time algorithm called ALGORITHM B that finds for $\overline{G}$ such a path cover.

Recall that in ALGORITHM A for computing a path cover that contains the minimum number of 0-paths, an alternating path saving a singleton v_0 starts from the singleton v_0 and reaches a vertex v_{2i+1} on a path of $\mathcal{P}_{\geq 3}$ or on a cycle of $\mathcal{C}$ (see Fig. 1). If v_{2i+1} is on a cycle, then the last vertex v_{2i+2} can be any one of the two neighbors of v_{2i+1} on the cycle. If v_{2i+1} is on a k-path, then the last vertex v_{2i+2} is a *non-endpoint* neighbor of v_{2i+1} on the path (the existence is guaranteed by $k \geq 3$); and the reason why v_{2i+2} cannot be an endpoint is obvious since otherwise v_{2i+2} would be left as a new singleton after the edge swapping. In the current variant we want to minimize the total number of 0-paths and 1-paths; clearly v_{2i+2} cannot be an endpoint either and cannot even be the vertex adjacent to an endpoint, for the latter case because the edge swapping saves v_0 but leaves a new 1-path. To guarantee the existence of such vertex v_{2i+2}, the k-path must have $k \geq 4$, and if $k = 4$ then v_{2i+1} cannot be the middle vertex of the 4-path.

ALGORITHM B is in spirit similar to but in practice slightly more complex than ALGORITHM A, mostly because the definition of an alternating path saving a singleton or a 1-path is different, and slightly more complex.

In the first step of ALGORITHM B, we apply any polynomial time algorithm to find a maximum 2-matching M in $\overline{G}$. Let $\mathcal{P}_0$ ($\mathcal{P}_1$, $\mathcal{P}_2$, $\mathcal{P}_3$, $\mathcal{P}_4$, $\mathcal{P}_{\geq 5}$, $\mathcal{C}$, respectively) denote the sub-collection of 0-paths (1-paths, 2-paths, 3-paths, 4-paths, paths of length at least 5, cycles, respectively) in M. We also let $\mathcal{P}_{0,1} = \mathcal{P}_0 \cup \mathcal{P}_1$ denote the collection of all 0-paths (called *singletons*) and 1-paths in M.

Let $e_0 = (v_0, u_0)$ be an edge in M. In the sequel when we say e_0 is *adjacent* to a vertex v_1 in the graph $\overline{G}$, we mean v_1 is a different vertex (from v_0 and u_0) and at least one of v_0 and u_0 is adjacent to v_1; if both v_0 and u_0 are adjacent to v_1, then pick one (often arbitrarily) for the subsequent purposes. This way, we unify our treatment on singletons and 1-paths, for the reasons to be seen in the following. For ease of presentation, we use an *object* to refer to a vertex or an edge. Like in the last subsection, an ending vertex of a k-path with $k \geq 1$ or an ending edge of a k-path with $k \geq 2$ is called an *end-object* for simplicity.

Let v_0 be a singleton or $e_0 = (v_0, u_0)$ be a 1-path in M. In the underlying graph $\overline{G}$, if v_0 is adjacent to a vertex v_1 on a cycle of $\mathcal{C}$, or on a path of $\mathcal{P}_{\geq 5}$, or on a 4-path such that v_1 is not the middle vertex, then we may delete a certain edge incident at v_1 from M while add the edge (v_0, v_1) to M to achieve another maximum 2-matching with one less singleton if v_0 is a singleton or with one less 1-path. In either of the three cases, assume the edge deleted from M is (v_1, v_2); then we say the *alternating* path v_0-v_1-v_2 *saves* the singleton v_0 or the 1-path $e_0 = (v_0, u_0)$.

Analogously as in the last subsection, in the general setting, in the underlying graph $\overline{G}$, v_0 is adjacent to a vertex v_1 of a path $P_1 \in \mathcal{P}_{2,3,4}$ (if P_1 is a 4-path then v_1 has to be the middle vertex). Note that this vertex v_1 basically separates the two end-objects of the path P_1 — an analogue to the role of the middle vertex of a 2-path that separates the two endpoints of the 2-path. We say "an end-object of P_1 is adjacent to v_1 via v_2", to mean that if the end-object is a vertex then it is v_2, or if the end-object is an edge, then it is (v_2, u_2), with the edge (v_1, v_2) on the path P_1 either way (see an illustration in Fig. 3).

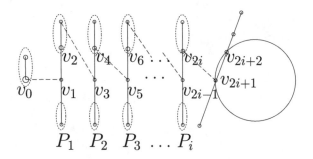

Fig. 3. An alternating path v_0-v_1-v_2-$\ldots$-v_{2i}-v_{2i+1}-v_{2i+2} that saves the singleton v_0 or the 1-path $e_0 = (v_0, u_0)$, where the last two vertices are on a cycle of $\mathcal{C}$, or on a path of $\mathcal{P}_{\geq 5}$, or on a 4-path such that v_{2i+1} is not the middle vertex. In the figure, solid edges are in the maximum 2-matching M, dashed edges are outside of M, and a dotted circle contains an object which is either a vertex or an edge.

Suppose one end-object of P_1, which is adjacent to v_1 via v_2, is adjacent to a vertex v_3 of another $P_2 \in \mathcal{P}_{2,3,4}$ (the same, if P_2 is a 4-path then v_3 has to be the middle vertex); one end-object of P_2, which is adjacent to v_3 via v_4, is adjacent to a vertex v_5 of another $P_3 \in \mathcal{P}_{2,3,4}$ (the same, if P_3 is a 4-path then v_5 has to be the middle vertex); and so on; one end-object of P_{i-1}, which is adjacent to v_{2i-3} via v_{2i-2}, is adjacent to a vertex v_{2i-1} of another $P_i \in \mathcal{P}_{2,3,4}$ (the same, if P_i is a 4-path then v_{2i-1} has to be the middle vertex); one end-object of P_i, which is adjacent to v_{2i-1} via v_{2i}, is adjacent to a vertex v_{2i+1} of a cycle of $\mathcal{C}$, or of a path of $\mathcal{P}_{\geq 5}$, or of a 4-path such that v_{2i+1} is not the middle vertex (see an illustration in Fig. 3), on which a certain edge (v_{2i+1}, v_{2i+2}) is to be deleted. Then we may delete the edges $\{(v_{2j+1}, v_{2j+2}) \mid j = 0, 1, \ldots, i\}$ from M while add the edges $\{(v_{2j}, v_{2j+1}) \mid j = 0, 1, \ldots, i\}$ to M to achieve another maximum

2-matching with one less singleton if v_0 is a singleton or with one less 1-path. We say the *alternating* path v_0-v_1-v_2-$\ldots$-v_{2i}-v_{2i+1}-v_{2i+2} *saves* the singleton v_0 or the 1-path $e_0 = (v_0, u_0)$. It is important to note that in this alternating path, the vertex v_2 "*represents*" the end-object of P_1, meaning that when the end-object is an edge, it is treated very the same as the vertex v_2.

The second step of the algorithm is to iteratively find a simple alternating path to save an object of $\mathcal{P}_{0,1}$; it terminates when no alternating path is found. The resulting maximum 2-matching is still denoted as M.

In the last step, we break the cycles in M by deleting one edge per cycle to produce a path cover. A high-level description of ALGORITHM B is similar to the one for ALGORITHM A shown in Fig. 2, replacing a singleton by an object of $\mathcal{P}_{0,1}$. We will prove in Theorem 2 that the path cover produced by ALGORITHM B contains the minimum total number of 0-paths and 1-paths.

Theorem 2. ALGORITHM B *is an* $O(m^2 \log n)$*-time algorithm for computing a path cover with the minimum total number of* 0*-paths and* 1*-paths in the agreement graph* $\overline{G} = (V, \overline{E})$.

Remark 1. The path cover produced by ALGORITHM B has the minimum total number of 0-paths and 1-paths in the agreement graph $\overline{G} = (V, \overline{E})$. One may run ALGORITHM A at the end of the second step of ALGORITHM B to achieve a path cover with the minimum total number of 0-paths and 1-paths, and with the minimum number of 0-paths. During the execution of ALGORITHM A, a singleton trades for a 1-path.

2.3   Approximation Algorithms for $F2 \mid G = (V, E), p_{ij} = 1 \mid C_{\max}$

Given an instance of the problem $F2 \mid G = (V, E), p_{ij} = 1 \mid C_{\max}$, where there are n unit jobs $V = \{J_1, J_2, \ldots, J_n\}$ to be processed on the two-machine flow-shop, with their conflict graph $G = (V, E)$, we want to find a schedule with a provable makespan.

For a k-path in the agreement graph $\overline{G} = (V, \overline{E})$, where $k \geq 0$, for example $P = J_1$-J_2-$\ldots$-J_k-J_{k+1}, we compose a sub-schedule π_P in which the machine M_1 continuously processes the jobs $J_1, J_2, \ldots, J_{k+1}$ in order, and the machine M_2 in one unit of time after M_1 continuously processes these jobs in the same order. The sub-makespan for the flow-shop to complete these $k + 1$ jobs is thus $k + 2$ (units of time). Let $M - \{P_1, P_2, \ldots, P_\ell\}$ be a path cover of size ℓ in the agreement graph $\overline{G}$. For each path P_i we use $|P_i|$ to denote its length and construct the sub-schedule π_{P_i} as above that has a sub-makespan of $|P_i| + 2$. We then concatenating these ℓ sub-schedules (in an arbitrary order) into a full schedule π, which clearly has a makespan $C_{\max}^\pi = \sum_{i=1}^{\ell}(|P_i| + 2) = n + \ell$.

On the other hand, given a schedule π, if two jobs J_{j_1} and J_{j_2} are processed concurrently on the two machines, then they have to be agreeing to each other and thus adjacent in the agreement graph $\overline{G}$; we select this edge (J_{j_1}, J_{j_2}). Note that one job can be processed concurrently with at most two other jobs as there are only two machines. Therefore, all the selected edges form into a number of vertex-disjoint paths in $\overline{G}$ (due to the flow-shop, no cycle is formed); these paths together with the vertices outside of the paths, which are the 0-paths, form a path cover for $\overline{G}$. Assuming without loss of generality that two machines cannot both idle at any time point, the makespan of the schedule is exactly the sum of the number of jobs and the number of paths.

We state this relationship between a feasible schedule and a path cover in the agreement graph $\overline{G}$ into the following lemma.

Lemma 2 [32]. *A feasible schedule π for the problem $F2 \mid G = (V, E), p_{ij} = 1 \mid C_{\max}$ one-to-one corresponds to a path cover M in the agreement graph $\overline{G}$, and $C^{\pi}_{\max} = n + |M|$, where n is the number of jobs in the instance.*

Theorem 3. *The problem $F2 \mid G = (V, E), p_{ij} = 1 \mid C_{\max}$ admits an $O(m^2 \log n)$-time 4/3-approximation algorithm, where $n = |V|$ and $m = |\overline{E}|$.*

Remark 2. If ALGORITHM A is used in the proof of Theorem 3 to compute a path cover with the minimum number of 0-paths and subsequently to construct a schedule π, then we have $C^{\pi}_{\max} \leq \frac{3}{2} C^{*}_{\max}$. That is, we have an $O(m^2 \log n)$-time 3/2-approximation algorithm based on ALGORITHM A.

When the agreement graph $\overline{G}$ consists of k vertex-disjoint triangles such that a vertex of the i-th triangle is adjacent to a vertex of the $(i + 1)$-st triangle, for $i = 1, 2, \ldots, k-1$, and the maximum degree is 3, ALGORITHM B could produce a path cover containing k 2-paths, while there is a Hamiltonian path in the graph. This suggests that the approximation ratio 4/3 is asymptotically tight.

3 Approximating $F2 \mid G = K_{\ell} \cup K_{n-\ell}, p_{ij} \mid C_{\max}$

In this section, we present a 3/2-approximation algorithm for the weakly NP-hard problem $F2 \mid G = K_{\ell} \cup K_{n-\ell}, p_{ij} \mid C_{\max}$ for arbitrary jobs with a conflict graph that is the union of two disjoint cliques. Note that the agreement graph $\overline{G} = K_{\ell, n-\ell}$ is a complete bipartite graph. Without loss of generality, let the job set of K_{ℓ} be $A = \{J_1, J_2, \ldots, J_{\ell}\}$ and the job set of $K_{n-\ell}$ be $B = \{J_{\ell+1}, J_{\ell+2}, \ldots, J_n\}$.

For the job set A, we merge all its jobs (in the sequential order with increasing indices) to become a single "*aggregated*" job denoted as J_A, with its processing time on the machine M_1 being $P^1_A = \sum_{j=1}^{\ell} p_{1j}$ and its processing time on the machine M_2 being $P^2_A = \sum_{j=1}^{\ell} p_{2j}$. Likewise, for the job set B, we merge all its jobs (in the sequential order with increasing indices) to become a single aggregated job denoted as J_B, with its two processing times being $P^1_B = \sum_{j=\ell+1}^{n} p_{1j}$ and $P^2_B = \sum_{j=\ell+1}^{n} p_{2j}$. We now have an instance of the classical two-machine flow-shop scheduling problem consisting of only two aggregated jobs J_A and J_B, and we may apply Johnson's algorithm [23] to obtain a schedule denoted as π. From π we obtain a schedule for the original instance of the problem $F2 \mid G = K_{\ell} \cup K_{n-\ell}, p_{ij} \mid C_{\max}$, which is also denoted as π as there is no major difference. We call this algorithm as ALGORITHM C.

Theorem 4. ALGORITHM C *is an $O(m)$-time $3/2$-approximation algorithm for the problem $F2 \mid G = K_\ell \cup K_{n-\ell}, p_{ij} \mid C_{\max}$, where m is the number of edges in the conflict graph G.*

In the schedule produced by ALGORITHM C, one sees that when the jobs of Λ are processed on the machine M_1, the other machine M_2 is left idle. This is certainly disadvantageous. For instance, when the jobs are all unit jobs and $|A| = |B| = \frac{1}{2}n$, the makespan of the produced schedule is $\frac{3}{2}n$, while the agreement graph is Hamiltonian and thus the optimal makespan is only $n + 1$. This huge gap suggests that one could probably design a better approximation algorithm and we leave it as an open question.

4   Concluding Remarks

In this paper, we investigated approximation algorithms for the two-machine flow-shop scheduling problem with a conflict graph. In particular, we considered two special cases of all unit jobs and of a conflict graph that is the union of two disjoint cliques, that is, $F2 \mid G = (V, E), p_{ij} = 1 \mid C_{\max}$ and $F2 \mid G = K_\ell \cup K_{n-\ell}, p_{ij} \mid C_{\max}$. For the first problem we studied the graph theoretical problem of finding a path cover with the minimum total number of 0-paths and 1-paths, and presented a polynomial time exact algorithm. This exact algorithm leads to a 4/3-approximation algorithm for the problem $F2 \mid G = (V, E), p_{ij} = 1 \mid C_{\max}$. We also showed that the performance ratio 4/3 is asymptotically tight. For the second problem $F2 \mid G = K_\ell \cup K_{n-\ell}, p_{ij} \mid C_{\max}$, we presented a 3/2-approximation algorithm.

Designing approximation algorithms for $F2 \mid G = (V, E), p_{ij} = 1 \mid C_{\max}$ with a performance ratio better than 4/3 is challenging, since one way or the other, one has to deal with longer paths in a path cover or has to deal with the original PATH COVER problem. Nevertheless, better approximation algorithms for $F2 \mid G = K_\ell \cup K_{n-\ell}, p_{ij} \mid C_{\max}$ can be expected.

Acknowledgements. This research is partially supported by the NSFC Grants 11571252, 11771114 and 61672323, the China Scholarship Council Grants 201508330054 and 201706315073, NSERC Canada and the Fundamental Research Funds for the Central Universities (Grant No. 20720160035).

References

1. Asdre, K., Nikolopoulos, S.D.: A linear-time algorithm for the k-fixed-endpoint path cover problem on cographs. Networks **50**, 231–240 (2007)
2. Asdre, K., Nikolopoulos, S.D.: A polynomial solution to the k-fixed-endpoint path cover problem on proper interval graphs. Theoret. Comput. Sci. **411**, 967–975 (2010)
3. Baker, B.S., Coffman, E.G.: Mutual exclusion scheduling. Theoret. Comput. Sci. **162**, 225–243 (1996)

4. Bendraouche, M., Boudhar, M.: Scheduling jobs on identical machines with agreement graph. Comput. Oper. Res. **39**, 382–390 (2012)
5. Bendraouche, M., Boudhar, M.: Scheduling with agreements: new results. Int. J. Prod. Res. **54**, 3508–3522 (2016)
6. Błażewicz, J., et al.: Scheduling Under Resource Constraints - Deterministic Models (1986)
7. Błażewicz, J., et al.: Scheduling unit - time tasks on flow - shops under resource constraints. Ann. Oper. Res. **16**, 255–266 (1988)
8. Blazewicz, J., et al.: Scheduling subject to resource constraints: classification and complexity. Discrete Appl. Math. **5**, 11–24 (1983)
9. Bodlaender, H.L., Jansen, K.: Restrictions of graph partition problems. Part I. Theoret. Comput. Sci. **148**, 93–109 (1995)
10. Chen, B., et al.: A new heuristic for three-machine flow shop scheduling. Oper. Res. **44**, 891–898 (1996)
11. Even, G., et al.: Scheduling with conflicts: online and offline algorithms. J. Sched. **12**, 199–224 (2009)
12. Gabow, H.N.: An efficient reduction technique for degree-constrained subgraph and bidirected network flow problems. In: STOC 1983, pp. 448–456 (1983)
13. Garey, M.R., Graham, R.L.: Bounds for multiprocessor scheduling with resource constraints. SIAM J. Comput. **4**, 187–200 (1975)
14. Garey, M.R., Johnson, D.S.: Complexity results for multiprocessor scheduling under resource constraints. SIAM J. Comput. **4**, 397–411 (1975)
15. Garey, M.R., Johnson, D.S.: Computers and Intractability: A Guide to the Theory of NP-Completeness. W. H. Freeman and Company, San Francisco (1979)
16. Garey, M.R., et al.: The complexity of flowshop and jobshop scheduling. Math. Oper. Res. **1**, 117–129 (1976)
17. Garey, M.R., et al.: The planar hamiltonian circuit problem is NP-complete. SIAM J. Comput. **5**, 704–714 (1976)
18. Golumbic, M.C.: Algorithmic Graph Theory and Perfect Graphs. Elsevier, New York (2004)
19. Gonzalez, T., Sahni, S.: Flowshop and jobshop schedules: complexity and approximation. Oper. Res. **26**, 36–52 (1978)
20. Graham, R.L., et al.: Optimization and approximation in deterministic sequencing and scheduling: a survey. Ann. Discrete Math. **5**, 287–326 (1979)
21. Hall, L.A.: Approximability of flow shop scheduling. Math. Program. **82**, 175–190 (1998)
22. Halldórsson, M.M., et al.: Multicoloring trees. Inf. Comput. **180**, 113–129 (2003)
23. Johnson, S.M.: Optimal two- and three-machine production schedules with setup times included. Naval Res. Logist. **1**, 61–68 (1954)
24. Müller, H.: Hamiltonian circuits in chordal bipartite graphs. Discrete Math. **156**, 291–298 (1996)
25. Pao, L.L., Hong, C.H.: The two-equal-disjoint path cover problem of matching composition network. Inf. Process. Lett. **107**, 18–23 (2008)
26. Petrank, E.: The hardness of approximation: gap location. Comput. Complex. **4**, 133–157 (1994)
27. Rizzi, R., et al.: On the complexity of minimum path cover with subpath constraints for multi-assembly. BMC Bioinform. **15**, S5 (2014)
28. Röck, H.: Scheduling unit task shops with resource constraints and excess usage costs. Technical report, Fachbereich Informatik, Technical University of Berlin, Berlin (1983)

29. Röck, H.: Some new results in flow shop scheduling. Zeitschrift für Oper. Res. **28**, 1–16 (1984)
30. Süral, H., et al.: Scheduling unit-time tasks in renewable resource constrained flowshops. Zeitschrift für Oper. Res. **36**, 497–516 (1992)
31. Tellache, N.E.H., Boudhar, M.: Two-machine flow shop problem with unit-time operations and conflict graph. Int. J. Prod. Res. **55**, 1664–1679 (2017)
32. Tellache, N.E.H., Boudhar, M.: Flow shop scheduling problem with conflict graphs. Ann. Oper. Res. **261**, 339–363 (2018)
33. Williamson, D.P., et al.: Short shop schedules. Oper. Res. **45**, 288–294 (1997)

On Contact Representations of Directed Planar Graphs

Chun-Hsiang Chan and Hsu-Chun Yen[(✉)]

Department of Electrical Engineering, National Taiwan University, Taipei, Taiwan
kennyhchan@gmail.com, hcyen@ntu.edu.tw

Abstract. Prior work on contact representations of planar graphs deals with undirected graphs only. We introduce a notion of point-side contact representations for directed planar graphs. We show every outerplanar digraph of out-degree at most three to enjoy a point-side triangle contact representation. The result is generalized to outerplanar digraphs of out-degree at most n, which are shown to have convex n-gon (i.e., n-sided polygon) point-side contact representations. Our result is tight is the sense that there exists a 2-outerplanar digraph that does not have a point-side triangle contact representation. For maximal outerplanar digraphs of out-degree at most three, an efficient constructive procedure is designed to yield their point-side triangle contact representations. For general planar digraphs of degree d, they are shown to admit $2d$-gon point-side contact representations.

1 Introduction

With potential applications to floorplanning, cartography, and more, the study of contact representations of graphs from an algorithmic viewpoint has received increasing attention in computational geometry. A *contact graph representation* refers to a drawing style in which vertices are represented by interior-disjoint geometric objects with edges corresponding to contacts between those objects. A classical example of a contact graph representation is illustrated in Koebe's circle packing theorem [9], saying that every planar graph can be drawn as touching circles. By altering the contact style (point vs. side contact, for instance) and the object shape (circle, triangle, . . . , etc), a wide variety of contact representations have been proposed and studied over the years.

As triangles represent the simplest form of convex polygons, much work along the line of contact representations of graphs has focused on representing vertices by triangles, and edges by point- or side-contacts of triangles. For instance, [6,8] studied a drawing style called a *proper touching triangle graph* (proper-TTG) representation, which is a side-contact representation that forms a triangular tiling of a triangle. Touching triangle representations without boundary constraints can be found in [7]. For contact representations of graphs using polygons, the reader is referred to, e.g., [1,3,5]. In particular, it was shown in [1]

H.-C. Yen—Research supported in part by Ministry of Science and Technology, Taiwan, under grant MOST 106-2221-E-002-036-MY3.

© Springer International Publishing AG, part of Springer Nature 2018
L. Wang and D. Zhu (Eds.): COCOON 2018, LNCS 10976, pp. 218–229, 2018.
https://doi.org/10.1007/978-3-319-94776-1_19

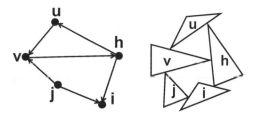

Fig. 1. (Left) A directed planar graph G; (Right) A point-side triangle contact representation of G.

that any planar graph can be represented by touching hexagons, and there are planar graphs that cannot be represented by pentagons.

To our knowledge, all the previous work in the literature on contact representations of graphs deals with undirected graphs. In this paper, we propose for the first time contact representations for directed graphs (i.e., digraphs) based on the notion of *point-side contacts*. In our setting, a planar digraph $G = (V, E)$ is displayed in a way that each vertex $v \in V$ is drawn as a polygon, and each directed edge $e = (v, u) \in E$ corresponds to a contact of a side of the polygon associated with u and a point of the polygon associated with v. Figure 1 displays an example of a point-side contact representation of a planar digraph. It is worth noting that the conventional notion of a point-contact in, e.g., [2], is in fact a contact between a point and an edge; however, such a point-side contact applies to an edge with no direction (i.e., an edge in an undirected graph).

We first show that every outerplanar digraph of out-degree at most 3 admits a point-side triangle contact representation. Our result is tight in the sense that there exists a 2-outerplanar digraph of out-degree at most 3 that does not have a point-side triangle contact representation. For maximal outerplanar digraphs of out-degree at most 3, we are able to come up with a procedure to construct point-side triangle contact representations in linear time. As a natural extension from triangles to general polygons, we also study the use of n-gons (i.e., n-sided polygons) to represent vertices. Using a strategy that parallels the triangular case, we show every outerplanar digraph of out-degree at most n to admit a point-side convex n-gon contact representation. Finally, for general planar digraphs of degree (in-degree + out-degree) d, they are shown to admit $2d$-gon point-side contact representations.

2   Preliminaries

A graph is *planar* iff it can be drawn in the Euclidean plane without crossings. A *plane graph* is a planar graph with a fixed combinatorial embedding and a designated outer face. An *outerplanar graph* is a graph for which there exists a planar embedding with all vertices of the graph belonging to the outer face. We call such an embedding of an outerplanar graph an *outerplanar embedding*. We define a *k-outerplanar embedding* recursively as follows. If k equals 1, a 1-outerplanar embedding of a graph is just the outerplanar embedding. A k-outerplanar embedding, $k > 1$, is that the removal of the vertices on the outer face results in a $(k-1)$-outerplanar embedding. A graph exhibiting a k-outerplaner embedding is said to be *k-outerplanar*. *Directed outerplanar graphs* (DOPGs, for short) are directed versions of outerplanar graphs. *Directed k-outerplanar graphs* (k-DOPGs, for short) can be defined similarly. We write DOPG_d $(d \geq 1)$ to denote the class of DOPGs, for which each vertex v of the graph has out-degree $deg^+(v) \leq d$.

A *point-side triangle contact representation* (psTCR, for short) of a planer digraph $G = (V, E)$ is a drawing meeting the following conditions:

1. each vertex $v \in V$ is drawn as a triangle, and
2. each directed edge $e = (v, u) \in E$ corresponds to a contact of a side of the triangle associated with u and a point of the triangle associated with v.

Throughout this paper, we assume that for $G = (V, E)$, if $(v, u) \in E$, then $(u, v) \notin E$; otherwise, it is obvious that G has no psTCR. See Fig. 1 for an example of a psTCR corresponding to a planar digraph.

Fig. 2. A contact system of pseudo-segments with extremal points marked in red. (Color figure online)

One of our main results shows that every graph in DOPG_3 admits a psTCR. To show this result, we require the notion and results of the so-called *contact systems of pseudo-segments* [4]. A *contact system of pseudo-segments* (or a *contact system*, for short) is a set of non-crossing Jordan arcs where any two of them intersect in at most one point, and each intersecting point is internal to at most one arc. If a contact system is *stretchable*, then there exists a homeomorphism transforming the contact system into a drawing where each arc is a straight line. Stretchable contact systems were characterized in [4] based on the notion of *extremal points*. A point p is an extremal point of a contact system S if the following three conditions are satisfied:

1. p is an endpoint of a pseudo-segment in S,
2. p is not interior to any pseudo-segment in S, and
3. p is incident to the unbounded region of S.

Figure 2 displays an example of a contact system. The following result characterizes necessary and sufficient conditions for a contact system to be stretchable, based on extremal points. Throughout the rest of this paper, we let $ext(S)$ denote the number of extremal points of a contact system S.

Theorem 1 *(Theorem 38 in [4]). A contact system S of pseudo-segments is stretchable iff each of its subsystems S' of cardinality greater than 1 has at least 3 extremal points (i.e., $ext(S') \geq 3$).*

3   Point-Side Contact Representations of Outerplanar Digraphs

To prove our main result, we associate outerplanar embeddings of graphs in $DOPG_3$ with a special class of contact systems of pseudo-segments called *trinity contact systems* of pseudo-segments (*TCSs*, for short), showing that a psTCR of an outerplanar graph in $DOPG_3$ is exactly a TCS that is stretched. Thus, if all graphs in $DOPG_3$ have stretchable TCSs, then $DOPG_3$ enjoys having psTCRs.

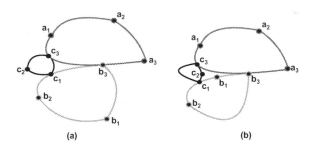

(a)   (b)

Fig. 3. (a) A minimal TCS; (b) A TCS that is not minimal.

TCSs are defined based on the so-called *unit of pseudo-segments* (or simply *unit*, for short). For convenience of representation, we write $s = \{a, a'\}$ to denote that pseudo-segment s has end points a and a'. A set of pseudo-segments U is a *unit* if U contains exactly three pseudo-segments $s_1 = \{a_1, a_2\}$, $s_2 = \{a_2, a_3\}$, and $s_3 = \{a_3, a_1\}$, for some a_1, a_2, and a_3. The set of the three end points associated with a unit U is denoted by P_U.

A TCS S is a set of pseudo-segments which can be partitioned into units such that for each pair of units U_1, U_2 in S,

1. $U_1 \cap U_2 = \emptyset$ and $P_{U_1} \cap P_{U_2} = \emptyset$.
2. U_1 and U_2 intersect in at most one point in S, and each intersecting point is internal to at most one segment.

A end point x in a unit U is a *free point* if x is not a touching point between U and a segment of another unit. For example, c_2 in Fig. 3(a) is a free point.

Our strategy of showing that every graph G in $DOPG_3$ has a psTCR relies on relating G to a TCS S, and then showing S to be stretchable. To this end, we take advantage of the following well-known result in graph theory:

Theorem 2 (Koebe's circle packing theorem [9]). *Every connected simple planar graph G has a circle packing in the plane whose intersection graph is isomorphic to G.*

A TCS S is said to correspond to a planar digraph $G = (V, E)$ in $DOPG_3$ if there is a 1-to-1 correspondence between S and G such that:

1. each vertex $v \in V$ maps to a unit U_v in S, and
2. if there exists a directed edge $e = (v, u) \in E$, then one of the end point $p \in P_{U_v}$ must touch one of the segments $s \in U_u$.

Note that a graph in $DOPG_3$ might correspond to several TCSs, as shown in Fig. 3. As Theorem 1 suggests, the number of extremal points in a contact system plays a key role as to whether the system is stretchable or not. In other words, if a graph has several corresponding TCSs, it is beneficial, as far as stretchability is concerned, to consider the one with the possibly maximal number of extremal points, i.e., keeping the internal points as few as possible. A point in a contact system is said to be *internal* if it is not along the outer boundary of the drawing. A TCS is said to be *minimal* (called an *mTCS*) if none of its points can be relocated to form another TCS corresponding to the same graph but with fewer internal points. Figure 3(b) is not minimal as b_1 and c_2 can be moved to the outer boundary as displayed in Fig. 3(a) which is minimal.

Lemma 1. *Every graph G in $DOPG_3$ has an mTCS.*

Proof. (Sketch) First apply Theorem 2 to yield a circle packing of G. Then divide the circumference of each circle into three segments in the following way. If $(v, u) \in E$, the touching point between the circles associated with v and u becomes an end point of the unit associated with v. Finally, for a circle with fewer than 3 end points, add the remaining end points, if necessary, to the outer boundary of that circle.   □

Lemma 1 shows a transformation from a graph G in $DOPG_3$ to an mTCS S. To show G to have a psTCR, it remains to show S to be stretchable. Lemma 2, which leads to the main result, is the key statement in this section. The lemma is proved by induction, based on the fact that every outerplanar graph has *degeneracy* at most two.

A *k-degenerate* graph is an undirected graph in which every subgraph has a vertex of degree at most k, i.e., some vertex in the subgraph has k or fewer incident edges in the subgraph. The *degeneracy* of a graph is the smallest value of k for which it is k-degenerate. Take the class of outerplanar graphs for example. It is known that outerplanar graphs are of degeneracy 2. As a result, given an outerplanar graph $G = (V, E)$, there exists an ordering $v_1, v_2, ..., v_n$ of V such that deleting vertices of V in the above order requires the removal of no more than two edges in each step.

Lemma 2. *Let G be a graph in $DOPG_3$. Every mTCS S of G is stretchable.*

Proof. Based on Theorem 1, the lemma is proved by induction on the number of vertices of G.

First, the base case (i.e., a graph with one vertex and zero edge) is trivial. Now assume that all the mTPSs corresponding to an arbitrary outerplanar digraph with n or fewer vertices are stretchable. Now consider a graph $G = (V, E)$ in $DOPG_3$ with $|V| = n + 1$. Let v be a vertex of G with one or two incident edges, guaranteed by the fact that G has degeneracy at most two. The case when v has one incident edge is simpler, and is omitted here. Let G' be the graph resulting from removing v and its incident edges from G. Clearly G' is in $DOPG_3$.

Let S be a mTCS of G. Based on our earlier discussion, v corresponds to a unit of segments U in S. Let S' be the TCS obtained from S by removing U. Clearly, S' is an mTCS of G'. Also note that among the three segments in U, only one segment (say r) touches S' at two points, say a and b. Let the other two segments of U be r' and r''.

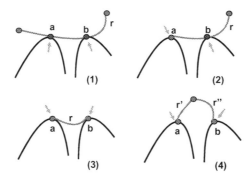

Fig. 4. Some cases of segments of U touching S_1. (Color figure online)

For any subsystem S_1 of S, let $S_1 - U = S_1'$. Figures 4(1)–(4) display some cases in which r (r' and r'') of U touches S_1'. The red (resp., blue) dots are points belonging to U (resp., S_1') and the arrows indicate the directions of the directed edges in the original graph. We show in the following that $ext(S_1) \geq 3$. Consider the following three cases:

- (Case 1): none of the extremal points of S_1' vanishes. In this case, $ext(S_1) \geq ext(S_1') \geq 3$.
- (Case 2): the number of regions of S_1 is the same as that of S_1'. In this case, if a segment $\bar{r} \in (S_1 \cap U)$ cancels an extremal point of S_1', then a new extremal point must be introduced in S_1. See Figs. 4(1) and (2). In Fig. 4(1), the two blue points are extremal points of S_1' canceled in S_1 (because of r). However, S_1 has two additional extremal points (i.e., the two red points). The case in Fig. 4(2) is similar, in which one extremal point of S_1' is canceled but a new one is introduced in S_1. Note that in Figs. 4(3) and (4), none of the extremal points disappears. Hence, $ext(S_1) \geq ext(S_1') \geq 3$ holds.

– (Case 3): some extremal points of S_1' disappear because of becoming internal in S_1. We call such points as *changing points*. In this case, the introduction of one or more segments of U forms a new internal face (called R) enclosing those changing points. First note that none of the changing points could lie on a segment of U, otherwise, the degree of v is at least 3 – a contradiction. See Fig. 6(2). Now we add to S_1' all the segments of S' in region R, and call the resulting system S_1''. See Figs. 5(1) and (2). It is easy to see that those changing points cannot be free points in S_1''; otherwise, they are also free (internal) points in S' contradicting the fact that the underlying TCS is minimal (see point x in Fig. 6(1)). As S_1'' contains at least three extremal points and none of which comes from points in region R, S_1' must have at least three extremal points excluding those changing points.

Now the only way that one of more of the above three extremal points could disappear due to the insertion of $U \cap S_1$ is shown in Figs. 4(1) and (2). In both cases, additional extremal points belonging to $U \cap S_1$ will be provided so that $ext(S_1) \geq 3$ remains true.

In view of the above, S is stretchable.   □

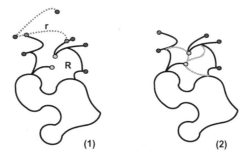

Fig. 5. (1) A subsystem S_1 of S. S_1' is obtained by removing the two red segments (which belong to U); (2) Subsystem S_1'' resulting from filling out missing segments in R. (Color figure online)

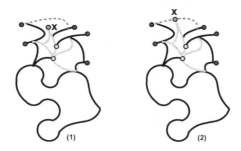

Fig. 6. (1) A TCS that is not minimal; (2) the corresponding graph is not in $DOPG_3$.

Lemmas 1 and 2, in conjunction with the fact that stretching the three segments in a unit of an mTPS forms a triangle, give rise to the following result.

Theorem 3. *Every outerplanar digraph in $DOPG_3$ admits a point-side triangle contact representation.*

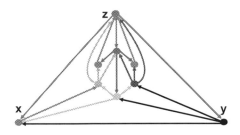

Fig. 7. Graph G, a maximal planar as well as a 2-outerplanar digraph, has no psTCR.

As our next result shows, the stretchability property may not hold as we go beyond the class $DOPG_3$.

Lemma 3. *There exists a 2-outerplanar digraph (which is also maximal) of maximal out-degree at most 3 that does not admit a psTCR.*

It is not known whether those triangles used in the psTCR of Theorem 3 are homothetic or not. For trees of out-degree at most 3 (which are trivially outerplanar), we have the following result:

Theorem 4. *Every directed tree with out-degree at most 3 admits an equilateral triangle contact representation such that the ratio between the lengths of edges of a child triangle and its parent triangle is bounded by $min\{\frac{1}{8}, \frac{1}{2d}\}$, where d is the in-degree of the parent node w.r.t. its children.*

4   A Linear-Time Procedure for Constructing psTCRs of Maximal Outerplanar Diagraphs

In Sect. 3, every outerplanar graph in $DOPG_3$ was shown to admit a psTCR. The proof, based on relating the problem to stretchable contact systems of pseudo-segments, is in a sense not a typical constructive approach. In this section, we show that if the underlying outerplanar graph is also maximal, i.e., no more edges can be added to the graph while preserving outerplanarity, then a constructive procedure for yielding the psTCR is available, and the procedure runs in linear time. We write $mDOPG_3$ to denote the class of outerplanar graphs in $DOPG_3$ which are also maximal. Notice that every maximal outerplanar graph is biconnected.

As each of the simple cycles in a maximal outerplanar graph $G = (V, E)$ is a triangle, our procedure takes advantage of the existence of an ordering $\pi = v_1, v_2, ..., v_n$ of V such that $(v_1, v_2) \in E$ and for all $3 \leq i \leq n$, with respect to the subgraph induced by $v_1, v_2, ..., v_{i-1}$, v_i is connected to exactly two vertices $v_j, v_k, 1 \leq j < k \leq i - 1$ and $(v_j, v_k) \in E$.

A vertex p of a triangle in a psTCR of an outerplanar graph is said to be *external* if p is located on the outer boundary of the psTCR, and p is not a contact point with other triangle; otherwise, p is called *internal*. For instance, in the left figure of Fig. 8, $p_1, ..., p_7$ are external, while the rest are internal. Let $p_1, p_2, ..., p_k$ be a clock-wise ordering of external points of a psTCR R. R is said to form a *convex group* if the k-sided polygon with endpoints $p_1, p_2, ..., p_k$ is convex and no three points are co-linear. In our subsequent discussion, we let $I(p_x, p_y)$ denote the internal region enclosed by $\overline{p_x p_y}$ and the edges of psTCR, where p_x and p_y are external points. The left figure of Fig. 8 is a convex group in which $I(p_1, p_2)$ is the internal region enclosed in the region specified by p_1, p_2 and p'. Notice that $I(p_1, p_2)$ might be empty if $\overline{p_x p_y}$ is an edge of a triangle. The right figure of Fig. 8, on the other hand, is not a convex group.

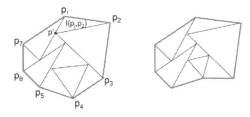

Fig. 8. (Left): A convex group; (Right): A non-convex group

Theorem 5. *Given a plane graph $G = (V, E)$ in $mDOPG_3$, there is an algorithm which can construct a psTCR of G in linear time.*

Proof. Given any planar embedding of $G = (V, E)$ in $mDOPG_3$, let $\pi = v_1, v_2, ..., v_n$ be an ordering of V mentioned in the beginning of this section. We construct a psTCR T of G in a greedy fashion in the order specified by π. Let t_i denote the triangle corresponding to v_i in T, G_i denote the subgraph induced by $v_1, v_2, ..., v_i$, and T_i denote the psTCR with respect to subgraph G_i. As we shall see later, in each step i of the algorithm, psTCR T_i is always a convex group.

(Initial step). We start with a psTCR corresponding to a subgraph induced by v_1 and v_2. If $(v_1, v_2) \in E$, then a corner of t_1 touches an edge of t_2; otherwise, a corner of t_2 touches an edge of t_1. The convexity of T_2 is easy to enforce.

(Iteration steps). From $i = 3$ to $|V|$, we add a triangle t_i to T_{i-1} and keep the psTCR T_i a convex group. Since v_i is connected to exactly two neighboring vertices among $v_1, ..., v_{i-1}$, there are three cases to be considered.

(a) (v_i has two out-going edges) See Fig. 9.

A T_i can be formed by placing the external point p'_j of t_i in region O (excluding the boundary, see Fig. 9), and the two internal points in $I(p_j, p_{j+1})$ touching the edges of the two triangles corresponding to v_i's two neighbors. Recall that in a convex group, no three points are co-linear, and hence, region O is not empty.

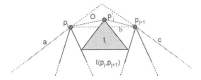

Fig. 9. Region O is enclosed by edge b, and the extensions of edges a and c.

(b) (v_i has two incoming edges) See Fig. 10.

A T_i can be formed by placing two external points p'_j and p'_{j+2} of t_i in a way shown in the figure (i.e., edge $\overline{p'_j p'_{j+2}}$ extends $\overline{p_j p_{j+1}}$). The third external point p'_{j+1} is placed in region O to form a triangle $\{p'_j, p'_{j+1}, p'_{j+2}\}$. It is not hard to see that the new region $\{p_1, ..., p_{j-1}, p'_j, p'_{j+1}, p'_{j+2}, ..., p_k\}$ is a convex polygon, for some k.

(c) (v_i has one incoming edge and one outgoing edge) See Fig. 11.

A T_i can be formed by placing an edge $\overline{p'_j q}$ in a way shown in the figure, with the external point p'_j lying outside the boundary, and an internal point q in $I(p_j, p_{j+1})$ such that $\overline{p'_j q}$ goes through point p_j (serving as a touching point of an incoming edge) and touches an edge of a triangle capturing an outgoing edge connecting to a triangle t' at point q. Notice that the figure only shows a case of t'; the rest are similar. Then we place the last point p'_{j+1} in region O to form a triangle $\{p'_j, q, p'_{j+1}\}$. The new region $\{p_1, ..., p_{j-1}, p'_j, p'_{j+1}, p_{j+1}, ..., p_k\}$ is a convex polygon, for some k.

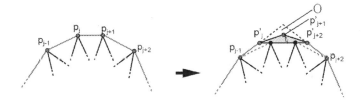

Fig. 10. (Left): The original convex group of T_{i-1}, where red edges denote the boundary; (Right): After the new triangle's edge $\overline{p'_j p'_{j+2}}$ is added to T_{i-1}, the originally external points p_j and p_{j+1} become internal. (Color figure online)

Finally, it is easy to see that the procedure runs in time linear in the size of the given graph.   □

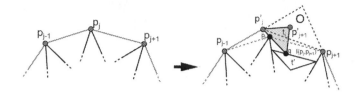

Fig. 11. (Left): The original convex group of T_{i-1}, where red edges denote the boundary; (Right): After the new triangle's edge $\overline{p'_j q}$ is added to T_{i-1}, the originally external point p_j becomes internal. (Color figure online)

Recall that in graph theory, a *bridge* e in a connected graph G is an edge which is not along any cycle of the graph (i.e., the removal of e disconnects G). Graph G is *2-edge-connected* if it has no bridges. A *2-edge-connected component* of G is a maximal subgraph which is 2-edge-connected. We let $bcDOPG_3$ be the class of graphs in $DOPG_3$ such that each 2-edge-connected component of a graph is in $mDOPG_3$, i.e., graphs formed by connecting maximal outerplanar graphs by bridges. We have the following result:

Theorem 6. *Given a plane graph $G = (V, E)$ in $bcDOPG_3$, there is an algorithm which can construct a psTCR of G in linear time.*

5   Contact Representations of Planar Digraphs Using n-gons

To deal with digraphs of out-degree greater than 3, we extend the definition of a psTCR to a *point-side n-gon contact representation* ($psCR_n$, for short) as follows. A $psCR_n$ of a planar digraph $G = (V, E)$ is a drawing meeting the following conditions: (1) each vertex $v \in V$ is drawn as an n-side polygon, and (2) each directed edge $e = (v, u) \in E$ corresponds to a contact of a side of the polygon associated with u and a point of the polygon associated with v.

In Sect. 3, we show that every outerplanar digraph of out-degree at most 3 admits a psTCR. Following a proof that parallels that of Theorem 3 based on contact systems of pseudo-segments, we are able to establish the following result:

Theorem 7. *Every outerplanar digraph in $DOPG_n$ admits a convex $psCR_n$.*

Recall from Lemma 3 that there is a 2-outerplanar graph that does not have a psTCR. If one examines Fig. 7 (i.e., a witnessing example) carefully, the three units (or triangles) associated with nodes x, y and z (of out-degrees 1, 3, and 3, resp.) along the outer boundary could have at most 2 extremal points (corresponding to the two "free corners" of x). It is therefore natural to ask whether allowing the number of "corners" of a polygon to exceed its out-degree, it is possible to have a contact representation for every planar digraph. The following theorem answers the above affirmatively.

Theorem 8. *Every planar digraph of degree (i.e., in-degree + out-degree) d admits a $psCR_{2d}$.*

References

1. Aerts, N., Felsner, S.: Straight line triangle representations. Discrete Comput. Geom. **57**(2), 257–280 (2017)
2. Alam, M.J., Biedl, T., Felsner, S., Kaufmann, M., Kobourov, S.G.: Proportional contact representations of planar graphs. In: van Kreveld, M., Speckmann, B. (eds.) GD 2011. LNCS, vol. 7034, pp. 26–38. Springer, Heidelberg (2012). https://doi.org/10.1007/978-3-642-25878-7_4
3. Chang, Y.-J., Yen, H.-C.: A new approach for contact graph representations and its applications. In: Dehne, F., Sack, J.-R., Stege, U. (eds.) WADS 2015. LNCS, vol. 9214, pp. 166–177. Springer, Cham (2015). https://doi.org/10.1007/978-3-319-21840-3_14
4. de Fraysseix, H., de Mendez, P.O.: Barycentric systems and stretchability. Discrete Appl. Math. **155**, 1079–1095 (2007)
5. Duncan, C., Gansner, E., Hu, Y., Kaufmann, M., Kobourov, S.: Optimal polygonal representation of planar graphs. Algorithmica **63**(3), 672–691 (2012)
6. Fowler, J.J.: Strongly-connected outerplanar graphs with proper touching triangle representations. In: Wismath, S., Wolff, A. (eds.) GD 2013. LNCS, vol. 8242, pp. 155–160. Springer, Cham (2013). https://doi.org/10.1007/978-3-319-03841-4_14
7. Gansner, E.R., Hu, Y., Kobourov, S.G.: On touching triangle graphs. In: Brandes, U., Cornelsen, S. (eds.) GD 2010. LNCS, vol. 6502, pp. 250–261. Springer, Heidelberg (2011). https://doi.org/10.1007/978-3-642-18469-7_23
8. Kobourov, S.G., Mondal, D., Nishat, R.I.: Touching triangle representations for 3-connected planar graphs. In: Didimo, W., Patrignani, M. (eds.) GD 2012. LNCS, vol. 7704, pp. 199–210. Springer, Heidelberg (2013). https://doi.org/10.1007/978-3-642-36763-2_18
9. Koebe, P.: Kontaktprobleme der konformen Abbil-dung. Ber. Verh. Sachs. Akademie der Wissenschaften Leipzig, Math.-Phys. Klasse, vol. 88, pp. 141–164 (1936)

Computation and Growth of Road Network Dimensions

Johannes Blum[✉] and Sabine Storandt

Institut für Informatik, Julius-Maximilians-Universität Würzburg,
Würzburg, Germany
{blum,storandt}@informatik.uni-wuerzburg.de

Abstract. There is a plethora of route planning techniques which work remarkably well on real-world road networks. To explain their good performance, theoretical bounds in dependency of road network parameters as the highway dimension h or the skeleton dimension k were investigated. For example, for the hub label technique, query times in the order of $\mathcal{O}(p \log D)$ and a space consumption of $\mathcal{O}(np \log D)$ were proven for both $p = h$ and $p = k$, with D denoting the graph diameter and n the number of nodes in the network. But these bounds are only meaningful when the dimension values h or k are known. While it was conjectured that h and k grow polylogarithmically in n, their true nature was not thoroughly investigated before – primarily because of a lack of efficient algorithms for their computation. For the highway dimension, this is especially challenging as it is NP-hard to decide whether a network has highway dimension at most h. We describe the first efficient algorithms to lower bound the highway dimension and to compute the skeleton dimension exactly, even in huge networks. This allows us to formulate new conjectures about their growth behavior. Somewhat surprisingly, our results imply that h and k scale very differently. While k turns out to be a constant, we expect h to grow superpolylogarithmically in n. These observations have implications for the future design and analysis of route planning techniques.

1 Introduction

To accelerate the computation of shortest paths in road networks, several pre-processing-based techniques have been developed, as contraction hierarchies (CH) [11], transit node routing (TN) [4] or hub labels (HL) [2]. While shortest path planning with Dijkstra's algorithm takes seconds on large road networks, these acceleration techniques allow for query answering in the order of milliseconds or even microseconds. Furthermore, the preprocessing time and space consumption is usually moderate. The empirical justification stems from the investigation of real-world road networks, e.g. extracted from OpenStreetMap (Germany about 20 million, Europe 174 million nodes) or TIGER data (USA 24 million nodes).

But for all listed acceleration techniques, one can construct artificial input networks on which they perform unsatisfactorily. This inspired the question what

© Springer International Publishing AG, part of Springer Nature 2018
L. Wang and D. Zhu (Eds.): COCOON 2018, LNCS 10976, pp. 230–241, 2018.
https://doi.org/10.1007/978-3-319-94776-1_20

characteristics of real road networks enable their great performance. As a result, new parameters were designed which try to capture the essence of road networks: In [3], it was conjectured that road networks exhibit a low highway dimension h and h-dependent bounds for CH, TN and HL were proven. In [12], the skeleton dimension k was suggested as an alternative network parameter for the analysis of HL. While both h and k are assumed to grow at most polylogarithmically in the size n of the network, grid instances with $h, k \in \Theta(\sqrt{n})$ are known. As grid substructures are ubiquitous in real-world road networks, computing h and k on multiple instances is necessary to gain insights in their real dependency on n.

In this paper, we devise algorithms that are efficient enough to compute (lower bounds for) the highway dimension and the skeleton dimension in road networks with millions of nodes and edges. This enables the first practical study of their growth behavior.

1.1 Related Work

The treewidth t – a classical network parameter – was also investigated for theoretical analyses of shortest path planning techniques. For CH, query times in $\mathcal{O}(t \log n)$ and a space consumption of $\mathcal{O}(nt \log n)$ were proven [5]. It is NP-complete to decide whether a network has treewidth at most t. Nevertheless, the recent Parameterized Algorithms and Computational Experiments Challenge (PACE '17) focused on the search for practical algorithms that compute t [6]. The approach presented in the winning paper [13] was able to compute the exact treewidth for over 100 different benchmark networks. Upper bounds of t for selected road networks were reported in [8], e.g. $t \leq 479$ for the network of Europe. In [5], it was proven that for a given network G, there exist edge lengths such that $h(G) \geq (pw(G)-1)/(\log_{3/2} n+2)$ where $pw(G)$ denotes the pathwidth of G. As $pw(G) \geq t(G)$, this inequality also relates h and t.

For graphs with bounded maximum degree, $k \in \mathcal{O}(h)$ was proven [12]. Furthermore, in a carefully weighted grid graph, it was shown that there can be an exponential gap between the highway and the skeleton dimension ($h \in \Omega(\sqrt{n})$ and $k \in \mathcal{O}(\log n)$). In other grid networks, though, the skeleton dimension is also in $\Omega(\sqrt{n})$. Hence it is not a priori clear how these two parameters behave with respect to each other in road networks. Preliminary results for h and k on real networks are available for New York (which contains about 200,000 nodes). There $h > 173$ and $k = 73$ holds [12]. But these results were obtained using naive algorithms which unfortunately are impractical for larger networks with respect to running time or solution quality (as will be shown in our experiments).

1.2 Contributions

On the theoretical side, we establish the first relationship between the skeleton dimension and the pathwidth of a network.

On the practical side, we first discuss a new approach to lower bound the highway dimension h in real-world road networks by combining a greedy algorithm and an ILP relaxation. In addition, we describe a new multi-phase algorithm for computing the exact skeleton dimension k which leverages contraction hierarchies and transit node routing. Equipped with these new algorithms, we compute $h^* \leq h$ and k in real-world road networks with up to 24 million nodes, cut from Germany and the US. Based on the respective results, we conjecture that the highway dimension grows worse in the size n of the network than previously thought, while the value of the skeleton dimension seems not to depend on n but instead on the densest region within the network. We support our new conjectures with additional experiments. These results provide new insights in the structure of road networks. They also imply that the theoretical running time bounds in dependency of k are much stronger than the ones for h. As at the moment such bounds are only available for HL, this motivates to look for bounds depending on k also for CH, TN and other graph algorithms.

2   Preliminaries

Throughout the paper, we consider a weighted, undirected graph $G(V, E)$ which represents a road network. We assume that the shortest path between any two nodes is unique in G (which can be enforced e.g. by symbolic perturbation). The distance of the shortest path from $s \in V$ to $t \in V$ is denoted by $d_s(t)$.

2.1   Road Network Dimensions

For a node v let $B_r(v) = \{w \in V \mid d_v(w) \leq r\}$ be the ball around v with radius r, and let $S_r(v)$ be the set of all shortest paths which intersect $B_r(v)$ and have a length in the range $(r/2, r]$. The *highway dimension* [1] is the smallest $h \in \mathbb{N}$ such that for all r and for all $v \in V$, there exists a hitting set H for $S_{2r}(v)$ of size at most h (i.e. $H \subseteq V$ and for every path $P \in S_{2r}(v)$, we have $H \cap P \neq \emptyset$).

Let T_s be the shortest path tree of a node s and let $\tilde{T}_s$ be its geometric realization, i.e. every edge is seen as infinitely many nodes of degree two that are incident to infinitely small edges. The distance of a node v to its furthest descendant in $\tilde{T}_s$ is denoted by $Reach_s(v)$. The skeleton T_s^* is defined as the subtree of $\tilde{T}_s$ induced by all nodes v satisfying $d_s(v) \leq 2 \cdot Reach_s(v)$ from s. Intuitively, we obtain T_s^* by cutting every branch of T_s at two thirds of its length. The width of T_s^* is the maximum over all $Cut_r(T_s^*)$ where $Cut_r(T_s^*)$ denotes the number of nodes in T_s^* at distance r from s. The *skeleton dimension* k of G is defined as the maximum width of all skeletons T_v^* in G [12].

2.2   Contraction Hierarchies

The idea behind CH is to augment the road network with so called shortcut edges, which allow to reduce the number of necessary edge relaxations in a Dijkstra run significantly.

The CH graph construction relies on the node contraction operation. Here, a node v is removed from the graph, and shortcut edges are inserted between the neighbors of v if they are necessary to preserve their pairwise shortest path distances. The preprocessing phase of CH consists of contracting all nodes one-by-one until the graph is gone. In the end, a new graph $G^+(V, E \cup E^+)$ is constructed, with E^+ being the set of shortcut edges that were inserted during the contraction process.

The position of a node in the order of contraction is called $rank(v)$. An edge $e = (u, v)$ is called upwards from u if $rank(u) < rank(v)$ and downwards otherwise. A path is called upwards/downwards if it consists of upwards/downwards edges only. The graph induced by all upwards paths emerging from a node v is referred to as $G^\uparrow(v)$. For answering an s-t-query, a bidirectional Dijkstra run is used but restricted to $G^\uparrow(s)$ and $G^\uparrow(t)$. It was proven that both runs will settle the node that was contracted last on the original shortest path from s to t in G. Hence identifying p such that $d_s(p) + d_t(p)$ is minimized leads to correct query answering [11]. As the graphs $G^\uparrow(s)$ and $G^\uparrow(t)$ are usually concise, a bidirectional Dijkstra run in these graphs is several orders of magnitudes faster than in the original graph.

2.3 Transit Node Routing

The TN algorithm [4] relies on the observation that all shortest paths from some small region to faraway destinations (for some notion of far) pass through a small set of so-called access nodes (e.g. on-ramps or important crossings). The union of all these access nodes then forms the transit node set T. For all pairs of nodes in T, the shortest path distance is precomputed and stored in a look-up table. Additionally, for every node $v \in V$ the set of access nodes $AN(v)$ together with the respective shortest path distances are stored.

A 'long' shortest path is defined as one that contains an access node from s and one of t. The distance of such a shortest path can be determined by looking up the distances from s to all nodes in $AN(s)$, the distances from $AN(s)$ to the nodes in $AN(t)$ and the distances from $AN(t)$ to t, and keeping track of the shortest concatenated path. As all of the considered distances are precomputed, the query time is $|AN(s)| \cdot |AN(t)|$. On real road networks, a transit node set T of size $\approx \sqrt{n}$ can be chosen such that the access node sizes are less than 10 on average, allowing for very fast query answering (order of microseconds). For 'short' queries, however, the transit node approach might report a too long distance. In that case, a (local) Dijkstra run is used as backup.

3 Skeleton Dimension and Pathwidth

Bauer et al. showed that for any unweighted graph G there exist edge weights such that the highway dimension is at least $(pw(G) - 1)/(\log_{3/2} n + 2)$ where $pw(G)$ is the pathwidth of G [5]. With a slight adaptation of their proof we can show that there is a similar relation for the skeleton dimension.

We consider the edge weights that maximize the skeleton dimension of a given unweighted graph G. We first show that G can be decomposed into smaller graphs by recursively removing at most k nodes.

Lemma 1. *Let G be a unweighted connected graph, let k be the maximum skeleton dimension of G over all possible edge weights and let G' be a connected subgraph of G with $n' \geq 2k + 2$ nodes. Then G' can be separated into at least two connected components of size at most $\lfloor n'/2 \rfloor$ by removing at most k nodes.*

This can be used to show the following theorem (similar to [5]).

Theorem 1. *Let G be an undirected graph. There exist edge weights on G such that for the skeleton dimension k and the pathwidth pw of G we have $k \geq (pw - 1)/(\log_2 n + 2)$.*

4   Lower Bounds for the Highway Dimension

Deciding whether a network exhibits a highway dimension of at most h is NP-hard [10]. Also, computing h exactly seems to be hardly possible in large networks, as this would require to extract and solve for each $v \in V$ and for each $r \in [1, D]$ the respective hitting set instance on $S_{2r}(v)$. Even just the extraction of the instances – requiring Dijkstra computations from all $w \in B_{2r}(v)$ and storing all shortest paths in $S_{2r}(v)$ – is too demanding for larger r to be practical. But we observe that for any node $v \in V$, for any r, and for any subcollection S of the paths in $S_{2r}(v)$, any lower bound on the hitting set (HS) size for these paths is also a lower bound for h. Solving the relaxation of the respective standard HS-ILP gives a lower bound for the HS instance. Based on this observation, we propose the following algorithm to lower bound the highway dimension.

With h^* we denote the best lower bound for h found so far. Initially, we have $h^* = 0$. For a source node s, chosen randomly from a dense region of the graph, we consider a subset $S \subseteq S_{2r}(v)$ that is constructed as follows. We compute the shortest path tree of s up to a distance of $3r$, backtrack the paths of all leaves up to a distance of r (provided they are not too close to s) and add them to S. This is supposed to select paths that intersect $B_{2r}(s)$ only in a few nodes. Then we select a fixed number of nodes from $B_{2r}(s)$ (we use 500 in our experiments) and run Dijkstra computations up to a radius of $r + \delta$ from them, where δ is the maximum length of all edges incident to the root. Again we backtrack leave paths of sufficient lengths and add them to S.

Then we first run the standard greedy algorithm for HS. If the greedy solution does not exceed h^*, we discard the instance S, otherwise we consider the respective ILP. To reduce the size of the ILP (as otherwise a practical solution would be too time-consuming), we let P_i be the set of paths hit by node i selected in the greedy algorithm, and choose at most 100 paths from each P_i. With examples from each P_i considered in the ILP, we hope to preserve the essence of the instance. If the solution of the ILP relaxation exceeds h^*, we choose more nodes from $B_{2r}(s)$, perform Dijktra runs with radius $r + \delta$ from them and add the resulting paths to the ILP, still considering at most 100 paths from each P_i. We iterate this process as long as the solution of the extended ILP relaxation exceeds the previous solution. If the last solution exceeds h^*, we update h^* accordingly. Then we proceed and choose a new source s.

To choose the radius r in the experiments, we first apply the above described algorithm for a small number of selected source nodes and for a variety of radii in every network. We then select the radius that led to the largest lower bound for h in this test set for each network individually. For the source nodes that resulted in the best bounds in every network, we tried to improve h^* further by choosing different radii.

5   Skeleton Dimension Computation

In contrast to t and h, the skeleton dimension k can be computed in polynomial time by determining the skeleton of all shortest path trees in G and the maximum width therein. But for large networks, computing all shortest path trees is far too time consuming. We will now propose a significantly more efficient approach for determining k which takes the structure of road networks into account.

5.1   Computing the Width of a Tree

The width of a given rooted tree T can be computed by iterating over the nodes by increasing distance r from the root and storing the target nodes of the edges cut at radius r in a priority queue. In every step we pop all nodes from the priority queue that have the same distance label as the top element and push their direct descendants. The width of the tree is the maximal size of the priority queue during this process, which takes $O(n \log n)$ time for $n = |T|$.

5.2   Naive Algorithm

Naively, one would simply run Dijkstra's algorithm from every node, determine the distance to the furthest successor in every resulting shortest path tree for every contained node and construct the skeletons by pruning nodes whose furthest descendant is too close. As the skeleton is based on the geometric realization, it might also be required to insert an additional node $\tilde{w}$ at the end of every branch that satisfies $d_v(\tilde{w}) = 2Reach_v(\tilde{w})$. Then the widths can be computed. In order to compute for every node v the distance to its furthest descendant in a shortest path tree T, one can iterate over the nodes of T in reverse topological order and propagate the distance of the furthest descendant of every node to its predecessor. But as running a one-to-all Dijkstra computation alone is already expensive on large networks (order of multiple seconds), and the computation of the furthest descendants adds on that, one cannot afford to repeat this procedure for millions of nodes.

5.3   Faster Processing with PHAST

To perform efficient one-to-all shortest path computations we use the PHAST algorithm [7]. It requires a CH data structure G^+. To determine the shortest path distances from a node s to all other nodes, it executes a Dijkstra run in the CH upwards graph $G^{\uparrow}(s)$ before iterating over all edges of G^+ in the inverse contraction order of the source nodes. Whenever a shorter path to some node is discovered during this edge sweep, its distance and predecessor labels are updated. Eventually, every node is labeled with the correct distance. The predecessor labels correspond however only to a shortest path tree in the CH graph (which includes shortcuts) and the algorithm does not construct any explicit topological order. This disables the naive propagation of furthest descendant distance labels as described above.

But as every resulting shortest path is unimodal (i.e., it consists of a sequence of upwards edges followed by a sequence of downwards edges) , we can compute $Reach_s(v)$ for every node v as follows. After running a PHAST query, we iterate over all nodes in contraction order (i.e. nodes of low rank first) and propagate the distance of the furthest descendant of every node to its predecessor. When this step is finished, every node v is labeled with the distance to the furthest descendant x, for which the path from v to x consists only of downwards edges. In the next step, we iterate over the nodes in $G^{\uparrow}(s)$ in reverse topological order and propagate the descendant labels again to the predecessors. Now, every node is labeled with the distance of its furthest descendant in the shortest path tree in the CH graph. To propagate the correct value also to the nodes, that are shortcut on the shortest path from the root to their furthest descendant, we sweep over all edges of G^+ as in the PHAST query. For every edge (v, w) we check if it is contained in the actual shortest path tree, which is the case if $d_s(v) + \ell(v, w) = d_s(w)$ where $\ell(v, w)$ denotes the edge cost. If this holds, we propagate the descendant label of v to w. With this approach, we can compute the value $Reach_s(v)$ for every node v wrt some root s in the time required for one Dijkstra run in the upwards graph and two linear sweeps over the edges in the CH graph (creating only a mild overhead of one edge sweep over PHAST). But as each PHAST run still requires about 1 s on a continental-sized network (without using SSE instructions, GPUs, or further advanced optimization) [7], we now describe multiple steps which allow to spare a lot computations.

5.4   Upper Bounds

We observe that given a supergraph T' of a tree T, the width of T' is an upper bound on the width of T. To compute small supergraphs of the shortest path tree skeletons in G and hence to obtain good upper bounds on the skeleton dimension, we make use of the following lemma.

Lemma 2. *Let v be a descendant of u in the shortest path tree of s. Then we have $Reach_s(v) \leq Reach_u(v)$.*

Assume now that for some nodes $a_1, \ldots, a_\ell$ the distances $Reach_{a_i}(u)$ are already known for all $i \in \{1, \ldots, \ell\}$ and all $u \in V$. Then we can compute a supergraph of the skeleton T_s^* by performing a Dijkstra search from s and keeping track of the first a_i encountered on every branch of the search tree. Whenever we scan a node v such that $d_s(v) > 2 \cdot Reach_{a_i}(v)$ for the corresponding preceding

a_i, it follows from Lemma 2 that no descendant of v is contained in T_s^* an we can prune the current branch. When the algorithm terminates, it has explored a supergraph of the skeleton.

But if none of the nodes $a_1, \ldots, a_\ell$ was encountered on some branch, it has been explored entirely and can still be pruned. Therefore, we iterate over all nodes v in such a branch in reverse topological order and determine the value $Reach_s(v)$ before discarding all nodes v with $d_s(v) > 2 \cdot Reach_s(v)$ and adding boundary nodes $\tilde{v}$ that satisfy $d_s(\tilde{v}) = 2 \cdot Reach_s(\tilde{v})$. After this step, we obtain a smaller supergraph of T_s^* and compute its width $\hat{k}(s)$, which is an upper bound on the width $k(s)$ of T_s^*.

5.5   Exploiting Transit Nodes

The idea of our algorithm is to compute a transit node set of the given network and to use the access nodes $AN(v)$ as the nodes $a_1, \ldots, a_\ell$ of every node v. Usually, for every node $v \in V$ there is a whole set of nodes $v_1, \ldots, v_c$ that satisfy $AN(v) = AN(v_i)$ for $i = 1, \ldots, c$. We call such a set $\{v_1, \ldots, v_c\}$ also a *cell*. For every cell $\{v_1, \ldots, v_c\}$ and every access node $a_j \in AN(v_1)$, we need to compute $Reach_{a_j}(u)$ only once in order to compute $\hat{k}(v_i)$ for all $v_i \in \{v_1, \ldots, v_c\}$ as described previously. The whole network is processed by iterating over all cells via a depth-first search. As adjacent cells are very likely to share some of their access nodes, we store the computed values for $Reach_{a_j}(u)$ for the most recent access nodes in a least recently used cache, where the values of the least recently considered access node gets evicted when the cache is full. By doing so we can avoid computing the values of $Reach_{a_j}(u)$ several times for some access nodes.

5.6   Pruning

During the computation some nodes can further be pruned based on the following lemma.

Lemma 3. *A path $v_1, \ldots, v_p$ is called a* chain *with end nodes v_1 and v_p, if $v_2, \ldots, v_{p-1}$ have degree 2 and v_1, v_n do not. Let $v_1, \ldots, v_p$ be a chain. Then it yields:*

> *i. We have $k(v_1) \leq k(v_n)$ if $deg(v_1) = 1$.*
> *ii. For all $i \in \{2, \ldots, p-1\}$ we have $k(v_i) \leq \max\{2, k(v_n)\}$ if $deg(v_1) = 1$.*
> *iii. For all $i \in \{2, \ldots, p-1\}$ we have $k(v_i) \leq k(v_1) + k(v_n)$.*

Note that replacing $k(v_1)$ and $k(v_n)$ on the right hand side of the inequalities by some upper bound $\hat{k}(v_1)$ and $\hat{k}(v_n)$ does not invalidate the same. Provided that the network contains some node of degree at least 3 (which implies $k \geq 3$), we can therefore skip all nodes of degree 1 or less in our algorithm (isolated nodes do not contribute to the skeleton dimension at all). Consider now a node u with $deg(u) = 2$ that lies on a chain with end nodes v and v'. Then we can simply choose $\hat{k}(u) = \max\{2, \hat{k}(v)\}$ if $deg(v') = 1$, $\hat{k}(u) = \max\{2, \hat{k}(v')\}$

if $deg(v) = 1$ and $\hat{k}(u) = \hat{k}(v) + \hat{k}(v')$ otherwise. This requires however that $\hat{k}(v)$ and $\hat{k}(v')$ are already known. In every cell we consider therefore all non degree 2 nodes before processing the degree 2 nodes. It may however happen that the bound $\hat{k}(u) = \hat{k}(v) + \hat{k}(v')$ is not very tight. At the beginning of our algorithm we compute therefore a lower bound $\check{k}$ on the skeleton dimension k by computing the width of a few skeletons and choosing $\check{k}$ as the maximum of this widths. Then we use the bound $\hat{k}(u) \leq \hat{k}(v) + \hat{k}(v')$ for a degree 2 node u only if $\hat{k}(v) + \hat{k}(v') \leq \check{k}$, otherwise we compute a better bound based on the access nodes of u.

5.7 Computing Exact Values

When all cells have been processed, we have an upper bound $\hat{k}(u)$ on the width of every skeleton T_u^*. In order to compute the exact value of k, we iterate over all nodes u sorted descending by $\hat{k}(u)$ and compute the actual skeleton T_u^* and its width. During this process we keep updating the lower bound $\check{k}$, the maximum width of all skeletons computed so far. If at some point for the currently considered node u we have $\hat{k}(u) \leq \check{k}$, it follows that $k = \check{k}$. Provided that the bounds $\hat{k}(u)$ are not too bad, this last step involves considerably fewer complete one-to-all shortest path computations than the naive approach.

6 Experiments

We implemented the proposed algorithms for computing bounds on the highway dimension and the skeleton dimension in C++. We used the GNU C++ compiler 5.4.0 with optimization level 3. Experiments were conducted on a AMD Ryzen Threadripper 1950X CPU (16 cores and 32 threads; clocked at 2.2 GHz) with 128 GB main memory, running Ubuntu 16.04.3 (kernel 4.13.0).

Experiments were executed on the OSM road network of Germany (22.9 million nodes; 24.6 million undirected edges) and the TIGER/Line data of the US (24.3 million nodes; 29.5 million undirected edges). Shortest paths were computed wrt travel time. For the TIGER/Line data, the travel time is the spatial distance divided by a factor between 0.4 and 1.0 depending on the road category.

We applied our algorithms for computing (bounds on) the highway and skeleton dimension to the aforementioned networks and several subnetworks. Some of the results are shown in Table 1. Note that the computed values are slightly implementation-dependent, as in the considered networks some shortest paths are ambiguous.

We observe that the skeleton dimension is significantly smaller than the highway dimension for all considered networks. For the highway dimension, it seems safe to assume that due to the nature of our lower bound construction the obtained results are far from being tight (and the more so the larger the network). Still, there seems to be a clear correlation between the size of the network and the highway dimension (cf. Fig. 1). For the complete US (24 million nodes) we computed a lower bound of 512, whereas for Michigan (739,000 nodes) and

Table 1. Highway and skeleton dimensions of different networks and the corresponding radii r in seconds/time units. $\sqrt{n}$ and $\log_2(n)$ are provided for comparison.

Network	n	$\sqrt{n}$	$\log_2(n)$	h	r	k	r
Germany	22,919,324	4,787	24	≥ 512	1,000	114	186
US	24,278,285	4,927	25	≥ 562	2,826	92	650
Michigan	673,534	821	19	≥ 321	22,829	92	650
Washington, D.C.	9,559	98	13	≥ 156	17,806	42	8,903

Washington, D.C. (10,000 nodes) we obtained $h \geq 321$ and $h \geq 156$, respectively. Taking the looseness of our bounds into account, the results suggest that the dependency between h and n might be even superlogarithmic (cf. Table 1). But of course, the model $h \leq a \cdot \log^b n$ can not be declared void based on our results as there might be suitable choices of a and b that also hold for larger networks. For more conclusive results, better lower bound techniques (in terms of running time and solution quality) need to be investigated in future work.

Computing the exact skeleton dimension was possible on all benchmark networks. For the US, the computation required 10 days runtime (using 18 threads on average) and 110 GB RAM. We used $50\sqrt{n}$ transit nodes and dedicated 50 GB memory to simultaneously store the results of 1,200 PHAST runs. In the last step slightly less than one million exact computations were required to close the gap between upper and lower bounds. More than 99% of the degree 2 nodes could be pruned. On these instances we estimated that the naive algorithm based on Dijkstra computations would have required more than 6 years CPU time, and simply using CPU-based PHAST runs followed by computing the distances to the furthest descendants without our improvements and pruning strategies a bit less than 3 years.

We observe for k that the radius where the maximum width in the skeleton is assumed is small in most considered networks. For the network of Germany we obtained $k = 114$. This value is assumed at a radius of only 186 s; for larger radii the skeleton width drops significantly (cf. Fig. 2). The relevant part of the network which contains a skeleton of width 114 has only about 16,000 nodes (cf. Fig. 3). This indicates that the value of k is indeed not dependent on the total size of the network but rather on the densest cluster therein (e.g. the largest city). This is also reflected in our results on the networks of the whole US and its individual states. There we have $k < 100$ for all instances (cf. Fig. 1) and the whole US induces the same skeleton dimension value as Michigan (cf. Table 1). Moreover, the maximum skeleton width is almost always assumed within a metropolis. But the shape and the distribution of cities within the network also have an influence on the road network dimensions and the respective radii.

In [12], bounds of $\mathcal{O}(k \log D)$ and $\mathcal{O}(kn \log D)$ were proven for the search spaces and space consumption of hub labels, respectively. Our results imply that search spaces are in $\mathcal{O}(\log D)$ and the space consumption in $\mathcal{O}(n \log D)$.

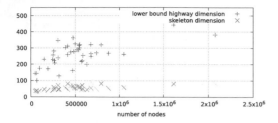

Fig. 1. Distribution of the highway dimension h and the skeleton dimension k in dependency of the road network size for all states of the US.

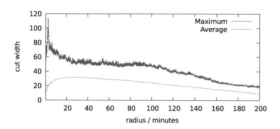

Fig. 2. The skeleton width in dependency of the radius in the road network of Germany

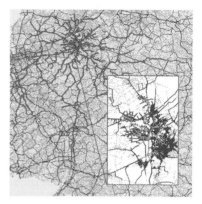

Fig. 3. A skeleton of width 114 (green), the 114 shortest paths cut at a radius of 186 s (yellow) and the relevant ball of radius 279 s (blue) (Color figure online)

7   Conclusions and Future Work

We showed that computing the exact skeleton dimension of large road networks is possible when applying suitable speed-up schemes and pruning techniques. The obtained experimental results strongly imply that the skeleton dimension depends on the local graph density rather than the network size. As the computed values are small for all considered benchmark networks, the skeleton dimension seems to be an excellent parameter for conducting practically useful theoretical analyses.

The highway dimension turned out to be larger than the skeleton dimension in all networks, and the more the bigger the network. The growth is at least logarithmic in the size of the network but possibly way larger. Besides the listed route planning techniques, also many other algorithms have been analyzed based on the assumption that the highway dimension of road networks is small, see e.g. [9,10]. It might be worthwhile to investigate whether the bounds obtained in these analyses also hold when replacing h with k, as the respective results would be much stronger.

References

1. Abraham, I., Delling, D., Fiat, A., Goldberg, A.V., Werneck, R.F.: VC-dimension and shortest path algorithms. In: Aceto, L., Henzinger, M., Sgall, J. (eds.) ICALP 2011. LNCS, vol. 6755, pp. 690–699. Springer, Heidelberg (2011)
2. Abraham, I., Delling, D., Goldberg, A.V., Werneck, R.F.: A hub-based labeling algorithm for shortest paths in road networks. In: Pardalos, P.M., Rebennack, S. (eds.) SEA 2011. LNCS, vol. 6630, pp. 230–241. Springer, Heidelberg (2011)
3. Abraham, I., Fiat, A., Goldberg, A.V., Werneck, R.F.: Highway dimension, shortest paths, and provably efficient algorithms. In: Proceedings of the 21st Annual ACM-SIAM Symposium on Discrete Algorithms (SODA), pp. 782–793. SIAM (2010)
4. Bast, H., Funke, S., Sanders, P., Schultes, D.: Fast routing in road networks with transit nodes. Science **316**(5824), 566 (2007)
5. Bauer, R., Columbus, T., Rutter, I., Wagner, D.: Search-space size in contraction hierarchies. In: Fomin, F.V., Freivalds, R., Kwiatkowska, M., Peleg, D. (eds.) ICALP 2013. LNCS, vol. 7965, pp. 93–104. Springer, Heidelberg (2013). https://doi.org/10.1007/978-3-642-39206-1_9
6. Dell, H., Komusiewicz, C., Talmon, N., Weller, M.: The pace 2017 parameterized algorithms and computational experiments challenge: the second iteration. In: Proceedings of the 12th International Symposium on Parameterized and Exact Computation (IPEC), Leibniz International Proceedings in Informatics (LIPIcs), Dagstuhl, Germany, pp. 30:1–30:12. Schloss Dagstuhl-Leibniz-Zentrum für Informatik (2017)
7. Delling, D., Goldberg, A.V., Nowatzyk, A., Werneck, R.F.: PHAST: hardware-accelerated shortest path trees. In: Proceedings of the 25th IEEE International Symposium on Parallel and Distributed Processing (IPDPS), pp. 921–931. IEEE (2011)
8. Dibbelt, J., Strasser, B., Wagner, D.: Customizable contraction hierarchies. In: Gudmundsson, J., Katajainen, J. (eds.) SEA 2014. LNCS, vol. 8504, pp. 271–282. Springer, Cham (2014)
9. Feldmann, A.E.: Fixed parameter approximations for k-center problems in low highway dimension graphs. In: Halldórsson, M.M., Iwama, K., Kobayashi, N., Speckmann, B. (eds.) ICALP 2015. LNCS, vol. 9135, pp. 588–600. Springer, Heidelberg (2015)
10. Feldmann, A.E., Fung, W.S., Könemann, J., Post, I.: A $(1 + \varepsilon)$-embedding of low highway dimension graphs into bounded treewidth graphs. In: Halldórsson, M.M., Iwama, K., Kobayashi, N., Speckmann, B. (eds.) ICALP 2015. LNCS, vol. 9134, pp. 469–480. Springer, Heidelberg (2015). https://doi.org/10.1007/978-3-662-47672-7_38
11. Geisberger, R., Sanders, P., Schultes, D., Vetter, C.: Exact routing in large road networks using contraction hierarchies. Transp. Sci. **46**(3), 388–404 (2012)
12. Kosowski, A., Viennot, L.: Beyond highway dimension: small distance labels using tree skeletons. In: Proceedings of the 28th Annual ACM-SIAM Symposium on Discrete Algorithms (SODA), pp. 1462–1478. SIAM (2017)
13. Tamaki, H.: Positive-instance driven dynamic programming for treewidth. In: Proceedings of the 25th Annual European Symposium on Algorithms (ESA), LIPIcs, vol. 87, pp. 68:1–68:13. Schloss Dagstuhl - Leibniz-Zentrum für Informatik (2017)

Car-Sharing Between Two Locations: Online Scheduling with Flexible Advance Bookings

Kelin Luo[1,2(✉)], Thomas Erlebach[2(✉)], and Yinfeng Xu[1]

[1] School of Management, Xi'an Jiaotong University, Xi'an, China
luokelin@stu.xjtu.edu.cn
[2] Department of Informatics, University of Leicester, Leicester, UK
te17@leicester.ac.uk

Abstract. We study an on-line scheduling problem that is motivated by applications such as car-sharing. Users submit ride requests, and the scheduler aims to accept requests of maximum total profit using a single server (car). Each ride request specifies the pick-up time and the pick-up location (among two locations, with the other location being the destination). The scheduler has to decide whether or not to accept a request immediately at the time when the request is submitted (booking time). We consider two variants of the problem with respect to constraints on the booking time: In the fixed booking time variant, a request must be submitted a fixed amount of time before the pick-up time. In the variable booking time variant, a request can be submitted at any time during a certain time interval that precedes the pick-up time. We present lower bounds on the competitive ratio for both variants and propose a greedy algorithm that achieves the best possible competitive ratio.

1 Introduction

In a car-sharing system, a company offers cars to customers for a period of time. Customers can pick up a car in one location, drive it to another location, and return it there. Car booking requests arrive on-line, and the goal is to maximize the profit obtained from satisfied requests. We consider a setting where all driving routes go between two fixed locations, but can be in either direction. For example, the two locations could be a residential area and a nearby shopping mall or central business district. Other applications that provide motivation for the problems we study include taxi dispatching and boat rental for river crossings.

In real life, customer requests for car bookings usually arrive over time, and the decision about each request must be made immediately, without knowledge of future requests. This gives rise to an on-line problem that bears some resemblance to interval scheduling, but in which additionally the pick-up and drop-off locations play an important role: The server that serves a request must be at the pick-up location at the start time of the request and will be located at the

L. Wang and D. Zhu (Eds.): COCOON 2018, LNCS 10976, pp. 242–254, 2018.
https://doi.org/10.1007/978-3-319-94776-1_21

drop-off location at the end time of the request. A server can serve two consecutive requests only if the drop-off location of the first request is the same as the pick-up location of the second request, or if there is enough time to travel between the two locations otherwise. (We allow 'empty movements' that allow a server to be moved from one location to another while not serving a request. Such empty movements could be implemented by having company staff drive a car from one location to another, or in the future by self-driving cars.)

An important aspect of the problem is the relation between the *booking time*, i.e., the time when the request is submitted, and the *start time* of the request, i.e., the time when the customer picks up the car at the pick-up location. Constraints on the booking time (also called the *reservation window* in the context of advance reservation systems) can affect the performance of a system [7]. There are generally two types of bookings, current and advance. Current bookings are requests that are released and must be served immediately. Advance bookings are requests that are released before the start time. In this paper we consider advance bookings. More specifically, we study two variants of advance bookings: In the *fixed booking time* variant, the amount of time between the booking time of a request and its start time is a fixed value, independent of the request. In the *variable booking time* variant, the booking time of a request must lie in a certain time interval (called the *booking horizon*) before the start time of the request.

We assume that every request is associated with a profit that is obtained if the request is accepted. When a server moves from one location to another while not serving a request, a certain cost is incurred. The goal is to maximize the total profit, which is the sum of the profits of the accepted requests minus the costs incurred for moving servers while not serving a request. In this paper, we focus on the special case of a single server.

1.1 Related Work

The car sharing problem considered in this paper belongs to the class of dynamic pickup and delivery problems surveyed by Berbeglia et al. [2]. The problem that is closest to our setting is the on-line dial-a-ride problem (OLDARP) that has been studied widely. In OLDARP, transportation requests between locations in a metric space arrive over time, but typically it is assumed that requests want to be served 'as soon as possible' rather than at a specific time as in our problem. Known results for OLDARP include on-line algorithms for minimizing the makespan [1,3] or the maximum flow time [8]. Work on versions of OLDARP where not all requests can be served includes competitive algorithms for requests with deadlines where each request must be served before its deadline or rejected [9], and for settings with a given time limit where the goal is to maximize the revenue from requests served before the time limit [6]. In contrast to existing work on OLDARP, in this paper we consider requests that need to be served at a specific time that is specified by the request when it is released.

Off-line versions of car-sharing problems are studied by Böhmová et al. [4]. They show that if all customer requests for car bookings are known in advance, the problem of maximizing the number of accepted requests can be solved in polynomial time using a minimum-cost network flow algorithm. Furthermore, they consider the problem variant with two locations where each customer requests two rides (in opposite directions) and the scheduler must accept either both or neither of the two. They prove that this variant is NP-hard and APX-hard. In contrast to their work, we consider the on-line version of the problem.

1.2 Problem Description and Preliminaries

We consider a setting with only two locations (denoted by 0 and 1) and a single server. The travel time from 0 to 1 is the same as the travel time from 1 to 0 and is denoted by t. Let R denote a sequence of requests that are released over time. The i-th request is denoted by $r_i = (\tilde{t}_{r_i}, t_{r_i}, p_{r_i})$ and is specified by the *booking time* or *release time* $\tilde{t}_{r_i}$, the *start time* t_{r_i}, and the pick-up location $p_{r_i} \in \{0, 1\}$. Requests with the same release time arrive one by one in arbitrary order, and each request must be processed by the algorithm before the next request arrives. If r_i is accepted, the server must pick up the customer at p_{r_i} at time t_{r_i} and drop off the customer at location $\dot{p}_{r_i} = 1 - p_{r_i}$ at time $\dot{t}_{r_i} = t_{r_i} + t$, the *end time* of the request. We say that the request r_i *starts* at time t_{r_i}. For an interval $[b, d)$, we say that r_i starts in the interval if $t_{r_i} \in [b, d)$.

The server can only serve one request at a time. Serving a request yields profit $r > 0$. The server is initially located at location 0. If the pick-up location p_{r_i} of a request r_i is different from the current location of the server and if at least t time units remain before the start time of r_i, the server can move from its current location to p_{r_i}. We refer to such moves (which do not serve a request) as *empty* moves. An empty move takes time t and incurs a cost of c, $0 \leq c \leq r$, and we say that r_i is accepted *with cost* in this case. If the server is already located at p_{r_i}, we say that r_i is accepted *without cost*. We forbid 'unprompted' moves, i.e., the algorithm is allowed to make an empty move to the other location only if it does so in order to serve a request that was accepted before the current time and whose pick-up location is the other location. If two requests are such that they cannot both be served by one server, we say that the requests are *in conflict*.

We denote the requests accepted by an algorithm by R', and the i-th request in R', in order of request start times, is denoted by r_i'. We say that request r_i' is accepted *without cost* if $i = 1$ and $p_{r_i'} = 0$ or if $i > 1$ and $p_{r_i'} = \dot{p}_{r_{i-1}'}$. Otherwise, r_i' is accepted *with cost*. We denote the profit of serving the requests in R' by $P_{R'}$. If R_c' denotes the subset of R' consisting of the requests that are accepted with cost, we have $P_{R'} = r \cdot |R'| - c \cdot |R_c'|$.

The goal is to accept a set of requests R' that maximizes the profit $P_{R'}$. The problem for one server and two locations for the fixed booking time variant in which $t_{r_i} - \tilde{t}_{r_i} = a$ for all requests r_i, where $a \geq 0$ is a constant, is called the *1S2L-F problem*. For the variable booking time variant, the booking time $\tilde{t}_{r_i}$ of any request r_i must satisfy $t_{r_i} - b_u \leq \tilde{t}_{r_i} \leq t_{r_i} - b_l$, where b_l and b_u are constants, with $b_l \leq b_u$, that specify the minimum and maximum length, respectively, of the time interval between booking time and start time. The problem for one server and two locations for the variable booking time variant is called the *1S2L-V problem*. If $b_l = b_u$, the 1S2L-V problem turns into the 1S2L-F problem.

The performance of an algorithm for 1S2L-F or 1S2L-V is measured using competitive analysis (see [5]). For any request sequence R, let P_{R^A} denote the objective value produced by an on-line algorithm A, and P_{R^*} that obtained by an optimal scheduler OPT that has full information about the request sequence in advance. Like for the algorithm, we also require that OPT does not make unprompted moves, i.e., OPT is allowed to make an empty move starting at time t_0 only if there is an accepted request r_i with $\tilde{t}_{r_i} \le t_0$ and $t_{r_i} \ge t_0 + t$ whose pick-up location is the other location. Without this restriction on OPT, it would not be possible to achieve finite competitive ratio in cases where a request can be booked less than t units of time before its starting time.

The competitive ratio of A is defined as $\rho_A = \sup_R \frac{P_{R^*}}{P_{R^A}}$. We say that A is ρ-competitive if $P_{R^*} \le \rho \cdot P_{R^A}$ for all request sequences R. Let ON be the set of all on-line algorithms for a problem. A value β is a *lower bound* on the best possible competitive ratio if $\rho_A \ge \beta$ for all A in ON. We say that an algorithm A is optimal if there is a lower bound β with $\rho_A = \beta$.

1.3 Paper Outline

In Sect. 2, we study the 1S2L-F problem. We give lower bounds and propose a greedy algorithm that achieves the best possible competitive ratio. In Sect. 3, we study the 1S2L-V problem. Although variable booking times provide much greater flexibility to customers, we show that our greedy algorithm is still optimal. Some proofs are omitted due to space restrictions. An overview of our results is shown in Table 1.

Table 1. Lower and upper bounds for the car sharing problem

Problem	Booking constraint	$0 \le c < r$		$c = r$	
		LB	UB	LB	UB
1S2L-F	$0 \le a < t$	1	1	1	1
1S2L-F	$t \le a$	$\frac{2r}{r-c}$	$\frac{2r}{r-c}$	1	1
1S2L-V	$0 < b_u < t$	3	3	3	3
1S2L-V	$b_u = t$	$\max\{\frac{2r}{r-c}, 3\}$	$\max\{\frac{2r}{r-c}, 3\}$	3	3
1S2L-V	$t < b_u$	$\frac{3r-c}{r-c}$	$\frac{3r-c}{r-c}$	$1 + 2\lceil \frac{b_u - b_l}{2t} \rceil$	$1 + 2\lceil \frac{b_u - b_l}{2t} \rceil$

2   Car Sharing with Fixed Booking Times

In this section, we study the 1S2L-F problem. First, we present a lower bound. We use ALG to denote any on-line algorithm and OPT to denote an optimal scheduler. We refer to the server of ALG and OPT as s' and s^*, respectively. The set of requests accepted by ALG is referred to as R', and the set of requests accepted by OPT as R^*.

Theorem 1. *For $0 \leq c < r$ and $a \geq t$, no deterministic on-line algorithm for 1S2L-F can achieve competitive ratio smaller than $\frac{2r}{r-c}$.*

Proof. Initially, the adversary releases a request $r_1 = (0, a, 1)$. We distinguish two cases.

Case 1: ALG accepts r_1 (with cost). The adversary releases requests $r_2 = (\varepsilon, a + \varepsilon, 0)$ and $r_3 = (\varepsilon + t, a + \varepsilon + t, 1)$, where $0 < \varepsilon < t$. OPT accepts r_2 and r_3 without cost, but ALG cannot accept either of these requests as they are in conflict with r_1. We have $P_{R^*} = 2r$ and $P_{R'} = r - c$, and hence $\frac{P_{R^*}}{P_{R'}} = \frac{2r}{r-c}$.

Case 2: ALG does not accept request r_1. In this case, OPT accepts r_1 and we have $P_{R^*} = r - c$ and $P_{R'} = 0$, and hence $\frac{P_{R^*}}{P_{R'}} = \infty$. □

Algorithm 1. Greedy Algorithm (GA)

Input: one server, requests arrive over time.

Step: When request r_i arrives, accept r_i if r_i is acceptable and $P_{R_i^{GA} \cup \{r_i\}} - P_{R_i^{GA}} > 0$;

Note 1. R_i^{GA} is the set of requests accepted by GA before r_i is released.

Note 2. r_i is acceptable if and only if $\forall r_j' \in R_i^{GA}, |t_{r_i} - t_{r_j'}| \geq 2t$ if $p_{r_i} = p_{r_j'}$, and $|t_{r_i} - t_{r_j'}| \geq t$ if $p_{r_i} \neq p_{r_j'}$, and $t_{r_i} - \tilde{t}_{r_i} \geq t$ if s' is at location $p_{s'} \in \{0, 1\}$ at time $\tilde{t}_{r_i}$ and $p_{r_i} = 1 - p_{s'}$.

We propose a Greedy Algorithm (GA) for the 1S2L-F problem, shown in Algorithm 1. For an arbitrary request sequence $R = \{r_1, r_2, r_3, \ldots, r_n\}$, note that we have $t_{r_i} \leq t_{r_{i+1}}$ for $1 \leq i < n$ because $t_{r_i} - \tilde{t}_{r_i} = a$ is fixed. Denote the requests accepted by OPT by $R^* = \{r_1^*, r_2^*, \ldots, r_{k^*}^*\}$ and the requests accepted by GA by $R' = \{r_1', r_2', \ldots, r_k'\}$ indexed in order of non-decreasing start times.

Theorem 2. *Algorithm GA is 1-competitive for 1S2L-F if $c = r$, or if $0 \leq c < r$ and $0 \leq a < t$.*

Proof. If $0 \leq c < r$ and $0 \leq a < t$, GA and OPT only accept requests without cost because the release time of a request is too late for the server to be able to serve it with cost (recall that we forbid unprompted moves by OPT). Observe that this means that both GA and OPT accept requests with alternating pick-up location, starting with a request with pick-up location 0.

We claim that R^* can be transformed into R' without reducing its profit, thus showing that $P_{R^*} = P_{R'}$. As GA accepts the first request r_j with $p_{r_j} = 0$, it is clear that $t_{r_1'} \leq t_{r_1^*}$. If $r_1' \neq r_1^*$, we can replace r_1^* by r_1' in R^*, and R^* is still a valid solution with the same profit. Now assume, that R' and R^* are identical with respect to the first i requests, and that s' and s^* are at location $p \in \{0, 1\}$ at time $\tilde{t}_{r_i'}$. If there is a request r_{i+1}^*, there must also be a request r_{i+1}' with $t_{r_{i+1}'} \leq t_{r_{i+1}^*}$, as GA could accept r_{i+1}^*. We can replace r_{i+1}^* by r_{i+1}' in R^*. The claim thus follows by induction.

If $c = r$, accepting a request with cost yields profit $r - c = 0$. Without loss of generality, we can therefore assume that both GA and OPT only accept requests without cost. The arguments of the previous paragraph can then be applied to this case as well. □

Theorem 3. *Algorithm GA is $\frac{2r}{r-c}$-competitive for 1S2L-F if $0 \leq c < r$ and $a \geq t$.*

Proof. We partition the time horizon $[0, \infty)$ into intervals (periods) that can be analyzed independently. Period i, for $1 < i < k$, is the interval $[t_{r'_i}, t_{r'_{i+1}})$. Period 1 is $[0, t_{r'_2})$, and period k is $[t_{r'_k}, \infty)$. (If $k = 1$, there is only a single period $[0, \infty)$.) Exactly one request in R', namely r'_i, starts in period i, for $1 \leq i \leq k$. We define $R'_i = \{r'_i\}$ for $1 \leq i \leq k$. Let R^*_i denote the set of requests accepted by OPT that start in period i, for $1 \leq i \leq k$.

For $1 < i \leq k$, r'_i starts at time $t_{r'_i}$ and the first request of R^*_i starts during the interval $[t_{r'_i}, t_{r'_{i+1}})$ (or the interval $[t_{r'_i}, \infty)$ if $i = k$). Furthermore, r'_1 is the first acceptable request in R, and so the first request of R^*_1 cannot start before r'_1. Hence, for all $1 \leq i \leq k$, the first request in R^*_i cannot start before the request r'_i.

We bound the competitive ratio of GA by analyzing each period independently. As $R' = \bigcup_i R'_i$ and $R^* = \bigcup_i R^*_i$, it is clear that $P_{R^*}/P_{R'} \leq \alpha$ follows if we can show that $P_{R^*_i}/P_{R'_i} \leq \alpha$ for all i, $1 \leq i \leq k$.

For all $1 \leq i \leq k$, as $R'_i = \{r_i\}$, we have $P_{R'_i} \in \{r, r - c\}$. It suffices to show $P_{R^*_i}/P_{R'_i} \leq 2r/(r - c)$ to prove the theorem. We claim that R^*_i contains at most two requests. Assume that R^*_i contains at least three requests. Let r_j be the third request (in order of start time) in R^*_i. As the first request in R^*_i does not start before $t_{r'_i}$, we have $t_{r_j} \geq t_{r'_i} + 2t$. This means that r_j would be acceptable to GA after it has accepted r'_i. Therefore, GA accepts either r_j or another request starting before t_{r_j}, and that request becomes r'_{i+1}. Hence, there cannot be such a request r_j that starts in period i.

As we have shown that R^*_i contains at most two requests, we get that $P_{R^*_i} \leq 2r$. Since $P_{R'_i} \geq r - c$, we have $P_{R^*_i}/P_{R'_i} \leq 2r/(r - c)$. The theorem follows. □

3 Car Sharing with Variable Booking Times

In this section, we study the 1S2L-V problem. Recall that the booking time of a request r_i must satisfy $t_{r_i} - b_u \leq \tilde{t}_{r_i} \leq t_{r_i} - b_l$. First, we present three lower bound results, one for the case $c = r$ and two for the case $c < r$.

Theorem 4. *No deterministic algorithm for 1S2L-V can have competitive ratio smaller than $1 + 2\lceil \frac{b_u - b_l}{2t} \rceil$ if $c = r$. In particular, the lower bound is 3 if $0 < b_u < t$.*

Proof. Let ALG be an arbitrary on-line algorithm, and let OPT be an optimal scheduler. We distinguish two cases based on the value of $b_u - b_l$.

Case 1: $0 < b_u - b_l \leq 2t$. We need to show that the competitive ratio is at least 3. Define four requests as follows: $r_1 = (\frac{b_u + b_l}{2} + t, b_u + b_l + t, 0)$, $r_2 = (\frac{2b_u + b_l}{3} + t, b_l + \frac{2b_u + b_l}{3} + t, 0)$, $r_3 = (\frac{2b_u + b_l}{3} + 2t, b_l + \frac{2b_u + b_l}{3} + 2t, 1)$, $r_4 =$

$(\frac{2b_u+b_l}{3} + 3t, b_l + \frac{2b_u+b_l}{3} + 3t, 0)$. Note that a server can accept either r_1, or all of r_2, r_3, r_4. Furthermore, r_2 is released after r_1 but starts earlier.

Initially, the adversary releases r_1. There are two sub-cases.

Case 1.1: ALG accepts r_1. The adversary releases r_2, r_3 and r_4. OPT accepts r_2, r_3, r_4 without cost, so we have $P_{R^*} = 3r$ and $P_{R'} = r$, showing that $P_{R^*}/P_{R'} = 3$.

Case 1.2: ALG does not accept request r_1. OPT accepts r_1. We have $P_{R^*} = r$ and $P_{R'} = 0$, and hence $P_{R^*}/P_{R'} = \infty$.

The lower bound of 3 follows.

Case 2: $2t < b_u - b_l$. Let $n = \lceil \frac{b_u-b_l}{2t} \rceil - 1$. Choose values ε_i for $1 \le i \le n+2$ satisfying $0 \le \varepsilon_1 < \varepsilon_2 < \cdots < \varepsilon_{n+1} < \varepsilon_{n+2} < \min\{t, b_u - b_l - 2tn\}$.

Initially, the adversary releases the request sequence R_1 consisting of the following requests: $r_1 = (\varepsilon_1, b_l + \varepsilon_{n+2} + t, 1), r_2 = (\varepsilon_2, b_l + \varepsilon_{n+2} + 3t, 1), \ldots, r_i = (\varepsilon_i, b_l + \varepsilon_{n+2} + (2i - 1)t, 1), \ldots, r_n = (\varepsilon_n, b_l + \varepsilon_{n+2} + (2n - 1)t, 1)$ and $r_{n+1} = (\varepsilon_{n+1}, b_u + \varepsilon_{n+1}, 0)$. Note that $b_l \le b_l + \varepsilon_{n+2} + (2i - 1)t \le b_u$ for all $1 \le i \le n$. There are three sub-cases.

Case 2.1: ALG rejects all the requests of R_1. In this case, OPT accepts the request r_{n+1}. We have $P_{R^*} = r$ and $P_{R'} = 0$, yielding $P_{R^*}/P_{R'} = \infty$.

Case 2.2: The first request accepted by ALG is r_i for some i with $1 \le i \le n$. In this case, the adversary does not release the remaining requests of R_1. Instead, it releases only one final request $r_f = (\varepsilon_{i+1}, b_l + (2i - 1)t, 0)$. ALG cannot accept r_f as it is in conflict with r_i. OPT accepts r_f. We have $P_{R^*} = r$ and $P_{R'} = r - c = 0$, hence $P_{R^*}/P_{R'} = \infty$.

Case 2.3: The first request accepted by ALG is r_{n+1}. The adversary then releases the request sequence R_2 consisting of the following requests: $r_{n+1+1} = (\varepsilon_{n+2}, b_l + \varepsilon_{n+2}, 0), r_{n+1+2} = (\varepsilon_{n+2}, b_l + \varepsilon_{n+2} + 2t, 0), \ldots, r_{n+1+i} = (\varepsilon_{n+2}, b_l + \varepsilon_{n+2} + 2(i-1)t, 0), \ldots, r_{n+1+n} = (\varepsilon_{n+2}, b_l + \varepsilon_{n+2} + 2(n-1)t, 0)$. After this, the adversary releases the request sequence R_3 consisting of three more requests: $r_{2n+1+1} = (\varepsilon_{n+2}, b_u, 0), r_{2n+1+2} = (\varepsilon_{n+2} + t, b_u + t, 1), r_{2n+1+3} = (\varepsilon_{n+2} + 2t, b_u + 2t, 0)$. ALG cannot accept any requests of R_3 as they all conflict with r_{n+1}. ALG can accept any number of requests of R_2, but since they all have pick-up location 0 (as does r_{n+1}), its total profit will be $P_{R'} = r$. OPT accepts all requests of R_1 except r_{n+1}, and all requests of R_2 and R_3. We have $P_{R^*} = (2n + 3)r = (2\lceil \frac{b_u-b_l}{2t} \rceil + 1)r$. Hence, $P_{R^*}/P_{R'} = 1 + 2\lceil \frac{b_u-b_l}{2t} \rceil$.

The claimed lower bound of $1 + 2\lceil \frac{b_u-b_l}{2t} \rceil$ follows. □

Theorem 5. *No deterministic algorithm for 1S2L-V can have competitive ratio smaller than $\frac{3r-c}{r-c}$ if $0 \le c < r$ and $b_u > t$.*

Proof. Initially, the adversary releases the request $r_1 = (0, b_u, 1)$. We distinguish two cases.

Case 1: ALG accepts r_1 (with cost). The adversary releases requests $r_2 = (\varepsilon, t_{r_1} - \varepsilon, 1), r_3 = (\varepsilon + t, t_{r_2} + t, 0)$ and $r_4 = (\varepsilon + 2t, t_{r_3} + t, 1)$, where $0 < \varepsilon < \min\{t, \frac{b_u-b_l}{2}\}$. OPT accepts r_2, r_3 and r_4. As they are in conflict with r_1, ALG cannot accept any of them, see also Fig. 1. We have $P_{R^*} = 3r - c$ and $P_{R'} = r - c$. Hence, $P_{R^*}/P_{R'} = \frac{3r-c}{r-c}$.

ALG: accepts the first request   OPT: accepts three requests

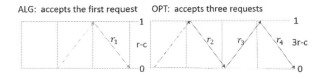

Fig. 1. Case 1: $\frac{P_{R^*}}{P_{R'}} = \frac{3r-c}{r-c}$

Case 2: ALG does not accept request r_1. In this case, OPT accepts r_1. We have $P_{R^*} = r - c$, $P_{R'} = 0$, and hence $P_{R^*}/P_{R'} = \infty$.   □

Theorem 6. *No deterministic algorithm for 1S2L-V can have competitive ratio smaller than 3 if $0 \le c < r$ and $b_u \le t$.*

Proof. Initially, the adversary releases the request $r_1 = (0, b_u, 0)$. We distinguish two cases.

Case 1: ALG accepts r_1 (without cost). The adversary releases requests $r_2 = (\varepsilon, t_{r_1} - \varepsilon, 0)$, $r_3 = (\varepsilon + t, t_{r_2} + t, 1)$ and $r_4 = (\varepsilon + 2t, t_{r_3} + t, 0)$, where $0 < \varepsilon < \min\{t, \frac{b_u - b_l}{2}\}$. OPT accepts r_2, r_3 and r_4. As they are in conflict with r_1, ALG cannot accept any of them. We have $P_{R^*} = 3r$ and $P_{R'} = r$. Hence, $P_{R^*}/P_{R'} = 3$.

Case 2: ALG does not accept request r_1. In this case, OPT accepts r_1. We have $P_{R^*} = r$, $P_{R'} = 0$, and hence $P_{R^*}/P_{R'} = \infty$.   □

From Theorems 1 and 6 we can conclude that no deterministic algorithm for 1S2L-V can have competitive ratio smaller than $\max\{\frac{2r}{r-c}, 3\}$ if $0 \le c < r$ and $b_u = t$.

We now turn to upper bounds and analyze Algorithm GA (which was presented as Algorithm 1 in Sect. 2) for the 1S2L-V problem. Denote the set of requests accepted by OPT by R^* and the set of requests accepted by GA as R'. The server of OPT is referred to as s^*, and the server of GA as s'. Let $R' = \{r'_1, \ldots, r'_k\}$, with the requests indexed in order of increasing start time. For $1 \le i \le k$, let $R'_i = \{r'_i\}$. We partition the time horizon $[0, \infty)$ into intervals (periods) that can be analyzed independently. The partition differs for GA and OPT, so we refer to *GA periods* and *OPT periods*. GA period i is the interval $[0, t_{r'_2})$ if $i = 1$, the interval $[t_{r'_k}, \infty)$ if $i = k$, and the interval $[t_{r'_i}, t_{r'_{i+1}})$ if $1 < i < k$. Note that R'_i consists of the only request in R' that starts in GA period i.

For $1 \le i \le k$, define $\hat{t}_{r'_i}$ to be the first time when the optimal server s^* is at location $\dot{p}_{r'_i}$ at or after time $\dot{t}_{r'_i}$, or ∞ if s^* never reaches $\dot{p}_{r'_i}$ from time $\dot{t}_{r'_i}$ onward. Now, define *OPT period* i to be the interval $[0, \hat{t}_{r'_1})$ if $i = 1$, the interval $[\hat{t}_{k-1}, \infty)$ if $i = k$, and the interval $[\hat{t}_{r'_{i-1}}, \hat{t}_{r'_i})$ if $1 < i < k$. For $1 \le i \le k$, let R^*_i be the set of requests accepted by OPT that start during *OPT period* i. If $\hat{t}_{r'_i} \le \hat{t}_{r'_{i-1}}$, *OPT period* i is empty and $R^*_i = \emptyset$. The jth (in order of start times) request of R^*_i is denoted by $R^*_i(j)$.

We will compare the profit $P_{R'_i}$ that GA accrues in GA period i with the profit $P_{R^*_i}$ that OPT accrues in OPT period i. We can again analyze each period independently: If we can show that $P_{R^*_i}/P_{R'_i} \leq \alpha$ for all i, this implies that $P_{R^*}/P_{R'} \leq \alpha$. We first state two observations.

Observation 1. *For all i, s' is at $\dot{p}_{r'_i}$ at time $\dot{t}_{r'_i}$, and s^* is at $\dot{p}_{r'_i}$ at time $\hat{t}_{r'_i}$, and $\hat{t}_{r'_i} \geq \dot{t}_{r'_i}$.*

Observation 2. *If $c = r$, by definition of Algorithm GA (Algorithm 1), for $1 < i \leq k$ we have that $\hat{t}_{r'_{i-1}} \leq \tilde{t}_{r'_i}$ (otherwise, $P_{R^{GA}_i \cup \{r'_i\}} - P_{R^{GA}_i} = 0$ and r'_i would be rejected).*

Theorem 7. *Algorithm GA is 3-competitive for 1S2L-V if $0 < b_u < t$ and $0 \leq c < r$.*

Proof. Due to $0 < b_u < t$, all requests of R^*_i and R'_i must be accepted without cost because the request arrival is too late to serve a request with cost (recall that we forbid unprompted moves by OPT).

First, consider period $i = 1$. OPT cannot accept any request that is released during the time interval $[0, \tilde{t}_{r'_1})$, because otherwise such a request accepted by OPT could be accepted by GA instead of r'_1. Thus $\tilde{t}_{R^*_1(1)} \geq \tilde{t}_{r'_1}$, and hence $t_{R^*_1(1)} \geq \tilde{t}_{r'_1} \geq t_{r'_1} - b_u$. By Observation 1 and because OPT does not accept any request with cost, $\dot{p}_{R^*_1(1)} = \dot{p}_{r'_1}$, $\dot{p}_{R^*_1(2)} = p_{r'_1}$, and $\dot{p}_{R^*_1(3)} = \dot{p}_{r'_1}$. Hence, s^* is at $\dot{p}_{r'_1}$ at time $\dot{t}_{R^*_1(3)}$, which is not before $\hat{t}_{r'_1}$ (because $t_{R^*_1(3)} \geq t_{R^*_1(1)} + 2t \geq t_{r'_1} - b_u + 2t > t_{r'_1} + t$). Therefore, $|R^*_1| \leq 3$. Thus, $P_{R^*_1}/P_{R'_1} \leq 3$.

For $1 < i \leq k$, we distinguish the following cases in order to bound $P_{R^*_i}/P_{R'_i}$. As $R'_i = \{r'_i\}$, $P_{R'_i} = r$. We need to show that $P_{R^*_i} \leq 3r$.

Case 1: $\hat{t}_{r'_{i-1}} < \hat{t}_{r'_i}$ and $\hat{t}_{r'_{i-1}} < t_{r'_i}$. Assume that $|R^*_i| \geq 3$. (Otherwise, there is nothing to show.) By Observation 1 and because OPT does not accept any request with cost, $\dot{p}_{R^*_i(1)} = \dot{p}_{r'_i}$, $\dot{p}_{R^*_i(2)} = p_{r'_i}$, and $\dot{p}_{R^*_i(3)} = \dot{p}_{r'_i}$. If $t_{R^*_i(1)} \geq t_{r'_i}$, we have $\hat{t}_{r'_i} = \dot{t}_{R^*_i(1)}$ and thus $|R^*_i| = 1$. Therefore, we must have $t_{R^*_i(1)} < t_{r'_i}$. Observe that OPT can only accept requests with pick-up location $p_{r'_i}$ that start before $t_{r'_i}$ if they start after $t_{r'_i} - b_u > t_{r'_i} - t$. Otherwise, such a request accepted by OPT would arrive before r'_i and so it would be accepted by GA instead of r'_i. So we have $t_{R^*_i(3)} \geq t_{R^*_i(1)} + 3t \geq t_{r'_i} - b_u + 3t > \dot{t}_{r'_i}$. When OPT finishes serving the third request of R^*_i, s^* is at $\dot{p}_{r'_i}$, and this happens at a time after $\hat{t}_{r'_i}$ (see Fig. 2 for an illustration). Therefore, this time is $\hat{t}_{r'_i}$ and thus the end of OPT period i, and R^*_i cannot contain any further requests. (If $i = k$, the end of OPT period i is ∞, but any further request accepted by OPT could also be accepted by GA, a contradiction.) We have shown $|R^*_i| \leq 3$, as required. Hence $P_{R^*_i}/P_{R'_i} \leq 3$.

Case 2: $\hat{t}_{r'_{i-1}} < \hat{t}_{r'_i}$ and $\hat{t}_{r'_{i-1}} \geq t_{r'_i}$. We claim that $|R^*_i| \leq 1$ and argue as follows. Because $\hat{t}_{r'_{i-1}} \geq t_{r'_i}$, OPT can accept at most one request in OPT period i: When s^* finishes serving the first request in R^*_i, s^* is located at $\dot{p}_{r'_i}$, and the time when this happens becomes $\hat{t}_{r'_i}$ and thus the end of OPT period i. (If

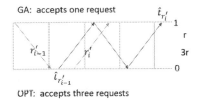

Fig. 2. Example configuration for R_i^* and R_i' in Case 1

$i = k$, we can argue as in Case 1 that OPT cannot accept any further requests.)
Hence, $P_{R_i^*}/P_{R_i'} \leq 1 < 3$.

Case 3: $\hat{t}_{r_{i-1}'} \geq \hat{t}_{r_i'}$. As OPT period i is empty by definition, we have $R_i^* = \emptyset$
and hence $P_{R_i^*} = 0$. Thus, $P_{R_i^*}/P_{R_i'} < 3$.

Because $P_{R_i^*}/P_{R_i'} \leq 3$ holds for all $1 \leq i \leq k$, we have $P_{R^*}/P_{R'} \leq 3$. This
proves the theorem. □

Theorem 8. *Algorithm GA is* $\max\{\frac{2r}{r-c}, 3\}$*-competitive for 1S2L-V if* $0 \leq c < r$
and $b_u = t$.

Theorem 9. *Algorithm GA is* $\frac{3r-c}{r-c}$*-competitive for 1S2L-V if* $0 \leq c < r$ *and*
$b_u > t$.

The proofs of Theorems 8 and 9 are omitted due to space restrictions.

Theorem 10. *Algorithm GA has competitive ratio at most* $1 + 2\lceil \frac{b_u - b_l}{2t} \rceil$ *for*
1S2L-V if $c = r$. *In particular, it is 3-competitive if* $0 < b_u < t$.

Proof. We prove the theorem by induction over the size of R'.

Base Case: $|R'| = 1$. We can show that $|R^*| \leq 1 + 2 \cdot \lceil \frac{b_u - b_l}{2t} \rceil$. As $P_{R'} = r$
and $P_{R^*} = |R^*| \cdot r$, we get $P_{R^*}/P_{R'} \leq 1 + 2\lceil \frac{b_u - b_l}{2t} \rceil$. The arguments are similar
to those used in the Induction Step below and are omitted here.

Induction Step: We assume that $\frac{P_{R^*}}{P_{R'}} \leq 1 + 2\lceil \frac{b_u - b_l}{2t} \rceil$ holds for all instances
with $|R'| \leq i$ and show that then $\frac{P_{R^*}}{P_{R'}} \leq 1 + 2\lceil \frac{b_u - b_l}{2t} \rceil$ also holds for all instances
with $|R'| = i + 1$.

Consider an instance of 1S2L-V given by a request sequence R where GA
accepts $i + 1$ requests. As GA accepts requests in order of increasing arrival
time by Observation 2, GA accepts i requests before time $\tilde{t}_{r_{i+1}'}$. Let $\bar{R}$ be the
sub-instance of R that contains all requests in R except r_{i+1}' and all requests
that are released after r_{i+1}' (i.e., released at time $\tilde{t}_{r_{i+1}'}$ and processed after r_{i+1}',
or released after time $\tilde{t}_{r_{i+1}'}$) and that GA could accept instead of r_{i+1}'. By the
inductive hypothesis, OPT can achieve profit at most $i \cdot (1 + 2\lceil \frac{b_u - b_l}{2t} \rceil) \cdot r$ on
the request sequence $\bar{R}$. The increase in profit that OPT can achieve on request
sequence R compared to $\bar{R}$ must be due to requests accepted without cost that
start in the interval $[\tilde{t}_{r_{i+1}'} + b_l, \infty)$ as all requests that start earlier were presented
before time $\tilde{t}_{r_{i+1}'}$ and are thus contained in $\bar{R}$. Let Q be a largest set of requests

in $R \setminus \bar{R}$ with start times in $[\tilde{t}_{r'_{i+1}} + b_l, \infty)$ that can be accepted without cost by OPT. Clearly, $P_{R^*} \leq P_{\bar{R}^*} + r \cdot |Q|$, where $\bar{R}^*$ denotes the requests accepted by an optimal solution for the instance $\bar{R}$.

We claim that the first request in Q must have pick-up location $p_{r'_{i+1}}$ as otherwise that request would have to be contained in $\bar{R}$, a contradiction to Q being a subset of $R \setminus \bar{R}$. To see this, assume that the request with earliest start time after $\tilde{t}_{r'_{i+1}} + b_l$ in Q has pick-up location $\dot{p}_{r'_{i+1}}$. Denote that request by r_j. If r_j was presented before r'_{i+1}, it is clearly contained in $\bar{R}$. If r_j was presented after r'_{i+1}, it is also contained in $\bar{R}$ because GA cannot accept it instead of r'_{i+1} (as s' is at location $p_{r'_{i+1}}$ after serving r'_i and GA accepts requests in order of increasing start times by Observation 2).

Before time $\tilde{t}_{r'_{i+1}}$, OPT can accept requests with pick-up location $\dot{p}_{r'_{i+1}}$ that start no later than $\tilde{t}_{r'_{i+1}} + b_u$. After time $\tilde{t}_{r'_{i+1}}$, OPT can only accept requests with pick-up location $\dot{p}_{r'_{i+1}}$ that start strictly before $t_{r'_{i+1}} + t$, because s' arrives at $\dot{p}_{r'_{i+1}}$ at time $t_{r'_{i+1}} + t$ and could serve any request with pick-up location $\dot{p}_{r'_{i+1}}$ from that time onward.

First, consider the case that $\tilde{t}_{r'_{i+1}} + b_u \geq t_{r'_{i+1}} + t$. The last request that OPT can accept with pick-up location $\dot{p}_{r'_{i+1}}$ starts no later than $\tilde{t}_{r'_{i+1}} + b_u$. After that request, OPT can accept at most one more request with pick-up location $p_{r'_{i+1}}$. To bound the size of Q, we bound the maximum number of pairs of requests (one with pick-up location 0 and the next with pick-up location 1) that OPT can accept. As the last request with pick-up location $\dot{p}_{r'_{i+1}}$ that OPT can accept has start time at most $\tilde{t}_{r'_{i+1}} + b_u$, the start time of the first request of the last pair that OPT accepts is at most $\tilde{t}_{r'_{i+1}} + b_u - t$. As the start times of consecutive pairs differ by at least $2t$, the number of pairs is bounded by $1 + \lfloor ((\tilde{t}_{r'_{i+1}} + b_u - t) - (\tilde{t}_{r'_{i+1}} + b_l))/2t \rfloor = 1 + \lfloor (b_u - b_l - t)/2t \rfloor$. If $b_u - b_l - t$ is a multiple of $2t$, this bound is equal to $1 + (b_u - b_l - t)/2t = \lceil (b_u - b_l)/2t \rceil$. Otherwise, the bound is equal to $\lceil (b_u - b_l - t)/2t \rceil \leq \lceil (b_u - b_l)/2t \rceil$. After the last pair, OPT can accept at most one more request with pick-up location $p_{r'_{i+1}}$. Therefore, $|Q| \leq 1 + 2 \cdot \lceil \frac{b_u - b_l}{2t} \rceil$.

Now, consider the case that $\tilde{t}_{r'_{i+1}} + b_u < t_{r'_{i+1}} + t$. Again, we consider the maximum number of pairs of requests (one with pick-up location $p_{r'_{i+1}}$ and the next with pick-up location $\dot{p}_{r'_{i+1}}$) that OPT can accept. As the last request with pick-up location $\dot{p}_{r'_{i+1}}$ that OPT can accept must have start time strictly smaller than $t_{r'_{i+1}} + t \leq \tilde{t}_{r'_{i+1}} + b_u + t$, the start time of the first request of the last pair that OPT accepts is strictly smaller than $\tilde{t}_{r'_{i+1}} + b_u$. The start times of consecutive pairs differ by at least $2t$. If $b_u - b_l$ is a multiple of $2t$, the number of pairs is bounded by $((\tilde{t}_{r'_{i+1}} + b_u) - (\tilde{t}_{r'_{i+1}} + b_l))/2t = (b_u - b_l)/2t = \lceil (b_u - b_l)/2t \rceil$. If $b_u - b_l$ is not a multiple of $2t$, the number of pairs is bounded by $1 + \lfloor (b_u - b_l)/2t \rfloor = \lceil (b_u - b_l)/2t \rceil$. In any case, the number of pairs is at most $\lceil (b_u - b_l)/2t \rceil$. After the last pair, OPT can accept at most one more request with pick-up location $p_{r'_{i+1}}$. Therefore, $|Q| \leq 1 + 2 \cdot \lceil \frac{b_u - b_l}{2t} \rceil$.

In either case, $|Q| \leq 1 + 2\lceil \frac{b_u - b_l}{2t} \rceil$. Thus, $P_{R^*} \leq P_{\bar{R}^*} + r \cdot |Q| \leq i(1 + 2\lceil \frac{b_u - b_l}{2t} \rceil)r + (1 + 2\lceil \frac{b_u - b_l}{2t} \rceil)r = (i + 1)(1 + 2\lceil \frac{b_u - b_l}{2t} \rceil)r$. As $P_{R'} = (i + 1)r$, we get $P_{R^*}/P_{R'} \leq 1 + 2\lceil \frac{b_u - b_l}{2t} \rceil$. $\qquad\square$

4 Conclusion

We have studied an on-line problem with one server and two locations that is motivated by applications such as car sharing and taxi dispatching. In particular, we have analyzed the effects of different constraints on the booking time of requests on the competitive ratio that can be achieved. For all variants of booking time constraints and costs for empty server movements we have given matching lower and upper bounds on the competitive ratio. The upper bounds are all achieved by the same greedy algorithm (GA). Interestingly, the size of the booking horizon does not affect the competitive ratio if $0 \leq c < r$, but the competitive ratio increases as $b_u - b_l$ increases if $c = r$.

A number of directions for future work arise from this work. In particular, it would be interesting to extend our results to the case of more than one server and more than two locations. It would be interesting to determine how the constraints on the booking time affect the competitive ratio for the general car-sharing problem with k servers and m locations.

Acknowledgments. This work was partially supported by the China Postdoctoral Science Foundation (Grant No. 2016M592811), and the China Scholarship Council (Grant No. 201706280058).

References

1. Ascheuer, N., Krumke, S.O., Rambau, J.: Online dial-a-ride problems: minimizing the completion time. In: Reichel, H., Tison, S. (eds.) STACS 2000. LNCS, vol. 1770, pp. 639–650. Springer, Heidelberg (2000)
2. Berbeglia, G., Cordeau, J., Laporte, G.: Dynamic pickup and delivery problems. Eur. J. Oper. Res. **202**(1), 8–15 (2010)
3. Bjelde, A., Disser, Y., Hackfeld, J., Hansknecht, C., Lipmann, M., Meißner, J., Schewior, K., Schlöter, M., Stougie, L.: Tight bounds for online TSP on the line. In: SODA 2017, pp. 994–1005. SIAM (2017)
4. Böhmová, K., Disser, Y., Mihalák, M., Šrámek, R.: Scheduling transfers of resources over time: towards car-sharing with flexible drop-offs. In: Kranakis, E., Navarro, G., Chávez, E. (eds.) LATIN 2016. LNCS, vol. 9644, pp. 220–234. Springer, Heidelberg (2016)
5. Borodin, A., El-Yaniv, R.: Online Computation and Competitive Analysis. Cambridge University Press, Cambridge (1998)
6. Christman, A., Forcier, W., Poudel, A.: From theory to practice: maximizing revenues for on-line dial-a-ride. J. Comb. Optim. **35**(2), 512–529 (2018)
7. Kaushik, N.R., Figueira, S.M., Chiappari, S.A.: Flexible time-windows for advance reservation scheduling. In: MASCOTS 2006, pp. 218–225. IEEE Computer Society (2006)

8. Krumke, S.O., de Paepe, W.E., Poensgen, D., Lipmann, M., Marchetti-Spaccamela, A., Stougie, L.: On minimizing the maximum flow time in the online dial-a-ride problem. In: Erlebach, T., Persinao, G. (eds.) WAOA 2005. LNCS, vol. 3879, pp. 258–269. Springer, Heidelberg (2006)
9. Yi, F., Tian, L.: On the Online Dial-A-Ride Problem with Time-Windows. In: Megiddo, N., Xu, Y., Zhu, B. (eds.) AAIM 2005. LNCS, vol. 3521, pp. 85–94. Springer, Heidelberg (2005). https://doi.org/10.1007/11496199_11

Directed Path-Width and Directed Tree-Width of Directed Co-graphs

Frank Gurski and Carolin Rehs[✉]

Algorithmics for Hard Problems Group, Institute of Computer Science,
Heinrich-Heine-University Düsseldorf, 40225 Düsseldorf, Germany
carolin.rehs@hhu.de

Abstract. In this paper we consider the directed path-width and directed tree-width of directed co-graphs. As an important combinatorial tool, we show how the directed path-width and the directed tree-width can be computed for the disjoint union, series composition, and order composition of two directed graphs. These results imply the equality of directed path-width and directed tree-width for directed co-graphs and a linear-time solution for computing the directed path-width and directed tree-width of directed co-graphs, which generalizes the known results for undirected co-graphs of Bodlaender and Möhring.

Keywords: Directed path-width · Directed tree-width
Directed co-graphs

1 Introduction

Tree-width is a well-known graph parameter [20]. Many NP-hard graph problems admit polynomial-time solutions when restricted to graphs of bounded tree-width using the tree-decomposition. The same holds for path-width since a path-decomposition can be regarded as a special case of a tree-decomposition. Computing both parameters is hard even for bipartite graphs and complements of bipartite graphs [1], while for co-graphs it has been shown [6,7] that the path-width equals the tree-width and how to compute this value in linear time.

During the last years, width parameters for directed graphs have received a lot of attention [11]. Among these are directed path-width and directed tree-width [16]. We show useful properties of directed path-decompositions and directed tree-decompositions, such as bidirectional complete subdigraph and bidirectional complete bipartite subdigraph lemmas. These results allow us to show how the directed path-width and directed tree-width can be computed for the disjoint union, series composition, and order composition of two directed graphs. Our proofs are constructive, i.e. a directed path-decomposition and a directed tree-decomposition can be computed from a di-co-tree. We show that the

---

C. Rehs—The work of the second author was supported by the German Research Association (DFG) grant GU 970/7-1.

L. Wang and D. Zhu (Eds.): COCOON 2018, LNCS 10976, pp. 255–267, 2018.
https://doi.org/10.1007/978-3-319-94776-1_22

directed path-width and directed tree-width are equal for directed co-graphs and give a linear-time solution for computing this value of directed co-graphs. Since for complete bioriented digraphs the directed path-width equals the (undirected) path-width of the corresponding underlying undirected graph and the directed tree-width equals the (undirected) tree-width of the corresponding underlying undirected graph our results generalize the known results from [6,7].

2   Preliminaries

Co-graphs have been introduced in the 1970s by a number of authors under different notations. Co-graphs can be characterized as the set of graphs without an induced path with four vertices [8]. From an algorithmic point of view a recursive definition based on the following operations is very useful.

Let $G_1 = (V_1, E_1), \ldots, G_k = (V_k, E_k)$ be k vertex-disjoint graphs.

- The *disjoint union* of $G_1, \ldots, G_k$, denoted by $G_1 \cup \ldots \cup G_k$, is the graph with vertex set $V_1 \cup \ldots \cup V_k$ and edge set $E_1 \cup \ldots \cup E_k$.
- The *join composition* of $G_1, \ldots, G_k$, denoted by $G_1 \times \ldots \times G_k$, is defined by their disjoint union plus all possible edges between vertices of G_i and G_j for all $1 \leq i, j \leq k$, $i \neq j$.

Definition 1 (Co-graphs). *The class of* co-graphs *is recursively defined as follows.*

(i) *Every graph on a single vertex* $(\{v\}, \emptyset)$, *denoted by* •, *is a co-graph.*
(ii) *If* $G_1, \ldots, G_k$ *are vertex-disjoint co-graphs, then*
   (a) *the disjoint union* $G_1 \cup \ldots \cup G_k$ *and*
   (b) *the join composition* $G_1 \times \ldots \times G_k$ *are co-graphs.*

By this definition every co-graph can be represented by a tree structure, denoted as *co-tree*. The leaves of the co-tree represent the vertices of the graph and the inner nodes of the co-tree correspond to the operations applied on the subexpressions defined by the subtrees. For every graph G one can decide in linear time, whether G is a co-graph and in the case of a positive answer construct a co-tree for G, see [14]. Using the co-tree a lot of hard problems have been shown to be solvable in polynomial time when restricted to co-graphs. Such problems are clique, independent set, partition into independent sets (chromatic number), partition into cliques, hamiltonian cycle, isomorphism [8].

We recall the definition of directed co-graphs from [9]. The following operations have already been considered in [3]. Let $G_1 = (V_1, E_1), \ldots, G_k = (V_k, E_k)$ be k vertex-disjoint digraphs.

- The *disjoint union* of $G_1, \ldots, G_k$, denoted by $G_1 \oplus \ldots \oplus G_k$, is the digraph with vertex set $V_1 \cup \ldots \cup V_k$ and arc set $E_1 \cup \ldots \cup E_k$.
- The *series composition* of $G_1, \ldots, G_k$, denoted by $G_1 \otimes \ldots \otimes G_k$, is defined by their disjoint union plus all possible arcs between vertices of G_i and G_j for all $1 \leq i, j \leq k$, $i \neq j$.

– The *order composition* of $G_1, \ldots, G_k$, denoted by $G_1 \oslash \ldots \oslash G_k$, is defined by
  their disjoint union plus all possible arcs from vertices of G_i to vertices of G_j
  for all $1 \leq i < j \leq k$.

Definition 2 (Directed co-graphs, [9]). *The class of* directed co-graphs *is
recursively defined as follows.*

(i) Every digraph on a single vertex $(\{v\}, \emptyset)$, *denoted by* •, *is a directed co-
    graph.*
(ii) If $G_1, \ldots, G_k$ *are vertex-disjoint directed co-graphs, then*
    (a) the disjoint union $G_1 \oplus \ldots \oplus G_k$,
    (b) the series composition $G_1 \otimes \ldots \otimes G_k$, *and*
    (c) the order composition $G_1 \oslash \ldots \oslash G_k$ *are directed co-graphs.*

By the definition we conclude that for every directed co-graph $G = (V, E)$
the underlying undirected graph $u(G)$, which is defined by $u(G) = (V, E_u)$,
$E_u = \{\{u, v\} \mid (u, v) \in E \text{ or } (v, u) \in E\}$ is a co-graph, but not vice versa.

Similar as undirected co-graphs by the P_4, also directed co-graphs can be
characterized by excluding eight forbidden induced subdigraphs [9].

Obviously for every directed co-graph we can define a tree structure, denoted
as *di-co-tree*. The leaves of the di-co-tree represent the vertices of the graph and
the inner nodes of the di-co-tree correspond to the operations applied on the
subexpressions defined by the subtrees. For every directed co-graph one can
construct a di-co-tree in linear time, see [9]. The following lemma shows that it
suffices to consider binary di-co-trees.

Lemma 1 ($\bigstar^1$). *Every di-co-tree T can be transformed into an equivalent
binary di-co-tree T', such that every inner vertex in T' has exactly two sons.*

In [12] the relation of directed co-graphs to the set of graphs of directed
NLC-width 1 and to the set of graphs of directed clique-width 2 is analyzed.

3   Directed Path-Width of Directed Co-graphs

According to Barát [2], the notation of directed path-width was introduced by
Reed, Seymour, and Thomas around 1995 and relates to directed tree-width
introduced by Johnson et al. in [16].

Definition 3 (directed path-width). *A directed path-decomposition of a
digraph $G = (V, E)$ is a sequence $(X_1, \ldots, X_r)$ of subsets of V, called* bags,
such that the following three conditions hold true.

(dpw-1) $X_1 \cup \ldots \cup X_r = V$.
(dpw-2) For each $(u, v) \in E$ there is a pair $i \leq j$ such that $u \in X_i$ and $v \in X_j$.
*(dpw-3) If $u \in X_i$ and $u \in X_j$ for some $u \in V$ and two indices i, j with $i \leq j$,
    then $u \in X_\ell$ for all indices ℓ with $i \leq \ell \leq j$.*

---

[1] The proofs of the results marked with a ★ are omitted due to space restrictions.

The width *of a directed path-decomposition* $\mathcal{X} = (X_1, \ldots, X_r)$ *is*

$$\max_{1 \leq i \leq r} |X_i| - 1.$$

The directed path-width *of G, d-pw(G) for short, is the smallest integer w such that there is a directed path-decomposition of G of width w.*

Lemma 2 ([22]). *Let G be some digraph, then $d\text{-}pw(G) \leq pw(u(G))$.*

Lemma 3 ([2]). *Let G be some complete bioriented digraph, then $d\text{-}pw(G) = pw(u(G))$.*

Determining whether the (undirected) path-width of some given (undirected) graph is at most some given value w is NP-complete even for bipartite graphs, complements of bipartite graphs [1], chordal graphs [13], and planar graphs with maximum vertex degree 3 [18]. Lemma 3 implies that determining whether the directed path-width of some given digraph is at most some given value w is NP-complete even for digraphs whose underlying graphs lie in the mentioned classes. On the other hand, determining whether the (undirected) path-width of some given (undirected) graph is at most some given value w is polynomial for permutation graphs [5], circular arc graphs [21], and co-graphs [7].

While undirected path-width can be solved by an FPT-algorithm [4], the existence of such an algorithm for directed path-width is still open. The directed path-width of a digraph $G = (V, E)$ can be computed in time $\mathcal{O}(\frac{|E| \cdot |V|^{2\text{d-pw}(G)}}{(\text{d-pw}(G)-1)!})$ by [17]. This leads to an XP-algorithm for directed path-width w.r.t. the standard parameter and implies that for each constant w, it is decidable in polynomial time whether a given digraph has directed path-width at most w.

In order to prove our main results we show some properties of directed path-decompositions. Similar results are known for undirected path-decompositions and are useful within several places.

Lemma 4 ([22]). *Let G be some digraph and H be an induced subdigraph of G, then $d\text{-}pw(H) \leq d\text{-}pw(G)$.*

Lemma 5 (★, **Bidirectional complete subdigraph**). *Let $G = (V, E)$ be some digraph, $G' = (V', E')$ with $V' \subseteq V$ be a bidirectional complete subdigraph, and $(X_1, \ldots, X_r)$ a directed path-decomposition of G. Then there is some i, $1 \leq i \leq r$, such that $V' \subseteq X_i$.*

Lemma 6 (★). *Let $G = (V, E)$ be a digraph and $(X_1, \ldots, X_r)$ a directed path-decomposition of G. Further let $A, B \subseteq V$, $A \cap B = \emptyset$, and $\{(u, v), (v, u) \mid u \in A, v \in B\} \subseteq E$. Then there is some i, $1 \leq i \leq r$, such that $A \subseteq X_i$ or $B \subseteq X_i$.*

Lemma 7 (★). *Let $\mathcal{X} = (X_1, \ldots, X_r)$ be a directed path-decomposition of some digraph $G = (V, E)$. Further let $A, B \subseteq V$, $A \cap B = \emptyset$, and $\{(u, v), (v, u) \mid u \in A, v \in B\} \subseteq E$. If there is some i, $1 \leq i \leq r$, such that $A \subseteq X_i$ then there are $1 \leq i_1 \leq i_2 \leq r$ such that*

1. for all i, $i_1 \leq i \leq i_2$ is $A \subseteq X_i$,
2. $B \subseteq \cup_{i=i_1}^{i_2} X_i$, and
3. $\mathcal{X}' = (X'_{i_1}, \ldots, X'_{i_2})$ where $X'_i = X_i \cap (A \cup B)$ is a directed path-decomposition of the digraph induced by $A \cup B$.

Theorem 1. Let $G = (V_G, E_G)$ and $H = (V_H, E_H)$ be two vertex-disjoint digraphs, then the following properties hold.

1. $d\text{-}pw(G \oplus H) = \max\{d\text{-}pw(G), d\text{-}pw(H)\}$
2. $d\text{-}pw(G \oslash H) = \max\{d\text{-}pw(G), d\text{-}pw(H)\}$
3. $d\text{-}pw(G \otimes H) = \min\{d\text{-}pw(G) + |V_H|, d\text{-}pw(H) + |V_G|\}$

Proof. 1. In order to show d-pw$(G \oplus H) \leq \max\{$d-pw$(G),$ d-pw$(H)\}$ we consider a directed path-decomposition $(X_1, \ldots, X_r)$ for G and a directed path-decomposition $(Y_1, \ldots, Y_s)$ for H. Then $(X_1, \ldots, X_r, Y_1, \ldots, Y_s)$ leads to a directed path-decomposition of $G \oplus H$.

Since G and H are induced subdigraphs of $G \oplus H$, by Lemma 4 the directed path-width of both digraphs leads to a lower bound on the directed path-width for the combined graph.

2. By the same arguments as used for (1.).
3. In order to show d-pw$(G \otimes H) \leq$ d-pw$(G) + |V_H|$ let $(X_1, \ldots, X_r)$ be a directed path-decomposition of G. Then we obtain by $(X_1 \cup V_H, \ldots, X_r \cup V_H)$ a directed path-decomposition of $G \otimes H$. In the same way a directed path-decomposition of H leads to a directed path-decomposition of $G \otimes H$ which implies that d-pw$(G \otimes H) \leq$ d-pw$(H) + |V_G|$. Thus d-pw$(G \otimes H) \leq \min\{$d-pw$(G) + |V_H|,$ d-pw$(H) + |V_G|\}$.

For the reverse direction let $\mathcal{X} = (X_1, \ldots, X_r)$ be a directed path-decomposition of $G \otimes H$. By Lemma 6 we know that there is some i, $1 \leq i \leq r$, such that $V_G \subseteq X_i$ or $V_H \subseteq X_i$. We assume that $V_G \subseteq X_i$. We apply Lemma 7 using $G \otimes H$ as digraph, $A = V_G$ and $B = V_H$ in order to obtain a directed path-decomposition $\mathcal{X}' = (X'_{i_1}, \ldots, X'_{i_2})$ for $G \otimes H$ where for all i, $i_1 \leq i \leq i_2$, it holds $V_G \subseteq X_i$ and $V_H \subseteq \cup_{i=i_1}^{i_2} X_i$. Further $\mathcal{X}'' = (X''_{i_1}, \ldots, X''_{i_2})$, where $X''_i = X'_i \cap V_H$ leads to a directed path-decomposition of H. Thus there is some i, $i_1 \leq i \leq i_2$, such that $|X_i \cap V_H| \geq$ d-pw$(H) + 1$. Since $V_G \subseteq X_i$, we know that $|X_i \cap V_H| = |X_i| - |V_G|$ and thus $|X_i| \geq |V_G| +$ d-pw$(H) + 1$. Thus the width of directed path-decomposition $(X_1, \ldots, X_r)$ is at least d-pw$(H) + |V_G|$. If we assume that $V_H \subseteq X_i$ it follows that the width of directed path-decomposition $(X_1, \ldots, X_r)$ is at least d-pw$(G) + |V_H|$. $\square$

Lemma 8 (★). Let G and H be two directed co-graphs, then $pw(u(G \oslash H)) > d\text{-}pw(G \oslash H)$.

Corollary 1 (★). Let G be some directed co-graph, then $d\text{-}pw(G) = pw(u(G))$ if and only if there is an expression for G without any order operation. Further $d\text{-}pw(G) = 0$ if and only if there is an expression for G without any series operation.

4   Directed Tree-Width of Directed Co-graphs

An *acyclic* digraph (*DAG* for short) is a digraph without any cycles as subdigraph. An out-tree is a digraph with a distinguished root such that all arcs are directed away from the root. For two vertices u, v of an out-tree T the notation $u \leq v$ means that there is a directed path on ≥ 0 arcs from u to v and $u < v$ means that there is a directed path on ≥ 1 arcs from u to v.

Let $G = (V, E)$ be some digraph and $Z \subseteq V$. A vertex set $S \subseteq V$ is *Z-normal*, if there is no directed walk in $G - Z$ with first and last vertices in S that uses a vertex of $G - (Z \cup S)$. That is, a set $S \subseteq V$ is Z-normal, if every directed walk which leaves and again enters S must contain only vertices from $Z \cup S$.[2]

Definition 4 (directed tree-width, [16]). *A (arboreal) tree-decomposition of a digraph $G = (V_G, E_G)$ is a triple $(T, \mathcal{X}, \mathcal{W})$. Here $T = (V_T, E_T)$ is an out-tree, $\mathcal{X} = \{X_e \mid e \in E_T\}$ and $\mathcal{W} = \{W_r \mid r \in V_T\}$ are sets of subsets of V_G, such that the following two conditions hold true.*

(dtw-1) $\mathcal{W} = \{W_r \mid r \in V_T\}$ *is a partition of V_G into nonempty subsets.*[3]
(dtw-2) For every $(u, v) \in E_T$ the set $\bigcup \{W_r \mid r \in V_T, v \leq r\}$ is $X_{(u,v)}$-normal.

The width *of a (arboreal) tree-decomposition $(T, \mathcal{X}, \mathcal{W})$ is*

$$\max_{r \in V_T} |W_r \cup \bigcup_{e \sim r} X_e| - 1.$$

Here $e \sim r$ means that r is one of the two vertices of arc e. The directed tree-width *of G, $d\text{-}tw(G)$ for short, is the smallest integer k such that there is a (arboreal) tree-decomposition $(T, \mathcal{X}, \mathcal{W})$ of G of width k.*

Lemma 9 ([16]). *Let G be some digraph, then $d\text{-}tw(G) \leq tw(u(G))$.*

Lemma 10 ([16]). *Let G be some complete bioriented digraph, then $d\text{-}tw(G) = tw(u(G))$.*

Determining whether the (undirected) tree-width of some given (undirected) graph is at most some given value w is NP-complete even for bipartite graphs and complements of bipartite graphs [1]. Lemma 10 implies that determining whether the directed tree-width of some given digraph is at most some given value w is NP-complete even for digraphs whose underlying graphs lie in the mentioned classes.

The results of [16] lead to an XP-algorithm for directed tree-width w.r.t. the standard parameter which implies that for each constant w, it is decidable in polynomial time whether a given digraph has directed tree-width at most w.

In order to show our main results we show some properties of directed tree-decompositions.

---

[2] Please note that our definition of Z-normality differs slightly from the definition in [16] but this trivially makes no difference for the directed tree-width.
[3] A remarkable difference to the undirected tree-width [20] is that the sets W_r have to be disjoint and non-empty.

Lemma 11 ([16])**.** *Let G be some digraph and H be an induced subdigraph of G, then $d\text{-}tw(H) \leq d\text{-}tw(G)$.*

Lemma 12 (★)**.** *Let G be some digraph, then $d\text{-}tw(G) \leq d\text{-}pw(G)$.*

Lemma 13 (★, **Bidirectional complete subdigraph**)**.** *Let $(T, \mathcal{X}, \mathcal{W})$, $T = (V_T, E_T)$, where r_T is the root of T, be a directed tree-decomposition of some digraph $G = (V, E)$ and $G' = (V', E')$ with $V' \subseteq V$ be a bidirectional complete subdigraph. Then $V' \subseteq W_{r_T}$ or there is some $(r, s) \in E_T$, such that $V' \subseteq W_s \cup X_{(r,s)}$.*

Lemma 14 (★)**.** *Let $G = (V, E)$ be some digraph, $(T, \mathcal{X}, \mathcal{W})$, $T = (V_T, E_T)$, where r_T is the root of T, be a directed tree-decomposition of G. Further let $A, B \subseteq V$, $A \cap B = \emptyset$, and $\{(u, v), (v, u) \mid u \in A, v \in B\} \subseteq E$. Then $A \cup B \subseteq W_{r_T}$ or there is some $(r, s) \in E_T$, such that $A \subseteq W_s \cup X_{(r,s)}$ or $B \subseteq W_s \cup X_{(r,s)}$.*

Lemma 15. *Let G be a digraph of directed tree-width at most k. Then there is a directed tree-decomposition $(T, \mathcal{X}, \mathcal{W})$, $T = (V_T, E_T)$, of width at most k for G such that $|W_r| = 1$ for every $r \in V_T$.*

Proof. Let $G = (V, E)$ be a digraph and $(T, \mathcal{X}, \mathcal{W})$, $T = (V_T, E_T)$, be a directed tree-decomposition of G. Let $r \in V_T$ such that $W_r = \{v_1, \ldots, v_k\}$ for some $k > 1$. Further let p be the predecessor of r in T and $s_1, \ldots, s_\ell$ be the successors of r in T. Let $(T', \mathcal{X}', \mathcal{W}')$ be defined by the following modifications of $(T, \mathcal{X}, \mathcal{W})$: We replace vertex r in T by the directed path $P(r) = (\{r_1, \ldots, r_k\}, \{(r_1, r_2), \ldots, (r_{k-1}, r_k)\})$ and replace arc (p, r) by (p, r_1) and the ℓ arcs (r, s_j), $1 \leq j \leq \ell$, by the ℓ arcs (r_k, s_j), $1 \leq j \leq \ell$ in T'. We define the sets $W'_{r_j} = \{v_j\}$ for $1 \leq j \leq k$. Further we define the sets $X'_{(p, r_1)} = X_{(p, r)}$, $X_{(r_k, s_j)} = X_{(r, s_j)}$, $1 \leq j \leq \ell$, and $X'_{(r_j, r_{j+1})} = X_{(p, r)} \cup \{r_1, \ldots, r_j\}$, $1 \leq j \leq k-1$.

By our definition $\mathcal{W}'$ leads to a partition of V into nonempty subsets. Further for every new arc (r_{i-1}, r_i), $1 < i \leq k$, the set $\bigcup\{W'_{r'} \mid r' \in V_{T'}, r_i \leq r'\}$ is $X'_{(r_{i-1}, r_i)}$-normal since $\bigcup\{W_{r'} \mid r' \in V_T, r \leq r'\}$ is $X_{(p, r)}$-normal and $X'_{(r_{i-1}, r_i)} = X_{(p, r)} \cup \{r_1, \ldots, r_{i-1}\}$. The property is fulfilled for arc (p, r_1) and (v_k, s_j), $1 \leq j \leq \ell$ since the considered vertex sets of G did not change. Thus triple $(T', \mathcal{X}', \mathcal{W}')$ is a directed tree-decomposition of G.

The width of $(T', \mathcal{X}', \mathcal{W}')$ is at most the width of $(T, \mathcal{X}, \mathcal{W})$ since for every r_j, $1 \leq j \leq k$, the following holds: $|W'_{r_j} \cup \bigcup_{e \sim r_j} X'_e| \leq |W_r \cup \bigcup_{e \sim r} X_e|$.

If we perform this transformation for every $r \in V_T$ such that $|W_r| > 1$, we obtain a directed tree-decomposition of G which fulfills the properties of the lemma. □

Lemma 16. *Let $G = (V, E)$ be a digraph of directed tree-width at most k, such that $V_1 \cup V_2 = V$, $V_1 \cap V_2 = \emptyset$, and $\{(u, v), (v, u) \mid u \in V_1, v \in V_2\} \subseteq E$. Then there is a directed tree decomposition $(T, \mathcal{X}, \mathcal{W})$, $T = (V_T, E_T)$, of width at most k for G such that for every $e \in E_T$ holds $V_1 \subseteq X_e$ or for every $e \in E_T$ holds $V_2 \subseteq X_e$.*

Proof. Let $G = (V, E)$ be a digraph of directed tree-width at most k and $(T, \mathcal{X}, \mathcal{W})$, $T = (V_T, E_T)$, be a directed tree-decomposition of width at most k for G. By Lemma 15 we can assume that holds: $|W_r| = 1$ for every $r \in V_T$.

We show the claim by traversing T in a bottom-up order. Let t' be a leaf of T, t be the predecessor of t' in T and $W_{t'} = \{v\}$ for some $v \in V_1$. Then the following holds: $V_2 \subseteq X_{(t,t')}$ since $(v, v') \in E$ and $(v', v) \in E$ for every $v' \in V_2$.

If t' is a non-leaf of T and there is a successor t'' of t' in T such that $V_1 \subseteq X_{(t',t'')}$ and there is a successor t''' of t' in T such that $V_2 \subseteq X_{(t',t''')}$. Then the width of $(T, \mathcal{X}, \mathcal{W})$ is $|V_1| + |V_2| - 1$ which allows us to insert V_1 into every set X_e as well as V_2 into every set X_e.

Otherwise let t' be a non-leaf of T and $V_2 \subseteq X_{(t',t'')}$ for every successor t'' of t'. Let t be the predecessor of t' and s be the predecessor of t in T. We distinguish the following two cases.

– Let $V_1 \subseteq \cup_{t' \leq \tilde{t}} W_{\tilde{t}}$. We replace $X_{(t,t')}$ by $X_{(t,t')} \cup V_2$ in order to meet our claim for edge (t, t').

  We have to show that this does not increase the width of the obtained directed tree-decomposition at vertex t' and at vertex t.

  The value of $|W_{t'} \cup \bigcup_{e \sim t'} X_e|$ does not change, since $V_2 \subseteq X_{(t',t'')}$ by induction hypothesis and $(t', t'') \sim t'$.

  Since $V_1 \subseteq \cup_{t \leq \tilde{t}} W_{\tilde{t}}$ by (dtw-2) we can assume that $V_1 \cap X_{(s,t)} = \emptyset$. Since all W_r have size one we know that $|W_t \cup \bigcup_{e \sim t} X_e| \leq |W_{t'} \cup \bigcup_{e \sim t'} X_e|$.

– Let $V_1 \not\subseteq \cup_{t' \leq \tilde{t}} W_{\tilde{t}}$. We distinguish the following two cases.

  • Let $V_2 \cap \cup_{t' \leq \tilde{t}} W_{\tilde{t}} = \emptyset$, then $W_{t'} = \{v\}$ for some $v \in V_1$ and thus $V_2 \subseteq X_{(t,t')}$ since $(v, v') \in E$ and $(v', v) \in E$ for every $v' \in V_2$.

  • Let $V_2 \cap \cup_{t' \leq \tilde{t}} W_{\tilde{t}} \neq \emptyset$. Since $\{(u,v), (v,u) \mid u \in V_1, v \in V_2\} \subseteq E$ the following is true:

$$V - \bigcup_{t' \leq \tilde{t}} W_{\tilde{t}} = (V_1 \cup V_2) - \bigcup_{t' \leq \tilde{t}} W_{\tilde{t}} \subseteq X_{(t,t')}. \tag{1}$$

  That is, all vertices of G which are not of one of the sets $W_{\tilde{t}}$ for all successors $\tilde{t}$ of t' are in set $X_{(t,t')}$.

  We define $X_{(t,t')} = (V - \cup_{t' \leq \tilde{t}} W_{\tilde{t}}) \cup V_2$ in order to meet our claim for edge (t, t').

  We have to show that this does not increase the width of the obtained directed tree-decomposition at vertex t' and and vertex t.

  The value of $|W_{t'} \cup \bigcup_{e \sim t'} X_e|$ does not change, since $V_2 \subseteq X_{(t',t'')}$ by induction hypothesis and $(t', t'') \sim t'$ and by (1).

  Further (1) implies that $X_{(s,t)} \subseteq X_{(t,t')}$ and thus $|W_t \cup \bigcup_{e \sim t} X_e| \leq |W_{t'} \cup \bigcup_{e \sim t'} X_e|$.

Thus if T has a leaf t' such that $W_{t'} = \{v\}$ for some $v \in V_1$ we obtain a directed tree-decomposition $(T, \mathcal{X}, \mathcal{W})$, $T = (V_T, E_T)$, such that $V_2 \subseteq X_e$ for every $e \in E_T$. And if T has a leaf t' such that $W_{t'} = \{v\}$ for some $v \in V_2$ we obtain a directed tree-decomposition $(T, \mathcal{X}, \mathcal{W})$, $T = (V_T, E_T)$, such that $V_1 \subseteq X_e$ for every $e \in E_T$. □

Theorem 2. *Let* $G = (V_G, E_G)$ *and* $H = (V_H, E_H)$ *be two vertex-disjoint digraphs, then the following properties hold.*

1. $d\text{-}tw(G \oplus H) = \max\{d\text{-}tw(G), d\text{-}tw(H)\}$
2. $d\text{-}tw(G \oslash H) = \max\{d\text{-}tw(G), d\text{-}tw(H)\}$
3. $d\text{-}tw(G \otimes H) = \min\{d\text{-}tw(G) + |V_H|, d\text{-}tw(H) + |V_G|\}$

Proof. Let $G = (V_G, E_G)$ and $H = (V_H, E_H)$ be two vertex-disjoint digraphs. Further let $(T_G, \mathcal{X}_G, \mathcal{W}_G)$ be a directed tree-decomposition of G such that r_G is the root of $T_G = (V_{T_G}, E_{T_G})$ and $(T_H, \mathcal{X}_H, \mathcal{W}_H)$ be a directed tree-decomposition of H such that r_H is the root of $T_H = (V_{T_H}, E_{T_H})$.

1. We define a directed tree-decomposition $(T_J, \mathcal{X}_J, \mathcal{W}_J)$ for $J = G \oplus H$. Let ℓ_G be a leaf of T_G. Let T_J be the disjoint union of T_G and T_H with an additional arc (ℓ_G, r_H). Further let $\mathcal{X}_J = \mathcal{X}_G \cup \mathcal{X}_H \cup \{X_{(\ell_G, r_H)}\}$, where $X_{(\ell_G, r_H)} = \emptyset$ and $\mathcal{W}_J = \mathcal{W}_G \cup \mathcal{W}_H$. Triple $(T_J, \mathcal{X}_J, \mathcal{W}_J)$ satisfies (dtw-1) since the combined decompositions satisfy (dtw-1). Further $(T_J, \mathcal{X}_J, \mathcal{W}_J)$ satisfies (dtw-2) since additionally in J there is no arc from a vertex of H to a vertex of G. This shows that $d\text{-}tw(G \oplus H) \le \max\{d\text{-}tw(G), d\text{-}tw(H)\}$. Since G and H are induced subdigraphs of $G \oplus H$, by Lemma 11 the directed tree-width of both leads to a lower bound on the directed tree-width for the combined graph.

2. The same arguments lead to $d\text{-}tw(G \oslash H) = \max\{d\text{-}tw(G), d\text{-}tw(H)\}$.

3. In order to show $d\text{-}tw(G \otimes H) \le d\text{-}tw(G) + |V_H|$ let T_J be the disjoint union of a new root r_J and T_G with an additional arc (r_J, r_G). Further let $\mathcal{X}_J = \mathcal{X}'_G \cup \{X_{(r_J, r_G)}\}$, where $\mathcal{X}'_G = \{X_e \cup V_H \mid e \in E_{T_G}\}$ and $X_{(r_J, r_G)} = V_H$ and $\mathcal{W}_J = \mathcal{W}_G \cup \{W_{r_H}\}$, where $W_{r_J} = V_H$. Then we obtain by $(T_J, \mathcal{X}_J, \mathcal{W}_J)$ a directed tree-decomposition of width at most $d\text{-}tw(G) + |V_H|$ for $G \otimes H$. In the same way a new root r_J and T_H with an additional arc (r_J, r_H), $\mathcal{X}'_H = \{X_e \cup V_G \mid e \in E_{T_H}\}$, $X_{(r_J, r_H)} = V_G$, $W_{r_J} = V_G$ lead to a directed tree-decomposition of width at most $d\text{-}tw(H) + |V_G|$ for $G \otimes H$. Thus $d\text{-}tw(G \otimes H) \le \min\{d\text{-}tw(G) + |V_H|, d\text{-}tw(H) + |V_G|\}$.

   For the reverse direction let $(T_J, \mathcal{X}_J, \mathcal{W}_J)$, $T_J = (V_T, E_T)$, be a directed tree-decomposition of minimal width for $G \otimes H$. By Lemma 16 we can assume that $V_G \subseteq X_e$ for every $e \in E_T$ or $V_H \subseteq X_e$ for every $e \in E_T$. Further by Lemma 15 we can assume that $|W_t| = 1$ for every $t \in V_T$.

   We assume that $V_G \subseteq X_e$ for every $e \in E_T$. We define $(T'_J, \mathcal{X}'_J, \mathcal{W}'_J)$, $T'_J = (V'_T, E'_T)$, by $X'_e = X_e \cap V_H$ and $W'_s = W_s \cap V_H$. Whenever this leads to an empty set W'_s where t is the predecessor of s in T'_J we remove vertex s from T'_J and replace every arc (s, t') by (t, t') with the corresponding set $X_{(t, t')} = X_{(s, t')} \cap V_H$.

   Then $(T'_J, \mathcal{X}'_J, \mathcal{W}'_J)$ is a directed tree-decomposition of H as follows.
   - $\mathcal{W}'_J$ is a partition of V_H into nonempty sets.
   - Let e be an arc in T'_J which is also in T_J. Since $e \sim s$ implies $W_s = W'_s = \{v\}$ for some $v \in V_H$ normality condition remains true.
     Arcs (t, t') in T'_J which are not in T_J are obtained by two arcs (t, s) and (s, t') from T_J. If $\cup\{W_r \mid r \in V_T, t' \le r\}$ is $X_{(s, t')}$-normal, then $\cup\{W_r \mid r \in V'_T, t' \le r\}$ is $X_{(t, t')}$-normal since $X_{(t, t')} = X_{(s, t')} \cap V_H$.

The width of $(T'_J, \mathcal{X}'_J, \mathcal{W}'_J)$ is at most d-tw$(G \otimes H) - |V_G|$ as follows.

– Let s be a vertex in T'_J such that $W_t \cap V_H \neq \emptyset$ for all (s,t) in T_J.

$$
\begin{aligned}
|W'_s \cup \bigcup_{e \sim s} X'_e| &= |(W_s \cap V_H) \cup \bigcup_{e \sim s} (X_e \cap V_H)| && \text{by definition} \\
&= |(W_s \cup \bigcup_{e \sim s} X_e) \cap V_H| && \text{factor out } V_H \\
&= |W_s \cup \bigcup_{e \sim s} X_e| - |V_G| && \text{since } V_G \subseteq X_e
\end{aligned}
$$

– Let s be a vertex in T'_J such that there is (s,t) in T_J with $W_t \cap V_H = \emptyset$.

$$
\begin{aligned}
|W'_s \cup \bigcup_{e \sim s} X'_e| = &\, |(W_s \cap V_H) \cup (X_{(t'',s)} \cap V_H) \cup \bigcup_{\substack{(s,t) \in E_T \\ W_t \cap V_H = \emptyset}} (X_{(t,t')} \cap V_H) \\
&\cup \bigcup_{\substack{(s,t) \in E_T \\ W_t \cap V_H \neq \emptyset}} (X_{(s,t)} \cap V_H) |
\end{aligned} \tag{2}
$$

In order to bound this value we observe that for $W_t \cap V_H = \emptyset$ the following is true: $W_t = \{v\}$ for $v \in V_G$. Then $X_{(s,t)} = ((V_G \cup V_H) - \cup_{t \leq \tilde{t}} W_{\tilde{t}}) \cup V_G$ by Lemma 16. That is, $X_{(s,t)}$ consists of all vertices from V_G and all vertices which are not of one of the sets $W_{\tilde{t}}$ for all successors $\tilde{t}$ of t. Applying this argument to $X_{(t,t')}$ we only can have v as an additional vertex. But since $v \in V_G$ we know that $v \in X_{(s,t)}$ by our assumption. This implies

$$
X_{(t,t')} \subseteq X_{(s,t)} \text{ for all arcs } (s,t) \text{ in } T_J \text{ such that } W_t \cap V_H = \emptyset \tag{3}
$$

which allows the following estimations:

$$
\begin{aligned}
|W'_s \cup \bigcup_{e \sim s} X'_e| &= |(W_s \cap V_H) \cup \bigcup_{e \sim s} (X_e \cap V_H)| && \text{by (2) and (3)} \\
&= |(W_s \cup \bigcup_{e \sim s} X_e) \cap V_H| && \text{factor out } V_H \\
&= |W_s \cup \bigcup_{e \sim s} X_e| - |V_G| && \text{since } V_G \subseteq X_e
\end{aligned}
$$

Thus the width of $(T'_J, \mathcal{X}'_J, \mathcal{W}'_J)$ is at most d-tw$(G \otimes H) - |V_G|$ and since $(T'_J, \mathcal{X}'_J, \mathcal{W}'_J)$ is a directed tree-decomposition of H it follows d-tw$(H) \leq$ d-tw$(G \otimes H) - |V_G|$
If we assume that $V_H \subseteq X_e$ for every $e \in E_T$ it follows that d-tw$(G) \leq$ d-tw$(G \otimes H) - |V_H|$. $\qquad\square$

The proof of Theorem 2 even shows that for any directed co-graph there is a tree-decomposition $(T, \mathcal{X}, \mathcal{W})$ of minimal width such that T is a path.

Similar to the path-width results, we conclude the following results.

Corollary 2. *Let G be some directed co-graph, then d-tw$(G) = $ tw$(u(G))$ if and only if there is an expression for G without any order operation. Further d-tw$(G) = 0$ if and only if there is an expression for G without any series operation.*

5   Directed Tree-Width and Directed Path-Width

Theorem 3. *For every directed co-graph G, it holds that $d\text{-}pw(G) = d\text{-}tw(G)$.*

Proof. Let $G = (V, E)$ be some directed co-graph. We show the result by induction on the number of vertices $|V|$. If $|V| = 1$, then $d\text{-}pw(G) = d\text{-}tw(G) = 0$. If $G = G_1 \oplus G_2$, then by Theorems 1 and 2 follows:

$$d\text{-}pw(G) = \max\{d\text{-}pw(G_1), d\text{-}pw(G_2)\} = \max\{d\text{-}tw(G_1), d\text{-}tw(G_2)\} = d\text{-}tw(G).$$

For the other two operations a similar relation holds.   □

By Lemmas 3 and 10 our results generalize the known results from [6,7] but can not be obtained by the known results.

Theorem 4. *For every directed co-graph $G = (V, E)$ which is given by a binary di-co-tree the directed path-width and directed tree-width can be computed in time $\mathcal{O}(|V|)$.*

Proof. The statement follows by the algorithm given in Fig. 1, Theorems 1 and 2. The necessary sizes of the subdigraphs defined by subtrees of di-co-tree T_G can be precomputed in time $\mathcal{O}(|V|)$.   □

Algorithm DIRECTED PATH-WIDTH(v)

---

if v is a leaf of di-co-tree T_G
    then $d\text{-}pw(G[T_v]) = 0$
    else {
        Directed Path-width(v_ℓ) ▶ v_ℓ *is the left successor of* v
        Directed Path-width(v_r) ▶ v_r *is the right successor of* v
        if v corresponds to a $\oplus$ or a $\oslash$ operation
            then $d\text{-}pw(G[T_v]) = \max\{d\text{-}pw(G[T_{v_\ell}]), d\text{-}pw(G[T_{v_r}])\}$
            else $d\text{-}pw(G[T_v]) = \min\{d\text{-}pw(G[T_{v_\ell}]) + |V_{G[T_{v_r}]}|, d\text{-}pw(G[T_{v_r}]) + |V_{G[T_{v_\ell}]}|\}$
    }

Fig. 1. Computing the directed path-width of G for every vertex of a di-co-tree T_G.

For general digraphs $d\text{-}pw(G)$ leads to a lower bound for $pw(u(G))$ and $d\text{-}tw(G)$ leads to a lower bound for $tw(u(G))$, see [2,16]. For directed co-graphs we obtain a closer relation as follows.

Corollary 3 (★). *Let G be a directed co-graph and $\overleftrightarrow{\omega}(G)$ be the size of a largest bioriented clique of G. It then holds that*

$$\overleftrightarrow{\omega}(G) = d\text{-}pw(G) - 1 = d\text{-}tw(G) - 1 \leq pw(u(G)) - 1 = tw(u(G)) - 1 = \omega(u(G)).$$

All values are equal if and only if G is a complete bioriented digraph.

6 Conclusion and Outlook

In this paper we could generalize the equivalence of path-width and tree-width of co-graphs which is known from [6,7] to directed graphs. Our results also hold for more general directed tree-width definitions such as allowing empty sets W_r in [15].

Related definitions given in [19] using sets W_r of size one only for the leaves of T and in [10, Chap. 6] using strong components within (dtw-2) should be considered in future work. Further research directions should extend the shown results to larger classes as well as consider related width parameters.

References

1. Arnborg, S., Corneil, D.G., Proskurowski, A.: Complexity of finding embeddings in a k-tree. SIAM J. Algebr. Discrete Methods **8**(2), 277–284 (1987)
2. Barát, J.: Directed pathwidth and monotonicity in digraph searching. Graphs Comb. **22**, 161–172 (2006)
3. Bechet, D., de Groote, P., Retoré, C.: A complete axiomatisation for the inclusion of series-parallel partial orders. In: Comon, H. (ed.) RTA 1997. LNCS, vol. 1232, pp. 230–240. Springer, Heidelberg (1997). https://doi.org/10.1007/3-540-62950-5_74
4. Bodlaender, H.L.: A linear-time algorithm for finding tree-decompositions of small treewidth. SIAM J. Comput. **25**(6), 1305–1317 (1996)
5. Bodlaender, H., Kloks, T., Kratsch, D.: Treewidth and pathwidth of permutation graphs. In: Lingas, A., Karlsson, R., Carlsson, S. (eds.) ICALP 1993. LNCS, vol. 700, pp. 114–125. Springer, Heidelberg (1993). https://doi.org/10.1007/3-540-56939-1_66
6. Bodlaender, H.L., Möhring, R.H.: The pathwidth and treewidth of cographs. In: Gilbert, J.R., Karlsson, R. (eds.) SWAT 1990. LNCS, vol. 447, pp. 301–309. Springer, Heidelberg (1990). https://doi.org/10.1007/3-540-52846-6_99
7. Bodlaender, H.L., Möhring, R.H.: The pathwidth and treewidth of cographs. SIAM J. Disc. Math. **6**(2), 181–188 (1993)
8. Corneil, D.G., Lerchs, H., Stewart-Burlingham, L.: Complement reducible graphs. Discrete Appl. Math. **3**, 163–174 (1981)
9. Crespelle, C., Paul, C.: Fully dynamic recognition algorithm and certificate for directed cographs. Discrete Appl. Math. **154**(12), 1722–1741 (2006)
10. Dehmer, M., Emmert-Streib, F. (eds.): Quantitative Graph Theory: Mathematical Foundations and Applications. CRC Press Inc., New York (2014)
11. Ganian, R., Hlinený, P., Kneis, J., Meisters, D., Obdrzálek, J., Rossmanith, P., Sikdar, S.: Are there any good digraph width measures? J. Combin. Theory Ser. B **116**, 250–286 (2016)
12. Gurski, F., Wanke, E., Yilmaz, E.: Directed NLC-width. Theor. Comput. Sci. **616**, 1–17 (2016)
13. Gusted, J.: On the pathwidth of chordal graphs. Discrete Appl. Math. **45**(3), 233–248 (1993)
14. Habib, M., Paul, C.: A simple linear time algorithm for cograph recognition. Discrete Appl. Math. **145**, 183–197 (2005)
15. Johnson, T., Robertson, N., Seymour, P.D., Thomas, R.: Addendum to "Directed tree-width" (2001)

16. Johnson, T., Robertson, N., Seymour, P.D., Thomas, R.: Directed tree-width. J. Comb. Theory Ser. B **82**, 138–155 (2001)
17. Kitsunai, K., Kobayashi, Y., Komuro, K., Tamaki, H., Tano, T.: Computing directed pathwidth in $O(1.89^n)$ time. Algorithmica **75**, 138–157 (2016)
18. Monien, B., Sudborough, I.H.: Min cut is NP-complete for edge weighted trees. Theor. Comput. Sci. **58**, 209–229 (1988)
19. Reed, B.: Introducing directed tree width. Electron. Notes Discrete Math. **3**, 222–229 (1999)
20. Robertson, N., Seymour, P.D.: Graph minors II. Algorithmic aspects of tree width. J. Algorithms **7**, 309–322 (1986)
21. Suchan, K., Todinca, I.: Pathwidth of circular-arc graphs. In: Brandstädt, A., Kratsch, D., Müller, H. (eds.) WG 2007. LNCS, vol. 4769, pp. 258–269. Springer, Heidelberg (2007). https://doi.org/10.1007/978-3-540-74839-7_25
22. Yang, B., Cao, Y.: Digraph searching, directed vertex separation and directed pathwidth. Discrete Appl. Math. **156**(10), 1822–1837 (2008)

Generalized Graph k-Coloring Games

Raffaello Carosi[1]([⊠]) and Gianpiero Monaco[2]([⊠])

[1] Gran Sasso Science Institute, L'Aquila, Italy
raffaello.carosi@gssi.it
[2] DISIM - University of L'Aquila, L'Aquila, Italy
gianpiero.monaco@univaq.it

Abstract. We investigate pure Nash equilibria in *generalized graph k-coloring games* where we are given an edge-weighted undirected graph together with a set of k colors. Nodes represent players and edges capture their mutual interests. The strategy set of each player consists of k colors. The utility of a player v in a given state or coloring is given by the sum of the weights of edges $\{v, u\}$ incident to v such that the color chosen by v is different than the one chosen by u, plus the profit gained by using the chosen color. Such games form some of the basic payoff structures in game theory, model lots of real-world scenarios with selfish players and extend or are related to several fundamental class of games.

We first show that generalized graph k-coloring games are potential games. In particular, they are convergent and thus Nash Equilibria always exist. We then evaluate their performance by means of the widely used notions of price of anarchy and price of stability and provide tight bounds for two natural and widely used social welfare, i.e., utilitarian and egalitarian social welfare.

1 Introduction

We consider *generalized graph k-coloring games*. These are played on edge-weighted undirected graphs where nodes correspond to players and edges identify social connections or relations between players. The strategy set of each player is a set of k available colors (we assume that the colors are the same for each player). When players select a color they induce a k-coloring or simply a coloring. Each player has a *profit function* that expresses how much a player likes a color. Given a coloring, the *utility* (or *payoff*) of a player v colored i is the sum of the weights of edges $\{v, u\}$ incident to v, such that the color chosen by v is different than the one chosen by u, plus the profit deriving from choosing color i. This class of games forms some of the basic payoff structures in game theory, and can model lots of real-life scenarios. Consider, for example, a set of companies that have to decide which product to produce in order to maximize their revenue. Each company has its own competitors (for example the ones that are in its same region), and it is reasonable to assume that each company wants to minimize the number of competitors that produce the same product. However, this is not their only concern. Indeed, different products may guarantee different profits

© Springer International Publishing AG, part of Springer Nature 2018
L. Wang and D. Zhu (Eds.): COCOON 2018, LNCS 10976, pp. 268–279, 2018.
https://doi.org/10.1007/978-3-319-94776-1_23

to a company according to many economic factors like the expected profit, the sponsorship revenue, and so on and so forth. Another possible scenario is the one with miners deciding which land to drill for resources. To a miner it is surely important to choose the land in which the number of rivals is minimized, but also the land itself is important: maybe there is a land with no miners that is very poor in resources, while another land that has been chosen by many miners may be very rich in resources. Thus, for a player is important to find a compromise between her neighbors' decisions and her own strategy choice. Other interesting applications can be found in [13, 18, 21].

Since players are assumed to be selfish, a well-known solution concept for this kind of setting is the Nash Equilibrium. Formally, a coloring is a (pure) Nash equilibrium if no player can improve her utility by unilaterally deviating from her actual strategy. We stress that in our setting it is not required that edges are properly colored, that is, in a Nash equilibrium, we can have edges whose two endpoints use the same color.

Nash equilibrium is one of the most important concepts in game theory and it provides a stable solution that is robust to deviations of single players. However, selfishness may cause loss of social welfare, that is, a stable solution is not always good with respect to the well-being of the society. We consider two natural and widely used notions of welfare. Given a coloring, the *utilitarian* social welfare is defined as the sum of the utilities of the players in the coloring, while the *egalitarian* social welfare is defined as the minimum utility among all the players in the coloring. Two used way of measuring the goodness of a Nash equilibrium with respect to a social welfare are the price of anarchy [20] and the price of stability [3]. We adopt such measures and study the quality of the worst (resp. best) Nash stable outcome and refer to the ratio of its social welfare to the one of the socially optimum one as to the price of anarchy (resp. stability). Roughly speaking, the price of anarchy says, in the worst case, how the efficiency of a system degrades due to selfish behavior of its players, while the price of stability has a natural meaning of stability, since it is the optimal solution among the ones which can be accepted by selfish players.

Our aim is to study the existence and the performance of Nash equilibria in generalized graph k-coloring games. We focus only on undirected graphs since for directed graphs even the problem of deciding whether an instance admits a Nash equilibria is an hard problem, and there exist instances for which a Nash equilibrium does not exist at all [21] (see Sect. 6 for further discussion about directed graphs).

Our Results. We first show that generalized graph k-coloring games are potential games. In particular, they are convergent games and thus Nash Equilibria always exist. Moreover, if the graph is unweighted then it is possible to compute a Nash equilibrium in polynomial time. This is different from weighted undirected graph, for which the problem of computing a Nash equilibrium is PLS-complete even for $k = 2$ [25], since the max cut game is a special case of our game. We then evaluate the goodness of Nash equilibria by means of the widely used notions of price of anarchy and price of stability and show tight bounds for two natural and widely used social welfare, i.e., utilitarian and egalitarian social welfare. Moreover, we provide tight results for the egalitarian social welfare related to the case in which players have no personal preference on colors, that is, all color profits are set to 0. Our results are illustrated in Table 1 (our original results are marked with an $*$). Due to space constraints some proofs have been omitted.

Table 1. Overview of the results about the graph k-coloring games. New results are marked with a "$*$", while the other ones are obtained from [21].

	Utilitarian SW		Egalitarian SW	
	PoA	PoS	PoA	PoS
Graph k-coloring without profits	$\frac{k}{k-1}$	1	$\frac{k}{k-1}$	$\frac{k}{k-1}$ $*$
Generalized graph k-coloring	2 $*$	$\frac{3}{2}$ $*$	2 $*$	2 $*$

Related Work. The graph k-coloring games (also called Max k-Cut games and anticoordination games), that is the special case of the generalized graph k-coloring games where the colors profits are set to zero, have been studied in [18,21]. They consider the game applied to both undirected and directed graphs (in the last case, each player is interested only in her outgoing neighbors). They show that the graph k-coloring game is a potential game [22] in case of undirected graphs and therefore a Nash equilibrium always exists. They only consider the utilitarian social welfare and give a tight bound for the price of anarchy, which is $\frac{k}{k-1}$, and show that any optimum is a Nash equilibrium (i.e., the price of stability is 1). Conversely, they show that even deciding whether an unweighted directed graph admits a Nash equilibrium is NP-Hard, for any number of colors $k \geq 2$. As far as concerns graph k-coloring games in edge-weighted undirected graphs, computing a Nash equilibrium is PLS-Complete [25], while for unweighted undirected graphs the problem becomes polynomially solvable.

A more complex payoff function is considered in [19], where the utility of a player is equal to the sum of the distance of her color to the color of each of her neighbors, applying a non-negative, real-valued, concave function.

Apt et al. [4] consider a coordination game in which, given a graph, players are nodes and each player has to select a color so that the number of neighbors with her same color is maximized. Here, each player has her own set of colors. When the graph is undirected, the game converges to a Nash equilibria in polynomial time. Instead, for directed graphs computing a Nash equilibria is NP-Complete [5]. Feldman and Friedler [15] study the strong price of anarchy [2] of graph k-coloring games, that is, the ratio of the social optimum to the worst strong equilibrium [6], which is a Nash equilibrium resilient to deviations by group of players.

To the best of our knowledge there is no paper that considers the price of anarchy and stability for the graph k-coloring games under the egalitarian social welfare. However, we stress that the egalitarian social welfare has been studied in many other settings, like e.g., congestion games [14], hedonic games [7], and fair division problems [23].

The graph k-coloring games are strictly related to many fundamental games in the scientific literature. For instance, they are strictly related to the unfriendly partition problem [1,9]. Moreover, they can be seen as a particular hedonic game (see [8] for an introduction to the topic), in which nodes with the same color belong to the same coalition, and the utility of each player is equal to her degree minus the number of neighbors that are in her own coalition. Nash equilibria in hedonic games have been largely investigated [11,12,16,17] (just to cite a few).

Finally, there are other related games on graphs that involve coloring. In [24] the authors study a game in which players are nodes of a graph. Each player has to choose a color among k available ones, and her utility is defined as follows: if no neighbor has chosen her same color then her utility is equal to the number of players (not in her neighborhood) that have chosen her same color, otherwise it is zero. They prove that this is a potential game and a Nash equilibria can be found in polynomial time. Moreover, they show that any pure equilibrium is a proper coloring.

2   Preliminaries

We are given an undirected simple graph $G = (V, E, w)$, where $|V| = n$, $|E| = m$, and $w : E \rightarrow \mathbb{R}_{\geq 0}$ is the edge-weight function that associates a positive weight to each edge. When weights are omitted they are assumed to be 1. We denote by $\delta^v(G) = \sum_{u \in V:\{v,u\} \in E} w(\{v, u\})$ the sum of the weights of all the edges incident to v. The set of nodes with which a node v has an edge in common is called v's neighborhood. We will omit to specify (G) when clear from the context. An instance of the generalized graph k-coloring game is a tuple (G, K, P). $G = (V, E, w)$ is an undirected weighted graph without self loops, in which each node $v \in V$ is a selfish player (in the following we will use node and player interchangeably). K is a set of k available colors (we assume that $k \geq 2$). The strategy set of each player is given by the k available colors, that is, the players have the same set of actions. We denote with $P : V \times K \rightarrow \mathbb{R}_{\geq 0}$, the color profit function, that defines how much a player likes a color, that is, if player v chooses to use color i, then she gains $P_v(i)$. For each player v, we define P_v^M as the greatest profit that v can gain from a color, namely, $P_v^M = max_{i=1,...,k} P_v(i)$. When $P_v(i) = 0$ $\forall v \in V$ and $\forall i \in k$, that is the case without profits for the chosen color, then the game is equivalent to the one analysed in [18,21], and we refer to it as graph k-coloring game. A state of the game $c = \{c_1, \ldots, c_n\}$ is a k-coloring, or simply a coloring, where c_v is the color (i.e., a number from 1 to k) chosen by player v. In a certain coloring c, the payoff (or the utility) of a player v is the sum of the weights of edges $\{v, u\}$ incident to v, such that the color chosen by v is different than the one chosen by u, plus the profit gained by using the chosen color. Formally, for a coloring c, a player v's payoff $\mu_c(v) = \sum_{u \in V:\{v,u\} \in E \wedge c_v \neq c_u} w(\{v, u\}) + P_v(c_v)$. From now on, when an edge

$\{v, u\}$ provides utility to its endpoints in a coloring c, that is, when $c_v \neq c_u$ we say that such edge is *proper*. We also say that an edge $\{v, u\}$ is *monochromatic* in a coloring c when $c_v = c_u$.

Let (c_{-v}, c'_v) denote the coloring obtained from c by changing the strategy of player v from c_v to c'_v. Given a coloring $c = \{c_1, \ldots, c_n\}$, an *improving move* of player v in the coloring c is a strategy c'_v such that $\mu_{(c_{-v}, c'_v)}(v) > \mu_c(v)$. A state of the game is a pure Nash or stable equilibrium if and only if no player can perform an improving move. Formally, $c = \{c_1, \ldots, c_n\}$ is a NE if $\mu_c(v) \geq \mu_{(c_{-v}, c'_v)}(v)$ for any possible color c'_v and for any player $v \in V$. An *improving dynamics* (shortly *dynamics*) is a sequence of improving moves. A game is said to be *convergent* if, given any initial state c, any sequence of improving moves leads to a Nash equilibrium.

Given a coloring c, we define the *utilitarian social welfare function* (denoted with $SW_{UT}(c)$) and the *egalitarian social welfare* (denoted with $SW_{EG}(c)$) as follows:

$$SW_{UT}(c) = \sum_{v \in V} \mu_c(v) = \sum_{v \in V} P_v(c_v) + \sum_{\{v,u\} \in E : c_v \neq c_u} 2w(\{v, u\}) \tag{1}$$

$$SW_{EG}(c) = \min_{v \in V} \mu_c(v) \tag{2}$$

Let us denote C the set of all the possible colorings, and let Q be the set of all the stable colorings. Given a social welfare function SW, we define the Price of Anarchy (PoA) of the generalized graph k-coloring game as the ratio of the maximum social welfare among all the possible colorings over the minimum social welfare among all the possible stable colorings. Formally, $PoA = \frac{max_{c \in C} SW(c)}{min_{c' \in Q} SW(c')}$. We further define the Price of Stability (PoS) of the generalized graph k-coloring game as the ratio of the maximum social welfare among all the possible colorings over the maximum social welfare among all the possible stable colorings. Formally, $PoS = \frac{max_{c \in C} SW(c)}{max_{c' \in Q} SW(c')}$. Intuitively, the PoA (resp. PoS) says us how much worse is the social welfare at a worst (resp. best) Nash equilibrium, relative to the social welfare of a centralized enforced optimum. We refer to the utilitarian price of anarchy and the egalitarian price of anarchy when we are dealing with utilitarian social welfare and egalitarian social welfare, respectively, and likewise the same is for the utilitarian price of stability and the egalitarian price of stability.

3   Existence of Nash Equilibria

We first show that the generalized graph k-coloring game is convergent, and therefore, the Nash equilibria always exists. In fact for each coloring c we define the potential function $\Phi(c)$ as the sum of the weights of proper colored edges plus the profits that each player v gains for using color c_v. Formally:

$$\Phi(c) = \sum_{\{v,u\} \in E : c_v \neq c_u} w(\{v, u\}) + \sum_{v \in V} P_v(c_v) \tag{3}$$

Proposition 1. *For all* k, *any finite generalized graph* k-*coloring game* (G, K, P) *is convergent.*

We notice that, on the one hand, if the graph is unweighted the dynamics starting from the coloring in which each player v selects the color giving her the maximum possible profit, that is, the color i such that $P_v(i) = P_v^M$, converges to a Nash equilibrium in at most $|E|$ improving moves. On the other hand, if the graph is weighted, computing a Nash Equilibrium is PLS-complete. It follows from the fact that, when $k = 2$, our game is a generalization of the Cut Games that is one of the first problem proved to be PLS-complete [25].

4   Utilitarian Social Welfare

In this section we focus on the utilitarian social welfare. We show tight bounds both for utilitarian price of anarchy and stability.

4.1   Price of Anarchy

We recall that in the case with no color profits, the utilitarian price of anarchy is exactly $\frac{k}{k-1}$ [21]. Here we prove that for generalized graph k-coloring games the utilitarian price of anarchy is equal to 2, that is, it is independent from the number of colors. We start by showing that the utilitarian price of anarchy is at most 2.

Theorem 2. *The utilitarian price of anarchy of the generalized graph* k-*coloring games is at most* 2.

Proof. It is easy to see that, for any possible coloring, the utility of any player $v \in V$ cannot exceed $P_v^M + \delta^v$. Let c^* be the coloring that maximizes the social welfare, and c a stable coloring. We notice that, a player v, in any equilibrium, has utility at least $max\{\frac{k-1}{k}\delta^v, P_v^M\}$. In fact, on the one hand, by the Pigeonhole principle, there always exists a color i such that, $\sum_{u:c_u=i} w(\{v,u\}) \leq \frac{1}{k}\delta^v$. On the other hand, player v can always select the color that maximizes her profit function. We now consider the two following cases:

Case 1: $P_v^M \geq \delta^v$
In this case it holds that:

$$\frac{\mu_{c^*}(v)}{\mu_c(v)} \leq \frac{P_v^M + \delta^v}{P_v^M} \leq \frac{2P_v^M}{P_v^M} = 2 \tag{4}$$

Case 2: $P_v^M < \delta^v$
Let $P_v^M = \delta^v - x$, where $0 < x \leq \delta^v$. Let $Y_v(c)$ be the set of v's neighbors whose color in c is different from the one, say i, that maximizes the color profit of v, namely $u \in Y_v(c)$ if u is a neighbor of v and $c_u \neq i$, where $i \in K$ and $P_v(i) = P_v^M$. Let $y = \sum_{u \in Y_v(c)} w(\{v,u\})$. Thus, in any Nash equilibrium the v's

payoff is at least $max\{P_v^M + y, \delta^v - y\}$. Therefore, both the following inequalities hold:

$$\frac{\mu_{c^*}(v)}{\mu_c(v)} \leq \frac{P_v^M + \delta^v}{\delta^v - x + y} = \frac{2\delta^v - x}{\delta^v - x + y} \tag{5}$$

$$\frac{\mu_{c^*}(v)}{\mu_c(v)} \leq \frac{P_v^M + \delta^v}{\delta^v - y} = \frac{2\delta^v - x}{\delta^v - y} \tag{6}$$

Inequality (5) is true because, if v chooses color i, then she earns at least P_v^M (that is equal to $\delta^v - x$) plus all the edges that are proper (and they are exactly y). Inequality (6) holds because if v chooses any other color than i, she earns at least $\delta^v - y$. Notice that the ratio is upper-bounded by the minimum between these two values, namely:

$$\frac{\mu_{c^*}(v)}{\mu_c(v)} \leq min\left\{\frac{2\delta^v - x}{\delta^v - x + y}, \frac{2\delta^v - x}{\delta^v - y}\right\}, \tag{7}$$

and thus it is maximized when $\frac{2\delta^v - x}{\delta^v - x + y} = \frac{2\delta^v - x}{\delta^v - y}$, that is, when $x = 2y$.

By applying it to inequality (5), we obtain:

$$\frac{\mu_{c^*}(v)}{\mu_c(v)} \leq \frac{2\delta^v - 2y}{\delta^v - 2y + y} = \frac{2(\delta^v - y)}{\delta^v - y} = 2$$

Therefore, we have that for any player v, $\frac{\mu_{c^*}(v)}{\mu_c(v)} \leq 2$. By summing over all the players, the theorem follows.   □

We now show that the utilitarian price of anarchy is at least 2 even for the special case of unweighted star graphs.

Theorem 3. *The utilitarian price of anarchy of the generalized graph k-coloring games is at least 2, even for the special case of unweighted star graphs.*

4.2   Price of Stability

We now turn our attention to the utilitarian price of stability. We recall that in the case with no color profits, the utilitarian price of stability is 1 [21]. Here we start by showing that the utilitarian price of stability for the generalized graph k-coloring games is at least $\frac{3}{2} - \epsilon$, for any $\epsilon > 0$, even for the special case of unweighted star graphs.

Theorem 4. *The utilitarian price of stability of the generalized graph k-coloring games is at least $\frac{3}{2} - \epsilon$, for any $\epsilon > 0$, even for the special case of unweighted star graphs.*

We now show that the utilitarian price of stability for the generalized graph k-coloring games is at most $\frac{3}{2}$.

Theorem 5. *The utilitarian price of stability of the generalized graph k-coloring games is at most $\frac{3}{2}$.*

Proof. In order to prove the upper bound for the utilitarian price of stability we first need some preliminary definitions. Given a coloring c we define the following two variables: let $A(c) = \sum_{v \in V} P_v(c_v)$ be the sum of the color profits gained by the players in the coloring c, and let $B(c) = \sum_{\{v,u\} \in E : c_v \neq c_u} w(\{v, u\})$ be the sum of the weights of the properly colored edges in c. Thus, we can rewrite the Eqs. (1) and (3) for the utilitarian social welfare and the potential function with respect to a given coloring c as $SW_{UT}(c) = A(c) + 2B(c)$ and $\Phi(c) = A(c) + B(c)$, respectively.

Let c^* be the coloring that maximizes the social welfare, and let N be the Nash equilibrium that is reached by a dynamics starting from c^*. By the potential function argument, since every improving move increases the potential value, it must hold that:

$$A(N) + B(N) > A(c^*) + B(c^*) \tag{8}$$

On the other hand, since c^* is the coloring that maximizes the social welfare, it holds that:

$$A(c^*) + 2B(c^*) \geq A(N) + 2B(N) \tag{9}$$

In order to prove that $PoS \leq \frac{3}{2}$ it is sufficient to show that the ratio between $SW_{UT}(c^*)$ and $SW_{UT}(N)$ is less or equal than $\frac{3}{2}$. In fact, if such inequality holds then, if N^* is the best Nash equilibrium, $SW_{UT}(N) \leq SW_{UT}(N^*)$ and, consequently, $\frac{SW_{UT}(c^*)}{SW_{UT}(N^*)} \leq \frac{SW_{UT}(c^*)}{SW_{UT}(N)} \leq \frac{3}{2}$.

We want to prove that:

$$\frac{SW_{UT}(c^*)}{SW_{UT}(N)} \leq \frac{3}{2} \tag{10}$$

and, by using the above defined $A(c)$ and $B(c)$, we get that this is true if and only if:

$$A(c^*) + 2B(c^*) \leq \frac{3}{2} A(N) + 3B(N) \tag{11}$$

For the remainder of the proof we are going to show that:

$$B(c^*) \leq \frac{1}{2} A(N) + 2B(N). \tag{12}$$

Indeed, if inequality (12) holds, then by summing it with inequality (8) we get that (11) holds, and this ends the proof.

Given N, for every player v, let $B_v(N) = \sum_{u \in V : N_v \neq N_u} w(\{v, u\})$ be the sum of the weights of properly colored edges incident to v. Thus, $B(N) = \frac{1}{2} \sum_{v \in V} B_v(N)$. Similarly, let $\bar{B}_v(N) = \sum_{u \in V : N_v = N_u} w(\{v, u\}) = \delta_v - B_v(N)$ be the sum of the weights of monochromatic edges incident to v, and let $\bar{B}(N) = \sum_{v \in V} \delta_v - B(N) = \frac{1}{2} \sum_{v \in V} \bar{B}_v(N)$ be the sum of the weights of all monochromatic edges. Since N is a Nash equilibrium, it must hold that

$$P_v(N(v)) + B_v(N) \geq \bar{B}_v(N) \quad \forall v \in V, \tag{13}$$

otherwise, player v could switch to any other color, thus improving her utility. We can calculate $A(N)$ as follows:

$$A(N) = \sum_{v \in V} P_v(N(v)) \geq \sum_{v \in V} \left(\bar{B}_v(N) - B_v(N) \right) \tag{14}$$

$$= \sum_{v \in V} \left(\bar{B}_v(N) - B_v(N) \right) + 2B(N) - 2B(N) \tag{15}$$

$$= 2B(N) + \sum_{v \in V} \bar{B}_v(N) - \sum_{v \in V} B_v(N) - 2B(N) \tag{16}$$

$$= 2 \sum_{v \in V} \delta_v - 2B(N) - 2B(N) = 2 \sum_{v \in V} \delta_v - 4B(N)$$

In (14) we apply inequality (13), then in (15) we add and remove $2B(N)$. Since each edge is either proper or monochromatic, it appears in only one of the two summations in (16). Moreover, each edge is counted exactly twice in a summation (one for endpoint), that is, $\sum_{v \in V} \bar{B}_v(N) = 2\bar{B}(N)$ and $\sum_{v \in V} B_v(N) = 2B(N)$. By summing $2\bar{B}(N)$ and $2B(N)$, we get $2 \sum_{v \in V} \delta_v$. Thus, $A(N) \geq 2 \sum_{v \in V} \delta_v - 4B(N)$, that is, $A(N) + 4B(N) \geq 2 \sum_{v \in V} \delta_v \geq 2B(c^*)$. Dividing by 2, we get that inequality (12) holds, and this concludes the proof. □

5   Egalitarian Social Welfare

In this section we focus on the egalitarian social welfare. We show tight bounds both for egalitarian price of anarchy and stability.

5.1   Price of Anarchy

For the graph k-coloring game (i.e., without color profits), a lower bound on the egalitarian price of anarchy is provided by the instance in [21] (Sect. 3, Proposition 2). In such instance, the optimal solution is such that each player v has utility equal to her degree δ^v, however there exists a stable coloring in which each player v has utility equal to $\frac{k-1}{k} \delta^v$, for any number of colors $k \geq 2$. This result together with the fact that by the pigeonhole principle, in any stable coloring each player v achieves utility at least $\frac{k-1}{k} \delta^v$, imply that the egalitarian price of anarchy is exactly $\frac{k}{k-1}$ for the graph k-coloring games.

Therefore, we now consider the generalized graph k-coloring games. We first notice that the instance defined in Theorem 3 gives us a lover bound of 2 to the egalitarian price of anarchy. In fact, in the optimal coloring the minimum utility is 2, and there exists a stable coloring in which the minimum utility is 1. Moreover, we point out that the proof of Theorem 2 basically shows that for any player i, her utility in any stable outcome is at least half the payoff that she has in the optimum. Therefore we easily get the following theorem which says that the egalitarian price of anarchy of the generalized graph k-coloring games is 2.

Theorem 6. *The egalitarian price of anarchy of the generalized graph k-coloring games is 2.*

5.2   Price of Stability

We now turn our attention to the egalitarian price of stability. We start by showing that for the graph k-coloring games (i.e., without color profits), the egalitarian price of stability is exactly $\frac{k}{k-1}$.

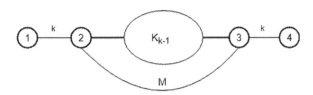

Fig. 1. Instance used in Theorem 7.

Theorem 7. *The egalitarian price of stability of the graph k-coloring games is* $\frac{k}{k-1}$.

Proof. Consider the weighted graph of Fig. 1 with $k \geq 2$ colors, where K_{k-1} stands for a clique of size $k-1$, and the bold arrows indicate that there is complete incidence between the two adjacent subgraphs, i.e., there is an edge of weight 1 between any node of the clique K_{k-1} and node 2, and any node of the clique K_{k-1} and node 3. Moreover, there is an edge from 2 to 3 of weight M, where M is an integer greater or equal than $2k$. The only way to get a coloring that maximizes the egalitarian social welfare is the following: nodes 2 and 3 pick the same color, the nodes in the clique choose the remaining $k-1$ colors, one per vertex, and players 1 and 4 pick any color different from the one chosen by 2 and 3. By coloring the nodes in this way, the resulting egalitarian social welfare is equal to k. We notice that this is the unique possible optimal coloring, since the maximum possible utility that each node in the clique K_{k-1} can achieve is exactly k, and the only way to get it is by assigning the same color to nodes 2 and 3. However, this coloring is not stable, since player 2 or 3 can switch to a different available color in order to improve her utility, passing from $2k-1$ to an utility of at least M. Indeed, it is easy to see that in any Nash equilibrium nodes 2 and 3 must choose different colors, and by doing in this way there is at least a node in the clique, say i, that has the same color as 2 or 3, thus achieving an utility of at most $k-1$. Moreover, player i cannot increase her utility by changing color because she has exactly one neighbor colored h, for each color $h = 1, \ldots, k$. Thus, we get that the egalitarian price of stability of the graph k-coloring games is at least $\frac{k}{k-1}$.

The theorem follows from the fact that the egalitarian price of anarchy in the graph k-coloring games is at most $\frac{k}{k-1}$.   □

We now consider the egalitarian price of stability of the generalized graph k-coloring games and show a tight bound of 2.

Theorem 8. *The egalitarian price of stability of the generalized graph k-coloring games is* 2.

6   Future Work

A possible future research direction is the study of other types of equilibria for the generalized graph k-coloring game. For instance, Feldman and Friedler [15] prove that, when strong Nash equilibria exist, the strong price of anarchy in the graph k-coloring game without profits depends on the number of colors and its maximum value is $\frac{3}{2}$ when $k = 2$. It would be interesting to study how much worse the strong price of anarchy gets for the generalized graph k-coloring games.

We also believe that questions related to the computational complexity of (approximate) Nash equilibria for the generalized graph k-coloring games deserve investigation. Indeed, already for graph k-coloring games, if the graph is unweighted and directed the problem of deciding whether an instance admits a Nash equilibria is NP-hard, and there exist instances for which a Nash equilibrium does not exist at all [21]. Moreover, for the case of weighted undirected graphs, even if Nash equilibria always exist, computing them is PLS-complete [25] even for $k = 2$, since the max cut game is a special case. A γ-Nash equilibrium is a coloring such that no player can improve her payoff by a (multiplicative) factor of γ by changing color. To the best of our knowledge, there are few papers which deal with the problem of computing approximate Nash equilibria for graph k-coloring games. In [10], the authors show that it is possible to compute in polynomial time a $(3 + \epsilon)$-Nash equilibrium, for any $\epsilon > 0$, for max cut games, while in [13] the authors present a randomized polynomial time algorithm that computes a constant approximate Nash equilibrium for a large class of directed unweighted graphs.

References

1. Aharoni, R., Milner, E., Prikry, K.: Unfriendly partitions of a graph. J. Comb. Theory Ser. B **50**(1), 1–10 (1990)
2. Andelman, N., Feldman, M., Mansour, Y.: Strong price of anarchy. Games Econ. Behav. **65**(2), 289–317 (2009)
3. Anshelevich, E., Dasgupta, A., Kleinberg, J.M., Tardos, É., Wexler, T., Roughgarden, T.: The price of stability for network design with fair cost allocation. In: Proceedings of the 45th Symposium on Foundations of Computer Science, FOCS, pp. 295–304 (2004)
4. Apt, K.R., Rahn, M., Schäfer, G., Simon, S.: Coordination games on graphs (extended abstract). In: Liu, T.-Y., Qi, Q., Ye, Y. (eds.) WINE 2014. LNCS, vol. 8877, pp. 441–446. Springer, Cham (2014). https://doi.org/10.1007/978-3-319-13129-0_37
5. Apt, K.R., Simon, S., Wojtczak, D.: Coordination games on directed graphs. In: Proceedings Fifteenth Conference on Theoretical Aspects of Rationality and Knowledge, TARK, pp. 67–80 (2015)
6. Aumann, R.J.: Acceptable points in general cooperative N-person games. In: Luce, R.D., Tucker, A.W. (eds.) Contribution to the Theory of Game IV. Annals of Mathematical Study, vol. 40, pp. 287–324 (1959)
7. Aziz, H., Gaspers, S., Gudmundsson, J., Mestre, J., Täubig, H.: Welfare maximization in fractional hedonic games. In: Proceedings of the Twenty-Fourth International Joint Conference on Artificial Intelligence, IJCAI, pp. 461–467 (2015)
8. Aziz, H., Savani, R., Moulin, H.: Hedonic Games, pp. 356–376. Cambridge University Press, Cambridge (2016)

9. Bazgan, C., Tuza, Z., Vanderpooten, D.: Satisfactory graph partition, variants, and generalizations. Eur. J. Oper. Res. **206**(2), 271–280 (2010)
10. Bhalgat, A., Chakraborty, T., Khanna, S.: Approximating pure nash equilibrium in cut, party affiliation, and satisfiability games. In: Proceedings 11th ACM Conference on Electronic Commerce, EC, pp. 73–82 (2010)
11 Bilò, V., Fanelli, A., Flammini, M., Monaco, G., Moscardelli, L.: Nash stability in fractional hedonic games. In: Liu, T.-Y., Qi, Q., Ye, Y. (eds.) WINE 2014. LNCS, vol. 8877, pp. 486–491. Springer, Cham (2014). https://doi.org/10.1007/978-3-319-13129-0_44
12. Bilò, V., Fanelli, A., Flammini, M., Monaco, G., Moscardelli, L.: On the price of stability of fractional hedonic games. In: Proceedings of the 2015 International Conference on Autonomous Agents and Multiagent Systems, AAMAS, pp. 1239–1247 (2015)
13. Carosi, R., Flammini, M., Monaco, G.: Computing approximate pure nash equilibria in digraph k-coloring games. In: Proceedings of the 16th Conference on Autonomous Agents and MultiAgent Systems, AAMAS, pp. 911–919 (2017)
14. Christodoulou, G., Koutsoupias, E.: The price of anarchy of finite congestion games. In: Proceedings of the 37th Annual ACM Symposium on Theory of Computing, STOC, pp. 67–73 (2005)
15. Feldman, M., Friedler, O.: A unified framework for strong price of anarchy in clustering games. In: Halldórsson, M.M., Iwama, K., Kobayashi, N., Speckmann, B. (eds.) ICALP 2015. LNCS, vol. 9135, pp. 601–613. Springer, Heidelberg (2015). https://doi.org/10.1007/978-3-662-47666-6_48
16. Feldman, M., Lewin-Eytan, L., Naor, J.S.: Hedonic clustering games. ACM Trans. Parallel Comput. **2**(1), 4:1–4:48 (2015)
17. Gairing, M., Savani, R.: Computing stable outcomes in hedonic games. In: Kontogiannis, S., Koutsoupias, E., Spirakis, P.G. (eds.) SAGT 2010. LNCS, vol. 6386, pp. 174–185. Springer, Heidelberg (2010). https://doi.org/10.1007/978-3-642-16170-4_16
18. Hoefer, M.: Cost sharing and clustering under distributed competition. Ph.D. thesis, University of Konstanz (2007)
19. Kliemann, L., Sheykhdarabadi, E.S., Srivastav, A.: Price of anarchy for graph coloring games with concave payoff. J. Dyn. Games **4**(1), 41–58 (2017)
20. Koutsoupias, E., Papadimitriou, C.: Worst-Case equilibria. In: Meinel, C., Tison, S. (eds.) STACS 1999. LNCS, vol. 1563, pp. 404–413. Springer, Heidelberg (1999). https://doi.org/10.1007/3-540-49116-3_38
21. Kun, J., Powers, B., Reyzin, L.: Anti-coordination games and stable graph colorings. In: Vöcking, B. (ed.) SAGT 2013. LNCS, vol. 8146, pp. 122–133. Springer, Heidelberg (2013). https://doi.org/10.1007/978-3-642-41392-6_11
22. Monderer, D., Shapley, L.S.: Potential games. Games Econ. Behav. **14**(1), 124–143 (1996)
23. Nguyen, N., Nguyen, T.T., Roos, M., Rothe, J.: Complexity and approximability of social welfare optimization in multiagent resource allocation. In: International Conference on Autonomous Agents and Multiagent Systems, AAMAS, pp. 1287–1288 (2012)
24. Panagopoulou, P.N., Spirakis, P.G.: A game theoretic approach for efficient graph coloring. In: Hong, S.-H., Nagamochi, H., Fukunaga, T. (eds.) ISAAC 2008. LNCS, vol. 5369, pp. 183–195. Springer, Heidelberg (2008). https://doi.org/10.1007/978-3-540-92182-0_19
25. Schäffer, A.A., Yannakakis, M.: Simple local search problems that are hard to solve. SIAM J. Comput. **20**(1), 56–87 (1991)

On Colorful Bin Packing Games

Vittorio Bilò[1]($\boxtimes$), Francesco Cellinese[2], Giovanna Melideo[3],
and Gianpiero Monaco[3]

[1] University of Salento, Lecce, Italy
vittorio.bilo@unisalento.it
[2] Gran Sasso Science Institute, L'Aquila, Italy
francesco.cellinese@gssi.it
[3] University of L'Aquila, L'Aquila, Italy
{giovanna.melideo,gianpiero.monaco}@univaq.it

Abstract. We consider colorful bin packing games in which a set of
items, each one controlled by a selfish player, are to be packed into a
minimum number of unit capacity bins. Each item has one of $m \geq 2$
colors and no items of the same color may be adjacent in a bin. All bins
have the same unitary cost which is shared among the items it contains,
so that players are interested in selecting a bin of minimum shared cost.
We adopt two standard cost sharing functions, i.e., the egalitarian and
the proportional ones. Although, under both cost functions, these games
do not converge in general to a (pure) Nash equilibrium, we show that
Nash equilibria are guaranteed to exist. We also provide a complete characterization of the efficiency of Nash equilibria under both cost functions
for general games, by showing that the prices of anarchy and stability
are unbounded when $m \geq 3$, while they are equal to 3 when $m = 2$. We
finally focus on the subcase of games with uniform sizes (i.e., all items
have the same size). We show a tight characterization of the efficiency of
Nash equilibria and design an algorithm which returns Nash equilibria
with best achievable performance.

1 Introduction

A classical problem in combinatorial optimization is the one-dimensional bin
packing problem, in which items with different sizes in $[0, 1]$ have to be packed
into the smallest possible number of unit capacity bins. This problem is known to
be NP-hard (see [8] for a survey). The study of bin packing in a game theoretical
context has been introduced in [4]. In such a setting, items are handled by selfish
players and the unitary cost of each bin is shared among the items it contains.
In the literature, two natural cost sharing functions have been considered: the
egalitarian cost function, which equally shares the cost of a bin among the items
it contains (see [9,18]), and the *proportional* cost function, where the cost of a
bin is split among the items proportionally to their sizes. Namely, each player
is charged with a cost according to the fraction of the used bin space her item
requires (see [4,10]). We stress that for games with uniform sizes, where all the

© Springer International Publishing AG, part of Springer Nature 2018
L. Wang and D. Zhu (Eds.): COCOON 2018, LNCS 10976, pp. 280–292, 2018.
https://doi.org/10.1007/978-3-319-94776-1_24

items have the same size s, the two cost functions coincide. Each player would prefer to choose a strategy that minimizes her own cost, where the strategy is the bin chosen by the player. Pure Nash equilibria, i.e. packings in which no player can lower her cost by changing the selected bin in favor of a different one, are mainly considered as natural stable outcomes for these games. The social cost function that we aim to minimize is the number of open bins (a bin is open if it stores at least one item). Bin packing games can model many practical scenarios, like bandwidth allocations problems, packet scheduling problems (see [4,13]).

In this paper we consider *colorful bin packing* games, a generalization of the bin packing games where we are given a set of n selfish players, a set of $m \geq 2$ colors, and a set of n unit capacity bins. The special case of games with two colors is called *black and white* bin packing games. Each player controls an indivisible colored item of size in $[0, 1]$. Each item needs to be packed into a bin without exceeding its capacity and in such a way that no item is misplaced, that is no item is adjacent to another one of the same color in the bin. We use both the egalitarian and the proportional cost functions, where we set the cost of any misplaced item as infinite, and adopt Nash equilibria as stable outcomes of the games. We notice that in any Nash equilibrium no player can be charged with an infinite cost since any player can move to an empty bin and getting cost 1. We notice that, if all the items have different colors, these games correspond to the bin packing ones. However, when there are items with the same color, the stable outcomes of the games are structurally different than bin packing ones. In fact, we show that in our games, Nash equilibria perform very differently than in bin packing ones. Colorful bin packing games can model many practical scenarios (see [5,6,11]) like television and radio stations which schedule a set of programs of various genre on different channels, or displays on websites that alternate between different types of information and advertisements, or a software which renders user-generated content and assigns it to columns which are to be displayed.

Our Contribution. We first focus on the existence of Nash equilibria. We show that colorful bin packing games may not converge to Nash equilibria even for special cases in which games have only two colors and uniform sizes (Proposition 1). However, in Theorems 1 and 2, we show that, under both cost functions, if one allows the players to perform only improving deviations towards bins in which no item is misplaced, then any game possesses the finite improving path property and therefore Nash equilibria are guaranteed to exist under both cost functions. We also show a very natural and simple algorithm, Algorithm 1 (a similar approach was already considered in [10]), that computes a Nash equilibrium whose running time is polynomial under the egalitarian cost function and pseudo-polynomial for a constant number of colors under the proportional one (Theorem 5). We then measure the quality of Nash equilibria using the standard notions of PoA (price of anarchy) and PoS (price of stability), that are defined as the worst/best case ratio between the social cost of a Nash equilibrium and the cost of a social optimum, which corresponds to the minimum number of open bin needed to feasibly pack all colored items. We provide a complete characterization of the efficiency of Nash equilibria by showing that, under both cost functions, the PoA and the PoS are unbounded (we consider the absolute approximation ratio), when $m \geq 3$ (Theorems 3 and 4), while they are equal to 3 when $m = 2$ (Theorems 6, 7, 8). We also consider the basic setting in which all items have the same size s and again provide a complete picture of the efficiency of Nash equilibria which happens to depend on the parity of the number $\kappa = \lfloor 1/s \rfloor$ of items that can be packed into a bin without exceeding its capacity. In particular, we show that, when κ is even, the price of stability is 2 for any $m \geq 2$ (Theorems 9 and 10), while the price of anarchy is 2 for $m = 2$ (Theorem 13), and unbounded for $m \geq 3$ (Theorem 11). When κ is odd, the price of stability is 1 for any $m \geq 2$ (Theorem 10), while the price of anarchy is 3 for $m = 2$ (Theorems 6 and 12), and unbounded for $m \geq 3$ (Theorem 11). We also design an algorithm (Algorithm 2) which returns a Nash equilibrium which is socially optimal when κ is odd and 2-approximates the social optimal when κ is even.

Due to space constraints, some proofs have been omitted.

Related Work. The classical one-dimensional bin packing problem has been widely studied (see [8] for a general survey). Bin packing games under the proportional cost function have been introduced in [4]. The author proved the existence of Nash equilibria by showing that the best-response dynamics converge in finite time. He also established that there is always a Nash equilibrium with minimal number of bins, i.e., the PoS is 1, but that finding such a good equilibrium is NP-hard. Finally, he presented constant upper and lower bounds on the PoA. Nearly tight bounds on the PoA have been later shown in [13]. Yu and Zhang [19] have designed a polynomial time algorithm which returns a Nash equilibrium. Bin packing games under the egalitarian cost function were considered in [18]. They showed constant tight bounds on the PoA and the PoS and design a polynomial time algorithm for computing a Nash equilibrium. In [9], the authors provided tight bounds on the exact worst-case number of steps needed to reach a Nash equilibrium. Other types of equilibria (like for instance strong equilibria) and other bin packing games were also considered in [1,7,10,12–14,16]. The offline version of the black and white bin packing problem was considered in [3]. Most of the literature on colorful bin packing is about the online version of the problem. Competitive algorithms for the online colorful bin packing problem were presented in [11]. The special case black and white was considered in [2], while, the one where all items have size 0, was considered in [5]. All such results on the online version of the problem were improved in [6]. Related colorful bin packing problems have been also considered. For instance, in the bin coloring [17], the problem is to pack colored items into bins, such that the maximum number of different colors per bin is minimized. The bin coloring games were considered in [15], where each player controls a colored item and aims at packing its item into a bin with as few different colors as possible. To the best of our knowledge, this is the first paper dealing with Nash equilibria in colorful bin packing games.

2   Model and Preliminaries

In a *colorful bin packing game* $G = (N, C, \mathcal{B}, (s_i)_{i \in N}, (c_i)_{i \in N})$ we have a set of n players $N = \{1, \ldots, n\}$, a set of $m \geq 2$ colors $C = \{1, \ldots, m\}$ and a set of n unit capacity bins $\mathcal{B} = \{B_1 \ldots, B_n\}$. Each player $i \in N$ controls an indivisible item, denoted for convenience as x_i (i.e., we denote by $X = \{x_1, \ldots, x_n\}$ the set of items), having size $s_i \in [0, 1]$ and color $c_i \in C$ which needs to be packed into one bin in $\mathcal{B}$ without exceeding its capacity. Game G has *uniform sizes* if $s_i = s_j$ for every $i, j \in N$. The special case in which G has $m = 2$ colors is called the *black and white bin packing game*; we shall define color 1 as black, color 2 as white and denote by $\#B$ and $\#W$ the number of black and white items in G, respectively.

A strategy profile is modeled by an n-tuple $\boldsymbol{\sigma} = (\sigma_1, \ldots, \sigma_n)$ such that, for each $i \in N$, $\sigma_i \in \mathcal{B}$ is the bin chosen by player i. We denote by $B_j(\boldsymbol{\sigma}) \doteq \{x_i \in X : \sigma_i = B_j\}$ the set of items packed into B_j according to the strategy profile $\boldsymbol{\sigma}$. Similarly, we also write $\sigma_i(\boldsymbol{\sigma}) = \{x_l \in X : \sigma_l = \sigma_i\}$ to indicate the set of items packed in the same bin as x_i (i.e., the bin chosen by player i), according to $\boldsymbol{\sigma}$. Given any bin B_j, we assume to pack the items in a fixed internal order, going from bottom to top, that is the sequential order in which players have chosen the bin B_j as strategy. This is what basically happens in queue systems. Namely, for any pair of items x_i and x_l in $B_j(\boldsymbol{\sigma})$, we say that x_i precedes x_l inside the bin B_j, and we write $x_i \prec_\sigma x_l$, if player i chose bin B_j before l. Formally, given any strategy profile $\boldsymbol{\sigma}$, each item x_i occupies a precise position p_i in the sequential order of items in bin σ_i, counting from bottom to top, computed as $p_i(\boldsymbol{\sigma}) = 1 + |\{x_l \in \sigma_i(\boldsymbol{\sigma}) : x_l \prec_\sigma x_i\}|$. We notice that with such packing, the last player (say i), choosing the bin σ_i, occupies the top position in σ_i.

Denoted by $\ell_{B_j}(\boldsymbol{\sigma}) = \sum_{x_i \in B_j(\boldsymbol{\sigma})} s_i$ the total size of items packed into $B_j(\boldsymbol{\sigma})$, we always assume that $\ell_{B_j}(\boldsymbol{\sigma}) \leq 1$, so that every strategy profile induces a packing of items in $\mathcal{B}$ and vice versa. We say that an item x_i is *misplaced* if there exists an item x_l with $c_i = c_l$ such that $\sigma_i = \sigma_l$ and $|p_i - p_l| = 1$, that is, x_i is adjacent to an item of the same color in the bin. A bin is *feasible* if it stores no misplaced items. In particular, an empty bin is feasible. A strategy profile is feasible if so are all of its bins. For games with uniform sizes $s_i = s$ for every $i \in N$, we denote by $\kappa = \lfloor \frac{1}{s} \rfloor$ the maximum number of items that can be packed into any (even non-feasible) bin. We only consider the cases in which $\kappa > 1$ as, otherwise, the game is trivial.

We shall denote by $cost_i(\boldsymbol{\sigma})$ the cost that player $i \in N$ pays in the strategy profile $\boldsymbol{\sigma}$ and each player aims at minimizing it. We consider two different cost functions: the *egalitarian cost function* and the *proportional cost function*. We have $cost_i(\boldsymbol{\sigma}) = \infty$ under both cost functions when x_i is a misplaced item, while, for non-misplaced ones, we have $cost_i(\boldsymbol{\sigma}) = \frac{1}{|\sigma_i(\boldsymbol{\sigma})|}$ under the egalitarian cost function and $cost_i(\boldsymbol{\sigma}) = \frac{s_i}{\ell_{\sigma_i}(\boldsymbol{\sigma})}$ under the proportional one. Note that, for games with uniform sizes, the two cost functions coincide. For a fixed strategy profile $\boldsymbol{\sigma}$, we say that a bin is a *singleton* bin if it stores only one item. Moreover, when considering the egalitarian (resp. proportional) cost function, we denote by $\overline{B}_\sigma$ the bin storing the maximum number of items (resp. the fullest bin) in the packing corresponding to σ, breaking ties arbitrarily.

A *deviation* for a player i in a strategy profile σ is the action of changing the selected bin σ_i in favor of another bin, say B_j, such that $\ell_{B_j}(\sigma) + s_i \leq 1$. We shall denote as (σ_{-i}, B_j) the strategy profile realized after the deviation. Formally, $\sigma' = (\sigma_{-i}, B_j) = (\sigma'_1,, \ldots, \sigma'_n)$ is defined as follows: $\sigma'_i = B_j$ and $\sigma'_l = \sigma_l$ for each player $l \neq i$. In this paper, we consider deviations of the following form: x_i is removed from σ_i and packed on top of B_j, consistently with the sequential order of items in a bin. An *improving deviation* for a player i in a strategy profile σ is a deviation towards a bin B_j such that $cost_i(\sigma_{-i}, B_j) < cost_i(\sigma)$. Fix a feasible strategy profile σ. Under the egalitarian cost function, player i admits an improving deviation in σ if there exists a bin $B_j \in \mathcal{B} \setminus \{\sigma_i\}$ such that (i) the item on top of B_j has a color different than c_i and (ii) $|\sigma_i(\sigma)| \leq |B_j(\sigma)|$. Under the proportional cost function, player i admits an improving deviation in σ if there exists a bin $B_j \in \mathcal{B} \setminus \{\sigma_i\}$ such that (i) the item on top of B_j has a color different than c_i and (ii) $\ell_{\sigma_i}(\sigma) < \ell_{B_j}(\sigma) + s_i$. Conversely, when a strategy profile σ is unfeasible, under both cost functions, a player controlling a misplaced item x always possesses an improving deviation, for instance, by moving x to an empty bin which is always guaranteed to exist as there are n items, n bins and the bin storing x is non-singleton. We note that, as a side-effect of an improving deviation, (σ_{-i}, B_j) may be unfeasible even if σ is feasible: this happens when x_i separates two items of the same color. We say that an improving deviation is *valid* whenever the destination bin is feasible before the deviation.

A strategy profile σ is a (pure) Nash equilibrium if $cost_i(\sigma) \leq cost_i(\sigma_{-i}, B_j)$ for each $i \in N$ and $B_j \in \mathcal{B}$, that is, no player has an improving deviation in σ. Let $\mathsf{NE}(G)$ denote the set of Nash equilibria of game G. It is easy to see that any Nash equilibrium is a feasible strategy profile. This implies that $m = 1$ would force each item to be packed into a different bin: this justifies our choice of $m \geq 2$. A game G has the *finite improvement path property* if it does not admit an infinite sequence of improving deviations. Clearly, if G enjoys the finite improvement path property, it follows that $\mathsf{NE}(G) \neq \emptyset$. Given a strategy profile σ, let $\bar{\mathcal{B}}(\sigma) \subseteq \mathcal{B}$ be the set of open bins in σ, where a bin is *open* if it stores at least one item. Let $\mathsf{F}(\sigma)$ be the number of open bins in σ, i.e., $\mathsf{F}(\sigma) = |\bar{\mathcal{B}}(\sigma)|$. We shall denote with $\sigma^*(G)$ the *social optimum*, that is, any strategy profile minimizing function F. It is easy to see that any social optimum is a feasible strategy profile.

The *price of anarchy* of G is defined as $\mathsf{PoA}(G) = \max_{\sigma \in \mathsf{NE}(G)} \frac{\mathsf{F}(\sigma)}{\mathsf{F}(\sigma^*(G))}$, while the *price of stability* of G is defined as $\mathsf{PoS}(G) = \min_{\sigma \in \mathsf{NE}(G)} \frac{\mathsf{F}(\sigma)}{\mathsf{F}(\sigma^*(G))}$. Given a class of colorful bin packing games $\mathcal{C}$, the prices of anarchy and stability of $\mathcal{C}$ are defined as $\mathsf{PoA}(\mathcal{C}) = \sup_{G \in \mathcal{C}} \mathsf{PoA}(G)$ and $\mathsf{PoS}(\mathcal{C}) = \sup_{G \in \mathcal{C}} \mathsf{PoS}(G)$. Let $\mathcal{G}_m$ denote the set of all colorful bin packing games with m colors and $\mathcal{U}_m^{odd}$ (resp. $\mathcal{U}_m^{even}$) denote the set of all colorful bin packing games with m colors and uniform sizes for which κ is odd (resp. even). Finally, denote $\mathcal{U}_m = \mathcal{U}_m^{even} \cup \mathcal{U}_m^{odd}$.

3 Existence and Efficiency of Nash Equilibria in General Games

In this section, we first show that, without any particular restriction on the type of improving deviations performed by the players, even games with uniform sizes and only two colors may not admit the finite improvement path property (Proposition 1). However, if one allows the players to perform only valid improving deviations, then any game possesses the finite improving path property under both cost functions (Theorems 1 and 2). These two theorems, together with the fact that in any strategy profile which is not a Nash equilibrium there always exists a valid improving deviation, imply the existence of Nash equilibria for colorful bin packing games under both cost functions.

Proposition 1. *There exists a black and white bin packing game with uniform sizes not possessing the finite improvement path property.*

Theorem 1. *If players are restricted to perform only valid improving deviations, then each colorful bin packing game under the egalitarian cost function admits the finite improvement path property.*

Proof. To prove the claim, we define a suitable potential function which strictly increases each time a player performs a valid improving deviation. Given a strategy profile $\boldsymbol{\sigma}$, consider the potential function

$$\Phi(\boldsymbol{\sigma}) = \sum_{B_j \in \bar{\mathcal{B}}(\boldsymbol{\sigma}):B_j \text{ is feasible}} |B_j(\boldsymbol{\sigma})|^{|B_j(\boldsymbol{\sigma})|}$$

and assume that a player i performs a valid improving deviation by moving x_i onto bin $B_j \neq \sigma_i$. We distinguish between two cases.

If bin σ_i is not feasible since both $B_j(\boldsymbol{\sigma})$ and $B_j(\boldsymbol{\sigma}_{-i}, B_j)$ are feasible, we obtain $\Phi(\boldsymbol{\sigma}_{-i}, B_j) - \Phi(\boldsymbol{\sigma}) \geq (|B_j(\boldsymbol{\sigma})| + 1)^{|B_j(\boldsymbol{\sigma})|+1} - |B_j(\boldsymbol{\sigma})|^{|B_j(\boldsymbol{\sigma})|} > 0$ when $B_j(\boldsymbol{\sigma})$ is open and $\Phi(\boldsymbol{\sigma}_{-i}, B_j) - \Phi(\boldsymbol{\sigma}) \geq 1^1 - 0 > 0$ otherwise.

If σ_i is feasible, under the egalitarian cost function, this implies that $|B_j(\boldsymbol{\sigma})| \geq |\sigma_i(\boldsymbol{\sigma})|$. Observe that, since σ_i is feasible, $cost_i(\boldsymbol{\sigma}) \leq 1$, which implies that it must be $|B_j(\boldsymbol{\sigma})| \geq 1$. For the ease of notation, set $|\sigma_i(\boldsymbol{\sigma})| = \alpha$ and $|B_j(\boldsymbol{\sigma})| = \beta$, so that $\beta \geq \alpha$ and $\beta \geq 1$; we get $\Phi(\boldsymbol{\sigma}_{-i}, B_j) - \Phi(\boldsymbol{\sigma}) \geq (\beta+1)^{\beta+1} - \alpha^\alpha - \beta^\beta = (\beta+1)(\beta+1)^\beta - \alpha^\alpha - \beta^\beta \geq 2(\beta+1)^\beta - \alpha^\alpha - \beta^\beta > 2(\beta+1)^\beta - 2\beta^\beta > 0$, where the second inequality comes from $\beta \geq 1$ and the third one comes from $\beta \geq \alpha$. Thus, in any case, $\Phi(\boldsymbol{\sigma}_{-i}, B_j) > \Phi(\boldsymbol{\sigma})$ which, since the number of possible strategy profiles is finite, implies the claim. $\square$

The next proof uses a different function than the one used above.

Theorem 2. *If players are restricted to perform only valid improving deviations, then each colorful bin packing game under the proportional cost function admits the finite improvement path property.*

In the following we give a tight characterization of the efficiency of Nash equilibria in colorful bin packing games under both cost functions. For games with at least three colors, Theorems 3 and 4 show that, under both cost functions, the PoS can be unbounded, thus, in the worst-case, no efficient Nash equilibria are guaranteed to exist.

Theorem 3. *Under the egalitarian cost function,* $\mathsf{PoS}(\mathcal{G}_m)$ *is unbounded for each* $m \geq 3$.

Theorem 4. *Under the proportional cost function,* $\mathsf{PoS}(\mathcal{G}_m)$ *is unbounded for each* $m \geq 3$.

We conclude the section by presenting a simple algorithm, namely Algorithm 1, for computing a Nash equilibrium in colorful bin packing games under both cost functions. In particular, we shall prove that its running time is polynomial for the egalitarian cost function and pseudo-polynomial for the proportional one for the special case of constant number of colors. Algorithm 1 is based on the computation of a solution for the following two optimization problems.

Max Cardinality Colorful Packing: Given a set of items $X = \{x_1, \ldots, x_n\}$, where each item x_i has size $s_i \in [0, 1]$ and color c_i, compute a set of items of maximum cardinality which can be packed into a feasible bin without exceeding its capacity.

Colorful Subset Sum: Given a set of items $X = \{x_1, \ldots, x_n\}$, where each item x_i has size $s_i \in [0, 1]$ and color c_i, compute a set of items, of maximum total size, which can be packed into a feasible bin without exceeding its capacity.

Algorithm 1. It takes as input a colorful bin packing game G

 1: $X \leftarrow \{x_1, \ldots, x_n\}$
 2: **while** ($X \neq \emptyset$) **do**
 3:     **if** (G is defined under the egalitarian cost function) **then**
 4:         Let B be a solution to Max Cardinality Colorful Packing(X)
 5:     **else**
 6:         Let B be a solution to Colorful Subset Sum(X)
 7:     **end if**
 8:     Open a new bin and assign it the set of items B
 9:     $X \leftarrow X \setminus B$
10: **end while**
11: **return** the strategy profile induced by the set of open bins

---

Next lemma shows the correctness of the algorithm.

Lemma 1. *Algorithm 1 computes a Nash equilibrium for any colorful bin packing game* G.

Lemma 2. Max Cardinality Colorful Packing *can be solved in polynomial time.*

Lemma 3. Colorful Subset Sum *can be solved in pseudo-polynomial time as long as the number of colors is constant.*

As a consequence of Lemmas 1, 2 and 3, we obtain the following result.

Theorem 5. *A Nash equilibrium for colorful bin packing games can be computed in polynomial time under the egalitarian cost function and in pseudo-polynomial time for a constant number of colors under the proportional one.*

4   Efficiency of Nash Equilibria in Black and White Games

For black and white bin packing games, things get much more interesting, as we show an upper bound of 3 on the PoA and a corresponding lower bound on the PoS. To address this particular case, given a black and white bin packing game G, we make use of the following additional notation. Given a strategy profile σ, we denote by $S_b(\sigma)$ the set of singleton bins storing a black item, by $S_w(\sigma)$ the set of singleton bins storing a white item, by $M_b(\sigma)$ the set of non-singleton bins having a black item on top, and by $M_w(\sigma)$ the set of non-singleton bins having a white item on top.

The following lemma relates the set of open bins of a feasible strategy profile with that of a social optimum.

Lemma 4. *Fix a feasible strategy profile σ and a social optimum σ^* for a black and white bin packing game G. Then, $|S_b(\sigma)| - |S_w(\sigma)| - |M_w(\sigma)| \leq \mathsf{F}(\sigma^*)$.*

The following theorem gives an upper bound on the PoA of black and white bin packing games under both cost functions.

Theorem 6. *Under both cost functions, $\mathsf{PoA}(\mathcal{G}_2) \leq 3$.*

Proof. Given a black and white bin packing game G under a certain cost function, fix a Nash equilibrium σ and a social optimum σ^*. Let $S = \sum_{i \in N} s_i$ be the sum of the sizes of all the items. Notice that $\mathsf{F}(\sigma^*) \geq \lceil S \rceil$. Assume without loss of generality that $|S_b(\sigma)| \geq |S_w(\sigma)|$ (if this is not the case, we simply swap the two colors).

Let P be the set of pairs of bins constructed as follows: each bin in $S_w(\sigma)$ is paired with a bin in $S_b(\sigma)$, each remaining bin in $S_b(\sigma)$ is paired with a bin in $M_w(\sigma)$, finally, all the remaining bins in $M_w(\sigma)$ and all the bins in $M_b(\sigma)$ are joined into pairs until possible. It is easy to check that, for each created pair of bins (B_j, B_k), it must be

$$\ell_{B_j}(\sigma) + \ell_{B_k}(\sigma) > 1 \tag{1}$$

under both cost functions, otherwise the hypothesis that σ is a Nash equilibrium would be contradicted. Moreover, by (1), it follows that $S > |P|$ which implies that $\mathsf{F}(\sigma^*) \geq \lceil S \rceil \geq |P| + 1$.

Now two cases may occur:

- no bin in $S_b(\sigma)$ is left unmatched by P, which implies that $|S_b(\sigma)|+|S_w(\sigma)|+|M_b(\sigma)|+|M_w(\sigma)| \leq 2|P|+1$, as at most one bin from the set $M_w(\sigma) \cup M_b(\sigma)$ may remain unmatched. Thus, we obtain

$$\mathsf{F}(\sigma) = |S_b(\sigma)| + |S_w(\sigma)| + |M_b(\sigma)| + |M_w(\sigma)| \leq 2|P| + 1 < 2\mathsf{F}(\sigma^*);$$

- at least one bin in $S_b(\sigma)$ is unmatched by P, which implies that $|S_b(\sigma)| + |S_w(\sigma)| + |M_b(\sigma)| + |M_w(\sigma)| \leq 2|P| + 1 + |S_b(\sigma)| - |S_w(\sigma)| - |M_w(\sigma)|$. Thus, we obtain

$$\begin{aligned}
\mathsf{F}(\sigma) &= |S_b(\sigma)| + |S_w(\sigma)| + |M_b(\sigma)| + |M_w(\sigma)| \\
&\leq 2|P| + 1 + |S_b(\sigma)| - |S_w(\sigma)| - |M_w(\sigma)| \\
&\leq 2|P| + 1 + \mathsf{F}(\sigma^*) < 2\mathsf{F}(\sigma^*) + \mathsf{F}(\sigma^*) = 3\mathsf{F}(\sigma^*),
\end{aligned}$$

where the second inequality comes from Lemma 4.   □

In the next two theorems, we show a matching lower bound on the PoS of black and white bin packing games under both cost functions.

Theorem 7. *Under the egalitarian cost function, $\mathsf{PoS}(\mathcal{G}_2) \geq 3$.*

Theorem 8. *Under the proportional cost function, $\mathsf{PoS}(\mathcal{G}_2) \geq 3$.*

5   Efficiency of Nash Equilibria in Games with Uniform Sizes

In this section, we provide a complete picture of the efficiency of Nash equilibria for games with uniform sizes. We remind the reader that, in this setting, the egalitarian and proportional cost functions are equivalent. For the sake of simplicity, we say that a bin is *full* if it contains κ items.

First, we give a lower bound of 2 on the PoS for games with any number of colors under the hypothesis that κ is an even number.

Theorem 9. *For each $m \geq 2$, $\mathsf{PoS}(\mathcal{U}_m^{even}) \geq 2$.*

We show that, unlike the case of general games considered in the previous section, under the hypothesis of uniform sizes, efficient Nash equilibria are always guaranteed to exist for any number of colors. In particular, we design an algorithm which, given a colorful bin packing game G with uniform sizes, returns a Nash equilibrium σ such that $\mathsf{F}(\sigma) \leq 2\mathsf{F}(\sigma^*(G))$ when κ is even and $\mathsf{F}(\sigma) = \mathsf{F}(\sigma^*(G))$ when κ is odd. Given the result on the price of stability of Theorem 9, these are the best achievable performance.

Theorem 10. *For each $m \geq 2$, $\mathsf{PoS}(\mathcal{U}_m^{even}) \leq 2$ and $\mathsf{PoS}(\mathcal{U}_m^{odd}) = 1$. Moreover, for any game $G \in \mathcal{U}_m$, a Nash equilibrium σ such that $\mathsf{F}(\sigma) \leq 2\mathsf{F}(\sigma^*(G))$ if $G \in \mathcal{U}_m^{even}$ and such that $\mathsf{F}(\sigma) = \mathsf{F}(\sigma^*(G))$ if $G \in \mathcal{U}_m^{odd}$ can be computed in pseudo-polynomial time.*

Algorithm 2. It takes as input a colorful bin packing game with uniform sizes G

1: $X \leftarrow \{x_1, \ldots, x_n\}$
2: $i \leftarrow 1$
3: $c_{old} \leftarrow 0$
4: **while** $(X \neq \emptyset)$ **do**
5:    **if** $(|B_i| < \kappa)$ && $(\exists x_j \in X$ s.t. $c_j \neq c_{old}))$ **then**
6:        $c \leftarrow$ most frequent color among the items in X having color other than c_{old}
7:        Select an item x_j of color c
8:        $X \leftarrow X \setminus \{x_j\}$
9:        $c_{old} \leftarrow c$
10:        $\sigma_j \leftarrow B_i$
11:    **else**
12:        $i \leftarrow i + 1$
13:        $c_{old} \leftarrow 0$
14:    **end if**
15: **end while**
16: **return** σ

Proof. Fix an integer $m \geq 2$ and a game $G \in \mathcal{U}_m$. We prove the claim by showing that Algorithm 2 computes a Nash equilibrium σ for G such that $\mathsf{F}(\sigma) \leq 2\mathsf{F}(\sigma^*(G))$ if $G \in \mathcal{U}_m^{even}$ and such that $\mathsf{F}(\sigma) = \mathsf{F}(\sigma^*(G))$ if $G \in \mathcal{U}_m^{odd}$.

We start by showing that the strategy profile σ returned by Algorithm 2 is a Nash equilibrium for G. Let us partition σ into three sets, namely Γ, Δ, Θ, where Γ contains all the full bins, Δ contains all the non-full and non-singleton bins and Θ contains all the singleton bins. It is not difficult to see that, by the definition of Algorithm 2, Δ and Θ are such that *(i)* Δ is either empty or contains only one bin, *(ii)* all items stored into bins belonging to Θ have the same color, denoted as c_Θ, *(iii)* the item on top of the bin in Δ (if any) has color c_Θ.

Now assume, by way of contradiction, that there exists a player j possessing an improving deviation in σ towards a bin B_i. Clearly, this can only be possible if x_j is packed into a singleton bin and $B_i \in \Delta \cup \Theta$, but properties *(ii)* and *(iii)* above imply a contradiction. So, σ is a Nash equilibrium.

Let $n_z(c)$ be the number of items of color c belonging to X at the zth iteration of Algorithm 2.

Lemma 5. *If either $|\Theta| \geq 2$ or $|\Delta| = |\Theta| = 1$, then the color of each item occupying an odd position in a bin belonging to $\Gamma \cup \Delta$ is c_Θ.*

By the previous lemma, we get the following corollary which gives us the number of items of color c_Θ and the number of items of color different that c_Θ.

Corollary 1. *If either $|\Theta| \geq 2$ or $|\Delta| = |\Theta| = 1$, then each bin in $B_j \in \bar{\mathcal{B}}(\sigma)$ contains $\lceil |B_j(\sigma)|/2 \rceil$ items of color c_Θ.*

Let $\#c_\Theta$ be the number of items having color c_Θ. We conclude by showing that $\mathsf{F}(\sigma) \leq 2\mathsf{F}(\sigma^*(G))$ when κ is even and that $\mathsf{F}(\sigma) = \mathsf{F}(\sigma^*(G))$ when κ is odd. Towards this end, we use Corollary 1 together with the simple basic fact.

Fact 1. $2\#c_\Theta \le n + \mathsf{F}(\sigma^*(G))$.

Let us start with the cases not covered by Corollary 1, that is, $|\Theta| = 0$ and $|\Theta| = 1 \wedge |\Delta| = 0$. In both cases, we have $\mathsf{F}(\sigma) = \mathsf{F}(\sigma^*(G))$ independently of the parity of κ, as $\bar{\mathcal{B}}(\sigma)$ contains at most one non-full bin. Thus, in the remaining of the proof, we can assume that Corollary 1 holds. Let $\delta \in \{0, \ldots, \kappa - 1\}$ be the number of items stored into the bin belonging to Δ ($\delta = 0$ models the case in which this bin does not exist).

For the case in which κ is odd, by Corollary 1, we have $\#c_\Theta = |\Gamma|\frac{\kappa+1}{2} + \lceil\frac{\delta}{2}\rceil + |\Theta|$ and $n = |\Gamma|\kappa + \delta + |\Theta|$. Assume, by way of contradiction, that $\mathsf{F}(\sigma^*(G)) < \mathsf{F}(\sigma)$, that is, $\mathsf{F}(\sigma^*(G)) \le |\Gamma| + |\Delta| + |\Theta| - 1$. By Fact 1, we obtain

$$2\left\lceil\frac{\delta}{2}\right\rceil \le \delta + |\Delta| - 1. \tag{2}$$

Now observe that, for $|\Delta| = 0$, which implies $\delta = 0$, (2) is not satisfied. Hence, it must be $|\Delta| = 1$ which, as $|\Theta| \ge 1$ (recall that we are under the hypothesis in which Corollary 1 holds), implies that $\delta > 1$. Now, if δ is even, by Corollary 1, the item on top of the unique bin in Δ has color different than c_Θ. This means that a player controlling an item packed into any bin in Θ has an improving deviation by migrating to the unique bin in Δ, thus contradicting the fact that σ is a Nash equilibrium. Thus, under the hypothesis of $|\Delta| = 1$ and δ odd, (2) is again not satisfied, thus rising a contradiction. Hence, it follows that $\mathsf{F}(\sigma^*(G)) = \mathsf{F}(\sigma)$.

For the case in which κ is even, by Corollary 1, we have $\#c_\Theta = |\Gamma|\frac{\kappa}{2} + \lceil\frac{\delta}{2}\rceil + |\Theta|$ and $n = |\Gamma|\kappa + \delta + |\Theta|$. As $\mathsf{F}(\sigma^*(G)) \ge |\Gamma| + |\Delta|$, if $|\Gamma| \ge |\Theta|$, it follows $\mathsf{F}(\sigma) = |\Gamma| + |\Delta| + |\Theta| \le 2|\Gamma| + |\Delta| \le 2\mathsf{F}(\sigma^*(G))$. Thus, in the remaining of the proof, we assume that $|\Theta| > |\Gamma|$. Assume now, by way of contradiction, that $\mathsf{F}(\sigma^*(G)) < \mathsf{F}(\sigma)/2$, which implies $\mathsf{F}(\sigma^*(G)) \le \frac{|\Gamma|+|\Delta|+|\Theta|}{2} - \frac{1}{2}$. By Fact 1, we obtain

$$2\left\lceil\frac{\delta}{2}\right\rceil + \frac{|\Theta|}{2} \le \delta + \frac{|\Gamma| + |\Delta| - 1}{2}. \tag{3}$$

Using the hypothesis that $|\Theta| > |\Gamma|$ within (3), we obtain

$$2\left\lceil\frac{\delta}{2}\right\rceil \le \delta + \frac{|\Delta|}{2} - 1 \tag{4}$$

which is never satisfied, thus rising a contradiction. Hence, it follows that $\mathsf{F}(\sigma) \le 2\mathsf{F}(\sigma^*(G))$.

We now argue the complexity of Algorithm 2. We first notice that, for uniform sizes, the compact representation of the input has size $\Omega(m + \log n)$. Moreover, it is easy to see that Algorithm 2 has complexity $O(n)$. It turns out that when, for instance, $m = \Omega(n^{\frac{1}{h}})$, for some constant h, the algorithm has polynomial time complexity. However, when $m = O(\log n)$, the complexity is pseudo-polynomial. □

Theorems 9 and 10 completely characterize the PoS of colorful bin packing games with uniform sizes. For what concerns the PoA, we also obtain a complete picture by means of the following results.

For the case of at least three colors, the PoA can be arbitrarily high.

Theorem 11. *For each $m \geq 3$, both $PoA(\mathcal{U}_m^{odd})$ and $PoA(\mathcal{U}_m^{even})$ are unbounded.*

For the case of black and white bin packing games, we show a lower bound of 3 on the PoA of games for which κ is an odd number, thus matching the upper bound showed in Theorem 6 which holds for general sizes.

Theorem 12. $PoA(\mathcal{U}_2^{odd}) \geq 3$.

For the leftover case of black and white bin packing games for which κ is even, we show that the upper bound on the PoA drops to 2 which matches the lower bound given in Theorem 9 for the PoS.

Theorem 13. $PoA(\mathcal{U}_2^{even}) \leq 2$.

References

1. Adar, R.: Selfish bin packing with cardinality constraints. Theor. Comput. Sci. **495**, 66–80 (2013)
2. Balogh, J., Békési, J., Dósa, G., Epstein, L., Kellerer, H., Tuza, Z.: Online results for black and white bin packing. Theory Comput. Syst. **56**(1), 137–155 (2015)
3. Balogh, J., Békési, J., Dósa, G., Epstein, L., Kellerer, H., Levin, A., Tuza, Z.: Offline black and white bin packing. Theor. Comput. Sci. **596**, 92–101 (2015)
4. Bilò, V.: On the packing of selfish items. In: 20th International Parallel and Distributed Processing Symposium - IPDPS, IEEE (2006)
5. Böhm, M., Sgall, J., Veselý, P.: Online colored bin packing. In: Bampis, E., Svensson, O. (eds.) WAOA 2014. LNCS, vol. 8952, pp. 35–46. Springer, Cham (2015). https://doi.org/10.1007/978-3-319-18263-6_4
6. Böhm, M., Dósa, G., Epstein, L., Sgall, J., Veselý, P.: Colored bin packing: online algorithms and lower bounds. Algorithmica **80**(1), 155–184 (2018)
7. Cao, Z., Yang, X.: Selfish bin covering. Theor. Comput. Sci. **412**(50), 7049–7058 (2011)
8. Coffman Jr., E.-G., Garey, M.-R., Johnson, D.-S.: Approximation algorithms for bin packing: a survey. In: Approximation Algorithms for NP-hard Problems. PWS Publishing Co. (1996)
9. Dósa, G., Epstein, L.: The Convergence time for selfish bin packing. In: Lavi, R. (ed.) SAGT 2014. LNCS, vol. 8768, pp. 37–48. Springer, Heidelberg (2014). https://doi.org/10.1007/978-3-662-44803-8_4
10. Dósa, G., Epstein, L.: Generalized selfish bin packing. CoRR, abs/1202.4080 (2012)
11. Dósa, G., Epstein, L.: Colorful bin packing. In: Ravi, R., Gørtz, I.L. (eds.) SWAT 2014. LNCS, vol. 8503, pp. 170–181. Springer, Cham (2014). https://doi.org/10.1007/978-3-319-08404-6_15
12. Epstein, L.: Bin packing games with selfish items. In: Chatterjee, K., Sgall, J. (eds.) MFCS 2013. LNCS, vol. 8087, pp. 8–21. Springer, Heidelberg (2013). https://doi.org/10.1007/978-3-642-40313-2_2

13. Epstein, L., Kleiman, E.: Selfish bin packing. Algorithmica **60**(2), 368–394 (2011)
14. Epstein, L., Kleiman, E.: Selfish vector packing. In: Bansal, N., Finocchi, I. (eds.) ESA 2015. LNCS, vol. 9294, pp. 471–482. Springer, Heidelberg (2015). https://doi.org/10.1007/978-3-662-48350-3_40
15. Epstein, L., Krumke, S.-O., Levin, A., Sperber, H.: Selfish bin coloring. J. Comb. Optim. **22**(4), 531–548 (2011)
16. Fernandes, C.-G., Ferreira, C.-E., Miyazawa, F.-K., Wakabayashi, Y.: Selfish square packing. Electron. Notes Discrete Math. **37**(1), 369–374 (2011)
17. Krumke, S.-O., de Paepe, W., Rambau, J., Stougie, L.: Bincoloring. Theor. Comput. Sci. **407**(1–3), 231–241 (2008)
18. Ma, R., Dósa, G., Han, X., Ting, H., Ye, D., Zhang, Y.: A note on a selfish bin packing problem. J. Global Optim. **56**(4), 1457–1462 (2013)
19. Yu, G., Zhang, G.: Bin packing of selfish items. In: Papadimitriou, C., Zhang, S. (eds.) WINE 2008. LNCS, vol. 5385, pp. 446–453. Springer, Heidelberg (2008). https://doi.org/10.1007/978-3-540-92185-1_50

Nonbipartite Dulmage-Mendelsohn Decomposition for Berge Duality

Nanao Kita[✉]

Tokyo University of Science, 2641 Yamazaki,
Noda, Chiba 278-8510, Japan
kita@rs.tus.ac.jp

Abstract. The *Dulmage-Mendelsohn decomposition* is a classical *canonical decomposition* in matching theory applicable for bipartite graphs and is famous not only for its application in the field of matrix computation, but also for providing a prototypal structure in matroidal optimization theory. The Dulmage-Mendelsohn decomposition is stated and proved using the two color classes of a bipartite graph, and therefore generalizing this decomposition for nonbipartite graphs has been a difficult task. In this paper, we obtain a new canonical decomposition that is a generalization of the Dulmage-Mendelsohn decomposition for arbitrary graphs using a recently introduced tool in matching theory, the *basilica decomposition*. Our result enables us to understand all known canonical decompositions in a unified way. Furthermore, we apply our result to derive a new theorem regarding *barriers*. The duality theorem for the maximum matching problem is the celebrated *Berge formula*, in which dual optimizers are known as barriers. Several results regarding maximal barriers have been derived by known canonical decompositions; however, no characterization has been known for general graphs. In this paper, we provide a characterization of the family of *maximal barriers* in general graphs, in which the known results are developed and unified.

1 Introduction

We establish the Dulmage-Mendelsohn decomposition for general graphs. The *Dulmage-Mendelsohn decomposition* [2–4], or the *DM decomposition* in short, is a classical canonical decomposition in matching theory [17] applicable for bipartite graphs. This decomposition is famous for its application for combinatorial matrix theory, especially for providing an efficient solution for a system of linear equations [1,4] and is also important in matroidal optimization theory.

Canonical decompositions of a graph are fundamental tools in matching theory [17]. A canonical decomposition partitions a given graph in a way uniquely determined for the graph and describes the structure of maximum matchings using this partition. The classical canonical decompositions are the *Gallai-Edmonds* [5,6] and *Kotzig-Lovász* decompositions [13–15] in addition to the DM

N. Kita—Supported partly by JSPS KAKENHI Grant Number 15J09683.

L. Wang and D. Zhu (Eds.): COCOON 2018, LNCS 10976, pp. 293–304, 2018.
https://doi.org/10.1007/978-3-319-94776-1_25

decomposition. The DM and Kotzig-Lovász decompositions are applicable for bipartite graphs and *factor-connected graphs*, respectively. The Gallai-Edmonds decomposition partitions an arbitrary graph into three parts: that is, the so-called $D(G)$, $A(G)$, and $C(G)$ parts. Comparably recently, a new canonical decomposition was proposed: the *basilica decomposition* [8–10]. This decomposition is applicable for arbitrary graphs and contains a generalization of the Kotzig-Lovász decomposition and a refinement the Gallai-Edmonds decomposition. (The $C(G)$ part can be decomposed nontrivially.)

In this paper, we establish an analogue of the DM decomposition for general graphs using the basilica decomposition. Our results accordingly provide a paradigm that enables us to handle any graph and understand the known canonical decompositions in a unified way. In the original theory of DM decomposition, the concept of the *DM components* of a bipartite graph is first defined, and then it is proved that these components form a poset with respect to a certain binary relation.

This theory depends heavily on the two color classes of a bipartite graph and cannot be easily generalized for nonbipartite graphs. In our generalization, we first define a generalization of the DM components using the basilica decomposition. To capture the structure formed by these components in nonbipartite graphs, we introduce a slightly more complex concept: *posets with a transitive forbidden relation*. We then prove that the generalized DM components form a poset with a transitive forbidden relation for certain binary relations. We also show that this structure can be computed in strongly polynomial time.

Furthermore, we apply our generalized DM decomposition to derive a characterization of the family of *maximal barriers* in general graphs. The *Berge formula* is a combinatorial min-max theorem in which maximum matchings are the optimizers of one hand, and the optimizers of the other hand are known as *barriers* [17]. That is, barriers are the dual optimizers of the maximum matchings problem. Barriers are heavily employed as a tool for studying matchings. However, not as much is known about barriers themselves [17]. Aside from several observations that are derived rather easily from the Berge formula, several substantial results about (inclusion-wise) maximal barriers have been provided by canonical decompositions.

Our result for maximal barriers proves that our generalization of the DM decomposition has a reasonable consistency with the relationship between each known canonical decomposition and maximal barriers. Each known canonical decomposition can be used to state the structure of maximal barriers. The original DM decomposition provides a characterization of the family of maximal barriers in a bipartite graph in terms of ideals in the poset; minimum vertex covers in bipartite graphs are equivalent to maximal barriers. The Gallai-Edmonds decomposition derives a characterization of the intersection of all maximal barriers (that is, the $A(G)$ part) [17]; this characterization is known as the *Gallai-Edmonds description*. The Kotzig-Lovász decomposition is used for characterizing the family of maximal barriers in factor-connected graphs [17]; this result is known as *Lovász's canonical partition theorem* [16,17]. The basilica decomposition provides the structure of a given maximal barrier in general graphs, which contains a common generalization of the Gallai-Edmonds description and Lovász's canonical partition theorem. Hence, a generalization of the DM decomposition would be reasonable if it can characterize the family of maximal barriers, and our generalization attains this in a way analogical to the classical DM decomposition, that is, in terms of ideals in the poset with a transitive forbidden relation.

Our results imply a new possibility in matroidal optimization theory. Submodular function theory [18] is a systematic field of study that captures many well-solved problems in terms of submodular functions and generalizations. In this theory, the bipartite maximum matching problem is an important exemplary problem. The DM decomposition is essential in the relationship between bipartite matchings and submodular functions, because it corresponds to the structural characterization of the family of minimizers of a submodular function. The nonbipartite maximum matching problem is also an important well-solved problem in combinatorial optimization, and is even referred to as the archetype of well-solved problems [17,18]. However, the nonbipartite maximum matching problem and its duality shown by the Berge formula are not included in submodular function theory today and nor in any of its generalizations. Our nonbipartite DM decomposition may provide a clue to a new epoch of submodular function theory that can be brought in by capturing these concepts.

The remainder of this paper is organized as follows: In Sect. 2, we provide the basic definitions. In Sect. 3, we present the preliminary results from the basilica decomposition theory. In Sect. 4, we introduce the new concept of posets with a transitive forbidden relation. In Sect. 5, we provide our main result, the nonbipartie DM decomposition. In Sect. 6, we use the generalized DM decomposition to characterize the family of maximal barriers. In Sect. 7, we show how our results contain the original DM decomposition for bipartite graphs.

2 Basic Preliminaries

2.1 General Definitions

For basic notation for sets, graphs, and algorithms, we mostly follow Schrijver [18]. In this section, we explain exceptions or nonstandard definitions. In Sect. 2, unless otherwise stated, let G be a graph. The vertex set and the edge set of G are denoted by $V(G)$ and $E(G)$, respectively. We treat paths and circuits as graphs. For a path P and vertices x and y from P, xPy denotes the subpath of P between x and y. We often treat a graph as the set of its vertices.

In the remainder of this section, let $X \subseteq V(G)$. The subgraph of G induced by X is denoted by $G[X]$. The graph $G[V(G) \setminus X]$ is denoted by $G - X$. The contraction of G by X is denoted by G/X. Let $F \subseteq E(G)$. The graph obtained by deleting F from G without removing vertices is denoted by $G - F$. Let H be a subgraph of G. The graph obtained by adding F to H is denoted by $H + F$. Regarding these operations, we identify vertices, edges, subgraphs of the newly created graph with the naturally corresponding items of old graphs.

A *neighbor* of X is a vertex from $V(G) \setminus X$ that is adjacent to some vertex from X. The neighbor set of X is denoted by $N_G(X)$. Let $Y \subseteq V(G)$. The set of edges joining X and Y is denoted by $E_G[X, Y]$. The set $E_G[X, V(G) \setminus X]$ is denoted by $\delta_G(X)$.

A set $M \subseteq E(G)$ is a *matching* if $|\delta_G(v) \cap M| \leq 1$ holds for each $v \in V(G)$. For a matching M, we say that M *covers* a vertex v if $|\delta_G(v) \cap M| = 1$; otherwise, we say that M *exposes* v. A matching is *maximum* if it consists of the maximum number of edges. A matching is *perfect* if it covers every vertex. A graph is *factorizable* if it has a perfect matchings. A graph is *factor-critical* if, for each vertex v, $G - v$ is factorizable. A graph with only one vertex is defined to be factor-critical. The number of edges in a maximum matching is denoted by $\nu(G)$. We define $\mathrm{def}(G) := |V(G)| - 2\nu(G)$.

Let $M \subseteq E(G)$. A circuit or path is said to be *M-alternating* if edges in M and not in M appear alternately. The precise definition is the following: A circuit C of G is M-alternating if $E(C) \cap M$ is a perfect matching of C. We define the three types of M-alternating paths. Let P be a path with ends s and t. We say that P is *M-forwarding* from s to t if $M \cap E(P)$ is a matching of P that covers every vertex except for t. A path with exactly one vertex is defined to be an M-forwarding path. For P with odd number of edges, we say that P is *M-saturated* (resp. *M-exposed*) between s and t if $M \cap E(P)$ (resp. $E(P) \setminus M$) is a perfect matching of P.

A path P with $|E(P)| \geq 1$ is an *ear* relative to X if the internal vertices of P are disjoint from X, whereas the ends are in X. A circuit C is an *ear* relative to X if exactly one vertex of C is in X; for simplicity, we call the vertex in $X \cap V(C)$ the *end* of the ear C. We call an ear P relative to X an *M-ear* if $\delta_P(X) \cap M = \emptyset$ holds and $P - X$ is empty or an M-saturated path.

2.2   Barriers, Gallai-Edmonds Family, and Factor-Components

We now explain the Berge Formula and the definition of barriers. An *odd component* (resp. *even component*) of a graph is a connected component with an odd (resp. even) number of vertices. The number of odd components of $G - X$ is

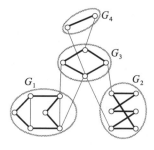

Fig. 1. The factor-components of a graph G: bold lines indicate allowed edges. This graph has four factor-components $G_1, \ldots, G_4$.

denoted by $q_G(X)$. The set of vertices from odd components (resp. even components) of $G - X$ is denoted by D_X (resp. C_X).

Theorem 1 (Berge Formula [17]). *For a graph G, $\mathrm{def}(G)$ is equal to the maximum value of $q_G(X) - |X|$, where X is taken over all subsets of $V(G)$.*

The set of vertices that attains the maximum value in this relation is called a *barrier*. That is, a set of vertices X is a *barrier* if $\mathrm{def}(G) = q_G(X) - |X|$.

The set of vertices that can be exposed by maximum matchings is denoted by $D(G)$. The neighbor set of $D(G)$ is denoted by $A(G)$, and the set $V(G) \setminus D(G) \setminus A(G)$ is denoted by $C(G)$. The following statement about $D(G)$, $A(G)$, and $C(G)$ is the celebrated *Gallai-Edmonds structure theorem* [5,6,17].

Theorem 2 (Gallai-Edmonds Structure Theorem). *For any graph G,*

(i) $A(G)$ *is a barrier for which* $D_{A(G)} = D(G)$ *and* $C_{A(G)} = C(G)$;
(ii) *each odd component of* $G - A(G)$ *is factor-critical; and,*
(iii) *any edge in* $E_G[A(G), D(G)]$ *is allowed.*

An edge is *allowed* if it is contained in some maximum matching. Two vertices are *factor-connected* if they are connected by a path whose edges are allowed. A subgraph is *factor-connected* if any two vertices are factor-connected. A maximal factor-connected subgraph is called a *factor-connected component* or *factor-component*. A graph consists of its factor-components and edges joining them that are not allowed. The set of factor-components of G is denoted by $\mathcal{G}(G)$.

A factor-component C is *inconsistent* if $V(C) \cap D(G) \neq \emptyset$. Otherwise, C is said to be *consistent*. We denote the sets of consistent and inconsistent factor-components of G by $\mathcal{G}^+(G)$ and $\mathcal{G}^-(G)$, respectively. The next property is easily confirmed from the Gallai-Edmonds structure theorem.

Fact 3. *A subgraph C of G is an inconsistent factor-component if and only if C is a connected component of $G[D(G) \cup A(G)]$. Any consistent factor-component has the vertex set contained in $C(G)$.*

That is, the structure of inconsistent factor-components are rather trivial under the Gallai-Edmonds structure theorem.

3   Basilica Decomposition of Graphs

In this section, we now introduce the basilica decomposition of graphs [9,10]. The theory of basilica decomposition is made up of the three central concepts: a canonical partial order between factor-components (Theorem 4), the general Kotzig-Lovász decomposition (Theorem 5), and an interrelationship between the two (Theorem 6). These theorems can be found in Kita [9,10]. In the following, let G be a graph unless otherwise stated.

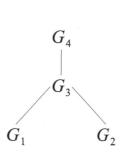

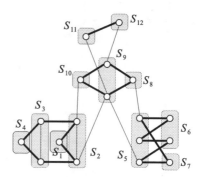

Fig. 2. The Hasse diagram of the poset $(\mathcal{G}(G), \lhd)$.

Fig. 3. The general Kotzig-Lovász decomposition of G: $\mathcal{P}(G)$ has 12 members $S_1, \ldots, S_{12}$.

Definition 1. *A set $X \subseteq V(G)$ is said to be* separating *if there exist $H_1, \ldots, H_k \in \mathcal{G}(G)$, where $k \geq 1$, such that $X = V(H_1) \cup \cdots \cup V(H_k)$. For $G_1, G_2 \in \mathcal{G}(G)$, we say $G_1 \lhd G_2$ if there exists a separating set $X \subseteq V(G)$ with $V(G_1) \cup V(G_2) \subseteq X$ such that $G[X]/G_1$ is a factor-critical graph.*

Theorem 4. *For a graph G, the binary relation $\lhd$ is a partial order over $\mathcal{G}(G)$.*

Definition 2. *For $u, v \in V(G) \setminus D(G)$, we say $u \sim_G v$ if u and v are identical or if u and v are factor-connected and satisfy $\mathrm{def}(G - u - v) > \mathrm{def}(G)$.*

Theorem 5. *For a graph G, the binary relation $\sim_G$ is an equivalence relation.*

We denote as $\mathcal{P}(G)$ the family of equivalence classes determined by $\sim_G$. This family is known as the *general Kotzig-Lovász decomposition* or just the *Kotzig-Lovász decomposition* of G. From the definition of $\sim_G$, for each $H \in \mathcal{G}(G)$, the family $\{S \in \mathcal{P}(G) : S \subseteq V(H)\}$ forms a partition of $V(H) \setminus D(G)$. We denote this family by $\mathcal{P}_G(H)$.

Let $H \in \mathcal{G}(G)$. The sets of strict and nonstrict upper bounds of H are denoted by $\mathcal{U}_G(H)$ and $\mathcal{U}_G^*(H)$, respectively. The sets of vertices $\bigcup\{V(I) : I \in \mathcal{U}_G(H)\}$ and $\bigcup\{V(I) : I \in \mathcal{U}_G^*(H)\}$ are denoted by $U_G(H)$ and $U_G^*(H)$, respectively.

Theorem 6. *Let G be a graph, and let $H \in \mathcal{G}(G)$. Then, for each connected component K of $G[U_G(H)]$, there exists $S \in \mathcal{P}_G(H)$ such that $N_G(K) \cap V(H) \subseteq S$.*

Under Theorem 6, for $S \in \mathcal{P}_G(H)$, we denote by $\mathcal{U}_G(S)$ the set of factor-components that are contained in a connected component K of $G[U_G(H)]$ with $N_G(K) \cap V(H) \subseteq S$. The set $\bigcup\{V(I) : I \in U_G(H)\}$ is denoted by $U_G(S)$. We denote $U_G(H) \setminus S \setminus U_G(S)$ by $^\top U_G(S)$.

Theorem 6 integrates the two structures given by Theorems 4 and 5 into a structure of graphs that is reminiscent of an architectural building. We call this integrated structure the *basilica decomposition* of a graph. See Figs. 1, 2, and 3 for an example of the basilica decomposition.

Inconsistent factor-components in a graph have a trivial structure regarding the basilica decomposition. The next statement is easily confirmed from Fact 3 and the Gallai-Edmonds structure theorem.

Fact 7. *Let G be a graph. Any inconsistent component is minimal in the poset $(\mathcal{G}(G), \lhd)$. For any $H \in \mathcal{G}^-(G)$, if $V(H) \cap A(G) \neq \emptyset$, then $\mathcal{P}_G(H) = \{V(H) \cap A(G)\}$; otherwise, $\mathcal{P}_G(H) = \emptyset$.*

For simplicity, even for $H \in \mathcal{G}^-(G)$ with $V(H) \cap A(G) = \emptyset$, we treat as if $V(H) \cap A(G)$ is a member of $\mathcal{P}(G)$. That is, we let $\mathcal{P}_G(H) = \{V(H) \cap A(G)\}$ and $^\top U_G(V(H) \cap A(G)) = {}^\top U_G(\emptyset) = V(H) \cap D(G) = V(H)$. Under Fact 7, the substantial information provided by the basilica decomposition lies in the consistent factor-components.

We now present some properties of the basilica decomposition that are used in later sections. The next lemma can be found in Kita [11].

Lemma 1. *Let G be a graph, and let M be a maximum matching of G. Let $H \in \mathcal{G}^+(G)$, $S \in \mathcal{P}_G(H)$, and $s \in S$.*

(i) *For any $t \in S$, there is an M-forwarding path from s to t, whose vertices are contained in $S \cup {}^\top U_G(S)$; however, there is no M-saturated path between s and t.*

(ii) *For any $t \in {}^\vdash U_G(S)$, there exists an M-saturated path between s and t whose vertices are contained in $S \cup {}^\top U_G(S)$.*

(iii) *For any $t \in U_G(S)$, there is an M-forwarding path from t to s, whereas there is no M-forwarding path from s to t or M-saturated path between s and t.*

The first part of the next lemma is provided in Kita [12], and the second part can be easily proved from Lemma 1.

Lemma 2. *Let G be a graph, and let M be a maximum matching of G. Let $S \in \mathcal{P}(G)$. If there is an M-ear relative to $S \cup {}^\top U_G(S)$ that has internal vertices, then the ends of this ear are contained in S.*

4   Poset with Transitive Forbidden Relation

We now introduce the new concept of *posets with a transitive forbidden relation*, which serves as a language to describe the nonbipartite DM decomposition.

Definition 3. *Let X be a set, and let $\preceq$ be a partial order over X. Let $\smile$ be a binary relation over X such that,*

(i) *for each $x, y, z \in X$, if $x \preceq y$ and $y \smile z$ hold, then $x \smile z$ holds (transitivity);*

(ii) *for each $x \in X$, $x \smile x$ does not hold (nonreflexivity); and,*

(iii) *for each $x, y \in X$, if $x \smile y$ holds, then $y \smile x$ also holds (symmetry).*

We call this poset endowed with this additional binary relation a poset with a transitive forbidden relation or TFR poset in short, and denote this by $(X, \preceq, \smile)$. We call a pair of two elements x and y with $x \smile y$ forbidden.

Let $(X, \preceq, \smile)$ be a TFR poset. For two elements $x, y \in X$ with $x \smile y$, we say that $x \overset{\star}{\smile} y$ if, there is no $z \in X \setminus \{x, y\}$ with $x \preceq z$ and $z \smile y$. We call such a forbidden pair of x and y *immediate*. A TFR poset can be visualized in a similar way to an ordinary posets. We represent $\preceq$ just in the same way as the Hasse diagrams and depict $\smile$ by indicating every immediate forbidden pairs.

Definition 4. *Let P be a TFR poset $(X, \preceq, \smile)$. A lower or upper ideal Y of P is* legitimate *if no elements $x, y \in Y$ satisfy $x \smile y$. Otherwise, we say that Y is* illegitimate. *Let Y be a consistent lower or upper ideal, and let Z be the subset of $X \setminus Y$ such that, for each $x \in Z$, there exists $y \in Y$ with $x \smile y$. We say that Y is* spanning *if $Y \cup Z = X$.*

5   DM Decomposition for General Graphs

We now provide our new results of the DM decomposition for general graphs. In this section, unless otherwise stated, let G be a graph.

Definition 5. *A Dulmage-Mendelsohn component, or a DM component in short, is a subgraph of the form $G[S \cup {}^{\top}U_G(S)]$, where $S \in \mathcal{P}(G)$, endowed with S as an attribute known as the* base. *For a DM component C, the base of C is denoted by $\pi(C)$. Conversely, for $S \in \mathcal{P}(G)$, $K(S)$ denotes the DM components whose base is S. We denote by $\mathcal{D}(G)$ the set of DM components of G.*

Hence, distinct DM components can be equivalent as a subgraph of G. Each member from $\mathcal{P}(G)$ serves as an identifier of a DM component.

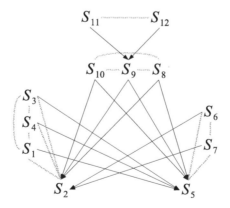

Fig. 4. The nonbipartite Dulmage-Mendelsohn decomposition of G: for each immediate compatible pair, an arrow points from the lower element to the upper element. The two elements from each immediate forbidden pair are connected by a gray broken line.

Definition 6. *A DM component C is said to be* inconsistent *if $\pi(C) \in \mathcal{P}_G(H)$ for some $H \in \mathcal{G}^-(G)$; otherwise, C is said to be* consistent. *The sets of consistent and inconsistent DM components are denoted by $\mathcal{D}^+(G)$ and $\mathcal{D}^-(G)$, respectively.*

Definition 7. *We define binary relations $\preceq^\circ$ and $\preceq$ over $\mathcal{D}(G)$ as follows: for $D_1, D_2 \in \mathcal{D}(G)$, we let $D_1 \preceq^\circ D_2$ if $D_1 = D_2$ or if $N_G(^\top U_G(S_1)) \cap S_2 \neq \emptyset$; we let $D_1 \preceq D_2$ if there exist $C_1, \dots, C_k \in \mathcal{D}(G)$, where $k \geq 1$, such that $\pi(C_1) = \pi(D_1)$, $\pi(C_k) = \pi(D_2)$, and $C_i \preceq^\circ C_{i+1}$ for each $i \in \{1, \dots, k\} \setminus \{k\}$.*

Also, we define binary relations $\smallsmile^\circ$ and $\smallsmile$ over $\mathcal{D}(G)$ as follows: for $D_1, D_2 \in \mathcal{D}(G)$, we let $D_1 \smallsmile^\circ D_2$ if $\pi(D_2) \subseteq V(D_1) \setminus \pi(D_1)$ holds; we let $D_1 \smallsmile D_2$ if there exists $D' \in \mathcal{D}(G)$ with $D_1 \preceq D'$ and $D' \smallsmile^\circ D_2$.

In the following, we provide some lemmas and prove that $(\mathcal{D}(G), \preceq, \smallsmile)$ is a TFR poset in Theorem 8. First, note the next lemma, which can be easily observed from Facts 3 and 7.

Lemma 3. *If C is an inconsistent DM component of a graph G, then there is no $C' \in \mathcal{D}(G) \setminus \{C\}$ with $C \preceq C'$ or $C \smallsmile C'$.*

Lemma 4. *Let G be a graph, let M be a maximum matching of G, and let $D_1, \dots, D_k \in \mathcal{D}(G)$, where $k \geq 1$, be DM components with $D_1 \preceq^\circ \cdots \preceq^\circ D_k$ no two of which share vertices and for which $D_k \in \mathcal{D}^+(G)$ holds. Then, for any $s \in \pi(D_1)$ and any $t \in \pi(D_k)$ (resp. $t \in V(D_k) \setminus \pi(D_k)$), there is an M-forwarding path from s to t (resp. M-saturated path between s and t) whose vertices are contained in $V(D_1) \dot\cup \cdots \dot\cup V(D_k)$.*

Proof. A desired path can be obtained by concatenating alternating paths given by Lemma 1 and edges joining distinct DM components. □

Lemma 5. *Let G be a graph, let M be a maximum matching of G, and let $D_1, \dots, D_k$, where $k \geq 2$, be DM components with $D_1 \preceq^\circ \cdots \preceq^\circ D_k$ such that $\pi(D_i) \neq \pi(D_{i+1})$ for any $i \in \{1, \dots, k-1\}$. Then, for any $i, j \in \{1, \dots, k\}$ with $i \neq j$, $V(D_i) \cap V(D_j) = \emptyset$.*

Proof. Suppose that the claim fails. Then, there exist $p, q \in \{1, \dots, k-1\}$ with $p \leq q$ such that $D_p, \dots, D_{q+1}$ are mutually disjoint except that $V(D_p) \cap V(D_{q+1}) \neq \emptyset$. Then, Lemma 3 implies $D_p, \dots, D_{q+1} \in \mathcal{D}^+(G)$. If $\pi(D_{q+1}) \subseteq V(D_p)$ holds, then let $t_{q+1} \in {}^\top U_G(\pi(D_{q+1}))$; otherwise, let $t_{q+1} \in {}^\top U_G(\pi(D_{q+1})) \cap V(D_p)$. Let $t_p \in {}^\top U_G(\pi(D_p))$ and $s_{p+1} \in \pi(D_{p+1})$ be vertices with $t_p s_{p+1} \in E(G)$, and let Q be an M-saturated path Q between s_{p+1} and t_q taken under Lemma 4. Then, $t_p s_{p+1} + Q + t_q s_{q+1} + s_{q+1} P t_{q+1}$ contains an M-ear relative to D_p one of whose ends is t_p. This contradicts Lemma 2. □

Lemma 6. *Let G be a graph, and let M be a maximum matching of G. Let $s, t \in V(G)$, and let S and T be the members from $\mathcal{P}(G)$ with $s \in S$ and $t \in T$, respectively. Then, there is an M-saturated path between s and t if and only if $K(S) \smallsmile K(T)$ holds.*

Proof. By combining Lemmas 4 and 5, the necessity is easily proved. For proving the sufficiency, let P be an M-saturated between s and t. We proceed by induction on the number of edges in P. Lemma 2 implies that $P - E(K(S))$ is an M-exposed path; let x be the end of $P - E(K(S))$ other than t. Let $y \in V(P)$ be the vertex with $xy \in E(P)$, and let $R \in \mathcal{P}(G)$ with $y \in R$. By Lemma 1, we have $K(S) \preceq^\circ K(R)$. The subpath yPt is M-saturated between y and t. Therefore, the induction hypothesis implies $K(R) \smile K(T)$, and proves the claim.   □

We are now ready to prove Theorem 8:

Theorem 8. *For a graph G, the triple $(\mathcal{D}(G), \preceq, \smile)$ is a TFR poset.*

Proof. Lemmas 4 and 5, it is clear that $\preceq$ is a partial order over $\mathcal{D}(G)$. Nonreflexivity and transitivity of $\smile$ are obvious from the definition. Symmetry is proved by Lemma 6.   □

For a graph G, the TFR poset $(\mathcal{D}(G), \preceq, \smile)$ is uniquely determined and is denoted by $\mathcal{O}(G)$. We call this canonical structure the *nonbipartite Dulmage-Mendelsohn (DM) decomposition* of G. We show in Sect. 7 that this is a generalization of the classical DM decomposition for bipartite graphs.

As mentioned previously, a DM component is identified by its base. Therefore, the nonbipartite DM decomposition is essentially the relations between the members of $\mathcal{P}(G)$. In Fig. 4, we provide an example of the nonbipartite DM decomposition for the graph G from Figs. 1, 2, and 3. Our result is distinct from the result by Iwata [7]. This can also be confirmed from the example graph G in Fig. 1.

Given a graph G, its basilica decomposition can be computed in $O(|V(G)| \cdot |E(G)|)$ time [9,10]. Hence, the next thereom can be stated.

Theorem 9. *Given a graph G, the TFR poset $\mathcal{O}(G)$ can be computed in $O(|V(G)| \cdot |E(G)|)$ time.*

6   Characterization of Maximal Barriers

In this section, we derive the characterization of the family of maximal barriers in general graphs using the nonbipartite DM decomposition. A barrier is *maximal* if it is inclusion-wise maximal. In this section, unless otherwise stated, let G be a graph.

Definition 8. *For $\mathcal{I} \subseteq \mathcal{D}(G)$, the* normalization *of $\mathcal{I}$ is the set $\mathcal{I} \cup \mathcal{D}^-(G)$. A set $\mathcal{I}' \subseteq \mathcal{D}(G)$ is said to be* normalized *if $\mathcal{I}' = \mathcal{I} \cup \mathcal{D}^-(G)$ for some $\mathcal{I} \subseteq \mathcal{D}(G)$.*

Note that the normalization of an upper ideal is an upper ideal; the normalization of a legitimate upper ideal is legitimate.

We present several known facts to derive Theorem 11.

Proposition 1 (see Lovász and Plummer [17]). *Let G be a graph, and let $X \subseteq V(G)$ be a barrier. Then, X is a maximal barrier if and only if every connected component of $G - X$ are odd and factor-critical.*

The next statements is a generalization of *Lovász's canonical partition theorem* [11,16,17].

Theorem 10 (Kita [11]). *Let G be a graph and $X \subseteq V(G)$ be a maximal barrier of G. Then, there exist $S_1, \ldots, S_k \in \mathcal{P}(G)$, where $k \geq 1$, such that $X = S_1 \dot\cup \cdots \dot\cup S_k$ and $D_X = {}^\top U_G(S_1) \dot\cup \cdots \dot\cup {}^\top U_G(S_k)$. The odd components of $G - X$ are the connected components of $G[{}^\top U_G(S_i)]$, where i is taken over all $\{1, \ldots, k\}$.*

From Proposition 1 and Theorem 10, the next theorem that characterizes the family of maximal barriers can be derived rather easily.

Theorem 11. *Let G be a graph. A set of vertices $X \subseteq V(G)$ is a maximal barrier if and only if there exists a spanning legitimate normalized upper ideal $\mathcal{I}$ of the TFR poset $\mathcal{O}(G)$ such that $X = \bigcup\{\pi(C) : C \in \mathcal{I}\}$.*

7   Original DM Decomposition for Bipartite Graphs

In this section, we explain the original DM decomposition for bipartite graphs, and prove this from our result in Sect. 5. In the remainder of this section, unless stated otherwise, let G be a bipartite graph with color classes A and B, and let $W \in \{A, B\}$.

Definition 9. *The binary relations $\leq_W^\circ$ and $\leq_W$ over $\mathcal{G}(G)$ are defined as follows: for $G_1, G_2 \in \mathcal{G}(G)$, let $G_1 \leq_W^\circ G_2$ if $G_1 = G_2$ or if $E_G[W \cap V(G_2), V(G_1) \setminus W] \neq \emptyset$; let $G_1 \leq_W G_2$ if there exist $H_1, \ldots, H_k \in \mathcal{G}(G)$, where $k \geq 1$, such that $H_1 = G_1$, $H_k = G_2$, and $H_1 \leq_W^\circ \cdots \leq_W^\circ H_k$.*

Note that $G_1 \leq_A G_2$ holds if and only if $G_2 \leq_B G_1$ holds.

Theorem 12 (Dulmage and Mendelsohn [2–4,17]). *Let G be a bipartite graph with color classes A and B, and let $W \in \{A, B\}$. Then, the binary relation $\leq_W$ is a partial order over $\mathcal{G}(G)$.*

We call the poset $(\mathcal{G}(G), \leq_W)$ proved by Theorem 12 the *Dulmage-Mendelsohn decomposition* of a bipartite graph G. It is easily confirmed, e.g., from the Gallai-Edmonds structure theorem that $\mathcal{G}_A^-(G) \cap \mathcal{G}_B^-(G) = \emptyset$ and that any $C \in \mathcal{G}_B^-(G)$ is minimal with respect to $\leq_A$.

Additionally, bipartite graphs have a trivial structure regarding the basilica decomposition:

(i) For each $H \in \mathcal{G}^+(G)$, $\mathcal{P}_G(H) = \{V(H) \cap A, V(H) \cap B\}$. For each $H \in \mathcal{G}_W^-(G)$, $\mathcal{P}_G(H) = \{V(H) \cap W\}$.

(ii) For any $H_1, H_2 \in \mathcal{G}(G)$ with $H_1 \neq H_2$, $H_1 \lhd H_2$ does not hold.

Under these properties, we define $\mathcal{D}^W(G)$ as the set $\{C \in \mathcal{D}(G) : \pi(C) \subseteq W\}$.

Define a mapping $f_W : \mathcal{G}^+(G) \cup \mathcal{G}_W^-(G) \to \mathcal{D}^W(G)$ as $f_W(C) := K(V(C) \cap W)$ for $C \in \mathcal{G}^+(G)$. The next statement is now obvious.

Observation 13. *The mapping f_W is a bijection; and, for any $C_1, C_2 \in \mathcal{G}(G)$, $C_1 \leq_W C_2$ holds if and only if $f(C_1) \preceq f(C_2)$ holds.*

According to Theorem 8 and Observation 13, the system $(\mathcal{G}^+(G) \cup \mathcal{G}_W^-(G), \leq_W)$ is a poset. Thus, this proves Theorem 12.

References

1. Duff, I.S., Erisman, A.M., Reid, J.K.: Direct Methods for Sparse Matrices. Clarendon Press, Oxford (1986)
2. Dulmage, A.L., Mendelsohn, N.S.: Coverings of bipartite graphs. Can. J. Math. **10**(4), 516–534 (1958)
3. Dulmage, A.L., Mendelsohn, N.S.: A structure theory of bi-partite graphs. Trans. Roy. Soc. Can. Sect. 3 **53**, 1–13 (1959)
4. Dulmage, A.L., Mendelsohn, N.S.: Two algorithms for bipartite graphs. J. Soc. Ind. Appl. Math. **11**(1), 183–194 (1963)
5. Edmonds, J.: Paths, trees and flowers. Can. J. Math. **17**, 449–467 (1965)
6. Gallai, T.: Maximale systeme unabhängiger kanten. A Magyer Tudományos Akadémia: Intézetének Közleményei **8**, 401–413 (1964)
7. Iwata, S.: Block triangularization of skew-symmetric matrices. Linear Algebra Appl. **273**(1–3), 215–226 (1998)
8. Kita, N.: New canonical decomposition in matching theory (under review). arXiv preprint arXiv:1708.01051
9. Kita, N.: A partially ordered structure and a generalization of the canonical partition for general graphs with perfect matchings. In: Chao, K.-M., Hsu, T., Lee, D.-T. (eds.) ISAAC 2012. LNCS, vol. 7676, pp. 85–94. Springer, Heidelberg (2012). https://doi.org/10.1007/978-3-642-35261-4_12
10. Kita, N.: A partially ordered structure and a generalization of the canonical partition for general graphs with perfect matchings. CoRR abs/1205.3 (2012)
11. Kita, N.: Disclosing barriers: a generalization of the canonical partition based on Lovász's formulation. In: Widmayer, P., Xu, Y., Zhu, B. (eds.) COCOA 2013. LNCS, vol. 8287, pp. 402–413. Springer, Cham (2013). https://doi.org/10.1007/978-3-319-03780-6_35
12. Kita, N.: A graph theoretic proof of the tight cut lemma (2015). arXiv preprint arXiv:1512.08870. In: The Proceedings of the 11th Annual International Conference on Combinatorial Optimization and Applications (COCOA 2017) (2017). https://doi.org/10.1007/978-3-319-71150-8_20
13. Kotzig, A.: Z teórie konečných grafov s lineárnym faktorom. I (in Slovak). Math. Slovaca **9**(2), 73–91 (1959)
14. Kotzig, A.: Z teórie konečných grafov s lineárnym faktorom. II (in Slovak). Math. Slovaca **9**(3), 136–159 (1959)
15. Kotzig, A.: Z teórie konečných grafov s lineárnym faktorom. III (in Slovak). Math. Slovaca **10**(4), 205–215 (1960)
16. Lovász, L.: On the structure of factorizable graphs. Acta Math. Hung. **23**(1–2), 179–195 (1972)
17. Lovász, L., Plummer, M.D.: Matching Theory, vol. 367. American Mathematical Society, Providence (2009)
18. Schrijver, A.: Combinatorial Optimization: Polyhedra and Efficiency, vol. 24. Springer, New York (2002)

The Path Set Packing Problem

Chenyang Xu and Guochuan Zhang[⊠]

College of Computer Science, Zhejiang University, Hangzhou, China
{xcy1995,zgc}@zju.edu.cn

Abstract. In this paper, we study a variant of set packing, in which a set P of paths in a graph $G = (V, E)$ is given, the goal is to find a maximum number of edge-disjoint paths of P. We show that the problem is NP-hard even if each path in P contains at most three edges, while it is hard to approximate within $O(|E|^{1/2-\epsilon})$ for the general case unless $NP = ZPP$. In the positive aspect, a parameterized algorithm relying on the maximum degree and the tree-width of G is derived. For tree networks, we present a polynomial time optimal algorithm.

1 Introduction

Recent years have witnessed the great development of an emerging network architecture named Software-Defined Networking (SDN). Different from the traditional network architecture, SDN is a central network, meaning that the state of the whole network and the instructions the data plane needs are all managed by a centralized controller. Actually, the central network [1] was first proposed in the 1980s. However, at that time, not much research was focused on the study of this architecture. Until recently the cloud computing becomes more and more important, more and more attention is paid to studying this architecture since it is ideal for the high-bandwidth, dynamic nature of today's computing.

The failure of network nodes or links may cause network service interruptions. In the SDN architecture, the controller can recompute the network state upon a failure, but the computation will take a lot of time and incur high processing delays if a large number of requests are interrupted. Fortunately, computing edge-disjoint routing paths for each request in the network can often resolve this problem [2]. Upon arrival of a request, the SDN controller computes two paths from its source node to its destination node, one is used in the current network, and the other serves as the backup path. Once a breakdown occurs and this request is affected, the SDN controller will arrange the backup path for this request if it fits. Since this method avoids recomputing for a large number of affected requests, high processing delays will not happen. However, those backup paths may conflict with each other in links. It means that not all affected requests can be recovered by using their backup paths. In order to maximize the throughput of the network, the controller needs to decide whether a request should be rerouted by its backup path or not.

This work was partially supported by NSFC Grant 11531014.

The scenario motivates the path set packing problem: Given some paths in a network, our goal is to find as many paths as possible such that those paths do not conflict with each other in links. Actually, many applications have the path set packing structure, like wireless network design, VLSI circuits and so on. Compared with the well-known set packing problem, the path set packing problem is more closely related to these applications' structure.

The paper is organized as follows: Sect. 2 provides the basic notations and related work. In Sect. 3, we show the problem is NP-complete even when each given path contains at most three edges. In addition, we prove that the problem is hard to approximate within $\mathcal{O}(m^{\frac{1}{2}-\epsilon})$ unless $NP = ZPP$ (where m is the number of edges). In Sect. 4, we consider the case that the given graph G is a tree, and design a polynomial time algorithm. Section 5 presents a parameterized algorithm for this problem, which takes the maximum degree and the treewidth of G as parameters. This paper is concluded in Sect. 6.

2   Basic Notations and Related Work

We shall use some standard graph-theoretic notations throughout the paper. Considering a graph G, we denote its sets of vertices and edges by $V(G)$ and $E(G)$, respectively. For a vertex v in G, we use $d(v)$ to represent its degree in G and let $\Delta(G) := \max_{v \in V(G)} d(v)$. A graph is a simple graph if there is at most one edge between any two vertices and a path is a simple path if any vertex in the path occurs only once. The graphs and paths mentioned in this paper are all simple. Let p be a path in G. By $Len(p)$ we denote the length of path p and by $E(p)$ we denote the set of all edges in path p. A *path set packing* is a set of paths $\mathcal{P} = \{p_1, \ldots, p_k\}$ where no two paths have an edge in common.

The Path Set Packing Problem: Given an undirected graph G, a set of paths $\mathcal{P} = \{p_i | i = 1, 2, \ldots, L\}$ in G and a non-negative integer k, the goal is to determine if there exists a path set packing $\mathcal{Q} \subseteq \mathcal{P}$, where $|\mathcal{Q}| \geq k$.

In contrast to the above decision problem, the maximum path set packing problem is to find a path set packing with the maximum size. The path set packing problem can be seen as a special case of the set packing problem, which is one of Karp's 21 NP-complete problems. The set packing problem is NP-complete even if the size of the sets in the collection is no more than 3 [3]. In 1996, Håstad [4] pointed out that the clique problem is hard to approximate within $n^{1-\epsilon}$ unless $NP = ZPP$. Thus, the set packing problem has an $\Omega(m^{\frac{1}{2}-\epsilon})$ lower bound for any $\epsilon > 0$ (where m is the size of the fundamental set). Later, Halldórsson et al. [5] proposed a greedy algorithm which obtains an approximation near the best possible. Their algorithm simply chooses a set with the minimum cardinality and then removes the sets that conflict with the selected set. Repeat this process until all sets have been considered. This algorithm can approximate the set packing problem within a factor of $\mathcal{O}(\sqrt{m})$. Due to the lower bound mentioned above, this approximation is near the best possible. The k-set packing problem is also an interesting topic, where all input sets have no more than k elements. Any maximal solution of this problem is k-approximated.

To see this, each set in the maximal solution conflicts with at most k sets in an optimal solution, since each set in the optimal solution must share one or more common elements with a set in the maximal solution and the cardinality of any set is no more than k. Halldórsson [6] showed that this approximation can be improved using local search. By trying to add a constant number of sets to a feasible solution while removing a constant number of conflicting sets, we can obtain a $(\frac{k}{2} + \epsilon)$-approximation algorithm. Hurkens and Schrijver [7] pointed out that the analysis of such an algorithm is tight. Additionally, Halldórsson [8] gave a quasipolynomial time algorithm with $(\frac{k+2}{3})$-approximation. In 2013, Cygan et al. [9] approximated this problem within $\frac{k+1+\epsilon}{3}$ in quasipolynomial time, successfully improving on Halldórsson's result. In the same year, Sviridenko and Ward [10] also improved upon Halldórsson's result. They showed that the k-set packing problem can be approximated within $\frac{k+2}{3}$ in polynomial time, using a large neighborhood local search.

Besides approximation algorithms, the parameterized complexity and parameterized algorithms for set packing also attract a lot of attention. Taking the optimal value s of the set packing problem as a parameter, we can obtain an $\mathcal{O}(n^s m)$-time parameterized algorithm by checking each sub-collections of s sets, where n is the total number of the given sets. Since this problem parameterized by its optimal value has been proved W[1]-complete [11], we can not expect a much better parameterized algorithm. But if the size of each set is bounded by a constant, this problem has an $\mathcal{O}(f(s)poly(n))$-time algorithm [11]. In other words, the parameterized k-set packing problem is fixed-parameter tractable if we take both k and the optimal value s as its parameters. Jia et al. [12] presented an $\mathcal{O}(k^s(g(k,s)^{ks} + s^2 k^2 n))$-time algorithm, where $g(k,s)$ is linear in ks. One year later, Koutis [13] gave an $\mathcal{O}(2^{\mathcal{O}(ks)} nm \log m)$-time algorithm, improving the above algorithm. These two papers both first proposed a randomized algorithm and then did derandomization. Different from this method, Fellows et al. [14] put forward an $\mathcal{O}(n + 2^{\mathcal{O}(ks)})$-time kernelization algorithm. They found a kernel of the parameterized k-set packing problem whose order is $\mathcal{O}(s^k)$. Their result was improved by Abu-Khzam [15], where an algorithm that can produce kernels with order $\mathcal{O}(s^{k-1})$ was proposed.

Our problem is a special case of the set packing problem, therefore, all algorithms designed for set packing can be applied to the path set packing problem. It is worth mentioning that the approximation of the algorithm proposed in [5] is also near the best possible in our problem, which we will show in Sect. 3. Different from the set packing problem, there are some features of the input graph in our problem that can be taken advantage while designing algorithms. We will show how to make use of this point in Sects. 4 and 5.

3   Hardness

If each path has a length of at most two, the path set packing problem can be solved in polynomial time. Without loss of generality, assume that no two paths are identical. Pick up all paths of single edge and remove those of two edges which conflict with the selected paths. Now consider the remaining paths of two edges. It becomes a maximum matching problem since such a path is chosen as long as its two edges are matched.

However, when paths are allowed to be longer than 2, this problem becomes hard. In the following, we prove that the path set packing problem is NP-complete even when the maximum length of the given paths is 3. For simplicity, denote this special case as the 3-length-path set packing problem. We construct a reduction from a variant of the 3-dimensional matching problem introduced in [16].

The variant of the 3-dimensional matching problem (3DM$'$) is defined as following: Given a positive integer k, three disjoint sets of elements $W = \{w_i, \bar{w}_i | i = 1, \ldots, 3n\}$, $X = \{s_i, a_i | i = 1, \ldots, 3n\}$, $Y = \{s'_i, b_i | i = 1, \ldots, 3n\}$, and two sets of triples $T_1 \subseteq \{(w_i, s_j, s'_j), (\bar{w}_i, s_j, s'_j) | w_i \in W, s_j \in X, s'_j \in Y\}$, $T_2 = \{(w_i, a_i, b_i), (\bar{w}_i, a_i, b_{\zeta(i)}) | i = 1, \ldots, 3n\}$ where ζ is defined as $\zeta(3m+1) = 3m+2$, $\zeta(3m + 2) = 3m + 3$ and $\zeta(3m + 3) = 3m + 1$ for $m = 0, \ldots, n - 1$, the goal is to determine if there exist at least k triples in $T_1 \cup T_2$ such that no two of them share an element.

Given an instance $I_{3DM'}$ of 3DM$'$, we construct an instance I_{3PSP} of the 3-length-path set packing problem such that it has a path set packing whose size is no less than k if and only if $I_{3DM'}$ has at least k disjoint triples.

Define a node for each element in $I_{3DM'}$ and add an extra node g. Namely, we have a node set $V = W \cup X \cup Y \cup \{g\}$. The edge set E consists of five kinds of edges: (1) $E_1 = \{(w_i, a_i), (\bar{w}_i, a_i) | i = 1, \ldots, 3n\}$, (2) $E_2 = \{(a_i, g) | i = 1, \ldots, 3n\}$, (3) $E_3 = \{(g, b_i) | i = 1, \ldots, 3n\}$, (4) $E_4 = \{(a_i, s_j) | i = 1, \ldots, 3n, j = 1, \ldots, 3n\}$, (5) $E_5 = \{(s_i, s'_i) | i = 1, \ldots, 3n\}$. Figure 1 illustrates the construction with $n = 1$.

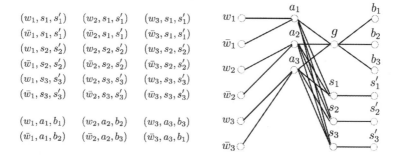

Fig. 1. An illustration of the first reduction. The left is the given triples and the right is the graph we constructed.

For each triple $(w_i(\bar{w}_i), s_j, s'_j)$ in T_1, we define a path $(w_i(\bar{w}_i), a_i, s_j, s'_j)$ and put it into the path set $\mathcal{P}$. For each triple (w_i, a_i, b_i) (or $(\bar{w}_i, a_i, b_{\zeta(i)})$) in T_2, we define a path (w_i, a_i, g, b_i) (or $(\bar{w}_i, a_i, g, b_{\zeta_i})$) and put it into $\mathcal{P}$.

Two triples in $I_{3DM'}$ share w_i (or $\bar{w}_i$) if and only if their corresponding paths in I_{3PSP} have the common edge (w_i, a_i) (or $(\bar{w}_i, a_i)$). They share (s_i, s'_i) if and only if their corresponding paths share the edge (s_i, s_i'). They have the common element a_i (or b_i) if and only if their corresponding paths share the edge (a_i, g) (or (g, b_i)). So it is easy to see that $\mathcal{P}$ has a path set packing whose size is no less than k if and only if we can choose no less than k disjoint triples from $T_1 \cup T_2$, which gives the following theorem.

Theorem 1. *The path set packing problem is NP-complete even if the length of any given path is no more than 3.*

Now we turn to the general case and prove that the maximum path set packing problem is hard to approximate within $\mathcal{O}(|E(G)|^{\frac{1}{2}-\epsilon})$ unless $NP = ZPP$ (G is the given graph). To this end, we construct a reduction from the independent set problem. An independent set is a set of vertices in a graph, no two of which are adjacent. The maximum independent set problem is to find the largest independent set in a given graph.

Given an instance I_{IS} of the maximum independent set problem, the instance I_{PSP} of the maximum path set packing problem is constructed as following. Use $G = (V, E)$ to represent the graph of I_{IS} and $G' = (V', E_1' \cup E_2')$ to represent the graph of I_{PSP}. We index the edges in E. For each edge $e_i = (u, v) \in E$, define two nodes $u^i, v^i \in V'$ and an edge $e_i' = (u^i, v^i) \in E_1'$.

We further define a function $f : V' \rightarrow V'$, where $f(u^i)$ denotes the other endpoint of e_i' other than u^i. Namely, $f(u^i) = v^i$ if $e_i = (u, v) \in E$. Now, there exists an edge in E_1' corresponding to each edge in E, and for each vertex $v \in V$, we have $d(v)$ nodes $\{v^{k_1}, v^{k_2}, \ldots, v^{k_{d(v)}}\}$ in V' accordingly. Let $k_1 < k_2 < \ldots < k_{d(v)}$. Connect $v^{k_{i-1}}$ to $f(v^{k_i})$ for $2 \leq i \leq d(v)$. The set of edges constructed in this way is denoted by E_2'. Finally, we construct the given path set: for each vertex $v \in V$, we can have a path $(f(v^{k_1}), v^{k_1}, f(v^{k_2}), \ldots, f(v^{k_{d(v)}}), v^{k_{d(v)}})$ in G'. Let all these paths be the given paths in I_{PSP}.

Take an arbitrary given path $p = (f(v^{k_1}), v^{k_1}, f(v^{k_2}), \ldots, f(v^{k_{d(v)}}), v^{k_{d(v)}})$ in I_{PSP}. It is easy to see that any edge in $\{(v^{k_{i-1}}, f(v^{k_i})) | i = 2, \ldots, d(v)\}$ can not appear in other paths. If the path p conflicts with some other path p', assuming that the edge $(f(v^{k_i}), v^{k_i})$ is their common edge and $f(v^{k_i}) = u^{k_i}$, p' must correspond to the vertex u in I_{IS}. The edge $(f(v^{k_i}), v^{k_i})$ suggests that u and v are adjacent in I_{IS}. If the vertex u and the vertex v are neighbors in I_{IS}, assuming $(u, v) = e_j$, two paths corresponding to them must contain the edge e_j', implying that they can not co-exist in a path set packing. Thus, two paths share a common edge in I_{PSP} if and only if two nodes corresponding to them in I_{IS} arc adjacent. The size of a maximum independent set in I_{IS} is equal to the size of a maximum path set packing in I_{PSP} (see Fig. 2 as an illustration).

Due to the inapproximability of Clique proposed in [4], the independent set problem is hard to approximate within $m^{\frac{1}{2}-\epsilon}$ unless $NP = ZPP$, where m is the number of edges in the graph. We conclude the following theorem.

Theorem 2. *The maximum path set packing problem is hard to approximate within $\mathcal{O}(m^{\frac{1}{2}-\epsilon})$ unless $NP = ZPP$.*

Additionally, we can obtain a $\sqrt{m}$-approximation algorithm simply applying the greedy method mentioned in [6].

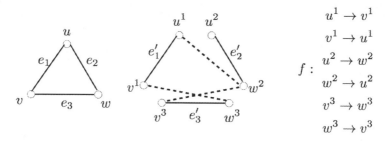

Fig. 2. An illustration of the second reduction. The left graph is G and the right is G', where E_1' is drawn by real edges and E_2' is drawn by dotted edges. We also show how the function f maps the vertices. u corresponds to the path (v^1, u^1, w^2, u^2), v corresponds to the path (u^1, v^1, w^3, v^3), and w corresponds to the path (u^2, w^2, v^3, w^3), respectively.

4   The Case of Trees

Although the path set packing problem is NP-complete even when the maximum length of the given path is 3, we find that if restricted to trees, this problem can be solved in polynomial time. In this section, we propose a polynomial time dynamic programming on trees.

Given a tree T, we root T at an arbitrary node r and let $T(v)$ denote the subtree rooted at v. In addition, for each child u of v, we use $T(v \setminus u)$ to represent the subgraph if removing the edge (v, u) and $T(u)$ from $T(v)$. For a subgraph $G' \subseteq T$, we denote with $\mathcal{P}(G')$ the set of the paths in $\mathcal{P}$ whose edges are all in G' and with $\mathcal{M}(G')$ a maximum path set packing in $\mathcal{P}(G')$. Thus, $\mathcal{M}(T(r))$ provides an optimal solution.

Without loss of generality, denote a path p by a sequence of vertices $(w_1, \ldots, w_s, u_i, v, u_j, \bar{w}_1, .., \bar{w}_t)$, where v is the closest vertex to r in p and $s, t \in N$. u_i and u_j are children of v (if v is an endpoint of p, p only contains one child of v). $w_1, \ldots, w_s$ are the vertices in $T(u_i)$ and $\bar{w}_1, \ldots, \bar{w}_t$ are the vertices in $T(u_j)$. We say a path p is *free* if

$$|\mathcal{M}(T(u_i))| = |\mathcal{M}(T(u_i \setminus w_s)) \cup \mathcal{M}(T(w_s \setminus w_{s-1})) \cup ... \cup \mathcal{M}(T(w_1))|$$

and

$$|\mathcal{M}(T(u_j))| = |\mathcal{M}(T(u_j \setminus \bar{w}_1)) \cup \mathcal{M}(T(\bar{w}_1 \setminus \bar{w}_2)) \cup ... \cup \mathcal{M}(T(\bar{w}_t))|.$$

We compute all $\mathcal{M}(T(v))$ and $\mathcal{M}(T(v \setminus u))$ for each child u of v in a bottom-to-top way, starting at the leaves of $T(r)$. If v is a leaf, $\mathcal{M}(T(v)) = 0$. If v is not a leaf, we use Algorithm 1 to compute $\mathcal{M}(T(v))$.

Algorithm 1.

1: P and M are initially empty
2: **for** each path p in $\mathcal{P}(T(v))$ **do**
3:    **if** p is a *free* path containing v **then**
4:        add p into P
5:    **end if**
6: **end for**
7: compute a maximum path set packing $\hat{P}$ in P /* to be specified later in Theorem 4*/
8: **for** each path p in $\hat{P}$ **do**
9:    /* Denote p by $(w_1, ..., w_s, u_i, v, u_j, \bar{w}_1, ..., \bar{w}_t), s, t \in N$ */
10:    $M \leftarrow M \cup \mathcal{M}(T(u_i \setminus w_s)) \cup \mathcal{M}(T(w_s \setminus w_{s-1})) \cup ... \cup \mathcal{M}(T(w_1)) \cup \mathcal{M}(T(u_j \setminus \bar{w}_1)) \cup \mathcal{M}(T(\bar{w}_1 \setminus \bar{w}_2)) \cup ... \cup \mathcal{M}(T(\bar{w}_t))$
11: **end for**
12: **for** each child u of v **do**
13:    **if** no path in $\hat{P}$ contains u **then**
14:        $M \leftarrow M \cup \mathcal{M}(T(u))$
15:    **end if**
16: **end for**
17: return $M \cup \hat{P}$

Theorem 3. *The path set returned by Algorithm 1 is a maximum path set packing in $\mathcal{P}(T(v))$.*

Proof. The path set packing M is computed in the subgraph which is obtained by removing all edges in $\hat{P}$, so no path in M conflicts with paths in $\hat{P}$, showing that $M \cup \hat{P}$ is a path set packing. Next, we will show that it is actually a maximum path set packing in $\mathcal{P}(T(v))$.

Consider an arbitrary maximum path set packing P_0 in $\mathcal{P}(T(v))$. If it contains a path $p = (w_1, \ldots, w_s, u_i, v, u_j, \bar{w}_1, \ldots, \bar{w}_t)$ which is not *free* but passes v, we replace this path and paths in $P_0 \cap (\mathcal{P}(T(u_i)) \cup \mathcal{P}(T(u_j)))$ with $\mathcal{M}(T(u_i)) \cup \mathcal{M}(T(u_j))$ and obtain a new path set packing P_0'.

Note that

$$|P_0 \cap (\mathcal{P}(T(u_i)) \cup \mathcal{P}(T(u_j)))| \leq |\mathcal{M}(T(u_i \setminus w_s)) \cup \mathcal{M}(T(w_s \setminus w_{s-1})) \cup \\ ... \cup \mathcal{M}(T(w_1)) \cup \mathcal{M}(T(u_j \setminus \bar{w}_1)) \cup \mathcal{M}(T(\bar{w}_1 \setminus \bar{w}_2)) \cup ... \cup \mathcal{M}(T(\bar{w}_t))|. \quad (1)$$

It is easy to see that $|P_0'| \geq |P_0|$ since p is not *free*. Thus, this new path set packing is also a maximum path set packing in $\mathcal{P}(T(v))$.

We do the above until all paths passing v are *free*. Denote the final path set packing by $\hat{P}_0$. Let $\hat{P}_0 = \hat{P}_1 \cup \hat{P}_2$, where $\hat{P}_1$ is the set of all paths which contain v and $\hat{P}_2$ is the set of remaining paths. Namely, $\hat{P}_2$ is a path set packing in $\bigcup_{\forall \text{ child } u \text{ of } v} \mathcal{P}(T(u))$. Thus,

$$|\hat{P}_2| \leq |\bigcup_{\forall \text{ child } u \text{ of } v} \mathcal{M}(T(u))| = |M|. \quad (2)$$

Recall P and $\hat{P}$ computed in Algorithm 1. Since $\hat{P}_1$ is a path set packing in P, we know

$$|\hat{P}_1| \leq |\hat{P}|. \tag{3}$$

Combining 2 and 3, we have

$$|\hat{P}_0| \leq |M \cup \hat{P}|. \tag{4}$$

So $M \cup \hat{P}$ is a maximum path set packing in $\mathcal{P}(T(v))$.

Theorem 4. *Algorithm 1 is a polynomial time algorithm.*

Proof. Denote the number of the vertices and edges by n and m, respectively. The number of the given paths is denoted by L. It takes $\mathcal{O}(n)$ time to check if a path is *free*. Thus, the first loop costs $\mathcal{O}(Ln)$ time.

For any path $(..., u_i, v, u_j, ...)$ (or $(..., u_i, v)$) in P, it shares common edges with other paths if and only if they have the common edge (u_i, v) or (v, u_j) due to the structure of trees. So when computing $\hat{P}$, we can simply use (u_i, v, u_j) to represent each path. Namely, computing $\hat{P}$ is a 2-length-path set packing problem. Recall that the 2-length-path set packing problem can be solved in polynomial time. This step takes $\mathcal{O}(\sqrt{m}L)$ time as the maximum matching problem is solvable in $\mathcal{O}(\sqrt{|V|}|E|)$ time [17].

Because the last two loops combine at most n states, their running time is $\mathcal{O}(n)$. Overall, Algorithm 1 takes $\mathcal{O}(Ln)$ time.

It is easy to see that Algorithm 1 can also be used to compute $\mathcal{M}(T(v \setminus u))$ for each child u of v. The number of states in the whole dynamic programming is $(m + n)$. Hence, we can solve the maximum path set packing on trees in $\mathcal{O}(Ln^2)$ time.

5 A Parameterized Algorithm

Inspired by the parameterization of the maximum independent set problem, we consider the maximum path set packing problem parameterized by $\Delta(G)$ and t, where t is the treewidth of the given graph G, with which we propose a parameterized algorithm.

Given a graph G, denote a nice tree decomposition by a pair $(\{X_i | i \in I\}, T = (I, F))$. Note that if T is a rooted nice tree decomposition, it has three properties [18]: (1) each node of T has at most 2 children. (2) if a node i has two children j and k, then $X_i = X_j = X_k$ and this node is called a join node. (3) if a node i has one child j, then it is either an introduce node or a forget node, where i is an introduce node if $|X_i| = |X_j| + 1$ and $X_j \subset X_i$ and is a forget node if $|X_i| = |X_j| - 1$ and $X_i \subset X_j$. Root T at an arbitrary node r and let $T(i)$ be the subtree rooted at i. Different from Sect. 4, by P_i we denote the set of all paths in $\mathcal{P}$ that contain one or more nodes in X_i. $P_{T(i)}$ is the set of all paths in $\mathcal{P}$ that contain one or more nodes in $\bigcup_{j \in T(i)} X_j$.

Lemma 1. *If node j lies on the i-k path in T, then $P_i \cap P_k \subseteq P_j$.*

Proof. Assume that there exists a path $p \in P_i \cap P_k$ and $p \notin P_j$. Let i' be the node closest to j in the i-j path, where $p \in P_{i'}$. Analogously, k' is the node closest to j in the j-k path, where $p \in P_{k'}$. In other words, for any intermediate h other than i' and k' in the i'-k' path, P_h does not contain p.

Since $X_{i'} \cap X_{k'} \subseteq X_j$, $X_{i'}$ and $X_{k'}$ can not contain a common node on the path p. Assume that $X_{i'}$ contains $v_{i'}$ in p and $X_{k'}$ contains $v_{k'}$ in p. Without loss of generality, any node between $v_{i'}$ and $v_{k'}$ in p does not belong to $X_{i'}$ or $X_{k'}$. If $(v_{i'}, v_{k'})$ is an edge in G, there must exist an X_h that contains the two nodes $v_{i'}$ and $v_{k'}$. According to the property of the tree decomposition, h has to lie on the i'-k' path, which conflicts with the definition of i' and j'. If $(v_{i'}, v_{k'})$ is not an edge in G, for any X_h that contains the nodes between $v_{i'}$ and $v_{k'}$ on p, h must all lie on either (i')'s side or (k')'s side in the tree decomposition. Thus, there must exist an edge in p that is not included in any node set in X, which conflicts with the property of the tree decomposition. It implies the non-existence of the path p.

Lemma 2. *For any path set packing $\hat{P}_i \subseteq P_i$ and $\hat{P}_k \subseteq P_k$, if node j lies on the i-k path in T and $\hat{P}_i \cap P_j = \hat{P}_k \cap P_j$, $\hat{P}_i \cup \hat{P}_k$ is a path set packing.*

Proof. Assume that there exist two paths $p_1 \in \hat{P}_i$ and $p_2 \in \hat{P}_k$ such that p_1 and p_2 share a common edge e. Denote by X_h the set containing e. It is easy to see that either j lies on the i-h path or j lies on the k-h path. Due to Lemma 1, P_j contains at least one path in $\{p_1, p_2\}$. Without loss of generality, let $p_1 \in P_j$. Since $p_1 \in \hat{P}_i$, hence $p_1 \in \hat{P}_i \cap P_j = \hat{P}_k \cap P_j$. But if p_1 and p_2 share a common edge, they can not co-exist in $\hat{P}_k$, which implies the non-existence of these two paths.

For any $i \in I$ and any path set packing $P \subseteq P_i$, we denote with $\mathcal{M}(i, P)$ a maximum path set packing in $P_{T(i)}$ such that $\mathcal{M}(i, P) \cap P_i = P$. It is easy to obtain an optimal solution if $\mathcal{M}(r, P)$ is known for each path set packing $P \subseteq P_r$. We compute each $\mathcal{M}(i, P)$ in a bottom-to-top way, starting at the leaves of the nice tree decomposition $T(r)$. If i is a leaf, $\mathcal{M}(i, P) = P$. If i is a forget node and has a child j, $\mathcal{M}(i, P) = \mathcal{M}(j, P)$ for any path set packing $P \subseteq P_i$. If i is an introduce node and has a child j, it is easy to see that $\mathcal{M}(i, P) = P \cup \mathcal{M}(j, P \cap P_j)$ due to Lemma 2. If i is a join node and has two children j and k, $\mathcal{M}(i, P) = \mathcal{M}(j, P) \cup \mathcal{M}(k, P)$ holds also because of Lemma 2. Thus, the time of computing an $\mathcal{M}(i, P)$ is $\mathcal{O}(1)$. In the following, we count the total number of states in this dynamic programming.

For simplicity, we use Δ to represent $\Delta(G)$ and n to present $|V(G)|$. Recall that L is the number of the given paths and t is the treewidth of G. For each node $i \in I$, the nodes in X_i connect to at most $t\Delta$ edges in G. Since each edge can only be used at most once in a path set packing, the number of path set packings $P \subseteq P_i$ is at most $L^{t\Delta}$. Thus, the total number of states is $\mathcal{O}(L^{t\Delta}n)$, which implies that our parameterized algorithm runs in $\mathcal{O}(L^{t\Delta}n)$ time.

6   Conclusion

In this paper, we have introduced the path set packing problem. This problem is NP-complete even if the length of each path is at most 3 and it is hard to approximate within $\mathcal{O}(m^{\frac{1}{2}-\epsilon})$ unless $NP = ZPP$. When restricted to trees, this problem can be solved in polynomial time. In addition, we propose a parameterized algorithm for the general case. It is interesting to consider other special graphs, such as bipartite graphs, grid graphs and so on. Whether this problem is fixed-parameter tractable on the treewidth is still open. Note that if we consider vertex-disjoint paths instead of edge-disjoint paths, the path set packing problem in a general graph is exactly the same as the set packing problem. But for tree networks, this problem can still be solved polynomially with the similar technique in this paper.

References

1. Horing, S., Menard, J., Staehler, R., Yokelson, B.: Stored program controlled network: overview. Bell Syst. Tech. J. **61**(7), 1579–1588 (1982)
2. Kuipers, F.A.: An overview of algorithms for network survivability. ISRN Commun. Netw. **2012**, 24 (2012)
3. Michael, R.G., David, S.J.: Computers and Intractability: A Guide to the Theory of NP-Completeness, pp. 90–91. W. H. Freeman and Company, San Francisco (1979)
4. Håstad, J.: Clique is hard to approximate within $n^{1-\epsilon}$. In: Acta Mathematica, pp. 627–636 (1996)
5. Halldórsson, M.Ú.M., Kratochvíl, J., Telle, J.A.: Independent sets with domination constraints. In: Larsen, K.G., Skyum, S., Winskel, G. (eds.) ICALP 1998. LNCS, vol. 1443, pp. 176–187. Springer, Heidelberg (1998). https://doi.org/10.1007/BFb0055051
6. Halldórsson, M.Ú.M.: Approximations of independent sets in graphs. In: Jansen, K., Rolim, J. (eds.) APPROX 1998. LNCS, vol. 1444, pp. 1–13. Springer, Heidelberg (1998). https://doi.org/10.1007/BFb0053959
7. Hurkens, C.A.J., Schrijver, A.: On the size of systems of sets every t of which have an SDR, with an application to the worst-case ratio of heuristics for packing problems. SIAM J. Discrete Math. **2**(1), 68–72 (1989)
8. Halldórsson, M.Ú.M.: Approximating discrete collections via local improvements. In: Proceedings of the Sixth Annual ACM-SIAM Symposium on Discrete Algorithms, pp. 160–169 (1995)
9. Cygan, M., Grandoni, F., Mastrolilli, M.: How to sell hyperedges: the hypermatching assignment problem. In: Proceedings of the Twenty-Fourth Annual ACM-SIAM Aymposium on Discrete Algorithms, SIAM, pp. 342–351 (2013)
10. Sviridenko, M., Ward, J.: Large neighborhood local search for the maximum set packing problem. In: Fomin, F.V., Freivalds, R., Kwiatkowska, M., Peleg, D. (eds.) ICALP 2013. LNCS, vol. 7965, pp. 792–803. Springer, Heidelberg (2013). https://doi.org/10.1007/978-3-642-39206-1_67
11. Fellows, M.R., Downey, R.G.: Parameterized Complexity. Springer, New York (1999). https://doi.org/10.1007/978-1-4612-0515-9
12. Jia, W., Zhang, C., Chen, J.: An efficient parameterized algorithm for m-set packing. J. Algorithms **50**(1), 106–117 (2004)

13. Koutis, I.: A faster parameterized algorithm for set packing. Inf. Process. Lett. **94**(1), 7–9 (2005)
14. Fellows, M.R., Knauer, C., Nishimura, N., Ragde, P., Rosamond, F., Stege, U., Thilikos, D.M., Whitesides, S.: Faster fixed-parameter tractable algorithms for matching and packing problems. In: Albers, S., Radzik, T. (eds.) ESA 2004. LNCS, vol. 3221, pp. 311–322. Springer, Heidelberg (2004). https://doi.org/10.1007/978-3-540-30140-0_29
15. Abu-Khzam, F.N.: An improved kernelization algorithm for r-set packing. Inf. Process. Lett. **110**(16), 621–624 (2010)
16. Chen, L., Ye, D., Zhang, G.: An improved lower bound for rank four scheduling. Oper. Res. Lett. **42**(5), 348–350 (2014)
17. Micali, S., Vazirani, V.V.: An $O(\sqrt{|V|}|E|)$ algorithm for finding maximum matching in general graphs. In: Proceedings of the Twenty-First Annual Symposium on Foundations of Computer Science, pp. 17–27 (1980)
18. Bodlaender, H.L., Kloks, T.: Efficient and constructive algorithms for the pathwidth and treewidth of graphs. J. Algorithms **21**(2), 358–402 (1996)

Manipulation Strategies
for the Rank-Maximal Matching Problem

Pratik Ghosal$^{(\boxtimes)}$ and Katarzyna Paluch

University of Wrocław, Wrocław, Poland
pratikghosal20082@gmail.com, abraka@cs.uni.wroc.pl

Abstract. We consider manipulation strategies for the rank-maximal matching problem. Let $G = (A \cup P, \mathcal{E})$ be a bipartite graph such that A denotes a set of applicants and P a set of posts. Each applicant $a \in A$ has a preference list over the set of his neighbours in G, possibly involving ties. A matching M is any subset of edges from $\mathcal{E}$ such that no two edges of M share an endpoint. A *rank-maximal* matching is one in which the maximum number of applicants is matched to their rank one posts, subject to this condition, the maximum number of applicants is matched to their rank two posts and so on. A central authority matches applicants to posts in G using one of rank-maximal matchings. Let a_1 be the sole manipulative applicant, who knows the preference lists of all the other applicants and wants to falsify his preference list, so that, he has a chance of getting better posts than if he were truthful, i.e., than if he gave a true preference list.

We give three manipulation strategies for a_1 in this paper. In the first problem 'best nonfirst', the manipulative applicant a_1 wants to ensure that he is never matched to any post worse than the most preferred post among those of rank greater than one and obtainable, when he is truthful. In the second strategy 'min max' the manipulator wants to construct a preference list for a_1 such that the worst post he can become matched to by the central authority is best possible or in other words, a_1 wants to minimize the maximal rank of a post he can become matched to. To be able to carry out strategy 'best nonfirst', a_1 only needs to know the most preferred post of each applicant, whereas putting into effect 'min max' requires the knowledge of whole preference lists of all applicants. The last manipulation strategy 'improve best' guarantees that a_1 is matched to his most preferred post at least in some rank-maximal matchings.

1 Introduction

We consider manipulation strategies for the rank-maximal matching problem. In the rank-maximal matching problem, we are given a bipartite graph $G = (A \cup P, \mathcal{E})$ where A denotes a set of applicants and P a set of posts. Each applicant $a \in A$ has a preference list over the set of his neighbours in G, possibly

Partly supported by Polish National Science Center grant UMO-2013/11/B/ST6/01748.

L. Wang and D. Zhu (Eds.): COCOON 2018, LNCS 10976, pp. 316–327, 2018.
https://doi.org/10.1007/978-3-319-94776-1_27

involving ties. Preference lists are represented by ranks on the edges - an edge (a, p) has rank i, denoted as $rank(a, p) = i$, if post p belongs to one of a's i-th choices. An applicant a prefers a post p to a post p' if $rank(a, p) < rank(a, p')$. In this case, we say that (a, p) has higher rank than (a, p'). If a is indifferent between p and p', then $rank(a, p) = rank(a, p')$. Posts most preferred by an applicant a have rank one in his preference list. A *matching* M is any subset of edges $\mathcal{E}$ such that no two edges of M share an endpoint. A matching is called a *rank-maximal* matching if it matches the maximum number of applicants to their rank one posts and subject to this condition, the maximum number of applicants to their rank two posts, and so on. A rank-maximal matching can be computed in $O(\min(c\sqrt{n}, n)m)$ time, where n is the number of applicants, m the number of edges and c the maximum rank of an edge in an optimal solution [20].

A central authority matches applicants to posts by using the rank-maximal matching algorithm. Since there may be more than one rank-maximal matching of G, we assume that the central authority may choose any one of them arbitrarily. Let a_1 be a manipulative applicant, who knows the preference lists of all the other applicants and wants to falsify his preference list, so that, he has a chance of getting better posts than if he were truthful, i.e., than if he gave a true preference list. We can always assume that a_1 does not get his most preferred post in every rank-maximal matching when he is truthful, otherwise, a_1 does not have any incentive to cheat. Also, we can notice that it is usually advantageous for a_1 to truncate his preference list. Let H_p denote the graph, in which a_1's preference list consists of only one post p. Then as long as no rank-maximal matching of H_p leaves a_1 unmatched, he is guaranteed to always get the post p. To cover the worst case situation for a_1, our strategies require a_1 to provide a full preference list that includes every post from P. Also, a_1 could make the posts, he does not want to be matched to, appear very far in his preference list. Thus, we assume that a_1 does not have any gap in his preference list, i.e., it cannot happen that in a_1's preference list there are a rank i and rank $(i + 2)$ posts but none of rank $(i + 1)$.

Our Contribution: Our contribution consists in developing manipulation strategies for the rank-maximal matching problem. Given a graph instance with the true preference list of every applicant, we introduce three manipulation strategies for a_1. We consider the case where a_1 is the sole manipulator in G.

Our first manipulation strategy named 'best nonfirst' is described in Sect. 3. The strategy may not provide an optimal improvement for a_1, but it is simple and fast. This strategy guarantees that a_1 is never matched to any post worse than the second best post he can be matched to in a rank-maximal matching, when he is truthful. In other words, if a_1 is matched to a post p when he is truthful and p is not his most preferred post, then the strategy 'best nonfirst' ensures that he is never matched to any post ranked worse than p in any rank-maximal matching. The advantage of this strategy is that a_1 does not need to know full preference lists of the other applicants. He only needs to know the most preferred post of each applicant to be able to successfully execute the strategy.

Next, in Sect. 4.2 we propose the strategy 'min max'. The strategy minimizes the maximal rank of a post a_1 can become matched to. Thus it optimally improves the worst post of a_1 that is obtainable from the central authority. What is more, the strategy has the property that by using it, a_1 always gets matched to p_1, which is the best among worst posts he can be matched to. Moreover, we prove that there does not exist a strategy that simultaneously guarantees that a_1 never gets a post worse than p_1 and sometimes gets a post better than p_1.

Last but not least, we have studied the manipulation strategy 'improve best'. The previous two manipulation strategies improve the worst post a_1 can be matched to in a rank-maximal matching. Hence, these strategies may not match a_1 to his most preferred post in any rank maximal matching. In this manipulation strategy, a_1 has a different goal - he wants to be matched to his most preferred post in some rank-maximal matchings. Note that it is not possible for him to ensure that he always gets his most preferred post. Due to space constraints, we have included the description of this strategy in the full version of the paper.

Previous and Related Work. The rank-maximal matching problem belongs to the class of matching problems with preferences. In the problems with one-sided preferences, the considered graph is bipartite and each vertex of only one set of the bipartition expresses preferences over the set of its neighbours. Apart from rank-maximal matchings, other types of matchings from this class include pareto-optimal [1,5,28], popular [3] and fair [15] matchings among others. In the problems with two-sided preferences, the underlying graph is also bipartite but vertices from both sides of the bipartition express preferences over their neighbours. The most famous example of a matching problem with two-sided preferences is that of a stable matching known also as the stable marriage problem. Since the seminal paper by Gale and Shapley [8], it has been studied very intensively, among others in [12,18,27]. In the non-bipartite matching problems with preferences each vertex from the graph ranks all of its neighbours. The stable roommate problem [17] is a counterpart of the stable marriage problem in the non-bipartite setting.

The rank-maximal matching problem was first introduced by Irving [19]. A rank-maximal matching can be found via a relatively straightforward reduction to the maximum weight matching problem. The already mentioned [20] gives a combinatorial algorithm that runs in $O(\min(n, c\sqrt{n})m)$ time. The capacitated and weighted versions were considered, respectively, in [21,25]. A switching graph characterization of the set of all rank-maximal matchings is described in [11]. Finally, the dynamic version of the rank-maximal matching problem was considered in [10,24].

A matching problem with preferences is called strategy-proof if it is in the best interest of each applicant to provide their true preference list. An example of a strategy-proof mechanism among matching problems with one-sided preferences is that of a pareto optimal matching. The strategyproofness of a pareto optimal matching has applications in house allocation [2,16,22,30] and kidney exchange [4,29]. Regarding the stable matching problem, if a stable matching algorithm produces a men-optimal stable matching, then it is not possible for men to gain any advantage by changing or contracting their preference lists and then the best strategy for them is to keep their true preference lists [7,26].

In the context of matching with preferences cheating strategies were mainly studied for the stable matching problem. Gale and Sotomayor [9] showed that women can shorten their preference lists to force an algorithm, that computes the men-optimal stable matching, to produce the women-optimal stable matching. Teo et al. [31] considered a cheating strategy, where women are required to give a full preference list and one of the women is a manipulator. Huang [13] explored the versions, in which, men can make coalitions. Manipulation strategy in the stable roommate problem was also considered by Huang [14]. For a matching problem with one-sided preferences, Nasre [23] studied manipulation strategies for the popular matching problem.

2   Background

A matching M is said to be *maximum (in a graph G)* if, among all matchings of G, it has the maximum number of edges. A path P is said to be *alternating with respect to matching M* or *M-alternating* if its edges belong alternately to M and $\mathcal{E} \setminus M$. A vertex v is *unmatched* or *free* in M if it is not incident to any edge of M. An M-alternating path P such that both its endpoints are unmatched in M, is said to be *M-augmenting* (or augmenting with respect to M). It was proved by Berge [6] that a matching M is maximum if and only if there exists no M-augmenting path.

We state the following well-known properties of maximum matchings in bipartite graphs. Let $G = (A \cup P, \mathcal{E})$ be a bipartite graph and let M be a maximum matching in G. The matching M defines a partition of the vertex set $A \cup P$ into three disjoint sets. A vertex $v \in A \cup P$ is even (resp. odd) if there is an even (resp. odd) length alternating path with respect to M from an unmatched vertex to v. A vertex v is unreachable if there is no alternating path from an unmatched vertex to v. The even, odd and unreachable vertices are denoted by E, O and U respectively. The following lemma is well known in matching theory. The proofs can be found in [20].

Lemma 1. *Let E, O and U be the sets of vertices defined as above by a maximum matching M in G. Then,*

1. *E, O and U are pairwise disjoint, and independent of the maximum matching M in G.*
2. *In any maximum matching of G, every vertex in O is matched with a vertex in E, and every vertex in U is matched with another vertex in U. The size of a maximum matching is $|O| + |U|/2$.*
3. *G contains no edge between a vertex in E and a vertex in $E \cup U$.*

2.1   Rank-Maximal Matchings

Next we review an algorithm by Irving et al. [20] for computing a rank-maximal matching. Let $G = (A \cup P, \mathcal{E})$ be an instance of the rank-maximal matching problem. Every edge $e = (a, p)$ has a rank reflecting its position in the preference list of applicant a. $\mathcal{E}$ is the union of disjoint sets $\mathcal{E}_i$, i.e., $\mathcal{E} = \mathcal{E}_1 \cup \mathcal{E}_2 \cup \mathcal{E}_3 \ldots \cup \mathcal{E}_r$, where $\mathcal{E}_i$ denotes the set of edges of rank i and r denotes the lowest rank of an edge in G.

Definition 1. *The signature of a matching M is defined as an r-tuple $\rho(M) = (x_1, \ldots, x_r)$ where, for each $1 \leq i \leq r$, x_i is the number of applicants who are matched to their i-th rank post in M.*

Let M and M' be two matchings of G, with the signatures $sig(M) = (x_1, \ldots, x_r)$ and $sig(M') = (y_1, \ldots, y_r)$. We say $M \succ M'$ if there exists k such that $x_i = y_i$ for each $1 \leq i < k \leq r$ and $x_k > y_k$.

Definition 2. *A matching M of a graph G is called rank-maximal if and only if M has the best signature under the ordering $\succ$ defined above.*

Let us denote $G_i = (A \cup P, \mathcal{E}_1 \cup \mathcal{E}_2 \cup \ldots \cup \mathcal{E}_i)$ as a subgraph of G that only contains edges of rank at most i. We define G'_1 as G_1 and G'_{i+1} as the subgraph of G_{i+1} occurring at the end of phase i of the algorithm for computing a rank-maximal matching [20]. G' is also called the reduced graph of G.

Lemma 2. *Let $G = (A \cup P, \mathcal{E})$ and $G' = (A \cup P, \mathcal{E}')$ be two bipartite graphs with ranks on the edges. Suppose that $\mathcal{E}' \subseteq \mathcal{E}$. Also, every edge $e \in \mathcal{E}'$ has the same rank in G and G'. Then any rank-maximal matching M of G such that $M \subseteq \mathcal{E}'$ is also a rank-maximal matching of G'.*

3   Properties of a Preference List and Strategy 'best nonfirst'

Here we note down some properties of the preference list of any applicant. Let us assume that the preference list of a_1 in G has the form $(P_1, P_2, P_3, \ldots, P_i, \ldots, P_t)$, where P_i denotes the set of posts of rank i in the preference list of a_1. $G \setminus \{a_1\}$ denotes the graph obtained from G after the removal of the vertex a_1 from G. We define an f-post of G in a similar way as in the popular matching problem [3].

Definition 3. *A post is called an f-post of G if and only if it belongs to $O(G_1 \setminus \{a_1\})$ or $U(G_1 \setminus \{a_1\})$, where $G_1 = (A \cup P, \mathcal{E}_1)$. The remaining posts of G are called non-f-posts.*

Lemma 3. *If P_1 contains a post that is a non-f-post, then a_1 is always matched to one of such posts in a rank-maximal matching of G and thus to one of his first choices.*

Next lemma shows that if a_1 is not matched to a rank one post in some rank-maximal matching of G, then an f-post may be defined in an alternative way that takes into account the whole graph G_1. This property is needed during the construction of strategy 'min max'.

Lemma 4. *Let us assume that a_1 is not matched to a rank one post in some rank-maximal matching of G. Then a post is an f-post if and only if it belongs to $O(G_1)$ or $U(G_1)$.*

The lemma below characterises the set of potential posts a_1 can be matched to, if he provides his true preference list.

Lemma 5. *Let G be a bipartite graph and i be the rank of the highest ranked non-f-post in the preference list of a_1. If a_1 is not matched to a rank one post, then a_1 can only be matched to a post of rank i or greater than i in any rank-maximal matching of G.*

Proof. Let M be a rank-maximal matching of G. While computing a rank-maximal matching of G we start by finding a maximum matching of G_1. Since a_1 is not matched to a rank one post in every rank-maximal matching of G, by Lemma 4, the set of f-posts contains every vertex from $O(G_1)$ and $U(G_1)$. Hence, we delete every edge, that has rank bigger than 1, incident to an f-post. Thus, every edge $e = (a_1, p)$ such that p is an f-post and $rank(a, p) > 1$ gets deleted after the first iteration of the algorithm. Therefore, no such edge can belong to a rank-maximal matching and a_1 can only be matched to a post of rank i or worse.   □

The above lemmas provide us with an easy method of manipulation that guarantee that a_1 can always be matched to the best non-f-post in his true preference list. Lemma 5 shows that the most preferred non-f-post of a_1 is ranked not worse than the second most preferred post he can be matched to, when he is truthful. We assume that a_1 is not matched to a rank one post in every rank-maximal matching of G. Otherwise, the manipulator has no incentive to cheat. Let $p_i \in P_i$ be a highest ranked non-f post in the true preference list of a_1. We put p_i as a rank 1 post in the falsified preference list of a_1. Next, we fill the falsified preference list of a_1 arbitrarily. This completes the description of strategy 'best nonfirst'.

Algorithm 1. Strategy 'best nonfirst'

1: $p_i \leftarrow$ a highest ranked non-f-post in the true preference list of a_1.
2: $p_i \leftarrow$ the rank one post in the falsified preference list of a_1 in H
3: Fill the rest of the preference list of a_1 in an arbitrary order
4: Output H

Theorem 1. *The graph H computed by Algorithm 1 is a strategy 'best nonfirst'.*

The correctness of Algorithm 1 follows from Lemma 3.

4   Strategy 'min max'

The strategy 'best nonfirst' may not provide an optimal solution. An example illustrating this fact can be found in the full version of the paper. In this section we introduce the strategy 'min max' that optimizes the worst post a_1 can be matched to in a rank-maximal matching.

4.1   Critical Rank

The notion that is going to be very useful while constructing a preference list is that of a critical rank.

Definition 4. *Let $G = (A \cup P, \mathcal{E})$ be a bipartite graph with ranks on the edges belonging to $\{1, 2, \ldots, r\}$. Suppose that $a \in A, p \in P$ and (a, p) does not belong to $\mathcal{E}$. Let $H = (A \cup P, \mathcal{E} \cup \{(a, p)\})$. We define a critical rank of (a, p) in H as follows.*

If there exists a natural number $1 \leq i \leq r$ such that $a \in O(G_i') \cup U(G_i')$ or $p \in O(G_i') \cup U(G_i')$, then the critical rank of (a, p) in H is equal to $\min\{i : (O(G_i') \cup U(G_i')) \cap \{a, p\} \neq \emptyset\}$. Otherwise, the critical rank of (a, p) is defined as $r + 1$.

The next lemma reveals an interesting property of the critical rank of an edge (a, p).

Lemma 6. *Let G, H and (a, p) be as in Definition 4. Then the critical rank of (a, p) is c if and only if*

1. *for every $1 \leq i < c$, the edge (a, p) belongs to every rank-maximal matching of H_i, in which (a, p) has rank i, and*
2. *for every $c < i \leq r$, the edge (a, p) does not belong to any rank-maximal matching of H_i, in which (a, p) has rank i, and*
3. *there exists a rank-maximal matching M of H_c, in which (a, p) has rank c such that (a, p) is not contained in M.*

Corollary 1. *The critical rank of an edge incident to an f-post of G is 1 in G.*

The next two lemmas explain the change of the critical rank of (a, p) when we add an f-post p' as a rank 1 post to the preference list of a.

Lemma 7. *Let G be a bipartite graph and a be an applicant. Let p be the only post in the preference list of a. Suppose that the critical rank of (a, p) is c in G. Let $\hat{G} = G \cup \{(a, p')\}$ where p' is a rank 1, f-post in the preference list of a. Then the critical rank of (a, p) is at most c in $\hat{G}$.*

Lemma 8. *Let $G = (A \cup P, \mathcal{E})$ be a bipartite graph, in which a has two neighbors p and p' such that apart from (a, p), each edge has a rank belonging to $\{1, 2, \ldots, r\}$. Additionally, p' is a rank one f-post in the preference list of a. Let $G' = (A \cup P, \mathcal{E} \setminus \{(a, p)\})$ and $G'' = (A \cup P, \mathcal{E} \setminus \{(a, p')\})$. Suppose that a becomes unreachable in G_i' and the critical rank of (a, p) is c in G''. Then the critical rank of (a, p) in the graph G is equal to, correspondingly:*

1. *c if $c \leq i$,*
2. *i if $c > i$.*

Proof. We can prove that for every $j < i$ there exists a rank-maximal matching of G_j' that contains (a, p') and there exists a rank-maximal matching of G_j' that does not contain (a, p').

Claim. Let us suppose that (a, p) has rank $c' < i$ in G. Then, the existence of a rank-maximal matching M of G that does not contain (a, p) implies that the critical rank of (a, p) in G'' is at most c'.

Proof. By Lemma 2, M is a rank-maximal matching of G' and hence, every rank-maximal matching of G' is also rank-maximal in G. We know that there exists a rank-maximal matching M' of G' that does not contain (a, p'). Thus a is unmatched in M'. By Fact 2, M' is also a rank-maximal matching of G''. M' does not contain (a, p), which shows that the critical rank of (a, p) is at most c' in G''.   □

First we assume that $c \le i$. Since G'' is a subgraph of G, by Lemma 7, the critical rank of $(a, p) \le c$ in G. Suppose the critical rank of $(a, p) = c' < c$ in G. Let us consider a graph G, in which (a, p) has rank c'. Since the critical rank of (a, p) is equal to c', there exists a rank-maximal matching M of G that does not contain (a, p). By the above claim, the critical rank of (a, p) in G'' is at most $c' < c$ - a contradiction. We conclude that the critical rank of (a, p) remains c in G if $c \le i$.

Let us consider now the case when $c > i$. First we show that the critical rank of $(a, p) \le i$ in G. Let us consider the graph G'. The vertex a becomes unreachable in G' after iteration i. We know that $G = G' \cup \{(a, p)\}$. Hence, the edge (a, p) is deleted in the graph G if the rank of $(a, p) > i$ in G. Therefore, the critical rank of $(a, p) \le i$ in G.

Next we show that the critical rank of $(a, p) = i$ in G. Suppose the critical rank of (a, p) equals $i' < i$ in G. Since the critical rank of (a, p) is equal to $i' < i$, there exists a rank-maximal matching M of G, in which (a, p) has rank i' that does not contain (a, p). Again by the claim, the critical rank of (a, p) in G'' is at most $i' < c$ - a contradiction. Therefore we have proved that the critical rank of (a, p) is equal to i in G.   □

Corollary 2. *Let G be a bipartite graph such that p' is a rank one, f-post in the preference list of a. Suppose that a becomes unreachable after iteration i. Let $\hat{G} = G \cup \{(a, p)\}$ with the rank of the edge (a, p) being c. Then (a, p) is never matched in a rank- maximal matching of $\hat{G}$, if $c > i$.*

The next lemma is useful while building a falsified preference list of a using the strategy 'min max'. This lemma basically combines two short preference lists of a into a longer preference list.

Lemma 9. *Let us consider two bipartite graphs G_1 and G_2 such that p is a rank one, f-post in the preference list of a in both graphs. Also a has only two neighbors in each of the graphs. In G_1, a has p_1 as a rank i post. In G_2, a has p_2 as a rank j post. We assume that G_3 is the union of graphs G_1 and G_2. Then a is matched to p in every rank-maximal matching of both G_1 and G_2 if and only if a is matched to p in every rank-maximal matching of G_3.*

Proof. Assume that a is matched to p_1 in a rank-maximal matching M of G_3. M is also a matching in the graph G_1. Since G_1 is a subgraph of G_3, from the fact 2, M is a rank-maximal matching of G_1. This means that a is matched to p_1 in some rank-maximal matchings of G_1, which is a contradiction.

Conversely, let a be matched to p in every rank-maximal matching of G_3. Let M_3 be a rank-maximal matching of G_3. Without loss of generality, suppose that a is matched to p_1 in a rank-maximal matching M_1 of G_1. Since G_1 is a subgraph of G_3, from fact 2, M_3 is a rank-maximal matching of G_1. Thus, M_1 and M_3 have the same signature. Therefore, M_1 is a rank-maximal matching of G_3, which is a contradiction.   □

4.2   Algorithm for Strategy 'min max'

In this section, we give an algorithm that computes a graph H by using the strategy 'min max' for the applicant a_1. We recall that strategy 'min max' consists in finding a full preference list for a_1 such that the maximal rank of a post he can obtain is minimized. Since we have assumed that a_1 is not always matched to his first choice when he is truthful and since strategy 'best nonfirst' ensures that a_1 always gets the highest ranked non-f-post, it remains to check if it is possible for a_1 to get one of the f-posts in every rank-maximal matching. For a given f-post p we want to verify if a_1 can construct a full preference list that guarantees that a becomes matched to p in every rank-maximal matching of the resultant graph. From all such f-posts, we want to choose that of the highest rank in the true preference list of a_1. Below we show that this way we indeed compute the strategy 'min max'.

Let p be an f-post that a_1 wants to be matched to in every rank-maximal matching of H_p, where H_p contains a full falsified preference list of a_1. How do we construct such H_p? Let $\hat{H}_p$ denote the graph, in which a_1 is incident only to p and (a_1, p) has rank one. By Lemma 9, we know that in order to obtain H_p, it suffices to find a certain number of graphs H_{p,p_j} such that p and p_j are the only posts in the preference list of a_1, p has rank 1, p_j has rank $j > 1$ and every rank-maximal matching of H_{p,p_j} matches a_1 to p. Then we can combine those graphs into one graph H_p. In fact, it suffices to fill the preference list of a_1 only till rank k, where k is the rank, when a_1 becomes an unreachable vertex in $\hat{H}'_{p,i}$. This follows from Corollary 2, which says that no rank-maximal matching of $H_{p,p'}$ such that (a_1, p') has rank $i > k$ contains (a_1, p'). Therefore, the ranks greater than k in the preference list of a_1 may be filled with arbitrary posts not occurring previously.

Suppose that we want to find a "good" post for rank $i < k$ in the preference list of a_1. First, we check if there is any available post p' such that the critical rank of (a_1, p') is smaller than i in $H_{p,p'}$. If we find such a post, then by Lemma 6, the edge (a_1, p') never occurs in a rank-maximal matching of $H_{p,p'}$ in which (a_1, p') has rank i. Therefore, we may add p' to the preference list of a_1 as a rank i post. Otherwise, we consider a post p'' with critical rank i in the graph $H_{p,p''}$. We verify if p is matched to a_1 in every rank-maximal matching, when we add (a_1, p'') as a rank i edge to the graph $\hat{H}_p$. If yes, then we put p'' as an ith choice in H_p. If not, we check another post with critical rank i. If we are unable to find any post for rank i, Algorithm 2 outputs that there does not exist any preference list that matches a_1 to p in every rank-maximal matching of H_p.

The algorithm that computes a graph H_p, if it exists, is given below as Algorithm 2. The thing that still requires explanation is how we verify if (a, p) belongs to every rank-maximal matching of $H_{p,p'}$. For this, we need the reduced graph of $II_{p,p'}$ from phase r, which we can obtain by either applying the standard rank-maximal matching algorithm [20] or we can use one of the dynamic algorithms [10, 24] if we want to have a faster algorithm. Once we have access to this reduced graph of $H_{p,p'}$ we can use the following lemma.

Algorithm 2. Construction of H_p

1: $C_i \leftarrow \{p' \in P :$ the critical rank of (a_1, p') in $H_{p,p'}$ equals $i\}$
2: $L \leftarrow$ an empty list - L is the falsified preference list of a_1 that is going to have the
    form $(p_1, p_2, ..., p_n)$, where p_i denotes the rank i post in L.
3: add p to L – this is the rank 1 post in the preference list L of a_1
4: $k \leftarrow$ the number of phase when a_1 becomes unreachable in $\hat{H}_p$, i.e., $a_1 \in U(\hat{H}'_{p,k})$
    and $a_1 \in E(\hat{H}'_{p,i})$ for every $i < k$, where $\hat{H}'_{p,i}$ is the i-th reduced graph of $\hat{H}_p$.
5: $C \leftarrow C_1$
6: **for** $i = 2, \ldots, k$ **do**
7:     **if** $C \neq \emptyset$ **then**   (there exists a post p' in C)
8:         add p' as a rank i post to the falsified preference list L of a_1
9:         $C \leftarrow C \setminus \{p'\}$
10:    **else** $(C = \emptyset)$
11:        $SEARCH \leftarrow TRUE$
12:        **while** $\exists p'$ with critical rank of (a, p') equal to i in $\hat{H}_p$ and $SEARCH$ **do**
13:            **if** (a, p) belongs to every rank-maximal matching of $H_{p,p'}$ (Lemma 10)
    then
14:                add p' as a rank i post to the falsified preference list L of a_1
15:                $SEARCH \leftarrow FALSE$
16:        **if** SEARCH **then** Break
17:        $C \leftarrow C \cup C_i$
18: **if** L is a full preference list **then**
19: **return** H_p
20: **else**
21: **return** p is not a feasible f-post

Lemma 10. *Let G be an instance of the rank-maximal matching problem, in which the maximal rank of an edge is r. Also, we assume that M is a fixed rank-maximal matching of G that matches an edge (a, p). Let us consider the switching graph of the matching M in G. Then the edge (a, p) belongs to every rank-maximal matching of G if there does not exist any switching path or switching cycle in the switching graph of M that contains the vertex p.*

Proof. Let us fix a rank-maximal matching M of G that matches the edge (a, p). Theorem 1 from [11] states that every rank-maximal matching G can be obtained from M by applying some vertex-disjoint switching paths and switching cycles in the switching graph of M. If there does not exist any switching path or switching cycle containing the vertex p, p has the same partner in every rank-maximal matching of G. Therefore, (a, p) is matched in every rank-maximal matching of G.   □

Definition 5. *We say that an f-post p is* feasible *if there exists a graph H_p such that every rank-maximal matching of H_p matches a_1 to p.*

In the lemma below we prove the correctness of Algorithm 2.

Lemma 11. *If Algorithm 2 outputs a graph H_p, then every rank-maximal matching of H_p matches a_1 to p. Otherwise, there does not exist a graph H_p, in which a_1 gives a full preference list such that every rank-maximal matching of H_p matches a_1 to p.*

Theorem 2. *Let p be the highest ranked feasible f-post in the true preference list of a_1. Then H_p output by Algorithm 2 is a strategy 'min max'. Moreover, each graph H that is a strategy 'min max' has the property that each rank-maximal matching of H matches a_1 to p.*

References

1. Abdulkadiroğlu, A., Sönmez, T.: Random serial dictatorship and the core from random endowments in house allocation problems. Econometrica **66**(3), 689–701 (1998)
2. Abraham, D.J., Cechlárová, K., Manlove, D.F., Mehlhorn, K.: Pareto optimality in house allocation problems. In: Deng, X., Du, D.-Z. (eds.) ISAAC 2005. LNCS, vol. 3827, pp. 1163–1175. Springer, Heidelberg (2005). https://doi.org/10.1007/11602613_115
3. Abraham, D.J., Irving, R.W., Kavitha, T., Mehlhorn, K.: Popular matchings. SIAM J. Comput. **37**(4), 1030–1045 (2007)
4. Ashlagi, I., Fischer, F.A., Kash, I.A., Procaccia, A.D.: Mix and match: a strategyproof mechanism for multi-hospital kidney exchange. Games Econ. Behav. **91**, 284–296 (2015)
5. Aziz, H., Brandt, F., Harrenstein, P.: Pareto optimality in coalition formation. Games Econ. Behav. **82**, 562–581 (2013)
6. Berge, C.: Two theorems in graph theory. Proc. Nat. Acad. Sci. **43**(9), 842–844 (1957)
7. Dubins, L.E., Freedman, D.A.: Machiavelli and the Gale-Shapley algorithm. Am. Math. Mon. **88**(7), 485–494 (1981)
8. Gale, D., Shapley, L.S.: College admissions and the stability of marriage. Am. Math. Mon. **69**(1), 9–15 (1962)
9. Gale, D., Sotomayor, M.: Ms. Machiavelli and the stable matching problem. Am. Math. Mon. **92**(4), 261–268 (1985)
10. Ghosal, P., Kunysz, A., Paluch, K.E.: The dynamics of rank-maximal and popular matchings. CoRR, abs/1703.10594 (2017)

11. Ghosal, P., Nasre, M., Nimbhorkar, P.: Rank-maximal matchings–structure and algorithms. In: Ahn, H.-K., Shin, C.-S. (eds.) ISAAC 2014. LNCS, vol. 8889, pp. 593–605. Springer, Cham (2014). https://doi.org/10.1007/978-3-319-13075-0_47

12. Gusfield, D., Irving, R.W.: The Stable Marriage Problem: Structure and Algorithms. Foundations of Computing. MIT Press, Cambridge (1989)

13. Huang, C.-C.: Cheating by men in the Gale-Shapley stable matching algorithm. In: Azar, Y., Erlebach, T. (eds.) ESA 2006. LNCS, vol. 4168, pp. 418–431. Springer, Heidelberg (2006). https://doi.org/10.1007/11841036_39

14. Huang, C.-C.: Cheating to get better roommates in a random stable matching. In: Thomas, W., Weil, P. (eds.) STACS 2007. LNCS, vol. 4393, pp. 453–464. Springer, Heidelberg (2007). https://doi.org/10.1007/978-3-540-70918-3_39

15. Huang, C.-C., Kavitha, T., Mehlhorn, K., Michail, D.: Fair matchings and related problems. Algorithmica **74**(3), 1184–1203 (2016)

16. Hylland, A., Zeckhauser, R.: The efficient allocation of individuals to positions. J. Polit. Econ. **87**(2), 293–314 (1979)

17. Irving, R.W.: An efficient algorithm for the stable roommates problem. J. Algorithms **6**(4), 577–595 (1985)

18. Irving, R.W.: Matching medical students to pairs of hospitals: a new variation on a well-known theme. In: Bilardi, G., Italiano, G.F., Pietracaprina, A., Pucci, G. (eds.) ESA 1998. LNCS, vol. 1461, pp. 381–392. Springer, Heidelberg (1998). https://doi.org/10.1007/3-540-68530-8_32

19. Irving, R.W.: Greedy matchings. Technical report, University of Glasgow, Tr-2003-136 (2003)

20. Irving, R.W., Kavitha, T., Mehlhorn, K., Michail, D., Paluch, K.E.: Rank-maximal matchings. ACM Trans. Algorithms (TALG) **2**(4), 602–610 (2006)

21. Kavitha, T., Shah, C.D.: Efficient algorithms for weighted rank-maximal matchings and related problems. In: Asano, T. (ed.) ISAAC 2006. LNCS, vol. 4288, pp. 153–162. Springer, Heidelberg (2006). https://doi.org/10.1007/11940128_17

22. Krysta, P., Manlove, D., Rastegari, B., Zhang, J.: Size versus truthfulness in the house allocation problem. In: Proceedings of ACM Conference on Economics and Computation, EC, pp. 453–470 (2014)

23. Nasre, M.: Popular matchings: structure and strategic issues. SIAM J. Discrete Math. **28**(3), 1423–1448 (2014)

24. Nimbhorkar, P., Rameshwar, V.A.: Dynamic rank-maximal matchings. In: Cao, Y., Chen, J. (eds.) COCOON 2017. LNCS, vol. 10392, pp. 433–444. Springer, Cham (2017). https://doi.org/10.1007/978-3-319-62389-4_36

25. Paluch, K.: Capacitated rank-maximal matchings. In: Spirakis, P.G., Serna, M. (eds.) CIAC 2013. LNCS, vol. 7878, pp. 324–335. Springer, Heidelberg (2013). https://doi.org/10.1007/978-3-642-38233-8_27

26. Roth, A.E.: The economics of matching: stability and incentives. Math. Oper. Res. **7**(4), 617–628 (1982)

27. Roth, A.E.: The evolution of the labor market for medical interns and residents: a case study in game theory. J. Polit. Econ. **92**(6), 991 (1984)

28. Roth, A.E., Postlewaite, A.: Weak versus strong domination in a market with indivisible goods. J. Math. Econ. **4**(2), 131–137 (1977)

29. Roth, A.E., Sönmez, T., Ünver, M.U.: Pairwise kidney exchange. J. Econ. Theor. **125**(2), 151–188 (2005)

30. Shapley, L., Scarf, H.: On cores and indivisibility. J. Math. Econ. **1**(1), 23–37 (1974)

31. Teo, C.P., Sethuraman, J., Tan, W.P.: Gale-Shapley stable marriage problem revisited: strategic issues and applications. Manag. Sci. **47**(9), 1252–1267 (2001)

Finding Maximal Common Subgraphs via Time-Space Efficient Reverse Search

Alessio Conte[1], Roberto Grossi[2(✉)], Andrea Marino[2], and Luca Versari[2]

[1] National Institute of Informatics, Tokyo, Japan
conte@nii.ac.jp
[2] Università di Pisa, Pisa, Italy
{grossi,marino,luca.versari}@di.unipi.it

Abstract. For any two given graphs, we study the problem of finding isomorphisms that correspond to inclusion-maximal common induced subgraphs that are *connected*. While common (induced or not) subgraphs can be easily listed using some well known reduction and state-of-the-art algorithms, they are not guaranteed to be connected. To meet the connectivity requirement, we propose an algorithm that revisits the paradigm of reverse search and guarantees polynomial time per solution (delay) and linear space, on top of showing good practical performance.

1   Introduction

The problem of finding common subgraphs, as studied in this paper, has been introduced and investigated in the practical setting of proteins [5,13,14], and can be employed to mine significant information in many domains, for example identifying compound similarity and structural relationships between biological molecules [9]. These patterns find motivation in the increasing amount of structured data arising from X-ray crystallography and nuclear magnetic resonance. For these reasons, the bioinformatics community has repeatedly expressed its interest in the detection of common subgraphs (see for instance [9,12,14,21]).

From a computational point of view, the problem has been studied as one of the application examples of an algorithmic framework [7] to efficiently enumerate maximal subgraphs satisfying a given property (e.g. being a clique, a cut, a cycle, a matching, etc.), also known as set systems [16]. In this paper we are interested to design efficient algorithms for the following scenario.

For any two given input graphs H and F, a subgraph S of H is *in common* with F if S is isomorphic to a subgraph of F: it is maximal if there is no other common subgraph that strictly contains it, and maximum if it is the largest. The *maximum* common subgraph problem asks for the maximum ones, or simply for their size. The *maximal* common subgraph (MCS) problem further requires discovering all the MCS's of H and F. The MCS problem can be constrained to *connected* and *induced* subgraphs (MCCIS) [3,13,14], where the latter means that all the edges of H between nodes in the MCS are mapped to edges of F, and vice versa: considering induced subgraphs reduces the search

© Springer International Publishing AG, part of Springer Nature 2018
L. Wang and D. Zhu (Eds.): COCOON 2018, LNCS 10976, pp. 328–340, 2018.
https://doi.org/10.1007/978-3-319-94776-1_28

Table 1. Comparison of polynomial space algorithms: running time of Koch's [13] algorithm vs our BC-ENUM and its parallel implementation. The first two rows are two pairs of random Erdos-Renyi graphs, and the last one a pair of graphs representing proteins from the Protein Data Bank (1ald and 1gox) with a time limit of two hours. As for the notation, n, m, and σ are the number of nodes, edges, and node labels of the graphs, Δ_H and Δ_F their maximum degrees, and q the size of the largest found MCCIS.

GRAPH H			GRAPH F			σ	q	Koch [13]		BC-ENUM		par.BC-ENUM	
n	m	Δ_H	n	m	Δ_F			TIME	#sol	TIME	#sol	TIME	#sol
200	235	5	200	234	7	5	12	28 s	6691	0.2 s	6691	0.04 s	6691
100	122	7	100	119	5	4	22	11 s	3654	0.6 s	3654	0.1 s	3654
2763	9488	14	2629	9059	12	12	68	2 h	1998	2 h	33874	2 h	887293

space [3], and their connectivity further alleviates the explosion of the number of solutions [13,14], as otherwise each permutation of a maximal independent set corresponds to a different maximal isomorphism.

MCCIS Problem. Given any two graphs H and F, list all (isomorphisms corresponding to) maximal common connected induced subgraphs (MCCIS's) between H and F in polynomial time per solution and total polynomial space.

Actually, MCS and MCCIS will refer to isomorphisms corresponding to MCS or MCCIS. Note that solving the above MCCIS problem is computationally more demanding than listing just maximal common induced subgraphs (i.e. relaxing the connectivity constraint) as we will comment later in the state of the art.

Contributions. We present algorithm BC-ENUM, which lists the (isomorphisms corresponding to) MCCIS's with polynomial delay and using linear total space. Given any two graphs H and F, let Δ_H and Δ_F be their maximum degree, respectively. For each reported MCCIS, letting q be its number of nodes, we pay $O(q^4 \Delta_H^2 \Delta_F^2)$ time using $O(q)$ space: the time complexity gives the *delay*, which is the worst-case time between any two consecutively reported MCCIS's. Note that a strength of these bounds is that they are parameterized by the solution size q and independent of the sizes of H and F (just their maximum degree).

Table 1 reports the running time of a sequential and parallel implementation[1] of BC-ENUM in C++, compared to the state-of-the-art algorithm by Koch [13]. Experiments were executed on a 12-core machine with two Intel Xeon E5-2620 CPUs and 128 gigabytes of RAM, with a time limit of two hours, showing that on top of giving theoretical guarantees, BC-ENUM is also fast in practice.

---

[1] Code available at https://github.com/veluca93/parallel_enum/tree/bccliques as part of a parallel enumeration framework.

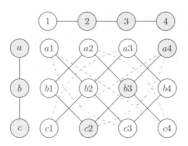

Fig. 1. An example of MCCIS ($\{a, b, c\}$ to $\{4, 3, 2\}$, in this order), with the corresponding BC-clique $\{a4, b3, c2\}$. White edges are represented as dashed lines.

Clarification on Maximum vs Maximal Common Subgraphs. As it is clear, *maximal* and *maximum* subgraphs are inherently different problems: listing *all* maximal ones can potentially find an exponential number of solutions, while finding the maximum connected ones corresponds to just the *single* largest one, and is in practice much faster (e.g. [19]). As pointed out in [5,13,14], however, a maximum common subgraph does *not* always contain all the relevant/large common structures, which motivates the MCCIS problem.

Converting the MCCIS Problem to a Maximal Clique Problem. Clique-based methods are widely employed on the product graph G, which transforms common subgraphs of H and F into maximal cliques in G, as proved in [17].

As in [13], we define the *product graph* between H and F as follows. (i) any pair of nodes $(x, i) \in H \times F$ is a node of G iff they have the same label; (ii) there is a *black edge* between (x, i) and (y, j) iff $(x, y) \in E(H)$ and $(i, j) \in E(F)$; (iii) there is a *white edge* between (x, i) and (y, j) iff $x \neq y$, $i \neq j$, $(x, y) \notin E(H)$ and $(i, j) \notin E(F)$, where $E(\cdot)$ denotes the edge set.

The key property is that MCCIS's between H and F correspond to cliques in G spanned by black edges [13], which we will call BC-*cliques*. An example is shown in Fig. 1.

Role of the Reverse Search. Reverse search is a powerful enumeration technique, introduced by Avis and Fukuda [1], that applies to a wide range of problems (e.g. [5,18]). If we try to apply it to BC-cliques, a number of obstacles appear along the road and thus this paper proposes a novel, restructured, way to use reverse search on BC-cliques: Cao et al. [3] observe that materializing the product graph G can be memory-wise expensive. BC-ENUM does *not* materialize G, but navigates the huge solution space of the BC-cliques by navigating G implicitly using H and F, just requiring $O(q)$ additional space (e.g. for H and F in the last row of Table 1, G would contain millions of nodes whereas $q = 68$). This simultaneously improves memory usage and running time, as detailed in Sect. 5.

State of the Art and Related Work. Common subgraphs problems have been studied for decades [3,10], with the great majority of the results dealing with *maximum* common subgraphs, rather than MCCIS's as we do. Previous work can be roughly classified into the following categories: backtracking methods [15], techniques based on special classes of graphs [11], clique-based methods [10,14, 19], methods which are applications of a generic framework [4], and restricted to trees [8]. Among of them, Koch [13] considers MCCIS's and employs a modified version of the Bron-Kerbosch algorithm [2] to work on explicit product graphs: it is still the state of the art [22], greatly used in practice, even in the very last years (e.g. [23]). Further methods have relaxed the definition of MCCIS to improve the practical performance, at the price of loosing some solutions [5]. Unfortunately, the aforementioned algorithms, when applied to listing all the MCCIS's, do not give any guaranteed polynomial bound on space or time per solution. Interestingly, a couple of other roads can be pursued successfully.

The framework presented in [4] uses the formulation of the reverse search on restricted problems and introduces new techniques for a class of set systems satisfying the connected hereditary property. The space is proportional to the number of solutions found, so it can be exponential, and thus space efficiency is one of the open problems posed there.

Along these lines, the framework presented in [7] provides new techniques for the class of set systems called commutable, and requires total polynomial space independently of the number of solutions found, thus answering to the question posed in [4].

We observe that BC-cliques can fit both frameworks, with polynomial time per solution. In this paper, we focus on the latter to provide polynomial bounds on the delay and space, while the implementation of the former in practice deserves further investigation in future work to evaluate the impact of the higher space usage. Compared to the bounds polynomial in the graph size from [7], which is a general theoretical framework for which BC-cliques are just an instance, BC-ENUM aims at specializing and parameterizing these bounds for the MCCIS problem, so they are polynomial in the max degree of the graphs (rather than in the size of the product graph), and at providing practical performance.

2   Using Reverse Search for Finding BC-cliques

As in [13], we reduce the problem of finding MCCIS's to finding BC-cliques in the product graph. Hence, in this section, we focus on the problem of listing BC-cliques in a graph G, whose edges are colored black or white, where a BC-clique is a maximal clique whose black edges connect all the nodes. To this aim, we employ reverse search, which can be successfully used when a suitable parent-child relationship between solutions is defined (see for instance [1,6,18]). Here we restructure the technique to deal with the more challenging BC-cliques. We keep the schema very simple for the sake of description, and hide the technical complexity in the definition of the parent-child relationship between solutions, which is the difficult part and it will be described in the following sections.

Algorithm 1. BC-ENUM: Enumerate all maximal BC-cliques of G

Function SPAWN (K)
  foreach $S \in$ CHILDREN(K) **do**
  $\quad$ SPAWN(S);
  Output K;

foreach $R \in$ ROOTS(G) **do**
$\quad$ SPAWN(R)

General Scheme. As is the case of reverse-search algorithms, our algorithm will implicitly define a rooted forest among all solutions, where some solutions are the *roots*, and from each solution we can find all its *children* in the forest. As we can identify all the roots and recursively visit the children of each solution, Algorithm 1 will not miss any solution.

In the following, let P be the *parent* of a solution S, denoted as P(S), if S is a child of P in this rooted forest-like structure. Note that every solution has exactly one parent, except the roots who have none.

Lemma 1. *Algorithm 1 lists all maximal* BC*-cliques when the following conditions are all met.*

1. *Each solution is either a* root, *or has exactly one* parent.
2. *The edges* P$(S) \rightarrow S$ *induce a forest* $\mathcal{Z}$ *whose sources are the* roots.
3. *The generic function* CHILDREN(P) *computes the set* $\{S : \text{P}(S) = P\}$.

The proof of Lemma 1 is straightforward as Algorithm 1 corresponds to a recursive traversal of the trees composing the forest $\mathcal{Z}$ induced by the parent-child relationship. We will design the latter so that the properties in Lemma 1 are satisfied. We remark that a tree traversal, rather than a graph traversal, does not require keeping track of visited nodes so far, and can be done without storing any information other than the current node and the previously visited one. We use this property to define an equivalent algorithm, which we call "stateless", that uses just $O(q)$ space and has the same complexity. We give further discussion in Sect. 5, and refer the reader to [6].

3   Canonical Representation and Operations

We consider G to have an arbitrary ordering of the nodes $\langle v_1 \ldots v_n \rangle$, and we consider each node v_i to have label i. A node v_i is *smaller* than v_j if $i < j$. For convenience, we call G_B the subgraph of G induced by the black edges. We introduce the representation for BC-cliques in G at the base of our approach. For a BC-clique K, we call the smallest node in K the *head* of K, and define the notion of *black-edge distance* as follows.

Definition 1 (black-edge distance). *The black-edge distance* $\beta_K(v)$ *of a node* $v \in K$ *is the distance in the induced subgraph* $G_B[K]$ *between* v *and the head of* K. *If* $v \notin K$ *but* $K \cup \{v\}$ *is a* BC*-clique,* $\beta_K(v)$ *is similarly defined on* $G_B[K \cup \{v\}]$.

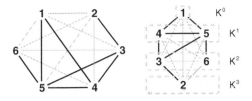

Fig. 2. A BC-clique K (left) and its canonical form $K^0 \ldots K^3$ (right)

When K is clear from the context, we will omit the subscript and just write $\beta(v)$. (When v is the head, $\beta(v) = 0$.) Moreover, let us define the canonical order of K.

Definition 2 (canonical order). *Given a* BC-*clique K, the* canonical order *of K is $\langle k_1, \ldots, k_{|K|} \rangle$, where elements of K are ordered in increasing lexicographical order of the pairs $(\beta(k_i), k_i)$.*

This order is a specialized version of the layer-based canonical order used in [7], which is the key to bound the running time to parameters of the two original input graphs H and F, rather than that of the larger G.

We will also refer to K^i as the set of nodes v with $\beta(v) = i$. This order essentially corresponds to the visiting order of a *breadth-first* search of $G_B[K]$, starting from the head k_1, where ties in the distance from k_1 are broken by taking the node with smallest label. As prefixes will be used extensively in our approach, we define $K_{<k_i}$ as the prefix $k_1, \ldots, k_{i-1}$ of K. It can be easily seen from the above how any prefix of K is a (non maximal) BC-clique.

An example of a BC-clique K in canonical order is shown in Fig. 2 with K^i's ordered by node label, where the head is node 1 and, for instance, $\beta_K(3) = 2$. When levels are relevant in the context, we represent K as a sequence of sets, which corresponds to the K^i's in increasing order: the clique in the example would be represented as $K = \{1\}, \{4, 5\}, \{3, 6\}, \{2\}$. As an example of prefix we have $K_{<3} = \langle 1, 4, 5 \rangle$.

We define the lexicographical order on BC-cliques using our canonical form. For any two pairs of integers (a, b) and (c, d), we write $(a, b) < (c, d)$ if the former pair is lexicographically smaller than the latter.

Definition 3 (lexicographical order). *Given two distinct maximal* BC-*cliques K and J, in their canonical orders $\langle k_1, \ldots, k_{|K|} \rangle$ and $\langle j_1, \ldots, j_{|J|} \rangle$, we say that K is lexicographically smaller than J, denoted as $K < J$, iff $(\beta_K(k_i), k_i) < (\beta_J(j_i), j_i)$, where i is the smallest index for which $(\beta_K(k_i), k_i) \neq (\beta_J(j_i), j_i)$.*

We also define a *forced* order of K with respect to $x \in K$, which is obtained by the same process as the canonical ordering, but computing the black-edge distances β with respect to x rather than the head of K.[2] A prefix of K *with*

---

[2] Thus the forced order of K with respect to its head is indeed the canonical order.

Algorithm 2. HEADS, CHILDREN functions that make Algorithm 1 working for BC-cliques

---

1 **Function** COMPLETE(K)
2   **while** $A \leftarrow \{v \in N(K) : v \text{ has a black neighbour in } K\} \neq \emptyset$ **do**
3   add $\text{argmin}_{x \in A}\{(\beta_k(x), x)\}$ to K
4   **return** K

5 **Function** PI(K)
6   **return** $\text{argmax}_{v \in K}\{(\beta(v), v) : \text{COMPLETE}(K_{<v}) \neq K\}$

7 **Function** P(K)
8     $v \leftarrow$ PI(K)
9   **return** COMPLETE($K_{<v}$)

10 **Function** ROOTS(G)
11   **return** $\{\text{COMPLETE}(v) : v = \min\{\text{COMPLETE}(\{v\})\}\}$

12 **Function** CAND(K)
13   **return** $\bigcup_{u \in K} N_B(u)$

14 **Function** CHILDREN(K)
15   **foreach** $v \in$ CAND(K) **do**
16       $K'_v \leftarrow$ unique maximal BC-clique containing v in $G_B[K \cap N(v) \cup \{v\}]$
17   **foreach** $h \in K'_v$ **do**
18          $K''_v \leftarrow$ prefix of K'_v with respect to h, truncated at v
19          $D \leftarrow$ COMPLETE(K''_v)
20   **if** $K =$ P(D) $\wedge h = \min D \wedge v =$ PI(D) **then yield** D

---

respect to x corresponds to a prefix of the forced order of K with respect to x. Clearly, this kind of prefix also corresponds to a BC-clique.

We now introduce the function COMPLETE(), which will be a key component as in [6,18]. Given a BC-clique K' which may or may not be maximal, COMPLETE(K') returns a *maximal* BC-clique K such that $K' \subseteq K$. This is achieved by recursively and greedily adding to K' the node $x \notin K'$ that minimizes $(\beta_{K'}(x), x)$, among all x for which $K' \cup \{x\}$ is a BC-clique. It is important to notice that the head of K' changes (as well as the values of $\beta_{K'}()$) whenever an element with label smaller than the current head is added. The COMPLETE operation is detailed in Algorithm 2. Finally, we remark that the properties of the COMPLETE function defined in [6], that make the reverse search algorithm work for cliques, are NP-hard to obtain in the context of BC-cliques [7].

4 Main Algorithm

In this section we describe the proposed algorithm BC-ENUM, using the definitions given in Sect. 3, to obtain the desired reverse search structure described in Sect. 2.

Firstly, using the function COMPLETE() we can easily identify the solutions which will be the roots of the forest $\mathcal{Z}$ induced by the parent-child relationship.

Definition 4 (root). *Let K be a maximal BC-clique and $h = \min K$ its head. Then, K is a root if and only if* COMPLETE($\{h\}$) $= K$.

This definition implies that the number of roots is at most n, and each can be identified by performing COMPLETE(v) on some node v. We now give definitions of P and CHILDREN as detailed in Algorithm 2.

The parent P(K) of K is defined as the result of applying COMPLETE to the longest prefix $K_{<v}$ such that this operation does not yield K. We call the element v of K that immediately follows this prefix $K_{<v}$ the *parent index* of K, PI(K). The definitions of P and root are consistent with Definition 4.

Lemma 2. P(K) $=$ NULL *if and only if $K \in$ ROOTS.*

We give a definition of CHILDREN, whose correctness will be proven in Sect. 4.1. Given a BC-clique K, let CAND(K) be the set of nodes that do not belong to K, but are neighbors of some node of K in G_B. For each such node v, we compute the largest BC-clique K'_v that contains v and is contained in $K \cup \{v\}$ (which corresponds to the connected component containing v in $G_B[K \cap N(v)]$). Then, for each $h \in K'_v$, we consider the *forced* order $\langle k_1, \ldots, k_{|K'_v|} \rangle$ with respect to h (noting that $k_1 := h$), and compute K''_v as the prefix $k_1, \ldots, k_i$ of this order truncated at $k_i = v$.

Finally, we compute $D =$ COMPLETE(K''_v) and control if D satisfies the check at line 20, which is required to ensure that the parent of D is indeed K and that we did not generate D multiple times from K itself.

4.1   Correctness

In order to prove the correctness of Algorithm 1 using the routines defined in Algorithm 2, we prove that the conditions listed in Lemma 1 are met, recalling that the directed graph $\mathcal{Z}$ induced by the parent function P has the arcs from P(K) to K for each solution K. By definition of P and by Lemma 2 we get Condition 1. Lemma 3 focuses on Condition 2, and Lemma 4 focuses on Condition 3.

Lemma 3. *The directed graph $\mathcal{Z}$ induced by P is a forest rooted in* ROOTS.

Lemma 4. *If $K = $ P(S), then $S \in$ CHILDREN(K).*

Proof. Consider an execution of CHILDREN(K), referring to its implementation in Algorithm 2, and let S be an arbitrary maximal BC-clique with P(S) $= K$. We need to prove that at some point in the execution Algorithm 2 will choose $v =$ PI(S) and $h = \min(S)$ in lines 15 and 17 respectively, and that this will give $D = S$ on line 19, which means D is yielded in line 20.

Consider the prefix $S_{<v}$ of S. As $K = $ P(S) $=$ COMPLETE($S_{<v}$), clearly $S_{<v} \subset K$ and, since S has a parent, it is not a root and so $S_{<v} \neq \emptyset$. By definition

of prefix, $S_{<v} \cup \{v\}$ is a BC-clique, so there must be a black edge between v and a node in $S_{<v}$, and thus to a node in K since $S_{<v} \subset K$, meaning that $v \in$ CAND and v is considered on line 15.

Consider now the execution of lines 16–20 when $v =$ PI(S). We have that $S_{<v} \cup \{v\}$ must be a subset of K'_v, the maximal BC-clique in $G_B[K \cap N(v) \cup \{v\}]$ containing v, since $S_{<v} \cup \{v\}$ is a BC-clique containing v and contained in $K \cup \{v\}$. Since $\min(S) \in S_{<v} \subseteq K'_v$, h will be chosen as $\min(S)$ in some iteration of line 17.

Finally, we need to prove that $S_{<v} \cup \{v\}$ is exactly K''_v, i.e., a prefix with respect to h of K'_v. If this were not the case, let d be the earliest element in K'_v (according to the forced ordering with respect to h) that is not in $S_{<v} \cup \{v\}$. Then $S_{<v} \cup \{v, d\}$ is still a BC-clique (as d must have a backwards black edge and all nodes before d are in $S_{<v}$); moreover, $(\beta_{K'_v}(d), d) < (\beta_{K'_v}(v), v)$. Thus d could be chosen by COMPLETE$(S_{<v} \cup \{v\})$, and since COMPLETE$(S_{<v} \cup \{v\}) = S$, this would mean that d is in S and occurs before v in its canonical ordering, implying $d \in S_{<v}$, which is a contradiction. $\square$

As a result, we obtain the correctness of BC-ENUM.

Theorem 1. *Algorithm 1 implemented with the methods from Algorithm 2 finds all and only maximal BC-cliques exactly once.*

5   Complexity and Implicit Product Graph

In this section, we give the complexity of BC-ENUM, taking into account that G is not a generic graph with white and black edges, but an implicit product graph between H and F that we do not want to materialize, whose size and features depend on H and F.

Recall that each node of G corresponds to a mapping between two nodes of H and F. For any given $v \in V(G)$, let these nodes be respectively $v_H \in V(H)$ and $v_F \in V(F)$. Further recall that Δ_H and Δ_F are the maximum node degree in H and F, while Δ_B is the maximum degree in G_B. By construction of the product graph we have $\Delta_B \leq \Delta_H \cdot \Delta_F$. For brevity, we define Δ as $\Delta_H + \Delta_F$. These parameters are all significantly smaller than the size of G, which has $|V(H)| \cdot |V(F)|$ nodes, and $O(|V(H)|^2 \cdot |V(F)|^2)$ edges, either black or white.

Let X be a BC-clique in G. We denote as X_H and X_F respectively the set of nodes of H and F mapped in X. We keep a dictionary between the nodes of X and those of X_H and X_F, allowing us to retrieve v_H and v_F from v, or vice versa, in $O(1)$ time.[3]

Lemma 5. *Let X be a BC-clique in G and v a node in $V(G)$. Testing whether $X \cup \{v\}$ is a BC-clique takes $O(\min(|X|, \Delta))$ time and $O(|X|)$ space.*

---

[3] This data structure will be built at the beginning of a COMPLETE call. As building it takes $O(|X|)$ time and space, it will not affect the final complexity.

Proof. As X is a BC-clique in G, in order to check that $X \cup \{v\}$ is a BC-clique in G we need to check that $\{v\}$ is connected to a node in X through a black edge, and to all the others through either white or black edges. This can trivially be done in $O(|X|)$ time by checking adjacency with the nodes of X one by one.

However, a faster solution is possible if we focus on the edges that are *not* in G: for a given node $x \in X$, corresponding to a mapping between $x_H \in V(H)$ and $x_F \in V(F)$, there is *no* edge in G between v and x if either $\{v_H, x_H\} \in E(H)$ and $\{v_F, x_F\} \notin E(F)$, or $\{v_H, x_H\} \notin E(H)$ and $\{v_F, x_F\} \in E(F)$. Otherwise, there is either a black or white edge between v and x.

To check the presence of missing edges between v and nodes of X we can simply iterate over all $x_H \in N_H(v_H) \cap X_H$, and check that each is mapped by X in a node $x_F \in N_F(v_F) \cap X_F$. Then, similarly, iterate over all $x_F \in N_F(v_F) \cap X_F$ and check that they are mapped in some $x_H \in N_H(v_H) \cap X_H$. This can be done in $O(|N_H(v_H)| + |N_F(v_F)|) = O(\Delta)$ time. If no missing edge exists then $X \cup \{v\}$ is a clique in G. As a byproduct, this process finds all black edges between v and X, thus we may check at the same time that there is at least one, and thus that $X \cup \{v\}$ is a BC-clique. $\square$

Lemma 6. *For any* BC-*clique* X *in* G, *computing* $\beta_X(v)$ *for all* $v \in X$ *takes* $O(|X| \cdot \min(|X|, \Delta_H, \Delta_F))$ *time and* $O(|X|)$ *space.*

Proof. The values of $\beta_X(v)$ corresponds to their distance from the head x of X in $G_B[X]$. This can be done via a BFS of $G_B[X]$ rooted at x. As $G_B[X]$ has $|X|$ nodes, the trivial bound for this traversal is $|X|^2$. Once again, we can exploit the fact that G is the product graph of H and F: indeed, each node v of X corresponds to a mapping of a node v_H of H into *one* node v_F of F. For this reason, while v can have up to Δ_B neighbors in G_B, v_H may have at most $|N_H(v_H)|$ neighbors in X_H.

We can thus iterate on the neighborhood of v in $O(\min(\Delta_H, \Delta_F))$ time by iterating on the neighbors of either v_H in H or v_F in F and then retrieve the corresponding nodes in X. In total, we process $|X|$ nodes, each in $O(\min(\Delta_H, \Delta_F))$ time, or in $O(|X|)$ time using the trivial version of the BFS. The cost follows. $\square$

Lemma 7. COMPLETE(X) *takes* $O(q(q + \Delta_B)\Delta) = O(q^2\Delta + q\Delta_B\Delta)$ *time and* $O(q)$ *space.*

Proof. In order to perform COMPLETE(X), we iterate over all nodes that can be added to X, adding the lexicographically smallest, with respect to X and its head x, first. For each node v in X (including those that are added during the procedure), we keep an iterator which will scan in increasing order its black neighbors. Clearly, each node must be considered after the smallest ones, and once it is considered it is either added to X or discarded, thus it does not need to be considered as a candidate anymore.

Given a node $c \notin X$, that has a black neighbor in X, we can see that $\beta_X(c) = \beta_X(v) + 1$, where v is the black neighbor of c in X that minimizes this value. Hence, to select the lexicographically smallest node, we must first consider the black neighbors of the nodes v that minimize $\beta_X(v)$. We thus order the nodes in a priority queue by value of $\beta_X(v)$, breaking ties by the value of the smallest black neighbor yet to consider, so that the first node in the priority queue is the smallest candidate to consider for addition to X.

As X will contain $|X| = O(q)$ nodes, and we will iterate on the $O(\Delta_B)$ black neighbors of each node exactly once, the total cost of this iteration is $O(q\Delta_B)$ time, and will yield up to $q\Delta_B$ nodes. Since by Lemma 5 testing a candidate takes $O(\min(q, \Delta))$ time, the total cost is $O(q\Delta_B \min(q, \Delta))$.

Furthermore, we need to account for the cost of changing *heads*: after we add a node x to X, this becomes the new head of X if its label is smaller than that of the previous head. In this case, we need to update both the values of $\beta_X(v)$, and the priority queue of candidate nodes. By Lemma 6, this can be done in $O(q \min(q, \Delta_H, \Delta_F))$ time. We pay this cost at most q times as we add up to q nodes, for a cost of $O(q^2 \min(q, \Delta_H, \Delta_F))$ which is upper bounded by $O(q^2\Delta)$.

The total cost is thus $O(q\Delta_B \min(q, \Delta) + q^2\Delta) = O(q^2(\Delta_B + \Delta))$ time.   □

Lemma 8. CHILDREN(K) *takes* $O(q^4\Delta_B(\Delta_B + \Delta)) = O(q^4\Delta_H^2\Delta_F^2)$ *time and* $O(q)$ *space.*

Proof. The cost of CHILDREN(K) is bounded by the cost of lines 19 and 20, times the number of nodes CAND(K), times the number of nodes in K'_v.

Nodes in CAND(K) are at most $|K|\Delta_B = O(q\Delta_B)$ (line 13 of Algorithm 2) and K'_v size is bounded by $O(q)$. In the following, we prove that the time cost of line 20 of Algorithm 2 is $O(q(q + \Delta_B)\Delta)$. Let $\langle d_1, \ldots, d_{|D|} \rangle$ be the canonical ordering of D. By definition, $d_i = \text{PI}(D)$ is the latest element in the canonical order of D such that COMPLETE$(D_{<d_i}) \neq D$. By the proof of Lemma 3, we have that COMPLETE$(D_{<d_i}) \leq D$ and COMPLETE$(D_{<d_j}) = D$ for any $j > i$. To check that v is indeed the parent index of D, we thus simply need to check that COMPLETE$(D_{<d_i}) \neq D$ and COMPLETE$(D_{<d_{i+1}}) = D$. Furthermore, if this is the case, COMPLETE$(D_{<d_i})$ also gives us the parent P(D) of D. Checking that h is the node of smallest label in D does not affect the cost, thus the total cost is that of calling the COMPLETE$()$ function twice. As Δ_B and Δ are bounded by $\Delta_H \cdot \Delta_F$, the statement follows.   □

Looking at Algorithm 1, we can see that the complexity of BC-ENUM is bounded by the cost of the function CHILDREN(K). Furthermore, as shown in Lemmas 5, 6, and 7, the space required is always $O(q)$. By turning the recursion into a stateless iteration (see [6]), no more space is needed as we do not need to store the recursion stack. We also address the *delay* of the algorithm, that is, the maximum elapsed time between two consecutive outputs, by applying the *alternative output* technique in [20]: for each recursive call on K in the recursion tree of Algorithm 1, we output solution K at the *beginning* of the call if its depth is even, and at the *end* if it is odd. In this way, the delay is equal to the cost per solution. We can thus state the main result.

Theorem 2. *Given two graphs H and F, BC-ENUM lists all their (isomorphisms corresponding to) MCCIS's in $O(q^4\Delta_H^2\Delta_F^2)$ delay and $O(q)$ space.*

Acknowledgements. Alessio Conte is supported by JST CREST, grant number JPMJCR1401, Japan, and Roberto Grossi, Andrea Marino and Luca Versari are supported by MIUR, Italy.

References

1. Avis, D., Fukuda, K.: Reverse search for enumeration. Discret. Appl. Math. **65**(1), 21–46 (1996)
2. Bron, C., Kerbosch, J.: Finding all cliques of an undirected graph (algorithm 457). Commun. ACM **16**(9), 575–576 (1973)
3. Cao, Y., Jiang, T., Girke, T.: A maximum common substructure-based algorithm for searching and predicting drug-like compounds. Bioinformatics **24**(13), i366–i374 (2008)
4. Cohen, S., Kimelfeld, B., Sagiv, Y.: Generating all maximal induced subgraphs for hereditary and connected-hereditary graph properties. J. Comput. Syst. Sci. **74**(7), 1147–1159 (2008)
5. Conte, A., Grossi, R., Marino, A., Tattini, L., Versari, L.: A fast algorithm for large common connected induced subgraphs. In: Figueiredo, D., Martín-Vide, C., Pratas, D., Vega-Rodríguez, M.A. (eds.) AlCoB 2017. LNCS, vol. 10252, pp. 62–74. Springer, Cham (2017). https://doi.org/10.1007/978-3-319-58163-7_4
6. Conte, A., Grossi, R., Marino, A., Versari, L.: Sublinear-space bounded-delay enumeration for massive network analytics: maximal cliques. In: ICALP, pp. 148:1–148:15 (2016)
7. Conte, A., Grossi, R., Marino, A., Versari, L.: Listing Maximal Subgraphs in Strongly Accessible Set Systems, March 2018. arXiv e-prints arXiv:1803.03659
8. Droschinsky, A., Heinemann, B., Kriege, N., Mutzel, P.: Enumeration of maximum common subtree isomorphisms with polynomial-delay. In: Ahn, H.-K., Shin, C.-S. (eds.) ISAAC 2014. LNCS, vol. 8889, pp. 81–93. Springer, Cham (2014). https://doi.org/10.1007/978-3-319-13075-0_7
9. Ehrlich, H.C., Rarey, M.: Maximum common subgraph isomorphism algorithms and their applications in molecular science: a review. Wiley Interdiscip. Rev.: Comput. Mol. Sci. **1**(1), 68–79 (2011)
10. Gardiner, E.J., Artymiuk, P.J., Willett, P.: Clique-detection algorithms for matching 3-dimensional molecular structures. J. Mol. Graph. Model. **15**, 245–253 (1997)
11. Gupta, A., Nishimura, N.: Finding largest subtrees and smallest supertrees. Algorithmica **21**(2), 183–210 (1998)
12. Huang, X., Lai, J., Jennings, S.: Maximum common subgraph: some upper bound and lower bound results. BMC Bioinform. **7**(Suppl 4), S6 (2006)
13. Koch, I.: Enumerating all connected maximal common subgraphs in two graphs. Theor. Comp. Sci. **250**(1), 1–30 (2001)
14. Koch, I., Lengauer, T., Wanke, E.: An algorithm for finding maximal common subtopologies in a set of protein structures. J. Comput. Biol. **3**(2), 289–306 (1996)
15. Krissinel, E.B., Henrick, K.: Common subgraph isomorphism detection by backtracking search. Softw.: Pract. Exp. **34**(6), 591–607 (2004)
16. Lawler, E.L., Lenstra, J.K., Rinnooy Kan, A.H.G.: Generating all maximal independent sets: NP-hardness and polynomial-time algorithms. SIAM J. Comput. **9**(3), 558–565 (1980)
17. Levi, G.: A note on the derivation of maximal common subgraphs of two directed or undirected graphs. CALCOLO **9**(4), 341–352 (1973)

18. Makino, K., Uno, T.: New algorithms for enumerating all maximal cliques. In: Hagerup, T., Katajainen, J. (eds.) SWAT 2004. LNCS, vol. 3111, pp. 260–272. Springer, Heidelberg (2004). https://doi.org/10.1007/978-3-540-27810-8_23
19. Suters, W.H., Abu-Khzam, F.N., Zhang, Y., Symons, C.T., Samatova, N.F., Langston, M.A.: A new approach and faster exact methods for the maximum common subgraph problem. In: Wang, L. (ed.) COCOON 2005. LNCS, vol. 3595, pp. 717–727. Springer, Heidelberg (2005). https://doi.org/10.1007/11533719_73
20. Uno, T.: Two general methods to reduce delay and change of enumerationalgorithms. NII Technical report NII-2003-004E, Tokyo, Japan (2003)
21. Van Berlo, R.J.P., Winterbach, W., De Groot, M.J.L., Bender, A., Verheijen, P.J.T., Reinders, M.J.T., de Ridder, D.: Efficient calculation of compound similarity based on maximum common subgraphs and its application to prediction of gene transcript levels. Int. J. Bioinform. Res. Appl. **9**(4), 407–432 (2013)
22. Welling, R.: A performance analysis on maximal common subgraph algorithms. In: 15th Twente Student Conference on IT. University of Twente, The Netherlands (2011)
23. Yuan, Y., Wang, G., Chen, L., Wang, H.: Graph similarity search on large uncertain graph databases. VLDB J. **24**(2), 271–296 (2015)

An FPT Algorithm for Contraction to Cactus

R. Krithika[1], Pranabendu Misra[2], and Prafullkumar Tale[1(✉)]

[1] The Institute of Mathematical Sciences, HBNI, Chennai, India
{rkrithika,pptale}@imsc.res.in
[2] Department of Informatics, University of Bergen, Bergen, Norway
Pranabendu.Misra@uib.no

Abstract. For a collection $\mathcal{F}$ of graphs, given a graph G and an integer k, the $\mathcal{F}$-Contraction problem asks whether we can contract k edges in G to obtain a graph in $\mathcal{F}$. $\mathcal{F}$-Contraction is well studied and known to be C-complete for several classes $\mathcal{F}$. Heggerners et al. [Algorithmica (2014)] were the first to explicitly study contraction problems in the realm of parameterized complexity. They presented FPT algorithms for Tree-Contraction and Path-Contraction. In this paper, we study contraction to a class larger than trees, namely, *cactus graphs*. We present an FPT algorithm for Cactus-Contraction that runs in $c^k n^{\mathcal{O}(1)}$ time for some constant c.

1 Introduction

For a collection $\mathcal{F}$ of graphs, $\mathcal{F}$-Modification problem is to determine if an input graph G can be converted to some graph in $\mathcal{F}$ using at most k modifications. $\mathcal{F}$-Modification is an abstraction of practically well motivated problems like Vertex Cover, Feedback Vertex Set, Odd Cycle Transversal, Minimum Fill-In, to name a few. In recent times, there has been increasing interest in the study of Edge Contraction problems where the modification operation allowed is edge contraction. These problems generally turn out to be more difficult compared to their vertex/edge deletion/addition counterparts. For example, even determining whether a given graph G can be contracted to a path of length four turns out to be NP-complete [3]. Formally, for a collection $\mathcal{F}$ of graphs, the $\mathcal{F}$-Contraction problem is to determine if an input graph G can be contracted to some graph in $\mathcal{F}$ using at most k edge contractions. For several choices of $\mathcal{F}$, early papers by Watanabe et al. and Asano and Hirata showed that $\mathcal{F}$-Edge Contraction is NP-complete even for several simple and well structured graph classes such as paths, stars, trees [2,3,14,15].

Graph contraction problems have received a lot of attention in parameterized complexity. It turns out that graph contraction problems are harder than their vertex/edge deletion/addition counterparts even in this setting. One of the

---

Due to space constraints, the proofs of results marked with ⋆ are omitted. These proof can be found in full version of the paper.

© Springer International Publishing AG, part of Springer Nature 2018
L. Wang and D. Zhu (Eds.): COCOON 2018, LNCS 10976, pp. 341–352, 2018.
https://doi.org/10.1007/978-3-319-94776-1_29

intuitive reasons is that the classical *branching technique* does not work even for graph classes $\mathcal{F}$ that have a finite forbidden structure characterization. In case of vertex deletion or edge deletion/addition operations, to destroy a structure which forbids the input graph from being in $\mathcal{F}$, one needs to include at least one vertex (or edge) from that structure into the solution. This is not necessarily true in the case of contractions. Indeed, a forbidden structure may be destroyed by contracting edges which are not contained in the structure. Despite this inherent difficulty, there are several fixed-parameter tractability results known when the parameter is the solution size, i.e., the maximum number k of edges that can be contracted. To best of our knowledge, Heggernes et al. [10] were the first to explicitly study edge contraction problems in the realm of parameterized complexity. They presented a $4^k n^{\mathcal{O}(1)}$ algorithm for TREE CONTRACTION and a $2^{k+o(k)} n^{\mathcal{O}(1)}$ algorithm for PATH CONTRACTION. When $\mathcal{F}$ is the set of graphs whose minimum degree is at least d, $\mathcal{F}$ is known to be FPT when parameterized by both k and d [8]. Golovach et al. proved that PLANAR CONTRACTION is FPT [7]. BIPARTITE CONTRACTION has been proved to be FPT by Heggernes et al. [11] and a faster algorithm was presented by Guillemot and Marx [9]. Cai and Guo [4] showed that CLIQUE CONTRACTION is FPT. On the negative side, it is known that $\mathcal{F}$-CONTRACTION is W[2]-hard when $\mathcal{F}$ is either the family of $P_{\ell+1}$-free graphs or the family of C_ℓ-free cycles for some $\ell \geq 4$ [4,13]. Recently, Agrawal et al. [1] proved that SPLIT CONTRACTION is W[1]-hard.

In this paper, we present an algorithm for CACTUS CONTRACTION, adding it to the small list of graph classes for which FPT algorithms for contraction problems are known. A graph is called a *cactus* if every edge is a part of at most one simple cycle. Formally, the problem can be stated as follows.

CACTUS CONTRACTION   **Parameter:** k
Input: A graph G and an integer k
Question: Does there exist $F \subseteq E(G)$ of size at most k whose contraction results in a cactus?

It is easy to verify that the problem is in NP and its NP-completeness follows from [12]. As a cactus has treewidth at most 2, it follows that if a graph is k-contractible to a cactus, then its treewidth is at most $k + 2$. Therefore, the problem is FPT by the celebrated result of Courcelle [5], as it is expressible in Monadic Second Order Logic. However, this approach yields an impractical algorithm whose running time involves a large function of k. The main contribution of this work is a $c^k n^{\mathcal{O}(1)}$ algorithm for CACTUS CONTRACTION, where c is a fixed constant. Our algorithm builds upon ideas presented in [10], but requires a more involved structural analysis of the graph.

Outline of the Algorithm: We can think of graph contraction problem as partition problem. The task is to find a partition where each partition, called *witness set*, is connected and contracting all witness set to a point leads to desired graph. The idea is to color the graph with a small number of colors to "highlight" certain portions of the graph that contain the desired solution. This solution is then extracted via the structural properties of the graph. In first phase, we color $V(G)$ using three colors $\{1, 2, 3\}$ with the hope that all vertices of a *big witness set* (set with at least two vertices) receive the same color and that two distinct big witness sets with certain properties are "separated". We then identify some vertices that are not part of any big witness sets and recolor them using new colors 4 and 5. For instance, we identify certain induced paths that do not intersect with any minimal solution and are "adjacent" to only one big witness set (Lemma 3). The vertices of such paths are colored 4. After this we identify vertices that are not part of any big witness set and lie on a path between two big witness sets (Lemma 4) and color them using color 5. This completes the first phase. In the second phase, we extract the big witness sets from the components highlighted in the first phase. For this purpose, we define the notion of a *connected core* (Definition 4) which can be thought of as generalization of connected vertex cover. For every monochromatic component colored with $\{1, 2, 3\}$ by the first phase, we find connected vertex cover containing certain boundary vertices. The desired solution is the set of edges of a spanning forest of the corresponding connected cores.

The paper is organized as follows. In Sect. 2 we review some graph theoretic preliminaries. We present the properties of solution in Sect. 3 which are used in proving the correctness of algorithm. Following the approach of [10], we first give a randomized algorithm for the problem on 2-connected graphs, which is then used to give an algorithm in general graphs. Algorithm can be divided into two phases viz coloring phase (Sect. 4) and extracting a solution from colored graph (Sect. 5). Finally, in Sect. 6 we present overall algorithm and illustrate how this algorithm can be derandomized via (n, k)-universal sets. We remark that the main goal of this paper is to provide a $c^k n^{\mathcal{O}(1)}$ algorithm for CACTUS CONTRACTION, where c is a fixed constant. For the sake of simplicity, we have not attempted to optimize the running time.

2   Preliminaries

For graph theoretic terms and notation which are not explicitly defined here, we refer the reader to the book by Diestel [6]. An *undirected graph* is a pair consisting of a set V of vertices and a set E of edges where $E \subseteq V \times V$. An edge uv between vertices u and v is specified as an unordered pair of vertices. For a graph G, $V(G)$ and $E(G)$ denote the set of vertices and edges, respectively. Two vertices u, v are said to be *adjacent* if there is an edge uv in the graph. The *neighbourhood* of a vertex v, denoted by $N_G(v)$, is the set of vertices adjacent to v. The *degree* $d_G(v)$ of a vertex v is $|N_G(v)|$. The subscript in the notation for neighbourhood and degree are omitted if the graph under consideration is clear. For a set of edges F, $V(F)$ denotes the set of endpoints of edges in F. For a set $S \subseteq V(G)$, $G - S$ denotes the graph obtained by deleting S from G and $G[S]$ denotes the subgraph of G induced on the set S. For sets $X, Y \subseteq V(G)$, $E(X, Y)$ denotes the set of edges with one endpoint in X and the other endpoint in Y. Similarly, $E(X)$ denotes the set of edges whose both endpoints are in X.

A *path* $P = (v_1, \ldots, v_l)$ is a sequence of distinct vertices in which there is an edge between any pair of consecutive vertices. The vertex set of P is the set $\{v_1, \ldots, v_l\}$ and is denoted by $V(P)$. The path P is called as a *cycle* if v_1 and v_l are adjacent. An induced path (or cycle) is a path (or a cycle) in which no two non-consecutive vertices are adjacent. An induced path $P = (v_1, v_2, \ldots, v_\ell)$ in G with $v_1 \neq v_\ell$ is called a *simple path* if $N_G(v_i) = \{v_{i-1}, v_{i+1}\}$ for each $2 \leq i \leq \ell - 1$. We define the neighborhood of such a path P as the set $N_G(P) = (N_G(v_1) \cup N_G(v_\ell)) \setminus V(P)$. We say that a set $X \subseteq V(G)$ is a simple path if there is an ordering of vertices in X that is a simple path. A graph is *connected* if there is a path between every pair of its vertices and it is *disconnected* otherwise. A set $S \subseteq V(G)$ is a *connected set of vertices* if $G[S]$ is connected. A *component* of a disconnected graph G is a maximal connected subgraph of G. A *cut-vertex* of a connected graph G is a vertex v such that $G - \{v\}$ is disconnected. A connected graph that has no cut-vertex is called *2-connected*. The operation of *subdividing* an edge uv results in the graph obtained by deleting uv and adding a new vertex w adjacent to both u and v. The operation of *short-circuiting* a degree two vertex v with neighbors u and w results in the graph obtained by deleting v and then adding the edge uw if it is not already present. A graph is called a *cactus* if every edge is a part of at most one cycle. Following properties of cactus are direct consequence of the definition.

Observation 1 *[12]. The following statements hold for a cactus T.*

1. *The vertices of T can be properly colored using 3 colors.*
2. *Every vertex of degree at least 3 is a cut-vertex.*
3. *The graph obtained from T by subdividing any edge is a cactus.*
4. *The graph obtained from T by short-circuiting any degree 2 vertex is a cactus.*

The *contraction* operation of an edge $e = uv$ in G results in the deletion of u and v and the addition of a new vertex w adjacent to vertices that were adjacent to either u or v. The resulting graph is denoted by G/e. Formally, $V(G/e) = V(G) \cup \{w\} \setminus \{u, v\}$ and $E(G/e) = \{xy \mid x, y \in V(G) \setminus \{u, v\}, xy \in E(G)\} \cup \{wx \mid x \in N_G(u) \cup N_G(v)\}$. For a set of edges $F \subseteq E(G)$, G/F denotes the graph obtained from G by contracting the edges in F (in an arbitrary order). It is easy to see that G/F is oblivious to the contraction sequence.

A graph G is *contractible* to a graph T, if T can be obtained from G by a sequence of edge contractions. For graphs G and T with $V(T) = \{t_1, \ldots, t_l\}$, G is said to have a *T-witness structure* $\mathcal{W}$ if $\mathcal{W}$ is a partition of $V(G)$ into l sets and there is a bijection $W : V(T) \mapsto \mathcal{W}$ such that the following properties hold.

– For each $t_i \in V(T)$, $G[W(t_i)]$ is connected.
– For a pair $t_i, t_j \in V(T)$, $t_i t_j \in E(T)$ if and only if there is an edge between a vertex in $W(t_i)$ and a vertex in $W(t_j)$ in G.

The sets $W(t_1), \ldots, W(t_l)$ in $\mathcal{W}$ are called *witness sets* or *bags*. The bags $W(t)$ which contain a single vertex are called *small bags*, while the bags with more than one vertex are called *big bags*. For the sake of brevity, we omit curly brackets while denoting a singleton set. We associate a set $F \subseteq E(G)$ with a T-witness structure $\mathcal{W}$ of G, where F is the union of the set of edges of a spanning tree of the $G[W]$ for each $W \in \mathcal{W}$. Observe that $G/F = T$ and we say that G is $|F|$-contractible to T. Note that there is a unique T-witness structure of G corresponding to a set F of edges. There are at most $|F|$ many big witness sets. Also the number of vertices which are contained in a big witness set is upper bounded by $|F| + 1$.

3   Key Properties of a Solution

In this section, we start with a simplifying assumption that let us concentrate on 2-connected graphs.

Proposition 1 *[12]. A graph is k-contractible to a cactus if and only if each of its 2-connected components is contractible to a cactus using at most k edges in total.*

Subsequently, we assume that the input graph G is 2-connected.

Observation 2 ($\star$). *For a cactus T, let $\mathcal{W}$ be a T-witness structure of 2-connected graph G. If t is a cut-vertex in T, then $|W(t)| > 1$.*

Every big witness set need not be a cut vertex in T. We now define certain structures (or subgraphs) in T with respect to witness structure $\mathcal{W}$.

Definition 1 (Internal-Cactus). *The subgraph T_I of T obtained by removing any vertex which does not lie on a path between two distinct vertices in T corresponding to big bags is called as internal-cactus of T.*

We see that vertices of G which are not contained in witness sets corresponding to vertices in internal-cactus are easy to identify. For a given cactus T and its leaf t, if t does not correspond to a big witness set then it can not be part of its internal cactus. We can say similar thing for cycles in T which have only one vertex which corresponds to one big witness set.

Definition 2 (Pendant Cycle). *A cycle in T is called as pendant cycle if there is exactly one vertex in cycle for which corresponds to a big witness set.*

If t is a unique vertex in cycle which corresponds to a big witness set, we say that pendant cycle is incident on t. To obtain an internal cactus, we need to delete all but one vertices in any pendant cycle. By Observation 2, every cut vertex in T corresponds to a big bag and hence it is a part of internal-cactus. In following observation, we bound the cardinality of neighborhood of such cut vertices in internal-cactus.

Observation 3 ($\star$). *Let C_T be the set of cut-vertices in T. The number of neighbors of C_T in internal-cactus is at most $4|C_T|$. In other words, the number of vertices in $N_T(C_T)$ that are neither leaves nor part of a pendant cycle is at most $4|C_T|$.*

We end this section with following lemma which resolves a special instance of CACTUS CONTRACTION in polynomial time.

Lemma 1 ($\star$). *If G is a 2-connected graph such that $V(G)$ can be partitioned into two simple paths P and Q in G, then we can solve the instance (G, k) of* CACTUS CONTRACTION *in polynomial time.*

4   Phase 1: The Coloring Phase

In coloring phase, we start with assigning uniformly at random one of colors $\{1, 2, 3\}$ to vertices of input graph G. Once we have obtained this coloring, we identify certain vertices of G which are contained in small witness sets. We re-color them using new colors $\{4, 5\}$ and move on to Phase 2 of algorithm to extract a solution from components of G which are colored $1, 2$ or 3.

We need notion of *compatible coloring* to argue the correctness of this coloring step. Let cactus T can be obtained from graph G by contracting edges F in G. Also, let $\mathcal{W}$ be T-witness structure of graph G. We determine whether a given coloring is compatible or not with respect to this witness structure. Informally speaking, for each big bag, a compatible coloring colors every vertex in this big bag with same color. It separates two big witness sets which shares an edge among them. If two big witness sets are connected by a path in G than the coloring gives different color to end points to this path.

Definition 3 (Compatible Coloring). *We say ϕ is compatible with $\mathcal{W}$ if the following three conditions are satisfied.*

- *For all $W(t) \in \mathcal{W}$, $W(t)$ is monochromatic.*
- *For all $t_x, t_y \in V(T)$ such that $|W(t_x)|, |W(t_y)| > 1$ and there is an edge in T between t_x and t_y, we have $\phi(W(t_x)) \neq \phi(W(t_y))$.*
- *For all $t_x, t_y \in V(T)$, such that $|W(t_x)|, |W(t_y)| > 1$ and there exists a simple path $P = (t_x, t_1, t_2, \ldots, t_q, t_y)$ in T such that $|W(t_i)| = 1$ for all $1 \leq i \leq q$, we have $\phi(W(t_x)) \neq \phi(W(t_1))$ and $\phi(W(t_y)) \neq \phi(W(t_q))$.*

We say that ϕ is compatible with solution F if ϕ is compatible with the witness structure $\mathcal{W}$ associated with F. We later argue that if (G, k) is an YES instance of CACTUS CONTRACTION than any random 3-coloring is compatible coloring with respect to an optimum solution with high probability. For this section, we assume that we are given a 3-coloring ϕ of G which is compatible with some optimum solution. Notice that we are not given the optimum solution. It is possible that same coloring can be compatible with different optimum solutions. In this section we prune coloring components and re-color them in order to move closer to obtain one of the optimum solution.

4.1   Properties of Coloring

We derive some structural properties of ϕ in G and use those properties to compute a solution. A set $X \subseteq V(G)$ is called a *colored component of ϕ*, if

X is a maximal connected set of vertices that have the same color in ϕ. Let $\mathcal{X}$ be the set of all components of ϕ. Since $\mathcal{X}$ is a T-compatible partition and contracting an edge in a cactus graph results in another cactus graph, $\mathcal{X}$ is the witness structure of some cactus. For every color component X in $\mathcal{X}$, either all vertices of X are in small bags in $\mathcal{W}$ or X contains exactly one big witness set in $W(t) \in \mathcal{W}$ and the remaining vertices $X \setminus W(t)$ are in small bags. Given a coloring ϕ, we are only interested in finding an optimum solution which is compatible with this coloring. Hence, for any two components X, Y of ϕ, no edge uv in $E(X, Y)$ is in optimum solution.

We start with simple case when a connected component X in $\mathcal{X}$ is simple path in G. Lemma 2 states that X is either one big witness set or all vertices in X are singleton sets. The proof of the lemma is based on the observation that if two adjacent bags have only one edge crossing them then this edge is not incident vertex which has degree two.

Lemma 2 ($\star$). *If colored component X in $\mathcal{X}$ is a simple path in G then either all vertices of X are in small bags or X is a big witness set in $\mathcal{W}$.*

4.2   Identifying Vertices in Pendant Cycles and Leaves

We now specify the criteria to identify vertices in G that are contained in pendant cycles in T or are leaves in T. We can not identify all such vertices in this phase.

Re-coloring I: For any colored component X in $\mathcal{X}$, if $G - X$ contains a vertex or a simple path as its connected component then recolor vertices in that connected component with color 4.

The re-coloring signifies that these vertices are part of pendant cycles or they are leaves in T. Notice that since v is not included in X and it is adjacent with X, initially vertex v had different color than X. We can say similar things for end points of path P. We argue that if vertices and simple paths in G are adjacent to only one colored component then they are either part of pendant cycles or leaves in T.

Lemma 3 ($\star$). *For a colored component X in $\mathcal{X}$, let P be a connected component of $G - X$. If P is a simple path in G whose neighborhood is contained in X then P is either a part of a pendant cycle or it is a leaf in T.*

Notice that above Lemma also holds when P contains only one vertex. For a colored component X in $\mathcal{X}$, suppose there is an isolated vertex v which is connected component of $G - X$. Since ϕ is compatible with optimum solution, all big witness sets are monochromatic. This implies v can not be part of any big witness set and remains as singleton witness set. As it can have path to at most one big witness set, it is either part of some pendant cycle in T or it is a leaf.

4.3   Identifying Vertices in Simple Paths

We now identify vertices in G that correspond to paths in T that are between two big witness sets. Recall that in simple path no internal vertex is adjacent to any vertex outside this path. A simple path is *maximal* if it is not contained in any other simple path. In other words, in *maximal simple path* every internal vertex has degree exactly two and end points have degree strictly greater than two. We color vertices which are in maximal simple path which has neighbors in two different colored components.

Re-coloring II: For any two colored component Y, Z in $\mathcal{X}$, if $G - (Y \cup Z)$ contains a vertex or a maximal simple path as its connected component then recolor vertices in that connected component with color 5.

The re-coloring signifies that these vertices are part of simple paths in internal cactus of T. As in case of Re-coloring-I, a vertex and end points of paths have different color than either Y or Z. We prove the correctness of this coloring in following Lemma. We state this lemma when P is maximal simple path but it holds for a vertex.

Lemma 4 ($\star$). *For two colored components Y, Z in $\mathcal{X}$, let P be a connected component of $G - (Y \cup Z)$. If P is a maximal simple path in G then no optimum solution contains a solution edge incident on vertices in P. Furthermore, both Y and Z contain big witness sets.*

4.4   Properties of Recoloring

By definition of compatible coloring, every colored component contains at most one big witness set. Before re-coloring, any colored component may or may not contain big witness set. In Lemma 5, we argue that after re-coloring, all colored components colored with $\{1, 2, 3\}$ must contains a big witness set. We can think of Lemma 5 as *(partial) completeness* part for Lemmas 3 and 4. In other words, in Lemma 3 (in Lemma 4) we argue that vertices in G which satisfy some criteria are contained in witness sets which are part of pendent cycles or are leaves (in simple paths) of cactus T. In Lemma 5, we claim that *all* vertices in colored component which do not contain a big witness set and are part of pendent cycles or are leaves (simple paths) of cactus T satisfies the premise of Lemma 3 (Lemma 4).

Lemma 5 ($\star$). *If a colored component X in $\mathcal{X}$ is monochromatic with color from $\{1, 2, 3\}$ after exhaustive application of two re-coloring rules then X contains a big witness set.*

5   Phase 2: Identifying Big Witness Sets

At the start of Phase 2, we have identified colored component which must contains big witness set. For a colored component X in $\mathcal{X}$, let $W(t)$ is a big witness set contained in X. Our objective in this section is to find subset X' of X which is *at least as good as* $W(t)$. Informally speaking, this means we can replace edges in spanning tree of $G[W(t)]$ by edges in spanning tree of $G[X']$ in any optimum solution F and we get another optimum solution F'. We examine what properties $W(t)$ has in graph $G[X]$. In fact, we consider a superset $\hat{X}$ of X and examine the properties of $W(t)$ with respect to graph $G[\hat{X}]$.

Let $\hat{X}$ be the superset of X which contains vertices in the connected components of $G - X$ that are either isolated vertices or a simple path in G whose neighborhood is contained in X. We now define the notion of connected core.

Definition 4 (Core). *A* core *of a graph G is a set $Z \subseteq V(G)$ such that every connected component of $G - Z$ is either an isolated vertex or a simple path whose neighborhood is contained in Z. If a core Z is a connected set in G, then we call it a* connected core *of G.*

Notice that any superset of a connected-core which induces a connected subgraph is also a connected core. We postpone discussion on how to find a connected core of given graph which contains specified vertex set and is of minimum size to Subsect. 5.1. We claim that $W(t)$ is a connected core of graph $G[\hat{X}]$.

Lemma 6 ($\star$). *For a colored component X in $\mathcal{X}$, if $W(t)$ is the big witness set contained in X then $W(t)$ is a connected core of $G[\hat{X}]$.*

We point out that it is possible that there exists a proper superset of $W(t)$ which is a connected core of $G[\hat{X}]$. In other words, every vertex in $W(t)$ has at least one of the two responsibility: it is a part of connected core of $G[X]$ or it is in $W(t)$ because of external constraints. We introduce Marking Scheme 1 to mark vertices which are in $W(t)$ because of external constraints. Once we mark vertices which are present in big witness set because of external constraints, we can find *any* connected core of minimum cardinality which contains these vertices and this connected core is *as good as* $W(t)$ for our purposes. Marking scheme is as follows.

Marking Scheme 1. *For a colored component X in $\mathcal{X}$,*

1. *If there exists y in $N(X)$ such that $\phi(y) = 5$ then mark all the vertices in $N(y) \cap X$.*
2. *For a colored component X in $\mathcal{X}$ which contains a big witness set, mark all vertices in $N(X') \cap X$.*

We now prove the soundness of this marking scheme. Lemmas 7 and 8 argue that if X contains a big witness set $W(t)$ then all the vertices marked by marking scheme are contained in $W(t)$.

Lemma 7 ($\star$). *If there exists v in $N_G(X)$ such that v is colored 5 then $N_G(v) \cap X$ is contained in a big witness set of X.*

Lemma 8 ($\star$). *Let X, Y be two colored component in $\mathcal{X}$ which contain big witness sets, say, W_X and W_Y, respectively. Then, $N(X) \cap Y \subseteq W_Y$ and $N(Y) \cap X \subseteq W_X$.*

In the following Lemma we prove *completeness* of the marking scheme. We argue that all vertices which are present in big witness set because of external constraints has been marked by Marking Scheme 1. It is sufficient to argue that if t_1 is neighbor of t in internal-cactus of T_I then all vertices in $N_G(W(t_1)) \cap X$ has been marked. Completeness of Marking Scheme 1.1 and 1.2 follows when $|W(t_1)|$ is one and strictly greater than one, respectively.

Lemma 9 ($\star$). *For a colored component X in $\mathcal{X}$ let $W(t)$ be the big witness set contained in X. If t_1 is a neighbor of t in the internal cactus T_I of T then all the vertices in $N_G(W(t_1)) \cap X$ has been marked by Marking Scheme 1.*

We now prove how this marking scheme and connected core help us to identify a set in X which is as good as $W(t)$.

Pruning Operation: For a given collection of colored component X, consider another set $\mathcal{X}'$ obtained by performing following operations. For every colored-component Y in $\mathcal{X}$ of cardinality at least 2, if a vertex in v got recolored to 4 or 5, remove Y from $\mathcal{X}$ and add $Y \setminus \{v\}$ and $\{v\}$ to $\mathcal{X}$. For a colored component X in $\mathcal{X}$ which contains a big witness set, let M_X be set of marked vertices in X by Marking Scheme 1. Let Z_X be a connected core of $G[\hat{X}]$ of minimum cardinality which contains set M_X. For every colored component X in $\mathcal{X}$, if Z_X is proper subset of X then remove X and add Z_X to $\mathcal{X}$. For every vertex v in $\hat{X} \setminus Z_X$, add a singleton set $\{v\}$ to $\mathcal{X}$.

We stop the pruning operation when no colored component is replaced in $\mathcal{X}$. Notice that this pruning operation stops in polynomial time with respect to number of vertices in graph. As final lemma in this section, we argue that if we start applying pruning operation on set of colored classes obtained from compatible coloring ϕ, we end up with a witness structure corresponding with an optimum solution. Recall that F is a minimum set of edges such that G/F is a cactus and $\mathcal{W}$ is the G/F witness structure of G. Also, ϕ is coloring of $V(G)$ which is compatible coloring with respect to $\mathcal{W}$. Set $\mathcal{X}$ is collection of colored components of ϕ.

Lemma 10 ($\star$). *Let set $\mathcal{X}'$ be obtained from $\mathcal{X}$ by exhaustive application of Pruning Operations. If F^* be a union of spanning trees of graph induced on colored component in $\mathcal{X}^*$ then G/F^* is a cactus and $|F'| = |F|$.*

5.1   Finding Connected Cores

Recall that a connected-core of a graph G is subset Z of vertices such that, $G[Z]$ is connected and each connected component of $G - Z$ is either an isolated vertex or a simple path whose both end points have neighbors in Z. Here, we present a simple branching algorithm that determines if G has a connected core of size at most k or not. We use algorithm for STEINER TREE problem as subroutine. In STEINER TREE problem, we are given a graph G and set of vertices, called *terminals*, and a positive integer ℓ. The goal is to determine whether there is a tree with at most ℓ edges that connects all the terminals.

Lemma 11 ($\star$). *There is an algorithm that given a connected graph G and a subset X of its vertices, computes a minimum connected core of G which has at most k vertices and contains X in $\mathcal{O}^*(6^k)$ time if one such exists in the graph.*

6 Putting It All Together: The Overall Algorithm

The pseudo-code of the algorithm is presented as Algorithm 6.1 and Theorem 1 formally states our result.

Algorithm 6.1: Randomized Algorithm for Cactus Contraction

Input: A 2-connected graph G and an integer k
Output: A set F of k edges in G such that G/F is a cactus

1 Generate random coloring $\phi : V(G) \to \{1,2,3\}$ and construct $\mathcal{X}$.
2 **for** *each* $X \in \mathcal{X}$ **do**
3   **if** *P is a simple path or a isolated vertex in $G - X$* **then**
4   **for** all $u \in P$: set color of u to 4

5 **for** *each pair* $X_1, X_2 \in \mathcal{X}$ **do**
6   **if** *P is a simple path or a isolated vertex in $G - (X_1 \cup X_2)$* **then**
7   **for** all $u \in P$: set color of u to 5

8 **for** *each* $X \in \mathcal{X}$ **do**
9   Apply Marking Scheme to obtain the set of marked vertices $Y_X \subseteq X$
10     $Z_X \leftarrow$ minimum connected core of $(G[\hat{X}], Y_X)$
11 Construct $\mathcal{X}^*$ from $\mathcal{X}$ and $\{Z_X \,|\, X \in \mathcal{X}\}$.
12 **if** *a spanning forest F^* of $\mathcal{X}^*$ has $\leq k$ edges* **then**
13   return F^*

14 **else**
15   return NO

Theorem 1 ($\star$). *There is an one-sided error Monte Carlo algorithm with false negatives which solves* CACTUS CONTRACTION *in time $c^k n^{\mathcal{O}(1)}$ on 2-connected graphs. It returns correct answer with constant probability.*

We apply the arguments presented in [10] to extend above theorem to solve CACTUS CONTRACTION on general graphs.

Theorem 2 ($\star$). *There is an one-sided error Monte Carlo algorithm with false negatives which solves* CACTUS CONTRACTION *in time $c^k n^{\mathcal{O}(1)}$. It returns correct answer with constant probability.*

We can derandomize our algorithms by constructing a family of coloring function, that is derived from a perfect hash family. The details of the same are deferred to the full version of paper. This leads to the following result.

Theorem 3. CACTUS CONTRACTION *can be solved in* $c^k n^{\mathcal{O}(1)}$ *time.*

Acknowledgements. We would like to thank Prof. Saket Saurabh for invaluable advice and several helpful suggestions.

References

1. Agrawal, A., Lokshtanov, D., Saurabh, S., Zehavi, M.: Split contraction: the untold story. In: Symposium on Theoretical Aspects of Computer Science, STACS (2017)
2. Asano, T., Hirata, T.: Edge-contraction problems. J. Comput. Syst. Sci. **26**(2), 197–208 (1983)
3. Brouwer, A.E., Veldman, H.J.: Contractibility and NP-completeness. J. Graph Theory **11**(1), 71–79 (1987)
4. Cai, L., Guo, C.: Contracting few edges to remove forbidden induced subgraphs. In: Gutin, G., Szeider, S. (eds.) IPEC 2013. LNCS, vol. 8246, pp. 97–109. Springer, Cham (2013). https://doi.org/10.1007/978-3-319-03898-8_10
5. Courcelle, B.: The monadic second-order logic of graphs. I. Recognizable sets of finite graphs. Inf. Comput. **85**(1), 12–75 (1990)
6. Diestel, R.: Graph Theory. Springer, Heidelberg (2006). https://doi.org/10.1007/978-3-662-53622-3
7. Golovach, P.A., van 't Hof, P., Paulusma, D.: Obtaining planarity by contracting few edges. Theoret. Comput. Sci. **476**, 38–46 (2013)
8. Golovach, P.A., Kamiński, M., Paulusma, D., Thilikos, D.M.: Increasing the minimum degree of a graph by contractions. Theoret. Comput. Sci. **481**, 74–84 (2013)
9. Guillemot, S., Marx, D.: A faster FPT algorithm for Bipartite contraction. Inf. Process. Lett. **113**(22–24), 906–912 (2013)
10. Heggernes, P., van 't Hof, P., Lévêque, B., Lokshtanov, D., Paul, C.: Contracting graphs to paths and trees. Algorithmica **68**(1), 109–132 (2014)
11. Heggernes, P., van 't Hof, P., Lokshtanov, D., Paul, C.: Obtaining a bipartite graph by contracting few edges. SIAM J. Discret. Math. **27**(4), 2143–2156 (2013)
12. Krithika, R., Misra, P., Rai, A., Tale, P.: Lossy kernels for graph contraction problems. In: 36th IARCS Annual Conference on Foundations of Software Technology and Theoretical Computer Science, FSTTCS 2016, pp. 23:1–23:14 (2016)
13. Lokshtanov, D., Misra, N., Saurabh, S.: On the hardness of eliminating small induced subgraphs by contracting edges. In: Gutin, G., Szeider, S. (eds.) IPEC 2013. LNCS, vol. 8246, pp. 243–254. Springer, Cham (2013). https://doi.org/10.1007/978-3-319-03898-8_21
14. Watanabe, T., Ae, T., Nakamura, A.: On the removal of forbidden graphs by edge-deletion or by edge-contraction. Discret. Appl. Math. **3**(2), 151–153 (1981)
15. Watanabe, T., Ae, T., Nakamura, A.: On the NP-hardness of edge-deletion and contraction problems. Discret. Appl. Math. **6**(1), 63–78 (1983)

An Approximation Framework for Bounded Facility Location Problems

Wenchang Luo[1,2(✉)], Bing Su[3], Yao Xu[2], and Guohui Lin[2,3(✉)]

[1] Faculty of Science, Ningbo University, Ningbo, China
luowenchang@163.com
[2] Department of Computing Science, University of Alberta, Edmonton, Canada
{xu2,guohui}@ualberta.ca
[3] School of Economics and Management, Xi'an Technological University,
Xi'an, China
subing684@sohu.com

Abstract. We study the bounded metric uncapacitated facility location (bUFL) problem and its two variants, the bounded fault-tolerant facility location (bFTFL) problem and the bounded fault-tolerant facility placement (bFTFP) problem. We propose a unified approximation framework built on the state-of-the-art approximation algorithms for the three unbounded counterparts, leading to a $(2.488+\epsilon)$-approximation algorithm for the bUFL problem in the Euclidean plane, a $(1.488+H(n))$-approximation algorithm for the bUFL problem, a $(1.725 + H(n))$-approximation algorithm for the bFTFL problem, and a $(1.515+H(n))$-approximation algorithm for the bFTFP problem in a general metric space. We also prove an inapproximability result for all the three bounded facility location problems in a general metric space.

Keywords: Approximation algorithm
Bounded uncapacitated facility location
Weighted dominating set in unit disk graphs · Weighted set multi-cover

1 Introduction

The facility location problem is a classical optimization problem that finds various applications in the fields of operations research and management science [20]. In the classical *uncapacitated facility location* (UFL) problem, we are given a set of facilities $\mathcal{F}$ and a set of clients $\mathcal{C}$. Between each facility $i \in \mathcal{F}$ and every client $j \in \mathcal{C}$, the non-negative distance d_{ij} represents the connection cost for the facility i serving the client j. Opening the facility $i \in \mathcal{F}$ incurs a non-negative cost f_i, and only if i is open then it may serve clients and it can serve any number of clients. The UFL problem is to find a subset of facilities that should be opened, and to assign every client to an open facility such that the total cost of opening facilities and connection between clients and open facilities is minimized. Clearly, once the open facilities are decided, each client is assigned to the nearest open facility or one such.

L. Wang and D. Zhu (Eds.): COCOON 2018, LNCS 10976, pp. 353–364, 2018.
https://doi.org/10.1007/978-3-319-94776-1_30

The UFL problem, and its variants, have been studied for decades. The UFL problem is NP-hard; Hochbaum presented a greedy approximation algorithm with a performance guarantee of $O(\log n)$ [9], where $n = |\mathcal{C}|$ is the number of clients. This $O(\log n)$-approximation is the best possible from the approximability perspective for the general UFL problem, unless NP $\subseteq$ DTIME($n^{\log \log n}$).

If for any quadruple $i, i' \in \mathcal{F}$ and $j, j' \in \mathcal{C}$, we have $d_{ij} \leq d_{ij'} + d_{i'j'} + d_{i'j}$, then we say that the distances satisfy the triangle inequality and refer the problem to as the *metric* UFL problem. In the sequel we discuss only the metric versions of the UFL problem and its variants, without explicitly pointing out "metric". Shmoys et al. presented the first constant ratio approximation algorithm for the UFL problem [23]; the algorithm is based on linear programming followed by rounding and has a performance ratio of 3.16. Afterwards, a series of improvements were made by multiple groups of researchers [1,3,5,10,11,15,18,19], with the current best known 1.488-approximation algorithm due to Li [18].

The UFL with penalties (UFLwP) is a variant of the UFL problem, where we pay a penalty c_j for not serving the client $j \in \mathcal{C}$, with the objective to minimize the total cost of opening facilities, connection costs, and penalties. Charikar et al. gave the first constant ratio approximation algorithm for the UFLwP problem, with a performance ratio 3 [4]; Jain and Vazirani presented an improved 2-approximation [12]. The current best known result is a 1.853-approximation algorithm by Xu and Xu [27].

Among other variants of the UFL problem discussed in the literature, many real life applications motivate the *bounded* UFL (bUFL) problem, in which an open facility can serve only those clients within a certain distance D. For example, when an open facility represents a fire department, it is required that the fire crew must be able to arrive at the point of accident within a pre-specified time; and thus the distance from the accident point to the fire department has to be bounded by a given threshold. The goal of the bUFL problem is to open a subset of facilities and to assign every client to an open facility within a given distance threshold D, such that the total cost of opening facilities and connection between clients and open facilities is minimized. We note that we may assume without loss of generality that every client is within the distance threshold D to at least one facility, since otherwise no feasible solution would exist. Krysta and Solis-Oba studied the bUFL problem and presented some bi-criteria approximation algorithms by relaxing the distance constraint [16]; Weng presented a $(6.853 + \varepsilon)$-approximation algorithm for the bUFL problem in the Euclidean plane, for any positive ϵ [26].

In this paper, we study the bUFL problem and we present a $(1.488 + H(n))$-approximation algorithm, where $n = |\mathcal{C}|$ is the number of clients and $H(n)$ is the n-th harmonic number; when the bUFL problem is in the Euclidean plane, we present a $(2.488 + \varepsilon)$-approximation algorithm, improving the result by Weng [26].

In some applications of the UFL problem, *fault-tolerant* solutions are required to safeguard against facility failures. Typically, every client $j \in \mathcal{C}$ has a positive integral connection requirement $r_j \in \mathbb{Z}^+$, and in a fault-tolerant solution the client j needs to be served by r_j open facilities. Two variants of fault-tolerant solutions have been studied in the literature: in the (more general) *fault-tolerant facility placement* (FTFP) problem multiple copies of a facility $i \in \mathcal{F}$ can be open, each costs f_i; while in the (more restricted) *fault-tolerant facility location* (FTFL) problem at most one facility can be open at each facility site. Both the FTFP and the FTFL problems seek to find a subset of sites to open facilities, possibly multiple facilities at the same site in FTFP, and to assign each client $j \in \mathcal{C}$ to r_j open facilities, such that the total cost of opening facilities and the connection cost is minimized. Clearly, when $r_j = 1$ for all the clients $j \in \mathcal{C}$, both FTFP and FTFL reduce to the classical UFL problem.

The FTFL problem was first studied by Jain and Vazirani [13], who presented a primal-dual algorithm achieving a performance guarantee that is logarithmic in $\max_{j \in \mathcal{C}} r_j$—the largest connection requirement. Guha et al. proposed the first constant factor approximation algorithm with an approximation ratio of 2.408 [8], which was later improved to 2.076 by Swamy and Shmoys [24]. The current best known result is a 1.725-approximation proposed by Byrka et al. [2].

The FTFP problem was first studied by Xu and Shen [28], and then by Yan and Chrobak [29] who obtained the first constant factor approximation algorithm with a performance ratio of 3.16, which was later improved to 1.575 [30]. The current best known result is a 1.515-approximation proposed by Rybicki and Byrka [22].

In this paper, we study the *bounded* FTFL (bFTFL) and the *bounded* FTFP (bFTFP) problems, that is, every client has to be served by r_j open facilities within the given distance threshold D, from the approximation algorithm perspective. Together with the bUFL problem, we propose a unified approximation framework for all three bounded problems, built on the best known approximation algorithms for the unbounded counterparts. This leads to a $(1.488 + H(n))$-approximation for the bUFL problem, a $(1.725 + H(n))$-approximation for the bFTFL problem, and a $(1.515 + H(n))$-approximation for the bFTFP problem, where $H(n)$ is the n-th harmonic number and $n = |\mathcal{C}|$ is the number of clients.

The rest of the paper is organized as follows. In Sect. 2 we consider the bUFL problem; we first present a $(2.488 + \epsilon)$-approximation algorithm for the bUFL problem in the Euclidean plane, for any positive ϵ, which improves the best result by Weng [26]; we then present the unified approximation framework, leading to a $(1.488 + H(n))$-approximation algorithm for the general metric bUFL problem, where $H(n)$ is the n-th harmonic number and $n = |\mathcal{C}|$ is the number of clients. In Sect. 3, we present a $(1.725 + H(n))$-approximation algorithm for the bFTFL problem and a $(1.515 + H(n))$-approximation algorithm for the bFTFP problem, respectively. We prove an inapproximability result for all the three bounded facility location problems in Sect. 4. Lastly, we conclude the paper in Sect. 5.

2   The bUFL Problem

We mentioned in the introduction that the current best known approximation algorithm for the (metric) UFL problem has a performance ratio of 1.488 and is due to Li [18]. In fact, the algorithm is based on a combination of the algorithm due to Byrka [1] parameterized by $\gamma \in [0, \infty)$ and the algorithm proposed by Jain et al. [11], and 1.488 is the expected ratio as γ is chosen from a probability distribution. Let y_i be the indicator variable denoting whether the facility $i \in \mathcal{F}$ is open, and x_{ij} be the indicator variable denoting whether the client $j \in \mathcal{C}$ is assigned to the facility $i \in \mathcal{F}$. The algorithm of Byrka first solves the following natural linear programming relaxation (LP1) for the UFL problem, then scales the y-variables up by γ to open facility i with probability γy_i, followed by greedily assigning each client to the closest open facility.

$$\text{minimize} \sum_{i \in \mathcal{F}} f_i y_i + \sum_{i \in \mathcal{F}} \sum_{j \in \mathcal{C}} d_{ij} x_{ij} \qquad \text{(LP1)}$$

$$\text{subject to} \qquad \sum_{i \in \mathcal{F}} x_{ij} = 1, \forall j \in \mathcal{C} \qquad (1)$$

$$y_i - x_{ij} \geq 0, \forall i \in \mathcal{F}, j \in \mathcal{C} \qquad (2)$$

$$x_{ij} \geq 0, y_i \geq 0, \qquad \forall i \in \mathcal{F}, j \in \mathcal{C} \qquad (3)$$

In (LP1), basically the constraint group (1) ensures that each client is assigned to a facility, the constraint group (2) ensures that a facility been assigned with some client must be open, and the constraint group (3) relaxes the integral constraints $x_{ij}, y_i \in \{0, 1\}$ in the UFL problem.

Since we only use the 1.488-approximation algorithm as a black-box to generate a starting point for the bUFL problem, we refer the interested readers to [18] for more technical details of the distribution of γ and the performance analysis. Let $\{\hat{x}_{ij}, \hat{y}_i\}$ denote the (integral) solution for the UFL problem provided by the 1.488-approximation algorithm, which in general is infeasible to the bUFL problem, since the following distance constraints (4) could be violated:

$$x_{ij}(D - d_{ij}) \geq 0, \ \forall i \in \mathcal{F}, j \in \mathcal{C}, \qquad (4)$$

that is, if the client $j \in \mathcal{C}$ is assigned to the facility $i \in \mathcal{F}$ then the distance d_{ij} must be no greater than D.

We next use $\{\hat{x}_{ij}, \hat{y}_i\}$ to construct a feasible solution to the bUFL problem. To this purpose, let $\mathcal{F}_1 = \{i \in \mathcal{F} \mid \hat{y}_i = 1\}$ be the subset of open facilities and $\hat{\mathcal{C}} = \{j_1, j_2, \ldots, j_q\} \subseteq \mathcal{C}$ be the set of clients each is assigned to an open facility at distance greater than D in the solution $\{\hat{x}_{ij}, \hat{y}_i\}$. It follows that each client of $\hat{\mathcal{C}}$ needs to be re-assigned in order to achieve a feasible solution.

We address first a special case where the bUFL problem is in the Euclidean plane, that is, the locations of the facilities and the clients are points in the two-dimensional Euclidean space. We will then deal with the general metric bUFL problem by presenting a unified approximation framework, which can be used to construct feasible solutions for the other two variants of the bUFL problem.

2.1   bUFL in the Euclidean Plane

Previously, Weng presented a $(6.853 + \epsilon)$-approximation for the bUFL problem in the Euclidean plane [26], for any positive ϵ. Their algorithm has two phases. In the first phase, the algorithm constructs an instance of the *weighted dominating set problem in unit disk graphs* (WDS in UDG) in which the ground set is $\mathcal{C}$ and a unit disk is centered at a facility $i \in \mathcal{F}$ with its weight f_i and its radius D; the algorithm calls the $(5 + \epsilon)$-approximation due to Dai and Yu [6] for the WDS in UDG to compute a set $\mathcal{F}'$ to cover all clients and let C_j be the connection cost for the client $j \in \mathcal{C}$. In the second phase, using the C_j as the penalty for not serving the client $j \in \mathcal{C}$, the algorithm constructs an instance of the UFLwP problem, and calls the 1.853-approximation due to Xu and Xu [27] to compute a set $\mathcal{F}''$ to serve some clients and pay penalties for the others. The final set of open facilities is $\mathcal{F}' \cup \mathcal{F}''$, and each client is assigned to its closest open facility.

Since the optimal objective function values in both the WDS in UDG and the UFLwP are lower bounds on the optimum in our bUFL problem, this two-phase algorithm is a $(6.853 + \epsilon)$-approximation.

We remark that for the WDS in UDG, Li and Jin presented a polynomial time approximation scheme (PTAS) in 2015 [17], after Weng's [26]. Therefore, the $(5 + \epsilon)$-approximation in the first phase can be replaced by a $(1+\epsilon)$-approximation due to Li and Jin; this way, Weng's algorithm becomes a $(2.853 + \epsilon)$-approximation for the bUFL problem in the Euclidean plane, for any positive ϵ.

In our algorithm, we take a different route, by keeping $\mathcal{F}_1 = \{i \in \mathcal{F} \mid \hat{y}_i = 1\}$ the set of open facilities that can serve the clients of $\mathcal{C} - \hat{\mathcal{C}}$, and to open some more facilities to serve the clients of $\hat{\mathcal{C}}$. Deciding these new open facilities is done via a reduction to the WDS in UDG, where similarly a unit disk is centered at a facility $i \in \mathcal{F}$ with its weight f_i and its radius D, but the ground set is $\hat{\mathcal{C}}$ not the whole set of clients $\mathcal{C}$. Using the PTAS due to Li and Jin [17], we find a nearly optimal subset $\mathcal{F}_2 \subseteq \mathcal{F}$ of open facilities to serve all the clients of $\hat{\mathcal{C}}$. The final set of open facilities in our algorithm is $\mathcal{F}_1 \cup \mathcal{F}_2$, and then each client is assigned to its nearest open facility. We denote our algorithm as $\mathcal{A}_1$, of which a high-level description is shown in Fig. 1.

Theorem 1. *The algorithm $\mathcal{A}_1$ is a $(2.488 + \epsilon)$-approximation algorithm for the bUFL problem in the Euclidean plane, for any $\epsilon > 0$.*

Proof. Let us fix an optimal solution OPT to the bUFL problem, let $\mathcal{F}^*$ be the set of facilities selected by the optimal solution, and let $\text{OPT}_f = \sum_{i \in \mathcal{F}^*} f_i$.

Clearly $\mathcal{F}^*$ is a feasible solution to the constructed instance of the WDS in UDG; we conclude that the total opening cost of the facilities of $\mathcal{F}_2$ is

$$F_2 = \sum_{i \in \mathcal{F}_2} f_i \le (1 + \epsilon)\text{OPT}_f,$$

for any $\epsilon > 0$.

Let OPT_c denote the connection cost in the optimal solution OPT.

For the solution $\{\hat{x}_{ij}, \hat{y}_i\}$ to the corresponding UFL problem obtained by the 1.488-approximation, let $F_1 = \sum_{i \in \mathcal{F}_1} f_i$ and C_1 be the connection cost for

---

Algorithm $\mathcal{A}_1$:

Step 1. Invoke the 1.488-approximation for the UFL problem to obtain a solution $\{\hat{x}_{ij}, \hat{y}_i\}$. Let $\mathcal{F}_1 = \{i \in \mathcal{F} \mid \hat{y}_i = 1\}$ be the subset of open facilities by this solution; let $\hat{\mathcal{C}} \subseteq \mathcal{C}$ be the subset of clients each is assigned to an open facility at distance greater than D.

Step 2. Construct an instance of WDS in UDG by setting a unit disk of radius D at each facility $i \in \mathcal{F} - \mathcal{F}_1$ and of a weight f_i, and setting $\hat{\mathcal{C}}$ as the ground set. Invoke the PTAS due to Li and Jin for WDS in UDG to find a subset $\mathcal{F}_2 \subseteq \mathcal{F}$ of unit disks covering all the points of $\hat{\mathcal{C}}$.

Step 3. Open the facilities in $\mathcal{F}_1 \cup \mathcal{F}_2$ and assign each client to its nearest open facility.

Fig. 1. A high-level description of $\mathcal{A}_1$ for the bUFL problem.

assigning all the clients in $\mathcal{C}$ to open facilities. Since the optimal solution OPT to the bUFL problem is a feasible solution to the UFL problem, we have $F_1 + C_1 \leq 1.488(\text{OPT}_f + \text{OPT}_c)$.

Re-assigning the clients in $\hat{\mathcal{C}}$ to the open facilities of $\mathcal{F}_2$ only decreases their respective connection costs to the open facilities of $\mathcal{F}_1$. Therefore, the total connection cost in the solution by the algorithm $\mathcal{A}_1$ is no greater than C_1.

Putting these together, the total cost of the open facilities and the connection cost in the solution by the algorithm $\mathcal{A}_1$ is at most

$$F_1 + F_2 + C_1 \leq 1.488(\text{OPT}_f + \text{OPT}_c) + (1 + \epsilon)\text{OPT}_f \leq (2.488 + \epsilon)(\text{OPT}_f + \text{OPT}_c).$$

That is, $\mathcal{A}_1$ is a $(2.488 + \epsilon)$-approximation for the bUFL problem.   □

2.2 bUFL in General Metrics

For the bUFL problem in general metrics, we take the same route as in $\mathcal{A}_1$, to keep $\mathcal{F}_1 = \{i \in \mathcal{F} \mid \hat{y}_i = 1\}$ the set of open facilities that can serve the clients of $\mathcal{C} - \hat{\mathcal{C}}$, and to open some more facilities to serve the clients of $\hat{\mathcal{C}}$. Deciding these new open facilities in a general metric space is done via a reduction to the *weighted set cover* problem. We construct an instance of the weighted set cover problem as follows: We set the ground set to be $\hat{\mathcal{C}}$, and for each facility $i \in \mathcal{F} - \mathcal{F}_1$, we define a set $S_i = \{j \in \hat{\mathcal{C}} \mid d_{ij} \leq D\}$ which has a weight $w(S_i) = f_i$. The goal is to find a minimum weight collection of sets to cover all the elements in the ground set, which corresponds to a minimum total cost of open facilities in $\mathcal{F} - \mathcal{F}_1$ to serve all the clients in $\hat{\mathcal{C}}$ satisfying the distance constraint. By invoking the well-known greedy $H(n)$-approximation for the weighted set cover problem [14], we obtain a subset $\mathcal{F}_2 \subseteq \mathcal{F} - \mathcal{F}_1$ of open facilities with a total opening cost no greater than $H(n)$ times of the minimum, where $n = |\mathcal{C}|$. The final set of open facilities in our algorithm is $\mathcal{F}_1 \cup \mathcal{F}_2$, and then each client is assigned to its nearest open facility. We denote our algorithm as $\mathcal{A}_2$, of which a high-level description is shown in Fig. 2.

Algorithm $\mathcal{A}_2$ (the unified approximation framework):

Step 1. Invoke the 1.488-approximation for the UFL problem to obtain a solution $\{\hat{x}_{ij}, \hat{y}_i\}$. Let $\mathcal{F}_1 = \{i \in \mathcal{F} \mid \hat{y}_i = 1\}$ be the subset of open facilities by this solution; let $\hat{\mathcal{C}} \subseteq \mathcal{C}$ be the subset of clients each is assigned to an open facility at distance greater than D.

Step 2. Construct an instance of the weighted set cover problem by defining a set of weight f_i for each facility $i \in \mathcal{F} - \mathcal{F}_1$ and setting $\hat{\mathcal{C}}$ as the ground set. Invoke the $H(n)$-approximation due to Johnson for the weighted set cover problem to find a subset $\mathcal{F}_2 \subseteq \mathcal{F}$ of open facilities that can serve all the clients of $\hat{\mathcal{C}}$.

Step 3. Open the facilities in $\mathcal{F}_1 \cup \mathcal{F}_2$ and assign each client to its nearest open facility.

Fig. 2. A high-level description of $\mathcal{A}_2$ for the bUFL problem.

Theorem 2. *The algorithm $\mathcal{A}_2$ is a $(1.488 + H(n))$-approximation algorithm for the bUFL problem in general metrics, where $H(n) = 1 + \frac{1}{2} + \frac{1}{3} + \ldots + \frac{1}{n}$ is the n-th harmonic number and $n = |\mathcal{C}|$ is the number of clients.*

Proof. Let us fix an optimal solution OPT to the bUFL problem, let $\mathcal{F}^*$ be the set of facilities selected by the optimal solution, and let $\text{OPT}_f = \sum_{i \in \mathcal{F}^*} f_i$.

Clearly $\mathcal{F}^*$ is a feasible solution to the constructed instance of the weighted set cover problem; we conclude that the total opening cost of the facilities of $\mathcal{F}_2$ is

$$F_2 = \sum_{i \in \mathcal{F}_2} f_i \leq H(n) \text{OPT}_f.$$

Let OPT_c denote the connection cost in the optimal solution OPT.

For the solution $\{\hat{x}_{ij}, \hat{y}_i\}$ to the corresponding UFL problem obtained by the 1.488-approximation, let $F_1 = \sum_{i \in \mathcal{F}_1} f_i$ and C_1 be the connection cost for assigning all the clients in $\mathcal{C}$ to open facilities. Since the optimal solution OPT to the bUFL problem is a feasible solution to the UFL problem, we have $F_1 + C_1 \leq 1.488\text{OPT}$.

Re-assigning the clients in $\hat{\mathcal{C}}$ to the open facilities of $\mathcal{F}_2$ only decreases their respective connection costs to the open facilities of $\mathcal{F}_1$. Therefore, the total connection cost in the solution by the algorithm $\mathcal{A}_2$ is no greater than C_1.

Putting these together, the total cost of the open facilities and the connection cost in the solution by the algorithm $\mathcal{A}_2$ is at most

$$F_1 + F_2 + C_1 \leq 1.488\text{OPT} + H(n)\text{OPT}_f \leq (1.488 + H(n))\text{OPT}.$$

That is, $\mathcal{A}_2$ is a $(1.488 + H(n))$-approximation for the bUFL problem. $\quad\square$

3   The bFTFL and bFTFP Problems

In this section, we present an approximation algorithm for the bFTFL problem and the bFTFP problem, respectively. We will adopt the framework of the algorithm $\mathcal{A}_2$, using the current best known approximation for the unbounded counterpart in the first phase, and reducing the remainder problem to a variant of the weighted set cover problem in the second phase, for which we invoke the current best known $H(n)$-approximation algorithm.

Recall that in both the bFTFL and the bFTFP problems, each client $j \in \mathcal{C}$ is required to be served by r_j open facilities within the distance threshold D. Correspondingly, in the *weighted set multi-cover* problem, we are given a ground set $U = \{u_1, u_2, \ldots, u_n\}$, a collection $\mathcal{S} = \{S_1, S_2, \ldots, S_m\}$ of which each set S_i is associated with a weight f_i, and the goal is to find a minimum weight sub-collection of sets such that each element $u_j \in U$ is covered by r_j sets. Including the set S_i in the sub-collection each time incurs a weight f_i, whether or not multiple times are allowed.

For the bFTFL problem, we first invoke the current best known 1.725-approximation for the FTFL problem, due to Byrka et al. [2], to obtain a solution denoted as $\{\hat{x_{ij}}, \hat{y_i}\}$, where x_{ij} is a variable recording whether there is a connection between the client $j \in \mathcal{C}$ and the facility $i \in \mathcal{F}$ and y_i is a variable recording whether the facility $i \in \mathcal{F}$ is open. We then keep $\mathcal{F}_1 = \{i \in \mathcal{F} \mid \hat{y_i} = 1\}$ the set of open facilities and let $\hat{\mathcal{C}} \subseteq \mathcal{C}$ be the subset of clients each has not yet assigned to r_j open facilities of $\mathcal{F}_1$. For each client $j \in \hat{\mathcal{C}}$, let r'_j denote the remaining connection requirement, that is, the client j still needs to be served by r'_j open facilities within the distance threshold D.

In the second phase, we construct an instance of the weighted set multi-cover problem as follows: We set the ground set to be $\hat{\mathcal{C}}$, of which each element j needs to be covered by at least r'_j sets, and for each facility $i \in \mathcal{F} - \mathcal{F}_1$, we define a set $S_i = \{j \in \hat{\mathcal{C}} \mid d_{ij} \le D\}$ which has a weight $w(S_i) = f_i$. The goal is to find a minimum weight collection of sets to cover each element in the ground set the required number of times, which corresponds to a minimum total cost of open facilities in $\mathcal{F} - \mathcal{F}_1$ to serve all the clients in $\hat{\mathcal{C}}$ satisfying the distance constraint and the required connections. We remark that in this weighted set multi-cover problem, each set S_i can be included in the solution at most once; the problem is also referred to as the *constrained* weighted set multi-cover problem and admits an $H(n)$-approximation [21,25], where $n = |\hat{\mathcal{C}}|$. We thus invoke the $H(n)$-approximation for the constrained weighted set multi-cover problem to obtain a subset $\mathcal{F}_2 \subseteq \mathcal{F} - \mathcal{F}_1$ of open facilities with a total opening cost no greater than $H(n)$ times of the minimum. The final set of open facilities in our algorithm is $\mathcal{F}_1 \cup \mathcal{F}_2$, and then each client is assigned to its r_j nearest open facilities. We denote our algorithm as $\mathcal{A}_3$, of which a high-level description is very the same as shown in Fig. 2, except that we replace "1.488-approximation for the UFL problem" by "1.725-approximation for the FTFL problem", and replace "instance of the weighted set cover problem" by "instance of the constrained weighted set multi-cover problem" (where the connection requirement of each client $j \in \mathcal{C}$ is changed from 1 to r'_j).

The following theorem can be proved very the same as the proof of Theorem 2, and we omit its details.

Theorem 3. *The algorithm $\mathcal{A}_3$ is a $(1.725 + H(n))$-approximation algorithm for the bFTFL problem in a general metric space, where $H(n) = 1 + \frac{1}{2} + \frac{1}{3} + \ldots + \frac{1}{n}$ and $n = |\mathcal{C}|$ is the number of clients.*

In the bFTFP problem, we are allowed to open multiple copies of the facility at each facility site $i \in \mathcal{F}$. We first invoke the current best known 1.515-approximation for the FTFP problem, due to Rybicki and Byrka [22], to obtain a solution denoted as $\{\hat{x_{ij}}, \hat{y_i}\}$, where x_{ij} is a variable recording the number of connections between the client $j \in \mathcal{C}$ and the facility $i \in \mathcal{F}$ and y_i is a variable recording the number of facility copies open at site $i \in \mathcal{F}$. We then keep $\mathcal{F}_1 = \{i \in \mathcal{F} \mid \hat{y_i} \geq 1\}$ the set of open facilities as well as their multiple copies, and let $\hat{\mathcal{C}} \subseteq \mathcal{C}$ be the subset of clients each has not yet assigned to r_j open facilities of $\mathcal{F}_1$. For each client $j \in \hat{\mathcal{C}}$, let r'_j denote the remaining connection requirement, that is, the client j still needs to be served by r'_j open facilities within the distance threshold D.

In the second phase, we construct an instance of the weighted set multi-cover problem as follows: We set the ground set to be $\hat{\mathcal{C}}$, of which each element j needs to be covered by at least r'_j sets, and for each facility $i \in \mathcal{F}$, we define a set $S_i = \{j \in \hat{\mathcal{C}} \mid d_{ij} \leq D\}$ which has a weight $w(S_i) = f_i$. The goal is to find a minimum weight collection of multi-sets to cover each element in the ground set the required number of times, which corresponds to a minimum total cost of open facilities in $\mathcal{F}$ to serve all the clients in $\hat{\mathcal{C}}$ satisfying the distance constraint and the required connections. We remark that in this weighted set multi-cover problem, each set S_i can be included in the solution however number of times possible. This version of the weighted set multi-cover problem also admits an $H(n)$-approximation [21,25], where $n = |\hat{\mathcal{C}}|$. We thus invoke the $H(n)$-approximation for the weighted set multi-cover problem to obtain a subset $\mathcal{F}_2 \subseteq \mathcal{F}$ of open facilities with a total opening cost no greater than $H(n)$ times of the minimum. The final set of open facilities is $\mathcal{F}_1 \cup \mathcal{F}_2$ and for each facility $i \in \mathcal{F}_1 \cap \mathcal{F}_2$ its total number of open copies is the sum of the number determined in the solution $\{\hat{x_{ij}}, \hat{y_i}\}$ to the FTFP problem and the number determined in the solution to the weighted set multi-cover problem. Lastly, each client is assigned to its r_j nearest open facilities. We denote our algorithm as $\mathcal{A}_4$, of which a high-level description is also very the same as shown in Fig. 2.

The following theorem can be proved very the same as the proof of Theorem 2, and we omit its details.

Theorem 4. *The algorithm $\mathcal{A}_4$ is a $(1.515 + H(n))$-approximation algorithm for the bFTFP problem in a general metric space, where $H(n) = 1 + \frac{1}{2} + \frac{1}{3} + \ldots + \frac{1}{n}$ and $n = |\mathcal{C}|$ is the number of clients.*

4 An Inapproximability Result

In this section, we prove a lower bound on the approximation ratios for all three bounded facility location problems in general metrics. Recall that for the bUFL, bFTFL and bFTFP problems, we have presented a $(1.488 + H(n))$-approximation, a $(1.725 + H(n))$-approximation and a $(1.515 + H(n))$-approximation, respectively, where n is the number of clients. All three performance ratios are in $\Theta(\ln n)$ since $H(n) \in \Theta(\ln n)$.

Theorem 5. *The metric bUFL problem cannot be approximated within a factor* $\rho < (1 - \epsilon) \ln n$, *for any* $\epsilon > 0$, *unless* $P = NP$.

Proof. Dinur and Steurer have proved that approximating the set cover problem within a factor of $(1 - \epsilon) \ln n$ is NP-hard, for any $\epsilon > 0$ [7].

We prove the theorem by contradiction to construct a reduction from the set cover problem to the bUFL problem, which shows that a ρ-approximation for bUFL, where $\rho < (1 - \epsilon) \ln n$, implies a $(1 - \epsilon) \ln n$-approximation for the set cover problem.

Given an instance $(U, \mathcal{S})$ of the set cover problem, in which U is the ground set and $\mathcal{S}$ is a collection of sets, we construct an instance of the bUFL problem as follows. Suppose $n = |U|$. Let $\mathcal{C} = U$ be the set of clients; for each set $S_i \in \mathcal{S}$, there is a facility i with a uniform opening cost $f_i = n^2 \ln n$; let $\mathcal{F}$ be the set of facilities. Set the distance threshold $D = 1$. If the element $u_j \in U$ belongs to a set S_i, then the distance between the client j and the facility i is $d_{ij} = \frac{2}{3}$; if the element $u_j \in U$ doesn't belong to a set S_i, then $d_{ij} = \frac{4}{3}$. One may verify that the triangle inequality is satisfied across the board.

Clearly, in any feasible solution to the bUFL problem, each client will be served by an open facility at distance $\frac{2}{3}$; therefore, the total connection cost is always $2n/3$. From the fact that every facility has the uniform opening cost of $n^2 \ln n$, we conclude that the quality of a solution depends solely on the number of open facilities. More precisely, a set cover of cardinality k one-to-one corresponds to a set of k open facilities that are able to serve all the clients satisfying the distance constraint, which being a solution to the bUFL problem has an objective function value $kn^2 \ln n + 2n/3$. Let k^* denote the number of sets in the minimum set cover. It follows that the objective function value of an optimal solution to the bUFL problem is $k^* n^2 \ln n + 2n/3$.

Next suppose there is a ρ-approximation $\mathcal{A}$ for the bUFL problem with $\rho < (1 - \epsilon) \ln n$, for some positive ϵ. Running the algorithm $\mathcal{A}$ on the constructed instance of the bUFL problem gives a solution with the objective function value less than

$$(1 - \epsilon) \ln n (k^* n^2 \ln n + 2n/3).$$

We can then estimate the number of open facilities inside the solution to be less than

$$\frac{(1 - \epsilon) \ln n (k^* n^2 \ln n + 2n/3) - 2n/3}{n^2 \ln n} = ((1 - \epsilon) \ln n) k^* + O(\frac{1}{n}).$$

This implies a $(1 - \epsilon) \ln n$-approximation for the set cover problem, a contradiction. □

Corollary 1. *The metric bFTFL and the metric bFTFP problems cannot be approximated within a factor* $\rho < (1 - \epsilon) \ln n$, *for any* $\epsilon > 0$, *unless* $P = NP$.

5   Conclusion

Motivated by the real life applications, we studied the distance bounded version of three uncapacitated facility location problems, bUFL, bFTFL, and bFTFP. We proposed a unified approximation framework in which first the current best known approximation algorithm for their unbounded counterpart [2,18,22] is invoked to obtain a solution and then to construct an instance of a variant of the weighted set (multi-)cover problem to take care of the clients not yet served, subsequently to invoke the current best known approximation algorithm for the variant of the weighted set (multi-)cover problem to open some more new facilities. This approximation framework, together with the current best known approximation algorithms for the various ingredient problems, gives rise to a $(1.488 + H(n))$-approximation algorithm for the bUFL problem, a $(1.725 + H(n))$-approximation algorithm for the bFTFL problem, and a $(1.515 + H(n))$-approximation algorithm for the bFTFP problem, in general metrics. For a special case of the bUFL problem in the Euclidean plane, we presented a $(2.488 + \epsilon)$-approximation algorithm for any positive ϵ, which improves the previous best $(2.853 + \epsilon)$-approximation due to Weng [26]. We also proved an approximability lower bound of $\ln n$ for these three bounded facility location problems.

Acknowledgements. This research is partially supported by NSERC Canada, NNSF of China (Grant No. 61672323), the K. C. Wong Magna Foundation of Ningbo University, the China Scholarship Council (Grant No. 201408330402), and the Ningbo Natural Science Foundation (Grant No. 2016A610078).

References

1. Byrka, J.: An optimal bifactor approximation algorithm for the metric uncapacitated facility location problem. In: Charikar, M., Jansen, K., Reingold, O., Rolim, J.D.P. (eds.) APPROX/RANDOM -2007. LNCS, vol. 4627, pp. 29–43. Springer, Heidelberg (2007). https://doi.org/10.1007/978-3-540-74208-1_3

2. Byrka, J., Srinivasan, A., Swamy, C.: Fault-tolerant facility location: a randomized dependent LP-rounding algorithm. In: Eisenbrand, F., Shepherd, F.B. (eds.) IPCO 2010. LNCS, vol. 6080, pp. 244–257. Springer, Heidelberg (2010). https://doi.org/10.1007/978-3-642-13036-6_19

3. Charikar, M., Guha, S.: Improved combinatorial algorithms for the facility location and k-median problems. In: FOCS 1999, pp. 378–388 (1999)

4. Charikar, M., Khuller, S., Mount, D.: Algorithms for facility location problems with outliers. In: SODA 2001, pp. 642–651 (2001)

5. Chudak, F.A., Shmoys, D.B.: Improved approximation algorithms for the uncapacitated facility location problem. SIAM J. Comput. **33**, 1–25 (2004)

6. Dai, D., Yu, C.: A $(5 + \epsilon)$-approximation algorithm for minimum weighted dominating set in unit disk graph. Theoret. Comput. Sci. **410**, 756–765 (2009)

7. Dinur, I., Steurer, D.: Analytical approach to parallel repetition. In: STOC 2014, pp. 624–633 (2014)

8. Guha, S., Meyerson, A., Munagala, K.: A constant factor approximation algorithm for the fault-tolerant facility location problem. J. Algorithms **48**, 429–440 (2003)

9. Hochbaum, D.S.: Heuristics for the fixed cost median problem. Math. Program. **22**, 148–162 (1982)

10. Jain, K., Mahdian, M., Markakis, E., Saberi, A., Vazirani, V.V.: Greedy facility location algorithms analyzed using dual fitting with factor-revealing LP. J. ACM **50**, 795–824 (2003)

11. Jain, K., Mahdian, M., Saberi, A.: A new greedy approach for facility location problems. In: STOC 2002, pp. 731–740 (2002)

12. Jain, K., Vazirani, V.V.: Approximation algorithms for metric facility location and k-median problems using the primal-dual schema and Lagrangian relaxation. J. ACM **48**, 274–296 (2001)

13. Jain, K., Vazirani, V.V.: An approximation algorithm for the fault tolerant metric facility location problem. Algorithmica **38**, 433–439 (2004)

14. Johnson, D.S.: Approximation algorithms for combinatorial problems. J. Comput. Syst. Sci. **9**, 256–278 (1974)

15. Korupolu, M.R., Plaxton, C.G., Rajaraman, R.: Analysis of a local search heuristic for facility location problems. J. Algorithms **37**, 146–188 (2000)

16. Krysta, P., Solis-Oba, R.: Approximation algorithms for bounded facility location problems. J. Comb. Optim. **5**, 233–247 (2001)

17. Li, J., Jin, Y.: A PTAS for the weighted unit disk cover problem. In: Halldórsson, M.M., Iwama, K., Kobayashi, N., Speckmann, B. (eds.) ICALP 2015. LNCS, vol. 9134, pp. 898–909. Springer, Heidelberg (2015). https://doi.org/10.1007/978-3-662-47672-7_73

18. Li, S.: A 1.488 approximation algorithm for the uncapacitated facility location problem. Inf. Comput. **222**, 45–58 (2013)

19. Mahdian, M., Ye, Y., Zhang, J.: Approximation algorithms for metric facility location problems. SIAM J. Comput. **36**, 411–432 (2006)

20. Mirchandani, P.B., Francis, R.L.: Discrete Location Theory. Wiley, New York (1990)

21. Rajagopalan, S., Vazirani, V.V.: Primal-dual RNC approximation algorithms for set cover and covering integer programs. SIAM J. Comput. **28**, 525–540 (1998)

22. Rybicki, B., Byrka, J.: Improved approximation algorithm for fault-tolerant facility placement. In: Bampis, E., Svensson, O. (eds.) WAOA 2014. LNCS, vol. 8952, pp. 59–70. Springer, Cham (2015). https://doi.org/10.1007/978-3-319-18263-6_6

23. Shmoys, D.B., Tardos, É., Aardal, K.: Approximation algorithms for facility location problems (extended abstract). In: STOC 1997, pp. 265–274 (1997)

24. Swamy, C., Shmoys, D.B.: Fault-tolerant facility location. ACM Trans. Algorithms **4**, 1–27 (2008)

25. Vazirani, V.V.: Approximation Algorithms. Springer, Berlin (2003). https://doi.org/10.1007/978-3-662-04565-7

26. Weng, K.: Approximation algorithm for uniform bounded facility location problem. J. Comb. Optim. **26**, 284–291 (2013)

27. Xu, G., Xu, J.: An improved approximation algorithm for uncapacitated facility location problem with penalties. J. Comb. Optim. **17**, 424–436 (2008)

28. Xu, S., Shen, H.: The fault-tolerant facility allocation problem. In: Dong, Y., Du, D.-Z., Ibarra, O. (eds.) ISAAC 2009. LNCS, vol. 5878, pp. 689–698. Springer, Heidelberg (2009). https://doi.org/10.1007/978-3-642-10631-6_70

29. Yan, L., Chrobak, M.: Approximation algorithms for the fault-tolerant facility placement problem. Inf. Process. Lett. **111**, 545–549 (2011)

30. Yan, L., Chrobak, M.: LP-rounding algorithms for the fault-tolerant facility placement problem. J. Discret. Algorithms **33**, 93–114 (2015)

Reconfiguration of Satisfying Assignments and Subset Sums: Easy to Find, Hard to Connect

Jean Cardinal[1], Erik D. Demaine[2], David Eppstein[3], Robert A. Hearn[4], and Andrew Winslow[5(✉)]

[1] Université libre de Bruxelles (ULB), Brussels, Belgium
`jcardin@ulb.ac.be`
[2] Massachusetts Institute of Technology, Cambridge, MA, USA
`edemaine@mit.edu`
[3] University of California, Irvine, CA, USA
`eppstein@ics.uci.edu`
[4] Portola Valley, CA, USA
`bob@hearn.to`
[5] University of Texas Rio Grande Valley, Edinburg, TX, USA
`andrew.winslow@utrgv.edu`

Abstract. We consider the computational complexity of *reconfiguration* problems, in which one is given two combinatorial configurations satisfying some constraints, and is asked to transform one into the other using elementary transformations, while satisfying the constraints at all times. Such problems appear naturally in many contexts, such as model checking, motion planning, enumeration and sampling, and recreational mathematics. We provide hardness results for problems in this family, in which the constraints and operations are particularly simple.

More precisely, we prove the PSPACE-completeness of the following decision problems:

- Given two satisfying assignments to a planar monotone instance of Not-All-Equal 3-SAT, can one assignment be transformed into the other by single variable "flips" (assignment changes), preserving satisfiability at every step?
- Given two subsets of a set S of integers with the same sum, can one subset be transformed into the other by adding or removing at most three elements of S at a time, such that the intermediate subsets also have the same sum?
- Given two points in $\{0,1\}^n$ contained in a polytope P specified by a constant number of linear inequalities, is there a path in the n-hypercube connecting the two points and contained in P?

These problems can be interpreted as reconfiguration analogues of standard problems in NP. Interestingly, the instances of the NP problems that appear as input to the reconfiguration problems in our reductions

A preprint of this paper containing three omitted proofs is available on arXiv: https://arxiv.org/abs/1805.04055.

L. Wang and D. Zhu (Eds.): COCOON 2018, LNCS 10976, pp. 365–377, 2018.
https://doi.org/10.1007/978-3-319-94776-1_31

can be shown to lie in P. In particular, the elements of S and the coefficients of the inequalities defining P can be restricted to have logarithmic bit-length.

Keywords: Boolean satisfiability · Subset sum
Combinatorial reconfiguration · PSPACE-completeness

1   Introduction

Many computational problems consist of deciding the existence of a combinatorial object subject to constraints expressible in algebraic or logical terms. We consider *reconfiguration* problems, in which one is given two objects satisfying a set of constraints, and the goal is to transform one into the other using simple reconfiguration moves such that all the constraints remain satisfied at every intermediate step. Such problems find applications in dynamic environments or reactive systems, in which solutions are required or designed to evolve, in accessibility problems in model checking, as well as in enumeration and sampling problems, in which connectivity of the search space plays a major role.

We focus on reconfiguration problems that are naturally derived from standard NP-complete problems. This line of inquiry seems to have begun with the Sliding Tokens problem, a reconfiguration version of Independent Set, by Hearn and Demaine [11], and has gained momentum with publications such as the extension of Schaefer's dichotomy to the connectivity of Boolean satisfiability due to Gopalan et al. [9], and an overview of the complexity of reconfiguration problems by Ito et al. [16]. In the canonical example of Boolean satisfiability, one is given two satisfying assignments to connect by a sequence of variable assignment flips, such that the formula remains satisfied at every step. The study of this type of question also benefits from the interest of puzzle designers and recreational mathematicians; token-sliding problems, for instance, are related to the famous 15-puzzle, popular in the late 19th century [29]. Combinatorial reconfiguration now constitutes a quickly developing field with dedicated research groups and workshops (such as the combinatorial reconfiguration workshop held in Banff in January 2017). For a more thorough survey and history of this family of problems, we refer to van den Heuvel [13].

Reconfiguration of independent sets in graphs is among the most studied problem in this vein (see [4,6,8,14] for recent results) and is relevant to our findings. In these problems, one is given a graph G and two independent sets of G of the same size k, and the goal is to transform one into the other using elementary operations, preserving independence at every step. The operations consist either of "token slides", in which a vertex in the independent set is replaced by one of its neighbors [11], or of vertex additions and removals such that the size of the independent set is either k or $k - 1$ [16]. A third, related, model is that of "token jumping", in which a vertex is replaced by another, so that the size remains unchanged [20]. In general, these reconfiguration problems are known to be PSPACE-complete.

Reconfiguration problems for graph colorings followed, and have a large dedicated body of results as well [1,3,5,7,17]. Again, many such problems are known to be PSPACE-complete. Reconfiguration problems for shortest paths [2,19], vertex covers [18], dominating sets [10], and Steiner trees [24] have also been considered.

As discussed by van den Heuvel [13], the question of the relation between the complexity of the existence problem (of a satisfying assignment, for instance) and that of the reconfiguration problem is intriguing. In many early examples, reconfiguration problems in P are obtained from existence problems that are in P, and many PSPACE-hardness proofs follows the lines of the NP-hardness proof of the corresponding satisfiability problem. In the Schaefer-type dichotomy theorem established by Gopalan et al. [9], all satisfiability problems in P yield a reconfiguration problem in P as well. In some cases, the satisfiability problem is NP-complete while the reconfiguration problem is in P. (For example, this is the case for 1-in-3 SAT, whose reconfiguration problem is trivial.) Examples in which the existence problem is in P, but the reconfiguration problem is PSPACE-complete can also be found. Prominent examples are reconfiguration of shortest paths [2] and reconfiguration of 4-colorings of bipartite and planar graphs [3]. Our results provide further examples of such a situation.

Our Results. We give hardness results for reconfiguration problems involving solutions of special families of Boolean satisfiability problems, subset sum and knapsack problems, and, more generally, 0–1 linear programming problems.

In Sect. 2, we prove that the problem of reconfiguring satisfying assignments to a planar monotone instance of Not-All-Equal 3-SAT by single variable flips is PSPACE-complete. Interestingly, the planar Not-All-Equal 3-SAT problem is in P. If we further restrict to monotone instances, the reconfiguration problem is equivalent to reconfiguration of 2-colorings of 3-uniform hypergraphs with planar vertex-edge incidence graphs.

In Sect. 3, we consider the Subset Sum reconfiguration problem, that is, reconfiguration of subsets of a set of integers with the same sum. For this, we need to be able to perform elementary moves involving three elements of the set. We show that this problem is again PSPACE-complete.

Finally, in Sect. 4, we prove the PSPACE-completeness of the problem of finding a path between two points of the hypercube that is constrained to lie within a polytope. We show that the hardness result holds even if the number of inequalities defining the polytope is $O(1)$, and the coefficients involved are polynomial.

2 Planar NAE 3-SAT Reconfiguration

In this section, we give new results on the reconfiguration problems for a variant of Boolean satisfiability.

Definition 1 (Boolean Satisfiability Reconfiguration Problem). *Given an instance of a Boolean satisfiability problem and two satisfying assignments s* *and t, does there exist a sequence of satisfying assignments $s_1, s_2, \ldots, s_k$ such that $s_1 = s$, $s_k = t$, and for all $i \in [k-1]$, s_{i+1} can be obtained from s_i by a single variable flip?*

Such problems (also referred to as the $s-t$-*connectivity* problems for Boolean satisfiability) have been considered extensively before [9, 22, 23, 26–28]. Here we investigate the complexity of the reconfiguration versions of Boolean satisfiability problems in which the *variable-clause incidence graph* is planar. The variable-clause incidence graph of a CNF formula is a bipartite graph in whose set of vertices is the union of the set of clauses and the set of variables of the formula, and a variable vertex is adjacent to a clause vertex if the variable appears in the clause, in either positive or negative form. The *planar 3-SAT problem* is the 3-SAT problem restricted to instances with a planar variable-clause incidence graph. It has long been known that planar 3-SAT is NP-complete [21].

In the *NAE 3-SAT problem*, satisfying assignments are forbidden from containing clauses in which all literals have the same value. Hence in a satisfying assignment, every clause has exactly two literals with the same value. In an instance of *Monotone NAE 3-SAT*, all literals appearing in the clauses are positive.

Monotone NAE 3-SAT is equivalent to 2-coloring 3-uniform hypergraphs, and known to be NP-complete from Schaefer's dichotomy theorem. We consider instances of *Planar NAE 3-SAT*, where the variable-clause incidence graph is planar. In 1988, Moret proved the surprising result that Planar NAE 3-SAT is in P by reducing the problem to that of finding a maximum cut in a planar graph [25]. We prove:

Theorem 1. *Planar Monotone NAE 3-SAT Reconfiguration is* PSPACE-*complete.*

It is interesting to observe that the problem is PSPACE-complete despite the satisfiability problem lying in P. The proof relies on the Nondeterministic Constraint Logic framework of Hearn and Demaine [11, 12].

Nondeterministic Constraint Logic (NCL). In nondeterministic constraint logic, a *constraint graph* is an edge- and node-weighted graph. A *configuration* of such a graph is an orientation of its edges, and an orientation is *legal* provided that the sum of the weights of edges pointing to a node is at least the weight of this node. In what follows, we will further restrict to graphs in which all node weights equal 2, and edges have weights either 1 or 2. The latter are referred to as red and blue edges, respectively. Furthermore, we only have two types of nodes: AND nodes with one blue and two red incident edges, and OR nodes with three blue incident edges. It was proved that the framework retains all of its expressive power, even under these restrictions [11]. The names of the two node types come from the interpretation of the incoming weight constraint: a configuration is legal if and only if (i) for all AND nodes, the blue edge is not outgoing unless both red edges are incoming, (ii) for all OR nodes, at least one edge is incoming.

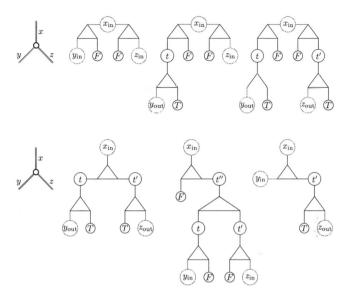

Fig. 1. Monotone NAE clauses implementing the AND (top) and OR (bottom) nodes in a constraint graph. Only one of the three versions of the gadget is used, depending on the encoding used for the edge orientations. (Color figure online)

Definition 2 (C2C Problem). *Given a constraint graph and two legal configurations C_1 and C_2, can C_2 be obtained from C_1 by flipping one edge at a time, so that all intermediate configurations are also legal?*

Theorem 2 ([11]). *The C2C problem is* PSPACE-*complete, even if the constraint graph is restricted to be planar.*

Sketch of Proof of Theorem 1. As a warmup, we first consider the known reduction from the planar C2C problem to planar 3-SAT reconfiguration. Given a planar constraint graph, we define one Boolean variable per edge. When considering an edge x incident to a node, we denote by x_{in} the literal corresponding to the orientation of x towards the node, and the opposite literal by x_{out}. For a given AND node with a blue incident edge x and two red incident edges y and z, we add the two clauses $(x_{in} \lor y_{in})$ and $(x_{in} \lor z_{in})$, forcing both y_{in} and z_{in} to be true whenever x_{in} is false. For a given OR node with three incident blue edges x, y, and z, we add the single clause $(x_{in} \lor y_{in} \lor z_{in})$. The resulting variable-clause incidence graph is planar whenever the initial constraint graph is. This reduction is due to Sarah Eisenstat,[1] and is also alluded to by Gopalan et al. [9].

The hardness proof for the planar monotone NAE 3-SAT reconfiguration problem is more involved. The AND and OR nodes are also translated into

---

[1] MIT Course 6.890, "Algorithmic Lower Bounds: Fun with Hardness Proofs" (Fall '14), Lecture 17.

monotone NAE clauses involving variables representing edges of the constraint graph, but also other additional variables. Furthermore, the translation may use different sets of clauses depending on whether the variable representing the edge corresponds to the incoming or outgoing orientation. The reduction is summarized on Fig. 1. In this figure, triangles represent monotone NAE clauses. The symbols F and T represent variables whose values have been fixed by a "rigid" gadget. The complete proof is provided in the full version of the paper on arXiv.

3   Subset Sum Reconfiguration

We now consider the reconfiguration problem for the well-known subset sum problem.

Definition 3 (Subset Sum Problem). *Given an integer x and a set of integers $S = \{a_1, a_2, \ldots, a_n\}$, does there exist a subset $A \subseteq [n]$ such that $\sum_{i \in A} a_i = x$?*

If we restrict our reconfiguration steps to involve only a single element of S, the reconfiguration problem is trivial, as no single such move can maintain the same sum. We therefore consider more general reconfiguration steps. We say that a set of integers A_1 can be *k-move reconfigured* into a second set of integers A_2 whenever the symmetric difference of A_1 and A_2 has cardinality at most k.

Definition 4 (k-move Subset Sum Reconfiguration Problem). *Given two solutions A_1 and A_2 to an instance of the subset sum problem, can A_2 be obtained by repeated k-move reconfiguration, beginning with A_1, so that all intermediate subsets are also solutions?*

The problem remains trivial for $k = 2$, since any removed element must be replaced by itself. For $k = 3$, we prove the following theorem.

Theorem 3. *The 3-move subset sum reconfiguration problem is strongly* PSPACE-*complete.*

The problem is *strongly* PSPACE-complete, meaning that it remains PSPACE-complete when the input integer set is given in unary. The corresponding instances of the subset sum problem can be solved in polynomial time using dynamic programming. This is another example of a reconfiguration problem that is PSPACE-complete, despite the underlying decision problem lying in P. We first note that the problem is contained in PSPACE. A simple proof can be found in the full version of the paper on arXiv. As for hardness, the reduction is done in two steps. First, from the Sliding Tokens problem to the Exact Cover reconfiguration problem (Lemma 1), then to the 3-move Subset Sum reconfiguration problem (Theorem 3).

Definition 5 (Token Slide Reconfiguration). *Given two independent sets I_1, I_2 of a graph $G = (V, E)$, I_1 can be reconfigured into I_2 via a token slide provided $(I_1 - I_2) \cup (I_2 - I_1) = \{v_1, v_2\}$ and $\{v_1, v_2\} \in E$.*

Observe that a token slide corresponds to changing the selection of a vertex $v_1 \in I_1$ to a neighboring vertex $v_2 \in I_2$, possible exactly when v_1 is the only vertex in I_1 among v_1, v_2, and their neighbors.

Definition 6 (Sliding Tokens Problem). *Given two independent sets I_1, I_2, can I_1 be reconfigured into I_2 via repeated token slides?*

An *exact cover* is a set cover that covers every element exactly once.

Definition 7 (Exact Cover Split and Merge Reconfiguration). *Given a set S of subsets of a set U, and two exact covers $C_1, C_2 \subseteq S$, C_1 can be reconfigured into C_2 via a split (and C_2 can be reconfigured into C_1 via a merge) provided that there exist $S_1, S_2, S_3 \subseteq S$ with $C_1 - C_2 = S_1$ and $C_2 - C_1 = \{S_2, S_3\}$.*

Since C_1, C_2 are exact covers, $S_1 = S_2 \cup S_3$ and $S_2 \cap S_3 = \varnothing$.

Definition 8 (Exact Cover Reconfiguration Problem). *Given a set S of subsets of a set U, can C_1 be reconfigured into C_2 via repeated splits and merges?*

Recall that a set S of subsets of a set U can be considered as a hypergraph $G = (U, S)$, where each element of U is a vertex and each element of S is a hyperedge. We say that a hypergraph is *k-colorable* whenever we can assign one of k colors to each vertex such that no two vertices in a hyperedge have the same color.

Lemma 1. *The exact cover reconfiguration problem is* PSPACE-*hard for instances that are 23-colorable hypergraphs.*

Proof. The proof of Theorem 23 of [11] establishes that the sliding tokens problem is PSPACE-hard on 3-regular graphs (see Sect. 3.2 of [3] for further discussion). A trivial modification of the proof suffices to prove that a *labeled* variant of the sliding tokens problem, where each token has a unique label, is also PSPACE-hard. The reduction is from this variant. The following describes an input instance of the labeled sliding tokens problem:

- $G = (V, E)$, a 3-regular graph.
- T, a set of labeled tokens.
- $p_1 : T \to V$, a function mapping each labeled token to a vertex placement in the starting configuration.
- $p_2 : T \to V$, a function mapping each labeled token to a vertex placement in the ending configuration.

Also, $I_1 = \{p_1(t) : t \in T\}$ and $I_2 = \{p_2(t) : t \in T\}$ are independent sets of size $|T| \leq |V|$.

Output U and S. The output exact cover instance has a set U consisting of two types of elements: *vertices* $v_1, v_2, \ldots, v_{|V|}$ and *tokens* $t_1, t_2, \ldots, t_{|T|}$. That is, $U = \{v_1, v_2, \ldots, v_{|V|}\} \cup \{t_1, t_2, \ldots, t_{|T|}\}$.

For each pair of adjacent vertices $v_i, v_j \in V$, the set consisting of these two vertices and their neighbors is called a *slide set*, denoted $S_{i,j}$. The output set S of subsets of U contains the following subsets for every pair of adjacent vertices v_i, v_j and token t_k:

- All subsets of $S_{i,j} - \{v_i\}$ and $S_{i,j} - \{v_j\}$.
- $\{v_i, t_k\}$ and $\{v_j, t_k\}$.
- $S_{i,j} \cup \{t_k\}$.

Output C_1 and C_2. The starting configuration C_1 is the union of $\{\{v_i\} : v_i \in V - I_1\}$ and, for every $v_i \in I_1$, a set $\{v_i, t_k\}$ with a distinct t_k. Similarly, the ending configuration C_2 is the union of $\{\{v_i\} : v_i \in V - I_2\}$ and, for every $v_i \in I_2$, a set $\{v_i, t_k\}$ with a distinct t_k.

23-colorability of $(U, \mathcal{S})$. Since G is 3-regular, G^3 has degree at most 21. So G can be 22-colored such that no two vertices of distance at most 3 (i.e. in a common slide set) have the same color. Such a coloring ensures that no pair of vertices in a common set in $\mathcal{S}$ share a color. Coloring the tokens in T a distinct (23rd) color then gives a coloring of U such that no pair of elements of a common set share the same color.

High-level Idea. The subsets containing exactly one vertex and token (e.g., $\{v_i, t_k\}$) represent the presence of the token t_k on vertex v_i. Subsets consisting of a slide set and token (e.g., $S_{i,j} \cup \{t_k\}$) represent the presence of a "mid-slide" token between v_i and v_j.

Sliding a token t_k from v_i to v_j is simulated by first merging $\{v_i, t_k\}$ and $S_{i,j} - \{v_i\}$ into $S_{i,j} \cup \{t_k\}$, and then splitting this set into $S_{i,j} - \{v_j\}$ and $\{v_j, t_k\}$. This sequence enforces the absence of tokens on neighbors of v_i and v_j, and the presence of a token on v_i or v_j, but not both. Before a merge-split sequence, additional splits and merges of token-less sets may be needed to obtain $S_{i,j} - \{v_i\}$.

Bijection Between Configurations. Call a configuration C of the output Exact Cover Reconfiguration instance *maximally split* if every C in $\mathcal{C}$ contains exactly one vertex and up to one token. The following defines a function f_{red} from token arrangements to maximally split covers:

- Each token-less vertex corresponds to a set $\{v_i\}$ in the cover.
- Each token t_k placed at v_i corresponds to a set $\{v_j, t_k\}$ in the cover.

Notice that f_{red} is a bijection and $f_{\text{red}}(p_1) = C_1$, $f_{\text{red}}(p_2) = C_2$.

Reduction Structure. The remainder of the proof is devoted to proving the following claim: a token arrangement p' is reachable from a token arrangement p if and only if $f_{\text{red}}(p')$ is reachable from $f_{\text{red}}(p)$ via splits and merges.

Both directions are proved inductively. That is, we consider only "adjacent" configurations. We also assume that the starting token arrangement $p : T \to V$ has $\{p(t) : t \in T\}$ independent.

Sliding tokens reachability $\Rightarrow$ exact cover reachability. Let p be a token arrangement that can be reconfigured into p' via a token slide from v_i to v_j. Then $f_{\text{red}}(p')$ can be reached from $f_{\text{red}}(p)$ via the following sequence of merges and splits:

1. Repeatedly merge token-less vertex sets to form $S_{i,j} - \{v_i\}$.
2. Merge $S_{i,j} - \{v_i\}$ and $\{v_i, t_k\}$ into $S_{i,j} \cup \{t_k\}$.
3. Split $S_{i,j} \cup \{t_k\}$ into $S_{i,j} - \{v_j\}$ and $\{v_j, t_k\}$.
4. Repeatedly split the token-less vertex set $S_{i,j} - \{v_j\}$ into single vertex sets.

Exact cover reachability $\Rightarrow$ *sliding tokens reachability.* For each exact cover configuration C in the output instance, at least one maximally split configuration is reachable from C via a sequence of splits. Call the set of all such configurations the *sploot set* of C, denoted $\mathrm{sploot}(C)$.

Let C, C' be maximally split configurations such that C can be reconfigured into C' and C_{inter} is the first configuration reached such that $\mathrm{sploot}(C_{\mathrm{inter}}) \neq \{C\}$. By induction, assume $C' \in \mathrm{sploot}(C_{\mathrm{inter}})$.

Since splits and token-less merges do not add elements to a sploot set, C_{inter} is obtained by merging two sets, one of which contains a token. Since the only token-containing sets that can be merged are those of the form $\{v_i, t_k\}$, C_{inter} is obtained by merging $\{v_i, t_k\}$ and $S_{i,j} - \{v_i, t_k\}$ to obtain $S_{i,j} \cup \{t_k\}$ for some v_i, v_j, and t_k. Notice that it may be the case that $S_{i,j} = S_{i',j'}$ for other pairs i', j'.

Such a merge allows two kinds of splits:

- Splitting $S_{i,j}$ into $S_{i,j} - \{v_i, t_k\}$ (to obtain the previous configuration, with sploot set $\{C\}$).
- Splitting $S_{i,j}$ into $S_{i',j'} - \{v'_j, t_k\}$, where $S_{i,j} = S_{i',j'}$ (to obtain a new configuration with sploot set $\{C'\}$, where C' is identical to C, except that C' contains $\{v'_j, t_k\}, \{v'_i\}$ instead of $\{v_i, t_k\}, \{v_j\}$).

Since $S_{i,j} - \{v_i\}, \{v_j, t_k\} \in C$, the token arrangement p with $f_{\mathrm{red}}(p) = C$ has no tokens on vertices in $S_{i,j}$ except for token t_k on v_i. Since $S_{i,j} \cup S_{i',j'} = S_{i,j}$ contains all neighbors of v_i, v_j, v'_i, v'_j, the token arrangement obtained by moving the location of t_k in p from v_i to v_j, v'_i, or v'_j is an independent set.

So all that remains is to prove that there are a sequence of slides moving t_k from v_i to v'_j via vertices in $\{v_i, v_j, v'_i, v'_j\}$. Since $S_{i,j} = S_{i',j'}$, $v'_i, v'_j \in S_{i,j}$ and so either $v_i \in \{v'_i, v'_j\}$, or there is an edge $\{v_i, v'_i\}$ or $\{v_i, v'_j\} \in E$. So t_k can slide from v_i to either v'_i or v'_j (via 0 or 1 slides), and then from v'_i or v'_j to v'_j (via 0 or 1 slides). □

We are now ready to prove the main result of this section.

Theorem 3. *The 3-move subset sum reconfiguration problem is strongly* PSPACE-*complete.*

Proof. The reduction is from the Exact Cover reconfiguration problem for instances that are 23-colorable induced hypergraphs, proved PSPACE-hard by Lemma 1. Observe that every 3-move subset sum reconfiguration is either a *merge*, where a_i and a_j are replaced by $a_i + a_j$, or a *split*, where $a_i + a_j$ is replaced by a_i and a_j. Each set split or merge will correspond to a 3-move split or merge, respectively, in the output instance.

Output Numbers and Sum. A function $f : U \to \mathbb{N}$ maps each element of the universe U of the input exact cover reconfiguration problem to a positive integer, and the numbers in the output 3-move subset sum reconfiguration instance are $\{\sum_{a \in S} f(a) : S \in \mathcal{S}\}$ and the output target sum is $\sum_{a \in U} f(a)$.

Elements of U are partitioned according to their colors $1, 2, \ldots, 23$ and (arbitrarily) labeled $a_1, a_2, \ldots, a_{|U|}$. The function f maps a color-j element a_i to $i \cdot 2^{100j\lceil \log_2(|U|)\rceil}$. In binary, this mapping consists of the binary encoding of i followed by by $100j\lceil \log_2(|U|)\rceil$ zeros.

Output Size. The output instance consists of $|\mathcal{S}|$ numbers, each between 0 and $|U| \cdot 2^{100 \cdot 23\lceil \log_2(|U|)\rceil} = O(|U|^2)$. So the output sum is $O(|U|^3)$. Thus the output instance, encoded in unary, has length $O(|\mathcal{S}||U|^2 + |U|^3)$, i.e. polynomial in the input instance.

Correctness. A reconfiguration in both the exact cover and 3-move subset sum problems involves splitting or merging elements. Thus it suffices to prove that the function f yields a one-to-one mapping $g : \mathcal{S} \to \mathbb{N}$ given by $g(S) = \sum_{a \in S} f(a)$.

Recall that the function f maps each element $a_i \in U$ to a value based upon the color of a_i. The sums of the outputs of f for all elements of all colors 1 to $j - 1$ is at most $2^{100(j-1)\lceil \log_2(|U|)\rceil} \cdot |U|^2 \leq 2^{(100j-98)\lceil \log_2(|U|)\rceil}$ while the output of f for any element of any color j or larger is at least $2^{100j\lceil \log_2(|U|)\rceil} \geq 2^{98} \cdot 2^{(100j-98)\lceil \log_2(|U|)\rceil}$.

Thus if a pair of sets $S_1, S_2 \subseteq \mathcal{S}$ have $S_1 \neq S_2$, then their color-j elements differ, this difference cannot be made up by adding or removing elements of colors 1 to $j - 1$ (values too small) or colors $j + 1$ to 23 (values too large). Thus if $S_1 \neq S_2$, then $g(S_1) \neq g(S_2)$. $\square$

4   Reconfiguration Problems and Paths in Hypercubes

The n-hypercube is the graph with vertex set $\{0, 1\}^n$ such that two vertices are adjacent whenever their coordinates differ by exactly one component. In this section, we consider the following abstraction of reconfiguration problems involving subsets.

Definition 9 (Constrained Hypercube Path). *Given two vertices s, t of the n-hypercube, both contained in a polytope $P := \{x \in \mathbb{R}^n : Ax \leq b\}$ for some $A = (a_{ij}) \in \mathbb{Z}^{d \times n}$ and $b \in \mathbb{Z}^d$, does there exist a path from s to t in the hypercube, all vertices of which lie in P?*

The constrained hypercube path problem can be seen as a reconfiguration analogue of the *0–1 integer linear programming* (0–1 ILP) satisfiability problem, which simply asks for the existence of a 0–1 point in the inside P, and is a standard NP-complete problem from Karp's list. (Note that this problem is distinct from the 0–1 ILP Reconfiguration problem defined in Ito et al. [16]: in the latter, a solution must optimize some objective function, while we are only concerned with satisfiability.)

The subset sum problem is the question of the existence of a 0–1 point in a polytope consisting of a subspace of dimension $n - 1$, hence defined by two linear constraints with the same coefficients. Similarly, the knapsack (decision) problem involves exactly two linear constraints, and the Knapsack reconfiguration problem can be cast as a special case of the constrained hypercube path problem where $d = 2$. The definitions are as follows.

Definition 10 (Knapsack Problem.) *Given integers ℓ and u and two sets of integers $S = \{a_1, a_2, \ldots, a_n\}$ and $W = \{w_1, w_2, \ldots, w_n\}$, does there exist a subset $A \subseteq [n]$ such that $\sum_{i \in A} a_i \geq \ell$ and $\sum_{i \in A} w_i \leq u$?*

Definition 11 (Knapsack Reconfiguration Problem). *Given two solutions A_1 and A_2 to an instance of the knapsack problem, can A_2 be obtained by repeated 1-move reconfiguration, beginning with A_1, so that all intermediate subsets are also solutions?*

Demaine and Ito considered the knapsack reconfiguration problem in the case where $S = W$ [15]. They proved that the problem was NP-hard, and gave an approximation algorithm for finding a reconfiguration sequence in which the intermediate steps satisfy one of the constraints only up to some multiplicative factor. Whether the knapsack reconfiguration problem is PSPACE-complete is a tantalizing open question. Characterizing the complexity of the knapsack reconfiguration problem implies understanding the complexity of the constrained hypercube path problem for bounded values of d. We do not settle the former question, but provide an answer to the latter. The proof of Theorem 4 uses techniques from the proof of Theorem 3, and can be found in full version of the paper on arXiv.

Theorem 4. *The Constrained Hypercube Path problem is PSPACE-complete, even when $d = O(1)$.*

Acknowledgements. This work was initiated at the 32nd Bellairs Winter Workshop on Computational Geometry, January 27–February 3, 2017. We thank the other participants of the workshop for a productive and positive atmosphere.

References

1. Bonamy, M., Johnson, M., Lignos, I., Patel, V., Paulusma, D.: Reconfiguration graphs for vertex colourings of chordal and chordal bipartite graphs. J. Comb. Optim. **27**(1), 132–143 (2014)
2. Bonsma, P.S.: The complexity of rerouting shortest paths. Theor. Comput. Sci. **510**, 1–12 (2013)
3. Bonsma, P.S., Cereceda, L.: Finding paths between graph colourings: PSPACE-completeness and superpolynomial distances. Theor. Comput. Sci. **410**(50), 5215–5226 (2009)
4. Bonsma, P., Kamiński, M., Wrochna, M.: Reconfiguring independent sets in claw-free graphs. In: Ravi, R., Gørtz, I.L. (eds.) SWAT 2014. LNCS, vol. 8503, pp. 86–97. Springer, Cham (2014). https://doi.org/10.1007/978-3-319-08404-6_8

5. Bonsma, P., Mouawad, A.E., Nishimura, N., Raman, V.: The complexity of bounded length graph recoloring and CSP reconfiguration. In: Cygan, M., Heggernes, P. (eds.) IPEC 2014. LNCS, vol. 8894, pp. 110–121. Springer, Cham (2014). https://doi.org/10.1007/978-3-319-13524-3_10

6. Bonsma, P.S.: Independent set reconfiguration in cographs and their generalizations. J. Graph Theory **83**(2), 164–195 (2016)

7. Cereceda, L., van den Heuvel, J., Johnson, M.: Finding paths between 3-colorings. J. Graph Theory **67**(1), 69–82 (2011)

8. Demaine, E.D., Demaine, M.L., Fox-Epstein, E., Hoang, D.A., Ito, T., Ono, H., Otachi, Y., Uehara, R., Yamada, T.: Linear-time algorithm for sliding tokens on trees. Theor. Comput. Sci. **600**, 132–142 (2015)

9. Gopalan, P., Kolaitis, P.G., Maneva, E.N., Papadimitriou, C.H.: The connectivity of Boolean satisfiability: computational and structural dichotomies. SIAM J. Comput. **38**(6), 2330–2355 (2009)

10. Haddadan, A., Ito, T., Mouawad, A.E., Nishimura, N., Ono, H., Suzuki, A., Tebbal, Y.: The complexity of dominating set reconfiguration. Theor. Comput. Sci. **651**, 37–49 (2016)

11. Hearn, R.E., Demaine, E.D.: PSPACE-completeness of sliding-block puzzles and other problems through the nondeterministic constraint logic model of computation. Theor. Comput. Sci. **343**(1–2), 72–86 (2005)

12. Hearn, R.E., Demaine, E.D.: Games, Puzzles, & Computation. A. K. Peters/CRC Press, Boca Raton (2009)

13. van den Heuvel, J.: The complexity of change. In: Blackburn, S.R., Gerke, S., Wildon, M. (eds.) Surveys in Combinatorics. London Mathematical Society Lecture Note Series (2013)

14. Hoang, D.A., Uehara, R.: Sliding tokens on a cactus. In: 27th International Symposium on Algorithms and Computation (ISAAC), pp. 37:1–37:26 (2016)

15. Ito, T., Demaine, E.D.: Approximability of the subset sum reconfiguration problem. J. Comb. Optim. **28**(3), 639–654 (2014)

16. Ito, T., Demaine, E.D., Harvey, N.J.A., Papadimitriou, C.H., Sideri, M., Uehara, R., Uno, Y.: On the complexity of reconfiguration problems. Theor. Comput. Sci. **412**, 154–165 (2011)

17. Ito, T., Kaminski, M., Demaine, E.D.: Reconfiguration of list edge-colorings in a graph. Discrete Appl. Math. **160**(15), 2199–2207 (2012)

18. Ito, T., Nooka, H., Zhou, X.: Reconfiguration of vertex covers in a graph. IEICE Trans. **99–D**(3), 598–606 (2016)

19. Kaminski, M., Medvedev, P., Milanic, M.: Shortest paths between shortest paths. Theor. Comput. Sci. **412**(39), 5205–5210 (2011)

20. Kaminski, M., Medvedev, P., Milanic, M.: Complexity of independent set reconfigurability problems. Theor. Comput. Sci. **439**, 9–15 (2012)

21. Lichtenstein, D.: Planar satisfiability and its uses. SIAM J. Comput. **11**, 329–343 (1982)

22. Makino, K., Tamaki, S., Yamamoto, M.: On the Boolean connectivity problem for Horn relations. Discrete Appl. Math. **158**(18), 2024–2030 (2010)

23. Makino, K., Tamaki, S., Yamamoto, M.: An exact algorithm for the Boolean connectivity problem for k-CNF. Theor. Comput. Sci. **412**(35), 4613–4618 (2011)

24. Mizuta, H., Ito, T., Zhou, X.: Reconfiguration of Steiner trees in an unweighted graph. In: Mäkinen, V., Puglisi, S.J., Salmela, L. (eds.) IWOCA 2016. LNCS, vol. 9843, pp. 163–175. Springer, Cham (2016). https://doi.org/10.1007/978-3-319-44543-4_13

25. Moret, B.M.E.: Planar NAE3SAT is in P. ACM SIGACT News **19**(2), 51–54 (1988)

26. Mouawad, A.E., Nishimura, N., Pathak, V., Raman, V.: Shortest reconfiguration paths in the solution space of Boolean formulas. In: 42nd International Colloquium on Automata, Languages, and Programming (ICALP), pp. 985–996 (2015)
27. Scharpfenecker, P.: On the structure of solution-graphs for Boolean formulas. In: Kosowski, A., Walukiewicz, I. (eds.) FCT 2015. LNCS, vol. 9210, pp. 118–130. Springer, Cham (2015) https://doi.org/10.1007/978-3-319-22177-9_10
28. Schwerdtfeger, K.W.: A computational trichotomy for connectivity of Boolean satisfiability. JSAT 8(3/4), 173–195 (2014)
29. Slocum, J., Sonneveld, D.: The 15 puzzle: how it drove the world crazy. The Slocum Puzzle Foundation (2006)

Solving the Gene Duplication Feasibility Problem in Linear Time

Alexey Markin, Venkata Sai Krishna Teja Vadali, and Oliver Eulenstein[✉]

Department of Computer Science, Iowa State University, Ames, IA 50011, USA
{amarkin,vvadali,oeulenst}@iastate.edu

Abstract. The gene duplication model, which has been pioneered by Goodman et al. nearly 40 years ago, is widely-used for resolving the discordance between the evolutionary history of a gene family (gene tree), and the species tree through which this family has evolved. This discordance is explained by *reconciling* the gene tree with postulated gene duplications that have occurred while the gene tree has evolved along the edges of the species tree, such that the reconciled tree can be embedded into the species tree. Today, for many gene families lower bounds on the number of gene duplications that have occurred along each edge in the species tree can be derived, for example, from known genome duplications. Here, we augment the gene duplication model by using a species tree for the reconciliation whose edges are decorated with such lower bounds, called a *(duplication) scenario*. A scenario is *feasible* for a gene family under consideration if there exists a reconciled gene tree for this family whose embedding into the species tree satisfies the lower bounds of the scenario. Non-feasibility of a credible scenario for a gene family can provide a strong indication that this family might not be well-resolved, and identifying well-resolved gene families is a challenging task in evolutionary biology. Here, we provide a linear time algorithm that decides whether a scenario is not feasible when provided a gene family.

1 Introduction

Tree reconciliation is a fundamental approach for analyzing discordant evolutionary relationships among the family histories of genes when contemplated with the histories of the species in which they have evolved. This approach has become common practice in many biological oriented research disciplines, such as molecular biology, microbiology, and biotechnology [16]. For example, gene tree reconciliation is one of the most comprehensive ways to describe the dynamics of gene family evolution [8,15], and it is also a widely-used approach to differentiate between orthologous and paralogous genes [1,2], an elementary task in the functional determination of genes [14]. Tree reconciliation can be performed using different biological models under which discordant relationships can be explained. Here we focus on the gene duplication model that has been pioneered by Goodman et al. nearly 40 years ago [12] and has laid the groundwork for tree reconciliation [7,9].

© Springer International Publishing AG, part of Springer Nature 2018
L. Wang and D. Zhu (Eds.): COCOON 2018, LNCS 10976, pp. 378–390, 2018.
https://doi.org/10.1007/978-3-319-94776-1_32

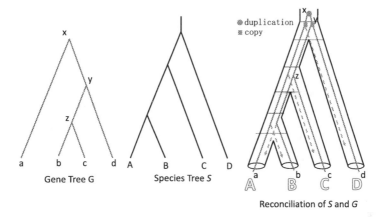

Fig. 1. An example of a gene tree - species tree reconciliation with inferred gene duplication events. The reconciliation is based on the least common ancestor mapping – it is not difficult to note that gene tree nodes x and y map to the root node of the species tree based on LCAs; hence, the duplication is at the root edge. (Color figure online)

The gene duplication model takes the following pair of rooted and full binary trees: (i) a *gene (family) tree* that represents the family history of a set of genes, and (ii) its *(corresponding) species tree* that is the evolutionary history of the species hosting these genes. Discordance between a gene tree and a species tree is often caused by complex histories of gene duplication events [16,18], but can also originate from other evolutionary events like deep coalescence or lateral transfer [17]. The gene duplication model is reconciling the gene tree with its species tree under the assumption that discordance is only caused by gene duplication events. Following the parsimony principle, the reconciliation process under the gene duplication model seeks an embedding, called *reconciliation*, of the gene tree into the species tree that infers the minimum number of duplication events. The resulting embedding is the *reconciled (gene) tree* that can reveal complex histories of gene duplication events, elucidating the evolution of function and discriminating between orthologous and paralogous genes. Figure 1 depicts an example for such a reconciliation. For a more detailed treatment of the gene duplication model, the interested reader is referred to [7,9].

A gene tree - species tree reconciliation infers a *duplication scenario* on the species tree, which can be characterized as the number of gene duplications that occurred along each edge of the species tree (note that we consider species trees to be *planted*, i.e., having an auxiliary edge connected to the root node). Formally, we define a duplication scenario as a function that maps each species tree edge to an integer that specifies the *lower bound* on the number duplications that occurred along that edge for the given gene family. While the exact number of duplications might be a more natural choice, the lower bounds are much easier to obtain in practice, for example, using histories of *whole genome duplications* [5,19]. The phylogenetic inference of gene trees has never been subjected

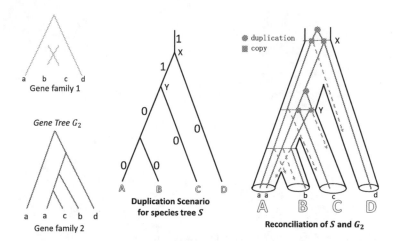

Fig. 2. An example of a duplication scenario (in the center). The duplication lower bounds are shown as edge annotations. Observe that edges going into nodes X and Y in S have duplication lower bounds of 1, which requires that a proper gene tree reconciliation will show at least 1 duplication occurring at each of these edges. On the left hand side there are given two gene families, namely, $\{a, b, c, d\}$ and $\{a, a, b, c, d\}$, which belong to the respective capital letter species. There does not exist any gene tree for the first family that will induce the required reconciliation. On the other hand, the second gene family allows to build the gene tree G_2 that satisfies the given duplication scenario. (Color figure online)

to known duplication scenarios, which in turn could lead to the inference of more biologically informed and accurate trees.

Here, we set out to reveal the space of *feasible duplication scenarios* for a specified species tree topology and a gene family. We call a duplication scenario *feasible*, if there exists a gene family tree and a corresponding reconciliation (under the Goodman et al. duplication model) that satisfies the provided lower bounds on the number of duplications. Figure 2 demonstrates an example of a duplication scenario. Note that satisfiability of the lower bounds depends on the gene family provided (namely, the number of gene copies for each leaf species). Consequently, we introduce the *Feasibility of a Duplication Scenario (FDS)* problem that decides whether a duplication scenario is feasible, and describe a linear time algorithm for this problem. In addition, the augmentation of this algorithm provides a smallest (most parsimonious) gene tree satisfying the duplication scenario. Software implementing the FDS algorithm is freely available from the web-page http://genome.cs.iastate.edu/ComBio/software.htm.

Related Work. Gene duplication is a major and frequently occurring evolutionary process that is known to cause discordance between gene trees themselves and gene trees and their corresponding species trees [18]. An efficient approach to identify such discordance is the gene duplication model from Goodman et al. [12]. This approach takes a gene tree and its corresponding species tree (both of them are rooted and full binary), and is essentially embedding the gene tree into the species tree by possibly introducing gene duplication events. The left side of Fig. 1, depicts an example of discordance between gene tree G and species tree S. To explain this discordance in the absence of phylogenetic inference error let us temporarily direct our attention only to the species tree S. The right side of the figure depicts one out of infinitely many scenarios of how a gene tree evolves within the species tree S using blue edges (solid and dashed). Initially, gene x, represented by a red circle, duplicates into two copies that are each represented by a red square. Then, each of these copies evolves along the topology of the species tree by speciation events and losses (lost gene lineages are displayed as dashed blue arrows). The resulting gene tree scenario is inferred from species tree S using exactly one duplication event. As shown in Fig. 1, gene tree G can be embedded (solid blue edges) into the gene tree scenario (solid and dashed blue edges). Thus, the gene tree scenario reconciles gene tree G by invoking one duplication, offering an explanation for the discordance between G and S. However, there can be infinitely many such scenarios, each of them invoking some number of gene duplications. Following the parsimony principle, the gene duplication model explains the incongruence with a smallest scenario, which is *unique*, invoking the minimum number of duplications that can be specified through a mapping that relates each gene in the gene tree to its *host species* [6,9,10].

The host species in the duplication model are defined based on the *least common ancestor mapping*. Formally, M is a function mapping gene tree nodes to species tree nodes, such that for each gene g that is a leaf $M(g)$ is the leaf species from which g was sampled. Further, for an internal gene node x with children y and z, $M(x)$ is defined as the *least common ancestor (LCA)* of $M(y)$ and $M(z)$ in the species tree; that is, the furthest from the root node s, which is the ancestor of both $M(y)$ and $M(z)$.

A gene in the gene tree is a *gene duplication* when it has a child with the same host species. Visually, we say that such gene duplication happened on the edge connecting the host species to its ancestor (see Fig. 1). The mapping and the gene duplications are linear time computable [20]. There is a rich literature of extensions and variants of the gene duplication model, which can, in most cases, be efficiently computed [7,9]. While computationally highly complex, probabilistic models for gene/species tree reconciliation, as well as gene sequence evolution, have also been developed [1,3].

Contribution. We present a linear time algorithm for solving the Feasibility of a Duplication Scenario (FDS) problem. The algorithm is based on dynamic programming that became possible through intrinsic properties of the gene duplication model formulated and proven in this work. In particular, a simple, but powerful property is that the *caterpillar* substructure of a gene tree is a minimum substructure allowing a gene tree to satisfy a lower bound on duplications in the given duplication scenario. Further, the majority of our analysis builds on the here introduced concept of *gene forests* that proved to be effective for establishing feasibility conditions. The algorithm utilizes the dynamic bottom-up strategy computing maximum gene forests at each step. Further, an augmentation of this algorithm can be used to produce an example gene tree that satisfies the duplication scenario. Such gene tree, as we prove, will have the property that it is smallest in size among all gene trees satisfying the duplication scenario; hence, it represents the most parsimonious way to "explain" this duplication scenario.

Applying the presented FDS algorithm, practitioners will now be able to verify the feasibility for various gene families of interest using established duplication scenarios.

2   Basics and Preliminaries

We only consider full binary rooted trees where each leaf is identified with a taxon, which we refer to as *(phylogenetic) trees*. Adhering to the standard notation, given a tree T, we denote its root, node set, edge set and leaf set by $\mathsf{Rt}(T), V(T), E(T)$, and $\mathsf{L}(T)$, respectively. The sibling and the parent of each non-root node $v \in V(T)$ are denoted by $\mathsf{Sb}(v)$ and $\mathsf{Pa}(v)$, respectively. If a tree is *planted*, then the root has a parent node as well. The set of children of each internal node $v \in V(T)$ is denoted by $\mathsf{Ch}(v)$. Further, we let $T(v)$ be a subtree of T rooted at $v \in V(T)$. A set of leaves $\mathsf{L}(T(v))$ is called a *cluster* of the node v and is denoted by C_v.

We define a partial order $\preceq_T$ on the node set $V(T)$, such that $u \preceq v$, if v is a node on the path from u to $\mathsf{Rt}(T)$. Additionally, we say $u \prec v$, if $u \preceq v$ and $u \neq v$. The *least common ancestor (LCA)* of a set of nodes $\{u_1, \ldots, u_k\}$, $\mathsf{lca}_T(u_1, \ldots, u_k)$, is the furthest from the root node, w, such that $u_i \preceq w \; \forall i \in \{1, \ldots, k\}$. A *species tree* is a planted tree with leaves referring to species names. *Gene tree*, G, is a tree that is defined by a set of species X, such that there exists a *labeling* (function) $\Lambda_G \colon \mathsf{L}(G) \to X$.

LCA Mapping. Let S be a species tree, and G be a gene tree over $\mathsf{L}(S)$. An *LCA mapping* $M \colon V(G) \to V(S)$ is a function such that for each leaf node $g \in V(G)$, $M(g) := \Lambda_G(g)$, and for each internal node g with children u and w, $M(g) := \mathsf{lca}_S(\{M(u), M(w)\})$. Observe that the mapping function M is *monotone*, implying that for $g_1 \preceq g_2$, $M(g_1) \preceq M(g_2)$.

A node g with children u and w is a *duplication node* if either $M(g) = M(u)$ or $M(g) = M(w)$. For a species tree node $s \in V(S)$, $\boldsymbol{\xi(G, s)}$ denotes the number of duplication nodes $g \in V(G)$, such that $M(g) = s$.

Duplication Scenario. Given a species tree S, a *duplication scenario* (described in the introduction) is defined by a function $\delta \colon V(S) \to \mathbb{N}_0$. We say that a gene tree G over $\mathsf{L}(S)$ *satisfies* a duplication scenario $\langle S, \delta \rangle$ if $\forall s \in V(S) : \xi(G, s) \geq \delta(s)$. Note that, while in the introduction the duplication scenario function was defined on edges of a species tree, here for later convenience we define it on the nodes of a species tree (which is identical, since each node uniquely defines its ancestral edge, $(\mathsf{Pa}(v), v)$, in planted trees).

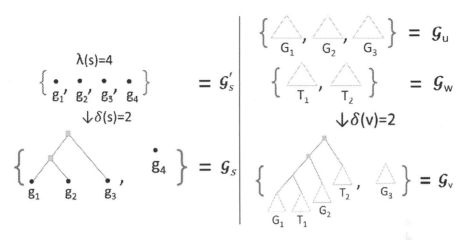

Fig. 3. *Left*: an example of a species leaf s with four gene copies in the provided gene family (i.e., four genes within the family are hosted by s). $\mathcal{G}'_s$ represents the maximum gene forest (consisting of each gene copy individually) if we are not taking the duplication scenario into the account. Gene forest $\mathcal{G}_s$ represents a maximum gene forest, when the duplication scenario puts the lower bound $\delta(s) = 2$ on the edge incident to s. Yellow squares indicate the duplication nodes. *Right*: an example of a gene forest construction for an internal species node v with children u and w. $\mathcal{G}_v$ is a maximum gene forest for the node v constructed using $\mathcal{G}_u$ and $\mathcal{G}_w$, representing maximum gene forests of nodes u and w respectively. Observe that the tree satisfying $\delta(v) = 2$ was assembled from subtrees as a caterpillar (assembly of a profile $\{G_1, T_1, G_2, T_2\}$). (Color figure online)

3   Feasibility of Duplication Scenarios

In this section, we analyze the problems of the feasibility of duplication scenarios as motivated in the introduction.

3.1   General Feasibility

Given a duplication scenario and a gene family, we would like to know whether there exists a gene tree for the family whose reconciliation is satisfying the scenario, i.e., the scenario is feasible for the gene family.

A gene family is characterized by the number of gene copies for each extant species. That is, assume, we are given a species tree, S, and for each leaf-species we know the number of gene copies, given by a function $\lambda \colon \mathsf{L}(S) \to \mathbb{N}$. We say that a gene tree G *satisfies* $\langle S, \delta, \lambda \rangle$ if G satisfies $\langle S, \delta \rangle$, and G contains at most $\lambda(s)$ taxa labeled with s for all $s \in \mathsf{L}(S)$.

Problem FDS. Feasibility of a Duplication Scenario
Instance: Duplication scenario with a gene copy function $\langle S, \delta, \lambda \rangle$
Question: Does there exist a gene tree G over $\mathsf{L}(S)$, such that G satisfies $\langle S, \delta, \lambda \rangle$

Informal Solution Description. We consider a bottom up construction of a gene tree G satisfying $\langle S, \delta, \lambda \rangle$ (if one exists). At each node of the tree S we maintain certain tree structures (parts of future tree G) that jointly satisfy all the duplication lower bounds, given by δ, below that node. These structures can be treated as building blocks (subtrees) for G. The algorithm seeks to maximize the number of such building blocks at each node in order to supply them further to the parent node and so on. These building blocks are later formally defined as *gene forests*. The maximization is of importance, since in case that here are no available tree structures at the root of S, then G does not exist and the scenario is not feasible.

At the leaf level of species tree S the number of building blocks available is simply the number of gene copies for a species. Indeed, if a species taxon s requires $\delta(s) > 0$ duplications mapped into that node, then we need to use the available building blocks (gene copies) to generate that mapping. Similarly, for intermediate nodes. Note that maximizing the number of building blocks at each node entails minimizing the number of blocks needed to satisfy the duplications lower bound at that node. An important observation here is that if at node v the lower bound is $\delta(v) > 0$, then to create a structure (new block) satisfying that duplication count at least $\delta(v) + 1$ blocks are needed, when v is a leaf, and $\delta(v) + 2$ blocks are required otherwise. Figure 3 illustrates that observation.

Applying these ideas, Algorithm 1 below checks feasibility of a given duplication scenario.

3.2   Proof of Correctness of Algorithm 1

As was mentioned above, we introduce the concept of *gene forests*. We define $\mathcal{G}$ to be a gene forest *over* $\langle S, \lambda \rangle$ if the following properties hold.

(F1) $\mathcal{G}$ is a set of phylogenetic trees over $\mathsf{L}(S)$.

Algorithm 1. Feasibility of Duplication Scenario $\langle S, \delta, \lambda \rangle$

```
 1: function MaxForestSize(Node v)
 2:     if v is a leaf then
 3:         if δ(v) > 0 then  // The scenario requires > 0 duplications mapped to v
 4:             return max(0, λ(v) − (δ(v) + 1) + 1)
 5:         else return λ(v)
 6:         end if
 7:     else  // v has two children
 8:         l := MaxForestSize(Ch(v).left);   r := MaxForestSize(Ch(v).right)
 9:         if l > 0 and r > 0 then
10:             if δ(v) > 0 then  return max(0, l + r − (δ(v) + 2) + 1)
11:             else return l + r
12:             end if
13:         end if
14:         return 0  // either l or r is 0
15:     end if
16: end function
17: return (MaxForestSize(Rt(S)) > 0)  // returns TRUE if the scenario is feasible
```

(F2) Let $\Lambda_G^{-1}(s)$ be a set of leaves in G labeled by s, then $\sum_{G \in \mathcal{G}} |\Lambda_G^{-1}(s)| \leq \lambda(s)$.

(F3) $\bigcup_{G \in \mathcal{G}} \Lambda_G(\mathsf{L}(G)) = \mathsf{L}(S)$.

We say that a gene forest $\mathcal{G}$ *satisfies* $\langle S, \delta, \lambda \rangle$ if $\mathcal{G}$ is over $\langle S, \lambda \rangle$ and for all $s \in V(S)$ we have $\sum_{G \in \mathcal{G}} \xi(G, s) \geq \delta(s)$. Further, given a gene forest $\mathcal{G}$, we define $\overline{\mathcal{G}}$ to be a gene tree obtained by assembling the trees in $\mathcal{G}$ into a single tree. While there could be many ways to assemble the trees from $\mathcal{G}$, we are interested in a *caterpillar* structure that will be used later in the analysis.

Definition 1. *A gene tree G_C represents a **caterpillar assembly** of a profile of trees $G_1, \ldots, G_k$ if it is obtained as follows. First, set $G_C := G_1$, then*

> **for** $i = 2, \ldots, k$ **do**
> > *join the trees G_C and G_i by introducing a new root node v_i*
> > *and attaching subtrees G_C and G_i to v_i as children.*
> > *Let G_C denote the resulting tree.*
> **end for**

Given a duplication scenario, $\langle S, \delta, \lambda \rangle$, and a node $v \in V(S)$, we say that forest $\mathcal{G}_v$ satisfies $\langle S, \delta, \lambda \rangle|_v$ implying that $\mathcal{G}_v$ satisfies the duplication-scenario restricted to subtree $S(v)$, i.e., $\langle S(v), \delta|_{V(S(v))}, \lambda|_{C_v} \rangle$. Let $\alpha(v)$ denote the *maximum* size of a gene forest (in terms of a number of trees) that satisfies $\langle S, \delta, \lambda \rangle|_v$. The following observation then explains Line 17 of Algorithm 1.

Observation 1 (Feasibility for a given gene family). *An instance $\langle S, \delta, \lambda \rangle$ of FDS is a yes-instance if and only if $\alpha(\mathsf{Rt}(S)) > 0$.*

Observation 2 and Lemma 1 summarize the core properties needed for the proof of correctness. Note that Observation 2 was informally described above.

Observation 2. *Consider $v \in V(S)$ such that $\delta(v) > 0$.*

(I) If v is a leaf, then at least $\delta(v)+1$ gene copies for v are required to construct a gene tree with $\delta(v)$ duplication nodes mapping into v. Note that any binary tree with $\delta(v)+1$ leaves mapping into v will induce exactly $\delta(v)$ duplications on v (for example, a caterpillar assembly of $\delta(v) + 1$ gene copies – see Fig. 3).

(II) If v has children u and w, then at least $\delta(v) + 2$ gene trees from maximum gene forests of u and w are required to construct a gene tree with $\delta(v)$ duplication nodes mapping into v. Further, it is always possible to use exactly $\delta(v) + 2$ gene trees via a caterpillar assembly (see Fig. 3).

Lemma 1. *Let $\mathcal{G}_v$ be a maximum gene forest satisfying $\langle S, \delta, \lambda \rangle|_v$ for some $v \in V(S)$. Then*

(MF4) If $\mathcal{G}_v$ contains a tree G with a duplication vertex mapping into v, then no other tree in the forest can have a node mapping into v. That is, all duplications for a specific node are localized within the same tree.

(MF5) *For $v \notin L(V)$, if $\delta(v) = 0$, then for each tree $G \in \mathcal{G}_v$, the root of G maps below v. That is, $M(\mathsf{Rt}(G)) \prec v$.*

Proof.

(MF4) Let $\mathcal{G}_v$ be a maximum gene forest satisfying $\langle S, \delta, \lambda \rangle|_v$ and G be a tree in that forest with a duplication node mapping into v. For the purpose of contradiction, assume that $\mathcal{G}_v$ also contains another tree G' with a node mapping into v. Let $\mathcal{G}'$ be a forest consisting of subtrees of G' obtained by removing all internal nodes mapping into v from G' (this will split G' into at least two subtrees). If G' induced k duplication nodes on v (k could be 0), then, by Observation 2, $\mathcal{G}'$ will contain at least $k + 2$ subtrees (or $k + 1$ if v is a leaf). Let us enumerate any $k + 1$ trees in $\mathcal{G}'$ as $G'_1, G'_2, \ldots, G'_k, G'_{k+1}$. Consider now an augmentation of G, tree G_a, obtained by a caterpillar assembly of a profile $(G, G'_1, G'_2, \ldots, G'_k)$. Note that G_a induces k more duplications onto v than G. Hence, the gene forest $\mathcal{G}_v - \{G\} \cup \{G_a, G'_{k+1}\}$ satisfies the duplication scenario $\langle S, \delta, \lambda \rangle|_v$ and is of size larger than $\mathcal{G}_v$ – contradiction.

(MF5) Assume (for contradiction) a maximum forest $\mathcal{G}_v$ contains a tree G, such that $M(\mathsf{Rt}(G)) = v$. Removing the root of G will split it into two subtrees and increase the size of the gene forest by one (we denote the new gene forest by $\mathcal{G}'_v$). Since $\delta(v) = 0$, this operation will maintain that $\mathcal{G}'_v$ satisfies $\langle S, \delta, \lambda \rangle|_v$. Hence, $\mathcal{G}_v$ is not maximum – contradiction.

Lemma 2 then proves the correctness of Algorithm 1 (the lemma's proof is omitted for brevity).

Lemma 2. *Function* MAXFORESTSIZE *from Algorithm 1 given a node $v \in V(S)$ returns $\alpha(v)$.*

3.3   Gene Trees for Feasible Scenarios

Algorithm 1 is designed to solve the feasibility problem. However, in addition, this algorithm can be modified to construct an example gene tree satisfying the given duplication scenario (if one exists). This modification would require maintaining the maximum gene forests themselves, instead of only keeping track of the number of trees in maximum gene forests. For species tree nodes v with $\delta(v) > 0$ the algorithm will need to join a subset of the available subtrees in a caterpillar that would satisfy that lower bound on duplications.

At the root of the species tree the trees in the maximum forest (if it is non-empty) should be joined together to produce an example gene tree satisfying the duplication scenario.

Time Complexity. Let m be the size of the gene family, i.e., $m = \sum_{s \in L(S)} \lambda(s)$. Then the complexity of constructing the gene tree as outlined above is $\Theta(m)$. Representing gene forests as linked lists allows us to join two forests in $O(1)$ time. Further, when assembling trees in a caterpillar, the linked lists will allow to use $O(1)$ time per each new node created. Hence, the overall time complexity is bounded by the size of the gene tree, which is $\Theta(m)$.

Minimum Gene Trees. Such gene tree building procedure, while of potential value on its own, can be further altered to produce a *smallest* gene tree satisfying the given duplication scenario. Such minimality can be achieved by not employing some of the gene copies present in the family, if it is not necessary.

For a node $v \in V(S) \setminus L(S)$ with $\delta(v) > 0$ and children u and w, let $\mathcal{G}'_v$ denote the forest of available gene trees at v; that is, the union of the maximum gene forests of the children of v. We will call a gene tree *trivial*, if it contains exactly one node; i.e., it represents a single gene copy.

The original algorithm will proceed by assembling $\delta(v) + 2$ trees from $\mathcal{G}'_v$ in a caterpillar structure. However, since we are interested in using as few gene copies as possible, we would like to use as few trivial trees for the caterpillar assembly as possible. Formally, consider all profiles of trees of the type $\{G_1, G_2, \ldots, G_{\delta(v)+2}\}$, such that $G_1 \in \mathcal{G}_u, G_2 \in \mathcal{G}_w, G_3, \ldots, G_{\delta(v)+2} \in \mathcal{G}_u \cup \mathcal{G}_w$ (where $\mathcal{G}_u$ and $\mathcal{G}_w$ are maximum forests for u and w, respectively). Let $P(\mathcal{G}_u, \mathcal{G}_w)$ be such a profile with the *minimum number of trivial trees*. Then let the algorithm for the minimum gene tree construction will use such a profile for the caterpillar assembly.

This constitutes a greedy strategy for the minimum gene tree construction: at each step use the minimum number of trivial trees for the caterpillar assembly. Then let $\mathcal{G}_p$ be a maximum gene forest satisfying $\langle S, \delta, \lambda \rangle$ constructed that way.

Theorem 1. *A minimum (most parsimonious) gene tree is obtained by removing all the trivial trees from $\mathcal{G}_p$ and joining the rest together (e.g., as a caterpillar). The time complexity for constructing a minimum gene tree is $O(m)$. That is, Algorithm 2 constructs a minimum gene tree satisfying $\langle S, \delta, \lambda \rangle$ in $O(m)$.*

Proof. The proof of correctness is omitted for brevity. Here we argue the time-complexity. The algorithm can be implemented efficiently by representing gene forests as two separate linked lists: one list for trivial trees and the other for non-trivial trees (see Algorithm 2). Then joining two forests (as needed for internal nodes v with $\delta(v) = 0$) encompasses joining two pairs of linked lists, which can be done in constant time (by maintaining a link to the last element of each list). Further, for internal nodes v with $\delta(v) > 0$ and children u and w, the algorithm has to construct a caterpillar assembly of a minimum profile $P(\mathcal{G}_u, \mathcal{G}_w)$. Let $\mathcal{G}_u^t$ and $\mathcal{G}_u^{nt}$ denote the linked lists containing trivial and non-trivial trees of $\mathcal{G}_u$ respectively (similarly, for w). Then $P(\mathcal{G}_u, \mathcal{G}_w)$ can be obtained in $O(\delta(v))$ time as demonstrated in function MTRIVFORESTSPECIAL of Algorithm 2.

The caterpillar assembly of the profile also takes $O(\delta(v))$ time and it produces $\delta(v)$ new nodes contributing to the resulting gene tree. Hence, the algorithm spends constant time per each node created. Finally, observe that for each leaf node s of the species tree the algorithm spends $O(\lambda(s))$ time. Hence, the total time spent for all leaves is $O(m)$. Overall, the time complexity bounded by $O(m)$.

Algorithm 2. Minimum Gene Tree Satisfying $\langle S, \delta, \lambda \rangle$

1: **function** MTRIVFORESTSPECIAL(Internal node v with $\delta(v) > 0$)
2:    $\mathcal{G}_u^t, \mathcal{G}_u^{nt} :=$MTRIVFOREST(Ch($v$).$left$); $\mathcal{G}_w^t, \mathcal{G}_w^{nt} :=$MTRIVFOREST(Ch($v$).$right$)
3:    $P = ()$ // *empty linked list representing* $P(\mathcal{G}_u, \mathcal{G}_w)$
4:    **if** $\mathcal{G}_u^{nt}.size > 0$ **then**
5:        $P.add(\mathcal{G}_u^{nt}.pop())$ // *remove the head of* $\mathcal{G}_u^{nt}$ *and add it to P.*
6:    **else** $P.add(\mathcal{G}_u^t.pop())$ // *assume .add() adds an element to the end of the list*
7:    **end if**
8:    **if** $\mathcal{G}_w^{nt}.size > 0$ **then** $P.add(\mathcal{G}_w^{nt}.pop())$
9:    **else** $P.add(\mathcal{G}_w^t.pop())$
10:    **end if**
11:    $\mathcal{G}_v^t = join(\mathcal{G}_u^t, \mathcal{G}_w^t); \mathcal{G}_v^{nt} = join(\mathcal{G}_u^{nt}, \mathcal{G}_w^{nt})$
12:    **while** $P.size < \delta(v) + 2$ **do** // *use as many non-trivial trees as possible*
13:        **if** $\mathcal{G}_v^{nt}.size > 0$ **then** $P.add(\mathcal{G}_v^{nt}.pop())$
14:        **else** $P.add(\mathcal{G}_v^t.pop())$
15:        **end if**
16:    **end while**
17:    $\mathcal{G}_v^{nt}.add(\overline{P})$ // *add a caterpillar assembly of P to non-trivial trees of v*
18:    **return** $\mathcal{G}_v^t, \mathcal{G}_v^{nt}$
19: **end function**
20: **function** MTRIVFOREST(Node v)
21:    **if** v is a leaf **then**
22:        $\mathcal{G}_v^t = (g_i : \forall i \in [\lambda(v)]); \mathcal{G}_v^{nt} = ();$
23:        **if** $\delta(v) > 0$ **then** $P :=$ first $\delta(v) + 1$ trees from $\mathcal{G}_v^t; \mathcal{G}_v^{nt} := (\overline{P}); \mathcal{G}_v^t := \mathcal{G}_v^t \setminus P;$
24:        **end if**
25:        **return** $\mathcal{G}_v^t, \mathcal{G}_v^{nt}$
26:    **else** // *v has two children*
27:        **if** $\delta(v) > 0$ **then return** MTRIVFORESTSPECIAL(v)
28:        **else return** joined trivial and non-trivial lists from
29:            MTRIVFOREST(Ch(v).$left$) and MTRIVFOREST(Ch(v).$right$)
30:        **end if**
31:    **end if**
32: **end function**
33: Call Algorithm 1 to verify that a tree exists. If exists:
34: $\mathcal{G}^t, \mathcal{G}^{nt} =$ MTRIVFOREST(Rt(S)); **return** caterpillar assembly of $\mathcal{G}^{nt}$.

---

4   Conclusion and Discussion

Gene trees play a crucial role in the inference of species trees and networks, in the systematic analysis of protein function, and other related areas [4,13, 15]. Refining the credibility of gene trees is thereof one of the central topics in phylogenetics for many years.

Here we propose a novel framework, where the evidence of gene duplications collected throughout an abundance of biological studies can be utilized to improve on the accuracy of gene trees. In this framework we define a duplication scenario as an augmentation of the species tree with localized evidence of duplication events, and introduce a linear time algorithm for determining whether a duplication scenario is feasible for a particular gene family; that is, whether there exists an evolutionary history of the gene family whose reconciliation is in agreement with the postulated duplication events in the species tree. In addition, following the phylogenetic parsimony paradigm [11], this algorithm can construct a smallest gene tree that will satisfy the given duplication scenario.

The presented work is laying the foundation for practitioners to assess, aggregate, and study various duplication scenarios that can be inferred from the existing studies of gene families and their evolution. Our algorithm has the ability to support a much broader range of applications beyond the feasibility question, e.g., pointing out where the additional (lost) gene lineages might have existed for duplication scenarios failing the most parsimonious duplication model.

References

1. Akerborg, O., Sennblad, B., Arvestad, L., Lagergren, J.: Simultaneous Bayesian gene tree reconstruction and reconciliation analysis. Proc. Natl. Acad. Sci. USA **106**(14), 5714 5719 (2009)
2. Altenhoff, A.M., Dessimoz, C.: Inferring Orthology and Paralogy, pp. 259–279. Humana Press, Totowa (2012)
3. Arvestad, L., Berglund, A.C., Lagergren, J., Sennblad, B.: Bayesian gene/species tree reconciliation and orthology analysis using MCMC. Bioinformatics **19**(suppl1), 7–15 (2003)
4. Bininda-Emonds, O.R. (ed.): Phylogenetic Supertrees: Combining Information to Reveal the Tree of Life. Computational Biology, vol. 4. Springer, Heidelberg (2004). https://doi.org/10.1007/978-1-4020-2330-9
5. Blanc, G., Wolfe, K.H.: Widespread paleopolyploidy in model plant species inferred from age distributions of duplicate genes. Plant Cell **16**(7), 1667–1678 (2004)
6. Bonizzoni, P., Della Vedova, G., Dondi, R.: Reconciling a gene tree to a species tree under the duplication cost model. Theor. Comp. Sci. **347**, 36–53 (2005)
7. Chauve, C., El-Mabrouk, N., Guéguen, L., Semeria, M., Tannier, E.: Duplication, Rearrangement and Reconciliation: A Follow-Up 13 Years Later. In: Chauve, C., El-Mabrouk, N., Tannier, E. (eds.) Models and Algorithms for Genome Evolution. Computational Biology, vol. 19, pp. 47–62. Springer, London (2013). https://doi.org/10.1007/978-1-4471-5298-9_4
8. Chen, K., Durand, D., Farach-Colton, M.: Notung: a program for dating gene duplications and optimizing gene family trees. J. Comput. Biol. **7**(3–4), 429–447 (2000)
9. Eulenstein, O., Huzurbazar, S., Liberles, D.: Reconciling phylogenetic trees. In: Evolution After Gene Duplication. John Wiley (2010)
10. Eulenstein, O.: Vorhersage von Genduplikationen und deren Entwicklung in der Evolution. Ph.D. thesis, Rheinische Friedrich-Wilhelms-Universität Bonn, Bonn, Germany (1998)
11. Felsenstein, J.: Inferring Phylogenies. Sinauer Associates, Inc., Sunderland (2004)
12. Goodman, M., Czelusniak, J., Moore, G., Romero-Herrera, A., Matsuda, G.: Fitting the gene lineage into its species lineage, a parsimony strategy illustrated by cladograms constructed from globin sequences. Syst. Zool. **28**(2), 132–163 (1979)
13. Huson, D.H., Scornavacca, C.: A survey of combinatorial methods for phylogenetic networks. Genome Biol. Evol. **3**, 23–35 (2011)
14. Ihara, K., Umemura, T., Katagiri, I., Kitajima-Ihara, T., Sugiyama, Y., Kimura, Y., Mukohata, Y.: Evolution of the archaeal rhodopsins: evolution rate changes by gene duplication and functional differentiation. J. Mol. Biol. **285**(1), 163–74 (1999)

15. Kamneva, O.K., Knight, S.J., Liberles, D.A., Ward, N.L.: Analysis of genome content evolution in PVC bacterial super-phylum: assessment of candidate genes associated with cellular organization and lifestyle. Genome Biol. Evol. **4**(12), 1375–1390 (2012)
16. Kamneva, O.K., Ward, N.L.: Reconciliation approaches to determining HGT, duplications, and losses in gene trees, Chap. 9. In: Michael Goodfellow, I.S., Chun, J. (eds.) New Approaches to Prokaryotic Systematics, Methods in Microbiology, vol. 41, pp. 183–199. Academic Press (2014)
17. Maddison, W.P.: Gene trees in species trees. Syst. Biol. **46**, 523–536 (1997)
18. Page, R.D., Cotton, J.: Vertebrate phylogenomics: reconciled trees and gene duplications. In: Pacific Symposium on Biocomputing, pp. 536–547 (2002)
19. Renny-Byfield, S., Wendel, J.F.: Doubling down on genomes: polyploidy and crop plants. Am. J. Bot. **101**(10), 1711–1725 (2014)
20. Zhang, L.: On a Mirkin-Muchnik-Smith conjecture for comparing molecular phylogenies. J. Comput. Biol. **4**(2), 177–187 (1997)

An Efficiently Recognisable Subset of Hypergraphic Sequences

Syed M. Meesum[✉]

Institute of Computer Science, University of Wrocław, Wrocław, Poland
meesum.syed@gmail.com

Abstract. The problem of efficiently characterizing degree sequences of simple hypergraphs (without repeated hyper-edges) is a fundamental long-standing open problem in Graph Theory. Several results are known for restricted versions of this problem. This paper adds to the list of sufficient conditions for a degree sequence to be *hypergraphic* and proposes a polynomial time algorithm which correctly identifies at least $2^{\frac{(n-1)(n-2)}{2}}$ *hypergraphic* sequences. For comparison, the number of hypergraphic sequences on n vertices is at most $2^{n \cdot (n-1)}$.

1 Introduction

For a given list of positive integers $D = (d_1, \ldots, d_n) \in \mathbb{Z}_+^n$, the *graph realization* problem asks if there exists a simple graph[1] G_D on n vertices whose vertex degrees are given by the list D. If such a graph exists then the degree sequence D is said to be *realizable*. A hypergraph is a k-hypergraph if every edge has k vertices. A k-hypergraph is called as a simple hypergraph if none of its edges are repeated. Simple graphs can be seen to correspond to the class of 2-hypergraphs. Generally, given a degree sequence and a positive integer k, the k-HYPERGRAPHIC SEQUENCE problem asks if there exists a k-hypergraph *realizing* the given sequence. When $k = 2$, Erdös-Gallai Theorem [11] gives necessary and sufficient conditions that must be satisfied by D for it to be *realizable*. For this case, another criteria in terms of smaller degree sequences were given by Havel [14] and Hakimi [12]. Further, Hoogeveen and Sierksma [19] listed seven criteria and gave a unifying proof for them.

Recently, for $k \geq 3$, k-HYPERGRAPHIC SEQUENCE was shown to be NP-Complete for any fixed value of k by Deza et al. [9]. In the past, a characterization was given by Dewdney [8] for all $k \geq 3$. Some sharp sufficient conditions for *realizability* of a degree sequence based on a sequence's length and degree sum were given in [2]. Colbourn et al. [7] proved that several other problems related to 3-graphic sequences are NP-Complete. Achuthan et al. [1], Billington [4] and Choudum [6] gave several necessary conditions for 3-hypergraphs, however Achuthan et al. [1] also showed that none of these conditions are sufficient.

S. M. Meesum—Supported by the NCN grant number 2015/18/E/ST6/00456. This work was partially done at the Institute of Mathematical Sciences, HBNI, India.

[1] A loopless graph without repeated edges.

© Springer International Publishing AG, part of Springer Nature 2018
L. Wang and D. Zhu (Eds.): COCOON 2018, LNCS 10976, pp. 391–402, 2018.
https://doi.org/10.1007/978-3-319-94776-1_33

There are many surveys available for this problem [13,17,20–22], for a recent survey on related problems see [10].

If we give up the restriction on sizes of edges to be k, we get the HYPER-GRAPHIC SEQUENCE problem which, along with its variants, was first considered by Boonyasombat in [5]. In particular, the HYPERGRAPHIC SEQUENCE problem, given a list of degrees D, asks if there exists a simple hypergraph which is a *realization* of D. This problem appears to be harder than the class of k-hypergraph problems in the sense that it is not even known to be in NP. It can be easily seen to be in PSPACE. Several restricted versions of this problem have been studied in the past. Bhave et al. [3] gave an Erdös-Gallai type characterization of degree sequences of loopless linear hypergraphs, where a linear hypergraph is one in which any two edges have at most one common vertex. In another direction, characterization for a partial Steiner triple system (PSTS), which is a linear 3-hypergraph, were given by Keranen et al. [15]. The results in [3] and [15] were recently generalized by Khan [16] using partial (n, k, λ)-systems. For greedy based approaches to this problem refer to [18]. This paper provides an efficiently checkable sufficient condition for a degree sequence to be *realizable* by a simple hypergraph.

2 Preliminaries

The set of non-negative integers is denoted using the symbol $\mathbb{Z}_+$. The set of integers $\{1, \ldots, n\}$ is denoted by $[n]$. An n-tuple $L = (\ell_1, \ell_2, \ldots, \ell_n)$, also simply referred to as a list is an ordered collection of elements. We refer to its ith element ℓ_i as $L(i)$. We index any list or tuple with natural numbers starting with 1. The notation $a_{\times m}$ denotes the tuple $(a, \ldots, a)$ consisting of a repeated m times. We use $L_1 \cdot L_2$ to denote the list obtained by the operation of concatenating two lists L_1 and L_2. A table $T = [L_1, L_2, \ldots, L_m]$ is a collection of lists of equal size. Pictorially, the lists $L_1, \ldots, L_m$ are arranged as columns in the table T. If each list L_i is of size n, then the table T is said to be of size $n \times m$. We refer to the ith row of a table T using the notation $T(i)$ and to an $(i, j)^{th}$ entry using the symbol $T(i, j)$.

For ease of notation, the sum of the entries in a list L will be denoted by $\sum L = \sum_{i \in [|L|]} L(i)$. The sum of $L = (\ell_1, \ldots, \ell_n)$ and $L' = (\ell'_1, \ldots, \ell'_n)$, denoted by $L + L'$, is the n-tuple $(\ell_1 + \ell'_1, \ldots, \ell_n + \ell'_n)$.

For $k \geq 0$, a permutation π is called as a cyclic permutation of order k, for each $i \in [n]$ it maps $i \mapsto 1 + ((i + k - 1) \bmod n)$.

A simple hypergraph H is a pair $([n], \mathcal{F})$, where $\mathcal{F}$ is a family of subsets of $[n]$. In general, a hypergraph may have sets repeated in $\mathcal{F}$, in which case $\mathcal{F}$ is a multiset. The degree of a vertex $v \in [n]$ is equal to $|\{F \in \mathcal{F} : v \in F\}|$. We will be working with an equivalent version, which can be stated in terms of co-ordinate wise sum of binary sequences of length n.

2.1   An Equivalent Version of Hypergraphic Sequences

For a given positive integer n, consider the set $S_n = \{0,1\}^n$ consisting of all binary tuples of length n. The binary tuples of length n will be used to encode the incidence vector of an hyperedge. The elements of S_n will also be referred to as binary sequences. The set S_n can be seen to be the complete hypergraph, having all the possible hyperedges in it. Any simple hypergraph is easily seen to be a subset of S_n. Finally, note that adding together all the incidence vectors corresponding to hyperedges in a hypergraph S gives us the degree sequence of hypergraph S. Therefore, the set $H_n = \{\sum_{x \in S} x : \varnothing \subseteq S \subseteq S_n\}$ is the set of all the hypergraphic sequences. Note that for the empty set $\varnothing$ the corresponding sum $\sum_{x \in \varnothing} x$ is defined to be $0_{\times n}$. By construction, each element of H_n is *realized* by some simple hypergraph and the degree sequence of every simple hypergraph is contained in H_n. Each element of H_n is said to be *representable* or is said to admit a *representation*. Given this setting the *realizability* problem for simple hypergraphs can be restated as follows.

HYPERGRAPHIC SEQUENCE

Input: A tuple $w \in \mathbb{Z}_+^n$ which is provided as a binary input.

Question: Is $w \in H_n$?

As $\sum S_n = 2_{\times n}^{n-1}$, the maximum possible value of any entry in w, for $w \in H_n$, is 2^{n-1}. Thus, if any entry of w is outside the range $\{0, \ldots, 2^{n-1}\}$, then it is not a member of H_n. Even though the number of subsets of S_n is 2^{2^n}, the cardinality of hypergraphic sequences on n vertices is at most $2^{n \cdot (n-1)}$ if every vertex is assumed to appear in at least one hyper-edge.

Table 1. n-Bit-Table for $n = 3$

×	$c_{3,3}$	$c_{2,3}$	$c_{1,3}$
0	0	0	0
1	0	0	1
2	0	1	0
3	0	1	1
4	1	0	0
5	1	0	1
6	1	1	0
7	1	1	1

Our Results: We define a notion of cyclic permutations of the columns of a binary table (for example see Table 1) and use it to find an efficiently computable sufficient condition for a given degree sequence to be *realizable* by a simple hypergraph. In this paper we prove the following.

1. Firstly, we prove that for any set of cyclic permutations acting individually on each column of a binary table, the resulting table has all of its 2^n rows distinct.
2. Next, we define a notion of *cyclic hyper degrees* viz. the degree sequences which are the sum of contiguous rows in a binary table in which each column has been permuted by different cyclic permutations. As cyclic permutations of binary tables have every row distinct, a *cyclic hyper degree* sequence is *realizable* for some simple hypergraph.
3. Exploiting the special structure of the columns of a binary table, we give an efficient algorithm which checks if a given degree sequence is a *cyclic hyper degree*.
4. Finally, we provide a lower bound of $2^{\frac{(n-1)(n-2)}{2}}$ for the number of *cyclic hyper degree* sequences. This also gives us a lower bound on the number of *hypergraphic* sequences.

3   Cyclically Permuting Binary Tables

In this section, we will be working with binary tables of size $2^n \times n$ and study the action of cyclic permutations on the columns of the table. We first define the notion of a binary table formally and give its structural properties which would be used later. For a given number $n \in \mathbb{Z}_+$, list out the binary expansion of numbers in increasing order from $\{0, \ldots, 2^n - 1\}$ as rows in a table. We pad the binary expansion with sufficient numbers of zeros on the left to make the length of each row exactly equal to n. For example, when $n = 3$, the table is as given in Table 1.

To state it formally we need the following definitions. Given a number m we denote the ith bit in its binary representation by $\mathrm{bin}(m, i)$. If the most significant bit in the binary expansion occurs at the sth-position in the binary expansion of m, then for all values of $i > s$ the value of $\mathrm{bin}(m, i)$ is zero. For example, $\mathrm{bin}(4, 2) = 1$ and $\mathrm{bin}(4, i) = 0$ for every $i \geq 3$.

For a given n, we construct n lists $c_{1,n}, \ldots, c_{n,n}$, with each list $c_{i,n}$ having length equal to 2^n. For $n = 2$, we have $c_{1,2} = (0, 1, 0, 1)$ and $c_{2,2} = (0, 0, 1, 1)$. For $n = 3$, the lists $c_{1,3}, c_{2,3}, c_{3,3}$ correspond to the columns of the Table 1. Formally, the lists are defined as follows.

Definition 1 (n-Bit-Lists). *For a given n, we define n-Bit-List to consist of n lists $c_{1,n}, \ldots, c_{n,n}$. For $j \in [2^n]$, the value of $c_{i,n}(j)$ is equal to $\mathrm{bin}(j - 1, i)$.*

Definition 2 (n-Bit-Table). *For a positive integer n, the n-Bit-Table T_n is defined to be a size $2^n \times n$ table with $T_n = [c_{n,n}, c_{n-1,n}, \ldots, c_{1,n}]$.*

The lists $c_{1,n}, \ldots, c_{n,n}$ have a nice recursive structure and can be generated in an alternative way by concatenation. Given a positive integer n, the base case of $n = 1$ is one list $c_{1,1} = (0, 1)$. For $n \geq 2$, the tuple $c_{n,n} = 0_{\times 2^{n-1}} \cdot 1_{\times 2^{n-1}}$ and for $j \in [n-1]$, the list $c_{j,n}$ is equal to the concatenated list $c_{j,n-1} \cdot c_{j,n-1}$. Thus, we get the following observation about the lists.

Observation 1. *For $n \in \mathbb{Z}_+$ and $i \in [n]$, we have $c_{i,n} = (0_{\times 2^{i-1}} \cdot 1_{\times 2^{i-1}})_{\times 2^{n-i}}$.*

Let $\mathcal{C}_n$ be the set of all cyclic permutations on an 2^n length list. Let $\Pi \subseteq \mathcal{C}_n$ be a multi-set consisting of n arbitrary cyclic permutations $\pi_1, \ldots, \pi_n$. Let $\Pi(T_n) = [\pi_n(c_{n,n}), \ldots, \pi_1(c_{1,n})]$ be the new table obtained from T_n. For clarity we will change the notation slightly, let $\Pi(T_n, i)$ denote row i of $\Pi(T_n)$, it is a length n binary tuple $(c_{n,n}(\pi_n(i)), c_{n-1,n}(\pi_{n-1}(i)), \ldots, c_{1,n}(\pi_1(i)))$, for $i \in [2^n]$.

We would be looking at the action of cyclic groups on the columns of a binary table. We next state a general lemma whose proof follows by induction over the order of a cyclic permutation.

Lemma 1. *Let L be a list and π be any cyclic permutation, then*

$$\pi(L \cdot L) = \pi(L) \cdot \pi(L).$$

Recall that for $j \in [n-1]$, the tuple $c_{j,n}$ consists of two copies of $c_{j,n-1}$ concatenated together. Combining this fact with Lemma 1, we obtain the following.

Corollary 1. *Given a positive integer n and a cyclic permutation $\pi \in \mathcal{C}_n$. For $j \in [n-1]$, the tuple $\pi(c_{j,n})$ is equal to $\pi(c_{j,n-1}) \cdot \pi(c_{j,n-1})$.*

We next prove that for any set of cyclic permutations Π, any two rows in $\Pi(T_n)$ will never become equal. This simple looking invariance property of the binary table, when combined with the periodic structure of the binary table columns gives as a way to give short certificates of membership of many degrees sequences in H_n.

Theorem 1. *For any set of n cyclic permutations Π, the rows of $\Pi(T_n)$ are pair-wise distinct.*

Proof. We prove this by induction on n. For the base case $n = 1$, the statement is trivially true. For the rest of the proof, assume that the list of cyclic permutations is $\Pi = (\pi_1, \ldots, \pi_n)$ and $\Pi_{\overline{n}}$ is used to denote the list $(\pi_1, \ldots, \pi_{n-1})$.

The table T_n can be constructed recursively by taking two copies of T_{n-1} and appending the rows of one below the other, after that we add $0_{\times 2^{n-1}} \cdot 1_{\times 2^{n-1}}$ as the first column. Consider the table

$$\begin{aligned} T_{\overline{n}} &= [c_{n-1,n}, \ldots, c_{1,n}] \\ &= [c_{n-1,n-1} \cdot c_{n-1,n-1}, \ldots, c_{1,n-1} \cdot c_{1,n-1}]. \end{aligned}$$

Apply the list of permutations $\Pi_{\overline{n}}$ on $T_{\overline{n}}$ to get

$$\begin{aligned} \Pi_{\overline{n}}(T_{\overline{n}}) &= [\pi_{n-1}(c_{n-1,n}), \ldots, \pi_1(c_{1,n})] \\ &= [\pi_{n-1}(c_{n-1,n-1}) \cdot \pi_{n-1}(c_{n-1,n-1}), \ldots, \pi_1(c_{1,n-1}) \cdot \pi_1(c_{1,n-1})], \end{aligned}$$

where the second equality follows from Corollary 1. By the induction hypothesis, the first row-wise half of $\Pi_{\overline{n}}(T_{\overline{n}})$, which is the same as $\Pi_{\overline{n}}(T_{n-1})$, consists of distinct rows. Therefore, the table $\Pi_{\overline{n}}(T_{\overline{n}})$ consists of rows which are repeated exactly twice. For $i < j$, the rows $\Pi_{\overline{n}}(T_{\overline{n}}, i)$ and $\Pi_{\overline{n}}(T_{\overline{n}}, j)$ are equal when $j = i + 2^{n-1}$. Therefore, it suffices to prove that the rows $\Pi(T_n, i)$ and $\Pi(T_n, i+2^{n-1})$ are distinct. Observe that to obtain the table $\Pi(T_n)$, we need to append $\pi_n(c_{n,n})$ as the first column in $\Pi_{\overline{n}}(T_{\overline{n}})$. As $c_{n,n}$ is equal to $0_{\times 2^{n-1}} \cdot 1_{\times 2^{n-1}}$, we have $c_{n,n}(i) \neq c_{n,n}(i + 2^{n-1})$, this implies that $c_{n,n}(\pi_n(i)) \neq c_{n,n}(\pi_n(i + 2^{n-1}))$. $\quad\square$

Given a binary table, if we sum a sequence of contiguous rows we would get a hypergraphic sequence. However, the number of such hypergraphic sequences would be at most $\mathcal{O}(2^{2n})$ as there are at most 2^n choices for both the starting row and ending row. The additional operation of permuting the columns allows for creating a lot of different binary tables with distinct rows. In Sect. 5 we show that the number of distinct contiguous sums for permuted binary tables is at least $2^{(n-1)(n-2)/2}$. Theorem 1 proved above can also be seen as a way to argue about the membership certificates of a subset of hypergraphic sequences. We call that subset as *cyclic hyper degrees*, which is defined as follows.

Definition 3 (Cyclic Hyper Degree). *Let Π be a multi-set of n cyclic permutations. A tuple $d \in \mathbb{Z}_+^n$ is said to be a cyclic hyper degree if there exist $i, N \in [2^n]$ such that $d = \sum_{k=i}^{N} \Pi(T_n, k)$.*

As the rows of $\Pi(T_n)$ are distinct, their contiguous sum is hypergraphic and is in H_n by definition. Also, any permutation of a hypergraphic sequence is hypergraphic as well, by a relabelling of the realising hypergraph's vertices. This gives us the following theorem.

Theorem 2. *If $w \in \mathbb{Z}_+^n$ is a cyclic hyper degree and π is a permutation over $[n]$, then $\pi(w) = (w_{\pi(1)}, \ldots, w_{\pi(n)}) \in H_n$.*

We note that $(4, 1, 1, 1)$ is a *realizable* hypergraph degree sequence but it is not a *cyclic hyper degree* sequence. So the reverse direction of the theorem above is false. If it was true this would have given us a polynomial time algorithm for the HYPERGRAPHIC SEQUENCE problem. In the next section, we will show how to efficiently check if a given sequence d is a *cyclic hyper degree*.

4   Efficiently Recognizing *Cyclic Hyper Degrees*

Observe that the elements in the columns of T_n do not change their relative position after application of a cyclic permutation when seen as a cyclic list. Suppose we are given the number of edges N in a hypergraph which realises a given cyclic hyper degree sequence. If the given input sequence is a cyclic hyper degree sequence then there must be a set of cyclic permutations Π such that the first N rows of $\Pi(T_n)$ sum up to it. As we will see in Theorem 3, we can actually list out all the possible cyclic hyper degrees having N edges. Using Theorem 3 we can efficiently compute the range of values taken by contiguous sum of N elements in a list $c_{i,n}$, for any i. In the second part we will show how to find the value of N using Theorem 3. Finally, combining it with Lemma 6 we show how to efficiently check if the input degree sequence is a cyclic hyper degree.

Definition 4 (Contiguous Sum). *Given a list L of length m, the contiguous sum of N elements in L starting at the index $i \in [m]$ is defined to be*

$$\mathcal{S}(L, i, N) := \sum_{j=0}^{N-1} L(1 + ((i + j - 1) \bmod m)).$$

The summation above treats the list L as a cyclic list. Next, we prove that the contiguous sum function is a 'continuous' function, this property is very important as it will allow us to specify the range of sum by stating the minimum and the maximum value taken by it. Note that if L is a 0-1 list, for any index $\ell \in [m]$, we have $|\mathcal{S}(L, \ell, N) - \mathcal{S}(L, \ell + 1, N)| \in \{0, 1\}$. This fact gives us the following property.

Observation 2 (Continuity of Range). *Let L be a size m list having 0-1 entries and $N \in \mathbb{Z}_+$. If $v_i = \mathcal{S}(L, i, N)$ and $v_j = \mathcal{S}(L, j, N)$, for some $i, j \in [m]$, then for every $v \in \mathbb{Z}_+$ contained between v_i and v_j there exists a $k \in [m]$ such that $\mathcal{S}(L, k, N) = v$.*

As the lists $c_{i,n}$ are over 0-1 we get an easy relation between the maximum and minimum values taken by the contiguous sum as follows.

Lemma 2. *Let $j \in \{0, \ldots, n\}$, $i \in [n]$ and $N \in [2^n]$. The minimum of the sum of N contiguous bits in a bit list $c_{i,n}$ is m if and only if its maximum is $N - m$.*

Proof. Let $\bar{c}_{i,n}$ be the bit list obtained from the list $c_{i,n}$ by flipping each zero to one and vice versa. Let $\sigma_{2^{i-1}}$ be an order 2^{i-1} cyclic permutation, observe that $\bar{c}_{i,n}$ is equal to $\sigma_{2^{i-1}}(c_{i,n})$. If the minimum value is obtained at the contiguous segment which starts at the index j in $c_{i,n}$, then the value $N - m$ can be obtained by the contiguous sum starting at index $\sigma_{2^{i-1}}(j)$. Finally, note that m is the minimum value if and only if $N - m$ is the maximum value. $\square$

Combining Observation 2 and Lemma 2 we get the following.

Lemma 3. *Let $N \in [2^n]$ and $m = \min_{j \in [2^n]} \mathcal{S}(c_{i,n}, j, N)$. For every value v in the range $\{m, \ldots, N - m\}$ there exists a $j \in [2^n]$ such that $\mathcal{S}(c_{i,n}, j, N) = v$.*

The lemma above allows us to find the range of values taken by the contiguous sum by just finding the minimum value taken by it. Next we prove a simpler lemma about the range of values taken. Using that, in Theorem 3, we will find the range of values taken by the contiguous sum of N elements in any list $c_{i,n}$.

Lemma 4. *For $j \in \{0, \ldots, n\}$ and $i \in [n]$, the sum of 2^j contiguous bits in a bit list $c_{i,n}$ takes the following values.*

1. *If $j \leq (i - 1)$, then the range is $\{0, \ldots, 2^j\}$, and*
2. *If $j \geq i$, then the sum is exactly 2^{j-1}.*

Proof. By Lemma 3, it suffices to find the minimum value of contiguous sum function. Notice that we have, $c_{i,n} = (0_{\times 2^{i-1}} \cdot 1_{\times 2^{i-1}})_{\times 2^{n-i}}$, by Observation 1.

1. When $j \leq (i-1)$, we can pick a block of 2^j zeros giving a total of zero, which is the minimum possible value.

2. When $j \geq i$, let L_k be a list of 2^j contiguous bits of $c_{i,n}$ starting at the index k in $c_{i,n}$. To prove that $\sum L_k = \sum L_{k+1}$, it suffices to show that $c_{i,n}(k) = c_{i,n}(k + 2^j)$. Rewriting $c_{i,n} = ((0_{\times 2^{i-1}} \cdot 1_{\times 2^{i-1}})_{\times 2^{j-i}})_{\times 2^{n-j}}$ shows that any two indices with difference equal to 2^j store the same value. As the choice of k was arbitrary, the contiguous sum is equal to 2^{j-1}.

$\square$

Theorem 3. *For $i \in [n]$, $N \in [2^n]$ and $p = 2^i$, the sum of N contiguous bits in a bit list $c_{i,n}$ takes values in the range,* $\mathrm{range}(i, N) \triangleq$

$$\left\{ \left\lfloor \frac{N}{p} \right\rfloor \frac{p}{2} + \max\left((N \bmod p) - \frac{p}{2}, 0\right), \cdots, \left\lfloor \frac{N}{p} \right\rfloor \frac{p}{2} + \min\left(N \bmod p, \frac{p}{2}\right) \right\}.$$

Proof. For a fixed $i \in [n]$ consider the list $c_{i,n}$. Assuming that the minimum value of the range is as claimed above, by Lemma 3, the maximum value is

$$\max_{j \in [2^n]} \mathcal{S}(c_{i,n}, j, N) = N - \min_{j \in [2^n]} \mathcal{S}(c_{i,n}, j, N)$$

$$= N - \left(\left\lfloor \frac{N}{p} \right\rfloor \frac{p}{2} + \max\left((N \bmod p) - \frac{p}{2}, 0\right) \right)$$

$$= \left\lfloor \frac{N}{p} \right\rfloor \frac{p}{2} + (N \bmod p) - \max\left((N \bmod p) - \frac{p}{2}, 0\right)$$

$$= \left\lfloor \frac{N}{p} \right\rfloor \frac{p}{2} + \min\left(\frac{p}{2}, N \bmod p\right).$$

As proved in case 2 of Lemma 4, the sum of $\lfloor \frac{N}{p} \rfloor p$ contiguous bits is equal to $\lfloor \frac{N}{p} \rfloor \frac{p}{2}$ irrespective of the starting index. Therefore, it suffices to find the minimum sum of $R = (N \bmod 2^i)$ contiguous bits. Next, we find the starting index for achieving the minimum possible sum of R values. Let L_k be a list of R bits occurring contiguously in $c_{i,n}$ starting at index k. If the first bit of L_k is 1, then $\sum L_{k+1} \leq \sum L_k$. Therefore, we can keep on increasing the value of k until the first bit is zero, without increasing the value of the contiguous sum. On the other hand, if $c_{i,n}(k-1) = 0$, then $\sum L_{k-1} \leq \sum L_k$. Therefore, we can keep on decreasing the value of k one at a time until $c_{i,n}(k-1) = 1$, without increasing the value of the contiguous sum. Thus the minimum value of the contiguous sum is achieved when the index k points to the start of any block $0_{\times 2^{i-1}}$ contained in $c_{i,n}$. Thus, the value of minimum is equal to $\max(R - \frac{p}{2}, 0)$, as the ones start appearing after $\frac{p}{2}$ indices from the start of the list $0_{\times 2^{i-1}} \cdot 1_{\times 2^{i-1}}$. Adding it to $\lfloor \frac{N}{p} \rfloor \frac{p}{2}$ gives the required minimum value. $\square$

Theorem 4. *A list $w = \{w_1, \ldots, w_n\} \in \mathbb{Z}_+^n$ is a cyclic hyper degree if and only if there exist $N \in [2^n]$ and a permutation π over n, such that for each $i \in [n]$, $w_{\pi(i)} \in \mathrm{range}(i, N)$.*

Proof. Forward direction is a direct consequence of Theorem 4.

Using Definition 3 and Theorem 4, we get that there exist numbers $s_1, \ldots, s_n \in [2^n]$ such that for each $i \in [n]$, we have $w_{\pi(i)} = \mathcal{S}(c_{i,n}, s_i, N)$. Let $\Pi^{-1} = (\sigma_{s_1}^{-1}, \ldots, \sigma_{s_n}^{-1})$ be the list of cyclic permutations, where for each $i \in [n]$, $\sigma_{s_i}^{-1}$ is the inverse of the cyclic permutation of order s_i. Consider the table $\Pi^{-1}(T_n)$, by Theorem 1, all its rows are distinct. In particular, the first N rows are distinct and their sum is $\pi(w)$. Finally, note that $w \in H_n$ if and only if $\pi(w) \in H_n$. □

Theorem 3 gives us a way to efficiently find the number of bits in a contiguous sum of N bits. If we know the number of distinct bit sequences that can sum up to a given vector $w \in \mathbb{Z}_+^n$, then using Theorem 3 we can generate all the possible ranges of values which can be taken by each coordinate of the sum. Finally, we need to check if each coordinate of w is contained in different ranges, this corresponds to finding the permutation π in Theorem 4. In the next lemma, we will find the number of possible distinct bit-sequences which can sum up to a given w using cyclic permutations of binary tables, this corresponds to finding N in Theorem 4.

Lemma 5. *If $w = \{w_1, \ldots, w_n\} \in \mathbb{Z}_+^n$ is a cyclic hyper degree, then the number of bit sequences which sum up to w is an element of the set*

$$\mathcal{N}_w \triangleq \{2w_i + j \; : \; i \in [n], j \in \{-1, 0, 1\}\}.$$

Proof. As one of the coordinates of w, say w_k, is the contiguous sum of $c_{1,n}$, we need to find the number of bits which sum up to w_k. From the structure of $c_{1,n}$, it is easily seen that there are just three values viz. $2w_k - 1, 2w_k, 2w_k + 1$ which contain w_k in their range of sums. Conversely, for any number x not contained in $\{2w_i + j \; : \; i \in [n], j \in \{-1, 0, 1\}\}$, the sum of x contiguous bits $c_{1,n}$ will not contain any of w_i, for $i \in [n]$. □

Lemma 6 (Embedding integer sequence in integer intervals). *Given $w \in \mathbb{Z}_+^n$ and a list of integer intervals $R_1, \ldots, R_n \subset \mathbb{Z}_+^2$. There exists an algorithm running in time polynomial in n which correctly answers if there exists a permutation π such that for each $i \in [n]$, $w_{\pi(i)} \in R_i$.*

Proof. Construct a bipartite graph $G = (A, B, E)$ on $2n$ vertices. Let $A = B = [n]$ and $(i, j) \in E$ if and only if $w_i \in R_j$. Using any of the classical polynomial time algorithms one can find if there exists a perfect matching in G. If there is a perfect matching then the answer is YES, otherwise it is NO. □

We note that there is a greedy algorithm which can find the correct permutation π, if it exists. However, our purpose in this paper is to show that this step can be done efficiently, so we will content ourselves with the simpler to describe perfect matching algorithm.

Theorem 5. *There is a polynomial time algorithm in n which decides if a given $w \in \mathbb{Z}_+^n$ is a cyclic hyper degree.*

Proof. For an element $N \in \mathcal{N}_w$ given by Lemma 5, for each $i \in [n]$ compute range(i, N) as given by Theorem 3. Now, use Lemma 6 on these ranges of numbers and decide if w is a *cyclic hyper degree*, if it is not then try the next number from the set $\mathcal{N}_w$. If it succeeds for at least one element of $\mathcal{N}_w$, we answer YES, otherwise we answer NO. Finally, note that $|\mathcal{N}_w| \leq 3n$ and all the other steps can be performed in time which is a polynomial function of n.   □

5   Lower Bound on the Number of *Cyclic Hyper Degrees*

In this section we will lower bound the number of *cyclic hyper degrees* for a given value of n. For this, given N we will find the size of range of contiguous sum, denoted by $R_{i,N}$, for each value of i as given by Theorem 3. The number of distinct *cyclic degree sequences* which are the sum of N bit sequences is then $\prod_{i \in [n]} R_{i,N}$. This is so because we have $R_{i,N}$ choices for the ith coordinate. We proceed to find $R_{i,N}$, the size of each range, as a corollary of Theorem 3 as follows.

Corollary 2. *For* $i \in [n]$, $N \in [2^n]$ *and* $p = 2^i$, *the sum of* N *contiguous bits in a bit list* $c_{i,n}$ *takes* $1 + \min(N \bmod p, p - (N \bmod p))$ *distinct values.*

Proof. The number of values is $1 + \max_{j \in [2^n]} \mathcal{S}(c_{i,n}, j, N) - \min_{j \in [2^n]} \mathcal{S}(c_{i,n}, j, N)$

$$= 1 + N - 2 \min_{j \in [2^n]} \mathcal{S}(c_{i,n}, j, N)$$

$$= 1 + N - \left\lfloor \frac{N}{p} \right\rfloor p - 2 \max \left((N \bmod p) - \frac{p}{2}, 0 \right)$$

$$= 1 + N \bmod p - 2 \max \left((N \bmod p) - \frac{p}{2}, 0 \right)$$

$$= 1 + \min \left(N \bmod p - 2(N \bmod p) + p, N \bmod p \right)$$

$$= 1 + \min \left(p - (N \bmod p), N \bmod p \right).$$

□

Lemma 7. *For* $n \in \mathbb{Z}_+$, *the number of* cyclic hyper degrees *on* n *vertices is at least* $2^{\frac{(n-1)(n-2)}{2}}$.

Proof. Given a fixed number n, we are going to count the number of *cyclic hyper degrees* which are the sum of exactly M bit-sequences, where $M = \sum_{j=0}^{\lfloor \frac{n}{2} \rfloor} 2^{2j}$. By Corollary 2, the range of values possible for i^{th} bit is $B_i = 1 + \min(2^i - M \bmod 2^i, M \bmod 2^i) = 1 + \min(2^i - \sum_{j=1}^{\lfloor \frac{i}{2} \rfloor} 2^{2j}, \sum_{j=1}^{\lfloor \frac{i}{2} \rfloor} 2^{2j})$. Depending on whether i is even or odd, it can be broken into two cases, but in both the cases, for $i \geq 2$ we have $B_i \geq 2^{i-2}$. The number of representable bit sequences possible with these ranges of coordinates is

$$\prod_{k \in [n]} B_k \geq \prod_{k=2}^{n} 2^{k-2} = \prod_{k=0}^{n-2} 2^k = 2^{\frac{(n-1)(n-2)}{2}}.$$

□

6   Conclusion

We looked at the HYPERGRAPHIC SEQUENCE problem and gave a sufficient condition for hypergraphic sequences. We proved that for a list $w \in \mathbb{Z}_+^n$, there is an algorithm $\mathcal{A}$, which given w as input, runs in time polynomial in n and if it answers YES then $w \in II_n$. If $\mathcal{A}$ answers NO then w may or may not be a member of H_n. However, if $w \notin H_n$, then the algorithm answers correctly. It would be interesting to look at the instances of hypergraphic sequences for which our algorithm fails. These instances may be helpful in finding hardness reductions for it.

References

1. Achuthan, N., Achuthan, N., Simanihuruk, M.: On 3-uniform hypergraphic sequences. J. Comb. Math. Comb. Comput **14**, 3–13 (1993)
2. Behrens, S., Erbes, C., Ferrara, M., Hartke, S.G., Reiniger, B., Spinoza, H., Tomlinson, C.: New results on degree sequences of uniform hypergraphs. Electron. J. Comb. **20**(4), P14 (2013)
3. Bhave, N.S., Bam, B., Deshpande, C.: 13 a characterization of degree sequences of linear hypergraphs. J. Indian Math. Soc. **75**(1), 151 (2008)
4. Billington, D.: Conditions for degree sequences to be realisable by 3-uniform hypergraphs. J. Comb. Math. Comb. Comput. **3**, 71–91 (1988)
5. Boonyasombat, V.: Degree sequences of connected hypergraphs and hypertrees. In: Koh, K.M., Yap, H.P. (eds.) Graph Theory Singapore 1983. LNM, vol. 1073, pp. 236–247. Springer, Heidelberg (1984). https://doi.org/10.1007/BFb0073123
6. Choudum, S.A.: On graphic and 3-hypergraphic sequences. Discrete Math. **87**(1), 91–95 (1991)
7. Colbourn, C.J., Kocay, W.L., Stinson, D.R.: Some NP-complete problems for hypergraph degree sequences. Discrete Appl. Math. **14**(3), 239–254 (1986)
8. Dewdney, A.K.: Degree sequences in complexes and hypergraphs. Proc. Am. Math. Soc. **53**(2), 535–540 (1975)
9. Deza, A., Levin, A., Meesum, S.M., Onn, S.: Optimization over degree sequences. arXiv preprint arXiv:1706.03951 (2017)
10. Ferrara, M.: Some problems on graphic sequences. Graph Theory Notes N. Y. **64**, 19–25 (2013)
11. Gallai, P.E.T., Erdos, P.: Graphs with prescribed degree of vertices (Hungarian), Mat. Lapok **11**, 264–274 (1960)
12. Hakimi, S.L.: On realizability of a set of integers as degrees of the vertices of a linear graph. I. J. Soc. Ind. Appl. Math. **10**(3), 496–506 (1962). SIAM
13. Hakimi, S.L., Schmeichel, E.F.: Graphs and their degree sequences: a survey. In: Alavi, Y., Lick, D.R. (eds.) Theory and applications of graphs, pp. 225–235. Springer, Heidelberg (1978). https://doi.org/10.1007/BFb0070380
14. Havel, V.: A remark on the existence of finite graphs. Casopis Pest. Mat. **80**(477–480), 1253 (1955)
15. Keranen, M.S., Kreher, D., Kocay, W., Li, P.C.: Degree sequence conditions for partial steiner triple systems. Bull. Inst. Comb. Appl **57**, 71–73 (2009)
16. Khan, M.: Characterizing the degree sequences of hypergraphs, August 2015. https://open.library.ubc.ca/cIRcle/collections/48630/items/1.0228065

17. Rao, S.B.: A survey of the theory of potentially P-graphic and forcibly P-graphic degree sequences. In: Rao, S.B. (ed.) Combinatorics and Graph Theory. LNM, vol. 885, pp. 417–440. Springer, Heidelberg (1981). https://doi.org/10.1007/BFb0092288

18. Sahakyan, H., Aslanyan, L.: Evaluation of greedy algorithm of constructing (0, 1)-matrices with different rows. Int. J. Inf. Technol. Knowl. **5**(1), 55–66 (2011)

19. Sierksma, G., Hoogeveen, H.: Seven criteria for integer sequences being graphic. J. Graph theory **15**(2), 223–231 (1991)

20. Tyshkevich, R., Chernyak, A., Chernyak, Z.A.: Graphs and degree sequences. i. Cybernetics **23**(6), 734–745 (1987)

21. Tyshkevich, R., Chernyak, A., Chernyak, Z.A.: Graphs and degree sequences: a survey. iii. Cybernetics **24**(5), 539–548 (1988)

22. Tyshkevich, R., Chernyak, A., Chernyak, Z.A.: Graphs and degree sequences. ii. Cybernetics **24**(2), 137–152 (1988)

Partial Homology Relations - Satisfiability in Terms of Di-Cographs

Nikolai Nøjgaard[1,2]($\boxtimes$), Nadia El-Mabrouk[3], Daniel Merkle[2], Nicolas Wieseke[4], and Marc Hellmuth[1,5]

[1] Institute of Mathematics and Computer Science, University of Greifswald, Walther-Rathenau-Strasse 47, 17487 Greifswald, Germany
nojgaard@imada.sdu.dk, mhellmuth@mailbox.org
[2] Department of Mathematics and Computer Science, University of Southern Denmark, Campusvej 55, 5230 Odense M, Denmark
[3] Department of Computer Science and Operational Research, University of Montreal, CP 6128 succ Centre-Ville, Montreal, Canada
[4] Swarm Intelligence and Complex Systems Group, Department of Computer Science, University of Leipzig, Augustusplatz 10, 04109 Leipzig, Germany
[5] Center for Bioinformatics, Saarland University, Building E 2.1, 66041 Saarbrücken, Germany

Abstract. Directed cographs (di-cographs) play a crucial role in the reconstruction of evolutionary histories of genes based on homology relations which are binary relations between genes. A variety of methods based on pairwise sequence comparisons can be used to infer such homology relations (e.g. orthology, paralogy, xenology). They are *satisfiable* if the relations can be explained by an event-labeled gene tree, i.e., they can simultaneously co-exist in an evolutionary history of the underlying genes. Every gene tree is equivalently interpreted as a so-called cotree that entirely encodes the structure of a di-cograph. Thus, satisfiable homology relations must necessarily form a di-cograph. The inferred homology relations might not cover each pair of genes and thus, provide only partial knowledge on the full set of homology relations. Moreover, for particular pairs of genes, it might be known with a high degree of certainty that they are not orthologs (resp. paralogs, xenologs) which yields forbidden pairs of genes. Motivated by this observation, we characterize (partial) satisfiable homology relations with or without forbidden gene pairs, provide a quadratic-time algorithm for their recognition and for the computation of a cotree that explains the given relations.

Keywords: Directed cographs · Partial relations
Forbidden relations · Recognition algorithm · Homology
Orthology · Paralogy · Xenology

1 Introduction

Directed cographs (di-cographs) are a well-studied class of graphs that can uniquely be represented by an ordered rooted tree (T, t), called cotree, where

© Springer International Publishing AG, part of Springer Nature 2018
L. Wang and D. Zhu (Eds.): COCOON 2018, LNCS 10976, pp. 403–415, 2018.
https://doi.org/10.1007/978-3-319-94776-1_34

each inner vertex gets a unique label "1", "$\overrightarrow{1}$" or "0" [6,8,10,13]. In particular, di-cographs have been shown to play an important role for the reconstruction of the evolutionary history of genes or species based on genomic sequence data [12,18,19,21–23,25]. Genes are the molecular units of heredity holding the information to build and maintain cells. During evolution, they are mutated, duplicated, lost and passed to organisms through speciation or horizontal gene transfer (HGT), which is the exchange of genetic material among co-existing species. A gene family comprises genes sharing a common origin. Genes within a gene family are called *homologs*.

The history of a gene family is equivalent to an event-labeled gene tree, the leaves correspond to extant genes, internal vertices to ancestral genes and the label of an internal vertex highlighting the event at the origin of the divergence leading to the offspring, namely *speciation-*, *duplication-* or *HGT-events* [14]. Equivalently, the history of genes is described by an event-labeled rooted tree for which each inner vertex gets a unique label "1" (for speciation), "0" (for duplication) or "$\overrightarrow{1}$" (for HGT). In other words, any gene tree is also a cotree. The type of event "1", "0" and "$\overrightarrow{1}$" of the lowest common ancestor of two genes gives rise to one of three distinct *homology relations* respectively, the *orthology-relation* R_1, the *paralogy-relation* R_0 and the *xenology-relation* $R_{\overrightarrow{1}}$. The orthology-relation R_1 on a set of genes forms an undirected cograph [18]. In [19] it has been shown that the graph G with arc set $E(G) = R_1 \cup R_{\overrightarrow{1}}$ must be a di-cograph (see [20] for a detailed discussion).

In practice, these homology relations are often estimated from sequence similarities and synteny information, without requiring any *a priori* knowledge on the topology of either the gene tree or the species tree (see e.g. [2–4,7,11,26–28,31,32,34,35]). The starting point of this contribution is a set of estimated relations R_1, $R_{\overrightarrow{1}}$ and R_0. In particular, we consider so-called *partial* and *forbidden* relations: In fact, similarity-based homology inference methods often depend on particular threshold parameters to determine whether a given pair of genes is in one of R_0, R_1 or $R_{\overrightarrow{1}}$. Gene pairs whose sequence similarity falls below (or above) a certain threshold cannot be unambiguously identified as belonging to one of the considered homology relations. Hence, in practice one usually obtains partial relations only, as only subsets of these relations may be known. Moreover, different homology inference methods may lead to different predictions. Thus, instead of a yes or no orthology, paralogy or xenology assignment, a confidence score can rather be assigned to each relation [12]. A simple way of handling such weighted gene pairs is to set an upper threshold above which a relation is predicted, and a lower threshold under which a relation is rejected, leading to partial relations with forbidden gene pairs, i.e., gene pairs that are not included in any of the three relations but for which it is additionally known to which relations they definitely *not* belong to.

In this contribution, we generalize results established by Lafond and El-Mabrouk [24] and characterize satisfiable partial relations with and without forbidden relations. We provide a recursive quadratic-time algorithm testing whether the considered relations are satisfiable, and if so reconstructing a corresponding cotree. This, in turn, allows us to extend satisfiable partial relations to full relations. Finally, we evaluate the accuracy of the designed algorithm on large-scaled simulated data sets. As it turns out, it suffices to have only a very few but correct pairs of relations to recover most of them.

Note, the results established here may also be of interest for a broader scientific community. Di-cographs play an important role in computer science because many combinatorial optimization problems that are NP-complete for arbitrary graphs become polynomial-time solvable on di-cographs [6,9,15]. However, the cograph-editing problem is NP-hard [29]. Thus, an attractive starting point for heuristics that edit a given graph to a cograph may be the knowledge of satisfiable parts that eventually lead to partial information of the underlying di-cograph structure of the graph of interest.

2   Preliminaries

Basics. In what follows, we always consider *binary* and *irreflexive* relations $R \subseteq V_{irr}^{\times} := V \times V \setminus \{(v,v) \mid v \in V\}$ and we omit to mention it each time. If we have a non-symmetric relation R, then we denote by $R^{sym} = R \cup \{(x,y) \mid (y,x) \in R\}$ the *symmetric extension* of R and by $\overleftarrow{R}$ the set $\{(x,y) \mid (y,x) \in R\}$. For a subset $W \subseteq V$ and a relation R, we define $R[W] = \{(x,y) \in R \mid x,y \in W\}$ as the *sub-relation of R that is induced by W*. Moreover, for a set of relations $\mathcal{R} = \{R_1, \ldots, R_n\}$ we set $\mathcal{R}[W] = \{R_1[W], \ldots, R_n[W]\}$.

A directed graph (digraph) $G = (V,E)$ has vertex set $V(G) = V$ and arc set $E(G) = E \subseteq V_{irr}^{\times}$. Given two disjoint digraphs $G = (V,E)$ and $H = (W,F)$, the digraphs $G \cup H = (V \cup W, E \cup F)$, $G \oplus H = (V \cup W, E \cup F \cup \{(x,y),(y,x) \mid x \in V, y \in W\})$ and $G \oslash H = (V \cup W, E \cup F \cup \{(x,y) \mid x \in V, y \in W\})$ denote the *union*, *join* and *directed join* of G and H, respectively. For a given subset $W \subseteq V$, the *induced subgraph* $G[W] = (W,F)$ of $G = (V,E)$ is the subgraph for which $x,y \in W$ and $(x,y) \in E$ implies that $(x,y) \in F$. We call $W \subseteq V$ a *(strongly) connected component* of $G = (V,E)$ if $G[W]$ is a *maximal* (strongly) connected subgraph of G.

Given a digraph $G = (V,E)$ and a partition $\{V_1, V_2, \ldots, V_k\}$ of its vertex set V, the *quotient digraph* $G/\{V_1, V_2, \ldots, V_k\}$ *has as vertex set* $\{V_1, V_2, \ldots, V_k\}$ and two distinct vertices V_i, V_j form an arc (V_i, V_j) in $G/\{V_1, \ldots, V_k\}$ if there are vertices $x \in V_i$ and $y \in V_j$ with $(x,y) \in E$.

An acyclic digraph is called *DAG*. It is well-known that the vertices of a DAG can be *topologically ordered*, i.e., there is an ordering of the vertices as $v_1, \ldots, v_n$ such that $(v_i, v_j) \in E$ implies that $i < j$. To check whether a digraph G contains no cycles one can equivalently check whether there is a topological order, which can be done via a depth-first search in $O(|V(G)| + |E(G)|)$ time.

Furthermore, we consider a *rooted tree* $T = (W,E)$ *(on V)* with root $\rho_T \in W$ and leaf set $V \subseteq W$ such that the root has degree ≥ 2 and each vertex $v \in W \setminus V$ with $v \neq \rho_T$ has degree ≥ 3. We write $x \succeq_T y$, if x lies on the path from ρ_T to y. The *children* of a vertex x are all adjacent vertices y for which $x \succeq_T y$. Given two leaves $x, y \in V$, their lowest common ancestor $lca(x,y)$ is the first vertex that lies on both paths from x to the root and y to the root. We say that rooted trees $T_1, \ldots, T_k$, $k \geq 2$ *are joined under a new root in the tree* T if T is obtained by the following procedure: add a new root ρ_T and all trees $T_1, \ldots, T_k$ to T and connect the root ρ_{T_i} of each tree T_i to ρ_T with an edge (ρ_T, ρ_{T_i}).

Di-cographs. Di-cographs generalize the notion of undirected cographs [6,8, 10,13] and are defined recursively as follows: The single vertex graph K_1 is a di-cograph, and if G and H are di-cographs, then $G \cup H$, $G \oplus H$, and $G \oslash H$ are di-cographs [8,16]. Each Di-cograph $G = (V, E)$ is associated with a unique ordered least-resolved tree $T = (W, F)$ (called *cotree*) with leaf set $L = V$ and a labeling function $t : W \setminus L \to \{0, 1, \overrightarrow{1}\}$ defined by

$$
t(\mathrm{lca}(x, y)) = \begin{cases} 0, & \text{if } (x, y), (y, x) \notin E \\ 1, & \text{if } (x, y), (y, x) \in E \\ \overrightarrow{1}, & \text{otherwise.} \end{cases}
$$

Since the vertices in the cotree T are ordered, the label $\overrightarrow{1}$ on some $\mathrm{lca}(x, y)$ of two distinct leaves $x, y \in L$ means that there is an arc $(x, y) \in E$, while $(y, x) \notin E$, whenever x is placed to the left of y in T.

Some important properties of di-cographs that we need for later reference are given now.

Lemma 1 ([8,16,30]). *A digraph $G = (V, E)$ is a di-cograph if and only if each induced subgraph of G is a di-cograph.*

Determining whether a digraph is a di-cograph, and if so, computing the corresponding cotree can be done in $O(|V| + |E|)$ time.

3   Problem Statement

As argued in the introduction and explained in more detail in [19], the evolutionary history of genes is equivalently described by an ordered rooted tree $\mathcal{T} = (T, t)$ where the leaf set of T are genes and t is a map that assigns to each non-leaf vertex a unique label $0, 1$ or $\overrightarrow{1}$. The labels correspond to the classical evolutionary events that act on the genes through evolution, namely *speciation* (1), *duplication* (0) and *horizontal gene transfer (HGT)* ($\overrightarrow{1}$). The tree T is ordered to represent the inherently asymmetric nature of HGT events with their unambiguous distinction between the vertically transmitted "original" gene and the horizontally transmitted "copy".

Therefore, a given gene tree $\mathcal{T} = (T, t)$ is a cotree of some di-cograph $G(\mathcal{T}) = (V, E)$. In particular, $\mathcal{T}$ gives rise to the following three well-known *homology relations* between genes:

the *orthology-relation* : $R_1(\mathcal{T}) = \{(x, y) \mid t(\mathrm{lca}(x, y)) = 1\}$,
the *paralogy-relation* : $R_0(\mathcal{T}) = \{(x, y) \mid t(\mathrm{lca}(x, y)) = 0\}$, and
the *xenology-relation* : $R_{\overrightarrow{1}}(\mathcal{T}) = \{(x, y) \mid t(\mathrm{lca}(x, y)) = \overrightarrow{1}$ and x is left of y in $T\}$.

Equivalently, $R_1(\mathcal{T}) = \{(x,y) \mid (x,y),(y,x) \in E\}$, $R_0(\mathcal{T}) = \{(x,y) \mid (x,y),(y,x)$ $\notin E\}$, $R_{\overrightarrow{1}}(\mathcal{T}) = \{(x,y) \mid (x,y) \in E, (y,x) \notin E\}$. By construction, $R_1(\mathcal{T})$ and $R_0(\mathcal{T})$ are symmetric relations, while $R_{\overrightarrow{1}}(\mathcal{T})$ is an anti-symmetric relation.

In practice, however, one often has only empirical estimates R_0, R_1 and $R_{\overrightarrow{1}}$ of some "true" relations $R_0(\mathcal{T}), R_1(\mathcal{T})$ and $R_{\overrightarrow{1}}(\mathcal{T})$, respectively. Moreover, it is often the case that for two distinct leaves x, y none of the pairs (x,y) and (y,x) is contained in the estimates R_0, R_1 and $R_{\overrightarrow{1}}$.

In what follows we always assume that R_0, R_1 and $R_{\overrightarrow{1}}$ are subsets of $V_{\text{irr}}^{\times}$ and pairwise disjoint. Furthermore R_0 and R_1 are always symmetric relations while $R_{\overrightarrow{1}}$ is always an anti-symmetric relation.

Definition 1 (Full and Partial Relations). *A set $\mathcal{H} = \{R_0, R_1, R_{\overrightarrow{1}}\}$ of relations is* full *if $R_0 \cup R_1 \cup R_{\overrightarrow{1}}^{\text{sym}} = V_{\text{irr}}^{\times}$ and* partial *if $R_0 \cup R_1 \cup R_{\overrightarrow{1}}^{\text{sym}} \subseteq V_{\text{irr}}^{\times}$.*

The definition allows considering full relations as partial. In other words, all results that will be obtained for partial relations will also be valid for full relations.

The question arises under which conditions the given partial relations R_0, R_1 and $R_{\overrightarrow{1}}$ are satisfiable, i.e., there is a cotree $\mathcal{T} = (T,t)$ such that $R_1 \subseteq R_1(\mathcal{T})$, $R_0 \subseteq R_0(\mathcal{T})$ and $R_{\overrightarrow{1}} \subseteq R_{\overrightarrow{1}}(\mathcal{T})$, or equivalently, there is a di-cograph $G^{\star} = (V, E^{\star})$ such that $(R_1 \cup R_{\overrightarrow{1}}) \subseteq E^{\star}$ and $R_0 \cap E^{\star} = \overleftarrow{R_{\overrightarrow{1}}} \cap E^{\star} = \emptyset$.

Definition 2 (Satisfiable Relations). *A* full *set $\mathcal{H} = \{R_0, R_1, R_{\overrightarrow{1}}\}$ is satisfiable, if there is a cotree $\mathcal{T} = (T,t)$ such that $R_1 = R_1(\mathcal{T})$, $R_0 = R_0(\mathcal{T})$ and $R_{\overrightarrow{1}} = R_{\overrightarrow{1}}(\mathcal{T})$.*

A partial *set $\mathcal{H} = \{R_0, R_1, R_{\overrightarrow{1}}\}$ is satisfiable, if there is a full set $\mathcal{H}^{\star} = \{R_0^{\star}, R_1^{\star}, R_{\overrightarrow{1}}^{\star}\}$ that is satisfiable such that $R_0 \subseteq R_0^{\star}$, $R_1 \subseteq R_1^{\star}$, and $R_{\overrightarrow{1}} \subseteq R_{\overrightarrow{1}}^{\star}$.*

In this case, we say that $\mathcal{H}$ can be extended to a satisfiable full set $\mathcal{H}^{\star}$ and that $\mathcal{T}$ explains $\mathcal{H}$ and $\mathcal{H}^{\star}$.

It is easy to see that a full set $\mathcal{H}$ is satisfiable if and only if the graph $G(R_1, R_{\overrightarrow{1}}) = (V, R_1 \cup R_{\overrightarrow{1}})$ is a di-cograph. The latter result has already been observed in [19] and is summarized below.

Theorem 1 ([19]). *The full set $\mathcal{H} = \{R_0, R_1, R_{\overrightarrow{1}}\}$ is satisfiable if and only if $G(R_1, R_{\overrightarrow{1}}) = (V, R_1 \cup R_{\overrightarrow{1}})$ is a di-cograph.*

Due to errors and noise in the data, the graph $G(R_1, R_{\overrightarrow{1}})$ is often not a di-cograph. However, in case that $\mathcal{H}$ is partial, it might be possible to extend $G(R_1, R_{\overrightarrow{1}})$ to a di-cograph. Moreover, in practice, one often has additional knowledge about the unknown parts of the relations, that is, one may know that a pair (x,y) is *not* in relation R_i for some $i \in \{0, 1, \overrightarrow{1}\}$. To be able to model such forbidden pairs, we introduce the concept of satisfiability in terms of forbidden relations.

Definition 3 (Satisfiable Relations w.r.t. Forbidden Relations). *Let* $\mathcal{F} = \{F_0, F_1, F_{\overrightarrow{1}}\}$ *be a partial set of relations. We say that a full set* $\mathcal{H} = \{R_0, R_1, R_{\overrightarrow{1}}\}$ *is satisfiable w.r.t.* $\mathcal{F}$ *if* $\mathcal{H}$ *is satisfiable and*

$$R_0 \cap F_0 = R_1 \cap F_1 = R_{\overrightarrow{1}} \cap F_{\overrightarrow{1}} = \emptyset.$$

On the other hand, a partial set $\mathcal{H} = \{R_0, R_1, R_{\overrightarrow{1}}\}$ *is satisfiable w.r.t.* $\mathcal{F}$ *if* $\mathcal{H}$ *can be extended to a full set* $\mathcal{H}^\star$ *that is satisfiable w.r.t.* $\mathcal{F}$.

Equivalently, a partial set $\mathcal{H} = \{R_0, R_1, R_{\overrightarrow{1}}\}$ is satisfiable w.r.t. $\mathcal{F}$, if there is a cotree $\mathcal{T} = (T, t)$ such that $R_0 \subseteq R_0(\mathcal{T})$, $R_1 \subseteq R_1(\mathcal{T})$, $R_{\overrightarrow{1}} \subseteq R_{\overrightarrow{1}}(\mathcal{T})$ and $R_0(\mathcal{T}) \cap F_0 = R_1(\mathcal{T}) \cap F_1 = R(\mathcal{T})_{\overrightarrow{1}} \cap F_{\overrightarrow{1}} = \emptyset$.

4   Satisfiable Relations

In what follows, we consider the problem of deciding whether a partial set $\mathcal{H}$ of relations is satisfiable, and if so, finding an extended full set $\mathcal{H}^\star$ of $\mathcal{H}$ and the respective cotree that explains $\mathcal{H}$. *Due to space limitation, all proofs are omitted and can be found in* [1].

Based on results provided by Böcker and Dress [5], the following result has been established for the HGT-free case.

Theorem 2 ([18,24]). *Let* $R_{\overrightarrow{1}} = \emptyset$ *,* $F_0 = F_1 = \emptyset$ *and* $F_{\overrightarrow{1}} = V_{irr}^\times$. *A full set* $\mathcal{H} = \{R_0, R_1, R_{\overrightarrow{1}}\}$ *is satisfiable w.r.t.* $\mathcal{F} = \{F_0, F_1, F_{\overrightarrow{1}}\}$ *if and only if the graph* $G = (V, R_1)$ *is an undirected cograph.*

A partial set $\mathcal{H} = \{R_0, R_1, \emptyset\}$ *is satisfiable w.r.t.* $\mathcal{F}$ *if and only if at least one of the following statements is satisfied:*

1. $G = (V, R_1)$ *is disconnected and each of its connected components is satisfiable.*
2. $G = (V, R_0)$ *is disconnected and each of its connected components is satisfiable.*

To generalize the latter result for non-empty relations $R_{\overrightarrow{1}}, F_0$ and F_1 and to allow pairs to be added to $R_{\overrightarrow{1}}$, i.e., $F_{\overrightarrow{1}} \neq V_{irr}^\times$, we need the following result.

Lemma 2. *A partial set* $\mathcal{H} = \{R_0, R_1, R_{\overrightarrow{1}}\}$ *is satisfiable w.r.t. the set of forbidden relations* $\mathcal{F} = \{F_1, F_0, F_{\overrightarrow{1}}\}$ *if and only if for any partition* $\{C_1, \ldots, C_k\}$ *of* V *the set* $\mathcal{H}[C_i]$ *is satisfiable w.r.t.* $\mathcal{F}[C_i]$, $1 \leq i \leq k$.

Lemma 2 characterizes satisfiable partial sets $\mathcal{H} = \{R_0, R_1, R_{\overrightarrow{1}}\}$ in terms of a partition $\{C_1, \ldots, C_k\}$ of V and the induced sub-relations in $\mathcal{H}[C_i]$. In what follows, we say that *a component* C *of a graph is satisfiable* if the set $\mathcal{H}[C]$ is satisfiable w.r.t. $\mathcal{F}[C]$.

We are now able to state the main result.

Theorem 3. *Let* $\mathcal{H} = \{R_0, R_1, R_{\overrightarrow{1}}\}$ *be a partial set. Then,* $\mathcal{H}$ *is satisfiable w.r.t. the set of forbidden relations* $\mathcal{F} = \{F_0, F_1, F_{\overrightarrow{1}}\}$ *if and only if* $R_0 \cap F_0 = R_1 \cap F_1 = R_{\overrightarrow{1}} \cap F_{\overrightarrow{1}} = \emptyset$ *and at least one of the following statements hold:*

Rule (0): $|V| = 1$.

Rule (1): *(a)* $G_0 := (V, R_1 \cup R_{\overrightarrow{1}} \cup F_0)$ *is disconnected and (b) each connected component of* G_0 *is satisfiable.*

Rule (2): *(a)* $G_1 := (V, R_0 \cup R_{\overrightarrow{1}} \cup F_1)$ *is disconnected and (b) each connected component of* G_1 *is satisfiable.*

Rule (3): *(a)* $G_{\overrightarrow{1}} := (V, R_0 \cup R_1 \cup R_{\overrightarrow{1}} \cup \overleftarrow{F_{\overrightarrow{1}}})$ *contains more than one strongly connected component, and (b) each strongly connected component of* $G_{\overrightarrow{1}}$ *is satisfiable.*

Note, the notation G_0, G_1 and $G_{\overrightarrow{1}}$ in Theorem 3 is chosen because if G_0 (resp. G_1 and $G_{\overrightarrow{1}}$) satisfies Rule (1) (resp. (2) and (3)), then the root of the cotree that explains $\mathcal{H}$ is labeled "0" (resp. "1" and "$\overrightarrow{1}$").

In the absence of forbidden relations, Theorem 3 immediately implies

Corollary 1. *The partial set* $\mathcal{H} = \{R_0, R_1, R_{\overrightarrow{1}}\}$ *is satisfiable if and only if at least one of the following statements hold*

Rule (0): $|V| = 1$

Rule (1): *(a)* $G_0 := (V, R_1 \cup R_{\overrightarrow{1}})$ *is disconnected and (b) each connected component of* G_0 *is satisfiable.*

Rule (2): *(a)* $G_1 := (V, R_0 \cup R_{\overrightarrow{1}})$ *is disconnected and (b) for each connected component of* G_1 *is satisfiable.*

Rule (3): *(a)* $G_{\overrightarrow{1}} := (V, R_0 \cup R_1 \cup R_{\overrightarrow{1}})$ *contains more than one strongly connected component, and (b) each strongly connected component of* $G_{\overrightarrow{1}}$ *is satisfiable.*

Theorems 1 and 3 together with Corollary 1 imply the following characterization of di-cographs.

Corollary 2. $G = (V, E)$ *is a di-cograph if and only if either*

(0) $|V| = 1$

(1) G *is disconnected and each of its connected components are di-cographs.*

(2) $\overline{G}$ *is disconnected and each of its connected components are di-cographs.*

(3) G *and* $\overline{G}$ *are connected, but* G *contains more than one strongly connected component, each of them is a di-cograph.*

Theorem 3 gives a characterization of satisfiable partial sets $\mathcal{H} = \{R_0, R_1, R_{\overrightarrow{1}}\}$ with respect to some forbidden sets $\mathcal{F} = \{F_0, F_1, F_{\overrightarrow{1}}\}$. In the appendix [1], it is shown that the order of applied rules has no influence on the correctness of the algorithm. Clearly, Theorem 3 immediately provides an algorithm for the recognition of satisfiable sets $\mathcal{H}$, which is summarized in Algorithm 1. Figure 1 shows an example of the application of Theorem 3 and Algorithm 1.

Theorem 4. *Let* $\mathcal{H} = \{R_0, R_1, R_{\overrightarrow{1}}\}$ *be a partial set, and* $\mathcal{F} = \{F_0, F_1, F_{\overrightarrow{1}}\}$ *a forbidden set. Additionally, let* $n = |V|$ *and* $m = |R_0 \cup R_1 \cup R_{\overrightarrow{1}} \cup F_0 \cup F_1 \cup F_{\overrightarrow{1}}|$. *Then, Algorithm 1 runs in* $O(n^2 + nm)$ *time and either:*

(i) outputs a cotree (T, t) *that explains* $\mathcal{H}$; *or*

(ii) outputs the statement "$\mathcal{H}$ is not satisfiable w.r.t. $\mathcal{F}$".

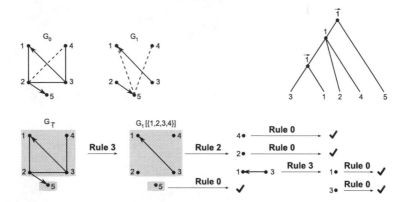

Fig. 1. Consider the partial relations $R_0 = \emptyset$, $R_1 = \{[1,2],[2,3],[3,4]\}$, and $R_{\vec{1}} = \{(3,1),(2,5)\}$, and the forbidden relations $F_0 = \{[2,4]\}$, $F_1 = \{[1,5],[4,5]\}$ and $F_{\vec{1}} = \emptyset$. Here $[x,y] \in R$ means that both (x,y) and (y,x) are contained in R. Left, the three graphs G_0, G_1 and $G_{\vec{1}}$ as defined in Theorem 3 are shown (arrows are omitted for symmetric arcs and dashed arcs correspond to forbidden pairs). Both G_0 and G_1 are connected. However, $G_{\vec{1}}$ contains the two strongly connected components $C_1 = \{1,2,3,4\}$ and $C_2 = \{5\}$ (highlighted with gray rectangles). Thus, Rule (3) can be applied. The graphs $G_0[C_1]$ and $G_{\vec{1}}[C_1]$ are connected. However, since $G_1[C_1]$ is disconnected one can apply Rule (2). For the graph $G[C_2]$ only Rule (0) can be applied. The workflow in the lower part shows stepwise application of allowed rules on the respective components. The final cotree (T,t) that explains the full set $\mathcal{H}^* = \{R_0^* = \emptyset, R_1 = \{[1,2],[2,3],[3,4],[1,4],[2,4]\}, R_{\vec{1}} = \{(3,1),(1,5),(2,5),(3,5),(4,5)\}$ is shown in the upper right part.

Algorithm 1 provides a cotree; $(T;t)$, explaining a full satisfiable set; $\mathcal{H}^* = \{R_0^*, R_1^*, R_{\vec{1}}^*\}$ extended from a given partial set $\mathcal{H}$, such that $\mathcal{H}^*$ is satisfiable w.r.t. a forbidden set $\mathcal{F}$. Nevertheless, it can be shown that $\mathcal{H}^*$ can easily be reconstructed from a given cotree $(T;t)$ in $O(|V|^2)$ time (see Appendix [1]).

5   Experiments

In this section, we investigate the accuracy of the recognition algorithm and compare recovered relations with known full sets that are obtained from simulated cotrees. The intended practical application that we have in mind, is to reconstruct estimated homology relations. In this view, sampling random trees would not be sufficient, as the evolutionary history of genes and species tend to produce fairly balanced trees. Therefore, we used the DendroPy uniform_pure_birth_tree model [17,33] to simulate 1000 binary gene trees for each of the three different leaf sizes $|L| \in \{25, 50, 100\}$. In addition, we randomly labeled the inner vertices of all trees as "0", "1" or "$\vec{1}$" with equal probability.

Each cotree $\mathcal{T} = (T;t)$ then represents a full set $\mathcal{H}^*(\mathcal{T}) = \{R_0^*(\mathcal{T}), R_1^*(\mathcal{T}), R_{\vec{1}}^*(\mathcal{T})\}$. For each of the full sets $\mathcal{H}^*(\mathcal{T})$ and any two vertices $x, y \in V$, the

Algorithm 1. Recognition of satisfiable partial sets $\mathcal{H}$ w.r.t. forbidden sets $\mathcal{F}$ and reconstruction of a cotree (T, t) that explains $\mathcal{H}$.

Input: Partial sets $\mathcal{H} = \{R_0, R_1, R_{\overrightarrow{1}}\}$ and $\mathcal{F} = \{F_0, F_1, F_{\overrightarrow{1}}\}$

Output: A cotree $(T; t)$ that explains $\mathcal{H}$, if one exists and $R_0 \cap F_0 = R_1 \cap F_1 = R_{\overrightarrow{1}} \cap F_{\overrightarrow{1}} = \emptyset$
    or the statement "$\mathcal{H}$ is not satisfiable w.r.t. $\mathcal{F}$"

1: **if** $R_0 \cap F_0 = R_1 \cap F_1 = R_{\overrightarrow{1}} \cap F_{\overrightarrow{1}} = \emptyset$ **then** Call BuildCotree$(V, \mathcal{H}, \mathcal{F})$
2: **else**
3:     Halt and output: "$\mathcal{H}$ is not satisfiable w.r.t. $\mathcal{F}$"

4: **function** BuildCotree$(V, \mathcal{H} = \{R_0, R_1, R_{\overrightarrow{1}}\}, \mathcal{F} = \{F_0, F_1, F_{\overrightarrow{1}}\}$)
                                     ▷ G_0, G_1 and $G_{\overrightarrow{1}}$ are defined as in Thm. 3 for given $\mathcal{H}$ and $\mathcal{F}$
5:     **if** $|V| = 1$ **then return** the cotree $((V, \emptyset), \emptyset)$
6:     **else if** G_0 (resp. G_1) is disconnected **then**
7:         $\mathcal{C} :=$ the set of connected components $\{C_1, \ldots, C_k\}$ of G_0 (resp. G_1)
8:         $\mathcal{T} := \{\text{BuildCotree}(V[C_i], \mathcal{H}[C_i], \mathcal{F}[C_i]) \mid C_i \in \mathcal{C}\}$
9:         **return** the cotree from joining the cotrees in $\mathcal{T}$ under a new root labeled 0 (resp. 1)
10:    **else if** $G_{\overrightarrow{1}}$ has more than one strongly connected component **then**
11:        $\mathcal{C} :=$ the set of strongly connected components $\{C_1, \ldots, C_k\}$ of $G_{\overrightarrow{1}}$
12:        $\pi :=$ a topological order on the quotient $G/\{C_1, \ldots, C_k\}$
13:        $\mathcal{T} := \{\text{BuildCotree}(V[C_i], \mathcal{H}[C_i], \mathcal{F}[C_i]) \mid \text{ for all } C_i, \ i = 1, \ldots, k\}$
14:        **return** the cotree (T, t) obtained by joining the cotrees in $\mathcal{T}$ under a new root
                  with label $\overrightarrow{1}$, where T_i is placed left from T_j whenever $\pi(C_i) < \pi(C_j)$
15:    **else**
16:        Halt and output: "$\mathcal{H}$ is not satisfiable w.r.t. $\mathcal{F}$"

---

corresponding gene pairs (x, y) and (y, x) is removed from $R_0^\star(\mathcal{T}) \cup R_1^\star(\mathcal{T}) \cup R_{\overrightarrow{1}}^\star(\mathcal{T})$ with a fixed probability $p \in \{0.1, 0.2, \ldots, 1\}$. Hence, for each $p \in \{0.1, 0.2, \ldots, 1\}$ and each fixed leaf size $|L| \in \{25, 50, 100\}$, we obtain 1000 partial sets $\mathcal{H} = \{R_0, R_1, R_{\overrightarrow{1}}\}$. We then use Algorithm 1 on each partial set $\mathcal{H}$, to obtain a cotree $\widetilde{\mathcal{T}} = (\widetilde{T}; \tilde{t})$ explaining the full set $\mathcal{H}^\star(\widetilde{\mathcal{T}}) = \{R_0^\star(\widetilde{\mathcal{T}}), R_1(\widetilde{\mathcal{T}}), R_{\overrightarrow{1}}(\widetilde{\mathcal{T}})\}$.

Figure 2(left) shows the average relative difference of the original full set $\mathcal{H}^\star(\mathcal{T})$ and the recovered full sets $\mathcal{H}^\star(\widetilde{\mathcal{T}})$ for each instance. The dashed line in

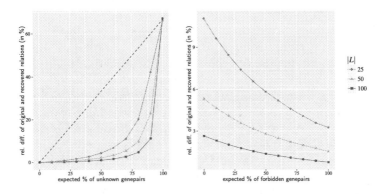

Fig. 2. Shown are the average relative differences of original and recovered relations depending on the size of unknowns (left) and the size of additional forbidden relations (right).

the plots of Fig. 2(left) shows the expected relative difference when each unknown gene pair is assigned randomly to one of the relations in the partial set $\mathcal{H}$. As expected, the relative differences increases with the number of unassigned leaf pairs. Somewhat surprisingly, even if 80% of pairs of leaves are expected to be unassigned in $\mathcal{H}$, it is possible to averagely recover 79.8%–95.3% of the original relations. Moreover, the plot in Fig. 2(left) also suggests that the accuracy of recovered homology relations increases with the input size, i.e., the number of leaves. To explain this fact, observe that the number of constraints given by the full set of homology relations on some leaf set L is $O(|L|^2)$. Conversely, the number of inner vertices in a tree only increases linearly with L, $O(|L|)$. Hence, on average the number of constraints given on the labeling of an internal vertex in the gene tree is $O(|L|^2)/O(|L|) = O(|L|)$. Note, Algorithm 1 constructs cotrees and hence, if there are more leaves, then there are also more constraints for the rules (labeling of the inner vertices) that are allowed to be applied. Therefore, with an increasing number of leaves the correct relation between unassigned pairs of leaves in $\mathcal{H}$ are already determined.

Figure 2(right) shows the impact of additional forbidden relations for the instances where we have removed pairs (x, y) with probability $p = 0.7$. For each of the partial sets $\mathcal{H}$, we have chosen two vertices x and y where neither (x, y) nor (y, x) is contained in $\mathcal{H}$ with probability $p' \in \{0.1, 0.2, \dots, 1\}$ and assigned $(x, y), (y, x)$ to a forbidden relation F_i such that $(x, y), (y, x) \in R_j^\star(\mathcal{T})$ with $i, j \in \{0, 1, \overrightarrow{1}\}$ implies that $i \neq j$. In other words, if $(x, y), (y, x)$ are assigned to F_i with $i \in \{0, 1, \overrightarrow{1}\}$ then $(x, y), (y, x)$ were not in the original set $R_j^\star(\mathcal{T})$. The latter is justified to ensure satisfiability of the partial relations w.r.t. the forbidden relations. Again, we compared the relative difference of the original full sets $\mathcal{H}^\star(\mathcal{T})$ and the recovered full sets $\mathcal{H}^\star(\widetilde{\mathcal{T}})$ computed with Algorithm 1. The plot shows that, with an increasing number of forbidden leaf pairs, the relative difference decreases. Clearly, the more leaf pairs are forbidden the more of such leaf pairs are not allowed to be assigned to one of the relations. Therefore, the degree of freedom for assigning a relation to an unassigned pair decreases with an increase of the number of forbidden pairs.

One factor that may affect the results of the plots shown in Fig. 2 is the order in which rules are chosen when more than one rule is satisfied. By construction, Algorithm 1 fixes the order of applied rules as follows: first Rule (1), then Rule (2), then Rule (3). In other words, when possible the trees for the satisfiable (strongly) connected components are first joined by a common root labeled "1"; if this does not apply, then with common root labeled "0", and "$\overrightarrow{1}$" otherwise. To investigate this issue in more details, we modified Algorithm 1 so that either a different fixed rule order or a random rule order is applied.

Figure 3(left) shows the plot for the partial relations for the fixed leaf size $|L| = 50$. The rule orders are shown in the legend of Fig. 3. Here, $X/Y/Z$, with $X, Y, Z \in \{1 \,\widehat{=}\, \text{Rule}(2), 0 \,\widehat{=}\, \text{Rule}(1), \overrightarrow{1} \,\widehat{=}\, \text{Rule}(3)\}$ being distinct; indicates that first rule X is checked, then rule Y and, if both are not applicable, then rule Z is used. $RAND$ means that each of the allowed rules are chosen with equal probability. As one can observe, the rule order does not have a significant impact

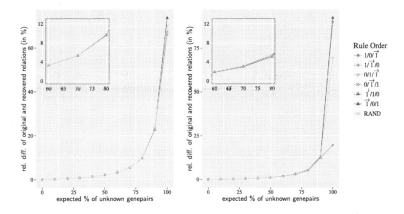

Fig. 3. Shown are the average relative differences of original and recovered relations, depending on the percentage of unassigned pairs and the rule order for cotrees with uniform (left) and skewed (right) label distribution.

on the quality of the recovered full sets. This observation might be explained by the fact that we have used a random assignment of events $0, 1$ and $\vec{1}$ for the vertices of the initial simulated trees T.

To investigate this issue in more detail, we additionally used 1000 unlabeled simulated trees T with $|L| = 50$ and assigned to each vertex v a label $t(v) = 1$ with a probability $p = 0.8$, label $t(v) = \vec{1}$ with $p = 0.1$ and label $t(v) = 0$ with $p = 0.1$ Again, each resulting cotree $\mathcal{T} = (T; t)$ represents a full set $\mathcal{H}^\star(\mathcal{T}) = \{R_0^\star(\mathcal{T}), R_1^\star(\mathcal{T}), R_{\vec{1}}^\star(\mathcal{T})\}$ from which we obtain partial sets $\mathcal{H} = \{R_0, R_1, R_{\vec{1}}\}$ as for the other instances. Figure 3(right) shows the resulting plots. As it can be observed, even a quite skewed distribution of labels in the cotrees and the choice of rule order does not have an effect on the quality of the recovered full sets.

In summary, the results show that it suffices to have only a very few but correct pairs of the original relations to reconstruct most of them.

Acknowledgment. This contribution is supported in part by the Independent Research Fund Denmark, Natural Sciences, grant DFF-7014-00041.

References

1. https://arxiv.org/abs/1711.00504v3
2. Altenhoff, A.M., Dessimoz, C.: Phylogenetic and functional assessment of orthologs inference projects and methods. PLoS Comput. Biol. **5**, e1000262 (2009)
3. Altenhoff, A.M., Gil, M., Gonnet, G.H., Dessimoz, C.: Inferring hierarchical orthologous groups from orthologous gene pairs. PLoS ONE **8**(1), e53786 (2013)
4. Altenhoff, A.M., Škunca, N., Glover, N., Train, C.M., Sucki, A., Piližota, I., Gori, K., Tomiczek, B., Müller, S., Redestig, H., Gonnet, G.H., Dessimoz, C.: The OMA orthology database in 2015: function predictions, better plant support, synteny view and other improvements. Nucleic Acids Res. **43**(D1), D240–D249 (2015)

5. Böcker, S., Dress, A.W.M.: Recovering symbolically dated, rooted trees from symbolic ultrametrics. Adv. Math. **138**, 105–125 (1998)
6. Brandstädt, A., Le, V.B., Spinrad, J.P.: Graph Classes: A Survey. Society for Industrial and Applied Mathematics, Philadelphia (1999)
7. Chen, F., Mackey, A.J., Stoeckert, C.J., Roos, D.S.: OrthoMCL-db: querying a comprehensive multi-species collection of ortholog groups. Nucleic Acids Res. **34**(S1), D363–D368 (2006)
8. Corneil, D.G., Lerchs, H., Steward Burlingham, L.: Complement reducible graphs. Discret. Appl. Math. **3**, 163–174 (1981)
9. Corneil, D.G., Perl, Y., Stewart, L.K.: A linear recognition algorithm for cographs. SIAM J. Comput. **14**, 926–934 (1985)
10. Crespelle, C., Paul, C.: Fully dynamic recognition algorithm and certificate for directed cographs. Discret. Appl. Math. **154**, 1722–1741 (2006)
11. Dessimoz, C., Margadant, D., Gonnet, G.H.: DLIGHT – lateral gene transfer detection using pairwise evolutionary distances in a statistical framework. In: Vingron, M., Wong, L. (eds.) RECOMB 2008. LNCS, vol. 4955, pp. 315–330. Springer, Heidelberg (2008). https://doi.org/10.1007/978-3-540-78839-3_27
12. Dondi, R., El-Mabrouk, N., Lafond, M.: Correction of weighted orthology and paralogy relations - complexity and algorithmic results. In: Frith, M., Storm Pedersen, C.N. (eds.) WABI 2016. LNCS, vol. 9838, pp. 121–136. Springer, Cham (2016). https://doi.org/10.1007/978-3-319-43681-4_10
13. Engelfriet, J., Harju, T., Proskurowski, A., Rozenberg, G.: Characterization and complexity of uniformly nonprimitive labeled 2-structures. Theor. Comp. Sci. **154**, 247–282 (1996)
14. Fitch, W.M.: Homology: a personal view on some of the problems. Trends Genet. **16**, 227–231 (2000)
15. Gao, Y., Hare, D.R., Nastos, J.: The cluster deletion problem for cographs. Discret. Math. **313**(23), 2763–2771 (2013)
16. Gurski, F.: Dynamic programming algorithms on directed cographs. Stat. Optim. Inf. Comput. **5**(1), 35–44 (2017)
17. Hartmann, K., Wong, D., Stadler, T.: Sampling trees from evolutionary models. Syst. Biol. **59**(4), 465–476 (2010)
18. Hellmuth, M., Hernandez-Rosales, M., Huber, K.T., Moulton, V., Stadler, P.F., Wieseke, N.: Orthology relations, symbolic ultrametrics, and cographs. J. Math. Biol. **66**(1–2), 399–420 (2013)
19. Hellmuth, M., Stadler, P.F., Wieseke, N.: The mathematics of xenology: Dicographs, symbolic ultrametrics, 2-structures and tree- representable systems of binary relations. J. Math. Biol. **75**(1), 199–237 (2017)
20. Hellmuth, M., Wieseke, N.: From sequence data including orthologs, paralogs, and xenologs to gene and species trees. In: Pontarotti, P. (ed.) Evolutionary Biology, pp. 373–392. Springer, Cham (2016). https://doi.org/10.1007/978-3-319-41324-2_21
21. Hellmuth, M., Wieseke, N.: On tree representations of relations and graphs: symbolic ultrametrics and cograph edge decompositions. J. Comb. Optim. 1–26 (2017)
22. Hellmuth, M., Wieseke, N., Lechner, M., Lenhof, H.P., Middendorf, M., Stadler, P.F.: Phylogenomics with paralogs. Proc. Natl. Acad. Sci. **112**(7), 2058–2063 (2015)
23. Lafond, M., Dondi, R., El-Mabrouk, N.: The link between orthology relations and gene trees: a correction perspective. Algorithms Mol. Biol. **11**(1), 1 (2016)
24. Lafond, M., El-Mabrouk, N.: Orthology and paralogy constraints: satisfiability and consistency. BMC Genomics **15**(6), S12 (2014)

25. Lafond, M., El-Mabrouk, N.: Orthology relation and gene tree correction: complexity results. In: Pop, M., Touzet, H. (eds.) WABI 2015. LNCS, vol. 9289, pp. 66–79. Springer, Heidelberg (2015). https://doi.org/10.1007/978-3-662-48221-6_5
26. Lawrence, J.G., Hartl, D.L.: Inference of horizontal genetic transfer from molecular data: an approach using the bootstrap. Genetics **131**(3), 753–760 (1992)
27. Lechner, M., Findeiß, S., Steiner, L., Marz, M., Stadler, P.F., Prohaska, S.J.: Proteinortho: detection of (co-)orthologs in large-scale analysis. BMC Bioinform. **12**, 124 (2011)
28. Lechner, M., Hernandez-Rosales, M., Doerr, D., Wiesecke, N., Thevenin, A., Stoye, J., Hartmann, R.K., Prohaska, S.J., Stadler, P.F.: Orthology detection combining clustering and synteny for very large datasets. PLoS ONE **9**(8), e105015 (2014)
29. Liu, Y., Wang, J., Guo, J., Chen, J.: Complexity and parameterized algorithms for cograph editing. Theor. Comput. Sci. **461**, 45–54 (2012)
30. McConnell, R.M., De Montgolfier, F.: Linear-time modular decomposition of directed graphs. Discret. Appl. Math. **145**(2), 198–209 (2005)
31. Östlund, G., Schmitt, T., Forslund, K., Köstler, T., Messina, D.N., Roopra, S., Frings, O., Sonnhammer, E.L.: InParanoid 7: new algorithms and tools for eukaryotic orthology analysis. Nucleic Acids Res. **38**(suppl 1), D196–D203 (2010)
32. Ravenhall, M., Škunca, N., Lassalle, F., Dessimoz, C.: Inferring horizontal gene transfer. PLoS Comput. Biol. **11**(5), e1004095 (2015)
33. Sukumaran, J., Holder, M.T.: Dendropy: a python library for phylogenetic computing. Bioinformatics **26**(12), 1569–1571 (2010)
34. Tatusov, R.L., Galperin, M.Y., Natale, D.A., Koonin, E.V.: The COG database: a tool for genome-scale analysis of protein functions and evolution. Nucleic Acids Res. **28**(1), 33–36 (2000)
35. Trachana, K., Larsson, T.A., Powell, S., Chen, W.H., Doerks, T., Muller, J., Bork, P.: Orthology prediction methods: a quality assessment using curated protein families. BioEssays **33**(10), 769–780 (2011)

Improved Algorithm for Finding the Minimum Cost of Storing and Regenerating Datasets in Multiple Clouds

Yingying Wang, Kun Cheng, and Zimao Li[✉]

College of Computer Science, South-Central University for Nationalities,
182 Minyuan Road, Wuhan 430074, Hubei Province, People's Republic of China
2279457429@qq.com, 820714776@qq.com, lizm@mail.scuec.edu.cn

Abstract. This paper studies intermediate datasets storage problem with linear dataflow in multiple clouds. The proliferation of cloud computing allows users to flexibly store, re-compute or transfer large generated datasets with multiple cloud service providers. However, due to the pay-as-you-go model, the total cost of using cloud services depends on the consumption of storage, computation and bandwidth resources. Given cloud service providers with different pricing models on their resources, users can flexibly choose a cloud service to store a generated dataset, or delete it and then regenerate it when needed, or transfer it to another cloud service in order to reduce the total cost for datasets storage and re-computation. The current best algorithm for finding an optimal strategy of a linear dataflow in multiple clouds takes $O\left(m^4 n^3\right)$, where m is the number of the clouds and n is the number of datasets in a dataflow. In this paper, we present an improved algorithm for the linear dataflow with time complexity $O\left(m^3 n^3\right)$.

Keywords: Multiple clouds · Intermediate datasets
Storage strategies · Time complexity

1 Introduction

A scientific workflow is data intensive running in a cloud computing environment [1]. In a scientific workflow, there are usually a large number of datasets, including initial dataset, output dataset and a large volume of intermediate datasets generated during the execution [2]. The intermediate datasets often contain valuable intermediate results, thus which would be frequently traced back for re-analyzing or re-using [5]. In the process of the datasets, all used resources need to be paid for. As indicated in [1], for a scientific cloud workflow system, storing all the intermediate datasets generated during workflow executions may cause a high storage cost. In contrast, if we delete all the intermediate datasets and re-generate them whenever they are needed, the computation cost of the system may well be very high too. A possible solution is to store some datasets and delete the rest of datasets for minimizing the total cost of the cloud workflow system [3,6]. This leads to the intermediate datasets storage problem (Abbr. for IDSP). In one cloud, we may need to design an optimal strategy to find the best trade-off between the re-generation cost and the storage cost. Hence, in multiple clouds, this problem comes to be more complex, and we may need to design an optimal strategy to find the best trade-off between the re-generation cost, the storage cost and the bandwidth cost. In this paper, we focus on the IDSP in multiple clouds.

In [4], Yuan et al. proposed the newest algorithm with time complexity $O\left(m^4 n^3\right)$ for the linear IDSP in multiple clouds. This paper addresses a new algorithm for linear IDSP, which can improve the time complexity to $O\left(m^3 n^3\right)$, where m is the number of the clouds and the n is the number of the intermediate datasets.

The rest of the paper is organized as follows. Section 2 introduces the intermediate datasets storage problem and related notions. Section 3 describes the algorithm proposed by Yuan et al. Our improved algorithm is in Sect. 4. Section 5 concludes and prospects future work.

2   Related Notions and Notations

2.1   Data Dependency Graph

The intermediate datasets and their relationship can be demonstrated on a data dependency graph (Abbr. for DDG). A DDG is a directed acyclic graph, where the vertexes and directed arcs represent the datasets and the dependencies, respectively. For example, Fig. 1 depicts a simple workflow. d_i points to d_j represents that regenerating d_j needs d_i. In the same way, multiple nodes point to d_j represents that regenerating d_j needs multiple nodes.

To better describe the relationships of datasets in DDG, we define a symbol:$\rightarrow$, which denotes the transitive and dependency relationships between the datasets. In Fig. 1, for example, $d_1 \rightarrow d_2$ means that d_1 is the precursor of d_2 in DDG, moreover, $d_2 \rightarrow d_3$, $d_3 \rightarrow d_5$, $d_4 \rightarrow d_5$. $((d_2 \rightarrow d_3) \cup (d_3 \rightarrow d_5)) \Rightarrow (d_2 \rightarrow d_5)$, etc.

2.2   Datasets Storage Cost, Computing Cost and Bandwidth Cost Model

In cloud environment, the IDSP in multiple clouds is complex, including the IDSP in linear workflow with multiple clouds and the IDSP in non-linear workflow with multiple clouds. In [4], Yuan et al. for the first time proposed a solution for linear DDG with multiple clouds. Facing the complex situation, they simplified the cost model. In this cost model, there are three basic resources involving the $Storage_C ost, Computing_C ost$, and $Bandwidth_C ost$.

In this paper, we consider the algorithm solutions based on this cost model, as follows.

$$Cost = Storage_C ost + Computing_C ost + Bandwidth_C ost \tag{1}$$

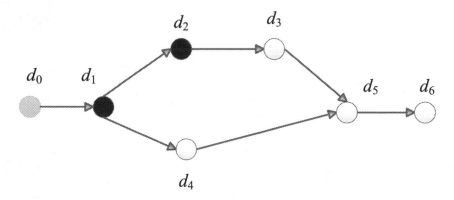

Fig. 1. An example of a workflow.

In this cost model, the Cost is the sum of storage cost, and computing cost, and bandwidth cost, and it is the sum of total cost of storage resources and total cost of computation resources and total cost of the bandwidth resources adopted for transferring intermediate datasets via network.

To utilize the cost model, we present the following assumptions and denotations and definitions.

Assumption [4]: we assume that the datasets are transferred between the m Cloud Service Providers(CSP). And CSP = $\{c_1, c_2, \ldots, c_m\}$.

Denotations [4]: the regeneration cost, the storage cost and the bandwidth cost are represented as X, Y and Z, respectively. In multiple clouds, the three items are related to the cloud service providers. So the three items are denoted as follows:

$X_{d_i}^{c_j}$ denotes the regeneration cost of regenerating d_i from its direct stored precursor in DDG with the cloud service provider c_j

$Y_{d_i}^{c_j}$ denotes the storage cost of d_i with the cloud provider c_j

$Z_{d_i}^{c_j,c_k}$ denotes the bandwidth cost of transferring the d_i from the cloud provider c_j to the cloud provider c_k

In multiple clouds, we aim to find a generic best trade-off among computing and storage and bandwidth (Abbr. GT-CSB) in linear DDG.

2.3   Cost Transitive Tournament Graph for Linear DDG

When considering a linear DDG (See Fig. 2), we generally need to construct a cost transitive tournament graph (Abbr. CTG). For example, in Fig. 2, the linear DDG is constructed by six datasets with determined status. Hence, the total cost of the linear DDG is sum of all datasets storage cost or regeneration cost based on their status. However, in multiple clouds, the dataset storage cost and the regeneration cost are all with the cloud service providers. We can construct the CTG of Fig. 2 by the three steps (See Figs. 3 and 4).

Step 1: Add vertices to CTG As indicated in Fig. 3, we can add m vertices to CTG for each dataset in DDG, denoted as $V_{d_i} = \{ver_{d_i}^{c_1}, ver_{d_2}^{c_2}, \ldots, ver_{d_i}^{c_m}\}$. In

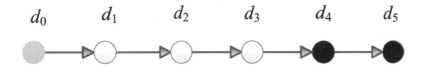

Fig. 2. An example of a linear DDG

order to utilize the Dijkstra algorithm, we need to add the extra vertices called start node and end node, denoted as $ver_{d_0}^{c_0}$ and $ver_{d_{n+1}}^{c_{m+1}}$ without cost.

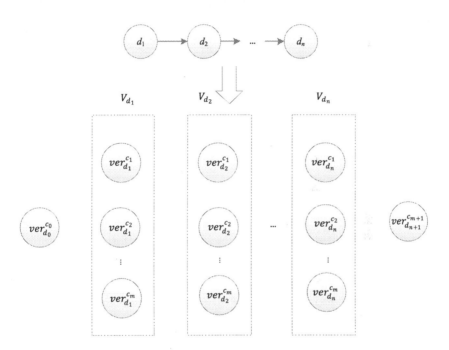

Fig. 3. Create vertices for CTG

Step 2: Add edges to CTG As indicated in Fig. 4, we add $O(m^2n^2)$ edges to construct CTG (See Theorem 1).

Theorem 1. *A complete CTG needs to be added $O(m^2n^2)$ edges, where m is the number of clouds and n is the number of datasets.*

Proof. In Fig. 4, we can see that there are $m*n+2$ vertices, including the secondary nodes such as $ver_{d_0}^{c_0}$ and $ver_{d_{n+1}}^{c_{m+1}}$. According to induction, we can prove Theorem 1, as follows.

(1) For $ver_{d_0}^{c_0}$, there are $m*n+1$ edges created.
(2) For $ver_{d_1}^{c_1}$, there are $m*(n-1)+1$ edges created.

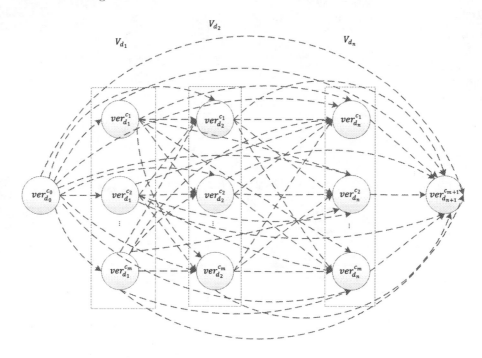

Fig. 4. Create edges for CTG

(3) For $ver_{d_1}^{c_m}$, there are $m * (n - 1) + 1$ edges created.

(4) Apply broadly, for $ver_{d_2}^{c_1}$, there are $m * (n - 1) + 1$ edges created.

(5) For $ver_{d_n}^{c_1}$, there are $m * (n - 2) + 1$ edges created.

(6) Apply broadly, for $ver_{d_{n+1}}^{c_{m+1}}$, there is one edge created.

(7) For $ver_{d_0}^{c_0}$, there is not any edge created.

According to (2)(3)(4), we can calculate the total edges as $m * (m * (n - 1) + 1)$ for one dataset with m clouds. According to (2)(5)(6), we can calculate the total edges as $(m * (n - 1) + 1) + (m * (n - 2) + 1) + \ldots + 1$ for all datasets with one cloud. So, we define the total edges as t_edges, as follows.

$t_edges = (m * n + 1) + (m * (m * (n - 1) + 1)) + (m * (m * (n - 2) + 1)) + (m * (m * (n - 3) + 1)) + \ldots + (m * 1)$

$= (m * n + 1) + m * (n - 1) + m^2 * [(n - 1) + (n - 2) + \ldots + 2 + 1 + 0]$

$= 2mn + 1 - m + m^2 \left(\frac{n^2 - n}{2} \right)$

$= \frac{m^2 n^2}{2} - \frac{m^2 n}{2} + 2mn - m + 1$. This ends the proof.

In this cost transitive graph, we guarantee that the paths in the CTG (from start vertex to end vertex) have one-to-one mapping to the storage strategies of datasets in the DDG. (See Fig. 4)

In one word, for any two vertices, $ver_{d_i}^{c_s}, ver_{d_{i'}}^{c_{s'}} \in$ CTG belonging to different datasets vertex sets (i.e. $V_{d_i} \neq V_{d_{i'}}$), we create an edge between them. Formally,

$$(ver_{d_i}^{c_s}, ver_{d_i'}^{c_{s'}} \in CTG) \cap (d_i \to d_{i'}) \Rightarrow e\langle ver_{d_i}^{c_s}, ver_{d_{i'}}^{c_{s'}} \rangle \tag{2}$$

Lemma 1 [4]. *The storage strategies for a DDG have one-to-one mapping to the paths from $ver_{d_0}^{c_0}$ to $ver_{d_{n+1}}^{c_{m+1}}$ in the CTG.*

Step 3: Set weight to CTG

Definition 1 [4]. *We define the value of a vertex $ver_{d_i}^{c_s} \in$ as the minimum regeneration cost of d_i with cloud service provider c_s from its direct stored predecessors, as follows.*

$$\begin{cases} ver_{d_{i+1}}^{c_k} = Z_{d_i}^{c_s, c_k} + X_{d_{i+1}}^{c_k} \\ ver_{d_j}^{c_k} = min_{h=1}^{m} \left\{ ver_{d_{j-1}}^{c_h} + Z_{d_{j-1}}^{c_h, c_k} \right\} + X_{d_j}^{c_k} \end{cases} \tag{3}$$

Where $d_j \in$ DDG and $d_{i+1} \to d_j \to d_{i'}$, $c_k \in \{c_1, c_2, ..., c_m\}$. And the cost of $ver_{d_0}^{c_0}$ or $ver_{d_{n+1}}^{c_{m+1}}$ is zero.

Definition 2 [4]. *For an edge $e\langle ver_{d_i}^{c_s}, ver_{d_{i'}}^{c_{s'}} \rangle$, the weight of the edge is the sum of $d_{i'}$ storage cost and the datasets' regeneration cost between d_i and $d_{i'}$, supposing that the only d_i and $d_{i'}$ is stored with cloud service provider c_s and $c_{s'}$ respectively and the rest of the datasets are all deleted with cloud service provider c_h $(h = 1 \ldots m)$.*

Based on above definition, the weight of $e\langle ver_{d_i}^{c_s}, ver_{d_{i'}}^{c_{s'}} \rangle$ is as follows:

$$e\langle ver_{d_i}^{c_s}, ver_{d_{i'}}^{c_{s'}} \rangle = Y_{d_{i'}}^{c_{s'}} + \sum_{d_i \to d_j \to d_{i'}} mim_{h=1}^{m}\{ver_{d_j}^{c_h}\} \tag{4}$$

Lemma 2 [4]. *The length of every path from $ver_{d_0}^{c_0}$ to $ver_{d_{n+1}}^{c_{m+1}}$ in the CTG equals to the total cost of the corresponding storage strategy and the minimum cost regeneration strategy for the deleted datasets.*

From the steps above, we can create a weighted directed acyclic graph. Based on the graph theory, we can utilize the Dijkstra algorithm to find the shortest path. We denote the shortest path as $P_{min}\langle ver_{d_0}^{c_0}, ver_{d_{n+1}}^{c_{m+1}} \rangle$.

Theorem 2 [4]. *Given a linear DDG with n datasets $\{d_1, d_2, \ldots, d_n\}$ and m cloud services $\{c_1, c_2, \ldots, c_m\}$ for storage, the length of $P_{min}\langle ver_{d_0}^{c_0}, ver_{d_{n+1}}^{c_{n+1}} \rangle$ of its CTG is the minimum cost of storing and regenerating datasets of the DDG in clouds.*

3 Original GT-CSB Algorithm

In [4], based on the CTG, Yuan et al. proposed the detailed algorithm to find the generic best trade-off among computing and storage and bandwidth (GT-CSB). The pseudo code is as follows. Seen from Fig. 5, we can know that the

Step 1 is demonstrated on line 1–4 and Step 2 is demonstrated on line 5–7. For the linear DDG with n datasets and m cloud service providers, we need to create $m * n + 2$ vertexes (line 1–4) and $O(m^2 n^2)$ edges (line 5–7). Step 3 is demonstrated on line 8–16. To calculate the weight of an edge in the CTG, we need to calculate at most $m * n$ vertices' values (line 11–12). Because of the iterations, the calculated vertices' values can be reused in the next iteration. Hence the time complexity for calculating the value of one vertex is $O(m)$ (line 13). Hence, the time complexity of calculating the weight of an edge is $O(m^2 n)$. In Step 4 (line 17), the time complexity of Dijkstra algorithm is $O(m^2 n^2)$. Hence, the time complexity of calculating weights of all edges in the CTG is $O(m^4 n^3)$.

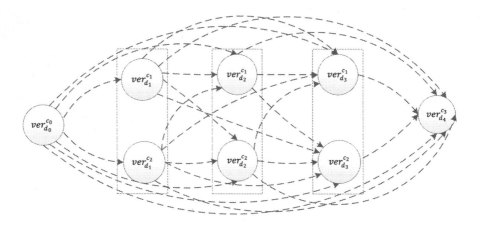

Fig. 5. Pseudo code of GT-CSB algorithm

Based on the above analysis, the total time complexity of the GT-CSB algorithm is $O(m^4 n^3)$. The space complexity of the GT-CSB algorithm is $O(m^2 n^2)$. Because of the iterations, the calculated vertices' values can be reused in the next iteration.

4   Improved Algorithm

4.1   Weakness of Proposed Algorithm

In Yuan's algorithm, they re-calculated the values of vertices calculated by them before. For example, As indicated in Fig. 6, they calculate the weight of $e\langle ver_{d_1}^{c_1}, ver_{d_3}^{c_1} \rangle$ in Yuan's algorithm by the steps.

(1) Calculate the values of $ver_{d_1}^{c_1}$ and $ver_{d_1}^{c_2}$, as follows:

$$ver_{d_1}^{c_1} = Z_{d_0}^{c_0, c_1} + X_{d_1}^{c_1} = X_{d_1}^{c_1}$$
$$ver_{d_1}^{c_2} = Z_{d_0}^{c_0, c_2} + X_{d_1}^{c_2} = X_{d_1}^{c_2}$$

(2) Calculate the values of $ver_{d_2}^{c_1}$ and $ver_{d_2}^{c_1}$, as follows:

$$ver_{d_2}^{c_1} = min_{h=1}^{2}\{ver_{d_1}^{c_h} + Z_{d_1}^{c_h,c_1}\} + X_{d_2}^{c_1}$$
$$ver_{d_2}^{c_2} = min_{h=1}^{2}\{ver_{d_1}^{c_h} + Z_{d_1}^{c_h,c_2}\} + X_{d_2}^{c_2}$$

(3) Calculate the value of $e\langle ver_{d_1}^{c_1}, ver_{d_3}^{c_1}\rangle$, as follows:

$$e\langle ver_{d_1}^{c_1}, ver_{d_3}^{c_1}\rangle = min\{ver_{d_2}^{c_1}, ver_{d_2}^{c_2}\} + Y_{d_3}^{c_1}$$

For instance again, they calculate the weight of $e\langle ver_{d_1}^{c_1}, ver_{d_4}^{c_3}\rangle$ in Yuan's algorithm by the steps.

(1) Calculate the values of $ver_{d_1}^{c_1}$ and $ver_{d_1}^{c_2}$, as follows:

$$ver_{d_1}^{c_1} = Z_{d_0}^{c_0,c_1} + X_{d_1}^{c_1} = X_{d_1}^{c_1}$$
$$ver_{d_1}^{c_2} = Z_{d_0}^{c_0,c_2} + X_{d_1}^{c_2} = X_{d_1}^{c_2}$$

(2) Calculate the values of $ver_{d_2}^{c_1}$ and $ver_{d_2}^{c_2}$, as follows:

$$ver_{d_2}^{c_1} = min_{h=1}^{2}\{ver_{d_1}^{c_h} + Z_{d_1}^{c_h,c_1}\} + X_{d_2}^{c_1}$$
$$ver_{d_2}^{c_2} = min_{h=1}^{2}\{ver_{d_1}^{c_h} + Z_{d_1}^{c_h,c_2}\} + X_{d_2}^{c_2}$$

(3) Calculate the values of $ver_{d_3}^{c_1}$ and $ver_{d_3}^{c_2}$ as follows:

$$ver_{d_3}^{c_1} = min_{h=1}^{2}\{ver_{d_2}^{c_h} + Z_{d_2}^{c_h,c_1}\} + X_{d_3}^{c_1}$$
$$ver_{d_3}^{c_2} = min_{h=1}^{2}\{ver_{d_2}^{c_h} + Z_{d_2}^{c_h,c_2}\} + X_{d_3}^{c_2}$$

(4) Calculate the value of $e\langle ver_{d_1}^{c_1}, ver_{d_4}^{c_3}\rangle$, as follows:

$$e\langle ver_{d_1}^{c_1}, ver_{d_4}^{c_3}\rangle = min_{h=1}^{2}\{ver_{d_2}^{c_h}\} + min_{h=1}^{2}\{ver_{d_3}^{c_h}\}$$

From the examples, we can see that when they calculate the weight of $e\langle ver_{d_1}^{c_1}, ver_{d_3}^{c_1}\rangle$, they calculate the values of $min_{h=1}^{2}\{ver_{d_1}^{c_h}\}$ and $min_{h=1}^{2}\{ver_{d_2}^{c_h}\}$. However, when they calculate the weight of $e\langle ver_{d_1}^{c_1}, ver_{d_4}^{c_3}\rangle$, they did it again, even though they calculated them while calculating the weight of $e\langle ver_{d_1}^{c_1}, ver_{d_4}^{c_3}\rangle$.

We can firstly calculate the weight of $e\langle ver_{d_1}^{c_1}, ver_{d_4}^{c_3}\rangle$ and then calculate the edges with left endpoint $ver_{d_1}^{c_1}$ and arbitrary right endpoint. So we can utilize the values of the nodes that the path passes from $ver_{d_1}^{c_1}$ to $ver_{d_4}^{c_3}$. In this paper, we find a way to reduce calculation redundancies so as to decrease the time complexity.

Definition 3 *If the left endpoint of one edge and the left endpoint of another edge are the same, we define the two edges are the SameClassification (Abbr. for SClass). Formulation:*

$$SClass = \begin{cases} \text{all edges whose the left endpoints are the same} \\ \text{and right endpoints are arbitrary, in other words,} \\ e\langle ver_{d_i}^{c_j}, ver_{d_{i'}}^{c_{j'}}\rangle \cong e\langle ver_{d_i}^{c_j}, ver_{d_{i''}}^{c_{j''}}\rangle \cap \left((d_{i'} \neq d_{i''})\, or\, \left(c_{j'} \neq c_{j''}\right)\right) \end{cases}$$

(5)

```
Algorithm: GT-CSB
Input: A linear DDG {d₁, d₂,…,dₙ};
        Cloud service providers {c₁,c₂,…,cₘ};
Output: The minimum cost
```

1. Create $\text{ver}_{d_0}^{c_0}, \text{ver}_{d_{n+1}}^{c_{m+1}}$;

2. For every dataset d_i in DDG //create vertices for CTG

3.   Create $V_{d_i} \leftarrow \{\text{ver}_{d_i}^{c_1}, \text{ver}_{d_i}^{c_2}, \text{ver}_{d_i}^{c_3}, \cdots, \text{ver}_{d_i}^{c_m}\}$;

4. Add vertices $V_{d_1} \cup V_{d_2} \cup \cdots \cup \{\text{ver}_{d_0}^{c_0}, \text{ver}_{d_{n+1}}^{c_{m+1}}\}$ to CTG ;

5. For every $\text{ver}_{d_i}^{c_s} \in$ CTG;//create edges for CTG

6.   For every $\left(\text{ver}_{d_i'}^{c_s'} \in \text{CTG}\right) \cap \left((d_i, d_{i'} \in \text{CTG}) \cap (d_i \rightarrow d_{i'})\right)$

7.   Create $e \langle \text{ver}_{d_i}^{c_s}, \text{ver}_{d_i}^{c_s'} \rangle$;//create an edge

8.   For every $\text{ver}_{d_{i+1}}^{c_k} \in V_{d_{i+1}}$ //calculate the edge weight

9. $\text{ver}_{d_{i+1}}^{c_k} \leftarrow Z_{d_i}^{c_s, c_k} + X_{d_{i+1}}^{c_k}$;

10.   weight$\leftarrow \min_{h=1}^{m} \{\text{ver}_{d_{i+1}}^{c_h}\}$

11.   For every $\left(d_j \in \text{DDG}\right) \cap \left(d_{i+1} \rightarrow d_j \rightarrow d_{i'}\right)$

12.   For every $\text{ver}_{d_j}^{c_k} \in V_{d_j}$

13. $\text{ver}_{d_j}^{c_k} \leftarrow \min_{h=1}^{m} \left\{\text{ver}_{d_{j-1}}^{c_h} + Z_{d_{j-1}}^{c_h, c_k}\right\} + X_{d_j}^{c_k}$

14.   weight$:=$weight$+\min_{h=1}^{m} \left\{\text{ver}_{d_j}^{c_h}\right\}$

15.   weight$:=$weight$+Y_{d_{i'}}^{c_s'}$

16.   Set $e \langle \text{ver}_{d_i}^{c_s}, \text{ver}_{d_i}^{c_s'} \rangle =$weight;//Set edge weight

17. P:=Dijkstra$(\text{ver}_{d_0}^{c_0}, \text{ver}_{d_{n+1}}^{c_{m+1}}, \text{CTG})$;//Find the shortest path

18. Return P;//the minimum cost is the length of P

Fig. 6. CTG for three datasets with two cloud service providers

Specially, in the *SameClassification*, there is a longest edge whose right end-point is the end node.

To avoid the repeated calculations, among the edges in the $SClass$, we only calculate the longest edge, denoted as $e\langle ver_{d_i}^{c_j}, ver_{d_{n+1}}^{c_{m+1}}\rangle (i \in \{1, 2, \ldots, n\}, j \in \{1, 2, \ldots, m\})$.

Hence, we can directly utilize the calculated values of vertices that the longest path passed to set the weight of the edges in the $SClass$.

After calculating weight of the longest path corresponding to $e\langle ver_{d_i}^{c_j},$ $ver_{d_{n+1}}^{c_{m+1}}\rangle$, we set weight of the edge in the $SClass$ by rules.

(1) The right endpoint is the end node. a. The left endpoint is next to the end node. The weight of the edge is zero. b. The left endpoint is not next to the end node. The weight of the edge is the sum of the values of the vertices that the path passed.
(2) The right endpoint is not the end node. a. The left endpoint is next to the right endpoint. The weight of the edge is the storage cost of the right endpoint. b. The left endpoint is not next to the right endpoint. The weight of the edge is the sum of the values of the storage cost of the right endpoint and the values of the vertices that the path passed.

4.2   Detailed Algorithm

We list the detailed steps, as follows:

Step 1: Adding vertices, which is the same as the above Step 1 in the Sect. 3.
Step 2: Set weight of edges.
Step 2.1: Calculate the value of the next vertice for next iteration (line 5–7);
Step 2.2: Calculate the values of the vertices that the edge passes from the vertice with value of $min_{h=1}^{m}\{ver_{d_{i+1}}^{c_h}\}$ to the $ver_{d_{n+1}}^{c_{m+1}}$ (line 8–10);
Step 3: Create edge and set weight to the edge (line 11–27);
Step 4: Find the shortest path of the CTG (line 28).

Based on the analysis, we demonstrate the pseudo code and make the time complexity analysis, as follows.

From Fig. 7, we can see that Step 1 is demonstrated on line 1–4, where we need to create $m * n$ vertices. Step 2 is demonstrated on line 5–10. In line 5–7, we calculated the values of m adjacent vertices. In line 8–10, we calculated the most $m*n$ values of vertices that the path passed from $ver_{d_{i+1}}^{c_k}$ to the end vertice $ver_{d_{n+1}}^{c_{m+1}}$. Step 3 is demonstrated on line 11–27. In line 11–27, we set weight of $O(m^2n^2)$ edges. The path corresponding to edge passes m*n vertices at most. In conclusion, the time complexity of this algorithm is $O\left(m^3n^3\right)$, compared to $O\left(m^4n^3\right)$ of Yuan, obviously decreased. Also, the space complexity of this algorithm is $O\left(m^2n^2\right)$ same as that of Yuan.

```
Algorithm: GT-CSB
Input: A liner DDG{d₁, d₂, ... , dₙ};
        Cloud service providers {c₁, c₂, ... , cₘ};
Output: The minimum cost benchmark;
```

1. Create $\text{ver}_{d_0}^{c_0}, \text{ver}_{d_{n+1}}^{c_{m+1}}$; //indicating start node and end node

2. For every dataset d_i in DDG //Create vertices for CTG

3.   Create $V_{d_i} \leftarrow \{\text{ver}_{d_i}^{c_1}, \text{ver}_{d_i}^{c_2}, \text{ver}_{d_i}^{c_3}, \cdots, \text{ver}_{d_i}^{c_m}\}$;

4. Add vertices $V_{d_1} \cup V_{d_2} \cup \cdots \cup V_{d_n} \cup \{\text{ver}_{d_0}^{c_0}, \text{ver}_{d_{n+1}}^{c_{m+1}}\}$ to CTG

5. For every $\text{ver}_{d_i}^{c_s} \in$ CTG

6.   For every $\text{ver}_{d_{i+1}}^{c_k} \in V_{d_{i+1}}$ //calculate the adjacent nodes

7. $\text{Ver}_{d_{i+1}}^{c_k} \leftarrow Z_{d_i}^{c_s, c_k} + X_{d_{i+1}}^{c_k}$

8.   For every $(d_j \in DDG) \cap (d_{i+1} \rightarrow d_j \rightarrow d_{n+1})$

9.   For every $\text{ver}_{d_j}^{c_k} \in V_{d_j}$

10. $\text{ver}_{d_j}^{c_k} \leftarrow \min_{h=1}^{m} \{\text{ver}_{d_{j-1}}^{c_h} + Z_{d_{j-1}}^{c_h, c_k}\} + x_{d_j}^{c_k}$

11.   For every $\left(\text{ver}_{d_{i'}}^{c_s'} \in CTG\right) \cap (d_i \rightarrow d_{i'})$

12. $e\langle\text{ver}_{d_i}^{c_s}, \text{ver}_{d_{i'}}^{c_s'}\rangle \leftarrow 0$; //initialize edge

13.   If $i' := n+1$ // right endpoint is the end node

14.   If $i := i'-1$ // left endpoint is next to end node

15. $e\langle\text{ver}_{d_i}^{c_s}, \text{ver}_{d_{i'}}^{c_s'}\rangle := 0$;

16.   Else// left endpoint is not next to the end node

17.   For every $d_j \in DDG \cap d_i \rightarrow d_j \rightarrow d_{i'}$

18. $e\langle\text{ver}_{d_i}^{c_s}, \text{ver}_{d_{i'}}^{c_s'}\rangle := e\langle\text{ver}_{d_i}^{c_s}, \text{ver}_{i'}^{c_s'}\rangle + \min_{h=1}^{m}\{\text{ver}_{d_j}^{c_h}\}$

19.   Else// right endpoint is not the end node

20.   If $i := i'-1$

21. $e\langle\text{ver}_{d_i}^{v_s}, \text{ver}_{d_{i'}}^{c_s'}\rangle := Y_{d_{i'}}^{c_s'}$;

22.   Else

23.   For $(d_j \in DDG) \cap (d_i \rightarrow d_j \rightarrow d_{i'})$

24. $e\langle\text{ver}_{d_i}^{c_s}, \text{ver}_{d_{i'}}^{c_s'}\rangle := e\langle\text{ver}_{d_i}^{c_s}, \text{ver}_{d_{i'}}^{c_s'}\rangle + \min_{h=1}^{m}\{\text{ver}_{d_j}^{c_h}\}$

25. $e\langle\text{ver}_{d_i}^{c_s}, \text{ver}_{d_{i'}}^{c_s'}\rangle := e\langle\text{ver}_{d_i}^{c_s}, \text{ver}_{d_{i'}}^{c_s'}\rangle + Y_{d_{i'}}^{c_s'}$

26. P=Dijkstra($\text{ver}_{d_0}^{c_0}, \text{ver}_{d_{n+1}}^{c_{m+1}}$, CTG);

27. Return P;

Fig. 7. Pseudo code of improved algorithm

5   Conclusions and Future Work

In this paper, we have designed a pseudo algorithm with $O\left(m^3n^3\right)$ for the linear workflow in multiple clouds, which improves upon the best known exited algorithm. Looking forward to future, it is more important and interesting to design more efficient algorithms for parallel intermediate datasets storage problem in multiple clouds.

Acknowledgement. The author thanks reviewers for their constructive suggestions.

References

1. Yuan, D., Yang, Y., Liu, X., et al.: On-demand minimum cost benchmarking for intermediate data storage in scientific cloud workflow systems. J. Parallel Distrib. Comput. **71**(2), 316–332 (2011)
2. Cheng, J., Zhu, D., Zhu, B.: Improved algorithms for intermediate dataset storage in a cloud-based dataflow. Theor. Comput. Sci. **657**, 48–53 (2017)
3. Yuan, D., Yang, Y., Liu, X., et al.: A data dependency based strategy for intermediate data storage in scientific cloud workflow systems. Concurr. Comput.: Pract. Exp. **24**(9), 956–976 (2010)
4. Yuan, D., Cui, L., Li, W., et al.: An algorithm for finding the minimum cost of storing and regenerating datasets in multiple clouds. IEEE Trans. Cloud Comput. (99), 1 (2015)
5. Deelman, E., Chervenak, A.: Data management challenges of data-intensive scientific workflows. In: IEEE International Symposium on Cluster Computing and the Grid (CCGrid 2008), Lyon, France, pp. 687–692 (2008). https://doi.org/10.1109/CCGRID.2008.24
6. Adams, I., Long, D.D.E., Miller, E.L., et al.: Maximizing efficiency by trading storage for computation. In: Workshop on Hot Topics in Cloud Computing (HotCloud 2009), San Diego, CA, pp. 1–5 (2009)

Reconfiguring Spanning and Induced Subgraphs

Tesshu Hanaka[1], Takehiro Ito[2], Haruka Mizuta[2(✉)], Benjamin Moore[3],
Naomi Nishimura[3], Vijay Subramanya[3], Akira Suzuki[2],
and Krishna Vaidyanathan[3]

[1] Chuo University, Tokyo, Japan
hanaka.91t@g.chuo-u.ac.jp
[2] Tohoku University, Sendai, Japan
{takehiro,a.suzuki}@ecei.tohoku.ac.jp,
haruka.mizuta.s4@dc.tohoku.ac.jp
[3] University of Waterloo, Waterloo, Canada
{brmoore,nishi,v7subram,kvaidyan}@uwaterloo.ca

Abstract. SUBGRAPH RECONFIGURATION is a family of problems focusing on the reachability of the solution space in which feasible solutions are subgraphs, represented either as sets of vertices or sets of edges, satisfying a prescribed graph structure property. Although there has been previous work that can be categorized as SUBGRAPH RECONFIGURATION, most of the related results appear under the name of the property under consideration; for example, independent set, clique, and matching. In this paper, we systematically clarify the complexity status of SUBGRAPH RECONFIGURATION with respect to graph structure properties.

1 Introduction

Combinatorial reconfiguration [5], [4], [10] studies the reachability/connectivity of the solution space formed by feasible solutions of an instance of a search problem. More specifically, consider a graph such that each node in the graph represents a feasible solution to an instance of a search problem P, and there is an edge between nodes representing any two feasible solutions that are "adjacent," according to a prescribed *reconfiguration rule* $\mathcal{A}$; such a graph is called the *reconfiguration graph* for P and $\mathcal{A}$. In the *reachability problem* for P and $\mathcal{A}$, we are given *source* and *target* solutions to P, and the goal is to determine whether or not there is a path between the two corresponding nodes in the reconfiguration graph for P and $\mathcal{A}$. We call a desired path a *reconfiguration sequence* between source and target solutions, where a *reconfiguration step* from one solution to another corresponds to an edge in the path.

This work is partially supported by JST ERATO Grant Number JPMJER1201, JST CREST Grant Number JPMJCR1402, and JSPS KAKENHI Grant Numbers JP16K00004 and JP17K12636, Japan. Research by Canadian authors is supported by the Natural Science and Engineering Research Council of Canada.

Table 1. Subgraph representations and variants

Subgraph representations	Variant names	Known reachability problems
Edge subset	Edge	Spanning tree [5]
		Matching [5,9], and b-matching [9]
Vertex subset	Induced	Clique [6]
		Independent set [5,7]
		Induced forest [8]
		Induced bipartite [8]
		Induced tree [11]
	Spanning	Clique [6]

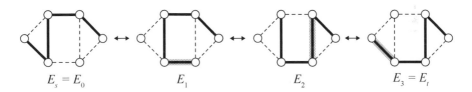

$E_s = E_0$ E_1 E_2 $E_3 = E_t$

Fig. 1. A reconfiguration sequence $\langle E_0, E_1, E_2, E_3 \rangle$ in the edge variant under the TJ rule (also under the TS rule) with the property "a graph is a path," where the edges forming solutions are depicted by thick lines.

1.1 Subgraph Reconfiguration

In this paper, we use the term SUBGRAPH RECONFIGURATION to describe a family of reachability problems that take subgraphs (more accurately, vertex subsets or edge subsets of a given graph) as feasible solutions. Each of the individual problems in the family can be defined by specifying the node set and the edge set of a reconfiguration graph, as follows. (We use the terms *node* for reconfiguration graphs and *vertex* for input graphs.)

Nodes of a Reconfiguration Graph. The set of feasible solutions (i.e., subgraphs) can be defined in terms of a specified graph structure property Π which subgraphs must satisfy; for example, "a graph is a tree," "a graph is edgeless (an independent set)," and so on. By the choice of how to represent subgraphs, each specific problem in the family can be categorized into one of three variants. (See also Table 1.) If a subgraph is represented as an edge subset, which we will call the *edge variant*, then the subgraph formed (induced) by the edge subset must satisfy Π. For example, Fig. 1 illustrates four subgraphs represented as edge subsets, where Π is "a graph is a path." On the other hand, if a subgraph is represented as a vertex subset, we can opt either to require that the subgraph induced by the vertex subset satisfies Π or that the subgraph induced by the vertex subset contains at least one spanning subgraph that satisfies Π; we will refer

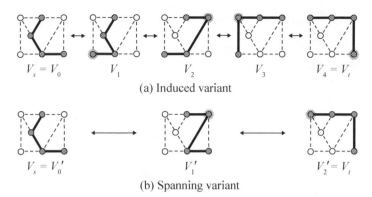

(a) Induced variant

(b) Spanning variant

Fig. 2. Reconfiguration sequences $\langle V_0, V_1, V_2, V_3, V_4 \rangle$ in the induced variant under the TJ rule and $\langle V_0', V_1', V_2' \rangle$ in the spanning variant under the TJ rule with the property "a graph is a path," where the vertices forming solutions are depicted by colored circles, and the subgraphs satisfying the property by thick lines.

to these as the *induced variant* and *spanning variant*, respectively. For example, if Π is "a graph is a path," then in the induced variant, the vertex subset must induce a path, whereas in the spanning variant, the vertex subset is feasible if its induced subgraph contains at least one Hamiltonian path. Figure 2 illustrates feasible vertex subsets of the induced variant and spanning variant. In the figure, the vertex subset V_1' is feasible in the spanning variant, but is not feasible in the induced variant, because it contains a spanning path but does not induce a path. As can be seen by this simple example, in the spanning variant, we need to pay attention to the additional complexity of finding a spanning subgraph and the complications resulting from the fact that the subgraph induced by the vertex subset may contain more than one spanning subgraph which satisfies Π.

Edges of a Reconfiguration Graph. Since we represent a feasible solution by a set of vertices (or edges) in any variant, we can consider that tokens are placed on each vertex (resp., edge) in the feasible solution. Then, in this paper, we mainly deal with the two well-known reconfiguration rules, called the token-jumping (TJ) [7] and token-sliding (TS) rules [2,3,7]. In the former, a token can move to any other vertex (edge) in a given graph, whereas in the latter it can move only to an adjacent vertex (adjacent edge, that is sharing a common vertex.) For example, Figs. 1 and 2 illustrate reconfiguration sequences under the TJ rule for each variant. Note that the sequence in Fig. 1 can also be considered as a sequence under the TS rule. In the reconfiguration graph, two nodes are adjacent if and only if one of the two corresponding solutions can be obtained from the other one by a single move of one token that follows the specified reconfiguration rule. Therefore, all nodes in a connected component of the reconfiguration graph represent subgraphs having the same number of vertices (edges).

We note in passing that since in most cases we wish to retain the same number of vertices and/or edges, we rarely use the token-addition-and-removal (TAR) rule [5,7], where we can add or remove a single token at a time, for SUBGRAPH RECONFIGURATION problems.

1.2 Previous Work

Although there has been previous work that can be categorized as SUBGRAPH RECONFIGURATION, most of the related results appear under the name of the property Π under consideration. Accordingly, we can view reconfiguration of independent sets [5,7] as the induced variant of SUBGRAPH RECONFIGURATION such that the property Π is "a graph is edgeless." Other examples can be found in Table 1. We here explain only known results which are directly related to our contributions.

Reconfiguration of cliques can be seen as both the spanning and the induced variant; the problem is PSPACE-complete under any rule, even when restricted to perfect graphs [6]. Indeed, for this problem, the rules TAR, TJ, and TS have all been shown to be equivalent from the viewpoint of polynomial-time solvability. It is also known that reconfiguration of cliques can be solved in polynomial time for several well-known graph classes [6].

Wasa et al. [11] considered the induced variant under the TJ and TS rules with the property Π being "a graph is a tree." They showed that this variant under each of the TJ and TS rules is PSPACE-complete, and is W[1]-hard when parameterized by both the size of a solution and the length of a reconfiguration sequence. They also gave a fixed-parameter algorithm when parameterized by both the size of a solution and the maximum degree of an input graph, under both the TJ and TS rules. In closely related work, Mouawad et al. [8] considered the induced variants of SUBGRAPH RECONFIGURATION under the TAR rule with the properties Π being either "a graph is a forest" or "a graph is bipartite." They showed that these variants are W[1]-hard when parameterized by the size of a solution plus the length of a reconfiguration sequence.

1.3 Our Contributions

In this paper, we study the complexity of SUBGRAPH RECONFIGURATION under the TJ and TS rules. (Our results are summarized in Table 2, together with known results, where an (i,j)-biclique is a complete bipartite graph with the bipartition of i vertices and j vertices.) As mentioned above, because we consider the TJ and TS rules, it suffices to deal with subgraphs having the same number of vertices or edges. Subgraphs of the same size may be isomorphic for certain properties Π, such as "a graph is a path" and "a graph is a clique," because there is only one choice of a path or a clique of a particular size. On the other hand, for the property "a graph is a tree," there are several choices of trees of a particular size. (We will show an example in Sect. 3 with Fig. 3.)

As shown in Table 2, we systematically clarify the complexity of SUBGRAPH RECONFIGURATION for several fundamental graph properties. In particular, we show that the edge variant under the TJ rule is computationally intractable for the property "a graph is a path" but tractable for the property "a graph is a tree." This implies that the computational (in)tractability does not follow directly from the inclusion relationship of graph classes required as the properties Π; one possible explanation is that the path property implies a specific graph, whereas the tree property allows several choices of trees, making the problem easier.

We omitted proofs for the claims marked with (*) from this extended abstract due to the page limitation.

Table 2. Previous and new results

Property Π	Edge variant	Induced variant	Spanning variant
Path	NP-hard (TJ) [Theorem 2]	PSPACE-c. (TJ, TS) [Theorems 7, 9]	PSPACE-c. (TJ, TS) [Theorems 7, 9]
Cycle	P (TJ, TS) [Theorem 3]	PSPACE-c. (TJ, TS) [Theorems 8, 9]	PSPACE-c. (TJ, TS) [Theorems 8, 9]
Tree	P (TJ) [Theorem 6]	PSPACE-c. (TJ, TS) [11]	P (TJ) PSPACE-c. (TS) Theorems [11], [10]
(i, j)-biclique	P (TJ, TS) [Theorem 5]	PSPACE-c. for $i = j$ (TJ) PSPACE-c. for fixed i (TJ) [Corollary 1, Theorem 12]	NP-hard for $i = j$ (TJ) P for fixed i (TJ) [Theorems 13, 14]
Clique	P (TJ, TS) [Theorem 4]	PSPACE-c. (TJ, TS) [6]	PSPACE-c. (TJ, TS) [6]
Diameter two		PSPACE-c. (TS) [Theorem 15]	PSPACE-c. (TS) [Theorem 15]
Any property	XP for solution size (TJ, TS) [Theorem 1]	XP for solution size (TJ, TS) [Theorem 1]	XP for solution size (TJ, TS) [Theorem 1]

1.4   Preliminaries

Although we assume throughout the paper that an input graph G is simple, all our algorithms can be easily extended to graphs having multiple edges with small modifications. We denote by (G, V_s, V_t) an instance of a spanning variant or an induced variant whose input graph is G and source and target solutions are vertex subsets V_s and V_t of G. Similarly, we denote by (G, E_s, E_t) an instance of the edge variant. We may assume without loss of generality that $|V_s| = |V_t|$ holds for the spanning and induced variants, and $|E_s| = |E_t|$ holds for the edge variant; otherwise, the answer is clearly no since under both the TJ and TS rules, all solutions must be of the same size.

2   General Algorithm

In this section, we give a general XP algorithm when the size of a solution (that is, the size of a vertex or edge subset that represents a subgraph) is taken as the parameter. For notational convenience, we simply use *element* to represent a vertex (or an edge) for the spanning and induced variants (resp., the edge variant), and *candidate* to represent a set of elements (which does not necessarily satisfy the property Π.) Furthermore, we define the *size* of a given graph as the number of elements in the graph.

Theorem 1. *Let Π be any graph structure property, and let $f(k)$ denote the time to check if a candidate of size k satisfies Π. Then, all of the spanning, induced, and edge variants under the TJ or TS rules can be solved in time $O(n^{2k}k + n^k f(k))$, where n is the size of a given graph and k is the size of a source (and target) solution. Furthermore, a shortest reconfiguration sequence between source and target solutions can be found in the same time bound, if it exists.*

Proof. Our claim is that the reconfiguration graph can be constructed in the stated time. Since a given source solution is of size k, it suffices to deal only with candidates of size exactly k. For a given graph, the total number of possible candidates of size k is $O(n^k)$. For each candidate, we can check in time $f(k)$ whether it satisfies Π. Therefore, we can construct the node set of the reconfiguration graph in time $O(n^k f(k))$. We then obtain the edge set of the reconfiguration graph. Since there are $O(n^k)$ nodes in the reconfiguration graph, the number of pairs of nodes is $O(n^{2k})$. Since each node corresponds to a set of k elements, we can check if two nodes are adjacent or not in $O(k)$ time. Therefore, we can find all pairs of adjacent nodes in time $O(n^{2k}k)$.

In this way, we can construct the reconfiguration graph in time $O(n^{2k}k + n^k f(k))$ in total. The reconfiguration graph consists of $O(n^k)$ nodes and $O(n^{2k})$ edges. Therefore, by breadth-first search starting from the node representing a given source solution, we can determine in time $O(n^{2k})$ whether or not there exists a reconfiguration sequence between two nodes representing the source and target solutions. Notice that if a desired reconfiguration sequence exists, then the breadth-first search finds a shortest one. $\square$

3   Edge Variants

In this section, we study the edge variant of SUBGRAPH RECONFIGURATION for the properties of being paths, cycles, cliques, bicliques, and trees.

We first consider the property "a graph is a path" under the TJ rule.

Theorem 2 (*). *The edge variant of* SUBGRAPH RECONFIGURATION *under the TJ rule is NP-hard for the property "a graph is a path."*

We now consider the property "a graph is a cycle," as follows.

Theorem 3. *The edge variant of* SUBGRAPH RECONFIGURATION *under each of the TJ and TS rules can be solved in linear time for the property "a graph is a cycle."*

Proof. Let (G, E_s, E_t) be a given instance. We claim that the reconfiguration graph is edgeless, in other words, no feasible solution can be transformed at all. Then, the answer is yes if and only if $E_s = E_t$ holds; this condition can be checked in linear time.

Let E' be any feasible solution of G, and consider a replacement of an edge $e^- \in E'$ with an edge e^+ other than e^-. Let u, v be the endpoints of e^-. When we remove e^- from E', the resulting edge subset $E' \setminus \{e\}$ forms a path whose ends are u and v. Then, to ensure that the candidate forms a cycle, we can choose only $e^- = uv$ as e^+. This contradicts the assumption that $e^+ \neq e^-$. □

The same arguments hold for the property "a graph is a clique," and we obtain the following theorem. We note that, for this property, both induced and spanning variants (i.e., when solutions are represented by vertex subsets) are PSPACE-complete under any rule [6].

Theorem 4. *The edge variant of* SUBGRAPH RECONFIGURATION *under each of the TJ and TS rules can be solved in linear time for the property "a graph is a clique."*

We next consider the property "a graph is an (i, j)-biclique," as follows.

Theorem 5. *The edge variant of* SUBGRAPH RECONFIGURATION *under each of the TJ and TS rules can be solved in polynomial time for the property "a graph is an (i, j)-biclique" for any pair of positive integers i and j.*

Proof. We may assume without loss of generality that $i \leq j$ holds. We prove the theorem in the following three cases: **Case 1:** $i = 1$ and $j \leq 2$; **Case 2:** $i, j \geq 2$; and **Case 3:** $i = 1$ and $j \geq 3$.

We first consider **Case 1**, which is the easiest case. In this case, any $(1, j)$-biclique has at most two edges. Therefore, by Theorem 1 we can conclude that this case is solvable in polynomial time.

We then consider **Case 2**. We show that (G, E_s, E_t) is a yes-instance if and only if $E_s = E_t$ holds. To do so, we claim that the reconfiguration graph is edgeless, in other words, no feasible solution can be transformed at all. To see this, because $i, j \geq 2$, notice that the removal of any edge e in an (i, j)-biclique results in a bipartite graph with the same bipartition of i vertices and j vertices. Therefore, to obtain an (i, j)-biclique by adding a single edge, we must add back the same edge e.

We finally deal with **Case 3**. Notice that a $(1, j)$-biclique is a star with j leaves, and its center vertex is of degree $j \geq 3$. Then, we claim that (G, E_s, E_t) is a yes-instance if and only if the center vertices of stars represented by E_s and E_t are the same. The if direction clearly holds, because we can always move edges in $E_s \setminus E_t$ into ones in $E_t \setminus E_s$ one by one. We thus prove the only-if direction; indeed, we prove the contrapositive, that is, the answer is no if the center vertices of stars represented by E_s and E_t are different. Consider such a star T_s formed by E_s. Since T_s has j (≥ 3) leaves, the removal of any edge in E_s results in a star having $j - 1$ (≥ 2) leaves. Therefore, to ensure that each intermediate solution is a star with j leaves, we can add only an edge of G which is incident to the center of T_s. Thus, we cannot change the center vertex.   □

In this way, we have proved Theorem 5. Although **Case 1** takes non-linear time, our algorithm can be easily improved (without using Theorem 1) so that it runs in linear time.

We finally consider the property "a graph is a tree" under the TJ rule. As we have mentioned in the introduction, for this property, there are several choices of trees even of a particular size, and a reconfiguration sequence does not necessarily consist of isomorphic trees (see Fig. 3). This "flexibility" of subgraphs may yield the contrast between Theorem 2 for the path property and the following theorem for the tree property.

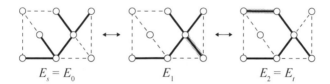

$$E_s = E_0 \qquad E_1 \qquad E_2 = E_t$$

Fig. 3. Reconfiguration sequence $\langle E_0, E_1, E_2 \rangle$ in the edge variant under the TJ rule with the property "a graph is a tree."

Theorem 6 (*). *The edge variant of* SUBGRAPH RECONFIGURATION *under the TJ rule can be solved in linear time for the property "a graph is a tree."*

4   Induced and Spanning Variants

In this section, we deal with the induced and spanning variants where subgraphs are represented as vertex subsets. Most of our results for these variants are hardness results, except for Theorems 11 and 14.

4.1   Path and Cycle

In this subsection, we show that both induced and spanning variants under the TJ or TS rules are PSPACE-complete for the properties "a graph is a path" and "a graph is a cycle." All proofs in this subsection make use of reductions that employ almost identical constructions. Therefore, we describe the detailed proof for only one case, and give proof sketches for the other cases.

We give polynomial-time reductions from the SHORTEST PATH RECONFIGU- RATION problem, which can be seen as a SUBGRAPH RECONFIGURATION prob- lem, defined as follows [1]. For a simple, unweighted, and undirected graph G and two distinct vertices s, t of G, SHORTEST PATH RECONFIGURATION is the induced (or spanning) variant of SUBGRAPH RECONFIGURATION under the TJ rule for the property "a graph is a shortest st-path." Notice that there is no difference between the induced variant and the spanning variant for this property, because any shortest path in a simple graph forms an induced subgraph. This problem is known to be PSPACE-complete [1].

Let d be the (shortest) distance from s to t in G. For each $i \in \{0, 1, \ldots, d\}$, we denote by $L_i \subseteq V(G)$ the set of vertices that lie on a shortest st-path at distance i from s. Therefore, we have $L_0 = \{s\}$ and $L_d = \{t\}$. We call each L_i a *layer*. Observe that any shortest st-path contains exactly one vertex from each layer, and we can assume without loss of generality that G has no vertex which does not belong to any layer.

We first give the following theorem.

Theorem 7. *For the property "a graph is a path," the induced and spanning variants of* SUBGRAPH RECONFIGURATION *under the TJ rule are both PSPACE-complete on bipartite graphs.*

Proof. Observe that these variants are in PSPACE. Therefore, we construct a polynomial-time reduction from SHORTEST PATH RECONFIGURATION.

Let (G, V_s, V_t) be an instance of SHORTEST PATH RECONFIGURATION. Since any shortest st-path contains exactly one vertex from each layer, we can assume without loss of generality that G has no edge joining two vertices in the same layer, that is, each layer L_i forms an independent set in G. Then, G is a bipartite graph. From (G, V_s, V_t), we construct a corresponding instance (G', V'_s, V'_t) for the induced and spanning variants; note that we use the same reduction for both variants. Let G' be the graph obtained from G by adding four new vertices s_1, s_2, t_1, t_2 which are connected with four new edges $s_2 s_1$, $s_1 s$, $t t_1$, $t_1 t_2$. Note that G' is also bipartite. We then set $V'_s = V_s \cup \{s_1, s_2, t_1, t_2\}$ and $V'_t = V_t \cup \{s_1, s_2, t_1, t_2\}$. Since each of V_s and V_t induces a shortest st-path in G, each of V'_s and V'_t is a feasible solution to both variants. This completes the polynomial-time construction of the corresponding instance.

We now give the key lemma for proving the correctness of our reduction.

Lemma 1 (*). *Let $V' \subseteq V(G')$ be any solution for the induced or spanning variant which is reachable by a reconfiguration sequence from V'_s (or V'_t) under the TJ rule. Then, V' satisfies the following two conditions:*

(a) $s_2, s_1, s, t, t_1, t_2 \in V'$; and
(b) V' contains exactly one vertex from each layer of G.

Consider any vertex subset $V'' \subseteq V(G')$ which satisfies conditions (a) and (b) of Lemma 1; note that V'' is not necessarily a feasible solution. Then, these conditions ensure that $V'' \setminus \{s_2, s_1, t_1, t_2\}$ forms a shortest st-path in G if and only if the subgraph represented by V'' induces a path in G'. Thus, an instance (G, V_s, V_t) of SHORTEST PATH RECONFIGURATION is a **yes**-instance if and only if the corresponding instance (G', V'_s, V'_t) of the induced or spanning variant is a **yes**-instance. This completes the proof of Theorem 7.   □

Similar arguments give the following theorem.

Theorem 8. *Both the induced and spanning variants of* SUBGRAPH RECONFIG-URATION *under the TJ rule are PSPACE-complete for the property "a graph is a cycle."*

Proof. Our reduction is the same as in the proof of Theorem 7 except for the following point: instead of adding four new vertices, we connect s and t by a path of length three with two new vertices s_1 and t_1. Then, the same arguments hold as the proof of Theorem 7.   □

4.2   Path and Cycle Under the TS Rule

We now consider the TS rule. Notice that, in the proofs of Theorems 7 and 8, we exchange only vertices contained in the same layer. Since any shortest st-path in a graph G contains exactly one vertex from each layer, we can assume without loss of generality that each layer L_i of G forms a clique. Then, the same reductions work for the TS rule, and we obtain the following theorem.

Theorem 9. *Both the induced and spanning variants of* SUBGRAPH RECONFIG-URATION *under the TS rule are PSPACE-complete for the properties "a graph is a path" and "a graph is a cycle."*

4.3   Tree

Wasa et al. [11] showed that the induced variant under the TJ and TS rules is PSPACE-complete for the property "a graph is a tree." In this subsection, we show that the spanning variant for this property is also PSPACE-complete under the TS rule, while it is linear-time solvable under the TJ rule.

We first note that our proof of Theorem 9 yields the following theorem.

Theorem 10. *The spanning variant of* SUBGRAPH RECONFIGURATION *under the TS rule is PSPACE-complete for the property "a graph is a tree."*

Proof. We claim that the same reduction as in Theorem 9 applies. Let $V' \subseteq V(G')$ be any solution which is reachable by a reconfiguration sequence from V'_s (or V'_t) under the TS rule, where (G', V'_s, V'_t) is the corresponding instance for the spanning variant, as in the reduction. Then, the TS rule ensures that $s_2, s_1, s, t, t_1, t_2 \in V'$ holds, and V' contains exactly one vertex from each layer of G. Therefore, any solution forms a path even for the property "a graph is a tree," and hence the theorem follows.   □

In contrast to Theorem 10, the spanning variant under the TJ rule is solvable in linear time. We note that the reduction in Theorem 10 does not work under the TJ rule, because the tokens on s_2 and t_2 can move (jump) and hence there is no guarantee that a solution forms a path for the property "a graph is a tree."

Theorem 11. *The spanning variant of* SUBGRAPH RECONFIGURATION *under the TJ rule can be solved in linear time for the property "a graph is a tree."*

Suppose that (G, V_s, V_t) is a given instance. We assume that $|V_s| = |V_t| \geq 2$ holds; otherwise it is a trivial instance. Then, Theorem 11 can be obtained from the following lemma.

Lemma 2 (*). (G, V_s, V_t) with $|V_s| = |V_t| \geq 2$ is a yes-*instance if and only if* V_s *and* V_t *are contained in the same connected component of* G.

4.4   Biclique

For the property "a graph is an (i, j)-biclique," we show that the induced variant under the TJ rule is PSPACE-complete even if $i = j$ holds, or i is fixed. On the other hand, the spanning variant under the TJ rule is NP-hard even if $i = j$ holds, while it is polynomial-time solvable when i is fixed.

We first give the following theorem for a fixed $i \geq 1$.

Theorem 12. *For the property "a graph is an (i, j)-biclique," the induced variant of* SUBGRAPH RECONFIGURATION *under the TJ rule is PSPACE-complete even for any fixed integer $i \geq 1$.*

Proof. We give a polynomial-time reduction from the MAXIMUM INDEPENDENT SET RECONFIGURATION problem [12], which can be seen as a SUBGRAPH RECONFIGURATION problem. The MAXIMUM INDEPENDENT SET RECONFIGURATION problem is the induced variant for the property "a graph is edgeless" such that two given independent sets are maximum. Note that, because we are given maximum independent sets, there is no difference between the TJ and TS rules for this problem. This problem is known to be PSPACE-complete [12].

Suppose that (G, V_s, V_t) is an instance of MAXIMUM INDEPENDENT SET RECONFIGURATION. We now construct a corresponding instance (G', V_s', V_t') of the induced variant under the TJ rule for the property "a graph is an (i, j)-biclique," where i is any fixed positive integer. Let L and R be distinct sets of new vertices such that $|L| = i$ and $|R| = 1$. The vertex set of G' is defined as $V(G') = V(G) \cup L \cup R$, and the edge set of G' as $E(G') = E(G) \cup \{uv \mid u \in V(G), v \in L\} \cup \{vw \mid v \in L, w \in R\}$, that is, new edges are added so that there are edges between each vertex of L and each vertex of $V(G) \cup R$. Let $V_s' = V_s \cup L \cup R$ and $V_t' = V_t \cup L \cup R$. Since L, R, V_s and V_t are all independent sets in G', both V_s' and V_t' form (i, j)-bicliques, where $i = |L|$ and $j = |V_s \cup R| = |V_t \cup R|$. We have now completed the construction of our corresponding instance, which can be accomplished in polynomial time. We omit the correctness proof of our reduction from this extended abstract. $\square$

The corresponding instance (G', V_s', V_t') constructed in the proof of Theorem 12 satisfies $i = j$ if we set $i = |V_s| + 1 = |V_t| + 1$. Therefore, we can obtain the following corollary.

Corollary 1. *For the property "a graph is an (i, j)-biclique," the induced variant of* SUBGRAPH RECONFIGURATION *under the TJ rule is PSPACE-complete even if $i = j$ holds.*

We next give the following theorem.

Theorem 13 (*). *For the property "a graph is an (i,j)-biclique," the spanning variant of* SUBGRAPH RECONFIGURATION *under the TJ rule is NP-hard even if $i = j$ holds.*

We now give a polynomial-time algorithm solving the spanning variant for a fixed constant $i \geq 1$.

Theorem 14 (*). *For the property "a graph is an (i,j)-biclique," the spanning variant of* SUBGRAPH RECONFIGURATION *under the TJ rule is solvable in polynomial time when $i \geq 1$ is a fixed constant.*

4.5   Diameter-Two Graph

In this subsection, we consider the property "a graph has diameter at most two." Note that the induced and spanning variants are the same for this property.

Theorem 15 (*). *Both induced and spanning variants of* SUBGRAPH RECON-FIGURATION *under the TS rule are PSPACE-complete for the property "a graph has diameter at most two."*

5   Conclusions and Future Work

The work in this paper initiates a systematic study of SUBGRAPH RECONFIGURA-TION. Although we have identified graph structure properties which are harder for the induced variant than the spanning variant, it remains to be seen whether this pattern holds in general. For the general case, questions of the roles of diameter and the number of subgraphs satisfying the property are worthy of further investigation. Another obvious direction for further research is an investigation into the fixed-parameter complexity of SUBGRAPH RECONFIGURATION.

A natural extension of SUBGRAPH RECONFIGURATION is the extension from isomorphism of graph structure properties to other mappings, such as topological minors.

References

1. Bonsma, P.: The complexity of rerouting shortest paths. Theor. Comput. Sci. **510**, 1–12 (2013)
2. Bonsma, P., Cereceda, L.: Finding paths between graph colourings: PSPACE-completeness and superpolynomial distances. Theor. Comput. Sci. **410**(50), 5215–5226 (2009)
3. Hearn, R.A., Demaine, E.D.: PSPACE-completeness of sliding-block puzzles and other problems through the nondeterministic constraint logic model of computation. Theor. Comput. Sci. **343**(1–2), 72–96 (2005)
4. van den Heuvel, J.: The complexity of change. In: Surveys in Combinatorics, London Mathematical Society Lecture Note Series, vol. 409, pp. 127–160. Cambridge University Press (2013)

5. Ito, T., Demaine, E.D., Harvey, N.J.A., Papadimitriou, C.H., Sideri, M., Uehara, R., Uno, Y.: On the complexity of reconfiguration problems. Theor. Comput. Sci. **412**(12–14), 1054–1065 (2011)
6. Ito, T., Ono, H., Otachi, Y.: Reconfiguration of cliques in a graph. In: Jain, R., Jain, S., Stephan, F. (eds.) TAMC 2015. LNCS, vol. 9076, pp. 212–223. Springer, Cham (2015). https://doi.org/10.1007/978-3-319-17142-5_19
7. Kamiński, M., Medvedev, P., Milanič, M.: Complexity of independent set reconfigurability problems. Theor. Comput. Sci. **439**, 9–15 (2012)
8. Mouawad, A.E., Nishimura, N., Raman, V., Simjour, N., Suzuki, A.: On the parameterized complexity of reconfiguration problems. Algorithmica **78**(1), 274–297 (2017)
9. Mühlenthaler, M.: Degree-constrained subgraph reconfiguration is in P. In: Italiano, G.F., Pighizzini, G., Sannella, D.T. (eds.) MFCS 2015. LNCS, vol. 9235, pp. 505–516. Springer, Heidelberg (2015). https://doi.org/10.1007/978-3-662-48054-0_42
10. Nishimura, N.: Introduction to reconfiguration. Algorithms **11**(4), 52 (2018)
11. Wasa, K., Yamanaka, K., Arimura, H.: The complexity of induced tree reconfiguration problems. In: Dediu, A.-H., Janoušek, J., Martín-Vide, C., Truthe, B. (eds.) LATA 2016. LNCS, vol. 9618, pp. 330–342. Springer, Cham (2016). https://doi.org/10.1007/978-3-319-30000-9_26
12. Wrochna, M.: Reconfiguration in bounded bandwidth and tree-depth. J. Comput. Syst. Sci. **93**, 1–10 (2018)

Generalizing the Hypergraph Laplacian via a Diffusion Process with Mediators

T.-H. Hubert Chan[✉] and Zhibin Liang

Department of Computer Science, The University of Hong Kong,
Pokfulam, Hong Kong
hubert@cs.hku.hk, liangzb@connect.hku.hk

Abstract. In a recent breakthrough STOC 2015 paper, a continuous diffusion process was considered on hypergraphs (which has been refined in a recent JACM 2018 paper) to define a Laplacian operator, whose spectral properties satisfy the celebrated Cheeger's inequality. However, one peculiar aspect of this diffusion process is that each hyperedge directs flow only from vertices with the maximum density to those with the minimum density, while ignoring vertices having strict in-beween densities.

In this work, we consider a generalized diffusion process, in which vertices in a hyperedge can act as mediators to receive flow from vertices with maximum density and deliver flow to those with minimum density. We show that the resulting Laplacian operator still has a second eigenvalue satisfying the Cheeger's inequality.

Our generalized diffusion model shows that there is a family of operators whose spectral properties are related to hypergraph conductance, and provides a powerful tool to enhance the development of spectral hypergraph theory. Moreover, since every vertex can participate in the new diffusion model at every instant, this can potentially have wider practical applications.

1 Introduction

Spectral graph theory, and specifically, the well-known Cheeger's inequality give a relationship between the edge expansion properties of a graph and the eigenvalues of some appropriately defined matrix [1,2]. Loosely speaking, for a given graph, its edge expansion or *conductance* gives a lower bound on the ratio of the number of edges leaving a subset S of vertices to the sum of vertex degrees in S. It is natural that graph conductance is studied in the context of graph partitioning or clustering [10,12,13], whose goal is to minimize the weight of edges crossing different clusters with respect to intra-cluster edges. The reader can refer to the standard references [6,9] for an introduction to spectral graph theory.

The full version of this paper is available online [3].

T.-H. H. Chan—This work was partially supported by the Hong Kong RGC under the grant 17200817.

© Springer International Publishing AG, part of Springer Nature 2018
L. Wang and D. Zhu (Eds.): COCOON 2018, LNCS 10976, pp. 441–453, 2018.
https://doi.org/10.1007/978-3-319-94776-1_37

Recent Generalization to Hypergraphs. In an edge-weighted hypergraph $H = (V, E, w)$, an edge $e \in E$ is a non-empty subset of V. The edges have positive weights indicated by $w : E \to \mathbb{R}_+$. The weight of each vertex $v \in V$ is its weighted degree $w_v := \sum_{e \in E : v \in e} w_e$. A subset S of vertices has weight $w(S) := \sum_{v \in S} w_v$, and the edges it cuts is $\partial S := \{e \in E : e \text{ intersects both } S \text{ and } V \setminus S\}$.

The conductance of $S \subseteq V$ is defined as $\phi(S) := \frac{w(\partial S)}{w(S)}$. The conductance of H is defined as:

$$\phi_H := \min_{\emptyset \subsetneq S \subsetneq V} \max\{\phi(S), \phi(V \setminus S)\}. \tag{1}$$

Until recently, it was an open problem to define a spectral model for hypergraphs. In a breakthrough STOC 2015 paper, Louis [11] considered a continuous diffusion process on hypergraphs (which has been refined in a recent JACM paper [4]), and defined an operator $\mathsf{L}_w f := -\frac{df}{dt}$, where $f \in \mathbb{R}^V$ is some appropriate vector associated with the diffusion process. As in classical spectral graph theory, L_w has non-negative eigenvalues, and the all-ones vector $\mathbf{1}$ is an eigenvector with eigenvalue 0. Moreover, the operator L_w has a second eigenvalue γ_2, and the Cheeger's inequality can be recovered[1] for hypergraphs:

$$\frac{\gamma_2}{2} \leq \phi_H \leq 2\sqrt{\gamma_2}.$$

Limitation of the Existing Diffusion Model [4,11]. Suppose at some instant, each vertex has some *measure* that is given by a measure vector $\varphi \in \mathbb{R}^V$. A corresponding *density* vector $f \in \mathbb{R}^V$ is defined by $f_u := \frac{\varphi_u}{w_u}$, for each $u \in V$. Then, at this instant, each edge $e \in E$ will cause measure to flow from vertices $S_e(f) := \mathrm{argmax}_{s \in e} f_s$ having the maximum density to vertices $I_e(f) := \mathrm{argmin}_{i \in e} f_i$ having the minimum density, at a rate of $w_e \cdot \max_{s,i \in e}(f_s - f_i)$. Observe that there can be more than one vertex achieving the maximum or the minimum density in an edge, and a vertex can be involved with multiple number of edges. As shown in [4], it is non-trivial to determine the net rate of incoming measure for each vertex.

One peculiar aspect of this diffusion process is that each edge e only concerns its vertices having the maximum or the minimum density, and ignores the vertices having strict in-between densities. Even though this diffusion process leads to a theoretical treatment of spectral hypergraph properties, its practical use is somehow limited, because it would be considered more natural if vertices having intermediate densities in an edge also take part in the diffusion process.

For instance, in a recent work on semi-supervised learning on hypergraphs [15], the diffusion operator is used to construct an update vector that changes only the solution values of vertices attaining the maximum or the minimum in hyperedges. Therefore, we consider the following open problem in this work:

---

[1] In fact, as shown in the full version [3], a stronger upper bound holds: $\phi_H \leq \sqrt{2\gamma_2}$.

Is there a diffusion process on hypergraphs that involves all vertices in every edge at every instant such that the resulting operator still retains desirable spectral properties?

1.1 Our Contribution and Results

Generalized Diffusion Process with Mediators. We consider a diffusion process where for each edge e, a vertex $j \in e$ can act as a *mediator* that receives flow from vertices in $S_e(f)$ and delivers flow to $I_e(f)$. Formally, we denote $[e] := e \cup \{0\}$, where 0 is a special index that does not refer to any vertex. Each edge e is equipped with non-negative constants $(\beta_j^e : j \in [e])$ such that $\sum_{j \in [e]} \beta_j^e = 1$. Intuitively, for $j = 0$, β_0^e refers to the effect of flow going directly from $S_e(f)$ to $I_e(f)$; for each vertex $j \in e$, β_j^e refers to the significance of j as a mediator between $S_e(f)$ and $I_e(f)$. The complete description of the diffusion rules is in Definition 1. Here are some interesting special cases captured by the new diffusion model.

- For each $e \in E$, $\beta_0^e = 1$. This is the existing model in [4,11].
- For each $e \in E$, there is some $j_e \in e$ such that $\beta_{j_e}^e = 1$, i.e., each edge has one special vertex that acts as its mediator who regulates all flow within the edge.
- For each $e \in E$, for each $j \in e$, $\beta_j^e = \frac{1}{|e|}$, i.e., every vertex in an edge are equally important as mediators.

Theorem 1 (Recovering Cheeger's Inequality via Diffusion Process with Mediators). *Given a hypergraph $H = (V, E, w)$ and mediator constants $(\beta_j^e : e \in E, j \in [e])$, the diffusion process in Definition 1 defines an operator $L_w f := -\frac{df}{dt}$ that has a second eigenvalue γ_2 satisfying $\frac{\gamma_2}{2} \leq \phi_H \leq 2\sqrt{\gamma_2}$, where ϕ_H is the hypergraph conductance defined in (1).*

Impacts of New Diffusion Model. Our generalized diffusion model shows that there is a family of operators whose spectral properties are related to hypergraph conductance. On the theoretical aspect, this provides a powerful tool to enhance the development of spectral hypergraph theory.

On the practical aspect, as mentioned earlier, in the context of semi-supervised learning [8,15], the following minimization convex program is considered: the objective function is $Q(f) := \langle f, L_w f \rangle_w$, and the f values of labeled vertices are fixed. For an iterative method to solve the convex program, our new diffusion model can possibly lead to an update vector that modifies every coordinate in the current solution, thereby potentially improving the performance of the solver.

1.2 Related Work

Other Works on Diffusion Process and Spectral Graph Theory. Apart from the most related aforementioned works [4,11] that we have already mentioned, similar diffusion models (without mediators) have been considered for directed normal graphs [14] and directed hypergraphs [5] to define operators whose spectral properties are analyzed.

2   Preliminaries

We consider an edge-weighted hypergraph $H = (V, E, w)$. Without loss of generality, we assume that the weight $w_i := \sum_{e \in E : i \in e} w_e$ of each vertex $i \in V$ is positive, since any vertex with zero weight can be removed. We use $\mathsf{W} \in \mathbb{R}^{V \times V}$ to denote the diagonal matrix whose (i, i)-th entry is the vertex weight w_i; we let I_n denote the identity matrix.

We use $\mathbb{R}^V$ to denote the set of column vectors. Given $f \in \mathbb{R}^V$, we use f_u or $f(u)$ to indicate the coordinate corresponding to $u \in V$. We use A^T to denote the transpose of a matrix A.

We use $\mathbf{1} \in \mathbb{R}^V$ to denote the vector having 1 in every coordinate. For a vector $x \in \mathbb{R}^V$, we define its support as the set of coordinates at which x is non-zero, i.e. $\mathsf{supp}(x) := \{i : x_i \neq 0\}$.

We use $\chi_S \in \{0, 1\}^V$ to denote the indicator vector of the set $S \subset V$, i.e., $\chi_S(v) = 1$ *iff* $v \in S$.

Recall that the conductance ϕ_H of a hypergraph H is defined in (1). We drop the subscript whenever the hypergraph is clear from the context.

Generalized Quadratic Form. For each edge $e \in E$, we denote $[e] := e \cup \{0\}$, where 0 is a special index that does not correspond to any vertex. Then, each edge e is associated with non-negative constants $(\beta_j^e : j \in [e])$ such that $\sum_{j \in [e]} \beta_j^e = 1$. The *generalized quadratic form* is defined for each $f \in \mathbb{R}^V$ as:

$$\mathsf{Q}(f) := \sum_{e \in E} w_e \{ \beta_0^e \max_{s,i \in e} (f_s - f_i)^2 \\ + \sum_{j \in e} \beta_j^e [(\max_{s \in e} f_s - f_j)^2 + (f_j - \min_{i \in e} f_i)^2] \}.$$

For each non-zero $f \in \mathbb{R}^V$, its *discrepancy ratio* is defined as $\mathsf{D}_w(f) := \frac{\mathsf{Q}(f)}{\sum_{u \in V} w_u f_u^2}$.

Remark. Observe that for each $S \subseteq V$, the corresponding indicator vector $\chi(S) \in \{0, 1\}^V$ satisfies $\mathsf{Q}(\chi(S)) = w(\partial S)$. Hence, we have $\mathsf{D}_w(\chi(S)) = \phi(S)$.

Special Case. We denote $\mathsf{Q}^0(f) := \sum_{e \in E} w_e \max_{s,i \in e} (f_s - f_i)^2$ for the case when $\beta_0^e = 1$ for all e, which was considered in [4]. As we shall see later, for $j \in e$, the weight β_j^e denotes the significance of vertex j as a "mediator" in the diffusion process to direct measure from vertices of maximum density to those with minimum density. As in [4], we consider three isomorphic spaces as follows.

Density Space. This is the space associated with the quadratic form Q. For $f, g \in \mathbb{R}^V$, the inner product is defined as $\langle f, g \rangle_w := f^\mathsf{T} \mathsf{W} g$, and the associated norm is $\|f\|_w := \sqrt{\langle f, f \rangle_w}$. We use $f \perp_w g$ to denote $\langle f, g \rangle_w = 0$.

Normalized Space. Given $f \in \mathbb{R}^V$ in the density space, the corresponding vector in the normalized space is $x := \mathsf{W}^{\frac{1}{2}} f$. The normalized discrepancy ratio is $\mathcal{D}(x) := \mathsf{D}_w(\mathsf{W}^{-\frac{1}{2}} x) = \mathsf{D}_w(f)$.

In the normalized space, the usual ℓ_2 inner product and norm are used. Observe that if x and y are the corresponding normalized vectors for f and g in the density space, then $\langle x, y \rangle = \langle f, g \rangle_w$.

Towards Cheeger's Inquality. Using the inequality $a^2 + b^2 \leq (a+b)^2 < 2(a^2 + b^2)$ for non-negative a and b, we conclude that $\mathsf{Q}(f) < \mathsf{Q}^0(f) \leq 2\mathsf{Q}(f)$ for all $f \in \mathbb{R}^V$. This immediately gives a partial result of Theorem 1.

Lemma 1 (Cheeger's Inequality for Quadratic Form). *Suppose* $\gamma_2 := \min_{0 \neq f \perp_w 1} \frac{\mathsf{Q}(f)}{\|f\|_w^2}$. *Then, we have* $\frac{\gamma_2}{2} \leq \phi_H \leq 2\sqrt{\gamma_2}$, *where* ϕ_H *is the hypergraph conductance defined in (1).*

Proof. Denote $\gamma_2^0 := \min_{0 \neq f \perp_w 1} \frac{\mathsf{Q}^0(f)}{\|f\|_w^2}$. Then, the result from [4] and an improved upper bound in the full version [3] give: $\frac{\gamma_2^0}{2} \leq \phi_H \leq \sqrt{2\gamma_2^0}$. Finally, $\mathsf{Q} \leq \mathsf{Q}^0 \leq 2\mathsf{Q}$ implies that $\gamma_2 \leq \gamma_2^0 \leq 2\gamma_2$. Hence, the result follows. $\square$

Goal of This Paper. In view of Lemma 1, the most technical part of the paper is to define an operator[2] $\mathsf{L}_w : \mathbb{R}^V \to \mathbb{R}^V$ such that $\langle f, \mathsf{L}_w f \rangle_w = \mathsf{Q}(f)$, and show that γ_2 defined in Lemma 1 is indeed an eigenvalue of L_w. To achieve this, we shall consider a diffusion process in the following measure space.

Measure Space. Given a density vector $f \in \mathbb{R}^V$, multiplying each coordinate with its corresponding weight gives the measure vector $\varphi := \mathsf{W}f$. Observe that a vector in the measure space can have negative coordinates. We do not consider inner product explicitly in this space, and so there is no special notation for it.

Transformation between Different Spaces. We use the Roman letter f for vectors in the density space, x for vectors in the normalized space, and Greek letter φ for vectors in the measure space. Observe that an operator defined on one space induces operators on the other two spaces. For instance, if L is an operator defined on the measure space, then $\mathsf{L}_w := \mathsf{W}^{-1}\mathsf{L}\mathsf{W}$ is the corresponding operator on the density space and $\mathcal{L} := \mathsf{W}^{-\frac{1}{2}}\mathsf{L}\mathsf{W}^{\frac{1}{2}}$ is the one on the normalized space. Moreover, all three operators have the same eigenvalues. Recall that the Rayleigh quotients are defined as $\mathsf{R}_w(f) := \frac{\langle f, \mathsf{L}_w f \rangle_w}{\langle f, f \rangle_w}$ and $\mathcal{R}(x) := \frac{\langle x, \mathcal{L}x \rangle}{\langle x, x \rangle}$. For $\mathsf{W}^{\frac{1}{2}}f = x$, we have $\mathsf{R}_w(f) = \mathcal{R}(x)$.

3   Diffusion Process with Mediators

Intuition. Given an edge-weighted hypergraph $H = (V, E, w)$, suppose at some instant, each vertex has some measure given by the vector $\varphi \in \mathbb{R}^V$, whose corresponding density vector is $f = \mathsf{W}^{-1}\varphi$. The idea of a diffusion process is that within each edge $e \in E$, measure should flow from vertices with higher densities to those with lower densities, and the rate of flow has a positive correlation with

---

[2] In the literature, the weighted Laplacian is actually $\mathsf{W}\mathsf{L}_w$ in our notation. Hence, to avoid confusion, we restrict the term Laplacian to the normalized space.

the difference in densities and the strength of the edge e given by w_e. If the diffusion process is well-defined, then an operator on the density space can be defined as $\mathsf{L}_w f := -\frac{df}{dt}$. This induces the Laplacian operator $\mathcal{L} := \mathsf{W}^{\frac{1}{2}} \mathsf{L}_w \mathsf{W}^{-\frac{1}{2}}$ on the normalized space.

In previous work [4], within an edge, measure only flows from vertices $S_e(f) := \mathrm{argmax}_{s \in e} f_s \subseteq e$ having the maximum density to those $I_e(f) := \mathrm{argmin}_{i \in e} f_i$ having minimum densities, where the rate of flow is $w_e \cdot \mathrm{max}_{s,i \in e}(f_s - f_i)$. If all f_u's for an edge e are equal, then we use the convention that $I_e(f) = S_e(f) = e$. Note that vertices $j \in e \setminus (S_e(f) \cup I_e(f))$ with strict in-between densities do not participate due to edge e at this instant.

Generalized Diffusion Process with Mediators. In some applications as mentioned in Sect. 1, it might be more natural if every vertex in an edge e plays some role in diverting flow from $S_e(f)$ to $I_e(f)$. In our new diffusion model, each edge e is associated with constants $(\beta_j^e : j \in [e])$ such that $\sum_{j \in [e]} \beta_j^e = 1$.

Here, 0 is a special index and the parameter β_0^e corresponds to the significance of measure flowing directly from $S_e(f)$ to $I_e(f)$. For $j \in e$, β_j^e indicates the significance of vertex j as a "mediator" to receive measure from $S_e(f)$ and deliver measure to $I_e(f)$. The formal rules are given as follows.

Definition 1 (Rules of Diffusion Process). *Suppose at some instant the system is in a state given by the density vector $f \in \mathbb{R}^V$, with measure vector $\varphi = \mathsf{W}f$. Then, at this instant, measure is transferred between vertices according to the following rules. For $u \in e$ and $j \in [e]$, the pair (e, j) imposes some rules on the diffusion process; let $\varphi_u'(e, j)$ be the net rate of measure flowing into vertex u due to the pair (e, j).*

R(0) For each vertex $u \in V$, the density changes according to the net rate of incoming measure divided by its weight:

$$w_u \frac{df_u}{dt} = \varphi_u' := \sum_{e \in E : u \in e} \sum_{j \in [e]} \varphi_u'(e, j).$$

R(1) We have $\varphi_u'(e, j) < 0$ and $u \neq j$ implies that $u \in S_e(f)$.
Similarly, $\varphi_u'(e, j) > 0$ and $u \neq j$ implies that $u \in I_e(f)$.
R(2) Each edge $e \in E$ and $j \in [e]$, the rates of flow satisfy the following.
For $j = 0$, the rate of flow from $S_e(f)$ to $I_e(f)$ due to $(e, 0)$ is:

$$-\sum_{u \in S_e(f)} \varphi_u'(e, 0) = w_e \cdot \beta_0^e \cdot \mathrm{max}_{s,i \in e}(f_s - f_i) = \sum_{u \in I_e(f)} \varphi_u'(e, 0).$$

For $j \in e$, the rate of flow from $S_e(f)$ to j due to (e, j) is:

$$-\sum_{u \in S_e(f)} \varphi_u'(e, j) = w_e \cdot \beta_j^e \cdot (\mathrm{max}_{s \in e} f_s - f_j);$$

the rate of flow from j to $I_e(f)$ due to (e, j) is:

$$\sum_{u \in I_e(f)} \varphi_u'(e, j) = w_e \cdot \beta_j^e \cdot (f_j - \mathrm{min}_{i \in e} f_i).$$

Then the net rate of flow received by j due to (e, j) is:

$$w_e \cdot \beta_j^e \cdot (\max_{s \in e} f_s + \min_{i \in e} f_i - 2f_j) = \varphi_j'(e, j).$$

Existence of Diffusion Process. The diffusion rules in Definition 1 are much more complicated than those in [4]. It is not immediately obvious whether such a process is well-defined. However, the techniques in [5] can be employed. Intuitively, by repeatedly applying the procedure described in Sect. 4, all higher-order derivatives of the density vector can be determined, which induce an equivalence relation on V such that vertices in the same equivalence class will have the same density in infinitesimal time. This means the hypergraph can be reduced to a simple graph, in which the diffusion process is known to be well-defined. However, to argue this formally is non-trivial, and the reader can refer to the details in [5].

As in [4], if we define an operator using the diffusion process in Definition 1, then the resulting Rayleigh quotient coincides with the discrepancy ratio. The proof of the following lemma is deferred to the full version [3].

Lemma 2 (Rayleigh Quotient Coincides with Discrepancy Ratio).
Suppose L_w on the density space is defined as $\mathsf{L}_w f := -\frac{df}{dt}$ by the rules in Definition 1. Then, the Rayleigh quotient associated with L_w satisfies that for any f in the density space, $\mathsf{R}_w(f) = \mathsf{D}_w(f)$. By considering the isomorphic normalized space, we have for each x, $\mathcal{R}(x) = \mathcal{D}(x)$.

4 Computing the First Order Derivative in the Diffusion Process

In Sect. 3, we define a diffusion process, whose purpose is to define an operator $\mathsf{L}_w f := -\frac{df}{dt}$, where $f \in \mathbb{R}^V$ is in the density space. In this section, we show that the diffusion rules uniquely determine the first order derivative vector $\frac{df}{dt}$; moreover, we give an algorithm to compute it.

Infinitesimal Considerations. In Definition 1, if a vertex u is losing measure due to the pair (e, j) and $u \neq j$, then u must be in $S_e(f)$. However, u must also continue to stay in $S_e(f)$ in infinitesimal time; otherwise, if u is about to leave $S_e(f)$, then u should no longer lose measure due to (e, j). Hence, the vertex u should have the maximum first-order derivative of f_u among vertices in $S_e(f)$. A similar rule should hold when u is gaining measure due to (e, j) and $u \neq j$. This is formalized as the first-order variant of (R1):

Rule (R3) First-Order Derivative Constraints:
If $\varphi_u'(e, j) < 0$ and $u \neq j$, then $u \in \mathsf{argmax}_{s \in S_e(f)} \frac{df_s}{dt}$.
If $\varphi_u'(e, j) > 0$ and $u \neq j$, then $u \subset \mathsf{argmin}_{i \in I_e(f)} \frac{df_i}{dt}$.

Considering Each Equivalence Class U Independently. As in [4], we consider the equivalence relation induced by $f \in \mathbb{R}^V$, where two vertices u and v are in the same equivalence class *iff* $f_u = f_v$. For vertices in some equivalence class U, their current f values are the same, but their values could be about to be separated because their first derivatives might be different.

Subset with the Largest First Derivative: Densest Subset. Suppose $X \subseteq U$ are the vertices having the largest derivative in U. Then, these vertices should receive or contribute rates of measure in each of the following cases.

1. The subset X receives measure due to edges $I_X := \{e \in E : I_e(f) \subseteq X\}$, because the corresponding vertices in X continue to have minimum f values in these edges; we let $c_e^I \geq 0$ be the rate of measure received by $I_e(f)$ due to (e, j) for $j \notin I_e(f)$.
2. The subset X contributes measure due to edges $S_X := \{e \in E : S_e(f) \cap X \neq \emptyset\}$, because the corresponding vertices in X continue to have maximum f values in these edges; we let $c_e^S \geq 0$ be the rate of measure delivered by $S_e(f)$ due to (e, j) for $j \notin S_e(f)$.
3. Each $j \in X$ receives or contributes measure due to all (e, j)'s such that $e \in E$ and $j \in e$; we let $c_j \in \mathbb{R}$ be the net rate of measure received by vertex j due to (e, j) for all $e \in E$ such that $j \in e$.

Hence, the net rate of measure received by X is

$$\mathfrak{C}(X) := \sum_{e \in I_X} c_e^I - \sum_{e \in S_X} c_e^S + \sum_{j \in X} c_j.$$

Therefore, given an instance (U, I_U, S_U), the problem is to find a maximal subset $P \subseteq U$ with the largest density $\delta(P) := \frac{\mathfrak{C}(P)}{w(P)}$, which will be the $\frac{df}{dt}$ values for the vertices in P. For the remaining vertices in U, the sub-instance $(U \setminus P, I_U \setminus I_P, S_U \setminus S_P)$ is solved recursively. The procedure and the precise parameters are given in Fig. 1. Efficient algorithms for this densest subset problem are described in [4,7].

The next lemma shows that the procedure in Fig. 1 returns a vector $r \in \mathbb{R}^V$ that coincides with the first-order derivative $\frac{df}{dt}$ of the density vector obeying rules (R0) to (R3). This implies that these rules uniquely determine the first-order derivative. Given $f \in \mathbb{R}^V$ and $r = \frac{df}{dt}$, we denote $r_S(e) := \max_{u \in S_e(f)} r_u$ and $r_I(e) := \min_{u \in I_e(f)} r_u$.

Lemma 3 (Densest Subset Problem Determines First-Order Deriative). *Given a density vector $f \in \mathbb{R}^V$, rules (R0) to (R3) uniquely determine $r = \frac{df}{dt} \in \mathbb{R}^V$, which can be found by the procedure described in Fig. 1. Moreover, $\sum_{e \in E} c_e^I \cdot r_I(e) - \sum_{e \in E} c_e^S \cdot r_S(e) + \sum_{j \in V} c_j \cdot r_j = \sum_{u \in V} \varphi'_u r_u = \|r\|_w^2.$*

Proof. Using the same approach as in [4], we consider each equivalence class U in Fig. 1, where all vertices in a class have the same f values.

For each such equivalence class $U \subset V$, define $I_U := \{e \in E : U \cap I_e(f) \neq \emptyset\}$, $S_U := \{e \in E : U \cap S_e(f) \neq \emptyset\}$. Notice that each e can only be in exactly one of I_U and S_U.

Given a hypergraph $H = (V, E, w)$ and a vector $f \in \mathbb{R}^V$ in the density space, define an equivalence relation on V such that u and v are in the same equivalence class *iff* $f_u = f_v$. We consider each such equivalence class $U \subseteq V$ and define the $r = \frac{df}{dt}$ values for vertices in U as follows.

1. Denote $E_U := \{e \in E : U \cap [I_e(f) \cup S_e(f)] \neq \emptyset\}$.
   For $e \in E$, define
   $c_e^I := w_e \cdot [\beta_0^e \cdot \max_{s,i \in e}(f_s - f_i) + \sum_{j \in e} \beta_j^e \cdot (f_j - \min_{i \in e} f_i)]$,
   $c_e^S := w_e \cdot [\beta_0^e \cdot \max_{s,i \in e}(f_s - f_i) + \sum_{j \in e} \beta_j^e \cdot (\max_{s \in e} f_s - f_j)]$;
   for $j \in V$, define $c_j := \sum_{e \in E : j \in e} \beta_j^e \cdot w_e \cdot (\max_{s \in e} f_s + \min_{i \in e} f_i - 2f_j)$.
   For $X \subseteq U$, define $I_X := \{e \in E_U : I_e(f) \subseteq X\}$, $S_X := \{e \in E_U : S_e(f) \cap X \neq \emptyset\}$.
   Denote $\mathfrak{C}(X) := \sum_{e \in I_X} c_e^I - \sum_{e \in S_X} c_e^S + \sum_{j \in X} c_j$ and $\delta(X) := \frac{\mathfrak{C}(X)}{w(X)}$.
2. Find $P \subseteq U$ such that $\delta(P)$ is maximized. For all $u \in P$, set $r_u := \delta(P)$.
3. Recursively, find the r values for the remaining vertices in $U' := U \setminus P$ using $E_{U'} := E_U \setminus (I_P \cup S_P)$.

Fig. 1. Procedure to compute $r = \frac{df}{dt}$

Considering Each Equivalence Class U. Suppose T is the set of vertices within U that have the maximum first-order derivative $r = \frac{df}{dt}$. It suffices to show that T is the maximal densest subset in the densest subset instance $(U, I_U \cup S_U)$ defined in Fig. 1.

Because of rule (R3), the rate of net measure received by T is $\mathfrak{C}(T)$. Hence, all vertices $u \in T$ have $r_u = \frac{\mathfrak{C}(T)}{w(T)}$.

Next, suppose P is the maximal densest subset found in Fig. 1. Observe that the net rate of measure entering P is at least $\mathfrak{C}(P)$. Hence, there exists some vertex $v \in P$ such that $\frac{\mathfrak{C}(P)}{w(P)} \leq r_v \leq \frac{\mathfrak{C}(T)}{w(T)}$, where the last inequality follows from the definition of T.

Since P is the maximal densest subset, it follows that in the above inequality, actually all equalities hold and all vertices in P have the same r value. In general, the maximal densest subset contains all densest subsets, and it follows that $T \subseteq P$. Since all vertices in P have the maximum r value within U, we conclude that $P = T$.

Recursive Argument. Hence, it follows that the set T can be uniquely identified in Fig. 1 as the set of vertices having maximum r values, which is also the unique maximal densest subset. Then, the argument can be applied recursively for the smaller instance with $U' := U \setminus T$, $I_{U'} := I_U \setminus I_T$, $S_{U'} := S_U \setminus S_T$.

Claim. $\sum_{e \in E} c_e^I \cdot r_I(e) - \sum_{e \in E} c_e^S \cdot r_S(e) + \sum_{j \in V} c_j \cdot r_j = \sum_{u \in V} \varphi_u' r_u = \|r\|_w^2$.
Consider some T defined above with $\delta := \delta(T) = r_u$, for $u \in T$.

Observe that

$$\sum_{u \in T} \varphi'_u r_u = \left(\sum_{e \in I_T} c_e^I - \sum_{e \in S_T} c_e^S + \sum_{j \in T} c_j \right) \cdot \delta$$

$$= \sum_{e \in I_T} c_e^I \cdot \min_{i \in I_e} r_i - \sum_{e \in S_T} c_e^S \cdot \max_{s \in S_e} r_s + \sum_{j \in T} c_j \cdot r_j$$

where the last equality is due to rule (R3).

Observe that every $u \in V$ will be in exactly one such T, and every $e \in E$ will be accounted for exactly once in each of I_T and S_T, ranging over all T's. Hence, summing over all T's gives the result. $\square$

5   Spectral Properties of Laplacian

A classical result in spectral graph theory is that for a 2-graph whose edge weights are given by the adjacency matrix A, the parameter $\gamma_2 := \min_{0 \neq x \perp W^{\frac{1}{2}} 1} \mathcal{D}(x)$ is an eigenvalue of the normalized Laplacian $\mathcal{L} := I_n - W^{-\frac{1}{2}} A W^{-\frac{1}{2}}$, where a corresponding minimizer x_2 is an eigenvector of $\mathcal{L}$. Observe that γ_2 is also an eigenvalue on the operator $L_w := I_n - W^{-1} A$ induced on the density space.

In this section, we generalize the result to hypergraphs. Observe that any result for the normalized space has an equivalent counterpart in the density space, and vice versa.

Theorem 2 (Eigenvalue of Hypergraph Laplacian).   *For a hypergraph with edge weights w, there exists a normalized Laplacian $\mathcal{L}$ such that the normalized discrepancy ratio $\mathcal{D}(x)$ coincides with the corresponding Rayleigh quotient $\mathcal{R}(x)$. Moreover, the parameter $\gamma_2 := \min_{0 \neq x \perp W^{\frac{1}{2}} 1} \mathcal{D}(x)$ is an eigenvalue of $\mathcal{L}$, where any minimizer x_2 is a corresponding eigenvector.*

Before proving Theorem 2, we first consider the spectral properties of the normalized Laplacian $\mathcal{L}$ induced by the diffusion process defined in Sect. 4.

Lemma 4 (First-Order Derivatives).   *Consider the diffusion process satisfying rules (R0) to (R3) on the measure space with $\varphi \in \mathbb{R}^V$, which corresponds to $f = W^{-1} \varphi$ in the density space. Suppose L_w is the induced operator on the density space such that $\frac{df}{dt} = -L_w f$. Then, we have the following derivatives.*

1. $\frac{d\|f\|_w^2}{dt} = -2 \langle f, L_w f \rangle_w.$
2. $\frac{d\langle f, L_w f \rangle_w}{dt} = -2 \| L_w f \|_w^2.$
3. *Suppose $R_w(f)$ is the Rayleigh quotient with respect to the operator L_w on the density space. Then, for $f \neq 0$, $\frac{dR_w(f)}{dt} = -\frac{2}{\|f\|_w^4} \cdot (\|f\|_w^2 \cdot \|L_w f\|_w^2 - \langle f, L_w f \rangle_w^2) \leq 0$, by the Cauchy-Schwarz inequality on the $\langle \cdot, \cdot \rangle_w$ inner product, where equality holds iff $L_w f \in \mathrm{span}(f)$.*
   By considering a transformation to the normalized space, for any $x \neq 0$, $\frac{d\mathcal{R}(x)}{dt} \leq 0$, where equality holds iff $\mathcal{L} x \in \mathrm{span}(x)$.

Proof. For the first statement, $\frac{d\|f\|_w^2}{dt} = 2\langle f, \frac{df}{dt}\rangle_w = -2\langle f, \mathsf{L}_w f\rangle_w$.

For the second statement, from the proof of Lemma 2 we have $\langle f, \mathsf{L}_w f\rangle_w = \sum_{e \in E} w_e \{\beta_0^e \max_{s,i \in e} (f_s - f_i)^2 + \sum_{j \in e} \beta_j^e [(\max_{s \in e} f_s - f_j)^2 + (f_j - \min_{i \in e} f_i)^2]\}$.

Hence, by the Envelope Theorem,

$$\frac{d\langle f, \mathsf{L}_w f\rangle_w}{dt}$$

$$= 2\sum_{e \in E} w_e \left[\beta_0^e \max_{s,i \in e} (f_s - f_i) \left(\max_{s \in S_e} \frac{df_s}{dt} - \min_{i \in I_e} \frac{df_i}{dt}\right)\right.$$

$$+ \sum_{j \in e} \beta_j^e (\max_{s \in e} f_s - f_j) \left(\max_{s \in S_e} \frac{df_s}{dt} - \frac{df_j}{dt}\right)$$

$$\left. + \sum_{j \in e} \beta_j^e (f_j - \min_{i \in e} f_i) \left(\frac{df_j}{dt} - \min_{i \in I_e} \frac{df_i}{dt}\right)\right]$$

$$= 2\sum_{e \in E} w_e \left\{\left[\beta_0^e \max_{s,i \in e} (f_s - f_i) + \sum_{j \in e} \beta_j^e (\max_{s \in e} f_s - f_j)\right] \max_{s \in S_e} \frac{df_s}{dt}\right.$$

$$- \left[\beta_0^e \max_{s,i \in e} (f_s - f_i) + \sum_{j \in e} \beta_j^e (f_j - \min_{i \in e} f_i)\right] \min_{i \in I_e} \frac{df_i}{dt}$$

$$\left. + \sum_{j \in e} \beta_j^e (2f_j - \max_{s \in e} f_s - \min_{i \in e} f_i) \frac{df_j}{dt}\right\}.$$

$$= 2\left(\sum_{e \in E} c_e^I \cdot \max_{s \in S_e} r_s - \sum_{e \in E} c_e^S \cdot \max_{i \in I_e} r_i - \sum_{j \in V} c_j \cdot r_j\right)$$

where c_e^I, c_e^S, c_j are defined in Fig. 1. From Lemma 3, this equals $-2\|r\|_w^2 = -2\|\mathsf{L}_w f\|_w^2$.

Finally, for the third statement, we have $\frac{d}{dt} \frac{\langle f, \mathsf{L}_w f\rangle_w}{\langle f, f\rangle_w} = \frac{1}{\|f\|_w^4}(\|f\|_w^2 \cdot \frac{d\langle f, \mathsf{L}_w f\rangle_w}{dt} - \langle f, \mathsf{L}_w f\rangle_w \cdot \frac{d\|f\|_w^2}{dt}) = -\frac{2}{\|f\|_w^4} \cdot (\|f\|_w^2 \cdot \|\mathsf{L}_w f\|_w^2 - \langle f, \mathsf{L}_w f\rangle_w^2)$, where the last equality follows from the first two statements.   □

We next prove some properties of the normalized Laplacian $\mathcal{L}$ with respect to orthogonal projection in the normalized space.

Lemma 5 (Laplacian and Orthogonal Projection). *Suppose $\mathcal{L}$ is the normalized Laplacian. Moreover, denote $x_1 := \mathsf{W}^{\frac{1}{2}}\mathbf{1}$, and let Π denote the orthogonal projection into the subspace that is orthogonal to x_1. Then, for all x, we have the following:*

1. $\mathcal{L}(x) \perp x_1$,
2. $\langle x, \mathcal{L}x\rangle = \langle \Pi x, \mathcal{L}\Pi x\rangle$.
3. *For all real numbers a and b, $\mathcal{L}(ax_1 + bx) = b\mathcal{L}(x)$.*

Proof. For the first statement, observe that since the diffusion process is defined on a closed system, the total measure given by $\sum_{u \in V} \varphi_u$ does not change. Therefore, $0 = \langle \mathbf{1}, \frac{d\varphi}{dt}\rangle = \langle \mathsf{W}^{\frac{1}{2}}\mathbf{1}, \frac{dx}{dt}\rangle$, which implies that $\mathcal{L}x = -\frac{dx}{dt} \perp x_1$.

For the second statement, observe that from Lemma 2, we have $\langle x, \mathcal{L}x\rangle = \sum_{e \in E} w_e \{\beta_0^e \max_{s,i \in e}(\frac{x_s}{\sqrt{w_s}} - \frac{x_i}{\sqrt{w_i}})^2 + \sum_{j \in e} \beta_j^e [(\max_{s \in e} \frac{x_s}{\sqrt{w_s}} - \frac{x_j}{\sqrt{w_j}})^2 + (\frac{x_j}{\sqrt{w_j}} - \min_{i \in e} \frac{x_i}{\sqrt{w_i}})^2]\} = \langle (x + \alpha x_1), \mathcal{L}(x + \alpha x_1)\rangle$, where the last equality holds for all real numbers α. Observe that $\Pi x = x + \alpha x_1$, for some suitable real α.

For the third statement, it is more convenient to consider transformation into the density space $f = \mathsf{W}^{-\frac{1}{2}}x$. It suffices to show that $\mathsf{L}_w(a\mathbf{1} + bf) = b\mathsf{L}_w(f)$.

Observe that in the diffusion process, only pairwise difference in densities among vertices matters. Hence, we immediately have $\mathsf{L}_w(a\mathbf{1} + bf) = \mathsf{L}_w(bf)$.

For $b \geq 0$, observe that all the rates are scaled by the same factor b. Hence, we have $\mathsf{L}_w(bf) = b\mathsf{L}_w(f)$.

Finally, if we reverse the sign of every coordinate of f, then the roles of $S_e(f)$ and $I_e(f)$ are switched. Moreover, the direction of every component of the measure flow is reversed with the same magnitude. Hence, $\mathsf{L}_w(-f) = -\mathsf{L}_w(f)$, and the result follows. $\qquad\square$

Proof of Theorem 2. This follows the same argument as in [4]. Suppose $\mathcal{L}$ is the normalized Laplacian induced by the diffusion process in Lemma 3. Let $\gamma_2 := \min_{0 \neq x \perp \mathsf{W}^{\frac{1}{2}}\mathbf{1}} \mathcal{R}(x)$ be attained by some minimizer x_2. We use the isomorphism between the three spaces: $\mathsf{W}^{-\frac{1}{2}}\varphi = x = \mathsf{W}^{\frac{1}{2}}f$.

The third statement of Lemma 4 can be formulated in terms of the normalized space, which states that $\frac{d\mathcal{R}(x)}{dt} \leq 0$, where equality holds *iff* $\mathcal{L}x \in \text{span}(x)$.

We claim that $\frac{d\mathcal{R}(x_2)}{dt} = 0$. Otherwise, suppose $\frac{d\mathcal{R}(x_2)}{dt} < 0$. From Lemma 5, we have $\frac{dx}{dt} = -\mathcal{L}x \perp \mathsf{W}^{\frac{1}{2}}\mathbf{1}$. Hence, it follows that at this moment, the current normalized vector is at position x_2, and is moving towards the direction given by $x' := \frac{dx}{dt}|_{x=x_2}$ such that $x' \perp \mathsf{W}^{\frac{1}{2}}\mathbf{1}$, and $\frac{d\mathcal{R}(x)}{dt}|_{x=x_2} < 0$. Therefore, for sufficiently small $\epsilon > 0$, it follows that $x_2' := x_2 + \epsilon x'$ is a non-zero vector such that $x_2' \perp \mathsf{W}^{\frac{1}{2}}\mathbf{1}$ and $\mathcal{R}(x_2') < \mathcal{R}(x_2) = \gamma_2$, contradicting the definition of x_2.

Hence, it follows that $\frac{d\mathcal{R}(x_2)}{dt} = 0$, which implies that $\mathcal{L}x_2 \in \text{span}(x_2)$. Since $\gamma_2 = \mathcal{R}(x_2) = \frac{\langle x_2, \mathcal{L}x_2 \rangle}{\langle x_2, x_2 \rangle}$, it follows that $\mathcal{L}x_2 = \gamma_2 x_2$, as required. $\qquad\square$

References

1. Alon, N.: Eigenvalues and expanders. Combinatorica **6**(2), 83–96 (1986)
2. Alon, N., Milman, V.D.: $\lambda 1$, isoperimetric inequalities for graphs, and supercon- centrators. J. Comb. Theory Ser. B **38**(1), 73–88 (1985)
3. Chan, T.-H.H., Liang, Z.: Generalizing the hypergraph Laplacian via a diffusion process with mediators. arXiv e-prints (2018)
4. Chan, T.-H.H., Louis, A., Tang, Z.G., Zhang, C.: Spectral properties of hypergraph Laplacian and approximation algorithms. J. ACM **65**(3), 15:1–15:48 (2018)
5. Chan, T.-H.H., Tang, Z.G., Wu, X., Zhang, C.: Diffusion operator and spectral analysis for directed hypergraph Laplacian. CoRR, abs/1711.01560 (2017)
6. Chung, F.R.K.: Spectral Graph Theory, vol. 92. American Mathematical Society (1997)
7. Danisch, M., Chan, T.-H.H., Sozio, M.: Large scale density-friendly graph decom- position via convex programming. In: WWW, pp. 233–242. ACM (2017)
8. Hein, M., Setzer, S., Jost, L., Rangapuram, S.S.: The total variation on hypergraphs - learning on hypergraphs revisited. In: NIPS, pp. 2427–2435 (2013)
9. Hoory, S., Linial, N., Wigderson, A.: Expander graphs and their applications. Bull. Am. Math. Soc. **43**(4), 439–561 (2006)

10. Kannan, R., Vempala, S., Vetta, A.: On clusterings: good, bad and spectral. J. ACM **51**(3), 497–515 (2004)
11. Louis, A.: Hypergraph Markov operators, eigenvalues and approximation algorithms. In: STOC, pp. 713–722. ACM (2015)
12. Makarychev, K., Makarychev, Y., Vijayaraghavan, A.: Correlation clustering with noisy partial information. In: COLT. JMLR Workshop and Conference Proceedings, vol. 40, pp. 1321–1342. JMLR.org (2015)
13. Peng, R., Sun, H., Zanetti, L.: Partitioning well-clustered graphs: spectral clustering works! In: COLT. JMLR Workshop and Conference Proceedings, vol. 40, pp. 1423–1455. JMLR.org (2015)
14. Yoshida, Y.: Nonlinear Laplacian for digraphs and its applications to network analysis. In: Proceedings of the Ninth ACM International Conference on Web Search and Data Mining, pp. 483–492. ACM (2016)
15. Zhang, C., Hu, S., Tang, Z.G., Chan, T.-H.H.: Re-revisiting learning on hypergraphs: confidence interval and subgradient method. In: ICML. Proceedings of Machine Learning Research, vol. 70, pp. 4026–4034. PMLR (2017)

Efficient Enumeration of Bipartite Subgraphs in Graphs

Kunihiro Wasa$^{(\boxtimes)}$ and Takeaki Uno

National Institute of Informatics, Tokyo, Japan
{wasa,uno}@nii.ac.jp

Abstract. Subgraph enumeration problems ask to output all subgraphs of an input graph that belong to a specified graph class or satisfy a given constraint. These problems have been widely studied in theoretical computer science. So far, many efficient enumeration algorithms for the fundamental substructures such as spanning trees, cycles, and paths, have been developed. This paper addresses the enumeration problem of bipartite subgraphs. Even though bipartite graphs are quite fundamental and have numerous applications in both theory and practice, their enumeration algorithms have not been intensively studied, to the best of our knowledge. We propose the first non-trivial algorithms for enumerating all bipartite subgraphs in a given graph. As the main results, we develop two efficient algorithms: the one enumerates all bipartite induced subgraphs of a graph with degeneracy k in $\mathcal{O}(k)$ time per solution. The other enumerates all bipartite subgraphs in $\mathcal{O}(1)$ time per solution.

Keywords: Graph algorithms · Subgraph enumeration
Bipartite graphs · Constant delay · Binary partition method
Degeneracy

1 Introduction

A *subgraph enumeration problem* is, for given a graph G and a constraint R, to output all subgraphs of G that satisfy R once for each and without duplication. An example is to enumerate all the trees in the given graph, and all the subgraphs whose minimum degree is at least k. The complexity and polynomiality of the subgraph enumeration have been intensively studied in theoretical computer science in the terms of both output size sensitivity and input size sensitivity. Compared to optimization approach, enumeration has an advantage on exploring and investigating all possibilities and all aspects of the data, thus is widely studied from a practical point of view, e.g. in Bioinformatics [1], machine learning [18], and data mining [22,26]. We say that an enumeration algorithm is efficient if the algorithm is *output sensitive* [11]. Especially, we say that $\mathcal{A}$ runs in *polynomial amortized time*, if the total running time of an enumeration algorithm $\mathcal{A}$ is $\mathcal{O}(N \cdot poly(n))$ time, where N is the number of solutions, n is the size of the input, and $poly$ is a polynomial function. That is, $\mathcal{A}$ enumerates

© Springer International Publishing AG, part of Springer Nature 2018
L. Wang and D. Zhu (Eds.): COCOON 2018, LNCS 10976, pp. 454–466, 2018.
https://doi.org/10.1007/978-3-319-94776-1_38

all solutions in $poly(n)$ time per solution. Such algorithms have been considered to be efficient, and one of our research goals is to develop efficient enumeration algorithms. So far, there have been studied enumeration algorithms for many fundamental graph structures such as spanning trees [17,19], st-paths [7,17], cycles [2,8,17], maximal cliques [3,6,15], minimal dominating sets [12], and so on. See the comprehensive list in [23] of this area. Recently, Uno [21] developed a technique for a fine-grained analysis of enumeration algorithms.

A bipartite graph which is a fundamental structure is a graph containing no cycle of odd length, that is, whose vertex set can be partitioned into two disjoint independent sets. Bipartite graphs widely appear in real-world graphs such as itemset mining [22,26], chemical information [13], Bioinformatics [27], and so on. Further, enumeration problems for matchings [9,10,20] and biclique [4,15] in bipartite graphs are well studied. However, to the best of our knowledge, there has been proposed no non-trivial enumeration algorithm for bipartite subgraphs.

In this paper, we propose efficient enumeration algorithms for bipartite induced subgraphs and bipartite subgraphs. For enumerating both substructures, we employ a simple binary partition method and develop a data structure for efficiently updating the candidates that are called *child generators*. Intuitively speaking, child generators are vertices or edges such that adding them to a current solution generates another solution. For bipartite induced subgraph, we look at the *degeneracy* [14] of a graph. The degeneracy of a graph is the upper bound of the minimum degree of any its subgraph, so the graph is sparse when the degeneracy is small. It is widely considered as a sparsity measure [3,5,24,25]. There are several graph classes have constant degeneracies, e.g., forests, grid graphs, planar graphs, bounded treewidth graphs, H-minor free graphs with some fixed H, and so on [14]. In addition, Real-world graphs such as road networks, social networks, and internet networks are said to often have small degeneracies, or do so after removing a small part of vertices. Our algorithm utilizes a *good* ordering on the vertices called a *degeneracy ordering* [16], that achieves $\mathcal{O}(k)$ amortized time per solution, where k is the degeneracy of an input graph. This implies that when we restrict the class of input graphs, such as planar graphs, the algorithm runs in constant time per solution and is optimal in the sense of time complexity. Next, for developing an algorithm for bipartite induced subgraph, we show that we can avoid redundant edge additions and removal to obtain a solution from another solution. As the main result, we give an optimal enumeration algorithm, that is, the algorithm runs in constant time per solution. These algorithms are quite simple, but by giving non trivial analysis, we show the algorithms are efficient. These are the first non-trivial efficient enumeration algorithms for bipartite subgraphs.

2   Preliminaries

Let $G = (V, E)$ be an *undirected graph* with vertex set $V = \{1, \ldots, n\}$ and edge set $E = \{e_1, \ldots, e_m\} \subseteq V \times V$. An edge is denoted by $e = (u, v)$. We say that u and v are *endpoints* of $e = (u, v)$, and u is *adjacent* to v if $(u, v) \in E$. When the graph is undirected, $(u, v) = (v, u)$. Two edges are said to be adjacent to each other if a vertex is an endpoint of both edges. The set of *neighbors* of v is the set of vertices that are adjacent to v and is denoted by $N(v)$. For any vertex subset S of V, $E[S] = E \cap (S \times S)$, that is, $E[S]$ is the set of edges whose both endpoints are in S. For any edge subset F of E, $V[F] = \{v \in V \mid \exists e \in F (v \in e)\}$, that is, $V[F]$ is the collection of endpoints of edges in F. The *induced graph* of G by S is $(S, E[S])$ and is denoted by $G[S]$. $G[F] = (V[F], F)$ is a *subgraph* of G by F. We denote by $G \setminus S = G[V \setminus S]$. Since $G[S]$ (resp. $G[F]$) is uniquely determined by S (resp. F), we identify S with $G[S]$ (resp. F with $G[F]$) if no confusion arises.

We say that a sequence $\pi = (v = w_1, \ldots, w_\ell = u)$ of vertices in V is a *path* of G between v and u if for each $i = 1, \ldots, \ell - 1$, $(w_i, w_{i+1}) \in E$, and each vertex in π appears exactly once. We denote by the *length* of a path the number of edges in the path. π is a cycle if $v = u$ and the length of π is at least three. The *distance* $dist(u, v)$ between u and v is the We say G is *connected* if there is a path between any pair of vertices in G. G is *bipartite* if G has no cycle with odd length. For a vertex subset $S \subseteq V$ (resp. an edge subset $F \subseteq E$) such that $G[S]$ (resp. $G[F]$) is bipartite, we say S (resp. F) a *bipartite vertex set* (resp. a *bipartite edge set*). For any bipartite vertex set S, if $G[S]$ is connected, we say S a *connected bipartite vertex set*. We also say a bipartite edge set F is a *connected bipartite edge set* if $G[F]$ is connected. Let $\mathcal{B}^V(G)$ and $\mathcal{B}^E(G)$ be the collection of connected bipartite vertex sets and connected bipartite edge sets, respectively. We call $\mathcal{B}^V(G)$ (resp. $\mathcal{B}^E(G)$) the *solution space* for Problem 1 (resp. for Problem 2). Since we only focus on connected ones, we simply call a connected bipartite vertex (resp. edge) set a bipartite vertex (resp. edge) set. In what follows, we assume that G is connected and simple. We now define the enumeration problems of this paper as follows:

Problem 1 (Bipartite induced subgraph enumeration). For given a graph G, output all vertex sets in $\mathcal{B}^V(G)$ without duplication.

Problem 2 (Bipartite subgraph enumeration). For given a graph G, output all subgraphs in $\mathcal{B}^E(G)$ without duplication.

3   Enumeration of Bipartite Induced Subgraphs

In this paper, we propose two enumeration algorithms for Problems 1 and 2, and this section describes the algorithm for Problem 1. The pseudocode of the algorithm is described in Algorithm 1. We employ *binary partition method* for constructing the algorithms. The algorithm outputs the minimal solution to be output, and partitions the set of remaining solutions to be output into two or more disjoint subsets. Then, the algorithm recursively solves the problems for each subset, by generating recursive calls. We call this dividing step excluding recursive calls (Line 11 in Algorithm 1) an *iteration*.

For any pair X and Y of iterations, X is the *parent* of Y if Y is called from X and Y is a *child* of X if X is the parent of Y.

Algorithm 1. Enumeration algorithm based on binary method

```
 1 Procedure Main(G = (V, E))
 2 │   foreach v ∈ V do
 3 │   │   Rec(G, {v}, N(v));
 4 │   │   G ← G \ {v};
 5 Subprocedure Rec(G, S, C (S, G))
   │   // C (S, G): the set of child generators of G.
 6 │   output (S);
 7 │   while C (S, G) ≠ ∅ do
 8 │   │   u ← the smallest child generator in C (S, G);
 9 │   │   C (S, G) ← C (S, G) \ {u};
10 │   │   S' ← S ∪ {u};
11 │   │   Rec(G, S', ComputeChildGen(C (S, G), u, G));
12 │   │   G ← G \ {u};
13 Subprocedure ComputeChildGen(C (S, G), u, G)
   │   // Compute the set of child generators by Lemma 3.
14 │   if u ∈ C_L (S, G) then
15 │   │   C (S ∪ {u}, G) ← C (S, G) \ (C_L (S, G) ∩ N(u));
16 │   else if u ∈ C_R (S, G) then
17 │   │   C (S ∪ {u}, G) ← C (S, G) \ (C_R (S, G) ∩ N(u));
18 │   C (S ∪ {u}, G) ← C (S ∪ {u}, G) ∪ Γ (S, u, G);
19 │   return C (S ∪ {v}, G);
```

For any bipartite vertex set S, we say that S' is a *child* of S if there exists a vertex u such that $S' = S \cup \{u\}$. A vertex $v \notin S$ is a *child generator* of S for G if $S \cup \{v\}$ is a bipartite vertex set in G. That is, the proposed algorithm enumerates all bipartite vertex sets by recursively adding a child generator to a current bipartite vertex set S. We denote by $C(S, G)$ the set of child generators of S in G. Suppose that r is the smallest vertex in S. Let $L(S) = \{u \in S \mid dist(u, r) \bmod 2 = 0\}$ and $R(S) = \{u \in S \mid dist(u, r) \bmod 2 = 1\}$. For any vertex v in G, any descendant iteration of Rec($G, \{v\}, N(v)$) does not output a bipartite vertex set including vertices less than v. Hence, no vertex will never move to the other side in any descendant bipartite vertex set. Let $C_L(S, G) = \{u \in C(S, G) \mid u \in L(S \cup \{u\})\}$ and

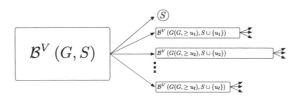

Fig. 1. Example of the partitioning the solution space. Algorithm 1 recursively partitions the solution space $\mathcal{B}^V(G, S)$ into smaller disjoint solution spaces, according to $C(S, G) = \{u_1, \dots, u_\ell\}$.

$C_R(S,G) = \{u \in C(S,G) \mid u \in R(S \cup \{u\})\}$. Note that $C(S,G) = C_L(S,G) \sqcup C_R(S,G)$, where $A \sqcup B$ is the disjoint union of A and B. We denote by $\mathcal{B}^V(G,S) = \{S' \in \mathcal{B}^V(G) \mid S \subseteq S'\}$ the collection of bipartite vertex sets which include S. Note that $\mathcal{B}^V(G) = \mathcal{B}^V(G, \emptyset)$. From now on, we fix a graph G and a bipartite vertex set S of G. By the following lemma, the algorithm divides $\mathcal{B}^V(G,S)$ according to $C(S,G)$ (Fig. 1). For an edge $u \in C(S,G)$, we define $G(S, \geq u)$ by $G \setminus \{v \in C(S,G) \mid v < u\}$.

Lemma 1. $\mathcal{B}^V(G(S, \geq u), S \cup \{u\}) \cap \mathcal{B}^V(G(S, \geq v), S \cup \{v\}) = \emptyset$ for any $u \neq v$ of $C(S,G)$.

Lemma 2. $\mathcal{B}^V(G,S) = \{S\} \cup \bigsqcup_{u \in C(S,G)}^{\ell} \mathcal{B}^V(G(S, \geq u), S \cup \{u\})$.

Next, we consider the correctness of `ComputeChildGen`. For brevity, we introduce some notations: Let u be a child generator in $C(S,G)$. $\Gamma(S,u,G) = \{w \in N(u) \mid w \notin N[S]\}$ is the set of vertices that are adjacent to only u in $S \cup \{u\}$. Note that $C(S,G) \cap \Gamma(S,u,G) = \emptyset$. $\Delta(S,G,u) = C_L(S,G) \cap N[u]$ if $u \in C_L(S,G)$; $\Delta(S,G,u) = C_R(S,G) \cap N[u]$ if $u \in C_R(S,G)$. Intuitively, $\Gamma(S,u,G)$ and $\Delta(S,G,u)$ are the set of vertices that are added to and removed from $C(S,G)$ to compute $C(S \cup \{u\}, G(G, \geq u))$, respectively. The following lemma shows the sufficient and necessary conditions for computing $C(S \cup \{u\}, G(G, \geq u))$.

Lemma 3. $C(S \cup \{u\}, G(S, \geq u)) = ((C(S,G) \setminus \Delta(S,G,u)) \sqcup \Gamma(S,u,G)) \setminus \{v \in C(S,G) \mid v < u\}$.

Proof. We let $C_* = ((C(S,G) \setminus \Delta(S,G,u)) \sqcup \Gamma(S,u,G)) \setminus \{v \in C(S,G) \mid v < u\}$. Suppose that $x \in C(S \cup \{u\}, G(S, \geq u))$. Without loss of generality, we can assume that $u \in C_L(S,G)$. From the definition of $G(S, \geq u)$, $x \notin \{v \in C(S,G) \mid v < u\}$. If $x \notin N[S]$, then since x can be added to $S \cup \{u\}$, x is adjacent to only one vertex u in $S \cup \{u\}$. Hence, $x \in \Gamma(S,u,G)$. If $x \in N[S]$, then since $x \in C(S \cup \{u\}, G(S, \geq u))$, $x \in C(S,G)$. Moreover, if x is in $C_L(S,G) \cap N(u)$, then $S \cup \{u,x\}$ has an odd cycle. Hence, the statement holds.

Suppose that $x \in C_*$. Without loss of generality, we can assume that $u \in C_L(S,G)$. Since $x \in C(S,G) \sqcup \Gamma(S,u,G)$, $S' = S \cup \{u,x\}$ is connected. Suppose that S' has an odd cycle C_o. Since $S \cup \{u\}$ is bipartite, C_o must contain x. This implies that x has neighbors both in $L(S')$ and $R(S')$. If $x \in \Gamma(S,u,G)$, then x has exactly one neighbor in S' since $u \in L(S \cup \{u\})$. Hence, $x \in C(S,G)$. This implies that either (I) $N(x) \cap (S \cup \{x\}) \subseteq L(S)$ or (II) $N(x) \cap (S \cup \{x\}) \subseteq R(S)$. If (I) holds, then x has neighbors only in $L(S \cup \{u\})$ on S' since $u \in C_L(S,G)$ and $x \in C_R(S,G)$. If (II) holds, then x has neighbors only in $R(S \cup \{u\})$ on S' since $x \notin N(u)$. Both cases contradict that x in C_o. Hence, $x \in C(S \cup \{u\}, G(S, \geq u))$ and the statement holds.   □

From the above discussion, we can show the correctness of our algorithm.

Lemma 4. *Algorithm 1 correctly enumerates all bipartite vertex sets in G.*

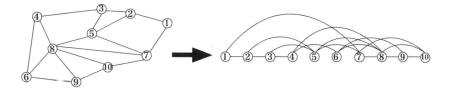

Fig. 2. An example of a degeneracy ordering of G. The degeneracy of G is two even though there is a vertex with degree six. The right-hand side shows a degenerate ordering of G. In the figure, if a vertex u is larger than a vertex v, then u is placed at the right v. Each vertex has at most two larger neighbors.

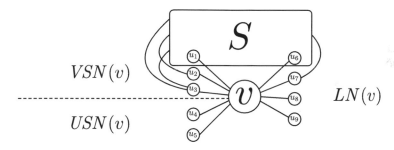

Fig. 3. Three types of neighbors of v. Here, $VSN(v, S) = \{u_1, u_2, u_3\}$, $USN(v, S) = \{u_4, u_5\}$, and $LN(v, S) = \{u_6, \ldots, u_9\}$. All of vertices in $VSN(v, S)$ are adjacent to some vertices in S. Thus, for any $S \subseteq S'$, if u_i is in $VSN(v, S)$, then u_i is also in $VSN(v, S')$. In addition, Algorithm 1 also stores $VSN_C(v, S)$ and $VSN_{\overline{C}}(v, S)$.

3.1 Update of Child Generators

In this section, we consider the time complexity for the maintenance of the sets of child generators. If we naïvely use Lemma 3 for ComputeChildGen, we cannot achieve $\mathcal{O}(k)$ amortized time per solution. To overcome this, we use a *degeneracy ordering* on vertices. G is a *k-degenerate* graph [14] if for any induced graph S of G, S has a vertex whose degree is at most k (Fig. 2). The *degeneracy* of G is the smallest k such that G is k-degenerate. Every k-degenerate graph G has a *degeneracy ordering* on V. The definition of a degeneracy ordering is that for any vertex v in G, the number of neighbors of v that are larger than v is at most k. By recursively removing a vertex with the minimum degree, we can obtain this ordering in linear time [16]. Note that there are many degeneracy orderings for a graph. In what follows, we pick one of degeneracy orderings of G and then fix it as the vertex ordering of G. For any two vertices u, v in G, we write $u < v$ if u is smaller than v in the ordering. We can easily see that if u is the smallest child generator, then u has at most k neighbors in $G[C(S, G)]$ since $G[C(S, G)]$ is k-degenerate. Therefore, Lemma 3 implies that we can compute the child generators of $S \sqcup \{u\}$ by removing at most k vertices and adding some vertices that generate some grandchildren of S.

Next, we define the three types of neighbors of a vertex v in S, *larger neighbors*, *visited smaller neighbors*, and *unvisited smaller neighbors*: For any vertex $u \in N(v)$, (1) u is a larger neighbor of v if $v < u$, (2) u is a visited smaller neighbor of v if $u \in N[S]$ and $u < v$, and (3) u is an unvisited smaller neighbor otherwise (Fig. 3). Intuitively, u is a visited smaller neighbor if one of its neighbors is already picked in some ancestor iteration of X which receives S. We denote by $LN(v, S)$, $VSN(v, S)$, and $USN(v, S)$ the sets of larger neighbors, visited smaller neighbors, and unvisited smaller neighbors of v, respectively. In addition, the algorithm divides $VSN(v, S)$ into two disjoint parts $VSN_C(v, S)$ and $VSN_{\overline{C}}(v, S)$; $VSN_C(v, S) \subseteq C(S, G)$ and $VSN_{\overline{C}}(v, S) \cap C(S, G) = \emptyset$. We omit S if no confusion arises.

We now consider the data structure for the algorithm. For each vertex v, the algorithm stores $LN(v)$, $VSN_C(v)$, $VSN_{\overline{C}}(v)$, and $USN(v)$ in doubly linked lists. $C(S, G)$ is also stored in a doubly linked list and sorted by the degeneracy ordering. The algorithm needs $\mathcal{O}(m) = \mathcal{O}(kn)$ space for storing these data structures. The algorithm also records the modification when an iteration X calls a child iteration Y. Let S_X (resp. S_Y) be bipartite vertex sets received by X (resp. Y). Note that for each neighbor w of a vertex v, if w moves from $USN(v, S_X)$ to $VSN(v, S_Y)$, then w will never move from the list in any descendant of Y. Moreover, when w moves to $VSN(v, S_Y)$, w becomes a child generator of Y, and thus, $w \in VSN_C(v, S_Y)$. Initially, for all smaller neighbors of v is in $USN(v, \emptyset)$. In addition, v will be never added to the set in any descendant of S if v is not a child generator of S. Hence, the algorithm totally needs $\mathcal{O}(m)$ space for storing the modification history. When the algorithm backtracks to X from Y, the algorithm can completely restore the data structure in the same complexity as the transition from X to Y. Now, we consider the time complexity for the transition from X to Y. Suppose that when we remove a vertex v from $C(S, G)$ or add v to S, for each larger neighbor w of v, we give a flag which represents w is not a child generator of the child of S. This can be done in $\mathcal{O}(k)$ time per vertex because of the degeneracy. The next technical lemma shows the number of the larger neighbors which are checked for updating the set of child generators.

Lemma 5. *S has at most one larger neighbor of v for any vertex v in $C(S, G)$.*

Proof. Suppose that two or more neighbors of v are in S. Let x and y be two of them such that y is added after x, and $S' \subseteq S$ be an ancestor bipartite vertex set of S for some graph G' such that $x \in S'$ and $y \notin S'$. Without loss of generality, we can assume that v and y are child generators of S'. We can also assume that v is added after y. Then, from Lemma 3, When y is added to S', v is not in $G(G', \geq y)$ since $v < y$. This contradicts, and thus, the statement holds. □

Lemma 6. *Let u and v be two vertices in $C(S, G)$ such that $u < v$ and $\nexists w \in C(S, G)(u < w < v)$. $C(S \cup \{v\}, G(S, \geq v))$ can be computed from $C(S \cup \{u\}, G(S, \geq u))$ in $\mathcal{O}(k|C(S \cup \{u\}, G(S, \geq u))| + k|C(S \cup \{v\}, G(S, \geq v))|)$ time.*

Proof. From Lemma 3, only the neighbors of v or u may be added to or removed from $C\left(S \cup \{u\}, G(S, \geq u)\right)$ to obtain $C\left(S \cup \{v\}, G(S, \geq v)\right)$. Let w be a vertex in $LN(v)$. We consider the following cases: (L.1) $w \in C\left(S \cup \{u\}, G(S, \geq u)\right) \cap C\left(S \cup \{v\}, G(S, \geq v)\right)$ or $w \notin C\left(S \cup \{u\}, G(S, \geq u)\right) \cup C\left(S \cup \{v\}, G(S, \geq v)\right)$. In this case, there is nothing to do. (L.2) $w \in C\left(S \cup \{u\}, G(S, \geq u)\right) \setminus C\left(S \cup \{v\}, G(S, > v)\right)$. For each larger neighbor x of w, we need to move w from $VSN_C\left(x, S \cup \{u\}\right)$ to $VSN_{\overline{C}}\left(x, S \cup \{v\}\right)$. The number such x is at most $k \left| C\left(S \cup \{u\}, G(S, \geq u)\right) \right|$. (L.3) $w \in C\left(S \cup \{v\}, G(S, \geq v)\right) \setminus C\left(S \cup \{u\}, G(S, \geq u)\right)$. For each larger neighbor x of w, we need to move w from $VSN_{\overline{C}}\left(x, S \cup \{u\}\right)$ to $VSN_C\left(x, S \cup \{v\}\right)$. The number of such x is at most $k \left| C\left(S \cup \{v\}, G(S, \geq v)\right) \right|$. Note that for each vertex, at most one larger its neighbor is in S from Lemma 5. Thus, the above three conditions can be checked in constant time for each w by checking whether or not w is in the same partition as v. Therefore, the larger part can be done in $\mathcal{O}\left(k + k \left| C\left(S \cup \{u\}, G(S, > u)\right) \right| + k \left| C\left(S \cup \{v\}, G(S, \geq v)\right) \right|\right)$ time.

Next, let w be a vertex in $VSN_C\left(v, S\right)$. From Lemma 3, such w does not belong to $C\left(S \cup \{v\}, G(S, \geq v)\right)$. Moreover, since u and v are consecutive on $C(S, G)$, such w is also not in $C\left(S \cup \{u\}, G(S, \geq u)\right)$. Thus, this case can be done in constant time by skipping such vertices. For each vertex w in $VSN_{\overline{C}}\left(v, S\right)$, w cannot be added to both $S \cup \{u\}$ and $S \cup \{v\}$. Hence, we skip them. In addition, we need to remove v from $G(S, \geq u)$. This takes $\mathcal{O}(k)$ time since we only need to update larger neighbors of v. The same procedure needs for updating the neighbors of u. Hence, the statement holds.   □

Roughly speaking, by ignoring neighbors of u or v such that they cannot be added to both $S \cup \{u\}$ and $S \cup \{v\}$, we can compute $C\left(S \cup \{v\}, G(S, \geq v)\right)$ from $C\left(S \cup \{u\}, G(S, \geq u)\right)$, efficiently. In addition, other neighbors have corresponding bipartite vertex sets with size $|S| + 2$, that is, grandchildren of S. This implies that we can amortize the cost for these neighbors as follows.

Lemma 7. *Let u be a vertex in $C(S, G)$ and $T(S, u)$ be the computation time for $C\left(S \cup \{u\}, G(S, \geq u)\right)$. The total computation time for all the sets of child generators of S's children and recording the modification history is*
$$\sum_{u \in C(S,G)} T(S, u) = \mathcal{O}\left(k \left| C(S, G) \right| + \sum_{u \in C(S,G)} k \left| C\left(S \cup \{u\}, G(S, \geq u)\right) \right|\right)$$
time.

Proof. From Lemma 3, we need $\mathcal{O}\left(k \left| C(S, G) \right| + k \left| C\left(S \cup \{u_*\}, G(S, \geq u_*)\right) \right|\right)$ time for computing $C\left(S \cup \{u_*\}, G(S, \geq u_*)\right)$, where u_* is the smallest child generator in $C(S, G)$. From Lemma 6, we can compute all the sets of child generators for children of S except for $S \cup \{u_*\}$ in $\mathcal{O}\left(\sum_{u \in C(S,G)} k \left| C\left(S \cup \{u\}, G(S, \geq u)\right) \right|\right)$ time in total. Moreover, recording the modification history can be done in the same time complexity in above. Hence, the statement holds.   □

Theorem 1. *Given a graph G with degeneracy k, Algorithm 1 enumerates all solutions in $\mathcal{O}\left(k \left| \mathcal{B}^V(G) \right|\right)$ total time, that is, $\mathcal{O}(k)$ time per solution with $\mathcal{O}(m) = \mathcal{O}(kn)$ space and preprocessing time.*

Proof. From Lemma 7, we can see the larger neighbors of u are always checked. Thus, Line 12 can be done in $\mathcal{O}(k)$ time since the algorithm does not need to remove edges whose endpoints are u and a smaller neighbor of u. Moreover, Line 9 can be done in $\mathcal{O}(1)$ time. In addition, in the preprocessing, we need to initialize the data structure and compute the degeneracy ordering. Both need $\mathcal{O}(kn)$ time and space since the number of edges is at most kn. From Lemma 7 and the above this discussion, the algorithm runs in $\mathcal{O}\left(\sum_{S \in \mathcal{B}^V(G)} \sum_{u \in C(S,G)} T(S,u)\right)$ time.

Now, $\mathcal{O}\left(\sum_{S \in \mathcal{B}^V(G)} \left(|C(S,G)| + \sum_{u \in C(S,G)} |C(S \cup \{u\}, G(S, \geq u))|\right)\right) = \mathcal{O}\left(|\mathcal{B}^V(G)|\right)$. Hence, the statement holds. $\square$

Algorithm 2. Enumeration algorithm for bipartite edges sets

---

1 **Procedure** Main$(G = (V, E))$
2   **foreach** $e \in E$ **do** // Pick the smallest edge in E.
3   Rec$(G, \{e\}, N(e))$;
4    $G \leftarrow G \setminus \{e\}$;
5 **Subprocedure** Rec$(G, F, N'(F, G))$
6   output(F);
7   **while** $N'(F, G) \neq \emptyset$ **do**
8    $e \leftarrow$ the smallest child generator in $N'(F)G$;
9    $F' \leftarrow F \cup \{e\}$;
10    $G' \leftarrow G(F, \geq e) \setminus E(B(G(F, \geq e), F'))$;
11    $N'(G', F') \leftarrow$
    $(N'(G, F) \cup N_+(G, F, e)) \setminus (N_-(G, F, e) \cup \{f \in E \mid f \leq e\})$;
12   Rec$(G', F', N'(G', F'))$;
13    $N'(F, G) \leftarrow N'(F, G) \setminus \{e\}$;

Corollary 1. *All bipartite induced subgraphs in graphs with constant degeneracy, such as planar graphs, can be listed in $\mathcal{O}(1)$ time per solution with $\mathcal{O}(n)$ space and preprocessing time.*

4 Enumeration of Bipartite Subgraphs

In this section, we describe our algorithm for Problem 2. For a graph G and a bipartite edge set F of G, let $B(G, F)$ be the set of edges e of G such that $F \cup \{e\}$ is not bipartite, i.e., $F \cup \{e\}$ has an odd cycle that includes e. Let $\mathcal{B}^E(G, F) = \{F' \in \mathcal{B}^E(G) \mid F \subseteq F'\}$. We can see that $\mathcal{B}^E(G(F), F) = \mathcal{B}^E(G(F) \setminus B(G, F), F)$. For an edge e of G, we define $N'(G, e) = \{f \in E \setminus \{e\} \mid f \text{ is adjacent to } e\}$, $N'(G, F) = \bigcup_{e \in F} N'(e) \setminus F$, and we also define $G(F, \geq e)$ by $G \setminus \{f \in N'(G, F) \mid f < e\}$.

The framework of the algorithm is the same as the algorithm for Problem 1. The algorithm starts from the empty edge set, and add edges recursively so that the edge sets generated are always connected and bipartite, and no duplication occurs. For given a graph G and a bipartite edge set F of G, the algorithm first removes edges of $B(G, F)$ from G, and outputs F as a solution. Then for each $e \in N'(G, F)$, the algorithm generates the problems of enumerating all bipartite subgraphs that include $F \cup \{e\}$ but no edge $f < e, f \in N'(G, F)$, that is, bipartite subgraphs in $G(F, \geq e)$ that includes $F \cup \{e\}$. Before generating the recursive call, the algorithm computes the edges of $B(G(F, \geq e), F \cup \{e\})$ and removes them from $G(F, \geq e)$ so that the computation of the iteration will be accelerated. The correctness of our strategy for the enumeration is as follows.

Lemma 8. $\mathcal{B}^E(G(F, \geq e), F \cup \{e\}) \cap \mathcal{B}^E(G(F, \geq f), F \cup \{f\}) = \emptyset$ for any $e \neq f$ of $N'(G, F)$.

Lemma 9. $\mathcal{B}^E(G, F) = \{F\} \cup \bigsqcup_{e \in N'(G,F)} \mathcal{B}^E(G(F, \geq e), F \cup \{e\})$,

For the efficient computation, our algorithm always keeps $N'(G, F)$ in the memory, and update and passes it to the recursive calls. For the efficient update of $N'(G, F)$, we keep the graph $G \setminus F$ in the memory since edges of F never be added to $N'(G, F)$, until the completion of the iteration. We also put a label of "1" or "2" to each vertex in $V(F)$ and update so that each edge of F connects vertices of different labels, that is always possible since F is bipartite. For a vertex v of G that is not in $V(F)$, let $N_1(G, F, v)$ (resp., $N_2(G, F, v)$) be the set of edges f of $N(v) \setminus F$ such that the endpoint of f other than v has label "1" (resp., "2"). We also keep and update $N_1(G, F, v)$ and $N_2(G, F, v)$. For an edge $(u, v) \in N'(G, F)$ such that $u \notin V(F)$, we define $N_+(G, F, (u, v))$ by $N_1(G, F, v)$ and $N_-(G, F, (u, v))$ by $N_2(G, F, v)$ if the label of v is "1", and $N_+(G, F, (u, v))$ by $N_2(G, F, v)$ and $N_-(G, F, (u, v))$ by $N_1(G, F, v)$ otherwise. We define $N_1(G, F, (u, v))$ and $N_2(G, F, (u, v))$ by the empty set if both u and v are in $V(F)$

For an edge $e \in N'(G, F)$, let $F' = F \cup \{e\}$ and G' be the graph obtained from $G(F, \geq e)$ by removing edges of $B(G(F, \geq e), F')$.

Lemma 10. Suppose that $B(G, F) = \emptyset$ and $e = (u, v)$. Then, $B(G', F') = N_-(G, F, v)$.

Proof. Since $B(G, F) = \emptyset$, any edge f in $B(G', F')$ must share one of its endpoint with e, and the endpoint is adjacent to no edge of F. Further, the edge is not included in F. The addition of f to F generates an odd cycles if and only if the label of both endpoints of f are the same. Therefore the statement holds. □

The following lemma shows that the computation of $N'(G', F')$ from $N'(G, F)$ can be also done in $\mathcal{O}(|N_+(G, F, e)| + 1)$ time.

Lemma 11. $N'(G', F') = (N'(G, F) \cup N_+(G, F, e)) \setminus (N_-(G, F, e) \cup \{f \in E \mid f \leq e\})$.

Proof. We first prove $N'(G', F') \subseteq (N'(G, F) \cup N_+(G, F, e)) \setminus (N_-(G, F, e) \cup \{f \in E \mid f \leq e\})$. Let f be an edge in $N'(G', F')$. Then, f is in either $N'(G, F)$ or $N_+(G, F, e)$. From the definition of $G(F, \geq e)$ and $B(G', F')$, f

is not in $B(G', F')$. Further, $f > e$ from $f \in G'$. This implies that $f \in (N'(G, F) \cup N_+(G, F, e)) \setminus (N_-(G, F, e) \cup \{f \in E \mid f \le e\})$.

We next prove $(N'(G, F) \cup N_+(G, F, e)) \setminus (N_-(G, F, e) \cup \{f \in E \mid f \le e\}) \supseteq N'(G', F')$. Suppose that $f \in (N'(G, F) \cup N_+(G, F, e)) \setminus (N_-(G, F, e) \cup \{f \in E \mid f \le e\})$. Then, f is not in F', and adjacent to an edge of F'. Further, $f > e$ and the addition of f to F' generates no odd cycle. Thus, $f \in N'(G', F')$. □

When we generate $G(F, \ge e)$ for each $e \in N'(G, F)$ one by one in increasing order, the total computation time is $\mathcal{O}(|N'(G, F)|)$. Computation of $G' \setminus F'$ is at most the time to compute G'. From this together with these lemmas, we can see that an iteration of the algorithm spends $\mathcal{O}\left(|N'(G, F)| + \sum_{e \in N'(G, F)} |N_+(G, F, e)|\right)$ time. The following lemma bound this complexity in another way. Let G_e is the graph obtained from $G(F \cup \{e\}, \ge e)$ by removing edges of $B(G(F \cup \{e\}, \ge e), F \cup \{e\})$.

Lemma 12. *Suppose that e' is next to e in the edge ordering in $N'(G, F)$. The computation of $N'(G_{e'}, F \cup \{e'\})$ from $N'(G_e, F \cup \{e\})$ can be done in $\mathcal{O}(|N'(G_{e'}, F \cup \{e'\})| + |N'(G_e, F \cup \{e\})|)$ time.*

Proof. The computation is to recover $N'(G, F) \setminus \{f \in E \mid f \le e\}$ from $N'(G_e, F \cup \{e\})$ and construct $N'(G_{e'}, F \cup \{e'\})$ from it. From Lemma 11, its time is linear in $|N_+(G, F, e) \setminus \{f \in E \mid f \le e\}| + |N_-(G, F, e) \setminus \{f \in E \mid f \le e\}| + |N_+(G, F, e') \setminus \{f \in E \mid f \le e\}| + |N_-(G, F, e') \setminus \{f \in E \mid f \le e\}|$. We see that $N_+(G, F, e) \setminus \{f \in E \mid f \le e\} \subseteq N'(G_e, F \cup \{e\})$, and $N_+(G, F, e') \setminus \{f \in E \mid f \le e'\} \subseteq N'(G_{e'}, F \cup \{e'\})$. When $N_-(G, F, e) \ne N_-(G, F, e')$, we have $N_-(G, F, e) \cap N_-(G, F, e') = \emptyset$, thus $N_-(G, F, e) \setminus \{f \in E \mid f \le e\} \subseteq N'(G_{e'}, F \cup \{e'\})$, and $N_-(G, F, e') \setminus \{f \in E \mid f \le e'\} \subseteq N'(G_e, F \cup \{e\})$ thus the statement holds. When $N_-(G, F, e) = N_-(G, F, e')$, they are canceled out and no need of taking care in the computation, thus the statement also holds. □

Lemma 13. *For any iteration inputting G and F such that $B(G, F) = \emptyset$, its computation time is at most proportional to one plus the number of its children and the grandchildren.*

Proof. For the first recursive call with respect to an edge e, we pay computation time of $\mathcal{O}(|N'(G, F)| + |N'(G_e, F \cup \{e\})|)$. For the remaining recursive calls, as we see in Lemma 12, the computation time is linear in the number of grandchildren generated in the recursive call, and that generated just before. Thus, the statement holds. □

Sine any iteration requires at most $\mathcal{O}(|V| + |E|)$ space. When the iteration generates a recursive call, the graphs and variants which the iteration is using have to be recovered, just after the termination of the recursive call. This can be done by just keeping the vertices and edges that are removed to make the input graph of the recursive call. Thus, the total accumulated space spent by all its ancestors is at most $\mathcal{O}(|V| + |E|)$. Therefore, we obtain the following theorem.

Theorem 2. *All bipartite subgraphs in a graph* $G = (V, E)$ *can be listed in* $\mathcal{O}\left(\left|\mathcal{B}^E(G)\right|\right)$ *total time, that is,* $\mathcal{O}(1)$ *time per solution with* $\mathcal{O}(|V| + |E|)$ *space.*

References

1. Ahmed, N.K., Neville, J., Rossi, R.A., Duffield, N.: Efficient graphlet counting for large networks. In: ICDM 2015, pp. 1–10 (2015)
2. Birmelé, E., Ferreira, R., Grossi, R., Marino, A., Pisanti, N., Rizzi, R., Sacomoto, G.: Optimal listing of cycles and st-paths in undirected graphs. In: SODA 2012, pp. 1884–1896 (2012)
3. Conte, A., Grossi, R., Marino, A., Versari, L.: Sublinear-space bounded-delay enumeration for massive network analytics: maximal cliques. In: ICALP 2016. LIPIcs, vol. 55, pp. 148:1–148:15 (2016)
4. Dias, V.M., de Figueiredo, C.M., Szwarcfiter, J.L.: Generating bicliques of a graph in lexicographic order. Theor. Comput. Sci. **337**(1–3), 240–248 (2005)
5. Eppstein, D., Löffler, M., Strash, D.: Listing all maximal cliques in sparse graphs in near-optimal time. In: Cheong, O., Chwa, K.-Y., Park, K. (eds.) ISAAC 2010. LNCS, vol. 6506, pp. 403–414. Springer, Heidelberg (2010). https://doi.org/10.1007/978-3-642-17517-6_36
6. Eppstein, D., Strash, D.: Listing all maximal cliques in large sparse real-world graphs. In: Pardalos, P.M., Rebennack, S. (eds.) SEA 2011. LNCS, vol. 6630, pp. 364–375. Springer, Heidelberg (2011). https://doi.org/10.1007/978-3-642-20662-7_31
7. Ferreira, R., Grossi, R., Rizzi, R.: Output-sensitive listing of bounded-size trees in undirected graphs. In: Demetrescu, C., Halldórsson, M.M. (eds.) ESA 2011. LNCS, vol. 6942, pp. 275–286. Springer, Heidelberg (2011). https://doi.org/10.1007/978-3-642-23719-5_24
8. Ferreira, R., Grossi, R., Rizzi, R., Sacomoto, G., Sagot, M.-F.: Amortized $\tilde{O}(|V|)$-delay algorithm for listing chordless cycles in undirected graphs. In: Schulz, A.S., Wagner, D. (eds.) ESA 2014. LNCS, vol. 8737, pp. 418–429. Springer, Heidelberg (2014). https://doi.org/10.1007/978-3-662-44777-2_35
9. Fukuda, K., Matsui, T.: Finding all the perfect matchings in bipartite graphs. Appl. Math. Lett. **7**(1), 15–18 (1994)
10. Gély, A., Nourine, L., Sadi, B.: Enumeration aspects of maximal cliques and bicliques. Discrete Appl. Math. **157**(7), 1447–1459 (2009)
11. Johnson, D.S., Yannakakis, M., Papadimitriou, C.H.: On generating all maximal independent sets. Inform. Process. Lett. **27**(3), 119–123 (1988)
12. Kanté, M.M., Limouzy, V., Mary, A., Nourine, L.: Enumeration of minimal dominating sets and variants. In: Owe, O., Steffen, M., Telle, J.A. (eds.) FCT 2011. LNCS, vol. 6914, pp. 298–309. Springer, Heidelberg (2011). https://doi.org/10.1007/978-3-642-22953-4_26
13. Koichi, S., Arisaka, M., Koshino, H., Aoki, A., Iwata, S., Uno, T., Satoh, H.: Chemical structure elucidation from 13c NMR chemical shifts: efficient data processing using bipartite matching and maximal clique algorithms. J. Chem. Inf. Model. **54**(4), 1027–1035 (2014)
14. Lick, D.R., White, A.T.: k-degenerate graphs. Can. J. Math. **22**(5), 1082–1096 (1970)
15. Makino, K., Uno, T.: New algorithms for enumerating all maximal cliques. In: Hagerup, T., Katajainen, J. (eds.) SWAT 2004. LNCS, vol. 3111, pp. 260–272. Springer, Heidelberg (2004). https://doi.org/10.1007/978-3-540-27810-8_23

16. Matula, D.W., Beck, L.L.: Smallest-last ordering and clustering and graph coloring algorithms. J. ACM **30**(3), 417–427 (1983)
17. Read, R.C., Tarjan, R.E.: Bounds on backtrack algorithms for listing cycles, paths, and spanning trees. Networks **5**(3), 237–252 (1975)
18. Ruggieri, S.: Enumerating distinct decision trees. In: ICML 2017. Proceedings of Machine Learning Research, vol. 70, pp. 2960–2968 (2017)
19. Shioura, A., Tamura, A., Uno, T.: An optimal algorithm for scanning all spanning trees of undirected graphs. SIAM J. Comput. **26**(3), 678–692 (1997)
20. Uno, T.: Algorithms for enumerating all perfect, maximum and maximal matchings in bipartite graphs. In: Leong, H.W., Imai, H., Jain, S. (eds.) ISAAC 1997. LNCS, vol. 1350, pp. 92–101. Springer, Heidelberg (1997). https://doi.org/10.1007/3-540-63890-3_11
21. Uno, T.: Constant time enumeration by amortization. In: Dehne, F., Sack, J.-R., Stege, U. (eds.) WADS 2015. LNCS, vol. 9214, pp. 593–605. Springer, Cham (2015). https://doi.org/10.1007/978-3-319-21840-3_49
22. Uno, T., Kiyomi, M., Arimura, H.: LCM ver. 2: efficient mining algorithms for frequent/closed/maximal itemsets. In: FIMI 2004 (2004)
23. Wasa, K.: Enumeration of enumeration algorithms. CoRR, abs/1605.05102 (2016)
24. Wasa, K., Arimura, H., Uno, T.: Efficient enumeration of induced subtrees in a K-degenerate graph. In: Ahn, H.-K., Shin, C.-S. (eds.) ISAAC 2014. LNCS, vol. 8889, pp. 94–102. Springer, Cham (2014). https://doi.org/10.1007/978-3-319-13075-0_8
25. Xu, Y., Cheng, J., Fu, A.W.-C.: Distributed maximal clique computation and management. IEEE T. Serv. Comput. **9**(1), 1 (2015)
26. Zaki, M.J.: Scalable algorithms for association mining. IEEE Tans. Knowl. Data. Eng. **12**(3), 372–390 (2000)
27. Zhang, Y., Phillips, C.A., Rogers, G.L., Baker, E.J., Chesler, E.J., Langston, M.A.: On finding bicliques in bipartite graphs: a novel algorithm and its application to the integration of diverse biological data types. BMC Bioinform. **15**(1), 110 (2014)

Bipartite Graphs of Small Readability

Rayan Chikhi[1], Vladan Jovičić[2], Stefan Kratsch[3], Paul Medvedev[4],
Martin Milanič[5], Sofya Raskhodnikova[6], and Nithin Varma[6(✉)]

[1] CNRS, UMR 9189, Lille, France
rayan.chikhi@univ-lille1.fr
[2] ENS Lyon, Lyon, France
vladan94.jovicic@gmail.com
[3] Institut für Informatik, Humboldt-Universität zu Berlin, Berlin, Germany
kratsch@informatik.hu-berlin.de
[4] The Pennsylvania State University, State College, USA
pashadag@cse.psu.edu
[5] IAM and FAMNIT, University of Primorska, Koper, Slovenia
martin.milanic@upr.si
[6] Boston University, Boston, USA
{sofya,nvarma}@bu.edu

Abstract. We study a parameter of bipartite graphs called readability, introduced by Chikhi et al. (*Discrete Applied Mathematics* 2016) and motivated by applications of overlap graphs in bioinformatics. The behavior of the parameter is poorly understood. The complexity of computing it is open and it is not known whether the decision version of the problem is in NP. The only known upper bound on the readability of a bipartite graph (Braga and Meidanis, *LATIN* 2002) is exponential in the maximum degree of the graph. Graphs that arise in bioinformatic applications have low readability. In this paper we focus on graph families with readability $o(n)$, where n is the number of vertices. We show that the readability of n-vertex bipartite chain graphs is between $\Omega(\log n)$ and $\mathcal{O}(\sqrt{n})$. We give an efficiently testable characterization of bipartite graphs of readability at most 2 and completely determine the readability of grids, showing in particular that their readability never exceeds 3. As a consequence, we obtain a polynomial-time algorithm to determine the readability of induced subgraphs of grids. One of the highlights of our techniques is the appearance of Euler's totient function in the proof of the upper bound on the readability of bipartite chain graphs. We also develop a new technique for proving lower bounds on readability, which is applicable to dense graphs with a large number of distinct degrees.

1 Introduction

In this work we further the study of *readability* of bipartite graphs initiated by Chikhi et al. [6]. Given a bipartite graph $G = (V_s, V_p, E)$, an *overlap labeling* of G is a mapping from vertices to strings, called labels, such that for all $u \in V_s$

The full version of this paper is available online [5].

© Springer International Publishing AG, part of Springer Nature 2018
L. Wang and D. Zhu (Eds.): COCOON 2018, LNCS 10976, pp. 467–479, 2018.
https://doi.org/10.1007/978-3-319-94776-1_39

and $v \in V_p$ there is an edge between u and v if and only if the label of u *overlaps* with the label of v (i.e., a non-empty suffix of u's label is equal to a prefix of v's label). The *length* of an overlap labeling of G is the maximum length (i.e., number of characters) of a label. The *readability* of G, denoted $r(G)$, is the smallest nonnegative integer r such that there is an overlap labeling of G of length r. In this definition, no restriction is placed on the alphabet. One could also consider variants of readability parameterized by the size of the alphabet. A result of Braga and Meidanis [4] implies that these variants are within constant factors of each other, where the constants are logarithmic in the alphabet sizes.

The notion of readability arises in the study of overlap digraphs. Overlap digraphs constructed from DNA strings have various applications in bioinformatics.[1] Most of the graphs that occur as the overlap graphs of genomes have low readability. Chikhi et al. [6] show that the readability of overlap digraphs is asymptotically equivalent to that of balanced bipartite graphs: there is a bijection between overlap digraphs and balanced bipartite graphs that preserves readability up to (roughly) a factor of 2. This motivates the study of bipartite graphs with low readability. In this work we derive several results about bipartite graphs with readability sublinear in the number of vertices.

For general bipartite graphs, the only known upper bound on readability is implicit in a paper on overlap digraphs by Braga and Meidanis [4]. As observed by Chikhi et al. [6], it follows from [4] that the readability of a bipartite graph is well defined and at most $2^{\Delta+1} - 1$, where Δ is the maximum degree of the graph. Chikhi et al. [6] showed that almost all bipartite graphs with n vertices in each part have readability $\Omega(n/\log n)$. They also constructed an explicit graph family (called Hadamard graphs) with readability $\Omega(n)$.

For trees, readability can be defined in terms of an extremal question on certain integer functions on the edges, without any reference to strings or their overlaps [6]. In this work, we reveal another connection to number theory, through Euler's totient function, and use it to prove an upper bound on the readability of bipartite chain graphs.

So far, our understanding of readability has been hindered by the difficulty of proving lower bounds. Chikhi et al. [6] developed a lower bound technique for graphs where the overlap between the neighborhoods of any two vertices is limited. In this work, we add another technique to the toolbox. Our technique is applicable to dense graphs with a large number of distinct degrees. We apply this technique to obtain a lower bound on readability of bipartite chain graphs.

We give a characterization of bipartite graphs of readability at most 2 and use this characterization to obtain a polynomial-time algorithm for checking if a graph has readability at most 2. This is the first nontrivial result of this kind: graphs of readability at most 1 are extremely simple (disjoint unions of

---

[1] In the context of genome assembly, variants of overlap digraphs appear as either de Bruijn graphs [11] or string graphs [18,21] and are the foundation of most modern assemblers (see [17,19] for a survey). Several graph-theoretic parameters of overlap digraphs have been studied [1–3,9,15,16,20,23], with a nice survey in [14].

complete bipartite graphs, see [6]), whereas the problem of recognizing graphs of readability k is open for all $k \geq 3$.

We also give a formula for the readability of grids, showing in particular that it never exceeds 3. As a corollary, we obtain a polynomial-time algorithm to determine the readability of induced subgraphs of grids.

1.1   Our Results and Structure of the Paper

Preliminaries are summarized in Sect. 2; here we only state some of the most important technical facts. All missing proofs can be found in the full version [5].

To study readability, it suffices to consider bipartite graphs that are connected and *twin-free*, i.e., no two nodes in the same part have the same sets of neighbors [6]. As connected bipartite graphs have a unique bipartition up to swapping the two parts, we state some of our results without specifying the bipartition.

Bounds on the Readability of Bipartite Chain Graphs (Sect. 3). Bipartite chain graphs are the bipartite analogue of a family of digraphs that occur naturally as subgraphs of overlap graphs of genomes. In a *bipartite chain graph* $G = (V_s, V_p, E)$, the vertices in V_s (or V_p) can be linearly ordered with respect to inclusion of their neighborhoods. That is, we can write $V_s = \{v_1, \ldots, v_k\}$ so that $N(v_1) \subseteq \ldots \subseteq N(v_k)$ (where $N(u)$ denotes the set of u's neighbors). A twin-free connected bipartite chain graph must have the same number of vertices on either side. For each $n \in \mathbb{N}$, there is, up to isomorphism, a unique connected twin-free bipartite chain graph with n vertices in each part, denoted $C_{n,n}$. The graph $C_{n,n}$ is (V_s, V_p, E) where $V_s = \{s_1, \ldots, s_n\}, V_p = \{p_1, \ldots, p_n\}$, and $E = \{(s_i, p_j) \mid 1 \leq i \leq j \leq n\}$. The graph $C_{4,4}$ is shown in Fig. 1. We prove an upper and a lower bound on the readability of $C_{n,n}$.

Fig. 1. The graph $C_{4,4}$

Theorem 1. *For all $n \in \mathbb{N}$, the graph $C_{n,n}$ has readability $\mathcal{O}(\sqrt{n})$, with labels over an alphabet of size 3.*

We prove Theorem 1 by giving an efficient algorithm that constructs an overlap labeling of $C_{n,n}$ of length $\mathcal{O}(\sqrt{n})$ using strings over an alphabet of size 3.

Theorem 2. *For all $n \in \mathbb{N}$, the graph $C_{n,n}$ has readability $\Omega(\log n)$.*

Characterization of Bipartite Graphs with Readability at Most 2 (Sect. 4). Let C_t for $t \in \mathbb{N}$ denote the simple cycle with t vertices. The *domino* is the graph obtained from the cycle C_6 by adding an edge between two diametrically opposite vertices. For a graph G and a set $U \subseteq V(G)$, let $G[U]$ denote the subgraph of G induced by U.

Every bipartite graph with readability at most 1 is a disjoint union of complete bipartite graphs (also called bicliques) [6]. The characterization in the following theorem extends our understanding to graphs of readability at most 2.

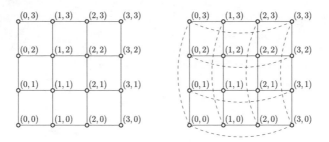

Fig. 2. The 4×4 grid $G_{4,4}$ and toroidal grid $TG_{4,4}$.

Theorem 3. *A twin-free bipartite graph G has readability at most 2 iff G has a matching M such that the graph $G' = G - M$ satisfies the following properties:*

1. *G' is a disjoint union of complete bipartite graphs.*
2. *For $U \subseteq V(G)$, if $G[U]$ is a C_6, then $G'[U]$ is the disjoint union of three edges.*
3. *For $U \subseteq V(G)$, if $G[U]$ is a domino, then $G'[U]$ is the disjoint union of a C_4 and an edge.*

Theorem 3 expresses a condition on vertex labels of a bipartite graph in purely graph theoretic terms, reducing the problem of deciding if a graph has readability at most 2 to checking the existence of a matching with a specific property.

An Efficient Algorithm for Readability 2 (in the Full Version). It is unknown whether computing the readability of a given bipartite graph is NP-hard. In fact, it is not even known whether the decision version of the problem is in NP, as the only upper bound on the readability of a bipartite graph with n vertices in each part is $\mathcal{O}(2^n)$ [4]. We make some progress on this front by showing that for readability 2, the decision version is polynomial-time solvable.

Theorem 4. *There exists an algorithm that, given a bipartite graph G, decides in polynomial time whether G has readability at most 2. Moreover, if the answer is "yes", the algorithm can also produce an overlap labeling of length at most 2.*

Readability of Grids and Their Induced Subgraphs (Sect. 5). We fully characterize the readability of grids. A *(two-dimensional) grid* is a graph $G_{m,n}$ with vertex set $\{0, 1, \ldots, m - 1\} \times \{0, 1, \ldots, n - 1\}$ such that there is an edge between two vertices iff the L_1-distance between them is 1. An example is shown in Fig. 2. Our next theorem fully settles the question of readability of grids.

Theorem 5. *For any two positive integers m, n with $m \leq n$,*

$$r(G_{m,n}) = \begin{cases} 3, & \text{if } m \geq 3 \text{ and } n \geq 3; \\ 2, & \text{if } (m = 2 \text{ and } n \geq 3) \text{ or } (m = 1 \text{ and } n \geq 4); \\ 1, & \text{if } (m, n) \in \{(1, 2), (1, 3), (2, 2)\}; \\ 0, & \text{if } m = n = 1. \end{cases}$$

Theorem 5 has an algorithmic implication for the readability of grid graphs. A *grid graph* is an induced subgraph of a grid. Several problems are NP-hard on the class of grid graphs, including Hamiltonicity problems [12], various layout problems [8], and others (see, e.g., [7]). We show that, unless P = NP, this is not the case for the readability problem.

Corollary 1. *The readability of a given grid graph can be computed in polynomial time.*

1.2   Technical Overview

We now give a brief description of our techniques. The key to proving the upper bound on the readability of bipartite chain graphs is understanding the combinatorics of the following process. We start with the sequence $(1, 2)$. The process consists of a series of rounds, and as a convention, we start at round 3: we write $3 \ (= 1 + 2)$ between 1 and 2 and obtain the sequence $(1, 3, 2)$. More generally, in round r, we insert r between all the consecutive pairs of numbers in the current sequence that sum up to r. Thus, we obtain $(1, 4, 3, 2)$ in round 4, then $(1, 5, 4, 3, 5, 2)$ in round 5, and so on. The question is to determine the length of the sequence formed in round r as a function of r. We prove that this length is $\frac{1}{2} \sum_{k=1}^{r} \varphi(k) = \Theta(r^2)$, where $\varphi(k)$ is the famous Euler's totient function denoting the number of integers in $\{1, \ldots, k\}$ that are coprime to k.

To prove our lower bound on the readability of bipartite chain graphs, we define a special sequence of subgraphs of the bipartite chain graph such that the number of graphs in the sequence is a lower bound on the readability. The sequence that we define has the additional property that if two vertices in the same part have the same set of neighbors in one of the graphs, then they have the same set of neighbors in all preceding graphs in the sequence. If the readability is very small, then we cannot simultaneously cover all the edges incident with two large-degree nodes as well as have their degrees distinct. The only properties of the connected twin-free bipartite chain graph that our proof uses are that it is dense and all vertices in the same part have distinct degrees. Hence, this technique is more broadly applicable to any class of dense graphs with a large number of distinct degrees.

Our characterization of graphs of readability at most 2, roughly speaking, states that a twin-free bipartite graph has readability at most 2 iff the graph can be decomposed into two subgraphs G_1 and G_2 such that G_1 is a disjoint union of bicliques and G_2 is a matching satisfying some additional properties. For $i \in \{1, 2\}$, the edges in G_i model overlaps of length exactly i. The heart of the proof lies in observing that for each pair of bicliques in the first subgraph, there can be at most one matching edge in the second subgraph that has its left endpoint in the first biclique and the right endpoint in the second biclique.

To derive a polynomial-time algorithm for recognizing graphs of readability two, we first reduce the problem to connected twin-free graphs of maximum degree at least three. For such graphs, we show that the constraints from our characterization of graphs of readability at most 2 can be expressed with a 2SAT formula having variables on edges and modeling the selection of edges forming a matching to form the graph G_2 of the decomposition.

In order to determine the readability of grids, we establish upper and lower bounds and in both cases use the fact that readability is monotone under induced subgraphs (that is, the readability of a graph is at least the readability of each of its induced subgraphs). The upper bound is derived by observing that every grid is an induced subgraph of some $4n \times 4n$ toroidal grid (see Fig. 2) and exploiting the symmetric structure of such toroidal grids to show that their readability is at most 3. This is the most interesting part of our proof and involves partitioning the edges of the $4n \times 4n$ toroidal grid into three sets and coming up with labels of length at most 3 for each vertex based on the containment of the four edges incident with the vertex in each of these three parts. Our characterization of graphs of readability at most 2 is a helpful ingredient in proving the lower bound on the readability of grids, where we construct a small subgraph of the grid for which our characterization easily implies that its readability is at least 3.

2   Preliminaries

For a string x, let $\mathrm{pre}_i(x)$ (respectively, $\mathrm{suf}_i(x)$) denote the prefix (respectively, suffix) of x of length i. A string x *overlaps* another string y if there exists an i with $1 \leq i \leq \min\{|x|, |y|\}$ such that $\mathrm{suf}_i(x) = \mathrm{pre}_i(y)$. If $1 \leq i < \min\{|x|, |y|\}$, we say that x *properly overlaps* with y. For a positive integer k, we denote by $[k]$ the set $\{1, \ldots, k\}$. Let $G = (V, E)$ be a (finite, simple, undirected) graph. If G is a connected bipartite graph, then it has a unique bipartition (up to the order of the parts). In this paper, we consider bipartite graphs $G = (V, E)$. If the bipartition $V = V_s \cup V_p$ is specified, we denote such graphs by $G = (V_s, V_p, E)$. Edges of a bipartite graph G are denoted by $\{u, v\}$ or by (u, v) (which implicitly implies that $u \in V_s$ and $v \in V_p$). We respect bipartitions when we perform graph operations such as taking an induced subgraph and disjoint union. For example, we say that a bipartite graph $G_1 = (V_s^1, V_p^1, E_1)$ is an *induced subgraph* of a bipartite graph $G_2 = (V_s^2, V_p^2, E_2)$ if $V_s^1 \subseteq V_s^2$, $V_p^1 \subseteq V_p^2$, and $E_1 = E_2 \cap \{(x, y) : x \in V_s^1, y \in V_p^1\}$. The *disjoint union* of two vertex-disjoint bipartite graphs $G_1 = (V_s^1, V_p^1, E_1)$ and $G_2 = (V_s^2, V_p^2, E_2)$ is the bipartite graph $(V_s^1 \cup V_s^2, V_p^1 \cup V_p^2, E_1 \cup E_2)$.

The path on n vertices is denoted by P_n. Given two graphs F and G, graph G is said to be *F-free* if no induced subgraph of G is isomorphic to F. Two vertices u, v in a bipartite graph are called *twins* if they belong to the same part of the bipartition and have the same neighbors (that is, if $N(u) = N(v)$). Given a bipartite graph $G = (V_s, V_p, E)$, its *twin-free reduction* $TF(G)$ is the graph with vertices being the equivalence classes of the twin relation on $V(G)$ (that is, $x \sim y$ iff x and y are twins in G), and two classes X and Y are adjacent iff $(x, y) \in E$ for some $x \in X$ and $y \in Y$. For graph theoretic terms not defined here, we refer to [24]. We now state some basic results for later use.

Lemma 1. *Let G and H be two bipartite graphs.*

Then:

(a) If G is an induced subgraph of H, then $r(G) \leq r(H)$.
(b) If F is the disjoint union of G and H, then $r(F) = \max\{r(G), r(H)\}$.
(c) The readability of G is the same for all bipartitions of $V(G)$.
(d) $r(G) = r(TF(G))$.

Lemma 1(b) shows that the study of readability reduces to the case of connected bipartite graphs. By Lemma 1(c), the readability of a bipartite graph is well defined even if a bipartition is not given in advance. Lemma 1(d) further shows that to understand the readability of connected bipartite graphs, it suffices to study the readability of connected twin-free bipartite graphs.

3   Readability of Bipartite Chain Graphs

In this section, we prove an upper bound on the readability of twin-free bipartite chain graphs, $C_{n,n}$, and prove Theorem 1. The lower bound on their readability (Theorem 2) is proved in the full version. Recall that the graph $C_{n,n}$ is (V_s, V_p, E) where $V_s = \{s_1, \ldots, s_n\}$, $V_p = \{p_1, \ldots, p_n\}$, and $E = \{(s_i, p_j) \mid 1 \leq i \leq j \leq n\}$.

3.1   Upper Bound

To prove Theorem 1, we construct a labeling ℓ of length $\mathcal{O}(\sqrt{n})$ for $C_{n,n}$ that satisfies (1) $\ell(s_i) = \ell(p_i)$ for all $i \in [n]$, and (2) $\ell(s_i)$ properly overlaps $\ell(s_j)$ iff $i < j$. It is easy to see that such an ℓ will be a valid overlap labeling of $C_{n,n}$. As the labels on either side of the bipartition are equal, we will just come up with a sequence of n strings to be assigned to one of the sides of $C_{n,n}$ such that the strings satisfy condition (2) above.

Definition 1. *A sequence of strings $(s_1, \ldots, s_t)$ is* forward-matching *if*

- *$\forall i \in [t]$, string s_i does not have a proper overlap with itself and*
- *$\forall i, j \in [t]$, string s_i overlaps string s_j iff $i \leq j$.*

Given an integer $r \geq 2$, we will show how to construct a forward-matching sequence S_r with $\Theta(r^2)$ strings, each of length at most r, over an alphabet of size 3. This will imply an overlap labeling of length $\mathcal{O}(\sqrt{n})$ for $C_{n,n}$, proving Theorem 1. The following lemma is crucial for this construction.

Lemma 2. *For all integers $t \geq 2$ and all $i \in [t-1]$, if $(s_1, \ldots, s_t)$ is forward-matching, so is $(s_1, \ldots, s_i, s_i s_{i+1}, s_{i+1}, \ldots, s_t)$.*

Proof. For the purposes of notation, let A be an arbitrary string from $s_1, \ldots, s_{i-1}$ (if it exists), let $B = s_i$, $C = s_{i+1}$, and let D be an arbitrary string from $s_{i+2}, \ldots, s_t$ (if it exists). The reader can easily verify that A and B overlap with the new string BC, and BC overlaps with C and D, as desired. What remains to show is that there are no undesired overlaps. Suppose for the sake of contradiction that BC overlaps B, and let i be the length of any such overlap. If $\text{suf}_i(BC)$ only includes characters from C, then C overlaps B; if it includes

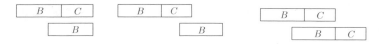

(a) BC does not overlap B.   (b) BC has no proper overlap with itself.

Fig. 3. Overlaps in the proof of Lemma 2

characters from B (and the entire C) then B has a proper overlap with itself (see Fig. 3a). In either case, we reach a contradiction. So, BC does not overlap B. By a symmetric argument, C does not overlap BC.

Next, suppose for the sake of contradiction that BC overlaps A, and let i be the length of any such overlap. If $\mathrm{suf}_i(BC)$ only includes characters from C, then C overlaps A; if it includes characters from B (and the entire C) then B overlaps A. In either case, we reach a contradiction. So, BC does not overlap A. By a symmetric argument, D does not overlap BC.

Finally, suppose for the sake of contradiction that BC has a proper overlap with itself, and let i be the length of any such overlap. Since C does not overlap BC, it follows that $\mathrm{suf}_i(BC)$ must include characters from B and the entire C. But then B has a proper overlap with B, a contradiction (see Fig. 3b). So, BC does not have a proper overlap with itself, completing the proof.   □

Now, we show how to construct a forward-matching sequence S_r. For the base case, we let $S_2 = (20, 0, 01)$. It can be easily verified that S_2 is forward-matching. Inductively, let S_r for $r > 2$ denote the sequence obtained from S_{r-1} by applying the operation in Lemma 2 to all indices i such that $s_i s_{i+1}$ is of length r, that is, add all obtainable strings of length r. Let B_r, for all integers $r \geq 2$, be the sequence of lengths of strings in S_r. We can obtain B_r directly from B_{r-1} by performing the following operation: for each consecutive pair of numbers x, y in B_{r-1}, if $x + y = r$ then insert r between x and y. Note that there is a mirror symmetry to the sequences with respect to the middle element, 1. The right sides of the first 6 sequences B_r, starting from the middle element, are as follows:

$$
\begin{array}{r|l}
r = 2 & 1\ 2 \\
r = 3 & 1\ 3\ 2 \\
r = 4 & 1\ 4\ 3\ 2 \\
r = 5 & 1\ 5\ 4\ 3\ 5\ 2 \\
r = 6 & 1\ 6\ 5\ 4\ 3\ 5\ 2 \\
r = 7 & 1\ 7\ 6\ 5\ 4\ 7\ 3\ 5\ 7\ 2
\end{array}
$$

It turns out that $|B_r|$, and, by extension, $|S_r|$, is closely related to the totient summatory function [22], also called the partial sums of Euler's totient function. This is the function $\Phi(r) = \sum_{k=1}^{r} \varphi(k)$, where $\varphi(k)$ is the number of integers in $[k]$ that are coprime to k. The asymptotic behavior of $\Phi(r)$ is well known: $\Phi(n) = \frac{3n^2}{\pi^2} + \mathcal{O}(n \log n)$ [10, p. 268]. The following lemma therefore implies $|S_r| = |B_r| = \Theta(r^2)$, completing the proof of Theorem 1.

Lemma 3. *For all integers $r \geq 2$, the length of the sequence B_r is $\Phi(r) + 1$.*

Proof. For the base case, observe that $|B_2| = 3 = \Phi(2) + 1$. In general, consider the case of $r \geq 3$.

Definition 2. *Two elements of B_r are called* neighbors in B_r *if they appear in two consecutive positions in B_r.*

We will show that any two neighbors are coprime (Claim 7) and any pair (i, j) of coprime positive integers that sum up to r appears exactly once as a pair of ordered neighbors in B_r (Claim 8). Together, these claims show that the neighbor pairs in B_{r-1} that sum up to r are exactly the pairs of coprime positive integers that sum up to r.

Fact 6. *If i and j are coprime then each of them is coprime with $i + j$ and with $i - j$.*

By this fact, there is a bijection between pairs (i, j) of coprime positive integers that sum up to r and integers $i \in [r]$ that are coprime to r. Hence, the number of neighbor pairs in B_{r-1} that sum up to r is $\varphi(r)$. Therefore, B_r contains $\varphi(r)$ occurrences of r. By induction, it follows that $|B_r| = |B_{r-1}| + \varphi(r) = \Phi(r-1) + 1 + \varphi(r) = \Phi(r) + 1$, proving the Lemma. □

We now prove the necessary claims.

Claim 7. *For all $r \geq 2$, if two numbers are neighbors in B_r, they are coprime.*

Proof. We prove the claim by induction. For the base case of $r = 2$, the claim follows from the fact that 1 and 2 are coprime. For the general case of $r \geq 3$, recall that B_r was obtained from B_{r-1} by inserting an element r between all neighbors i and j in B_{r-1} that summed to r. By the induction hypothesis, $gcd(i, j) = 1$, and, hence, by Fact 6, $gcd(i, r) = gcd(i, i+j) = 1$ and $gcd(r, j) = gcd(i+j, j) = 1$. Therefore, any two neighbors in B_r must be coprime. □

Claim 8. *For all $r \geq 3$, every ordered pair (i, j) of coprime positive integers that sum to r occurs exactly once as neighbors in B_{r-1}.*

Proof. We prove the claim by strong induction. The reader can verify the base case (when $r = 3$). For the inductive step, suppose the claim holds for all $k \leq r-1$ for some $r \geq 4$. Consider an ordered pair (i, j) of coprime positive integers that sum to r. Assume that $i > j$; we know that $i \neq j$, and the case of $i < j$ is symmetric. Since $r \geq 4$, we have that $i \geq 3$. In the recursive construction of the sequences $\{B_k\}$, the elements i are added to the sequence B_i when B_i is created from B_{i-1}. Since $j < i$, all the elements j are already present in B_{i-1}. By Fact 6, since $gcd(i, j) = 1$, we get that $gcd(i - j, j) = 1$. By the inductive hypothesis, pair $(i - j, j)$ appears exactly once as an ordered pair of neighbors in B_{i-1}. Consequently, (i, j) must appear exactly once as an ordered pair of neighbors in B_i. No new elements i, j are added to the sequence in later stages, when $k > i$. Also, no new elements are inserted between i and j when $i + 1 \leq k \leq i + j - 1 = r - 1$. Therefore, the ordered neighbor pair (i, j) appears exactly once in B_{r-1}. □

4   A Characterization: Graphs with Readability at Most 2

We characterize bipartite graphs with readability at most 2 by proving Theorem 3. By Lemma 1, it is enough to obtain such a characterization for connected twin-free bipartite graphs. We use this characterization (in the full version) to develop a polynomial-time algorithm for recognizing graphs of readability at most 2 and also (in Sect. 5) to prove a lower bound on the readability of general grids. Recall that a *domino* is the graph obtained from C_6 by adding an edge between two vertices at distance 3. We first define the notion of a feasible matching, which is implicitly used in the statement of Theorem 3.

Definition 3 (Feasible Matching). *A matching M in a bipartite graph G is feasible if the following conditions are satisfied:*

1. *The graph $G' = G - M$ is a disjoint union of bicliques (equivalently: P_4-free).*
2. *For $U \subseteq V(G)$, if $G[U]$ is a C_6, then $G'[U]$ is the disjoint union of three edges.*
3. *For $U \subseteq V(G)$, if $G[U]$ is a domino, then $G'[U]$ is the disjoint union of a C_4 and an edge.*

In the full version, we prove Theorem 3 by showing that a bipartite graph G has readability at most 2 iff G has a feasible matching. The following corollary of Theorem 3 is used in Sect. 5.

Corollary 2. *Every bipartite graph G of maximum degree at most 2 has readability at most 2.*

Proof. If G is a connected twin-free bipartite graph of maximum degree at most 2, then G is a path or an (even) cycle. In this case, the edge set of G can be decomposed into two matchings M_1 and M_2. Both M_1 and M_2 are feasible matchings. Thus, by Theorem 3, G has readability at most 2.   □

5   Readability of Grids and Their Induced Subgraphs

In this section, we determine the readability of grids by proving Theorem 5. We first look at toroidal grids, which are closely related to grids. For positive integers $m \geq 3$ and $n \geq 3$, the *toroidal grid* $TG_{m,n}$ is obtained from the grid $G_{m,n}$ by adding edges $((i, 0), (i, n-1))$ and $((0, j), (m-1, j))$ for all $i \in \{0, \ldots, m-1\}$ and $j \in \{0, \ldots, n-1\}$ (See Fig. 2 for an example.). The graph $TG_{m,n}$ is bipartite iff m and n are both even. In this case, a bipartition can be obtained by setting $V(TG_{m,n}) = V_s \cup V_p$ where $V_s = \{(i, j) \in V(TG_{m,n}) : i + j \equiv 0 \pmod{2}\}$ and $V_p = \{(i, j) \in V(TG_{m,n}) : i + j \equiv 1 \pmod{2}\}$.

Lemma 4. *For all integers $n > 0$, we have $r(TG_{4n,4n}) \leq 3$.*

We now prove Theorem 5, about the readability of $G_{m,n}$. We first recall the following simple observation (which follows, e.g., from [6, Theorem 4.3]).

Lemma 5. *A bipartite graph G has: (i) $r(G) = 0$ iff G is edgeless, and (ii) $r(G) \leq 1$ iff G is P_4-free (equivalently: a disjoint union of bicliques).*

Proof (of Theorem 5). First, by Lemma 5, $r(G_{m,n})$ is 0 if $m = n = 1$ and positive, otherwise. Second, when $(m, n) \subset \{(1, 2), (1, 3), (2, 2)\}$, the graphs $G_{m,n}$ are isomorphic to $K_{1,1}, K_{1,2}$, and $K_{2,2}$, respectively. Thus, by Lemma 5, their readability is 1.

Third, when $m + n \geq 5$, the grid $G_{m,n}$ contains an induced P_4, implying that $r(G_{m,n}) \geq 2$. By Theorem 3, a twin-free bipartite graph G has readability at most 2 iff G has a feasible matching. (See Definition 3.) When $m + n \geq 5$, the grid $G_{m,n}$ is twin-free. If $m = 2$ and $n \geq 3$, then $M = \{((i, j), (i, j + 1)) \mid i \in \{0, 1\}$ and $j \in \{0, \ldots, n - 2\}$ is even$\}$ is a feasible matching in $G_{m,n}$, so $r(G_{m,n}) = 2$. If $m = 1$ and $n \geq 4$, then $G_{m,n}$ is isomorphic to a path of length at least three. Since its maximum degree is 2, we get $r(G_{m,n}) \leq 2$, by Corollary 2. Thus, $r(G_{m,n}) = 2$.

To show that $r(G_{m,n}) \leq 3$ for $m \geq 3$ and $n \geq 3$, we observe that $G_{m,n}$ (for $m \leq n$) is an induced subgraph of $TG_{4n,4n}$. By Lemmas 1(a) and 4, we have that $r(G_{m,n}) \leq r(TG_{4n,4n}) \leq 3$. The proof that $r(G_{m,n}) \geq 3$ can be found in the full version. $\qquad\square$

6   Conclusion

In this work, we gave several results on families of n-vertex bipartite graphs with readability $o(n)$. The results were obtained by developing new or applying a variety of known techniques to the study of readability. These include a graph theoretic characterization in terms of matchings, a reduction to 2SAT, an explicit construction of overlap labelings analyzed via number theoretic notions, and a new lower bound applicable to dense graphs with a large number of distinct degrees. One of the main specific questions left open by our work is to close the gap between the $\Omega(\log n)$ lower bound and the $\mathcal{O}(\sqrt{n})$ upper bound on the readability of n-vertex bipartite chain graphs. In the context of general bipartite graphs, it would be interesting to determine the computational complexity of determining whether the readability of a given bipartite graph is at most k, where k is either part of input or a constant greater than 2, to study the parameter from an approximation point of view, and to relate it to other graph invariants. For instance, for a positive integer k, what is the maximum possible readability of a bipartite graph of maximum degree at most k? Another interesting direction would be to study the complexity of various computational problems on graphs of low readability.

Acknowledgments. The result of Sect. 3.1 was discovered with the help of The On-Line Encyclopedia of Integer Sequences Ⓡ [22]. This work has been supported in part by NSF awards DBI-1356529, CCF-1439057, IIS-1453527, and IIS-1421908 to P.M. and by the Slovenian Research Agency (I0-0035, research program P1-0285 and research projects N1-0032, J1-6720, and J1-7051) to M.M. The authors S.R. and N.V. were supported by NSF grant CCF-1422975 to S.R. The author N.V. was also supported by Pennsylvania State University College of Engineering Fellowship, PSU Graduate Fellowship, and by NSF grant IIS-1453527 to P.M. The main idea of the proof of Lemma 4 was developed by V.J. in his undergraduate final project paper [13] at the University of Primorska.

References

1. Błażewicz, J., Formanowicz, P., Kasprzak, M., Kobler, D.: On the recognition of de Bruijn graphs and their induced subgraphs. Discrete Math. **245**(1), 81–92 (2002)
2. Błażewicz, J., Formanowicz, P., Kasprzak, M., Schuurman, P., Woeginger, G.J.: DNA sequencing, eulerian graphs, and the exact perfect matching problem. In: Goos, G., Hartmanis, J., van Leeuwen, J., Kučera, L. (eds.) WG 2002. LNCS, vol. 2573, pp. 13–24. Springer, Heidelberg (2002). https://doi.org/10.1007/3-540-36379-3_2
3. Blazewicz, J., Hertz, A., Kobler, D., de Werra, D.: On some properties of DNA graphs. Discrete Appl. Math. **98**(1), 1–19 (1999)
4. Braga, M.D.V., Meidanis, J.: An algorithm that builds a set of strings given its overlap graph. In: Rajsbaum, S. (ed.) LATIN 2002. LNCS, vol. 2286, pp. 52–63. Springer, Heidelberg (2002)
5. Chikhi, R., Jovicic, V., Kratsch, S., Medvedev, P., Milanic, M., Raskhodnikova, S., Varma, N.: Bipartite graphs of small readability. CoRR (2018). http://arxiv.org/abs/1805.04765
6. Chikhi, R., Medvedev, P., Milanič, M., Raskhodnikova, S.: On the readability of overlap digraphs. Discrete Appl. Math. **205**, 35–44 (2016)
7. Clark, B.N., Colbourn, C.J., Johnson, D.S.: Unit disk graphs. Discrete Math. **86**(1–3), 165–177 (1990)
8. Díaz, J., Penrose, M.D., Petit, J., Serna, M.J.: Approximating layout problems on random geometric graphs. J. Algorithms **39**(1), 78–116 (2001)
9. Gevezes, T.P., Pitsoulis, L.S.: Recognition of overlap graphs. J. Comb. Optim. **28**(1), 25–37 (2014)
10. Hardy, G.H., Wright, E.M.: An Introduction to the Theory of Numbers, 5th edn. The Clarendon Press, Oxford University Press, New York (1979)
11. Idury, R.M., Waterman, M.S.: A new algorithm for DNA sequence assembly. J. Comput. Biol. **2**(2), 291–306 (1995)
12. Itai, A., Papadimitriou, C.H., Szwarcfiter, J.L.: Hamilton paths in grid graphs. SIAM J. Comput. **11**(4), 676–686 (1982)
13. Jovičić, V.: Readability of digraphs and bipartite graphs (2016), final project paper. University of Primorska, Faculty of Mathematics, Natural Sciences and Information Technologies, Koper, Slovenia (2016). https://arxiv.org/abs/1612.07113
14. Kasprzak, M.: Classification of de Bruijn-based labeled digraphs. Discrete Appl. Math. **234**, 86–92 (2016)
15. Li, X., Zhang, H.: Characterizations for some types of DNA graphs. J. Math. Chem. **42**(1), 65–79 (2007)
16. Li, X., Zhang, H.: Embedding on alphabet overlap digraphs. J. Math. Chem. **47**(1), 62–71 (2010)
17. Miller, J.R., Koren, S., Sutton, G.: Assembly algorithms for next-generation sequencing data. Genomics **95**(6), 315–327 (2010)
18. Myers, E.W.: The fragment assembly string graph. In: ECCB/JBI, p. 85 (2005)

19. Nagarajan, N., Pop, M.: Sequence assembly demystified. Nat. Rev. Genet. **14**(3), 157–167 (2013)
20. Pendavingh, R., Schuurman, P., Woeginger, G.J.: Recognizing DNA graphs is difficult. Discrete Appl. Math. **127**(1), 85–94 (2003)
21. Simpson, J.T., Durbin, R.: Efficient de novo assembly of large genomes using compressed data structures. Genome Res. **22**, 549–556 (2011)
22. Sloane, N.J.A.: The On-Line Encyclopedia of Integer Sequences (2016). https://oeis.org
23. Tarhio, J., Ukkonen, E.: A greedy approximation algorithm for constructing shortest common superstrings. Theoret. Comput. Sci. **57**(1), 131–145 (1988)
24. West, D.B.: Introduction to Graph Theory. Prentice Hall Inc., Upper Saddle River (1996)

Maximum Colorful Cliques
in Vertex-Colored Graphs

Giuseppe F. Italiano[2], Yannis Manoussakis[1], Nguyen Kim Thang[3],
and Hong Phong Pham[1(✉)]

[1] LRI, University Paris-Saclay, Orsay, France
phongph.hut@gmail.com
[2] University of Rome Tor Vergata, Rome, Italy
[3] IBISC, University Paris-Saclay, Evry, France

Abstract. In this paper we study the problem of finding a maximum colorful clique in vertex-colored graphs. Specifically, given a graph with colored vertices, we wish to find a clique containing the maximum number of colors. Note that this problem is harder than the maximum clique problem, which can be obtained as a special case when each vertex has a different color. In this paper we aim to give a dichotomy overview on the complexity of the maximum colorful clique problem. We first show that the problem is NP-hard even for several cases where the maximum clique problem is easy, such as complement graphs of bipartite permutation graphs, complement graphs of bipartite convex graphs, and unit disk graphs, and also for properly vertex-colored graphs. Next, we provide a XP parameterized algorithm and polynomial-time algorithms for classes of complement graphs of bipartite chain graphs, complete multipartite graphs and complement graphs of cycle graphs, which are our main contributions.

1 Introduction

In this paper we deal with vertex-colored graphs, which are useful in various applications. For instance, the Web graph may be considered as a vertex-colored graph where the color of a vertex represents the content of the corresponding site (i.e., green for corporate sites, red for blogs, yellow for ecommerce sites, etc.) [2]. In a biological population, vertex-colored graphs can be used to represents the connections and interactions between species where different species have different colors. Other applications of vertex-colored graphs arise also in bioinformatics (Multiple Sequence Alignment Pipeline or for multiple Protein-Protein Interaction networks) [5], and in scheduling problems [11].

Given a vertex-colored graph, a *tropical subgraph* is a subgraph where each color of the initial graph appears at least once. Many graph properties, such as the domination number, the vertex cover number, independent sets, connected

---

N. Kim Thang—Research supported by the ANR project OATA n° ANR-15-CE40-0015-01, Hadamard PGMO and DIM RFSI.

components, shortest paths, matchings, etc. can be studied in their tropical version. Tropical subgraphs find applications in many scenarios: in a (biological) population, for instance, a tropical subgraph fully represents the (bio-)diversity of the population. There are many cases, however, where tropical subgraphs do not necessarily exist. Hence, one might be interested in the more general question of finding a *maximum colorful* subgraph, i.e., a subgraph with the maximum possible number of colors. For instance in biology this would represent a subgraph with the most diverse population. As a special case, a maximum colorful subgraph is tropical if it contains all colors. The notion of colorful subgraph is close to, but somewhat different from the *colorful* concept considered in [1,9,10], where neighbor vertices must have different colors. It is also related to the concepts of *color patterns* or *colorful* used in bio-informatics [6]. Note that in a *colorful* subgraph considered in our paper, two adjacent vertices may have the same color, i.e., the subgraph is not necessarily properly colored.

In this paper we are interested in finding maximum colorful cliques in vertex-colored graphs. Throughout, we let $G = (V, E)$ denote a simple undirected graph. Given a set of colors $\mathcal{C}$, $G^c = (V, E)$ denotes a vertex-colored graph whose vertices are (not necessarily properly) colored by one of the colors in $\mathcal{C}$. The number of colors of G^c is $|\mathcal{C}|$. Given a subset of vertices $U \subseteq V$, the set of colors of vertices in U is denoted by $\mathcal{C}(U)$. Moreover, we denote by $c(v)$ the color of vertex v and by $v(H, c)$ the number of vertices of H whose color is c. The set of neighbors of v is denoted by $N(v)$. More formally, in this paper we study the following:

Maximum Colorful Clique Problem (MCCP). Given a vertex-colored graph $G^c = (V, E)$, find a clique with the maximum number of colors of the original graph.

Related Work. In the special case where each vertex has a distinct color, MCCP reduces to the maximum clique problem. The maximum clique problem has been widely studied in the literature and it is known to be NP-complete for general graphs, fixed-parameter intractable and hard to approximate. However, the maximum clique problem can be efficiently solved in polynomial time for several special classes of graphs, such as complement graphs of bipartite graphs, permutation graphs, comparability graphs, chordal graphs. All those are perfect graphs, and the maximum clique problem can be solved in polynomial time also in other non-perfect graphs such as circle graphs and unit disk graphs.

Another related problem, which has also been widely considered in the literature, is listing all maximal cliques in a graph. Clearly, if one can list all maximal cliques, one can also find a maximum clique, since a maximum clique must be maximal. In a similar vein, one can also find the maximum colorful clique, since any maximum colorful clique can be extended to a maximal clique. Therefore, MCCP is easy for all classes of graphs for which we can list in polynomial time all maximal cliques. Those graphs include chordal graphs, complete graphs, triangle-free graphs, interval graphs, graphs of bounded boxicity, and planar graphs.

Tropical subgraph and maximum colorful subgraph problems in vertex-colored graphs have been studied only recently. In particular, tropical subgraph problems in vertex-colored graphs such as tropical connected subgraphs, tropical dominating sets have been investigated in [7]. The maximum colorful matching problem [3], the maximum colorful path problem [4] and the maximum colorful cycles problem [8] have been studied, and several hardness results and polynomial-time algorithms were shown for different classes of graphs.

Our Contributions. In this paper, we aim to give a dichotomy overview on the complexity of MCCP. First, we show that MCCP is NP-hard even for several cases where the maximum clique problem is known to be easy, such as complement graphs of bipartite permutation graphs, complement graphs of convex bipartite graphs, and unit disk graphs. Also, we show that MCCP is NP-hard for properly vertex-colored graphs. Next, we present polynomial-time algorithms for several classes of graphs. First, we prove that MCCP belongs to the class of XP parameterized algorithms. Second, we show that MCCP can be solved in polynomial time for complement graphs of bipartite chain graphs, which is a special case of complement graphs of bipartite permutation graphs (for which MCCP is NP-hard).

Our main contribution is polynomial-time algorithms for complete multipartite graphs and complement graphs of cycle graphs. A graph is called multipartite if its vertices can be partitioned into different independent sets. In a complete multipartite graph any two vertices in different independent sets are adjacent. To solve MCCP on complete multipartite graphs, we proceed as follows. We start with a maximum clique, by picking one vertex from each independent set. To compute a maximum colorful clique, we iterate through different maximum cliques by increasing at each step the number of colors, without decreasing the number of vertices. To do this efficiently, we define a special structure, called k-colorful augmentation, which might be of independent interest. The running time of our algorithm is $O(|\mathcal{C}|M(m + n, n))$, where $|\mathcal{C}|$ is the total number of colors and $M(m, n)$ is the time required for finding a maximum matching in a general graph with m edges and n vertices (Currently, $M(m, n) = O(\sqrt{n}m)$ [12]).

A cycle graph is a graph that consists of a single cycle. Similar to the algorithm for complete multipartite graphs, we also investigate a special structure to obtain another better clique from the current cliques in a complement graph of a cycle graph and this yields a polynomial algorithm in this case.

Due to space limit, some results, proofs and details are omitted from this extended abstract and are deferred to the appendix.

2   Hardness Results for MCCP

In this section, we present several NP-hardness results for MCCP. Specifically, we show that MCCP is NP-hard for the complement graphs of biconvex bipartite graphs and the complement graphs of permutation bipartite graphs. Note that the maximum clique problem can be solved in polynomial time for the complement graphs of bipartite graphs based on König's Theorem, and therefore the maximum clique problem can be also efficiently solved for the complement graphs of biconvex bipartite graphs and the complement graphs of permutation bipartite graphs. Next, we also show that MCCP is NP-hard for unit disk graphs, which are also easy cases for the maximum clique problem. We also prove that MCCP is NP-hard for properly vertex-colored graphs. The following lemma shows that MCCP is NP-hard for the complement graphs of bipartite permutation graphs. Recall that a graph is a bipartite permutation graph, if it is both bipartite and a permutation graph.

Lemma 1. *The maximum colorful clique problem is NP-hard for the complement graphs of permutation bipartite graphs.*

Proof. We reduce from the MAX-3SAT problem. Consider a boolean expression B in CNF with variables $X = \{x_1, \ldots, x_s\}$ and clauses $B = \{b_1, \ldots, b_t\}$. In addition, suppose that B contains exactly 3 literals per clause (actually, we may also consider clauses of arbitrary size). We show how to construct a vertex-colored graph G^c associated with any such formula B, such that, there exists a truth assignment to the variables of B satisfying t' clauses if and only if G^c contains a clique with t' distinct colors. Suppose that $\forall i, 1 \leq i \leq s$, the variable x_i appears in clauses $b_{i1}, b_{i2}, \ldots, b_{i\alpha_i}$ and $\overline{x_i}$ appears in clauses $b'_{i1}, b'_{i2}, \ldots, b'_{i\beta_i}$ in which $b_{ij} \in B$ and $b'_{ik} \in B$. Now a vertex-colored permutation bipartite graph G^c is constructed as follows. We give geometrical definition as the intersection graphs of line segments whose endpoints lie on two parallel lines L_1 and L_2.

We create first $\alpha_1 + \beta_1$ endpoints for two parallel lines as follows. Firstly, from left to right, we let α_1 endpoints of L_1 corresponding to pairs (x_1, b_{11}), $(x_1, b_{12}), \ldots, (x_1, b_{1\alpha_1})$, and next β_1 endpoints of L_1 will be $(\overline{x_1}, b'_{11}), (\overline{x_1}, b'_{12})$ $\ldots, (\overline{x_1}, b'_{1\beta_1})$. Conversely, on L_2, from left to right, we let first β_1 endpoints including $(\overline{x_1}, b'_{11}), (\overline{x_1}, b'_{12}) \ldots, (\overline{x_1}, b'_{1\beta_1})$, and next α_1 endpoints of L_2 will be $(x_1, b_{11}), (x_1, b_{12}), \ldots, (x_1, b_{1\alpha_1})$. This way, the segment of (x_1, b_{1i}) and the segment (x_1, b_{1j}) are in parallel and they do not intersect each other for $\forall i, j, 1 \leq i \neq j \leq \alpha_1$; similarly for each pair $(\overline{x_1}, b'_{1i})$ and $(\overline{x_1}, b'_{1j})$, for $\forall i, j, 1 \leq i \neq j \leq \beta_1$. However, it is easy to see that the pair of segment (x_1, b_{1i}) and $(\overline{x_1}, b'_{1j})$ intersect each other. Now we finish arrangement for clauses corresponding to the variable x_1 and $\overline{x_1}$. Next, we similarly create and arrange $\alpha_2 + \beta_2$ endpoints corresponding to the variable x_2 and $\overline{x_2}$ on L_1 and L_2. The segments of this section are separated with the segments of x_1 and $\overline{x_1}$ but they still guarantee the properties of intersection: the segment of (x_2, b_{2i}) and the segment (x_2, b_{2j}) are in parallel, similarly for $(\overline{x_2}, b'_{2i})$ and $(\overline{x_2}, b'_{2j})$ but the pair of segment (x_2, b_{2i}) and $(\overline{x_2}, b'_{2j})$ intersect each other.

By doing so, we created a permutation bipartite graph. In fact, it is a bipartite graph since edges are between (x_i, b_{ij}) and $(\overline{x_i}, b'_{ik})$. It is also a permutation graph since it is the intersection graph of line segments whose endpoints lie on two parallel lines. Now we color vertices as follows. We use color c_l for the vertex (x_i, b_{ij}) and $(\overline{x_i}, b'_{ik})$ if b_{ij} or b'_{ik} is the clause b_l of B. From this vertex-colored permutation bipartite graph, we focus on its complement graph. Clearly the obtained graph is the complement graph of a vertex-colored permutation bipartite graphs, denoted it by G^c. Now we claim that there exists a truth assignment to the variables of B satisfying t' clauses if and only if G^c contains a clique with t' distinct colors. Observe that in the complement graph G^c, there are no edges between (x_i, b_{ij}) and $(\overline{x_i}, b'_{ik})$ for $\forall i, 1 \leq i \leq s$ but there are all edges between (x_i, b_{ij}) and (x_i, b_{ik}), between (x_i, b_{ij}) and $(x_{i'}, b_{i'k})$, between (x_i, b_{ij}) and $(\overline{x_{i'}}, b_{i'k})$ in which $\forall i, i', 1 \leq i \neq i' \leq s$.

Now we extract a subgraph from a truth assignment to the variables of B satisfying t' clauses, as follows. For each $\forall i, 1 \leq i \leq s$, in the case that x_i is assigned true, then we choose all vertices (x_i, b_{ij}), $1 \leq j \leq \alpha_i$. Otherwise, we choose all vertices $(\overline{x_i}, b_{ij})$, $1 \leq j \leq \beta_i$. It is possible to see that this is a clique and this clique contains all vertices corresponding to satisfied clauses, so the number of colors of this cliques is equal to t'. Conversely, from a clique with t' colors we obtain an assignment as follows. Note that in this clique we can ot have both vertices (x_i, b_{ij}) and $(\overline{x_i}, b'_{ik})$. Therefore, it is possible to assign x_i as true if this clique contains vertices (x_i, b_{ij}), otherwise x_i is assigned as false. Clearly, this assignment is consistent. Since this clique has t' colors, it must contain vertices corresponding to t' different clauses. Thus, it is not difficult to see that the corresponding assignment leads to t' satisfied clauses. This complete our proof. $\qquad\square$

Next, we show that MCCP is also NP-hard for the complement graphs of biconvex bipartite graphs. Recall that a bipartite graph (X, Y, E) is biconvex if there is an ordering of X and Y that fulfills the adjacency property, i.e., for every vertex $y \in Y$ (resp., $x \in X$), $N(y)$ (resp., $N(x)$) consists of vertices that are consecutive in the sorted ordering of X (resp., Y).

Lemma 2. *The maximum colorful clique problem is NP-hard for the complement graphs of biconvex bipartite graphs.*

Proof. We use the same notation as in Lemma 1, and reduce again from MAX-3SAT, as follows. We first create a vertex-colored biconvex bipartite graph (X, Y, E) from an instance of the MAX-3SAT problem such that, there exists a truth assignment to the variables of B satisfying t' clauses if and only if G^c contains a clique with t' distinct colors. We also assume that $\forall i, 1 \leq i \leq s$, the variable x_i appears in clauses $b_{i1}, b_{i2}, \ldots, b_{i\alpha_i}$ and $\overline{x_i}$ appears in clauses $b'_{i1}, b'_{i2}, \ldots, b'_{i\beta_i}$ in which $b_{ij} \in B$ and $b'_{ik} \in B$. Now each vertex of X represents a pair of (x_i, b_{ij}) with $1 \leq j \leq \alpha_i$ (x_i appears in the clause b_{ij}). Similarly, each vertex of Y represents a pair of $(\overline{x_i}, b'_{ik})$ with $1 \leq k \leq \beta_i$ ($\overline{x_i}$ appears in the clause b'_{ik}). Next, we sort an ordering over X from left to right as follows. The vertices of X are sorted as (x_1, b_{11}), (x_1, b_{12}), $\ldots$, $(x_1, b_{1\alpha_1})$, (x_2, b_{21}), (x_2, b_{22}), $\ldots$, $(x_2, b_{2\alpha_2})$, $\ldots$, (x_s, b_{s1}), (x_s, b_{s2}), $\ldots$, $(x_s, b_{s\alpha_s})$. Similarly, vertices over Y are sorted from left to right as follows: $(\overline{x_1}, b'_{11})$, $(\overline{x_1}, b'_{12})$, $\ldots$, $(\overline{x_1}, b'_{1\beta_1})$, $(\overline{x_2}, b'_{21})$, $(\overline{x_2}, b'_{22})$, $\ldots$, $(\overline{x_2}, b'_{2\beta_2})$, $\ldots$, $(\overline{x_s}, b'_{s1})$, $(\overline{x_s}, b'_{s2})$, $\ldots$, $(\overline{x_s}, b'_{s\beta_s})$. Now edges between X and Y are created by connecting each vertex (x_i, b_{ij}) to each vertex $(\overline{x_i}, b'_{ik})$ for $\forall j, 1 \leq j \leq \alpha_i$ and $\forall k, 1 \leq k \leq \beta_i$. Clearly the obtained graph is a biconvex bipartite graph since for each vertex (x_i, b_{ij}), all its neighbors, i.e., $(\overline{x_i}, b'_{i1})$, $(\overline{x_i}, b'_{i2})$, $\ldots$, $(\overline{x_i}, b'_{i\beta_i})$, are consecutive by the sorted ordering over Y and vice versa for each vertex $(\overline{x_i}, b'_{ik})$. Similar to the case of complement graphs of permutation bipartite graphs, we use the color c_l for the vertex (x_i, b_{ij}) and $(\overline{x_i}, b'_{ik})$ if b_{ij} or b'_{ik} is the clause b_l of B. Now we focus on the complement graph of this vertex-colored biconvex bipartite graph, denoted by G^c. It is not difficult to see that this graph has the same structure as the complement graphs of bipartite permutation graphs in Lemma 1. In other words, there are no edges between (x_i, b_{ij}) and $(\overline{x_i}, b'_{ik})$ for all $1 \leq i \leq s$ but there are all edges between (x_i, b_{ij}) and (x_i, b_{ik}), between (x_i, b_{ij}) and $(x_{i'}, b_{i'k})$, between (x_i, b_{ij}) and $(\overline{x_{i'}}, b_{i'k})$ in which $\forall i, i', 1 \leq i \neq i' \leq s$. So, exactly the same arguments used in Lemma 1 yield the proof. $\qquad\square$

Now we prove that MCCP is NP-hard also for unit disk graphs and properly vertex-colored graphs, as shown in the following lemmas (proofs in the appendix).

Lemma 3. *The maximum colorful clique problem is NP-hard for unit disk graphs.*

Lemma 4. *The maximum colorful clique problem is NP-hard for properly vertex-colored graphs.*

3   Efficient Algorithms for MCC

In this section we present several efficient algorithms for the maximum colorful clique problem. We start by proving that MCPP belongs to the class *XP* parameterized problems. Next, we show that MCCP can be efficiently solved for complement graphs of bipartite chain graphs. This is in contrast with the case of complement graphs of bipartite permutation graphs and complement graphs of biconvex bipartite graphs, for which we have shown in Sect. 2 that MCCP is NP-hard. Finally, we present our polynomial-time algorithms for MCCP in complete multipartite graphs and complement graphs of cycle graphs.

3.1   A XP Parameterized Algorithm for MCCP

Our algorithm is based on the following observation: each maximum colorful clique can be reduced to another maximum colorful clique in which each color appears at most once. Indeed, if a color c appears more than once in maximum colorful clique, we can maintain only one vertex of this color. By doing so, we can keep all colors of the original clique, and obtain a new maximum colorful clique where each color appears only once. From this observation, we can list all cases by trying each vertex from the set of vertices of a color for each subset of the original set of colors. Let $\mathcal{C} = \{c_1, c_2, \ldots, c_{|\mathcal{C}|}\}$ be the set of colors of the original graph G^c. For each color c_i, let n_i be the number of vertices of the graph with the color c_i and let denote $V(c_i)$ be the set of vertices of G^c of color c_i. Our *XP* parameterized algorithm for MCCP is as follows.

The following theorem shows the correctness of our *XP* parameterized algorithm for computing a maximum colorful cycle in a vertex-colored graph G^c.

Algorithm 1. Maximum colorful clique in vertex-colored graphs.

1: $max \leftarrow 0$ /* the number of colors of a maximum colorful clique */
2: $H^* \leftarrow \emptyset$ /* the maximum colorful clique returned */
3: **for** $\{c_{i_1}, c_{i_2}, \ldots, c_{i_j}\} \subseteq \{c_1, c_2, \ldots, c_{|\mathcal{C}|}\}$ **do** # Consider all subsets of colors
    of $\mathcal{C}$
4:    **for** $\{v_1, v_2, \ldots, v_j\} \subseteq \{V(c_{i_1}) \times V(c_{i_2}) \times \ldots \times V(c_{i_j})\}$ **do**
5:       $H \leftarrow$ the induced graph of the vertices $\{v_1, v_2, \ldots, v_j\}$
6:       **if** H is a clique and $max < |\mathcal{C}(H)|$ **then**
7:          $H^* \leftarrow H$
8:       **end if**
9:    **end for**
10: **end for**
11: **return** H^* as a maximum colorful clique

Theorem 1. *Algorithm 1 computes a maximum colorful clique of G^c in time $O((\frac{n}{|\mathcal{C}|})^{|\mathcal{C}|})$ where $|\mathcal{C}|$ is the number of colors in G^c and n is the number of vertices of G^c.*

Proof. The algorithm first consider all subsets of colors of the set $\mathcal{C}$. This takes $2^{|\mathcal{C}|}$ times in the outer loop. Inside each of these iterations (corresponding to the set $\{c_{i_1}, c_{i_2}, \ldots, c_{i_j}\}$), the algorithm considers all subsets of vertices of $\{V(c_{i_1}) \times V(c_{i_2}) \times \ldots \times V(c_{i_j})\}$ (each vertex from a set), to check the existence of clique. It is easy to see that this takes $O(|V(c_{i_1})| \times |V(c_{i_2})| \times \ldots \times |V(c_{i_j})|)$. By the Cauchy inequality, with $j \leq |\mathcal{C}|$ we have that $|V(c_{i_1})| \times |V(c_{i_2})| \times \ldots \times |V(c_{i_j})| \leq (\frac{\sum |V(c_i)|}{|\mathcal{C}|})^{|\mathcal{C}|}$. Therefore, the complexity of this algorithm is $O((\frac{n}{|\mathcal{C}|})^{|\mathcal{C}|})$. $\qquad\square$

3.2 An Algorithm for MCCP for Complement Graphs of Bipartite Chain Graphs

In this section, we show a polynomial algorithm for MCCP for complement graphs of bipartite chain graphs. Recall that a bipartite graph $G = (X, Y, E)$ is said to be a *chain bipartite* graph if its vertices of X can be linearly ordered such that $N(x_1) \supseteq N(x_2) \supseteq \ldots \supseteq N(x_{|X|})$. As a consequence, we also immediately obtain a similar linear ordering over Y such that $N(y_1) \supseteq N(y_2) \supseteq \ldots \supseteq N(y_{|Y|})$. It is known that these orderings over X and Y can be computed in $O(n)$ time. Here we will look for a maximum colorful clique in the complement graph of a vertex-colored bipartite chain graph $G^c = (X, Y, E)$. Let us denote this complement graphs by $\overline{G^c}$. First, we observe that in $\overline{G^c}$, we have that $N(x_1) \subseteq N(x_2) \subseteq \ldots \subseteq N(x_{|X|})$ and $N(y_1) \subseteq N(y_2) \subseteq \ldots \subseteq N(y_{|Y|})$. Let K be a maximum colorful clique of $\overline{G^c}$ in which the set of vertices of X and Y of K are denoted by X_K and Y_K, respectively. Then, we can convert K to another maximum colorful clique by exploiting the following lemma.

Lemma 5. *Let K be a maximum colorful clique of $\overline{G^c}$, let i and j be the minimum numbers such that $x_i \in X_K$ and $y_j \in Y_K$. Then there exists another maximum colorful clique K' where $V(K') = \{x_i, x_{i+1}, \ldots, x_{|X|}\} \cup \{y_j, y_{j+1}, \ldots, y_{|Y|}\}$.*

Proof. Let $x_{i'}$ be a vertex such that $x_{i'} \notin X_K$ and $i < i' < |X|$. Since $N(x_i) \subseteq N(x_{i'})$ and K is a clique, adding the vertex $x_{i'}$ yields a larger clique which contains the old clique. Similarly, it is possible to add vertices $y_{j'}$ to our clique such that $y_{j'} \notin Y_K$ and $j < j' < |Y|$. As a consequence, there exists another maximum colorful clique K' where $V(K') = \{x_i, x_{i+1}, \ldots, x_{|X|}\} \cup \{y_j, y_{j+1}, \ldots, y_{|Y|}\}$. □

The following algorithm, which computes a maximum colorful clique in complement graphs of vertex-colored bipartite chain graphs, is based directly on Lemma 5.

Algorithm 2. Maximum colorful clique in complement graphs of vertex-colored bipartite chain graphs.

1: $max \leftarrow 0$ /* the number of colors of a maximum colorful clique */
2: $H^* \leftarrow \emptyset$ /* the maximum colorful clique returned */
3: **for** $1 \le i \le |X|$ **do**
4:    **for** $1 \le j \le |Y|$ **do**
5:       $H \quad \leftarrow \quad$ the induced graph of the vertices $\{x_i, x_{i+1}, \ldots, x_{|X|}\} \cup \{y_j, y_{j+1}, \ldots, y_{|Y|}\}$
6:       **if** H is a clique and $max < |\mathcal{C}(H)|$ **then**
7:          $H^* \leftarrow H$
8:       **end if**
9:    **end for**
10: **end for**
11: **return** H^* as a maximum colorful clique

The proof of following theorem is immediate.

Theorem 2. *Algorithm 2 computes a maximum colorful clique of $\overline{G^c}$ in time $O(n^2)$.*

3.3   An Algorithm for MCCP for Complete Multipartite Graphs

In this section, we present our new algorithm for MCCP for vertex-colored complete multibipartite graphs. Recall that a k-partite graph is a graph whose vertices can be partitioned into k different independent sets. A k-partite graph is complete if there exists an edge from each vertex of an independent set to each vertex of another independent set. Observe that if we pick one vertex from each independent set we obtain a clique. Therefore, it is possible to reduce the original MCCP problem to the problem of choosing one vertex from each independent set such that the number of colors is maximized. Note that all independent sets have to be selected, otherwise one could add more vertices. From now on, we denote a set of such vertices as a maximum clique. Suppose that our complete multipartite graph has $\mathcal{N}$ independent sets and let us denote these independent sets by $I_1, I_2, \ldots, I_{\mathcal{N}}$. The main idea behind our algorithm is to create first a maximum clique K by randomly picking a vertex from each independent set. Next, we construct a maximum colorful clique from K by iteratively increasing the number of covered colors, without decreasing the number of vertices. To show how to accomplish this task, we need some further notation. Given a maximum clique K and a color c, let $v(K, c)$ be the total number of vertices of color c in K. We let c_0 be a color of the original graph which does not appear in the current maximum clique, i.e., $v(K, c_0) = 0$. Now a set of pairs of vertices $\{(v_1', v_1), (v_2', v_2), \ldots, (v_k', v_k)\}$ $(1 \leq k \leq \mathcal{N})$ is called a k-*colorful augmentation w.r.t. a maximum clique* K if it satisfies the following properties:

1. For all $i = 1, 2, \ldots, k$, v_i is covered by K and v_i' is not covered by K.
2. The color $c(v_1') = c_0$ is not presented in K, i.e., $v(K, c_0) = 0$.
3. For all $i = 1, 2, \cdots, k - 1$, $c(v_i) = c(v_{i+1}') = c_i$ and $v(K, c_i) = 1$. (Note that $c_i \neq c_j$, for all $1 \leq i, j \leq k - 1$.)
4. The color $c(v_k) = c_k$ such that $v(K, c_k) \geq 2$ and $c_k \neq c_i$ for all $i = 1, 2, \cdots, k - 1$.

Note that a k-colorful augmentation $\{(v_1', v_1), (v_2', v_2), \ldots, (v_k', v_k)\}$ w.r.t. K provides a better solution for our problem. Indeed, if we replace the vertices $\{v_1, v_2, \ldots, v_k\}$ by the set of vertices $\{v_1', v_2', \ldots, v_k'\}$, we obtain a new maximum clique, which includes a new color $c(v_1') = c_0$ and preserves the old colors $c(v_1), c(v_2), \ldots, c(v_k)$, i.e., the number of colors in the new clique increases by one. The following theorem is at the heart of our algorithm for finding a maximum colorful clique in a vertex-colored complete multipartite graph G^c.

Theorem 3. *Let K be a maximum clique in a vertex-colored complete multipartite graph G^c. Then, K admits a k-colorful augmentation w.r.t. K if and only if there exists another maximum clique K' such that $|\mathcal{C}(K')| > |\mathcal{C}(K)|$.*

Proof. First, assume that K admits a k-colorful augmentation. A k-colorful augmentation $\mathcal{P} = \{(v_1', v_1), (v_2', v_2), \ldots, (v_k', v_k)\}$ w.r.t. K can be used to transform K into another clique K' such that $|\mathcal{C}(K')| > |\mathcal{C}(K)|$. Replacing vertices in K by vertices not in K' inside this k-colorful augmentation yields another maximum clique K', which increases the number of distinct colors used.

Conversely, suppose that there exists a maximum colorful clique K' such $|\mathcal{C}(K')| > |\mathcal{C}(K)|$. Since K and K' are maximum cliques, $|K| = |K'| = \mathcal{N}$, i.e., both K and K' contains one vertex from each independent set. Let S be the set of independent sets in which the vertex chosen by K is equal to the vertex chosen by K'. Clearly, the set of colors of vertices of K in S is equal to the set of colors of vertices of K' in S. Thus, from now we can ignore all those vertices. We first focus on vertices v_1' of K' such that $c(v_1') = c_0$ is not present in K, i.e., $v(K, c_0) = 0$. Let us denote the set of those vertices by $V_1(K')$. From each vertex v_1' of $V_1(K')$, we extend to a sequence of pairs of vertices of K and K', namely $\{(v_1', v_1), (v_2', v_2), \ldots, (v_k', v_k)\}$ such that: (i) for all $i = 1, 2, \ldots, k$, $v_i \in K$ and $v_i' \in K'$, and (ii) for all $i = 1, 2, \cdots, k - 1$, $c(v_i) = c(v_{i+1}') = c_i$ and $v(K, c_i) = 1$. Clearly this extension is unique for each vertex v_1'. Note that the ending vertex in each of the extended sequences is the vertex v_k. Next, we will focus on those ending vertices.

- In the case that there exists an ending vertex v_k such that $v(K, c_k) \geq 2$ then this set of pair of vertices $\{(v'_1, v_1), (v'_2, v_2), \ldots, (v'_k, v_k)\}$ is *a k-colorful augmentation w.r.t. K.*
- Otherwise, for each v'_1 of $V_1(K')$ such that $v(K, c_0) = 0$ and $v(K', c_0) \geq 1$, we have that the ending vertex v_k satisfies $v(K, c_k) = 1$ and $v(K', c_k) = 0$. Note that $v(K', c_k) = 0$ since otherwise it is possible to extend this sequence. Clearly, except for the vertices of $V_1(K')$, then for each vertex v' of K' such that $v' \notin V_1(K')$ (denote this set of vertices by $V_2(K')$), we have that $c(v') \in \mathcal{C}(K)$, i.e., $v(K, c(v')) \geq 1$. Combining these properties of vertices in the sets $V_1(K')$, $V_2(K')$, we can deduce that $|\mathcal{C}(K')| \leq |\mathcal{C}(K)|$, a contradiction. This completes our proof.   □

In order to complete our algorithm, we need a polynomial sub-routine for finding colorful augmentations. This is achieved with the following lemma.

Lemma 6. *Let K be a maximum clique in a vertex-colored complete multipartite graph G^c. The problem of finding a k-colorful augmentation with respect to K in G^c can be reduced in polynomial time to the problem of finding an M-augmenting path in another graph G'^c w.r.t. K.*

Proof. Since K is a maximum clique, each vertex of K is in an independent set I_i of G^c. Let us denote these vertices by $v_1, v_2, \ldots, v_\mathcal{N}$. W.l.o.g. assume that $v_i \in I_i$. To look for a k-colorful augmentation with respect to K in G^c, we construct a new graph $G' = (V', E')$ and a matching M as follows. Its vertex-set is defined as $V(G') = \{v \in V(G^c) : v(M, c(v)) \neq 1\} \cup \{v_\eta : v(M, \eta) = 1 \text{ and } \eta \in \mathcal{C}\} \cup \{u_i | 1 \leq i \leq \mathcal{N}\} \cup \{z\}$, where u_i and z are new artificial vertices not in $V(G^c)$ and u_i is corresponding to the set I_i. Next the edge-set of G' is defined as $E(G') = \{(t(v), t(w)) : (v, w) \in E(G^c)\} \cup \{(u_i, v) : u_i \text{ and } v \text{ are covered by } I_i\} \cup \{(z, v) : v = t(v) \text{ and } v \text{ is covered by } K\}$, where $t(v) = v$ if $v(M, c(v)) \neq 1$, otherwise $t(v) = v_\eta$ if $v(M, \eta) = 1$ and $c(v) = \eta$.

Now we define our matching: $M = \{(u_1, v_1), (u_2, v_2), \ldots, (u_\mathcal{N}, v_\mathcal{N})\}$. Let M' be a subset of $E(G')$ such that $M' = \{(t(x), t(y)) : (x, y) \in M\}$. It is easy to see that M' is also a matching of G'.

Conversely, let $P = \{(v'_1, v_1), (v'_2, v_2), \ldots, (v'_k, v_k)\}$ be a k-colorful augmentation w.r.t. K, $1 \leq k \leq \mathcal{N}$, in which v_i and v'_i are in I_i, $1 \leq i \leq k$. From this k-colorful augmentation, we can define an M-augmenting path P' with respect to M in G' as $P = \{v'_1, u_1, v_{c_1}, u_2, v_{c_2}, u_3, \ldots, v_{c_{k-1}}, u_k, v_k, z\}$ in which $c_i = c(v_i)$. Recall that $v(K, c(v'_1)) = 0$ and $v(K, c(v_k)) \geq 2$. In conclusion, finding a k-colorful augmentation with respect to a maximum clique K in G^c is equivalent to finding an augmenting path P' with respect to M in G'.   □

Our algorithm for MCCP in vertex-colored complete multipartite graphs derives immediately from Theorem 3 and Lemma 6.

The following theorem shows that Algorithm 3 runs in polynomial time.

Theorem 4. *Let G^c be a vertex-colored complete multipartite graph. Algorithm 3 computes a maximum colorful clique of G^c in time $O(|\mathcal{C}|M(m + n, n))$, where $O(M(m, n))$ is the time required to find a maximum matching in a general graph with n vertices and m edges.*

Algorithm 3. Maximum colorful clique in vertex-colored complete multipartite graphs.

1: $K \leftarrow$ any maximum clique of G^c /* pick each vertex from each independent set */
2: **while** a k-colorful augmentation w.r.t. K $\mathcal{P}$ is found **do**
3:     $K \leftarrow K \oplus \mathcal{P}$ /* replace vertices of $\{v_i\}$ by vertices of $\{v_i'\}$ */
4: **end while**
5: **return** K as a maximum colorful clique

Proof. The initialization step of the algorithm is trivial. Next, after each iteration of the while loop in the algorithm, a new color is included into the new maximum clique. Thus, the maximum number of iterations of the while loop is $|\mathcal{C}|$ (the number of colors of G^c). Inside each iteration of the while loop, we can use Lemma 6 to look for a k-colorful augmentation, which requires $O(M(m, n))$ time. Note that the new graph G'^c has $O(n)$ vertices and $O(n+m)$ edges. In summary, the total running time of the algorithm is $O(|\mathcal{C}|M(m + n, n))$, as claimed.   □

3.4   An Algorithm for MCCP for Complement Graphs of Cycle Graphs

In the final section, we propose another algorithm for the case of complement graphs of cycle graphs. Related theorems and the algorithm can be found in the appendix.

References

1. Akbari, S., Liaghat, V., Nikzad, A.: Colorful paths in vertex coloring of graphs. Electron. J. Comb. **18**(1), P17 (2011)
2. Bruckner, S., Hüffner, F., Komusiewicz, C., Niedermeier, R.: Evaluation of ILP-based approaches for partitioning into colorful components. In: Bonifaci, V., Demetrescu, C., Marchetti-Spaccamela, A. (eds.) SEA 2013. LNCS, vol. 7933, pp. 176–187. Springer, Heidelberg (2013). https://doi.org/10.1007/978-3-642-38527-8_17
3. Cohen, J., Manoussakis, Y., Pham, H., Tuza, Z.: Tropical matchings in vertex-colored graphs. In: Latin and American Algorithms, Graphs and Optimization Symposium (2017)
4. Cohen, J., Italiano, G.F., Manoussakis, Y., Nguyen, K.T., Pham, H.P.: Tropical paths in vertex-colored graphs. In: Gao, X., Du, H., Han, M. (eds.) COCOA 2017. LNCS, vol. 10628, pp. 291–305. Springer, Cham (2017). https://doi.org/10.1007/978-3-319-71147-8_20
5. Corel, E., Pitschi, F., Morgenstern, B.: A min-cut algorithm for the consistency problem in multiple sequence alignment. Bioinformatics **26**(8), 1015–1021 (2010)
6. Fellows, M.R., Fertin, G., Hermelin, D., Vialette, S.: Upper and lower bounds for finding connected motifs in vertex-colored graphs. J. Comput. Syst. Sci. **77**(4), 799–811 (2011)
7. Foucaud, F., Harutyunyan, A., Hell, P., Legay, S., Manoussakis, Y., Naserasr, R.: Tropical homomorphisms in vertex-coloured graphs. Discrete Appl. Math. **229**, 1–168 (2017)

8. Italiano, G.F., Manoussakis, Y., Kim Thang, N., Pham, H.P.: Maximum colorful cycles in vertex-colored graphs. In: Fomin, F., Podolskii, V. (eds.) CSR 2018. LNCS, vol. 10846, pp. 106–117. Springer, Cham (2018). https://doi.org/10.1007/978-3-319-90530-3_10

9. Li, H.: A generalization of the Gallai–Roy theorem. Graphs and Combinatorics **17**(4), 681–685 (2001)

10. Lin, C.: Simple proofs of results on paths representing all colors in proper vertex-colorings. Graphs and Combinatorics **23**(2), 201–203 (2007)

11. Marx, D.: Graph colouring problems and their applications in scheduling. Periodica Polytech. Electr. Eng. **48**(1–2), 11–16 (2004)

12. Micali, S., Vazirani, V.V.: An $O(\sqrt{|V|}|E|)$ algorithm for finding maximum matching in general graphs. In: Proceedings of 21st Symposium on Foundations of Computer Science, pp. 17–27 (1980)

Partial Sublinear Time Approximation and Inapproximation for Maximum Coverage

Bin Fu[✉]

Department of Computer Science, University of Texas Rio Grande Valley,
Edinburg, TX 78539, USA
bin.fu@utrgv.edu

Abstract. We develop a randomized approximation algorithm for the classical maximum coverage problem, which given a list of sets $A_1, A_2, \cdots, A_m$ and integer parameter k, select k sets $A_{i_1}, A_{i_2}, \cdots, A_{i_k}$ for maximum union $A_{i_1} \cup A_{i_2} \cup \cdots \cup A_{i_k}$. In our algorithm, each input set A_i is a black box that can provide its size $|A_i|$, generate a random element of A_i, and answer the membership query ($x \in A_i$?) in $O(1)$ time. Our algorithm gives $(1 - \frac{1}{e})$-approximation for maximum coverage problem in $O(\text{poly}(k)m \cdot \log m)$ time, which is independent of the sizes of the input sets. No existing $O(p(m)n^{1-\epsilon})$ time $(1 - \frac{1}{e})$-approximation algorithm for the maximum coverage has been found for any function $p(m)$ that only depends on the number of sets, where $n = \max(|A_1|, \cdots, |A_m|)$ (the largest size of input sets). The notion of partial sublinear time algorithm is introduced. For a computational problem with input size controlled by two parameters n and m, a partial sublinear time algorithm for it runs in a $O(p(m)n^{1-\epsilon})$ time or $O(q(n)m^{1-\epsilon})$ time. The maximum coverage has a partial sublinear time $O(\text{poly}(m))$ constant factor approximation since $k \leq m$. On the other hand, we show that the maximum coverage problem has no partial sublinear $O(q(n)m^{1-\epsilon})$ time constant factor approximation algorithm.

1 Introduction

The maximum coverage problem is a classical NP-hard problem with many applications [7,13], and is directly related to set cover problem, one of Karp's twenty-one NP-complete problems [16]. The input has several sets and a number k. The sets may have some elements in common. You must select at most k of these sets such that the maximum number of elements are covered, i.e. the union of the selected sets has a maximum size. The greedy algorithm for maximum coverage chooses sets according to one rule: at each stage, choose a set which contains the largest number of uncovered elements. It can be shown that this algorithm achieves an approximation ratio of $(1 - \frac{1}{e})$ [7,14].

Inapproximability results show that the greedy algorithm is essentially the best-possible polynomial time approximation algorithm for maximum coverage [10]. The existing implementation for the greedy $(1 - \frac{1}{e})$-approximation

© Springer International Publishing AG, part of Springer Nature 2018
L. Wang and D. Zhu (Eds.): COCOON 2018, LNCS 10976, pp. 492–503, 2018.
https://doi.org/10.1007/978-3-319-94776-1_41

algorithm for the maximum coverage problem needs $\Omega(mn)$ time for a list of m sets $A_1, \cdots, A_m$ with $n = |A_1| = |A_2| = \cdots = |A_m|$ [14,23]. We have not found any existing $O(p(m)n^{1-\epsilon})$ time algorithm for the same ratio $(1 - \frac{1}{e})$ of approximation for any function $p(m)$ that only depends on the number of sets. The variant versions and methods for this problem have been studied in a series of papers [1,5,6,18,21].

This paper sticks to the original definition of the maximum coverage problem, and studies its complexity under several concrete models. In the first model, each set is accessed as a black box that only provides random elements and answers membership queries. When m input sets $A_1, A_2, \cdots, A_m$ are given, our model allows random sampling from each of them, and the cardinality $|A_i|$ (or approximation for $|A_i|$) of each A_i is also part of the input. The results of the first model can be transformed into other conventional models. A set could be a set of points in a geometric shape. For example, a set may be all lattice points in a d-dimensional rectangular shape. If the center position, and dimension parameters of the rectangle are given, we can count the number of lattice points and provide a random sample for them.

A more generalized maximum coverage problem was studied under the model of submodular set function subject to a matroid constraint [2,11,20], and has same approximation ratio $1 - \frac{1}{e}$. The maximum coverage problem in the matroid model has time complexity $O(r^3 m^2 n)$ [11], and $O(r^2 m^3 n + m^7)$ [2], respectively, according to the analysis in [11], where r is the rank of matroid, m is the number of sets, and n is the size of the largest set. The maximum coverage problem in the matroid model has the oracle query to the submodular function [2] and is counted $O(1)$ time per query. Computing the size of union of input sets is #P-hard if each input set as a black box is a set of high dimensional rectangular lattice points since #DNF is #P-hard [22]. The size of union of sets is approximated in this paper. This enable us to apply our model to this high dimensional space maximum coverage problem.

In this paper, we develop a randomized algorithm to approximate the maximum coverage problem. We show an approximation algorithm for maximum coverage problem with $(1 - \frac{1}{e})$-ratio. For an input list L of finite sets $A_1, \cdots, A_m$, an integer k, and parameter $\epsilon \in (0, 1)$, our randomized algorithm returns an integer z and a subset $H \subseteq \{1, 2, \cdots, m\}$ such that $|\cup_{j \in H} A_j| \geq (1 - \frac{1}{e})C^*(L, k)$ and $|H| = k$, and $(1 - \epsilon)|\cup_{j \in H} A_j| \leq z \leq (1 + \epsilon)|\cup_{j \in H} A_j|$, where $C^*(L, k)$ is the maximum union size for a solution of maximum coverage. Its complexity is $O(\frac{k^6}{\epsilon^2}(\log(\frac{3m}{k}))m)$ and its probability to fail is less than $\frac{1}{4}$.

Our computational time is independent of the size of each set if the membership checking for each input set takes one step. When each set A_i is already saved in an efficient data structure such as B-tree, we can also provide an efficient random sample, and make a membership query to each A_i in a $O(\log |A_i|)$ time. This model also has practical importance because B-tree is often used to collect a large set of data. Our algorithms are suitable to estimate the maximum coverage when there are multiple big data sets, and each data set is stored in a efficient data structure that can support efficient random sampling and membership query. The widely used B-tree in modern data base clearly fits our algorithm. Our model and algorithm are suitable to support online computation.

We apply the randomized algorithm to several versions of maximum coverage problem: 1. Each set contains the lattice points in a rectangle of d-dimensional space. It takes $O(d)$ time for a random element, or membership query. This gives an application to a #P-hard problem. 2. Each set is stored in a unsorted array. It takes $O(1)$ time for a random element, and $O(n)$ time for membership query. It takes $O(\log n)$ time for a random element, or membership query. 3. Each set is stored in a sorted array. 4. Each set is stored in a B-tree. It takes $O(\log n)$ time for a random element, or a membership query. Furthermore, B-tree can support online version of maximum coverage that has dynamic input. 5. Each set is stored in a hashing table. The time for membership query needs some assumption about the performance of hashing function. We show how the computational time of the randomized algorithm for maximum coverage depends on the these data structures.

Sublinear time algorithms have been found for many computational problems, such as checking polygon intersections [3], estimating the cost of a minimum spanning tree [4,8,9], etc.

The notion of partial sublinear time computation is introduced in this paper. It characterizes a class of computational problems that are sublinear in one of the input parameters, but not necessarily the other ones. For a function $f(.)$ that maps a list of sets to nonnegative integers, a $O(p(m)n^{1-\epsilon})$ time or $O(q(n)m^{1-\epsilon})$ time approximation to $f(.)$ is a partial sublinear time computation. The maximum coverage has a partial sublinear time constant factor approximation scheme. We prove that the special case of maximum coverage problem with equal size of sets, called *equal size maximum coverage*, is as hard as the general case. On the other hand, we show that the equal size maximum coverage problem has no partial sublinear $O(q(n)m^{1-\epsilon})$ constant factor approximation randomized algorithm in a randomized model. Thus, the partial sublinear time computation is separated from the conventional sublinear time computation via the maximum coverage problem.

2   Computational Model and Complexity

In this section, we show our model of computation, and the definition of complexity. For a finite set A, we use $|A|$, *cardinality* of A, to be the number of distinct elements in A. For a real number x, let $\lceil x \rceil$ be the least integer y greater than or equal to x, and $\lfloor x \rfloor$ be the largest integer z less than or equal to x. Let $N = \{0, 1, 2, \cdots\}$ be the set of nonnegative integers, $R = (-\infty, +\infty)$ be the set of all real numbers, and $R^+ = [0, +\infty)$ be the set of all nonnegative real numbers. An integer s is a $(1 + \epsilon)$-approximation for $|A|$ if $(1 - \epsilon)|A| \le s \le (1 + \epsilon)|A|$.

Definition 1. *Let parameters α_L and α_R be in $[0, 1)$. An (α_L, α_R)-biased generator RandomElement(A) for set A generates an element in A such that for each $y \in A$, $(1 - \alpha_L) \cdot \frac{1}{|A|} \le \text{Prob}(\text{RandomElement}(A) = y) \le (1 + \alpha_R) \cdot \frac{1}{|A|}$.*

Definition 2 gives the randomized computation model that allows biased random generators and approximate sizes for input sets. It is suitable to apply our algorithm for high dimensional geometric problems that may not give uniform random sampling or exact set size. For example, it is not trivial to count the number of lattice points or generate a random lattice point in a d-dimensional ball with its center not at a lattice point.

Definition 2. *The randomized computation model for our algorithm is defined below: Let real parameters $\alpha_L, \alpha_R, \delta_L, \delta_R$ be in $[0,1)$. An input L is a list of sets $A_1, A_2, \cdots, A_m$ that provide an approximate cardinality s_i of A_i with $(1-\delta_L)|A_i| \le s_i \le (1+\delta_R)|A_i|$ for $i = 1, 2, \cdots, m$, the largest approximate cardinality of input sets $s = \max\{s_i : 1 \le i \le m\}$, and support the following operations:*

1. *Function RandomElement(A_i) is a (α_L, α_R)-biased random generator for A_i for $i = 1, 2, \cdots, m$.*
2. *Function Query(x, A_i) returns 1 if $x \in A_i$, and 0 otherwise.*

The model is called type 0 model if $\alpha_L = \alpha_R = \delta_L = \delta_R = 0$, and type 1 model otherwise.

The main problem, which is called maximum coverage, is that given a list of sets $A_1, \cdots, A_m$ and an integer k, find k sets from $A_1, A_2, \cdots, A_m$ to maximize the size of the union of the selected sets in the computational model defined in Definition 2. For real number $a \in [0,1]$, an approximation algorithm is a $(1-a)$-approximation for the maximum coverage problem that has input of integer parameter k and a list of sets $A_1, \cdots, A_m$ if it outputs a sublist of sets $A_{i_1}, A_{i_2}, \cdots, A_{i_k}$ such that $|A_{i_1} \cup A_{i_2} \cup \cdots \cup A_{i_k}| \ge (1-a)|A_{j_1} \cup A_{j_2} \cup \cdots \cup A_{j_k}|$, where $A_{j_1}, A_{j_2}, \cdots, A_{j_k}$ is a solution with maximum size of union.

We use the triple $(T(.), R(.), Q(.))$ to characterize the computational complexity, where

- $T(.)$ is a function for the number of steps that each access to RandomElement(.) or Query(.) is counted one step,
- $R(.)$ is a function to count the number of random samples from A_i for $i = 1, 2, \cdots, m$. It is measured by the total number of times to access those functions RandomElement(A_i) for all input sets A_i, and
- $Q(.)$ is a function to count the number of queries to A_i for $i = 1, \cdots, m$. It is measured by the total number of times to access those functions Query(x, A_i) for all input sets A_i.

The parameters $\epsilon, \gamma, k, n, m$ can be used to determine the three complexity functions, where $n = \max(|A_1|, \cdots, |A_m|)$ (the largest cardinality of input sets), ϵ controls the accuracy of approximation, and γ controls the failure probability of a randomized algorithm. Their types could be written as $T(\epsilon, \gamma, k, m), R(\epsilon, \gamma, k, m)$, and $Q(\epsilon, \gamma, k, m)$. All of the complexity results of this paper at both model 0 and model 1 are independent of parameter n.

Definition 3. *For a list L of sets $A_1, A_2, \cdots, A_m$ and real $\alpha_L, \alpha_R, \delta_L, \delta_R \in [0, 1)$, it is called $((\alpha_L, \alpha_R), (\delta_L, \delta_R))$-list if each set A_i is associated with a number s_i with $(1 - \delta_L)|A_i| \leq s_i \leq (1 + \delta_R)|A_i|$ for $i = 1, 2, \cdots, m$, and the set A_i has a (α_L, α_R)-biased random generator* RandomElement(A_i).

3   Outline of Our Methods

For two sets A and B, we develop a randomized method to approximate the cardinality of the difference $B - A$. We approximate the size of $B - A$ by sampling a small number of elements from B and calculating the ratio of the elements in $B - A$ by querying the set A. The approximate $|A \cup B|$ is the sum of an approximation of $|A|$ and an approximation of $|B - A|$.

A greedy approach will be based on the approximate difference between a new set and the union of sets already selected. Assume that $A_1, A_2, \cdots, A_m$ is the list of sets for the maximum coverage problem. After $A_{i_1}, \cdots, A_{i_t}$ have been selected, the greedy approach needs to check the size $|A_j - (A_{i_1} \cup A_{i_2} \cup \cdots \cup A_{i_t})|$ before selecting the next set. Our method to estimate $|A_j - (A_{i_1} \cup A_{i_2} \cup \cdots \cup A_{i_t})|$ is based on randomization in order to make the time independent of the sizes of input sets. Some random samples are selected from set A_j.

The classical greedy approximation algorithm provides $1 - (1 - \frac{1}{k})^k$ ratio for the maximum coverage problem. The randomized greedy approach gives $1 - (1 - \frac{1}{k})^k - \xi$ ratio, where ξ depends on the accuracy of estimation to $|A_j - (A_{i_1} \cup A_{i_2} \cup \cdots \cup A_{i_t})|$. As $(1 - \frac{1}{k})^k$ is increasing and $\frac{1}{e} = (1 - \frac{1}{k})^k + \Omega(\frac{1}{k})$, we can let $(1 - \frac{1}{k})^k + \xi \leq \frac{1}{e}$ by using sufficient number of random samples for the estimation of set difference when selecting a new set. Thus, we control the accuracy of the approximate cardinality of the set difference so that it is enough to achieve the approximation ratio $1 - \frac{1}{e}$ for the maximum coverage problem.

During the accuracy analysis, Hoeffding Inequality [15] plays an important role. It shows how the number of samples determines the accuracy of approximation.

Theorem 1 ([15]). *Let $X_1, \ldots, X_s$ be s independent random 0-1 variables and $X = \sum_{i=1}^{s} X_i$.*

i. If X_i takes 1 with probability at most p for $i = 1, \ldots, s$, then for any $\epsilon > 0$, $\Pr(X > ps + \epsilon s) < e^{-\frac{1}{2}s\epsilon^2}$.

ii. If X_i takes 1 with probability at least p for $i = 1, \ldots, s$, then for any $\epsilon > 0$, $\Pr(X < ps - \epsilon s) < e^{-\frac{1}{2}s\epsilon^2}$.

We define the function $\mu(x)$ in order to simply the probability mentioned in Theorem 1

$$\mu(x) = e^{-\frac{1}{2}x^2} \tag{1}$$

Chernoff Bound (see [19]) is also used in the maximum coverage approximation when our main result is applied in some concrete model. It implies a similar result as Theorem 1.

Theorem 2. *Let* $X_1, \ldots, X_s$ *be* s *independent random 0-1 variables, where* X_i *takes 1 with probability at least* p *for* $i = 1, \ldots, s$. *Let* $X = \sum_{i=1}^{s} X_i$, *and* $\mu = E[X]$. *Then for any* $\delta > 0$, $\Pr(X < (1 - \delta)ps) < e^{-\frac{1}{2}\delta^2 ps}$.

Theorem 3. *Let* $X_1, \ldots, X_s$ *be* s *independent random 0-1 variables, where* X_i *takes 1 with probability at most* p *for* $i = 1, \ldots, s$. *Let* $X = \sum_{i=1}^{s} X_i$. *Then for any* $\delta > 0$, $\Pr(X > (1 + \delta)ps) < \left[\frac{e^\delta}{(1+\delta)^{(1+\delta)}} \right]^{ps}$.

A well known fact in probability theory called union bound is expressed by the inequality

$$\Pr(E_1 \cup E_2 \ldots \cup E_t) \leq \Pr(E_1) + \Pr(E_2) + \ldots + \Pr(E_t), \tag{2}$$

where $E_1, E_2, \ldots, E_t$ are t events that may not be independent. In the analysis of our randomized algorithm, there are multiple events such that the failure from any of them may fail the entire algorithm. We often characterize the failure probability of each of those events, and use the above inequality to show that the whole algorithm has a small chance to fail, after showing that each of them has a small chance to fail.

Our algorithm performance will depend on the initial accuracy of approximation to each set size, and how biased the random sample from each input set. This consideration is based on the applications to high dimensional geometry problems which may be hard to count the exact number of elements in a set, and is also hard to provide perfect uniform random source. We plan to release more applications to high dimensional geometry problems that need approximate counting and biased random sampling.

Overall, our method is an approximate randomized greedy approach for the maximum coverage problem. The numbers of random samples is controlled so that it has enough accuracy to derive the classical approximation ratio $1 - \frac{1}{e}$. The main results are stated at Theorem 4.

There is an existing analysis about the accuracy of approximation and the approximate gain when adding a new set to a partial solution [13]. If the gain $|A_j - (A_{i_1} \cup A_{i_2} \cup \cdots \cup A_{i_t})|$ for set A_j is at least $\beta |A_{j'} - (A_{i_1} \cup A_{i_2} \cup \cdots \cup A_{i_t})|$, then the approximation ratio is $1 - \frac{1}{e^\beta}$, where $A_{j'}$ has the maximal gain $|A_{j'} - (A_{i_1} \cup A_{i_2} \cup \cdots \cup A_{i_t})|$. Our method has a more careful control to the approximate gain when selecting a set in order to match the best possible approximation ratio $1 - \frac{1}{e}$.

Definition 4. *Let the maximum coverage problem have integer parameter* k, *and a list* L *of sets* $A_1, A_2, \cdots, A_m$ *as input. We always assume* $k \leq m$. *Let* $C^*(L, k) = |A_{t_1} \cup A_{t_2} \cup \cdots \cup A_{t_k}|$ *be the maximum union size of a solution* $A_{t_1}, \cdots, A_{t_k}$ *for the maximum coverage.*

Theorem 4. *Let* ρ *be a constant in* $(0, 1)$. *For parameters* $\epsilon, \gamma \in (0, 1)$ *and* $\alpha_L, \alpha_R, \delta_L, \delta_R \in [0, 1 - \rho]$, *there is an algorithm to give a* $(1 - \frac{1}{e^\beta})$ *approximation for the maximum cover problem, such that given a* $((\alpha_L, \alpha_R), (\delta_L, \delta_R))$-*list* L *of finite sets* $A_1, \cdots, A_m$ *and an integer* k, *with probability at least* $1 - \gamma$, *it returns an integer* z *and a subset* $H \subseteq \{1, 2, \cdots, m\}$ *that satisfy*

1. $|\cup_{j\in H} A_j| \geq (1 - \frac{1}{e^{\beta}})C^*(L,k)$ and $|H| = k$,
2. $((1-\alpha_L)(1-\delta_L) - \epsilon)|\cup_{j\in H} A_j| \leq z \leq ((1+\alpha_R)(1+\delta_R) + \epsilon)|\cup_{j\in H} A_j|$, and
3. Its complexity is $(T(\epsilon, \gamma, k, m), R(\epsilon, \gamma, k, m), Q(\epsilon, \gamma, k, m))$ with $T(\epsilon, \gamma, k, m) = O(\frac{k^5}{\epsilon^2}(k\log(\frac{3m}{k}) + \log\frac{1}{\gamma})m)$, $R(\epsilon, \gamma, k, m) = O(\frac{k^4}{\epsilon^2}(k\log(\frac{3m}{k}) + \log\frac{1}{\gamma})m)$, and $Q(\epsilon, \gamma, k, m) = O(\frac{k^5}{\epsilon^2}(k\log(\frac{3m}{k}) + \log\frac{1}{\gamma})m)$, where $\beta = \frac{(1-\alpha_L)(1-\delta_L)}{(1+\alpha_R)(1+\delta_R)}$.

4 Randomized Algorithm for Maximum Coverage

We give a randomized algorithm for approximating the maximum coverage. It is based on an approximation to the cardinality of set difference. The algorithms are described at type 1 model, and has corollaries for type 0 model.

4.1 Randomized Algorithm for Set Difference Cardinality

In this section, we develop a method to approximate the cardinality of $B - A$ based on random sampling. It will be used as a submodule to approximate the maximum coverage.

Definition 5. Let $R = x_1, x_2, \cdots, x_w$ be a list of elements from set B, and let L be a list of sets $A_1, A_2, \cdots, A_u$. Define $\text{test}(L, R) = |\{j : 1 \leq j \leq w, \text{and } x_j \notin (A_1 \cup A_2 \cup \cdots \cup A_u)\}|$.

The Algorithm ApproximateDifference(.) gives an approximation s for the size of $B - A$. It is very time consuming to approximate $|B - A|$ when $|B - A|$ is much less than $|B|$. The algorithm ApproximateDifference(.) returns an approximate value s for $|B - A|$ with a range in $[(1-\delta)|B - A| - \epsilon|B|, (1+\delta)|B - A| + \epsilon|B|]$, and will not lose much accuracy when it is applied to approximate the maximum coverage by controlling the two parameters δ and ϵ.

Algorithm 1 : RandomTest(L, B, w)

Input: L is a list of sets $A_1, A_2, \cdots, A_u$, B is another set with a random generator RandomElement(B), and w is an integer to control the number of random samples from B.

1.   For $i = 1$ to w let $x_i = \text{RandomElement}(B)$;
2.   For $i = 1$ to w
3.   Let $y_i = 0$ if $(x_i \in A_1 \cup A_2 \cup \cdots \cup A_u)$, and 1 otherwise;
4.   Return $t = y_1 + \cdots + y_w$;

End of Algorithm

Algorithm 2 : ApproximateDifference$(L, B, s_2, \epsilon, \gamma)$

Input: L is a list of sets $A_1, A_2, \cdots, A_u$, B is another set with a random generator RandomElement(B), integer s_2 is an approximation for $|B|$ with $(1 - \delta_L)|B| \leq s_2 \leq (1 + \delta_R)|B|$, and ϵ and γ are real parameters in $(0, 1)$, where $\delta \in [0, 1]$.

Steps:

1.   Let w be an integer with $\mu(\frac{\epsilon}{3})^w \leq \frac{\gamma}{4}$, where $\mu(x)$ is defined in equation (1).

2.   Let $t = $ RandomTest(L, B, w);

3.   Return $s = \frac{t}{w} \cdot s_2$

End of Algorithm

Lemma 1 shows how Algorithm ApproximateDifference(.) returns an approximation s for $|B - A|$ with a small failure probability γ, and its complexity depends on the accuracy ϵ of approximation and probability γ. Its accuracy is controlled for the application to the approximation algorithms for maximum coverage problem.

Lemma 1. *Assume that real number $\epsilon \in [0, 1]$, B is a set with (α_L, α_R)-biased random generator* RandomElement(B) *and an approximation s_2 for $|B|$ with $(1 - \delta_L)|B| \leq s_2 \leq (1 + \delta_R)|B|$, and L is a list of sets $A_1, A_2, \cdots, A_u$. Then*

1. *If $R = x_1, x_2, \cdots, x_w$ be a list of elements generated by* RandomElement(B), *and $\mu(\frac{\epsilon}{3})^w \leq \frac{\gamma}{4}$, then with probability at most γ, the value $s = \frac{t}{w} \cdot s_2$ fails to satisfy inequality (3)*

$$(1 - \alpha_L)(1 - \delta_L)|B - A| - \epsilon|B| \leq s \leq (1 + \alpha_R)(1 + \delta_R)|B - A| + \epsilon|B|, \quad (3)$$

   where $A = A_1 \cup A_2 \cup \cdots \cup A_u$ is the union of sets in the input list L.
2. *With probability at most γ, the returned value s by the algorithm ApproximateDifference(.) fails to satisfy inequality (3), and*
3. *If the implementation of RandomTest(.) in Algorithm 1 is used, then the complexity of ApproximateDifference(.) is $(T_D(\epsilon, \gamma, u), R_D(\epsilon, \gamma, u), Q_D(\epsilon, \gamma, u))$ with $T_D(\epsilon, \gamma, u) = O(\frac{u}{\epsilon^2} \log \frac{1}{\gamma})$, $R_D(\epsilon, \gamma, u) = O(\frac{1}{\epsilon^2} \log \frac{1}{\gamma})$, and $Q_D(\epsilon, \gamma, u) = O(\frac{u}{\epsilon^2} \log \frac{1}{\gamma})$.*

4.2   A Randomized Algorithm for Set Union Cardinality

We describe a randomized algorithm for estimating the cardinality for set union. It will use the algorithm for set difference developed in Sect. 4.1. Karp et al. [17] developed a $(1 + \epsilon)$ randomized approximation algorithm to improve the running time for approximating the number of distinct elements in the union $A_1 \cup \cdots \cup A_m$ to linear $O((1 + \epsilon)m/\epsilon^2)$ time. Their algorithm requires each set provides exact size and a uniform random generator. Their algorithm cannot be directly used for the maximum coverage problem algorithm in this paper as we use biased random generator and approximate size from each set. Furthermore, the greedy approach for the maximum coverage needs to estimate gain, which is based on Lemma 1 in this paper, for adding each new set to a partial solution. The following lemma gives an approximation for the size of sets union. Its accuracy is enough when it is applied in the approximation algorithms for maximum coverage problem.

Lemma 2. *Assume* $\epsilon, \delta_L, \delta_R, , \delta_{2,L}, \delta_{2,R}, \alpha_L, \alpha_R \in [0,1]$, $(1 - \delta_L) \leq (1 - \alpha_L)(1 - \delta_{2,L})$ *and* $(1 + \delta_R) \geq (1 + \alpha_R)(1 + \delta_{2,R})$. *Assume that* L *is a list of sets* $A_1, A_2, \cdots, A_u$, *and* X_2 *is set with an* (α_L, α_R)-*biased random generator* RandomElement(X_2). *Let integers* s_1 *and* s_2 *satisfy* $(1 - \delta_L)|X_1| \leq s_1 \leq (1 + \delta_R)|X_1|$, *and* $(1 - \delta_{2,L})|X_2| \leq s_2 \leq (1 + \delta_{2,R})|X_2|$, *then*

1. *If* t *satisfies* $(1 - \alpha_L)(1 - \delta_{2,L})|X_2 - X_1| - \epsilon|X_2| \leq t \leq (1 + \alpha_R)(1 + \delta_{2,R})|X_2 - X_1| + \epsilon|X_2|$, *then* $s_1 + t$ *satisfies*

$$(1 - \delta_L - \epsilon)|X_1 \cup X_2| \leq s_1 + t \leq (1 + \delta_R + \epsilon)|X_1 \cup X_2|. \tag{4}$$

2. *If* $t =$ *ApproximateDifference*$(L, X_2, s_2, \epsilon, \gamma)$, *with probability at most* γ, $s_1 + t$ *does not have inequality (4)*,

where $X_1 = A_1 \cup A_2 \cup \cdots \cup A_u$.

4.3   Approximation to the Maximum Coverage Problem

In this section, we show that our randomized approach to the cardinality of set union can be applied to the maximum coverage problem. Lemma 3 gives the approximation performance of greedy method for the maximum coverage problem. It is adapted to a similar result [14] with our approximation accuracy to the size of set difference.

In [13], it showed that if each iteration selects a set A_j with $|A_j - (A_{i_1} \cup A_{i_2} \cup \cdots \cup A_{i_t})| \geq \beta|A_{j'} - (A_{i_1} \cup A_{i_2} \cup \cdots \cup A_{i_t})|$, then the approximation ratio is $1 - \frac{1}{e^\beta}$, where $A_{j'}$ has the maximal gain $|A_{j'} - (A_{i_1} \cup A_{i_2} \cup \cdots \cup A_{i_t})|$. When the accuracy for the approximate gain $|A_j - (A_{i_1} \cup A_{i_2} \cup \cdots \cup A_{i_t})|$ is controlled by a single factor β, it is not enough to obtain $(1 - \frac{1}{e})$-approximation ratio. The error control at the term $\epsilon|B|$ in inequality (3) enables us to achieve the $(1 - \frac{1}{e})$-approximation ratio.

Definition 6. *For a list* L *of sets* $A_1, A_2, \cdots, A_m$, *define its initial* h *sets by* $L(h) = A_1, A_2, \cdots, A_h$, *and the union of sets in* L *by* $U(L) = A_1 \cup A_2 \cup \cdots \cup A_m$.

Lemma 3. *Let* L' *be a sublist of sets* $A_{t_1}, A_{t_2}, \cdots, A_{t_k}$ *selected from the list* L *of sets* $A_1, A_2, \cdots, A_m$. *If each subset* $A_{t_{j+1}}(j = 0, 1, 2, \cdots, k - 1)$ *in* L' *satisfies* $|A_{t_{j+1}} - U(L'(j))| \geq \theta \cdot \frac{C^*(L,k) - |U(L'(j))|}{k} - \delta C^*(L, k)$, *then* $|U(L'(l))| \geq (1 - (1 - \frac{\theta}{k})^l)C^*(L, k) - l \cdot \delta C^*$ *for* $l = 1, 2, \cdots, k$.

Definition 7. *If* L' *is a list of sets* $B_1, B_2, \cdots, B_u$, *and* B_{u+1} *is another set, define* $Append(L', B_{u+1})$ *to be the list* $B_1, B_2, \cdots, B_u, B_{u+1}$, *which is to append* B_{u+1} *to the end of* L'.

In Algorithm ApproximateMaximumCover(.), there are several virtual functions including RandomSamples(.), ApproximateSetDifferenceSize(.), and ProcessSet(.), which have variant implementations and will be given in Virtual Function Implementations 1, 2 and 3. We use a virtual function ApproximateSetDifferenceSize$(L', A_i, s_i, \epsilon', \gamma, k, m)$ to approximate

$|A_i - \cup_{A_j \ is \ in \ L'} A_j|$. We will have variant implementations for this function, and get different time complexity. Another function ProcessSet(A_j) also has variant implementations. Its purpose is to process a newly selected set A_j to list L' of existing selected sets, and may sort it in one of the implementations. The function RandomSamples(.) is also virtual and will have two different implementations (see the full version of this paper [12] for details).

Algorithm 3 : ApproximateMaximumCover(L, k, ξ, γ)

Input: a list $((\alpha_L, \alpha_R), (\delta_L, \delta_R))$-list L of m sets $A_1, A_2, \cdots, A_m$, an integer parameter k, and two real parameters $\xi, \gamma \in (0, 1)$. Each A_i has a (α_L, α_R)-biased random generator RandomElement(A_i), and an approximation s_i for $|A_i|$.

Steps:

1. Let $H = \emptyset$, and list L' be empty;
2. Let $z = 0$;
3. Let $\epsilon' = \frac{\xi}{4k}$;
4. For $i = 1$ to m let R_i =RandomSamples(A_i, ξ, γ, k, m);
5. For $j = 1$ to k
6. {
7.   Let $s_j^* = -1$;
8.   For each A_i in L,
9.   {
10.   Let $s_{i,j}$ =ApproximateSetDifferenceSize($L', A_i, s_i, R_i, \epsilon', \gamma, k, m$);
11.   If ($s_{i,j} > s_j^*$) then let $s_j^* = s_{i,j}$ and $t_j = i$;
12.   }
13.   Let $H = H \cup \{t_j\}$;
14.   Let $z = z + s_{t_j, j}$;
15.   ProcessSet(A_{t_j});
16.   Let L' =Append(L', A_{t_j});
17.   Remove A_{t_j} from list L;
18. }
19. Return z and H;

End of Algorithm

The performance of the algorithm is stated in Theorem 4. It gives an $(1 - \frac{1}{e})$-approximation for the maximum coverage problem for $((0,0), (0,0))$-list according to Definition 3.

5 Inapproximability of Partial Sublinear Time

In this section, we introduce the concept of partial sublinear time computation. The maximal coverage has a partial sublinear constant factor approximation algorithm. On the other hand, we show that an inapproximability result for maximum coverage if the time is $q(n)m^{1-\epsilon}$, where m is the number of sets. This makes the notion of partial sublinear computation different from conventional sublinear computation.

The inapproximability result is derived on a randomized computational model that includes the one used in developing our randomized algorithms for the maximum coverage problem. The randomized model result needs some additional work than a deterministic model to prove the inapproximability. A deterministic algorithm with $q(n)m^{1-\epsilon}$ time let some set be never queried by the computation, but all sets can be queried in randomized algorithm with $q(n)m^{1-\epsilon}$ time as there are super-polynomial (of both n and m) many paths.

6   Conclusions

We developed a randomized greed approach for the maximum coverage problem. It obtains the same approximation ratio $(1 - \frac{1}{e})$ as the classical approximation for the maximum coverage problem, while its computational time is independent of the cardinalities of input sets under the model that each set answers query and generates one random sample in $O(1)$ time. It can be applied to find approximate maximum volume by selecting k objects among a list of objects such as rectangles in high dimensional space. Our approximation ratio depends on the how much the random sampling is biased, and the initial approximation accuracy for the input set sizes. The two accuracies are determined by the parameters $\alpha_L, \alpha_R, \delta_L$ and δ_R in a $((\alpha_L, \alpha_R), (\delta_L, \delta_R))$-list. It seems that our method can be generalized to deal with more general version of the maximum coverage problems, and it is expected to obtain more results in this direction. The notion of partial sublinear time algorithm will be used to characterize more computational problems than the sublinear time algorithm.

Acknowledgements. The author is grateful to Jack Snoeyink for his suggestions that improve the presentation of this paper. This author would like to thank the reviewers for their helpful comments. This research was supported in part by National Science Foundation Early Career Award 0845376 and Bensten Fellowship of the University of Texas Rio Grande Valley. The first draft of this paper was posted at https://arxiv.org/abs/1604.01421 (with date 5 Apr 2016). The readers can find the full version of this paper there [12].

References

1. Ageev, A., Sviridenko, M.: Pipage rounding: a new method of constructing algorithms with proven performance guarantee. J. Comb. Optim. **8**(3), 307–328 (2004)
2. Calinescu, G., Chekuri, C., Pal, M., Vondrak, J.: Maximizing a submodular set function subject to a matroid constraint. SIAM J. Comput. **40**(6), 1740–1766 (2011)
3. Chazelle, B., Liu, D., Magen, A.: Sublinear geometric algorithms. SIAM J. Comput. **35**, 627–646 (2005)
4. Chazelle, B., Rubinfeld, R., Trevisan, L.: Approximating the minimum spanning tree weight in sublinear time. SIAM J. Comput. **34**, 1370–1379 (2006)

5. Chekuri, C., Kumar, A.: Maximum coverage problem with group budget constraints and applications. In: Jansen, K., Khanna, S., Rolim, J.D.P., Ron, D. (eds.) APPROX/RANDOM -2004. LNCS, vol. 3122, pp. 72–83. Springer, Heidelberg (2004)
6. Cohen, R., Katzir, L.: The generalized maximum coverage problem. Inf. Process. Lett. **108**(1), 15–22 (2008)
7. Cornuejols, G., Fisher, M.L., Nemhauser, G.L.: Location of bank accounts to optimize float: an analytic study of exact and approximate algorithms. Manag. Sci. **23**, 789–810 (1977)
8. Czumaj, A., Ergun, F., Fortnow, L., Magen, A., Newman, I., Rubinfeld, R., Sohler, C.: Approximating of euclidean minimum spanning tree in sublinear time. SIAM J. Comput. **35**(1), 91–109 (2005)
9. Czumaj, A., Sohler, C.: Estimating the weight of metric minimum spanning trees in sublinear time. SIAM J. Comput. **39**(3), 904–922 (2009)
10. Feige, U.: A threshold of ln n for approximating set cover. J. ACM **45**(4), 634–652 (1998)
11. Filmus, Y., Ward, J.: The power of local search: maximum coverage over a matroid. In: Proceedings of the 29th International Symposium on Theoretical Aspects of Computer Science, pp. 601–612 (2012)
12. Fu, B.: Partial sublinear time approximation and inapproximation for maximum coverage. arXiv:1604.01421, 5 April 2016
13. Hochbaum, D.: Approximation Algorithms for NP-hard Problems. PWS Publishing Co., Boston (1997)
14. Hochbaum, D.S., Pathria, A.: Analysis of the greedy approach in problems of maximum k-coverage. Naval Res. Logist. (NRL) **45**(6), 615–627 (1998)
15. Hoeffding, W.: Probability inequalities for sums of bounded random variables. J. Am. Stat. Assoc. **58**(301), 13–30 (1963)
16. Karp, R.: Reducibility among combinatorial problems. In: Miller, R.E., Thatcher, J.W. (eds.) Complexity of Computer Computations. Plenum Press, New York (1972)
17. Karp, R.M., Luby, M., Madras, N.: Monte-carlo approximation algorithms for enumeration problems. J. Algorithms **10**(3), 429–448 (1989)
18. Khuller, S., Moss, A., Naor, J.: The budgeted maximum coverage problem. Inf. Process. Lett. **70**, 39–45 (1999)
19. Motwani, R., Raghavan, P.: Randomized Algorithms. Cambridge University Press, Cambridge (2000)
20. Nemhauser, G.L., Wolsey, L.A., Fisher, M.L.: An analysis of approximations for maximizing submodular set functions. Math. Program. **14**, 265–294 (1978)
21. Srinivasan, A.: Distributions on level-sets with applications to approximation algorithms. In: Proceedings of 42nd IEEE Symposium on Foundations of Computer Science, pp. 588–597 (2001)
22. Valiant, L.G.: The complexity of enumeration and reliability problems. SIAM J. Comput. **8**(3), 410–421 (1979)
23. Vazirani, V.V.: Approximation Algorithms. Computer Science Press/Springer, Rockville/Berlin (2001)

Characterizing Star-PCGs

Mingyu Xiao[1(✉)] and Hiroshi Nagamochi[2]

[1] School of Computer Science and Engineering,
University of Electronic Science and Technology of China, Chengdu, China
myxiao@gmail.com
[2] Department of Applied Mathematics and Physics, Graduate School of Informatics,
Kyoto University, Kyoto, Japan
nag@amp.i.kyoto-u.ac.jp

Abstract. A graph G is called a pairwise compatibility graph (PCG, for short) if it admits a tuple $(T, w, d_{\min}, d_{\max})$ of a tree T whose leaf set is equal to the vertex set of G, a non-negative edge weight w, and two non-negative reals $d_{\min} \leq d_{\max}$ such that G has an edge between two vertices $u, v \in V$ if and only if the distance between the two leaves u and v in the weighted tree (T, w) is in the interval $[d_{\min}, d_{\max}]$. The tree T is also called a witness tree of the PCG G. The problem of testing if a given graph is a PCG is not known to be NP-hard yet. To obtain a complete characterization of PCGs is a wide open problem in computational biology and graph theory. In the literature, most witness trees admitted by known PCGs are stars and caterpillars. In this paper, we give a complete characterization for a graph to be a star-PCG (a PCG that admits a star as its witness tree), which provides us the first polynomial-time algorithm for recognizing star-PCGs.

Keywords: Pairwise compatibility graph
Polynomial-time algorithm · Graph algorithm · Graph theory

1 Introduction

Pairwise compatibility graph is a graph class originally motivated from computational biology. In biology, the evolutionary history of a set of organisms is represented by a phylogenetic tree, which is a tree with leaves representing known taxa and internal nodes representing ancestors that might have led to these taxa through evolution. Moreover, the edges in the phylogenetic tree may be assigned weights to represent the evolutionary distance among species. Given a set of taxa and some relations among the taxa, we may want to construct a phylogenetic tree of the taxa. The set of taxa may be a subset of taxa from a large phylogenetic tree, subject to some biologically-motivated constraints. Kearney et al. [12] considered the following constraint on sampling based on the observation in [10]: the pairwise distance between any two leaves in the sample phylogenetic tree is between two given integers d_{min} and d_{max}. This motivates the introduction of pairwise compatibility graphs (PCGs). Given a phylogenetic

© Springer International Publishing AG, part of Springer Nature 2018
L. Wang and D. Zhu (Eds.): COCOON 2018, LNCS 10976, pp. 504–515, 2018.
https://doi.org/10.1007/978-3-319-94776-1_42

tree T with an edge weight w and two real numbers d_{min} and d_{max}, we can construct a graph G whose each vertex is corresponding to a leaf of T so that there is an edge between two vertices in G if and only if the corresponding two leaves of T are at a distance within the interval $[d_{min}, d_{max}]$ in T. The graph G is called the PCG of the tuple (T, w, d_{min}, d_{max}).

It is straightforward to construct a PCG from a given tuple (T, w, d_{min}, d_{max}). However, the inverse direction seems a considerably hard task. Few methods have been known for constructing a corresponding tuple (T, w, d_{min}, d_{max}) from a given graph G. The inverse problem attracts certain interests in graph algorithms, which may also have potential applications in computational biology. PCG has been extensively studied from many aspects after its introduction [3,6,7,9,18,19].

A natural question was whether all graphs are PCGs. This was proposed as a conjecture in [12], and was confuted in [18] by giving a counterexample of a bipartite graph with with 15 vertices. Later, a counterexample with eight vertices and a counterexample of a planar graph with 20 vertices were found [9]. It has been checked that all graphs with at most seven vertices are PCGs [3] and all bipartite graphs with at most eight vertices are PCGs [14]. In fact, it is even not easy to check whether a graph with a small constant number of vertices is a PCG or not. Whether recognizing PCGs is NP-hard or not is currently open. Some references conjecture the NP-hardness of the problem [7,9]. A generalized version of PCG recognition is shown to be NP-hard [9].

Several graph classes contained in PCG have been investigated. PCG contains the well-studied graph class of *leaf power graphs* (LPGs) as a subset of instances such that $d_{min} = 0$, which was introduced in the context of constructing phylogenies from species similarity data [8,13,15]. Another natural relaxation of PCG is to set $d_{max} = \infty$. This graph class is known as *min leaf power graph* (mLPG) [6], which is the complement of LPG. Several other known graph classes have been shown to be subclasses of PCG, e.g., disjoint union of cliques [2], forests [11], cacti [17,19], chordless cycles and single chord cycles [19], complete k-partite graphs [17], tree power graphs [18], threshold graphs [6], triangle-free outerplanar 3-graphs [16], some particular subclasses of split matrogenic graphs [6], Dilworth 2 graphs [5], the complement of a forest [11] and so on. It is also known that a PCG with a witness tree being a caterpillar also allows a witness tree being a centipede [4]. A method for constructing PCGs is derived [17], where it is shown that a graph G consisting two graphs G_1 and G_2 that share a vertex as a cut-vertex in G is a PCG if and only both G_1 and G_2 are PCGs.

How to recognize PCGs or construct a corresponding phylogenetic tree for a PCG becomes a wide open problem in this area. To make a step toward this open problem, we consider PCGs with a witness tree being a star in this paper, which we call star-PCGs. Note that in the literature, most of the witness trees of PCGs have simple graph structures, such as stars and caterpillars [7]. It is fundamental to consider the problem of characterizing subclasses of PCGs derived from a specific topology of trees. Although stars are trees with a rather simple topology, star-PCG recognition is not easy at all. It is known that threshold graphs are star-PCGs (even in star-LPG and star-mLPG) and the class of star-PCGs is nearly the class of three-threshold graphs, a graph class extended from the threshold graphs [6]. However, no complete characterization of star-PCGs and no polynomial-time recognition of star-PCGs are known. In this paper, we give a complete characterization for a graph to be a star-PCG, which provides us the first polynomial-time algorithm for recognizing star-PCGs.

The main idea of our algorithm is as follows. Without loss of generality, we always rank the leaves of the witness star T_V (and the corresponding vertices in the star-PCG G) according to the weight of the edges incident on it. When such an ordering of the vertices in a star-PCG G is given, we can see that all the neighbors of each vertex in G must appear consecutively in the ordering. This motivates us to define such an ordering to be "consecutive ordering". To check if a graph is a star-PCG, we can first check if the graph can have a consecutive ordering of vertices. Consecutive orderings can be computed in polynomial time by reducing to the problem of recognizing interval graphs. However, this is not enough to test star-PCGs. A graph may not be a star-PCG even if it has a consecutive ordering of vertices. We further investigate the structural properties of star-PCGs on a fixed consecutive ordering of vertices. We find that three cases of non-adjacent vertex pairs, called *gaps*, can be used to characterize star-PCGs. A graph is a star-PCG if and only if it admits a consecutive ordering of vertices that is gap-free (Theorem 3). Finally, to show that whether a given graph is gap-free or not can be tested in polynomial time (Theorem 4), we also use a notion of "contiguous orderings". All these together contribute to a polynomial-time algorithm for our problem.

The paper is organized as follows. Section 2 introduces some basic notions and notations necessary to this paper. Section 3 discusses how to test whether a given family $\mathcal{S}$ of subsets of an element set V admits a special ordering on V, called "consecutive" or "contiguous" orderings and proves the uniqueness of such orderings under some conditions on $\mathcal{S}$. Section 4 characterizes the class of star-PCGs $G = (V, E)$ in terms of an ordering σ of the vertex set V, called a "gap-free" ordering, and shows that given a gap-free ordering of V, a tuple $(T, w, d_{\min}, d_{\max})$ that represents G can be computed in polynomial time. Section 5 first derives structural properties on a graph that admits a "gap-free" ordering, and then presents a method for testing if a given graph is a star-PCG or not in polynomial time by using the result on contiguous orderings to a family of sets. Finally Sect. 6 makes some concluding remarks. Due to the space limitation, some proofs are omitted in the extended abstract.

2   Preliminaries

For two integers a and b, let $[a, b]$ denote the set of integers i with $a \leq i \leq b$. For a sequence σ of elements, let $\bar{\sigma}$ denote the reversal of σ. A sequence obtained by concatenating two sequences σ_1 and σ_2 in this order is denoted by (σ_1, σ_2).

Families of Sets. Let V be a set of $n \geq 1$ elements. We call a subset $S \in V$ *trivial* in V if $|S| \leq 1$ or $S = V$. We say that a set X has a common element with a set Y if $X \cap Y \neq \emptyset$. We say that two subsets $X, Y \subseteq V$ *intersect* (or X *intersects* Y) if three sets $X \cap Y$, $X \setminus Y$, and $Y \setminus X$ are all non-empty sets. A *partition* $\{V_1, V_2, \ldots, V_k\}$ of V is defined to be a collection of disjoint non-empty subsets V_i of V such that their union is V, where possibly $k = 1$.

Let $\mathcal{S} \subseteq 2^V$ be a family of m subsets of V. A total ordering $u_1, u_2, \ldots, u_n$ of elements in V is called *consecutive* to $\mathcal{S}$ if each non-empty set $S \in \mathcal{S}$ consists of elements with consecutive indices, i.e., S is equal to $\{u_i, u_{i+1}, \ldots, u_{i+|S|-1}\}$ for some $i \in [1, n - |S| - 1]$. A consecutive ordering $u_1, u_2, \ldots, u_n$ of elements in V to $\mathcal{S}$ is called *contiguous* if any two sets $S, S' \in \mathcal{S}$ with $S' \subseteq S$ start from or end with the same element along the ordering, i.e., $S' = \{u_j, u_{j+1}, \ldots, u_{j+|S'|-1}\}$ and $S = \{u_i, u_{i+1}, \ldots, u_{i+|S|-1}\}$ satisfy $j = i$ or $j + |S'| = i + |S|$.

Graphs. Let a graph stand for a simple undirected graph. A graph (resp., bipartite graph) with a vertex set V and an edge set E (resp., an edge set E between two vertex sets V_1 and $V_2 = V \setminus V_1$) is denoted by $G = (V, E)$ (resp., (V_1, V_2, E)). Let G be a graph, where $V(G)$ and $E(G)$ denote the sets of vertices and edges in a graph G, respectively. For a vertex v in G, we denote by $N_G(v)$ the set of neighbors of a vertex v in G, and define *degree* $\deg_G(v)$ to be the $|N_G(v)|$. We call a pair of vertices u and v in G a *mirror pair* if $N_G(v) \setminus \{u\} = N_G(u) \setminus \{v\}$. Let X be a subset of $V(G)$. Define $N_G(X)$ to be the set of neighbors of X, i.e., $N_G(X) = \{u \in N_G(v) \setminus X \mid v \in X\}$. Let $G - X$ denote the graph obtained from G by removing vertices in X together with all edges incident to vertices in X, where $G - \{v\}$ for a vertex v may be written as $G - v$. Let $G[X]$ denote the graph induced by X, i.e., $G[X] = G - (V(G) \setminus X)$.

Let T be a tree. A vertex v in T is called an *inner vertex* if $\deg_T(v) \geq 2$ and is called a *leaf* otherwise. Let $L(T)$ denote the set of leaves. An edge incident to a leaf in T is called a *leaf edge* of T. A tree T is called a *star* if it has at most one inner vertex.

Weighted Graphs. An edge-weighted graph (G, w) is defined to be a pair of a graph G and a non-negative weight function $w : E(G) \to \Re_+$. For a subgraph G' of G, let $w(G')$ denote the sum $\sum_{e \in E(G')} w(e)$ of edge weights in G'.

Let (T, w) be an edge-weighted tree. For two vertices $u, v \in V(T)$, let $d_{T,w}(u, v)$ denote the sum of weights of edges in the unique path of T between u and v.

PCGs. For a tuple $(T, w, d_{\min}, d_{\max})$ of an edge-weighted tree (T, w) and two non-negative reals $d_{\min}$ and $d_{\max}$, define $G(T, w, d_{\min}, d_{\max})$ to be the simple graph $(L(T), E)$ such that, for any two distinct vertices $u, v \in L(T)$, $uv \in E$ if and only if $d_{\min} \leq d_{T,w}(u, v) \leq d_{\max}$. Note that $G(T, w, d_{\min}, d_{\max})$ is not necessarily connected.

A graph G is called a *pairwise compatibility graph* (PCG, for short) if there exists a tuple $(T, w, d_{\min}, d_{\max})$ such that G is isomorphic to the graph $G(T, d_{\min}, d_{\max})$, where we call such a tuple a *pairwise compatibility representation* (PCR, for short) of G, and call a tree T in a PCR of G a *pairwise compatibility tree* (PCT, for short) of G. The tree T is called a *witness tree* of G. We call a PCG G a *star-PCG* if it admits a PCR $(T, w, d_{\min}, d_{\max})$ such that T is a star. Figure 1 illustrates examples of star-PCGs and PCRs of them. Although phylogenetic trees may not have edges with weight 0 or degree-2 vertices by some biological motivations [4], our PCTs do not have these constraints. This relaxation will be helpful for us to analyze structural properties of PCGs from graph theory. Furthermore, it is easy to get rid of edges with weight 0 or degree-2 vertices in a tree by contracting an edge.

Lemma 1. *Every PCG admits a PCR (T, w, d_{min}, d_{max}) such that $0 < d_{min} < d_{max}$ and $w(e) > 0$ for all edges $e \in E(T)$.*

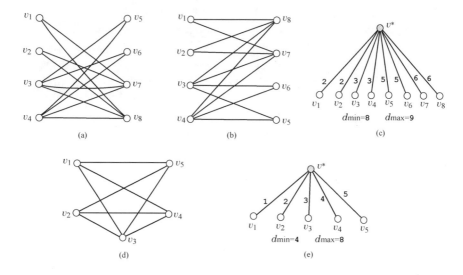

Fig. 1. Illustration of examples of star-PCG. (a) A connected and bipartite star-PCG $G_1 = (V_1, V_2, E)$. (b) The same graph G_1 in (a) with a different ordering of vertices (the two different orderings of vertices in (a) and (b) will be used to illustrate the concept of "gap-free" later). (c) A PCR $(T, w, d_{min} = 8, d_{max} = 9)$ of G_1. (d) A connected and non-bipartite star-PCG G_2. (e) A PCR $(T, w, d_{min} = 4, d_{max} = 8)$ of G_2 in (d).

3   Consecutive/Contiguous Orderings of Elements

Let $\mathcal{S} \subseteq 2^V$ be a family of m subsets of a set V of $n \geq 1$ elements in this section. Let $V(\mathcal{S})$ denote the union of all subsets in $\mathcal{S}$.

3.1   Consecutive Orderings of Elements

Observe that when $\mathcal{S}$ admits a consecutive ordering of $V(\mathcal{S})$, any subfamily $\mathcal{S}' \subseteq \mathcal{S}$ admits a consecutive ordering of $V(\mathcal{S}')$. We call a non-trivial set $C \subseteq V$ a *cut* to $\mathcal{S}$ if no set $S \in \mathcal{S}$ intersects C, i.e., each $S \in \mathcal{S}$ satisfies one of $S \supseteq C$, $S \subseteq C$ and $S \cap C = \emptyset$. We call $\mathcal{S}$ *cut-free* if $\mathcal{S}$ has no cut.

Theorem 1. *For a set V of $n \geq 1$ elements and a family $\mathcal{S} \subseteq 2^V$ of $m \geq 1$ sets, a consecutive ordering of V to $\mathcal{S}$ can be found in $O(nm^2)$ time if one exists. Moreover if $\mathcal{S}$ is cut-free, then a consecutive ordering of V to $\mathcal{S}$ is unique up to reversal.*

3.2 Contiguous Orderings of Elements

We call two elements $u, v \in V$ *equivalent* in $\mathcal{S}$ if no set $S \in \mathcal{S}$ satisfies $|\{u, v\} \cap S| = 1$. We call $\mathcal{S}$ *simple* if there is no pair of equivalent elements $u, v \in V$. Define $\mathcal{X}_{\mathcal{S}}$ to be the family of maximal sets $X \subseteq V$ such that any two vertices in X are equivalent and X is maximal subject to this property.

A non-trivial set $S \in \mathcal{S}$ is called a *separator* if no other set $S' \in \mathcal{S}$ contains or intersects S, i.e., each $S' \in \mathcal{S}$ satisfies $S' \subseteq S$ or $S' \cap S = \emptyset$. We call $\mathcal{S}$ *separator-free* in $\mathcal{S}$ if $\mathcal{S}$ has no separator.

Theorem 2. *For a set V of $n \geq 1$ elements and a family $\mathcal{S} \subseteq 2^V$ of $m \geq 1$ sets, a contiguous ordering of V to $\mathcal{S}$ can be found in $O(nm^2)$ time if one exists. Moreover, all elements in each set $X \in \mathcal{X}_{\mathcal{S}}$ appear consecutively in any contiguous ordering of V to $\mathcal{S}$, and if $\mathcal{S}$ is separator-free, then a contiguous ordering of V to $\mathcal{S}$ is unique up to reversal of the entire ordering and arbitrariness of orderings of elements in each set $X \in \mathcal{X}_{\mathcal{S}}$.*

4   Star-PCGs

Let $G = (V, E)$ be a graph with $n \geq 2$ vertices, not necessarily connected. Let M_G denote the set of mirror pairs $\{u, v\} \subseteq V$ in G, i.e., $N_G(u) \setminus \{v\} = N_G(v) \setminus \{u\}$, where u and v are not necessarily adjacent. Let T_V be a star with a center v^* and $L(T) = V$. An *ordering* of V is defined to be a bijection $\sigma : V \to \{1, 2, \ldots, n\}$, and we simply write a vertex v with $\sigma(v) = i$ with v_i. For an edge weight w in T_V, we simply denote $w(v^* v_i)$ by w_i. When G is a star-PCG of a tuple $(T_V, w, d_{\min}, d_{\max})$, there is an ordering σ of V such that $w_1 \leq w_2 \leq \cdots \leq w_n$. Conversely this section derives a necessary and sufficient condition for a pair (G, σ) of a graph G and an ordering σ of V to admit a PCR $(T_V, w, d_{\min}, d_{\max})$ of G such that $w_1 \leq w_2 \leq \cdots \leq w_n$.

For an ordering σ of V, a non-adjacent vertex pair $\{v_i, v_j\}$ with $i < j$ in G is called a *gap* (with respect to edges $e_1, e_2 \in E$) if there are edges $e_1, e_2 \in E$ that satisfy one of the following:

(g1) $e_1 = v_i v_{j'}$ and $e_2 = v_i v_{j''}$ such that $j' < j < j''$ (or $e_1 = v_{i'} v_j$ and $e_2 = v_{i''} v_j$ such that $i' < i < i''$), as illustrated in Fig. 2(a);

(g2) $e_1 = v_i v_{i'}$ and $e_2 = v_j v_{j'}$ such that $j' < i$ and $j < i'$, as illustrated in Fig. 2(b); and

(g3) $e_1 = v_i v_{i'}$ and $e_2 = v_j v_{j'}$ such that $i' < j$ and $i < j'$, as illustrated in Fig. 2(c).

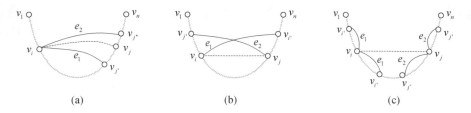

Fig. 2. Illustration of a gap $\{v_i, v_j\}$ in an ordered graph $G = (V = \{v_1, v_2, \ldots, v_n\}, E)$, where edges are denoted by solid lines and anti-edges are denoted by dashed lines: (a) $e_1 = v_i v_{j'}$ and $e_2 = v_i v_{j''}$ such that $j' < j < j''$, (b) $e_1 = v_i v_{i'}$ and $e_2 = v_j v_{j'}$ such that $j' < i$ and $j < i'$, (c) $e_1 = v_i v_{i'}$ and $e_2 = v_j v_{j'}$ such that $i' < j$ and $i < j'$, where possibly $j' \leq i'$ or $i' < j'$.

We call an ordering σ of V *gap-free* in G if it has no gap. Clearly the reversal of a gap-free ordering of V is also gap-free. We can test if a given ordering is gap-free or not in $O(n^4)$ time by checking the conditions (a)–(c) for each non-adjacent vertex pair $\{v_i, v_j\}$ in G.

Figure 1(a) and (b) illustrate the same graph G_1 with different orderings $\sigma_a = v_1, v_2, v_3, v_4, v_8, v_7, v_6, v_5$ and $\sigma_b = v_1, v_2, \ldots, v_8$, where σ_a is not gap-free while σ_b is gap-free.

We have the following result, which implies that a graph $G = (V, E)$ is a star-PCG if and only if it admits a gap-free ordering of V.

Theorem 3. *For a graph $G = (V, E)$, let σ be an ordering of V. Then there is a PCR $(T_V, w, d_{\min}, d_{\max})$ of G such that $w_1 \leq w_2 \leq \cdots \leq w_n$ if and only if σ is gap-free.*

The necessity of this theorem is relatively easy to prove. Next we consider the sufficiency of Theorem 3, which is implied by the next lemma.

Lemma 2. *For a graph $G = (V, E)$, let $\sigma = v_1, v_2, \ldots, v_n$ be an gap-free ordering of V. There is a PCR $(T_V, w, d_{\min}, d_{\max})$ of G such that $w_1 \leq w_2 \leq \cdots \leq w_n$. Such a set $\{w_1, w_2, \ldots, w_n, d_{\min}, d_{\max}\}$ of weights and bounds can be obtained in $O(n^3)$ time.*

A main technique used to prove this lemma is to color all edges and anti-edges in the graph (i.e., all edges in the complete graph $K_{|V|} = (V, E \cup \overline{E})$ on the vertex set V, where $\overline{E} = \binom{V}{2} \setminus E$) to indicate the range of the distance between the two endpoints of the edge or anti-edge in the PCT. Note that when two vertices u and v are not adjacent in a PCG G, i.e., there is anti-edge between u and v, there are two reasons: one is that the distance between them in the PCR $(T, w, d_{\min}, d_{\max})$ is smaller than $d_{\min}$, and the other is that the distance is larger than $d_{\max}$. We will use two different colors to distinguish these two kinds of anti-edges. It will become easier to set weights to edges and bounds after coloring all the anti-edges.

A *coloring* of a graph $G = (V, E)$ is a function $c : E \cup \overline{E} \to \{\mathtt{r}, \mathtt{g}, \mathtt{b}\}$. We call an edge e with $c(e) = \mathtt{r}$ (resp., $\mathtt{g}$ and $\mathtt{b}$) a *red* (resp., *green* and *blue*) edge and use red (resp., green and blue) to denote the sets of red (resp., green and blue) edges. We denote by $N_{\text{red}}(v)$ the set of neighbors of a vertex v via red edges. We define $N_{\text{green}}(v)$ and $N_{\text{blue}}(v)$ analogously.

For a graph G with an ordering $\sigma = v_1, v_2, \ldots, v_n$ of the vertices, a coloring c is *proper* to (G, σ) if it satisfies the following conditions: each $v_i \in V$ admits integers $a(i), b(i) \in [1, n]$ such that

$$N_{\text{red}}(v_i) = \{v_j \mid 1 \le j \le a(i) - 1\} \setminus \{v_i\}$$

and

$$N_{\text{blue}}(v_i) = \{v_j \mid b(i) + 1 \le j \le n\} \setminus \{v_i\},$$

where $a(i) = 1$ if $N_{\text{red}}(v_i) = \emptyset$; $b(i) = n$ if $N_{\text{blue}}(v_i) = \emptyset$; $N_{\text{green}}(v_i) = V \setminus (N_{\text{red}}(v_i) \cup N_{\text{blue}}(v_i) \cup \{v_i\})$, which is $\emptyset$ if $b(i) < a(i)$ and $\{v_j \mid a(i) \le j \le b(i)\} \setminus \{v_i\}$ otherwise.

We will show that for a graph G with an ordering σ of the vertices, if there is a coloring c proper to (G, σ), then we can find weights w_i and bounds $d_{\min}$ and $d_{\max}$ so that the next holds:

$w_1 \le w_2 \le \cdots \le w_n$;
$d_{\min} \le w_i + w_j \le d_{\max}$ for $(i, j) \in$ green;
$w_i + w_j < d_{\min}$ for $(i, j) \in$ red; and
$w_i + w_j > d_{\max}$ for $(i, j) \in$ blue.

All these are necessary for us to set the weights and bounds.

Recall that $a(i)$ and $b(i)$ are defined in the above definition of proper coloring. We define integers i_{red} and i_{blue} as follows.

$$i_{\text{red}} = \begin{cases} \text{the largest index } i \text{ such that } i < a(i) & \text{if red} \ne \emptyset, \\ 0 & \text{if red} = \emptyset, \end{cases}$$

$$i_{\text{blue}} = \begin{cases} \text{the smallest index } i \text{ such that } b(i) < i & \text{if blue} \ne \emptyset, \\ n + 1 & \text{if blue} = \emptyset. \end{cases}$$

In other words, $i_{\text{red}} \ne 0$ is the largest i with $(i, i + 1) \in$ red, and $i_{\text{red}} < n$, whereas $i_{\text{blue}} \ne n + 1$ is the smallest i with $(i - 1, i) \in$ blue, and $i_{\text{blue}} > 1$. Given a graph G, a gap-free ordering $\sigma = v_1, v_2, \ldots, v_n$ of V, and a coloring c proper to (G, σ), we can find the set $\{a(i), b(i) \mid i = 1, 2, \ldots, n\} \cup \{i_{\text{red}}, i_{\text{blue}}\}$ of indices in $O(n^2)$ time. We also compute the set M_G of all mirror pairs in $O(n^3)$ time. Equipped with above results, we can design an $O(n)$-time algorithm that assigns the right values to weights $w_1, w_2, \ldots, w_n$ in T_V. The details can be found in the full version.

To prove Lemma 2, we also need the following lemma.

Lemma 3. *For a graph $G = (V, E)$ and a gap-free ordering σ of V, there is a coloring c of G that is proper to (G, σ), which can be found in in $O(n^2)$ time.*

5   Recognizing Star-PCGS

Based on Theorem 3, we can test whether a graph $G = (V, E)$ is a star-PCG or not by generating all $n!$ orderings of V. In this section, we show that whether a graph has a gap-free ordering of vertices can be tested in polynomial time.

Theorem 4. *Whether a given graph $G = (V, E)$ with n vertices has a gap-free ordering of V can be tested in $O(n^6)$ time.*

Lemma 4. *For a graph $G = (V, E)$ with a gap-free ordering $\sigma = v_1, v_2, \ldots, v_n$ of V and a coloring c proper to (G, σ), let $V_1 = \{v_i \mid 1 \le i \le i_{\text{red}}\}$, $V_2 = \{v_i \mid i_{\text{blue}} \le i \le n\}$, and $V^* = \{v_i \mid i_{\text{red}} - 1 \le i \le i_{\text{blue}} + 1\}$. Then*

(i) *If two edges $v_i v_j$ and $v_{i'} v_{j'}$ with $i < j$ and $i' < j'$ cross (i.e., $i < i' < j < j'$ or $i' < i < j' < j$), then they belong to the same component of G;*

(ii) *It holds $i_{\text{red}} + 1 \le i_{\text{blue}} - 1$. The graph $G[V^*]$ is a complete graph, and $G - V^*$ is a bipartite graph between vertex sets V_1 and V_2;*

(iii) *Every two vertices $v_i, v_j \in V_1 \cap N_G(V^*)$ with $i < j$ satisfy $v_{i_{\text{blue}} - 1} \in N_G(v_i) \cap V^* \subseteq N_G(v_j) \cap V^* \subseteq V^* \setminus \{v_{i_{\text{red}} + 1}\}$; and*
Every two vertices $v_i, v_j \in V_2 \cap N_G(V^*)$ with $i < j$ satisfy $v_{i_{\text{red}} + 1} \in N_G(v_j) \cap V^* \subseteq N_G(v_i) \cap V^* \subseteq V^* \setminus \{v_{i_{\text{blue}} - 1}\}$.

We call the complete graph $G[V^*]$ in Lemma 4(ii) the *core* of G. Based on the next lemma, we can treat each component of a disconnected graph G separately to test whether G is a star-PCG or not. Figure 3 illustrates a disconnected PCG and a connected non-bipartite PCG.

Lemma 5. *Let $G = (V, E)$ be a graph with at least two components.*

(i) *If G admits a gap-free ordering of V, then each component of G admits a gap-free ordering of its vertex set, and there is at most one non-bipartite component in G;*

(ii) *Let $G' = (V_1', V_2', E')$ be a bipartite component of G, and $G'' = G - V(G')$. Assume that G' admits a gap-free ordering $v_1', v_2', \ldots, v_p'$ of $V_1' \cup V_2'$ and G'' admits a gap-free ordering $v_1, v_2, \ldots, v_q$ of V_2. Then there is an index k such that $\{\{v_1', v_2', \ldots, v_k'\}, \{v_{k+1}', v_{k+2}', \ldots, v_p'\}\} = \{V_1', V_2'\}$. Moreover, the ordering $v_1', v_2', \ldots, v_k', v_1, v_2, \ldots, v_q, v_{k+1}', v_{k+2}', \ldots, v_p'$ of V is gap-free to G.*

Proof. (i) Let G admit a gap-free ordering of V. Any induced subgraph G such as a component of G is a star-PCG, and a gap-free ordering of its vertex set by Theorem 3. By Lemma 4(i), at most one component H containing a complete graph with at least three vertices can be non-bipartite, and the remaining graph $G - V(H)$ must be a collection of bipartite graphs.

(ii) Immediate from the definition of gap-free orderings.   □

We first consider the problem of testing if a given connected bipartite graph is a star-PCG or not. We reduce this to the problem of finding contiguous ordering to a family of sets.

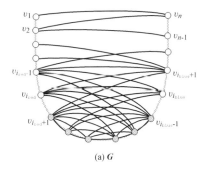

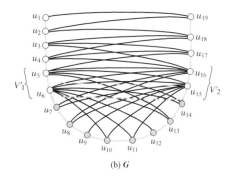

(a) G (b) G

Fig. 3. (a) A disconnected PCG, (b) A connected non-bipartite PCG, where the edges between two vertices in $V^* = \{u_7, u_8, u_9, u_{10}, u_{11}, u_{12}, u_{13}, u_{14}\}$ are not depicted.

For a bipartite graph $G = (V_1, V_2, E)$ and $i \in \{1, 2\}$, define $\mathcal{S}_i$ to be the family $\{N_G(v) \mid v \in V_j\}$ for the $j \in \{1, 2\} - \{i\}$, where even if there are distinct vertices $u, v \in V_j$ with $N_G(u) = N_G(v)$, $\mathcal{S}_i$ contains exactly one set $S = N_G(u) = N_G(v)$. For the example of a connected bipartite graph $G_1 = (V_1, V_2, E)$ in Fig. 1(a), we have $\mathcal{S}_1 = \{\{v_3, v_4\}, \{v_1, v_2, v_3, v_4\}\}$, and $\mathcal{S}_2 = \{\{v_5, v_6\}, \{v_5, v_6, v_7, v_8\}\}$.

Lemma 6. *Let* $G = (V_1, V_2, E)$ *be a connected bipartite graph with* $|E| \geq 1$. *Then family* $\mathcal{S}_i$ *is separator-free for each* $i = 1, 2$, *and* G *has a gap-free ordering of* V *if and only if for each* $i = 1, 2$, *family* $\mathcal{S}_i$ *admits a contiguous ordering* σ_i *of* V_i. *For any contiguous ordering* σ_i *of* V_i, $i = 1, 2$, *one of orderings* (σ_1, σ_2) *and* $(\sigma_1, \overline{\sigma_2})$ *of* V *is a gap-free ordering to* G.

Note that $|\mathcal{S}_1| + |\mathcal{S}_2| + |V(\mathcal{S}_1)| + |V(\mathcal{S}_2)| = O(n)$. By Theorem 2, a contiguous ordering of $V(\mathcal{S}_i)$ for each $i = 1, 2$ can be computed in $O(|V(\mathcal{S}_i)||\mathcal{S}_i|^2) = O(n^3)$ time.

Figure 1(a) illustrates an ordering $\sigma_a = v_1, v_2, v_3, v_4, v_8, v_7, v_6, v_5$ of $V(G_1)$ of a connected bipartite graph $G_1 = (V_1, V_2, E)$, where σ_a consists of a contiguous ordering $\sigma_1 = v_1, v_2, v_3, v_4$ of V_1 and a contiguous ordering $\sigma_2 = v_8, v_7, v_6, v_5$ of V_2. Although σ_a is not gap-free in G, the other ordering σ_b of $V(G_1)$ that consists of σ_1 and the reversal of σ_2 is gap-free, as illustrated in Fig. 1(b).

Finally we consider the case where a given graph G is a connected and non-bipartite graph. Figure 1(d) illustrates a connected and non-bipartite star-PCG whose maximum clique is not unique.

In a graph $G = (V, E)$, let E^{t} denote the union of edge sets of all cycles of length 3 in G, V^{t} denote the set of end-vertices of edges in E^{t}, and $N_G^{\mathrm{t}}(v)$ denote the set of neighbors $u \in N_G(v)$ of a vertex $v \in V$ such that $uv \in E^{\mathrm{t}}$.

Lemma 7. *For a connected non-bipartite graph* $G = (V, E)$ *with* $V^{\mathrm{t}} \neq \emptyset$, *and let* v_1^*, v_2^* *be two adjacent vertices in* V^{t}. *Let* $V^* = \{v_1^*, v_2^*\} \cup (N_G(v_1^*) \cap N_G(v_2^*))$, $V_1' = N_G(v_2^*) \backslash V^*$, *and* $V_2' = N_G(v_1^*) \backslash V^*$. *Assume that* G *has a gap-free ordering* σ *of* V *and a proper coloring* c *to* σ *such that* $v_1^* = v_{i_{\mathrm{red}}+1}$, $v_2^* = v_{i_{\mathrm{blue}}-1}$. *Then:*

(i) *A maximal clique $K_{v_1^*, v_2^*}$ of G that contains edge v_1^*, v_2^* is uniquely given as $G[V^*]$. The graph $G[V^*]$ is the core of the ordering σ, and $G - V^*$ is a bipartite graph (V_1, V_2, E');*

(ii) *Let S_i denote the family $\{N_G(v) \mid v \in V_j\}$ for $\{i, j\} = \{1, 2\}$, and $S = S_1 \cup S_2 \cup \{V^*\}$. Then S is a separator-free family that admits a contiguous ordering σ of V, and any contiguous ordering σ of V is a gap-free ordering to G.*

For example, when we choose vertices $v_1^* = u_7$ and $v_2^* = u_{14}$ in the connected non-bipartite graph $G = (V, E)$ in Fig. 3(b), we have $V^* = \{u_7, u_8, u_9, u_{10}, u_{11}, u_{12}, u_{13}, u_{14}\}$, $S_1 = \{\{u_1, u_2\}, \{u_2, u_3, u_4\}, \{u_3, u_4, u_5\}, \{u_3, u_4, u_5, u_6, u_7, u_8\}, \{u_5, u_6, u_7, u_8, u_9, u_{10}, u_{11}\}\}$, and $S_2 = \{\{u_{19}\}, \{u_{18}, u_{19}\}, \{u_{16}, u_{17}, u_{18}\}, \{u_{13}, u_{14}, u_{15}, u_{16}, u_{17}\}, \{u_{10}, u_{11}, u_{12}, u_{13}, u_{14}, u_{15}, u_{16}\}\}$.

For a fixed V^* in Lemma 7, we can test whether the separator-free family S in Lemma 7(ii) is constructed from V^* in $O(|V(S)||S|^2) = O(n^3)$ time by Theorem 2, since $|S| + |V(S)| = O(n)$ holds. It takes $O(n^4)$ time to check a given ordering is gap-free or not. To find the right choice of a vertex pair $v_1^* = v_{i_{\text{red}}+1}$ and $v_2^* = v_{i_{\text{blue}}-1}$ of some gap-free ordering σ of V, we need to try $O(n^2)$ combinations of vertices to construct V^* according to the lemma. Then we can find a gap-free ordering of a given graph, if one exists in $O(n^6)$ time, proving Theorem 4.

By Theorems 3 and 4, we conclude that whether a given graph with n vertices is a star-PCG or not can be tested in $O(n^6)$ time.

6   Concluding Remarks

Pairwise compatibility graphs were initially introduced from the context of phylogenetics in computational biology and later became an interesting graph class in graph theory. PCG recognition is a hard task and we are still far from a complete characterization of PCG. Significant progresses toward PCG recognition would be interesting from a graph theory perspective and also be helpful in designing sampling algorithms for phylogenetic trees. In this paper, we give the first polynomial-time algorithm to recognize star-PCGs. Although stars are trees of a simple topology, it is not an easy task to recognize star-PCGs. To do so, we need to develop several structural properties of this problem. We first show that three cases of non-adjacent vertex pairs (called gaps) under a fixed ordering of vertices can be used to characterize star-PCGs and the gaps in a graph can be tested in polynomial time. Then we show that we only need to test a polynomial number of orderings of vertices and thus we can get a polynomial time algorithm. For further study, it is an interesting topic to study the characterization of PCGs with witness trees of other particular topologies.

Acknowledgement. The work is supported by the National Natural Science Foundation of China, under grants 61772115 and 61370071.

References

1. Booth, S., Lueker, S.: Testing for the consecutive ones property, interval graphs, and graph planarity using PQ-tree algorithms. J. Comput. Syst. Sci. **13**, 335–379 (1976)
2. Brandstadt, A.: On leaf powers. Technical report, University of Rostock (2010)
3. Calamoneri, T., Frascaria, D., Sinaimeri, B.: All graphs with at most seven vertices are pairwise compatibility graphs. Comput. J. **56**(7), 882–886 (2013)
4. Calamoneri, T., Frangioni, A., Sinaimeri, B.: Pairwise compatibility graphs of caterpillars. Comput. J. **57**(11), 1616–1623 (2014)
5. Calamoneri, T., Petreschi, R.: On pairwise compatibility graphs having Dilworth number two. Theoret. Comput. Sci. **524**, 34–40 (2014)
6. Calamoneri, T., Petreschi, R., Sinaimeri, B.: On the pairwise compatibility property of some superclasses of threshold graphs. Discrete Math. Algorithms Appl. **5**(2) (2013). Article 360002
7. Calamoneri, T., Sinaimeri, B.: Pairwise compatibility graphs: a survey. SIAM Rev. **58**(3), 445–460 (2016)
8. Chen, Z.-Z., Jiang, T., Lin, G.: Computing phylogenetic roots with bounded degrees and errors. SIAM J. Comput. **32**, 864–879 (2003)
9. Durocher, S., Mondal, D., Rahman, M.S.: On graphs that are not PCGs. Theoret. Comput. Sci. **571**, 78–87 (2015)
10. Felsenstein, J.: Cases in which parsimony or compatibility methods will be positively misleading. Syst. Zool. **27**, 401–410 (1978)
11. Hossain, M.I., Salma, S.A., Rahman, M.S., Mondal, D.: A necessary condition and a sufficient condition for pairwise compatibility graphs. J. Graph Algorithms Appl. **21**(3), 341–352 (2017)
12. Kearney, P., Munro, J.I., Phillips, D.: Efficient generation of uniform samples from phylogenetic trees. In: Benson, G., Page, R.D.M. (eds.) WABI 2003. LNCS, vol. 2812, pp. 177–189. Springer, Heidelberg (2003). https://doi.org/10.1007/978-3-540-39763-2_14
13. Lin, G.-H., Kearney, P.E., Jiang, T.: Phylogenetic k-root and Steiner k-root. In: Goos, G., Hartmanis, J., van Leeuwen, J., Lee, D.T., Teng, S.-H. (eds.) ISAAC 2000. LNCS, vol. 1969, pp. 539–551. Springer, Heidelberg (2000). https://doi.org/10.1007/3-540-40996-3_46
14. Mehnaz, S., Rahman, M.S.: Pairwise compatibility graphs revisited. In: Proceedings of the 2013 International Conference on Informatics, Electronics Vision (ICIEV), pp. 1–6 (2013)
15. Nishimura, N., Ragde, P., Thilikos, D.M.: On graph powers for leaf-labeled trees. J. Algorithms **42**, 69–108 (2002)
16. Salma, S.A., Rahman, M.S., Hossain, M.I.: Triangle-free outerplanar 3-graphs are pairwise compatibility graphs. J. Graph Algorithms Appl. **17**, 81–102 (2013)
17. Xiao, M., Nagamochi, H.: Some reduction operations to pairwise compatibility graphs, abs/1804.02887, arXiv (2018)
18. Yanhaona, M.N., Bayzid, M.S., Rahman, M.S.: Discovering pairwise compatibility graphs. Discrete Math. Algorithms Appl. **2**(4), 607–624 (2010)
19. Yanhaona, M.N., Hossain, K.S.M.T., Rahman, M.S.: Pairwise compatibility graphs. J. Appl. Math. Comput. **30**, 479–503 (2009)

Liar's Dominating Set in Unit Disk Graphs

Ramesh K. Jallu, Sangram K. Jena, and Gautam K. Das$^{(\boxtimes)}$

Indian Institute of Technology, Guwahati, India
{j.ramesh,sangram,gkd}@iitg.ernet.in

Abstract. In this article, we study a variant of the dominating set problem known as the liar's dominating set problem on unit disk graphs. We prove that the liar's dominating set problem is NP-complete and admits a polynomial time approximation scheme in unit disk graphs.

Keywords: Dominating set · Liar's dominating set · Unit disk graph
Approximation scheme

1 Introduction

Given a simple undirected graph $G = (V, E)$, the open and closed neighborhoods of a vertex v_i are defined by $N_G(v_i) = \{v_j \in V \mid (v_i, v_j) \in E \text{ and } v_i \neq v_j\}$ and $N_G[v_i] = N_G(v_i) \cup \{v_i\}$, respectively. A *dominating set* D of G is a subset of V such that every vertex in $V \setminus D$ is adjacent to at least one vertex in D. That is, each vertex $v_i \in V$ is either in D or there exists a vertex $v_j \in D$ such that $(v_i, v_j) \in E$. Observe that for any dominating set $D \subseteq V$, $|N_G[v_i] \cap D| \geq 1$ for each $v_i \in V$. We say that a vertex v_i is dominated by v_j in G, if $v_j \in D$ and $(v_i, v_j) \in E$. The dominating set problem asks to find a dominating set of minimum size in a given graph. A set $D \subseteq V$ is a *k-tuple dominating set* in G, if each vertex $v_i \in V$ is dominated by at least k vertices in D. In other words, $|N_G[v_i] \cap D| \geq k$ for each $v_i \in V$. The minimum cardinality of a k-tuple dominating set of a graph G is called the *k-tuple domination number* of G.

A *liar's dominating set* (LDS) of a simple undirected graph, $G = (V, E)$, is a dominating set D having the following two properties: (i) for every $v_i \in V$, $|N_G[v_i] \cap D| \geq 2$, and (ii) for every pair of distinct vertices v_i and v_j in V, $|(N_G[v_i] \cup N_G[v_j]) \cap D| \geq 3$. For a given graph G, the problem of finding an LDS in G of minimum cardinality is known as the *minimum liar's dominating set* (MLDS) problem. The cardinality of an MLDS in a graph G is known as the liar's domination number of G. Every 3-tuple dominating set is a liar's dominating set as it satisfies both conditions, so the liar's domination number lies between 2-tuple and 3-tuple domination numbers.

L. Wang and D. Zhu (Eds.): COCOON 2018, LNCS 10976, pp. 516–528, 2018.
https://doi.org/10.1007/978-3-319-94776-1_43

2 Related Work

The MLDS problem is introduced by Slater [14]. He showed that the problem is NP-hard for general graphs, and gave a lower bound on the liar's domination number in case of trees by proving that the size of any liar's dominating set of a tree of order n is between $\frac{3}{4}(n + 1)$ and n. Later, Roden and Slater [13] characterized tree classes with liar's domination number equal to $\frac{3}{4}(n + 1)$. In the same paper, they also showed that the MLDS problem is NP-hard even for bipartite graphs. Panda and Paul [9] proved that the problem is NP-hard for split graphs and chordal graphs. They also proposed a linear time algorithm for computing an MLDS in case of trees.

Panda et al. [12] studied the approximability of the problem and presented an $O(\ln \Delta(G))$-factor approximation algorithm, where $\Delta(G)$ is the degree of the graph. Panda and Paul [10] considered the problem for proper interval graphs and proposed a linear time algorithm for computing a minimum cardinality liar's dominating set. The problem is also studied for bounded degree graphs, and p-claw free graphs [12]. Sterling [15] considered the problem on two-dimensional grid graphs and presented bounds on the liar's domination number.

Alimadadi et al. [1] provided the characterization of graphs and trees for which the liar's domination number is $|V|$ and $|V| - 1$, respectively. Panda and Paul [8,11] studied variants of liar's domination, namely, connected liar's domination and total liar's domination. A *connected liar's dominating set* (CLDS) is an LDS whose induced subgraph is connected. A *total liar's dominating set* (TLDS) is a dominating set L with the following two properties: (i) for every $v \in V$, $|N_G(v) \cap L| \geq 2$, and (ii) for every distinct pair of vertices u and v, $|(N_G(u) \cup N_G(v)) \cap L| \geq 3$, where $N_G(\cdot)$ is the open neighborhood of a vertex. The objective of both problems is to find CLDS and TLDS of minimum size, respectively. They proved that both problems are NP-hard and proposed $O(\ln \Delta(G))$-factor approximation algorithms. They also proved that the problems are APX-complete for graphs with maximum degree 4. Jallu and Das [6] studied the geometric version of the MLDS problem, and presented constant factor approximation algorithms.

2.1 Our Contribution

We study the MLDS problem on a geometric intersection graph model, particularly on unit disk graphs. A *unit disk graph* (UDG) is an intersection graph of disks of equal radii in the plane. Given a set $S = \{d_1, d_2, \ldots, d_n\}$ of n circular disks in the plane, each having diameter 1, the corresponding UDG $G = (V, E)$ is defined as follows: each vertex $v_i \in V$ corresponds to a disk $d_i \in S$, and there is an edge between two vertices if and only if the Euclidean distance between the respective disk centers is at most 1.

Our interest in this problem arises from the following scenario. Consider a graph in which each node is a possible location for an intruder such as a thief, or a saboteur. We would like to detect and report the intruder's location in the graph. A protection device such as a camera, or a sensor placed at a node can not only detect (and report) the intruder's presence at it, but also at its neighbors. Our objective is to place a minimum number of protection devices such that the intrusion of the intruder at any vertex is detected and reported. In this situation, one must place the devices at the vertices of a minimum dominating set of the graph to achieve the goal. The protection devices are prone to failure and hence certain degree of redundancy is needed in the solution. Also, some times the devices may misreport the intruder's location deliberately or due to transmission error. Assume that at most one protection device in the closed neighborhood of the intruder can lie (misreport). In this context, one must place the protection devices at the vertices of an MLDS of the graph to achieve the objective. The first property in the definition of LDS deals with single device fault-tolerance, while the second property deals with the case in which two distinct locations about the intruder are reported.

3   Hardness of the MLDS Problem on UDGs

In this section, we show that the MLDS problem on UDGs is NP-complete by reducing the *vertex cover* problem defined on planar graphs to it, which is known to be NP-complete [3]. The decision versions of both problems are defined below.

The MLDS problem on UDGs (LDS-UDG)
Instance: A unit disk graph $G = (V, E)$ and a positive integer k.
Question: Does there exist a liar's dominating set L in G such that $|L| \leq k$?.

The vertex cover problem on planar graphs (VC-PLA)
Instance: An undirected planar graph G with maximum degree 3 and a positive integer k.
Question: Does there exist a vertex cover D of G such that $|D| \leq k$?.

Lemma 1 ([16]). *A planar graph $G = (V, E)$ with maximum degree 4 can be embedded in the plane using $O(|V|^2)$ area in such a way that its vertices are at integer coordinates and its edges are drawn so that they are made up of line segments of the form $x = i$ or $y = j$, for integers i and j.*

This kind of embedding is known as *orthogonal drawing* of a graph. Biedl and Kant [2] gave a linear time algorithm that produces an orthogonal drawing of a given graph with the property that the number of bends along each edge is at most 2 (see Fig. 1).

Corollary 1. *A planar graph $G = (V, E)$ with maximum degree 3 and $|E| \geq 2$ can be embedded in the plane such that its vertices are at $(4i, 4j)$ and its edges are drawn as a sequence of consecutive line segments on the lines $x = 4i$ or $y = 4j$, for integers i and j.*

Lemma 2. *Let $G = (V, E)$ be an instance of VC-PLA with $|E| \geq 2$. An instance $G' = (V', E')$ of LDS-UDG can be constructed from G in polynomial-time.*

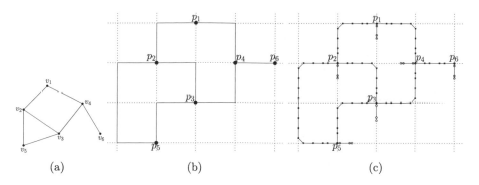

Fig. 1. (a) A planar graph G with maximum degree 3, (b) its embedding on a grid, and (c) a UDG construction from the embedding.

Proof. We construct G' in four phases.

Phase 1: Embedding of G into a grid of size 4×4

Embed the instance G in the plane as discussed previously using one of the algorithms in [4,5]. An edge in the embedding is a sequence of connected line segment(s) of length four units each. If the total number of line segments used in the embedding is ℓ, then the sum of the lengths of the line segments is 4ℓ as each line segment has length 4 units. We name the points in the embedding correspond to the vertices of G by *node points*.

Phase 2: Adding extra points to the embedding

Divide the set of line segments in the embedding into two categories, namely, proper and improper. We call a line segment *proper* if none of its end points correspond to a vertex in G. A line segment is *improper* if it is not a proper segment. For each edge (p_i, p_j) of length 4 units we add two points at distances 1 and 1.5 units of p_i and p_j, respectively (thus adding four points in total, see the edge (p_4, p_6) in Fig. 1(c)). For each edge of length greater than 4 units, we also add points as follows: for each improper line segment we add four points at distances 1, 1.5, 2.5, and 3.5 units from the end point corresponding to a vertex in G, and for each proper line segment we add four points at distances 0.5 and 1.5 units from its end points (see Fig. 1(c)). We name the points added in this phase *joint points*.

Phase 3: Adding extra line segments and points

Add a line segment of length 1.4 units (on the lines $x = 4i$ or $y = 4j$ for some integers i or j) for every point p_i, which corresponds to a vertex v_i in G, without coinciding with the line segments that had already been drawn. Observe that adding this line segment on the lines $x = 4i$ or $y = 4j$ is possible without losing the planarity as the maximum degree of G is 3. Now, add three points (say x_i, y_i, and z_i) on these line segments at distances 0.2, 1.2, and 1.4 units, respectively, from p_i. We name the points added in this phase *support points*.

Phase 4: Construction of UDG

For convenience, let us denote the set of node points, joint points, and support points by N, J, and S, respectively. Let $N = \{p_i \mid v_i \in V\}$, $J = \{q_1, q_2, \ldots, q_{4\ell}\}$, and $S = \{x_i, y_i, z_i \mid v_i \in V\}$. We construct a UDG $G' = (V', E')$, where $V' = N \cup J \cup S$ and there is an edge between two points in V' if and only if the Euclidean distance between the points is at most 1 (see Fig. 1(c)). Observe that, $|N| = |V|(= n)$, $|J| = 4\ell$, where ℓ is the total number of line segments in the embedding, and $|S| = 3|V|(= 3n)$. Hence, $|V'| = 4(n + \ell)$ and ℓ is bounded by a polynomial of n. Therefore G' can be constructed in polynomial-time.

Theorem 1. LDS-UDG *is NP-complete.*

Proof. For any given set $L \subseteq V$ and a positive integer k, we can verify whether L is a liar's dominating set of size at most k or not in polynomial-time.

   We prove the hardness of LDS-UDG by reducing VC-PLA to it. Let $G = (V, E)$ be an instance of VC-PLA. Construct an instance $G' = (V', E')$ of LDS-UDG as discussed in Lemma 2. We now prove the following claim: *G has a vertex cover of size at most k if and only if G' has a liar's dominating set of size at most $k + 3\ell + 3n$.*

Necessity: Let $D \subseteq V$ be a vertex cover of G such that $|D| \leq k$. Let $N' = \{p_i \in N \mid v_i \in D\}$, i.e., N' is the set of vertices in G' that correspond to the vertices in D. From each segment in the embedding we choose 3 vertices. The set of chosen vertices, say $J'(\subseteq J)$, together with N' and S will form an LDS of desired cardinality in G'. We now discuss the process of obtaining the set J'. Initially $J' = \emptyset$. As D is a vertex cover, every edge in G has at least one of its end vertices in D. Let (v_i, v_j) be an edge in G and $v_i \in D$ (the tie can be broken arbitrarily if both v_i and v_j are in D). Note that the edge (v_i, v_j) is represented as a sequence of line segments in the embedding. Start traversing the segments (of (v_i, v_j)) from p_i, where p_i corresponds to v_i, and add all the vertices to J' except the first one from each segment encountered in the traversal (see (p_2, p_5) in Fig. 2b. The big vertices are part of J' while traversing from p_2).

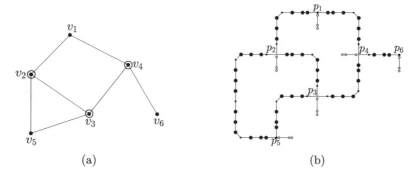

(a)   (b)

Fig. 2. (a) A vertex cover $\{v_2, v_3, v_4\}$ in G, and (b) the construction of J' in G' (the tie between v_2 and v_3, and v_3 and v_4 is broken by choosing v_3)

Apply the above process to each edge in G. Observe that the cardinality of J' is 3ℓ as we have chosen 3 vertices from each segment in the embedding. Let $L = N' \cup J' \cup S$. Now, we argue that L is a liar's dominating set in G'.

1. Each $p_i \in N$ is dominated by x_i in S. If $p_i \in N'$ (i.e., the corresponding vertex $v_i \in D$ in G), then $|N_{G'}[p_i] \cap L| \geq |\{p_i, x_i\}| = 2$. If $p_i \notin N'$, then there must exist at least one vertex q_j in J' dominating p_i. The existence of q_j is guaranteed by the way we constructed J'. Hence, $|N_{G'}[p_i] \cap L| \geq |\{q_j, x_i\}| = 2$. In either case every vertex in N is dominated by at least two vertices in L. It is needless to say that vertex in J is dominated by at least two vertices in $N' \cup J'$. Similarly, every vertex in S is dominated by itself, by its neighbor(s) in S, and, perhaps, by one vertex in N'. Therefore, every vertex in V' is double dominated by vertices in L.

2. Consider a pair of distinct vertices in V'. Of course, every pair of distinct vertices in S satisfy the liar's second condition. We prove that remaining pairs of distinct vertices also satisfy the liar's second condition by considering all possible cases.

   Case a. $p_i, p_j \in N$: If at least one of p_i, p_j belongs to N' (without loss of generality say $p_i \in N'$), then $|(N_{G'}[p_i] \cup N_{G'}[p_j]) \cap L| \geq |\{x_i, x_j, p_i\}| = 3$. If none of p_i, p_j belongs to N', then there must exists some $q_i, q_j \in J'$ such that q_i, q_j dominate p_i, p_j, respectively. Hence, $|(N_{G'}[p_i] \cup N_{G'}[p_j]) \cap L| \geq |\{x_i, x_j, q_i, q_j\}| = 4$.

   Case b. $q_i, q_j \in J$: If both $q_i, q_j \in J'$, then it is trivial that $|(N_{G'}[q_i] \cup N_{G'}[q_j]) \cap L| \geq 3$. Suppose one of q_i, q_j belongs to J' (without loss of generality say $q_i \in J'$). As every vertex in G' is double dominated, q_j must be dominated by two vertices in J' or by either some q_k in J' and some p_l in N'. In either case we get $|(N_{G'}[q_i] \cup N_{G'}[q_j]) \cap L| \geq 3$. A similar argument works even if none of q_i, q_j belong to J'.

   Case c. $p_i \in N$ **and** $q_j \in J$: If none of p_i and q_j belong to L, then the argument is trivial as each one is dominated by at least two vertices in L. If both belong to L, then $|(N_{G'}[p_i] \cup N_{G'}[q_j]) \cap L| \geq |\{p_i, x_i, q_j\}| = 3$. If $p_i \in L$ and $q_j \notin L$ (the other case is similar), then $|(N_{G'}[p_i] \cup N_{G'}[q_j]) \cap L| \geq 3$ holds as q_j is double dominated.

   Likewise, we can argue for other pair combinations too. Therefore, every pair of distinct vertices in V' is dominated by at least 3 vertices in L.

Thus, we have L is an LDS in G' and $|L| = |N'| + |J'| + |S| \leq k + 3\ell + 3n$.

Sufficiency: Let $L \subseteq V'$ be an LDS of size at most $k + 3\ell + 3n$. We prove that G has a vertex cover of size at most k with the aid of the following claims: (i) $S \subset L$, (ii) Every segment in the embedding must contribute at least 3 vertices to L and hence $|J \cap L| \geq 3\ell$, where ℓ is the total number of segments in the embedding, and (iii) If p_i and p_j correspond to end vertices of an edge (v_i, v_j) in G, and both p_i, p_j are not in L, then there must be at least $3\ell' + 1$ vertices in L form the segment(s) representing the edge (v_i, v_j), where ℓ' is the number of segments representing the edge (v_i, v_j) in the embedding.

We shall show that, by removing and/or replacing some vertices in L, a set of k vertices from N can be chosen such that the corresponding vertices in G is a vertex cover. The vertices in S account for $3n$ vertices in L (due to Claim (i)). Let $L = L \setminus S$ and $D = \{v_i \in V \mid p_i \in L \cap N\}$. If any edge (v_i, v_j) in G has none of its end vertices in D, then we do the following: consider the sequence of segments representing the edge (v_i, v_j) in the embedding. Since, both p_i and p_j are not in L, there must exist a segment having all its vertices in L (due to Claim (iii)). Consider the segment having its four vertices in L. Delete any one of the vertices on the segment and introduce p_i (or p_j). Update D and repeat the process till every edge has at least one of its end vertices in D. (due to Claim (ii)) D is a vertex cover in G with $|D| \leq k$. Therefore, LDS-UDG is NP-complete.

4   Approximation Scheme

We propose a PTAS for the MLDS problem on UDGs, i.e., for a given UDG $G = (V, E)$ and a parameter $\epsilon > 0$, we propose an algorithm which produces a liar's dominating set of size no more than $(1 + \epsilon)$ times the size of a minimum liar's dominating set in G. We use $\delta_G(u, v)$ to denote the number of edges on a shortest path between u and v in G. For $A, B \subseteq V$, $\delta_G(A, B)$ denotes the distance between A and B and is defined as $\delta_G(A, B) = \min_{u \in A, v \in B}\{\delta_G(u, v)\}$. For $A \subseteq V$, $LD(A)$ and $LD_{opt}(A)$ denote an LDS and an optimal (minimum size) LDS of A in G, respectively. We define the closed neighborhood of a set $A \subseteq V$ as $N_G[A] = \bigcup_{v \in A} N_G[v]$.

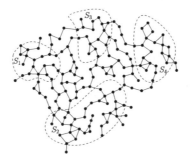

Fig. 3. A 4-separated collection $S = \{S_1, S_2, S_3, S_4\}$

The proposed PTAS is based on the concept of m-separated collection of subsets of V ($m \geq 4$). Let $G = (V, E)$ be a UDG. A collection $S = \{S_1, S_2, \ldots, S_k\}$ such that $S_i \subseteq V$ for $i = 1, 2, \ldots, k$, is said to be an m-separated collection, if $\delta_G(S_i, S_j) > m$, for $1 \leq i \leq k$ and $1 \leq j \leq k$ (see Fig. 3 for a 4-separated collection). Nieberg and Hurink [7] considered 2-separated collection to propose a PTAS for the minimum dominating set problem on unit disk graphs.

Lemma 3. *Let $S = \{S_1, S_2, \ldots, S_k\}$ be an m-separated collection. If $|S_i| \geq 3$ for $1 \leq i \leq k$, then $\sum_{i=1}^{k} |LD_{opt}(S_i)| \leq |LD_{opt}(V)|$.*

Proof. Observe that $N_G[S_i] \cap N_G[S_j] = \emptyset$ for $i \neq j$ and $1 \leq i, j \leq k$. Also, $LD_{opt}(S_i) \cap LD_{opt}(S_j) = \emptyset$ as S_i and S_j are m-separated. Let $S_i' = \{u \in V \mid v \in S_i$ and $\delta_G(u, v) \leq 2\}$ for $i = 1, 2, \ldots, k$. Observe that $S_i \subseteq S_i'$ and $S_i' \cap LD_{opt}(V)$ is a liar's dominating set of S_i for $i = 1, 2, \ldots, k$. Since, $\delta_G(S_i, S_j) > m(\geq 4)$ for $i \neq j$, implies $S_i' \cap S_j' = \emptyset$. Therefore, $(S_i' \cap LD_{opt}(V)) \cap (S_j' \cap LD_{opt}(V)) = \emptyset$ and $\bigcup_{i=1}^{k}(S_i' \cap LD_{opt}(V)) \subseteq LD_{opt}(V)$. Also, $LD_{opt}(S_i) \subseteq S_i' \cap LD_{opt}(V)$ for $i = 1, 2, \ldots, k$, $S_i' \cap LD_{opt}(V)$ is a liar's dominating set of S_i, and $LD_{opt}(S_i)$ is a minimum size liar's dominating set. Thus, $\bigcup_{i=1}^{k} LD_{opt}(S_i) \subseteq \bigcup_{i=1}^{k}(S_i' \cap LD_{opt}(V)) \subseteq LD_{opt}(V)$. Hence, the result of the lemma follows.

Lemma 4. *Let* $S = \{S_1, S_2, \ldots, S_k\}$ *be an* m-*separated collection, and* $N_1, N_2, \ldots, N_k$ *be subsets of* V *with* $S_i \subseteq N_i$ *for all* $i = 1, 2, \ldots, k$. *If there exists* $\rho \geq 1$ *such that* $|LD_{opt}(N_i)| \leq \rho|LD_{opt}(S_i)|$ *holds for all* $i = 1, 2, \ldots, k$, *and if* $\bigcup_{i=1}^{k} LD_{opt}(N_i)$ *is a liar's dominating set in* G, *then the value of* $\sum_{i=1}^{k} |LD_{opt}(N_i)|$ *is at most* ρ *times the size of a minimum liar's dominating set in* G.

In the proposed PTAS we have chosen the concept of 4-separated collection to make $LD_{opt}(N_i) \cap LD_{opt}(N_j) = \emptyset$ for $i \neq j$.

4.1   Algorithm

In this section, we discuss the construction of a 4-separated collection $S = \{S_1, S_2, \ldots, S_k\}$ and subsets $N_1, N_2, \ldots, N_k$ of V such that $S_i \subseteq N_i$ for all $i = 1, 2, \ldots, k$. The algorithm proceeds in an iterative manner. Initially $V_1 = V$. In the i-th iteration the algorithm computes S_i and N_i. For a given $\epsilon > 0$, the i-th iteration of the algorithm starts with an arbitrary vertex $v \in V_i$ and increases the value of $r(= 1, 2, \ldots)$ as long as $|LD(N_G^{r+4}[v])| > \rho|LD(N_G^r[v])|$ holds. Here, $LD(N_G^{r+4}[v])$ and $LD(N_G^r[v])$ are liar's dominating sets of $N_G^{r+4}[v]$ and $N_G^r[v]$, respectively, and $\rho = 1 + \epsilon$. The smallest r violating the above condition, say $\hat{r}$, is obtained. Set $S_i = N_G^{\hat{r}}[v]$ and $U_i = N_G^{\hat{r}+4}[v]$. Now, the removal of U_i from V_i may lead to some isolated (i) vertex $u \in V_i$, and/or (ii) connected component with two vertices $u, w \in V_i$. In case (i), for each such vertex u find $x, y \in U_i$ such that $\{u, x, y\}$ forms a connected component and update U_i as follows: $U_i = U_i \setminus \{x, y\}$. In case (ii), for each such pair of vertices u, w find $x \in U_i$ such that $\{u, w, x\}$ forms a connected component and update U_i as follows: $U_i = U_i \setminus \{x\}$. Set $N_i = U_i$ and $V_{i+1} = V_i \setminus N_i$. The process stops if $V_{i+1} = \emptyset$ and returns the sets S_is and N_is. The collection of the sets S_is is a 4-separated collection. The pseudo code is given in Algorithm 1.

The liar's dominating set of a r-th neighborhood of a vertex v, $LD(N_G^r[v])$, can be computed with respect to G as follows. We successively find maximal independent sets I_1, I_2 and I_3 such that $I_1 \cap I_2 \cap I_3 = \emptyset$. Now $I_1 \cup I_2 \cup I_3$ is a liar's dominating set for $N_G^r[v] \setminus I_1$ as every vertex not in I_1 either belongs to $I_2 \cup I_3$ or is adjacent to at least one vertex in each I_1, I_2, and I_3. To ensure the liar's domination conditions for the vertices in I_1, for each vertex u in I_1 we add two arbitrary vertices from the neighborhood of u, if they exist. If u

Algorithm 1. Liar's dominating set

Require: An undirected graph $G = (V, E)$ with $|V| \geq 3$ and an arbitrary small $\epsilon > 0$
Ensure: A liar's dominating set $\mathcal{D}$ of V

1: $i \leftarrow 0$ and $V_{i+1} \leftarrow V$
2: $\mathcal{D} \leftarrow \emptyset$ and $\rho \leftarrow 1 + \epsilon$
3: **while** $(V_{i+1} \neq \emptyset)$ **do**
4:     pick an arbitrary $v \in V_{i+1}$
5:     $N^0[v] \leftarrow v$ and $r \leftarrow 1$
6:     **while** $|(LD(N_G^{r+4}[v])| > \rho|LD(N_G^r[v])|$ **do**   ▷ call Algorithm 2
7:         $r \leftarrow r + 1$
8:     $\hat{r} \leftarrow r$ ▷ the smallest r violating while condition in step 6
9:     $i \leftarrow i + 1$ ▷ the index i keeps track of the number of iterations
10:    $S_i \leftarrow N_G^{\hat{r}}[v]$ and $U_i \leftarrow N_G^{\hat{r}+4}[v]$
11:    **if** $(V_{i+1} \setminus N_G^{\hat{r}+4}[v]$ contains isolated components of size 1 and/or 2) **then**
12:        **for** (each component, $\{u\}$, of size 1) **do**
13:            find $x, y \in U_i$ such that $\{u, x, y\}$ is a connected component
14:            $U_i \leftarrow U_i \setminus \{x, y\}$
15:        **for** (each component, $\{u, w\}$, of size 2) **do**
16:            find $x \in U_i$ such that $\{u, w, x\}$ is a connected component
17:            $U_i \leftarrow U_i \setminus \{x\}$
18:    $N_i \leftarrow U_i$
19:    $\mathcal{D} \leftarrow \mathcal{D} \cup LD(N_i)$ ▷ call Algorithm 2
20:    $V_{i+1} \leftarrow V_i \setminus N_i$
21: **return** $\mathcal{D}$

has only one neighbor, say u', then we add u' and one of its neighbors in the solution. The pseudo code is given in Algorithm 2. In summary, Algorithm 1 deals with obtaining an m-separated collection $S = \{S_1, S_2, \ldots, S_k\}$ and collection $N = \{N_1, N_2, \ldots, N_k\}$ such that $S_i \subseteq N_i \subseteq V$ and using Algorithm 2 (that deals with obtaining a liar's dominating set of the r-th neighborhood of a vertex) it computes a liar's dominating set for G.

Lemma 5. $LD(N_G^r[v])$ returned by Algorithm 2 is an LDS of $N_G^r[v]$ in G.

Proof. Let $u \in V$. We prove the first condition of liar's domination in the following cases.
Case 1. $u \in N_G^r[v] \setminus (I_1 \cup I_2 \cup I_3)$
Observe that $I_1 \cap I_2 \cap I_3 = \emptyset$. The vertex u is dominated by at least three vertices as in each round u is dominated by at least one vertex (see **for** loop in line number 2 in Algorithm 2). Thus, $|N_G[u] \cap LD(N_G^r[v])| \geq 3$.
Case 2. $u \in I_1 \cup I_2 \cup I_3$
We consider the following two sub-cases: (a) $u \in I_2 \cup I_3$: In this case $|N_G[u] \cap LD(N_G^r[v])| \geq 2$ holds as every vertex in $I_2 \cup I_3$ has at least one neighbor in I_1. (b) $u \in I_1$: For every vertex u in I_1, $LD(N_G^r[v])$ contains a neighbor of u or u's neighbor's neighbor (see line number 6 in Algorithm 2). Hence, $|N_G[u] \cap LD(N_G^r[v])| \geq 2$ holds.

Algorithm 2. Liar's dominating set in r-th neighborhood

Require: The r-th neighborhood of a vertex v i.e., $N_G^r[v]$
Ensure: A liar's dominating set $LD(N_G^r[v])$ of $N_G^r[v]$
1: $j \leftarrow 0$, $I_j \leftarrow \emptyset$, $LD(N_G^r[v]) \leftarrow \emptyset$ and $N \leftarrow N_G^r[v]$
2: **for** ($j = 1$ to 3) **do**
3:     **if** ($N \neq \emptyset$) **then**
4:         $I_j \leftarrow MIS(N \setminus I_{j-1})$ ▷ $MIS(\cdot)$ returns a maximal independent set
5:         $LD(N_G^r[v]) \leftarrow LD(N_G^r[v]) \cup I_j$; $N \leftarrow N \setminus I_j$
6: **for** every $u \in I_1$ **do**
7:     **if** ($|N_G[u]| > 2$) **then**
8:         add two arbitrary vertices from $N_G[u]$ to $LD(N_G^r[v])$
9:     **else**
10:        add w and its neighbor to $LD(N_G^r[v])$ ▷ w is the neighbor of u
11: **return** $LD(N_G^r[v])$

Now we prove the second condition of liar's domination. Let $u, w \in N_G^r[v]$. We prove the inequality $|(N_G[u] \cup N_G[w]) \cap LD(N_G^r[v])| \geq 3$ by considering the following cases.

Case 1. $u, w \in N_G^r[v] \setminus (I_1 \cup I_2 \cup I_3)$: Observe that both u and w have a neighbor in each I_1, I_2 and I_3. Hence $|(N_G[u] \cup N_G[w]) \cap LD(N_G^r[v])| \geq 3$.

Case 2. $u \in N_G^r[v] \setminus (I_1 \cup I_2 \cup I_3)$ and $w \in I_1 \cup I_2 \cup I_3$ (the other case proof is similar): In this case, the inequality $|(N_G[u] \cup N_G[w]) \cap LD(N_G^r[v])| \geq 3$ is true as $|N_G[u] \cap LD(N_G^r[v])| \geq 3$ and $|N_G[w] \cap LD(N_G^r[v])| \geq 2$.

Case 3. $u, w \in I_1 \cup I_2 \cup I_3$: The inequality $|(N_G[u] \cup N_G[w]) \cap LD(N_G^r[v])| \geq 3$ holds in this case as $u, w \in LD(N_G^r[v])$ and each of them has a neighbor (different from w, u, respectively) in $LD(N_G^r[v])$. This completes the proof of the lemma.

Lemma 6. $|LD(N_G^r[v])| \leq O(r^2)$.

Proof. Algorithm 2 computes $LD(N_G^r[v])$ by first computing maximal independent sets I_1, I_2, and I_3 subsequently. After computing I_1, I_2, and I_3, the algorithm adds two vertices for each vertex in I_1 to ensure that $LD(N_G^r[v])$ is a feasible solution. Without loss of generality we assume that $|I_1| \geq |I_2|, |I_3|$. Hence, $|LD(N_G^r[v])| = 3|I_1| + |I_2| + |I_3| \leq 5 \times |I_1| \leq 5 \times \frac{\pi(r+1)^2}{\pi(1)^2} = O(r^2)$. The latter inequality follows from the standard area argument, the number of non-intersecting unit disks can be packed in a larger disk of radius $r + 1$ centered at v.

Lemma 7. *In each iteration of Algorithm 1, there exists an r violating the condition $|(LD(N_G^{r+4}[v])| > \rho|LD(N_G^r[v])|$, where $\rho = 1 + \epsilon$.*

Proof. We prove the lemma by contradiction. Suppose there exists $v \in V$ such that $|(LD(N_G^{r+4}[v])| > \rho|LD(N_G^r[v])|$ for $r = 1, 2, \ldots$. Observe that $|LD(N_G^2[v])| \geq 3$ as there exists at least three vertices in G.
If r is even,

$(r+5)^2 \geq |(LD(N_G^{r+4}[v])| > \rho|LD(N_G^r[v])|) > \cdots > \rho^{\frac{r}{2}}|LD(N_G^2[v])| \geq 3\rho^{\frac{r}{2}}$, and
if r is odd,
$(r+5)^2 \geq |(LD(N_G^{r+4}[v])| > \rho|LD(N_G^r[v])|) > \cdots > \rho^{\frac{r-1}{2}}|LD(N_G^3[v])| \geq 3\rho^{\frac{r-1}{2}}$.
In both the cases the first inequality follows from Lemma 6. Hence,

$$(r+5) > \begin{cases} (\sqrt{\rho})^r, & \text{if } r \text{ is even} \\ (\sqrt{\rho})^{r-1}, & \text{if } r \text{ is odd} \end{cases} \tag{1}$$

The right hand part in inequality (1) is an exponential function in r and the left hand part is a polynomial in r, for arbitrarily large r none of the inequalities can be true. Hence we arrived at contradiction. Thus there exists an r violating the condition.

Lemma 8. *The smallest r violating inequality (1) is a constant and is bounded by $O(\frac{1}{\epsilon}\log\frac{1}{\epsilon})$.*

Proof. Lemma 7 suggests that the smallest r violating the condition cannot be sufficiently large but a constant. Let $\hat{r}$ be the smallest r violating the condition i.e., when $r = \hat{r}$ the inequalities in (1) are violated. Using the inequality $\log(1 + \epsilon) > \frac{\epsilon}{2}$ for $0 < \epsilon < 1$, we have $\hat{r} \leq O(\frac{1}{\epsilon}\log\frac{1}{\epsilon})$ as follows.

Consider the inequality $(1 + \epsilon)^x < x^2$ for a fixed $\epsilon > 0$. We prove that there exist ϵ' such that $(1 + \epsilon)^x < x^2 < (x + 5)^2$ for $0 < \epsilon < \epsilon'$. The latter inequality is trivial. Let $x = \frac{4}{\epsilon}\log\frac{1}{\epsilon}$. By taking the logarithm on both sides of the former inequality, we get, $\log 4 + \log\log\frac{1}{\epsilon} > 0$. Note that, we can always find an ϵ' such that $\log 4 + \log\log\frac{1}{\epsilon} > 0$ for $0 < \epsilon < \epsilon'$. Therefore, $(1 + \epsilon)^x < x^2$ holds for sufficiently smaller ϵ values and hence, $\hat{r} \leq \frac{4}{\epsilon}\log\frac{1}{\epsilon}$.

Lemma 9. *For a given $v \in V$, liar's dominating set $LD_{opt}(N_i)$ of N_i can be computed in polynomial time.*

Proof. Note that $N_i \subseteq N_G^{r+4}[v]$. The size of a liar's dominating set $LD(N_i)$ of N_i obtained by Algorithm 2 is bounded by $O(r^2)$ (by Lemma 6). Again, $r = O(\frac{1}{\epsilon}\log\frac{1}{\epsilon})$ by Lemma 8. Therefore, the size of the minimum size liar's dominating set $LD_{opt}(N_i)$ of N_i is bounded by a constant. The process of checking whether a given set is a liar's dominating set or not can be done in polynomial-time. Therefore, we can consider every subset of N_i as a possible liar's dominating set and check whether it is a liar's dominating set or not in polynomial-time. Finally, the minimum size liar's dominating set is reported. Thus the lemma.

Lemma 10. *For the collection of neighborhoods $\{N_1, N_2, \ldots, N_k\}$ created by Algorithm 1, the union $\mathcal{D} = \bigcup_{i=1}^{k} LD(N_i)$ is a liar's dominating set in G.*

Proof. We first prove that for every $v \in V, |N_G[v] \cap \mathcal{D}| \geq 2$. Observe that $\bigcup_{i=1}^{k} N_i = V$ as $V_{i+1} = V_i \setminus N_i$ and $N_i \subseteq V_i$. Thus, every vertex $v \in N_i$ for some $1 \leq i \leq k$. By Lemma 5, $|N_G[v] \cap \mathcal{D}| \geq 2$ is satisfied.

Now we prove the second condition. Consider two arbitrary vertices $u, v \in V$. The following cases may arise.

Case 1. $u, v \in N_i$ for some $1 \le i \le k$

Since $LD(N_i)$ is the liar's dominating set of N_i in G, we have, $|(N_G[u] \cup N_G[v]) \cap LD(N_i)| \ge 3$ for every $u, v \in N_i$. Hence, $|(N_G[u] \cup N_G[v]) \cap \mathcal{D}| \ge 3$ for every $u, v \in V$.

Case 2. $u \in N_i$ and $v \in N_j$ for some $i \ne j$ and $1 \le i, j \le k$

If u and v are not adjacent in G, the proof is trivial. Hence, we assume that $(u, v) \in E$ i.e., u and v are adjacent in G. Now the following sub-cases may arise.

(a) $u \in LD(N_i)$ and $v \in LD(N_j)$

Observe that $|N_G[u] \cap LD(N_i)| \ge 2$ and $|N_G[v] \cap LD(N_j)| \ge 2$ as $LD(N_i)$ and $LD(N_j)$ are liar's dominating sets of N_i and N_j, respectively. Hence, u has a neighbor, say w, in $LD(N_i)$, similarly v has also a neighbor, say x, in $LD(N_j)$. However, maybe $w = x$ or maybe not. In either case $|(N_G[u] \cup N_G[v]) \cap \mathcal{D}| \ge 3$ holds.

(b) $u \notin LD(N_i)$ and $v \in LD(N_j)$ (the other case proof similar)

Since $LD(N_i)$ is a liar's dominating set of N_i, we have $|N_G[u] \cap LD(N_i)| \ge 2$. Hence, $|(N_G[u] \cup N_G[v]) \cap \mathcal{D}| \ge 3$ is true as v is part of the solution.

(c) $u \notin LD(N_i)$ and $v \notin LD(N_j)$ (The proof is similar to the previous cases).

Corollary 2. *For the collection $N = \{N_1, N_2, \ldots, N_k\}$ created by Algorithm 1, the union $\mathcal{D}^* = \bigcup_{i=1}^{k} LD_{opt}(N_i)$ is a liar's dominating set.*

Theorem 2. *For a given UDG, $G = (V, E)$, and an $\epsilon > 0$, we can design a $(1 + \epsilon)$-factor approximation algorithm to find an LDS in G with running time $n^{O(c^2)}$, where $c = O(\frac{1}{\epsilon} \log \frac{1}{\epsilon})$.*

Proof. Note that Algorithm 1 generates the collection of sets $S = \{S_1, S_2, \ldots, S_k\}$ and $N = \{N_1, N_2, \ldots, N_k\}$ such that S is a 4-separated collection of V with $S_i \subseteq N_i$ for each $i \in \{1, 2, \ldots, k\}$ and $\bigcup_{i=1}^{k} N_i = V$ with $N_i \cap N_j = \emptyset$ for $i \ne j$. Corollary 2 suggests that $\mathcal{D}^* = \bigcup_{i=1}^{k} LD_{opt}(N_i)$ is a liar's dominating set of G. The approximation bound follows from Lemmas 3 and 4. Let $|N_i| = n_i$ for $1 \le i \le k$. By Lemma 9, an optimal liar's dominating set $LD_{opt}(N_i)$ of N_i can be computed in $n_i^{O(\frac{1}{\epsilon^2} \log \frac{1}{\epsilon})}$ time. Therefore, the total running time to compute $\mathcal{D}^*$ is $\sum_{i=1}^{k} n_i^{O(\frac{1}{\epsilon^2} \log \frac{1}{\epsilon})} \le n^{O(\frac{1}{\epsilon^2} \log \frac{1}{\epsilon})}$.

5   Conclusion

We studied the minimum liar's dominating set problem (MLDS) on unit disk graphs, showed that the MLDS problem is NP-complete and proposed a PTAS.

References

1. Alimadadi, A., Chellali, M., Mojdeh, D.A.: Liar's dominating sets in graphs. Discret. Appl. Math. **211**, 204–210 (2016)
2. Biedl, T., Kant, G.: A better heuristic for orthogonal graph drawings. Comput. Geom. **9**(3), 159–180 (1998)
3. Garey, M.R., Johnson, D.S.: Computers and Intractability: A Guide to the Theory of NP-completeness. Freeman, Dallas (1979)
4. Hopcroft, J., Tarjan, R.: Efficient planarity testing. J. ACM (JACM) **21**(4), 549–568 (1974)
5. Itai, A., Papadimitriou, C.H., Szwarcfiter, J.L.: Hamilton paths in grid graphs. SIAM J. Comput. **11**(4), 676–686 (1982)
6. Jallu, R.K., Das, G.K.: Liar's domination in 2D. In: Gaur, D., Narayanaswamy, N.S. (eds.) CALDAM 2017. LNCS, vol. 10156, pp. 219–229. Springer, Cham (2017). https://doi.org/10.1007/978-3-319-53007-9_20
7. Nieberg, T., Hurink, J.: A PTAS for the minimum dominating set problem in unit disk graphs. In: Erlebach, T., Persinao, G. (eds.) WAOA 2005. LNCS, vol. 3879, pp. 296–306. Springer, Heidelberg (2006). https://doi.org/10.1007/11671411_23
8. Panda, B.S., Paul, S.: Connected liar's domination in graphs: complexity and algorithms. Discret. Math. Algorithms Appl. **5**(04), 1350024 (2013)
9. Panda, B.S., Paul, S.: Liar's domination in graphs: complexity and algorithm. Discret. Appl. Math. **161**(7), 1085–1092 (2013)
10. Panda, B.S., Paul, S.: A linear time algorithm for liar's domination problem in proper interval graphs. Inf. Process. Lett. **113**(19), 815–822 (2013)
11. Panda, B.S., Paul, S.: Hardness results and approximation algorithm for total liar's domination in graphs. J. Comb. Optim. **27**(4), 643–662 (2014)
12. Panda, B.S., Paul, S., Pradhan, D.: Hardness results, approximation and exact algorithms for liar's domination problem in graphs. Theor. Comput. Sci. **573**, 26–42 (2015)
13. Roden, M.L., Slater, P.J.: Liar's domination in graphs. Discret. Math. **309**(19), 5884–5890 (2009)
14. Slater, P.J.: Liar's domination. Networks **54**(2), 70–74 (2009)
15. Sterling, C.: Liar's Domination in Grid Graphs. Ph.D. thesis, East Tennessee State University (2012)
16. Valiant, L.G.: Universality considerations in VLSI circuits. IEEE Trans. Comput. **100**(2), 135–140 (1981)

Minimum Spanning Tree of Line Segments

Sanjana Dey, Ramesh K. Jallu, and Subhas C. Nandy$^{(\boxtimes)}$

Indian Statistical Institute, Kolkata 700108, India
nandysc@isical.ac.in

Abstract. In this article, we study a variant of the geometric minimum spanning tree (MST) problem. Given a set $\mathcal{S}$ of n disjoint line segments in $I\!\!R^2$, we need to find a tree spanning one endpoint from each of the segments in $\mathcal{S}$. Note that, we have 2^n possible choices of such a set of endpoints, each being referred as an *instance*. Thus, our objective is to choose one among those instances such that the sum of the lengths of all the edges of the tree spanning the points of that instance is minimum. We show that finding such a spanning tree is NP-complete in general, and propose a $O(\log^2 n)$-factor approximation algorithm for the same.

Keywords: Minimum spanning tree · k-MST
Approximation algorithm · NP-complete

1 Introduction

The minimum spanning tree (MST) is well studied in both graph and geometric domain. In the context of an edge-weighted graph G, the objective is to find a tree spanning all the nodes in G such that the sum of the weights of all the tree edges is minimized. In the geometric setup, the nodes of the underlying graph correspond to a given set of objects, the graph is complete, and each edge of the graph is the distance (in some appropriate metric) between the objects corresponding to the nodes incident to that edge. If the objects are points in $I\!\!R^d$ and the distances are in the Euclidean metric, the problem is referred to as *Euclidean MST*. In $I\!\!R^2$ and $I\!\!R^3$, the problem can be solved in $O(n \log n)$ and $O((n \log n)^{4/3})$ time, respectively. In $I\!\!R^d$, the best-known algorithm runs in sub-quadratic time [1]. Using a well-separated pair decomposition, it is possible to produce a $(1 + \epsilon)$-approximation of the Euclidean MST in $I\!\!R^d$ ($d \geq 2$) in $O(n \log n)$ time [22]. For a survey on the Euclidean spanning tree of a point set in $I\!\!R^d$, refer to [7,11].

Several variations of minimum spanning tree problem are studied in the literature. In the k-MST problem, an edge-weighted graph with n vertices is given, and the objective is to compute a tree of minimum weight that spans at least k vertices. The problem is well studied in graphs as well as in the metric space. In the latter case, the underlying graph is a complete graph induced by a set of

© Springer International Publishing AG, part of Springer Nature 2018
L. Wang and D. Zhu (Eds.): COCOON 2018, LNCS 10976, pp. 529–541, 2018.
https://doi.org/10.1007/978-3-319-94776-1_44

points in the plane and the weight of an edge is the Euclidean distance between the pair of points defining that edge. For weighted undirected graph, the k-MST problem is NP-hard [23] for arbitrary k. A $O(\sqrt{k})$-factor approximation algorithm is proposed by Ravi et al. [21]. Later it was improved to $O(\log^2 k)$ [3], and then to $O(\log k)$ [19]. It is also shown that improving the approximation factor beyond $\frac{96}{95}$ is NP-hard [9]. The best known approximation result is 2 [14].

For the geometric k-MST problem, Ravi et al. [21] proposed a $O(k^{\frac{1}{4}})$-factor approximation algorithm, which is improved to $O(\log k)$ [15]. Blum et al. [4] gave a $O(1)$ approximation result where the constant involved in the O-notation is high. Mitchell [18] improved this factor to $2\sqrt{2}$. Finally, Arora [2] presented a polynomial time approximation scheme for this problem.

The MST problem for colored points is also studied in the literature. For a given set of n red and n blue points, the problem of computing an MST of $2n$ points whose edges are non-crossing, and each edge joins a red and a blue point, is studied in [5]. The problem is NP-hard. However, (i) if the points are in convex position, then the optimal tree can be obtained in $O(n^3)$ time, and (ii) if number of points of one color is bounded by a constant k, then the optimal tree can be computed in $n^{O(k^5)}$ time. Finally, they proposed a $O(\sqrt{n})$-factor approximation algorithm for the general problem. In [10] the MST problem with multicolor point set is studied, where n points of m different colors ($m \le n$) are given. The objective is to select m different colored points such that (i) the total edge length of spanning tree of those m points is minimum, (ii) the total edge length of a minimal spanning tree of those m points is as large as possible, and (iii) the perimeter of the convex hull of m different colored points is as small as possible. All the three problems are shown to be NP-complete. For problem (iii), the authors have proposed a $\sqrt{2}$-factor approximation algorithm.

The study on the Euclidean MST problem, where the objects are different from points, is relatively less. Bose and Toussaint [8] first considered the MST problem with disjoint line segments. Here, a set S of n disjoint straight-line segments in $\mathbb{R}^2$ is given, and the objective is to compute a minimum spanning tree $\mathcal{T}$ on $2n$ endpoints of these segments where each line-segment in S appears as an edge of $\mathcal{T}$ and the tree has no crossing edges. This kind of tree is known as *encompassing spanning tree*. A minimum encompassing spanning tree for a set of disjoint line segments can be computed in $O(n \log n)$ time [8]. Later Bose et al. [6] proved that every set of disjoint line segments admits an encompassing binary tree, i.e., the maximum vertex degree of the resulting tree is bounded by 3. Hoffman and Tóth [17] showed that the segment endpoint visibility graph[1] of a set S of n segments is Hamiltonian. Hoffman et al. [16] proposed an optimal $O(n \log n)$ time algorithm to compute an encompassing spanning tree of maximum degree three, such that at every vertex v all the edges of the tree that are incident to v lie in a half-plane bounded by the line through the input segment of which v is an endpoint. This kind of tree is known as *pointed binary encom-*

---

[1] Its vertices are the $2n$ segment endpoints; two vertices a and b are connected by an edge, if and only if the corresponding line segment ab is either in S or if the open segment ab does not intersect any (closed) segment from S.

passing tree. Rappaport et al. [20] showed that in $O(n \log n)$ time it is possible to test whether the convex hull of a set S of n segments has at least one endpoint of each segment on its boundary (as a vertex).

In this paper, we concentrate on a different variation of the MST problem for a set S of non-crossing line segments in $\mathbb{R}^2$. Here, the objective is to find a tree of n nodes with a minimum sum of edge costs such that each of its nodes corresponds to an endpoint of n distinct members of S (see Fig. 1). From now on, we call this problem the *minimum spanning tree of segments* (MSTS) problem. Surely, if the segments in S are of length zero, then the MSTS problem reduces to the standard Euclidean MST problem. We show that in the non-degenerate case, i.e., where not all the segments are of length zero, the MSTS problem is NP-hard. We also propose a $O(\log^2 n)$-factor approximation algorithm that runs in polynomial time. To the best of our knowledge, this version of the MST problem for line segment objects has not been studied in the literature.

The proposed MSTS problem may find its application in VLSI physical design. In global routing, the pins of the same net are routed in a hierarchical manner. In an intermediate stage, the objective is to connect a number of already routed segments using wires of minimum total length. In this scenario, a routing wire connecting two segments a and b may choose any point on the two segments a and b for the connection. Here, we consider a simplified version of the problem, where the chosen point of connection of a segment is restricted to one of its endpoints only.

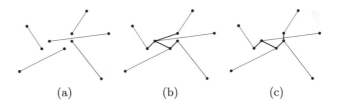

(a)   (b)   (c)

Fig. 1. (a) A set of line segments in the plane, (b) a spanning tree of the segments, and (c) a minimum spanning tree

2   The MSTS Problem is NP-hard

We prove the MSTS problem is NP-hard by a reduction from the MAX 2-SAT problem, defined below, and which is known to be NP-complete [13].

Instance: A boolean formula consisting of n variables and a set $\{C_1, C_2, \ldots, C_m\}$ of m clauses, each C_i is a disjunction of at most two literals, and an integer k, $1 \le k < m$.

Question: Is there a truth assignment to the variables that simultaneously satisfies at least k clauses?

Given an instance of MAX 2-SAT, we get an instance of the MSTS problem such that the given MAX 2-SAT formula satisfies k clauses if and only if the cost of the MSTS attains a specified value.

We represent a horizontal line segment s as $(l(s), r(s))$, where $l(s)$ and $r(s)$ are the left and right endpoints of s, respectively. Similarly, a vertical line segment s is represented as $(t(s), b(s))$, where $t(s)$ and $b(s)$ are the top and bottom endpoints of s, respectively. For a point p in the plane, we use $x(p)$ and $y(p)$ to denote the x and y-coordinate of p, respectively.

Our reduction is similar to [10], however, the gadgets we use in our reduction are entirely different. Let ψ be a 2-SAT formula having m clauses $C_1, C_2, \ldots, C_m$ and n variables $x_1, x_2, \ldots, x_n$. We use the following notation as used in [10]. Let $x_{i,j,k}$ (or $\neg x_{i,j,k}$) be the variable x_i that appears at the j-th literal in ψ from left to right such that x_i (including $\neg x_i$) appears $k-1$ times already in ψ before the current occurrence of x_i. For example in the boolean formula $(x_1 \vee \neg x_2) \wedge (x_2 \vee x_3)$, the literals $x_1, \neg x_2, x_2$, and x_3 are represented as $x_{1,1,1}, \neg x_{2,2,1}, x_{2,3,2}$, and $x_{3,4,1}$, respectively. We create gadgets for each variable x_i ($1 \le i \le n$) and for each literal $x_{i,j,k}$ (or $\neg x_{i,j,k}$). Each gadget consists of a set of horizontal/vertical line segments in the plane.

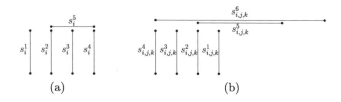

Fig. 2. (a) Variable gadget for x_i, and (b) literal gadget for $x_{i,j,k}$

Variable Gadget: For each variable x_i, five disjunct segments s_i^l ($1 \le l \le 5$) are considered. The first four are vertical and the last one is horizontal. The vertical segments are of equal length (say $\lambda > 0$) and their top endpoints are horizontally aligned with unit distance between consecutive endpoints. The horizontal segment s_i^5 spans as follows: $x(l(s_i^5)) = x(t(s_i^2)), y(l(s_i^5)) = y(t(s_i^2)) + \epsilon, x(r(s_i^5)) = x(t(s_i^4)), y(r(s_i^5)) = y(t(s_i^2)) + \epsilon$, where ϵ is a very small positive real number (see Fig. 2a).

Literal Gadget: For each literal $x_{i,j,k}$ (or $\neg x_{i,j,k}$), six disjunct segments $s_{i,j,k}^l$ ($1 \le l \le 6$) are considered. The first four segments are vertical, while the last two are horizontal (see Fig. 2b). Here also the vertical segments are of same length and their top endpoints are aligned as discussed in variable gadget. The two horizontal segments are above the vertical segments and are of different lengths. Their lengths depend on how many times its associated variable x_i appears in different clauses of ψ. The left endpoint of the horizontal segment $s_{i,j,k}^5$ (resp. $s_{i,j,k}^6$) is at $x(l(s_{i,j,k}^5)) = x(t(s_{i,j,k}^5)), y(l(s_{i,j,k}^5)) = y(t(s_{i,j,k}^5)) + \alpha$

(resp. $x(l(s^6_{i,j,k})) = x(t(s^4_{i,j,k})), y(l(s^6_{i,j,k})) = y(t(s^4_{i,j,k})) + \beta)$, where $\alpha = 2j\epsilon$ (resp. $\beta = (2j+1)\epsilon$). The two horizontal segments will be used to connect with the gadget of variable x_i. In the connection, the right endpoints of the horizontal segments are vertically aligned with s^2_i and s^4_i (see Fig. 3).

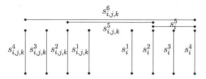

Fig. 3. Connecting $x_{i,j,k}$'s gadget to x_i's gadget

The basic idea of putting the segments in this manner is to get two minimum spanning trees of the segments for connecting the gadgets of x_i and $x_{i,j,k}$ depending on whether x_i is set to true or false respectively. If $x_i = false$, we choose the endpoints $F = \{l(s^6_{i,j,k}), t(s^4_{i,j,k}), t(s^3_{i,j,k}), t(s^2_{i,j,k}), t(s^1_{i,j,k}), t(s^1_i), t(s^2_i), t(s^3_i), t(s^4_i), l(s^5_i), r(s^5_{i,j,k})\}$, and if $x_i = true$, we choose the endpoints $T = \{t(s^4_{i,j,k}), t(s^3_{i,j,k}), t(s^2_{i,j,k}), t(s^1_{i,j,k}), t(s^1_i), t(s^2_i), t(s^3_i), t(s^4_i), l(s^5_{i,j,k}), r(s^5_i), r(s^6_{i,j,k})\}$. In Fig. 4 these two minimum spanning trees are shown using bold lines.

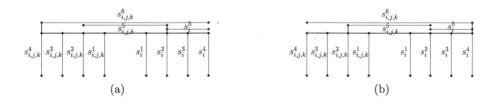

Fig. 4. (a) The MSTS when x_i is false, and (b) the MSTS when x_i true.

We now explain the arrangement of the gadgets according to the given formula ψ. The gadgets corresponding to the variables $x_1, x_2, \ldots, x_n$ are placed on the positive part of the x-axis in left to right order with the top endpoints of the vertical segments aligned with the x-axis. Similarly, the gadgets corresponding to the literals are arranged on the negative part of the x-axis from right to left (in the order they appear in ψ). Needless to mention, the y-coordinates of the top endpoint of the vertical segments (in both variable and literal gadgets) are zero. As mentioned earlier, the horizontal segments of a literal gadget establish a connection with its corresponding variable gadget[2]. Note that, the formula for the y-coordinates of the horizontal segments ensure that no two horizontal segments will overlap (refer Fig. 5 for the example $x_1 \vee x_2$).

---

[2] A variable gadget may be connected with multiple literal gadgets.

According to our construction, we get a set $\mathcal{S}$ of $5n + 12m$ segments ($5n$ segments corresponding to n variables, $12m$ segments corresponding to m clauses (since each clause contains two literals). Let $\mathcal{T}_{\mathcal{S}}$ be the MSTS over $\mathcal{S}$ and $\mathcal{W}(\mathcal{T}_{\mathcal{S}})$ be its weight. $\mathcal{T}_{\mathcal{S}}$ satisfies the following properties:

(i) In every (variable and literal) gadget, the top endpoints of all its vertical segments are part of it,

(ii) No edge in $\mathcal{T}_{\mathcal{S}}$ connects any two horizontal segments of literal gadgets,

(iii) No edge in $\mathcal{T}_{\mathcal{S}}$ connects horizontal segments of any two variable gadgets, and

(iv) The horizontal segments of any literal $x_{i,j,k}$ (or $\neg x_{i,j,k}$) gadget are either connected to its vertical segments or segments in its associated variable x_i's gadget using a vertical edge of $\mathcal{T}_{\mathcal{S}}$.

These properties justify the need of taking four equidistant vertical segments in each gadget as it prohibits from choosing non axis-parallel edges of $\mathcal{T}_{\mathcal{S}}$ to connect segments in the gadgets.

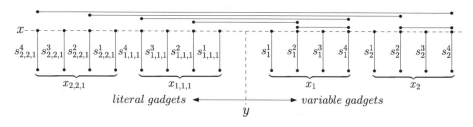

Fig. 5. Placement of variable and literal gadgets in the clause $x_1 \vee x_2$

So far, we have discussed the arrangement of variable and literal gadgets and the nature of the minimum spanning tree of those segments for a true/false assignment of the n variables. Now, we add one special horizontal segment[3] for each clause in ψ. These segments appear in the same vertical level in which the horizontal segments of the variable gadgets appear. Consider a clause $x_{i_1,j,k_1} \vee x_{i_2,j+1,k_2}$ in ψ. We put a segment $s^{\lceil \frac{j}{2} \rceil}$ that connects the gadgets corresponding to x_{i_1,j,k_1} and $x_{i_2,j+1,k_2}$, an endpoint representing a literal in the clause. Depending on whether the literal $x_{i,j,k}$ corresponds to the variable x_i or its negation $\neg x_i$, we put the corresponding endpoint of the segment $s^{\lceil \frac{j}{2} \rceil}$ on the top of $s^2_{i,j,k}$ or $s^4_{i,j,k}$, respectively. Figure 6(a) and (b) demonstrates the horizontal segments (with its endpoints represented as black squares) of clauses $x_1 \vee x_2$ and $x_1 \vee \neg x_2$, respectively. Figure 7 shows the MSTS instance for the formula $\psi = (\neg x_1 \vee x_2) \wedge (x_2 \vee \neg x_3)$. Let $\mathcal{S}' = \mathcal{S} \cup \{s^{\lceil \frac{j}{2} \rceil} \mid j = 1, 3, \ldots, 2m - 1\}$, and $\mathcal{T}$ be an MSTS of this extended set $\mathcal{S}'$ of $5n + 13m$ segments with weight $W(\mathcal{T})$.

---

[3] This segment corresponds to the binary relation *or*.

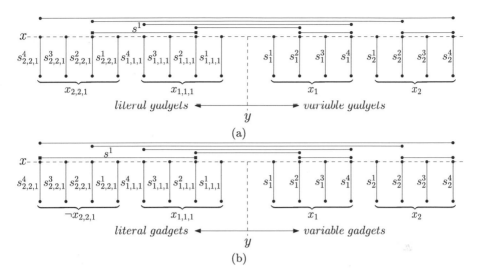

Fig. 6. Gadget for the clause (a) $x_1 \vee x_2$, and (b) $x_1 \vee \neg x_2$

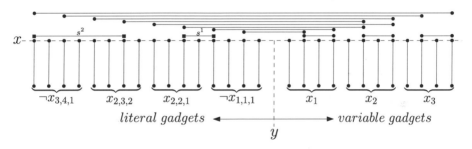

Fig. 7. Gadget for $\psi = (\neg x_1 \vee x_2) \wedge (x_2 \vee \neg x_3)$

Lemma 1. *From an MSTS $\mathcal{T}$ of $\mathcal{S}'$ one can get a conflict-free assignment of the variables.*

Proof. Without loss of generality, we assume that the horizontal segments of the literal and variable gadgets are connected to vertical segments in $\mathcal{T}$, in one of the ways shown in Fig. 4. If there is a connection of a pair of segments in $\mathcal{T}$ which is not in either of the forms as shown in Fig. 4 (see the curved edges in the left part of Fig. 8), then we can alter the tree without changing its weight (see the right part in Fig. 8).

For every variable x_i's gadget, we check which endpoint between $t(s_i^2)$ and $t(s_i^4)$ is chosen to connect to its horizontal segment s_i^5. Note that, both the endpoints cannot be connected to the horizontal segment simultaneously due to the feasibility[4] of the problem. By our construction of the gadgets, if $t(s_i^2)$ is connected then we set $x_i = false$, else set $x_i = true$. Also note that, if a

---

[4] Only one endpoint of a segment can participate in the tree.

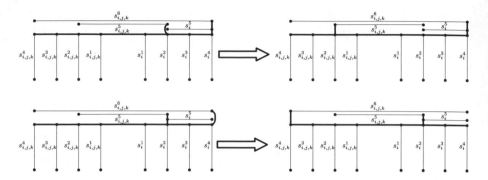

Fig. 8. The possible tree alterations in an MSTS

variable x_i gets an assignment it never changes. Because $\mathcal{T}$ is a spanning tree with minimum weight, if $t(s_i^2)$ (or $t(s_i^4)$) is chosen, then this choice forces to connect the horizontal segments' endpoints vertically aligned with it. Thus, we get a conflict-free assignment of the variables.

□

Lemma 2. *The* MAX 2-SAT *instance* ψ *satisfies at least* k *clauses simultaneously if and only if the weight of* $\mathcal{T}$ *is at most* $W(\mathcal{T}_S) + (m - k)\epsilon$, *where* m *is the number of clauses in* ψ, $\epsilon > 0$ *is a very small real number, and* $1 \leq k < m$.

Proof (**Necessity**). Let ψ satisfies at least k clauses simultaneously. For a literal $x_{i,j,k}$ (or $\neg x_{i,j,k}$), if variable x_i is true (or false), then T (or F) is chosen accordingly; this implies, one of the endpoints of the segment $s^{\lceil \frac{j}{2} \rceil}$ can be chosen to connect with $\mathcal{T}_S$ in MSTS $\mathcal{T}$ without adding any extra weight (see Fig. 9, where the blue edge implies $\neg x_1$ is *true*, i.e., x_1 is *false*). Therefore, if at least one literal is true in one clause, no extra weight will be added to $\mathcal{T}_S$ for that clause to get $W(\mathcal{T})$. If both the literals are false, then the extra weight of ϵ will be added to $W(\mathcal{T}_S)$. Therefore, if ψ satisfies at least k clauses simultaneously, then an extra weight of at most $(m - k) \times \epsilon$ will be added to $W(\mathcal{T}_S)$ (see the red edge in Fig. 9).

(**Sufficiency**). Let $\mathcal{T}$ be an MSTS over $\mathcal{S}'$ such that $W(\mathcal{T}) \leq W(\mathcal{T}_S) + (m - k)\epsilon$. We show that there is a truth assignment of variables such that at least k clauses are satisfied. In any gadget, all the top endpoints of its vertical segments are part of $\mathcal{T}$. Let in $\mathcal{T}$, the horizontal segments of literal and variable gadgets are connected to vertical segments in one of the ways as shown in Fig. 4 (if not, we can alter the tree as discussed in the proof of Lemma 1).

By Lemma 1, it is guaranteed that we can obtain a conflict-free assignment to the variables. Now consider the second term (i.e., $(m-k)\epsilon$) of $W(\mathcal{T})$. Each factor ϵ is due to the non-existence of an edge between some segment $s^{\lceil \frac{j}{2} \rceil}$ endpoints and any one of the horizontal segments' endpoints of the j-th and $(j+1)$-st literal gadgets. We assign *false* to both the literals in the clause associated with the

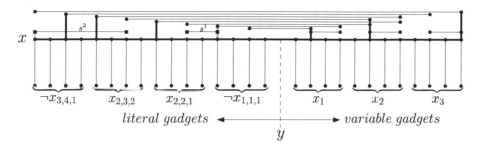

Fig. 9. The MSTS (shown in thick) obtained for the assignment $x_1 = \textit{false}, x_2 = \textit{false}$, and $x_3 = \textit{true}$

segment $s^{\lceil \frac{j}{2} \rceil}$, implying the clause corresponding to $s^{\lceil \frac{j}{2} \rceil}$ is not satisfied. Since at most $m - k$ clauses are not satisfied, the result follows.   □

Given a designated set of endpoints (one marked endpoint of each segment) of $\mathcal{S}$ and a parameter μ, in polynomial time it can be checked whether the sum of lengths of the edges of the MST of those points is less than or equal to μ. Thus, we have the following result:

Theorem 1. *The MSTS problem is NP-complete.*

Observe that, the MSTS problem remains NP-hard for the case where all the segments are horizontal. Here, the same reduction works by replacing each vertical segment by a segment of length zero.

3   Approximation Algorithm

We now propose a $O(\log^2 n)$-factor approximation algorithm for the MSTS problem. We use $d(p, q)$ to denote the Euclidean distance between the points p and q in the plane. The *lune* of two points p and q, denoted by $lune(p, q)$, is the intersection region of the two disks of radius $d(p, q)$ centered at p and q, respectively. Let $\mathcal{S} = \{s_1, s_2, \ldots, s_n\}$ be the given set of segments in the plane. The basic idea of our algorithm is as follows: for every pair of distinct segments s_i and s_j, we consider all four edges between the endpoints of s_i and s_j. Every edge, say (p_i, p_j), is processed separately by considering the lune formed by p_i and p_j. If $lune(p_i, p_j)$ contains at least one endpoint of each segment, then we judiciously choose n endpoints (one per segment) in $lune(p_i, p_j)$ having certain properties. Then we compute a minimum tree spanning those chosen points. We repeat the above process for all the four edges between s_i and s_j, and the spanning tree with the minimum cost is noted. Finally, we output a spanning tree with minimum cost considering all the pairs of segments. The pseudo-code of the algorithm is given in Algorithm 1.

Our algorithm finds a set of n endpoints that are close. The closeness of a point set is defined in terms of potential (see Definition 1) of the set. For a given point set P, potential on P is a function $f : 2^P \to \mathbb{R}^+$. Garg and Hochbaum [15] used the concept of potential to get a $O(\log k)$-factor approximation algorithm for the k-MST problem in the plane. We use their algorithm to propose a $O(\log^2 n)$-factor approximation result for our problem.

Let OPT be the cost of an optimal MSTS of $\mathcal{S}$, and P^* be the set of endpoints in the MSTS. Let $p_i, p_j \in P^*$ such that $d(p_i, p_j) = \delta = $ the diameter[5] of P^*. Note that all the points in P^* lie in $lune(p_i, p_j)$. If there exists a point $p_k \in P^*$, not in $lune(p_i, p_j)$, then either $d(p_i, p_k)$ or $d(p_j, p_k)$ is greater than δ, contradicting the fact that $d(p_i, p_j)$ is the diameter of the set P^*. Also, note that $OPT \geq \delta$.

Let S be the set of endpoints of the segments in $\mathcal{S}$ lying in $lune(p_i, p_j)$, and $P \subseteq S$ be the set of endpoints returned by our algorithm. Let Q be the smallest rectangle circumscribing $lune(p_i, p_j)$ (see Fig. 10). Needless to say, Q is of size $\sqrt{3}\delta \times \delta$. We will use some of the definitions and lemmas discussed in [15].

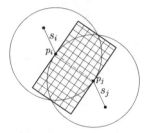

Fig. 10. The smallest rectangle circumscribing $lune(p_i, p_j)$.

Let G_0 be a grid having each cell of size $x_0 = \frac{\sqrt{3}\delta}{n}$. Let $G_1, \ldots, G_{\log n}$ be a family of grids defined on Q and constructed recursively such that each grid cell in G_i is composed of four cells in G_{i-1}. That is, if x_i is the size of a cell in G_i, then $x_i = 2x_{i-1} = 2^i x_0$. Since $x_{\log n} = nx_0$ for $i = \log n$, $G_{\log n}$ is the rectangle Q.

Definition 1. (a) The G_i-potential of a point set $P \subseteq S$, denoted by $G_i(P)$, is equal to $x_i \times t_i$, where t_i is the number of cells of G_i containing the points of P.

(b) The potential of a point set P is the sum of the G_i-potentials of P, i.e.,

$$\mathcal{P}(P) = \sum_{i=0}^{\log n - 1} G_i(P).$$

It is known that the potential of a point set is small if and only if the weight of the minimum spanning tree of the point set is small, and the authors used this fact to get a $O(\log k)$-factor approximation result for the k-MST problem. For our problem, the following two lemmas can be proved as in [15] using the fact that $\delta \leq OPT$.

Lemma 3 *[15]. The minimum tree spanning a set of points P has weight at most $\sqrt{2}\mathcal{P}(P)$.*

---

[5] The maximum possible distance between a pair of points.

Algorithm 1. Approximation algorithm

Require: A set $S = \{s_1, s_2, \ldots, s_n\}$ of line segments in the plane
Ensure: A spanning tree of S
 1: **for** each distinct pair of segments s_i and s_j in S **do**
 2:     **for** each edge (p_i, p_j) between s_i and s_j **do**
 3:         let $lune(p_i, p_j)$ be the lune of p_i and p_j
 4:         **if** ($lune(p_i, p_j)$ contains at least one endpoint from each segment) **then**
 5:             let Q be the smallest rectangle circumscribing the lune
 6:             Compute the grids $G_0, G_1, \ldots, G_{\log n}$ defined on Q as described earlier
 7:             let $l = n$ and $P \leftarrow \emptyset$
 8:             **while** ($l\ != 0$) **do**
 9:                 let S be the set of endpoints of the segments in S lying in $lune(p_i, p_j)$
10:                 $P' = \text{Min\_Potential}(S, G_0, G_1, \ldots, G_{\log n}, l)$
11:                                    ▷ $P' \subseteq S$ is of size l with minimum potential in Q [15]
12:                 **while** (there is a segment having both its endpoints in P') **do**
13:                     remove any one of its endpoints from P'
14:                 $P = P \cup P'$
15:                 remove every segment in S having at least one endpoint in P
16:                 $l = l - |P|$
17:             compute the MST of P
18: **return** spanning tree with the minimum cost and the set P corresponding to it

---

Lemma 4 $\mathcal{P}(P^*) \leq 8 \log n \times OPT$

Let p_i and p_j are the farthest pair of points in P^*. Let them correspond to the segments s_i and s_j. Consider the iteration in which p_i and p_j are picked by our algorithm (see line number 2 in Algorithm 1). As we are considering every distinct pair of segments and all the pairs of endpoints, we must have considered (p_i, p_j) in some iteration. Algorithm 1 returns a feasible solution by obtaining a set P of cardinality n iteratively. Initially $P = \emptyset$. The algorithm finds a set P' of size $n - |P|$ having minimum potential among the endpoints in $lune(p_i, p_j)$ by calling the subroutine $\text{Min\_Potential}()$ (see line number 10 in Algorithm 1). Note that P' need not be a feasible solution as there can be some segments having both endpoints in P', and some segments having no representation in P'. We update P' to get a (partial) feasible solution (see line numbers 12–13 in Algorithm 1) as follows: for every segment having both the endpoints in P' we delete any one of its endpoints arbitrarily. We also update S by deleting segments having at least one endpoint in P' (refer line number 15) to ensure the feasibility of the final solution returned by the algorithm. The same process is repeated with $P = P \cup P'$ until $|P| = n$.

Lemma 5 $\mathcal{P}(P) \leq \log n \times \mathcal{P}(P^*)$

Proof The factor $\log n$ in the bound is due to the number of iterations needed to find P in worst case[6]. Let us divide the set S of segments into disjoint

---

[6] In each iteration at least half of the segments are deleted from S.

subsets $\mathcal{S}_1, \mathcal{S}_2, \ldots, \mathcal{S}_{\log n}$, where $\mathcal{S}_i$ is the set of segments considered in i-th iteration by Algorithm 1. Analogously, we divide P and P^* into disjoint subsets $P_1, P_2, \ldots, P_{\log n}$ and $P_1^*, P_2^*, \ldots, P_{\log n}^*$, respectively, where P_i and P_i^* are the set of endpoints of the segments in $\mathcal{S}_i$ (one for each) included in the solution P and in the optimal solution P^*, respectively. By the way the algorithm chooses the sets P_i, it is clear that $\mathcal{P}(P_i) \leq \mathcal{P}(P_i^*)$ for $1 \leq i \leq \log n$. The lemma follows by taking the summation in both sides over $\log n$ iterations. $\qquad \square$

Theorem 2 *The minimum tree spanning P has weight at most $O(\log^2 n)OPT$.*

Proof The weight of the minimum tree spanning the points in P

$$\leq \sqrt{2}\mathcal{P}(P) \text{ (by Lemma 3)}$$
$$\leq \sqrt{2}\log n \times \mathcal{P}(P^*) \text{ (by Lemma 5)}$$
$$\leq 8\sqrt{2}\log^2 n \times OPT = O(\log^2 n)OPT \text{ (by Lemma 4)} \qquad \square$$

Algorithm 1 runs in time $O(n^2 \times \pi(n) \log n)$, where $\pi(n)$ is the running time of the sub-routine Min_Potential(). In [15], a $O(k^4)$ algorithm is proposed to find a set of size k with minimum potential in a given set P of n points. However, using the idea in [12], the running time can be improved to $O(nk \log k)$.

References

1. Agarwal, P.K., Edelsbrunner, H., Schwarzkopf, O., Welzl, E.: Euclidean minimum spanning trees and bichromatic closest pairs. Discret. Comput. Geom. **6**(1), 407–422 (1991)
2. Arora, S.: Polynomial time approximation schemes for euclidean traveling salesman and other geometric problems. J. ACM (JACM) **45**(5), 753–782 (1998)
3. Awerbuch, B., Azar, Y., Blum, A., Vempala, S.: New approximation guarantees for minimum-weight k-trees and prize-collecting salesmen. SIAM J. Comput. **28**(1), 254–262 (1998)
4. Blum, A., Chalasani, P., Vempala, S.: A constant-factor approximation for the k-MST problem in the plane. In: Proceedings of the Twenty-Seventh Annual ACM Symposium on Theory of Computing, pp. 294–302. ACM (1995)
5. Borgelt, M.G., Van Kreveld, M., Löffler, M., Luo, J., Merrick, D., Silveira, R.I., Vahedi, M.: Planar bichromatic minimum spanning trees. J. Discret. Algorithms **7**(4), 469–478 (2009)
6. Bose, P., Houle, M.E., Toussaint, G.T.: Every set of disjoint line segments admits a binary tree. Discret. Comput. Geom. **26**(3), 387–410 (2001)
7. Bose, P., Smid, M.: On plane geometric spanners: a survey and open problems. Comput. Geom. **46**(7), 818–830 (2013)
8. Bose, P., Toussaint, G.: Growing a tree from its branches. J. Algorithms **19**(1), 86–103 (1995)
9. Chlebík, M., Chlebíková, J.: The steiner tree problem on graphs: inapproximability results. Theoret. Comput. Sci. **406**(3), 207–214 (2008)
10. Daescu, O., Ju, W., Luo, J.: NP-completeness of spreading colored points. In: Wu, W., Daescu, O. (eds.) COCOA 2010. LNCS, vol. 6508, pp. 41–50. Springer, Heidelberg (2010). https://doi.org/10.1007/978-3-642-17458-2_5

11. Eppstein, D.: Spanning trees and spanners (1996). https://www2.cs.duke.edu/courses/spring07/cps296.2/papers/SpanningTrees.pdf
12. Eppstein, D.: Faster geometric k-point MST approximation. Comput. Geom. **8**(5), 231–240 (1997)
13. Garey, M.R., Johnson, D.S.: Computers and Intractability, vol. 29. W. H. Freeman, New York (2002)
14. Garg, N.: Saving an epsilon: a 2-approximation for the k-MST problem in graphs. In: Proceedings of the Thirty-Seventh Annual ACM Symposium on Theory of Computing, pp. 396–402. ACM (2005)
15. Garg, N., Hochbaum, D.S.: An $O(\log k)$ approximation algorithm for the k minimum spanning tree problem in the plane. In: Proceedings of the Twenty-Sixth Annual ACM Symposium on Theory of Computing, pp. 432–438. ACM (1994)
16. Hoffmann, M., Speckmann, B., Tóth, C.D.: Pointed binary encompassing trees: simple and optimal. Comput. Geom. **43**(1), 35–41 (2010)
17. Hoffmann, M., Tóth, C.D.: Segment endpoint visibility graphs are hamiltonian. Comput. Geom. **26**(1), 47–68 (2003)
18. Mitchell, J.S.: Guillotine subdivisions approximate polygonal subdivisions: a simple polynomial-time approximation scheme for geometric tsp, k-MST, and related problems. SIAM J. Comput. **28**(4), 1298–1309 (1999)
19. Rajagopalan, S., Vazirani, V.: Logarithmic approximation of minimum weight k trees. Unpublished Manuscript (1995)
20. Rappaport, D., Imai, H., Toussaint, G.T.: Computing simple circuits from a set of line segments. Discret. Comput. Geom. **5**(1), 289–304 (1990)
21. Ravi, R., Sundaram, R., Marathe, M.V., Rosenkrantz, D.J., Ravi, S.S.: Spanning trees - short or small. SIAM J. Discret. Math. **9**(2), 178–200 (1996)
22. Smid, M.: The well-separated pair decomposition and its applications (2016). https://people.scs.carleton.ca/~michiel/aa-handbook.pdf
23. Zelikovsky, A., Lozevanu, D.: Minimal and bounded trees. In: Tezele Cong. XVIII Acad. Romano-Americane, Kishniev, pp. 25–26 (1993)

Improved Learning of k-Parities

Arnab Bhattacharyya[1], Ameet Gadekar[2($\boxtimes$)], and Ninad Rajgopal[3]

[1] Department of Computer Science and Automation, Indian Institute of Science, Bengaluru, India
arnabb@iisc.ac.in
[2] Department of Computer Science, Aalto University, Helsinki, Finland
ameet.gadekar@aalto.fi
[3] Department of Computer Science, University of Oxford, Oxford, UK
ninad.rajgopal@cs.ox.ac.uk

Abstract. We consider the problem of learning k-parities in the online mistake-bound model: given a hidden vector $x \in \{0,1\}^n$ where the hamming weight of x is k and a sequence of "questions" $a_1, a_2, \cdots \in \{0,1\}^n$, where the algorithm must reply to each question with $\langle a_i, x \rangle \pmod{2}$, what is the best trade-off between the number of mistakes made by the algorithm and its time complexity? We improve the previous best result of Buhrman et al. [BGM10] by an $\exp(k)$ factor in the time complexity.

Next, we consider the problem of learning k-parities in the PAC model in the presence of random classification noise of rate $\eta \in (0, 1/2)$. Here, we observe that even in the presence of classification noise of non-trivial rate, it is possible to learn k-parities in time better than $\binom{n}{k/2}$, whereas the current best algorithm for learning noisy k-parities, due to Grigorescu et al. [GRV11], inherently requires time $\binom{n}{k/2}$ even when the noise rate is polynomially small.

1 Introduction

By now, the "Parity Problem" of Blum et al. [BKW03] has acquired widespread notoriety. The question is simple enough to be in our second sentence: in order to learn a hidden vector $x \in \{0,1\}^n$, what is the least number of random examples (a, ℓ) that need to be seen, where a is uniformly chosen from $\{0,1\}^n$ and $\ell = \sum_i a_i x_i \pmod{2}$ with probability at least $1 - \eta$? Information-theoretically, x can be recovered after only $O(n)$ examples, even if the noise rate η is close to $1/2$. But if we add the additional constraint that the running time of the learning algorithm be minimized, the barely sub-exponential running time of [BKW03]'s algorithm, $2^{O(n/\log n)}$ still holds the record of being the fastest known for this problem (for any distribution)!

A. Bhattacharyya—Supported in part by DST Ramanujan Fellowship.

A. Gadekar—Partially supported by European Research Council (ERC) under the European Unions Horizon 2020 research and innovation programme (grant agreement No. 759557). Work partially completed while at IISc.

N. Rajgopal—Work partially completed while at IISc.

L. Wang and D. Zhu (Eds.): COCOON 2018, LNCS 10976, pp. 542–553, 2018.
https://doi.org/10.1007/978-3-319-94776-1_45

Learning parities with noise, abbreviated as LPN, is a central problem in theoretical computer science. It has incarnations in several different areas of computer science, including coding theory as the problem of decoding random binary linear codes and cryptography as the "learning with errors" problem that underlies lattice-based cryptosystems [Reg09,BV11]. The learning with errors problem, often known as LWE, is in fact a generalization of LPN to finite fields of larger alphabets. Note that, LPN is a special case of LWE when the alphabet size is two.

In learning theory, the special case of the problem where the hidden vector x is known to be supported on a set of size k much smaller than n has great relevance. We refer to this problem as *learning k-parity with noise* or k-LPN. Feldman et al. [FGKP09] showed that learning k-juntas, as well as learning 2^k-term DNFs from uniformly random examples and variants of these problems in which the noise is adversarial instead of random, all reduce to the k-LPN problem. For the k-LPN problem, the current record is that of Grigorescu et al. [GRV11] who showed a learning algorithm that succeeds with constant probability, takes $\binom{n}{k/2}^{1+(2\eta)^2+o(1)}$ time and uses $\frac{k \log n}{(1-2\eta)^2} \cdot \omega(1)$ samples. When the noise rate η is close to $1/2$, this running time is improved by an algorithm due to Valiant [Val12] that runs in time $n^{0.8k} \cdot \text{poly}(\frac{1}{1-2\eta})$. It is a wide open challenge to find a polynomial time algorithm for k-LPN for growing k or to prove a negative result.

Another outstanding challenge in machine learning is the problem of learning parities without noise in the "attribute-efficient" setting [Blu96]. The algorithm is given access to a source of examples (a, ℓ) where a is chosen uniformly from $\{0,1\}^n$ and $\ell = \sum_i a_i x_i \pmod 2$ with no noise, and the question is to learn x using number of samples polynomial in the description length of x. Note that, the attribute-efficient setting is not an interesting question for general parities, as using $\text{poly}(n)$ many samples is attribute-efficient and this is enough for Gaussian elimination. We focus on the case where x has sparsity $k \ll n$. Information-theoretically, of course, $O(k \log n)$ examples should be sufficient, as each linearly independent example reduces the number of consistent k-parities by a factor of 2. But the fastest known algorithm requiring $O(k \log n)$ samples runs in time $\tilde{O}(\binom{n}{k/2})$ [KS06], and it is open whether there exists a polynomial time algorithm for learning parities that is attribute-efficient, i.e. it uses $\text{poly}(k \log n)$ samples. Buhrman et al. [BGM10] give the current best tradeoffs between the sample complexity and running time for learning parities in this noiseless setting and in fact, give an algorithm which achieves the current best running time for given sample complexity. Notice that with $O(n)$ samples, it is easy to learn the k-parity in polynomial time using Gaussian elimination.

1.1   Our Results

We first study the noiseless setting and consider the mistake-bound model [Lit89]. Our main technical result is an improved tradeoff between the sample complexity and runtime for learning parities.

We assume throughout, as in [BGM10], that $k, t : \mathbb{N} \to \mathbb{N}$ are two functions such that they are constructible in quadratic time. We also use the shorthand notation $k = k(n)$ and $t = t(n)$ henceforth. Finally, we use $f(n) \ll g(n)$ to mean that $f(n) = o(g(n))$.

Theorem 1. *Let $k, t : \mathbb{N} \to \mathbb{N}$ be two functions such that $\log \log n \ll k(n) \ll t(n) \ll n$. Then for every $n \in \mathbb{N}$, there is an algorithm that learns PAR(k) in the mistake-bound model, with mistake bound at most $(1 + o(1))\frac{kn}{t} + \log \binom{t}{k}$ and running time per round $e^{-k/4.01} \cdot \binom{t}{k} \cdot \tilde{O}\left((kn/t)^2\right)$.*

For comparison, let us quote the result of Buhrman et al.:

Theorem 2. *Let $k, t : \mathbb{N} \to \mathbb{N}$ be two functions satisfying $k(n) \leqslant t(n) \leqslant n$. For every $n \in \mathbb{N}$, there exists a deterministic algorithm that learns PAR(k) in the mistake-bound model, with the mistake bound $k\lceil \frac{n}{t} \rceil + \lceil \log \binom{t}{k} \rceil$ and running time per round $O\left(\binom{t}{k}(kn/t)^2\right)$.*

Thus, in the comparable regime, our Theorem 1 improves the runtime complexity of Theorem 2 by an $\exp(k)$ factor while its sample complexity remains the same upto constant factors. Note that as t approaches k, our algorithm requires $O(n)$ samples and takes poly(n) time which is the complexity of the Gaussian elimination approach. On the other hand, if $t = n/\log(n/k)$, our algorithm requires $O(k \log(n/k))$ samples and takes[1] $\exp(-k) \cdot \binom{n/k}{k}$ time (ignoring polynomial factors), compared to the trivial approach which explicitly keeps track of the subset of all the k-weight parities consistent with examples given so far and which requires $O(k \log(n/k))$ samples and takes $O(\binom{n}{k})$ time.

The mistake-bound model is stronger than the PAC model (in fact, strictly stronger assuming the existence of one-way functions [Blu94]). As a consequence, we can get a PAC learning algorithm from the above theorem. There are standard conversion techniques which can be used to transform any mistake-bound algorithm into a PAC learning algorithm (over arbitrary distributions):

Theorem 3 *[Ang88, Hau88, Lit89]. Any algorithm $\mathcal{A}$ that learns a concept class $\mathcal{C}$ in the mistake-bound model with mistake bound m and running time t per round can be converted into an algorithm $\mathcal{A}'$ that PAC-learns $\mathcal{C}$ with sample complexity $O(\frac{1}{\varepsilon}m + \frac{1}{\varepsilon}\log\frac{1}{\delta})$, running time $O(\frac{1}{\varepsilon}mt + \frac{t}{\varepsilon}\log\frac{1}{\delta})$, approximation parameter ε, and confidence parameter δ.*

Using Theorem 3, we directly obtain the following corollary. In fact, since Theorem 3 produces a PAC-learner over any distribution, a statement of the form of Corollary 1 also holds for examples obtained from any distribution.

Corollary 1. *Let $k, t : \mathbb{N} \to \mathbb{N}$ be two functions satisfying $\log \log n \ll k(n) \ll t(n) \ll n$. For any $\delta > 0$, there is an algorithm that learns the concept class of k-parities on n variables with confidence parameter δ, using $O(kn/t + \log \binom{t}{k} +$*

---

[1] By $\exp(\cdot)$, we mean $2^{O(\cdot)}$.

$\log(1/\delta))$ *uniformly random examples and* $e^{-k/4.01}\binom{t}{k}\cdot\text{poly}(n)\cdot\log(1/\delta)$ *running time*[2].

We next examine the noisy setting. Here, our contribution is a simple, general observation that does not seem to have been explicitly made before. Let $\text{PAR}(k)$ represent concept class of parities of Hamming weight k.

Theorem 4. *Given an algorithm $\mathcal{A}$ that learns $\text{PAR}(k)$ over the uniform distribution with confidence parameter δ using $s(\delta)$ samples and running time $t(\delta)$, there is an algorithm $\mathcal{A}'$ that solves the k-LPN problem with noise rate $\eta \in (0, 1/3)$, using $O(s(\delta/2)\log(4/\delta))$ examples and running time $\exp(O(H(3\eta/2) \cdot s(\delta/2))) \cdot \log(4/\delta) \cdot (t(\delta/2) + s(\delta/2))$ and with confidence parameter δ.*

In the above, $H : [0, 1] \rightarrow [0, 1]$ denotes the binary entropy function $H(p) = p\log_2\frac{1}{p} + (1 - p)\log_2\frac{1}{1-p}$. The main conceptual message carried by Theorem 4 is that improving the sample complexity for efficient learning of noiseless parity improves the running time for learning of noisy parity. For instance, if we use Spielman's algorithm as $\mathcal{A}$, reported in [KS06], that learns k-parity using $O(k\log n)$ samples and $O(\binom{n}{k/2})$ running time, we immediately get the following:

Corollary 2. *For any $\eta \in (0, 1/3)$ and constant confidence parameter, there is an algorithm for k-LPN with sample complexity $O(k\log n)$ and running time $O\left(\binom{n}{k/2}^{1+O(H(1.5\eta))}\right)$.*

For comparison, consider the current best result of [GRV11]:

Theorem 5 (Theorem 5 of [GRV11]). *For any $\varepsilon, \delta, \eta \in (0, 1/2)$, and distribution $\mathcal{D}$ over $\{0, 1\}^n$, the k-LPN problem over $\mathcal{D}$ with noise rate η can be solved using $\frac{k\log(n/\delta)\omega(1)}{\varepsilon^2(1-2\eta)^2}$ samples in time $\frac{\log 1/\delta}{\varepsilon^2(1-2\eta)^2}\cdot\binom{n}{k/2}^{1+(\frac{\eta}{\varepsilon+\eta-2\varepsilon\eta})^2+o(1)}$, where ε and δ are the approximation and confidence parameters respectively.*

This result has runtime $\binom{n}{k/2}^{1+4\eta^2+o(1)}$ and sample complexity $\omega(k\log n)$. In the regime under consideration, our algorithm's runtime has a worse exponent but an asymptotically better sample complexity.

The result of [GRV11] requires $\binom{n}{k/2}$ time regardless of how small η is. We show via Theorem 4 and Corollary 1 for the uniform distribution, that it is possible to break the $\binom{n}{k/2}$ barrier when η is a small enough function of n.

Corollary 3. *Suppose $k(n) = n/f(n)$ for some function $f : \mathbb{N} \rightarrow \mathbb{N}$ for which $f(n) \ll n/\log\log n$, and suppose $\eta(n) = o(\frac{1}{((f(n))^\alpha \log n)})$ for some $\alpha \in [1/2, 1)$. Then, for constant confidence parameter, there exists an algorithm for k-LPN with noise rate η with running time $O\left(e^{-k/4.01+o(k)} \cdot \binom{n}{k}^{1-\alpha} \cdot \text{poly}(n)\right)$ and sample complexity $O(k(f(n))^\alpha)$.*

We note that because of the results of Feldman et al. [FGKP09], the above results for k-LPN also extend to the setting where the example source adversarially mislabels examples instead of randomly but with the same rate η.

---

[2] The "4.01" can be replaced by any constant more than 4.

1.2   Our Techniques

We first give an algorithm to learn parities in the noiseless setting in the mistake bound model. We use the same approach as that of [BGM10] (which was itself inspired by [APY09]). The idea is to consider a family $\mathcal{S}$ of subsets of $\{0,1\}^n$ such that the hidden k-sparse vector is contained inside one of the elements of $\mathcal{S}$. We maintain this invariant throughout the algorithm. Now, each time an example comes, it specifies a halfspace H of $\{0,1\}^n$ inside which the hidden vector is lying. So, we can update $\mathcal{S}$ by taking the intersection of each of its elements with H. If we can ensure that the set of points covered by the elements of $\mathcal{S}$ is decreasing by a constant factor at every round, then after $O(\log \sum_{S \in \mathcal{S}} |S|)$ examples, the hidden vector is learned. The runtime is determined by the number of sets in $\mathcal{S}$ times the cost of taking the intersection of each set with a halfspace.

One can think of the argument of Buhrman et al. [BGM10] as essentially initializing $\mathcal{S}$ to be the set of all $\binom{n}{k}$ subspaces spanned by k standard basis vectors. The intersections of these subspaces with a halfspace can be computed efficiently by Gaussian elimination. Our idea is to reduce the number of sets in $\mathcal{S}$. Note that we can afford to make the size of each set in $\mathcal{S}$ larger by some factor C because this only increases the sample complexity by an additive $\log C$. Our approach is (essentially) to take $\mathcal{S}$ to be a random collection of subspaces spanned by αk standard basis vectors, where $\alpha > 1$ is a sufficiently large constant. We show that it is sufficient for the size of $\mathcal{S}$ to be smaller than $\binom{n/\alpha}{k}$ by a factor that is exponential in k, so that the running time is also improved by the same factor. Moreover, the sample complexity increases by only a lower-order additive term.

Our second main contribution is a reduction from noiseless parity learning to noisy parity learning. The algorithm is a simple exhaustive search which guesses the location of the mis-labelings, corrects those labels, applies the learner for noiseless parity and then verifies whether the output hypothesis matches the examples by drawing a few more samples. Surprisingly, this seemingly immediate algorithm allows us to devise the first algorithm which has a better running time than $\binom{n}{k/2}$ in the presence of a non-trivial amount of noise.

2   Preliminaries

Let $\mathrm{PAR}(k)$ be the class of all vectors $f \in \{0,1\}^n$ of Hamming weight k. So, $|\mathrm{PAR}(k)| = \binom{n}{k}$. With each vector $f \in \mathrm{PAR}(k)$, we associate a parity function $f : \{0,1\}^n \to \{0,1\}$ defined by $f(a) = \sum_{i=1}^n x_i a_i \pmod 2$.

Let $\mathcal{C}$ be a concept class of Boolean functions on n variables, such as $\mathrm{PAR}(k)$. We discuss two models of learning in this work. One is Littlestone's *online mistake bound* model [Lit89]. Here, learning proceeds in a series of rounds, where in each round, the learner is given an unlabeled boolean example $a \in \{0,1\}^n$ by an oracle and must predict the value $f(a)$ of an unknown target function $f \in \mathcal{C}$. Once the learner predicts the value of $f(a)$, the true value of $f(a)$ is revealed to the learner by the oracle. The *mistake bound* of a learning algorithm is the worst-case number of mistakes that the algorithm makes over all sequences of examples and all possible target functions $f \in \mathcal{C}$.

The second model of learning we consider is Valiant's famous *PAC model* [Val84] of learning from random examples. Here, for an unknown target function $f \in \mathcal{C}$, the learner has access to a source of examples $(a, f(a))$ where a is chosen independently from a distribution $\mathcal{D}$ on $\{0, 1\}^n$. A learning algorithm is said to PAC-learn $\mathcal{C}$ with *sample complexity s, running time t, approximation parameter ε* and *confidence parameter δ* if for all distributions $\mathcal{D}$ and all target functions $f \in \mathcal{C}$, the algorithm draws at most s samples from the example source, runs for time at most t and outputs a function f^* such that, with a probability at least $1 - \delta$:

$$\Pr_{a \leftarrow \mathcal{D}}[f(a) \neq f^*(a)] < \varepsilon$$

Often in this paper (e.g., all of the Introduction), we consider PAC-learning over the uniform distribution, in which case $\mathcal{D}$ is fixed to be uniform on $\{0, 1\}^n$. Notice that for learning $PAR(k)$ over the uniform distribution, we can take $\varepsilon = \frac{1}{2}$ because any two distinct parities differ on half of $\{0, 1\}^n$.

The k-LPN problem with noise rate η, introduced in Sect. 1, corresponds to the problem of PAC-learning $PAR(k)$ under the uniform distribution, when the example source can mislabel examples with a rate $\eta \in (0, 1/2)$. More generally, one can study the k-LPN *problem over* $\mathcal{D}$, an arbitrary distribution. In fact, Theorem 5 [GRV11] show their algorithm for any arbitrary distribution $\mathcal{D}$.

3   In the Absence of Noise

We re-state Theorem 1, which is the main result of this section.

Theorem 1. *Let $k, t : \mathbb{N} \to \mathbb{N}$ be two functions such that $\log \log n \ll k(n) \ll t(n) \ll n$. Then for every $n \in \mathbb{N}$, there is an algorithm that learns $PAR(k)$ in the mistake-bound model, with mistake bound at most $(1 + o(1))\frac{kn}{t} + \log \binom{t}{k}$ and running time per round $e^{-k/4.01} \cdot \binom{t}{k} \cdot \tilde{O}\left((kn/t)^2\right)$.*

For comparison, we quote the relevant result of [BGM10] in the mistake-bound model.

Theorem 2 (recalled) [Theorem 2.1 of [BGM10]]. *Let $k, t : \mathbb{N} \to \mathbb{N}$ be two functions such that $k(n) \leqslant t(n) \leqslant n$. Then for every $n \in \mathbb{N}$, there is a deterministic algorithm that learns $PAR(k)$ in the mistake-bound model, with mistake bound at most $k\lceil \frac{n}{t} \rceil + \log \binom{t}{k}$ and running time per round $\binom{t}{k} \cdot O((kn/t)^2)$.*

Note that their mistake bound is better by a lower-order term which we do not see how to avoid in our setup. This slack is not enough though to recover Theorem 1 from Theorem 2: dividing t by C roughly multiplies the sample complexity by C and divides the running time by C^k in [BGM10]'s algorithm, whereas in our algorithm, dividing t by C roughly multiplies the sample complexity by C and divides the running time by $(1.28C)^k$.

3.1   The Algorithm

Let $f \in \{0,1\}^n$ be the hidden vector of sparsity k that the learning algorithm is trying to learn. Let $e = \{e_1, e_2, \cdots, e_n\}$ be the set of standard basis of the vector space $\{0,1\}^n$.

Let α be a large constant we set later, and let $T = \alpha t$. Note that $T \ll n$. We define an arbitrary partition $\pi = C_1, C_2, \cdots, C_T$ on the set e into T parts, each of size at most $\lceil n/T \rceil$. Next, let $S_1, \ldots, S_m \subset [T]$ be m random subsets of $[T]$, each of size αk. We choose m to ensure the following:

Claim. If $m = \tilde{O}\left(\frac{\binom{T}{\alpha k}}{\binom{T-k}{\alpha k - k}}\right)$, then with nonzero probability, for every set $A \subset [T]$ of size k, $A \subset S_i$ for some $i \in [m]$.

Proof. This follows from the simple observation that for any fixed $i \in [m]$, $\mathbf{Pr}[A \subset S_i] = \binom{T-k}{\alpha k-k}/\binom{T}{\alpha k}$, and so,

$$\mathbf{Pr}[\forall i \in [m], A \not\subset S_i] = \left(1 - \binom{T-k}{\alpha k-k} \middle/ \binom{T}{\alpha k}\right)^m \leqslant e^{-m\binom{T-k}{\alpha k - k}/\binom{T}{\alpha k}}$$

Choosing $m = 2\frac{\binom{T}{\alpha k}}{\binom{T-k}{\alpha k-k}} \log \binom{T}{k}$ and applying the union bound finishes the proof.

We fix some choice of $S_1, \ldots, S_m \subset [T]$ that satisfies the conclusion of above claim for what follows. In fact, the rest is exactly [BGM10]'s algorithm, which we reproduce for completeness.

For every $i \in [m]$, let $M_i \subset \{0,1\}^n$ be the span of $\bigcup_{j \in S_i} C_j$. Note that $\left|\bigcup_{j \in S_i} C_j\right| \leqslant \alpha k \lceil n/T \rceil \leqslant \alpha k \cdot \left(\frac{n}{T} + 1\right) = \frac{kn}{t} + \alpha k = (1 + o(1))kn/t$, as $t \ll n$ and α is a constant. So, M_i is a linear subspace containing at most $2^{(1+o(1))kn/t}$ points.

Note that every $f \in \{0,1\}^n$ with $|f| = k$ is contained in some M_i. This is simply because every set of k standard basis vectors is contained in the union of at most k of the T parts in the partition π, and by Claim 3.1, every subset of $[T]$ of size k is contained in some S_i.

Initially, the unknown target vector f can be in any of the M_i's. Consider what happens when the learner sees an example $a \in \{0,1\}^n$ and a label $y \in \{0,1\}$. For $i \in [m]$, let $M_i(a, y) = \{v \in M_i : v(a) = y\}$. $M_i(a, y)$ may be of size 0, $|M_i|$ or $|M_i|/2$. Note that the size of $M_i(a, y)$ can be efficiently found using Gaussian elimination.

We are now ready to describe the algorithm:

- **Initialization:** The learning algorithm begins with a set of affine spaces $N_i, i \in [m]$ represented by a system of linear equations. Initialize the affine spaces $N_i = M_i$ for all $i \in [m]$.
- **On receiving an example** $a \in \{0,1\}^n$ **from the oracle:** Predict its label $\hat{y} \in \{0,1\}$ such that $\sum_{i \in [m]} |N_i(a, \hat{y})| \geqslant \sum_{i \in [m]} |N_i(a, 1 - \hat{y})|$.
- **On receiving the answer from the oracle** $y = f(a)$: Update N_i to $N_i(a, y)$ for each $i \in [m]$.
- **Termination:** The algorithm terminates when $|\bigcup_{i \in [m]} N_i| \leqslant 1$.

3.2   Analysis

Before we analyze the algorithm, we first establish a combinatorial claim that is the crux of our improvement:

Lemma 1. *If α is a large enough constant,*

$$\frac{\binom{T}{\alpha k}}{\binom{T-k}{\alpha k-k}} \leqslant e^{-k/4.01} \cdot \binom{t}{k}$$

Proof.

$$
\begin{aligned}
\frac{1}{\binom{t}{k}} \cdot \frac{\binom{T}{\alpha k}}{\binom{T-k}{\alpha k-k}} &= \prod_{i=0}^{k-1} \frac{k-i}{t-i} \cdot \frac{T-i}{\alpha k-i} \\
&= \prod_{i=0}^{k-1} \frac{\alpha t-i}{\alpha k-i} \cdot \frac{k-i}{t-i} \\
&= \prod_{i=1}^{k-1} \left(1 - \frac{i\left(1-\frac{1}{\alpha}\right)\left(\frac{1}{k-i}-\frac{1}{t-i}\right)}{1+\frac{i}{k-i}\left(1-\frac{1}{\alpha}\right)} \right) \\
&\leqslant \prod_{i=1}^{k-1} \left(1 - \frac{0.999}{1+\frac{k-i}{i}\frac{\alpha}{\alpha-1}} \right)
\end{aligned}
$$

where the equalities are routine calculation and the inequality is using that $k(n) \ll t(n)$. Each individual term in the product is strictly less than 1. So, the above is bounded by[3]:

$$
\begin{aligned}
&\leqslant \prod_{i=k/(2-\varepsilon)}^{k-1} \left(1 - \frac{0.999}{1+\frac{k-i}{i}\frac{\alpha}{\alpha-1}} \right) \\
&\leqslant \left(1 - \frac{0.999}{1+(1-\varepsilon)\frac{\alpha}{\alpha-1}} \right)^{\frac{1-\varepsilon}{2-\varepsilon}k} \\
&\leqslant \exp\left(-\lg e \cdot \frac{0.999(1-\varepsilon)}{(2-\varepsilon)(1+(1-\varepsilon)\frac{\alpha}{\alpha-1})}k \right) \leqslant e^{-k/4.01}
\end{aligned}
$$

for a small enough constant $\varepsilon > 0$ and large enough constant $\alpha > 1$.

Proof (Proof of Theorem 1). Fix α to be a constant that makes the conclusion of Lemma 1 true.

We first check that the invariant is maintained throughout the algorithm that $f \in \cup_{i \in [m]} N_i$. This holds at initiation by the argument given earlier. After that, obviously, if $f \in N_i$, then $f \in N_i(a, f(a))$ for any $a \in \{0,1\}^n$, and so the invariant holds. Therefore, if the algorithm terminates, it will find the hidden vector f and return it as the solution. The rate of convergence is precisely captured by the number of mistakes learning algorithm makes, which we describe next.

---

[3] Again, in the second last inequality, by $\exp(\cdot)$, we mean $2^{O(\cdot)}$.

Mistake Bound. Notice that when the algorithm begins, the sum of the sizes of all the affine spaces, $\sum_i |N_i| \leqslant \tilde{O}\left(\frac{\binom{T}{\alpha k}}{\binom{T-k}{\alpha k-k}}\right) 2^{(1+o(1))kn/t}$. Now whenever the learner makes a mistake by predicting $\hat{y} \neq y$, the size of all affine spaces $\sum_i |N_i|$ reduces by a factor of at least 2. This is due to the definition of $\hat{y}$ and the fact that $|N_i(a, \hat{y})| + |N_i(a, 1 - \hat{y})| = |N_i|$.

Hence, using Lemma 1, after at most

$$\log\left(\sum_i |N_i|\right) \leqslant \log\left[\tilde{O}\left(\frac{\binom{T}{\alpha k}}{\binom{T-k}{\alpha k-k}}\right) 2^{(1+o(1))kn/t}\right] \leqslant (1+o(1))kn/t + \log\binom{t}{k} - \Omega(k) + O\left(\log\log\binom{t}{k}\right)$$

mistakes, the size of $\cup_{i \in [m]} N_i$ will decrease to 1, which by the invariant above will imply that $\cup_{i \in [m]} N_i = \{f\}$, and hence the learner makes no more mistakes. Since we assume $k \gg \log\log n$ and $t \ll n$, we can bound the number of mistakes by: $(1 + o(1))kn/t + \log\binom{t}{k}$.

Running Time. We analyze the running time of the learner for each round. At each round, for a question $a \in \{0, 1\}^n$, we need to compute $|N_i(a, 0)|$ and $|N_i(a, 1)|$ as well as store a representation of the updated N_i. Now, since for each N_i is spanned by at most $\ell = (1 + o(1))kn/t$ basis vectors, we can treat each N_i as a linear subspace in $\{0, 1\}^\ell$. $N_i(a, 0)$ and $N_i(a, 1)$ can be computed by performing one step of Gaussian elimination on a system of linear equations involving ℓ variables, which takes $O(\ell^2)$ time. Thus, the total running time is $O(m\ell^2)$, which using Lemma 1 is exactly the bound claimed in Theorem 1.

4   In the Presence of Noise

Recall the k-LPN problem. In this section, we show a reduction from k-LPN to noiseless learning of PAR(k) and its applications.

4.1   The Reduction

We focus on the case when the noise rate η is bounded by a constant less than $\frac{1}{3}$.

Theorem 4 (recalled). *Given an algorithm $\mathcal{A}$ that learns PAR(k) over the uniform distribution with confidence parameter δ using $s(\delta)$ samples and running time $t(\delta)$, there is an algorithm $\mathcal{A}'$ that solves the k-LPN problem with noise rate $\eta \in (0, 1/3)$, using $O(s(\delta/2)\log(4/\delta))$ examples and running time $\exp(O(H(3\eta/2) \cdot s(\delta/2))) \cdot \log(4/\delta) \cdot (t(\delta/2) + s(\delta/2))$ and with confidence parameter δ.*

Let $\mathcal{A}(\delta)$ be a PAC-learning algorithm over the uniform distribution for PAR(k) of length n with confidence parameter δ that draws $s(\delta)$ examples and runs in time $t(\delta)$. Below is our algorithm NOISY for k-LPN. Here, H denotes the binary entropy function $p \mapsto p\log_2(1/p) + (1-p)\log_2(1/(1-p))$.

Algorithm . NOISY(δ, η)

1: $\mathcal{X} = \phi$
2: **for** $\rho = 1$ to $3 \log(4/\delta)$ **do**
3:     Draw $s' = s(\delta/2)$ random samples $(a_1, \ell_1), \ldots, (a_{s'}, \ell_{s'}) \in \{0,1\}^n \times \{0,1\}$.
4:     **for all** $S \subset [s'], |S| \leq \frac{3}{2}\eta s'$ **do**
5:         **for** $i \in [s']$ **do**
6:             **if** $i \in S$ **then** $\tilde{\ell}_i \leftarrow 1 - \ell_i$
7:             **else** $\tilde{\ell}_i \leftarrow \ell_i$
8:             **end if**
9:         **end for**
10:         $x_S \leftarrow \mathcal{A}(\delta/2)$ applied to examples $(a_1, \tilde{\ell}_1), \ldots, (a_{s'}, \tilde{\ell}_{s'})$.
11:         $\mathcal{X} = \mathcal{X} \cup x_S$
12:     **end for**
13: **end for**
14: Draw $s'' = 600\big(s' \cdot H(3\eta/2) + \log\big(\frac{24 \log(4/\delta)}{\delta}\big)\big)$ random samples $(b_1, m_1), \ldots, (b_{s''}, m_{s''}) \in \{0,1\}^n \times \{0,1\}$
15: $x_{S^*} \leftarrow \arg\max_{x_S \in \mathcal{X}} |\{i \in [s''] : \langle b_i, x_S \rangle = m_i\}|$
16: **return** x_{S^*}

Proof (Proof of Theorem 4*).*

Lemma 2. *The sample complexity of* NOISY *is* $s' \cdot 3\log(4/\delta) + s'' = O(s(\delta/2) \log(4/\delta))$.

Proof. Immediate.

Lemma 3. *The running time of* NOISY *is* $2^{O(H(3\eta/2)s(\delta/2))} \cdot \log(4/\delta) \cdot (t(\delta/2) + s(\delta/2))$.

Proof. We use the standard estimate $\sum_{i=0}^{\alpha x} \binom{x}{i} \leq 2^{H(\alpha)x}$ for $\alpha \leq \frac{1}{2}$. The bound is then immediate.

Lemma 4. *If x is the hidden vector and x^* is output by* NOISY(δ), *then with probability at least $1 - \delta$, $x^* = x$.*

Proof. Let $T = \{i \in [s'] : \langle a_i, x \rangle \neq \ell_i\}$ be the subset of the s' samples drawn in line 3 that are mislabeled by the example source. By Markov's inequality:

$$\mathbf{Pr}[|T| > 3\eta s'/2] \leq 2/3$$

Thus, if we repeat step 3, $3\log(4/\delta)$ times, then with probability at least $1 - \delta/4$, it is true that $|T| \leq 3\eta s'/2$ in some round. If $|T| \leq 3\eta s'/2$, we have with probability at least $1 - \delta/2$, $x_T = x$. Thus, for any $i \in [s'']$, $\mathbf{Pr}_{b_i}[\langle x_T, b_i \rangle \neq m_i] \leq \eta$. On the other hand, for all $x_S \neq x_T$, $\mathbf{Pr}_{b_i}[\langle x_S, b_i \rangle \neq \langle x, b_i \rangle] = 1/2$, and so $\mathbf{Pr}_{b_i}[\langle x_S, b_i \rangle \neq m_i] = 1/2$ as the noise is random. Again, using Chernoff bounds,

$$Pr[\exists S \neq T \text{ s.t.} |\{i \in [s''] : \langle b_i, x_S \rangle \neq m_i\}| \leq 5s''/12] \leq 2^{H(3\eta/2)s'} \cdot 3\log(4/\delta) \cdot e^{-s''/450} < \frac{\delta}{8}$$

On the other hand, for x_T itself, $\mathbf{Pr}[|\{i \in [s''] : \langle b_i, x_T \rangle \neq m_i\}| > 5s''/12] < \frac{\delta}{8}$ by a similar use of Chernoff bounds. So, in all, with probability at least $1 - \delta$, x_T will be returned in step 15.

When the noise rate η is more than $1/3$, a similar reduction can be given by adjusting the parameters accordingly. For example, for any $\eta < 1/2$, performing the for loop of step 2, $\frac{\log(4/\delta)}{(1-2\eta)}$ times, with $|S| \leqslant s'/2$ in step 4, we would still have that with probability at least $1 - \delta/4$, it is true that $|T| \leqslant s'/2$. Further, when the distribution is arbitrary, $\mathcal{A}$ is invoked with a smaller approximation parameter than the one given to NOISY so that the filtering step in line 14 works.

4.2   Applications

An immediate application of Theorem 4 is obtained by letting $\mathcal{A}$ be the current fastest known attribute-efficient algorithm for learning PAR(k), the algorithm due to Spielman[4] [KS06] that requires $O(k \log n)$ samples and takes $O(\binom{n}{k/2})$ time (for constant confidence parameter δ). (We ignore the confidence parameter in this section for simplicity.)

Corollary 2 (recalled). *For any $\eta \in (0, 1/3)$ and constant confidence parameter, there is an algorithm for k-LPN with sample complexity $O(k \log n)$ and running time $O\left(\binom{n}{k/2}^{1+O(H(1.5\eta))}\right)$.*

Proof. Immediate from Theorem 4.

Our next application of Theorem 4 uses our improved PAR(k) learning algorithm from Sect. 3.

Corollary 3 (recalled). *Suppose $k(n) = n/f(n)$ for some function $f : \mathbb{N} \to \mathbb{N}$ for which $f(n) \ll n/\log \log n$, and suppose $\eta(n) = o(1/((f(n))^\alpha \log n))$ for some $\alpha \in [1/2, 1)$. Then, for constant confidence parameter, there exists an algorithm for k-LPN with noise rate η with running time $O\left(e^{-k/4.01+o(k)} \cdot \binom{n}{k}^{1-\alpha} \cdot \text{poly}(n)\right)$ and sample complexity $O(k(f(n))^\alpha)$.*

Proof. Let $\mathcal{A}$ be the algorithm of Corollary 1 with $t(n) = \lceil n/(f(n))^\alpha \rceil$. The running time of $\mathcal{A}$ is $e^{-k/4.01} \cdot \binom{n}{k}^{1-\alpha} \cdot \text{poly}(n)$ and its sample complexity is $O(k \cdot f(n))^\alpha)$. Now, applying Theorem 4, we see that since $H(1.5\eta) = o((f(n))^{-\alpha})$, the running time for NOISY is only a $2^{o(k)}$ factor times the running time of $\mathcal{A}$. This yields our desired result.

---

[4] Though a similar algorithm was also proposed by Hopper and Blum [HB01].

References

[Ang88] Angluin, D.: Queries and concept learning. Mach. Learn. **2**(4), 319–342 (1988)

[APY09] Alon, N., Panigrahy, R., Yekhanin, S.: Deterministic approximation algorithms for the nearest codeword problem. In: Dinur, I., Jansen, K., Naor, J., Rolim, J. (eds.) APPROX/RANDOM -2009. LNCS, vol. 5687, pp. 339–351. Springer, Heidelberg (2009). https://doi.org/10.1007/978-3-642-03685-9_26

[BGM10] Buhrman, H., García-Soriano, D., Matsliah, A.: Learning parities in the mistake-bound model. Inform. Process. Lett. **111**(1), 16–21 (2010)

[BKW03] Blum, A., Kalai, A., Wasserman, H.: Noise tolerant learning, the parity problem, and the statistical query model. J. ACM **50**(4), 506–519 (2003)

[Blu94] Blum, A.: Separating distribution-free and mistake-bound learning models over the boolean domain. SIAM J. Comput. **23**(5), 990–1000 (1994)

[Blu96] Blum, A.: On-line algorithms in machine learning. In: Fiat, A., Woeginger, G.J. (eds.) Online Algorithms. LNCS, vol. 1442, pp. 306–325. Springer, Heidelberg (1998). https://doi.org/10.1007/BFb0029575

[BV11] Brakerski, Z., Vaikuntanathan, V.: Efficient fully homomorphic encryption from (standard) LWE. In: Proceedings of 52nd Annual IEEE Symposium on Foundations of Computer Science, pp. 97–106 (2011)

[FGKP09] Feldman, V., Gopalan, P., Khot, S., Ponnuswami, A.K.: On agnostic learning of parities, monomials, and halfspaces. SIAM J. Comput. **39**(2), 606–645 (2009)

[GRV11] Grigorescu, E., Reyzin, L., Vempala, S.: On noise-tolerant learning of sparse parities and related problems. In: Kivinen, J., Szepesvári, C., Ukkonen, E., Zeugmann, T. (eds.) ALT 2011. LNCS (LNAI), vol. 6925, pp. 413–424. Springer, Heidelberg (2011). https://doi.org/10.1007/978-3-642-24412-4_32

[Hau88] Haussler, D.: Space efficient learning algorithms. Technical report UCSC-CRL-88-2, University of California at Santa Cruz (1988)

[HB01] Hopper, N.J., Blum, M.: Secure human identification protocols. In: Boyd, C. (ed.) ASIACRYPT 2001. LNCS, vol. 2248, pp. 52–66. Springer, Heidelberg (2001). https://doi.org/10.1007/3-540-45682-1_4

[KS06] Klivans, A.R., Servedio, R.A.: Toward attribute efficient learning of decision lists and parities. J. Mach. Learn. Res. **7**, 587–602 (2006)

[Lit89] Littlestone, N.: From on-line to batch learning. In: Proceedings of 2nd Annual ACM Workshop on Computational Learning Theory, pp. 269–284 (1989)

[Reg09] Regev, O.: On lattices, learning with errors, random linear codes, and cryptography. J. ACM **56**(6), 1–40 (2009)

[Val84] Valiant, L.G.: A theory of the learnable. Comm. Assoc. Comp. Mach. **27**(11), 1134–1142 (1984)

[Val12] Valiant, G.: Finding correlations in subquadratic time, with applications to learning parities and juntas. In: Proceedings of 53rd Annual IEEE Symposium on Foundations of Computer Science, pp. 11–20. IEEE (2012)

On a Fixed Haplotype Variant
of the Minimum Error Correction
Problem

Axel Goblet, Steven Kelk, Matúš Mihalák, and Georgios Stamoulis[✉]

Department of Data Science and Knowledge Engineering, Maastricht University,
Maastricht, The Netherlands
a.goblet@student.maastrichtuniversity.nl
{steven.kelk,matus.mihalak,georgios.stamoulis}@maastrichtuniversity.nl

Abstract. Haplotype assembly is the problem of reconstructing the two parental chromosomes of an individual from a set of sampled DNA-sequences. A combinatorial optimization problem that models haplotype assembly is the *Minimum Error Correction* problem (MEC). This problem has been intensively studied in the computational biology literature and is also known in the clustering literature: essentially we are required to find two cluster centres such that the sum of distances to the nearest centre, is minimized. We introduce here the problem *Fixed haplotype-Minimum Error Correction* (FH-MEC), a new variant of MEC which corresponds to instances where one of the haplotypes/centres is already given. We provide hardness results for the problem on various restricted instances. We also propose a new and very simple 2-approximation algorithm for MEC on binary input matrices.

1 Introduction

Humans have, genetically speaking, an extremely high degree of similarity: at the vast majority of positions in our DNA sequence we share the same DNA symbol. The relatively few positions at which we differ are known as Single Nucleotide Polymorphisms (SNP) [5]. In most cases the variation observed at a given position involves two nucleotides (as opposed to three or four). For this reason the SNPs of an individual can be summarized as a string over a binary alphabet, also known as a *haplotype*.

A classical computational challenge in the genomic era is to efficiently infer such haplotypes from a set of overlapping, aligned haplotype fragments which have been obtained by sequencing the DNA at different intervals. This problem is complicated by the fact that humans (and diploid organisms in general) actually have two haplotypes (chromosomes): one inherited from the mother, and one from the father. We do not know easily which haplotype fragment originated from which of the two haplotypes, so the goal is to construct *two* haplotypes and to map the fragments to these two haplotypes. This is the haplotype assembly problem [11]. The *Minimum Error Correction (MEC)* model imposes the following

© Springer International Publishing AG, part of Springer Nature 2018
L. Wang and D. Zhu (Eds.): COCOON 2018, LNCS 10976, pp. 554–566, 2018.
https://doi.org/10.1007/978-3-319-94776-1_46

objective function on the selection of the haplotypes: find two haplotypes such that, summing over all the input fragments, the Hamming distance (interpreted as 'errors corrected') from the fragment to its nearest haplotype, is minimized. (When a fragment contains no information about a given position, a 'wildcard' character is used which does not contribute to the Hamming distance). The haplotypes can be thought of as cluster centres. This problem, originally introduced in 2001 [10], has been intensively studied in the computational biology literature: we refer to articles [2–4,7] and the references therein for a comprehensive overview. We assume that the optimal solution has always cost greater that zero since, otherwise, it is a trivial task to find the optimal solution.

Without restrictions on the use of wildcards it is straightforward to show NP-hardness, and the problem remains hard under a number of natural restrictions. For many years, however, it was unclear whether the problem is NP-hard if there are *no* wildcards in the input: this is the BINARYMEC problem. This was finally settled in 2014 by Feige, who showed that the equivalent *Hypercube 2-segmentation problem* is NP-hard [8].

Here we present a new variant of the problem: *Fixed-haplotype Minimum Error Correction* (FH-MEC). In this version of the problem one fixed haplotype (not necessarily the optimal one) is given as part of the input, and we are asked to find the other that minimizes the total error correction. This is a quite natural variation which models the situation when one of the haplotypes has already been determined. Fast algorithms for FH-MEC could also be used to heuristically explore the space of solutions to MEC, and thus to provide warm-start upper bounds for MEC algorithms.

It is straightforward (by simply adding many copies of the fixed haplotype to the input) to reduce FH-MEC *to* MEC in an approximation-preserving way (under preservation of common restrictions on the use of wildcards), so the Binary Fixed-Haplotype variant of MEC, BINARYFH-MEC, inherits the PTAS that via the clustering literature was already known to exist for BINARYMEC [4,9,12]. Determining the complexity of FH-MEC is, however, a more involved task since there is no obvious reduction in the opposite direction. We show in this article that FH-MEC is APX-hard by providing an L-reduction from the MAXCUT problem on cubic graphs. Our central result is a proof that BINARYFH-MEC is NP-hard. This is a non-trivial adaption of the elegant proof by Feige [8]. Feige, who reduces from MAXCUT, works purely with ℓ_1-norms, but unfortunately this option is not open to us due to the presence of the fixed haplotype, which cannot be interpreted this way. Another difficulty posed is that we also have to explicitly identify a fixed haplotype sufficient to induce hardness, and deal with a number of subtle technicalities concerning the way Hadamard (sub)matrices, and submatrices encoding the endpoints of graph edges (from the MAXCUT instance), are divided between the fixed haplotype and the variable haplotype. Although the NP-hardness of BINARYFH-MEC implies the NP-hardness of UNGAPPEDFH-MEC (where each haplotype fragment covers a *contiguous* interval of positions), we show an alternative NP-hardness reduction for this problem which is much simpler and potentially easier to manipulate into stronger forms of hardness, and thus of independent interest.

Ending the article on a positive note, we return to BINARYMEC. We follow the trend towards simplification given in [3] and provide another very simple polynomial-time 2-approximation algorithm for this problem. Our algorithm has, compared to [3], lower polynomial dependency on the length of the haplotypes we are constructing (at the expense of higher dependency on the number of fragments).

Definitions and Notations: A *fragment matrix* F is a matrix with n rows and m columns, every entry of which is in $\{-1, 1, *\}$. A $*$ entry is called a *hole*, and encodes an unknown value. F is *binary* if F contains no holes. F is *ungapped* if, for every row $r \in F$, there exists no hole in r such that there is a non-hole entry somewhere to the left and somewhere to the right of it.

Let r_i, r_j be two distinct rows of F. By $r_i[k]$ we denote the kth entry of r_i. Given two vectors r_i, r_j of the same dimension their (generalized) Hamming distance is defined as

$$d(r_i, r_j) = |\{k : r_i[k], r_j[k] \in \{-1, 1\}, r_i[k] \neq r_j[k]\}|. \tag{1}$$

i.e., $d(r_i, r_j)$ counts in how many positions the two vectors differ, where $*$ characters in one vector induce no errors, no matter what is the corresponding entry of the other vector. The Minimum Error Correction (MEC) problem is defined as follows.

Problem: MEC
Input: An $n \times m$ fragment matrix.
Output: Two m-dimensional vectors $h_1, h_2 \in \{-1, 1\}^m$, such that the following sum over all rows of F is minimized:

$$\sum_{r_i \in F} \min \left\{ d(h_1, r_i), d(h_2, r_i) \right\}.$$

In other words, the goal is to find two haplotypes h_1 and h_2 minimizing the sum of (generalized) Hamming distances of each row of F to its closest haplotype. This creates a bipartition of the rows into two groups, where rows that share the same closest haplotype are in the same group or *partition*. Ties can be broken arbitrarily. To make a row equal to its closest haplotype, the differing positions (errors) would have to be corrected. By minimizing the sum of these error corrections, the most likely parental haplotypes are found. Observe that a bipartition of the rows immediately induces two haplotypes by a simple majority voting rule on the rows within the same bipartition. Two variants of this problem, UNGAPPEDMEC and BINARYMEC, minimize the same function, but take as input an ungapped and binary fragment matrix, respectively.

In the Fixed-Haplotype MEC (FH-MEC) problem, one of the haplotypes h_1, h_2 is fixed and part of the input:

Problem: FH-MEC
Input: An $n \times m$ fragment matrix F and an m-dimensional vector $h_1 \in \{-1, 1\}^m$ (the fixed centre).
Output: An m-dimensional vector $h_2 \in \{-1, 1\}^m$, such that the following is minimized:

$$\sum_{r_i \in F} \min \left\{ d(h_1, r_i), d(h_2, r_i) \right\}.$$

For FH-MEC, binary and ungapped variants exist as well. In this paper, the haplotypes h_1 and h_2 are sometimes described as the *fixed* and *variable* haplotypes, respectively.

Given a minimization problem Π, we say that an algorithm $\mathcal{A}$ is a ρ-approximation algorithm if for any given instance I for Π (i) $\mathcal{A}$ runs in polynomial time in the size of I, and (ii) it outputs a solution $sol(I)$ with value at most $\rho \cdot opt(I)$. Here $opt(I)$ corresponds to the optimal solution value for I. Note that $\rho \geq 1$.

2 APX-Hardness of FH-MEC

In this section we will prove that FH-MEC is APX-hard by showing that the CUBICMAXCUT problem, where the input graph is *cubic*, L-reduces [13] to our problem. A cubic graph is a graph where every vertex has exactly three adjacent vertices (i.e., the degree of each vertex is exactly three). MAXCUT is APX-hard, even for cubic graphs [1]. Moreover, the value of MAXCUT on cubic graphs has a lower bound of 2/3 of the number of the edges [4], which will be used to prove that the proposed reduction is indeed an L-reduction.

Theorem 1. FH-MEC *is APX-hard.*

Proof. Let $G = (V, E)$ be an arbitrary, cubic, connected graph corresponding to an input to the CUBICMAXCUT problem. Let F be a $|V| \times 2|E|$ fragment matrix to be constructed as follows: Every edge $e \in E$ is represented by a block of two columns of F, and every vertex $v \in V$ is represented by a row of F. F is constructed as follows: First arbitrarily orient the edges of the graph. For every edge $e = (u, v)$, set its corresponding columns in F to $(1 \ -1)$ in row u, and to $(-1 \ 1)$ in row v and set its corresponding columns in the other rows to $(* \ *)$.

A simple example for the cycle graph C_3 on vertices $\{1, 2, 3\}$ with orientations $(1, 2), (2, 3), (1, 3)$ is given below.

$$F = \begin{pmatrix} 1 & -1 & * & * & 1 & -1 \\ -1 & 1 & 1 & -1 & * & * \\ * & * & -1 & 1 & -1 & 1 \end{pmatrix}$$

Now, let the fixed haplotype h_1 be the all -1 vector. We will first prove that MAXCUT$(G) = c$ if and only if UNGAPPEDFH-MEC$(F, h_1) = 2|E| - c$. Then, we will show that the conditions of an L-reduction are satisfied. There are 2 cases to consider:

Two vertices connected by an edge are in the same partition: The rows containing $(-1\ 1)$ and $(1\ -1)$ will be in the same partition. If both rows are closest to h_1 (having value $(-1\ -1)$ in the two corresponding columns), then these columns will contribute 2 to the error correction. If both rows are closest to the variable haplotype h_2, any values on the corresponding columns of h_2 will contribute 2 to the error correction.

Two vertices connected by an edge are *not* in the same partition: In every pair of columns representing an edge, there are only two rows filled with numbers. If these rows are placed in separate partitions, the row closest to h_1 will increase the error correction by 1. h_2 can be set to the other row without contributing to the error correction.

For vertices that are not connected by an edge, there is no column pair that is not already covered in the two cases discussed above. For every column pair covered in cases 1 and 2, the rows corresponding to these vertices will be $(*\ *)$. Thus, they do not contribute to the error correction.

For every edge, either case 1 or case 2 will hold. Edges that are split over the partitions (i.e., cut-edges) will contribute 1 to the error correction (case 2). Edges that are not split over the partitions will contribute 2 to the error correction (case 1). The minimum error correction is found by maximizing the number of cut edges. There are $|E|$ edges in G. Therefore, c edges are cut iff the error correction of a solution is $2|E| - c$.

To complete the proof, we will show that the conditions of an L-reduction are satisfied. Let G be an instance of CUBICMAXCUT. Let $R(G) = (F, h_1 = \{-1\}^{2|E|})$ be the instance of FH-MEC that is constructed from G. Clearly, $R(G)$ can be constructed in polynomial time. Let $Opt(G)$ be the value of the maximum cut of G. Let $Opt(R(G))$ be the minimum error correction of $R(G)$. Lastly, let s be a feasible solution of $R(G)$ and $S(s)$ the corresponding solution for G and let $c(s)$ and $c(S(s))$ be their respective costs. According to the definition of an L-reduction [13], two conditions need to be satisfied:

$$Opt(R(G)) \leq \alpha Opt(G) \tag{2}$$

$$|Opt(G) - c(S(s))| \leq \beta|Opt(R(G)) - c(s)| \tag{3}$$

where α, β are positive constants. We have that $Opt(G) \geq \frac{2}{3}|E|$ for any cubic graph G. We showed that $Opt(R(G)) = 2|E| - Opt(G) \leq 2Opt(G)$. Therefore, taking $\alpha = 2$ will be sufficient to satisfy Eq. (2). For any bipartition s of F of cost $c(s)$, the cost of the corresponding cut is $c(S(s)) = 2|E| - c(s)$. Thus, $Opt(G) - c(S(s)) = Opt(G) - 2|E| + c(s)$, and $Opt(R(G)) - c(s) = 2|E| - Opt(G) - c(s)$. This shows that $Opt(G) - c(S(s)) = -(Opt(R(G)) - c(s))$. Therefore, taking $\beta = 1$ will satisfy Eq. 3) and this completes the proof. □

3   NP-Hardness of BINARYFH-MEC

NP-hardness for a variant of BINARYMEC was proven by a reduction from MAXCUT [8]. In that variant the objective is to maximize the sum of the ℓ_1

norms of the vector sums of the rows of each bipartition, rather than to minimize the error correction as we are interested in this paper. Here, we will prove NP-hardness for BINARYFH-MEC using a reduction inspired by [8]. A vanilla approach does not work and we need to resolve several technicalities that arise from the difference in the objective function and from the presence of a fixed center. In the following we will show that optimizing the one objective function is equivalent to optimizing the other by showing how to translate one objective function value to another. The following allows the conversion of an ℓ_1 norm to an error correction.

Lemma 1. *Let P be a subset of n_p rows of a binary fragment matrix F. The ℓ_1 norm of the sum of the rows of P is l, if and only if the contribution to the error correction of the rows is $(n_p m - l)/2$.*

Proof. Let n_{maj} be the *number* of bits in P that belong to the majority bit of their column. Let n_{min} be the bits that belong to minority bit of their column. If a column contains equal numbers of -1's and 1's, the majority bit can be chosen arbitrarily. Let l be the ℓ_1 norm of the sum of the rows of P. The contribution of a column to l is the absolute difference between the number of majority and minority bits in that column. Therefore, $l = n_{\mathrm{maj}} - n_{\mathrm{min}}$.

The total number of bits in P is $n_p m = n_{\mathrm{maj}} + n_{\mathrm{min}}$. The contribution of P to the error correction is equal to the number of minority bits n_{min} in P. We have that $n_{\mathrm{min}} = n_p m - n_{\mathrm{maj}} = n_p m - n_{\mathrm{min}} - l$ from which we immediately get that $n_{\mathrm{min}} = (n_p m - l)/2$. □

Corollary 1. *Given a bipartition of the rows of a binary matrix, the task of maximizing the ℓ_1 norm of the sum of the rows of each partition is equivalent to minimizing the error correction.*

The reduction involves the use of *Hadamard* matrices. We recall that an M-dimensional Hadamard matrix is a set of M row vectors in $\{-1, 1\}^M$, such that the vectors are pairwise orthogonal. This means that every pair of vectors will differ in exactly $M/2$ positions. Hadamard matrices can be constructed recursively [14] as follows: Let $H_1 = (1)$. From here, we can construct H_{2M} from H_M by using

$$H_{2M} = \begin{pmatrix} H_M & H_M \\ H_M & -H_M \end{pmatrix}. \tag{4}$$

This construction is also known as Sylvester's construction (James Joseph Sylvester, 1867). Using this recursive construction, M will be a power of 2. In the proof of our reduction, we use the fact that all columns of a recursively constructed Hadamard matrix H_M contain $M/2$ 1's:

Lemma 2. *Let H_M be an M-dimensional Hadamard matrix that is constructed as above. In each column of H_M, except for the first one, the number of 1's in that column is $M/2$.*

Proof. Since the first row contains only 1's, and the rows of H_M are pairwise orthogonal, all other rows contain $M/2$ 1's. Due to the recursive construction by Eq. (4), $H_M = H_M^T$. Therefore, all columns except the first column contain $M/2$ 1's. □

Feige showed an upper bound on the ℓ_1-norm of an arbitrary subset of q vectors of a Hadamard matrix (Proposition 2 in [8]):

Lemma 3 ([8]). *Consider an arbitrary set of q distinct vectors from an arbitrary Hadamard matrix H_M. Then, the ℓ_1 norm of their sum is at most $\sqrt{q}M$.*

Theorem 2. BINARYFH-MEC *is NP-hard.*

Proof. Given a graph $G = (V, E)$, an instance to the MAXCUT problem, an $M|V| \times M|E|$ fragment matrix F is constructed where M will be fixed later on. Every vertex $v \in V$ is represented by a block of M rows, and every edge $e \in E$ is represented by a block of M columns. Arbitrarily orient all edges $e \in E$ so every edge is now an ordered pair of vertices. For every edge $e = (u, v)$, in the block of columns representing e, set the block representing vertex u to all 1's, and set the block representing vertex v to all -1's. Set each block representing one of the remaining vertices (not incident to e) to the M-dimensional Hadamard matrix H_M, constructed recursively as shown in Eq. (4).

The fixed haplotype $h_1 \in \{-1, 1\}^{M|E|}$ is set as follows: for each block of columns representing an edge, the corresponding coordinates of h_1 are set to $M/2$ 1's followed by $M/2$ -1's.

We will first discuss the case where an optimum solution to BINARYFH-MEC on the matrix F and fixed haplotype h_1 never splits the rows belonging to a single vertex block to two different parts of the bipartition. Such a block of rows will be assigned to either the fixed haplotype h_1 or the variable haplotype h_2. For each block of columns representing an edge, there are 4 options to consider. Each of these options shows the possible values for the error correction of cut and uncut edges. After that we will calculate the contribution of the blocks of Hadamard matrices to the error correction. In the following, when we say that an edge is cut we mean that the block that corresponds to one if its vertices is assigned to one haplotype but the block corresponding to the other vertex to the other haplotype.

The edge is cut and the block of -1's is closest to h_2. The block of 1's corresponding to one of the vertices incident to the edge, will contribute $M^2/2$ to the error correction. If the block of -1's is the only block closest to h_2, h_2 can be set to all -1's and the block of -1's will not contribute to the error correction. If other rows, that include blocks with Hadamard matrices are included in h_2, the first column of h_2 will be set to 1, and the block of -1's will contribute M to the error correction.

The edge is cut and the block of 1's is closest to h_2. The block of -1's will cause an error correction of $M^2/2$. h_2 can be set to all 1's on the corresponding entries without contributing to the error correction.

The edge is not cut and both blocks are closest to h_2. If both blocks are assigned to h_2, the first column among the rows assigned to h_2 will always have 1 as a majority bit. In every other column, there will be no unique majority bit, making the haplotype choice unimportant. The two blocks together will contribute M^2 to the total error correction.

The edge is not cut and both blocks are closest to h_1. In the first $M/2$ columns, the block of 1's will not contribute to the error correction and the block of -1's will contribute $M^2/2$. In the second $M/2$ columns, the block of -1's will not contribute to the error correction and the block of 1's will contribute $M^2/2$. The total contribution to the error correction for the two blocks will be M^2.

From the cases above it is clear that, when ignoring the Hadamard blocks, a cut edge will contribute to the error correction either $M^2/2$ or $M^2/2 + M$, while an uncut edge will have contribution of M^2. Note that every Hadamard block will have no error in the first column and an error correction of exactly $M/2$ in each one of the remaining columns, regardless which haplotype is assigned to, yielding a total error correction of $M(M-1)/2$. Note that for each edge there are $(|V| - 2)$ Hadamard blocks in that column block representing that edge, thus in total we have $(|V| - 2)|E|$ Hadamard blocks.

Summing up the terms of (i) the c cut edges each one contributing either $M^2/2$ or $M^2/2 + M$, (ii) the $|E| - c$ uncut edges each one with contribution of M^2, and (iii) $(|V| - 2)|E|$ Hadamard matrices each one contributing $M(M-1)/2$, we see that a cut of size c will have an error correction ec in the interval

$$\left[\frac{((|V||E| - c)M^2 - |E|M(|V| - 2))}{2}, \frac{((|V||E| - c)M^2 - M((|V| - 2)|E| + 2c))}{2} \right]$$

It is straightforward to see that the difference in error correction between cuts of size c and $c + 1$ is at least $M^2/2 - (c + 1)M$. By taking $M \geq 2|V|^2|E|^2$ and since $c \leq |E|$, knowing ec, it is always possible to distinguish between cuts of size c and $c + 1$.

On the other hand, it could be possible that splitting rows belonging to a block (representing a vertex) could give us a lower error correction as opposed to not splitting a block. If this potential decrease is less than $M^2/2 - (c + 1)M$, it is still possible to distinguish between cuts of size c and $c + 1$.

Assume a Hadamard block is split, and q of its rows are closest to h_2. By Lemma 3, the ℓ_1 norm of these q rows is at most $\sqrt{q}M$. By Lemma 1, the contribution to the error correction of this subset is at least $(qM - \sqrt{q}M)/2$. The crucial observation is that a fixed haplotype h_1 is equal to one of the rows of H_M i.e., there is a row in H_M that has $M/2$ 1's followed by $M/2$ -1's. This fact can be derived from the recursive construction shown in Eq. (4). Abusing slightly notation, we say that that row is equal to h_1. Thus, all rows, except the one row equal to h_1, will contribute $M/2$ to the error correction. Summing up, this will be $(M - q - 1)M/2$. From here it follows that the contribution of a split Hadamard codematrix to the error correction is at least $(M^2 - (\sqrt{q} + 1)M)/2$. Since $q < M$

(since we assume splitting of the Hadamard blocks), splitting a Hadamard block can decrease the total error correction by at most $M^{3/2}$, by Lemma 3. There are $(|V| - 2)|E|$ Hadamard blocks in F, so taking $M \gg O(|V|^2|E|^2)$, the total decrease is at most $M^2/2 - (c + 1)M$.

Lastly, we investigate whether partially cutting edges can close the gap between cuts of size c and $c + 1$. Assume an edge $e = (u, v)$ is cut partially, and its corresponding blocks of 1's and -1's are distributed over both partitions. Let x_u, x_v be the fractions of rows corresponding to vertices u, v, respectively, which are closest to h_2 (and so $(1 - x_u), (1 - x_v)$ fractions of rows are assigned to h_1). In this bipartition, by majority voting, the contribution to the error correction will be at least $\min(x_u, x_v)M^2$. For the rows closest to h_1, the contribution will be $(2 - x_u - x_v)M^2/2$. Thus, the total contribution tc of the edges to the error correction is

$$
\begin{aligned}
tc &= \min(x_u, x_v)M^2 + \left(2 - \min(x_u, x_v) - \max(x_u, x_v)\right)\frac{M^2}{2} \\
&= M^2 - \left(\max(x_u, x_v) - \min(x_u, x_v)\right)\frac{M^2}{2} \\
&= M^2 - |x_u - x_v|\frac{M^2}{2} \\
&= M^2 - y_e\frac{M^2}{2}.
\end{aligned}
$$

In the above expression, y_e is the extent to which e is cut. Summing up over all edges, the total contribution of the blocks of 1's and -1's to the error correction is $|E|M^2 - M^2/2\sum y_e$. The term $\sum y_e$ can be bounded by observing that local search can always change a fractional cut into an integer cut, which is at least as large. Indeed, within a connected set of fractional vertices we can always either increase them all by some amount of decrease them all by some amount such that in each case at least one of the fractional vertices becomes 1 or 0 respectively. Since we change all the fractional values at the same time by the same amount, this does not alter (i.e., worsen) the value of the term $y_e = |x_u - x_v|$. Hence, $\sum y_e \leq c$, the value of the cut. Thus, the edges contribute at least $(|E| - c/2)M^2$ to the error correction, allowing a maximum decrease of cM, which cannot close the gap between cuts of size c and $c + 1$.   □

4   NP-Hardness of UNGAPPEDFH-MEC

The NP-hardness of UNGAPPEDFH-MEC is implicitly proven by Theorem 2. The proof that follows does not involve the use of Hadamard matrices, and is therefore less technical and more intuitive. Since little is known yet about the approximability of both UNGAPPEDMEC and UNGAPPEDFH-MEC, this more straightforward proof might see potential use in future research.

Theorem 3. UNGAPPEDFH-MEC *is NP-hard.*

Proof. Let $G = (V, E)$ be an arbitrary, connected graph. Let F be a $(M + |V|) \times 4|E|$ ungapped fragment matrix. Here, M will be a sufficiently large number. Every edge $e \in E$ is represented by a block of four columns of F, and every vertex $v \in V$ is represented by one of the rows of F. F is constructed as follows: As usual, each edge is oriented arbitrarily. For every edge $e = (u, v)$, set its corresponding columns in F to $(1\ 1\quad 1\quad 1)$ in row u, and to $(\ 1\quad 1\ 1\ 1)$ in row v. Set the corresponding block of columns in all other rows corresponding to vertices to $(1\ -1\ 1\ -1)$. Now, set its corresponding columns in $M/2|E|$ of the last M rows to $(1\ 1\ -1\ -1)$, and in $M/2|E|$ rows to $(-1\ -1\ 1\ 1)$. In the remaining blocks of columns, set all values in these rows to $*$. Lastly, let the fixed haplotype h_1 be -1 in all positions.

A simple example of the above construction for the simple triangle graph C_3 with vertices $\{1, 2, 3\}$ and oriented edges $\{(1, 2), (2, 3), (1, 3)\}$ is given below.

$$
F = \left(
\begin{array}{cccccccccccc}
1 & 1 & -1 & -1 & 1 & -1 & 1 & -1 & 1 & 1 & -1 & -1 \\
-1 & -1 & 1 & 1 & 1 & 1 & -1 & -1 & 1 & -1 & 1 & -1 \\
1 & -1 & 1 & -1 & -1 & -1 & 1 & 1 & -1 & -1 & 1 & 1 \\
1 & 1 & -1 & -1 & * & * & * & * & * & * & * & * \\
-1 & -1 & 1 & 1 & * & * & * & * & * & * & * & * \\
* & * & * & * & 1 & 1 & -1 & -1 & * & * & * & * \\
* & * & * & * & -1 & -1 & 1 & 1 & * & * & * & * \\
* & * & * & * & * & * & * & * & 1 & 1 & -1 & -1 \\
* & * & * & * & * & * & * & * & -1 & -1 & 1 & 1
\end{array}
\right) \left.\begin{array}{c} \\ \\ \\ \\ \\ \\ \end{array}\right\} M/(2|E|) \text{ copies}
\tag{5}
$$

For every set of columns representing an edge, the variable haplotype h_2 will either be $(1\ 1\ -1\ -1)$ or $(-1\ -1\ 1\ 1)$. By doing this, half of the last M rows will not contribute to the error correction. The other half can simply be assigned to h_1, contributing M to the error correction. Setting h_2 to a value other than $(1\ 1\ -1\ -1)$ or $(-1\ -1\ 1\ 1)$ will increase the error correction among the last M rows by at least $M/2$, since $M/2$ rows contribute 0 to the error correction in these configurations. Since there are only $4|E||V|$ values in the upper $|V|$ rows, the decrease in error correction of the new configuration can be at most $4|E||V|$. Therefore, when setting $M > 8|E||V|$, no different configuration will yield an optimum result.

Any row of the first $|V|$ rows that is assigned to h_1, will contribute 2 to the error correction for every edge, since the possible values $(1\ 1\ -1\ -1)$, $(-1\ -1\ 1\ 1)$ and $(1\ -1\ 1\ -1)$ all have a hamming distance of 2 to $(-1\ -1\ -1\ -1)$.

Let $e = (u, v)$ be an edge that is part of a maximum cut of G. If u is assigned to h_2, then the h_2 can be set to $(1\ 1\ -1\ -1)$, causing row u not to contribute to the error correction in the columns corresponding to e. If v is also assigned to the h_2, it will contribute 4 to the error correction in the columns corresponding to e. When assigning v to h_1 instead, the contribution will be 2. It follows that assigning two vertices that are connected by an edge to the same haplotype will contribute 4 to the error correction in the columns corresponding to that edge. Assigning the vertices to different haplotypes will contribute 2 to the error correction. Since $c = \text{MAXCUT}(G)$ edges can be split up this way, the total error correction for the edges is $4|E| - 2c$. Sequences that do not encode an edge will contribute 2 to the error correction in any haplotype. There are $|E|(|V| - 2)$ of these sequences, yielding an error of $2|E|(|V| - 2)$. Combining the errors of the edges, the values that do not encode an edge, and the last M rows, the total error correction is equal to $2|E||V| + M - 2c$. Since M, $|V|$ and $|E|$ are known for any instance, the maximum cut can always be determined based on the minimum error correction. □

5   A Simple 2-Approximation for BINARYMEC

For the BINARYMEC polynomial time approximation schemata (PTAS) are known [9,12]. In [3] a simple and fast 2-approximation algorithm was shown. The algorithm follows the simple observation that given a "conflict-free" matrix M, then any heterozygous *column* (i.e., not all 1 or not all −1 column) of M naturally induces a bipartition of the rows of M^1. Their algorithm tries to built from any binary matrix M a conflict free matrix M' that is induced by a column of M.

Here we give an even simpler 2-approximation algorithm for BINARYMEC. We show that it is enough to work directly with rows and, in particular, we show that there always exists a pair of rows of M that when considered as the two haplotypes h_1 and h_2 induce a 2-approximate solution. The algorithm simply iterates over all pairs of rows and picks the pair inducing the smallest error correction.

Theorem 4. *For every binary matrix F, there exists a pair of rows r_1, r_2 such that taking r_1, r_2 as the haplotypes will yield a 2-approximation to BINARYMEC.*

Proof. Let h_1 and h_2 be the haplotypes of an optimum solution to BINARYMEC. Let R_1 and R_2 be the partition of rows induced by h_1 and h_2, respectively. Thus, the cost of the optimum solution is $\sum_{r \in R_1} d(r, h_1) + \sum_{r \in R_2} d(r, h_2) =:$ opt. For $i = 1, 2$, let $r_i \in R_i$ be the row from R_i that is closest to h_i (in the Hamming distance). Then, by triangle inequality, for every row $r \in R_i$, $d(r, r_i) \leq d(r, h_i) + d(h_i, r_i) \leq 2d(r, h_i)$.

Let R_1^*, R_2^* be the partition of the rows induced by considering r_1 and r_2 as the haplotypes. The cost of this partition is minimum among all partitions, and thus at most the cost induced by the partition R_1, R_2, which is $\sum_{r \in R_1} d(r, r_1) +$ $\sum_{r \in R_2} d(r, r_2) \leq \sum_{r \in R_1} 2d(r, h_1) + \sum_{r \in R_2} 2d(r, h_2) = 2opt$. □

The running time is $\mathcal{O}(n^2 m)$ since we loop over pairs of rows (n^2 pairs of rows in total on instances with n rows) and for each pair we compute the error correction which takes $\mathcal{O}(m)$ time. The algorithm of [3] runs in time $\mathcal{O}(m^2 n)$.

---

[1] A binary matrix M is conflict-free if the rows of M can be bipartitioned into two sets such that the corresponding entries on each set are identical.

Thus, the new algorithm is quicker for inputs where n is much smaller than m. Moreover, if the optimum solution to BINARYMEC is k, then (after collapsing identical rows) the number of rows in the input will also be at most $k + 2$ – yet the number of columns could still be large. Hence our algorithm might have a role in *parameterized* approaches to solve or approximate BINARYMEC [3,6].

To see that this is tight, let $F = \begin{pmatrix} I \\ 1 - I \end{pmatrix}$, where I is the identity matrix. We further replace each 0 in F by -1. The two optimum haplotypes will be $\{-1\}^m$ and $\{1\}^m$. It is easy to see that each row will contribute 1 to the error correction, for a total error correction of n on instances of n rows. The approximation algorithm, on the other hand, will pick one row of I and one of $1 - I$ as haplotypes. The two chosen rows will not contribute to the error correction. For the $n - 2$ remaining rows, the contribution will be 2 for a total error of $2(n - 2)$.

Acknowledgements. The last author acknowledges the support of an NWO TOP 2 grant.

References

1. Alimonti, P., Kann, V.: Hardness of approximating problems on cubic graphs. In: Bongiovanni, G., Bovet, D.P., Di Battista, G. (eds.) CIAC 1997. LNCS, vol. 1203, pp. 288–298. Springer, Heidelberg (1997). https://doi.org/10.1007/3-540-62592-5_80
2. Bansal, V., Bafna, V.: HapCUT: an efficient and accurate algorithm for the haplotype assembly problem. Bioinformatics **24**(16), i153–i159 (2008)
3. Bonizzoni, P., Dondi, R., Klau, G.W., Pirola, Y., Pisanti, N., Zaccaria, S.: On the minimum error correction problem for haplotype assembly in diploid and polyploid genomes. J. Comput. Biol. **23**(9), 718–736 (2016)
4. Cilibrasi, R., Van Iersel, L., Kelk, S., Tromp, J.: The complexity of the single individual SNP haplotyping problem. Algorithmica **49**(1), 13–36 (2007)
5. International HapMap Consortium, et al.: A haplotype map of the human genome. Nature **437**(7063), 1299 (2005)
6. Downey, R.G., Fellows, M.R.: Fundamentals of Parameterized Complexity, vol. 201. Springer, London (2016). https://doi.org/10.1007/978-1-4471-5559-1
7. Etemadi, M., Bagherian, M., Chen, Z.-Z., Wang, L.: Better ILP models for haplotype assembly. BMC Bioinform. **19**(1), 52 (2018)
8. Feige, U.: NP-hardness of hypercube 2-segmentation (2014). arXiv preprint arXiv:1411.0821
9. Jiao, Y., Xu, J., Li, M.: On the k-closest substring and k-consensus pattern problems. In: Sahinalp, S.C., Muthukrishnan, S., Dogrusoz, U. (eds.) CPM 2004. LNCS, vol. 3109, pp. 130–144. Springer, Heidelberg (2004). https://doi.org/10.1007/978-3-540-27801-6_10
10. Lancia, G., Bafna, V., Istrail, S., Lippert, R., Schwartz, R.: SNPs problems, complexity, and algorithms. In: auf der Heide, F.M. (ed.) ESA 2001. LNCS, vol. 2161, pp. 182–193. Springer, Heidelberg (2001). https://doi.org/10.1007/3-540-44676-1_15
11. Lippert, R., Schwartz, R., Lancia, G., Istrail, S.: Algorithmic strategies for the single nucleotide polymorphism haplotype assembly problem. Brief. Bioinform. **3**(1), 23–31 (2002)

12. Ostrovsky, R., Rabani, Y.: Polynomial-time approximation schemes for geometric min-sum median clustering. J. ACM **49**(2), 139–156 (2002)
13. Papadimitriou, C.H., Yannakakis, M.: Optimization, approximation, and complexity classes. J. Comput. Syst. Sci. **43**(3), 425–440 (1991)
14. Phelps, K.T., Rifa, J., Villanueva, M.: Rank and kernel of binary hadamard codes. IEEE Trans. Inf. Theory **51**(11), 3931–3937 (2005)

Non-monochromatic and Conflict-Free Coloring on Tree Spaces and Planar Network Spaces

Boris Aronov[1], Mark de Berg[2], Aleksandar Markovic[2(✉)], and Gerhard Woeginger[3]

[1] New York University, New York City, USA
[2] TU Eindhoven, Eindhoven, The Netherlands
a.markovic@tue.nl
[3] RWTH Aachen, Aachen, Germany

Abstract. It is well known that any set of n intervals in $\mathbb{R}^1$ admits a non-monochromatic coloring with two colors and a conflict-free coloring with three colors. We investigate generalizations of this result to colorings of objects in more complex 1-dimensional spaces, namely so-called tree spaces and planar network spaces.

1 Introduction

Conflict-free colorings, CF-colorings for short, were introduced by Even *et al.* [5] and Smorodinsky [9] to model frequency assignment to base stations in wireless networks. In the basic setting one is given a set S of objects in the plane—often disks are considered—and the goal is to assign a color to each object such that the following holds: for any point p in the plane such that the set $S_p := \{D \in S \mid p \in D\}$ of objects containing p is non-empty, S_p must contain an object whose color is different from the colors of the other objects in S_p. Even *et al.* proved, among other things, that any set of disks admits a CF-coloring with $O(\log n)$ colors. This bound is tight in the worst case. Since then many different geometric variants of CF-colorings have been studied. For example, Har-Peled and Smorodinsky [6] generalized the result to objects with near-linear union complexity, while Even *et al.* [5] considered the dual version of the problem. See the survey by Smorodinsky [11] for an overview. A restricted type of a CF-coloring is a *unique-maximum (UM) coloring*, in which the colors are identified with integers, and the maximum color in the set S_p is required to be unique. Another type of coloring, often used as an intermediate step to obtain a CF-coloring, is *non-monochromatic (NM)*. In an NM-coloring—sometimes called *a proper coloring*—we only require that, for any point p in the plane, if the set S_p

BA has been partially supported by NSF Grants CCF-11-17336, CCF-12-18791, and CCF-15-40656, and by BSF grant 2014/170.

MdB and AM are supported by the Netherlands' Organisation for Scientific Research (NWO) under project no. 024.002.003.

L. Wang and D. Zhu (Eds.): COCOON 2018, LNCS 10976, pp. 567–578, 2018.
https://doi.org/10.1007/978-3-319-94776-1_47

contains at least two elements, not all of them have the same color. Smorodinsky [10] showed that if an NM-coloring of k elements using $\beta(k)$ colors exists for every k, one can CF-color n elements with $O(\beta(n) \log n)$ colors.

CF- or NM-coloring objects in $\mathbb{R}^1$ is significantly easier than in the planar case. In $\mathbb{R}^1$ the objects become intervals, assuming we require the objects to be connected, and a folklore result states that any set of intervals in $\mathbb{R}^1$ can be CF-colored with three colors and NM-colored with two colors. (This is achieved by the *chain methods*, which we describe in the full version [2].) Thus, unlike in the planar case, the number of colors for a CF- or NM-coloring of intervals in $\mathbb{R}^1$ does not depend on the number of intervals to be colored.

We are interested in generalizations of this result to 1-dimensional spaces that have a more complex topology than $\mathbb{R}^1$. To this end we consider *network spaces*: 1-dimensional spaces with the topology of an arbitrary graph. It is convenient to view a network space $\mathcal{N}$ as being embedded in $\mathbb{R}^2$, although the embedding is actually immaterial. In this view the *nodes* of $\mathcal{N}$ are points in $\mathbb{R}^2$, and the *edges* are simple curves connecting pairs of nodes and otherwise disjoint. We let $d \colon \mathcal{N}^2 \to \mathbb{R}_+$ denote the geodesic distance on $\mathcal{N}$. In other words, for two points $p, q \in \mathcal{N}$—these points may lie in the interior of an edge—we let $d(p, q)$ denote the minimum Euclidean length of any path connecting p to q in $\mathcal{N}$. We consider two special types of network spaces, *tree spaces* and *planar network spaces*, whose topology is that of a tree and a planar graph, respectively.

The objective of our paper is to investigate the number of colors needed to CF- or NM-color a set $\mathcal{A}$ of n objects in a network space, where we consider various classes of connected objects. (Here CF- and NM-colorings are defined as above: in a CF-coloring, for any point $p \in \mathcal{N}$ the set $S_p := \{o \in \mathcal{A} \mid p \in o\}$ of objects containing p should have an object with a unique color when it is non-empty, and in an NM-coloring the set S_p should not be monochromatic when it consists of at least two objects.) In particular, we consider balls on $\mathcal{N}$—the *ball centered at* $p \in \mathcal{N}$ *of radius* r is defined as $B(p, r) := \{q \in \mathcal{N} \mid d(p, q) \leqslant r\}$—and, for tree spaces, we also consider arbitrary connected subsets as objects. Note that, if the given network space is a single curve, then our setting, both for balls and for connected subspaces, reduces to coloring intervals in $\mathbb{R}^1$. The main question we want to answer is: How does the maximum number of colors needed to NM- or CF-color a set $\mathcal{A}$ of objects in a network space depend on the complexity of the network space and of the objects to be colored?

Our Results. We assume without loss of generality that the nodes in our network space either have degree 1 or degree at least 3—there are no nodes of degree 2. Nodes of degree 1 are also called *leaves*, and nodes of degree at least 3 are also called *internal nodes*.

We start by considering colorings on a tree space, which we denote by $\mathcal{T}$. Let $\mathcal{A}$ be the set of n objects that we wish to color, where each object $T \in \mathcal{A}$ is a connected subset of $\mathcal{T}$. Note that each such object is itself also a tree. From now on we refer to the objects in $\mathcal{A}$ as "trees," and always use "tree space" when talking about $\mathcal{T}$. Observe that internal nodes of a tree are necessarily internal nodes of $\mathcal{T}$, but a tree leaf may lie in the interior of an edge of $\mathcal{T}$. We will investigate CF- and NM-chromatic number of trees on tree space as a function of the following parameters:

- k, the number of leaves of the tree space $\mathcal{T}$;
- ℓ, the maximum number of leaves of any tree in $\mathcal{A}$;
- n, the number of objects in $\mathcal{A}$.

We define the CF-chromatic number $X_{\mathrm{cf}}^{\mathrm{tree,trees}}(k,\ell;n)$ as the minimum number of colors sufficient to CF-color any set $\mathcal{A}$ of n trees of at most ℓ leaves each, in a tree space of at most k leaves. The NM-chromatic number $X_{\mathrm{nm}}^{\mathrm{tree,trees}}(k,\ell;n)$ is defined similarly. Rows 3 and 4 in Table 1 give our bounds on these chromatic numbers. Notice that the upper bounds do not depend on n. In other words, any set of trees in a tree space can be colored with a number of colors that depends only on the complexity of the tree space $\mathcal{T}$ and of the trees in $\mathcal{A}$. (Obviously the number of objects, n, is an upper bound on these chromatic numbers as well. To avoid cluttering the statements, we usually omit this trivial bound.) We also study balls in tree spaces. Here it turns out to be more convenient to not use k (the number of leaves) as the complexity measure of $\mathcal{T}$, but

- t, the number of internal nodes of $\mathcal{T}$.

Table 1. Overview of our results. The folklore result for intervals on the line (that is, in $\mathbb{R}^1$) is explained in the full version [2].

Space	Objects	Coloring	Upper bound	Lower bound	Reference
Line	Intervals	NM	2	2	Folklore
Line	Intervals	CF	3	3	Folklore
Tree	Trees	NM	$\min\left(\ell+1, 2\sqrt{6k}\right)$	$\min\left(\ell+1, \left\lfloor \frac{1+\sqrt{1+8k}}{2} \right\rfloor\right)$	Section 2
Tree	Trees	CF	$O(\ell \log k)$	$\lfloor \log_2 \min(k,n) \rfloor$	Section 2
Tree	Balls	NM	2	2	Section 3.1
Tree	Balls	CF	$\lceil \log t \rceil + 3$	$\lceil \log(t+1) \rceil$	Section 3.1
Planar	Balls	NM	4	4	Section 3.2
Planar	Balls	CF	$\lceil \log_{4/3} t \rceil + 3$	$\lceil \log(t+1) \rceil$	Section 3.2

We are interested in the chromatic numbers $X_{\mathrm{cf}}^{\mathrm{tree,balls}}(t;n)$ and $X_{\mathrm{nm}}^{\mathrm{tree,balls}}(t;n)$. Rows 5 and 6 of Table 1 state our bounds for these chromatic numbers.

After studying balls in tree spaces, we turn our attention to balls in planar network spaces. Rows 7 and 8 of Table 1 contain our bounds on the corresponding chromatic numbers $X_{\mathrm{cf}}^{\mathrm{planar,balls}}(t;n)$ and $X_{\mathrm{nm}}^{\mathrm{planar,balls}}(t;n)$. Due to space constraints, lower bounds and some proofs are deferred to the full version [2].

Related Results. Above we considered CF- and NM-colorings in a geometric setting, but they can also be defined more abstractly. A CF-coloring on a hypergraph $\mathcal{H} = (V, E)$ is a coloring of the vertex set V such that, for every (non-empty) hyperedge $e \in E$, there is a vertex in e whose color is different from that of the other vertices in e. In a NM-coloring any hyperedge with at least two vertices should not be monochromatic. Smorodinsky's survey [11] also gives an overview of results on CF-colorings in this abstract setting.

The basic geometric version mentioned above—coloring objects in $\mathbb{R}^2$ with respect to points—can be phrased in terms of hypergraphs by letting the objects be the node set V and, for each point p in the plane, creating a hyperedge $e := S_p$. Another avenue for constructing a hypergraph $\mathcal{H}$ to be colored is to start with a graph $\mathcal{N}$, let the vertices of $\mathcal{H}$ be the nodes of $\mathcal{N}$ and create hyperedges for (the sets of vertices of) certain subgraphs of $\mathcal{N}$. For example, Pach and Tardos [8] considered the case where hyperedges are all the node neighborhoods. For this case, Abel $et\ al.$ [1] recently showed that a planar graph can always be CF-colored with only three colors, if we allow some nodes to be uncolored. (Otherwise, we can use a dummy color, increasing the number of colors to four.) As another example, we let the hyperedges be induced by all the paths in the graph. This setting is equivalent to an older notion of $node\ ranking$ [3], or $ordered\ coloring$ [7]. Note that in the above results the goal is to color the nodes of a graph. We, on the other hand, do not want to color nodes, but objects (connected subsets) in a network space (which has a graph topology, but is a geometric object).

2   Trees on Tree Spaces

Overview of the Coloring Procedure. Let $\mathcal{T}$ be a tree space with k leaves and let $\mathcal{A}$ be a set of n trees in $\mathcal{T}$, each with at most ℓ leaves. We describe an algorithm that NM-colors $\mathcal{A}$ in two phases: first, we select a subset $\mathcal{C} \subseteq \mathcal{A}$ of size at most $6k - 12$ and color it with at most $\min\left(\ell + 1, 2\sqrt{6k}\right)$ colors. In the second phase we extend this coloring to the whole set $\mathcal{A}$ without using new colors.

An edge e of $\mathcal{T}$ is a $leaf\ edge$ if it is incident to a leaf; the remaining edges are $internal$. We define $\mathcal{C} \subseteq \mathcal{A}$ as the set of at most $6k - 12$ trees selected as follows. For every pair (e, v), where e is an edge of $\mathcal{T}$ and v is an endpoint of e that is not a leaf of $\mathcal{T}$, we choose two trees containing v and extending the furthest into e (if they exist), that is, trees T of $\mathcal{A}$ containing v for which length$(T \cap e)$ is maximal, and place them in $\mathcal{A}(e, v)$. If two or more trees of $\mathcal{A}$ fully contain e, then $\mathcal{A}(e, v)$ contains two of them, chosen arbitrarily. If a tree contains an internal edge e fully, it may be chosen by both endpoints. We now define $\mathcal{A}(e) := \mathcal{A}(e, u) \cup \mathcal{A}(e, v)$ for each internal edge $e = uv$, $\mathcal{A}(e) := \mathcal{A}(e, v)$ for each leaf edge $e = uv$ with non-leaf endpoint v, and $\mathcal{C} := \bigcup \mathcal{A}(e)$, with the union taken over all edges e of $\mathcal{T}$. Then $\mathcal{A}(e)$ contains at most four trees for any internal edge e and at most two trees for any leaf edge e. If $\mathcal{T}$ has at most k leaves, it has at most k leaf edges and at most $k - 3$ internal edges; recall that $\mathcal{T}$ has no degree-two nodes. Thus $|\mathcal{C}| \leqslant 6k - 12$, as claimed. We first explain how to color $\mathcal{C}$.

Coloring $\mathcal{C}$. We color $\mathcal{C}$ in two steps. Let $T \in \mathcal{C}$ be a tree. We define $E(T)$ to be the set of edges e of $\mathcal{T}$ with $T \in \mathcal{A}(e)$. Firstly, if $\ell > 2\sqrt{6k}$ we select all subtrees T with $|E(T)| \geqslant \sqrt{6k}$, and give each of them a unique color. Since $\sum_e |\mathcal{A}(e)| \leqslant$

$6k - 12$, there are at most $\sqrt{6k} - 1$ such trees, so we use at most $\sqrt{6k} - 1$ colors. For each uncolored $T \in \mathcal{C}$, we create a new tree T', defined as the smallest tree containing $\bigcup_{e \in E(T)} e \cap T$; see Fig. 1. T' has at most $\ell' := \min(\ell, \sqrt{6k})$ leaves because $|E(T)| < \sqrt{6k}$. Define $\mathcal{C}' := \{T' \mid T \in \mathcal{C}\}$.

Fig. 1. The original tree T (left), the set $\bigcup_{e \in E(T)} e \cap T$ (middle), and the new tree T' (right).

The second step is to color $\mathcal{C}'$. We need the following lemma, which shows that an NM-coloring of $\mathcal{C}'$ carries over to $\mathcal{C}$.

Lemma 1. *Any NM-coloring of $\mathcal{C}'$ corresponds to an NM-coloring of $\mathcal{C}$, that is, if we give each tree $T \in \mathcal{C}$ the color of the corresponding tree $T' \in \mathcal{C}'$ then we obtain an NM-coloring.*

Next we show how to NM-color $\mathcal{C}'$. Fix an arbitrary internal node r of $\mathcal{T}$ and treat $\mathcal{T}$ as rooted at r. Our coloring procedure for $\mathcal{C}'$ maintains the following invariant: any path from r to a leaf v of $\mathcal{T}$ consists of three disjoint consecutive subpaths (some possibly empty), in this order, as illustrated in Fig. 2:

- a *non-monochromatic* subpath containing the root on which at least two trees are colored with at least two different colors,
- a *singly-colored* subpath covered by exactly one colored tree, and
- an *uncolored* subpath containing the leaf on which no tree is colored.

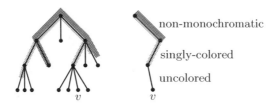

Fig. 2. A coloring of trees (left) and an illustration of the invariant for v (right). (Color figure online)

Observation 1. *Any set of trees containing r and satisfying the invariant described above is NM-colored if we disregard uncolored trees.*

We color the trees $T \in \mathcal{C}'$ that contain r in an arbitrary order, using $\ell' + 1$ colors, as follows: for each leaf v of T, we follow the path from v to the root r to find a singly-colored part. Note that if we find a singly-colored part—by the invariant there is at most one such part on the path from v to r—we cannot use that color for T. Since T has at most ℓ' leaves, this eliminates at most ℓ' colors. Hence, at least one color remains for T.

Lemma 2. *The procedure described above maintains the invariant and colors all trees of $\mathcal{C}'$ containing r with at most $\ell' + 1$ colors.*

Once all the trees containing r are colored we delete r from $\mathcal{T}$, that is, we consider the space $\mathcal{T} \setminus \{r\}$, and we take the closures of the resulting connected components. This creates a number of subspaces such that each uncolored tree in $\mathcal{C}'$ is contained in exactly one of them. Consider such a subspace $\mathcal{T}'$ and let r' be the neighbor of r in $\mathcal{T}'$. We now want to recursively color the uncolored trees in $\mathcal{T}'$, taking r' as the root of $\mathcal{T}'$. However, the invariant might not hold on the edge e from r' to the old root r: Since now r is considered a child of r', the order of the three parts might switch on e—see Fig. 3. Suppose this is the case, and let c_e be the color of the singly-colored part on the edge e. (If the singly-colored part is empty, we can cut the tree between the non-monochromatic and the uncolored part and recurse immediately, which maintains the invariant.) Note also that, for the order to switch, the non-monochromatic part needs to end on e, and therefore the only color used in any singly-colored part of the tree rooted at r' is c_e. We overcome this problem by carefully choosing the order in which we color the trees containing r'. Namely, we fist color the tree T extending the farthest into e. In this case, there is only one color forbidden, namely c_e. We can therefore easily color T. We can then trim the treespace $\mathcal{T}'$ to remove any non-monochromatic and singly-colored part and hence restore the invariant and continue with the coloring.

Fig. 3. When recursing on the subspace rooted at r' (leftmost), the invariant does not hold anymore (middle left), as the parts are switched on the edge between r and r'. To remedy this, we first color the tree extending the farthest into that edge (middle right), starting from r'. We then trim the tree to fix the invariant (rightmost).

Lemma 3. $\mathcal{C}$ *admits an NM-coloring with* $\min(\ell + 1, 2\sqrt{6k})$ *colors.*

Extending the Coloring from $\mathcal{C}$ to $\mathcal{A}$. Let $c\colon \mathcal{C} \to \mathbb{N}$ be an NM-coloring on $\mathcal{C}$. We extend the coloring to $\mathcal{A}$ as follows. We start by coloring all trees in $\mathcal{A} \setminus \mathcal{C}$ containing an internal node of $\mathcal{T}$ using an arbitrary color already used. We then treat all edges in an arbitrary order, coloring all trees contained in the edge as explained now.

Let $e = rr'$ be an arbitrary edge of $\mathcal{T}$ and $\mathcal{A}^*(e)$ be the set of uncolored trees contained in e. We color $\mathcal{A}^*(e)$ as follows. We first color the set of uncolored trees contained in e naively using the chain method. For this we use two new colors, which are used for all chains—we can re-use the same two colors for the chains, since trivially the chains in any two edges e, e' do not interact. However, we can avoid using two extra colors and re-use the colors from $\mathcal{C}$ as explained in the full version [2].

Theorem 2.

1. $X_{\text{nm}}^{\text{tree,trees}}(k, \ell; n) \leqslant \min\left(\ell + 1, 2\sqrt{6k}\right)$.
2. $X_{\text{cf}}^{\text{tree,trees}}(k, \ell; n) = O(\ell \log k)$.

Proof. For the NM-coloring part of the theorem, we use Lemma 3 and the coloring extension explained in the full version [2]. For the second part, if $\ell > 2\sqrt{6k}$ we again reduce $\mathcal{C}$ to $\mathcal{C}'$ using at most $\sqrt{6k} - 1$ colors. Then use the result by Smorodinsky [10] on the NM-coloring on $\mathcal{C}'$ provided by Lemma 2. Since this coloring uses at most $\ell' + 1$ colors and $|\mathcal{C}'| \leqslant 6k - 12$, the CF-coloring uses $O(\ell \log k)$ colors. We then extend the coloring to $\mathcal{A}$ using similar techniques as for the NM-coloring. This coloring uses $O(\sqrt{k} \log k)$ colors if $\ell > 2\sqrt{6k}$, which is in $O(\ell \log k)$, and directly $O(\ell \log k)$ colors otherwise. Note that a direct application of the result of Smorodinsty [10] would give a $O(\ell \log n)$ bound instead. □

3   Balls in Tree Spaces and on Planar Network Spaces

In this section we restrict the objects to balls. Let $\mathcal{N}$ be a network space, $d\colon \mathcal{N}^2 \to \mathbb{R}$ a distance function on $\mathcal{N}$, and let $\mathcal{A}$ be a set of balls on $\mathcal{N}$. We define the coverage $cov_x(B)$ of a node x by a ball $B = B(p, r)$ containing x as $cov_x(B) := r - d(p, x)$. Given a node x contained in at least one ball from $\mathcal{A}$, we define B_x as the ball maximizing the coverage of x, where we break ties using an arbitrary but fixed ordering on the balls. We say that B_x is *assigned* to x. Note that B_x does not exist if no ball contains x, and that a ball can be assigned to multiple nodes. We will regularly use the following lemma regarding the assigned balls.

Lemma 4. *Let x be an internal node of $\mathcal{N}$.*

(i) *Suppose $\mathcal{N}$ is a tree space, and let $\mathcal{T}_1, \ldots, \mathcal{T}_{\deg(x)}$ denote the subtrees resulting from removing x from $\mathcal{N}$ or, more precisely, the closures of the connected components of $\mathcal{T} \setminus \{x\}$. Let p be a point in some subtree $\mathcal{T}_i$ and suppose p is contained in a ball $B \in \mathcal{A}$ whose center lies in $\mathcal{T}_j$ with $j \neq i$. Then $p \in B_x$.*

(ii) *Suppose x is contained in at least one ball in $\mathcal{A}$. Let π be a shortest path from x to the center of B_x, and let y be a node on the path π. Then B_x is also assigned to y, that is, $B_x = B_y$.*

3.1 Tree Spaces

For balls on a tree space $\mathcal{T}$, the upper bounds from Theorem 2 with $\ell = k$ apply. Below we improve upon these bounds using the special structures of balls. Let $\mathcal{T}$ be a tree with t internal nodes. We present algorithms to NM-color balls on trees using two colors, and CF-color them with $\log t + 3$ colors.

Let $\mathcal{A}$ be a set of n balls on $\mathcal{T}$. Let also $\mathcal{C} := \{B = B(c, r) \mid \exists x : B = B_x\}$ be the set of balls assigned to at least one internal node. Recall that an internal node x is assigned the ball maximizing the coverage of x.

NM-Coloring. We first explain how to NM-color $\mathcal{A}$. We use a divide-and-conquer approach. If $t = 0$, that is $\mathcal{T}$ consists of a single node or a single edge, we use the chain method for NM-coloring with colors blue and red. If $t > 0$, then we proceed as follows. Let $e = uv$ be an edge of $\mathcal{T}$. Let $\mathcal{T}_u$, respectively $\mathcal{T}_v$, be the connected component of $\mathcal{T} \setminus e$ containing u, respectively v. Recall that B_u and B_v are the balls assigned to u and v, respectively. Note that we may assume that both B_u or B_v exist, for otherwise recursion is trivial. Also observe that B_u and B_v may coincide. We define

$$\mathcal{A}(u) := \{\text{balls } B \in \mathcal{A} \text{ whose center lies in } \mathcal{T}_u\} \cup \{B_u\},$$

We define $\mathcal{A}(v)$ similarly. We recursively color $\mathcal{A}(u)$ in $\mathcal{T}_u$ and $\mathcal{A}(v)$ in $\mathcal{T}_v$, obtaining colorings of $\mathcal{A}(u)$ and $\mathcal{A}(v)$ with colors blue and red. In the recursive calls on $\mathcal{A}(u)$, and similarly for $\mathcal{A}(v)$, we "clip" the balls to within $\mathcal{T}_u$. Note that the clipped balls are still balls in the space $\mathcal{T}_u$. This is clear for the balls whose center lies in $\mathcal{T}_u$. The center of B_u may not lie in $\mathcal{T}_u$, but in that case it behaves within $\mathcal{T}_u$ as a ball with center u and radius $cov_u(B_u)$.

Let $\mathcal{A}(e) := \mathcal{A} \setminus (\mathcal{A}(u) \cup \mathcal{A}(v))$ be the set of the remaining balls. In other words, $\mathcal{A}(e)$ contains the balls whose center is contained in e, except for B_u and B_v. We color $\mathcal{A}(e)$, possibly swapping colors in $\mathcal{A}(u)$ or $\mathcal{A}(v)$, as follows.

- If $B_u = B_v$, we first ensure that it gets the same color in both $\mathcal{A}(u)$ and $\mathcal{A}(v)$ by swapping colors in one of the two subsets if necessary. We then color all balls in $\mathcal{A}(e)$ blue if B_u is red, and red if B_u is blue.
- If $B_u \neq B_v$, let π be a longest simple path containing u and v. We color $\mathcal{A}(e) \cup \{B_u, B_v\}$ restricted to π using the non-monochromatic chain method. We then possibly swap colors in $\mathcal{A}(u)$ and $\mathcal{A}(v)$ so that B_u and B_v match the colors they were given by the chain method.

Both cases are illustrated in Fig. 4.

Theorem 3. $X_{nm}^{\text{balls,trees}}(t; n) = 2$.

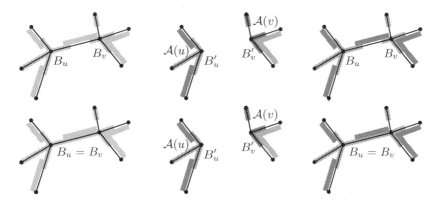

Fig. 4. On the left, we have the two different initial cases, i.e., on the top, $B_u \neq B_v$, on the bottom, $B_u = B_v$. In the middle, the recursive call is made. On the right, we use the two recursive colorings and swap colors if needed. (Color figure online)

Proof. The coloring obviously uses two colors. It remains to show it is non-monochromatic. We use induction on t. If $t = 0$, the coloring is non-monochromatic since it uses the chain method.

Suppose now that $t \geqslant 1$ and that the claim holds for any tree space with fewer than t internal nodes. Let p be a point contained in at least two balls.

If p is contained in balls only of $\mathcal{A}(v)$, only of $\mathcal{A}(u)$, or only of $\mathcal{A}(e)$, it is contained in at least two balls of different colors. Indeed, the colorings of $\mathcal{A}(v)$ and $\mathcal{A}(u)$ are non-monochromatic since they use the method on a tree with fewer than t internal nodes and we can use the induction hypothesis. Moreover $\mathcal{A}(e)$ is non-monochromatic due to the chain method.

It remains to consider the case where p is contained in balls from at least two of the sets $\mathcal{A}(u)$, $\mathcal{A}(v)$, and $\mathcal{A}(e)$. We distinguish two cases: p is contained in a ball of $\mathcal{A}(e)$ and p is not contained in a ball of $\mathcal{A}(e)$.

If p is contained in a ball B of $\mathcal{A}(e)$, we can assume without loss of generality that p is also contained in a ball of $\mathcal{A}(v)$. By Lemma 4(i), we have that $p \in B_v$.

If $B_u = B_v$ then all balls in $\mathcal{A}(e)$ are given a different color than B_v hence p is contained in two balls of different color. If $B_u \neq B_v$ then we use the chain method on π. Hence if $p \in \pi$, it is contained in two balls of different color. To show that if $p \notin \pi$ then p is still contained in two balls of different colors, it suffices to notice that for any subset of balls of $\mathcal{A}(e)$ in which p is contained, the point $p' \in \pi$ at distance $d(u, p)$ from u is contained in the same set of balls from $\mathcal{A}(e)$ as π is the longest path containing e.

On the other hand, if p is not contained in a ball of $\mathcal{A}(e)$, then it is contained in at least one ball from $\mathcal{A}(u)$ and one from $\mathcal{A}(v)$. By Lemma 4 we have that $p \in B_u \cap B_v$.

We then have two cases. If $B_u = B_v$, then p is contained in another ball of $\mathcal{A}(u)$ or $\mathcal{A}(v)$, and then the coloring is non-monochromatic by the induction hypothesis. Otherwise B_u and B_v are part of the chain $\mathcal{A}(e) \cup \{B_u, B_v\}$, and hence p is contained in at least two balls of different color. □

CF-Coloring. The second algorithm CF-colors $\mathcal{A}$ using $\lceil \log t \rceil + 3$ colors. As before, define $\mathcal{C} := \{B = B(c, r) \mid \exists x : B = B_x\}$. We explain how to color $\mathcal{C}$ and then extend the coloring to $\mathcal{A}$. Let r be a node whose removal results in subtrees each of at most $t/2$ internal nodes. We color B_r (if it exists) with color 1. Let $\mathcal{T}_1, \ldots, \mathcal{T}_{\deg(r)}$ be subtrees resulting from removing r, that is, the closures of the connected components of $\mathcal{T} \setminus \{r\}$. For each $i = 1, \ldots, \deg(r)$, we recurse on $\mathcal{T}_i$ with the balls from $\mathcal{C}$ whose centers lie in $\mathcal{T}_i$. In such a recursive call, we consider a node to be an internal node when it was an internal node in the original space $\mathcal{T}$ and when it has not yet been selected as a splitting node in a previous call. Hence, when $t = 0$ in a recursive call on a subtree $\mathcal{T}' \subset \mathcal{T}$, then $\mathcal{T}'$ must be a single edge both of whose endpoints have already been treated.

The recursion stops when there are no more balls left (which must be the case when we have a recursive call with $t = 0$). Note that the internal nodes are fixed from the beginning, hence at some point of the recursion, a leaf node might still be considered internal for the purposes of the recursion.

Lemma 5. *The above algorithm CF-colors $\mathcal{C}$ using $\lceil \log t \rceil$ colors.*

We now wish to extend the coloring to balls in $\mathcal{A} \setminus \mathcal{C}$. To this end, define $\mathcal{T}' := \mathcal{T} \setminus (\bigcup \mathcal{C})$ to be the part of $\mathcal{T}$ that remains after removing all points covered by the balls in $\mathcal{C}$.

We finish the coloring with three more colors (using the chain method for CF-colorings) as explained next, resulting in $\lceil \log t \rceil + 3$ colors. We use the following lemma to show that the remaining balls can be reduced to intervals on disjoint lines. Note that it does not use tree spaces and can hence be applied also for planar network spaces.

Lemma 6. *For any ball $B \notin \mathcal{C}$, we have $\{p \in B \mid p \notin \bigcup \mathcal{C}\} \subseteq e$, where e is the edge containing the center of B.*

Theorem 4. $X_{\mathrm{cf}}^{\mathrm{tree,balls}}(t; n) \leqslant \lceil \log t \rceil + 3.$

3.2   Planar Network Spaces

NM-Coloring. We first explain how to NM-color balls on a planar network space $\mathcal{N}$. Let again $\mathcal{C}$ be the set $\{B = B(c, r) \mid \exists x : B = B_x\}$. We create a graph $\mathcal{G}_{\mathcal{C}}$ whose node set is $\mathcal{C}$ and whose edge set is defined as follows: there is an edge between B and B' if and only if there is an edge vv' in $\mathcal{T}$ with $B_v = B$ and $B_{v'} = B'$. It follows from Lemma 4 that for any ball B, the set of nodes of $\mathcal{N}$ to which B is assigned, together with the edges between these nodes, is a connected set. Therefore, $\mathcal{G}_{\mathcal{C}}$ is planar as well since its nodes correspond to disjoint connected subspaces in the planar space $\mathcal{N}$. We now use the Four Color Theorem to color $\mathcal{G}_{\mathcal{C}}$ and we give each ball in $\mathcal{C}$ the same color as the corresponding node in $\mathcal{G}_{\mathcal{C}}$.

Lemma 7. *The coloring on $\mathcal{C}$ is non-monochromatic and uses at most four colors.*

We now wish to extend the coloring to balls in $\mathcal{A} \setminus \mathcal{C}$. To this end, define $\mathcal{N}' := \mathcal{N} \setminus (\bigcup \mathcal{C})$ to be the part of $\mathcal{N}$ that remains after removing all points covered by the balls in $\mathcal{C}$. The proof of the following lemma is similar to the proof of Lemma 6.

Lemma 8. *Consider a ball $B \in \mathcal{A} \setminus \mathcal{C}$, and let $B' := B \cap \mathcal{N}'$. Then B' is contained in a single edge of $\mathcal{N}'$.*

For each edge e of $\mathcal{N}'$, let $\mathcal{A}(e)$ denote the set of balls contained in e. Let u and v denote the endpoints of the edge in $\mathcal{N}$ containing e. We color the uncolored balls in e using the chain method with two colors not equal to $c(B_u)$ and $c(B_v)$. We have now colored the balls in $\mathcal{C}$ as well as the balls in $\mathcal{A} \setminus \mathcal{C}$ that lie at least partially in $\mathcal{N}'$. Next we explain how to color the remaining balls, which are fully covered by the balls in $\mathcal{C}$.

Lemma 9. *Any uncolored ball is contained in the union of at most three balls.*

Using this lemma, we can easily finish the NM-coloring.

Theorem 5. $X_{\mathrm{nm}}^{\mathrm{planar,balls}}(t; n) = 4$.

CF-Coloring. We now explain how to CF-color balls on a planar network. As before, define $\mathcal{C} := \{B = B(c, r) \mid \exists x : B = B_x\}$. We first CF-color $\mathcal{C}$ using the following recursive algorithm introduced by Smorodinsky [10]: we select a maximum independent set in $C_1 := \mathcal{C}$, we give it color 1, place all uncolored balls in C_2, and recurse. We claim that for all i, the Delauney graph $D_i := (C_i, E_i)$ on the balls in C_i is planar, where $E_i := \{\{B_1, B_2\} \mid \exists p \in \mathcal{N} : p \in B_1 \cap B_2 \text{ and } \forall B \notin \{B_1, B_2\} : p \notin B\}$.

Lemma 10. D_i *is planar.*

Using this lemma and the Four Color Theorem, we get a coloring on $\mathcal{C}$ using $\lceil \log_{4/3} t \rceil$ colors. Note that this method does not give an efficient algorithm because of the use of the Four Color Theorem. For a fast algorithm, we can use a linear-time algorithm [4] to find an independent set of size at least $n/5$, leading to $\lceil \log_{5/4} t \rceil$ colors.

We then color the balls in $\mathcal{A} \setminus \mathcal{C}$. Using Lemma 6, we have that for any such ball B, the set of points contained in B but not in any ball in $\mathcal{C}$ is contained in one edge of $\mathcal{N}$. Therefore, if we cut $\cup \mathcal{C}$ out of $\mathcal{N}$, the remaining space is a union of disjoint segments, and any object that is not colored is contained in at most one segment. We can therefore use the chain coloring on each segment with the two additional colors and the dummy one.

Finally, any point in $\cup \mathcal{C}$ is contained in a ball in $\mathcal{C}$ of unique color, and any point not in $\cup \mathcal{C}$, is contained in at most one ball of each of the two additional colors. Therefore, the coloring is conflict-free. This yields the following theorem.

Theorem 6. $X_{\mathrm{cf}}^{\mathrm{planar,balls}}(t; n) \leqslant \lceil \log_{4/3} t \rceil + 3$.

4   Concluding Remarks

We studied NM- and CF-colorings on network spaces, where the objects to be colored are connected regions of the network space. We showed that the number of colors can be bounded as a function of the complexity (which depends on the type of space and of objects) of the network space and the objects, rather than on the number of objects. All our bounds are tight up to some constants, except for $X_{\text{cf}}^{\text{tree,trees}}(k, \ell; n)$ where the upper bound is a factor ℓ away from the lower bound. Closing this gap remains an open problem. It would also be interesting to find bounds on general connected objects on any network space, or other settings where the number of colors depends on the complexity of the space and objects rather the number of objects.

References

1. Abel, Z., Alvarez, V., Demaine, E.D., Fekete, S.P., Gour, A., Hesterberg, A., Keldenich, P., Scheffer, C.: Three colors suffice: conflict-free coloring of planar graphs. In: Proceedings of the 28th Annual ACM-SIAM Symposium on Discrete Algorithms, pp. 1951–1963 (2017)
2. Aronov, B., de Berg, M., Markovic, A., Woeginger, G.: Non-monochromatic and conflict-free coloring on tree spaces and planar network spaces. CoRR (2018)
3. Bodlaender, H.L., Deogun, J.S., Jansen, K., Kloks, T., Kratsch, D., Müller, H., Tuza, Z.: Rankings of graphs. In: Mayr, E.W., Schmidt, G., Tinhofer, G. (eds.) WG 1994. LNCS, vol. 903, pp. 292–304. Springer, Heidelberg (1995). https://doi.org/10.1007/3-540-59071-4_56
4. Chiba, N., Nishizeki, T., Saito, N.: A linear 5-coloring algorithm of planar graphs. J. Algorithms **2**(4), 317–327 (1981)
5. Even, G., Lotker, Z., Ron, D., Smorodinsky, S.: Conflict-free colorings of simple geometric regions with applications to frequency assignment in cellular networks. SIAM J. Comput. **33**(1), 94–136 (2003)
6. Har-Peled, S., Smorodinsky, S.: Conflict-free coloring of points and simple regions in the plane. Discret. Comput. Geom. **34**(1), 47–70 (2005)
7. Katchalski, M., McCuaig, W., Seager, S.M.: Ordered colourings. Discret. Math. **142**(1–3), 141–154 (1995)
8. Pach, J., Tardos, G.: Conflict-free colourings of graphs and hypergraphs. Comb. Probab. Comput. **18**(5), 819–834 (2009)
9. Smorodinsky, S.: Combinatorial problems in computational geometry. Ph.D. thesis, Tel-Aviv University (2003)
10. Smorodinsky, S.: On the chromatic number of some geometric hypergraphs. In: Proceedings of the Seventeenth Annual ACM-SIAM Symposium on Discrete Algorithms, SODA 2006, Miami, Florida, USA, 22–26 January 2006, pp. 316–323 (2006)
11. Smorodinsky, S.: Conflict-free coloring and its applications. In: Bárány, I., Böröczky, K.J., Tóth, G.F., Pach, J. (eds.) Geometry—Intuitive, Discrete, and Convex. BSMS, vol. 24, pp. 331–389. Springer, Heidelberg (2013). https://doi.org/10.1007/978-3-642-41498-5_12

Amplitude Amplification for Operator Identification and Randomized Classes

Debajyoti Bera[✉]

Indraprastha Institute of Information Technology-Delhi (IIIT-D), New Delhi, India
dbera@iiitd.ac.in

Abstract. Amplitude amplification (AA) is tool of choice for quantum algorithm designers to increase the success probability of query algorithms that reads its input in the form of oracle gates. Geometrically speaking, the technique can be understood as rotation in a specific two-dimensional space. We study and use a generalized form of this rotation operator to design algorithms in a geometric manner. Specifically, we apply AA to algorithms that take their input in the form of input states and in which rotations with different angles and directions are used in a unified manner. We show that AA can be used to sequentially discriminate between two unitary operators, both without error and with bounded-error, in an asymptotically optimal manner. We also show how to reduce error probability in one and two-sided bounded error algorithms more efficiently than the usual parallel repetitions technique; in particular, errors can be completely eliminated from the exact error algorithms.

1 Introduction

Amplitude amplification (AA) is the engine that powers the "unordered quantum search" algorithm proposed by Grover in 1996 [1]. A lot of efficient quantum algorithms essentially ride this horse in some way or the other [2–5] and one wonders how much more can this idea deliver. It is now routine to apply AA for boosting the success probability of quantum algorithms. One reason behind this unmatched popularity is the black-box manner in which this technique can be applied. *Suppose $\mathcal{A}$ is a quantum algorithm without any intermediate measurement such that after measuring the output of $\mathcal{A}|00\ldots0\rangle$, we obtain a solution to $\mathcal{A}$ that may be "good" with some probability, say p. Then AA can be applied to $\mathcal{A}$ to generate an algorithm Q that basically calls $\mathcal{A}$ (and $\mathcal{A}^\dagger$) as black-boxes in an iterative manner.* Temptation to use AA becomes stronger due to the uniform nature of Q: $\mathcal{A}$ and $\mathcal{A}^\dagger$ are used as black-box here and the input state to $\mathcal{A}$ as well as the measurement operators at the end remain unchanged (maybe, extended). Therefore, it makes sense to apply this technique to a family of $\mathcal{A}$, e.g., to $\{\mathcal{A}_x\}_{x\in\{0,1\}^*}$ in which $\mathcal{A}_x$ uses an oracle gate to read bits of input string x. This is why AA has so far been applied in the query-complexity model in which $\mathcal{A}$ can read the "input" by making oracle queries.

© Springer International Publishing AG, part of Springer Nature 2018
L. Wang and D. Zhu (Eds.): COCOON 2018, LNCS 10976, pp. 579–591, 2018.
https://doi.org/10.1007/978-3-319-94776-1_48

The second reason behind the popularity of AA is the square-root promise that *the amplification algorithm Q makes $O(\sqrt{p})$ calls to A (and $A^\dagger$) and can guarantee a good solution with high probability*; this is in contrast to classical techniques that require $O(p)$ calls to A. There are also several improvements to workaround the requirement of knowing p beforehand [6,7].

This work is motivated by two other observations about AA. Amplitude amplification requires use of a diffusion operator that essentially depends upon the input state of the algorithm $\mathcal{A}$. Therefore, it is worth investigating if, and when, can amplitude amplification be applied to non-query algorithms, i.e., algorithms in which the input is supplied in the form of an input state. In this setting, we have a family of input states instead of a family of algorithms and therefore, we no longer have a uniform amplification circuit for different input states. We find that AA works in general, but with a subtlety for communication protocols.

Amplitude amplification can also be viewed as a rotation in a particular 2-dimensional space. Our second observation is that it is possible to mix-match rotations in different directions and by different angles but in a uniform manner across different instances – this we call as "differential amplification". This is an extension of the idea present in the original search algorithm by Grover that if $\mathcal{A}$ has no solution, then the amplified algorithm too will produce no solution— geometrically, the same amplification routine rotated different states differently.

Contribution: Sequential Operator Discrimination [Sect. 3]: A common manner of differentiating between output distributions of algorithms is to run them in parallel and statistically analyse the aggregate of the outcomes [8,9]. Differential amplification can be seen as a sequential technique for the same purpose. For instance, a recently proposed fault detection method for quantum circuits uses a classical repeated sampling of the output of a quantum circuit to distinguish between several output distributions, one for each type of faulty circuit [10]. Our technique can be used to replace the classical repetition by quantum amplification and we show a limited form of this in this work. Specifically, we design both exact and bounded-error sequential algorithms for discriminating between two unitary operators (given as black-box) without using any special input state for the operators, whereas, the existing parallel and sequential methods require preparation of a specific "optimum" state [9,11]. Moreover, if the optimum state is used, then our algorithm makes at most additional call compared to the optimum. In this process we also strengthen and generalize some known upper and lower bounds on sequential and parallel discrimination algorithms.

Contribution: Sequential Amplification of Bounded-Error Algorithms [Sect. 4]: Quantum algorithms that operate in the non-query mode, i.e., take input in the form of input states, appear sidelined in the crowd of quantum query algorithms. However, important problems like "Factoring" and "Discrete-logarithm" with eye-catching quantum algorithms, belong to the non-query **BQP** class. The current technique for boosting the success probability of **RQP** (one-sided error) and **BQP** (two-sided error) is by parallely and independently running the original algorithm [8, Chap. 6], [12]. We use differential amplification for reducing error of bounded-error algorithms faster compared to the parallel ones. We also show that one-sided and two-sided "exact" error quantum classes (**ERQP** and **EBQP**) can be improved to be included in **EQP** (a quantum version of the complexity class **P**) thus making **EQP** = **ERQP** = **EBQP**.

2   Grover Iterator Revisited

Brassard et al. [6] formalized the key technique of Grover's search algorithm as amplitude amplification (AA) and showed its use in general search problems. AA involves repeated application of an operator commonly known as the Grover iterator G. Traditionally G has been defined based on a quantum (oracle) algorithm $\mathcal{A}$ that on input $|00\ldots0\rangle$ searches a state space and outputs a superposition $|\Psi\rangle$ of "good" and "bad" solution states (in the standard basis) of some search problem (say, searching for 1 in an unordered array). Another operator $U_{\Psi_0} = (I - 2\sum_{x:good}|x\rangle\langle x|)$ is used to identify "good" solution states. Then, G is constructed as $G = -A(I - 2|00\ldots0\rangle\langle00\ldots0|)A^\dagger U_{\Psi_0} = (2|\Psi\rangle\langle\Psi| - I)U_{\Psi_0}$.

Soon after Grover proposed his quantum search algorithm, researchers observed that his algorithm, and the underlying amplitude amplification technique, has an elegant geometric interpretation of a rotation in a 2-dimensional state. Several extensions to Grover's search rely on this geometric interpretation, e.g., the generalization of Grover's search to handle arbitrary initial states [13,14]. The algorithms that we study are not search algorithms and we want to mix-and-match more than one generalizations of G. Even though such generalized Grover's iterator has been analyzed in the context of unordered quantum search [13,14], we did not find any independent characterization suitable for us.

Given a state $|\Psi\rangle$ and a two-outcome projective measurement $\mathbf{P} = \langle P^0, P^1\rangle$, we study the following operator family for any pair of angles $0 \le a, b < 2\pi$:

$$G_{a,b} = [(1 - e^{ia})|\Psi\rangle\langle\Psi| - I] \cdot [I - (1 - e^{ib})P^1]$$

It is easy to show that $G_{a,b}$ is a unitary operator for any a, b. These operators were used to amplify query algorithms in which they are applied to rotate certain types of states that are related to $|\Psi\rangle$ and P^1 [6,13,14]. Our motivation was to characterize the transformation and which all states can this be applied on.

Define angle $\theta \in [0, \pi/2]$ and orthogonal states $|\Psi_0\rangle$ and $|\Psi_1\rangle$ such that $P^0|\Psi\rangle = \cos\theta|\Psi_0\rangle$ and $P^1|\Psi\rangle = \sin\theta|\Psi_1\rangle$. Observe that $\sin^2\theta$ is the probability of observing outcome P^1 when $|\Psi\rangle$ is measured using $\mathbf{P}$. Denote the Hilbert space spanned by $|\Psi_0\rangle$ and $|\Psi_1\rangle$ by $\mathcal{H}$. If $P^0|\Psi\rangle = 0$ or $P^1|\Psi\rangle = 0$, then $\mathcal{H}$ is 1-dimensional, essentially spanned by $|\Psi\rangle$. In that case $G_{a,b} \stackrel{\varphi}{\simeq} I$; we use the notation $U \stackrel{\varphi}{\simeq} V$ to indicate that the two operators U and V are identical, except maybe for different global phases. So, henceforth, we will only consider the cases when $P^0|\Psi\rangle \ne 0 \ne P^1|\Psi\rangle$, and in that case, $\mathcal{H}$ is 2-dimensional.

We will use CP_ρ to denote the conditional phase-change unitary operator $P^0 + e^{i\rho}P^1$. Observe that $CP_\rho \stackrel{\varphi}{\simeq} I$ if $\mathcal{H}$ is one-dimensional. The following well-known theorem shows how to implement rotations in two-dimensional $\mathcal{H}$ and can be proved by observing the action of G and CP_ρ in $\mathcal{H}$. We will use R_x to denote rotation by angle x in $\mathcal{H}$ in the anti-clockwise direction from $|\Psi_0\rangle$ to $|\Psi_1\rangle$.

Theorem 1 (Proved by Høyer [15]). *Let $|\Psi\rangle$ denote a state and $\langle P^0, P^1 \rangle$ denote a two-outcome projective measurement. Let $\mathcal{H}$ be the space spanned by $\{P^0|\Psi\rangle, P^1|\Psi\rangle\}$ and let $(|\Psi_0\rangle, |\Psi_1\rangle)$ be a basis of $\mathcal{H}$ such that $\langle \Psi_0|P^0|\Psi\rangle = \cos\theta$ and $\langle \Psi_1|P^1|\Psi\rangle = \sin\theta$ for some $\theta \in [0, \pi/2]$. Let R_α denote rotation by angle α in $\mathcal{H}$ from $|\Psi_0\rangle$ towards $|\Psi_1\rangle$. If $\mathcal{H}$ is one-dimensional then $G_{\pi,\pi} \overset{\mathcal{L}}{\simeq} I$ and $CP_\rho \cdot G_{a,b} \cdot CP_\rho^\dagger \overset{\mathcal{L}}{\simeq} I$ for any ρ, a, b.*

On the other hand, if $\mathcal{H}$ is two-dimensional then, for any $0 \leq \theta' \leq 2\theta$, there exists angles $\rho, a, b \in [0, 2\pi]$ such that $R_{\theta'} \overset{\mathcal{L}}{\simeq} CP_\rho \cdot G_{a,b} \cdot CP_\rho^\dagger$. In particular, $R_{2\theta} = G_{\pi,\pi} = [2|\Psi\rangle\langle\Psi| - I] \cdot [I - 2P^1]$.

For any angle $\delta \in [0, \pi/2]$, R_δ can be implemented as $R_{\theta'} R_{2\theta}^k$ in which k is the largest integer such that $\delta = k \cdot 2\theta + \theta'$. The above theorem allows us to rotate *any* state in $\mathcal{H}$ by any angle and may be of independent interest.

Corollary 1. *Let $|\Phi\rangle$ denote some state in $\mathcal{H}$ of the form $\cos\phi|\Psi_0\rangle + \sin\phi|\Psi_1\rangle$ for some $\phi \in [0, \pi/2]$ and let δ be some angle. Then, $\cos(\delta+\phi)|\Psi_0\rangle + \sin(\delta+\phi)|\Psi_1\rangle$ can be obtained by executing $R_\delta|\Phi\rangle \overset{\mathcal{L}}{\simeq} CP_\rho G_{a,b} CP_\rho^\dagger G_{\pi,\pi}^k |\Phi\rangle$ for some angles ρ, a, b depending on δ, θ and $k = \lfloor \frac{\delta}{2\theta} \rfloor$.*

In particular, let $|\chi\rangle = \cos x|\Psi_0\rangle + \sin x|\Psi_1\rangle$ be some other state in $\mathcal{H}$. Define project measurement operators $\mathcal{P} = \langle P'^0 = I - |\chi\rangle\langle\chi|, P'^1 = |\chi\rangle\langle\chi| \rangle$. Then there exists ρ, a, b, k such that $\|P'^1 CP_\rho G_{a,b} CP_\rho^\dagger G_{\pi,\pi}^k |\Phi\rangle\|^2 = \|P'^1 R_{x-\phi}|\Phi\rangle\|^2 = \sin^2 x$.

Simpler rotation operators can surely be constructed for any Hilbert space. However, we shall see in the next two sections that the particular construction of R_δ allows us to *differentially amplify* different states in different manners.

3   Unitary Operator Discrimination

In the unitary operator discrimination problem, we are given a unitary operator $U \in \{U_1, U_2\}$ as a black-box with equal chance of picking either of the operators. The goal is to identify U. Let $\omega(U)$, for any unitary operator U, denote the angle of the smallest arc containing all the eigenvalues of U (on the unit circle). Let ω represent $\omega(U_1^\dagger U_2)$. It is known that $\frac{1}{2}(1 - \sin\frac{\omega}{2})$ is the minimum probability of error to discriminate between U_1 and U_2 by making only one call to U on an *appropriate input state* and using an appropriate measurement operator [9,11]. Thus, if $\omega \geq \pi$ then there exists a $|\gamma\rangle$ such that $U_1|\gamma\rangle$ and $U_2|\gamma\rangle$ are orthonormal and therefore, can be *perfectly distinguished*.

On the other hand, if $\omega < \pi$, then the optimal methods for exact discrimination require $k = \lceil \frac{\pi}{\omega} \rceil$ calls to U on a bespoke input state followed by a measurement in a suitable basis. These k calls may happen in parallel in which case the input state is a maximally entangled one over kd qubits [11,16] or may also happen sequentially in which the input state is a superposition of the eigenstates of $(U_1^\dagger U_2)$ [17]. Such bespoken input states may be difficult to create, all the more if k is large. It may be desirable to have a method that uses easy to construct input states (e.g., $|0\ldots0\rangle$) and it will be even better if one simple state can discriminate *any* U_1 and U_2. Our method requires a sequential application of U and can be applied to "any input state" (except a small subset).

To discriminate with a probability of error at most $1/3$, Kawachi et al. [9] reported a method that used parallel calls to U and an entangled state over kd qubits. They proved an upper bound of $\lceil\frac{\pi}{3\omega}\rceil$ calls and also showed that there exists operators that require at least $\lceil\frac{2}{3\omega}\rceil$ calls to U. Their method can be easily generalized for an arbitrary error ϵ and after doing that along with additional tightening (see full version [18]), we obtain an upper bound of $\lceil\frac{2}{\omega}\sin^{-1}(1-2\epsilon)\rceil$ calls and a lower bound of $\lceil\frac{1-2\epsilon}{\sin(\omega/2)}\rceil$ calls that almost matches their upper bound. However, even for their method a specific input state is required. Our method can be seen as an alternative sequential method but with fewer qubits.

Duan et al. gave a lower bound on the number of calls required in a sequential method for perfect discrimination [17]. Their method can also be easily generalized (see full version [18]) to arbitrary error and we obtained the same lower bound as that obtained from the generalization of Kawachi et al.'s result that was mentioned earlier. Duan et al. also gave a sequential algorithm for perfect discrimination (using a specific input state) but it was not immediately clear how to extend their algorithm for bounded-error discrimination. In any case, we would like to see our discrimination algorithm as an alternative sequential method that uses the idea of amplitude amplification, is independent of the input state and makes almost the same number of calls to the black-boxes as the currently known parallel discrimination method.

3.1 Separation Using Amplitude Amplification

Suppose that we want to use an input state $|\gamma\rangle$ which may be chosen optimally or may simply be available for use. We assume that we have access to the black-box $U \in \{U_1, U_2\}$ and its corresponding adjoint $U^\dagger$ as well. It should be noted that if U is implemented as a quantum circuit, then $U^\dagger$ is usually easy to implement. We will discuss both cases of error probability $\epsilon < 0.5$ and $\epsilon = 0$.

Let s be some phase and $\theta \in [0, \pi/4]$ be an angle such that $\langle\gamma|U_1^\dagger U_2|\gamma\rangle = \cos 2\theta e^{is}$; define $|\sigma_1\rangle = U_1|\gamma\rangle$ and $|\sigma_2\rangle = e^{-is}U_2|\gamma\rangle$ so that $\langle\sigma_1|\sigma_2\rangle = \cos 2\theta$ is real making it easier to apply Theorem 1. Given this $|\gamma\rangle$, the probability of error in discriminating between $U_1|\gamma\rangle$ and $U_2|\gamma\rangle$ can be expressed according to this well-known relationship: $\Pr[error] = \frac{1}{2}\left(1 - \sqrt{1 - |\langle\sigma_1|\sigma_2\rangle|^2}\right) = \frac{1-\sin 2\theta}{2}$.

Observe that if $\theta = \frac{\pi}{4}$, the states $|\sigma_1\rangle$ and $|\sigma_2\rangle$ are already orthogonal and so can be perfectly discriminated; on the other hand, if $\theta = 0$ (i.e., $|\sigma_1\rangle$ and $|\sigma_2\rangle$ differ only by a global phase), then they cannot be discriminated better than a random guess. Therefore, we will focus on the case when $\theta \in (0, \pi/4)$ and our strategy will be to devise a suitable projective measurement that allows us to use amplitude amplification (as Theorem 1) to identify between the states.

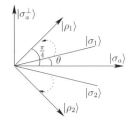

Fig. 1. The different states that are used in operator discrimination.

Construct an orthogonal basis for $\mathcal{H} = \mathcal{H}(|\sigma_1\rangle, |\sigma_2\rangle)$ by first defining $|\sigma_a\rangle = p|\sigma_1\rangle + p|\sigma_2\rangle$ and then defining an appropriate orthogonal state $|\sigma_a^\perp\rangle = p'|\sigma_1\rangle - p'|\sigma_2\rangle)$; it suffices to use $p = 1/(2\cos\theta)$ and $p' = 1/(2\sin\theta)$. It is now easy to represent $|\sigma_i\rangle$ in the above basis; $|\sigma_1\rangle = \langle\sigma_a|\sigma_1\rangle|\sigma_a\rangle + \langle\sigma_a^\perp|\sigma_1\rangle|\sigma_a^\perp\rangle = \cos\theta|\sigma_a\rangle + \sin\theta|\sigma_a^\perp\rangle$ and $|\sigma_2\rangle = \langle\sigma_a|\sigma_2\rangle|\sigma_a\rangle + \langle\sigma_a^\perp|\sigma_2\rangle|\sigma_a^\perp\rangle = \cos\theta|\sigma_a\rangle - \sin\theta|\sigma_a^\perp\rangle$. Define operators $Q_0 = |\sigma_a\rangle\langle\sigma_a|$, $Q_1 = |\sigma_a^\perp\rangle\langle\sigma_a^\perp|$ and define a projective measurement $\mathcal{Q} = \langle Q_0, Q_1\rangle$. It will be easier to visualize all rotations in the basis $(|\sigma\rangle, |\sigma^\perp\rangle)$.

Now define $|\rho_1\rangle = \frac{1}{\sqrt{2}}|\sigma_a\rangle + \frac{1}{\sqrt{2}}|\sigma_a^\perp\rangle$ and $|\rho_2\rangle = \frac{1}{\sqrt{2}}|\sigma_a\rangle - \frac{1}{\sqrt{2}}|\sigma_a^\perp\rangle$. Use this to define a two-outcome projective measurement operator $\mathcal{P} = \langle P_1, P_2\rangle$ in which $P_1 = |\rho_1\rangle\langle\rho_1|$ and $P_2 = I - P_1 = |\rho_2\rangle\langle\rho_2|$. The different states that were constructed are explained in Fig. 1.

Suppose $G_{a,b}^U$ denotes the operator $U[(1 - e^{ia})|\gamma\rangle\langle\gamma| - I]U^\dagger \cdot [I - (1 - e^{ib})Q_1]$ that uses the black-box U. Our first observation is that $G_{a,b}^{U_i}$ can also be written as $[(1 - e^{ia})|\sigma_i\rangle\langle\sigma_i| - I]U^\dagger \cdot [I - (1 - e^{ib})Q_1]$. Since the space spanned by $Q_0|\sigma_1\rangle$ and $Q_1|\sigma_1\rangle$ is $\mathcal{H}$ itself, and $\|Q^1|\sigma_1\rangle\|^2 = |\langle\sigma_a^\perp|\sigma_1\rangle|^2 = \sin^2\theta$, $G_{a,b}^{U_1}$ operators can be used in Theorem 1 for rotating any state in $\mathcal{H}$ in a counter-clockwise manner by some angle that is at most 2θ.

Our second observation arises from the fact that since $|\sigma_2\rangle = \cos\theta|\sigma_a\rangle + \sin\theta(-|\sigma_a^\perp\rangle)$, $G_{a,b}^{U_2}$ can still be used in Theorem 1 but the rotation will be from $|\sigma_a\rangle$ towards $-|\sigma_a^\perp\rangle$; that is, the rotation will be in a *clockwise* manner with everything else remaining the same as above (also illustrated in Fig. 1).

For discriminating with low probability of error, say ϵ, let $\varepsilon \in [0, \pi/2]$ be an angle such that $\sin^2\varepsilon = \epsilon$. Let $\phi = \pi/4 - \theta - \varepsilon$. Applying Corollary 1, one can calculate ρ, a, b and set $k = \lfloor\frac{\phi}{2\theta}\rfloor$ such that $CP_\rho G_{a,b}^{U_i} CP_\rho^\dagger [G_{\pi,\pi}^{U_i}]^k$ rotates in the following manner. Here, $|\bar{\sigma}_a^\perp\rangle$ denotes the state $-|\sigma_a^\perp\rangle$.

$$|\sigma_1\rangle = \cos\theta|\sigma_a\rangle + \sin\theta|\sigma_a^\perp\rangle \xrightarrow{CP_\rho G_{a,b}^{U_1} CP_\rho^\dagger [G_{\pi,\pi}^{U_1}]^k} \cos(\theta + \varepsilon)|\sigma_a\rangle + \sin(\theta + \varepsilon)|\sigma_a^\perp\rangle$$

$$|\sigma_2\rangle = \cos\theta|\sigma_a\rangle + \sin\theta|\bar{\sigma}_a^\perp\rangle \xrightarrow{CP_\rho G_{a,b}^{U_2} CP_\rho^\dagger [G_{\pi,\pi}^{U_2}]^k} \cos(\theta + \varepsilon)|\sigma_a\rangle + \sin(\theta + \varepsilon)|\bar{\sigma}_a^\perp\rangle$$

Let V^{U_i} denote the operator $CP_\rho G_{a,b}^{U_i} CP_\rho^\dagger [G_{\pi,\pi}^{U_i}]^k U_i$. Our discrimination procedure consists of first deriving the parameters (ρ, a, b, k), constructing a circuit for the operator V^U using the black-box U, executing $V^U|\gamma\rangle$ to obtain state $|\Psi\rangle$ and finally measuring $|\Psi\rangle$ in the basis $\mathcal{P}$. U is declared to be U_i if measurement outcome is $|\rho_i\rangle$. The probability of error can be calculated as $\|P_1 V^{U_2}|\gamma\rangle\|^2 = \|P_2 V^{U_1}|\gamma\rangle\|^2 = \sin^2 \varepsilon = \epsilon$, as desired.

Finally, we would like to discuss the query complexity of discrimination. Since each call to $G^U_{a,b}$ involves one call to U and one call to $U^\dagger$, the number of calls to $U^\dagger$ is $k+1$ while the number of calls to U is $k+2$ (including the initial call to $|\gamma\rangle$). Therefore, the total number of queries is $2k + 3$. Here $k = \lfloor \frac{\pi/4 - \theta - \sin^{-1}\sqrt{\epsilon}}{2\theta} \rfloor$ that can be simplified to $\mathsf{R}(\frac{\pi}{8\theta} - \frac{\sin^{-1}\sqrt{\epsilon}}{2\theta}) - 1$. We use $\mathsf{R}(f)$ to denote the nearest integer to any floating point number f ($\mathsf{R}(0.5)$ is set to 1).

Theorem 2. *The above algorithm can differentiate between two operators U_1 and U_2 for any input state $|\gamma\rangle$ with probability of error at most ϵ as long as $\theta = \cos^{-1}|\langle\gamma|U_1^\dagger U_2|\gamma\rangle| \neq 0$ and using $2 * \mathsf{R}\left(\frac{\pi}{8\theta} - \frac{\sin^{-1}\sqrt{\epsilon}}{2\theta}\right) + 1$ total calls to the black-boxes U and $U^\dagger$.*

Of course, $|\gamma\rangle$ can be chosen optimally to maximize θ. For the optimal $|\gamma\rangle$, $|\langle\gamma|U_1^\dagger U_2|\gamma\rangle|$ equals $\cos^2 \frac{\omega}{2}$ (therefore, use $\theta = \omega/4$) which leads to the following corollary about optimal discrimination between U_1 and U_2.

Corollary 2. *The above algorithm can differentiate between two operators U_1 and U_2 without any error using the optimal input state and a total of $2 \cdot \mathsf{R}\left(\frac{\pi}{2\omega}\right) + 1 \in \{\lceil \frac{\pi}{\omega} \rceil, \lceil \frac{\pi}{\omega} \rceil + 1\}$ calls to the black-boxes. The number of calls to discriminate with probability of error ϵ is at most $2 \cdot \mathsf{R}\left(\frac{\pi}{\omega} - \frac{2\sin^{-1}\sqrt{\epsilon}}{\omega}\right) + 1 = 2 \cdot \mathsf{R}\left(\frac{\sin^{-1}(1-2\epsilon)}{\omega}\right) + 1$ and in particular, with error $1/3$ is at most $2 \cdot \mathsf{R}(0.34/\omega) + 1$.*

For both exact and bounded-error algorithms, the query complexity of our algorithm using the optimal state is at most one more than current known bounds.

4   Randomized Non-query Classes

The current methods for improving success probability of bounded-error non-query quantum algorithms are parallel repetitions [12] that have the same complexity as classical methods. Intuitively, however, amplitude amplification ought to be applicable for such algorithms too. In fact, if the error is known and fixed (not simply bounded), then AA should be able to "precisely rotate" an output state to the basis states used for the (projective) measurement, thereby completely eliminating error. The current section discusses this idea in detail.

Consider the randomized complexity class $\mathbf{RQP}_\epsilon$. For any language $L \in \mathbf{RQP}_\epsilon$, there exists a corresponding uniform family of quantum circuits $\{C\}_n$, say, over $n + a$ qubits and an initial state $|\alpha\rangle$ over a ancilla qubits that may depend on n. As per standard practice, we assume that after C is applied the first qubit is measured in the standard basis, i.e., the output state is measured by the projective measurement operator $\mathcal{P} = \langle P^0 = |0\rangle\langle0|\otimes I, P^1 = |1\rangle\langle1|\otimes I\rangle$. This can be easily generalized to any other decision criterion that involves measuring the output state by a two-outcome projective measurement. We denote the output state $C_n(|x\rangle \otimes |\alpha\rangle)$ by $|\Psi_x\rangle$ in which n denotes $|x|$. Let $\theta \in (0, \pi/2)$ be an angle such that $\sin^2 \theta = \epsilon$; note that if $\theta = \pi/2$, then $p = 1$ and in that case we anyway have $L \in \mathbf{EQP}$. Since $L \in \mathbf{RQP}_\epsilon$, the following should hold for any $x \in \{0, 1\}^*$.

$$x \in L \implies \|P^1|\Psi_x\rangle\|^2 \geq \epsilon = \sin^2 \theta, \quad \text{and} \quad x \notin L \implies \|P^1|\Psi_x\rangle\|^2 = 0 = \sin^2 0$$

We also define the *exact one-sided error quantum class* $\mathbf{ERQP}_\epsilon$ by extending $\mathbf{RQP}_\epsilon$: for $x \in L$, $\|P^1|\Psi_x\rangle\|^2 = \epsilon$ and the probability is zero for $x \notin L$.

We can similarly define two-sided bounded-error quantum classes. For any $0 \leq \epsilon_2 < \frac{1}{2} < \epsilon_1 \leq 1$, define $\mathbf{BQP}_{\epsilon_1, \epsilon_2}$ as the class of languages L such that:

$$x \in L \implies \|P^1|\Psi_x\rangle\|^2 \geq \epsilon_2, \quad \text{and} \quad x \notin L \implies \|P^1|\Psi_x\rangle\|^2 \leq \epsilon_1 \quad \text{for any } x$$

Also, define its exact error version[1] $\mathbf{EBQP}_{\epsilon_1, \epsilon_2}$ which consists of languages with error probabilities that is exactly ϵ_1 if $x \in L$ and exactly ϵ_2 if $x \notin L$.

Furthermore, define $\mathbf{ERQP} = \bigcup_\epsilon \mathbf{ERQP}_\epsilon$ and $\mathbf{EBQP} = \bigcup_{\epsilon_1, \epsilon_2} \mathbf{EBQP}_{\epsilon_1, \epsilon_2}$. Obviously $\mathbf{EQP} = \mathbf{RQP}_0 = \mathbf{BQP}_{0,0}$; therefore, $\mathbf{EBQP} \supseteq \mathbf{EQP} \subseteq \mathbf{ERQP}$.

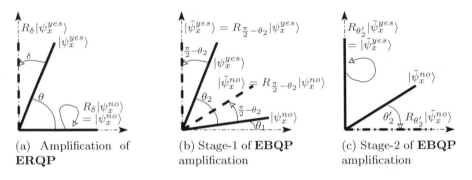

(a) Amplification of **ERQP**

(b) Stage-1 of **EBQP** amplification

(c) Stage-2 of **EBQP** amplification

Fig. 2. Amplification of exact-error classes can be seen as conditional rotations. Solid lines denote the states before rotation and dashed lines denotes the states after rotation.

4.1 One-Sided Exact Error Class: ERQP

Languages in $\mathbf{ERQP}_\epsilon$ can be amplified immediately by using Theorem 1. Consider any such L and any $x \in L$. Let $\mathcal{H}$ denote the space spanned by $P^1|\Psi_x\rangle$ and $P^0|\Psi_x\rangle$; clearly, $|\Psi_x\rangle \in \mathcal{H}$. Let $\delta = \pi/2 - \theta$. Construct rotation operator R_δ using $|\Psi_x\rangle$ and P^1. Applying this R_δ on $|\Psi_x\rangle$ gives us a state such that $\|P^1 R_\delta|\Psi_x\rangle\|^2 = 1$. On the other hand, if $x \notin L$, $\mathcal{H}$ is one-dimensional and in that case the constructed operator R_δ acts as the identity operator; therefore, $\|P^1 R_\delta|\Psi_x\rangle\|^2 = 0$. Figure 2a illustrates the action of R_δ for both the cases.

We only need to show that R_δ can be constructed in a uniform manner for a fixed L. R_δ will be constructed as $R_{\delta'} R_{2\theta}^k$ where $k = \lfloor \frac{\delta}{2\theta} \rfloor$ and $\delta' = \delta - k \cdot 2\theta$.

---

[1] A similar question on exact two-sided-error classical class was asked in http://cstheory.stackexchange.com/questions/20027.

However, the difficulty lies in constructing the $R_{\delta'}$ and $R_{2\theta}$ operators. Both of them involve some uniformly chosen gates (that depend upon $|\alpha\rangle$, P^1, δ and 2θ) but also operators of the type $[(1 - e^{i\gamma})|\Psi_x\rangle\langle\Psi_x| - I]$ that seem to be dependent on x. The key observation here is that

$$((1 - e^{i\gamma})|\Psi_x\rangle\langle\Psi_x| - I) - C \cdot ((1 - e^{i\gamma})|x\rangle\langle x| \otimes |\alpha\rangle\langle\alpha| - I) \cdot C^\dagger$$

and that a Fanout gate [19] can be used to make a copy of the input state $|x\rangle$ at the beginning which can be used later to implement $|x\rangle\langle x|$. The detailed construction of the above operator is given in the full version [18]. Essentially we are able to construct a uniform circuit R_δ that can completely eliminate any error in deciding strings in **ERQP** languages.

Theorem 3. EQP $= \bigcup_\epsilon$ **ERQP**$_\epsilon =$ **ERQP**.

We will now analyse the complexity of amplification. Let $s(n)$ be an upper bound on the size (number of gates) the circuits C_n and $C_n^\dagger$. Let C_n' denote the amplified zero-error circuit that calls C_n and $C_n^\dagger$ and let $s'(n)$ denote the size of C_n'. Assuming that A, B, E can be implemented without much overhead on size, $s'(n)/s(n)$ is proportional to $1 + 2(k+1) \approx 2(\frac{\pi}{4\sin^{-1}\sqrt{\epsilon}} - \frac{1}{2}) + 3 \leq \frac{1.6}{\sqrt{\epsilon}} + 2$. Now contrast this with the usual parallel scheme of running multiple copies of C_n in parallel (on copies of $|x\rangle \otimes |\alpha\rangle$, with $|x\rangle$ being copied using a Fanout gate). First of fall, such a parallel scheme cannot possible achieve 100% probability of success. Secondly, the size increases by a factor proportional to $\frac{1}{-\ln(1-\epsilon)} \approx \frac{1}{\epsilon}$ that is quadratic ally large compared to the amplified C_n' with zero-error.

4.2 Two-Sided Exact Error Class EBQP

Two-sided bounded-error classes can be treated similarly as their one-sided counterparts. We will leave out the details and only chalk the main ideas using the R_θ rotation operators for suitable θ. We have already explained in the earlier subsections how to implement R_θ in a uniform manner; this is sufficient to give us uniform circuits to decide languages with a higher probability of success.

For the two-sided exact error class **EBQP** and some language $L \in$ **EBQP**, consider the two-stage amplification process illustrated in Fig. 2. Consider any n-bit x and let C denote the corresponding circuit and $|\alpha\rangle$ denote the (uniformly generated) fixed-state ancilla register. Let $|\Psi_x\rangle$ denote $C|x\rangle \otimes |\alpha\rangle$.

Consider any $x^{yes} \in L$ and denote $C|x^{yes}\rangle \otimes |\alpha\rangle$ by $|\Psi_x^{yes}\rangle$. Similarly, if $x^{no} \notin L$, denote $C|x^{no}\rangle \otimes |\alpha\rangle$ by $|\Psi_x^{no}\rangle$. Furthermore, let $0 < \theta_1 < \theta_2 < \pi/2$ be angles such that $\sin^2\theta_1 = \epsilon_1$ and $\sin^2\theta_2 = \epsilon_2$.

In stage-1 (Fig. 2b), first $R_{\pi/2-\theta_2}$ is applied to $|\Psi_x\rangle$; let C_1 denote the circuit and $|\tilde{\Psi}_x\rangle = C_1|x\rangle \otimes |\alpha\rangle$ denote the state thus obtained. Here $R_{\pi/2-\theta_2}$ is constructed by using $|\Psi_x\rangle$ and P^1 and involves C and $C^\dagger$ similar to the construction in Subsect. 4.1. If $x \in L$, then $|\tilde{\Psi}_x\rangle$ is now aligned with $P^1|\Psi_x\rangle$ whereas if $x \notin L$, then $|P^1|\tilde{\Psi}_x\rangle| = \sin(\pi/2 - (\theta_2 - \theta_1))$.

Let θ_2' denote $\pi/2 - (\theta_2 - \theta_1)$. Observe that $|\tilde{\Psi}_x\rangle$ belongs to the same Hilbert space spanned by $P^1|\Psi_x\rangle$ and $P^0|\Psi_x\rangle$. In stage-2 (Fig. 2c), $\tilde{R}_{\theta_2'}$ is applied on $|\tilde{\Psi}_x\rangle$

but now $\tilde{R}_{\theta'_2}$ is constructed using $|\tilde{\Psi}_x\rangle$ and P^0 and involves C_1 and $C_1^\dagger$ similar to the construction in Subsect. 4.1; C_1 and $C_1^\dagger$ in turn calls C and $C^\dagger$.

First consider the case of $x \in L$. $P^0|\tilde{\Psi}_x\rangle = 0$ implies that $\tilde{R}_{\theta'_2}|\tilde{\Psi}_x\rangle$ is identical to $|\tilde{\Psi}_x\rangle$ (up to a global phase). Then consider the case of $x \notin L$. Since, $|P^0|\tilde{\Psi}_x\rangle| = \sin(\theta_2 - \theta_1)$, $\tilde{R}_{\theta'_2}|\tilde{\Psi}_x\rangle$ will be now aligned with P^0.

Thus we get the resultant circuit $C' = \tilde{R}_{\theta'_2}R_{\pi/2-\theta_2}C$ and the final state applying C' is measured using (P^0, P^1). If P^0 is observed, (correctly) decide that $x \notin L$ and otherwise, (correctly) decide that $x \in L$.

Theorem 4. EQP $= \bigcup_{\epsilon_1, \epsilon_2}$ EBQP$_{\epsilon_1, \epsilon_2}$ = EBQP.

We will now discuss the complexity of the amplified algorithm and compare it with the usual parallel repetition algorithm for **BQP** that outputs the majority. Like for the **EQP** case, we will use C and C' for the original circuit and the zero-error amplified circuit, respectively. We will focus only on overhead caused by the multiple calls to C and $C^\dagger$ hoping that the additional components in the R_δ gates can be implemented with a small number of gates.

For comparison with the parallel repetition algorithm, it will be convenient to assume that $\epsilon_2 = 1 - \epsilon_1$ and therefore, $\theta_2 = \pi/2 - \theta_1$ in which $\theta_2 > \pi/4$.

Apart from the initial call to C to generate $|\Psi_x\rangle$, notice that the number of calls in the first phase is at most 2 since $\frac{\pi/2-\theta_2}{2\theta_2} < 1$. The number of calls in the second phase is

$$\approx 2\left(1 + \frac{\pi/2-(\theta_2-\theta_1)}{2(\theta_2-\theta_1)}\right) = 1 + \frac{\pi}{2(\theta_2-\theta_1)} = 1 + \frac{\pi}{2(\sin^{-1}\sqrt{\epsilon_2}-\sin^{-1}\sqrt{\epsilon_1})}$$

$$= 1 + \frac{\pi/2}{\sin^{-1}(\sqrt{\epsilon_2(1-\epsilon_1)}-\sqrt{\epsilon_1(1-\epsilon_2)})} \le 1 + \frac{\pi/4}{\epsilon_2-1/2}. \text{ Let } n_s \text{ denote this upper bound.}$$

Contrast this with the parallel repetition method that takes the majority of several parallel executions of $C|x\rangle \otimes |\alpha\rangle$. Even though this method cannot collapse **EBQP** to **EQP**, suppose we are interested to improve the probability ϵ_2 to $\sigma \approx 1$. Applying the usual Chernoff's bound based analysis, the number of parallel executions necessary is $\frac{4(1-\epsilon_2)}{(\epsilon_2-1/2)^2}\ln\frac{1}{1-\sigma}$ that we denote by n_p.

Note that if $\epsilon_2 \ge 3/4$, i.e., $\theta_2 \ge \pi/3$ and $\theta_1 \le \pi/6$, then only 2 calls are necessary in the second phase. So, for comparison we consider $1/2 < \epsilon_2 < 3/4$. In that case, $n_p \ge \frac{1}{(\epsilon_2-1/2)^2}\ln\frac{1}{1-\sigma}$ and $n_s \approx \sqrt{n_p}$ (ignoring small constants).

4.3   Non-exact Classes RQP and BQP

First we address the amplification of non-exact **RQP**$_\epsilon$ languages. For such languages $1 - \epsilon$ is only an upper bound on the failure probability (when $x \in L$). Since amplitude amplification requires knowledge of the success probability, there has been several attempts to generalize amplitude amplification for the cases when this probability is not known. An often followed approach guesses the value of ϵ in an exponentially increasing manner until a solution is found [6] or time-out happens. Instead we suggest using the quantum "fixed-point" search techniques for **RQP**$_\epsilon$ languages, e.g., following the one proposed by Yoder et al. [7] gives us a quantum circuit that makes $O(\frac{1}{\sqrt{\epsilon}}\log\frac{2}{\sqrt{1-\delta}}) = O(\frac{1}{\sqrt{\epsilon}}\log\frac{1}{1-\delta})$ calls to C and $C^\dagger$ and is sufficient to increase the success probability from ϵ to δ. The gates in that circuit are either fixed for L or of the form $G_{a,b}$ that we showed how to construct in a uniform manner in the earlier subsection. Contrast this to classical techniques for amplifying probability of **RP** languages; if C was a classical algorithm, then $\frac{\ln(1-\delta)}{\ln(1-\epsilon)} \geq \frac{1}{\epsilon} \ln \frac{1}{1-\delta}$ calls to C are required which is almost square of the number of calls required for the quantum case.

Circuits for **BQP** languages can also be amplified using ideas presented here. However, we leave out the specific details from this paper.

4.4   Communication Protocols

Apart from black-box/query algorithms, quantum amplitude amplification has also been applied to quantum communication protocols [20] for reducing probability of error and for distributed leader election [4] but they do not involve protocols that use pre-shared entangled bits. One can observe that existing quantum communication complexity protocols can be applied to protocols in which the parties get their input in the form of input state and not as oracle gates.

In this context we want to point out that the subtle requirement that it is not possible to amplify protocols that use arbitrary shared entangled qubits as ancilla. This is in stark contrast to quantum circuits that may use ancilla in entangled states and yet, can be amplified.

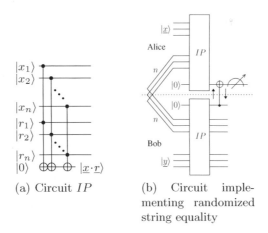

(a) Circuit IP

(b)   Circuit   implementing   randomized   string equality

Fig. 3. Circuit to detect if $\underline{x} = \underline{y}$ with probability at least $1/2$

Consider the quantum protocol illustrated in Fig. 3 that implements the well-known randomized algorithm for $\mathbf{EQ}(x, y)$ that asks whether two n-bit string x and y are identical. The protocol compares $x \cdot r$ and $y \cdot r$ in which r represents n random bits known to both parties. The circuit uses n EPR pairs to simulate n public random bits used in the randomized algorithm. It can be verified that if $x = y$, then the output qubit is always observed to be in $|0\rangle$ whereas if $x \neq y$, then the output qubit is observed in $|0\rangle$ or $|1\rangle$ with equal probability. If we could somehow apply amplitude amplification to this protocol, then Grover iterator would be applied only once, i.e., involving a total communication of 6 qubits (each Grover iterator involves a call to the circuit and a call to its inverse). That would invalidate the well-established lower bound that computing **EQ** by a communication protocol that involves pre-shared EPR pairs requires communication of at least $n/2$ qubits [21].

5   Conclusion

Amplitude amplification is commonly used to improve the probability of success of quantum query algorithms. We extend their usage to non-query algorithms by exploiting the fact that what they essentially do is increase the difference in probability of success between two cases. Based on this observation, we obtain efficient sequential algorithms for discrimination of unitary operators and for improving success probability of bounded error quantum algorithms.

References

1. Grover, L.K.: A fast quantum mechanical algorithm for database search. In: Proceedings of 28th STOC (1996)
2. Magniez, F., Santha, M., Szegedy, M.: Quantum algorithms for the triangle problem. SIAM J. Comput. **37**(2), 413–424 (2007)
3. Ozols, M., Roetteler, M., Roland, J.: Quantum rejection sampling. ACM Trans. Comput. Theory **5**(3), 1–33 (2013)
4. Kobayashi, H., Matsumoto, K., Tani, S.: Simpler exact leader election via quantum reduction. Chic. J. Theor. Comput. Sci. **2014**(10) (2014)
5. Berry, D.W., Childs, A.M., Cleve, R., Kothari, R., Somma, R.D.: Exponential improvement in precision for simulating sparse Hamiltonians. In: Proceedings of the 46th STOC (2014)
6. Brassard, G., Høyer, P., Mosca, M., Tapp, A.: Quantum amplitude amplication and estimation. Contemp. Math. **305**, 53–74 (2002)
7. Yoder, T.J., Low, G.H., Chuang, I.L.: Fixed-point quantum search with an optimal number of queries. Phys. Rev. Lett. **113**, 210501 (2014). https://journals.aps.org/prl/abstract/10.1103/PhysRevLett.113.210501
8. Lipton, R.J., Regan, K.W.: Quantum Algorithms via Linear Algebra: A Primer. The MIT Press, Cambridge (2014)
9. Kawachi, A., Kawano, K., Le Gall, F., Tamaki, S.: Quantum query complexity of unitary operator discrimination. In: Cao, Y., Chen, J. (eds.) COCOON 2017. LNCS, vol. 10392, pp. 309–320. Springer, Cham (2017). https://doi.org/10.1007/978-3-319-62389-4_26
10. Bera, D.: Detection and diagnosis of single faults in quantum circuits. IEEE Trans. Comput.-Aided Des. Integr. Circuits Syst. **37**(3), 587–600 (2018)
11. Acin, A.: Statistical distinguishability between unitary operations. Phys. Rev. Lett. **87**(17), 177901 (2001)
12. Bennett, C.H., Bernstein, E., Brassard, G., Vazirani, U.: Strengths and weaknesses of quantum computing. SIAM J. Comput. **26**(5), 1510–1523 (1997)

13. Biham, E., Biham, O., Biron, D., Grassl, M., Lidar, D.A.: Grover's quantum search algorithm for an arbitrary initial amplitude distribution. Phys. Rev. A **60**(4), 2742 (1999)
14. Biham, E., Kenigsberg, D.: Grover's quantum search algorithm for an arbitrary initial mixed state. Phys. Rev. A **66**(6), 062301 (2002)
15. Høyer, P.: Arbitrary phases in quantum amplitude amplification. Phys. Rev. A **62**(5), 052304 (2000)
16. D'Ariano, G.M., Presti, P.L., Paris, M.G.: Using entanglement improves the precision of quantum measurements. Phys. Rev. Lett. **87**(27), 270404 (2001)
17. Duan, R., Feng, Y., Ying, M.: Entanglement is not necessary for perfect discrimination between unitary operations. Phys. Rev. Lett. **98**(10), 100503 (2007)
18. Bera, D.: Amplitude amplification for operator identification and randomized classes. Technical report TR14-151. Electronic Colloquium on Computational Complexity (2018)
19. Bera, D., Green, F., Homer, S.: Small depth quantum circuits. ACM SIGACT News **38**(2), 35–50 (2007)
20. Høyer, P., de Wolf, R.: Improved quantum communication complexity bounds for disjointness and equality. In: Alt, H., Ferreira, A. (eds.) STACS 2002. LNCS, vol. 2285, pp. 299–310. Springer, Heidelberg (2002). https://doi.org/10.1007/3-540-45841-7_24
21. Buhrman, H., de Wolf, R.: Communication complexity lower bounds by polynomials. In: Proceedings of the 16th CCC (2001)

Reconstruction of Boolean Formulas in Conjunctive Normal Form

Evgeny Dantsin$^{(\boxtimes)}$ and Alexander Wolpert

Department of Computer Science, Roosevelt University, 430 S. Michigan Av.,
Chicago, IL 60605, USA
{edantsin,awolpert}@roosevelt.edu

Abstract. A long-standing open problem in graph theory is to prove or disprove the *graph reconstruction conjecture* proposed by Kelly and Ulam in the 1940s. This conjecture roughly states that every graph on at least three vertices is uniquely determined by its vertex-deleted subgraphs. We adapt the idea of reconstruction for Boolean formulas in conjunctive normal form (CNFs) and formulate the *reconstruction conjecture for CNFs*: every CNF with at least four clauses is uniquely determined by its clause-deleted subformulas. Our main results can be summarized as follows. First, we prove that our conjecture is equivalent to a well-studied variation of the graph reconstruction conjecture, namely, the edge-reconstruction conjecture for hypergraphs. Second, we prove that the number of satisfying assignments of a CNF is reconstructible, i.e., this number can be computed from the clause-deleted subformulas. Third, we show that every CNF with m clauses over n variables is reconstructible if $2^{m-1} > 2^n \cdot n!$.

1 Introduction

We consider *reconstruction* of combinatorial objects from their subobjects, for example, graphs from their subgraphs, or Boolean formulas from their subformulas. Suppose that an object A is uniquely determined by its subobjects $A_1, \ldots, A_n$. The problem arises when the subobjects are given "up to isomorphism". Can we reconstruct A, up to isomorphism, from the isomorphism classes of $A_1, \ldots, A_n$?

The Graph Reconstruction Conjecture. A typical example of reconstruction is reconstruction of a graph from its vertex-deleted subgraphs. Let G be a simple graph on vertices $v_1, \ldots, v_n$. Let $G - v_1, \ldots, G - v_n$ denote subgraphs, where $G - v_i$ is obtained from G by removing v_i and its incident edges. These subgraphs are called *vertex-deleted subgraphs* of G. The *deck* of G consists of n cards that represent the vertex-deleted subgraphs: the ith card shows $G - v_i$ up to isomorphism, i.e., the vertices on the card have no names. Can we reconstruct G up to isomorphism from its deck? Equivalently, does the deck determine G uniquely?

The *graph reconstruction conjecture*, proposed by Kelly and Ulam in the 1940s, also known as the *Kelly-Ulam conjecture* or *Ulam's conjecture*, gives a positive answer to these questions:

© Springer International Publishing AG, part of Springer Nature 2018
L. Wang and D. Zhu (Eds.): COCOON 2018, LNCS 10976, pp. 592–601, 2018.
https://doi.org/10.1007/978-3-319-94776-1_49

Every simple graph on at least three vertices is uniquely determined up to isomorphism by its deck.

The conjecture is stated in terms of a deck of cards, which was proposed by Harary in [Har64]. The following equivalent form is given in terms of two graphs [Kel57, Ula60]. Let G be a graph on vertices $v_1, \ldots, v_n$ and H be a graph on vertices $u_1, \ldots, u_n$. We say that H is a *reconstruction* of G if there is a permutation σ on $[1..n]$ such that $G - v_i$ is isomorphic to $H - u_{\sigma(i)}$ for all $i \in [1..n]$. A graph G is called *reconstructible* if every reconstruction of G is isomorphic to G. In these terms, the graph reconstruction conjecture states that all simple graphs with at least three vertices are reconstructible.

While it is an open problem whether all graphs are reconstructible, many well known classes of graphs are shown to be reconstructible, for example, regular graphs, Eulerian graphs, trees, disconnected graphs [Kel57]. Computerized verification shows that all graphs with up to 11 vertices are reconstructible [McK97]. Considering reconstruction for randomly chosen graphs, almost all graphs are reconstructible [Mül76]. Moreover, almost all graphs can be uniquely determined up to isomorphism using only three cards from their decks [Bol90].

Properties or parameters of a graph G are called *invariants* if they are preserved under isomorphisms. An invariant is called *reconstructible* if it is uniquely determined by the deck of G. For example, the number of edges of G is reconstructible (note that if G has n vertices then each edge of G appears in exactly $n - 2$ cards). Another simple example is the degree sequence of G which can be computed from the number of edges of G and the numbers of edges of the cards.

Reconstruction of the number of edges is a special case of Kelly's lemma [Kel57] stating that for every graph H with fewer vertices than in G, the number of subgraphs of G isomorphic to H is reconstructible. Other examples of reconstructible properties and parameters include the Tutte polynomial, the chromatic number, the number of Hamiltonian cycles, and planarity, see a survey in [LS16].

A number of variations of the graph reconstruction conjecture have been proposed and studied. Some of them have been disproved, for example, for directed graphs [Sto77] and for hypergraphs [Koc87]. Other variations have still been neither proved nor disproved, for example, the *set reconstruction conjecture* proposed by Harary in [Har64]: every simple graph with at least four vertices is uniquely determined up to isomorphism by the set (not the multiset) of its cards.

Edge Reconstruction. An *edge-card* of a graph G is its subgraph obtained from G by removing an edge and viewed up to isomorphism. The *edge-deck* of G is the multiset of all edge-cards of G. Harary proposed the *edge reconstruction conjecture* [Har64]: every simple graph with at least four edges is uniquely determined up to isomorphism by its *edge-deck*.

It was shown in [Gre71] that the deck of a graph with at least four edges and no isolated vertices can be uniquely determined from the graph's edge-deck. Therefore, if the Kelly-Ulam conjecture is true, then the edge reconstruction conjecture is true. Using the inclusion-exclusion principle, Lovász showed in

[Lov72] that a graph on n vertices with m edges is reconstructible from its deck if $m > \frac{1}{2}\binom{n}{2}$. Müller improved this result to $2^{m-1} > n!$; this result also follows from the sufficient condition for edge reconstruction given by Nash-Williams in [NW78].

Extension for Hypergraphs. Berge extended the notion of edge reconstruction to hypergraphs [Ber72]. The edge-deck for a hypergraph is defined in literally the same way as for graphs. The *edge reconstruction conjecture for hypergraphs* states:

> Every hypergraph with at least four edges is uniquely determined up to isomorphism by its edge-deck.

It is still open whether this conjecture is true or not. A number of hypergraph invariants were recently shown to be reconstructible from the edge-deck, for example, multivariate chromatic polynomials [Whi11], edge-induced and vertex-induced sub-hypergraph polynomials [Tad15].

Alon et al. extended edge reconstruction from graphs and hypergraphs to more general combinatorial structures in [ACKR89], see Sect. 5 for more details. Müller's inequality mentioned above follows from their results on reconstructible structures.

What is Done in this Paper. We adapt the idea of graph reconstruction to CNFs, Boolean formulas in conjunctive normal form. The motivation for this adaptation is natural. CNFs can be characterized using graph structures, for example, using their primal graphs, dual graphs, or incidence graphs [SS09]. Properties and parameters of the corresponding graphs, like bandwidth or treewidth, are used in satisfiability-testing algorithms and their analysis, see for example [AR11]. It would be interesting to see if results on graph reconstruction could help in learning more about CNF structures. It would also be interesting to see if reconstruction techniques could help in computing reconstructible invariants like the number of satisfying assignments.

In Sect. 2, we define the notion of deck for CNFs and we formulate the *reconstruction conjecture for CNFs*: every CNF with at least four clauses is reconstructible from its deck. It is shown in Sect. 3 that this conjecture is true if and only if the edge conjecture for hypergraphs is true. In Sect. 4, we use a combination of the inclusion-exclusion principle with the idea of Kelly's lemma [Kel57] to show that the number of satisfying assignments of a CNF can be computed from its deck. In Sect. 5, we apply the results of [ACKR89] about combinatorial structures to prove that a CNF on n variables with m clauses is reconstructible if $2^{m-1} > 2^n \cdot n!$.

2 Reconstruction of CNFs: Definitions

Terminology and Notation for CNFs. A *literal* is a Boolean variable x or its negation $\neg x$. The literals x and $\neg x$ are called *complementary*; each of them is the *complement* of the other. A *clause* is a finite set of literals that contains no pair of complements. The *width* of a clause C is $|C|$, i.e. the number of literals in C. A *formula in conjunctive normal form*, or a *CNF* for short, is a finite set of clauses.

For a clause C, we write $var(C)$ to denote the set of all variables appearing in C (with or without negation). A clause C is viewed as the disjunction of its literals: C is *satisfied* by an assignment of truth values to the variables in $var(C)$ if at least one literal in C is true under this assignment. For a CNF ϕ consisting of clauses $C_1, \ldots, C_m$, let $var(\phi)$ denote $\bigcup_{i=1}^{m} var(C_i)$. An *assignment* for ϕ is an assignment of truth values to all variables of $var(\phi)$. An assignment *satisfies* ϕ if all clauses of ϕ are satisfied by this assignment. The number of assignments satisfying of ϕ is denoted by $\#\phi$.

Isomorphism of CNFs. Consider a CNF ϕ and a variable $x \in var(\phi)$. We change ϕ by *flipping* x if

- all occurrences of x are replaced with $\neg x$;
- all occurrences of $\neg x$ are replaced with x.

Let y be a variable that may or may not belong to $var(\phi)$. We change ϕ by *renaming* x to y if

- all occurrences of x are replaced with y;
- all occurrences of y are replaced with x;
- all occurrences of $\neg x$ are replaced with $\neg y$;
- all occurrences of $\neg y$ are replaced with $\neg x$.

Thus, if both x and y appear in ϕ then these variables swap their names, otherwise x is renamed to the new variable y.

We say that CNFs ϕ and ψ are *isomorphic*, written $\phi \simeq \psi$, if one of them can be obtained from the other by a composition of renamings and flippings. Note that an isomorphism between CNFs is sometimes defined as a composition of renamings only, see for example [SS09], yet in this paper we adhere to the more general version. For a CNF ϕ, we write $[\phi]$ to denote the *isomorphism class* of ϕ, i.e., the class of all CNFs isomorphic to ϕ. Thus, $[\phi]$ can be viewed as ϕ up to isomorphism.

The Reconstruction Conjecture for CNFs. Let ϕ be a CNF consisting of clauses $C_1, \ldots, C_m$ and let $n = var(\phi)$. Consider m CNFs obtained from ϕ by removing one clause:

$$\phi - C_1, \ \ldots, \ \phi - C_m.$$

Each of them is called a *clause-deleted* CNF of ϕ. The *cards* of ϕ are the isomorphism classes of all clause-deleted CNFs of ϕ:

$$[\phi - C_1], \ \ldots, \ [\phi - C_m].$$

The *deck* of ϕ is defined to be the following pair: the number n and the multiset of the cards of ϕ.

The following conjecture about reconstruction of CNFs has the form similar to the Kelly-Ulam conjecture about reconstruction of graphs.

Conjecture 1 (reconstruction of CNFs). For every CNF ϕ with at least four clauses, the isomorphism class of ϕ is uniquely determined by the deck of ϕ.

3   CNFs and Hypergraphs

We prove that the reconstruction conjecture for CNFs (Conjecture 1 above) is equivalent to the following conjecture known as the *edge-reconstruction conjecture for hypergraphs* [Ber72]:

Conjecture 2 (edge-reconstruction of hypergraphs). For every hypergraph H with at least four edges, the isomorphism class of H is uniquely determined by the edge-deck of H.

Theorem 1. *Conjecture 1 implies Conjecture 2.*

Proof. Let H be a hypergraph with edges $e_1, \ldots, e_m$. Let $v_1, \ldots, v_n$ be all non-isolated vertices of H, i.e.

$$\{v_1, \ldots, v_n\} = e_1 \cup \ldots \cup e_m.$$

We define the *CNF representation* of H to be the following CNF ϕ_H: it has n variables identified with $v_1, \ldots, v_n$ and m clauses identified with $e_1, \ldots, e_m$. Note that the clauses of ϕ_H contain no literals with negations. The following properties of CNF representations are easy to verify:

– Two hypergraphs are isomorphic if and only if they have the same number of isolated vertices and their CNF representations are isomorphic.
– If two hypergraphs have identical edge-decks, then their CNF representations have identical decks.

Therefore, the edge-deck of H uniquely determines the deck of ϕ_H. By Conjecture 1, the deck of ϕ_H uniquely determines the isomorphism class of ϕ_H. Also, the edge-deck of H determines the number of isolated vertices in H. The isomorphism class of ϕ_H together with the number of isolated vertices uniquely determine the isomorphism class of H. $\qquad\square$

Theorem 2. *Conjecture 2 implies Conjecture 1.*

Proof. Theorem 1 has been proved using the CNF representation of a hypergraph. In Theorem 2, we use a "dual" method: every CNF is represented by a hypergraph such that two CNFs are isomorphic if and only if the corresponding hypergraphs are isomorphic.

Let ϕ be a CNF consisting of clauses $C_1, \ldots, C_m$. The idea of representation of ϕ by a hypergraph is simple. The hypergraph's vertices are the literals of ϕ and it has two types of edges: some edges represent the clauses of ϕ and the others represent the pairs of complementary literals. It would be straightforward to think of $C_1, \ldots, C_m$ as edges of the first type and think of pairs of complements as edges of the second type. However, this implementation of the idea does not work because ϕ can have two-literal clauses and then two-vertex edges of different types could be mapped to each other under isomorphisms. To fix this issue, we represent every clause of width k, where $k \geq 2$, by an edge of size $k + 1$. Namely, a clause consisting of literals $a_1, \ldots, a_k$ is represented by an edge that consists of the vertices corresponding to $a_1, \ldots, a_k$ plus an additional, "dummy" vertex which appears only in this edge.

More exactly, the *hypergraph representation* of ϕ is a hypergraph H_ϕ defined as follows.

– *Vertices.* If $var(\phi) = \{x_1, \ldots, x_n\}$ then H_ϕ has $2n + m$ vertices: vertices $v_1, \ldots, v_n$ representing the variables $x_1, \ldots, x_n$, vertices $v_{n+1}, \ldots, v_{2n}$ that represent the complements $\neg x_1, \ldots, \neg x_n$, and *dummy* vertices $u_1, \ldots, u_m$. We refer to vertices v_i and v_{i+n} representing complements as *complementary* vertices.

– *Edges.* The hypergraph H_ϕ has n two-vertex edges representing pairs of complements: edges $\{v_i, v_{i+n}\}$ for all $i \in [1..n]$. Also, there are edges $e_1, \ldots, e_m$ representing the clauses $C_1, \ldots, C_m$: if $C_i = \{a_1, \ldots, a_k\}$ then e_i consists of the vertices representing the literals $a_1, \ldots, a_k$ and the dummy vertex u_i.

We need to show that hypergraph representations have the same property as CNF representations defined above: CNFs ϕ and ψ are isomorphic if and only if their hypergraph representations H_ϕ and H_ψ are isomorphic.

To prove the "only if" part, we consider an isomorphism α from ϕ to ψ, i.e., a bijection from the set of all variables of ϕ and their negations to the set of all variables of ψ and their negations. This bijection preserves both the clauses and complements. The isomorphism α induces a bijection β from the non-dummy vertices of H_ϕ to the non-dummy vertices of H_ψ. The bijection β preserves two-vertex edges and preserves the non-dummy parts of edges representing clauses: $\{v_{i_1}, \ldots, v_{i_k}, u_i\}$ is an edge in H_ϕ if and only if $\{\beta(v_{i_1}), \ldots, \beta(v_{i_k}), w_j\}$ is an edge in H_ψ. Next, it remains to make β an isomorphism from H_ϕ to H_ψ by extending β to the dummy vertices in an obvious way: u_i is mapped to w_j. The "if" part is even easier: we have an isomorphism β from H_ϕ to H_ψ, and the restriction of β to the non-dummy vertices induces an isomorphism between ϕ and ψ.

Suppose that Conjecture 2 is true. Then we can reconstruct every CNF ϕ from its deck as follows. Let H_ϕ be the hypergraph representation of ϕ. Each card of ϕ obtained by removing a clause C uniquely determines the card in the edge-deck of H_ϕ obtained by removing the edge that represents C and by adding, if needed, vertices that represent literals. By Conjecture 2, the edge-deck of Π_ϕ uniquely determines $[H_\phi]$. In turn, $[H_\phi]$ uniquely determines $[\phi]$. $\square$

4   Number of Satisfying Assignments

An *invariant* of a CNF is a parameter or a property of this CNF that is preserved under isomorphism. For example, the number of satisfying assignment of a CNF is an invariant because for all CNFs ϕ and ψ, if $\phi \simeq \psi$ then $\#\phi = \#\psi$. An invariant is called *reconstructible* if it can be computed from the deck of ϕ.

Theorem 3. *The number of satisfying assignments of a CNF is reconstructible.*

Proof. Let ϕ be a CNF with clauses $C_1, \ldots, C_m$. For every $i \in [1..m]$, let $\mathcal{F}(C_i)$ denote the set of assignments for ϕ that falsify the clause C_i. Then the set of assignments falsifying ϕ is the union

$$\mathcal{F}(C_1) \cup \ldots \cup \mathcal{F}(C_m)$$

and the cardinality of this union can be computed using the inclusion-exclusion principle:

$$|\mathcal{F}(C_1) \cup \ldots \cup \mathcal{F}(C_m)| = s_1 - s_2 + \ldots (-1)^{m+1} s_m$$

where

$$s_k = \sum_{1 \leq i_1 \leq \ldots \leq i_k \leq m} |\mathcal{F}(C_{i_1}) \cap \ldots \cap \mathcal{F}(C_{i_k})|. \tag{1}$$

If the numbers $s_1, \ldots, s_m$ are known, then $\#\phi$ can be computed by

$$\#\phi = 2^n - \sum_{k=1}^{m} (-1)^{k+1} s_k \tag{2}$$

where $n = |var(\phi)|$. Thus, the claim will be proved if we show that $s_1, \ldots, s_m$ are reconstructible, i.e., they can be computed from the deck of ϕ.

Consider a k-subset of clauses $\{C_{i_1}, \ldots, C_{i_k}\}$ occurring in equality (1) and notice that the number $|\mathcal{F}(C_{i_1}) \cap \ldots \cap \mathcal{F}(C_{i_k})|$ is an invariant, i.e., this number remains the same for all isomorphic copies of $\{C_{i_1}, \ldots, C_{i_k}\}$. Also notice that isomorphic copies of $\{C_{i_1}, \ldots, C_{i_k}\}$ appear in exactly $m - k$ cards of ϕ. Hence, if $k < m$, then the number s_k can be reconstructed using the idea of Kelly's Lemma [Kel57] as follows.

In every card of ϕ, we enumerate all k-subsets $\{A_1, \ldots, A_k\}$ of clauses and compute $|\mathcal{F}(A_1) \cap \ldots \cap \mathcal{F}(A_k)|$ for each of them:

$$|\mathcal{F}(A_1) \cap \ldots \cap \mathcal{F}(A_k)| = \begin{cases} 0 & \text{if } A \text{ contains complements} \\ 2^{n-|A|} & \text{otherwise} \end{cases}$$

where $A = A_1 \cup \ldots \cup A_k$. Note that $2^{n-|A|}$ can be computed because the deck of ϕ contains n. Next, we sum up all these numbers; let S_k denote their sum:

$$S_k = \sum_{\{A_1, \ldots, A_k\}} |\mathcal{F}(A_1) \cap \ldots \cap \mathcal{F}(A_k)|$$

where the sum is taken over all k-subsets $\{A_1, \ldots, A_k\}$ in all cards of ϕ. The key point is that isomorphic copies of $\{C_{i_1}, \ldots, C_{i_k}\}$ were counted $m - k$ times and, therefore, $s_k = S_k/(m - k)$. Thus, all $s_1, \ldots, s_{m-1}$ are found from the deck. As for s_m, there are only two options:

- ϕ contains a pair of complements and then $s_m = 0$;
- ϕ contains no pair of complements and then $s_m = 1$.

Obviously, it can be determined from the deck whether or not ϕ contains a pair of complements. Therefore, all $s_1, \ldots, s_m$ are known and $\#\phi$ can be computed using (2). □

5   Reconstructible CNFs

Combinatorial Structures. Following [LS16], we define a *structure* $\mathcal{S}$ to be a triple (D, Γ, E) where D is a finite set, Γ is a subgroup of the symmetric group on D, and E is a subset of D. The set D is called the *domain* and elements of E are called *edges* of $\mathcal{S}$. For example, the following structure (D, Γ, E) represents a hypergraph on vertices $v_1, \ldots, v_n$. The domain D is the set of all subsets of $\{v_1, \ldots, v_m\}$, E is the set of edges of the hypergraph, and Γ is the subgroup of the symmetric group on D induced by all permutations on $v_1, \ldots, v_n$.

Two structures $\mathcal{S}_1 = (D, \Gamma, E_1)$ and $\mathcal{S}_2 = (D, \Gamma, E_2)$ are called *isomorphic* if there is a permutation α in Γ such that for all $s \in D$, we have

$$s \in E_1 \ \Leftrightarrow \ \alpha(s) \in E_2.$$

The isomorphism class of a structure $\mathcal{S}$ is denoted $[\mathcal{S}]$.

Consider a structure $\mathcal{S} = (D, \Gamma, E)$ where $E = \{e_1, \ldots, e_m\}$. For every $i \in [1..m]$, we write $\mathcal{S} - e_i$ to denote the structure obtained from $\mathcal{S}$ by removing e_i from E. The *deck* of $\mathcal{S}$ is defined to be the multiset

$$[\mathcal{S} - e_1], \ldots, [\mathcal{S} - e_m].$$

The following theorem has in fact been proved by Alon et al. [ACKR89], see Corollary 2.4. It shows that every structure (D, Γ, E) is uniquely determined by its deck for a certain relationship between the size of E and the order of Γ.

Theorem 4 ([ACKR89, LS16]). *Let* $\mathcal{S} = (D, \Gamma, E)$ *be a structure such that*

$$2^{|E|-1} > ord(G).$$

Then $[\mathcal{S}]$ *is uniquely determined by the deck of* $\mathcal{S}$.

Proof. See Theorem 11.8 and its proof in [LS16]. □

Representation of CNFs by Structures. Let ϕ be a CNF with clauses $C_1, \ldots, C_m$ and let $var(\phi) = \{x_1, \ldots, x_n\}$. We define the *structure representation* of ϕ to be the following structure $S = (D, \Gamma, E)$. The domain D consists of all subsets of the set $\{1, \ldots, 2n\}$. We think of the integers in this set as representations of literals of ϕ: the integers $1, \ldots, n$ represent the variables $x_1, \ldots, x_n$ respectively and the integers $n+1, \ldots, 2n$ represent their complements. We refer to integers i and $n+i$ as *complements*. The set E consists of edges $e_1, \ldots, e_m$ that represent clauses:

- for every integer i between 1 and n, we have $i \in e_j$ if and only if $x_i \in C_j$;
- for every integer i between $n+1$ and $2n$, we have $i \in e_j$ if and only if $\neg x_i \in C_j$.

To define the group Γ, we first define the following group Π of permutations on $\{1, \ldots, 2n\}$. Consider two types of permutations on $\{1, \ldots, 2n\}$: *renamings* and *flippings*, where a *renaming* is induced by a permutation on $\{1, \ldots, n\}$ and a *flipping* is a swapping integers i and $n + i$. The group Π consists of all compositions of renamings and flippings. Clearly, the order of Π is $2^n \cdot n!$. The group Γ is the group of permutations on D induced by the group Π.

Note that such representations of CNFs by structures have the following properties:

- Two CNFs are isomorphic if and only if their structure representations are identical.
- Two CNFs have identical decks if and only if their structure representations have identical decks.

Theorem 5. *Let ϕ be a CNF with m clauses over n variables such that*

$$2^{m-1} > 2^n \cdot n!.$$

Then $[\phi]$ is uniquely determined by the deck of ϕ.

Proof. The deck of ϕ uniquely determines the deck of its structure representation $\mathcal{S}_\phi$. By Theorem 4, $\mathcal{S}_\phi$ is reconstructible if $2^{m-1} > 2^n \cdot n!$. $\square$

Thus, all CNFs such that $m > 1 + \mathrm{e} + n \log_2 n$ are reconstructible.

References

[ACKR89] Alon, N., Caro, Y., Krasikov, I., Roditty, Y.: Combinatorial reconstruction problems. J. Comb. Theory Ser. B **47**(2), 153–161 (1989)

[AR11] Alekhnovich, M., Razborov, A.: Satisfiability, branch-width and tseitin tautologies. Comput. Complex. **20**(4), 649–678 (2011)

[Ber72] Berge, C.: Isomorphism problems for hypergraphs. In: Berge, C., Ray-Chaudhuri, D. (eds.) Hypergraph Seminar. LNM, vol. 411, pp. 1–12. Springer, Heidelberg (1974). https://doi.org/10.1007/BFb0066174

[Bol90] Bollobás, B.: Almost every graph has reconstruction number three. J. Graph Theory **14**(1), 1–4 (1990)

[Gre71] Greenwell, D.L.: Reconstructing graphs. Proc. Am. Math. Soc. **30**(3), 431–433 (1971)

[Har64] Harary, F.: On the reconstruction of a graph from a collection of subgraphs. In: Theory of Graphs and Its Applications, pp. 47–52 (1964)

[Kel57] Kelly, P.J.: A congruence theorem for trees. Pac. J. Math. **7**(1), 961–968 (1957)

[Koc87] Kocay, W.L.: A family of nonreconstructible hypergraphs. J. Comb. Theory Ser. B **42**(1), 46–63 (1987)

[Lov72] Lovász, L.: A note on the line reconstruction problem. J. Comb. Theory Ser. B **13**, 309–310 (1972)

[LS16] Lauri, J., Scapellato, R.: Topics in Graph Automorphisms and Reconstruction. London Mathematical Society Lecture Note Series, 2nd edn. Cambridge University Press, Cambridge (2016)

[McK97] McKay, B.D.: Small graphs are reconstructible. Aust. J. Comb. **15**, 123–126 (1997)

[Mül76] Müller, V.: Probabilistic reconstruction from subgraphs. Commentationes Mathematicae Universitatis Carolinae **17**(4), 709–719 (1976)

[NW78] Nash-Williams, C.: The reconstruction problem. In: Selected Topics in Graph Theory, chap. 8, pp. 205–235. Academic Press (1978)

[SS09] Samer, M., Szeider, S.: Fixed-parameter tractability. In: Handbook of Satisfiability, chap. 13, pp. 425 456. IOS Press (2009)

[Sto77] Stockmeyer, P.K.: The falsity of the reconstruction conjecture for tournaments. J. Graph Theory **1**(1), 19–25 (1977)

[Tad15] Tadesse, Y.: Using edge-induced and vertex-induced subhypergraph polynomials. Mathematica Scandinavica **117**(3), 161–169 (2015)

[Ula60] Ulam, S.: A Collection of Mathematical Problems. Interscience Tracts in Pure and Applied Mathematics, vol. 8. Wiley, New York (1960)

[Whi11] White, J.: On multivariate chromatic polynomials of hypergraphs and hyperedge elimination. Electron. J. Comb. **18**(1), #P160 (2011)

A Faster FPTAS for the Subset-Sums Ratio Problem

Nikolaos Melissinos[1] and Aris Pagourtzis[2(✉)]

[1] School of Applied Mathematical and Physical Sciences, National Technical
University of Athens, Polytechnioupoli, 15780 Zografou, Athens, Greece
nmelissinos@corelab.ntua.gr
[2] School of Electrical and Computer Engineering, National Technical University of
Athens, Polytechnioupoli, 15780 Zografou, Athens, Greece
pagour@cs.ntua.gr

Abstract. The Subset-Sums Ratio problem (SSR) is an optimization
problem in which, given a set of integers, the goal is to find two subsets
such that the ratio of their sums is as close to 1 as possible. In this
paper we develop a new FPTAS for the SSR problem which builds on
techniques proposed by Nanongkai (Inf Proc Lett 113, 2013). One of the
key improvements of our scheme is the use of a dynamic programming
table in which one dimension represents the *difference* of the sums of
the two subsets. This idea, together with a careful choice of a scaling
parameter, yields an FPTAS that is several orders of magnitude faster
than the best currently known scheme of Bazgan et al. (J Comput Syst
Sci 64(2), 2002).

Keywords: Approximation scheme · Subset-Sums Ratio
Knapsack problems · Combinatorial optimization

1 Introduction

We study the optimization version of the following NP-hard decision problem
which given a set of integers asks for two subsets of equal sum (but, in contrast
to the Partition problem, the two subsets do not have to form a partition of the
given set):

Equal Sum Subsets Problem (ESS). *Given a set* $A = \{a_1, \ldots, a_n\}$ *of* n
positive integers, are there two nonempty and disjoint sets $S_1, S_2 \subseteq \{1, \ldots, n\}$
such that $\sum_{i \in S_1} a_i = \sum_{j \in S_2} a_j$ *?*

Our motivation to study the ESS problem and its optimization version comes
from the fact that it is a fundamental problem closely related to problems appearing in many scientific areas. Some examples are the Partial Digest problem, which
comes from molecular biology (see [2,3]), the problem of allocating individual
goods (see [8]), tournament construction (see [7]), and a variation of the Subset

ⓒ Springer International Publishing AG, part of Springer Nature 2018
L. Wang and D. Zhu (Eds.): COCOON 2018, LNCS 10976, pp. 602–614, 2018.
https://doi.org/10.1007/978-3-319-94776-1_50

Sum problem, namely the Multiple Integrated Sets SSP, which finds applications in the field of cryptography (see [10]).

The ESS problem has been proven NP-hard by Woeginger and Yu in [11] and several of its variations have been proven NP-hard by Cieliebak et al. in [4–6]. The corresponding optimization problem is:

Subset-Sums Ratio Problem (SSR). *Given a set $A = \{a_1, \ldots, a_n\}$ of n positive integers, find two nonempty and disjoint sets $S_1, S_2 \subseteq \{1, \ldots, n\}$ that minimize the ratio*

$$\frac{\max\{\sum_{i \in S_1} a_i, \sum_{j \in S_2} a_j\}}{\min\{\sum_{i \in S_1} a_i, \sum_{j \in S_2} a_j\}}.$$

The SSR problem was introduced by Woeginger and Yu [11]. In the same work they present an 1.324 approximation algorithm which runs in $O(n \log n)$ time. The SSR problem received its first FPTAS by Bazgan et al. in [1], which approximates the optimal solution in time no less than $O(n^5/\varepsilon^3)$; to the best of our knowledge this is still the faster scheme proposed for SSR. A second, simpler but slower, FPTAS was proposed by Nanongkai in [9].

The FPTAS we present in this paper makes use of some ideas proposed in [9], strengthened by certain key improvements that lead to a considerable acceleration: our algorithm approximates the optimal solution in $O(n^4/\varepsilon)$ time, several orders of magnitude faster than the best currently known scheme of [1].

2   Preliminaries

We will first define two functions that will allow us to simplify several of the expressions that we will need throughout the paper. We will use the convention $\sum_{i \in S} a_i = 0$ if $S = \emptyset$.

Definition 1 (Ratio of two subsets). *Given a set $A = \{a_1, \ldots, a_n\}$ of n positive integers and two sets $S_1, S_2 \subseteq \{1, \ldots, n\}$ we define $\mathcal{R}(S_1, S_2, A)$ as follows:*

$$\mathcal{R}(S_1, S_2, A) = \begin{cases} +\infty & if\ S_2 = \emptyset, \\ \dfrac{\sum_{i \in S_1} a_i}{\sum_{j \in S_2} a_j} & otherwise. \end{cases}$$

Definition 2 (Max ratio of two subsets). *Given a set $A = \{a_1, \ldots, a_n\}$ of n positive integers and two sets $S_1, S_2 \subseteq \{1, \ldots, n\}$ we define $\mathcal{MR}(S_1, S_2, A)$ as follows:*

$$\mathcal{MR}(S_1, S_2, A) = \max\{\mathcal{R}(S_1, S_2, A), \mathcal{R}(S_2, S_1, A)\}.$$

Note that, in cases where at least one of the sets is empty, the Max Ratio function will return ∞. Using these functions, the SSR problem can be rephrased as shown below.

Subset-Sums Ratio Problem (SSR) (Equivalent Definition). *Given a set $A = \{a_1, \ldots, a_n\}$ of n positive integers, find two disjoint sets $S_1, S_2 \subseteq \{1, \ldots, n\}$ such that the value $\mathcal{MR}(S_1, S_2, A)$ is minimized.*

In addition, from now on, whenever we have a set $A = \{a_1, \ldots, a_n\}$ we will assume that $0 < a_1 < a_2 < \ldots < a_n$ (clearly, if the input contains two equal numbers then the problem has a trivial solution).

The FPTAS proposed by Nanongkai [9] approximates the SSR problem by solving a restricted version.

Restricted Subset-Sums Ratio Problem. *Given a set $A = \{a_1, \ldots, a_n\}$ of n positive integers and two integers $1 \leq p < q \leq n$, find two disjoint sets $S_1, S_2 \subseteq \{1, \ldots, n\}$ such that $\{\max S_1, \max S_2\} = \{p, q\}$ and the value $\mathcal{MR}(S_1, S_2, A)$ is minimized.*

Inspired by this idea, we define a less restricted version. The new problem requires one additional input integer, instead of two, which represents the smallest of the two maximum elements of the sought optimal solution.

Semi-Restricted Subset-Sums Ratio Problem. *Given a set $A = \{a_1, \ldots, a_n\}$ of n positive integers and an integer $1 \leq p < n$, find two disjoint sets $S_1, S_2 \subseteq \{1, \ldots, n\}$ such that $\max S_1 = p < \max S_2$ and the value $\mathcal{MR}(S_1, S_2, A)$ is minimized.*

Let $A = \{a_1, \ldots, a_n\}$ be a set of n positive integers and $p \in \{1, \ldots, n-1\}$. Observe that, if S_1^*, S_2^* is the optimal solution of SSR problem of instance A and S_1^p, S_2^p the optimal solution of Semi-Restricted SSR problem of instance A, p then:

$$\mathcal{MR}(S_1^*, S_2^*, A) = \min_{p \in \{1, \ldots, n-1\}} \mathcal{MR}(S_1^p, S_2^p, A).$$

Thus, we can find the optimal solution of SSR problem by solving the SSR Semi-Restricted SSR problem for all $p \in \{1, \ldots, n-1\}$.

3   Pseudo-Polynomial Time Algorithm for Semi-Restricted SSR Problem

Let the A, p be an instance of the Semi-Restricted SSR problem where $A = \{a_1, \ldots, a_n\}$ and $1 \leq p < n$. For solving the problem we have to check two cases for the maximum element of the optimal solution. Let S_1^*, S_2^* be the optimal solution of this instance and $\max S_2^* = q$. We define $B = \{a_i \mid i > p, a_i < \sum_{j=1}^p a_j\}$ and $C = \{a_i \mid a_i \geq \sum_{j=1}^p a_j\}$ from which we have that either $a_q \in B$ or $a_q \in C$. Note that $A = \{a_1, \ldots, a_p\} \cup B \cup C$.

Case 1 $(a_q \in C)$. It is easy to see that if $a_q \in C$, then $a_q = \min C$ and the optimal solution will be $(S_1 = \{1, \ldots, p\}, S_2 = \{q\})$. We describe below a function that returns this pair of sets, thus computing the optimal solution if Case 1 holds.

Definition 3 (Case 1 solution). *Given a set $A = \{a_1, \ldots, a_n\}$ of n positive integers and an integer $1 \leq p < n$ we define the function $\mathcal{SOL}_1(A, p)$ as follows:*

$$\mathcal{SOL}_1(A, p) = \begin{cases} (\{1, \ldots, p\}, \{\min C\}) & if\ C \neq \emptyset, \\ (\emptyset, \emptyset) & otherwise, \end{cases}$$

where $C = \{a_i \mid a_i > \sum_{j=1}^p a_j\}$.

Case 2 $(a_q \in B)$. This second case is not trivial. Here, we define an integer $m = \max\{j \mid a_j \in A \setminus C\}$ and a matrix T, where $T[i, d]$, $0 \leq i \leq m, -2 \cdot \sum_{k=1}^p a_k \leq d \leq \sum_{k=1}^p a_k$, is a quadruple to be defined below. A cell $T[i, d]$ is nonempty if there exist two disjoint sets S_1, S_2 with sums sum_1, sum_2 such that $sum_1 - sum_2 = d, \max S_1 = p$, and $S_1 \cup S_2 \subseteq \{1, \ldots, i\} \cup \{p\}$; if $i > p$, we require in addition that $p < \max S_2$. In such a case, cell $T[i, d]$ consists of the two sets S_1, S_2, and two integers $\max(S_1 \cup S_2)$ and $sum_1 + sum_2$. A crucial point in our algorithm is that if there exist more than one pairs of sets which meet the required conditions, we keep the one that maximize the value $sum_1 + sum_2$; for convenience, we make use of a function to check this property and select the appropriate sets. The algorithm for this case (Algorithm 1) finally returns the pair S_1, S_2 which, among those that appear in some $T[m, d] \neq \emptyset$, has the smallest ratio $\mathcal{MR}(S_1, S_2, A)$.

Definition 4 (Larger total sum tuple selection). *Given two tuples $v_1 = (S_1, S_2, q, x)$ and $v_2 = (S_1', S_2', q', x')$ we define the function $\mathcal{LTST}(v_1, v_2)$ as follows:*

$$\mathcal{LTST}(v_1, v_2) = \begin{cases} v_2 & if\ v_1 = \emptyset\ or\ x' > x, \\ v_1 & otherwise. \end{cases}$$

Algorithm 1. Case 2 solution [$\mathcal{SOL}_2(A, p)$ function]

Input: a strictly sorted set $A = \{a_1, \ldots, a_n\}$, $a_i \in \mathbb{Z}^+$, and an integer p, $1 \le p < n$.
Output: the sets of an optimal solution for Case 2.

```
 1: S'_1 ← ∅, S'_2 ← ∅
 2: Q ← ∑_{i=1}^{p} a_i, m ← max{i | a_i < Q}
 3: if m > p then
 4:     for all i ∈ {0, ..., m}, d ∈ {-2·Q, ..., Q} do
 5:         T[i, d] ← ∅
 6:     end for
 7:     T[0, a_p] ← ({p}, ∅, p, a_p)                        ▷ p ∈ S_1 by problem definition
 8:     for i ← 1 to m do
 9:         if i < p then
10:             for all T[i − 1, d] ≠ ∅ do
11:                 (S_1, S_2, q, x) ← T[i − 1, d]
12:                 T[i, d] ← LTST(T[i, d], T[i − 1, d])
13:                 T[i, d + a_i] ← LTST(T[i, d + a_i], (S_1 ∪ {i}, S_2, q, x + a_i))
14:                 T[i, d − a_i] ← LTST(T[i, d − a_i], (S_1, S_2 ∪ {i}, q, x + a_i))
15:             end for
16:         else if i = p then                              ▷ p is already placed in S_1
17:             for all T[i − 1, d] ≠ ∅ do
18:                 T[i, d] ← T[i − 1, d]
19:             end for
20:         else
21:             for all T[i − 1, d] ≠ ∅ do
22:                 (S_1, S_2, q, x) ← T[i − 1, d]
23:                 if i > p + 1 then
24:                     T[i, d] ← LTST(T[i, d], T[i − 1, d])
25:                 end if
26:                 if d − a_i ≥ −2 · Q then
27:                     T[i, d − a_i] ← LTST(T[i, d − a_i], (S_1, S_2 ∪ {i}, i, x + a_i))
28:                 end if
29:             end for
30:             for all T[p, d] ≠ ∅ do
31:                 (S_1, S_2, q, x) ← T[p, d]
32:                 if d − a_i ≥ −2 · Q then
33:                     T[i, d − a_i] ← LTST(T[i, d − a_i], (S_1, S_2 ∪ {i}, i, x + a_i))
34:                 end if
35:             end for
36:         end if
37:     end for
38:     for d ← −2 · Q to Q do
39:         (S_1, S_2, q, x) ← T[m, d]
40:         if MR(S_1, S_2, A) < MR(S'_1, S'_2, A) then
41:             S'_1 ← S_1, S'_2 ← S_2
42:         end if
43:     end for
44: end if
45: return S'_1, S'_2
```

We next present the complete algorithm for Semi-Restricted SSR (Algorithm 2) which simply returns the best among the two solutions obtained by solving the two cases. Algorithm 2 runs in time polynomial in n and Q (where $Q = \sum_{i=1}^{p} a_i$), therefore it is a pseudo-polynomial time algorithm. More precisely, by using appropriate data structures we can store the sets in the matrix cells in $O(1)$ time (and space) per cell, which implies that the time complexity of the algorithm is $O(n \cdot Q)$.

Algorithm 2. Exact solution for Semi-Restricted SSR [$\mathcal{SOL}_{ex}(A, p)$ function]

Input: a strictly sorted set $A = \{a_1, \ldots, a_n\}$, $a_i \in \mathbb{Z}^+$, and an integer p, $1 \le p < n$.
Output: the sets of an optimal solution of Semi-Restricted SSR.
1: $(S_1, S_2) \leftarrow \mathcal{SOL}_1(A, p)$
2: $(S_1', S_2') \leftarrow \mathcal{SOL}_2(A, p)$
3: **if** $\mathcal{MR}(S_1, S_2, A) \le \mathcal{MR}(S_1', S_2', A)$ **then**
4:     **return** S_1, S_2
5: **else**
6:     **return** S_1', S_2'
7: **end if**

---

4   Correctness of the Semi-Restricted SSR Algorithm

In this section we will prove that Algorithm 2 solves exactly the Semi-Restricted SSR problem. Let S_1^*, S_2^* be the sets of an optimal solution for input $(A = \{a_1, \ldots, a_n\}, p)$.
Starting with Case 1 (where $\max S_2^* \in \{i \mid a_i \ge \sum_{j=1}^{p} a_j\}$), is easy to see that:

Observation 1. *The sets* $S_1^* = \{1, \ldots, p\}$, $S_2^* = \{\min\{i \mid a_i \ge \sum_{j=1}^{p} a_j\}\}$ *give the optimal ratio.*

These are the sets which the function $\mathcal{SOL}_1(A, p)$ returns.
For Case 2 (where $\max S_2^* \in \{i \mid i > p, a_i < \sum_{j=1}^{p} a_j\}$) we have to show that the cell $T[m, d]$ (where $d = \sum_{i \in S_1^*} a_i - \sum_{j \in S_2^*} a_j$) contains two sets S_1, S_2 with ratio equal to optimum. Before that we will show a lemma for the sums of the sets of the optimal solution.

Lemma 1. *Let* $Q = \sum_{i=1}^{p} a_i$ *then we have* $\sum_{i \in S_1^*} a_i \le Q$ *and* $\sum_{i \in S_2^*} a_i < 2 \cdot Q$.

Proof. Observe that $\max S_1^* = p$. This gives us $\sum_{i \in S_1^*} a_i \le \sum_{i=1}^{p} a_i$ so it remains to prove $\sum_{i \in S_2^*} a_i < 2 \cdot Q$. Suppose that $\sum_{i \in S_2^*} a_i \ge 2 \cdot Q$. We can define the set S_2 as $S_2^* \setminus \{\min S_2^*\}$. Note that, for all $i \in S_2^*$, we have that the $a_i < \sum_{i=1}^{p} a_i$. Because of that,

$$\sum_{i \in S_1^*} a_i \le \sum_{i=1}^{p} a_i < \sum_{i \in S_2} a_i < \sum_{i \in S_2^*} a_i$$

which means that the pair (S_1^*, S_2) is a feasible solution with smaller max ratio than the optimal, which is a contradiction.   □

The next two lemmas describe conditions which imply that certain cells of T are nonempty. Furthermore, they secure that appropriate sets are stored so that an optimal solution is returned.

Lemma 2. *If there exist two disjoint sets (S_1, S_2) such that*

- $\max S_2 < \max S_1 = p$
- $\sum_{i \in S_1} a_i - \sum_{j \in S_2} a_j = d$

then $T[i, d] \neq \emptyset$ for all $p \geq i \geq \max(S_1 \cup S_2 \smallsetminus \{p\})$. Furthermore for the sets (S_1', S_2') which are stored in $T[i, d]$ it holds that

$$\sum_{i \in S_1'} a_i + \sum_{j \in S_2'} a_j \geq \sum_{i \in S_1} a_i + \sum_{j \in S_2} a_j.$$

Proof. Note that, for all pairs (S_1, S_2) which meet the conditions, their sums are smaller than Q because $\max(S_1 \cup S_2) = p$ so for the value $d = \sum_{i \in S_1} a_i - \sum_{j \in S_2} a_j$ we have

$$-Q \leq d \leq Q.$$

The same clearly holds for every pair of subsets of S_1, S_2.

We will prove the lemma by induction on $q = \max(S_1 \cup S_2 \smallsetminus \{p\})$. For convenience if $S_1 \cup S_2 \smallsetminus \{p\} = \emptyset$ we let $q = 0$.

- $q = 0$ (base case).

The only pair which meets the conditions for $q = 0$ is the $(\{p\}, \emptyset)$. Observe that cell $T[0, a_p]$ is nonempty by the construction of the table and the same holds for $T[i, a_p]$, $1 \leq i \leq p$ (by line 12). In this case the pair of sets which meets the conditions and the pair which is stored are exactly the same, so the lemma statement is obviously true.

- Assume that the lemma statement holds for $q = k \leq p - 1$; we will prove it for $q = k + 1$ as well.

Let (S_1, S_2) be a pair of sets which meets the conditions. Either $q \in S_1$ or $q \in S_2$; therefore either $(S_1 \smallsetminus \{q\}, S_2)$ or $(S_1, S_2 \smallsetminus \{q\})$ (respectively) meets the conditions. By the inductive hypothesis, we know that

- either $T[q - 1, d - a_q]$ or $T[q - 1, d + a_q]$ (resp.) is nonempty
- in any of the above cases for the stored pair (S_1', S_2') it holds that:
  $\sum_{i \in S_1'} a_i + \sum_{j \in S_2'} a_j \geq \sum_{i \in S_1} a_i + \sum_{j \in S_2} a_j - a_q$

In particular, if $(S_1 \smallsetminus \{q\}, S_2)$ meets the conditions then $T[q - 1, d - a_q]$ is nonempty. In line 13 q is added to the first set and therefore $T[q, d]$ is nonempty and the stored pair is $(S_1' \cup \{q\}, S_2')$ (or some other with larger total sum). Hence, the total sum of the pair in $T[q, d]$ is at least

$$\sum_{i \in S_1'} a_i + \sum_{j \in S_2'} a_j + a_q \geq \sum_{i \in S_1} a_i + \sum_{j \in S_2} a_j.$$

If on the other hand $(S_1, S_2 \setminus \{q\})$ is the pair that meets the conditions then $T[q-1, d+a_q]$ is nonempty. In line 14 q is added to the second set and therefore $T[q, d]$ is nonempty and the stored pair is $(S_1', S_2' \cup \{q\})$ (or other with larger total sum). Hence, the total sum of the pair in $T[q, d]$ is at least

$$\sum_{i \in S_1'} a_i + \sum_{j \in S_2'} a_j + a_q \geq \sum_{i \in S_1} a_i + \sum_{j \in S_2} a_j.$$

The same holds for cells $T[i, d]$ with $q < i \leq p$ (due to line 12). This concludes the proof. □

A similar lemma can be proved for sets with maximum element index greater than p. Due to lack of space the proof is deferred to the full version of the paper.

Lemma 3. *If there exist two disjoint sets (S_1, S_2) such that*

- $\max S_2 = q > p = \max S_1$
- $\sum_{i \in S_1} a_i \leq Q$, $\sum_{j \in S_2} a_j < 2 \cdot Q$
- $\sum_{i \in S_1} a_i - \sum_{j \in S_2} a_j = d$

then $T[i, d] \neq \emptyset$ for all $i \geq q$. Furthermore for the sets (S_1', S_2') which are stored in $T[i, d]$ it holds that

$$\sum_{i \in S_1'} a_i + \sum_{j \in S_2'} a_j \geq \sum_{i \in S_1} a_i + \sum_{j \in S_2} a_j.$$

Now we can prove that in Case 2 the pair of sets which the algorithm returns and the pair of sets of an optimal solution have the same ratio.

Lemma 4. *If (S_1', S_2') is the pair of sets that Algorithm 1 returns, then:*

$$\mathcal{MR}(S_1', S_2', A) = \mathcal{MR}(S_1^*, S_2^*, A).$$

Proof. Let m be the size of the first dimension of the matrix T. Observe that for all i, $p+1 \leq i \leq m$, the sets S_1, S_2 of the nonempty cells $T[i, d]$ are constructed (lines 21–35 of Algorithm 1) such that $\max S_1 = p$ and $i \geq \max S_2 > p$. Therefore the pair (S_1', S_2') returned by the algorithm is a feasible solution. We can see that the sets S_1^*, S_2^* meet the conditions of Lemma 3 (the conditions for the sums are met because of Lemma 1) which give us that the cell $T[m, d]$ (where $d = \sum_{i \in S_1^*} a_i - \sum_{j \in S_2^*} a_j$) is non empty and contains two sets with total sum non less than $\sum_{i \in S_1^*} a_i + \sum_{j \in S_2^*} a_j$. Let S_1, S_2 be the sets which are stored to the cell $T[m, d]$. Then we have

$$\mathcal{MR}(S_1', S_2', A) \leq \mathcal{MR}(S_1, S_2, A) \leq \mathcal{MR}(S_1^*, S_2^*, A) \tag{1}$$

where the second inequality is because

$$\sum_{i \in S_1^*} a_i - \sum_{j \in S_2^*} a_j = \sum_{i \in S_1} a_i - \sum_{j \in S_2} a_j$$

and

$$\sum_{i \in S_1^*} a_i + \sum_{j \in S_2^*} a_j \leq \sum_{i \in S_1} a_i + \sum_{j \in S_2} a_j.$$

By (1) and because the S_1^*, S_2^* have the smallest Max Ratio we have

$$\mathcal{MR}(S_1', S_2', A) = \mathcal{MR}(S_1^*, S_2^*, A).$$

$\square$

As a corollary of Observation 1 and Lemma 4 we obtain the following.

Theorem 1. *Algorithm 2 returns an optimal solution for Semi-Restricted SSR.*

5   FPTAS for Semi-Restricted SSR and SSR

Algorithm 2, which we presented at Sect. 3, is an exact pseudo-polynomial time algorithm for the Semi-Restricted SSR problem. In order to derivee a $(1 + \varepsilon)$-approximation algorithm we will define a scaling parameter $\delta = \frac{\varepsilon \cdot a_p}{3 \cdot n}$ which we will use to make a new set $A' = \{a_1', \ldots, a_n'\}$ with $a_i' = \lfloor \frac{a_i}{\delta} \rfloor$. The approximation algorithm solves the problem optimally on input (A', p) and returns the sets of this exact solution. The ratio of those sets is a $(1 + \varepsilon)$-approximation of the optimal ratio of the original input.

Algorithm 3. FPTAS for Semi-Restricted SSR [$\mathcal{SOL}_{apx}(A, p, \varepsilon)$ function]

Input: a strictly sorted set $A = \{a_1, \ldots, a_n\}$, $a_i \in \mathbb{Z}^+$, an integer p, $1 \leq p < n$, and
an error parameter $\varepsilon \in (0, 1)$.
Output: the sets of a $(1 + \varepsilon)$-approximation solution for Semi-Restricted SSR.
1: $\delta \leftarrow \frac{\varepsilon \cdot a_p}{3 \cdot n}$
2: $A' \leftarrow \emptyset$
3: **for** $i \leftarrow 1$ to n **do**
4:     $a_i' \leftarrow \lfloor \frac{a_i}{\delta} \rfloor$
5:     $A' \leftarrow A' \cup \{a_i'\}$
6: **end for**
7: $(S_1, S_2) \leftarrow \mathcal{SOL}_{ex}(A', p)$
8: **return** S_1, S_2

Now, we will prove that the algorithm approximates the optimal solution by factor $(1 + \varepsilon)$. Our proof follows closely the proof of Theorem 2 in [9].

Let S_A, S_B be the pair of sets returned by Algorithm 3 on input $A = \{a_1, \ldots, a_n\}$, p and ε and (S_1^*, S_2^*) be an optimal solution to the problem.

Lemma 5. *For any* $S \in \{S_A, S_B, S_1^*, S_2^*\}$

$$\sum_{i \in S} a_i - n \cdot \delta \leq \sum_{i \in S} \delta \cdot a_i' \leq \sum_{i \in S} a_i, \tag{2}$$

$$n \cdot \delta \leq \frac{\varepsilon}{3} \cdot \sum_{i \in S} a_i. \tag{3}$$

Proof. For Eq. (2) notice that for all $i \in \{1, \ldots, n\}$ we define $a'_i = \lfloor \frac{a_i}{\delta} \rfloor$. This gives us

$$\frac{a_i}{\delta} - 1 \leq a'_i \leq \frac{a_i}{\delta} \Rightarrow a_i - \delta \leq \delta \cdot a_i \leq a_i.$$

In addition, for any $S \in \{S_A, S_B, S_1^*, S_2^*\}$ we have $|S| \leq n$, which means that

$$\sum_{i \in S} a_i - n \cdot \delta \leq \sum_{i \in S} \delta \cdot a'_i \leq \sum_{i \in S} a_i.$$

For Eq. (3) observe that $\max S \geq p$ for any $S \in \{S_A, S_B, S_1^*, S_2^*\}$. By this observation, we can show the second inequality

$$n \cdot \delta \leq n \cdot \frac{\varepsilon \cdot a_p}{3 \cdot n} \leq \frac{\varepsilon}{3} \cdot \sum_{i \in S} a_i.$$

$\square$

Lemma 6. $\mathcal{MR}(S_A, S_B, A) \leq \mathcal{MR}(S_A, S_B, A') + \frac{\varepsilon}{3}$

Proof.

$$\mathcal{R}(S_A, S_B, A) = \frac{\sum_{i \in S_A} a_i}{\sum_{j \in S_B} a_j} \leq \frac{\sum_{i \in S_A} \delta \cdot a'_i + \delta \cdot n}{\sum_{j \in S_B} a_j} \qquad \text{[by Eq. (2)]}$$

$$\leq \frac{\sum_{i \in S_A} a'_i}{\sum_{j \in S_B} a'_j} + \frac{\delta \cdot n}{\sum_{j \in S_B} a_j} \qquad \text{[by Eq. (2)]}$$

$$\leq \mathcal{MR}(S_A, S_B, A') + \frac{\varepsilon}{3} \qquad \text{[by Eq. (3)]}$$

The same way, we have

$$\mathcal{R}(S_B, S_A, A) \leq \mathcal{MR}(S_A, S_B, A') + \frac{\varepsilon}{3}$$

thus the lemma holds. $\square$

Lemma 7. *For any $\varepsilon \in (0, 1)$, $\mathcal{MR}(S_1^*, S_2^*, A') \leq (1 + \frac{\varepsilon}{2}) \cdot \mathcal{MR}(S_1^*, S_2^*, A)$.*

Proof. If $\mathcal{R}(S_1^*, S_2^*, A') \geq 1$, let $(S_1, S_2) = (S_1^*, S_2^*)$, otherwise $(S_1, S_2) = (S_2^*, S_1^*)$. That gives us

$$\mathcal{MR}(S_1^*, S_2^*, A') = \mathcal{R}(S_1, S_2, A') = \frac{\sum_{i \in S_1} a'_i}{\sum_{j \in S_2} a'_j}$$

$$\leq \frac{\sum_{i \in S_1} a_i}{\sum_{j \in S_2} a_j - n \cdot \delta} \qquad \text{[by Eq. (2)]}$$

$$= \frac{\sum_{i \in S_2} a_i}{\sum_{j \in S_2} a_j - n \cdot \delta} \cdot \frac{\sum_{i \in S_1} a_i}{\sum_{j \subset S_2} a_j}$$

$$= (1 + \frac{n \cdot \delta}{\sum_{j \in S_2} a_j - n \cdot \delta}) \cdot \frac{\sum_{i \in S_1} a_i}{\sum_{j \in S_2} a_j}.$$

Because $S_2 \in \{S_1^*, S_2^*\}$ by Eq. (3) it follows that

$$\mathcal{MR}(S_1^*, S_2^*, A') \leq (1 + \frac{1}{\frac{3}{\varepsilon} - 1}) \cdot \frac{\sum_{i \in S_1} a_i}{\sum_{j \in S_2} a_j}$$

$$= (1 + \frac{\varepsilon}{3 - \varepsilon}) \cdot \frac{\sum_{i \in S_1} a_i}{\sum_{j \in S_2} a_j}$$

$$\leq (1 + \frac{\varepsilon}{2}) \cdot \frac{\sum_{i \in S_1} a_i}{\sum_{j \in S_2} a_j} \qquad \text{[because } \varepsilon \in (0, 1)\text{]}$$

$$\leq (1 + \frac{\varepsilon}{2}) \cdot \mathcal{MR}(S_1^*, S_2^*, A).$$

This concludes the proof.   □

Now we can prove that Algorithm 3 is a $(1 + \varepsilon)$ approximation algorithm.

Theorem 2. *Let S_A, S_B be the pair of sets returned by Algorithm 3 on input $(A = \{a_1, \ldots, a_n\}, p, \varepsilon)$ and S_1^*, S_2^* be an optimal solution, then:*

$$\mathcal{MR}(S_A, S_B, A) \leq (1 + \varepsilon) \cdot \mathcal{MR}(S_1^*, S_2^*, A).$$

Proof. The theorem follows from a sequence of inequalities:

$$\mathcal{MR}(S_B, S_A, A) \leq \mathcal{MR}(S_A, S_B, A') + \frac{\varepsilon}{3} \qquad \text{[by Lemma 6]}$$

$$\leq \mathcal{MR}(S_1^*, S_2^*, A') + \frac{\varepsilon}{3}$$

$$\leq (1 + \frac{\varepsilon}{2}) \cdot \mathcal{MR}(S_1^*, S_2^*, A) + \frac{\varepsilon}{3} \qquad \text{[by Lemma 7]}$$

$$\leq (1 + \varepsilon) \cdot \mathcal{MR}(S_1^*, S_2^*, A).$$

□

It remains to show that the complexity of Algorithm 3 is $\mathcal{O}(poly(n, 1/\varepsilon))$. As mentioned in Sect. 3 the algorithm solves the Semi-Restricted SSR problem in $O(nQ)$ time (where $Q = \sum_{i=1}^p a_i'$). We have to bound the value of Q. By the definition of a_i' we have,

$$Q = \sum_{i=1}^p a_i' \leq n \cdot a_p' \leq \frac{n \cdot a_p}{\delta} = \frac{3 \cdot n^2}{\varepsilon}$$

which means that the time complexity of Algorithm 3 is $O(n^3/\varepsilon)$.
Clearly, it suffices to perform $n - 1$ executions of the FPTAS for Semi-Restricted SSR (Algorithm 3), and pick the best of the returned solutions, in order to obtain an FPTAS for the (unrestricted) SSR problem. Therefore, we obtain the following.

Theorem 3. *The above described algorithm is an FPTAS for SSR with $O(n^4/\varepsilon)$ time complexity.*

6 Conclusion

In this paper we provide an FPTAS for the Subset-Sums Ratio (SSR) problem that is much faster than the best currently known scheme of Bazgan et al. [1]. There are two novel ideas that provide this improvement.

The first comes from observing that in [9], the proof of correctness essentially relies only on the value of the smallest of the two maximum elements; this led to the idea to use only that information in order to solve the problem by defining and solving a new variation which we call Semi-Restricted SSR. Note that the idea of using a single parameter instead of two was mentioned in [9] as a hint for further improvement; however, the suggestion was to use the maximum of the two parameters, whereas we obtained our improvement by using the minimum of the two. In particular, we observed that a key ingredient in the technique of Nanongkai [9] is the use, in the scaling parameter δ, of a value smaller than the sums of the sets of both the optimal and the approximate solutions. We believe that this method can be used in several other partition problems, e.g. such as the ones described in [8,10].

The second idea was to use one dimension only, for the difference of the sums of the two sets, instead of two dimensions, one for each sum. This idea, combined with the observation that between two pairs of sets with the same difference, the one with the largest total sum has ratio closer to 1, is the key to obtain an optimal solution in much less time. It's interesting to see whether and how this technique could be used to problems that seek more than two subsets.

A natural open question is whether our techniques can be applied to obtain approximation results for other variations of the SSR problem [5,6].

References

1. Bazgan, C., Santha, M., Tuza, Z.: Efficient approximation algorithms for the subset-sums equality problem. J. Comput. Syst. Sci. **64**(2), 160–170 (2002). https://doi.org/10.1006/jcss.2001.1784
2. Cieliebak, M., Eidenbenz, S., Penna, P.: Noisy data make the partial digest problem NP-hard. In: Benson, G., Page, R.D.M. (eds.) WABI 2003. LNCS, vol. 2812, pp. 111–123. Springer, Heidelberg (2003). https://doi.org/10.1007/978-3-540-39763-2_9
3. Cieliebak, M., Eidenbenz, S.: Measurement errors make the partial digest problem NP-Hard. In: Farach-Colton, M. (ed.) LATIN 2004. LNCS, vol. 2976, pp. 379–390. Springer, Heidelberg (2004). https://doi.org/10.1007/978-3-540-24698-5_42
4. Cieliebak, M., Eidenbenz, S., Pagourtzis, A., Schlude, K.: Equal sum subsets: complexity of variations. Technical report 370, ETH Zürich, Department of Computer Science (2002). ftp://ftp.inf.ethz.ch/doc/tech-reports/3xx/370.pdf
5. Cieliebak, M., Eidenbenz, S., Pagourtzis, A.: Composing equipotent teams. In: Lingas, A., Nilsson, B.J. (eds.) FCT 2003. LNCS, vol. 2751, pp. 98–108. Springer, Heidelberg (2003). https://doi.org/10.1007/978-3-540-45077-1_10
6. Cieliebak, M., Eidenbenz, S., Pagourtzis, A., Schlude, K.: On the complexity of variations of equal sum subsets. Nord. J. Comput. **14**(3), 151–172 (2008)

7. Khan, M.A.: Some problems on graphs and arrangements of convex bodies. Ph.D. thesis, University of Calgary (2017). https://prism.ucalgary.ca/handle/11023/3765

8. Lipton, R.J., Markakis, E., Mossel, E., Saberi, A.: On approximately fair allocations of indivisible goods. In: Proceedings of the 5th ACM Conference on Electronic Commerce (EC 2004), 17–20 May 2004, New York, NY, USA, pp. 125–131 (2004)

9. Nanongkai, D.: Simple FPTAS for the subset-sums ratio problem. Inf. Process. Lett. **113**(19–21), 750–753 (2013)

10. Voloch, N.: MSSP for 2-D sets with unknown parameters and a cryptographic application. Contemp. Eng. Sci. **10**(19), 921–931 (2017)

11. Woeginger, G.J., Yu, Z.: On the equal-subset-sum problem. Inf. Process. Lett. **42**(6), 299–302 (1992). https://doi.org/10.1016/0020-0190(92)90226-L

A Linear-Space Data Structure
for Range-LCP Queries in
Poly-Logarithmic Time

Paniz Abedin[1], Arnab Ganguly[2], Wing-Kai Hon[3], Yakov Nekrich[4],
Kunihiko Sadakane[5], Rahul Shah[6,7], and Sharma V. Thankachan[1(✉)]

[1] Department of Computer Science, University of Central Florida, Orlando, USA
paniz@cs.ucf.edu, sharma.thankachan@ucf.edu
[2] Department of Computer Science, University of Wisconsin - Whitewater,
Whitewater, USA
gangulya@uww.edu
[3] Department of Computer Science, National Tsing Hua University, Hsinchu, Taiwan
wkhon@cs.nthu.edu.tw
[4] Cheriton School of Computer Science, University of Waterloo, Waterloo, Canada
yakov.nekrich@googlemail.com
[5] Department of Computer Science, The University of Tokyo, Tokyo, Japan
sada@mist.i.u-tokyo.ac.jp
[6] Department of Computer Science, Louisiana State University, Baton Rouge, USA
rahul@csc.lsu.edu
[7] National Science Foundation (NSF), Alexandria, USA

Abstract. Let $\mathsf{T}[1, n]$ be a text of length n and $\mathsf{T}[i, n]$ be the suffix starting at position i. Also, for any two strings X and Y, let $\mathsf{LCP}(X, Y)$ denote their longest common prefix. The range-LCP of T w.r.t. a range $[\alpha, \beta]$, where $1 \leq \alpha < \beta \leq n$ is

$$\mathsf{rlcp}(\alpha, \beta) = \max\{|\mathsf{LCP}(\mathsf{T}[i, n], T[j, n])| \mid i \neq j \text{ and } i, j \in [\alpha, \beta]\}$$

Amir et al. [ISAAC 2011] introduced the indexing version of this problem, where the task is to build a data structure over T, so that $\mathsf{rlcp}(\alpha, \beta)$ for any query range $[\alpha, \beta]$ can be reported efficiently. They proposed an $O(n \log^{1+\epsilon} n)$ space structure with query time $O(\log \log n)$, and a linear space (i.e., $O(n)$ words) structure with query time $O(\delta \log \log n)$, where $\delta = \beta - \alpha + 1$ is the length of the input range and $\epsilon > 0$ is an arbitrarily small constant. Later, Patil et al. [SPIRE 2013] proposed another linear space structure with an improved query time of $O(\sqrt{\delta} \log^\epsilon \delta)$. This poses an interesting question, whether it is possible to answer $\mathsf{rlcp}(\cdot, \cdot)$ queries in poly-logarithmic time using a linear space data structure. In this paper, we settle this question by presenting an $O(n)$ space data structure with query time $O(\log^{1+\epsilon} n)$ and construction time $O(n \log n)$.

A part of this work was done at NII Shonan Meeting No. 126: Computation over Compressed Structured Data.

L. Wang and D. Zhu (Eds.): COCOON 2018, LNCS 10976, pp. 615–625, 2018.
https://doi.org/10.1007/978-3-319-94776-1_51

1   Introduction and Related Work

The longest common prefix (LCP) is an important primitive employed in various string matching algorithms. By preprocessing a text $T[1, n]$ (over an alphabet set Σ) into a suffix tree data structure, we can compute the longest common prefix of any two suffixes of T, say $T[i, n]$ and $T[j, n]$, denoted by $LCP(T[i, n], T[j, n])$, in constant time. From now onwards, we use the shorthand notation $lcp(i, j)$ for the length of $LCP(T[i, n], T[j, n])$. Given its wide range of applicability, various generalizations of LCP has also been studied [1–3,9,15].

   In this paper, we focus on the "range" versions of this problem. The line of research was initiated by Cormode and Muthukrishnan. They studied the *Interval Longest Common Prefix* (Interval-LCP) problem in the context of data compression [6,12,13].

Definition 1 (Interval-LCP). *The Interval-LCP of a text* $T[1, n]$ *w.r.t a query* (p, α, β), *where* $p, \alpha, \beta \in [1, n]$ *and* $\alpha < \beta$ *is*

$$ilcp(p, \alpha, \beta) = \max\{lcp(p, i) \mid i \in [\alpha, \beta]\}$$

As observed by Keller et al. [12], any Interval-LCP query on T can be reduced to two orthogonal range successor/predecessor queries over n points in two dimensions (2D). Therefore, using the best known data structures for orthogonal range successor/predecessor queries [14], we can answer any Interval-LCP query on T in $O(\log^\epsilon n)$ time using an $O(n)$ space data structure, where $\epsilon > 0$ is an arbitrarily small positive constant.[1] Moreover, queries with $p \in [\alpha, \beta]$ can be answered in faster $O(\log^\epsilon \delta)$ time, where $\delta = \beta - \alpha + 1$ is the length of the input range [15].

   Another variation of LCP, studied by Amir et al. [1,2], is the following.

Definition 2 (Range-LCP). *The Range-LCP of a text* $T[1, n]$ *w.r.t a range* $[\alpha, \beta]$, *where* $1 \leq \alpha < \beta \leq n$ *is*

$$rlcp(\alpha, \beta) = \max\{lcp(i, j) \mid i \neq j \ and \ i, j \in [\alpha, \beta]\}$$

In order to efficiently solve the data structure version of this problem, Amir et al. [1,2] introduced the concept of **"bridges"** and **"optimal bridges"** and showed that any Range-LCP query on $T[1, n]$ can be reduced to an equivalent 2D range maximum query over a set of $O(n \log n)$ weighted points in 2D. Therefore, an $O(n \log^{1+\epsilon} n)$ space data structure with $O(\log \log n)$ query time is immediate from the best known result for 2D range maximum problem [4]. The construction time is $O(n \log^2 n)$. By choosing an alternative structure for 2D (2-sided) range maximum query[2], the space can be improved to $O(n \log n)$ with a slowdown in

---

[1] All results throughout this paper assume the standard unit-cost word RAM model, in which any standard arithmetic or boolean bitwise operation on word-sized operands takes constant time. The space is measured in words of $\log n$ bits unless specified otherwise.

[2] See Theorem 9 in [16] on sorted dominance reporting in 3D.

query time to $O(\log n)$. This sets an interesting question, whether it is possible to reduce the space further without sacrificing the poly-logarithmic query time. Unfortunately, the query times of the existing linear space solutions (listed below) are dependent on the parameter δ, which is $\Theta(n)$ in the worst case.

- $O(n)$ space and $O(\delta \log \log n)$ query time [1,2].
- $O(n)$ space and $O(\sqrt{\delta} \log^\epsilon \delta)$ query time [15]

To this end, we present our main contribution below. Our model of computation is the word RAM with word size $\Omega(\log n)$.

Theorem 1. *A text* $\mathsf{T}[1, n]$ *can be preprocessed into an* $O(n)$ *space data structure in* $O(n \log n)$ *time, such that any Range-LCP query on* T *can be answered in* $O(\log^{1+\epsilon} n)$ *time.*

Map. We start with some preliminaries (Sect. 2). We the briefly sketch the framework by Amir et al. [1] in Sect. 3. Sections 4 and 5 are dedicated for the details of our solution. The details of the construction of our data structure is deferred to Appendix.

2 Preliminaries

2.1 Predecessor/Successor Queries

Let $\mathcal{S}$ be a subset of $\{1, 2, \ldots, n\}$. Then, $\mathcal{S}$ can be preprocessed into an $O(|\mathcal{S}|)$ space data structure, such that for any query p, we can return $\mathsf{pred}(p, \mathcal{S})$ and $\mathsf{succ}(p, \mathcal{S})$ in $O(\log \log n)$ time [19], where

$$\mathsf{pred}(p, \mathcal{S}) = \max \{i \mid i \leq p \quad \textbf{and} \quad i \in \mathcal{S} \cup \{-\infty\}\}$$
$$\mathsf{succ}(p, \mathcal{S}) = \min \{i \mid i \geq p \quad \textbf{and} \quad i \in \mathcal{S} \cup \{\infty\}\}$$

2.2 Range Minimum Query

Let $A[1, n]$ be an array of length n. A range minimum query (RMQ) with an input range $[i, j]$ asks to report $\mathsf{rmq}(i, j) = \arg\min_k \{A[k] \mid k \in [i, j]\}$. By maintaining a data structure of size $2n + o(n)$ bits, any RMQ on A can be answered in $O(1)$ time [8] (even without accessing A).

2.3 2D Range Maximum Query

Let $\mathcal{S}$ be a set of m weighted points in a $[1, n] \times [1, n]$ grid. A 2D-RMQ with input (a, b, a', b') asks to return the highest weighted point in $\mathcal{S}$ within the orthogonal region corresponding to $[a, b] \times [a', b']$. Data structures with the following space-time trade-offs are known for this problem.

- $O(m)$ space, $O(m \log m)$ preprocessing time and $O(\log^{1+\epsilon} m)$ query time [5].
- $O(m \log^\epsilon n)$ space, $O(m \log m)$ preprocessing time and $O(\log \log n)$ query time [4].

2.4   Orthogonal Range Predecessor/Successor Queries in 2D

A set $\mathcal{P}$ of n points in an $[1, n] \times [1, n]$ grid can be preprocessed into a linear-space data structure, such that the following queries can be answered in $O(\log^\epsilon n)$ time [14].

- $\mathsf{ORQ}([x', x''], [-\infty, y'']) = \arg\max_j\{(i, j) \in \mathcal{P} \cap [x', x''] \times [-\infty, y'']\}$
- $\mathsf{ORQ}([-\infty, x''], [y', y'']) = \arg\max_i\{(i, j) \in \mathcal{P} \cap [-\infty, x''] \times [y', y'']\}$
- $\mathsf{ORQ}([x', x''], [y', +\infty]) = \arg\min_j\{(i, j) \in \mathcal{P} \cap [x', x''] \times [y', +\infty]\}$
- $\mathsf{ORQ}([x', +\infty], [y', y'']) = \arg\min_i\{(i, j) \in \mathcal{P} \cap [x', +\infty] \times [y', y'']\}$

2.5   Suffix Trees and Suffix Arrays

For a string $\mathsf{T}[1, n]$, the suffix array $\mathsf{SA}[1, n]$ is an array of length n, such that $\mathsf{SA}[i]$ denotes the starting position of the lexicographically ith smallest suffix among all suffixes of T. The suffix tree ST is a compact trie of all its suffixes [18]. The suffix tree consists of n leaves and at most $n - 1$ internal nodes. The edges are labeled with substrings of T. For any node u, $\mathsf{path}(u)$ is defined as the concatenation of edge labels on the path from the root of the suffix tree to u. Therefore, $\mathsf{path}(\ell_x) = \mathsf{T}[\mathsf{SA}[x], n]$, where ℓ_x is the xth leftmost leaf node. Moreover, $\mathsf{path}(\mathsf{lca}(\ell_x, \ell_y)) = \mathsf{LCP}(\mathsf{T}[\mathsf{SA}[x], n], \mathsf{T}[\mathsf{SA}[y], n])$, where $\mathsf{lca}(\cdot, \cdot)$ denotes the lowest common ancestor. The suffix tree of T occupies $O(n)$ space, can be constructed in $O(n)$ time and space, and for any two text positions i, j, we can compute $\mathsf{lcp}(i, j)$ in constant time. Also, define inverse suffix array $\mathsf{ISA}[1, n]$, such that $\mathsf{ISA}[i] = j$, where $\mathsf{SA}[j] = i$.

2.6   Heavy Path Decomposition

We define the heavy path decomposition [11,17] of a suffix tree ST as follows. First, we categorize the nodes in ST into light and heavy. The root node is *light* and for any internal node, exactly one child is heavy. Specifically, the child having the largest number of leaves in its subtree (ties are broken arbitrarily). When all incident edges to the light nodes are removed, the remaining edges of ST are decomposed into maximal downward paths, each starting from an internal light node and following a sequence of heavy nodes. We call each path a heavy path.

Lemma 1. *The number of heavy paths intersected by any root to leaf path is at most $\log_2 n$. Equivalently, the number of light nodes on any root to leaf path is at most $\log_2 n$.*

3   Amir et al.'s Framework

We start with some definitions.

Definition 3 (Bridges). *Let i and j and two distinct positions in the text T and let $h = \mathsf{lcp}(i, j)$ and $h > 0$. Then, we call the tuple (i, j, h) a bridge. Moreover, we call h its height, i its left leg and j its right leg, and $\mathsf{LCP}(\mathsf{T}[i, n], \mathsf{T}[j, n])$ its label.*

Let $\mathcal{B}_{all}$ be the set of all such bridges. Then clearly,

$$\mathsf{rlcp}(\alpha, \beta) = \max\{h \mid (i, j, h) \in \mathcal{B}_{all} \quad \textbf{and} \quad i, j \in [\alpha, \beta]\}$$

Therefore, by mapping each bridge $(i, j, h) \in \mathcal{B}_{all}$ to a 2D point (i, j) with weight h, the problem can be reduced to a 2D-RMQ problem (refer to Sect. 2.3). This yields an $O(|\mathcal{B}_{all}| \log^{\epsilon} n)$ space data structure with query time $O(\log \log n)$. Unfortunately, this is not a space efficient approach as the size of $\mathcal{B}_{all}$ is $\Theta(n^2)$ in the worst case. To circumvent this, Amir et al. [1] introduced the concept of optimal bridges.

Definition 4 (Optimal Bridges). *A bridge $(i, j, h) \in \mathcal{B}_{all}$ is optimal if there exists no other bridge (i', j', h'), such that $i', j' \in [i, j]$ and $h' \geq h$.*

Let $\mathcal{B}_{opt}$ be the set of all optimal bridges. Then, it is easy to observe that

$$\mathsf{rlcp}(\alpha, \beta) = \max\{h \mid (i, j, h) \in \mathcal{B}_{opt} \quad \textbf{and} \quad i, j \in [\alpha, \beta]\}.$$

Thus, to answer an rlcp query, it is sufficient to examine the bridges in $\mathcal{B}_{opt}$, instead of all the bridges in $\mathcal{B}_{all}$. The crux of Amir et al.'s [1] data structure is the following lemma.

Lemma 2 ([1]). *The size of $\mathcal{B}_{opt}$ is $O(n \log n)$.*

Therefore, by applying the above reduction (from Range-LCP to 2D-RMQ) on the bridges in $\mathcal{B}_{opt}$, they got an $O(|\mathcal{B}_{opt}| \log^{\epsilon} n) = O(n \log^{1+\epsilon} n)$ space data structure with query time $O(\log \log n)$. Additionally, they showed that there exist cases where the the size of $\mathcal{B}_{opt}$ is $\Omega(n \log n)$. For example, when T is a Fibonacci word (see Sect. 4 in [1] for its definition). This means that the bound on the number of optimal bridges is tight.

4   Our Framework

Firstly, we present a replacement for optimal bridges, called *special bridges*.

Definition 5 (Special Bridges). *A bridge $(i, j, h) \in \mathcal{B}_{all}$ is special if there exists no other bridge $(i', j', h') \in \mathcal{B}_{all}$, such that $i', j' \in [i, j]$ and*

$$\mathsf{LCP}(\mathsf{T}[i, n], \mathsf{T}[j, n]) = \mathsf{LCP}(\mathsf{T}[i', n], \mathsf{T}[j', n])$$

Let $\mathcal{B}_{spe}$ be the set of all special bridges. Clearly $\mathcal{B}_{opt} \subseteq \mathcal{B}_{spe}$, therefore

$$\mathsf{rlcp}(\alpha, \beta) = \max\{h \mid (i, j, h) \in \mathcal{B}_{spe} \quad \textbf{and} \quad i, j \in [\alpha, \beta]\}$$

From Lemma 3, $|\mathcal{B}_{spe}| = \Theta(|\mathcal{B}_{opt}|)$, the same space-time trade-off as in Amir et al. [1] can be obtained by employing special bridges instead of optimal bridges. However, the main advantage over optimal bridges is that special bridges can be encoded efficiently, in $O(1)$-bits per bridge.

Lemma 3. *The size of $\mathcal{B}_{spe}$ is $O(n \log n)$.*

Proof. Firstly, we show how to bound the number of special bridges with a fixed label P. Let u be the node in ST such that $\mathsf{path}(u) = P$. For any such bridge (i, j, h), the leaves (say $\ell_{\mathsf{ISA}[i]}$ and $\ell_{\mathsf{ISA}[j]}$) corresponding to the suffixes $\mathsf{T}[i, n]$ and $\mathsf{T}[j, n]$ must be under the subtree of u, but not under the subtree of the same child of u. This means, either $\ell_{\mathsf{ISA}[i]}$ or $\ell_{\mathsf{ISA}[j]}$ must be under a light child of u. Moreover, the starting position of the suffix corresponding to a leaf can be the left-leg (resp., right-leg) of at most one special bridge with label P. Therefore the number of special bridges with a fixed label P is at most twice the sum of subtree sizes of all light children of u. Hence, the total number of special bridges is at most twice the sum of subtree sizes of all light nodes in the suffix tree, which is bounded by $O(n \log n)$, since each leaf is under at most $O(\log n)$ light ancestors. □

We now present an overview of our solution.

4.1   An Overview of Our Data Structure

We start by defining two queries, which are weaker than Range-LCP.

Definition 6. *For a parameter $\Delta = \Theta(\log n)$, a query $\mathcal{E}_\Delta(\alpha, \beta)$ asks to return an estimate τ of $\mathsf{rlcp}(\alpha, \beta)$, such that*

$$\tau \leq \mathsf{rlcp}(\alpha, \beta) < \tau + \Delta$$

Definition 7. *A query $\mathcal{Q}(\alpha, \beta, h)$ asks to return* **YES** *if there exists special bridge (i, j, h), such that $i, j \in [\alpha, \beta]$. Otherwise, $\mathcal{Q}(\alpha, \beta, h)$ returns* **NO**.

The following two are the main components of our data structure.

1. A linear space structure for $\mathcal{E}_\Delta(\cdot, \cdot)$ queries in $O(\log^{1+\epsilon} n)$ time.
2. A linear space structure for $Q(\cdot, \cdot, \cdot)$ queries in $O(\log^\epsilon n)$ time.

   Our algorithm for computing $\mathsf{rlcp}(\alpha, \beta)$ is straightforward. First obtain $\tau = \mathcal{E}_\Delta(\alpha, \beta)$. Then, for $h = \tau, \tau + 1, \tau + 2, ..., \tau + \Delta - 1$, compute $\mathcal{Q}(\alpha, \beta, h)$. Then report

$$\mathsf{rlcp}(\alpha, \beta) = \max\{h \mid h \in [\tau, \tau + \Delta - 1] \quad \textbf{and} \quad \mathcal{Q}(\alpha, \beta, h) = \textbf{YES} \}$$

The time complexity is $\log^{1+\epsilon} n + \Delta \cdot \log^\epsilon n = O(\log^{1+\epsilon} n)$ and the space complexity is $O(n)$, as claimed. In what follows, we present the details of these two components of our data structure.

5   Details of the Components

We maintain the suffix tree ST of T and the linear space data structure for various 2D range successor/predecessor queries (in $O(\log^\epsilon n)$ time [14]) over the following set of n points.

$$\mathcal{P} = \{(i, \mathsf{SA}[i]) \mid i \in [1, n]\}$$

We rely on this structure for computing interval-LCP and left-leg/right-leg queries (to be defined next).

Lemma 4. *We can answer an interval-LCP query* $\mathsf{ilcp}(p, \alpha, \beta)$ *in time* $O(\log^\epsilon n)$.

Proof. Find the leaf $\ell_{\mathsf{ISA}[p]}$ first. Then find the rightmost leaf ℓ_x before $\ell_{\mathsf{ISA}[p]}$ and the leftmost leaf ℓ_y after $\ell_{\mathsf{ISA}[p]}$, such that $\mathsf{SA}[x], \mathsf{SA}[y] \in [\alpha, \beta]$. We can rely on the following queries for this:

$$x = \mathsf{ORQ}([-\infty, p-1], [\alpha, \beta]) \quad \textbf{and} \quad y = \mathsf{ORQ}([p+1, +\infty], [\alpha, \beta])$$

Clearly, $\mathsf{ilcp}(p, \alpha, \beta)$ is given by $\max\{\mathsf{lcp}(p, x), \mathsf{lcp}(p, y)\}$. This completes the proof. □

Definition 8. *Let* $(i, j, h) \in \mathcal{B}_{spe}$, *then define*

$$\mathsf{rightLeg}(i, h) = j \quad \textbf{and} \quad \mathsf{leftLeg}(j, h) = i$$

If there exists no j, *such that* $(i, j, h) \in \mathcal{B}_{spe}$, *then* $\mathsf{rightLeg}(i, h) = \infty$. *Similarly, if there exists no* i, *such that* $(i, j, h) \in \mathcal{B}_{spe}$, *then* $\mathsf{leftLeg}(j, h) = -\infty$. *If exists, then* $\mathsf{rightLeg}(i, h)$ *(resp.* $\mathsf{leftLeg}(j, h)$*) is unique.*

Lemma 5. *By maintaining a linear space data structure, we can answer* $\mathsf{rightLeg}(k, h)$ *and* $\mathsf{leftLeg}(k, h)$ *queries in* $O(\log^\epsilon n)$ *time.*

Proof. Find the ancestor u (if it exists) of $\ell_{\mathsf{ISA}[k]}$, such that $|\mathsf{path}(u)| = h$ via a weighted level ancestor query on ST^3. If u does not exist, then $\mathsf{rightLeg}(k, h) = +\infty$ and $\mathsf{leftLeg}(k, h) = -\infty$. Otherwise, let u' be the child of u, such that $\ell_{\mathsf{ISA}[k]}$ is under u'. Also, let $[x, y]$ and $[x', y']$ be the range of leaves under u and u', respectively. Then,

$$\mathsf{rightLeg}(k, h) = \min(\mathsf{ORQ}([x, y] \backslash [x', y'], [k+1, +\infty]), +\infty)$$

$$\mathsf{leftLeg}(k, h) = \max(\mathsf{ORQ}([x, y] \backslash [x', y'], [-\infty, k-1]), -\infty)$$

This completes the proof. □

The structures described in Lemma 4 and Lemma 5 are the building blocks of our main components, to be described next. The following observation is exploited in both.

Lemma 6. *Suppose that* $(i, j, h) \in \mathcal{B}_{spe}$. *Then,* $\forall k \in [1, h-1]$, *there exists* $(i+k, \cdot, h-k) \in \mathcal{B}_{spe}$ *such that* $\mathsf{rightLeg}(i+k, h-k) \in (i+k, j+k]$.

Proof. Given $\mathsf{lcp}(i, j) = h$, we have $\mathsf{lcp}(i+k, j+k) = (h-k)$. Clearly, $(i+k, j+k, h-k) \in \mathcal{B}_{all}$. This means, there exists a special bridge $(i+k, l_k, h-k)$, where l_k is the smallest integer after $i+k$, such that $\mathsf{lcp}(i+k, l_k) = h-k$. Equivalently, $l_k = \mathsf{rightLeg}(i+k, h-k)$. Clearly, $l_k \le j+k$, since $\mathsf{lcp}(i+k, j+k) = h-k$. This completes the proof. □

---

[3] Weighted level ancestor queries on suffix trees can be answered in $O(1)$ time using a linear space data structure [10] (also see [7]).

5.1   The Structure for Estimating Range-LCP

Let $\mathcal{B}_t$ denotes the set of all special bridges with height t. Also, for $f = 0, 1, 2, \ldots, (\Delta - 1)$, where $\Delta = \Theta(\log n)$, define $\mathcal{C}_f$: the set of all special bridges with its height divided by Δ leaving remainder f. Specifically,

$$\mathcal{C}_f = \bigcup_{k=0}^{\lfloor \frac{n-f}{\Delta} \rfloor} \mathcal{B}_{(f+k\Delta)}$$

Let $\mathcal{C}_\pi : \pi \in [0, \Delta - 1]$ be the smallest set among all $\mathcal{C}_f$'s. Its size can be bounded by $O((n \log n)/\Delta)$ (by pigeonhole principle), which is $O(n)$. We map each special bridge $(i, j, h) \in \mathcal{C}_\pi$ into a 2D point (i, j) with weight h and maintain the linear-space data structure over them for answering 2D-RMQ. We use the linear-space structure by Chazelle [5]. The space is $|\mathcal{C}_\pi| = O(n)$ words and the query time is $O(\log^{1+\epsilon} n)$.

Our Algorithm. Let (α^*, β^*, h^*) be the tallest special bridge, such that both $\alpha^*, \beta^* \in [\alpha, \beta]$. For computing $\mathcal{E}_\Delta(\alpha, \beta)$, we query on the 2D-RMQ structure over $\mathcal{C}_\pi$ and find the tallest bridge $(i', j', h') \in \mathcal{C}_\pi$, such that $i', j' \in [\alpha, \beta]$. Two possible scenarios are

1. $\beta^* \in (\alpha, \beta - \Delta]$: We claim that $h^* \in [h', h' + \Delta)$. Proof follows from Lemma 6.
2. $\beta^* \in (\beta - \Delta, \beta]$: We can rely on Interval-LCP queries. Specifically, $h^* = \max\{\mathsf{ilcp}(p, \alpha, \beta) \mid p \in (\beta - \Delta, \beta]\}$.

By combining both cases, we have

$$\mathcal{E}_\Delta(\alpha, \beta) = \max\left(\{\mathsf{ilcp}(p, \alpha, \beta) \mid p \in (\beta - \Delta, \beta]\} \cup \{h'\}\right)$$

The time complexity is proportional to that of one 2D-RMQ and at most Δ number of Interval-LCP queries. That is, $\log^{1+\epsilon} n + \Delta \cdot \log^\epsilon n = O(\log^{1+\epsilon} n)$.

5.2   The Structure for Handling $\mathcal{Q}(\alpha, \beta, h)$ Queries

Recall that $\mathcal{B}_t$ is the set of all special bridges with height t. Let L_t represent the sorted list of left-legs of all bridges in $\mathcal{B}_t$ in the form of a y-fast trie for fast predecessor search. Also, let R_t be another array, such that $R_t[k] = \mathsf{rightLeg}(L_t[k], t)$. In other words, for $k = 1, 2, ..., |\mathcal{B}_t|$, $L_t[k]$ (resp., $R_t[k]$) denotes the left-leg (resp., right-leg) of kth bridge among all bridges in $\mathcal{B}_t$ in the ascending order of left-leg. Also, let

$$S_\pi = \{\pi, \pi + \Delta, \pi + 2\Delta, \pi + 3\Delta, ..., (\pi + \lfloor (n - \pi)/\Delta \rfloor \Delta)\}$$

For each $t \in [1, n]$, we maintain a separate structure that can answer queries of the type $\mathcal{Q}(\cdot, \cdot, t)$. Based on whether h in the query $\mathcal{Q}(\alpha, \beta, h)$ is in S_π or not, we have two cases.

Case 1: $h \in S_\pi$ To handle this case, we maintain L_t and the succinct data structure for range minimum query (RMQ) on R_t for all $t \in S_\pi$. The total space is $|\mathcal{C}_\pi| = O(n)$ words. Therefore, any query $\mathcal{Q}(\alpha, \beta, h)$ with $h \in S_\pi$ can be answered using the following steps.

1. Find the smallest k, such that $L_h[k] \geq \alpha$ via a successor query.
2. Then, find the index k' corresponding to the smallest element in $R_h[k, |R_h|]$ using a range minimum query. Note that R_h is not stored.
3. Then, find rightLeg$(L_h[k'], h)$ and report "YES" if it is $\leq \beta$, and report "NO" otherwise.

The time complexity is $(\log \log n + \log^\epsilon n) = O(\log^\epsilon n)$. The correctness can be easily verified.

Case 2: $h \notin S_\pi$ We first show how to design a structure for a predefined h. Let $q = \text{pred}(h, S_\pi) = \pi + \Delta \cdot \lfloor (h - \pi)/\Delta \rfloor$ and $z = (h - q)$. Note that for each special bridge (i, j, h), there exists a special bridge $(i + z, \cdot, h - z) = (i + z, \cdot, q)$ (refer to Lemma 6). This implies the following.

$$\{L_h[k] \mid k \in [1, |\mathcal{B}_h|]\} \subseteq \{(L_q[k] - z) \mid k \in [1, |\mathcal{B}_q|]\} \tag{1}$$

Now, define an array R'_h of length $|\mathcal{B}_q|$, such that for any $k \in [1, |\mathcal{B}_q|]$, $R'_h[k] = $ rightLeg$((L_q[k] - z), h)$. Note that $R'_h[k] = \infty$ if there exists no special bridge with left-leg $(L_q[k] - z)$ and height h. Our data structure is a succinct range minimum query (RMQ) structure over R'_h. We now show how to answer an $Q(\alpha, \beta, h)$ query using R'_h and L_q (in Case 1). The steps are as follows.

1. Find the smallest k, such that $(L_q[k] - z) \geq \alpha$. We perform a successor query on L_q for this.
2. Then, find the index k' corresponding to the smallest element in $R'_h[k, |R_q|]$ using a range minimum query.
3. Then, find rightLeg$(L_h[k'] - z, h)$ and report "YES" if it is $\leq \beta$, and report "NO" otherwise.

The time complexity is $(\log \log n + \log^\epsilon n) = O(\log^\epsilon n)$. The correctness follows from the definition of R'_h and Eq. 1. The space complexity for a fixed h is $|\mathcal{B}_q|(2 + o(1))$ bits. Therefore, by maintaining the above structure for all values of h, we can answer $\mathcal{Q}(\alpha, \beta, h)$ for any α, β and h in $O(\log^\epsilon n)$ time. Total space (in bits) is:

$$(2 + o(1)) \sum_{h=1}^{n} |\mathcal{B}_{\pi + \Delta \cdot \lfloor (h - \pi)/\Delta \rfloor}| = (2 + o(1)) \Delta \sum_{q \in S_\pi} |\mathcal{B}_q| = O(n \log n).$$

In summary, any Range-LCP query on the text $T[1, n]$ can be answered in $O(\log^{1+\epsilon} n)$ time using a linear space data structure. We remark that our data structure can be constructed in $O(n \log n)$ time.

Acknowledgments. This research is supported in part by the U.S. NSF under the grants CCF-1703489 and CCF-1527435, and the Taiwan Ministry of Science and Technology under the grant 105-2221-E-007-040-MY3.

References

1. Amir, A., Apostolico, A., Landau, G.M., Levy, A., Lewenstein, M., Porat, E.: Range LCP. In: Asano, T., Nakano, S., Okamoto, Y., Watanabe, O. (eds.) ISAAC 2011. LNCS, vol. 7074, pp. 683–692. Springer, Heidelberg (2011). https://doi.org/10.1007/978-3-642-25591-5_70
2. Amir, A., Apostolico, A., Landau, G.M., Levy, A., Lewenstein, M., Porat, E.: Range LCP. J. Comput. Syst. Sci. **80**(7), 1245–1253 (2014)
3. Amir, A., Lewenstein, M., Thankachan, S.V.: Range LCP queries revisited. In: Iliopoulos, C., Puglisi, S., Yilmaz, E. (eds.) SPIRE 2015. LNCS, vol. 9309, pp. 350–361. Springer, Cham (2015). https://doi.org/10.1007/978-3-319-23826-5_33
4. Chan, T.M., Larsen, K.G., Patrascu, M.: Orthogonal range searching on the RAM, revisited. In: Symposium on Computational Geometry, pp. 1–10 (2011)
5. Chazelle, B.: A functional approach to data structures and its use in multidimensional searching. SIAM J. Comput. **17**(3), 427–462 (1988)
6. Cormode, G., Muthukrishnan, S.: Substring compression problems. In: Proceedings of the Sixteenth Annual ACM-SIAM Symposium on Discrete Algorithms, pp. 321–330. Society for Industrial and Applied Mathematics (2005)
7. Farach, M., Muthukrishnan, S.: Perfect hashing for strings: formalization and algorithms. In: Hirschberg, D., Myers, G. (eds.) CPM 1996. LNCS, vol. 1075, pp. 130–140. Springer, Heidelberg (1996). https://doi.org/10.1007/3-540-61258-0_11
8. Fischer, J., Heun, V.: Space-efficient preprocessing schemes for range minimum queries on static arrays. SIAM J. Comput. **40**(2), 465–492 (2011)
9. Gagie, T., Karhu, K., Navarro, G., Puglisi, S.J., Sirén, J.: Document listing on repetitive collections. In: Fischer, J., Sanders, P. (eds.) CPM 2013. LNCS, vol. 7922, pp. 107–119. Springer, Heidelberg (2013). https://doi.org/10.1007/978-3-642-38905-4_12
10. Gawrychowski, P., Lewenstein, M., Nicholson, P.K.: Weighted ancestors in suffix trees. In: Schulz, A.S., Wagner, D. (eds.) ESA 2014. LNCS, vol. 8737, pp. 455–466. Springer, Heidelberg (2014). https://doi.org/10.1007/978-3-662-44777-2_38
11. Harel, D., Tarjan, R.E.: Fast algorithms for finding nearest common ancestors. SIAM J. Comput. **13**(2), 338–355 (1984)
12. Keller, O., Kopelowitz, T., Feibish, S.L., Lewenstein, M.: Generalized substring compression. Theor. Comput. Sci. **525**, 42–54 (2014)
13. Lewenstein, M.: Orthogonal range searching for text indexing. In: Brodnik, A., López-Ortiz, A., Raman, V., Viola, A. (eds.) Space-Efficient Data Structures, Streams, and Algorithms. LNCS, vol. 8066, pp. 267–302. Springer, Heidelberg (2013). https://doi.org/10.1007/978-3-642-40273-9_18
14. Nekrich, Y., Navarro, G.: Sorted range reporting. In: Fomin, F.V., Kaski, P. (eds.) SWAT 2012. LNCS, vol. 7357, pp. 271–282. Springer, Heidelberg (2012). https://doi.org/10.1007/978-3-642-31155-0_24
15. Patil, M., Shah, R., Thankachan, S.V.: Faster range LCP queries. In: Kurland, O., Lewenstein, M., Porat, E. (eds.) SPIRE 2013. LNCS, vol. 8214, pp. 263–270. Springer, Cham (2013). https://doi.org/10.1007/978-3-319-02432-5_29
16. Patil, M., Thankachan, S.V., Shah, R., Nekrich, Y., Vitter, J.S.: Categorical range maxima queries. In: Proceedings of the 33rd ACM SIGMOD-SIGACT-SIGART Symposium on Principles of Database Systems, PODS 2014, 22–27 June 2014, Snowbird, UT, USA, pp. 266–277 (2014)

17. Sleator, D.D., Tarjan, R.E.: A data structure for dynamic trees. In: Proceedings of the 13th Annual ACM Symposium on Theory of Computing, 11–13 May 1981, Milwaukee, Wisconsin, USA, pp. 114–122 (1981)
18. Weiner, P.: Linear pattern matching algorithms. In: SWAT, pp. 1–11 (1973)
19. Willard, D.E.: Log-logarithmic worst-case range queries are possible in space theta(n). Inf. Process. Lett. **17**(2), 81–84 (1983)

Non-determinism Reduces Construction Time in Active Self-assembly Using an Insertion Primitive

Benjamin Hescott[1], Caleb Malchik[2], and Andrew Winslow[3]([envelope])

[1] Northeastern University, Boston, MA, USA
b.hescott@northeastern.edu
[2] Yale University, New Haven, CT, USA
caleb.malchik@yale.edu
[3] University of Texas Rio Grande Valley, Edinburg, TX, USA
andrew.winslow@utrgv.edu

Abstract. We consider efficient construction of DNA-based polymers in a model introduced by Dabby and Chen (SODA 2013) called *insertion systems*, where monomers insert themselves into the middle of a growing linear polymer. Specifically, we describe a new family of non-deterministic insertion systems that construct length-n polymers in $\Theta(\log^{3/2}(n))$ expected time, breaking the lower bound of $\Omega(\log^{5/3}(n))$ for deterministic construction. We also prove that this time is optimal for systems constructing finite polymers, and that the $\Theta(\log(n))$ monomer types used in the construction is optimal for this time.

1 Introduction

We study a theoretical model of DNA-based *algorithmic self-assembly* introduced by Dabby and Chen [6], in which simple particles called *monomers* aggregate to form long, complex chains called *polymers* via individual monomers asynchronously inserting themselves between adjacent monomers. This model shares similarities with other *active*[1] self-assembly models, e.g. the graph grammars of Klavins et al. [14,15] and the *nubots* model of Woods et al. [2–4,19], where structures undergo reconfiguration.

One appeal of active self-assembly models is that they allow formation of complex assemblies exponentially quickly by enabling insertion of new particles simultaneously throughout the assembly, a phenomenon observed in a wide range of biological systems [6,19]. In contrast, passive self-assembly models such as the *abstract Tile Assembly Model (aTAM)* of Winfree [18] are limited to only polynomially fast growth [12].

A preprint containing the omitted proof of Theorem 1 is available on arXiv: https://arxiv.org/abs/1411.0973.

[1] Not to be confused with *active* tile assembly models [8,9,11,13,16,17] in which bond states change.

© Springer International Publishing AG, part of Springer Nature 2018
L. Wang and D. Zhu (Eds.): COCOON 2018, LNCS 10976, pp. 626–637, 2018.
https://doi.org/10.1007/978-3-319-94776-1_52

Of the active self-assembly models, both graph grammars and nubots are capable of a topologically rich set of assemblies and reconfigurations, but rely on stateful particles forming complex bond arrangements. In contrast, insertion systems consist of stateless particles forming a single chain of bonds. Indeed, all insertion systems are captured as a special case of nubots in which a linear polymer is assembled via parallel insertion-like reconfigurations, as in Theorem 5.1 of [19]. The simplicity of insertion systems makes their implementation in matter a more immediately attainable goal; Dabby and Chen [5,6] describe experimental implementations in DNA.

1.1 Non-determinism in Insertion Systems

Previous work in insertion systems [6,10] only considered deterministic systems, where each location accepts at most one monomer type. In the equivalent [10] model of context-free grammars, deterministic systems are those in which each non-terminal symbol appears on the left-hand side of a unique production rule. Allowing insertion sites in which different monomer types may be inserted gives rise to *non-deterministic* insertion systems. As a result of such sites, non-deterministic insertion systems may construct many distinct polymers.

In deterministic systems, the expected time of assembling length-n polymers is known [10] to have a tight lower bound of $\Omega(\log^{5/3}(n))$. In contrast, a simple non-deterministic system of just two monomer types is easily shown (see Sect. 3) to reduce expected assembly time to $\Theta(\log n)$. The cost of this reduced assembly time is a loss of precision: the two-monomer system yields infinitely many polymers of arbitrarily large lengths (i.e. is "pumpable" in grammar terminology), making it useless for targeted construction of polymers of a given length.

1.2 Our Results

We consider whether non-determinism may be used to reduce assembly time without resulting in the complete loss of precision in the lengths of constructed polymers (i.e. without becoming "pumpable"). We answer this question in the affirmative, giving a non-deterministic insertion system that assembles length-n polymers in $\Theta(\log^{3/2}(n))$ expected time (Theorem 1) and a proof that this time is optimal systems that assemble a finite number of polymers (Theorem 2). However, the system still suffers from reduced precision: polymers of $\Theta(n)$ distinct lengths up to n are constructed (see Sect. 5 for further discussion).

The proof of Theorem 2 also implies a monomer type and time tradeoff for systems constructing a finite set of polymers: constructing a length-n polymer using k monomer types takes $\Omega(\log^2(n)/\sqrt{k})$ expected time (Lemma 4). This lemma implies that both our upper bound construction using $k = O(\log(n))$ monomer types (and assembling in $O(\log^{3/2}(n)$ expected time) and the upper bound construction of [10] using $k = O(\log^{2/3}(n))$ monomer types (and assembling in $O(\log^{3/2}(n)$ expected time) are optimal constructions at both ends of this tradeoff curve.

2   Definitions

An *insertion system* in the active self-assembly model of Dabby and Chen [6] carries out the construction of a linear *polymer* consisting of constant length *monomers*. A polymer grows incrementally by the insertion of a monomer at an *insertion site* between two existing monomers in the polymer, according to complementary bonding sites between the monomer and the insertion site.

An insertion system $\mathcal{S}$ is defined as a 4-tuple $\mathcal{S} = (\Sigma, \Delta, Q, R)$. The first element, Σ, is a set of symbols. Each symbol $s \in \Sigma$ has a *complement* s^*. We denote the complement of a symbol s as $\overline{s}$, i.e. $\overline{s} = s^*$ and $\overline{s^*} = s$. The set Δ is a set of *monomer types*, each assigned a *concentration*. Each monomer is specified by a quadruple $(a, b, c, d)^+$ or $(a, b, c, d)^-$, where $a, b, c, d \in \Sigma \cup \{s^* : s \in \Sigma\}$, and each concentration is a real number between 0 and 1. The sum of all concentrations in Δ must be at most 1. The two symbols $Q = (a, b)$ and $R = (c, d)$ are special two-symbol monomers that together form the *initiator* of $\mathcal{S}$. It is required that either $\overline{a} = d$ or $\overline{b} = c$. The *size* of $\mathcal{S}$ is $|\Delta|$, the number of monomer types in $\mathcal{S}$.

A *polymer* is a sequence of monomers $Qm_1m_2\ldots m_n R$ where $m_i \in \Delta$ such that for each pair of adjacent monomers $(w, x, a, b)(c, d, y, z)$, either $\overline{a} = d$ or $\overline{b} = c$. The *length* of a polymer is the number of monomers, including Q and R, it contains. Each pair of adjacent monomer ends $(a, b)(c, d)$ form an *insertion site*. Monomers can be inserted into an insertion site $(a, b)(c, d)$ according to the following rules (see Fig. 1):

1. If $\overline{a} = d$, then any monomer $(\overline{b}, e, f, \overline{c})^+$ can be inserted.
2. If $\overline{b} = c$, then any monomer $(e, \overline{a}, \overline{d}, f)^-$ can be inserted.[2]

A monomer is inserted after time t, where t is an exponential random variable with rate equal to the concentration of the monomer type. The set of all polymers *constructed* by an insertion system is recursively defined as any polymer constructed by inserting a monomer into a polymer constructed by the system, beginning with the initiator. Note that the insertion rules guarantee by induction that for every insertion site $(a, b)(c, d)$, either $\overline{a} = d$ or $\overline{b} = c$.

We say that a polymer is *terminal* if no monomer can be inserted into any insertion site in the polymer, and that an insertion system is *deterministic* if every polymer P constructed by the system is either P or is non-terminal and has length less than that of P (i.e. can become P).

3   Upper Bound for Assembly Time

It is natural to ask whether faster construction of polymers is possible in non-deterministic systems: systems that do not construct a single terminal polymer. A two-monomer-type insertion system consisting of the initiator $(s_1, s_2)(s_2^*, s_1^*)$

---

[2] In [6], this rule is described as a monomer $(\overline{d}, f, e, \overline{a})^-$ that is inserted into the polymer as $(e, \overline{a}, \overline{d}, f)$.

Inserting $(c, d^*, e^*, b^*)^+$ into $(a^*, c^*)(b, a)$
to yield $(a^*, c^*)(c, d^*, e^*, b^*)(b, a)$:

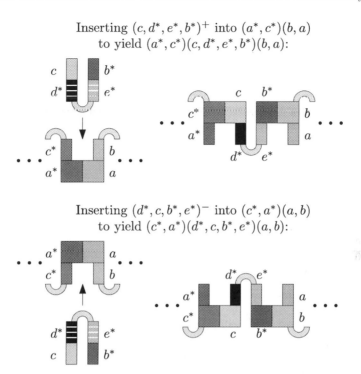

Inserting $(d^*, c, b^*, e^*)^-$ into $(c^*, a^*)(a, b)$
to yield $(c^*, a^*)(d^*, c, b^*, e^*)(a, b)$:

Fig. 1. A pictorial interpretation of the two insertion rules for monomers. Loosely based on Fig. 2 and corresponding DNA-based implementation of [6].

and monomer types $(s_2^*, s_1^*, s_1, s_2)^+$, and $(s_2^*, x, x, s_2)^+$ simultaneously constructs polymers of all lengths $n \geq 3$ in expected time $O(\log n)$ via balanced insertion sequences of logarithmic length. Moreover, any polymer in any system has $\Omega(\log n)$ expected construction time, since every insertion takes $\Omega(1)$ expected time, and constructing a polymer of length n requires an insertion sequence of length at least $\lfloor \log_2(n-2) \rfloor$. So if assembling anything is permitted, then this two-monomer-type system is asymptotically optimal.

This leads to the question considered in this paper: to what extent is assembly time reduction possible in systems that only construct finite number of polymers? Our next result proves that even this relaxation is sufficient to improve assembly. The key idea of the construction is allow large sets of monomer types to "compete" to insert into a common insertion site first. This competition increases the total concentration of insertable monomer types, reducing the expected insertion time, but results in a non-deterministic system.

Theorem 1. *For any positive, odd integer r, there exists an insertion system constructing a finite set of polymers with $O(r^2)$ monomer types that constructs a polymer of length $n = 2^{\Theta(r^2)}$ in $O(\log^{3/2}(n))$ expected time. Moreover, the expected time has an exponentially decaying tail probability.*

Due to space constraints, the proof of Theorem 1 is omitted; we give a brief sketch here. The construction uses insertion sites to store the (two) variable values of a double for-loop, with short insertion sequences used to either increment the inner variable or increment the outer variable and reset the inner variable to 0.

Non-determinism is used to speed up loop increments in the following way: loops with different variable values "share" insertable monomer types, increasing the concentration of insertable monomer types for all insertion sites to decrease the expected time for the increment to take place. This comes with the tradeoff that not all monomer types lead to successful incrementing; some instead cause the loop to "break", leading to assembly of additional polymers of length less than n.

4   Lower Bound for Assembly Time

Here we show that the construction in the previous section is the optimal in expected assembly time and, for the given assembly time, optimal in the number of monomer types used (Theorem 2). A collection of intervening lemmas are used to prove bounds on the number of monomer types and expected time to carry out an *insertion sequence*: a sequence of monomer insertions where each insertion is into a site created by the previous insertion.

Observe that if two monomer types of the same sign are insertable into a common site, then the set of sites each can be inserted into is equal. Nearly all of the lemmas involve consideration of not only monomer types, but *insertion sets*: maximal sets of same-signed monomer types sharing a common set of insertion sites each can be inserted into.

The first several lemmas of the section are used to prove Lemma 4, a lemma describing the trade-off between the number of monomer types and expected construction time for systems constructing finite polymer sets. This lemma is combined with extremal bounds on the minimum number of monomer types and insertion sets to prove the final result.

Lemma 1. *Any insertion sequence of length l with no repeated insertion sites has $\Theta(l)$ sites of the form $(a, b)(\overline{c}, \overline{a})$ with $b \neq c$.*

Proof. Insertion sites have one of three forms:

Positive: $(a, b)(\overline{c}, \overline{a})$ with $b \neq c$.
Mixed: $(a, b)(\overline{b}, \overline{a})$.
Negative: $(b, a)(\overline{a}, \overline{c})$ with $b \neq c$.

We prove that every sequence of four consecutive insertion sites has at least one positive site. Consider such a sequence of four sites (and the three intervening insertions). If the first site is positive, we're done. If the first site is mixed, then the first monomer type inserted may be negative or positive. If a negative monomer type is inserted:

$$(a, b) \diamond (\overline{b}, \overline{a})$$

$$(a, b)(\overline{c}, \overline{a}, a, d)(\overline{b}, \overline{a})$$

Since the sequence does not repeat sites, $b \neq c, d$ and either second site, $(a, b)(\overline{c}, \overline{a})$ or $(a, d)(\overline{b}, \overline{a})$, is positive. If a positive monomer type is inserted:

$$(a, b) \diamond (\overline{b}, \overline{a})$$

$$(a, b)(\overline{b}, \overline{c}, d, b)(\overline{b}, \overline{a})$$

As before, $a, d \neq c$ since sites cannot repeat. So the next insertion must use a negative monomer type. We assume the left site is used next in the sequence (a symmetric argument works if the right site is used instead). The entire insertion sequence has the form:

$$(a, b) \diamond (\overline{b}, \overline{a})$$

$$(a, b) \diamond (\overline{b}, \overline{c}, d, b)(\overline{b}, \overline{a})$$

$$(a, b) \diamond (\overline{b}, \overline{c})$$

$$(a, b)(\overline{e}, \overline{a}, c, f)(\overline{b}, \overline{c})$$

As before, $e, f \neq b$ since sites cannot repeat. So the next site, either $(a, b)(\overline{e}, \overline{a})$ with $b \neq e$ or $(c, f)(\overline{b}, \overline{c})$ with $f \neq b$ is positive. So the third site in the sequence is positive. Finally, if the initial site is negative then the first monomer type inserted is negative:

$$(b, a) \diamond (\overline{a}, \overline{c})$$

$$(b, a) \diamond (\overline{d}, \overline{b}, c, e)(\overline{a}, \overline{c})$$

$$(b, a) \diamond (\overline{d}, \overline{b})$$

We assume that the left site is used next in the sequence (a symmetric argument works if the right side is used instead). If $a \neq d$ then the second site is positive. Otherwise the second site is mixed, and by previous argument, at most two more insertions (a total of three) will take place until a positive site appears. So in the entire sequence of length l, a positive site appears at least once in every sequence of four consecutive sites.   □

Lemma 2. *Any insertion sequence with no repeated insertion sites using k monomer types forming m insertion sets has length $O(m\sqrt{k})$.*

Proof. Let $\mathcal{S} = (\Sigma, \Delta, Q, R)$ be the insertion system containing the sequence. Relabel the symbols in $\Sigma \cup \{s^* : s \in \Sigma\}$ as $s_1, s_2, \ldots, s_{4k}$, with some of these symbols possibly unused. Note that this implies that for every s_i, $\overline{s_i} = s_j$ for some $j \in 1, 2 \ldots, 4k$. Let l be the length of the sequence. By Lemma 1, $\Theta(l)$ sites are *positive*: they have the form $(s_a, s_b)(\overline{s_c}, \overline{s_a})$ with $b \neq c$.

A bound of $\sum_{i=1}^{4k} \min(|L_i|, |R_i|) \leq 3m$. Let L_i and R_i be the sets of monomer types of the forms $(\_, \_, s_i, \_)^{\pm}$ and $(\_, \overline{s_i}, \_, \_)^{\pm}$, respectively, used in the insertion sequence. Each positive site $(s_i, s_b)(\overline{s_c}, \overline{s_i})$ consists of a left monomer in L_i and right monomer in R_i. Every occurrence of a positive site in the sequence is followed by the use of the left or right resulting site, e.g.:

$$(s_i, s_b) \diamond (\overline{s_c}, \overline{s_i})$$

$$(s_i, s_b) \diamond (\overline{s_b}, \overline{s_d}, s_e, s_c)(\overline{s_c}, \overline{s_i})$$

$$(s_i, s_b) \diamond (\overline{s_b}, \overline{s_d})$$

It is the case that d is unique for c, i.e. no two insertions into positive sites using the left resulting sites both use monomers of the form $(\overline{s_b}, \overline{s_d}, \_, \_)^+$, since such a pair of monomers implies the sequence repeats the insertion site $(s_a, s_b)(\overline{s_b}, \overline{s_d})$. A similar claim holds for e and b in the case that the right site is used. So inserting into the resulting site requires a monomer from a distinct insertion set $\{(\_, \overline{s_i}, s_d, \_)^- \in \Delta\}$ or, in the special case that $i = d$, $\{(\overline{s_b}, \_, \_, s_b)^+ \in \Delta\}$.

The resulting sites require monomers from a number of distinct insertion sets equal to the sum of two values. First, the number of times the left side is used with a distinct c and a monomer is inserted into a site $(s_i, s_b)(\overline{s_b}, \overline{s_d})$ with d unique for c. Second, the number of times the right side is used with a distinct b and a monomer is inserted into a site $(s_e, s_c)(\overline{s_c}, \overline{s_i})$ with e unique for b. An assignment of left and right side usage that minimizes the number of distinct insertion sets needed is nearly equivalent to a minimum vertex covering of the following bipartite graph:

- A node $L_{(i,b)}$ for every site $(s_i, s_b)(\overline{s_c}, \overline{s_i})$ in the insertion sequence.
- A node $R_{(c,i)}$ for every site $(s_i, s_b)(\overline{s_c}, \overline{s_i})$ in the insertion sequence.
- An edge $(L_{(i,b)}, R_{(c,i)})$ for every site $(s_i, s_b)(\overline{s_c}, \overline{s_i})$ in the insertion sequence.

Selecting a vertex to cover a given edge corresponds to using the resulting left or right site of the edge's site, e.g. selecting $R_{(c,i)}$ to cover the edge $(L_{(i,b)}, R_{(c,i)})$ corresponds to using the resulting left site and inserting a monomer type of the form $(\_, \overline{s_i}, s_d, \_)^-$, where d is unique for c. By König's theorem (see [1,7]), since the graph is bipartite, the size of a minimum vertex covering is equal to the size of a maximum matching, which is bounded from above by $\sum_{i=1}^{4k} \min(|L_i|, |R_i|)$.

However, an insertion set $\{(\_, \overline{s_e}, s_d, \_)^- \in \Delta\}$ corresponds to selecting both $R_{(c,i)}$, where d is unique for c, and $L_{(j,b)}$, where e is unique for b. So the number of insertion sets needed may be as little as half the size of the vertex cover of the bipartite graph. Additionally, one site may not be inserted into. So $\sum_{i=1}^{4k} \min(|L_i|, |R_i|) - 1 \leq 2m$ and $\sum_{i=1}^{4k} \min(|L_i|, |R_i|) \leq 3m$.

Maximizing Insertion Sequence Length. Consider the number of positive sites y accepting some monomer type. We proved that $\Omega(l) = y$ and it is easily observed that $y \leq \sum_{i=1}^{4k} \min(m, |L_i| \cdot |R_i|)$. We also proved that $\sum_{i=1}^{4k} \min(|L_i|, |R_i|) \leq 3m$ and it is easily observed that $\sum_{i=1}^{4k} \max(|L_i|, |R_i|) \leq 2k$, since each monomer type is in at most one L_i and one R_i. This gives the following set of constraints:

1. $\Omega(l) = \sum_{i=1}^{4k} \min(m, |L_i| \cdot |R_i|)$.
2. $\sum_{i=1}^{4k} \min(|L_i|, |R_i|) \leq 3m$.
3. $\sum_{i=1}^{4k} \max(|L_i|, |R_i|) \leq 2k$.

Observe that $|L_i| \cdot |R_i| = \min(|L_i|, |R_i|) \cdot \max(|L_i|, |R_i|)$. Define two new variables $y_i = \min(|L_i|, |R_i|)$ and $z_i = \max(|L_i|, |R_i|)$ for an alternate formulation of the previous constraints:

1. $\Omega(l) = \sum_{i=1}^{4k} \min(m, y_i z_i)$.
2. $\sum_{i=1}^{4k} y_i \leq 3m$.
3. $\sum_{i=1}^{4k} z_i \leq 2k$.

Relax y_i, z_i to be real-valued and let $W = \{i : y_i z_i > 0\}$. If $0 < y_i z_i, y_j z_j < m$ for some $i \neq j$ and $y_i = \max(y_i, z_i, y_j, z_j)$, then $\min(m, y_i z_i) + \min(m, y_j z_j) < \min(m, y_i(z_i + \varepsilon)) + \min(m, y_j(z_j - \varepsilon))$ for sufficiently small $\varepsilon > 0$. More generally, if $0 < y_i z_i, y_j z_j < m$ for some $i \neq j$ then the values of y_i, z_i, y_j, z_j can be modified to increase $\sum_{i=1}^{4k} \min(m, y_i z_i)$. Therefore the maximum value is achieved when $m = y_i z_i$ for all but at most one $i \in W$.

We claim that it cannot be that $y_i z_i = m$ for $6\sqrt{k}$ distinct values of i. By contradiction, assume so. So $|W| \geq 6\sqrt{k}$ and the average value of y_i for $i \in W$ must be less than $3m/(6\sqrt{k}) = m/(2\sqrt{k})$. So for a subset $W' \subseteq W$ with $|W'| \geq |W|/2 \geq 3\sqrt{k}$, $y_i \leq 2 \cdot m/(2\sqrt{k}) = m/\sqrt{k}$ for all $i \in W'$. For every $i \in W'$, because $y_i \leq m/\sqrt{k}$ and $y_i z_i = m$, it must be the case that $z_i \geq \sqrt{k}$. So $\sum_{i=1}^{4k} z_i \geq |W'| \cdot \sqrt{k} \geq 3k$, a contradiction with the constraint that $\sum_{i=1}^{4k} z_i \leq 2k$.

So the maximum value is achieved when $m = y_i z_i$ for all but at most one $i \in W$, with $|W| + 1 < 6\sqrt{k} + 1 < 7\sqrt{k}$. So $\sum_{i=1}^{4k} \min(m, y_i z_i) \leq (|W| + 1)m < 7m\sqrt{k}$. So $\Omega(l) = 7m\sqrt{k}$ and $l = O(m\sqrt{k})$. □

Lemma 3. *An insertion sequence of length l using monomer types from m insertion sets with no repeated insertion sites takes $\Omega(ml)$ expected time.*

Proof. By linearity of expectation, the total expected time of the insertions is equal to the sum of the expected time for each insertion. By Lemma 1, $\Theta(l)$ sites are both *positive*, i.e. they have the form $(s_a, s_b)(\overline{s_c}, \overline{s_a})$ with $b \neq c$, and accept the monomer types of a positive, non-empty insertion set.

Let m be the number of insertion sets formed by the monomer types inserted into these $\Omega(l)$ sites. Let $c_1, c_2, \ldots, c_m$ be the sums of the concentrations of the monomer types in these sets, and $x_1, x_2, \ldots, x_m$ be the number of times a monomer from each set is inserted in the subsequence. Then the total expected time for all of the insertions in the subsequence is $\sum_{i=1}^{m} x_i/c_i$. Moreover, these variables are subject to the following constraints:

1. $\sum_{i=1}^{m} x_i = \Omega(l)$ (total number of insertions is $\Omega(l)$).
2. $\sum_{i=1}^{m} c_i \leq 1$ (total concentration is at most 1).

Minimizing Expected Time. We now consider minimizing the total expected time subject to these constraints, starting with proving that $x_i/c_i = x_j/c_j$ for all $1 \leq i, j \leq m$. That is, that the ratio of the number of sites that accept an insertion set to the total concentrations of the monomer types in the set is equal for all sets. Assume, without loss of generality, that $x_i/c_i > x_j/c_j$ and $c_i, c_j > 0$. Then it can be shown algebraically that the following two statements hold:

1. If $c_j \geq c_i$, then for sufficiently small $\varepsilon > 0$, $\frac{x_i}{c_i} + \frac{x_j}{c_j} > \frac{x_i}{c_i+\varepsilon} + \frac{x_j}{c_j-\varepsilon}$.
2. If $c_j < c_i$, then for sufficiently small $\varepsilon > 0$, $\frac{x_i}{c_i} + \frac{x_j}{c_j} > \frac{x_i}{c_i-\varepsilon} + \frac{x_j}{c_j+\varepsilon}$.

Since the ratios of every pair of monomer types are equal,

$$\frac{c_i}{1} \leq \frac{c_i}{\sum_{i=1}^{m} c_i} = \frac{x_i}{\sum_{i=1}^{m} x_i} = O(x_i/l)$$

So $\Omega(l) = x_i/c_i$ and $\Omega(ml) = \sum_{i=1}^{m} x_i/c_i$. □

Lemma 4. *Any polymer of length n constructed by an insertion system with k monomer types constructing a finite set of polymers takes $\Omega(\log^2(n)/\sqrt{k})$ expected time.*

Proof. By Lemma 2, $n = 2^{O(m\sqrt{k})}$. So $m = \Omega(\log n/\sqrt{k})$. Constructing any polymer of length n requires an insertion system of length $l = \Omega(\log n)$. Then by Lemma 3, the expected time to construct any polymer of length n is $\Omega(ml) = \Omega(\log^2(n)/\sqrt{k})$.

Before proving the final result, we prove a helpful lemma showing that the number of insertion sets cannot be too much smaller than the number of monomer types:

Lemma 5. *Any insertion sequence of length l with no repeated insertion sites using k monomer types forming m insertion sets has $m = \Omega(\sqrt{k})$.*

Proof. Notice that this bound can only be obtained by assuming the monomer types are used to carry out an insertion sequence, since it is possible to have an arbitrarily large set of monomer types belonging to a single insertion set. The number of monomer types used is at most the length of the insertion sequence ($k \leq l$), and the remainder of the proof is spent proving that the number of insertion sites in a system with m insertion sets is $O(m^2)$ ($l = O(m^2)$), giving the desired inequality.

Let $S = (\Sigma, \Delta, Q, R)$ be the insertion system containing the sequence. Relabel the symbols in $\Sigma \cup \{s^* : s \in \Sigma\}$ as $s_1, s_2, \ldots, s_{4k}$, with some of these symbols possibly unused. By Lemma 1, $\Omega(l)$ sites are *positive*: they have the form $(s_a, s_b)(s_c, \overline{s_a})$ with $b \neq c$.

Since the second monomer inserted to create the site must be negative, each positive site consists of at least one negative monomer type. Let L_i^- and R_i^- be the sets of monomer types of the forms $(\_, \_, s_i, \_)^-$ and $(\_, \overline{s_i}, \_, \_)^-$, respectively, used in the insertion sequence of length l. For a specific i, there exists a site of the form $(s_i, s_b)(s_c, \overline{s_i})$ only if $|L_i^-| + |R_i^-| > 0$. So the number of values of i such that a site of the form $(s_i, s_b)(s_c, \overline{s_i})$ exists is at most $\sum_{i=1}^{4k} |L_i^-| + \sum_{i=1}^{4k} |R_i^-|$. Since all monomer types of a negative insertion set belong to the same L_i^- and R_i^-, $\sum_{i=1}^{4k} |L_i^-| + \sum_{i=1}^{4k} |R_i^-| \leq 2m$.

Next, observe there are at most m sites of the form $(s_i, s_b)(s_c, \overline{s_i})$ that accept a monomer, since each site requires a monomer from a different positive insertion set. So the total number of positive sites that accept a monomer is at most

$2m \cdot m = 2m^2$. Since there are $\Omega(l)$ positive sites in the insertion sequence, $\Omega(l) = 2m^2$ and $l = O(m^2)$. □

Theorem 2. *Any polymer of length n constructed by an insertion system constructing a finite set of polymers takes $\Omega(\log^{3/2}(n))$ expected construction time. Moreover, constructing a polymer of length n in $\Theta(\log^{3/2}(n))$ expected time requires using $\Omega(\log n)$ monomer types.*

Proof. First, observe that constructing a polymer of length n in a system constructing a finite set of polymers involves an insertion sequence of length $\log_2(n) \leq l$ with no repeated sites. By Lemmas 2 and 5, $\log_2(n) = O(m^2)$ and so $m = \Omega(\sqrt{\log n})$ Then by Lemma 3, carrying out the insertion sequence and completing the construction of the polymer takes $\Omega(ml) = \Omega(\log^{3/2}(n))$ expected time and by Lemma 4, $k = \Omega(\log n)$. □

5   Open Problems

The results of in this paper, combined with those of [10] describe the landscape of efficient polymer construction using insertion systems:

– Trivial systems of just a few polymers can construct polymers of arbitrary length in optimal time, but with the caveat that the growth is uncontrolled and the systems construct infinite set of polymers.
– Deterministic construction a polymer of length n requires $\Omega(\log^{2/3}(n))$ monomer types and $\Omega(\log^{5/3}(n))$ expected time, and both of these are achievable simultaneously.
– The intermediate situation of constructing finite sets of polymers is more intricate – polymers can be constructed faster, but with the trade-off of using more monomer types *and* non-determinism.

In our system achieving $O(\log^{3/2}(n))$ expected construction time (Theorem 1), an exponential number $(2^{\Theta(n \log \log n)})$ of "junk" terminal polymers are constructed. Since achieving such speed requires significantly fewer insertion sets than monomer types, some junk is necessary – but how much? One approach to proving a lower bound is to prove that insertion sites accepting large insertion sets imply a large number of terminal polymers. We have been unable to prove such an implication even in the simplest case:

Conjecture 1. Every deterministic system with no unused monomer types has exclusively singleton insertion sets.

Since assembling a polymer in $o(\log^{5/3}(n))$ expected time requires that $\Omega(\log n)$ insertions along most insertion sequences are non-deterministic, the previous conjecture implies that any improvement in speed comes with an exponential number of junk terminal polymers:

Conjecture 2. Any system constructing a polymer of length n in $O(\log^{3/2}(n))$ expected time constructs a set of $2^{\Omega(n)}$ polymers.

Setting aside non-determinism, the trade-off between monomer types and construction time has a lower bound (Lemma 4) with matching upper bounds only at the extremes. Does there exist a parameterized system matching the lower bound across the entire range?

Conjecture 3. For every combination of n and k such that $\log_2^{2/3}(n) \leq k \leq \log_2(n)$, there exists a system with k monomer types that constructs a polymer of length n in $O(\log^2 n/\sqrt{k})$ time.

References

1. Bondy, J.A., Murty, U.S.R.: Graph Theory with Applications. Elsevier, New York (1976)
2. Chen, H.-L., Doty, D., Holden, D., Thachuk, C., Woods, D., Yang, C.-T.: Fast algorithmic self-assembly of simple shapes using random agitation. In: Murata, S., Kobayashi, S. (eds.) DNA 2014. LNCS, vol. 8727, pp. 20–36. Springer, Cham (2014). https://doi.org/10.1007/978-3-319-11295-4_2
3. Chen, M., Xin, D., Woods, D.: Parallel computation using active self-assembly. Nat. Comput. **14**, 225–250 (2015)
4. Chin, Y.-R., Tsai, J.-T., Chen, H.-L.: A minimal requirement for self-assembly of lines in polylogarithmic time. In: Brijder, R., Qian, L. (eds.) DNA 2017. LNCS, vol. 10467, pp. 139–154. Springer, Cham (2017). https://doi.org/10.1007/978-3-319-66799-7_10
5. Dabby, N.: Synthetic molecular machines for active self-assembly: prototype algorithms, designs, and experimental study. Ph.D. thesis, Caltech (2013)
6. Dabby, N., Chen, H.L.: Active self-assembly of simple units using an insertion primitive. In: Proceedings of 24th ACM-SIAM Symposium on Discrete Algorithms (SODA), pp. 1526–1536 (2013)
7. Diestel, R.: Graph Theory. Springer, Berlin (2005)
8. Gautam, V.K., Haddow, P.C., Kuiper, M.: Reliable self-assembly by self-triggered activation of enveloped DNA tiles. In: Dediu, A.-H., Martín-Vide, C., Truthe, B., Vega-Rodríguez, M.A. (eds.) TPNC 2013. LNCS, vol. 8273, pp. 68–79. Springer, Heidelberg (2013). https://doi.org/10.1007/978-3-642-45008-2_6
9. Hendricks, J., Padilla, J.E., Patitz, M.J., Rogers, T.A.: Signal transmission across tile assemblies: 3D static tiles simulate active self-assembly by 2D signal-passing tiles. In: Soloveichik, D., Yurke, B. (eds.) DNA 2013. LNCS, vol. 8141, pp. 90–104. Springer, Cham (2013). https://doi.org/10.1007/978-3-319-01928-4_7
10. Hescott, B., Malchik, C., Winslow, A.: Tight bounds for active self-assembly with an insertion primitive. Algorithmica **77**(2), 537–554 (2017)
11. Jonoska, N., Karpenko, D.: Active tile self-assembly, part 1: universality at temperature 1. Int. J. Found. Comput. Sci. **25**(2), 141–163 (2014)
12. Keenan, A., Schweller, R., Sherman, M., Zhong, X.: Fast arithmetic in algorithmic self-assembly. Technical report, arXiv (2013)
13. Keenan, A., Schweller, R., Zhong, X.: Exponential replication of patterns in the signal tile assembly model. In: Soloveichik, D., Yurke, B. (eds.) DNA 2013. LNCS, vol. 8141, pp. 118–132. Springer, Cham (2013). https://doi.org/10.1007/978-3-319-01928-4_9
14. Klavins, E.: Universal self-replication using graph grammars. In: Proceedings of International Conference on MEMS, NANO, and Smart Systems, pp. 198–204 (2004)

15. Klavins, E., Ghrist, R., Lipsky, D.: Graph grammars for self assembling robotic systems. In: Proceedings of the International Conference on Robotics and Automation (ICRA), vol. 5, pp. 5293–5300 (2004)

16. Majumder, U., LaBean, T.H., Reif, J.H.: Activatable tiles: compact, robust programmable assembly and other applications. In: Garzon, M.H., Yan, H. (eds.) DNA 2007. LNCS, vol. 4848, pp. 15–25. Springer, Heidelberg (2008). https://doi.org/10.1007/978-3-540-77962-9_2

17. Padilla, J.E., Liu, W., Seeman, N.C.: Hierarchical self assembly of patterns from the robinson tilings: DNA tile design in an enhanced tile assembly model. Nat. Comput. **11**(2), 323–338 (2012)

18. Winfree, E.: Algorithmic self-assembly of DNA. Ph.D. thesis, Caltech (1998)

19. Woods, D., Chen, H.L., Goodfriend, S., Dabby, N., Winfree, E., Yin, P.: Active self-assembly of algorithmic shapes and patterns in polylogarithmic time. In: Proceedings of 4th Conference on Innovations in Theoretical Compuer Science (ITCS), pp. 353–354 (2013)

Minimum Membership Hitting Sets
of Axis Parallel Segments

N. S. Narayanaswamy, S. M. Dhannya$^{(\boxtimes)}$, and C. Ramya

Department of Computer Science and Engineering,
Indian Institute of Technology Madras, Chennai, India
{swamy,dhannya,ramya}@cse.iitm.ac.in

Abstract. The Minimum Membership Set Cover (MMSC) problem is a well studied variant among set covering problems. We study the dual of MMSC problem which we refer to as *Minimum Membership Hitting Set* (MMHS) problem. Exact Hitting Set (EHS) problem is a special case of MMHS problem. In this paper, we show that EHS problem for hypergraphs induced by horizontal axis parallel segments intersected by vertical axis parallel segments is NP-complete. Our reduction shows that finding a hitting set in which the number of times any set is hit is minimized does not admit a $2 - \epsilon$ approximation. In the case when the horizontal segments are intersected by vertical lines (instead of vertical segments), we give an algorithm to optimally solve the MMHS problem in polynomial time. Clearly, this algorithm solves the EHS problem as well. Yet, we present a combinatorial algorithm for the special case of EHS problem for horizontal segments intersected by vertical lines because it provides interesting pointers to forbidden structures of intervals that have exact hitting sets. We also present partial results on such forbidden structures.

1 Introduction

A *set system* X is a pair $(S, \mathcal{C})$, where S is a set of elements and $\mathcal{C}$ is a collection of subsets of S. A *hitting set* of X is a set $S' \subseteq S$ that contains at least one element from every set in $\mathcal{C}$. The set S is trivially a hitting set for X. Given a set system X, finding a *minimum cardinality hitting set* of X is a fundamental computational problem known to be NP-complete [8]. This problem has several well-studied variants. In our paper, we study a slightly different hitting set problem, which we refer to as *minimum membership hitting set* (MMHS) problem.

The set $S' \subseteq S$ is an *exact hitting set* of X, if S' is a hitting set of X and every set in $\mathcal{C}$ contains exactly one element from S'. If X has an exact hitting set, then we refer to X as an *exactly hittable set system*. Given a set system X, the *Exact Hitting Set* (EHS) problem decides if X is exactly hittable. Given a set system X and a positive integer k, the MMHS problem seeks to find a hitting set $S' \subseteq S$ such that for every $\mathcal{C}_i \in \mathcal{C}$ the number of elements in set $S' \cap \mathcal{C}_i$ is atmost k. Observe that EHS is a special case of MMHS with $k = 1$. The MMHS problem is the dual of a well-studied set cover problem known as

L. Wang and D. Zhu (Eds.): COCOON 2018, LNCS 10976, pp. 638–649, 2018.
https://doi.org/10.1007/978-3-319-94776-1_53

Minimum Membership Set Cover (MMSC) problem. In the MMSC problem, given a set system $\mathcal{X} = (S, \mathcal{C})$, one must find if there exists a subset $\mathcal{C}'$ of $\mathcal{C}$ that covers all elements in S such that the maximum number of occurrences each element from S has in $\mathcal{C}'$ is at most k. MMSC problem was motivated by the need to reduce interference among transmitting base stations in cellular networks. Kuhn et al. [11] have addressed the problem of minimizing interference by assigning every base station a transmission power level such that the number of base stations covering any client is minimum. However, every client must be under the transmission range of at least one base station in order to maintain availability of the network. Formally, a base station is modelled as a set containing exactly all clients covered. The union of transmission ranges by all selected base stations is modelled as a collection of client sets. Kuhn et al. [11] showed that MMSC is NP-complete by a reduction from Minimum Set Cover problem. They also gave an $O(\ln n)$ approximation algorithm for MMSC and showed that there is no polynomial time approximation algorithm with ratio better than $\ln n$ unless NP $\subset$ TIME($n^{O(\log \log n)}$). MMSC problem has also been studied in contexts where the collection of sets have *consecutive ones property* (C1P) [4]. In a paper by Dom et al. [3], the authors studied the Red-Blue Hitting Set (RBHS) problem and special cases of RBHS where sets have C1P. RBHS problem is a generalization of MMSC problem. Given an n-element set S, two collections $\mathcal{C}_{red}$ and $\mathcal{C}_{blue}$ of subsets of S, and a non-negative integer k, RBHS problem seeks to find if there exists a subset $S' \subseteq S$ such that each set in $\mathcal{C}_{red}$ contains at least one element from S' and each set in $\mathcal{C}_{blue}$ contains at most k elements from S'. The authors showed that RBHS is NP-complete when either $\mathcal{C}_{red}$ or $\mathcal{C}_{blue}$ or both do not have C1P. However, when both $\mathcal{C}_{red}$ and $\mathcal{C}_{blue}$ obey C1P, the problem has been shown to admit a polynomial time solution. Observe that MMHS problem is exactly the same as RBHS problem when $\mathcal{C}_{red} = \mathcal{C}_{blue}$.

In this paper, we study the EHS and MMHS problems on geometric set systems obtained by intersection of segments on a two-dimensional plane. A set of line segments is called *axis parallel* (or *orthogonal*), if each line segment in the set is parallel either to the x-axis or to the y-axis on the plane. The line segments parallel to x-axis are called *horizontal* (denoted by set $\mathcal{H}$) and those line segments parallel to y-axis are called *vertical* (denoted by set $\mathcal{V}$). Katz et al. [10] studied geometric stabbing problems for axis parallel line segments. In particular, they studied the *Orthogonal Segment Dominating Set* (OSDS) problem to obtain a minimum cardinality dominating set for $\mathcal{H} \cup \mathcal{V}$ and *Orthogonal Segment Covering* (OSC) problem to find a subset of vertical segments of minimal size that intersects all horizontal segments in $\mathcal{H}$. They showed that, in general, both OSDS and OSC are NP-complete. For the special cases of OSC, where either the set of horizontal segments or the set of vertical segments or both are constituted by *rays*[1], they gave deterministic polynomial-time algorithm based on dynamic programming. Observe that OSDS and OSC focus on minimizing the cardinality of the hitting set, whereas in our paper, we focus on minimizing the number of times any set is hit.

---

[1] A ray is a line with one endpoint and extends infinitely in the other direction.

1.1   Our Results

We present the EHS problem and the MMHS problem as problems in hyper-graphs, which are graph theoretic representations of set systems. A *hypergraph* X is a pair $(\mathcal{U}, \mathcal{E})$ where $\mathcal{U}$ is a set of vertices and $\mathcal{E}$ is a set of *hyperedges* which are subsets of vertices in $\mathcal{U}$. In the problems we consider, the horizontal segments or the vertical segments are the vertices of the hypergraph. The set of vertical segments that intersect with horizontal segments are the hyperedges of the hypergraph. Formally, we consider the hypergraph $X = (\mathcal{V}, \{N(h)\}_{h \in \mathcal{H}})$ where $\mathcal{H}$ is a set of horizontal segments and $\mathcal{V}$ is a set of vertical segments in the plane and $N(h)$ denotes the set of vertical segments intersecting a horizontal segment h ($N(h)$ denotes the neighbourhood of segment h). Given a set $S \subseteq \mathcal{V}$, we say a vertex $v \in S$ *hits* a hyperedge $e \in \mathcal{E}$ if and only if hyperedge e contains the vertex v. We also consider hypergraphs whose vertex set is a set of vertical lines and not vertical segments. We clearly distinguish between the two cases in the following definitions:

1. MHSegments : *Minimum Membership Hitting Set for Horizontal Segments intersected by Vertical Segments*
   Let $\mathcal{H}$ be a set of horizontal segments and $\mathcal{V}$ be a set of vertical segments in the plane.
   Input : $X = (\mathcal{V}, \{N(h)\}_{h \in \mathcal{H}}), k \in \mathbb{Z}^+$
   Output : Does there exist a set $S \subseteq \mathcal{V}$ that hits every hyperedge such that maximum number of times each hyperedge in X is hit by S is at most k?
   We refer to the special case of $k = 1$ in MHSegments as EHSegments (*Exact Hitting Set for Horizontal Segments intersected by Vertical Segments*).
2. MHLines : *Minimum Membership Hitting Set for Horizontal Segments intersected by Vertical Lines*
   Let $\mathcal{H}$ be a set of horizontal segments and $\mathcal{V}_l$ be a set of vertical lines in the plane.
   Input : $X_l = (\mathcal{V}_l, \{N(h)\}_{h \in \mathcal{H}}), k \in \mathbb{Z}^+$
   Output : Does there exist a set $S \subseteq \mathcal{V}_l$ that hits every hyperedge such that maximum number of times each hyperedge in X_l is hit by S is at most k?
   When $k = 1$ in MHLines, we refer to it as EHLines (*Exact Hitting Set for Horizontal Segments intersected by Vertical Lines*).

Katz et al. [10] observed that the more there are "endpoints" for elements in sets $\mathcal{H}$ and $\mathcal{V}$, the harder the OSC problem becomes. Our results are similar wherein we show that MHSegments is harder than MHLines. In Sect. 2, we show that MHSegments is NP-complete through a reduction from PLANAR POSITIVE 1-IN-3 SAT. For arbitrary k, we show that MHSegments does not admit a $2 - \epsilon$ approximation for $\epsilon > 0$. For the case of MHLines, we give a polynomial time algorithm for arbitrary k. We do this by reducing MHLines to MHIntervals (MMHS problem in intervals, defined in Sect. 1.2). We also give a combinatorial algorithm for the special case when $k = 1$. However, we do not know if this algorithm can be extended to arbitrary k. In Sect. 4.1, we present combinatorial structures that are forbidden for exactly hittable interval hypergraphs. These forbidden structures give an insight into the complexity of recognizing exactly hittable interval hypergraphs.

In their paper, Katz et al. [10] have studied the minimum cardinality hitting set problem for segments intersected by rays and for rays intersected by segments. They showed that these problems can be solved in polynomial time using dynamic programming techniques. We believe that similar dynamic programming approaches can be devised for minimum membership variants as well.

1.2   Preliminaries

Definition 1 (Interval Hypergraph). *The hypergraph $H_n = (\mathcal{U}, \mathcal{I}_n)$, where $\mathcal{U} = \{1 \ldots n\}$ is a set of points and $\mathcal{I}_n = \{[i,j]|i \leqslant j, i, j \in [n]\}$ is a set of intervals, is known as the discrete intervals hypergraph [9] or complete interval hypergraph [1]. A hypergraph whose hyperedge set is a family of intervals $\mathcal{I} \subseteq \mathcal{I}_n$ is known as an interval hypergraph.*

Definition 2 (k-hitting set). *If S is a hitting set of hypergraph X that hits every hyperedge atmost k times, then we refer to S as a k-hitting set of X.*

Definition 3. MHIntervals *: Minimum Membership Hitting Set for Intervals*
Input *: An interval hypergraph $X = (\mathcal{V}, \mathcal{I})$ and $k \in \mathbb{Z}^+$*
Output *: Does there exist a set $S \subseteq \mathcal{V}$ that hits every interval in $\mathcal{I}$ such that maximum number of times each interval in X is hit by S is at most k?*

Definition 4 (Consecutive Ones Property (C1P)). *[12] A matrix is said to have Consecutive Ones Property for rows if it has a permutation of its columns such that 1's in every row are placed consecutively. A set family is said to have C1P for rows if the incidence matrix of the family (elements on the sets) has C1P.*

2   Minimum Membership Hitting Sets for Segments Intersecting Segments

In this section, we show that EHSegments is NP-complete. We prove the NP-hardness through a reduction from PLANAR POSITIVE 1-IN-3 SAT. Let φ be a boolean formula on n variables $X = \{x_1, x_2, \ldots, x_n\}$ having m clauses $C = \{C_1, C_2, \ldots, C_m\}$. Every boolean formula can be associated with a planar bipartite graph $G_\varphi = (C, X, E)$ where $X = \{x_1, x_2, \ldots, x_n\}, C = \{C_1, C_2, \ldots, C_m\}$ and $E = \{(x_i, C_j) \mid$ variable x_i appears in the clause $C_j\}$.
PP1in3SAT (PLANAR POSITIVE 1-IN-3 SAT):

Input : A positive 3-CNF Boolean formula $\varphi(X)$ such that G_φ is planar.

Output : Does there exist a satisfying assignment for φ that sets exactly one variable in every clause to **true**?
PP1in3SAT is known to be NP-complete [13]. We say a formula is 1-in-3 satisfiable if and only if there exists an assignment $\bar{a} = (a_1, a_2, \ldots, a_n) \in \{0, 1\}^n$ such that for every clause C_j in φ, there is exactly one variable $x_i \in C_j$ such that $a_i = 1$.

Theorem 1. EHSegments *is NP-complete.*

Proof. EHSegments is in NP. We show that EHSegments is NP-hard by reducing from PP1in3SAT. Let φ be a positive 3-CNF boolean formula such that G_φ is planar. Hartman et al. [6] have shown that every planar bipartite graph has a grid representation. That is, corresponding to every planar bipartite graph $G = (U, V, E)$, there is a set of horizontal segments U and vertical segments V on the plane such that for any $u_i \in U, v_j \in V$, edge (u_i, v_j) belongs to E if and only if the segments u_i and v_j intersect [6]. Since G_φ is a planar bipartite graph, $G_\varphi = (C, X, E)$ has a grid representation $(\mathcal{H}, \mathcal{V})$ where $\mathcal{H}$ and $\mathcal{V}$ are the horizontal and vertical segments in the grid representation of G_φ corresponding to C and X respectively. Such a grid representation can be obtained in polynomial time [2,6]. Let $X = (\mathcal{V}, \{N(h)\}_{h \in \mathcal{H}})$ be the hypergraph induced by segments in the grid representation $(\mathcal{H}, \mathcal{V})$. We now argue that φ is 1-in-3 satisfiable if and only if X is exactly hittable. Let $a = (a_1, a_2, \ldots, a_n)$ be a satisfying assignment of φ. Consider the set $S = \{v_i \mid a_i = 1, i \in [n]\}$. S is an exact hitting set for X. Now, suppose X is exactly hittable and let S be one of the exact hitting sets of X. Let $T = \{i \mid v_i \in S\}$. Consider the assignment $\bar{a} = (a_1, a_2, \ldots, a_n) \in \{0, 1\}^n$ defined by

$$a_i = \begin{cases} 1 \text{ if } i \in T \\ 0 \text{ therwise} \end{cases}$$

Note that φ is 1-in-3 satisfiable via the assignment $\bar{a}$.   □

Since EHSegments is a special case of MHSegments and EHSegments is NP-complete, it follows that MHSegments is also NP-complete.

2.1   Inapproximability of MHSegments

Theorem 2. *If there is a $2 - \epsilon$ approximation algorithm for* MHSegments *for some $\epsilon > 0$, then* P = NP.

Proof. Suppose there is a $2 - \epsilon$ approximation algorithm for MHSegments for some $\epsilon > 0$. Then there is an approximation algorithm A that outputs a set $S \subseteq \mathcal{V}$ such that S hits every hyperedge in X and that the maximum number of times each hyperedge in X is hit by S is atmost $(2 - \epsilon) \cdot OPT$ where OPT is the size of the optimum solution to MHSegments instance. We use algorithm A to decide EHSegments.

Let $A(\mathcal{J}) \triangleq |S|$ where S is the set returned by algorithm A on input $\mathcal{J}$. We use this to decide EHSegments.
Algorithm for EHSegments: On input $\mathcal{J}$, if $A(\mathcal{J}) \leq 1$ then return **yes**. Else if $A(\mathcal{J}) \geq 2$ then return **no**.

To argue the correctness of the algorithm for EHSegments, we argue:

(i) If $A(\mathcal{J}) \leq 1$, then $\mathcal{J}$ is a yes instance of EHSegments.
   Proof: If $A(\mathcal{J}) \leq 1$ then the set $S \subseteq \mathcal{V}$ returned by A hits every hyperedge in X atleast once and the maximum number of times each hyperedge is hit by S is atmost 1. Hence S is an exact-hitting set for X. $\mathcal{J}$ is a yes instance of EHSegments.

(ii) If $\mathcal{J}$ is a yes instance of EHSegments, then $A(\mathcal{J}) \leq 1$.
   Proof: If $\mathcal{J}$ is a yes instance, then we know $OPT = 1$. A outputs a set $S \subseteq \mathcal{V}$ that hits every hyperedge such that the maximum number of times each hyperedge in X is hit by S is atmost $(2 - \epsilon)$. Since the number of times a hyperedge is hit by S is an integer $(2 - \epsilon) \leq 1$ and S hits each hyperedge in X, it follows that $A(\mathcal{J}) \leq 1$.

Thus algorithm A decides EHSegments in polynomial time. By Theorem 1, EHSegments is NP-complete. Therefore, if there is a $2 - \epsilon$ approximation algorithm for MHSegments for some $\epsilon > 0$, then P $=$ NP.   □

3   Minimum Membership Hitting Sets for Lines Intersecting Segments

In this section, we show that MHLines can be solved in polynomial time. Given a set $\mathcal{H}$ of horizontal line segments and a set $\mathcal{V}_l$ of vertical lines in the plane, let $X_L = (\mathcal{V}_l, \{N(h)\}_{h \in \mathcal{H}})$ be a hypergraph, where $N(h)$ denotes the set of vertical lines intersecting horizontal segment h. Recall that the MHLines problem seeks to find if there exists a set $\mathcal{V}' \subseteq \mathcal{V}_l$ that hits every hyperedge $h \in \mathcal{H}$ such that maximum number of times each hyperedge in X_L is hit by $\mathcal{V}'$ is at most k.

We first reduce MHLines problem to MHIntervals (defined under preliminaries). Then we show that MHIntervals, and hence MHLines, can be optimally solved in polynomial time due to a result by Dom et al. [3]. They showed that an optimal solution to an integer linear program (say $\mathcal{B}$) for MHIntervals can be can be obtained in polynomial time. The coefficients of inequalities in $\mathcal{B}$ results in a binary matrix. Let $Ax \leqslant b$ represent the set of constraints in the linear programming relaxation corresponding to $\mathcal{B}$. If the underlying set system corresponds to an interval hypergraph, then the coefficient matrix A is known to have C1P [4]. Any matrix having C1P is known to be totally unimodular [5]. Since A is totally unimodular, then for all integral b, the polyhedron $P = \{x \mid Ax \leqslant b\}$ is known to be an integer polyhedron [7,14]. It follows that an optimal solution to MHIntervals can be obtained in polynomial time. Now, we present a reduction from MHLines to MHIntervals.

Lemma 1. MHLines $\leq_p$ MHIntervals.

Proof. Let $X_L = (\mathcal{V}_l, \{N(h)\}_{h \in \mathcal{H}})$ be an instance of MHLines. Let $X_I = (\mathcal{P}, \mathcal{I})$ be an interval hypergraph where $\mathcal{P} = \{1, 2, \ldots, n\}$ is a set of points on the integer line and $\mathcal{I}$ is a set of intervals which are obtained from X_L as follows.

Let $p : \mathcal{V}_l \to \{1, 2, \ldots, n\}$ be a function that maps every vertical line to a point in $\mathcal{P}$. For every vertical line $v_i \in \mathcal{V}_l$, let x_i be the point at which v_i intersects with the x-axis. Sort vertical lines according to their x_i value. Let $v_1 < v_2 < \ldots < v_n$ be the resulting order. Set $p(v_i)$ to i. In order to construct the intervals, for every horizontal segment h in $\mathcal{H}$, we add an interval $I(h) = [l(h), r(h)]$ to $\mathcal{I}$ as follows. Let $v_h(l)$ and $v_h(r)$ be the leftmost and rightmost vertical lines intersecting h. Set $l(h)$ to $p(v_h(l))$ and $r(h)$ to $p(v_h(r))$. For instance, Fig. 1 shows a set of vertical lines intersecting horizontal segments and the interval hypergraph obtained from it. Clearly, X_I can be obtained in polynomial time.

We now show the correctness of the reduction. First, we show that if X_L has a k-hitting set (defined under preliminaries) then X_I has a k-hitting set. Let $S_L \subseteq \mathcal{V}_l$ be a k-hitting set of X_L. Let $S_I = \{p(v) \mid v \in S_L\}$. We show that S_I is a k-hitting set of X_I. First, we show that S_I is a hitting set of X_I, that is, every hyperedge is hit atleast once. By construction, every interval in X_I corresponds to some horizontal segment in X_L. Assume for contradiction that there is an interval J that is not hit by S_I. Let h_J be a horizontal segment in X_L corresponding to J. By construction, $l(J)$ (and $r(J)$) corresponds to the leftmost (and rightmost) vertical segment intersecting h_J. It follows that if $J \cap S_I$ is empty, then $h_J \cap S_L$ is empty, a contradiction to our assumption that S_L is a hitting set of X_L. Now, we show that S_I is a k-hitting set of X_I. Again the proof is by contradiction. Let J' be an interval such that $|J' \cap S_I| > k$. Let $h_{J'}$ be a horizontal segment from which J' was obtained. Then, by construction, all vertical lines in the set $\{v \mid v = p^{-1}(x), \text{where } x \in J' \cap S_I\}$ intersect $h_{J'}$. It follows that $|S_L \cap h_{J'}| > k$, which is a contradiction.

Now, we show that if X_I has a k-hitting set then X_L has a k-hitting set. Let $S_I' \subseteq \mathcal{P}$ be a k-hitting set of X_I. Let $S_L' = \{v \mid v = p^{-1}(x), \text{where } x \in S_I'\}$. We show that S_L' is a k-hitting set of X_L. Let I be any interval in X_I. Then $|I \cap S_I'| \leq k$. Let h_I be a horizontal segment from which I was obtained. Then, $p^{-1}(I \cap S_I')$ belongs to S_L' and it also intersects with h_I. Since $|I \cap S_I'| \leq k$, there are atmost k vertical lines in S_L' that intersects with h_I. Thus X_L has a k-hitting set.   □

Theorem 3. MHLines *problem can be solved in polynomial time.*

Proof. Follows immediately from Lemma 1 and the fact that MHIntervals can be solved in polynomial time [3,5,7].   □

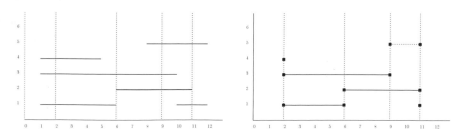

Fig. 1. A set of lines intersecting segments and the corresponding interval hypergraph

Recall that EHLines is a special case of MHLines when $k = 1$. Due to Theorem 3, EHLines problem is also solvable in polynomial time. However, we present a combinatorial algorithm for this problem via an algorithm to recognize exactly hittable interval hypergraphs. We have seen in Lemma 1 that an instance of EHLines can be reduced to an instance of an interval hypergraph such that if the EHLines instance has an exact hitting set, then the reduced interval hypergraph instance also has an exact hitting set and vice-versa. Effectively, if we have an algorithm that recognizes an exactly hittable interval hypergraph, then that algorithm can be made to recognize an yes instance of EHLines problem. In the next section, we present such an algorithm to recognize an exactly hittable interval hypergraph. This algorithm works by progressively pruning away vertices which cannot be elements of any exact hitting set. The algorithm returns a yes, if the set of intervals have an exact hitting set. Else, it returns a no. The feasibility of extending this algorithm to MHLines problem is an open question.

4 Algorithm to Recognize Exactly Hittable Interval Hypergraphs

We present an algorithm that recognizes an exactly hittable interval hypergraph X. Given an interval hypergraph $X = (\mathcal{V}, \mathcal{I})$, our problem seeks to find if there exists an exact hitting set for X. We define a function $c : \mathcal{V} \to \{B, W\}$, where B and W stand for colours black and white respectively. Initially, for every point v on the line (the vertex set of X), set $c(v)$ to W. As the algorithm proceeds, if v cannot belong to any exact hitting set, then $c(v)$ is set to black. We define another function $C : \mathcal{I} \to \{Y, N\}$ as follows.

$$C(I) = \begin{cases} N, & \text{if } c(v) = B, \forall v \in I \\ Y, & \text{otherwise} \end{cases}$$

If, for some I, $C(I)$ is set to N, then it means that no point in I can belong to any exact hitting set of X and hence I cannot be hit exactly once. We now present the algorithm.

Algorithm isEHS($X = (\mathcal{V}, \mathcal{I})$): If X is a proper interval hypergraph, then by Lemma 2, X is exactly hittable and the algorithm returns yes. Note that a proper interval hypergraph can be identified by checking if any interval shares its left endpoint (or right endpoint resp.) with the left endpoint (or right endpoint resp.) of another interval. If X is not a proper interval hypergraph, then we proceed as follows. Let $I_1, I_2 \ldots I_m$ be the intervals in $\mathcal{I}$. For all $v \in \mathcal{V}$, initialize $c(v)$ to W. For every pair of intervals I_i, I_j, if I_i contains I_j (that is $l(I_j) \geq l(I_i)$ and $r(I_j) \leq r(I_i)$), then for all $v \in I_i \setminus I_j$, set $c(v)$ to B. After the values are set, if there is any interval I for which $C(I)$ becomes N, the algorithm returns no. If not, construct a smaller hypergraph $X' = (\mathcal{V}', \mathcal{I}')$ as follows. The vertex set of X' is the set of points in X for which the colour has not been set to black. That is $\mathcal{V}' = \{v \mid v \in \mathcal{V} \land c(v) \neq B\}$. For each I_j in $\mathcal{I}$, we add a new hyperedge $I'_j = \{v \mid v \in \mathcal{V}' \cap I_j\}$ to $\mathcal{I}'$. If there are two intervals I_i and I_j in X' such that

$l(I_i) = l(I_j)$ and $r(I_i) = r(I_j)$, then retain either I_i or I_j but not both in X'. That is, if there are multiple intervals with the same left end points and same right end points, retain one among those intervals and discard the rest while constructing X'. We show in Lemma 3 that X' is indeed an interval hypergraph. Recurse on $X' = (\mathcal{V}', \mathcal{I}')$.

Lemma 2. *Let $X = (\mathcal{V}, \mathcal{I})$ be a proper interval hypergraph. Then X is exactly hittable.*

Proof. We prove the claim by constructing an exact hitting set S for X. Initialize S to $\emptyset$. Order intervals in $\mathcal{I}$ according to increasing order of their right end points. Since no interval is properly contained inside another interval, this ordering is well defined. Let this ordering be $I_1 < I_2 < \ldots < I_m$. Add $r(I_1)$ (which is the smallest right end point among all intervals) to set S. Remove every interval I such that $|r(I_1) \cap I| \neq \emptyset$. Recurse on the remaining set of intervals until all the intervals are hit by S. Clearly, S is a hitting set. We now show that S is an exact hitting set. Suppose it is not, then there exists an interval I such that $|I' \cap S| > 1$. Let $I' \cap S$ contain points p_1 and p_2 where $p_1 < p_2$. By construction, p_1 hits I' and p_2 is the right end point of some interval, say I'', that is not hit by p_1. Since $p_1 \in I'$, $p_1 < p_2$, $p_2 \in I'$, p_2 is the right end point of I'', and I'' is not hit by p_1, it follows that $I'' \subset I'$, contradicting the fact that X is an interval hypergraph in which no interval properly contains another. □

Lemma 3. *X' is an interval hypergraph.*

Proof. Clearly, every interval in X' is a subset of some interval in X. More importantly, every hyperedge in X' is an ordered subset of points of some interval in X. Let σ be the left to right ordering of points in $\mathcal{V}(H')$ on the line. Any consecutive subset of points in σ is an interval in X'. We show that every hyperedge in X' is a consecutive subset of points in σ. In X', the line has only those points which have not been set to B in X. Let I' be an interval in X' such that $I' \subseteq I$, where I is an interval in X. $I \setminus I'$ are the set of points in X that have been set to B and are absent in X'. So, the points in I' have the same order as in I and are consecutive in X'. Hence the hyperedges in X' correspond to intervals. □

Lemma 4. *Let $X = (\mathcal{V}, \mathcal{I})$ and $X' = (\mathcal{V}', \mathcal{I}')$ be interval hypergraphs that are as described in Algorithm isEHS. Then X is exactly hittable if and only if X' is exactly hittable.*

Theorem 4. *Let X be an interval hypergraph. Algorithm isEHS(X) decides, in polynomial time, if X is exactly hittable.*

With a slight modification, this algorithm can be made to output an exact hitting set of an exactly hittable interval hypergraph. However, when the set of intervals is not exactly hittable, it outputs a simple no. Can we answer more? Can we answer as to what makes it not exactly hittable? Our next section is an attempt to answer this question, at least partially.

4.1   Forbidden Structures in Exactly Hittable Interval Hypergraphs

We present sets of intervals that are forbidden configurations for exactly hittable interval hypergraphs. Table 1 shows instances of such forbidden structures. Row 1 and Row 2 of the table show F_0 and F_1 respectively, which are two basic forbidden structures from which infinite families of forbidden configurations can be constructed. It is easy to see that F_0 is not exactly hittable since any exact hitting set of I_2, I_3 must hit I_1 twice. Similarly F_1 is not exactly hittable because any exact hitting set of I_1, I_2, I_3, I_5 cannot hit I_4. We give a procedure to construct infinite families of forbidden structures as a combination of F_0s and/or F_1s. Rows 3 and 4 show two examples of such combinations. These forbidden structures will not have either F_0 or F_1 as an induced substructure. We start with F_0 comprising of intervals $\mathcal{I}_I = \{I_1, I_2, I_3\}$. Extend I_3 to the right such that the right endpoint of I_3 goes one point to the right of right endpoint of I_1. Now, $\mathcal{I}_I$ becomes exactly hittable. Then, append $\mathcal{I}_J = \{J_1, J_2, J_3\}$, which is another instance of F_0, to $\mathcal{I}_I$ as given below. We *fuse* I_3 and J_2 to become a single interval $I_3 J_2$ as shown in row 3 of the table. Also, $I_1 \cap J_1 \neq \emptyset$. We refer to this new set of intervals as $F_0 F_0$, to denote a contrived concatenation of two F_0s. Similarly, row 4 shows how to obtain $F_0 F_1$.

We now outline a general procedure to construct an infinite family of forbidden structures. We use $r(I)$ and $l(I)$ to denote the left and right endpoints, respectively, of interval I. Start with $F_i, i \in \{0, 1\}$. Extend the interval with the largest left endpoint, say I_r, to the right such that $r(I_r)$ is strictly larger than the right endpoint of every other interval in F_i. Let this new structure be F_i'. Observe that F_i' is exactly hittable due to the extended interval I_r. To obtain another forbidden structure, we place either an F_0 or an F_1 to the right of F_i'. We can think of this as *concatenating* or *attaching* two forbidden structures to obtain a bigger forbidden structure. Here, we show how to concatenate an F_0 to the right of F_i'. In F_i, let I be the interval which properly contained I_r before extending. In F_i', extend $r(I)$ such that $r(I)$ is strictly larger than the right endpoint of every other interval in F_i'. Add an interval I' such that $l(I') > r(I_r)$ and $r(I') < r(I)$. That is, I' is an interval properly contained inside I but is disjoint from I_r. This procedure concatenates an F_0 to the right of an F_i. We refer to this structure as $F_i F_0$. Since F_i' is exactly hittable due to the extended interval I_r, any exact hitting set of $F_i F_0$ hits the extended part of I_r. Consequently the concatenated F_0 cannot be exactly hit. Therefore, $F_i F_0$ is not exactly hittable.

By a similar procedure, we can concatenate an F_1 to the right of an F_i. We refer to the resulting structure as $F_i F_1$. Once we obtain a new forbidden structure, we can keep concatenating F_is to its right using exactly the same procedure as above. Observe that none of these forbidden structures have any F_i as a substructure. We believe that the infinite family of interval hypergraphs are the ones that are exactly the forbidden structures of exactly hittable interval hypergraphs.

Table 1. Instances of forbidden structures for exactly hittable interval hypergraphs

Name	Intervals	Construction / Restrictions	Figure
F_0	I_1, I_2, I_3	$I_2 \subset I_1, I_3 \subset I_1,$ $I_2 \cap I_3 = \emptyset.$	
F_1	$I_1, I_2, I_3,$ I_4, I_5	I_1 and I_2 overlap $I_4 \subseteq I_1 \cap I_2$ $I_3 \subset I_1, I_5 \subset I_2$	
$F_0 F_0$	$I_1, I_2, I_3 J_2,$ J_1, J_3	Start with 2 copies of F_0, extend I_3 to the right, append J_1, J_2, J_3 such that I_1, J_1 overlap, *fuse* I_3 and J_2 to become $I_3 J_2$	
$F_0 F_1$	$I_1, I_2,$ J_1, J_2, J_4, J_5 $I_3 J_3$	Start with an F_0 (I_1, I_2, I_3) and an F_1 $(J_1, J_2, J_3, J_4, J_5)$, extend I_3 to the right, append J_1, J_2, J_3, J_4, J_5 such that I_1, J_1 overlap, *fuse* I_3 and J_3 to become $I_3 J_3$	

References

1. Cheilaris, P., Gargano, L., Rescigno, A.A., Smorodinsky, S.: Strong conflict-free coloring for intervals. Algorithmica **70**(4), 732–749 (2014)
2. de Fraysseix, H., de Mendez, P.O., Pach, J.: A left-first search algorithm for planar graphs. Discrete Comput. Geom. **13**(3), 459–468 (1995)
3. Dom, M., Guo, J., Niedermeier, R., Wernicke, S.: Minimum membership set covering and the consecutive ones property. In: Arge, L., Freivalds, R. (eds.) SWAT 2006. LNCS, vol. 4059, pp. 339–350. Springer, Heidelberg (2006). https://doi.org/10.1007/11785293_32

4. Fulkerson, D., Gross, O.: Incidence matrices and interval graphs. Pac. J. Math. **15**(3), 835–855 (1965)
5. Ghouilahouri, A.: Programmes lineaires-caracterisation des matrices totalement unimodulaires. C. R. Hebd. Seances Acad. Sci. **254**(7), 1192 (1962)
6. Hartman, I.B.-A., Newman, I., Ziv, R.: On grid intersection graphs. Discrete Math. **87**(1), 41–52 (1991)
7 Hoffman, A., Kruskal, J.: Integral boundary points of convex polyhedra. In: Kuhn, H., Tucker, A. (eds.) Linear Inequalities and Related Systems, Ann. Math. Study **38**, 223–246 (1956)
8. Karp, R.M.: Reducibility among combinatorial problems. In: Miller, R.E., Thatcher, J.W., Bohlinger, J.D. (eds.) Complexity of Computer Computations. The IBM Research Symposia Series, pp. 85–103. Springer, Boston (1972). https://doi.org/10.1007/978-1-4684-2001-2_9
9. Katz, M.J., Lev-Tov, N., Morgenstern, G.: Conflict-free coloring of points on a line with respect to a set of intervals. Comput. Geom. **45**(9), 508–514 (2012)
10. Katz, M.J., Mitchell, J.S.B., Nir, Y.: Orthogonal segment stabbing. Comput. Geom. **30**(2), 197–205 (2005). Special Issue on the 19th European Workshop on Computational Geometry
11. Kuhn, F., von Rickenbach, P., Wattenhofer, R., Welzl, E., Zollinger, A.: Interference in cellular networks: the minimum membership set cover problem. In: Wang, L. (ed.) COCOON 2005. LNCS, vol. 3595, pp. 188–198. Springer, Heidelberg (2005). https://doi.org/10.1007/11533719_21
12. McConnell, R.M.: A certifying algorithm for the consecutive-ones property. In: Proceedings of the Fifteenth Annual ACM-SIAM Symposium on Discrete Algorithms, SODA 2004, Philadelphia, PA, USA, pp. 768–777. Society for Industrial and Applied Mathematics (2004)
13. Mulzer, W., Rote, G.: Minimum-weight triangulation is NP-hard. J. ACM **55**(2), 11:1–11:29 (2008)
14. Schrijver, A.: Theory of Linear and Integer Programming. Wiley, New York (1986)

Minimum Transactions Problem

Niranka Banerjee[1]([envelope]), Varunkumar Jayapaul[2], and Srinivasa Rao Satti[3]

[1] The Institute of Mathematical Sciences, HBNI, Chennai 600113, India
`nirankab@imsc.res.in`
[2] Université Libre De Bruxelles, Brussels, Belgium
`varunkumarj@gmail.com`
[3] Seoul National University, Seoul, South Korea
`ssrao@cse.snu.ac.kr`

Abstract. We are given a directed graph $G(V, E)$ on n vertices and m edges where each edge has a positive weight associated with it. The influx of a vertex is defined as the difference between the sum of the weights of edges entering the vertex and the sum of the weights of edges leaving the vertex. The goal is to find a graph $G'(V, E')$ such that the influx of each vertex in $G'(V, E')$ is same as the influx of each vertex in $G(V, E)$ and $|E'|$ is minimal. We show that

1. finding the optimal solution for this problem is NP-hard,
2. the optimal solution has at most $n - 1$ edges, and we give an algorithm to find one such solution with at most $n - 1$ edges in $O(m \log n)$ time, and
3. for one variant of the problem where we can delete as well as add extra edges to the graph, we can compute a solution that is within a factor $3/2$ from the optimal solution.

1 Introduction

Network flow problems have been studied in great depth in the literature. The maximum flow problem involves finding the maximum flow from a single source to a single sink in the graph. There have been many improvements down the years starting from the textbook algorithms of Ford-Fulkerson and Edmond-Karp [3] to the current state of the art result of James Orlin. Orlin [6] showed that maximum flow in a network can be found in $O(mn)$ time.

In a similar vein, Sleator and Tarjan [9] have talked about the acyclic flow problem. Given a flow from s to t in an arbitrary network, they reduce it to an acyclic flow by repeatedly finding a cycle in the graph and reducing the flow around the cycle to zero. In all these versions of flow problems we have a capacity associated with the edges which is an upper bound on the flow through that edge.

The problem of finding a subgraph G' of a graph G which maintains a specific property of G has also been studied extensively in the context of graph spanners [2,7,8], and in more recent years in its use in fault tolerant algorithms [1,4,5]. We define and study a new problem which is somewhat similar to these problems.

Problem Definition and Results. The input consists of a directed graph $G(V, E)$ where each edge has a positive weight. An edge $e \in E$ is a tuple (u, v, w),

© Springer International Publishing AG, part of Springer Nature 2018
L. Wang and D. Zhu (Eds.): COCOON 2018, LNCS 10976, pp. 650–661, 2018.
https://doi.org/10.1007/978-3-319-94776-1_54

which denotes that the edge is directed from u to v and its weight is $w > 0$. Let I_v be sum of weights of all edges entering v and O_v be sum of weights of all edges leaving v. We define the influx of a vertex $v \in V$ as $I_v - O_v$. The desired output is a graph $G'(V, E')$, such that for all $v_i \in V$ the influx of vertex v_i in G is same as the influx of vertex v_i in G'. Our aim is to minimize the size of $|E'|$ by redistributing the weights on the edges. We consider three variants of this problem depending on restrictions placed on the graph G':

1. G' is only allowed to have an edge (u, v, x) in E', if $(u, v, y) \in E$, i.e. one can only delete edges from G to obtain G'.
2. G' is only allowed to have an edge (u, v, x) in E', if (u, v, y) or $(v, u, y) \in E$, i.e., one can either delete or reverse edges in G to obtain G'.
3. G' is allowed to have any edge in E', which may or may not be present in E, i.e., one can either add or delete edges to/from G to obtain G'.

Henceforth, in the paper these variants are referred to as the first, second and third variant of the problem. We show the following results.

1. Finding the optimal solution of the problem is NP-hard for all three variants.
2. For all three versions, the optimal solution has at most $n - 1$ edges. We give an algorithm to find one (not necessarily optimal) solution with at most $n - 1$ edges in $O(m \log n)$ time for the first version, and $O(m + n)$ time for the second.
3. For the third version of the problem, we give a 3/2 approximation of the optimal solution in $O(m + n)$ time.

In Fig. 1, the influx of the vertices a, b, c, d are $0, 0, 5, -5$ respectively in the input graph G (graph on top). The three graphs below G show solutions to the three variants of the problem. The graph in Fig. 1(i) shows the desired output graph which has the minimum number of edges, provided edges have to be a subset of edges in the input graph. The graph in Fig. 1(ii) shows the desired output graph which has minimum number of edges provided the direction of edges (for example, of the edge between c and a) can be reversed in the final output. The graph in Fig. 1(iii) shows the case when the output can have edges (for example, edge from d to c) which are not present in the input graph.

A motivation comes from the following real-world application. Suppose the customers of n banks do wired transactions across these banks. Although the money is wired electronically, it is later complemented by a transfer of same amount of fiat currency. In such cases there is a cost involved with transfer of fiat currency. So the banks ideally want to reduce the number (and also the amount) of actual transfers of fiat currency.

Organization of the Paper. In Sects. 2.1 and 2.2, we define some notations and make a few elementary observations. Section 3 concludes that the minimum transactions problem is NP-hard. Section 4 gives a 3/2 approximation of the optimal solution for the third version of the problem. Section 5 gives algorithms for finding solutions which have at most $n - 1$ edges for the first and second version. Section 6 concludes with a few remarks and questions that warrant further research.

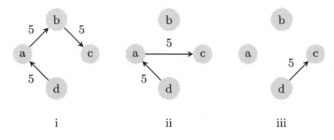

Fig. 1. The top graph shows the input graph $G(V, E)$ along with the transactions. The three graphs below the input graph show the different optimal solutions obtained from the different restrictions on G'.

2   Preliminaries

2.1   Notations

$G_U(V, E)$ denotes the undirected unweighted version of the directed graph $G(V, E)$. An edge in $G(V, E)$ from u to v is denoted by (u, v). $G_U(V, E)$ is assumed to be a connected graph without any loss of generality. An edge may also interchangeably be referred to as a transaction.

2.2   Elementary Observations

We make few elementary observations which apply to all the three variations of the Minimum Transactions problem mentioned earlier. The following lemma follows from the fact that the sum of the influxes of all the vertices in any graph is zero.

Lemma 1. *If $G_1(V, E_1)$ and $G_2(V, E_2)$ are two graphs with the same set of vertices and if there exist $n - 1$ vertices $\{v_1, v_2 \ldots v_{n-1}\}$, such that the influx of v_i ($\forall i \leq n - 1$) in G_1 is same as the influx of v_i in G_2, then influx of v_n in G_1 is also the same as the influx of v_n in G_2.*

Lemma 2. *Let $G(V, E')$ be an optimal solution to the Minimum Transactions problem for a given graph $G(V, E)$. Then the graph $G_U(V, E')$ is acyclic.*

Proof. Suppose, by contradiction, that $C = \{a_1, a_2, \ldots, a_c\}$ is a cycle in $G(V, E')$. Without loss of generality, assume that the edge having the smallest weight in cycle C is the edge (a_1, a_2) with weight w. The remaining edges have weights equal to or more than the weight of edge (a_1, a_2). Delete edge (a_1, a_2). Now a_2s' influx is reduced by w, and to compensate for that we need to increment/decrement the weight of edge between a_2 and a_3 by the value w depending on whether $(a_2, a_3) \in G(V, E)$ or $(a_3, a_2) \in G(V, E)$. We keep adjusting the weights of the edges in cycle C until we reach the vertex a_1. At this point all the vertices in cycle C (except a_1) have their influx adjusted to the influx in $G(V, E)$. The vertices which were not part of this cycle, have none of their corresponding edges changed in any way, so their influx also remains the same. The vertex a_1 does not have the option of adjusting the edge (a_1, a_2) since the edge has been deleted. Even though a_1 has no choice of edge to alter weights, we can still claim that a_1's influx is same as in $G(V, E)$ due to Lemma 1.

This shows that the undirected version of an optimal solution $(G_U(V, E'))$ does not have a cycle.   □

It is to be noted that the weight of the edges before and after a cycle is broken stays non-negative, since the weight added or subtracted from any edge is the smallest weight in that cycle.

Lemma 2 also shows the following result:

Corollary 1. *The optimal solution for any graph contains at most $n - 1$ edges.*

Consider the graph $G(V, E)$ such that $V = \{s, v_1, v_2 \ldots v_{n-1}\}$ and E consists of all edges of the form (s, v_i) ($\forall i \leq n - 1$). They all have arbitrary positive weights. We can easily see that the optimal solution is the graph $G(V, E)$ itself as each vertex has a non zero influx. Thus, there exists a problem instance whose optimal solution contains at least $n - 1$ edges.

The above observation can also be generalized to make the following claim.

Lemma 3. *In $G(V, E)$ if P is the number of vertices which have positive influx, and N is the number of vertices which have negative influx, then any optimal solution for G will have $\Omega(max(P, N))$ edges.*

Proof. Suppose $P > N$. Each of P vertices need at least one incoming edge to account for the positive influx in the optimal solution. Thus at least P edges are needed by the optimal solution. Similarly when $P \leq N$, we can see that at least N edges are present in the optimal solution.   □

It is to be noted that all the observations made in this subsection apply to all the three variants of the problem.

3   Hardness of the Problem

3.1   Reduction from Subset Sum

The Subset-Sum problem is known to be NP-complete when the weights are given in binary. We show that an instance of the subset-sum problem can be converted into an instance of Minimum Transactions problem. The Subset-Sum problem involves a sorted set (or multiset) of integers (given in binary) $S = \{d_1, d_2 \ldots d_n\}$ and a target value t. The goal is to find a set of integers $X \subseteq S$ (if it exists) such that sum of all integers in X is t.

First we show that an instance of Subset-Sum can be transformed into an instance of the Minimum Transactions problem where any directed edge can be added or deleted. Create a directed graph on $n + 3$ vertices $\{v_1 \ldots v_n, centre, v_t, v_{t'}\}$. The directed graph is created to have exactly $n + 2$ edges. The edges $(v_i, centre)$ ($\forall i \in [1, n]$) have weight d_i. The weight of edge $(centre, v_t)$ is t and weight of edge $(centre, v_{t'})$ is $\sum d_i - t$. This instance of minimum transactions problem is then given to an algorithm which solves it. If there exists a subset $X \subseteq S$ such that sum of all elements in X is t, then the algorithm will output a solution which has exactly n edges. The solution will consist of $|X|$ edges which are directed to v_t and $n - |X|$ edges which are directed to $v_{t'}$. Otherwise it will output a solution which has exactly $n+1$ edges, in which case all but one vertex in $\{v_1, v_2 \ldots v_n\}$ have exactly one edge to either v_t or $v_{t'}$ and one of the vertices has edges to both v_t and $v_{t'}$. This shows that any instance of Subset-Sum problem can be transformed into an instance of Minimum Transactions problem (where any edge can be added in the final solution) in polynomial time.

We now show that the Minimum Transactions problem where the final solution can only have edges present in the input is also NP-hard. Suppose $S = \{d_1, d_2 \ldots d_n\}$ denotes the sorted set of n positive numbers and we need to find whether a subset of $X \subseteq S$ exists such that the sum of the numbers in that subset is t. Create sets $S' = \{v_1, \ldots, v_n\}$ and $T' = \{v_t, v_t'\}$, as in the previous case. For each $i = 1, \ldots, n$, construct edges (v_i, v_t) and (v_i, v_t'), both of weight $d_i/2$. Thus, so far the influx of v_t and v_t' is $(\sum d_i)/2$ and the influx of v_i is $-d_i$ ($\forall i \in [1, n]$). Create an edge from v_t to v_t' or from v_t' to v_t, with weight chosen so that the influx of v_t is t and the influx of v_t' is $t' := (\sum d_i) - t$. Now check whether there is a solution with n edges. Such a solution must have an edge of weight d_i going out of each v_i. This edge goes either to v_t or v_t', and the edges going into v_t have to have total weight t so that the edge between v_t and v_t' does not need to be used. Otherwise, the solution consists of $n + 1$ edges and there will be an additional edge between v_t and v_t'. The same reduction also works for the case where the edges are allowed to be reversed in the final solution. All the edges are directed from S' to T'. If there exists a solution with exactly n edges, then all those n edges have to be directed from S' to T'. Thus the ability to reverse an existing edge does not help in any way. This proves that Minimum Transactions problem is NP-hard, irrespective of which of the above mentioned variants we are dealing with.

Theorem 1. *Solving the Minimum Transactions problem optimally is NP-hard for all the three versions.*

4   Approximation Algorithm for the Third Version

When the algorithm is allowed to add/delete any edge a simple algorithm exists which gives a near optimal solution. Compute the influx of all the vertices in $O(m+n)$ time (where m is number of edges and n is the number of vertices in the graph). Now partition the vertices into two sets P and N. P consists of vertices which have positive influx and N consists of vertices which have negative influx. The vertices with zero influx are neither added to P nor N. A vertex $v \in P$ is selected arbitrarily. The final solution consists of edges from all vertices in N to v and edges from v to all vertices in $P - \{v\}$. The weight of an edge (v_i, v) (where $v_i \in N$) is the absolute value of influx of vertex v_i in graph G. Similarly the weight of an edge (v, v_j) (where $v_j \in P$) is the absolute value of influx of vertex v_j in G. This solution has $|N|$ edges from N to v and $|P| - 1$ edges from v to P. Thus, this solution has $|P| + |N| - 1$ edges, which at most twice the optimal number of edges, since the optimal solution has at least $max(|P|, |N|)$ by Lemma 3.

This leads to the following theorem:

Theorem 2. *A solution with at most 2 times the optimal number of edges can be found for the Minimum Transactions problem, when the edges not present in the input are allowed to be added and the total time taken to find such a solution is* $O(m + n)$.

In fact, we can also show that a slightly modified version of the above mentioned algorithm can yield a solution which has at most 1.5 times the number of edges in the optimal solution.

In this pursuit, we show the following lemma:

Lemma 4. *Given an instance of Minimum Transactions problem, when the edges not present in the input are allowed to be added, if a vertex x has influx $t > 0$ and a vertex x' has influx $-t$, then there exists an optimal solution for which has a directed edge from x' to x with weight t .*

Proof. Suppose there exists an optimal solution, in which there is no edge from x' to x. In such a case, suppose the influx of x is satisfied by vertices $u_1, u_2 \ldots u_i$ and the influx of x' is satisfied by vertices $v_1, v_2 \ldots v_j$. These account for $(i + j)$ edges. These vertices may have other edges associated with them, but they will not matter in the analysis. Now we can construct a different optimal solution in the following manner. Construct a direct edge of weight t from x' to x. Now there are i positive influx vertices and j negative influx vertices, whose influx has not been satisfied. Here we use Theorem 2 and find a solution with $i + j - 1$ edges. The other edges in the new solution remain the same as in the old solution. Thus we can obtain a new solution which has a directed edge from x' to x and which has same number of edges as the optimal solution.   □

Now we will show how to use the above lemma to get a better solution. A matching pair is defined as a pair of vertices one of which has influx z and the other has influx $-z$.

As in the previous algorithm, compute the influx of all the vertices and ignore vertices which have zero influx. Separate the vertices into two groups of vertices P and N, where P denotes the set of vertices which have net positive influx, and N are the set of vertices which have net negative influx. Sort the vertices in P and N according to their influx in $O(n \lg n)$ time. Let k be the number of vertices in P which have a matching negative influx in N. These k vertices in P will have a directed edge to their respective matching vertex among the k vertices in N. Using the Lemma 4 we can remove these k vertices from P and k vertices from N and work on the remaining set of vertices. Thus, in the reduced problem, we have $P - k$ positive influx vertices, and $N - k$ negative influx vertices.

Since there are no matching pairs in the reduced problem, this implies that for every pair of vertices (x, y), where $x \in P$ and $y \in N$, either x has two edges incident on it or y has two edges incident on it. This implies that at least two edges are required to ensure that any 3 vertices have their influx correctly distributed. This implies that the optimal solution (for the reduced problem) has at least $2/3(P - k + N - k)$ and thus the optimal solution for the entire problem has at least $[2/3(P - k + N - k) + k = 2(P + N)/3 - k/3]$ edges.

Thus the new lower bound for the number of edges in the final solution is $\max(P, N, 2(P + N)/3 - k/3)$.

The improved algorithm involves finding these k matching pairs and then using the algorithm in Theorem 2 on the reduced problem, to find a solution of size $(P - k) + (N - k) - 1 = P + N - 2k - 1$. So the improved algorithm gives a solution with at most $P + N - k - 1$ edges for the entire problem. For all three possibilities of the lower bound, we analyse the quality of the solution.

1. $P = \max(P, N, 2(P + N)/3 - k/3)$.
   Then we get that $P \geq N$ and $P \geq 2P/3 + 2N/3 - k/3$.
   $\implies P/3 \geq 2N/3 - k/3$.
   $\implies (P + k)/2 \geq N$
   $\implies 3P/2 - k/2 - 1 \geq P + N - k - 1$

   Thus, in this case the lower bound on the number of edges in the optimal solution is P and the upper bound is $3P/2 - k/2 - 1$, which gives an approximation ratio of $3/2$.

2. $N = \max(P, N, 2(P + N)/3 - k/3)$.
   The analysis is analogous to one shown in case 1.

3. $2(P + N)/3 - k/3 = \max(P, N, 2(P + N)/3 - k/3)$.
   The number of edges in the solution is at most $(P + N - k - 1)$. Taking the ratio of number of edges in final solution to the lower bound on the number of edges required by the optimal solution, we get

$$\frac{P+N-k-1}{2/3(P+N-k/2)} = (1.5)\frac{P+N-k-1}{P+N-k/2} = 1.5\frac{P+N-k/2}{P+N-k/2} - \frac{3k/4+3/2}{P+N-k/2}$$

which gives a 3/2 approximation ratio of the solution.

This gives us the following theorem.

Theorem 3. *A solution with at most 3/2 times the optimal number of edges can be found for the Minimum Transactions problem, when the edges not present in input are allowed to be added and the total time taken to find such a solution is* $O(m + n \lg n)$.

5   Finding Solutions with at Most $n - 1$ Edges

5.1   Second Version: Case When Re-orienting an Edge Is Allowed

The algorithm assumes that $G(V, E)$ is a connected graph, otherwise disconnected components can be dealt with as separate disjoint problems.

We know that $G_U(V, E)$ has a single connected component, and thus it has a spanning tree. First we find an arbitrary spanning tree T on $G_U(V, E)$. If the edge directions can be changed in the final output, we will show that proper orientation of the edges in T is enough to find a valid solution with at most $n-1$ edges.

Let S denote the solution with at most $n - 1$ edges. Initially it is empty. Now we look at any leaf vertex (say u) and its neighbor (say v) in T. If u has a positive net flow w, then the edge (v, u) with the weight $|w|$ is added to S and net flow of v is decremented by $|w|$. If u has a negative net flow w, then the edge (u, v) with the weight $|w|$ is added to S and the net flow of v is incremented by $|w|$. If u has zero net flow, then do not add any edge to S. Irrespective of whether the net flow of u is positive, negative or zero, delete the edge uv from T. Repeat this process till T is empty. The pseudocode is given in the Appendix.

Correctness. The edges in final solution S are just orientated versions of the undirected edges in T. T is subgraph of G_U. This implies that if a directed edge (u, v) exists in S, then either (u, v) or (v, u) belongs to set E. When the first leaf vertex is deleted from T, the algorithm makes sure that the influx of deleted vertex (which is now added to S) is the same as the influx of that vertex in G. In a similar way, it can be shown that the influx of all but last two vertices being added to S is processed correctly. When T has only one edge, one of the vertices is correctly assigned the influx to match its influx in input graph, but the other vertex in the final edge does not have any option. Lemma 3 again helps in this case, and shows that if all but one pair of vertices have matching influx in both graphs, then the last pair of vertices also have the matching influx in both graphs.

This shows that the algorithm gives a correct solution with at most $n - 1$ edges.

Complexity. $O(m + n)$ time is spent to find compute the influx of each of the vertices and to find a spanning tree of the input. Another $O(n)$ time is enough to correctly assign the weights of the edges and their orientation.

This gives us the following theorem.

Theorem 4. *A solution with at most $n-1$ edges can be found for the Minimum Transactions problem, when the edges are allowed to be reversed and the total time taken is $O(m + n)$.*

5.2 First Version: Case When only Deletion of Edges Is Allowed

In this section, we deal with the hardest variant of the problem, where the final solution is required to be a subgraph of the input. We present two algorithms for the variant. One algorithm is simple to execute but takes $O(mn)$ time to find a solution with at most $n - 1$ edges. The other algorithm requires the use of Link-Cut trees but takes $O(m \lg n)$ time to find a solution with at most $n - 1$ edges.

5.2.1 $O(mn)$ Running Time Algorithm to Find G'

In this variant of the problem, we restrict the final solution to contain only those edges which are present in the input graph.

The algorithm for this variant processes the edges of the graph one by one and maintains a partial solution on edges which have been processed until that point. The algorithm initially has a solution $G'(V, E')$ where E' is empty. Then it processes edges one by one from E and adds them to E'. After adding an edge it checks if the undirected version of the graph has a cycle or not. If it has a cycle, then it is resolved using Lemma 2. Otherwise, if G' does not have a cycle, then the algorithm processes the next edge. This process continues until all edges in E have been processed.

The pseudocode is deferred to the full version.

Complexity. The process of detecting a cycle in G'_U is executed at most m times. At any point during the execution of the algorithm, the graph G'_U either does not have a cycle, or has exactly one cycle in G'_U. It cannot have more than one cycle, since adding an edge to an undirected tree/forest can create at most one cycle. The graph G'_U has at most n edges at any given time and $O(n)$ time is spent to find a cycle (if it exists), adjust the weights of the edges in this cycle and then delete an edge from the cycle. Thus, the total running time of the algorithm is $O(mn)$.

Correctness. We argue that the algorithm correctly maintains a partial solution for all the edges which have been processed. In other words, the graph G' after processing t edges in E, gives a valid solution for G if it consisted only of the edges processed till the tth iteration. Initially, when the first edge in G is processed, G' has only one edge and thus this claim is trivially true.

Suppose q edges have been processed and the partial solution G' is consistent with all the edges that have been processed till this point. When the $(q + 1)^{th}$

edge is processed either of two cases can happen: Either G'_U has an undirected cycle or not. In the former case, the edge is added to the cycle and some edge is removed from the cycle, by using Lemma 2. This ensures that the influx of vertices is consistent with the edges that have been processed. If the latter case happens, then a simple edge addition happens and this maintains the invariant that the influx of all the vertices in G' is consistent with the edges that have been processed.

Thus the algorithm always maintains a correct solution for the processed edges and the solution consists of at most $n - 1$ edges, which are sufficient to perform the transactions. This gives us the following theorem,

Theorem 5. *A solution with at most $n-1$ edges can be found for the Minimum Transactions problem, when the edges have to be a subset of the input graph and the total time taken to find this solution is $O(mn)$.*

5.2.2 $O(m \lg n)$ Running Time Algorithm to Find G'

The algorithm for an improved runtime bound uses the same algorithm as the previous one, but performs the book-keeping operations faster by the use of Link-Cut Trees. We use the Link-Cut trees in the cycle-detection phase, to efficiently detect if the edge being processed will create a cycle or not and if so make the necessary adjustments of weights along this cycle and then delete one edge from the newly formed cycle.

The Link-Cut trees [9] were designed for solving network flow problems in directed graphs. The Link-Cut trees maintain forests of trees that have edges e_i with non-zero weights w_i. We shall use the following operations supported by Link-Cut Trees which can be performed in $O(\lg n)$ worst case time.

1. Link (u, v, w) - Add edge between vertices u and v with weight w.
2. Cut(u, v) - Cut the edge between vertices u and v.
3. MakeRoot(v) - Restructure the tree around the vertex v to make it the root of the tree which it belongs to. While restructuring, it also negates the value of all the edges on the path from v to the previous root of the tree.
4. Add(v, w) - Adds a real value w to all edges on the path from the root to v.
5. Minima(v) - Returns the edge with smallest weight on path from root to v.

In our implementation, we slightly modify the implementation of Minima(v). If two edges have weight w_1 and w_2, then the smallest weight edge among the two is defined the edge with weight $min(|w_1|, |w_2|)$. Thus, when $Minima(v)$ is called, the data structure returns an edge with the smallest absolute weight on the path from the root to v and its weight.

We point out that not only the update of weights on the cycle is done by Link-Cut trees, but also the detection of the cycle itself (before it is about to be formed), as this was also an expensive operation in the naive algorithm.

It is also to be noted that the Link-Cut trees (which have rooted trees) do not store the direction of the edges. But we store the direction of the edges implicitly. For an edge uv in a tree $T_i \in F$, assume that u is nearer to the root than v, then the direction of edge is defined as follows. If the weight of the edge is positive, then the direction of the edge is from u to v, otherwise the direction of the edge in the original input graph (and final solution) is from v to u. When a vertex v which is not a root of its tree is made the root of its tree, then all the implicit directions of all the edges on the path from previous root of tree to v, have their directions reversed, which is then compensated by negating the value of weights of these edges. In this way it is ensured that the tree always captures the direction of the edge according to the original input graph.

The pseudocode of the algorithm is deferred to the full version.

The algorithm in this subsection, just like the previous algorithm, processes the edges one by one. We add these edges to the forest F. Denote the underlying undirected graph of F as F_U. If u_1 and v_1 belong to different trees, then adding an edge between them does not create a cycle(in F_U) and thus the edge is added "as is" between the two vertices. If u_1 and v_1 belong to the same tree, then adding an edge between them will create a single cycle. In order to avoid the cycle, the tree is pre-emptively adjusted, taking into account the edge which is being processed.

If the edge being processed is the edge which has the smallest absolute weight in that cycle (which would form in F_U, if (u_1, v_1) was added to F), then we can use Lemma 2 on this cycle and avoid adding that edge in the final solution. Similarly, if there is an edge on the path from u to v which has an absolute weight smaller than the weights of all the edges in that path and it is also smaller than the edge which is being processed, then this edge can be removed from the solution by the use of Lemma 2. The use of Lemma 2 for breaking cycles also ensures that if an edge (u, v) has positive weight in the final solution, then the edge (u, v) also had a positive weight in the input.

If T has negative weight for edge (u, v), it means that the input graph contains (v, u) as an edge. In this case, we check the directions of all edges in E', and add their corresponding edges in E in the final solution. The weight of an edge in the final solution is the absolute weight of that edge in E'. This takes care of adjusting the direction of edges in the final solution according to the input graph.

Complexity. The number of edges processed one by one is m and each edge while processed may create a cycle in G'_U. The detection of a cycle which is to be formed and necessary adjustments made to remove this cycle require a constant number of Minima(v), Makeroot(v), Add(v, w), Cut(u, v) and Link(u, v, w) operations. All these operations take $O(\lg n)$ amortized time when implemented using Link-Cut trees. Thus the algorithm takes $O(m \lg n)$ running time.

Correctness. Since the algorithm is essentially the same as the previous subsection, we do not argue its correctness separately.

Thus, we have the following theorem.

Theorem 6. *A solution with at most $n-1$ edges can be found for the Minimum Transactions problem, when the edges have to be a subset of the input graph and the total time taken to find this solution is $O(m \lg n)$.*

6   Conclusion

In this paper, we define a variation of the network flow problem and give various results on the possible lower bounds and upper bounds on the number of edges needed in the final solution. This network flow problem differs from most widely studied variations of network flow problem, as the edges do not have an upper limit on their capacity. The paper introduces this concept of unbounded edge flows. The following are the open research problems

- What is the fine-grained complexity of these problems?
- Are the algorithms efficient in practice?
- Can a better approximation be achieved for this problem?
- Given an instance of $G(V, E)$, a vertex v and an integer k, can we find a solution $G'(V, E')$ such that v has at most k edges in G'?
- Are there better lower bounds for the problem, when new edges cannot be added in the solution?

References

1. Baswana, S., Choudhary, K., Roditty, L.: Fault tolerant subgraph for single source reachability: generic and optimal. In: Proceedings of the 48th Annual ACM SIGACT Symposium on Theory of Computing, STOC 2016, 18–21 June 2016, Cambridge, MA, USA, pp. 509–518 (2016)
2. Chechik, S., Langberg, M., Peleg, D., Roditty, L.: Fault-tolerant spanners for general graphs. In: Proceedings of the 41st Annual ACM Symposium on Theory of Computing, STOC 2009, 31 May–2 June 2009, Bethesda, MD, USA, pp. 435–444 (2009)
3. Cormen, T.H., Leiserson, C.E., Rivest, R.L., Stein, C.: Introduction to Algorithms, 3rd edn. MIT Press, Cambridge (2009)
4. Gupta, M., Khan, S.: Multiple source dual fault tolerant BFS trees. In: 44th International Colloquium on Automata, Languages, and Programming, ICALP 2017, 10–14 July 2017, Warsaw, Poland, pp. 127:1–127:15 (2017)
5. Lengauer, T., Tarjan, R.E.: A fast algorithm for finding dominators in a flowgraph. ACM Trans. Program. Lang. Syst. **1**(1), 121–141 (1979)
6. Orlin, J.B.: Max flows in O(nm) time, or better. In: Symposium on Theory of Computing Conference, STOC 2013, 1–4 June 2013, Palo Alto, CA, USA, pp. 765–774 (2013)
7. Peleg, D., Schäffer, A.A.: Graph spanners. J. Graph Theor. **13**(1), 99–116 (1989)
8. Roditty, L., Thorup, M., Zwick, U.: Roundtrip spanners and roundtrip routing in directed graphs. In: Proceedings of the 13th Annual ACM-SIAM Symposium on Discrete Algorithms, 6–8 January 2002, San Francisco, CA, USA, pp. 844–851 (2002)
9. Sleator, D.D., Tarjan, R.E.: A data structure for dynamic trees. J. Comput. Syst. Sci. **26**(3), 362–391 (1983)

Heuristic Algorithms for the Min-Max Edge 2-Coloring Problem

Radu Stefan Mincu[1(✉)] and Alexandru Popa[1,2(✉)]

[1] Department of Computer Science, University of Bucharest, Bucharest, Romania
{mincu.radu,alexandru.popa}@fmi.unibuc.ro
[2] National Institute for Research and Development in Informatics,
Bucharest, Romania

Abstract. In multi-channel Wireless Mesh Networks (WMN), each node is able to use multiple non-overlapping frequency channels. Raniwala et al. (MC2R 2004, INFOCOM 2005) propose and study several such architectures in which a computer can have multiple network interface cards. These architectures are modeled as a graph problem named *maximum edge q-coloring* and studied in several papers by Feng et. al (TAMC 2007), Adamaszek and Popa (ISAAC 2010, JDA 2016). Later on Larjomaa and Popa (IWOCA 2014, JGAA 2015) define and study an alternative variant, named the *min-max edge q-coloring*.

The above mentioned graph problems, namely the maximum edge *q*-coloring and the min-max edge *q*-coloring are studied mainly from the theoretical perspective. In this paper, we study the min-max edge 2-coloring problem from a practical perspective. More precisely, we introduce, implement and test four heuristic approximation algorithms for the min-max edge 2-coloring problem. These algorithms are based on a *Breadth First Search* (BFS)-based heuristic and on *local search* methods like basic *hill climbing*, *simulated annealing* and *tabu search* techniques, respectively. Although several algorithms for particular graph classes were proposed by Larjomaa and Popa (e.g., trees, planar graphs, cliques, bi-cliques, hypergraphs), we design the first algorithms for general graphs.

We study and compare the running data for all algorithms on Unit Disk Graphs, as well as some graphs from the DIMACS vertex coloring benchmark dataset.

1 Introduction

MOTIVATION. In multi-channel Wireless Mesh Networks (WMN), each node is able to use multiple non-overlapping frequency channels. The use of many channels inside the same network can significantly improve overall performance. Interference from neighboring nodes can be decreased substantially when nodes do

A. Popa—This work was supported by the research programme PN 1819 "Advanced IT resources to support digital transformation processes in the economy and society - RESINFO-TD" (2018), project PN 1819-01-01 "New research in complex systems modelling and optimization with applications in industry, business and cloud computing", funded by the Ministry of Research and Innovation.

not need to use the same radio channel for every link. Multiple radio channels in the network imply that at least some of the nodes need to handle more than one channel at a time. In many proposed designs the multi-channel feature is achieved by packet-by-packet reconfiguration of the radio [6, 12, 15]. However, one of the drawbacks of this kind of continuous channel switching of a single radio interface is that it requires precise synchronization throughout the network.

An alternative approach would be to fit multiple radio interfaces to each node, thus allowing a more persistent channel allocation per interface. A couple of such multi-NIC (network interface card) architectures have been proposed by Raniwala et al. [13, 14]. Their simulation and testbed experiments show a promising improvement with only two NICs per node, compared to a single-channel WMN. Another appealing feature of these architectures is that they are based on readily available, commodity IEEE 802.11 interfaces, requiring only systems software modification.

The scenario of two or more NICs per node with fixed channels imposes some limitations to the assignment of channels on each interface. In order to set up a link between two nodes, both of them have to have at least one of their interfaces set to the same channel. On the other hand, links inside an interference range should use as many different channels as possible. Thus, the channels need to be assigned carefully in order to both keep every required link possible and maximize useful bandwidth throughout the network.

PROBLEM DEFINITION. The channel assignment problem can be modeled as a type of edge coloring problem: given a graph G, the edges have to be colored so that there are at most q different colors incident to each vertex. Here, vertices, edges and colors represent network nodes, links and channels, respectively. A coloring that satisfies this constraint, is called an *edge q-coloring*. Note, that the coloring constraint differs from the traditional coloring problems, where adjacent items are not allowed to have the same color. Also the goal is different; instead of minimizing, we want to maximize the number of different colors in an edge q-coloring.

Initially, the channel assignment was formulated as the *maximum edge q-coloring problem*, where the goal was to maximize the total number of colors in a q-coloring. The drawback of this model is that in an optimal solution the same color is assigned to many edges while other colors are used only once. We remind the reader that in the wireless mesh network setting, having the same color assigned to many edges is equivalent to having the same frequency used many times, and therefore, having interference. Since the goal of the application is to minimize the interference, max edge q-coloring is perhaps not the ideal theoretical formulation (although max edge q-coloring is still interesting as a combinatorial problem). Instead, it is more relevant for the network application to try to have the color components as balanced as possible. Thus, the *min-max edge q-coloring* had been introduced, where the goal is to minimize the maximum size of a color group. The formal definition of the min-max q-coloring follows.

Problem 1 (Min-max edge q-coloring). Given a graph $G = (V, E)$, find an edge q-coloring σ of G such that the amount $max_c|\{e \in E|\sigma(e) = c\}|$ is minimized.

In other words, find an edge coloring that minimizes the size of the largest set of edges with the same color.

PREVIOUS WORK. The problem of finding a maximum edge q-coloring of a given graph has been first studied by Feng et al. [2–4]. They provide a 2-approximation algorithm for $q = 2$ and a $(1 + \frac{4q-2}{3q^2-5q+2})$-approximation for $q > 2$. They show that the problem is solvable in polynomial time for trees and complete graphs in the case $q = 2$, but the complexity for general graphs has been left as an open problem. Later, Adamaszek and Popa [1] show that the problem is APX-hard and present a 5/3-approximation algorithm for graphs which have a perfect matching. The maximum edge q-coloring is also considered in combinatorics and is a particular case of the anti-Ramsey number. For a brief description of the connection of the two problems, the reader can refer to [1].

Larjomaa and Popa [8,9] introduce and study the min-max edge q-coloring problem. They prove that the problem is NP-hard for any $q \geq 2$ and show an exact polynomial time algorithm for trees, for $q = 2$. Moreover, Larjomaa and Popa [9] analyze the value of the optimal solution on special classes of graphs: cliques, bicliques and hypercubes. They provide the exact formulas of the optimal solutions for cliques. For bicliques they present a lower bound which is tight when both parts of the graph have an even number of vertices (and almost tight for the other cases). For a hypergraph Q_n they give a lower bound which is tight for even n, and similarly, almost tight for odd n. Although these classes of graphs have a very simple structure, finding lower bounds is much more difficult than in the case of the max edge q-coloring problem.

A good lower bound of the optimal solution is necessary in order to design approximation algorithms. For the min-max edge q-coloring problem, a trivial lower bound is half of the maximum degree. Larjomaa and Popa [9] show another lower bound in terms of the average degree of the graph. Larjomaa and Popa [9] also present an approximation algorithm for planar graphs which achieves a sublinear approximation ratio. The algorithm uses a theorem of Lipton and Tarjan [10] which says that a planar graph admits a small balanced separator.

OUR RESULTS. Although the min-max q-coloring problem has been studied for particular classes of graphs, little has been done for general graphs in the sense of an approximation algorithm. As such, we design, implement and analyze algorithms for the min-max 2-coloring problem for general graphs.

The paper is organized as follows. In Sect. 2 we show a Breadth First Search (BFS)-inspired approach to approximating min-max 2-coloring. In Sect. 3 we present min-max q-coloring as a local search problem in the context of combinatorial optimization. Subsequently, we build the necessary tools to tackle min-max edge 2-coloring as a local search problem (provide neighborhood structure, auxiliary objective function). After this framework is built, we construct algorithms to solve the problem using hill climbing (its basic nature led to omitting the full algorithm from this paper), simulated annealing (Subsect. 3.1) and tabu search (Subsect. 3.2) techniques. Finally, we reveal some experimental results in Sect. 4 and provide insight into the difficulty of the problem and the nature of the methods we employ to solve it. We reveal a simple design for a BFS-inspired algorithm

that yields good results while having the benefit of the linear time complexity
of BFS. We provide evidence that all of our local search algorithms successfully exploit the search space gradient in improving their working solutions as
shown by a linear decrease in the objective function. We show that a simple
hill climbing approach produces reasonably good solutions using a low number
of iterations over the initial solution. Algorithms 2 and 3 (based on simulated
annealing and tabu search techniques) take longer to complete but manage to
escape local optima and achieve better solutions.

In the Experimental Results (Sect. 4) we describe the testing dataset, analyze
the implementation of the local search algorithms and show the behavior of the
described algorithms on our selected dataset. The results are encouraging while
considering the upper bounds for the optimum solutions for a selection of the
input graphs that are obtained with an Integer Linear Program (ILP).

2   A BFS-Inspired Heuristic Algorithm

We show a simple algorithm for approximating the min-max edge 2-coloring
by using *Breadth First Search* (BFS). The idea is to color the uncolored edges
incident to each subsequent "level" in a BFS with a distinct color. The "levels"
denote the starting vertex, then its neighbors, then the neighbors of the neighbors and so on. The full algorithm is presented as Algorithm 1. The algorithm
takes time $O(n + m)$, same as BFS. We can improve the base algorithm by
coloring disconnected colored components with distinct colors as shown in step
5. By using a disjoint set forest data structure we may quickly determine these
disconnected components during the edge coloring step for only a small overhead
of $O(\alpha(m))$, α denoting the inverse of the Ackermann function.

Algorithm 1. input: graph $G = (V, E)$, an initial vertex v_0

---
1 : Let there be two sets $Q_1 \leftarrow \{v_0\}$ and $Q_2 \leftarrow \emptyset$. Mark v_0 as visited. Integer $c \leftarrow 1$.
2 : Color all uncolored edges (v_i, v_j) incident to each $v_i \in Q_1$ using integer color c.
3 : During the previous operation, add all unvisited v_j (neighbors of v_i) to set Q_2.
4 : Mark all these v_j as visited.
5 : (Improvement step) Consider the subgraph containing all the edges colored with
integer c. Color each disconnected component in this subgraph with a new color
obtained by incrementing c.
6 : Let $c \leftarrow c + 1$, $Q_1 \leftarrow Q_2$ and $Q_2 \leftarrow \emptyset$
7 : If $Q_1 = \emptyset$ then the algorithm terminates. Else, continue with step 2.

Theorem 1. *Algorithm 1 produces a valid 2-coloring.*

Proof. The colored subgraph G_i grows at each iteration i of the algorithm by
adding a new layer of previously uncolored edges. The vertices along the border
of G_i all have incident edges with the same color. At step $i + 1$ these vertices
may obtain a second incident color if they had any uncolored incident edges in

the main graph at iteration step i. Assume that $G_0 = (V, \emptyset)$ and at step i there is a 2-coloring using i colors in G_i, but in G_{i+1} we add a third incident color different from c_{i+1} to some vertex p (which has to be at the border of G_i). This third color comes from an edge that is incident to both p and a vertex q from the border of G_{i+1}. This edge can only be colored with c_{i+1}, contradiction.   □

3   Local Search Algorithms

Min-max edge q-coloring (including 2-coloring) can naturally be modeled as a combinatorial optimization problem:

- *a solution* ω is a color mapping from the edge set of the graph to a set of positive integers, for example.
- *the objective function* $f(\omega)$ used to evaluate the quality of the solution is the largest number of edges that share the same color. Our purpose is to minimize this amount, as such, it is a *minimization problem*.
- *the constraint* is that the set of edges incident to a vertex can contain edges that are colored with at most q (respectively, two for 2-coloring) different colors.
- *a feasible solution* will respect the constraint across all vertices while *an unfeasible solution* will not.

To solve this problem using local search, there are a few more requirements to fulfill:

- *some initial solution* ω_0 to start improving upon.
- *a neighborhood structure* $N(\omega)$ to provide slightly modified candidate colorings that we will evaluate with our objective function. If a neighbor is better in terms of the objective function then we select it as current solution (i.e. $\omega_{current} \leftarrow \omega_{best} \in N(\omega_{current})$).
- *some stopping criteria* to prevent the algorithm from looping.

For our neighborhood structure we choose *operations based neighborhood*, that is to say, we apply some local modifications or *moves* (i.e. color changes) to some components of the current solution (i.e. edges). The set of moves applied to every component of the solution ω will construct the neighborhood $N(\omega)$.

Notation 1. In the following we refer to the *color class of a vertex* as being the set of colors of its incident edges. We use $cc(v)$ to denote the color class of a vertex v. By definition, $cc(v) = \bigcup_{(v,v') \in E} \sigma((v, v'))$, where σ is an edge coloring.

We now consider a move set that can be applied only on feasible solutions (i.e. 2-colorings) and will also produce only feasible solutions.

The defined moves can only be applied in certain cases depending on the color classes of the endpoint vertices of the edge we operate on. Such scenarios are depicted in Fig. 1 but do not reveal all possible cases. The omitted cases are those that result in the removal of a color from either or both of the color classes for *exchange*, *connect* and *create*.

The effect of each move in our defined move set is detailed below:

1. *Exchange.* Applicable iff the color class of either endpoint is included in the other (or equal) and at least one of the endpoint vertices has a color class of cardinality 2: change the color of the edge to the other color in the endpoints' color classes.

$$\forall e = (v, v') \text{ if } \{col\} = cc(v) \cup cc(v') \setminus \{e_{color}\} : e_{color} \leftarrow col$$

2. *Connect.* Applicable iff the color classes of the endpoint vertices are both of cardinality 2 and not equal: repaint the edge color, as well as all the edges using the other two colors in the respective endpoints' color classes with a new, unified color.

$$\forall e = (v, v') if \{col_1, col_2\} = cc(v) \cup cc(v') \setminus \{e_{color}\} :$$

$$e_{color} \leftarrow col_{new}, \forall e' \in \left(\bigcup_{\sigma(e_1)=col_1} e_1 \right) \cup \left(\bigcup_{\sigma(e_2)=col_2} e_2 \right) : e'_{color} \leftarrow col_{new}$$

3. *Create.* Applicable iff the endpoints both have color classes of cardinality 1: assign a new color to the edge.

$$\forall e = (v, v') \text{ if } \emptyset = cc(v) \cup cc(v') \setminus \{e_{color}\} : e_{color} \leftarrow col_{new}$$

4. *Merge.* Essentially an operation that recolors two neighboring colored components with a new color. For consistency it is defined as operating on an edge like the other moves.

Theorem 2. *The move set defined above can only produce 2-colorings.*

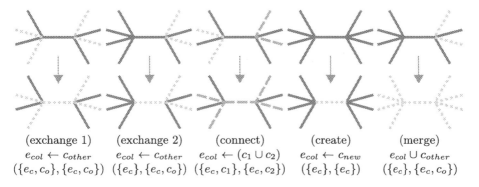

(exchange 1)	(exchange 2)	(connect)	(create)	(merge)
$e_{col} \leftarrow c_{other}$	$e_{col} \leftarrow c_{other}$	$e_{col} \leftarrow (c_1 \cup c_2)$	$e_{col} \leftarrow c_{new}$	$e_{col} \cup c_{other}$
$(\{e_c, c_o\}, \{e_c, c_o\})$	$(\{e_c\}, \{e_c, c_o\})$	$(\{e_c, c_1\}, \{e_c, c_2\})$	$(\{e_c\}, \{e_c\})$	$(\{e_c\}, \{e_c, c_o\})$

Fig. 1. Illustration of the considered move set in our local search algorithms. The central horizontal edge in all scenarios is the one that considers changing its color (e_{col}). The operation may affect the color of edges other than the horizontal one, as in *merge* and *connect*. Here, the $\cup$ operator stands for unifying two colors. The bottom row shows the vertex color classes where the move is applicable. (Color figure online)

Proof. All of the moves change the edge color and never add a third color to the edge endpoint vertices' color classes. We can observe that:

1. *Exchange case 1* does not add a new color to either color class. At most it can remove one from either or both.
2. *Exchange case 2* can at most add a color to a color class of cardinality 1.
3. *Connect* modifies the colors in the color classes but they remain of cardinality 2 (or may decrease). Other affected edges maintain their color class cardinality (or may decrease).
4. *Create* adds a color to color classes of cardinality 1. A color class may remain of cardinality 1 if the respective endpoint has degree 1.
5. *Merge* produces color classes of cardinality 1. Other affected edges maintain their color class cardinality (or may decrease).

Therefore any move applied on a 2-coloring will produce a 2-coloring.   □

Notation 2. We refer to an edge as being *color critical* if by removing this edge from a subgraph containing all of the edges that share its color will result in the number of connected components increasing in that subgraph.

In our algorithms, recoloring of a color critical edge will cause one of the resulting connected components to be colored with a new color.

Suppose that we have a solution ω and we operate on an edge which is colored with the most frequent color. Then, the moves defined above will affect $f(\omega)$, our objective function, in the following way:

1. *Exchange* will produce $f(\omega') \leftarrow f(\omega) - 1$ if the edge is not color critical. Otherwise the objective function can decrease by more than 1. However, if the other color present in the endpoints' color classes has the same frequency as the one on the edge we operate on, then $f(\omega') \leftarrow f(\omega) + 1$.
2. *Connect* will produce $f(\omega') \leftarrow f(\omega) - 1$ if the edge is not color critical. Otherwise the objective function can decrease by more than 1. However, if the sum of the frequencies of the other two colors in the endpoints' color classes is equal to or exceeds that of the edge we operate on, then $f(\omega')$ will increase by 1 or more.
3. *Create* will produce $f(\omega') \leftarrow f(\omega) - 1$ if the edge is not color critical. Otherwise the objective function can decrease by more than 1.
4. *Merge* will affect $f(\omega')$ if the number of colored edges that use the new color exceeds the previous objective function value. The new $f(\omega')$ can be no more than twice $f(\omega)$, just like with the *connect* case.

The cases when the moves affect a color critical edge of the most frequent color are tricky: to properly calculate the impact on the objective function one needs to perform for example a depth first search across all the neighboring edges to update the objective function. When we want to explore the entire neighborhood of a solution this becomes computationally expensive as we need to perform depth first searches for all edges in $O(|E|(|V| + |E|)$ for all iterations of our local search algorithms.

To avoid this, one can use probabilistic sampling of the neighborhood. In our implementations we prefer to discard this computation entirely, as it is certain that the objective function is decreased by at least 1 with all of the moves if we are careful about avoiding the special cases that worsen the value.

However, if the objective function can only decrease by 1 in all cases, there will not be sufficient information to drive the search to good solutions. As such, we use an auxiliary objective function in terms of defining an attractiveness value for each of the moves.

Notation 3. In the following we denote $count(c)$ to represent the number of edges that are colored with color $c \in C$. Formally, $count(c) = |\{e \in E | \sigma(e) = c\}|$.

Next, we define the attractiveness for each move (which must be *maximized*):

1. $att_{exchange}(e, \omega) = b_1 + w_1 \cdot count(e_{color}) \cdot \frac{f(\omega) - count(col_{other})}{f(\omega)}$
2. $att_{connect}(e, \omega) = b_2 + w_2 \cdot count(e_{color}) \cdot \frac{f(\omega) - count(col_1) - count(col_2) - 1}{f(\omega)}$
3. $att_{create}(e, \omega) = b_3 + w_3 \cdot count(e_{color}) \cdot \frac{1}{f(\omega)}$
4. $att_{merge}(e, \omega) = b_4 + w_4 \cdot \frac{f(\omega) - count(e_{color}) - count(col_{other})}{f(\omega)}$

The constants b_i, w_i are used for the fine tuning of the attractiveness values. Observe that for *connect* and *merge* the fraction part of the attractiveness will be 0 when the newly colored components reach exactly the size of $f(\omega)$ and negative if they exceed $f(\omega)$ (and thus worsen the new objective function value).

With all of the above we have all the elements required to build a simple hill climbing algorithm to approximate min-max 2-coloring by choosing the most attractive move at each iteration. We may improve upon this algorithm by using metaheuristics for local search such as simulated annealing and tabu search.

3.1   Simulated Annealing Algorithm (Algorithm 2)

Algorithm 2. input: graph $G = (V, E)$

---
1 : Let the working solution ω to be some initial 2-coloring.
2 : Set up some initial starting temperature of the annealing system: $temp \leftarrow temp_{initial}$
3 : Initialize $e_{chosen} \leftarrow nil$, $att_{chosen} \leftarrow 0$
4 : Cycle through edges $e \in E$:
  - (accept improving moves always:)
    if $att(e) > att_{chosen}$: $e_{chosen} \leftarrow e$, $att_{chosen} \leftarrow att(e)$
  - (accept worsening moves with temperature dependent probability:)
    else if $uniform(0, 1) < exp(\frac{att(e) - att_{chosen}}{temp})$: $e_{chosen} \leftarrow e$, $att_{chosen} \leftarrow att(e)$
5 : Perform the move on the working solution: $\omega \leftarrow move(e_{chosen}, \omega)$
6 : $temp \leftarrow temperature\_decrease\_schedule(temp)$
7 : Evaluate stopping criteria. If one of the criteria is met terminate the algorithm and output ω. Otherwise, continue with step 3.

In the simulated annealing setup we select an initial temperature for our system and we may accept worsening moves to our working solution with a probability p. This probability is affected by the temperature at a particular iteration step of the algorithm and by the loss in move attractiveness of our worsening operation. Lowering temperature causes p to decrease, while moves with low attractiveness also cause a small probability of acceptance.

3.2   Tabu Search Algorithm (Algorithm 3)

Algorithm 3. input: graph $G = (V, E)$

---

1 : Let the working solution ω to be some initial 2-coloring.

2 : Set up the frequency list to contain 0 for all edges, set up $TabuList \leftarrow nil$.

3 : Initialize $e_{chosen} \leftarrow nil$, $att_{chosen} \leftarrow 0$

4 : Cycle through edges $e \in E, e \notin TabuList$:

if $att(e) - k \cdot frequency(e) > att_{chosen}$: $e_{chosen} \leftarrow e$, $att_{chosen} \leftarrow att(e) - k \cdot frequency(e)$

5 : Perform the move on the working solution: $\omega \leftarrow move(e_{chosen}, \omega)$

6 : $TabuList \leftarrow TabuList \cup \{e\}$, $frequency(e) \leftarrow frequency(e) + 1$

7 : Evaluate stopping criteria. If one of the criteria is met terminate the algorithm and output ω. Otherwise, continue with step 3.

To explore the neighborhood of a solution in a more intelligent way we can employ memory to prevent cycling and drive the search to less explored areas of the search space. To solve the problem using tabu search techniques we use:

- a simple tabu list providing short-term memory that disallows a move on any edge recently changed;
- a frequency list on edge moves providing long-term memory. The frequency of an edge e increases by 1 each time it is used in an *exchange* operation, and move attractiveness values receive a penalty of $-k \cdot frequency(e)$ for some selected constant k.

Theorem 3. *Algorithms 2 and 3 produce valid 2-colorings.*

Proof. The algorithms take a feasible solution and apply a move set that only results in feasible solutions (2-colorings). □

4   Experimental Results

4.1   Testing Dataset Details

Our testing dataset includes computer generated Unit Disk Graphs and Quasi-Unit Disk Graphs (prefixed with "udg" and "qudg", respectively in Table 1) that are traditionally used to model Wireless Mesh Networks: two nodes can communicate only if they are within transmission range of each other.

In the testing setup, these two aforementioned types of graphs were generated by deploying 100, 500 and 1000 vertices with uniformly distributed coordinates over a square with the side measuring 2500 units. The maximum transmission range parameter is specified as a suffix (e.g. *udg500.140* is a Unit Disk Graph with 500 vertices and transmission range 140). The algorithms were tested on the largest connected component in each graph. The vertex count, edge count and maximum degree of the test graphs are presented in Table 1. The transmission range for the Quasi-Unit Disk Graphs varies uniformly between 50% and 100% of the maximum specified range. They are generated with the same random seed as their UDG counterparts so that the layout is identical excepting the absence of some edges from the qUDG cases.

The rest of our testing dataset contains graphs that are not themselves modelling wireless networks, for the sake of a more thorough analysis. These graphs are a part of the dataset for the DIMACS graph vertex coloring benchmarks and their high connectivity proves to be quite a challenge for our local search edge-coloring algorithms. Note that the following graphs featured in our experimental result showcase are geometric graphs, which are more relevant to the Wireless Mesh Network topology: *dsjr500.1c*, *dsjr500.5*, *r250.5*, *r1000.1c*.

4.2   Algorithms 2 and 3 Implementation Details

Our implementations are based on the JGraphT Java Graph Library and are made publicly available by means of a GitHub repository [11].

In all our local search algorithms we employ a disjoint set forest structure to keep track of colors when we use the *merge* and *connect* moves. The *create* move draws a new color by incrementing a static counter. Every so often, we renumber the colors because all moves except *create* can cause colors to disappear from the coloring. The vertex color classes are maintained inside a hash map structure. After every iteration it is necessary to perform a depth first search to recolor a potential new connected component that becomes disconnected when an edge changes color. Every iteration takes total time $O(|E|\alpha(|V|))$.

In our *simulated annealing algorithm* we have selected for our cooling schedule the exponential cooling scheme $T' \leftarrow kT$, with $k < 1$, close to 1, as first proposed by Kirkpatrik et al. [5].

In our *tabu search algorithm* we use hash map structures to keep track of the tabu moves and quickly determine if a move is tabu or not.

4.3   Running Data

Our experiments for the min-max 2-coloring approximation algorithms are performed on a selection of Unit Disk Graphs with increasing vertex density and transmission range, as well as on DIMACS benchmark graphs.

To compare the local search algorithms in terms of the quality of the produced solution the time required to obtain it, we plot the value of the objective function for the current solution at each iteration step in Fig. 2. To compute the values,

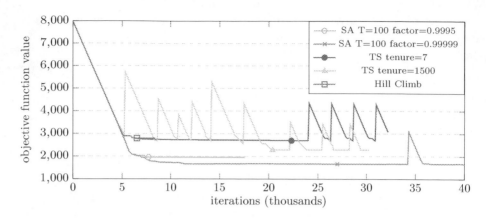

Fig. 2. A plot illustrating the quality of the incumbent solution at each iteration of the algorithms on a selected graph (7968 edges). Marks indicate best achieved solutions.

we start with an initial solution containing a single colored component and select graph *qudg1000.220* and the best combination of parameters we have discovered. The plot makes it easy to observe the linear drop in the objective function for 5000 units and between iterations 1 and 5000. This is a strong point of local search techniques as they exploit the gradient in the search space. Their weakness is that once they reach a local optimum it is hard to escape it as there are no more improving moves to be considered. Simulated annealing approaches this problem by adding randomness to the moves that are selected and we can see the result in the quality of the found solutions. Tabu search will run out of improving moves and will attempt worsening ones to escape the local optimum. The spikes in the objective value function correspond to applying the *merge* move which sometimes almost doubles the last best value.

For our local search algorithms we choose the initial solution to be either the solution given by Algorithm 1 or a single colored component. We stop the algorithms when they fail to produce an improvement for a set number of iterations.

We compare the algorithms with solutions obtained by running an ILP solver (Gurobi) on the linear program formulation from [7]. To obtain these solutions we had to limit the running time of the solver (3 h) and the maximum number of allowed colors in the linear program (which in turn decreases the number of variables). As such, the obtained linear program solutions are not the optimum solution and instead are an upper bound for each min-max 2-coloring on the respective graph.

Finally, we present the running data for our algorithms on a selection of graphs in Table 1. The results are encouraging for Unit Disk Graphs and their variants. Our tabu search heuristic applied to the solution of the BFS algorithm seems to consistently yield good results by improving (decreasing) the objective by up to 37% (21% on average). For some graphs, the BFS-inspired algorithm seems to create a harder to escape local optimum for the local search heuristic algorithms. This is where simulated annealing produces the best results starting from a blank (single color) initial solution.

Table 1. Algorithm running data. The first four columns display graph name, vertex count, edge count and maximum vertex degree. The next three columns represent the objective function value obtained by executing hill climbing, simulated annealing and tabu search with a blank (single color) initial solution. The following column is the solution for the BFS-based algorithm and then the solutions for the local search algorithms now starting with it as an initial solution. The last column gives an upper bound for the value of the optimum solution. The best solutions are highlighted.

Graph	$|V|$	$|E|$	Deg	HC	SA	TS	BFS	HC$'$	SA$'$	TS$'$	ILP
udg100.400	100	347	12	46	40	32	33	30	**27**	**27**	22
udg100.600	100	694	22	177	156	131	140	124	**113**	**113**	86
qudg100.400	100	232	9	29	19	19	21	19	**17**	**17**	12
qudg100.600	100	525	18	104	80	79	88	82	66	**64**	56
udg500.140	357	893	12	36	33	35	35	32	**28**	**28**	23
udg500.180	499	1862	16	198	156	120	77	70	59	**54**	63
udg500.220	500	2776	22	840	402	834	195	190	190	**148**	127
udg1000.140	1000	4641	20	359	173	183	163	141	133	**102**	—
udg1000.180	1000	7592	28	2218	1073	2218	579	579	579	**478**	—
udg1000.220	1000	11152	39	4132	3443	3854	**1058**	**1058**	**1058**	**1058**	—
qudg500.140	108	198	9	17	**12**	16	15	**12**	**12**	**12**	10
qudg500.180	480	1281	13	66	52	45	48	34	38	**31**	29
qudg500.220	500	1965	17	251	219	100	90	79	67	**64**	—
qudg1000.140	998	3305	14	120	135	78	78	65	60	**56**	—
qudg1000.180	1000	5427	21	614	542	586	295	255	220	**215**	—
qudg1000.220	1000	7968	31	2601	1727	2601	623	619	619	**614**	—
dsjc250.5	250	15668	147	7834	**7182**	7193	10148	7625	7625	7625	5234
dsjc500.1	500	12458	68	6123	**5558**	6123	9162	6084	6084	6084	5824
dsjc500.5	500	62624	286	31115	**30089**	31115	42946	30766	30766	33946	—
dsjr500.5	500	58862	388	29326	**28653**	29326	28724	28704	28704	28704	—
flat300_28_0	300	21695	162	10586	10781	10586	14604	**10551**	**10551**	**10551**	—
le450_25c	450	17343	179	8214	8416	8214	8614	8614	7549	**7286**	5781
le450_25d	450	17425	157	8339	7763	8339	8667	8667	7484	**7154**	5952
r250.5	250	14849	191	7425	6530	6806	7321	7321	**5813**	**5813**	4950

5   Conclusions and Future Work

The newly designed algorithms for the 2-coloring min-max problem offer a practical method of obtaining good solutions without resorting to more time consuming exact methods.

More techniques to approach the problem may be used, such as recombination heuristics. An idea is to attempt to find some coding for graphs with colored edges suitable for solving 2-coloring by using a genetic algorithm approach.

It would be interesting to find a constant factor approximation algorithm for min-max edge q-coloring.

References

1. Adamaszek, A., Popa, A.: Approximation and hardness results for the maximum edge q-coloring problem. In: Cheong, O., Chwa, K.-Y., Park, K. (eds.) ISAAC 2010. LNCS, vol. 6507, pp. 132–143. Springer, Heidelberg (2010). https://doi.org/10.1007/978-3-642-17514-5_12
2. Feng, W., Chen, P., Zhang, B.: Approximate maximum edge coloring within factor 2: a further analysis. In: ISORA, pp. 182–189 (2008)
3. Feng, W., Zhang, L., Qu, W., Wang, H.: Approximation Algorithms for maximum edge coloring problem. In: Cai, J.-Y., Cooper, S.B., Zhu, H. (eds.) TAMC 2007. LNCS, vol. 4484, pp. 646–658. Springer, Heidelberg (2007). https://doi.org/10.1007/978-3-540-72504-6_59
4. Feng, W., Zhang, L., Wang, H.: Approximation algorithm for maximum edge coloring. Theor. Comput. Sci. **410**(11), 1022–1029 (2009)
5. Kirkpatrick, S., Gelatt, C.D., Vecchi, M.P., et al.: Optimization by simulated annealing. Science **220**(4598), 671–680 (1983)
6. Kyasanur, P., Vaidya, N.H.: Routing and interface assignment in multi-channel multi-interface wireless networks. In: 2005 IEEE Wireless Communications and Networking Conference, vol. 4, pp. 2051–2056. IEEE (2005)
7. Larjomaa, T.: Improving bandwidth in wireless mesh networks. Master's thesis, School of Electrical Engineering, Aalto University, February 2013
8. Larjomaa, T., Popa, A.: The min-max edge q-coloring problem. In: Kratochvíl, J., Miller, M., Froncek, D. (eds.) IWOCA 2014. LNCS, vol. 8986, pp. 226–237. Springer, Cham (2015). https://doi.org/10.1007/978-3-319-19315-1_20
9. Larjomaa, T., Popa, A.: The min-max edge q-coloring problem. J. Graph Algorithms Appl. **19**(1), 507–528 (2015)
10. Lipton, R., Tarjan, R.: Applications of a planar separator theorem. SIAM J. Comput. **9**(3), 615–627 (1980)
11. Mincu, R.S.: Java implementation of heuristic algorithms for the maximum and min-max 2-coloring problems (2017). https://github.com/radusm/minmax
12. Muir, A., Garcia-Luma-Aceves, J.J.: A channel access protocol for multihop wireless networks with multiple channels. In: ICC 1998, vol. 3, pp. 1617–1621, June 1998
13. Raniwala, A., Chiueh, T.: Architecture and algorithms for an IEEE 802.11-based multi-channel wireless mesh network. In: INFOCOM, pp. 2223–2234 (2005)
14. Raniwala, A., Gopalan, K., Chiueh, T.: Centralized channel assignment and routing algorithms for multi-channel wireless mesh networks. Mobile Comput. Commun. Rev. **8**(2), 50–65 (2004)
15. So, J., Vaidya, N.H.: Multi-channel MAC for ad hoc networks: handling multi-channel hidden terminals using a single transceiver. In: MobiHoc 2004, pp. 222–233. ACM (2004)

Geometric Spanners in the MapReduce Model

Sepideh Aghamolaei[1]([⊠]), Fatemeh Baharifard[2], and Mohammad Ghodsi[1,2]

[1] Department of Computer Engineering,
Sharif University of Technology, Tehran, Iran
`aghamolaei@ce.sharif.edu, ghodsi@sharif.edu`
[2] School of Computer Science, Institute for Research in Fundamental Sciences (IPM),
Tehran, Iran
`f.baharifard@ipm.ir`

Abstract. A *geometric spanner* on a point set is a sparse graph that approximates the Euclidean distances between all pairs of points in the point set. Here, we intend to construct a geometric spanner for a massive point set, using a distributed algorithm on parallel machines. In particular, we use the MapReduce model of computation to construct spanners in several rounds with inter-communications in between. An algorithm in this model is called efficient if it uses a sublinear number of machines and runs in a polylogarithmic number of rounds. In this paper, we propose an efficient MapReduce algorithm for constructing a geometric spanner in a constant number of rounds, using linear amount of communication. The stretch factors of our spanner is $1 + \epsilon$, for any $\epsilon > 0$.

Keywords: Computational geometry · Parallel computation
Geometric spanners · MapReduce

1 Introduction

Space limitations are the main challenge in processing massive data i.e. data that do not fit inside the memory of a single machine. Given a bounded memory, an efficient algorithm has a low time complexity. Some space-bounded models allow a type of secondary slower memory or communication between multiple fast memories to reduce the running time of the algorithm. Allowing two types of memory, shifts the challenge in the algorithm design to data communication.

MapReduce is a framework for processing data in large scales in which a set of machines, each have a part of the input, run an algorithm in simultaneous rounds and after each round, they can communicate their data to each other. Efficient MapReduce algorithms have sublinear machines each with sublinear memory that run for polylogarithmic number of rounds and the number of machines used in the algorithm must be asymptotically as many as the input size.

An example of problems that has been discussed in MapReduce framework is Euclidean minimum spanning tree problem, which was studied by Andoni

© Springer International Publishing AG, part of Springer Nature 2018
L. Wang and D. Zhu (Eds.): COCOON 2018, LNCS 10976, pp. 675–687, 2018.
https://doi.org/10.1007/978-3-319-94776-1_56

et al. [3] who presented an algorithm with $O(1)$ round complexity and superlinear memory. Later, Yaroslavtsev and Vadapalli [23] proved lower bounds for this problem and proposed an algorithm for approximating each edge of the minimum spanning tree. Another example is computing the core-set for convex hull using the method proposed in [2], which also works in MapReduce model. Moreover, fixed-dimensional linear programming, 1-dimensional all nearest neighbors, 2-dimensional and 3-dimensional convex hull algorithms were solved in memory-bound MapReduce model [13] and practically proven algorithms for sky-line computation, merging two polygons, diameter and closest pair problems have been discussed in MapReduce model [8,10].

A network is called a *t-spanner*, if there is a short path between any pairs of nodes, within a guaranteed ratio t to the shortest paths between those nodes in an underlying base graph. Most of the time and in this paper, the complete graph which has $\theta(n^2)$ edges is considered as the underlying base graph.

Most efficient spanner construction algorithms are geometric and they find practical applications in areas such as terrain construction [12,21], metric space searching [19], broadcasting in communication networks [11] and solving approximately geometric problems like traveling salesman problem [20].

Recently, a divide and conquer algorithm for constructing geometric spanners has been studied as spanners merging problem in [4]. The size of the spanner created using this method is $O(n \log n)$, which requires $O(\log n)$ times more memory than the input. However, the proposed algorithm uses linear memory in its final merging steps, which is infeasible for MapReduce model. Moreover, in [6,14] the problems of well-separated pair decomposition on PRAM and 3D covex hull in MapReduce model were considered. Using the lifting transformation, 3D convex hull solves planar Delaunay triangulation. Using simulation of PRAM algorithms in MapReduce as discussed in [14] gives algorithms for the two spanners of [4,6]. Direct algorithms for Delaunay triangulation in MapReduce [5,18] are randomized and require a super-linear number of machines unlike the PRAM simulation. A summary of results on geometric spanners in MapReduce is shown in Table 1.

Contributions. Table 1 compares the previous results for geometric spanners with the one given in this paper.

Table 1. A summary of results on geometric spanners in MapReduce. $\frac{1}{\delta} = \log_m^n$, where m is the memory of each machine.

Spanner	$|E|$	Stretch factor	Rounds	Communication	Reference
WSPD	$O(n)$	$1 + \epsilon$	$O(\frac{\log n}{\delta})$	$O(\frac{n}{\delta \log n})$	Simulation [14], PRAM [6]
DT	$O(n)$	1.998	$O(\frac{1}{\delta})$	$O(\frac{n}{\delta})$	3D convex hull [14,22]
-	$O(\frac{n}{\epsilon})$	$1 + \epsilon$	$O(\frac{1}{\delta})$	$O(\frac{n}{\delta})$	this paper (Algorithm 4)

In this paper, we propose efficient algorithms for constructing a geometric spanner similar to Yao-graph and a special case of dynamic programming in the MapReduce model. Our algorithms run for a constant number of rounds and use linear memory.

2   Preliminaries

In this section we review basic knowledge required to understand the rest of the paper.

2.1   MapReduce Model

Different theoretical models for MapReduce has been introduced over the years [9,14,16]. In MapReduce class (MRC) model, for an input of size n, the following three conditions must be satisfied:

- the number of machines is sublinear: $L = o(n)$
- the memory of each machine is sublinear: $m = o(n)$
- the number of rounds is polylogarithmic: $O(polylog(n))$.

In MRC model, the input of each round is distributed among machines. Let S_i be the part of the input assigned to machine i in each round. Data in MapReduce are stored as (key, value) pairs. A MapReduce algorithm consists of three steps: *map*, *shuffle* and *reduce*:

- *map*: processes data into a set of (key, value) pairs.
- *shuffle*: sends data with the same key to the same machine.
- *reduce*: aggregates data with the same key.

Operations *map* and *reduce* are local, while *shuffle* distributes data between machines.

Two main parallel algorithms operations are semi-group and prefix sum:

- Semi-group: $x_1 \oplus x_2 \oplus \cdots \oplus x_n$, i.e. for a set S and a binary operation $\oplus$: $S \times S \to S$, the associative property holds: $\forall a, b, c \in S, (a \oplus b) \oplus c = a \oplus (b \oplus c)$.
- Prefix sum: $x_1 \oplus x_2 \oplus \cdots \oplus x_i, i = 1, \ldots, n$
- Diminished prefix sum: $x_1 \oplus x_2 \oplus \cdots \oplus x_{i-1}, i = 1, \ldots, n$

Both of these operations also take $O(\log_m^n)$ rounds and $O(n \log_m^n)$ computation in MapReduce [14]. Parallel algorithms in CRCW PRAM model can be simulated in MapReduce model by a factor 2 slow-down [14]. A class of functions that can be computed with minimum round and communication complexity are known as MRC-parallelizable functions [16]. An example of a MRC-parallelizable functions is computing the frequency of words in a set of documents which is known as word count algorithm [7].

Special cases of dynamic programming have been discuessed in MapReduce [15] with $(1 + \epsilon)$-approximation factor, $O(1)$ rounds and $\tilde{O}(n)$ communication, where $\tilde{O}$ ommits a $polylog(n)$ factor. The required conditions for this type of dynamic programming are monotonicity and decomposability. Monotonicity states that the cost of subproblems is less than the main problem, and decomposability states that the input can be decomposed into two-level laminar family of partial inputs where group is the higher level and block is the lower level, such that the solution to the main problem is a concatenation of the solutions to the subproblems and a nearly optimal solution can be constructed from $O(1)$ blocks.

Algorithms for constructing range searching data structures in MapReduce exist [1]. However, they lack the efficiency required for simultaneous queries.

2.2   Geometric Spanners

A geometric network G is a *t-spanner* for a point set P, if a $t > 1$ exists such that for each pair of points u and v in P, there is a path in G between u and v, whose length is less than or equal to the t times of the Euclidean distance between u and v. The minimum t such that G is a t-spanner of P is the *spanning ratio* of G.

Many spanner algorithms exist which excel in different quality measures. Here, we describe an algorithm for constructing a t-spanner of a set of points in Euclidean space which constructs a spanner with a linear number of edges in $O(n \log n)$ time.

Yao-Graph. One of the most common spanners is Yao-graph which is denoted by Y_k-graph. The Y_k-graph is constructed as follows. Given a set P of points in the plane, for each vertex $p \in P$, partition the plane into k disjoint cones (regions in the plane between two rays originating from the same point) with apex p, each defined by two rays at consecutive multiples of $\theta = \frac{2\pi}{k}$ radians from the negative y-axis and label the cones $C_0(p)$ through $C_{k-1}(p)$, in counterclockwise order around p. Then for each cone with apex p, connect p to its closest vertex q inside that cone.

It is proven that for any θ with $0 < \theta < \pi/3$, the Yao-graph with cones of angle θ, is a t-spanner of P for $t = \frac{1}{1-2\sin(\theta/2)}$ with $O(\frac{n}{\theta})$ edges, that can be constructed in $O(\frac{n \log n}{\theta})$ time [17].

3   Mergeable Dynamic Programming

In Algorithm 1 we solve the special dynamic programming, which is defined below, in MapReduce model. Actually, the idea behind dynamic programming in MapReduce is similar to the parallel prefix sum in PRAM.

Definition 1 (Mergeable DP). *A dynamic program (DP) with input set S and output set Q with a recurrence relation $f : S^k \to Q$ and a table T with a valid filling order $\Phi : T \to \mathbb{N}$ and size $|T| = n^d$, and mappings between $S \leftrightarrow T$ and $Q \leftrightarrow T$, is a mergeable DP if the following three conditions hold:*

– *Sparsity: The number of cells of T required for computing Q is $O(|S|)$.*

- *Neighbors: Computing f on each block requires data only from $O(1)$ previous blocks (in the order of Φ).*
- *Order Preserving: The value of each cell must only depend on the cells with smaller or equal index based on Φ and the order of each dimension of the table T.*
- *Parallelizable: Function f must be a semi-group function.*
- *Summarizable: There is an integer ℓ sublinear in $|S|$ ($\ell = o(|S|)$), such that using the last ℓ values of T in the order of Φ, it is possible to compute the rest of the table, i.e. function h exists such that $T[\Phi^{-1}(k)] = h(T[\Phi^{-1}(k-1)], \ldots, T[\Phi^{-1}(k-\ell)], S[\Phi^{-1}(k)])$.*

The sparsity and summarizability of the DP table allow us to summarize the computed part of the table into a sublinear subset of cells, which we call a frontier. A formal definition of frontier is given in Definition 2.

Definition 2 (Frontier). *The frontier is a subset of cells along with their indices F (denoted by F.cells and F.indices) of a DP table T, such that cells with indices $R = \{E(F) + 1, \ldots, |T|\}$ can be computed using cells with indices F.indices $\cup R$, where $E(F) = \max_{f \in F.indices} f$.*

Definition 3 (Frontier Merging). *Given two frontiers a, b, frontier merging operation creates a frontier c which is a frontier for cells from 1 to $k = \max\{E(a), E(b)\}$. Build a hypotetical table T_X similar to T but only store data from set $X = a.cells \cup b.cells$. Fill the table T_X from cell 1 to cell k. Using the summarizabilty property of the mergeable DP with table T_X and ordering Φ, there is a sequence $T_X[\Phi^{-1}(k)], \ldots, T_X[\Phi^{-1}(k - \ell - 1)]$ for each cell $T_X[\Phi^{-1}(k)]$. Report this as c.*

In Lemma 1, we show that a frontier of a mergeable DP can be constructed using a (diminished) parallel prefix algorithm.

Lemma 1. *The operation of Definition 3 builds a frontier for $\cup_{i=k}^{k-\ell} T[\Phi_{-1}(i)]$ and it is a semi-group function.*

Proof. Since T is a mergeable DP, T_X is also a mergeable DP. For two sets A and B, using the semi-group property of f, computing T_A and T_B and applying f on their results in the order of Φ, gives $T_{A \cup B}$. For $A = \{T[1], \ldots, T[E(a)]\}$ and $B = \{T[1], \ldots, T[E(b)]\}$, the result is $A \cup B = \{T[1], \ldots, T[k]\}$, $k = \max(E(a), E(b)) = E(c)$.

Assume three frontiers a, b and c of a DP table in the order of Φ with the set of indices denoted by S_a, S_b and S_c, respectively. Now, we prove the two possible orders of computation result in the same result. Since the frontiers in the computation follow the order of Φ:

$$a \oplus b = T_{S_a \cup S_b}[E(b)] \Rightarrow (a \oplus b) \oplus c = T_{S_a \cup S_b \cup S_c}[E(c)]$$

The other case can be proven similarly:

$$b \oplus c = T_{S_b \cup S_c}[E(c)] \Rightarrow a \oplus (b \oplus c) = T_{S_a \cup S_b \cup S_c}[E(c)]$$

which proves the lemma. $\square$

Now, we give an algorithm (Algorithm 1) for solving the mergeable DP (Definition 1) problems in MapReduce.

Algorithm 1. Mergeable DP in MapReduce

Input: A mergeable DP (S, Q, f, T, Φ)
Output: The output of DP
 1: Compute a valid ordering Φ on S and distribute the points by Algorithm 2.
 2: Map each cell $S[\Phi^{-1}(i)]$ to a ℓ-tuple $(\emptyset, \ldots, \emptyset, S[\Phi^{-1}(i)])$.
 3: Run the diminished parallel prefix algorithm with the frontier merging as the operation on ℓ-tuples.
 4: Compute T by applying f on ℓ-tuples and S and store the result.
 5: Use f to update the local values of T to compute Q.
 6: **return** Q

Theorem 1. *Algorithm 1 solves mergeable dynamic programming problems correctly.*

Proof. By Definition 1, for a mergeable dynamic programming, the order (Φ) in which the table is filled can be determined. Using this order, Algorithm 2 finds partitions that are ordered based on the order of the dynamic program (Theorem 4). Then, a parallel prefix computation can be used for computing the frontier in each cell (Lemma 1) and the minimum of the related cells gives the value of the cell in T. □

Theorem 2. *Algorithm 1 takes $O(\log_m^n)$ rounds and it has $O(n \log_m^n)$ communication complexity in MRC model, if $\ell = O(\sqrt[d+1]{n})$ for a d-dimensional DP.*

Proof. The algorithm consists of a parallel prefix computation and running Algorithm 2 once. The round and communication complexity of Algorithm 2 is $O(\log_m^n)$ and $O(n \log_m^n)$ respectively. The round complexity of parallel prefix algorithm is $O(\log_m^n)$ and the communication complexity of the algorithm is $O(\ell.\ell^d \log_m^n)$, since instead of data with $O(1)$ dimension, vectors of length $O(\ell)$ have been used, and there are $O(\ell^d)$ cells.

Using Theorem 4, the number of points in each row and column is $O(m)$, so the overall communication and space for sending data from a row and a column is also $O(m)$. Based on Definition 1, the amount of data from other cells is $O(m)$. Therefore, all the steps of the algorithm can be run using $O(n)$ communication per round. □

4   Application: Geometric Spanners in MapReduce

In this section, we present an efficient algorithm for constructing a geometric spanner in the MapReduce model.

4.1   A Balanced Grid in MapReduce

We can store a grid by its separating lines. A partitioning based on a regular grid and an indexing scheme is used to distribute data among machines. Algorithm 2 takes as input a set of point-sets and an ordering scheme and builds a regular grid and a partitioning of that grid based on the given ordering.

In Algorithm 2, $\mathbf{Sort}_f(\mathbf{S})$ means sorting based on function f of points in S in MapReduce model. Sorting points means they are distributed among L machines such that $i < j \Rightarrow \forall a \in S_i, b \in S_j, f(a) \le f(b)$. Sorting n points in this model can be done in $O(\log_m^n)$ rounds [14], which is $O(\frac{1}{\delta})$ for $m = n^\delta$ where δ is a constant $(0 \le \delta \le 1)$.

Algorithm 2. A Regular Grid in MapReduce

Input: a set of points set $S = \{S_1, \cdots, S_L\}$, an ordering function $f(.,.)$
Output: a space partitioning of S, the grid lines
  1: $\mathrm{Sort}_x(S)$
  2: $X_i \leftarrow \min_{(x,y)\in\cup_i S_i} x$
  3: send x_i to all other machines $(X = \cup_i X_i)$
  4: $\mathrm{Sort}_y(S)$
  5: $Y_i \leftarrow \min_{(x,y)\in\cup_i S_i} y$
  6: send Y_i to all other machines $(Y = \cup_i Y_i)$
  7: locally compute the index of each cell (i,j) using the ordering function $f(i,j)$
  8: $\mathrm{Sort}_{f(x,y)}(S)$
  9: Re-index the sets in the order of $\min_{(x,y)\in S_i} f(x,y)$
 10: **return** $S = \{S_1, \cdots, S_L\}$, X, Y

Here, we present some properties of Algorithm 2, which are used later.

Lemma 2. *The number of points in each cell of the grid in Algorithm 2 is at most m.*

Proof. Based on the sorting on x, the number of points between X_i and X_{i+1} is $O(m)$. Similarly, the number of points between Y_i and Y_{i+1} is $O(m)$. So, in the grid built on $X \times Y$, the number of points in each cell is $O(m)$. By indexing the points based on $f(x,y)$, the partitions lie inside cells of $X \times Y$. Since all equal keys in a MapReduce computation go to the same machine, there are no half-cells.

Theorem 3. *The round complexity of Algorithm 2 is $O(\log_m^n)$ and its communication complexity is $O(n \log_m^n)$.*

Proof. Each sorting takes $O(\log_m^n)$ rounds, which is constant for $m = O(n^\delta)$ and $O(\log_m^n)$ communication. Since the round and communication complexities of the algorithm are the sum of the complexities of these sorting steps, they are $O(\log_m^n)$ and $O(n \log_m^n)$ respectively.   □

Theorem 4. *There are at most $O(L)$ partitions in the output of Algorithm 2, each with $O(m)$ points.*

Proof. The last sorting step of the algorithm which sorts points based on the value of f, divides points into sets $S_1, \ldots, S_L$ such that $\forall (x, y) \in S_i, (x', y') \in S_j, i < j \Rightarrow f(x, y) \leq f(x', y')$. The size of the sets created using a sorting algorithm is $O(m)$ (Lemma 2) and the number of sets is $O(L)$, so the number of partitions is $O(L)$ and each of them has $O(m)$ points. $\square$

4.2   Simultaneous 2-Sided Queries

Now we present an offline algorithm for solving the 2-sided range queries in Algorithm 3, which can be extended to MapReduce in the same way as parallel prefix computation.

In Algorithm 3, the closest point to (X_i, Y_j) using distance function ℓ_1 that lies inside the 2-sided range $(x \leq X_i, y \leq Y_j)$ is computed. Also, $\|x\|_1$ denotes the length of vector x under ℓ_1.

Algorithm 3. Simultaneous 2-Sided Queries

Input: A point set S, a rectangular grid $\{X_i\}_{i=1}^{\ell} \times \{Y_j\}_{i=1}^{\ell}$
Output: The nearest neighbor to (X_i, Y_j) using points of S inside the 2-sided range
1: Run Algorithm 2 to build a $\ell \times \ell$ table T and index it using $\Phi(x, y) = x + y \times \ell$.
2: Run   Algorithm   1   with S as   point   set, T as   table, $T[i, j]$ $=$ $\arg \min_{t \in S[i,j] \cup T[i-1,j] \cup T[i,j-1]} \|t - (X_i, Y_j)\|_1$ as f and Φ as the ordering.
3: **return** T

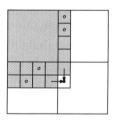

Fig. 1. For a 2D DP table which is filled in 2 directions (down and right), the data of the same row and column cells are needed in addition to C_i of the previous cells.

Lemma 3. *Algorithm 3 is a mergeable dynamic programming.*

Proof. The algorithm is mergeable since it satisfies the conditions of mergeable dynamic programming as defined in Definition 1:

– Sparsity: The number of cells required to answer 2-sided queries is $O(n)$, since we only need to know the value of cells which contain a point.

- Neighbors: The value of a cell (i, j) can be determined using the values of cells $T[i-1, j], T[i, j-1], S[i, j]$.
- Parallelizable: Minimum distance to the grid point corresponding to the corner of cell $[i, j]$ is a semi-group function, since minimum computations are semi-group.
- Summarizable: The anti-diagonal of T that passes through each cell is the frontier of that cell under 2-sided queries. So this problem is summarizable for $\ell = n$. □

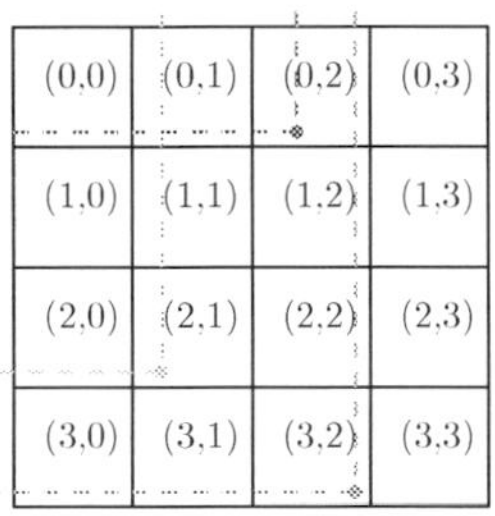

Fig. 2. Three 2-sided range queries for computing 3 nearest neighbors.

Lemma 4. *The nearest neighbor of a point inside a 2-sided range using ℓ_1 distance can be computed using Algorithm 3.*

Proof. Each 2-sided range query on the grid (Fig. 2), can be computed using Algorithm 1 for computing the recurrence relation. The nearest point to (X_i, Y_j) using ℓ_1 distance is either in cell $S[i, j]$ or in one of its neighbors: $T[i-1, j]$, $T[i, j-1]$ (Fig. 1). Since the algorithm checks all these values, it finds the exact nearest neighbor. □

4.3 A Geometric Spanner in MapReduce

To build a spanner similar to Yao-graph, we first solve simultaneous 2-sided range queries, using dynamic programming. These queries are then used in the spanner algorithm to find an approximate nearest neighbor in each cone.

Our algorithm for constructing a spanner (Algorithm 4) creates a grid and applies nearest neighbor search to find the edges. Algorithm 4 creates a set of oriented rhomboid grids with lines parallel to the ones creating cones around each point, as shown in Fig. 3.

In Algorithm 4, the distances computed in Algorithm 3 are ℓ_1 distances of the affine transformations of a grid, as shown in Fig. 4. Lemma 5 computes the approximation factor between this distance and the Euclidean distance.

Lemma 5. *The distance between two points on a grid with unit vectors that have an angle θ $(0 \le \theta \le \frac{\pi}{2})$ between them, is $1 + O(\theta)$ times the Euclidean distance between those points.*

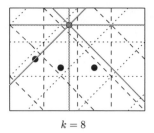

$$k = 8$$

Fig. 3. The overlay of two out of $k = 8$ square grids of Algorithm 4, indicated by dotted lines and dashed lines. The bold red lines denote the cones around a point. (Color figure online)

Algorithm 4. A Geometric Spanner in MapReduce

Input: A set of points set $S = \{S_1, \ldots, S_L\}$, an integer $k \geq 7$
Output: A spanner with k cones around each point
1: $\Theta = \{\frac{2\pi i}{k} | i = 1, \ldots, k\}$
2: locally create pairs of consecutive directions from Θ in clockwise order.
3: locally build a grid for each pair of directions from previous step.
4: repeat each point p once in each grid.
5: run Algorithm 3 with $d(.,.)$ defined as the ℓ_1 distance and $\cup_i S_i$ as the point set, using Algorithm 1 in each grid to find the nearest neighbor of each point inside its cone.
6: add an edge between each point and one of its nearest neighbors in each direction (cone).
7: **return** the edges of the spanner.

Proof. Assume w.l.o.g. that one of the points is $(0,0)$ and the other one is $p = (x, y), x, y > 0$. The angle between $\overrightarrow{op}$ and $\overrightarrow{i}$ is $\alpha \leq \theta$, since p lies inside the cone. Using basic trigonometry, the distance computed in our algorithm is $x + y$ and the Euclidean distance between these points is $\frac{y\sin(\theta)}{\sin(\alpha)}$. Also, $x = \cos(\alpha)\frac{y\sin(\theta)}{\sin(\alpha)} - y\cos(\theta)$. So, using Taylor series for cosine and Maclaurin series for $\frac{1}{1-X}$, the approximation factor is proved:

$$\frac{x+y}{|\overrightarrow{op}|} = \frac{\cos(\alpha)\frac{y\sin(\theta)}{\sin(\alpha)} - y\cos(\theta) + y}{\frac{y\sin(\theta)}{\sin(\alpha)}}$$

$$= \frac{y\cos(\alpha)\sin(\alpha) - y\cos(\theta)\sin(\alpha) + y\sin(\alpha)}{y\sin(\theta)} = \frac{\sin(\theta - \alpha) + \sin(\alpha)}{\sin(\theta)}$$

$$= \frac{2\sin(\frac{\theta}{2})\cos(\frac{\theta - 2\alpha}{2})}{2\sin(\frac{\theta}{2})\cos(\frac{\theta}{2})} = \frac{\cos(\frac{\theta - 2\alpha}{2})}{\cos(\frac{\theta}{2})} \leq \frac{1}{1 - \frac{\theta^2}{8}} = 1 + \frac{\theta^2}{8} + o(\theta^2).$$

$\square$

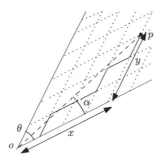

Fig. 4. A grid built inside a cone and the path on the grid compared to the distance between the point inside the cone and the apex.

Theorem 5. *The stretch factor of the spanner of Algorithm 4 is $1 + O(\theta)$.*

Proof. Applying Lemma 5 proves using ℓ_1 distance instead of the Euclidean distance (Algorithm 4) adds a factor $1 + O(\theta^2)$, and Lemma 4 proves Algorithm 3 computes the exact ℓ_1 distance. Using induction on the length of the path, similar to the proof of the stretch factor of Yao-graph, proves the approximation factor. ☐

Theorem 6. *Algorithm 4 has $O(k \log_m^n)$ round complexity and $O(nk \log_m^n)$ communication complexity.*

Proof. The algorithm solves one instance of range query per cone to compute the nearest neighbors simultaneously. Based on Lemma 3, Algorithm 3 is a mergeable dynamic program, and using Theorem 1, it takes $O(\log_m^n)$ rounds and $O(n \log_m^n)$ communications to solve it. Since there are k cones, the overall complexity of the algorithm is $O(k \log_m^n)$ rounds and $O(kn \log_m^n)$ communication. ☐

5   Conclusion

We introduced a $(1 + \epsilon)$-spanner in Euclidean plane and presented a MRC algorithm for constructing it in optimal round and communication complexities. However, the number of machines used in our algorithm is sub-quadratic. Finding algorithms that use fewer machines and algorithms for spanners with other geometric properties such as bounded degree spanners and bounded diameter spanners are also important.

We also proved conditions for parallelizable dynamic programming problems. Solving other problems in MapReduce using our method might also be interesting. Also, finding algorithms for other simultaneous queries can reduce the complexity of some MapReduce algorithms.

References

1. Agarwal, P.K., Fox, K., Munagala, K., Nath, A.: Parallel algorithms for constructing range and nearest-neighbor searching data structures. In: Proceedings of the 35th ACM SIGMOD-SIGACT-SIGAI Symposium on Principles of Database Systems, pp. 429–440. ACM (2016)
2. Agarwal, P.K., Har-Peled, S., Varadarajan, K.R.: Geometric approximation via coresets. Comb. Comput. Geom. **52**, 1–30 (2005)
3. Andoni, A., Nikolov, A., Onak, K., Yaroslavtsev, G.: Parallel algorithms for geometric graph problems. In: Proceedings of the 46th Annual ACM Symposium on Theory of Computing, pp. 574–583 (2014)
4. Bakhshesh, D., Farshi, M.: Geometric spanners merging and its applications. In: Proceedings of the 28th Canadian Conference on Computational Geometry, pp. 133–139 (2016)
5. Birn, M., Osipov, V., Sanders, P., Schulz, C., Sitchinava, N.: Efficient parallel and external matching. In: Wolf, F., Mohr, B., an Mey, D. (eds.) Euro-Par 2013. LNCS, vol. 8097, pp. 659–670. Springer, Heidelberg (2013). https://doi.org/10.1007/978-3-642-40047-6_66
6. Callahan, P.B.: Dealing with higher dimensions: the well-separated pair decomposition and its applications. Ph.D. thesis, Johns Hopkins University (1995)
7. Dean, J., Ghemawat, S.: MapReduce: simplified data processing on large clusters. Commun. ACM **51**, 107–113 (2008)
8. Eldawy, A., Li, Y., Mokbel, M.F., Janardan, R.: CG_Hadoop: computational geometry in MapReduce. In: Proceedings 21st ACM SIGSPATIAL International Conference on Advances in Geographic Information Systems, pp. 294–303 (2013)
9. Eldawy, A., Mokbel, M.F.: Communication steps for parallel query processing. In: Proceedings of the 32nd ACM SIGMOD-SIGACT-SIGAI Symposium on Principles of Database Systems, pp. 273–284 (2013)
10. Eldawy, A., Mokbel, M.F.: SpatialHadoop: A MapReduce framework for spatial data. In: Proceedings of the 31st International Conference on Data Engineering, pp. 1352–1363 (2015)
11. Farley, A.M., Proskurowski, A., Zappala, D., Windisch, K.: Spanners and message distribution in networks. Discrete Appl. Math. **137**, 159–171 (2004)
12. Ghodsi, M., Sack, J.: A coarse grained solution to parallel terrain simplification. In: Proceedings of 10th Canadian Conference on Computational Geometry (1998)
13. Goodrich, M.T.: Simulating parallel algorithms in the MapReduce framework with applications to parallel computational geometry. arXiv preprint arXiv:1004.4708 (2010)
14. Goodrich, M.T., Sitchinava, N., Zhang, Q.: Sorting, searching, and simulation in the MapReduce framework. In: Proceedings of the 22nd Annual International Symposium on Algorithms and Computation, pp. 374–383 (2011)
15. Im, S., Moseley, B., Sun, X.: Efficient massively parallel methods for dynamic programming. In: Proceedings of the 49th Annual ACM SIGACT Symposium on Theory of Computing, pp. 798–811. ACM (2017)
16. Karloff, H., Suri, S., Vassilvitskii, S.: A model of computation for MapReduce. In: Proceedings of the 21st ACM-SIAM Symposium on Discrete Algorithms, pp. 938–948 (2010)
17. Narasimhan, G., Smid, M.: Geometric Spanner Networks. Cambridge University Press, Cambridge (2007)

18. Nath, A., Fox, K., Munagala, K., Agarwal, P.K.: Massively parallel algorithms for computing TIN DEMs and contour trees for large terrains. In: Proceedings of the 24th ACM SIGSPATIAL International Conference on Advances in Geographic Information Systems (2016)

19. Navarro, G., Paredes, R., Chávez, E.: t-spanners as a data structure for metric space searching. In: Laender, A.H.F., Oliveira, A.L. (eds.) SPIRE 2002. LNCS, vol. 2476, pp. 298–309. Springer, Heidelberg (2002). https://doi.org/10.1007/3-540-45735-6_26

20. Rao, S.B., Smith, W.D.: Approximating geometrical graphs via "spanners" and "banyans". In: Proceedings of the 30th Annual ACM Symposium on Theory of Computing, pp. 540–550 (1998)

21. van Kreveld, M.: Algorithms for triangulated terrains. In: Plášil, F., Jeffery, K.G. (eds.) SOFSEM 1997. LNCS, vol. 1338, pp. 19–36. Springer, Heidelberg (1997). https://doi.org/10.1007/3-540-63774-5_95

22. Xia, G.: The stretch factor of the Delaunay triangulation is less than 1.998. SIAM J. Comput. **42**, 1620–1659 (2013)

23. Yaroslavtsev, G., Vadapalli, A.: Massively parallel algorithms and hardness for single-linkage clustering under ℓ_p-distances. arXiv preprint arXiv:1710.01431 (2017)

SDP Primal-Dual Approximation Algorithms for Directed Hypergraph Expansion and Sparsest Cut with Product Demands

T.-H. Hubert Chan$^{(\boxtimes)}$ and Bintao Sun

Department of Computer Science, The University of Hong Kong,
Pok Fu Lam, Hong Kong
hubert@cs.hku.hk, btsun@connect.hku.hk

Abstract. We give approximation algorithms for the edge expansion and sparsest cut with product demands problems on directed hypergraphs, which subsume previous graph models such as undirected hypergraphs and directed normal graphs.

Using an SDP formulation adapted to directed hypergraphs, we apply the SDP primal-dual framework by Arora and Kale (JACM 2016) to design polynomial-time algorithms whose approximation ratios match those of algorithms previously designed for more restricted graph models. Moreover, we have deconstructed their framework and simplified the notation to give a much cleaner presentation of the algorithms.

1 Introduction

The edge expansion of an edge-weighted graph gives a lower bound on the ratio of the weight of edges leaving any subset S of vertices to the sum of the weighted degrees of S. Therefore, this notion has applications in graph partitioning or clustering [10,13,14], in which a graph is partitioned into clusters such that, loosely speaking, the general goal is to minimize the number of edges crossing different clusters with respect to some notion of cluster weights.

The edge expansion and the sparsest cut problems [11] can be viewed as a special case when the graph is partitioned into two clusters. Even though the involved problems are NP-hard, approximation algorithms have been developed for them in various graph models and settings, such as undirected [4,5] or directed graphs [1,2], and uniform [2,5] or general demands [1,4] in the case of sparsest cut. Recently, approximation algorithms have been extended to the case of undirected hypergraphs [12]. In this paper, we consider these problems for the even more general class of directed hypergraphs.

The full version of this paper is available online [6].

T.-H. H. Chan—This work was partially supported by the Hong Kong RGC under the grant 17200817.

© Springer International Publishing AG, part of Springer Nature 2018
L. Wang and D. Zhu (Eds.): COCOON 2018, LNCS 10976, pp. 688–700, 2018.
https://doi.org/10.1007/978-3-319-94776-1_57

Directed Hypergraphs. We consider an edge-weighted *directed hypergraph* $H = (V, E, w)$, where V is the vertex set of size n and $E \subseteq 2^V \times 2^V$ is the set of m directed hyperedges; Each directed hyperedge $e \in E$ is denoted by $(\mathsf{T}_e, \mathsf{H}_e)$, where $\mathsf{T}_e \subseteq V$ is the *tail* and $\mathsf{H}_e \subseteq V$ is the *head*; we assume that both the tail and the head are non-empty, and we follow the convention that the direction is from tail to head. We denote $r := \max_{e \in E}(|\mathsf{T}_e| + |\mathsf{H}_e|)$.

The function $w : E \to \mathbb{R}_+$ assigns a non-negative weight to each edge. Note that T_e and H_e do not have to be disjoint. This notion of directed hypergraph was first introduced by Gallol et al. [8], who considered applications in propositional logic, analyzing dependency in relational database, and traffic analysis.

Observe that this model captures previous graph models: (i) an undirected hyperedge e is the special case when $\mathsf{T}_e = \mathsf{H}_e$, and (ii) a directed normal edge e is the special case when $|\mathsf{T}_e| = |\mathsf{H}_e| = 1$.

Directed Hyperedge Expansion. In addition to edge weights, each vertex $u \in V$ has weight $\omega_u := \sum_{e \in E : u \in \mathsf{T}_e \cup \mathsf{H}_e} w_e$ that is also known as its *weighted degree*. Given a subset $S \subseteq V$, denote $\overline{S} := V \setminus S$ and $\omega(S) := \sum_{u \in S} \omega_u$. Define the out-going cut $\partial^+(S) := \{e \in E : \mathsf{T}_e \cap S \neq \emptyset \wedge \mathsf{H}_e \cap \overline{S} \neq \emptyset\}$, and the in-coming cut $\partial^-(S) := \{e \in E : \mathsf{T}_e \cap \overline{S} \neq \emptyset \wedge \mathsf{H}_e \cap S \neq \emptyset\}$. The out-going edge expansion of S is $\phi^+(S) := \frac{w(\partial^+(S))}{\omega(S)}$, and the in-coming edge expansion is $\phi^-(S) := \frac{w(\partial^-(S))}{\omega(S)}$. The edge expansion of S is $\phi(S) := \min\{\phi^+(S), \phi^-(S)\}$. The edge expansion of H is

$$\phi_H := \min_{\emptyset \neq S \subset V : \omega(S) \leq \frac{\omega(V)}{2}} \phi(S).$$

Directed Sparsest Cut with Product Demands. As observed in previous works such as [5], we relate the expansion problem to the sparsest cut problem with *product demands*. For vertices $i \neq j \in V$, we assume that the demand between i and j is symmetric and given by the product $\omega_i \cdot \omega_j$. For $\emptyset \neq S \subsetneq V$, its *directed sparsity* is $\vartheta(S) := \frac{w(\partial^+(S))}{\omega(S) \cdot \omega(\overline{S})}$. The goal is to find a subset S to minimize $\vartheta(S)$.

Observe that $\omega(V) \cdot \vartheta(S)$ and $\frac{w(\partial^+(S))}{\min\{\omega(S), \omega(\overline{S})\}}$ are within a factor of 2 from each other. Therefore, the directed edge expansion problem on directed hypergraphs can be reduced (up to a constant factor) to the sparsest cut problem with product demands. Hence, for the rest of the paper, we just focus on the sparsest cut problem with product demands.

Vertex Weight Distribution. For the sparsest cut problem, the vertex weights $\omega : V \to \mathbb{R}_+$ actually do not have to be related to the edge weights. However, we do place restrictions on the *skewness* of the weight distribution. Without loss of generality, we can assume that each vertex has integer weight. For $\kappa \geq 1$, the weights ω are κ-skewed, if for all $i \in V$, $1 \leq \omega_i \leq \kappa$. In this paper, we assume $\kappa \leq n$.

Balanced Cut. For $0 < c < \frac{1}{2}$, a subset $S \subseteq V$ is c-balanced if both $\omega(S)$ and $\omega(V \setminus S)$ are at least $c \cdot \omega(V)$.

1.1   Our Contributions and Results

Our first observation is a surprisingly simple reduction of the problem from the more general directed hypergraphs to the case of directed normal graphs.

Fact 1 (Reduction to Directed Normal Graphs). *Suppose $H = (V, E)$ is a directed hypergraph with edge weights w and vertex weights ω. Then, transformation to a directed normal graph $\widehat{H} = (\widehat{V}, \widehat{E})$, where $|\widehat{V}| = n + 2m$ and $|\widehat{E}| = m + \sum_{e \in E}(|\mathsf{T}_e| + |\mathsf{H}_e|)$, is defined as follows.*

The new vertex set is $\widehat{V} := V \cup \{v_e^\mathsf{T}, v_e^\mathsf{H} : e \in E\}$, i.e., for each edge $e \in E$, we add two new vertices; the old vertices retain their original weights, and the new vertices have zero weight.

The new edge set is $\widehat{E} := \{(v_e^\mathsf{T}, v_e^\mathsf{H}) : e \in E\} \cup \{(u, v_e^\mathsf{T}) : e \in E, u \in \mathsf{T}_e\} \cup \{(v_e^\mathsf{H}, v) : e \in E, v \in \mathsf{H}_e\}$. An edge of the form $(v_e^\mathsf{T}, v_e^\mathsf{H})$ has its weight w_e derived from $e \in E$, while all other edges have large weights $\mathfrak{M} := n \sum_{e \in E} w_e$.

We overload the symbols for edge w and vertex ω weights. However, we use $\widehat{\partial}^+(\cdot)$ for out-going cut in $\widehat{H}$.

Given a subset $S \subseteq V$, we define the transformed subset $\widehat{S} := S \cup \{v_e^\mathsf{T} : S \cap \mathsf{T}_e \neq \emptyset\} \cup \{v_e^\mathsf{H} : \mathsf{H}_e \subseteq S\}$. Then, we have the following properties.

- *For any $S \subseteq V$, $\omega(S) = \omega(\widehat{S})$ and $w(\partial^+(S)) = w(\widehat{\partial}^+(\widehat{S}))$.*
- *For any $T \subseteq \widehat{V}$, $\omega(T \cap V) = \omega(T)$; moreover, if $w(\widehat{\partial}^+(T)) < \mathfrak{M}$, then $w(\partial^+(T \cap V)) = w(\widehat{\partial}^+(T))$.*

Fact 1 implies that for problems such as directed sparsest cut (with product demands), max-flow and min-cut, it suffices to consider directed normal graphs.

Semidefinite Program (SDP) Formulation. Arora et al. [5] formulated an SDP for the sparsest cut problem with uniform demands for undirected normal graphs. The SDP was later refined by Agarwal et al. [2] for directed normal graphs to give a rounding-based approximation algorithm. Since the method can be easily generalized to product demands with κ-skewed vertex weights by duplicating copies, we have the following corollary.

Corollary 1 (Approximation Algorithm for Directed Sparsest Cut with Product Demands). *For the directed sparsest cut problem with product demands (with κ-skewed vertex weights) on directed hypergraphs, there are randomized polynomial-time $O(\sqrt{\log \kappa n})$-approximate algorithms.*

Are We Done Yet? Unfortunately, solving an SDP poses a major bottleneck in running time. Alternatively, Arora and Kale [3] proposed an SDP primal-dual framework that iteratively updates the primal and the dual solutions.

Outline of SDP Primal-Dual Approach. The framework essentially performs binary search on the optimal SDP value. Each binary search step requires iterative calls to some ORACLE. Loosely speaking, given a (not necessarily feasible) primal solution candidate of the minimization SDP, each call of the ORACLE returns either (i) a subset $S \subset V$ of small enough sparsity $\vartheta(S)$, or (ii) a (not necessarily feasible) dual solution with large enough objective value to update the primal candidate in the next iteration. At the end of the last iteration, if a suitable subset S has not been returned yet, then the dual solutions returned in all the iterations can be used to establish that the optimal SDP value is large.

Disadvantage of Direct Reduction. For directed sparsest cut problem with uniform demands, the primal-dual framework gives an $O(\sqrt{\log n})$-approximate algorithm, which has running time[1] $\widetilde{O}(m^{1.5} + n^{2+o(1)})$. If we apply the reduction in Fact 1 directly, the resulting running time for directed hypergraphs becomes $\widetilde{O}((mr)^{1.5} + n^{2+o(1)} + m^{2+o(1)})$. The term $(mr)^{1.5}$ is due to a max-flow computation, which is not obvious how to improve. However, the extra $m^{2+o(1)}$ term is introduced, because the dimension of the primal domain is increased. Therefore, we think it is worthwhile to adapt the framework in [9] to directed hypergraphs to avoid the extra $m^{2+o(1)}$ term.

Other Motivations. We deconstruct the algorithm for directed normal graphs with uniform vertex weight in Kale's PhD thesis [9], and simplify the notation. The result is a much cleaner description of the algorithm, even though we consider more general directed hypergraphs and non-uniform vertex weights. As a by-product, we discover that since the subset returned by sparsest cut needs not be balanced, there should be an extra factor of $O(n^2)$ in the running time of their algorithm. We elaborate the details further as follows.

1. In their framework, they assume that in the SDP, there is some constraint on the trace $\mathbf{Tr}(\mathbf{X}) = \mathbf{I} \bullet \mathbf{X}$, which can be viewed as some dot-product with the identity matrix $\mathbf{I}$. The important property is that every non-zero vector is an eigenvector of $\mathbf{I}$ with eigenvalue 1. Therefore, if the smallest eigenvalue of $\mathbf{A}$ is at least $-\epsilon$ for some small $\epsilon > 0$, then the sum $\mathbf{A} + \epsilon\mathbf{I} \succeq 0$ has non-negative eigenvalues. This is used crucially to establish a lower bound on the optimal value of the SDP.

   However, for the SDP formulation of directed sparsest cut, the constraint loosely translates to $\mathbf{I} \cdot \mathbf{X} \leq O(\frac{n}{\omega(S) \cdot \omega(\overline{S})})$, where $S \subset V$ is some candidate subset. To achieve the claimed running time, one needs a good enough upper bound, which is achieved if the subset is balanced. However, for general S that is not balanced, there can be an extra factor of n in the upper bound, which translates to a factor of $O(n^2)$ in the final running time.

---

[1] After checking the calculation in [9] carefully, we conclude that there should actually be an extra factor of $O(n^2)$ in the running time. Through personal communication with Kale, we are told that it might be possible reduce a factor of $O(n)$, using the "one-sided width" technique in [9].

Instead, as we shall see, there is already a constraint $\mathbf{K} \bullet \mathbf{X} = 1$, where $\mathbf{K}$ is the Laplacian matrix of the complete graph. Since $\mathbf{K}$ is actually a scaled version of the identity operator on the space orthogonal to the all-ones vector $\mathbf{1}$, a more careful analysis can use this constraint involving $\mathbf{K}$ instead.

2. In capturing directed distance in an SDP [2], typically, one extra vector v_0 is added. However, in the SDP of [9], a different vector w_i is added for each $i \in V$, and constraints saying that all these w_i's are the same are added. At first glance, these extra vectors w_i's and constraints seem extraneous, and create a lot of dual variables in the description of the ORACLE. The subtle reason is that by increasing the dimension of the primal domain, the width of the ORACLE, which is measured by the spectral norm of some matrix, can be reduced.

   Observe that the matrix $\mathbf{K}$ does not involve any extra added vectors. If we do not use the trace bound on $\mathbf{Tr}(\mathbf{X})$ in the analysis, then we cannot add any extra vectors in the SDP. This can be easily rectified, because we can just label any vertex in V as 0 and consider two cases. In the first case, we formulate an SDP for the solution S to include 0; in the second case, we formulate a similar SDP to exclude 0 from the solution. The drawback is that now the width of the ORACLE increases by a factor of $O(n)$, which leads to a factor of $O(n^2)$ in the number of iterations.

   Therefore, in the end, we give a simpler presentation than [9], but the asymptotic running time is the same, although an improvement as mentioned in Footnote 1 might be possible.

3. For each simple path, they add a generalized ℓ_2^2-triangle inequality. This causes an exponential number of dual variables (even though most of them are zero). However, only triangle inequalities for triples are needed, because each triangle inequality for a long path is just a linear combination of inequalities involving only triples.

We summarize the performance of our modified primal-dual approach as follows.

Theorem 2 (SDP Primal-Dual Approximation Algorithm for Directed Sparsest Cut). *Suppose the vertex weights are κ-skewed. Each binary search step of the primal-dual framework takes $T := \widetilde{O}(\kappa^2 n^2)$ iterations. The running time of each iteration is $\widetilde{O}((rm)^{1.5} + (\kappa n)^2)$.*

The resulting approximation ratio is $O(\sqrt{\log \kappa n})$.

1.2   Related Work

As mentioned above, the most related work is the SDP primal-dual framework by Arora and Kale [3] used for solving various variants of the sparsest cut problems. The details for directed sparsest cut are given in Kale's PhD thesis [9]. We briefly describe the background of related problems as follows.

Edge Expansion and Sparsest Cut. Leighton and Rao [11] achieved the first $O(\log n)$-approximation algorithms for the edge expansion problem and the sparsest cut problem with general demands for undirected normal graphs. An SDP approach utilizing ℓ_2^2-representation was used by Arora et al. [5] to achieve $O(\sqrt{\log n})$-approximation for the special case of uniform demands; subsequently, $O(\sqrt{\log n} \cdot \log \log n)$-approximation has been achieved for general demands [4] via embeddings of n-point ℓ_2^2 metric spaces into Euclidean space with distortion $O(\sqrt{\log n} \cdot \log \log n)$. This embedding was also used to achieve $O(\sqrt{\log n \log r} \cdot \log \log n)$-approximation for the general demands case in undirected hypergraphs [7], where r is the maximum cardinality of an hyperedge.

More related works are given in the full version [6].

2   SDP Relaxation for Directed Sparsest Cut

We follow some common notation concerning sparsest cut (with uniform demands) in undirected [5] and directed [2] normal graphs.

Definition 1 (ℓ_2^2-Representation). *An ℓ_2^2-representation for a set of vertices V is an assignment of a vector v_i to each vertex $i \in V$ such that the ℓ_2^2-triangle inequality holds:*

$$\|v_i - v_j\|^2 \le \|v_i - v_k\|^2 + \|v_k - v_j\|^2, \qquad \forall i, j, k \in V.$$

Directed Distance [2]. We arbitrarily pick some vertex in V, and call it 0.

We first consider the case when 0 is always included in the feasible solution. Given an ℓ_2^2-representation $\{v_i\}_{i \in V}$, define the directed distance $d : V \times V \to \mathbb{R}_+$ by

$$d(i, j) := \|v_i - v_j\|^2 - \|v_i - v_0\|^2 + \|v_j - v_0\|^2.$$

It is easy to verify the directed triangle inequality: for all $i, j, k \in V$, $d(i, k) + d(k, j) \ge d(i, j)$.

For subsets $S \subseteq V, T \subseteq V$, we also denote $d(S, T) := \min_{i \in S, j \in T} \{d(i, j)\}$, $d(i, S) = d(\{i\}, S)$ and $d(S, i) = d(S, \{i\})$.

Interpretation. In an SDP-relaxation for directed sparsest cut, vertex 0 is always chosen in the solution $S \subseteq V$. For $i \in S$, v_i is set to v_0; for $i \in \overline{S} = V \setminus S$, v_i is set to $-v_0$. Then, it can be checked that $d(i, j)$ is non-zero *iff* $i \in S$ and $j \in \overline{S}$, in which case $d(i, j) = 8\|v_0\|^2$.

The Other Case. For the other case when 0 is definitely excluded from the solution S, it suffices to change the definition $d(i, j) := \|v_i - v_j\|^2 - \|v_i + v_0\|^2 + \|v_j + v_0\|^2$. For the rest of the paper, we just concentrate on the case that 0 is in the solution S.

We consider the following SDP relaxation (where $\{v_i : i \in V\}$ are vectors) for the directed sparsest cut problem with product demands on an edge-weighted hypergraph $H = (V, E, w)$ with vertex weights $\omega : V \to \{1, 2, \ldots, \kappa\}$. We denote $\mathfrak{W} := \sum_{i \in V} \omega_i$.

$$\text{SDP} \quad \min \quad \frac{1}{2} \sum_{e \in E} w_e \cdot d_e \tag{1}$$

$$\text{s.t.} \quad d_e \geq d(i,j), \qquad \forall e \in E, \forall (i,j) \in \mathsf{T}_e \times \mathsf{H}_e \tag{2}$$

$$\|v_i - v_j\|^2 \leq \|v_i - v_k\|^2 + \|v_k - v_j\|^2, \quad \forall i,j,k \in V \tag{3}$$

$$\sum_{\{i,j\} \in \binom{V}{2}} \omega_i \omega_j \|v_i - v_j\|^2 = 1, \tag{4}$$

$$d_e \geq 0, \quad \forall e \in E. \tag{5}$$

SDP Relaxation. To see that SDP is a relaxation of the directed sparsest cut problem, it suffices to show that any subset $S \subseteq V$ induces a feasible solution with objective function $\vartheta(S)$. We set v_0 to be a vector with $\|v_0\|^2 = \frac{1}{4\omega(S) \cdot \omega(\overline{S})}$. For each $i \in V$, we set $v_i := v_0$ if $i \in S$, and $v_i := -v_0$ if $i \in \overline{S}$. Then, the value of the corresponding objective is

$$\frac{1}{2} \sum_{e \in E} w_e \cdot d_e = \frac{1}{2} \sum_{e \in E} w_e \cdot \max_{(i,j) \in \mathsf{T}_e \times \mathsf{H}_e} \{d(i,j)\}$$

$$= \frac{1}{2} \sum_{e \in \partial^+(S)} w_e \cdot (\|v_0 + v_0\|^2 - \|v_0 - v_0\|^2 + \| - v_0 - v_0\|^2)$$

$$= \frac{w(\partial^+(S))}{\omega(S) \cdot \omega(\overline{S})} = \vartheta(S).$$

Trace Bound. We have $\sum_{i \in V} \|v_i\|^2 \leq \frac{n}{4\omega(S) \cdot \omega(\overline{S})} \leq O(\frac{\kappa n^2}{\mathfrak{W}^2})$. Note that if S is balanced, then the upper bound can be improved to $O(\frac{n}{\mathfrak{W}^2})$.

SDP Primal-Dual Approach [3]. Instead of solving the SDP directly, the SDP is used as a tool for finding an approximate solution. Given a candidate value α, the primal-dual approach either (i) finds a subset S such that $\vartheta(S) \leq O(\sqrt{\log n}) \cdot \alpha$, or (ii) concludes that the optimal value of the SDP is at least $\frac{\alpha}{2}$. Hence, binary search can be used to find an $O(\sqrt{\log n})$-approximate solution. This approach is described in Sect. 3.

3 SDP Primal-Dual Approximation Framework

We use the primal-dual framework by [3]. However, instead of using it just as a blackbox, we tailor it specifically for our problem to have a cleaner description.

Notation. We use a bold capital letter $\mathbf{A} \in \mathbb{R}^{V \times V}$ to denote a symmetric matrix whose rows and columns are indexed by V.

The sum of the diagonal entries of a square matrix $\mathbf{A}$ is denoted by the trace $\mathbf{Tr}(\mathbf{A})$. Given two matrices $\mathbf{A}$ and $\mathbf{B}$, let $\mathbf{A} \bullet \mathbf{B} := \mathbf{Tr}(\mathbf{A}^\top \mathbf{B})$, where $\mathbf{A}^\top$ is the transpose of $\mathbf{A}$. We use $\mathbf{1} \in \mathbb{R}^V$ to denote the all-ones vector.

Primal Solution. We use $\mathbf{X} \succeq 0$ to denote a positive semi-definite matrix that is associated with the vectors $\{v_i\}_{i \in V}$ such that $\mathbf{X}(i,j) = \langle v_i, v_j \rangle$.

We rewrite SDP (1) to an equivalent form as follows.

$$\text{SDP} \qquad \min \quad \frac{1}{2} \sum_{e \in E} w_e \cdot d_e \tag{6}$$

$$\text{s.t.} \quad d_e - \mathbf{A}_{ij} \bullet \mathbf{X} \geq 0, \qquad \forall e \in E, \forall (i,j) \in \mathsf{T}_e \times \mathsf{H}_e \tag{7}$$

$$\mathbf{T}_p \bullet \mathbf{X} \geq 0, \qquad \forall p \in \mathcal{T} \tag{8}$$

$$\mathbf{K} \bullet \mathbf{X} = 1, \tag{9}$$

$$\mathbf{X} \succeq 0; \quad d_e \geq 0, \quad \forall e \in E. \tag{10}$$

We define the notation used in the above formulation as follows:

- For $(i,j) \in V \times V$, $\mathbf{A}_{ij}$ is the unique symmetric matrix such that $\mathbf{A}_{ij} \bullet \mathbf{X} = d(i,j) = \|v_i - v_j\|^2 - \|v_i - v_0\|^2 + \|v_j - v_0\|^2$.

  Since we consider a minimization problem, we just use $\mathbf{X} \succeq 0$ to represent a primal solution, and automatically set $d_e := \max\{0, \max_{(i,j) \in \mathsf{T}_e \times \mathsf{H}_e} \mathbf{A}_{ij} \bullet \mathbf{X}\}$ for all $e \in E$. As we shall see, this implies that corresponding dual variable $y_{ij}^e \in \mathbb{R}$ can be set to 0.

  Moreover, we do not need the constraint $\mathbf{A}_{ij} \bullet \mathbf{X} \geq 0$, because we already have $d_e \geq 0$.

- The set $\mathcal{T}$ contains elements of the form $\begin{bmatrix} \{i,k\} \\ j \end{bmatrix} \in \binom{V}{2} \times V$, where i, j, k are distinct elements in V.

  They are used to specify the ℓ_2^2-triangle inequality.

  For $p = \begin{bmatrix} \{i,k\} \\ j \end{bmatrix}$, $\mathbf{T}_p$ is defined such that $\mathbf{T}_p \bullet \mathbf{X} = \|v_i - v_j\|^2 + \|v_j - v_k\|^2 - \|v_i - v_k\|^2$.

  Observe that in [9], a constraint is added for every path in the complete graph on V. However, these extra constraints are simply linear combinations of the triangle inequalities, and so, are actually unnecessary.

- As above, $\mathbf{K}$ is defined such that $\mathbf{K} \bullet \mathbf{X} = \sum_{\{i,j\} \in \binom{V}{2}} \omega_i \omega_j \|v_i - v_j\|^2$.

  Observe that any $\mathbf{X} \succeq 0$ can be re-scaled such that $\mathbf{K} \bullet \mathbf{X} = 1$.

- **Optional constraint.** In [9], an additional constraint is added, which in our notation[2] becomes:

$$-\mathbf{I} \bullet \mathbf{X} \geq -\Theta\left(\frac{n}{\mathfrak{W}^2}\right).$$

---

[2] In the original notation [9, p. 59], the claimed constraint is $\mathbf{Tr}(\mathbf{X}) \leq n$, but for general cut S, only the weaker bound $\mathbf{Tr}(\mathbf{X}) \leq \Theta(n^2)$ holds.

However, this holds only if the solution S is balanced. For general cut S, we only have the weaker bound: $-\mathbf{I} \bullet \mathbf{X} \geq -\Theta(\frac{\kappa n^2}{2\mathfrak{W}^2})$. As we shall see in the proof of Lemma 1, adding this weaker bound is less useful than the above constraint $\mathbf{K} \bullet \mathbf{X} = 1$.

The dual to SDP is as follows:

$$\text{Dual} \quad \max \quad z \tag{11}$$

$$\text{s.t.} \quad -\sum_{e \in E} \sum_{(i,j) \in T_e \times H_e} y_{ij}^e \mathbf{A}_{ij} + \sum_{p \in \mathcal{T}} f_p \mathbf{T}_p + z\mathbf{K} \preceq 0 \tag{12}$$

$$\sum_{(i,j) \in T_e \times H_e} y_{ij}^e \leq \frac{w_e}{2}, \quad \forall e \in E, \tag{13}$$

$$f_p \geq 0, \quad \forall p \in \mathcal{T}, \tag{14}$$

$$y_{ij}^e \geq 0, \quad \forall e \in E, \forall (i,j) \in T_e \times H_e. \tag{15}$$

Observe that, if we add the optional constraint $-\mathbf{I} \bullet \mathbf{X} \geq -b$ in the primal, then this will create a dual variable $x \geq 0$, which causes an extra term $-bx$ in the objective function and an extra term $-x\mathbf{I}$ on the left hand side of the constraint.

To use the primal-dual framework [3], we give a tailor-made version of the ORACLE for our problem.

Definition 2 (ORACLE for SDP). *Given $\alpha > 0$, ORACLE(α) has width ρ (which can depend on α) if the following holds. Given a primal candidate solution $\mathbf{X} \succeq 0$ (associated with vectors $\{v_i\}_{i \in V}$) such that $\mathbf{K} \bullet \mathbf{X} = 1$, it outputs either*

(i) a subset $S \subsetneq V$ such that its sparsity $\vartheta(S) \leq O(\sqrt{\log \kappa n}) \cdot \alpha$, or
(ii) some dual variables $(z, (f_p \geq 0 : p \in \mathcal{T}))$, where all y_{ij}^e's are implicitly 0, and a symmetric flow matrix $\mathbf{F} \in \mathbb{R}^{V \times V}$ such that all the following hold:

- $z \geq \alpha$
- $(\sum_{p \in \mathcal{T}} f_p \mathbf{T}_p + z\mathbf{K}) \bullet \mathbf{X} \leq \mathbf{F} \bullet \mathbf{X}$
- *For all feasible primal solution $\mathbf{X}^*$, $\mathbf{F} \bullet \mathbf{X}^* \leq \frac{1}{2} \sum_{e \in E} w_e d_e^*$, where $d_e^* := \max\{0, \max_{(i,j) \in T_e \times H_e} \mathbf{A}_{ij} \bullet \mathbf{X}^*\}$.*
- *For all $x \in \text{span}\{\mathbf{1}\}$, $\mathbf{F}x = 0$.*
- *The spectral norm $\|\sum_{p \in \mathcal{T}} f_p \mathbf{T}_p + z\mathbf{K} - \mathbf{F}\|$ is at most ρ.*

Using ORACLE in Definition 2, we give the primal-dual framework for one step of the binary search in Algorithm 1. As in [3], the running for each iteration is dominated by the call to the ORACLE.

Input: Candidate value $\alpha > 0$; ORACLE (α) with width ρ

1 T is chosen as in Lemma 1; $\eta \leftarrow \sqrt{\frac{\ln n}{T}}$;

2 $\mathbf{W}^{(1)} \leftarrow I \in \mathbb{R}^{V \times V}$;

3 **for** $t = 1, 2, \ldots, T$ **do**

4 $\quad$ $\mathbf{X}^{(t)} \leftarrow \frac{\mathbf{W}^{(t)}}{\mathbf{K} \bullet \mathbf{W}^{(t)}}$;

5 $\quad$ Run ORACLE (α) with $X^{(t)}$;

6 $\quad$ **if** ORACLE *returns some* $S \subset V$ **then**

7 $\quad$ $\quad$ **return** S and **terminate**.

8 $\quad$ **end**

9 $\quad$ Otherwise, the ORACLE returns some dual solution $(z^{(t)}, (f_p^{(t)} : p \in \mathcal{T}))$ and matrix $\mathbf{F}^{(t)}$ as promised in Definition 2.

10 $\quad$ $\mathbf{M}^{(t)} \leftarrow -\frac{1}{\rho} \left(\sum_{p \in \mathcal{T}} f_p^{(t)} \mathbf{T}_p + z^{(t)} \mathbf{K} - \mathbf{F}^{(t)} \right)$;

11 $\quad$ $\mathbf{W}^{(t+1)} \leftarrow \exp \left(-\eta \sum_{\tau=1}^{t} \mathbf{M}^{(\tau)} \right)$;

12 **end**

13 **if** *no subset S is returned yet* **then**

14 $\quad$ **report** the optimal value is at least $\frac{\alpha}{2}$.

15 **end**

Algorithm 1: Primal-Dual Approximation Algorithm for SDP

The following result is proved in [3, Corollary 3.2].

Fact 3 (Multiplicative Update). *Given any sequence of matrices* $\mathbf{M}^{(1)}$, $\mathbf{M}^{(2)}, \ldots, \mathbf{M}^{(T)} \in \mathbb{R}^{n \times n}$ *that all have spectral norm at most 1 and* $\eta \in (0, 1]$, *let* $\mathbf{W}^{(1)} = I$, $\mathbf{W}^{(t)} = \exp \left(-\eta \sum_{\tau=1}^{t-1} \mathbf{M}^{(\tau)} \right)$, *for* $t = 2, \ldots, T$; *let* $\mathbf{P}^{(t)} = \frac{\mathbf{W}^{(t)}}{\mathrm{Tr}(\mathbf{W}^{(t)})}$, *for* $t = 1, 2, \ldots, T$. *Then, we have*

$$\sum_{t=1}^{T} \mathbf{M}^{(t)} \bullet \mathbf{P}^{(t)} \leq \lambda_{\mathsf{min}} \left(\sum_{t=1}^{T} \mathbf{M}^{(t)} \right) + \eta T + \frac{\ln n}{\eta},$$

where $\lambda_{\mathsf{min}}(\cdot)$ *gives the minimum eigenvalue of a symmetric matrix.*

Lemma 1 (Correctness). *Set* $T := \lceil \frac{16\kappa^2 \rho^2 n^2 \ln n}{\alpha^2 \mathfrak{W}^4} \rceil$. *Suppose that in Algorithm 1, the* ORACLE *never returns any subset S in any of the T iterations. Then, the optimal value of SDP is at least* $\frac{\alpha}{2}$.

Proof. The proof follows the same outline as [3, Theorem 4.6], but we need to be more careful, depending on whether we use the constraint on $\mathbf{I} \bullet \mathbf{X}$.

For $t = 1, \ldots, T$, we use $\mathbf{M}^{(t)}$ as in Algorithm 1, and apply Fact 3. Definition 2 guarantees that $\mathbf{M}^{(t)} \bullet \mathbf{P}^{(t)} \geq 0$, because $\mathbf{X}^{(t)}$ is positively scaled from $\mathbf{P}^{(t)}$.

Hence, by Fact 3, we have $\lambda_{\mathsf{min}} \left(\sum_{t=1}^{T} \mathbf{M}^{(t)} \right) + \eta T + \frac{\ln n}{\eta} \geq 0$.

By setting $\eta := \sqrt{\frac{\ln n}{T}}$ and $\mathbf{Z} := \frac{\rho}{T} \sum_{t=1}^{T} \mathbf{M}^{(t)} = \frac{1}{T} \sum_{t=1}^{T} (\mathbf{F}^{(t)} - \sum_{p \in \mathcal{T}} f_p^{(t)} \mathbf{T}_p - z^{(t)} \mathbf{K})$, this is equivalent to $\lambda_{\mathsf{min}}(\mathbf{Z}) \geq -2\rho \cdot \sqrt{\frac{\ln n}{T}}$.

As in [3], we would like to add some matrix from the primal constraint to $\mathbf{Z}$ to make the resulting matrix positive semi-definite.

A possible candidate is $\mathbf{K}$, whose eigenvalues are analyzed as follows.

First, observe that for all $x \in \mathsf{span}\{\mathbf{1}\}$, it can be checked that $\mathbf{K}x = \mathbf{T}_p x = 0$, for all $p \in \mathcal{T}$. Furthermore, Definition 2 guarantees that $\mathbf{F}^{(t)}x = 0$, for all t. Hence, it follows that $\mathbf{Z}x = 0$, which implies that any negative eigenvalue of $\mathbf{Z}$ must be due to the space orthogonal to $\mathsf{span}\{\mathbf{1}\}$.

We next analyze the eigenvectors of $\mathbf{K}$ in this orthogonal space. Consider a unit vector $u \perp \mathsf{span}\{\mathbf{1}\}$, i.e., $\sum_{i \in V} u_i = 0$ and $\sum_{i \in V} u_i^2 = 1$.

Then, $u^\top K u = \frac{1}{2}\sum_{i \in V}\sum_{j \in V}\omega_i\omega_j(u_i - u_j)^2 = \mathfrak{W}^2 \cdot [\sum_{i \in V}\delta_i u_i^2 - (\sum_{i \in V}\delta_i u_i)^2]$, where $\delta_i := \frac{\omega_i}{\mathfrak{W}}$ can be interpreted as some probability mass function. Hence, this term can be interpreted as some variance.

Observe that the κ-skewness of the weights ω implies that for all $i \in V$, $\delta_i \geq \frac{1}{\kappa n}$. Therefore, Lemma 2 below implies that $u^\top K u \geq \mathfrak{W}^2 \cdot \frac{1}{\kappa n}$.

Hence, by enforcing $\epsilon \cdot \mathfrak{W}^2 \cdot \frac{1}{\kappa n} \geq 2\rho \cdot \sqrt{\frac{\ln n}{T}}$, we have $\lambda_{\min}(\mathbf{Z} + \epsilon\mathbf{K}) \geq 0$.

Next, suppose $\mathbf{X}^*$ (with induced d^*) is an optimal primal solution to SDP. Then, Definition 2 implies that $\frac{1}{2}\sum_{e \in E} w_e d_e^* \geq \frac{1}{T}\sum_{t=1}^T \mathbf{F}^{(t)} \bullet \mathbf{X}^*$.

Since $(\mathbf{Z} + \epsilon\mathbf{K}) \bullet \mathbf{X}^* \geq 0$, the optimal value is at least

$$\frac{1}{T}\sum_{t=1}^T\left(\sum_{p \in \mathcal{T}} f_p^{(t)}\mathbf{T}_p \bullet \mathbf{X}^*\right) + \frac{1}{T}\sum_{t=1}^T z^{(t)}\mathbf{K} \bullet \mathbf{X}^* - \epsilon\mathbf{K} \bullet \mathbf{X}^*$$

$$\geq 0 + \frac{1}{T}\sum_{t=1}^T z^{(t)} \cdot \mathbf{1} - \epsilon \cdot \mathbf{1}$$

$$\geq \alpha - \epsilon,$$

where the last two inequalities come from the properties of primal feasible $\mathbf{X}^*$ and ORACLE, respectively. Setting $\epsilon = \frac{\alpha}{2}$ gives the result. $\square$

Remark. One can see that in the proof of Lemma 1, if one uses the weaker bound $-\mathbf{I} \bullet \mathbf{X} \geq -\Theta(\frac{\kappa n^2}{\mathfrak{W}^2})$. Then, the proof continues by choosing $\nu = 2\rho \cdot \sqrt{\frac{\ln n}{T}}$, we have $\lambda_{\min}(\mathbf{Z} + \nu\mathbf{I}) \geq 0$.

Using the same argument, we conclude that the optimal value is at least $\alpha - \nu\mathbf{I} \bullet \mathbf{X}^* \geq \alpha - \nu \cdot \Theta(\frac{\kappa n^2}{\mathfrak{W}^2})$.

Setting $\frac{\alpha}{2} = \nu \cdot \Theta(\frac{\kappa n^2}{\mathfrak{W}^2})$ gives $T := \Theta(\frac{\kappa^2\rho^2 n^4 \ln n}{\alpha^2\mathfrak{W}^4})$ in this case, which has an extra factor of $O(n^2)$.

However, since we do not add any extra vectors in our primal domain, the width in our ORACLE in Theorem 4 has an extra $O(n)$ factor compared to that in [9], which brings back the $O(n^2)$ factor we have saved earlier.

Lemma 2. (Bounding the Variance). *For real numbers $u_1, u_2, \ldots, u_n$ and $\delta_0, \delta_1, \delta_2, \ldots, \delta_n$ such that $\sum_{i=1}^n u_i = 0$, $\sum_{i=1}^n u_i^2 = 1$, $\sum_{i=1}^n \delta_i = 1$ and $\delta_i \geq \delta_0 > 0$, $\forall i$, we have $\sum_{i=1}^n \delta_i u_i^2 - \left(\sum_{i=1}^n \delta_i u_i\right)^2 \geq \delta_0$.*

Moreover, we have $\sum_{i=1}^{n} \delta_i u_i^2 - \left(\sum_{i=1}^{n} \delta_i u_i\right)^2 \leq \max_i \delta_i$.

Proof. Let $u_1, \ldots, u_n$ be fixed and consider the function

$$q(\delta_1, \ldots, \delta_n) = \sum_{i=1}^{n} \delta_i u_i^2 - \left(\sum_{i=1}^{n} \delta_i u_i\right)^2$$

with domain $\{(\delta_1, \ldots, \delta_n) | \sum_{i=1}^{n} \delta_i = 1, \delta_i \geq \delta_0, \forall i\}$.

We claim that the minimum can be obtained at some point where at most one δ_i has value strictly greater than δ. Indeed, suppose there are two variables, say δ_1 and δ_2, whose value is strictly greater than δ_0. Consider $h(x) = g(x, s - x, \delta_3, \ldots, \delta_n)$ where $s = \delta_1 + \delta_2$. Simplifying it, we know that the coefficient associated to x^2 in h is $-(u_1 - u_2)^2 \leq 0$, which means that we can shift either δ_1 or δ_2 to δ_0 (and the other variable to $s - \delta_0$) without increasing the value of g.

Therefore, we only need to consider the case where there is at most one $\delta_i > \delta_0$. Without loss of generality, suppose $\delta_2 = \delta_3 = \cdots = \delta_n = \delta_0$ and thus $\delta_1 = (1 - n\delta_0) + \delta_0$. Then,

$$g(\delta_1, \ldots, \delta_n) = \delta_0 + (1 - n\delta_0)u_1^2 - ((1 - n\delta_0)u_1)^2 = \delta_0 + n\delta_0(1 - n\delta_0)u_1^2 \geq \delta_0,$$

since δ_0 cannot be greater than $\frac{1}{n}$.   □

Corollary 2. (Non-zero Eigenvalues of K). *All eigenvectors of* **K** *that are orthogonal to* **1** *has eigenvalues in the range* $[\frac{\mathfrak{W}^2}{\kappa n}, \frac{\kappa \mathfrak{W}^2}{n}]$.

Implementation of ORACLE. The construction is almost the same as [9], except that max-flow is computed on a directed hypergraph. Moreover, since we do not increase the dimension of the primal domain, there is an extra factor of $O(n)$ in the width of our ORACLE. For completeness, the proof of the following theorem is in the full version [6].

Theorem 4. *Given a candidate value* $\alpha > 0$ *and primal* $\mathbf{X} \succeq 0$ *such that* $\mathbf{K} \bullet \mathbf{X} = 1$, *the* ORACLE *returns one of the following:*

1. *A subset S with directed sparsity $\vartheta(S) = \frac{w(\partial^+(S))}{\omega(S)\omega(\overline{S})} = O(\sqrt{\log \kappa n}) \cdot \alpha$.*
2. *Dual variables $(z, (f_p : p \in T))$ and flow matrix* **F** *satisfying Definition 2. Moreover, the spectral norm satisfies* $\|\sum_p f_p \mathbf{T}_p + z\mathbf{K} - \mathbf{F}\| \leq O(\alpha \mathfrak{W}^2 \sqrt{\log \kappa n})$.

The running time is $\widetilde{O}((rm)^{1.5} + (\kappa n)^2)$, *where κ is the skewness of vertex weights, $m = |E|$ and $r = \max_{e \in E}(|\mathsf{T}_e| + |\mathsf{H}_e|)$.*

References

1. Agarwal, A., Alon, N., Charikar, M.: Improved approximation for directed cut problems. In: STOC, pp. 671–680. ACM (2007)
2. Agarwal, A., Charikar, M., Makarychev, K., Makarychev, Y.: O(sqrt(log n)) approximation algorithms for min UnCut, min 2CNF deletion, and directed cut problems. In: STOC, pp. 573–581. ACM (2005)
3. Arora, S., Kale, S.: A combinatorial, primal-dual approach to semidefinite programs. J. ACM **63**(2), 12:1–12:35 (2016)
4. Arora, S., Lee, J., Naor, A.: Euclidean distortion and the sparsest cut. J. Am. Math. Soc. **21**(1), 1–21 (2008)
5. Arora, S., Rao, S., Vazirani, U.: Expander flows, geometric embeddings and graph partitioning. J. ACM **56**(2), 1–37 (2009)
6. Chan, T.-H.H., Sun, B.: An SDP primal-dual approximation algorithm for directed hypergraph expansion and sparsest cut with product demands. ArXiv e-prints (2018)
7. Chan, T.-H.H., Louis, A., Tang, Z.G., Zhang, C.: Spectral properties of hypergraph Laplacian and approximation algorithms. J. ACM **65**(3), 15:1–15:48 (2018)
8. Gallo, G., Longo, G., Pallottino, S., Nguyen, S.: Directed hypergraphs and applications. Discrete Appl. Math. **42**(2), 177–201 (1993)
9. Kale, S.: Efficient algorithms using the multiplicative weights update method. Princeton University (2007)
10. Kannan, R., Vempala, S., Vetta, A.: On clusterings: good, bad and spectral. J. ACM **51**(3), 497–515 (2004)
11. Leighton, F.T., Rao, S.: Multicommodity max-flow min-cut theorems and their use in designing approximation algorithms. J. ACM **46**(6), 787–832 (1999)
12. Louis, A., Makarychev, K.: Approximation algorithm for sparsest k-partitioning. In: SODA, pp. 1244–1255. SIAM (2014)
13. Makarychev, K., Makarychev, Y., Vijayaraghavan, A.: Correlation clustering with noisy partial information. In: COLT JMLR Workshop and Conference Proceedings, vol. 40, pp. 1321–1342. JMLR.org (2015)
14. Peng, R., Sun, H., Zanetti, L.: Partitioning well-clustered graphs: spectral clustering works! In: COLT, JMLR Workshop and Conference Proceedings, vol. 40, pp. 1423–1455. JMLR.org (2015)

Lower Bounds for Special Cases of Syntactic Multilinear ABPs

C. Ramya[✉] and B. V. Raghavendra Rao

IIT Madras, Chennai, India
{ramya,bvrr}@cse.iitm.ac.in

Abstract. Algebraic Branching Programs (ABPs) are standard models for computing polynomials. Syntactic multilinear ABPs (smABPs) are restrictions of ABPs where every variable is allowed to occur at most once in every path from the start to terminal node. Proving lower bounds against syntactic multilinear ABPs remains a challenging open question in Algebraic Complexity Theory. The current best known bound is only quadratic [Alon,Kumar,Volk ECCC 2017].

In this article, we develop a new approach upper bounding the rank of the partial derivative matrix of syntactic multilinear ABPs: Convert the ABP to a syntactic multilinear formula with a super polynomial blow up in the size and then exploit the structural limitations of resulting formula to obtain a rank upper bound. Using this approach, we prove exponential lower bounds for special cases of smABPs and circuits namely, sum of Oblivious Read-Once ABPs, r-pass multilinear ABPs and sparse ROABPs. En route, we also prove super-polynomial lower bound for a special class of syntactic multilinear arithmetic circuits.

Keywords: Computational complexity
Algebraic complexity theory · Algebraic branching programs

1 Introduction

Algebraic Complexity Theory investigates the inherent complexity of computing polynomials with arithmetic circuit as the computational model. Arithmetic circuits introduced by Valiant [16] are standard models for computing polynomials over an underlying field. An *arithmetic formula* is a subclass of arithmetic circuits corresponding to arithmetic expressions. For circuits and formulas, the parameters of interest are *size* and *depth*, where size represents the number of nodes in the graph and depth the length of longest path in the graph. The arithmetic formulas are computationally weaker than circuits, a proper separation between them is not known. Nested in-between the computational power of formulas and circuits is yet another well-studied model for computing polynomials referred to as *Algebraic Branching Programs* (ABPs for short).

$$\text{Arithmetic Formula} \subseteq_P \text{ABP} \subseteq_P \text{Arithmetic Circuits.}$$

© Springer International Publishing AG, part of Springer Nature 2018
L. Wang and D. Zhu (Eds.): COCOON 2018, LNCS 10976, pp. 701–712, 2018.
https://doi.org/10.1007/978-3-319-94776-1_58

where the subscript P denotes the containment upto polynomial blow-up in size. Most of algebraic complexity theory revolves around understanding whether these containments are strict or not.

Separation of complexity classes of polynomials involves obtaining lower bound for specific polynomial against classes of arithmetic circuits.

For general classes of arithmetic circuits, Baur and Strassen [4] proved that any arithmetic circuit computing an explicit n-variate degree d polynomial must have size $\Omega(n \log d)$. In fact, this is the only super linear lower bound we know for general arithmetic circuits.

While the challenge of proving lower bounds for general classes of circuits still seems to be afar, recent research has focused on circuits with additional structural restrictions such as multilinearity, bounded read etc. We now look at some of the models based on these restrictions in more detail.

An arithmetic circuit (formula,ABP) is said to be *multilinear* if every gate (node) computes a multilinear polynomial. A seminal work of Raz [13] showed that multilinear formulas computing $\det_n$ or perm_n must have size $n^{\Omega(\log n)}$. Note that any multilinear ABP of $n^{O(1)}$ size computing f on n variables can be converted to a multilinear formula of size $n^{O(\log n)}$ computing f. In order to prove super-polynomial lower bounds for ABPs, it is enough to obtain a multilinear formula computing f of size $n^{o(\log n)}$ or prove a lower bound of $n^{\omega(\log n)}$ for multilinear formulas, both of which are not known.

Special cases of multilinear ABPs have been studied time and again. In this work, we focus on the class of Read-Once Oblivious Algebraic branching programs (ROABP for short). ROABPs are ABPs where every edge is labeled by a variable and every variable appears as edge labels in atmost one layer. There are explicit polynomials with $2^{\Omega(n)}$ ROABP size lower bound [7,8,10]. Also, ROABPs have been well studied in the context of polynomial identity testing algorithms (See e.g.,[6]).

In this article, we prove lower bounds against sum of multilinear ROABPs and other classes of restricted multilinear ABPs and circuits. Definitions of the models considered in this article can be found in Sect. 2.

Our Results. Let $X = \{x_1, \ldots, x_N\}$ and $\mathbb{F}$ be a field. Let g denote the family of N variate (for N even) defined by Raz and Yehudayoff [14]. (See Definition 5 for more details.) As our main result, we show that any sum of sub-exponential $(2^{o(N^\epsilon)})$ size ROABPs to represent g requires 2^{N^ϵ} many summands:

Theorem 1. *Let $f_1, \ldots f_m$ be polynomials computed by oblivious ROABPs such that $g = f_1 + \cdots + f_m$. Then, $m = \dfrac{2^{\Omega(N^{1/5})}}{s^{c \log N}}$, where c is a constant and $s = \max\{s_1, s_2, \ldots, s_m\}$, s_i is the size of the ROABP computing f_i.*

Further, we show that Theorem 1 extends to the case of r-pass multilinear ABPs (Theorem 3) for $r = o(\log n)$ and α-sparse multilinear ABPs (Theorem 4) for $1/1000 \le \alpha \le 1/2$.

Finally, we develop a refined approach to analyze syntactic multilinear formulas based on the central paths introduced by Raz [13]. Using this, we prove exponential lower bound against a class of $O(\log N)$ depth syntactic multilinear circuits (exact definition can be found in Sect. 4, Definition 8).

Theorem 2. *Let $\delta < N^{1/5}/10$ and $c = N^{o(1)}$. Any $O(\log N)$ depth (c, δ) variable close syntactically multilinear circuit computing the polynomial g requires size $2^{\Omega(N^{1/5}/\log N)}$.*

Our Approach. Our proofs are a careful adaptation of the rank argument developed by Raz [13]. This involves upper bounding the dimension of the partial derivative matrix (Definition 4) of the given model under a random partition of variables. However, upper bounding the rank of the partial derivative matrix of a syntactic multilinear ABP is a difficult task and there are no known methods for the same. To the best of our knowledge, there is no non-trivial upper bound on the rank of the partial derivative matrix of polynomials computed by ABPs (or special classes of ABPs) under a random partition.

Our crucial observation is, even though conversion of a syntactic multilinear ABP of size s into a syntactic multilinear formula blows the size to $s^{O(\log s)}$, the resulting formula is much simpler in structure than an arbitrary syntactic multilinear formula of size $n^{O(\log s)}$. For each of the special classes of multilinear ABPs (ROABPS, r-pass ABPs etc) considered in the article, we identify and exploit the structural limitations of the formula obtained from the corresponding ABP to prove upper bound on the rank of the partial derivative matrix under a random partition. Overall our approach to upper bound the rank can be summarized as follows:

1. Convert the given multilinear ABP P of size s to a multilinear formula Φ of size $s^{O(\log s)}$ (Lemmas 4, 7 and 9);
2. Identify structural limitations of the resulting formula Φ and exploit it to prove upper bound on the rank of the partial derivative matrix under a random partition (Lemmas 6, 2, 10 and 11);
3. Exhibit a hard polynomial that has full rank under all partitions. (Lemma 3)

Related Results. Anderson et al. [2] obtained exponential lower bound against oblivious read k branching programs. Kayal et al. [8] obtained a polynomial that can be written as sum of three ROABPs each of polynomial size such that any ROABP computing it has exponential size. Arvind and Raja [3] show that if permanent can be written as a sum of $N^{1-\epsilon}$ many ROABPs, then at least one of the ROABP must be of exponential size. Further, sum of read-once polynomials, a special class of oblivious ROABPs was considered by Mahajan and Tawari [9], independently by the authors [11]. Recently, Chillara et al. [5] show that any $o(\log N)$ depth syntactic multilinear circuit cannot a polynomial that is computable by width-2 ROABPs.

The existing lower bounds against ROABPs or sm-ABPs, implicitly restrict the number of different orders in which the variables can be read along any s to t path. In fact, the lower bound given in Arvind and Raja [3] allows only $N^{1-\epsilon}$ different ordering of the variables. To the best of our knowledge, this is the state of art with respect to the number of variable orders allowed in ABPs. Without any restriction on the orderings, the best known lower bound is only quadratic upto poly logarithmic factors [1]. In this light, our results in Theorems 1 and 3 can be seen as the first of the kind where the number of different orders allowed is sub-exponential. Certain proofs have been omitted due to space constraints.

2   Preliminaries

In this section we include necessary definitions and notations used. We begin with the formal definition of the models considered in this article.

An *arithmetic circuit* C over a field $\mathbb{F}$ and variables $X = x_1, \ldots, x_N$ is a directed acyclic graph with vertices of in-degree 0 or 2 and exactly one vertex of out-degree 0 called the output gate. The vertices of in-degree 0 are called input gates and are labeled by elements from $X \cup \mathbb{F}$. The vertices of in-degree 2 are labeled by either $+$ or $\times$. Every gate in C naturally computes a polynomial. The polynomial f computed by C is the polynomial computed by the output gate of the circuit. The *size* of an arithmetic circuit is the number of gates in C and *depth* of C is the length of the longest path from an input gate to the output gate in C. An *arithmetic formula* is an arithmetic circuit where the underlying undirected graph is a tree.

An *Algebraic Branching Program* P (ABP for short) is a layered directed acyclic graph with two special nodes, a start node s and a terminal node t. Each edge in P is labeled by either an $x_i \in X$ or $\alpha \in \mathbb{F}$. The size of p is the total number of nodes, width is the maximum number of nodes in any layer of P. Each path γ from s to t in P computes the product of the labels of the edges in γ which is a polynomial. The ABP P computes the sum over all s to t paths of such polynomials.

An ABP P is said to be *syntactic multilinear* (sm-ABP for short) if every variable occurs at most once in every path in P. An ABP is said to be *oblivious* if for every layer L in P there is at most one variable that labels edges from L.

Definition 1 (Read-Once Oblivious ABP). *An ABP P is said to be Read-Once Oblivious (ROABP for short) if P is an oblivious and each $x_i \in X$ appears as edge label in at most one layer.*

In any Oblivious ROABP, every variable appears in exactly one layer and all variables in a particular layer are the same. Hence, variables appear in layers from the start node to the terminal node in the *variable order* $x_{i_1}, x_{i_2}, \ldots, x_{i_n}$ where $(i_1, i_2, \ldots, i_n) \in S_n$ is a permutation on $[n]$. A natural generalization of ROABPs is the r-pass ABPs defined in [2]:

Definition 2 (r-pass multilinear ABP). *An oblivious sm-ABP P is said to be r-pass if there are permutations $\pi_1, \pi_2, \ldots, \pi_r \in S_n$ such that P reads the variables from s to t in the order $(x_{\pi_1(1)}, x_{\pi_1(2)}, \ldots, x_{\pi_1(n)}), \ldots, (x_{\pi_r(1)}, x_{\pi_r(2)}, \ldots, x_{\pi_r(n)})$.*

A polynomial $f \in \mathbb{F}[X]$ is s-sparse if it has atmost s monomials with non-zero coefficients.

Definition 3 (α-Sparse ROABP). *[6] An $d+1$ layer ABP P is said to be an α-sparse ROABP if there is a partition of X into $d = \Theta(N^\alpha)$ sets $X_1, X_2, \ldots, X_d$ with $|X_i| = N/d$ such that every edge label in layer L_i is an s-sparse multilinear polynomial in $\mathbb{F}[X_i]$ for $s = N^{O(1)}$.*

Let Ψ be a circuit over $\mathbb{F}$ with $X = \{x_1, \ldots, x_N\}$ as inputs. For a gate v in Ψ, let X_v denote the set of variables that appear in the sub-circuit rooted at v. The circuit Ψ is said to be *syntactic multilinear* (sm for short), if for every $\times$ gate $v = v_1 \times v_2$ in Ψ, we have $X_{v_1} \cap X_{v_2} = \emptyset$. By definition, every syntactic multilinear circuit is a multilinear circuit. In [13], it was shown that every multilinear formula can be transformed into a syntactic multilinear formula of the same size, computing the same polynomial.

Let Ψ be a circuit (formula) and v be a gate in Ψ. The *product-height* of v is the maximum number of $\times$ gates along any v to root path in Ψ.

We now review the partial derivative matrix of a polynomial introduced in [13]. Let $Y = \{y_1, \ldots, y_m\}$ and $Z = \{z_1, \ldots, z_m\}$ be disjoint sets of variables.

Definition 4 (Partial Derivative Matrix). *Let $f \in \mathbb{F}[Y, Z]$ be a polynomial. The partial derivative matrix of f (denoted by M_f) is a $2^m \times 2^m$ matrix defined as follows. For monic multilinear monomials p and q in variables Y and Z respectively, the entry $M_f[p, q]$ is the coefficient of the monomial pq in f.*

For a polynomial f, let $\mathsf{rank}(M_f)$ denote the rank of the matrix M_f over the field $\mathbb{F}$. $\mathsf{rank}(M_f)$ is known to satisfy sub-additivity and sub-multiplicativity.

Lemma 1 [13] (Sub-additivity, sub-multiplicativity). *Let $f, g \in \mathbb{F}[Y, Z]$. Then, we have that $\mathsf{rank}(M_{f+g}) \leq \mathsf{rank}(M_f) + \mathsf{rank}(M_g)$. Further, if $\mathsf{var}(f) \cap \mathsf{var}(g) = \emptyset$, then $\mathsf{rank}(M_{fg}) = \mathsf{rank}(M_f)\mathsf{rank}(M_g)$.*

Further, since row-rank of a matrix is equal to its column rank, we have:

Lemma 2 [13]. *For $f \in \mathbb{F}[Y_1, Z_1]$, $\mathsf{rank}(M_f) \leq 2^{\min\{|Y_1|, |Z_1|\}}$, where $Y_1 \subseteq Y, Z_1 \subseteq Z$.*

For $f \in \mathbb{F}[X]$, it may be noted that the matrix M_f is dependent on the partition of the variable set X into variables in $Y \cup Z$. In most cases, partition of the variable set is not apparent. In such cases, we need to consider a distribution over the set of all such partitions. We represent a partition as a bijective function $\varphi : X \to Y \cup Z$, where $|Y| = |Z| = |X|/2$.

Let $\mathcal{D}$ be the uniform distribution on the set of all partitions $\varphi : X \to Y \cup Z$, with $|Y| = |Z| = |X|/2$.

Now, we state a property of hypergeometric distribution that will be used later.

Proposition 1 [12,15] (Hypergeometric Distribution). *Let $M_1, M_2 \leq S$ be integers. Let $\mathcal{H}(M_1, M_2, S)$ denote the distribution of size of the intersection of a random set of size M_2 and a set of size M_1 in a universe of size S. Let χ be a random variable distributed according to $\mathcal{H}(M_1, M_2, S)$:*

1. If $S^{1/2} \leq M_1 \leq S/2$ and $S/4 \leq M_2 \leq 3S/4$ then $\Pr[\chi = a] \leq O(S^{-1/4})$.

2. If $0 \leq M_1 \leq 2S/3$ and $S/4 \leq M_2 \leq 3S/4$ then $\Pr[\chi = a] \leq O(M_1^{-1/2})$ for any $a \leq M_1$.

We consider the full rank polynomial g defined by Raz and Yehudayoff [14] to prove lower bounds for all models that arise in this work.

Definition 5 (Hard Polynomial). *Let $N \in \mathbb{N}$ be integer. Let $X = \{x_1, \ldots, x_N\}$ and $\mathcal{W} = \{w_{i,k,j}\}_{i,k,j \in [N]}$. For any two integers $i, j \in \mathbb{N}$, we define an interval $[i,j] = \{k \in \mathbb{N}, i \leq k \leq j\}$. Let $\|[i,j]\|$ be the length of the interval $[i,j]$. Let $X_{i,j} = \{x_p \mid p \in [i,j]\}$ and $W_{i,j} = \{w_{i',k,j'} \mid i', k, j' \in [i,j]\}$. Let $\mathbb{G} = \mathbb{F}(\mathcal{W})$, the rational function field. For every $[i,j]$ such that $\|[i,j]\|$ is even we define a polynomial $g_{i,j} \in \mathbb{G}[X]$ as $g_{i,j} = 1$ when $\|[i,j]\| = 0$ and if $\|[i,j]\| > 0$ then, $g_{i,j} \triangleq (1 + x_i x_j) g_{i+1,j-1} + \sum_k w_{i,k,j} g_{i,k} g_{k+1,j}$. where $x_k, w_{i,k,j}$ are distinct variables, $1 \leq k \leq j$ and the summation is over $k \in [i+1, j-2]$ such that $\|[i,k]\|$ is even. Let $g \triangleq g_{1,N}$.*

Lemma 3 *[14, Lemma 4.3]. Let $X = \{x_1, \ldots, x_N\}$ and $\mathcal{W} = \{w_{i,k,j}\}_{i,k,j \in [N]}$. Let $\mathbb{G} = \mathbb{F}(\mathcal{W})$ be the set of rational functions over field $\mathbb{F}$ and $\mathcal{W}$. Let $g \in \mathbb{G}[X]$ be the polynomial in Definition 5. Then for any $\varphi \sim \mathcal{D}$, $\mathsf{rank}(M_{g^\varphi}) = 2^{N/2}$.*

3   Lower Bounds for Special Cases of sm-ABPs

In this section, we obtain exponential lower bound for sum of ROABPs and related special classes of syntactic multilinear ABPs.

3.1   Sum of ROABPs: Proof of Theorem 1

Let P be an ROABP with $\ell + 1$ layers $L_0, L_1, L_2, \ldots, L_\ell$ computing a multilinear polynomial $f \in \mathbb{F}[x_1, x_2, \ldots, x_N]$. For every $i \in \{0, 1, \ldots, \ell - 1\}$, we say a layer L_i is a *constant* layer if every edge going out of a vertex in L_i is labeled by a constant from $\mathbb{F}$, else we call the layer L_i a *variable* layer. For any *variable* layer L_i denote by $\mathsf{var}(L_i)$ the variable in X that labels edges going out of vertices in L_i. For nodes u, v in P, we denote by $[u, v]$ the polynomial computed by the subprogram with u as the start node and v as the terminal node and let $X_{u,v}$ be the set of variables that occur in P between layers containing u and v respectively. We can assume without loss of generality that P does not have any two consecutive constant layers and that every ROABP P has exactly $2N$ layers by introducing dummy constant layers in between consecutive variable layers. Further, we assume that the variables occur in P in the order $x_1, \ldots x_N$, and hence indices of variables in $X_{u,v}$ is an interval $[i,j] = \{t \in \mathbb{N} \mid i \leq t \leq j\}$ for some $i < j$. (In case of a different order π for occurrence of variables, the interval would be $[i,j] = \{\pi(i), \pi(i+1), \ldots, \pi(j)\}$.)

Approach: In order to prove Theorem 1, we use $\mathsf{rank}(M_{f^\varphi})$ as a complexity measure, where $\varphi \sim \mathcal{D}$. The outline is as follows:

1. Convert the ROABP P into a multilinear formula Φ with a small (super polynomial) blow up in size (Lemma 4).

2. Obtain a partition $B_1, \ldots, B_t$ of the variable set with $O(\sqrt{N})$ parts of almost equal size, so that there is at least one set that is highly unbalanced under a random φ drawn from $\mathcal{D}$. (Observation 1 and Lemma 6.)
3. Using the structure of the formula Φ, show that if at least on of the B_i is highly unbalanced, then the formula Φ has low rank (Lemma 5).
4. Combining with Lemma 3 gives the required lower bound.

The following lemma lists useful properties of the straightforward conversion of an ROABP into a multilinear formula. The proof is a simple divide and conquer conversion of branching programs to formulas.

Lemma 4. *Let P be an ROABP of size s computing an $f \in \mathbb{F}[x_1, \ldots, x_N]$. Then f can be computed by a syntactic multilinear formula Φ of size $s^{O(\log N)}$ and depth $O(\log N)$ such that*

1. *Φ has an alternative of layers of $+$ and $\times$ gates; and*
2. *$\times$ gates have fan-in bounded by two; and*
3. *Every $+$ gate g in Φ computes a polynomial $[u, v]$ for some u, v in P; and*
4. *Every $\times$ gate computes a product $[u, v] \times [v, w]$, for some u, v and w in P.*
5. *The root of Φ is a $+$ gate.*

Let P be an ROABP and Φ be the syntactic multilinear formula obtained from P as in Lemma 4. Let g be a $+$ (respectively $\times$) gate in Φ computing $[u_g, v_g]$ (respectively $[u_g, v_g] \times [v_g, w_g]$) for some nodes u_g, v_g and w_g in P. Since P is an ROABP with variable order $x_1, x_2, \ldots x_N$, the set X_{u_g, v_g} (respectively $X_{u_g, v_g} \cup X_{v_g, w_g}$) corresponds to an interval I_g in $\{1, \ldots, N\}$. We call I_g the *interval associated with* g. By the construction of Φ in Lemma 4, the intervals have the following properties :

1. For any gate g in Φ at product-height i, $|I_g| \in [N/2^i - i, N/2^i + i]$.
2. For any $+$ gate g in Φ with children $g_1, \ldots, g_w$, we have $I_g = I_{g_1} = \cdots = I_{g_w}$.
3. Let $\mathcal{I}$ be the set of all distinct intervals associated with gates at product-height $\frac{\log N}{2}$ in Φ. The intervals in $\mathcal{I}$ are disjoint and $|\mathcal{I}| = \Theta(\sqrt{N})$. For any $I_j \in \mathcal{I}$, $\sqrt{N} - \log N \leq |I_j| \leq \sqrt{N} + \log N$.

We call the intervals in $\mathcal{I}$ as *blocks* $B_1, B_2, \ldots, B_t$ in Φ where $t = \Theta(\sqrt{N})$. For any block $B_\ell = [i_\ell, j_\ell]$, $X_\ell = \{x_{i_a} \mid i_\ell \leq i_a \leq j_\ell\} = \mathsf{var}(L_{i_\ell}) \cup \mathsf{var}(L_{i_\ell + 1}) \cup \cdots \cup \mathsf{var}(L_{j_\ell})$.

Let $\varphi : X \to Y \cup Z$ be a partition. We say a block B_ℓ is *k-unbalanced* with respect to φ iff $||Y \cap \varphi(X_\ell)| - |Z \cap \varphi(X_\ell)|| > k$. For any two intervals $I_1 = [i_1, j_1]$ and $I_2 = [i_2, j_2]$ we say $I_1 \subseteq I_2$ iff $i_2 \leq i_1 \leq j_1 \leq j_2$.

Observation 1. *Let P be an ROABP and Φ be the syntactic multilinear formula obtained from P and $B_1, \ldots, B_t$ be the blocks in Φ. Then, for any gate v in Φ,*

(1) If v is at a product-height $< \frac{\log N}{2}$ in Φ, then $B_i \subseteq I_v$ for some block B_i.
(2) If v is at product-height $> \frac{\log N}{2}$ in Φ, then for every $1 \leq i \leq t$, either $I_v \subseteq B_i$ or $B_i \cap I_v = \emptyset$.

(3) If v is at product-height $\frac{\log N}{2}$ in Φ, then for every $1 \leq i \leq t$, either $I_v = B_i$ or $B_i \cap I_v = \emptyset$.

We need the following before formalizing Step 3 in the approach outlined.

Definition 6 (k_B-hitting formula). Let $\varphi : X \to Y \cup Z$ be a partition and B be a k-unbalanced block in Φ with respect to φ. A gate v with product-height $\leq \frac{\log N}{2}$ in Φ is k_B-hitting if either

(i) $I_v = B$; Or
(ii) $B \subseteq I_v$ and,
   - If v is a sum gate with children $v_1, \ldots, v_w$, the gates $v_1, \ldots, v_w$ are k_B-hitting.
   - If v is a product gate with children v_1, v_2, then atleast one of v_1 or v_2 are k_B-hitting.

A formula Φ is k_B-hitting with respect to φ if the root r is k_B-hitting for some k-unbalanced block $B \in \{B_1, B_2, \ldots, B_t\}$ where $t = \Theta(\sqrt{N})$.

In the following, we note that the partial derivative matrix of k_B-hitting formulas have low rank. The proof is by induction on the structure of the formula.

Lemma 5. Let P be an ROABP computing f and Φ_P be the multilinear formula obtained from P computing f. Let $\varphi \sim \mathcal{D}$ such that block B is k-unbalanced in Φ with respect to φ. Let v be a gate in Φ that is k_B-hitting then $\mathrm{rank}(M_{f_v^\varphi}) \leq |\Phi_v| \cdot 2^{|X_v|/2 - k/2}$.

Observation 2. Let $\varphi : X \to Y \cup Z$ be a partition and B be a k-unbalanced block in Φ with respect to φ.

1. If $a + $ gate v in Φ with children $v_1, \ldots, v_w$ is not k_B-hitting then $I_{v_j} \cap B = \emptyset$ for some $j \in [w]$.
2. If $a \times$ gate v with children v_1, v_2 is not k_B-hitting then $I_{v_1} \cap B = \emptyset$ and $I_{v_2} \cap B = \emptyset$.

Further, we observe that, proving that a formula Φ is k_B-hitting with respect to a partition, is equivalent to showing existence of a k-unbalanced block among $B_1, \ldots, B_t$.

Observation 3. Let $B_1, \ldots, B_t$ be the blocks of the formula Φ obtained from an ROABP P. Let $B \in \{B_1, \ldots, B_t\}$ be a k-unbalanced block with respect to a partition φ. Then, Φ is k_B-hitting with respect to φ.

In the remainder of the section, we estimate the probability that at least one of the blocks among $B_1, \ldots, B_t$ is k-unbalanced. This is a straightforward application of the property of the hyper geometric distribution.

Lemma 6. Let P be an ROABP computing a polynomial $f \in \mathbb{F}[x_1, \ldots, x_N]$ and Φ_P be the syntactic multilinear formula computing f. Let $\varphi \sim \mathcal{D}$. Then, for any $k \leq N^{1/5}$, there exists a block B in Φ such that such that

$$\Pr_{\varphi \sim \mathcal{D}}[\Phi \text{ is } k_B\text{-hitting }] \geq 1 - 2^{-\Omega(\sqrt{N} \log N)}$$

Corollary 1. *Let P be an ROABP and Φ_P be the multilinear formula obtained from P computing f. Let $\varphi \sim \mathcal{D}$. Then with probability $1 - 2^{-\Omega(\sqrt{N}\log N)}$, $\mathsf{rank}(M_{f^\varphi}) \leq |\Phi| \cdot 2^{N/2 - N^{1/5}}$.*

Proof. Follows directly from Lemmas 5 and 6.

We are ready to combine the above to prove Theorem 1:

Proof (of Theorem 1). Suppose, f_i has an ROABP P_i of size s_i. Then, by Lemma 4, there is a multilinear formula Φ_i computing f_i. By Lemma 6, probability that Φ_i is not k_B-hitting is at most $2^{-\Omega(\sqrt{N}\log N)}$. Therefore, if $m < 2^{cN^{1/5}}$, there is a partition $\varphi \sim \mathcal{D}$ such that Φ_i is k_B-hitting for every $1 \leq i \leq m$. Therefore, by Lemma 5, there is a partition $\varphi \sim \mathcal{D}$ such that $\mathsf{rank}(M_{g^\varphi}) \leq m \cdot s^{O(\log N)} \cdot 2^{N/2 - k}$. If $m < 2^{c(N^{1/5})}/s^{\log N}$, we have $\mathsf{rank}(M_{g^\varphi}) < 2^{N/2}$, a contradiction to Lemma 3.

3.2 Lower Bound Against Multilinear r-pass ABPs

In this section, we extend Theorem 1 to the case of r-pass ABPs. Let P be a multilinear r-pass ABP of size s having ℓ layers. Let $\pi_1, \pi_2, \ldots, \pi_r$ be the r orders associated with P. Lemmas 7 and 8 show that techniques in Sect. 3.1 can be adapted to the case of r-pass sm-ABPs.

Lemma 7. *Let P be a multilinear r-pass ABP of size s having ℓ layers computing a polynomial $f \in \mathbb{F}[x_1, \ldots, x_N]$. Then there exists a syntactic multilinear formula $\Psi_P = \Psi_1 + \Psi_2 + \cdots + \Psi_t$, $t = s^{O(r)}$ where each Ψ_i is a syntactic multilinear formula obtained from an ROABP.*

Lemma 8. *Let P be a multilinear r-pass ABP computing a polynomial $f \in \mathbb{F}[x_1, \ldots, x_N]$ and $\Psi_P = \Psi_1 + \Psi_2 + \cdots + \Psi_t$, $t = s^{O(r)}$ be the syntactic multilinear formula computing f. Let $\varphi \sim \mathcal{D}$ and $k \leq N^{1/5}$. Then with probability $1 - 2^{-\Omega(\sqrt{N}\log N)}$, $\mathsf{rank}(M_f) \leq |\Psi| \cdot 2^{N/2 - k/2}$.*

Combining the above Lemmas with Lemma 3 we get the following.

Theorem 3. *Let $f_1, \ldots f_m$ be polynomials computed by multilinear r-pass ABPs of size $s_1, s_2, \ldots, s_m$ respectively such that $g = f_1 + \cdots + f_m$. Then, $m = \frac{2^{\Omega(N^{1/5})}}{s^{c(r + \log N)}}$, where c is a constant and $s = \max\{s_1, s_2, \ldots, s_m\}$.*

3.3 Lower Bound Against Sum of α-sparse ROABPs

In this section we prove lower bounds against sum of α-sparse ROABPs for $\alpha > 1/10$. We begin with a version of Lemma 4 for sparse ROABPs. The proof is similar to that of Lemma 4.

Lemma 9. *Let $\alpha \geq 1/10$ and P be an α-sparse ROABP of size s computing a polynomial $f \in \mathbb{F}[x_1, \ldots, x_N]$. Then f can be computed by a syntactic multilinear formula Φ of size $s^{O(\log d)}$ and depth $O(\log d)$ such that the leaves are labelled with sparse polynomials in X_i for some $1 \leq i \leq d$, where $d = \Theta(N^\alpha)$.*

Lemma 10. *Let P be an α-sparse ROABP computing $f \in \mathbb{F}[x_1, \ldots, x_N]$ and Φ be the syntactic multilinear formula computing f. Let $\varphi \sim \mathcal{D}$. Then, for any $k \leq N^{(1-\alpha)/4}$, there exists an $i \in [d]$ such that X_i is k-unbalanced with probability atleast $1 - 2^{\Omega(-N^{1/10} \log N/16)}$.*

Our first observation is that $X_1, \ldots, X_d$ can be treated as blocks $B_1, \ldots, B_d$ as in Sect. 3.1:

Observation 4. *If X_r is k-unbalanced , then Φ is k_B-hitting for $B = X_r$.*

Note that for any t-sparse polynomial f and any $\varphi \sim \mathcal{D}$, $\mathsf{rank}(M_{f\varphi}) \leq t$.

Corollary 2. *Let P be a α-sparse ROABP computing f and Φ be the multilinear formula obtained from P. Let $\varphi \sim \mathcal{D}$. Then with probability $1 - 2^{\Omega(-N^{1/10} \log N/16)}$, $\mathsf{rank}(M_{f\varphi}) \leq |\Phi| \cdot t \cdot 2^{N/2 - N^{9/40}}$, where t is the sparsity of the polynomials involved in the α-sparse ROABP computing f.*

Combining the above with Lemma 3, we get the following similar to Theorem 1:

Theorem 4. *Let $f_1, \ldots, f_m$ be polynomials computed by α-sparse ROABPs of size $s < 2^{N^{9/40}/\log N}$, for $\alpha > 1/10$ such that $g = f_1 + \cdots + f_m$. Then $m \geq 2^{N^{1/11}}$.*

4   Lower Bounds for Special Classes of Multilinear Circuits

In this section, we develop a framework for proving super polynomial lower bound against syntactic multilinear circuits and ABPs based on Raz [13]. Our approach involves a more refined analysis of central paths introduced by Raz [13].

Definition 7 (Central Paths). *Let Φ be a syntactic multilinear formula. For node v in Φ, let X_v denote the set of variables appearing in the sub-formula rooted at v. A leaf to root path $\rho = v_1, \ldots, v_\ell$ in Φ is said to be central, if $|X_{v_{i+1}}| \leq 2|X_{v_i}|$ for $1 \leq i \leq \ell - 1$.*

For a leaf to root path $\rho : v_1, \ldots, v_\ell$ in Φ, $X_{v_1} \subseteq \ldots \subseteq X_{v_\ell}$ is called the *signature* of the path ρ. A signature $X_{v_1} \subseteq \ldots \subseteq X_{v_\ell}$ is called central if $|X_{v_{i+1}}| \leq 2|X_{v_i}|$ for $1 \leq i \leq \ell - 1$. Let $\varphi : X \to Y \cup Z$ be a partition. A central signature $X_{v_1} \subseteq \ldots \subseteq X_{v_\ell}$ of a formula Φ is said to be k-unbalanced with respect to φ if for some $i \in [\ell]$, X_{v_i} is k-unbalanced with respect to φ , i.e., $|\varphi(X_{v_i}) \cap Y - \varphi(X_{v_i}) \cap Z| \geq k$. The formula Φ is said to be k-weak with respect to φ, if every central signature that terminates at the root is k-unbalanced. Our observation is that, we can replace central paths in [13] (Lemma 4.1) with central signatures and use the same arguments as in [13].

Observation 5. *Let $\varphi : X \to Y \cup Z$ be a partition of $X = \{x_1, \ldots, x_N\}$. Let Φ be any multilinear formula compuitng a polynomial $f \in \mathbb{F}[x_1, \ldots, x_N]$.*

1. If Φ is k-weak with respect to φ, then $\mathsf{rank}(M_{f\varphi}) \leq |\Phi| \cdot 2^{N/2-k}$.

2. Let $C : X_{v_1} \subseteq X_{v_2} \subseteq \cdots \subseteq X_{v_\ell}$ be a central signature in Φ such that $k < |X_{v_1}| \leq 2k$. Then $\mathsf{Pr}_{\varphi \sim \mathcal{D}}[C$ is not k-unbalanced$] = N^{-\Omega(\log N)}$.

Unfortunately, it can be seen that even when P is an ROABP the number of central signatures in a formula from an ROABP can be $N^{\Omega \log N}$. We show that a careful bound on the number of central signatures yields super-polynomial lower bounds for sum of ROABPs. Now, we consider a subclass of syntactic multilinear circuits where we can show that the equivalent formula obtained by duplicating nodes as and when necessary, has small number of central signatures. To start, we consider a refinement of the set of central signatures of a formula.

Let Φ be a syntactically multilinear formula of $O(\log N)$ depth. Two central paths ρ_1 and ρ_2 in Φ are said to *meet at* $\times$, if their first common node along leaf to root is labeled by $\times$. A set $\mathcal{T}$ of central paths in Φ is said to be $+$-*covering*, if for every central path $\rho \notin \mathcal{T}$, there is a $\rho' \in \mathcal{T}$ such that ρ and ρ' meet at $\times$. A *signature-cover* $\mathcal{C}$ of Φ is the set of all signatures of the $+$-*covering* set T of central paths in Φ.

Lemma 11. *Let Φ be a syntactic multilinear formula. Let φ be a partition. If there is a signature-cover $\mathcal{C}$ of Φ such that every signature in $\mathcal{C}$ is k-unbalanced with respect to φ, then $\mathsf{rank}(M_{f\varphi}) \leq |\Phi| \cdot 2^{N/2-k/2}$.*

Let $X_1, \ldots, X_r \subseteq X$, be subsets of variables. Let $\Delta(X_i, X_j)$ denote the Hamming distance between X_i and X_j, i.e, $\Delta(X_i, X_j) = |(X_i \setminus X_j) \cup (X_j \setminus X_i)|$. Let $C_1 : X_{11} \subseteq X_{12} \subseteq \cdots \subseteq X_{1\ell}$ and $C_2 : X_{21} \subseteq X_{22} \subseteq \cdots \subseteq X_{2\ell}$ be two central signatures in Φ. Define $\Delta(C_1, C_2) = \max_{1 \leq i \leq \ell} \Delta(X_{1i}, X_{2i})$. Let $\mathcal{C}$ be signature-cover in Φ. For $\delta > 0$, a δ-*cluster* of $\mathcal{C}$ is a set of signatures $C_1, \ldots, C_t \in \mathcal{C}$ such that for every $C \in \mathcal{C}$, there is a $j \in [t]$ with $\Delta(C, C_j) \leq \delta$. The following is immediate:

Observation 6. *Let $\mathcal{C}$ be a signature-cover, and $C_1, \ldots, C_t$ be a δ-cluster of $\mathcal{C}$. If φ is a partition of X such that for every $i \in [t]$, signature C_i is k-unbalanced, then for every $C \in \mathcal{C}$, signature C is $k - 2\delta$ unbalanced.*

We are ready to define the special class of sm-circuits where the above mentioned approach can be applied. For $X_1, \ldots, X_r \subseteq X$ and $\delta > 0$, a δ-*equivalence class* of $X_1, \ldots, X_r$, is a minimal set of indices $i_1, \ldots, i_t$ such that for $1 \leq i \leq r$, there is an $i_j, 1 \leq j \leq t$ such that $\Delta(X_i, X_{i_j}) \leq \delta$.

Definition 8. *Let $\delta \leq N \in \mathbb{N}$. Let Ψ be an sm-circuit with alternating layers of $+$ and $\times$ gates. Ψ is said to be (c, δ)-variable close, if for for every $+$ gate $v = v_{11} \times v_{12} + \cdots + v_{r1} \times v_{r2}$, there are indices $b_1, b_2, \ldots, b_r \in \{1, 2\}$ such that there is a δ-equivalence class of $X_{v_{1b_1}}, \ldots, X_{v_{rb_r}}$ with at most c different sets.*

We show (c, δ) close circuits have small number of signatures.

Lemma 12. *Let Ψ be a (c, δ)-variable close syntactic multilinear arithmetic circuit of size s and depth $O(\log N)$. Let Φ be the syntactic multilinear formula obtained by duplicating gates in Ψ as necessary. Then there is a signature-cover $\mathcal{C}$ for Φ such that $\mathcal{C}$ has a δ-cluster consisting of at most $c^{O(\log N)}$ sets.*

Proof (of Theorem 2). Let Ψ be a (c, δ) variable close circuit of depth $O(\log N)$ and Φ be formula obtained by duplicating nodes in Ψ. By Lemma 12 $\{C_1, \ldots, C_t\}$ be a δ-cluster of a signature-cover $\mathcal{C}$ of Φ for $t = N^{o(\log N)}$.By Observations 5 and 6, probability that there is a signature in $\mathcal{C}$ that is not $k - 2\delta$ unbalanced is at most $t \cdot N^{-\Omega(\log N)} < 1$ for $\varphi \sim \mathcal{D}$. Then, there is a φ such that every signature in $\{C_1, \ldots, C_t\}$ is $k - 2\delta$ unbalanced. By Lemma 11, there is φ with $\mathsf{rank}(M_{g^\varphi}) \leq |\Phi| \cdot 2^{N/2 - (k - 2\delta)} < 2^{N/2}$ for $s < 2^{k/10 \log N}$, a contradiction to Lemma 3.

References

1. Alon, N., Kumar, M., Volk, B.L.: An almost quadratic lower bound for syntactically multilinear arithmetic circuits. ECCC **24**, 124 (2017). https://eccc.weizmann.ac.il/report/2017/124
2. Anderson, M., Forbes, M.A., Saptharishi, R., Shpilka, A., Volk, B.L.: Identity testing and lower bounds for read-k oblivious algebraic branching programs. In: CCC, pp. 30:1–30:25 (2016). https://doi.org/10.4230/LIPIcs.CCC.2016.30
3. Arvind, V., Raja, S.: Some lower bound results for set-multilinear arithmetic computations. Chicago J. Theoret. Comput. Sci. (2016). http://cjtcs.cs.uchicago.edu/articles/2016/6/contents.html
4. Baur, W., Strassen, V.: The complexity of partial derivatives. Theoret. Comput. Sci. **22**, 317–330 (1983). https://doi.org/10.1016/0304-3975(83)90110-X
5. Chillara, S., Limaye, N., Srinivasan, S.: Small-depth multilinear formula lower bounds for iterated matrix multiplication, with applications. In: STACS (2018). http://arxiv.org/abs/1710.05481
6. Forbes, M.: Polynomial identity testing of read-once oblivious algebraic branching programs. Ph.D. thesis, Massachusetts Institute of Technology (2014)
7. Jansen, M.J.: Lower bounds for syntactically multilinear algebraic branching programs. In: Ochmański, E., Tyszkiewicz, J. (eds.) MFCS 2008. LNCS, vol. 5162, pp. 407–418. Springer, Heidelberg (2008). https://doi.org/10.1007/978-3-540-85238-4_33
8. Kayal, N., Nair, V., Saha, C.: Separation between read-once oblivious algebraic branching programs (ROABPs) and multilinear depth three circuits. In: STACS, pp. 46:1–46:15 (2016). https://doi.org/10.4230/LIPIcs.STACS.2016.46
9. Mahajan, M., Tawari, A.: Sums of read-once formulas: how many summands are necessary? Theoret. Comput. Sci. **708**, 34–45 (2018). https://doi.org/10.1016/j.tcs.2017.10.019
10. Nisan, N.: Lower bounds for non-commutative computation (extended abstract). In: STOC, pp. 410–418 (1991). https://doi.org/10.1145/103418.103462
11. Ramya, C., Rao, B.V.R.: Sum of products of read-once formulas. In: FSTTCS, pp. 39:1–39:15 (2016). https://doi.org/10.4230/LIPIcs.FSTTCS.2016.39
12. Raz, R.: Separation of multilinear circuit and formula size. Theory Comput. **2**(6), 121–135 (2006). https://doi.org/10.4086/toc.2006.v002a006
13. Raz, R.: Multi-linear formulas for permanent and determinant are of super-polynomial size. J. ACM **56**(2) (2009). https://doi.org/10.1145/1502793.1502797
14. Raz, R., Yehudayoff, A.: Balancing syntactically multilinear arithmetic circuits. Comput. Complex. **17**(4), 515–535 (2008). https://doi.org/10.1007/s00037-008-0254-0
15. Saptharishi, R.: A survey of lower bounds in arithmetic circuit complexity (2015). https://github.com/dasarpmar/lowerbounds-survey
16. Valiant, L.G.: Completeness classes in algebra. In: STOC, pp. 249–261 (1979). https://doi.org/10.1145/800135.804419

Approximation Algorithms on Multiple Two-Stage Flowshops

Guangwei Wu[1,2(✉)] and Jianer Chen[3,4]

[1] School of Information Science and Engineering, Central South University,
Changsha, People's Republic of China
will99031827@hotmail.com
[2] College of Computer and Information Engineering, Central South University
of Forestry and Technology, Changsha, People's Republic of China
[3] School of Computer Science & Education Software, Guangzhou University,
Guangzhou, People's Republic of China
[4] Department of Computer Science and Engineering, Texas A&M University,
College Station, USA

Abstract. This paper considers the problem of scheduling multiple two-stage flowshops that minimizes the makespan, where the number of flowshops is part of the input. We study the relationship between the problem and the classical MAKESPAN problem. We prove that if there exists an α-approximation algorithm for the MAKESPAN problem, then for the multiple two-stage flowshop scheduling problem, we can construct a 2α-approximation algorithm for the general case, and $(\alpha + 1/2)$-approximation algorithms for two restricted cases. As a result, we get a $(2 + \epsilon)$-approximation algorithm for the general case and a $(1.5 + \epsilon)$-approximation algorithm for the two restricted cases, which significantly improve the previous approximation ratios 2.6 and 11/6, respectively.

Keywords: Scheduling · Flowshops · Approximation algorithm
MAKESPAN

1 Introduction

This paper studies the scheduling problem of two-stage jobs on multiple two-stage flowshops. A job is a *two-stage job* if it consists of two operations: the *R-operation* and the *T-operation*. A flowshop is a *two-stage flowshop* if it contains two processors that can run in parallel: the *R-processor* and the *T-processor*. In this scheduling model, if a two-stage job is assigned to a two-stage flowshop, then both operations of the job are executed in the flowshop in such a way that the *T*-operation cannot be started by the *T*-processor of the flowshop until the *R*-operation has been finished by the *R*-processor of the same flowshop.

This work is supported by the National Natural Science Foundation of China under grants 61420106009, 61672536 and 61472449, Scientific Research Fund of Hunan Provincial Education Department under grant 16C1660.

L. Wang and D. Zhu (Eds.): COCOON 2018, LNCS 10976, pp. 713–725, 2018.
https://doi.org/10.1007/978-3-319-94776-1_59

Correspondingly, a *schedule* on a set of two-stage jobs includes an *assignment* for each job to a flowshop, and for each flowshop, an *execution order* of the R- and T-operations of the jobs assigned to that flowshop. Our objective is to construct a schedule of the two-stage jobs on m two-stage flowshops that minimizes the makespan, i.e., the completion time of the last operation.

The scheduling model derives from current research in data centers and cloud computing. A modern data center in cloud usually contains hundreds of thousands of servers [1]. Cloud computing is a paradigm for hosting and delivering services over network, which stores software and data as resources in the servers in data center, and allows clients to dynamically request these resources as services [17]. The resources are usually large in size thus are stored in second memory such as disks or flashes [2]. When a request for a resource from a client arrives at a server, the server needs to read the resource from the second memory to the main memory (i.e., the R-operation), and then send it to the client over the network (i.e., the T-operation). Therefore, each request has to pass through two stage operations in a server, and each server can be regarded as a two-stage flowshop. Given a set of requests, scheduling them on the servers in the data center to minimize the completion time of the last request, is exactly the scheduling model studied in this paper and is thus meaningful in practice.

Some characteristics should be taken into account when studying this scheduling model. The number of servers in data center is large and may vary frequently due to the issues such as economic factor and energy consumption factor [5,11]. Thus, it is natural to consider the number of servers as part of the input rather than a fixed constant in a scheduling algorithm. The servers in a data center are usually divided into clusters according to services they provide [2], and each request to a server in a cluster may be inclined towards one side: for some services requiring high reliability of data, the implementation of enterprise hard disk drives, which exceed the MTBF (mean time between failure) by sacrificing the transfer rate, makes the R-operation consume more time than the T-operation; on the other hand, for some services requiring a high I/O rate to deal with a huge requests, the use of cache system and SSDs (solid state drives) will make the R-operation less expensive than the T-operation.

Scheduling two-stage jobs on a single two-stage flowshop is the classical TWO-STAGE FLOWSHOP problem, that can be solved in polynomial time by sorting the jobs into Johnson's order [14]. Therefore, when scheduling two-stage jobs on multiple flowshops, if the job assignment to flowshops is given, an optimal schedule of the jobs on each flowshops can be easily obtained. Unfortunately, job assignment to flowshops is intractable, because the classical MAKESPAN problem can be viewed as a simpler case of our scheduling problem: each job is a one-stage job with its R-operation being 0, which is NP-hard even when the number m of machines is a fixed constant [3], and becomes strongly NP-hard when m is part of the input [7]. It is very likely that no algorithm can yield an optimal assignment in polynomial time.

As a result, it is of interest to study algorithms on this scheduling model that are guaranteed to yield, in polynomial time, a schedule that is close to the optimum. For a scheduling problem whose objective is to minimize the makespan, an α-*approximation algorithm*, achieving the *approximation ratio* α, is a polynomial time algorithm that for every instance of the problem produces a schedule with the makespan bounded by α times the minimum makespan [16, 20].

In the approximation algorithms proposed in this paper for scheduling two-stage jobs on multiple flowshops, the job assignment process is dealt with by approximation algorithms for the classical MAKESPAN problem, by transforming the two-stage jobs into one-stage jobs, then applying the algorithms for the MAKESPAN problem on these one-stage jobs, and finally constructing the assignment for the two-stage jobs from that for the one-stage jobs. The approximation algorithms then arrange the jobs assigned to each flowshop into Johnson's order. Therefore we will review some results in the scheduling literature about the classical MAKESPAN problem, and later the parallel two-stage flowshop problem.

The MAKESPAN problem has been extensively studied in the scheduling literature. Graham gave the well-known *ListRanking* algorithm for the problem with an approximation ratio $(2 - 1/m)$ [8], then, later further showed that if the jobs are sorted into an order such that the next job to be scheduled is the one with Longest Processing Time (i.e., LPT rule), then the approximation ratio of the *ListRanking* algorithm is $((4 - 1/m)/3)$ [9]. By studying the relationship between the MAKESPAN problem and the BIN-PACKING problem, Coffman *et al.* [4] gave their *MULTIFIT* algorithm and proved that this algorithm can achieve an approximation ratio 1.22. Hochbaum and Shmoys provided a polynomial-time approximation scheme [13] based on the approach of dual approximation algorithms. When the number m of machines is a fixed constant, there exist fully polynomial-time approximation schemes [9, 18].

Compared to the rich literature on the MAKESPAN problem, scheduling multiple two-stage flowshops had not received much research until recently, and most studies focused on the case when the number m of the flowshops is a fixed constant. Kovalyov seems the first to study this problem [15]. He *et al.* gave a mixed-integer programming formulation and a heuristic algorithm [12]. In order to cope with the hybrid flowshop problem, Vairaktarakis and Elhafsi proposed a formulation that leads to a pseudo-polynomial time algorithm for the problem when the number m is 2 [19]. Zhang and Velde provided two approximation algorithms with the approximation ratios $3/2$ and $12/7$, respectively, when the number m of flowshops equals 2 and 3 [24]. Based on a similar formulation to that in [19], Dong *et al.* [6] gave a pseudo-polynomial time algorithm for a fixed constant m, and constructed a fully polynomial-time approximation scheme under a fairly standard procedure. Wu *et al.* [21] proposed a new formulation totally different from that in [6, 19], which leads to a fully polynomial-time approximation scheme with improved running time for the problem on a fixed number m of flowshops.

Wu *et al.* also studied this scheduling problem when the number of flowshops is part of the input. The paper [22] dealt with two restricted cases of this model: for the case where each job has a more time-consuming R-operation, they proposed an online 2-competitive algorithm and an offline 11/6-approximation algorithm; for the second case where each job has a more time-consuming T-operation, they gave an online 5/2-competitive algorithm and an offline 11/6-approximation algorithm. They also provided a 2.6-approximation algorithm for the two-stage flowshops problem in the general case [23].

Our paper is organized as follows. A formal presentation of the scheduling model is given in Section 2, as well as some preliminary results. Section 3 gives an approximation algorithm for the scheduling model in the general case, which achieves an approximation ratio $2+\epsilon$ in time $O((2n/\epsilon)^{4/\epsilon^2}+n\log n)$ and thus significantly improves the ratio 2.6 of the algorithm in [23]. For the scheduling model in two restricted cases where each R-operation is more (less, respectively) expensive than the corresponding T-operation, Section 4 presents another approximation algorithm with the approximation ratio $1.5 + \epsilon$ and with the running time $O((n/\epsilon)^{1/\epsilon^2}+n\log n)$. It is also shown that the offline 11/6-approximation algorithms in [22] can be viewed as special cases of this approximation algorithm.

2   Preliminaries

For scheduling a given two-stage job set $G = \{J_1,\ldots,J_n\}$ on a system of m identical two-stage flowshops $\{M_1,\ldots,M_m\}$, we make some assumptions: each two-stage job $J_i = (r_i,t_i)$ of two integers consists of an R-operation and a T-operation, and each two-stage flowshop contains an R-processor and a T-processor, which can process, in parallel, the R- and T-operations of the jobs assigned to it; if a job is assigned to a flowshop, its R-operation and T-operation must be executed by the R-processor and the T-processor of this flowshop, respectively, with the cost R-time r_i and the cost T-time t_i, under the restriction that the T-operation cannot start unless the R-operation is finished; there is no precedence constraint among the jobs; and preemption is not allowed.

Given a schedule S for the two-stage job set on m two-stage flowshops, the *completion time* of a flowshop is the time when the flowshop finishes the execution of the last operation on it under the schedule S, and the *makespan C_{max}* is the maximum completion time over all flowshops. The objective of a schedule is to minimize the makespan. Using the three-field notation $\alpha|\beta|\gamma$ suggested by Graham *et al.* [10], we refer to the multiple two-stage flowshops scheduling problem as $P|2\mathrm{FL}|C_{\max}$ for general case, and as $P|2\mathrm{FL}_{R\geq T}|C_{\max}$ and $P|2\mathrm{FL}_{R\leq T}|C_{\max}$ for the two restricted cases respectively, when m is part of the input.

$P_1|2\mathrm{FL}_{R\geq T}|C_{\max}$ is the classical TWO-STAGE FLOWSHOP problem, which can be solved in time $O(n\log n)$ by sorting the given n jobs into Johnson's order [14]. The Johnson's order, which will be used as the permutation process of the schedules in the following sections, is given formally as follows:

A two-stage job sequence is in Johnson's order if the sequence is divided into two subsequences: the sequence G_1, which contains the jobs (r_f,t_f) with $r_f \leq t_f$, sorted in non-decreasing order by their R-times; and the following sequence G_2, which contains the rest jobs (r_s,t_s) with $r_s > t_s$, sorted in non-increasing order by their T-times.

The result also implies that if we are only concerned with the completion time of a flowshop, we only need to consider the schedules where the executions of the R-operations and the T-operations of the jobs on the flowshop follow the same order. Therefore in the current paper, a schedule for a flowshop M is expressed as a job sequence $\langle J_1, \ldots, J_t \rangle$, in which M executes the R-operations and the T-operations of these jobs in a way that strictly follows the ordered sequence. Let $\bar{\rho}_i$ and $\bar{\tau}_i$ denote the times when the R-operation and the T-operation of job J_i on a two-stage flowshop start, respectively. We have the following lemma.

Lemma 1 ([21]). *Let $\mathcal{S} = \langle J_1, J_2, \ldots, J_t \rangle$ be a two-stage job sequence scheduled on a single two-stage flowshop, where $J_i = (r_i, t_i)$, for $1 \leq i \leq t$. Then for all $i, 1 \leq i \leq t$, we can assume (with $\bar{\tau}_0 = t_0 = 0$): $\bar{\rho}_i = \sum_{k=1}^{i-1} r_k$; and $\bar{\tau}_i = \max\{\bar{\rho}_i + r_i, \bar{\tau}_{i-1} + t_{i-1}\}$.*

The lemma is intuitive. Given a schedule on a flowshop represented by a job sequence, the R-operation and the T-operation of a job always start as soon as they can if the objective of the schedule is to minimize the makespan. Thus, for a job, the execution of its R-operation starts once the previous R-operation in the sequence is finished. On the other hand, only when all previous T-operations and its own R-operation are all finished, the execution of the T-operation of a job can start. Therefore, the R-processor of a flowshop operates continuously, while the T-processor may have some idle times if there exists a T-operation of a job waiting for its own R-operation to finish. It is easy to see that after the R-processor finishes the R-operations of the given job sequence, the processing of the T-processor is continuous: the R-operations are finished, thus the remaining T-operations only need to wait for the previous T-operation to be finished.

We define some notations for the problem of scheduling n two-stage jobs on m identical two-stage flowshops. For each $1 \leq j \leq m$, let ρ_j and τ_j be the finishing times of the R-processor and the T-processor of the two-stage flowshop M_j, respectively, under the schedule for the jobs currently assigned to flowshops. Thus the *status* of a flowshop M_j is expressed as a pair (ρ_j, τ_j), which can be easily updated, by Lemma 1, after assigning a new job. Clearly, the finishing time τ_j after finishing all the jobs scheduled on M_j is also the completion time of M_j, and ρ_j equals the sum of the R-times of these jobs. Let ψ_j be the sum of the T-times of the current jobs assigned to M_j. We have a further observation.

Lemma 2. *For each two-stage flowshop M_j under a schedule, ρ_j plus ψ_j, where ρ_j and ψ_j are the sum of the R-times and the sum of the T-times of the current jobs on M_j by the schedule, respectively, is not larger than $2 \cdot \tau_j$, where τ_j is the finishing time of M_j under the schedule.*

3   Approximation Algorithms for $P|2FL|C_{\max}$

This section considers the relationship between the multiple two-stage flowshops problem in general case and the MAKESPAN problem, when the number of the flowshops is part of the input, and provides approximation algorithms based on these observations, which construct a schedule for the multiple two-stage flowshops problem.

Given a two-stage job $J_i = (r_i, t_i)$, we call a job J_i^p the $R\&T$-job of J_i if $J_i^p = (r_i + t_i, 0)$, i.e., the $R\&T$-job J_i^p is constructed from the job J_i by setting its R-time to the R-time plus the T-time of J_i and its T-time to 0. Let $G = \{J_1, \ldots, J_n\}$ be a two-stage job set, then its $R\&T$-job set is $G^p = \{J_1^p, \ldots, J_n^p\}$, where J_i^p is the $R\&T$-job of the job J_i for each $1 \leq i \leq n$. Note that the $R\&T$-job is actually a one-stage job with the T-time equaling 0, thus by Lemma 1, the execution order of the $R\&T$-jobs assigned to a flowshop will not affect the completion time of the flowshop, which always equals the sum of the R-times of these jobs. Let $Makespan(S)$ be the makespan C_{max} under a schedule S of scheduling a two-stage job set G on m two-stage flowshops, and denote by $Opt(G)$ the $Makespan(S)$ if S is an optimal schedule for the job set G. We have the following observation.

Lemma 3. *Given a schedule S for a two-stage job set G on m two-stage flow-shops, we construct a schedule $\bar{S}$ for the $R\&T$-job set G^p of G on m flowshops, in such a way that the job J_i^p is scheduled on a flowshop M_j if and only if J_i is scheduled on the same flowshop M_j under the schedule S. Then $Makespan(\bar{S})$ is not larger than $2 \cdot Makespan(S)$.*

Proof. Let M_h be the two-stage flowshop achieving the makespan $Makespan(\bar{S})$ denoted by $\bar{\tau}_h$, under the schedule $\bar{S}$ for the job set G^p, where $\langle J_1^p, J_2^p, \ldots, J_c^p \rangle$ is the job sequence on M_h. According to the construction of the schedule $\bar{S}$, that the job J_i^p is assigned to the flowshop M_h means that J_i is assigned to M_h under the schedule S for the job set G, thus the schedule on M_h under S is $\langle J_1, J_2, \ldots, J_c \rangle$. Combining this with the definition of the $R\&T$-job, where its R-time equals the sum of the R-time and the T-time of the original job and its T-time equals 0, it is easy to see that $\bar{\tau}_h$ equals ρ_h plus ψ_h, where ρ_h and ψ_h represent the sum of the T-times and the sum of the R-times of the jobs $\langle J_1, J_2, \ldots, J_c \rangle$, respectively. By Lemma 2, ρ_h plus ψ_h is bounded by $2 \cdot \tau_h$, where τ_h is the completion time of M_h under the schedule S, thus we have $\bar{\tau}_h \leq 2 \cdot \tau_h$ and further $\bar{\tau}_h \leq 2 \cdot Makespan(S)$, since τ_h is obviously not larger than $Makespan(S)$.   □

Suppose that S is an optimal schedule for a job set G. By Lemma 3, we can construct a schedule $\bar{S}$ for the $R\&T$-job set G^p of G with the makespan bounded by $2 \cdot Makespan(S)$. That is, there exists a schedule for the job set G^p, whose makespan is not larger than $2 \cdot Opt(G)$. Thus the above lemma can also be written as:

Theorem 1. *Given a two-stage job set G, let G^p be its corresponding $R\&T$-job set, then we have that $Opt(G^p)$ is not larger than $2 \cdot Opt(G)$.*

Now we show that by using Theorem 1, if there exists an α-approximation algorithm denoted by $AlgM_\alpha(G, m)$ for the classical MAKESPAN problem, we can get a 2α-approximation algorithm for the two-stage scheduling problem.

The 2α-approximation algorithm is presented in Fig. 1. Given a two-stage job set G and an integer m representing the number of the two-stage flowshops, we first construct the R&T-job set G^p of G. As we discussed, the jobs in G^p, which have their T-times being 0, can be considered as one-stage jobs, thus scheduling such jobs on flowshops to minimize the makespan is actually the classical MAKESPAN problem. As a result, for the R&T-job set G^p on m flowshops, applying $AlgM_\alpha$ based on the R-times of the jobs in G^p generates a schedule $\bar{S}^\alpha = \{\bar{S}_1^\alpha, \ldots, \bar{S}_m^\alpha\}$ for G^p with the makespan bounded by $\alpha \cdot \mathrm{Opt}(G^p)$, where $\bar{S}_j^\alpha$ is the job sequence scheduled on the flowshop M_j for all $1 \le j \le m$. Finally, we construct a schedule $S = \{S_1, \ldots, S_m\}$ for the original job set G from $\bar{S}^\alpha$, where the job sequence S_j for the flowshop M_j consists of the jobs whose corresponding R&T-job belongs to $\bar{S}_j^\alpha$ and is sorted in Johnson's order.

Algorithm ApproxProg-I
INPUT: a job set $G = \{J_1, \ldots, J_n\}$ of two-stage jobs, and an integer m
OUTPUT: a schedule S for G on m two-stage flowshops
1.   **for all** $1 \le i \le n$ **do** $r_i' = r_i + t_i$; $t_i' = 0$;
2.   let $G^p = \{J_1^p, \ldots, J_n^p\}$, where for each i, $J_i^p = (r_i', t_i')$;
3. $\bar{S}^\alpha = \{\bar{S}_1^\alpha, \ldots, \bar{S}_m^\alpha\}$ obtained from $AlgM_\alpha(G^p, m)$ on the job set G^p;
4.   **for all** $1 \le i \le n$ **do**
4.1   **if** J_i^p in $\bar{S}_j^\alpha$ **then** assign the job J_i to S_j;
5.   **for all** $1 \le j \le m$ **do**
5.1   sort the jobs in the job set S_j into Johnson's order;
6.   return $S = \{S_1, \ldots, S_m\}$;

Fig. 1. An approximation algorithm for $P|2\mathrm{FL}|C_{\max}$

In this algorithm, Steps 1–2 of constructing the R&T-job set of G take time $O(n)$. After calling $AlgM_\alpha(G^p, m)$, Step 4 of dividing the jobs in G into m job sets for m flowshops takes time $O(n)$. Step 5 sorts the jobs in each job set S_j into Johnson's order in time $O(n_1 \log n_1 + \ldots + n_m \log n_m)$, where n_j is the size of S_j for $1 \le j \le m$. Since $n_1 + \ldots + n_m = n$, the time complexity of this step is bounded by $O(n \log n)$. Therefore the time complexity of the algorithm excepting for the step of calling $AlgM_\alpha$ is $O(n \log n)$.

We give the relationship between the makespans of the schedules S and $\bar{S}^\alpha$.

Lemma 4. *Makespan(S) is bounded by Makespan($\bar{S}^\alpha$), where $\bar{S}^\alpha$ is a schedule of the job set G^p on m flowshops, and the schedule S of the job set G on m flowshops is constructed from $\bar{S}^\alpha$ according to Steps 4 and 5 of the algorithm* **ApproxProg-I**.

Thus, at the termination of the construction of the schedule S for G, we have

$$Makespan(S) \le Makespan(\bar{S}^a) \le \alpha \cdot \mathrm{Opt}(G^p) \le 2\alpha \cdot \mathrm{Opt}(G).$$

We conclude the discussion as the following theorem:

Theorem 2. *If the* MAKESPAN *problem has an α-approximation algorithm, then the multiple two-stage flowshops problem has a 2α-approximation algorithm.*

For the MAKESPAN problem when the number of machines is part of the input, Hochbaum *et al.* provided a polynomial-time approximation scheme [13], which runs in time $O((n/\epsilon)^{1/\epsilon^2})$ for each ϵ and has an approximation ratio $1 + \epsilon$. In addition, they presented more practical algorithms for $\epsilon = 1/5 + 2^{-k}$ and $\epsilon = 1/6 + 2^{-k}$, with the running times $O(n(k + \log n))$ and $O(n(km^4 + \log n))$, respectively. By Theorem 2, using this polynomial-time approximation scheme as $AlgM_\alpha$ makes the algorithm **ApproxProg-I** reach an approximation ratio $2 + \epsilon$ and run in time $O((2n/\epsilon)^{4/\epsilon^2} + n \log n)$, where the additional time $O(n \log n)$ is spent by the other steps except the step of calling $AlgM_\alpha$. This approximation ratio significantly improves the ratio 2.6 of the algorithm in the paper [23], which runs in time $O(n \log n)$. Similarly, when using the other two more practical algorithms as $AlgM_\alpha$, the algorithm **ApproxProg-I** achieves the approximation ratios $\alpha = 12/5 + 2^{-k+1}$ and $\alpha = 14/6 + 2^{-k+1}$, respectively, with more reasonable running times $O(n(k + \log n))$ and $O(n(km^4 + \log n))$.

4   Approximation Algorithms for $P|2\mathrm{FL}_{R \geq T}|C_{\max}$ and $P|2\mathrm{FL}_{R \leq T}|C_{\max}$

Motivated by the practical application in data centers and cloud computing, the paper [22] argues that if the servers in a cloud are divided into clusters based on the services they provide, then the requests to the servers in each cluster will be likely to be all with more expensive R-operations or all with more expensive T-operations. In this section, we consider the two-stage scheduling problem in these two restricted models, one assumes that for each job the R-time is not smaller than the T-time (i.e., $P|2\mathrm{FL}_{R \geq T}|C_{\max}$), while the other assumes that for each job the R-time is not larger than the T-time (i.e., $P|2\mathrm{FL}_{R \leq T}|C_{\max}$). By employing approximation algorithms for the MAKESPAN problem, we obtain approximation algorithms for the restricted cases, improving the results in [22].

We first give a simple but important lemma used in the following analysis.

Lemma 5 ([22]). *For each job J_i, $r_i + t_i$ is at most $\mathrm{Opt}(G)$. Consequently, the smaller of the values r_i and t_i is not larger than $1/2 \cdot \mathrm{Opt}(G)$.*

The lemma is obvious because $\mathrm{Opt}(G)$ cannot be smaller than the entire execution time of any single job $J_i = (r_i, t_i)$ in the job set G, which equals $r_i + t_i$ in this scheduling problem. We have a further observation holding in both cases.

Lemma 6. *In the models $P|2\mathit{FL}_{R \geq T}|C_{\max}$ and $P|2\mathit{FL}_{R \leq T}|C_{\max}$, the completion time of a flowshop under a schedule is bounded by the sum of the R/T-times of the jobs assigned to the flowshop plus $1/2 \cdot \mathrm{Opt}(G)$, where the R/T-time of a job denotes the cost of the more time-consuming operation of the job, i.e., the R/T-time of a job equals the R-time of the job in the case $P|2\mathit{FL}_{R \geq T}|C_{\max}$ and equals the T-time in the case $P|2\mathit{FL}_{R \leq T}|C_{\max}$.*

Proof. Without loss of generality, let $S = \langle J_1, \ldots, J_c \rangle$ be a two-stage job sequence scheduled on a single two-stage flowshop in any of the two restricted models. The status of the flowshop after scheduling by S for the two models is shown in Figs. 2 and 3, respectively. Let d be the minimum job index such that the T-operations of the jobs $J_d, J_{d+1}, \ldots, J_c$ are executed continuously by the flowshop. We denote by l the difference between the completion time τ of the flowshop and the sum of the R/T-times of the jobs on the flowshop. Let $a_0 = \sum_{i=1}^{d-1} r_i$ be the sum of the R-times of the jobs from J_1 to J_{d-1}, and $b_0 = \sum_{i=1}^{d-1} t_i$ be the sum of the T-times of these jobs. Notice that the T-operations of the jobs from J_1 to J_{d-1} do not need to be executed continuously though denoted by b_0 here. Similarly, let $a_1 = \sum_{i=d+1}^{c} r_i$ and $b_1 = \sum_{i=d+1}^{c} t_i$.

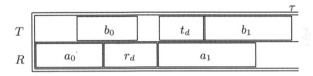

Fig. 2. The status of a flowshop after scheduling in $P|2\mathrm{FL}_{R \geq T}|C_{\max}$

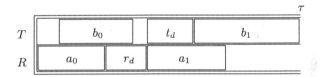

Fig. 3. The status of a flowshop after scheduling in $P|2\mathrm{FL}_{R \leq T}|C_{\max}$

We first show that in both models, the completion time τ of the flowshop can be expressed as $a_0 + r_d + t_d + b_1$. There are three reasons holding in both cases: (1) the T-operation of the job J_d starts right after the R-operation of J_d is finished: otherwise by Lemma 1, the T-operation of J_d must be waiting for the T-operation of J_{d-1} to be completed at that time, thus there would be no gap between the execution of the T-operations of J_{d-1} and J_d, contradicting the assumption of the minimality of the job index d, (2) by Lemma 1, the processing of the R-processor of any flowshop is continuous, and (3) by the definition of d, the T-operations of the jobs $J_d, J_{d+1}, \ldots, J_c$ are executed continuously.

Now consider the two models separately. In the case $P|2\mathrm{FL}_{R \geq T}|C_{\max}$ where the R-time is not smaller than the T-time for each job, we have that the R/T-time of a job equals the R-time of the job and b_1 is not larger than a_1, where a_1 and b_1 represent the sum of the R- and T-times of the same jobs from J_{d+1} to J_c, respectively (note that a_1 and b_1 could be 0). The difference l between the completion time τ and the sum of the R-times of the jobs is bounded as follows:

$$l = \tau - (a_0 + r_d + a_1) = (a_0 + r_d + t_d + b_1) - (a_0 + r_d + a_1) = t_d + b_1 - a_1 \leq t_d.$$

By Lemma 5, the assumption in this case that $r_i \geq t_i$ for each job J_i, implies that $t_d \leq 1/2 \cdot \text{Opt}(G)$, thus l is bounded by $1/2 \cdot \text{Opt}(G)$.

The analysis of the case $P|2\text{FL}_{R \leq T}|C_{\max}$ is similar. It is easy to see that the R/T-time of a job equals the T-time of the job and a_0 is not larger than b_0 (note that a_0 and b_0 could be 0). Therefore the difference l between the completion time τ and the sum of the T-times of the jobs is here:

$$l = \tau - (b_0 + t_d + b_1) = (a_0 + r_d + t_d + b_1) - (b_0 + t_d + b_1) = r_d + a_0 - b_0 \leq r_d.$$

Due to that r_d is not larger than $1/2 \cdot \text{Opt}(G)$ by Lemma 5, thus l is also bounded by $1/2 \cdot \text{Opt}(G)$ in this case. These complete the proof.   □

The lemma yields an interesting result that in both restricted cases, the completion time of a flowshop depends on the cost of the more time-consuming operation of the jobs assigned to it, and the difference between them is bounded by $1/2 \cdot \text{Opt}(G)$. Therefore when scheduling n two-stage jobs on m two-stage flowshops in these two cases, it should be more natural to be based on the scheduling strategy on the more time-consuming operation rather than the other operation, i.e., the R-operation in $P|2\text{FL}_{R \geq T}|C_{\max}$ and the T-operation in $P|2\text{FL}_{R \leq T}|C_{\max}$. We also remark that the lemma is not sensitive to the job order.

For convenience of the description of the algorithm, we define notations similar to that in the previous section. Call a job J_i^o a R/T-job of the job $J_i = (r_i, t_i)$ if J_i^o equals $(r_i, 0)$ in $P|2\text{FL}_{R \geq T}|C_{\max}$ or equals $(0, t_i)$ in $P|2\text{FL}_{R \leq T}|C_{\max}$, i.e., the job J_i^o is constructed from J_i by setting the cost of its less time-consuming operation to 0. For a two-stage job set $G = \{J_1, \ldots, J_n\}$, its corresponding R/T-job set is $G^o = \{J_1^o, \ldots, J_n^o\}$, where J_i^o is the R/T-job of job J_i for $1 \leq i \leq n$.

The algorithm for both restricted models is also similar to the algorithm **ApproxProg-I**. Given a two-stage job set $G = \{J_1, \ldots, J_n\}$ and an integer m, we first construct its R/T-job set $G^o = \{J_1^o, \ldots, J_n^o\}$. According to the definition, the job J_i^o in both cases is actually a one-stage job. Thus any α-approximation algorithm for the classical MAKESPAN problem, denoted by $AlgM_\alpha(G, m)$, can be used to schedule the R/T-job set G^o on m flowshops and will construct a schedule $\bar{S}^\alpha$ for these jobs with the makespan bounded by $\alpha \cdot \text{Opt}(G^o)$. Let $\bar{S}^\alpha = \{\bar{S}_1^\alpha, \ldots, \bar{S}_m^\alpha\}$ be the schedule for G^o, where $\bar{S}_j^\alpha$ is the job set on the flowshop M_j by $\bar{S}^\alpha$ for $1 \leq j \leq m$, then we construct a schedule $S = \{S_1, \ldots, S_m\}$ for the original job set G in such a way that J_i is scheduled on the two-stage flowshop M_j if and only if the corresponding R/T-job J_i^o is scheduled on M_j under schedule $\bar{S}^\alpha$ and for each flowshop M_j, the job set S_j is sorted into Johnson's order. The algorithm **ApproxProg-II** is shown in Fig. 4.

The approximation algorithm deals with the two cases almost the same way, except for the first step, which constructs the R/T-job for each job in G by setting the cost of each job's less time-consuming operation to 0. This step is also the only difference between the algorithm **ApproxProg-I** and the algorithm **ApproxProg-II**: the algorithm **ApproxProg-I** constructs the $R\&T$-job instead of the R/T-job for each job in the job set G. Thus the algorithm takes

---

Algorithm ApproxProg-II
INPUT: a job set $G = \{J_1, \ldots, J_n\}$ of two-stage jobs where each job has its
R-time no smaller than its T-time (i.e., the model $P|2\mathrm{FL}_{R \geq T}|C_{\max}$), or each
job has its R-time no larger than its T-time (i.e., the model $P|2\mathrm{FL}_{R \leq T}|C_{\max}$),
and an integer m
OUTPUT: a schedule S for G on m two-stage flowshops
1. **for all** $1 \leq i \leq n$
1.1   **do** $r_i' = r_i$; $t_i' = 0$ (in the model $P|2\mathrm{FL}_{R \geq T}|C_{\max}$)
1.1   or **do** $r_i' = 0$; $t_i' = t_i$ (in the model $P|2\mathrm{FL}_{R \leq T}|C_{\max}$);
2.   let $G^o = \{J_1^o, \ldots, J_n^o\}$, where for each i, $J_i^o = (r_i', t_i')$;
3. $\bar{S}^\alpha = \{\bar{S}_1^\alpha, \ldots, \bar{S}_m^\alpha\}$ obtained from $AlgM_\alpha(G^o, m)$ on the job set G^o;
4. **for all** $1 \leq i \leq n$ **do**
4.1   If J_i^o in S_j^α **then** assign the job J_i to S_j;
5. **for all** $1 \leq j \leq m$ **do**
5.1   sort the jobs in the job set S_j into Johnson's order;
6.   **return** $S = \{S_1, \ldots, S_m\}$;

Fig. 4. An approximation algorithm for $P|2\mathrm{FL}_{R \geq T}|C_{\max}$ and $P|2\mathrm{FL}_{R \leq T}|C_{\max}$

time $O(n \log n)$, excluding the step of calling $AlgM_\alpha(G, m)$. Based on Lemma 6, the following theorem holds for the algorithm **ApproxProg-II**.

Theorem 3. *If the* MAKESPAN *problem has an α-approximation algorithm, then the two restricted cases $P|2FL_{R \geq T}|C_{\max}$ and $P|2FL_{R \leq T}|C_{\max}$ of the multiple two-stage flowshops problem have an $(\alpha + 1/2)$-approximation algorithm.*

Proof. It is sufficient to verify the following inequality, where the schedule S and the schedule $\bar{S}^a$ are described in the algorithm **ApproxProg-II**.

$$Makespan(S) \leq Makespan(\bar{S}^a) + 1/2 \cdot \mathrm{Opt}(G)$$
$$\leq \alpha \cdot \mathrm{Opt}(G^o) + 1/2 \cdot \mathrm{Opt}(G) \leq (\alpha + 1/2) \cdot \mathrm{Opt}(G).$$

Therefore, the algorithm **ApproxProg-II** can construct a schedule for the job set G on flowshops with the makespan bounded by $(\alpha + 1/2) \cdot \mathrm{Opt}(G)$, when using an α-approximation algorithm for the classical MAKESPAN problem.   □

As a result, when using a polynomial-time approximation scheme from [13] as $AlgM_\alpha$, the algorithm **ApproxProg-II** is $(1.5 + \epsilon)$-approximation, running in time $O((n/\epsilon)^{1/\epsilon^2} + n \log n)$. This improves the results in [22], which provided 11/6-approximation algorithms for the two restricted cases in offline setting. One may argue that the algorithms in [22] have a more practical running time $O(n \log n)$. In fact, these algorithms can be viewed as the special cases of the algorithm **ApproxProg-II**. By Theorem 3, when using the *ListRanking* algorithm as $AlgM_\alpha$, whose approximation ratio is 4/3 and which runs in time $O(n \log n)$ [9], the approximation ratio of the algorithm **ApproxProg-II** is $4/3 + 1/2 = 11/6$, and the running time is $O(n \log n + n \log n) = O(n \log n)$, which are identical to that of the algorithms in [22].

References

1. Abts, D., Felderman, B.: A guided tour through data-center networking. Queue **10**(5), 10–23 (2012)
2. Barroso, L.A., Clidaras, J., Hölzle, U.: The Datacenter as a Computer: An Introduction to the Design of Warehouse-Scale Machines. Synthesis Lectures on Computer Architecture. Morgan & Claypool, San Rafael (2013)
3. Bruno, J., Coffman Jr., E.G., Sethi, R.: Scheduling independent tasks to reduce mean finishing time. Commun. ACM **17**(7), 382–387 (1974)
4. Coffman Jr., E.G., Garey, M.R., Johnson, D.S.: An application of bin-packing to multiprocessor scheduling. SIAM J. Comput. **7**(1), 1–17 (1978)
5. Dayarathna, M., Wen, Y., Fan, R.: Data center energy consumption modeling: a survey. IEEE Commun. Surv. Tutor. **18**(1), 732–794 (2016)
6. Dong, J., Tong, W., Luo, T., Wang, X., Hu, J., Xu, Y., Lin, G.: An FPTAS for the parallel two-stage flowshop problem. Theoret. Comput. Sci. **657**, 64–72 (2017)
7. Garey, M.R., Johnson, D.S.: Computers and Intractability: A Guide to the Theory of NP-Completeness. W.H. Freeman and Company, New York (1979)
8. Graham, R.L.: Bounds for certain multiprocessing anomalies. Bell Labs Tech. J. **45**(9), 1563–1581 (1966)
9. Graham, R.L.: Bounds on multiprocessing timing anomalies. SIAM J. Appl. Math. **17**(2), 416–429 (1969)
10. Graham, R.L., Lawler, E.L., Lenstra, J.K., Kan, A.R.: Optimization and approximation in deterministic sequencing and scheduling: a survey. Ann. Discret. Math. **5**, 287–326 (1979)
11. Greenberg, A., Hamilton, J., Maltz, D.A., Patel, P.: The cost of a cloud: research problems in data center networks. ACM SIGCOMM Comput. Commun. Rev. **39**(1), 68–73 (2008)
12. He, D.W., Kusiak, A., Artiba, A.: A scheduling problem in glass manufacturing. IIE Trans. **28**(2), 129–139 (1996)
13. Hochbaum, D.S., Shmoys, D.B.: Using dual approximation algorithms for scheduling problems theoretical and practical results. J. ACM (JACM) **34**(1), 144–162 (1987)
14. Johnson, S.M.: Optimal two-and three-stage production schedules with setup times included. Naval Res. Logist. Q. **1**(1), 61–68 (1954)
15. Kovalyov, M.Y.: Efficient epsilon-approximation algorithm for minimizing the makespan in a parallel two-stage system. In: Vesti Academii navuk Belaruskai SSR, Ser. Phiz.-Mat. Navuk, vol. 3, p. 119 (1985). (in Russian)
16. Pinedo, M.L.: Scheduling: Theory, Algorithms, and Systems. Springer Science, New York (2016). https://doi.org/10.1007/978-3-319-26580-3
17. Rittinghouse, J.W., Ransome, J.F.: Cloud Computing: Implementation, Management, and Security. CRC Press, New York (2016)
18. Sahni, S.K.: Algorithms for scheduling independent tasks. J. ACM (JACM) **23**(1), 116–127 (1976)
19. Vairaktarakis, G., Elhafsi, M.: The use of flowlines to simplify routing complexity in two-stage flowshops. IIE Trans. **32**(8), 687–699 (2000)
20. Vazirani, V.V.: Approximation Algorithms. Springer, New York (2013). https://doi.org/10.1007/978-3-662-045657
21. Wu, G., Chen, J., Wang, J.: On scheduling two-stage jobs on multiple two-stage flowshops. Technical report, School of Information Science and Engineering, Central South University (2016)

22. Wu, G., Chen, J., Wang, J.: On scheduling inclined jobs on multiple two-stage flowshops. Theoret. Comput. Sci. (2018). https://doi.org/10.1016/j.tcs.2018.04.005
23. Wu, G., Chen, J., Wang, J.: On scheduling multiple two-stage flowshops. Theoret. Comput. Sci. (2018). https://doi.org/10.1016/j.tcs.2018.04.017
24. Zhang, X., van de Velde, S.: Approximation algorithms for the parallel flow shop problem. Eur. J. Oper. Res. **216**(3), 544–552 (2012)

Constant Factor Approximation Algorithm for l-Pseudoforest Deletion Problem

Mugang Lin[1,2,3], Bin Fu[4], and Qilong Feng[1(✉)]

[1] School of Information Science and Engineering, Central South University, Changsha, China
{linmu718,csufeng}@csu.edu.cn
[2] School of Computer Science and Technology, Hengyang Normal University, Hengyang, China
[3] Hunan Provincial Key Laboratory of Intelligent Information Processing and Application, Hengyang, China
[4] Department of Computer Science, University of Texas Rio Grande Valley, Edinburg, TX 78539, USA
binfu@utrgv.edu

Abstract. An *l-pseudoforest* is a graph each of whose connected component is at most l edges away from being a tree. The *l-Pseudoforest Deletion problem* is to delete a vertex set P of minimum weight from a given vertex-weighted graph $G = (V, E)$ such that the remaining graph $G[V \setminus P]$ is an l-pseudoforest. The Feedback Vertex Set problem is a special case of the l-Pseudoforest Deletion problem with $l = 0$. In this paper, we present a polynomial time $4l$-approximation algorithm for the l-Pseudoforest Deletion problem with $l \geq 1$ by using the local ratio technique. When $l = 1$, we get a better approximation ratio 2 for the problem by further analyzing the algorithm, which matches the current best constant approximation factor for the Feedback Vertex Set problem.

Keywords: Approximation algorithm
l-Pseudoforest Deletion problem · Local ratio

1 Introduction

An *l-pseudoforest* is a graph which can be transformed into a forest by deleting at most l edges from each connected component. Let $G = (V, E)$ be an undirected graph with a weight function $w : V \rightarrow R_+$. A set $P \subseteq V$ is an *l-pseudoforest deletion set* (*l-PFDS*) of G if the induced graph $G[V \setminus P]$ is an l-pseudoforest. The *l-Pseudoforest Deletion problem* (*l-PFD*) is to find an l-PFDS of minimum weight

This work is supported by the National Natural Science Foundation of China under Grants (61772179, 61472449, 61420106009, 61402054), the Science and Technology Plan Project of Hunan Province under Grant (2016TP1020) and the Scientific Research Fund of Hunan Provincial Education Department under Grant (17C0222).

from graph G. When $l = 1$, the problem is called the *Pseudoforest Deletion problem* (PFD). Obviously, the classical *Feedback Vertex Set problem* (FVS) is the special case of the l-PFD problem with $l = 0$.

The FVS problem has been intensively studied for several decades, and a series of results have been obtained, such as, parameterized algorithms of running time $O^*(3.619^k)$ in deterministic setting [10] and $O^*(3^k)$ in randomized setting [6] parameterized by the size k of a feedback vertex set, an exact exponential algorithm with running time $O(1.7266^n)$ [16] and polynomial-time approximation algorithms of ratio 2 [1,3,5,8]. Philip et al. [13] introduced the l-PFD problem and studied the parameterized l-PDF problem which is to find an l-PFDS with at most k vertices in given graph $G = (V, E)$ where k is given non-negative integer. From the point of view of parameterized algorithm, the l-PFD problem has a kernel with $f(l)k^2$, and an algorithm with running time $c_l^k n^{O(1)}$ where the constant c_l depends only on l. For the PFD problem, Philip et al. [13] showed an explicit kernel with $O(k^2)$ vertices and proposed a deterministic algorithm with running time $O^*(7.5618^k)$ by applying the iterative compression technique. This was subsequently improved by Bodlaender et al. [4], who gave an $O^*(3^k k^{O(1)})$-time algorithm. Jansen et al. [9] showed the FVS problem has an $O(k^{10})$ kernel and a kernel lower bound of $\Omega(k^4)$ if parameterized by the size k of a pseudoforest deletion set of input instance. Majumdar [12] gave a kernel of $O(k^6)$ vertices. Related to the l-PFD problem, Rai and Saurabh [14] studied the Almost Forest Deletion problem which is to find a vertex set P of size at most k such that the induced subgraph $G[V \setminus P]$ is an l-forest, where an l-forest is a graph which can be transformed into a forest by deleting at most l edges. They gave a kernel of size $O(kl(k + l))$ and presented an FPT algorithm of running time $O^*(5.0024^{(k+l)})$ based on the iterative compression technique. Recently, we developed an improved parameterized algorithm with running time $O^*(5^k 4^l)$ for the Almost Forest Deletion problem [11].

A graph is a pseudoforest if and only if it does not have the butterfly and the diamond as minors. The class of l-pseudoforests has also finite forbidden minor characterization [15]. Based on the results given by Fomin et al. [7], it is easy to get that for the l-PFD problem, there exist a polynomial kernel (improved to quadratic by [13]), a constant factor randomized approximation algorithm with running time $O(nm)$, a randomized parameterized algorithm with running time $O(2^{O(k)}n)$ and a deterministic parameterized algorithms with running time $O(2^{O(k)}n log^2 n)$. In this paper, we generalize the results in [1,3,5,8] based on the local ratio technique [2] by presenting polynomial time constant factor approximation for all the l-PFD problem with fixed l. Our approximation ratio is still 2 if $l = 1$, and $4l$ for all $l \geq 2$.

2 Preliminaries

We assume that any graph $G = (V, E)$ in this paper is simple and undirected, with $|V| - n$ vertices and $|E| = m$ edges. For a subgraph G' of graph G, we denote the vertex set and the edge set of G' by $V(G')$ and $E(G')$, respectively.

If there exists an edge e between two vertices u and v, we say that u and v are adjacent, edge $e = (u,v)$ is incident to u and v, and u is a neighbor of v. If there exists an edge e between a vertex v and a vertex of subgraph C where $v \notin C$, we say v and C are adjacent by e. For vertex v, the degree of v in graph G, denoted by $deg_G(v)$, is the number of edges incident to v in G. The minimum degree of graph G is $\delta_G = min_{v \in G} deg_G(v)$. A *cut vertex* is a vertex removing which increases number of connected components. For any subset $W \subseteq V$, let $G[W]$ be the subgraph induced by W. For simplicity, we use the notation $G - w$ and $G - W$ for $G[V \setminus \{w\}]$ and $G[V \setminus W]$, respectively. For two subsets $P, Q \subseteq V$, $E(P,Q) = \{(u,v)|(u,v) \in E, u \in P, v \in Q\}$. A connected undirected graph T_l is called an *l-pseudotree* if it can be transformed into a tree by deleting l edges (i.e. $|E(T_l)| = |V(T_l)| + l - 1$). In graph G, an l-pseudotree T_l is *semidisjoint* if there exists at most one vertex $v \in V(T_l)$ not satisfying $deg_G(v) = deg_{T_l}(v)$, where T_l is an induced subgraph in G. Obviously, if there exists a vertex v with $deg_G(v) \neq deg_{T_l}(v)$, v must be a cut vertex of G.

For graph $G = (V,E)$, a vertex-weighted function w is a mapping $V \rightarrow R_+$. We use (G,w) to denote graph G with a vertex-weighted function w. For set $P \subseteq V$, $w(P) = \sum_{v \in P} w(v)$. Let $\{(G_i, w_i)\}$ be a collection of subgraphs of (G,w) that satisfies $\sum_i w_i(u) \leq w(u)$ for $\forall u \in V(G)$, $\{(G_i, w_i)\}$ is said to be a *decomposition* of (G,w). An l-PFDS of G is a set $P \subseteq V$ such that each connected component C_i of $G - P$ contains at most $|V(C_i)| - 1 + l$ edges. In graph G, an l-PFDS P is *minimal* if there exists no l-PFDS P' with $P' \subset P$; and an l-PFDS P is optimal if there exists no l-PFDS P' with $\sum_{v \in P'} w(v) < \sum_{v \in P} w(v)$.

3   Outline

In this section, we give the general idea of our method and a description of the approximation algorithm based on the local ratio technique for the l-PFD problem. The following lemma is to extend Theorem 2.1 in [1] to handle the l-PFD problem.

Lemma 1. *Let $\{(G_i, w_i)\}$ be a decomposition of (G,w), $opt(G,w)$ be an optimal l-PFDS of G with a weight function w, and P be any l-PFDS of G such that $w(P) = \sum_i w_i(P \cap V(G_i))$. Then $\frac{w(P)}{w(opt(G,w))} \leq \max\{\frac{w_i(P \cap V(G_i))}{w_i(opt(G_i, w_i))}\}$.*

Proof. Since $\{(G_i, w_i)\}$ is a decomposition of (G,w), $w(X) \geq \sum_i w_i(X \cap V(G_i))$ for any set $X \subseteq V(G)$. Thus, $w(opt(G,w)) \geq \sum_i w_i(opt(G,w) \cap V(G_i))$. As $opt(G_i, w_i)$ is the optimal l-PFDS for (G_i, w_i), $w_i(opt(G_i, w_i)) \leq w_i(opt(G,w) \cap V(G_i))$. Therefore, we have

$$\frac{w(P)}{w(opt(G,w))} \leq \frac{\sum_i w_i(P \cap V(G_i))}{\sum_i w_i(opt(G,w) \cap V(G_i))} \tag{1}$$

$$\leq \frac{\sum_i w_i(P \cap V(G_i))}{\sum_i w_i(opt(G_i, w_i))} \tag{2}$$

$$\leq \max\{\frac{w_i(P \cap V(G_i))}{w_i(opt(G_i, w_i))}\} \tag{3}$$

$\square$

Let (G, w) be an instance of given optimization problem, where G is a graph and w is a non-negative weight function defined on vertices of graph G. The idea of local ratio technique is to recursively find a decomposition $\{(G_i, w_i)\}$ of (G, w) such that w_i is a simple weight function which is suited for obtaining a good approximation solution to the optimal solution of (G_i, w_i). The local solution for each subgraph has a bounded approximation ratio. The approximation ratio of the algorithm is the maximum local ratio of all subgraphs in $\{(G_i, w_i)\}$ by Lemma 1. The key step in obtaining the desired approximation ratio is to find a special weight function w_i and prove the approximation ratio with respect to w_i. In this paper, we give a detailed analysis for the approximation ratio of those subgraphs derived in the decomposition by the algorithm.

Most of 2-approximation algorithms for the FVS problem are based on the similar degree observation that a vertex of larger degree has a larger chance to be in a minimum feedback vertex set. For the l-PFD problem, our algorithm is also based on the observation. In graph G with the minimum degree at least two, a larger degree vertex v can be in more cycles. If v is deleted, more cycles can be destroyed. Therefore, in our algorithm, we greedily choose some vertices with less weight and larger degree as candidate vertices. Our algorithm is described in Fig. 1. Alg-l-PFD(G, w, l) is recursive and works as follows. If G is an l-pseudoforest, then the algorithm stops and returns $\emptyset$ (line 1–line 3). Otherwise, it checks whether there exists $v \in V(G)$ with $w(v) = 0$ and determines whether vertex v with $w(v) = 0$ is in a minimal l-PFDS for graph G (line 4–line 13). If there exists a vertex $v \in V(G)$ with $w(v) = 0$, then recursively call Alg-l-PFD$(G - v, w', l)$. If the returned solution P of Alg-l-PFD$(G - v, w', l)$ for $G - v$ is an l-PFDS of graph G, then v is not in the minimal l-PFDS of graph G and Alg-l-PFD(G, w, l) returns P, otherwise returns $P \cup \{v\}$. If there does not exist any $v \in V(G)$ with $w(v) = 0$, graph G' is obtained by deleting all connected components with at most l cycles and all vertices of degree at most 1 from graph G (line 14–line 15). After that, the minimum degree of graph G' is at least two. Then, we extract a sequence of decomposition subgraphs of G' (line 16–line 25). If G' contains a semidisjoint $(l + 1)$-pseudotree T_{l+1}, then we choose a minimum vertex weight in T_{l+1} as a new weight function $\gamma = \min\{w(v) : v \in V(T_{l+1})\}$ of decomposition subgraph T_{l+1}, and call Alg-l-PFD$(G', w - \gamma, l)$. Otherwise, we greedily choose some vertex with less weight and larger degree, constructs a new weight function $\gamma = \min\{w(v)/(deg_{G'}(v) - 1) : v \in V(G)\}$ of decomposition subgraph G' and call Alg-l-PFD$(G', w - \gamma, l)$ to solve the problem.

In the following, we analyze the time complexity of Alg-l-PFD(G, w, l). Detecting an l-pseudoforest in line 1 and line 14 (or an l-PFDS P in line 8) can be executed in time $O(|V| + |E|)$ by checking whether every connected component C_i of G (or $G - P$) satisfies $|E(C_i)| < |V(C_i)| + l$. Deleting vertices of degree at most one in line 15 can also be executed in time $O(|V| + |E|)$. Detecting semidisjoint $(l + 1)$-pseudotrees in line 16 takes $O(|V| + |E|)$ time by finding and deleting all cut vertices and then determining every connected components C_i adjacent to at most one cut vertex v whether $|E(G[C_i \cup \{v\}])| = |V(C_i)| + 1 + l$. Computing the minimum weight $w(v)$ in line 17 (or $w(v)/deg_{G'}(v) - 1$ in line 22)

Algorithm Alg-l-PFD(G, w, l)
Input: an integer $l \geq 0$, graph $G = (V, E)$ with vertex weight $w : V(G) \to R_+$.
Output: an l-pseudoforest deletion set(l-PFDS) P.

```
1:  if G is an l-pseudoforest then
2:        return ∅;
3:  end if
4:  if there exists v ∈ V(G) with w(v) = 0 then
5:        G' ← G - v;
6:        w'(v) ← w(v) for v ∈ V(G');
7:        P ← Alg-l-PFD(G', w', l);
8:        if P is an l-PFDS of G then
9:              return P;
10:       else
11:             return P ∪ {v};
12:       end if
13: end if
14: for each connected component Cᵢ ∈ G with |E(Cᵢ)| ≤ |V(Cᵢ)|+l−1, G' ← G−Cᵢ;
15: while G' contains a vertex v with deg_{G'}(v) ≤ 1 do G' ← G' − {v};
16: if G' contains a semidisjoint (l + 1)-pseudotree T_{l+1} then
17:       let γ ← min{w(v) : v ∈ V(T_{l+1})};
18:       set w'(v) ← w(v) − γ, ∀v ∈ V(T_{l+1});
19:       set w'(v) ← w(v), ∀v ∈ V(G') \ V(T_{l+1});
20:       return Alg-l-PFD(G', w', l);
21: else
22:       let γ ← min{w(v)/(deg_{G'}(v) − 1) : v ∈ V(G')};
23:       set w'(v) ← w(v) − (deg_{G'}(v) − 1)γ, ∀v ∈ V(G');
24:       return Alg-l-PFD(G', w', l);
25: end if
```

Fig. 1. Approximation algorithm for the l-Pseudoforest Deletion Problem.

takes $O(|V|)$. The algorithm makes recursive calls at most $|V|$ times. Therefore, the whole running time of Alg-l-PFD(G, w, l) is $O(|V| \cdot (|E| + |V|))$.

4   Properties of Decomposition

In this section, we show some basic properties about the decomposition subgraphs of input graph after running Alg-l-PFD(G, w, l). Let P be an l-PFDS generated by Alg-l-PFD(G, w, l) and $G_i = (V_i, E_i)$ be i-th decomposition subgraph derived from Alg-l-PFD(G, w, l). In line 4–line 13 of Alg-l-PFD(G, w, l), we check whether vertices with zero weight are in a minimal l-PFDS, one by one, in the reverse order of their inclusion into P. Whenever a vertex is found to be extraneous, it is not in P. Therefore, this process ensures that P is a minimal l-PFDS for G. We introduce the following definitions and lemmas.

Definition 1 (type-1 subgraph and type-2 subgraph).
The subgraph G_i derived in the i-th decomposition is either a semidisjoint $(l + 1)$-pseudotree or G itself without semidisjoint $(l + 1)$-pseudotrees by Alg-

l-PFD(G, w, l). For a semidisjoint $(l + 1)$-pseudotree G_i in the decomposition, it is called type-1 subgraph. Otherwise, it is called type-2 subgraph.

Lemma 2. *Let P be a minimal l-PFDS of G generated by Alg-l-PFD(G, w, l) and $\{(G_i, w_i)\}$ be the decomposition of (G, w) derived by Alg-l-PFD(G, w, l). Then $w(P) = \sum_i w_i(P \cap V(G_i))$.*

Proof. Let (G_i, w_i) denote the decomposition subgraph derived in the i-th decomposition by Alg-l-PFD(G, w, l), and (G'_i, w'_i) denote the remaining graph right after the i-th extracting subgraph G_i. Assume that $v \in P$ is a vertex in (G_i, w_i) with $w'_{i-1}(v) - w_i(v) = 0$ in Alg-l-PFD(G, w, l) for some fixed integer $i \geq 1$. Since v is put into P in Alg-l-PFD(G, w, l) after its weight $w(v)$ is reduced to zero by a sequence subtractions, $w(v) = \sum_{j \leq i} w_j(v)$. Therefore, $w(P) = \sum_i w_i(P \cap V(G_i))$. $\qquad\square$

Lemma 3. *Let P be a minimal l-PFDS of G generated by Alg-l-PFD(G, w, l). For each type-1 subgraph (G_i, w_i) in the decomposition derived by Alg-l-PFD(G, w, l), $P \cap V_i$ contains an optimal l-PFDS for (G_i, w_i).*

Proof. Since type-1 subgraph $G_i = (V_i, E_i)$ is a semidisjoint $(l + 1)$-pseudotree, $|E_i| = |V_i| + l$. If $P \cap G_i$ does not contain any vertex in G_i, G is not an l-pseudoforest after completing Alg-l-PFD(G, w, l) as G contains G_i. It contradicts with the fact that P is a minimal l-PFDS of G. Thus, $P \cap G_i \neq \emptyset$. For each vertex $u \subset V_i$, $deg_{G_i}(u) \geq 2$ because the vertices of degree at most 1 are removed before detecting type-1 subgraph. For $v \in P \cap V_i$, we have

$$|E(G_i - v)| \leq |E_i| - 2 = (|V_i| - 1) + (l - 1) = |V(G_i - v)| + (l - 1) \quad (4)$$

By inequality (4) and $deg_{G_i - v}(u) \geq 1$ for each vertex $u \in V_i \setminus \{v\}$, we know that graph $G_i - v$ is an l-pseudoforest. Thus, v is a minimal l-PFDS of G_i. If $|P \cap V_i| \geq 2$, then since there is the same weight for every vertex of (G_i, w_i), v is an optimal l-PFDS for (G_i, w_i). Therefore, $P \cap V_i$ contains an optimal l-PFDS for (G_i, w_i). $\qquad\square$

Lemma 4. *Let P be a minimal l-PFDS of G generated by Alg-l-PFD(G, w, l). For each type-2 subgraph (G_i, w_i) in the decomposition $\{(G_i, w_i)\}$ derived by Alg-l-PFD(G, w, l), $P \cap V(G_i)$ is a minimal l-PFDS for G_i.*

Proof. Let (G_i, w_i) denote the decomposition subgraph derived in the i-th decomposition by Alg-l-PFD(G, w, l), and (G'_i, w'_i) denote the remaining graph right after the i-th extracting subgraph G_i. We know that G_i is a subgraph of G'_{i-1} by Alg-l-PFD(G, w, l). If G'_j is an l-pseudoforest for some j, Alg-l-PFD(G, w, l) stops. Hence, $G'_{j-1} = G_j$. After that, Alg-l-PFD(G, w, l) firstly checks whether the vertices satisfying $w'_{j-1}(v) - w_j(v) = 0$ are in a minimal l-PFDS of G'_{j-1}. When a vertex is found to be redundant, it is not in P. Otherwise, it is put into P. It is clearly seen that $P \cap V(G_j)$ is a minimal l-PFDS for G_j. Then, in the reverse order of recursive calls, Alg-l-PFD(G, w, l) examines whether every vertex satisfying $w'_{i-1}(v) - w_i(v) = 0$ is in a minimal l-PFDS of

G'_{i-1} and tests its redundancy where $i \leq j$. If it is redundant, it is not in P. Otherwise, it is put into P. Thus, $P \cap G'_{i-1}$ is a minimal l-PFDS of G'_{i-1}. For G_i is a type-2 subgraph, by Alg-l-PFD(G, w, l), we know that $G'_{i-1} = G_i$. Hence, $P \cap G_i$ is a minimal l-PFDS of G_i. $\square$

5   Approximation Ratio for the l-PFD Problem

In this section, we analyze the approximation ratio of Alg-l-PFD(G, w, l) which is the maximum local ratio of all subgraphs in $\{(G_i, w_i)\}$ by Lemma 1. Thus, we analyze the local ratio of type-1 subgraphs and type-2 subgraphs, respectively.

Lemma 5. *Let P be a minimal l-PFDS of G generated by Alg-l-PFD(G, w, l). For each type-1 subgraph (G_i, w_i) in the decomposition $\{(G_i, w_i)\}$ derived by Alg-l-PFD(G, w, l), $\frac{w_i(P \cap V(G_i))}{w_i(opt(G_i, w_i))} \leq l + 1$.*

Proof. For type-1 subgraph G_i, by Lemma 3, we know that $P \cap V(G_i)$ contains one vertex which is an optimal solution for (G_i, w_i). Since a semidisjoint $(l+1)$-pseudotree contains $l + 1$ cycles, $P \cap V(G_i)$ contains at most $l + 1$ vertices of G_i. In subgraph (G_i, w_i), each vertex has the same weight. Therefore, the local ratio of type-1 subgraph is that $\frac{w_i(P \cap V(G_i))}{w_i(opt(G_i, w_i))} \leq l + 1$. $\square$

For a type-2 subgraph in $\{(G_i, w_i)\}$, because the weight function of every type-2 subgraph G_i is degree proportional in Alg-l-PFD(G, w, l), it suffices to analyze this approximation ratio for the case when $w(v) = deg_{G_i}(v) - 1$ for every $v \in V(G_i)$. In the following argument, let P be a minimal l-PFDS for type-2 subgraph (G, w) without semidisjoint $(l+1)$-pseudotrees, where $w(v) = deg_G(v) - 1$ for every $v \in V(G)$. We give a lower bound and an upper bound for $w(P) = \sum_{v \in P} w(v)$ when $l \geq 1$.

Lemma 6. $w(P) \geq \frac{|E| - |V|}{2l}$.

Proof. In $G - P$, each connected component $C_i = (V_i, E_i)$ contains $|V_i| - 1 + l_i$ edges with $0 \leq l_i \leq l$ for $i = 1, \cdots, k$, where k is the number of connected components in $G - P$. We partition the connected components of graph $G - P$ into several parts by the partitioning operation (see Fig. 2.) such that each part which contains some connected components adjacent to vertex $v_i \in P$ is represented by $D(v_i)$.

By the partitioning operation, connected components in $G - P$ are partitioned into classes $D(v_1), \cdots, D(v_{|P|})$, where $P = \{v_1, v_2, \cdots, v_{|P|}\}$, and each connected component $C_i(V_i, E_i)$ of $G - P$ only belongs to a class $D(v_j)$. If $C_i \in D(v_j)$, there exists an edge $e = (v_j, v_h)$ that connects v_j and some $v_h \in V(G_i)$. Note that a class $D(v_j)$ cannot contain any connected component of $G - P$.

Consider a vertex $v \in P$ with $D(v) = \{C_1, \cdots, C_j\}$ $(j \geq 0)$. There are at most $\sum_{t=1}^{j}(|V(G_t)| - 1 + l)$ edges in $D(v)$. Let $E'(v) = E(C_1) \cup \cdots \cup E(C_j)$, $E''(v) = E(\{v\}, V(G) \setminus P)$, $E(v) = E'(v) \cup E''(v)$, and $V(v) = V(C_1) \cup \cdots \cup V(C_j)$.

Partitioning operation

1: let $P = \{v_1, v_2, \cdots, v_{|P|}\}$, and $C = \{C_1, \cdots, C_k\}$, where $C_i = (V_i, E_i)$ $(i = 1, \cdots, k)$ is a connected component in $G \setminus P$.
2: **for** each $v_i \in P$ **do**
3:     $D(v_i) = \emptyset$;
4:     **for** each $C_j \in C$ **do**
5:         **if** there exists an edge $e = (v_i, v_k)$ for v_i and some $v_k \in C_j$ **then**
6:             $D(v_i) = D(v_i) \cup C_j$;
7:         **end if**
8:     **end for**
9:     $C = C \setminus D(v_i)$;
10: **end for**

Fig. 2. Partition the connected components in $G - P$.

Case 1: $j = 0$. In this case, we have $E'(v) = \emptyset$, $V(v) = \emptyset$, and $|E''(v)| \geq 2$.

$$|E(v)| - (1 + |E'(v)|) \geq \frac{1}{2} \cdot |E(v)| \geq \frac{1}{l+1} \cdot (|E(v)| - |V(v)|) \qquad (5)$$

Case 2: $j = 1$. In this case, we have $E'(v) \neq \emptyset$, $V(v) \neq \emptyset$, $|E'(v)| - |V(v)| \leq l - 1$ and $|E''(v)| \geq 2$.

$$|E(v)| - (1 + |E'(v)|) \geq \frac{1}{l+1} \cdot (|E(v)| - |V(v)|) \qquad (6)$$

Case 3: $j \geq 2$. In this case, we have $E''(v) = j + h$ for some $h \geq 0$, $|E(C_i)| - |V(C_i)| \leq l - 1$ for $1 \leq i \leq j$. Hence we can obtain that

$$|E(v)| - |V(v)| \leq l \cdot j + h \qquad (7)$$

Thus, we have

$$|E(v)| - (1 + |E'(v)|) = |E''(v)| - 1 \qquad (8)$$
$$\geq (j - 1) + h \qquad (9)$$
$$\geq \frac{j}{2} + h \qquad (10)$$
$$\geq \frac{|E(v)| - |V(v)|}{2l} \qquad (11)$$

From inequality (10) to (11), we use inequality (7).

By the above inequality (5), (6), (11), and $l \geq 1$, we get the following conclusion.

$$|E| - |P| - |E(V \setminus P, V \setminus P)| \geq \frac{|E| - |V|}{2l} \qquad (12)$$

Therefore, we have

$$w(P) = \Sigma_{v \in P}(deg_G(v) - 1) = \Sigma_{v \in P}deg_G(v) - |P| \tag{13}$$
$$= 2|E(P, P)| + |E(P, V \setminus P)| - |P| \tag{14}$$
$$= (|E(P, P)| + |E(P, V \setminus P)|) + |E(P, P)| - |P| \tag{15}$$
$$= |E| - |E(V \setminus P, V \setminus P)| + |E(P, P)| - |P| \tag{16}$$
$$= |E| - (|E(V \setminus P, V \setminus P)| + |P|) + |E(P, P)| \tag{17}$$
$$\geq \frac{|E| - |V|}{2l} + |E(P, P)| \tag{18}$$
$$\geq \frac{|E| - |V|}{2l} \tag{19}$$

From inequality (17) to (18), we use inequality (12). □

Since P is a minimal l-PFDS of G obtained by Alg-l-PFD(G, w, l), each connected component $C_i = (V_i, E_i)$ in $G - P$ either is acyclic or contains at least one cycle and at most l cycles. If a connected component in $G - P$ is acyclic, it is called an *acyclic component*. Otherwise, it is called a *cyclic component*. Let k' be the number of acyclic components in $G - P$.

Lemma 7. $|P| + 2k' \leq |E(P, V \setminus P)|$.

Proof. For connected components in graph $G - P$, we can classify acyclic components and cyclic components. We consider the following cases.

Case 1: For an acyclic component C_i in $G - P$, if $|E(P, V(C_i)| \leq 1$, then $G[E(P, V(C_i)) \cup E(C_i)]$ is a acyclic graph and C_i is handled in line 14 of Alg-l-PFD(G, w, l). Thus, $|E(P, V(C_i)| \geq 2$, namely, $0+2\times 1$(one acyclic component)$\leq |E(P, V(C_i)|$.

Case 2: For a cyclic component C_i in $G - P$, if there exists a vertex $v \in P$ adjacent to C_i by edges, $|E(\{v\}, V(C_i)| \geq 1$. Hence, 1(one vertex v)$+2 \times 0 \leq |E(\{v\}, V(C_i)|$.

Case 3: For a vertex $v \in P$, assume that vertex v is adjacent to c' connected components $C_1, \cdots, C_{c'}$ of $G - P$. Let $C = \{C_1, \cdots, C_{c'}\}$, and c'' be the number of acyclic components of C.

Subcase 3.1: There exists at least a cyclic component $C_i \in C$. We can obtain that $1 + 2c'' \leq |E(\{v\}, V(C)|$ by Case 1 and Case 2.

Subcase 3.2: There exists at least one acyclic component $C_i \in C$ adjacent to v by at least three edges. $|E(\{v\}, V(C_i)| \geq 3$, namely, $1+2\times 1 \leq |E(\{v\}, V(C_i)|$. Thus, we also have $1 + 2c'' \leq |E(\{v\}, V(C)|$ by Case 1.

Subcase 3.3: $c' = 1$ and $C_{c'}$ is an acyclic component. Since P is a minimal l-PFDS for G, vertex v is adjacent to the acyclic component $C_{c'}$ by at least $l+2$ edges. It is easy to see that $1 + 2 \times 1 \leq l + 2 \leq |E(\{v\}, V(C_{c'}))|$ as $l \geq 1$.

Subcase 3.4: $c' \geq 2$ and each component in C is an acyclic component adjacent to v by at most two edge. Vertex v is adjacent to each acyclic component in C by at least one edge. Since P is a minimal l-PFDS of G, $G[\{v\} \cup V(C)]$ contains at least $l + 1$ cycles. If an acyclic component C_i is adjacent to v by only one edge, $G[\{v\} \cup V(C_i)]$ must be a tree. There exists a vertex $u \in P \setminus \{v\}$ adjacent to C_i, if not, C_i is handled in line 15 of Alg-l-PFD(G, w, l). We remove edge $E(\{v\}, C_i)$ and insert a new edge incident with u and C_i. That does not have impact on the cycles in $G[\{v\} \cup V(C)]$ and Case 1. Thus, for the subcase, we only consider the case that each component C_i in C is adjacent to v by two edges, namely $|F(\{v\}, C_i)| = 2$ $(1 \leq i < c')$ and $c' = c''$. If these components in C are not adjacent to vertices in $P \setminus \{v\}$, then there exists a semidisjoint l-pseudotree in $G[\{v\} \cup V(C)]$ which is handled in line 16–20 of Alg-l-PFD(G, w, l). Thus, there exists one component $C_i \in C$ adjacent to vertex $u \in P \setminus \{v\}$. If $|E(\{u\}, V(C_i))| = 1$, then remove edge $E(\{u\}, V(C_i))$ and insert a new edge incident with v and C_i. Thus, we have $|E(\{v\}, V(C_i))| = 3$, namely, $1 + 2 \times 1 \leq |E(\{v\}, V(C_i))|$. If $|E(\{u\}, V(C_i))| \geq 2$, then we have $|E(\{u, v\}, V(C_i))| \geq 4$, namely, $2 + 2 \times 1 \leq |E(\{u, v\}, V(C_i))|$. For $C \setminus \{C_i\}$, we obtain that $2 \times (c' - 1) \leq |E(\{v\}, V(C \setminus \{C_i\}))|$ by Case 1.

Therefore, by the above analysis, we have $|P| + 2k' \leq |E(P, V \setminus P)|$.   □

Lemma 8. $w(P) \leq 2(|E| - |V|)$.

Proof. Since graph $G - P$ is an l-pseudoforest and there are k' acyclic connected components in graph $G - P$, $|V| - |P| - k' \leq |E(V \setminus P, V \setminus P)|$.

$$w(P) = \Sigma_{v \in P}(deg_G(v) - 1) = \Sigma_{v \in P}deg_G(v) - |P| \tag{20}$$
$$= 2|E(P, P)| + |E(P, V \setminus P)| - |P| \tag{21}$$
$$= 2(|E(P, P)| + |E(P, V \setminus P)|) - |E(P, V \setminus P)| - |P| \tag{22}$$
$$= 2(|E| - |E(V \setminus P, V \setminus P)|) - |E(P, V \setminus P)| - |P| \tag{23}$$
$$\leq 2(|E| - (|V| - |P| - k')) - |E(P, V \setminus P)| - |P| \tag{24}$$
$$= 2|E| - 2|V| + |P| + 2k' - |E(P, V \setminus P)| \tag{25}$$
$$\leq 2(|E| - |V|) \tag{26}$$

□

Lemma 9. *Let P be a minimal l-PFDS of G generated by Alg-l-PFD(G, w, l). For each type-2 subgraph (G_i, w_i) in the decomposition $\{(G_i, w_i)\}$ derived by Alg-l-PFD(G, w, l), $\frac{w_i(P \cap V(G_i))}{w_i(opt(G_i, w_i))} \leq 4l$.*

Proof. For type-2 subgraph G_i, we obtain the approximation ratio with a lower bound of $w(P_i)$ by Lemma 6 and a upper bound of $w(P_i)$ by Lemma 8. Therefore, the local ratio of type-2 subgraph is that

$$\frac{w_i(P \cap V(G_i))}{w_i(opt(G_i, w_i))} \leq \frac{2(|E(G_i)| - |V(G_i)|)}{\frac{|E(G_i)| - |V(G_i)|}{2l}} = 4l \tag{27}$$

□

Thus, when $l \geq 1$, by Lemmas 1, 5 and 9, we get the following conclusion.

Theorem 1. *There is a $4l$-approximation polynomial time algorithm for the l-Pseudoforest Deletion problem where l is an integer with $l \geq 1$.*

6 Approximation Ratio for the PFD Problem

In this section, we further analyze the approximation ratio of Alg-l-PFD(G, w, l) for the PFD problem. Let P be a minimal 1-PFDS for type-2 subgraph (G, w) of the decomposition derived by Alg-l-PFD(G, w, l), and k' be the number of acyclic connected components of $G - P$, where $w(v) = deg_G(v) - 1$ for every $v \in V(G)$. Now we compute a lower bound of $w(P)$.

Lemma 10. $w(P) \geq |E| - |V|$.

Proof. Since every connected component in graph $G - P$ is a 1-pseudoforest and contains at most one cycle, $|E(V \setminus P, V \setminus P)| \leq |V| - |P|$. We have

$$w(P) = \Sigma_{v \in P}(deg_G(v) - 1) = \Sigma_{v \in P}deg_G(v) - |P| \tag{28}$$
$$= 2|E(P, P)| + |E(P, V \setminus P)| - |P| \tag{29}$$
$$= (|E(P, P)| + |E(P, V \setminus P)|) + |E(P, P)| - |P| \tag{30}$$
$$= |E| - |E(V \setminus P, V \setminus P)| + |E(P, P)| - |P| \tag{31}$$
$$\geq |E| - (|V| - |P|) + |E(P, P)| - |P| \tag{32}$$
$$\geq |E| - |V| \tag{33}$$

$\square$

Thus, when $l = 1$, by Lemmas 1, 5, 8 and 10, we get the following result.

Theorem 2. *There is a 2-approximation polynomial time algorithm for the PFD problem.*

7 Conclusions

We develop a $4l$-approximation algorithm with running time $O(|V| \cdot (|E| + |V|))$ for the l-PFD problem when $l \geq 1$. For any fixed l, the approximation ratio is fixed. The approximation ratio of the algorithm is 2 when $l = 1$. It seems that there is some chance to improve the ratio of approximation for $l \geq 2$, but needs some additional efforts for analysis about the structure of decomposition and local ratio. We do not know if there is a polynomial time reduction from the classical vertex cover problem to 1-PFD problem for a lower bound 2 of approximation ratio as it is for the FVS Problem (0-PFD).

References

1. Bafana, V., Berman, P., Fujito, T.: A 2-approximation algorithm for the undirected feedback vertex set problem. SIAM J. Discret. Math. **12**(3), 289–297 (1999)
2. Bar-Yehuda, R., Even, S.: A local-ratio theorem for approximating the weighted vertex cover problem. Ann. Discret. Math. **25**, 27–46 (1985)
3. Becker, A., Geiger, D.: Optimization of Pearl's method of conditioning and greedy-like approximation algorithms for the vertex feedback set problem. Artif. Intell. **83**(1), 167–188 (1996)

4. Bodlaender, H.L., Ono, H., Otachi, Y.: A faster parameterized algorithm for pseudo-oforest deletion. In: Guo, J., Danny, H. (eds.) IPEC 2016, LIPIcs, Dagstuhl, Germany, vol. 63, pp. 7:1–7:12 (2017). https://doi.org/10.4230/LIPIcs.IPEC.2016.7
5. Chudak, F.A., Goemans, M.X., Hochbaum, D.S., Williamson, D.P.: A primal-dual interpretation of two 2-approximation algorithms for the feedback vertex set problem in undirected graphs. Oper. Res. Lett. **22**(4–5), 111 118 (1998)
6. Cygan, M., Nederlof, J., Pilipczuk, M., Pilipczuk, M., Rooij, J.M.M., Wojtaszczyk, J.O.: Solving connectivity problems parameterized by treewidth in single exponential time. In: FOCS 2011, pp. 150–159. IEEE Press, New York (2011). https://doi.org/10.1109/FOCS.2011.23
7. Fomin, F.V., Lokshtanov, D., Misra, N., Saurabh, S.: Planar F-deletion: approximation, kernelization and optimal FPT algorithms. In: FOCS 2012, pp. 470–479. IEEE Press, New York (2012). https://doi.org/10.1109/FOCS.2012.62
8. Fujito, T.: A note on approximation of the vertex cover and feedback vertex set problems - unified approach. Inf. Process. Lett. **59**(2), 59–63 (1996)
9. Jansen, B.M., Raman, V., Vatshelle, M.: Parameter ecology for feedback vertex set. Tsinghua Sci. Technol. **19**(4), 387–409 (2014)
10. Kociumaka, T., Pilipczuk, M.: Faster deterministic feedback vertex set. Inf. Process. Lett. **114**(10), 556–560 (2014)
11. Lin, M., Feng, Q., Wang, J., Chen, J., Fu, B., Li, W.: An improved FPT algorithm for almost forest deletion problem. Inf. Process. Lett. **136**, 30–36 (2018)
12. Majumdar, D.: Structural parameterizations of feedback vertex set. In: Guo J., Danny H. (eds.) IPEC 2016, LIPIcs, Dagstuhl, Germany, vol. 63, pp. 21:1–21:16 (2017). https://doi.org/10.4230/LIPIcs.IPEC.2016.21
13. Philip, G., Rai, A., Saurabh, S.: Generalized pseudoforest deletion: algorithms and uniform kernel. In: Italiano, G.F., Pighizzini, G., Sannella, D.T. (eds.) MFCS 2015. LNCS, vol. 9235, pp. 517–528. Springer, Heidelberg (2015). https://doi.org/10.1007/978-3-662-48054-0_43
14. Rai, A., Saurabh, S.: Bivariate complexity analysis of ALMOST FOREST DELETION. In: Xu, D., Du, D., Du, D. (eds.) COCOON 2015. LNCS, vol. 9198, pp. 133–144. Springer, Cham (2015). https://doi.org/10.1007/978-3-319-21398-9_11
15. Rai, A.: Parameterized algorithms for graph modification problems. Ph.D. thesis, Homi Bhabha National Institute, Chennai, India (2016)
16. Xiao, M., Nagamochi, H.: An improved exact algorithm for undirected feedback vertex set. J. Comb. Optim. **30**(2), 214–241 (2015)

New Bounds for Energy Complexity
of Boolean Functions

Krishnamoorthy Dinesh[1], Samir Otiv[2], and Jayalal Sarma[1(✉)]

[1] Indian Institute of Technology Madras, Chennai, India
{kdinesh,jayalal}@cse.iitm.ac.in
[2] Maximl Labs, Chennai, India
samir.otiv@maximl.com

Abstract. For a Boolean function $f : \{0,1\}^n \to \{0,1\}$ computed by a circuit C over a finite basis $\mathcal{B}$, the *energy complexity* of C (denoted by $\mathsf{EC}_\mathcal{B}(C)$) is the maximum over all inputs $\{0,1\}^n$ the numbers of gates of the circuit C (excluding the inputs) that output a one. Energy complexity of a Boolean function over a finite basis $\mathcal{B}$ denoted by $\mathsf{EC}_\mathcal{B}(f) \stackrel{\text{def}}{=} \min_C \mathsf{EC}_\mathcal{B}(C)$ where C is a circuit over $\mathcal{B}$ computing f.

We study the case when $\mathcal{B} = \{\wedge_2, \vee_2, \neg\}$, the standard Boolean basis. It is known that any Boolean function can be computed by a circuit (with potentially large size) with an energy of at most $3n(1 + \epsilon(n))$ for a small $\epsilon(n)$(which we observe is improvable to $3n - 1$). We show several new results and connections between energy complexity and other well-studied parameters of Boolean functions.

- For all Boolean functions f, $\mathsf{EC}(f) \leq O(\mathsf{DT}(f)^3)$ where $\mathsf{DT}(f)$ is the optimal decision tree depth of f.
- We define a parameter *positive sensitivity* (denoted by psens), a quantity that is smaller than sensitivity and defined in a similar way, and show that for any Boolean circuit C computing a Boolean function f, $\mathsf{EC}(C) \geq \mathsf{psens}(f)/3$.
- Restricting the above notion of energy complexity to Boolean formulas, denoted $\mathsf{EC}^\mathsf{F}(f)$, we show that $\mathsf{EC}^\mathsf{F}(f) = \Theta(L(f))$ where $L(f)$ is the minimum size of a formula computing f.

We next prove lower bounds on energy for explicit functions. In this direction, we show that for the perfect matching function on an input graph of n edges, any Boolean circuit with bounded fan-in must have energy $\Omega(\sqrt{n})$. We show that any unbounded fan-in circuit of depth 3 computing the parity on n variables must have energy is $\Omega(n)$.

Keywords: Energy complexity · Boolean circuits · Decision trees

1 Introduction

For a Boolean function $f : \{0,1\}^n \to \{0,1\}$ computed by a circuit C over a basis $\mathcal{B}$, the *energy complexity* of C (denoted by $\mathsf{EC}_\mathcal{B}(C)$) is the maximum over all inputs $\{0,1\}^n$ the numbers of gates of the circuit C (excluding the inputs)

L. Wang and D. Zhu (Eds.): COCOON 2018, LNCS 10976, pp. 738–750, 2018.
https://doi.org/10.1007/978-3-319-94776-1_61

that outputs a one. The energy complexity of a Boolean function over a basis $\mathcal{B}$ denoted by $\mathsf{EC}_\mathcal{B}(f) \overset{\text{def}}{=} \min_C \mathsf{EC}_\mathcal{B}(C)$ where C is a circuit over $\mathcal{B}$ computing f. A particularly interesting case of this measure of Boolean function, is when the individual gates allowed in the basis $\mathcal{B}$ are threshold gates (with arbitrary weights allowed). In this case, the term energy in above model captures the number of neurons firing in the models of the human brain [10]. This motivated the study of upper and lower bounds [10] on various parameters of energy efficient circuits - in particular the question of designing threshold circuits which are efficient in terms of energy as well as size which computes various Boolean functions.

Indeed, irrespective of the recently discovered motivation mentioned above, the notion of energy complexity of Boolean functions, has been studied much before. Historically, the measure of energy complexity of Boolean functions[1] was first studied by Vaintsvaig [13] (under the name "power of a circuit"). Initial research was aimed at understanding the maximum energy needed to compute any n bit Boolean function for a *finite* basis $\mathcal{B}$ (denoted by $\mathsf{EC}_\mathcal{B}(n)$). Towards this end, Vaintsvaig [13] showed that for any finite basis $\mathcal{B}$, the value of $\mathsf{EC}_\mathcal{B}(n)$ is asymptotically between n and $\frac{2^n}{n}$. Refining this result further, Kasimzade [4] gave a complete characterization by showing the following remarkable trichotomy: for any finite complete basis $\mathcal{B}$, either $\mathsf{EC}_\mathcal{B}(n) = \Theta(2^n/n)$ or $\Omega(2^{n/2}) \le \mathsf{EC}_\mathcal{B}(n) \le O(\sqrt{n}2^{n/2})$ or $\Omega(n) \le \mathsf{EC}_\mathcal{B}(n) \le O(n^2)$.

An intriguing question about the above trichotomy is where exactly does the standard Boolean basis $\mathcal{B} = \{\wedge_2, \vee_2, \neg\}$ fits in. By an explicit circuit construction, Kasim-zade [4] showed that $\mathsf{EC}_\mathcal{B}(n) \le O(n^2)$. Recently, Lozhkin and Shupletsov [7] states (without proof) that the circuit construction by Kasim-zade [4] over the complete Boolean basis is of energy $4n$, thus deriving that $\mathsf{EC}_\mathcal{B}(n) \le 4n$. Lozhkin and Shupletsov improves it to $3n(1 + \epsilon(n))$ by constructing a circuit of size $\frac{2^n}{n}(1 + \epsilon(n))$ for an $\epsilon(n)$ tending to 0 for large n. We observe that, this bound can be further improved to be at most $3n - 1$ while size is $2^{O(n)}$ by carefully following the construction in [7] (see Proposition 1).

As mentioned in the beginning, in a more recent work, for the case when the basis is threshold gates[2], Uchizawa *et al.* [10] initiated the study of energy complexity for threshold circuits. More precisely, they defined the energy complexity of threshold circuits and gave some sufficient conditions for certain functions to be computed by small energy threshold circuits. In a sequence of works, Uchizawa *et al.* ([12] and references therein) related energy complexity of Boolean functions under the threshold basis to the other well-studied parameters like circuit size, depth for interesting classes of Boolean functions. In a culminating result, Uchizawa and Takimoto [11] showed that for constant depth thresholds circuits of unbounded weights with the energy restricted to $n^{o(1)}$ needs exponential sized circuits to compute the inner product function. This is also important in the con-

---

[1] We remark the notion of energy of Boolean circuits studied in this paper is very different from those studied in [2,3,5].

[2] With values of the weights and threshold being arbitrary rational numbers, notice that this basis is no longer finite and hence the bounds and the related trichotomy are not applicable.

text of circuit lower bounds, where it is an important open question to prove exponential lower bounds against constant depth threshold circuits in general (without the energy constraints) for explicit functions.

Our Results: Returning to the context of standard Boolean basis $\mathcal{B} = \{\wedge_2, \vee_2, \neg\}$, we show several new results and connections between energy complexity and other Boolean function parameters.

Relation to Parameters of Boolean Functions: As our first and main contribution, we relate energy complexity, $\mathsf{EC}(f)$ of Boolean functions to three parameters of Boolean functions that are not known to be related before, one in terms of a lower bound and the other in terms of an upper bound for the energy complexity $\mathsf{EC}(f)$. The third parameter characterizes the energy complexity of Boolean functions when restricted to formulas, in terms of optimal formula size itself.

For a function $f : \{0,1\}^n \to \{0,1\}$, let $\mathsf{DT}(f)$ denote the decision tree complexity of the Boolean function - the smallest depth of any decision tree computing the function f. We state our main result:

Theorem 1 (Main). *For any Boolean function f, $\mathsf{EC}(f) \leq O(\mathsf{DT}(f)^3)$.*

We remark that the size of the circuit constructed above is exponentially in $\mathsf{DT}(f)$. However, in terms of the energy of the circuit, this improves the bounds of [7] since it now depends only on $\mathsf{DT}(f)$. There are several Boolean functions, for which the decision trees are very shallow - a demonstrative example is the address function[3] where the decision tree is of depth $O(\log n)$. This gives a circuit computing the address function with $O(\log^3 n)$ energy.

On a related note, Uchizawa *et al.* [10], as a part of their main proof, showed a similar result for threshold decision trees which are decision trees where each internal node can query an arbitrary weighted threshold function on input variables. Let $\mathsf{EC}_{th}(f)$ denote the energy of the minimum energy threshold circuit computing f. They showed that $\mathsf{EC}_{th}(f) \leq 1 + \mathsf{DT}_{th}(f)$ where $\mathsf{DT}_{th}(f)$ denotes the depth of smallest depth threshold decision tree computing f. Since their construction produces a weighted threshold circuit, it does not directly give us a low energy Boolean circuit even for Boolean decision trees.

To obtain lower bounds on energy, we define a new parameter called the positive sensitivity (which is at most the sensitivity of the Boolean function). For a function $f : \{0,1\}^n \to \{0,1\}$ and an input $a \in \{0,1\}^n$, we define the *positive sensitivity* (denoted by $\mathsf{psens}(f)$) as the maximum over all inputs $a \in \{0,1\}^n$ - of the number of indices $i \in [n]$ such that $a_i = 1$ and $f(a \oplus e_i) \neq f(a)$. Here $e_i \in \{0,1\}^n$ has the ith bit alone set to 1. Using this parameter, we show the following.

Theorem 2. *For any Boolean function $f : \{0,1\}^n \to \{0,1\}$ computed by a circuit C over the Boolean basis, $\mathsf{EC}(C) \geq \mathsf{psens}(f)/3$.*

---

[3] $\mathsf{ADDR}_k(x_1, x_2, \ldots, x_k, y_0, y_1, \ldots, y_{2^k-1}) = y_{int(x)}$ where $int(x)$ is the integer representation of the binary string $x_1 x_2 \ldots x_k$.

The third parameter we relate to is the formula complexity when the computation is restricted to formulas. For a Boolean function f, let $L(f)$ denotes the number of leaves in the optimal bounded fan-in formula computing f, and $\mathsf{EC}^\mathsf{F}(f)$ denotes the minimum energy for any bounded fan-in formula computing f. We show the following tight relation.

Theorem 3. *For any Boolean function f, $\mathsf{EC}^\mathsf{F}(f) = \Theta(L(f))$*

This indeed shows that there are input settings that makes almost all the gates in a formula to output a one. Thus, bounded fan-in formulas are quite inefficient in terms of energy.

Energy Lower Bounds for Explicit Functions: We explore energy lower bounds for explicit functions for circuits over Boolean basis. In the first part, we show an $\Omega(\sqrt{n})$ lower bound for energy when the fan-in of the circuit is bounded and in the second we show an $\Omega(n)$ lower bound when the depth of the circuit is bounded by three and fan-in of the circuit is unbounded.

Theorem 4. *Let f be the perfect matching function of a graph on n edges. Then, any Boolean circuit with bounded fan-in, computing f will require energy at least $\Omega(\sqrt{n})$.*

To prove this result, we use the idea of continuous positive paths developed in proving Theorem 2, and show that the monotone Karchmer Wigderson games can be solved by exchanging at most $\mathsf{EC}(C) \log c$ where C is a circuit with fan-in at most c (see Lemma 3 for more details). On the depth restricted front, we show the following.

Theorem 5. *Let C be any unbounded fan-in circuit of depth 3 computing the parity function on n variables. Then, $\mathsf{EC}(C)$ is $\Omega(n)$.*

2   Preliminaries

Let $[n] \overset{\text{def}}{=} \{1, \ldots, n\}$. For $i \in [n]$, let e_i denote the n length Boolean vector with the ith entry alone as 1. For an $a \in \{0,1\}^n$, $a \oplus e_i$ denotes the input obtained by flipping the ith bit of a. The positive sensitivity of f on a, denoted by $\mathsf{psens}(f, a)$, is the number of $i \in [n]$ such that $a_i = 1$ and $f(a \oplus e_i) \neq f(a)$.

A Boolean circuit C over the basis $\mathcal{B} = \{\wedge_2, \vee_2, \neg\}$ is a directed acyclic graph (DAG) with a root node (of out-degree zero), input gates labeled by variables (of in-degree zero) and the non-input gates labeled by functions in $\mathcal{B}$. Define the size to be the number of non-input gates and, depth to be the length of the longest path from root to any input gate of the circuit C. A Boolean formula is a Boolean circuit where the underlying DAG is a tree. We call a negation gate that take input from a variable as a *leaf negation*. A circuit is said to be *monotone* if it does not use any negation gates. A function is monotone if it can be computed by a monotone circuit. Equivalently, a function f is monotone if $\forall\, x, y \in \{0,1\}^n$, $x \prec y \implies f(x) \leq f(y)$ where $x \prec y$ iff $x_i \leq y_i$ for all $i \in [n]$. For a circuit C, $\mathsf{negs}(C)$ denotes the number of negations in the circuit C.

For a monotone function $f : \{0,1\}^n \to \{0,1\}$, $x \in f^{-1}(1)$ and $y \in f^{-1}(0)$, define $S_f^+(x,y) = \{i \mid x_i = 1, y_i = 0, i \in [n]\}$. The monotone Karchmer Wigderson cost of f (denoted by $\mathsf{KW}^+(f)$) is the optimal communication cost of the problem where Alice has x, Bob has y and they have to find an $i \in [n]$ such that $i \in S_f^+(x,y)$. It is known that $\mathsf{KW}^+(f)$ equals the minimum depth monotone circuit computing f. For more details about this model, see [6]. For a Boolean circuit C, and an input a, the energy complexity of C on the input a (denoted by $\mathsf{EC}(C,a)$) is defined as the number of gates that output a 1 in C on the input a. Define the *energy complexity* of C (denoted $\mathsf{EC}(C)$) as $\max_a \mathsf{EC}(C,a)$. The energy complexity of a function f, (denoted by $\mathsf{EC}(f)$) to be the energy of the minimum energy circuit over the Boolean basis $\mathcal{B}$ computing f. As mentioned in the introduction, we observe the following about the construction of Lozhkin and Shupletsov [7].

Proposition 1. *For any $f : \{0,1\}^n \to \{0,1\}$, $\mathsf{EC}(f) \leq 3n - 1$.*

A *decision tree* is a rooted tree with all the non-leaf nodes labeled by variables and leaves labeled by a 0 or 1. Note that every assignment to the variable in the tree defines a unique path from root to leaf in the natural way. A Boolean function f is said to be computed by a decision tree if for every input a, the path from root to a leaf guided by the input is labeled by $f(a)$. Depth of a decision tree is the length of the longest path from root to any leaf. Define decision tree depth of f (denoted by $\mathsf{DT}(f)$) as the depth of the minimum depth decision tree computing f.

3   Energy Complexity as a Boolean Function Parameter

In this section, we show new techniques to obtain upper bounds and lower bounds on $\mathsf{EC}(f)$. In the upper bound front, we show that a Boolean function f whose $\mathsf{DT}(f)$ is low can be computed by a low energy circuit and also a weak converse of the statement (Sect. 3.1). We introduce a new parameter of Boolean function, called the *positive sensitivity*, and show that it forms a lower bound on $\mathsf{EC}(f)$ for any Boolean function f (Sect. 3.2) and conclude the section by a characterization of energy complexity of Boolean formulas (Sect. 3.3).

3.1   Energy Upper Bounds from Decision Trees

We know that any n bit function f can be computed by a circuit of energy at most $3n - 1$ (see Proposition 1). In this section, we identify the property of having low depth decision trees as a sufficient condition to guarantee energy efficient circuits. More precisely, we show that for any Boolean function f, $\mathsf{EC}(f) \leq O(\mathsf{DT}(f)^3)$.

One of the challenges in constructing a Boolean circuit is to use as few negation gates as possible. The reason is that non-leaf negation gates always contribute to the energy since either the gate or its input will always output a 1 on any input to the circuit. We achieve this in our construction via an idea inspired by the *connector circuit* introduced by Markov [8]. We now describe a circuit C computing f of energy at most $2\mathsf{DT}(f)^2$ with $\wedge$ gates having a large fan-in.

Lemma 1. *For any non-constant Boolean function f, there exists a circuit C computing f with, (1) all $\vee$ gates are of fan-in 2 and all $\wedge$ gates are of fan-in at most $\mathsf{DT}(f) + 2$, (2) no $\vee$ gate have a negation gate or a variable directly as its input, (3) $\mathsf{negs}(C) \leq \mathsf{DT}(f)$ and, (4) $\mathsf{EC}(C) \leq 2\mathsf{DT}(f)^2$.*

Proof. We first describe a circuit computing a function of the kind $f(x) = \neg x_1 \wedge f_0(x) \vee x_1 \wedge f_1(x)$ which is efficient in terms of the usage of negations. Using this, we obtain a C computing f satisfying (1) to (4).

Let f_0 (resp. f_1) be computed by a circuit C_0 (resp. C_1). We now construct a circuit C^* computing f with $1 + \max\{\mathsf{negs}(C_0), \mathsf{negs}(C_1)\}$ negations. We start with the circuit $A = \neg x_1 \wedge C_0(x) \vee x_1 \wedge C_1(x)$. Let g_0 (resp. g_1) be the lexicographically least gate that feeds into a negation in C_0 (resp. C_1). Let D_0' (resp. D_1') be the circuit C_0 with the negation gate that g_0 (resp. g_1) feeds into alone removed and let D_0 (resp. D_1) be the sub-circuit rooted at g_0 (resp. g_1). We construct a new circuit as shown in Fig. 1 with the *connector circuit* in the box. By construction, it can be argued that the resulting circuit correctly computes f while using one negation less that the circuit A. We repeat the previous steps restricted to gates in D_0' and D_1' as long as the negations in at least one of the circuits is exhausted. Hence the resulting circuit C^* (see Fig. 1 without the thinly dashed lines) has $1 + \mathsf{negs}(C_0) + \mathsf{negs}(C_1) - \min\{\mathsf{negs}(C_0), \mathsf{negs}(C_1)\} = 1 + \max\{\mathsf{negs}(C_0), \mathsf{negs}(C_1)\}$. We remark that though Markov [8] does shows the existence of such a circuit, their construction can be shown to have high energy making it unsuitable for our purpose.

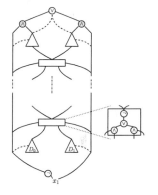

Fig. 1. Circuit C^* and the modifications

We describe the construction of the circuit C by an induction on $\mathsf{DT}(f)$. For f with $\mathsf{DT}(f) \leq 2$, the Boolean function f can be expressed as $(\neg x_1 \wedge \ell_1) \vee (x_1 \wedge \ell_2)$ where ℓ_i is a literal or a constant. By applying the construction of C^* outlined (if needed) the condition (3) can be ensured. It can be verified that for the resulting circuit, $(1), (2)$ and (4) holds for f. This completes the base case.

Let f be a Boolean function computed by a decision tree T of depth $\mathsf{DT}(f) \geq 3$. Let the root variable of T be x_1 and T_0 (resp. T_1) be the left (resp. right) subtree computing the function $f_0 = f|_{x_1=0}$ (resp. $f_1 = f|_{x_1=1}$). Since f_0 and f_1 are computed by decision trees of depth $\mathsf{DT}(f) - 1$, by induction, there exists circuits C_0 and C_1 computing f_0 and f_1, respectively, satisfying (1) to (4).

Observe that $f(x) = \neg x_1 \wedge f_0 \vee x_1 \wedge f_1$. Hence by our construction, there exists a circuit C^* computing f (Fig. 1 omitting the thinly dashed lines) with $\mathsf{negs}(C) = \max\{\mathsf{negs}(C_0), \mathsf{negs}(C_1)\} + 1$. We modify the circuit C^* as follows : for each $\wedge$ gate which was originally in C_0 (resp. C_1), we add $\neg x_1$ (resp. x_1) as input thereby increasing its fan-in by 1. We also remove the $\wedge$ gate (shaded in Fig. 1) feeding into the top $\vee$ gate and feed the output of the circuits directly to the top $\vee$ gate (shown as dashed in Fig. 1). Call the resulting circuit C' and the gates from C_0 as C_0' (the left part in Fig. 1) and the gates from C_1 as C_1' (the right part in Fig. 1).

For the circuit C', we observe that by construction and inductive hypothesis, the condition (1) holds. The removal of the shaded $\wedge$ gates never causes a variable or a negation to be fed to the top $\vee$ gate since f_0 and f_1 have a decision tree depth of at least 2 and hence the circuits of the respective functions have top gate as $\vee$ which is guaranteed by base case for depth 2 and by induction otherwise. Hence condition (2) holds. We now argue that C' correctly computes f. When $x_1 = 1$, all the $\wedge$ gates in C_0' evaluates to 0. Since no input variable or negation gate feeds into any $\vee$ gate in C_0' (condition (2)), all the $\vee$ gates and $\wedge$ gates output 0 irrespective of the remaining input bits. Hence the C_0' outputs 0. Since $x_1 = 1$, C_1' behaves exactly same as C_1. By construction of C^*, the circuit C_1 correctly computes f when $x_1 = 1$. Hence the circuit C' correctly computes f for $x_1 = 1$. The same argument with C_0 and C_1 interchanged shows that C' correctly computes f when $x_1 = 0$.

As no new negations are added in C', $\mathsf{negs}(C') = \mathsf{negs}(C^*)$, whence by induction hypothesis, condition (3) holds. We now show that condition (4) holds for C'. Let x be an input with $x_1 = 1$. We have already argued that when $x_1 = 1$, none of the $\wedge$ or $\vee$ gates of C_0' output a 1. Hence the gates that can output a 1 in C' are the negations in C_0', the gates that output 1 in C_1', the connector gates, the root gate and the negation gate for x_1 (recall that the shaded $\wedge$ gates are removed in C'). A similar argument holds for x with $x_1 = 0$. Hence, it can be argued that $\mathsf{EC}(C')$ is at most $\max\{\mathsf{EC}(C_0), \mathsf{EC}(C_1)\} + 3\max\{\mathsf{negs}(C_0), \mathsf{negs}(C_1)\} + 2$. By induction, we have $\mathsf{EC}(f) \leq \mathsf{EC}(C') \leq 2(\mathsf{DT}(f) - 1)^2 + 3(\mathsf{DT}(f) - 1) + 2$ which implies $\mathsf{EC}(f) \leq 2\mathsf{DT}(f)^2$ as f is non-constant. This completes the induction. $\qquad\square$

We now prove the main result of this section.

Proof (of Theorem 1). If f is constant, the result holds. Otherwise, applying Lemma 1, we have a circuit C' computing f with fan-in of $\vee$ gate being 2 and fan-in of $\wedge$ gate being at most $\mathsf{DT}(f) + 2$ of energy at most $2\mathsf{DT}(f)^2$. To obtain a bounded fan-in circuit from C', we replace the $\wedge$ gates by a tree of fan-in 2 $\wedge$ gates of $\mathsf{DT}(f) + 2$ leaves. Hence, $\mathsf{EC}(f) \leq \mathsf{EC}(C) \cdot (\mathsf{DT}(f) + 1) \leq 2\mathsf{DT}(f)^2 \cdot (\mathsf{DT}(f) + 1) = O(\mathsf{DT}(f)^3)$. $\qquad\square$

It can be argued that for any Boolean function f and a circuit C of size s and energy e computing f over an arbitrary finite basis $\mathcal{B}$, $s^e \geq \Omega(\mathsf{DT}(f))$ which can be seen as a weak converse of Theorem 1. The proof is deferred to the full version of this paper.

3.2 Energy Lower Bounds from Positive Sensitivity

In this section, we prove Theorem 2 from the introduction. We first describe an outline here. As a starting case, consider a monotone circuit C computing f evaluates to 1 on an input $a \in \{0, 1\}^n$. Let $i \in [n]$ be such that $a_i = 1$ and flipping a_i to 0 causes the circuit to evaluate to 0. We show that for such an index i on input a, there is a path from x_i to the root such that all the gates in the path outputs a 1. The latter already implies a weak energy lower bound. We then generalize this idea to non-monotone circuits as well and use it to prove energy lower bounds. This generalization also helps us to prove upper bounds for KW^+ games in Sect. 4.1.

To keep track of all input indices that are sensitive in the above sense, we introduce the measure of positive sensitivity denoted by $\mathsf{psens}(f)$ (as defined in Sect. 2). Let $\widetilde{\mathsf{psens}}(f, a)$ denote the set of positive sensitive indices on a.

Continuous Positive Paths: Let C be a Boolean circuit computing $f :$ $\{0, 1\}^n \rightarrow \{0, 1\}$. For an input $a \in \{0, 1\}^n$, we call a path of gates such that every gate in the path output 1 on a as a *continuous positive path* in C.

Fix an $a \in \{0, 1\}^n$. We argue that for every positive sensitive index i on a, either there is a continuous positive path from x_i to the root or it must be broken by a negation gate of the circuit. Using this we show that energy complexity of a function is lower bounded by its positive sensitivity.

Lemma 2. *Let $f : \{0, 1\}^n \rightarrow \{0, 1\}$ and $a \in \{0, 1\}^n$ be an input such that $\mathsf{psens}(f, a) \neq 0$ and $i \in \widetilde{\mathsf{psens}}(f, a)$. Let C be any circuit computing f. Then, either (1) there is a continuous positive path from x_i to root or (2) there is a continuous positive path from x_i to a gate which feeds into a negation gate of C.*

Proof. It suffices to prove the following stronger statement: for a Boolean function f and an $a \in \{0, 1\}^n$ with $\mathsf{psens}(f, a) \neq 0$ and $i \in \widetilde{\mathsf{psens}}(f, a)$, let C be any circuit such that $C(a) = f(a)$ and $C(a \oplus e_i) = f(a \oplus e_i)$. Then, either (1) there is a continuous positive path from x_i to root or (2) there is a continuous positive path from x_i to a gate which feeds into a negation gate of C. Proof is by induction on $\mathsf{negs}(C)$. Let C be any circuit such that $f(a) = C(a)$ and $f(a \oplus e_i) = C(a \oplus e_i)$.

Base Case: For the base case, $\mathsf{negs}(C) = 0$. As C is a monotone circuit and $i \in \mathsf{psens}(f, a)$, $C(a) = 1$. Hence, there must exist a series of gates all evaluating to 1 reaching some inputs. For any $i \in \mathsf{psens}(f, a)$, we show that (1) holds. Suppose not, then among all the paths from x_i to the root, collect all the gates that evaluate to 0 for the first time in the path and call this set as T. We fix all the variables except x_i to the values in a and view each of the gates in $g \in T$ as a function of x_i. Now, flipping x_i from $a_i = 1$ to 0 does not change the output of any $g \in T$ as they compute monotone functions and already evaluate to 0. Since all other values are fixed, the output of the root gate does not change by this flip which contradicts the fact that $i \in \mathsf{psens}(f, a)$.

Induction Step: Let C be a circuit with $\mathsf{negs}(C) \geq 1$. Let g be the first gate that feeds into a negation in the topologically sorted order of the gates of C.

We have the following two possibilities. In both the cases, we argue existence of continuous positive path in C from the variable x_i. The first case is, *on input a, flipping a_i change the output of g.* Denote the function computed at g as f_g.

Then f_g is monotone and $i \in \widetilde{\mathsf{psens}}(f_g, a)$ and is non-empty. Hence applying the argument in the base case to f_g and the monotone circuit rooted at g, we are guaranteed to get a continuous positive path from x_i to g. Since the circuit at g is a sub-circuit of C (that is, it appear as an induced subgraph), this gives a continuous positive path in C also. The second case is, *on input a, flipping* a_i *does not change the output of g*. In this case, we remove the negation gate that g feeds into and hard wire the output of this negation gate (on input a) in C to get a circuit C'. Note that all other gates in C are left intact. Observe that $C'(a) = f(a)$. Since flipping a_i did not change the output of g and as all other gates are left intact, $C'(a \oplus e_i) = f(a \oplus e_i)$. As $\mathsf{negs}(C') = \mathsf{negs}(C) - 1$, by induction, either (1) there is a continuous positive path from x_i to root or (2) there is a continuous positive path from x_i to a gate which feeds into a negation gate of C'. By construction, C' is same as C except for the negation gate. Hence a continuous positive path in C' is also a continuous positive path in C.   □

From Positive Sensitivity to Energy Lower Bounds: We call the negation gates and the root gate of a circuit as *target gates*. In Lemma 2, we have already shown the existence of continuous positive paths from a positive sensitive index up to a target gate. Using this, we show an energy lower bound for any circuit of bounded fan-in computing a Boolean function f in terms of $\mathsf{psens}(f)$. Since the fan-in of the circuit is limited, we exploit the idea that in a connected DAG, the number nodes with out degree at least 1 (internal nodes) is lower bounded by the number of source nodes.

Since every such positive sensitive index is reachable via a continuous positive path from a target gate, we obtain a lower bound on energy by applying this idea on an appropriate subgraph constructed from our circuit.

Proof (of Theorem 2). Without loss of generality assume that f is non-constant. Let C be any circuit computing f of fan-in 2 such that $\mathsf{EC}(C) = \mathsf{EC}(f)$. We prove that $\forall a \in \{0, 1\}^n, \mathsf{psens}(f, a) \leq 3\mathsf{EC}(C)$.

Let $a \in \{0, 1\}^n$ by any input. If $\mathsf{psens}(f, a) = 0$, the claim holds. Hence we can assume, $\mathsf{psens}(f, a) \neq 0$. Let T be the set of all target nodes in C. For every $i \in \widetilde{\mathsf{psens}}(f, a)$, by Lemma 2, there exists continuous positive paths starting from x_i to a gate $g \in T$. For every $g \in T$, let X_g be the set of all gates that lie in a continuous positive path from an x_i to g for some $i \in \widetilde{\mathsf{psens}}(f, a)$. Note that the subgraph induced by vertices in X_g is connected and does not include g. We now obtain a connected DAG with $\mathsf{psens}(f, a)$ leaves as follows. Let D be a full binary tree (with edges directed from child to parent) with $|T|$ many leaves and hence $|T| - 1$ internal nodes. For each $g \in T$ if it is a negation, we attach the gate feeding into g as a leaf of the D and if it is a root, we attach the root as a leaf of the D. Let H be the resulting DAG.

Since graph induced on X_g is connected for each g, this gives us a connected DAG on $\mathsf{psens}(f, a)$ many source nodes. Let $X = \cup_{g \in T} X_g$. Observe that the number of internal nodes is $|X| + (|T| - 1) + 1$ where the first term is the gates in X, the second term is the number of internal nodes of the tree and third term is due to the root. Since the target gates include negations and the root, $|T| = \mathsf{negs}(C) + 1$. Since the total number of negation gates in any circuit computing f is at most $\mathsf{EC}(f)$, we get that number of internal nodes of H is at most $|X| + |T| - 1 \le |X| + \mathsf{EC}(f) + 1 - 1 \le 2\mathsf{EC}(f)$. Since the resulting DAG is connected, the number of leaves, which is $\mathsf{psens}(f, a)$, is at most the number of internal nodes $+1$ which is at most $2\mathsf{EC}(f) + 1 \le 3\mathsf{EC}(f)$. $\square$

3.3   Formulas Size Characterizes Formula-Energy Complexity

Intuitively, Boolean formulas can take more energy than a circuit since we cannot "reuse" computation. Recall that $\mathsf{EC}^{\mathsf{F}}(f)$ as the energy of a minimum energy bounded fan-in formula computing f. In this section, we show that $\mathsf{EC}^{\mathsf{F}}(f) = \Theta(L(f))$ where $L(f)$ is the size of an optimal formula computing f.

Proof (of Theorem 3). Observe that since f can be computed by a fan-in 2 formula of size $L(f)$, there are at most $L(f) - 1$ internal nodes implying that the number of gates that fire cannot exceed $L(f) - 1$. Hence $\mathsf{EC}^{\mathsf{F}}(f) = O(L(f))$. We show that for any formula H of s leaves computing f, $\mathsf{EC}^{\mathsf{F}}(H) \ge s/32$ implying $\mathsf{EC}^{\mathsf{F}}(f) \ge \Omega(s)$ thereby completing the proof.

If at least half of the leaves of H read a negated variable, we have $\mathsf{EC}(H) \ge s/2$. Otherwise, at least $s/2$ of the leaves are unnegated and feeds directly to at least $s/4$ gates which are either $\wedge$ or $\vee$ gates. Of these, assume without loss of generality that $\wedge$ appears at least $s/8$ times. Then, on a random assignment, since $\wedge$ will be 1 with probability $1/4$, the expected number of gates that evaluate to 1 is at least $s/32$. Hence $\mathsf{EC}(H) \ge s/32$ for the input achieving the expectation. Similar argument for $\vee$ gives that $\mathsf{EC}(H) \ge 3s/32$. Hence $\mathsf{EC}^{\mathsf{F}}(H) \ge \Omega(s)$. Since we started with an arbitrary formula computing f, $\mathsf{EC}^{\mathsf{F}}(f) = \Omega(L(f))$. The above argument can be extended to show that for formulas of fan-in k, $\mathsf{EC}^{\mathsf{F}}(f) = \Omega(L(f)/2^k)$. $\square$

4   Energy Lower Bounds for Explicit Functions

In this section, we prove energy lower bounds for explicit functions. The first one is a lower bound for any Boolean circuit, and the second one is against unbounded fan-in depth three circuits.

4.1   Energy Lower Bounds from Karchmer-Wigderson Games

We use Lemma 2 and utilize the existence of continuous positive paths to design a KW^+ protocol of cost $O(\mathsf{EC}(C) \log \mathsf{fan\text{-}in}(C))$ (Lemma 3). Using this, we derive that any circuit C with constant fan-in computing the perfect matching function f_{PM} require energy at least $\Omega(\sqrt{n})$ since $\mathsf{KW}^+(f_{PM}) = \Omega(\sqrt{n})$ [9] thus proving Theorem 4 from the introduction.

Recall that $S_f^+(x, y) \stackrel{\mathrm{def}}{=} \{i \mid x_i = 1, y_i = 0, i \in [n]\}$. Also, we call the set of all negation gates, along with the root gate of C as the *target gates* of C.

Lemma 3. *For a non-constant monotone Boolean function f, let Alice and Bob hold inputs $a \in f^{-1}(1)$ and $b \in f^{-1}(0)$ respectively. Let C be any circuit computing f, and every gate in the circuit is either a $\wedge, \vee$ with fan-in of at most c or a negation gate. Then, $\mathsf{KW}^+(f) \leq \mathsf{EC}(C) \log c$.*

Proof. We argue that, without loss of generality it can be assumed that $\widetilde{\mathsf{psens}}(f, a) = \{i \mid a_i = 1\}$. Alice finds an $a' \prec a$ with $f(a') = f(a) = 1$ such that for any $a'' \prec a'$, $f(a'') = 0$. Observe that $a' \neq 0^n$ for otherwise, $f(0^n) = 1$ and since f is monotone, f must be a constant which is a contradiction. By construction, every bit in a' which is 1 is sensitive. Since $a' \prec a$, $S_f^+(a', b) \subseteq S_f^+(a, b)$, thereby it suffices to find an index in $S_f^+(a', b)$.

We now describe the protocol. Let $a \in f^{-1}(1)$ such that $\widetilde{\mathsf{psens}}(f, a) = \{i \mid a_i = 1\}$. Before the protocol begins, Alice does the following pre-computation. Let $\mathcal{P}$ be the collection of positive paths one each for every $i \in \widetilde{\mathsf{psens}}(f, a)$, which exists as per Lemma 2. Alice computes $\mathcal{P} = \bigcup_{g \in T} \mathcal{P}_g$ where $\mathcal{P}_g$ is the collection of all continuous positive paths ending at the target gate g. This ends the pre-processing. Now Alice and Bob fixes an ordering of the target gates. For each target gate $g \in T$ in the order, the following procedure is repeated. For each continuous positive path $p \in \mathcal{P}$, ending at g, Alice sends the address of the previous gate in the path p (using $\log c$ bits) until they trace back to an input index i. Now, Bob checks if $b_i = 0$, and if so, we have found $i \in S_f^+(a, b)$, else, they attempt on the next $p \in \mathcal{P}_g$.

We argue about the correctness of the protocol. Notice that the above protocol searches through all $i \in \widetilde{\mathsf{psens}}(f, a)$ by traversing through all $\mathcal{P}_g$, for $g \in T$. Since $\widetilde{\mathsf{psens}}(f, a) = \{i \mid a_i = 1\}$ and $S_f^+(a, b) \subseteq \widetilde{\mathsf{psens}}(f, a)$ the protocol correctly computes i such that $a_i = 1$ and $b_i = 0$. Since the protocol visits only those gates that output 1 on a, we have a protocol with communication cost $\leq \mathsf{EC}(C, a) \times \log(c) \leq \mathsf{EC}(C) \log c$. ☐

4.2   Energy Lower Bounds for Depth Three Circuits

We now turn to the energy complexity lower bounds for the constant depth circuits. While we are unable to prove strong lower bounds for circuits of depth d for an arbitrary constant d, we show that any depth $d = 3$ unbounded fan-in circuit computing the parity function requires large energy. For any Boolean function f, the trivial depth 2 circuit of unbounded fan-in computing f has an energy $n+2$ and it can be shown that any depth two circuit computing the parity on n bits require an energy of $n + 1$. We show that the same also holds for any depth 3 circuit computing parity thereby proving Theorem 5 from Introduction.

Razbarov showed that any circuit C of depth d of unbounded fan-in computing parity on n bits must be of size at least $2^{\Omega(n^{1/4d})}$ [1]. Using this result we show an energy lower bound of $\Omega(n)$ for any depth 3 circuit computing $\oplus_n$.

Proof (of Theorem 5). We call the root gate of the circuit as the "top" level and the two level immediately below as the "middle" and "bottom" levels respectively. Assume the circuit C does not have any redundant gates. Note that negations do not count towards the level.

Let there be i negated input variables and without loss of generality, assume $i < n$. We set these variables to 0 and let C' be the resulting circuit obtained. Let $g_1, g_2, \ldots, g_k$ be the k gates in the bottom layer that feeds to the layers above via negation gates. We set input variables to these k gates such that the output of the negations are fixed in the following way: for the gate g_i, consider any input variable, say x_j, that feeds into g_i and set it to 0 if g_i is $\wedge$ gate and 1 if g_i is $\vee$. We also remove the gates that have become a constant and hardwire their output to get the result circuit C'. Hence, all the gates at the bottom level are not fed negated to the level above.

In this process, we have eliminated the k negations leaving us with the circuit C'' where all the gates at bottom and middle layer computes some monotone function on the remaining $m = n - (i + j)$ for some $j \leq k$ variables. Since the resulting circuit must compute parity on m variables, by [1], $size(C'') \geq 2^{\Omega(m^{1/12})}$. Since C'' is of depth 3, the number of bottom and middle gates in C' must also be at least $2^{\Omega(m^{1/12})}$. As the gates in the bottom and middle level computes monotone function, there is a setting of input such that at least $i + k \geq i + j = n - m$ gates contributes an energy of 1 (since either the input to the negation or the negation gate itself will be 1) and $2^{\Omega(m^{1/12})}$ gates in C that evaluate to 1. Hence $\mathsf{EC}(C) \geq n - m + 2^{\Omega(m^{1/12})} = \Omega(n)$ irrespective of m. $\qquad\square$

References

1. Razborov, A.A.: Lower bounds on the size of constant-depth networks over a complete basis with logical addition. Mathemat. Zametki **41**(4), 598–607 (1987)
2. Antoniadis, A., Barcelo, N., Nugent, M., Pruhs, K., Scquizzato, M.: Energy-efficient circuit design. In: Proceedings of the 5th Conference on Innovations in Theoretical Computer Science, pp. 303–312 (2014)
3. Barcelo, N., Nugent, M., Pruhs, K., Scquizzato, M.: Almost all functions require exponential energy. Math. Found. Comput. Sci. **2015**, 90–101 (2015)
4. Kasim-zade, O.M.: On a measure of active circuits of functional elements. In: Mathematical Problems in Cybernetics "Nauka", vol. no. 4, pp. 218–228 (1992). (in Russian)
5. Kissin, G.: Measuring energy consumption in VLSI circuits: a foundation. In: Proceedings of the 14 Annual ACM Symposium on Theory of Computing (1982)
6. Kushilevitz, E., Nisan, N.: Communication Complexity, 2nd edn. Cambridge University Press, Cambridge (2006)
7. Lozhkin, S.A., Shupletsov, M.S.: Switching activity of boolean circuits and synthesis of boolean circuits with asymptotically optimal complexity and linear switching activity. Lobachevskii J. Math. **36**(4), 450–460 (2015)
8. Markov, A.A.: On the inversion complexity of a system of functions. J. ACM **5**(4), 331–334 (1958)
9. Raz, R., Wigderson, A.: Monotone circuits for matching require linear depth. J. ACM **39**(3), 736–744 (1992)
10. Uchizawa, K., Douglas, R.J., Maass, W.: On the computational power of threshold circuits with sparse activity. Neural Comput. **18**(12), 2994–3008 (2006)
11. Uchizawa, K., Takimoto, E.: Exponential lower bounds on the size of constant-depth threshold circuits with small energy complexity. Theoret. Comput. Sci. **407**(1–3), 474–487 (2008)

12. Uchizawa, K., Takimoto, E., Nishizeki, T.: Size-energy tradeoffs for unate cir-
    cuits computing symmetric boolean functions. Theoret. Comput. Sci. **412**, 773–782
    (2011)
13. Vaintsvaig, M.N.: On the power of networks of functional elements. In: Soviet
    Physics Doklady, vol. 6, p. 545 (1962)

Hitting and Covering Partially

Akanksha Agrawal[1], Pratibha Choudhary[2], Pallavi Jain[1],
Lawqueen Kanesh[1(✉)], Vibha Sahlot[1], and Saket Saurabh[1]

[1] Institute of Mathematical Sciences, HBNI, Chennai, India
akanksha.agrawal.2029@gmail.com,
{pallavij,lawqueen,vibhasahlot,saket}@imsc.res.in
[2] Indian Institute of Technology Jodhpur, Jodhpur, India
pratibhac247@gmail.com

Abstract. d-HITTING SET and d-SET COVER are among the classical
NP-hard problems. In this paper, we study variants of d-HITTING SET
and d-SET COVER, which are called PARTIAL d-HITTING SET (PAR-
TIAL d-HS) and PARTIAL d-EXACT SET COVER (PARTIAL d-EXACT SC),
respectively. In PARTIAL d-HS, given a universe U, a family $\mathcal{F}$, of sets
of size at most d over U, and integers k and t, the objective is to decide
if there exists a $S \subseteq U$ of size at most k such that S intersects with
at least t sets in $\mathcal{F}$. We obtain a kernel for PARTIAL d-HS in which the
size of the universe is bounded by $\mathcal{O}(dt)$ and the size of the family is
bounded by $\mathcal{O}(dt^2)$. Using this result, we obtain a kernel for PARTIAL
VERTEX COVER (PVC) with $\mathcal{O}(t)$ vertices, where t is the number of
edges to be covered. Next, we study the PARTIAL d-EXACT SC problem,
where, given a universe U, a family $\mathcal{F}$, of sets of size exactly d over U,
and integers k and t, the objective is to decide if there is $\mathcal{S} \subseteq \mathcal{F}$ of size at
most k, such that $\mathcal{S}$ covers at least t elements in U. We design a kernel
for PARTIAL d-EXACT SC in which sizes of the universe and the family
are bounded by $\mathcal{O}(k^{d+1})$. Finally, we study a special case of PARTIAL
d-HS, when $d = 2$, and design an exact exponential time algorithm with
running time $\mathcal{O}(1.731^n n^{\mathcal{O}(1)})$.

Keywords: Partial d-Hitting Set · Partial d-Set Cover
Partial Vertex Cover · k-Maximum Coverage · Kernel · Exact algorithm

1 Introduction

HITTING SET and SET COVER are among the most classic NP-hard problems [14]. Both these problems (and their variants) have received substantial
attention in algorithm design, and their studies have led to development of
many tools and techniques (see, for example [6,8,21]). Consider a universe U
and a family $\mathcal{F}$, of subsets of U. A set $X \subseteq U$ is said to "hit" a set $F \in \mathcal{F}$, if
$X \cap F \neq \emptyset$. Furthermore, a set $F \in \mathcal{F}$ is said to "cover" an element $u \in U$ if

---

P. Jain—The author acknowledges DST, India for SERB-NPDF fellowship
[PDF/2016/003508].

© Springer International Publishing AG, part of Springer Nature 2018
L. Wang and D. Zhu (Eds.): COCOON 2018, LNCS 10976, pp. 751–763, 2018.
https://doi.org/10.1007/978-3-319-94776-1_62

$u \in F$. The HITTING SET problem takes as an input a universe U, a family $\mathcal{F}$ of sets over U, and an integer k, and the objective is to test if there exists a set $S \subseteq U$ of size at most k such that S hits every set in $\mathcal{F}$. A problem equivalent to HITTING SET, is the SET COVER problem. SET COVER takes as an input a universe U, a family $\mathcal{F}$ of sets over U, and an integer k, and the objective is to test if there exists a set $\mathcal{S} \subseteq \mathcal{F}$ of size at most k, such that $\bigcup_{S \in \mathcal{S}} S$ covers every element in U.

HITTING SET and SET COVER have also been studied for the case when each set in the input family has size at most d (see, for example [3,13,19] and references therein). These variants of HITTING SET and SET COVER are called d-HITTING SET and d-SET COVER, respectively. Note that d-HITTING SET is NP-hard, even for $d = 2$. While SET COVER is polynomial time solvable for $d = 2$ [17], it becomes NP-hard for $d \geq 3$ [14,20]. A well studied variant of d-HITTING SET (d-SET COVER) is the PARTIAL d-HITTING SET (PARTIAL d-SET COVER) problem. The PARTIAL d-HITTING SET (PARTIAL d-SET COVER) takes as an input a universe U, a family $\mathcal{F}$ of sets, each of size at most d, over U, and integers k and t, the objective is to decide if there exists a $S \subseteq U$ ($\mathcal{S} \subseteq \mathcal{F}$) of size at most k, such that S ($\mathcal{S}$) hits (covers) at least t sets (elements) in $\mathcal{F}$ (U).

In this paper, we look at the d-PARTIAL HITTING SET and d-PARTIAL SET COVER problems. Firstly, we study these problems from the viewpoint of Kernelization complexity. *Kernelization* is one of the central notions in Parameterized Complexity. It mathematically models the efficiency of a preprocessing routine. The input to a parameterized problem is an instance of I, the classical problem, and an integer κ, which is called the *parameter*. A parameterized problem Π is said to admit an $f(\kappa)$-*kernel* if there exists a polynomial time algorithm (the degree of the polynomial is independent of κ), called a *kernelization* algorithm, that given an input (I, κ) of Π, outputs an equivalent instance (I', κ') such that $|I'| + \kappa' \leq f(\kappa)$. If the function $f(\cdot)$ is polynomial in κ, then we say that the problem admits a polynomial kernel. Secondly, we provide a *Fixed Parameter Tractable (FPT)* algorithm for PARTIAL VERTEX COVER. A problem is said to be *FPT* if it admits an algorithm that takes an input (I, κ), and correctly decides the problem in time $g(\kappa) \cdot |I|^c$, where $g(\cdot)$ is a computable function. For more details on Parameterized Complexity we refer to the books of Downey and Fellows [7], Flum and Grohe [9], Niedermeier [18], and Cygan et al. [6].

The d-HITTING SET problem admits a kernel of size $\mathcal{O}(k^{d-1})$ [1]. An equivalent formulation of PARTIAL 2-HITTING SET is the PARTIAL VERTEX COVER problem. PARTIAL VERTEX COVER is known not to admit a kernel (or an FPT algorithm) when parameterized by the solution size [12]. This implies that PARTIAL d-HITTING SET also does not admit a kernel, when parameterized by the solution size. Kneis et al. [16] gave a deterministic algorithm with running time $\mathcal{O}(1.396^t)$, and a randomized algorithm with running time $\mathcal{O}(1.2993^t)$ for PVC, where t is the number of edges to be covered. The PVC problem has also been studied for the case when the input is restricted to some special family of graphs (see for example [2,10]). The PARTIAL SET COVER problem is known not to admit an FPT algorithm, when parameterized by the solution size, but it admits an FPT algorithm (or an exponential kernel) when parameterized by t [5]. Weighted versions of PARTIAL VERTEX COVER have also been studied (see for example, [4,11,15]).

Our Results. Firstly, we study the PARTIAL d-HITTING SET problem, from the viewpoint of Kernelization Complexity. In the following, we formally define the problem PARTIAL d-HITTING SET.

PARTIAL d-HITTING SET (PARTIAL d-HS)   **Parameter:** t
Input: A universe U, a family $\mathcal{F}$ of sets of size at most d over U, and integers k and t.
Question: Is there a set $S \subseteq U$ of size at most k, for which there is $\mathcal{F}' \subseteq \mathcal{F}$ of size at least t such that for each $F \in \mathcal{F}'$, we have $S \cap F \neq \emptyset$?

In Sect. 3, we obtain a kernel for PARTIAL d-HS, where the size of the universe is bounded by $\mathcal{O}(dt)$ and the size of the family is bounded by $\mathcal{O}(dt^2)$. As a corollary to this result, we obtain a kernel for PVC (to be defined, shortly) with $\mathcal{O}(t)$ vertices and $\mathcal{O}(t^2)$ edges.

Next, we consider the problem PARTIAL d-EXACT SET COVER, which is formally defined below.

PARTIAL d-EXACT SET COVER (PARTIAL d-EXACT SC)   **Parameter:** t
Input: A universe U, a family $\mathcal{F}$ of sets of size exactly d over U, and integers k and t.
Question: Is there a set $\mathcal{S} \subseteq \mathcal{F}$ of size at most k, such that $|\cup_{F \in \mathcal{S}} F| \geq t$?

In Sect. 4, we obtain a kernel for PARTIAL d-EXACT SC, where the size of universe is bounded by $\mathcal{O}((k-1)^{d+1}d^{d+2})$ and size of the family is bounded by $\mathcal{O}((k-1)^{d+1}d^{d+1})$. Our kernelization algorithm is based on careful selection of a set that can be removed from the family, and once the number of sets in the family is bounded, we are able to obtain a bound on the size of the universe.

Finally, we consider the problem called PARTIAL VERTEX COVER, which is exactly the same as PARTIAL d-HS, for $d = 2$. The problem PARTIAL VERTEX COVER is formally defined below in a graph theoretic notation.

PARTIAL VERTEX COVER (PVC)   **Parameter:** t
Input: A (multi) graph G, and integers k and t.
Question: Is there a set $S \subseteq V(G)$ of size at most k, for which there is $E' \subseteq E(G)$ of size at least t, such that for each $uv \in E'$, we have $S \cap \{u, v\} \neq \emptyset$?

In Sect. 5, we give an exact exponential time algorithm for PVC. To the best of our knowledge, we do not know any exact exponential algorithm for the problem, apart from the trivial $2^n n^{\mathcal{O}(1)}$ algorithm, where n is the number of vertices in the input graph. We design an algorithm for PVC running in time $\mathcal{O}(2^{\omega n/3} n^{\mathcal{O}(1)}) \in \mathcal{O}(1.731^n n^{\mathcal{O}(1)})$. Here, ω is the exponent of matrix multiplication algorithm, for which the current best known bound is $\omega < 2.373$ [23]. Our algorithm for PVC is based on reducing the problem to finding maximum weighted triangle in a graph. Maximum weighted triangle in a graph can be found by using the algorithm given by Williams in [22].

2   Preliminaries

Sets and Functions. We denote the set of natural numbers and the set of integers by $\mathbb{N}$ and $\mathbb{Z}$, respectively. By $\mathbb{Z}^+$, we mean the set of positive integers. For $n \in \mathbb{N}$, we use $[n]$ to denote the sets $\{1, 2, \cdots, n\}$. We use ω to denote the exponent in running time of algorithm for matrix multiplication, the current best known bound for it is $\omega < 2.373$ [23].

In the following consider a set U and a family of sets $\mathcal{F}$, of subsets of U. We call U as the *universe* of $\mathcal{F}$, and $\mathcal{F}$ is a family over U. By 2^U, we denote the power set of U, i.e., $2^U = \{X \mid X \subseteq U\}$. For $U' \subseteq U$, by $\mathcal{F} - U'$, we denote the multi-set $\{F \setminus U' \mid F \in \mathcal{F}\}$. A set $X \subseteq U$ hits a set $F \in \mathcal{F}$, if $X \cap F \neq \emptyset$. Moreover, X is a *hitting set* for $\mathcal{F}$, if X hits each set in $\mathcal{F}$. For a set $\mathcal{F}' \subseteq \mathcal{F}$, $\mathcal{F}'$ *covers* an element $u \in U$, if there is an $F \in \mathcal{F}'$ such that $u \in F$. Moreover, $\mathcal{F}'$ is a *cover* of U if it covers every element in U.

Graphs. Consider a graph G. By $V(G)$ and $E(G)$ we denote the set of vertices and edges in G, respectively. For $X \subseteq V(G)$, $G[X]$ denotes the subgraph of G with vertex set X and edge set $\{uv \in E(G) \mid u, v \in X\}$.

Let G be a graph. A set $X \subseteq V(G)$, is said to cover an edge $uv \in E(G)$, if $\{u, v\} \cap X \neq \emptyset$. Moreover, X is a *vertex cover* in G if it covers every edge in G. By $d_G(v)$, we denote the degree of a vertex in graph G.

3   Kernel for PARTIAL d-HITTING SET

In this section, we design a kernelization algorithm for PARTIAL d-HITTING SET (PARTIAL d-HS for short) where the size of the universe is bounded by $\mathcal{O}(dt)$ and the size of the family is bounded by $\mathcal{O}(dt^2)$. Let $(U, \mathcal{F}, k, t)$ be an instance of PARTIAL d-HS. The algorithm starts by applying some reduction rules exhaustively, in the order in which they are stated. When none of the reduction rules are applicable, we argue that we have obtained an instance of the desired size. Note that, in this section we assume that $\mathcal{F}$ is a *multi-set*. This assumption is required for proving safeness of one of our reduction rules. Next, we state reduction rules that are used by the algorithm.

Reduction Rule 1. *If $k < 0$ then return that $(U, \mathcal{F}, k, t)$ is a* NO *instance of* PARTIAL d-HS.

The safeness of Reduction Rule 1 follows from the fact that the size of any set is at least 0.

Reduction Rule 2. *If $t \leq 0$ then return that $(U, \mathcal{F}, k, t)$ is a* YES *instance of* PARTIAL d-HS.

The safeness of Reduction Rule 2 follows from the fact that Reduction Rule 1 is not applicable and in this case an empty set is a solution to the given instance.

Reduction Rule 3. *If $t > 0$ and $k = 0$, then return that $(U, \mathcal{F}, k, t)$ is a* NO *instance of* PARTIAL d-HS.

The safeness of Reduction Rule 3 follows from the fact that we need at least one element from U in the solution in order to have a non-empty intersection with a set in $\mathcal{F}$. Hereafter, we assume that we have $k > 0$ and $t > 0$.

Next, we define some notations that will be used in upcoming reduction rules. For $u \in U$, $\mathcal{F}_u = \{F \in \mathcal{F} \mid u \in F\} \subseteq \mathcal{F}$, and for $X \subseteq U$, $\mathcal{F}_X = \{F \in \mathcal{F} \mid F \cap X \neq \emptyset\}$. The following reduction rule deals with elements in U that appear in many sets.

Reduction Rule 4. *If there exists $u \in U$ such that $|\mathcal{F}_u| \geq t$, then return that $(U, \mathcal{F}, k, t)$ is* YES *instance of* PARTIAL d-HS.

The safeness of Reduction Rule 4 follows from the fact that $u \in U$, with $|\mathcal{F}_u| \geq t$ will intersect at least t sets in $\mathcal{F}$, and since $k > 0$ (as Reduction Rule 1 to 3 are not applicable), $\{u\}$ is a solution to PARTIAL d-HS for the given instance. When Reduction Rule 4 is not applicable, an element of U can appear in at most $t - 1$ sets in $\mathcal{F}$.

To state our next reduction rule, we need the following notations. Let $\varphi = (u_1, \cdots, u_n)$ be a monotonically non increasing ordering of elements in U, based on $|\mathcal{F}_{u_i}|$, for $i \in [\ell]$, i.e. we arrange the elements of U based on non increasing order of their frequency of appearance in sets of $\mathcal{F}$. Let A be the set comprising of first $\min\{n, dt + 1\}$ elements in φ. We are now ready to state our final reduction rule.

Reduction Rule 5. *If there is $u \in U \setminus A$, then return $(U \setminus \{u\}, \mathcal{F} - \{u\}, k, t)$.*

Lemma 1. *Reduction Rule 5 is correct.*

Proof. Consider $u \in U \setminus A$. We show that $(U, \mathcal{F}, k, t)$ is a YES instance of PARTIAL d-HS if and only if $(U \setminus \{u\}, \mathcal{F} - \{u\}, k, t)$ is a YES instance of PARTIAL d-HS.

In the forward direction, let $(U, \mathcal{F}, k, t)$ be a YES instance of PARTIAL d-HS, and S be one of its solutions. If $u \notin S$, then clearly S is a solution to PARTIAL d-HS in $(U \setminus \{u\}, \mathcal{F} - \{u\}, k, t)$. Note that here we rely on the fact that $\mathcal{F} - \{u\}$ (and $\mathcal{F}$) is a multi-set. Next, we assume that $u \in S$, and construct some sets, which will be useful in constructing a solution to PARTIAL d-HS in $(U \setminus \{u\}, \mathcal{F} - \{u\}, k, t)$. Let $\mathcal{F}_S = \{F \in \mathcal{F} \mid F \cap S \neq \emptyset\}$, and $\widehat{\mathcal{F}}_S \subseteq \mathcal{F}_S$ be an arbitrarily chosen set of size t. Furthermore, let $\widehat{\mathcal{F}}_u = \{F \in \widehat{\mathcal{F}}_S \mid u \in F\}$, $\widetilde{\mathcal{F}} = \widehat{\mathcal{F}}_S \setminus \widehat{\mathcal{F}}_u$, and $\widetilde{U} = \cup_{F \in \widetilde{\mathcal{F}}} F$. Observe that $|\widetilde{U}| \leq dt$ and $|A| = dt + 1$ (since $U \setminus A \neq \emptyset$), and therefore, $A \setminus \widetilde{U} \neq \emptyset$. Consider $v \in A \setminus \widetilde{U}$, and let $S_v = (S \setminus \{u\}) \cup \{v\}$ and $\mathcal{F}_v = \{F \in \mathcal{F} \mid v \in F\}$. Notice that $\mathcal{F}_v \cap \widetilde{\mathcal{F}} = \emptyset$, as each set in $\mathcal{F}_v$ contains v and no set in $\widetilde{\mathcal{F}}$ contains v. Since v appears before u in the ordering φ of U, we have $|\mathcal{F}_v| \geq |\widehat{\mathcal{F}}_u|$. From the above arguments, together with the fact the $\widehat{\mathcal{F}}_S = \widehat{\mathcal{F}}_u \cup \widetilde{\mathcal{F}}$, we have $|\mathcal{F}_v \cup \widetilde{\mathcal{F}}| \geq |\widehat{\mathcal{F}}_S| = t$. Moreover, S_v is of size at most $|S| \leq k$, and S_v intersects each set in $\mathcal{F}_v \cup \widetilde{\mathcal{F}}$. Therefore, S_v is a solution to PARTIAL d-HS in $(U \setminus \{u\}, \mathcal{F} - \{u\}, k, t)$.

In the backward direction, let S be a solution to PARTIAL d-HS in $(U \setminus \{u\}, \mathcal{F} - \{u\}, k, t)$, and $\mathcal{F}'_S \subseteq \mathcal{F} - \{u\}$ be the set containing sets from $\mathcal{F} - \{u\}$ that have a non-empty intersection with S. Consider the (multi) set $\widehat{\mathcal{F}} = \{F \in \mathcal{F} \mid F \in \mathcal{F} - \{u\}\} \cup \{F \cup \{u\} \in \mathcal{F} \mid F \in \mathcal{F} - \{u\}\}$. Observe that $|\widehat{\mathcal{F}}| \geq |\mathcal{F}'_S| \geq t$, $S \subseteq U$, and each set in $\widehat{\mathcal{F}}$ has a non-empty intersection with S. Thus, S is a solution to PARTIAL d-HS in $(U, \mathcal{F}, k, t)$. $\qquad\square$

We are now ready to state the main lemma of this section.

Lemma 2. *Let $(U, \mathcal{F}, k, t)$ be an instance of PARTIAL d-HS. If Reduction Rules 1 to 5 are not applicable, then $|U| \in \mathcal{O}(dt)$ and $|\mathcal{F}| \in \mathcal{O}(dt^2)$.*

Proof. Let $(U, \mathcal{F}, k, t)$ be an instance of PARTIAL d-HS, where none of the Reduction Rules 1 to 5 are applicable. From safeness and non-applicability of reduction rules, (and particularly that of Reduction Rule 5) we have $|U| \in \mathcal{O}(dt)$. Since Reduction Rule 4 is not applicable, for each $u \in U$, we have $|\mathcal{F}_u| \leq t - 1$. Moreover, $\mathcal{F} = \cup_{u \in U} \mathcal{F}_u$. From the above arguments, we have $|\mathcal{F}| \in \mathcal{O}(dt^2)$. $\qquad\square$

Lemma 2 immediately implies the following theorem.

Theorem 1. PARTIAL d-HS *admits a kernel with $|U| \in \mathcal{O}(dt)$ and $|\mathcal{F}| \in \mathcal{O}(dt^2)$.*

As an immediate corollary to Theorem 1, we obtain the following result.

Corollary 1. PARTIAL VERTEX COVER *admits a kernel with $\mathcal{O}(t)$ vertices and $\mathcal{O}(t^2)$ edges, where t is the number of edges to be covered.*

4   Kernel for PARTIAL d-EXACT SET COVER

In this section, we design a kernelization algorithm for PARTIAL d-EXACT SET COVER (PARTIAL d-EXACT SC for short) where the size of the universe is bounded by $\mathcal{O}((k-1)^{d+1}d^{d+2})$, and the size of the family is bounded by $\mathcal{O}((k-1)^{d+1}d^{d+1})$. Let $(U, \mathcal{F}, k, t)$ be an instance of PARTIAL d-EXACT SC. The algorithm starts by applying some reduction rules exhaustively, in the order in which they are stated. When none of the reduction rules are applicable, we argue that we have an equivalent instance of the desired size. Next, we state reduction rules that are used by the algorithm.

Reduction Rule 6. *If $k < 0$, then return that $(U, \mathcal{F}, k, t)$ is a NO instance of* PARTIAL d-EXACT SC.

The safeness of Reduction Rule 6 follows from the fact that the size of any set is at least 0.

Reduction Rule 7. *If* $t \leq 0$, *then return that* $(U, \mathcal{F}, k, t)$ *is a* YES *instance of* PARTIAL d-EXACT SC.

The safeness of Reduction Rule 7 follows from the fact that Reduction Rule 6 is not applicable and in this case an empty set is a solution to the given instance.

Reduction Rule 8. *If* $t > 0$ *and* $k - 0$, *then return that* $(U, \mathcal{F}, k, t)$ *is a* NO *instance of* PARTIAL d-EXACT SC.

The safeness of Reduction Rule 8 follows from the fact that at least one set from $\mathcal{F}$ is required to cover an element of U. Hereafter, we assume that $k > 0$ and $t > 0$.

Reduction Rule 9. *If* $t \leq d$, *then return that* $(U, \mathcal{F}, k, t)$ *is a* YES *instance of* PARTIAL d-EXACT SC.

The safeness of Reduction Rule 9 follows from the fact that any set $F \in \mathcal{F}$ can cover d elements.

Reduction Rule 10. *If* $t > kd$, *then return that* $(U, \mathcal{F}, k, t)$ *is a* NO *instance of* PARTIAL d-EXACT SC.

The safeness of above reduction rule follows from the fact that any solution of size k can cover at most kd elements.

Let $T \subseteq U$ be a set of size l. We define $\mathcal{F}_T$ as a subset of family $\mathcal{F}$ such that sets in $\mathcal{F}_T$ contains T. In particular, $\mathcal{F}_T = \{F \in \mathcal{F} \mid T \subseteq F\}$. Next, the kernelization algorithm calls Algorithm 1 on instance $(U, \mathcal{F}, k, t)$.

Algorithm 1. Algo-PDSC$(U, \mathcal{F}, k, t)$

1: Fix an arbitrary ordering of sets in the family $\mathcal{F}$
2: **if** $|\mathcal{F}| \leq ((k-1)d)^{d+1}$ **then**
3:     **return** $(U, \mathcal{F}, k, t)$
4: **else**
5:     Let $\mathcal{F}^*$ be the family of first $(kd)^{d+1}$ sets in $\mathcal{F}$, $U^* = \{u \in F \mid F \in \mathcal{F}^*\}$.
6:     Apply Reduction Rules 11 and 12 exhaustively on instance $(U^*, \mathcal{F}^*, k, t)$. Let $(U^*, \mathcal{F}', k, t)$ be the reduced instance.
7:     **return** Algo-PDSC$(U, (\mathcal{F} \setminus \mathcal{F}^*) \cup \mathcal{F}', k, t)$
8: **end if**

Reduction Rule 11. *Let* $l_r = (k-1)d + 1$ *if* $r = 1$ *and* $l_r = l_{r-1}(k-1)d + 1$, *otherwise. Let* $r \in [d-1]$ *be the least integer such that there exists a set* $T \subseteq U^*$ *of size* $d - r$, *for which* $|\mathcal{F}_T^*| \geq l_r + 1$. *Then, delete an arbitrary set* $F \in \mathcal{F}_T^*$ *to generate a new instance* $(U^*, \mathcal{F}', k, t)$, *where* $\mathcal{F}' = \mathcal{F}^* \setminus F$.

Lemma 3. *Reduction Rule 11 is safe.*

Proof. We use induction on r to prove the lemma.

Base Step: $r = 1$. Let $T \subseteq U^*$ of size $d - 1$, for which $|\mathcal{F}_T^*| \geq (k-1)d + 2$. We delete an arbitrary set $F \in \mathcal{F}_T^*$ from $\mathcal{F}^*$ to generate a new instance $(U^*, \mathcal{F}' = \mathcal{F}^* \setminus F, k, t)$. Now, we prove that $(U^*, \mathcal{F}^*, k, t)$ is a YES instance of PARTIAL d-EXACT SC if and only if $(U^*, \mathcal{F}', k, t)$ is a YES instance of PARTIAL d-EXACT SC. In the forward direction, let $\mathcal{S}$ be a minimal solution of size at most k for PARTIAL d-EXACT SC in $(U^*, \mathcal{F}^*, k, t)$. If $\mathcal{S}$ does not contain F, then clearly $\mathcal{S}$ is also a solution for PARTIAL d-EXACT SC in $(U^*, \mathcal{F}', k, t)$. Now, suppose that $\mathcal{S}$ contains F and let $\mathcal{S}^* = \mathcal{S} \setminus F$. Let $U_{S*}^* = \{u \in S \mid S \in \mathcal{S}^*, u \notin T\}$. Observe that $|U_{S*}^*| \leq (k-1)d$. Since $U_{S*}^* \cap F = \emptyset$, $|X \cap S^*| \leq 1$, for any $X \in \mathcal{F}^*$. Hence, there can be at most $(k-1)d$ sets in $\mathcal{F}_T^* \setminus F$ which have non empty intersection with U_{S*}^*. Hence, there exist at least one set $F' \in \mathcal{F}_T^* \setminus F$ such that $F' \cap U_{S*}^* = \emptyset$. This implies that $\mathcal{S}' = F' \cup \mathcal{S}^*$ covers t elements and $\mathcal{S}'$ is a solution to PARTIAL d-EXACT SC in $(U^*, \mathcal{F}', k, t)$. For backward direction, since $\mathcal{F}' \subset \mathcal{F}^*$, therefore any solution $\mathcal{S}$ to $(U^*, \mathcal{F}', k, t)$ is also a solution to PARTIAL d-EXACT SC in $(U^*, \mathcal{F}^*, k, t)$.

Induction Hypothesis: Let us assume that Reduction Rule 11 is safe for $r \leq j - 1$. So when Reduction Rule 11 is applied for $r = j$, family $\mathcal{F}^*$ is already reduced by Reduction Rule 11 for $r \leq j - 1$.

Induction Step: $r = j$. For $r = j$, $l_j = l_{j-1}(k-1)d + 1$. Let $T \subseteq U^*$ of size $d - j$ and $|\mathcal{F}_T^*| \geq l_j + 1$. We delete an arbitrary set $F \in \mathcal{F}_T^*$ from $\mathcal{F}^*$ to obtain a new instance $(U^*, \mathcal{F}' = \mathcal{F}^* \setminus F, k, t)$. Now, we prove that $(U^*, \mathcal{F}^*, k, t)$ is a YES instance of PARTIAL d-EXACT SC if and only if $(U^*, \mathcal{F}', k, t)$ is a YES instance of PARTIAL d-EXACT SC. In the forward direction, let $\mathcal{S}$ be a minimal solution of size at most k for PARTIAL d-EXACT SC in $(U^*, \mathcal{F}^*, k, t)$. If $\mathcal{S}$ does not contain F, then clearly $\mathcal{S}$ is also a solution for PARTIAL d-EXACT SC in $(U^*, \mathcal{F}', k, t)$. Now, suppose that $\mathcal{S}$ contains F and let $\mathcal{S}^* = \mathcal{S} \setminus F$. Let $U_{S*}^* = \{u \in S \mid S \in \mathcal{S}^*, u \notin T\}$. Observe that $|U_{S*}^*| \leq (k-1)d$. We have that $|\mathcal{F}_T^* \setminus F| \geq l_j$. Since $\mathcal{F}^*$ is reduced by Reduction Rule 11 for $r \in [j-1]$, for a set T' of size $d - (j-1)$, $|\mathcal{F}_{T'}^*| \leq l_{j-1}$. This implies that for any element $u \in U_{S*}^*$, size of the family $\mathcal{F}_{T \cup u}^* \subseteq \mathcal{F}_T^*$ is at most l_{j-1} and $|\cup_{u \in U_{S*}^*} \mathcal{F}_{T \cup u}^*| \leq l_{j-1}(k-1)d$. Therefore, there exists at least one set $F' \in \mathcal{F}_T^* \setminus F$ such that, $F' \cap U_{S*}^* = \emptyset$. This implies that $\mathcal{S}' = F' \cup \mathcal{S}^*$ covers t elements and $\mathcal{S}'$ is a solution to PARTIAL d-EXACT SC in $(U^*, \mathcal{F}', k, t)$. For backward direction, since $\mathcal{F}' \subset \mathcal{F}^*$, therefore any solution $\mathcal{S}$ to $(U^*, \mathcal{F}', k, t)$ is also a solution to PARTIAL d-EXACT SC in $(U^*, \mathcal{F}^*, k, t)$.   □

Lemma 4. *Let* $(U^*, \mathcal{F}', k, t)$ *be reduced instance after applying Reduction Rule 11 exhaustively. Then, any element in* U^* *can be in at most* $((k-1)d)^d$ *sets in* $\mathcal{F}'$.

Proof. For $r \in [d-1]$, $l_r = (k-1)d + 1$ if $l = 1$ and $l_r = l_{r-1}(k-1)d + 1$, otherwise. When $r = 1$, the Reduction Rule 11 bounds the sets in family which shares subset of size $d - 1$ and when $r = j$ the Reduction Rule 11 bound the sets in family which shares subset of size $d - j$, and hence when $r = d - 1$ the Reduction Rule 11 bound the sets in family which shares an element.

$$l_{d-1} = l_{d-2}(k-1)d + 1$$
$$= 1 + (k-1)d + ((k-1)d)^2 + \cdots + ((k-1)d)^{d-1} \le ((k-1)d)^d$$

This completes the proof. $\square$

Let $(U^*, \mathcal{F}^*, k, t)$ be the reduced instance after applying Reduction Rule 11 exhaustively, we now give following reduction rule to bound size of family $\mathcal{F}^*$.

Reduction Rule 12. *If $|\mathcal{F}^*| \ge ((k-1)d)^{d+1} + 2$, then delete an arbitrary set $F \in \mathcal{F}^*$. Let the new instance be $(U^*, \mathcal{F}', k, t)$, where $\mathcal{F}' = \mathcal{F}^* \setminus F$.*

Lemma 5. *Reduction Rule 12 is safe.*

Proof. We prove that $(U^*, \mathcal{F}^*, k, t)$ is a YES instance of PARTIAL d-EXACT SC if and only if $(U^*, \mathcal{F}', k, t)$ is a YES instance of PARTIAL d-EXACT SC. In the forward direction, let $\mathcal{S}$ be a minimal solution of size at most k for PARTIAL d-EXACT SC in $(U^*, \mathcal{F}^*, k, t)$. If $\mathcal{S}$ does not contain F, then clearly $\mathcal{S}$ is also a solution for PARTIAL d-EXACT SC in $(U^*, \mathcal{F}', k, t)$. Now, suppose that $\mathcal{S}$ contains F and let $\mathcal{S}^* = \mathcal{S} \setminus F$. Let $U_{\mathcal{S}^*} = \{u \in S \mid S \in \mathcal{S}^*, u \notin F\}$. Observe that $|U_{\mathcal{S}^*}| \le (k-1)d$. We have that $|\mathcal{F}^* \setminus F| \ge ((k-1)d)^{d+1} + 1$. Since $\mathcal{F}^*$ is reduced by Reduction Rule 11, for an element $u \in U_{\mathcal{S}^*}$, $|\mathcal{F}_u^*| \le ((k-1)d)^d$ and $|\cup_{u \in U_{\mathcal{S}^*}} \mathcal{F}_u^*| \le ((k-1)d)^{d+1}$. Therefore, there exists at least one set $F' \in \mathcal{F}^* \setminus F$ such that $F' \cap U_{\mathcal{S}^*} = \emptyset$. This implies that $\mathcal{S}' = F' \cup \mathcal{S}^*$ covers t elements and $\mathcal{S}'$ is a solution to PARTIAL d-EXACT SC in $(U^*, \mathcal{F}', k, t)$. For backward direction, since $\mathcal{F}' \subset \mathcal{F}^*$, therefore any solution $\mathcal{S}$ to $(U^*, \mathcal{F}', k, t)$ is also a solution to PARTIAL d-EXACT SC in $(U^*, \mathcal{F}^*, k, t)$. $\square$

The correctness of Algorithm 1 follows from Lemmas 3, 4, and 5. Let $(U, \mathcal{F}, k, t)$ be the reduced instance returned by Algorithm 1. We now give the following reduction rule to bound size of universe U.

Reduction Rule 13. *Delete all the elements from U that are not present in any set in family $\mathcal{F}$.*

The proof of above reduction rule follows from that fact the elements not present in any set in $\mathcal{F}$ can not be covered. This completes the description of kernelization algorithm.

Now, we give the main lemma of this section.

Lemma 6. *Let $(U, \mathcal{F}, k, t)$ be an instance of PARTIAL d-EXACT SC. If none of the Reduction Rules 6 to 13 are applicable, then $|\mathcal{F}^*| \in \mathcal{O}(((k-1)d)^{d+1})$ and $|U| \in \mathcal{O}((k-1)^{d+1}d^{d+2})$.*

Proof. Let $(U, \mathcal{F}, k, t)$ be an instance of PARTIAL d-EXACT SC when none of the Reduction Rules 6 to 13 are applicable. From the safeness and exhaustive application of reduction rules, $|\mathcal{F}^*| \in \mathcal{O}(((k-1)d)^{d+1})$. Since each set in $|\mathcal{F}^*|$ has at most d elements, $|U| \in \mathcal{O}((k-1)^{d+1}d^{d+2})$. $\square$

Lemma 7. *Let $(U, \mathcal{F}, k, t)$ be an input instance of* PARTIAL d-EXACT SC, *then the kernelization algorithm can be implemented in $\mathcal{O}((kd)^{d^2}|\mathcal{F}|^{\mathcal{O}(1)})$ time.*

Proof. It is easy to see that Reduction Rules 6, 7, 8, 9, 10, and 13 can be applied in polynomial time. The Algorithm 1 applies Reduction Rule 11 on a family $\mathcal{F}^*$ of size $(kd)^{d+1}$ and universe U^* of size at most $(k^{d+1}d^{d+2})$. Reduction Rule 11 is applied on every subset of universe U^* of size at most $d - 1$, i.e. $\binom{k^{d+1}d^{d+2}}{d-1}$ and each application runs in $\mathcal{O}(|\mathcal{F}^*|)$ time. Reduction rule 12 can be applied in at most $\mathcal{O}(|\mathcal{F}|)$ time. Algorithm 1 has at most $|\mathcal{F}|$ many recursive calls. Thus running time of the algorithm is $\mathcal{O}((kd)^{\mathcal{O}(d^2)}|\mathcal{F}|^{\mathcal{O}(1)})$. $\square$

Lemma 6 along with Lemma 7 implies the following theorem.

Theorem 2. PARTIAL d-EXACT SC *admits a kernel with $|U| \in \mathcal{O}((k - 1)^{d+1}d^{d+2})$ and $|\mathcal{F}| \in \mathcal{O}((k - 1)^{d+1}d^{d+1})$.*

5   Exact Algorithm for PVC

In this section, we design an exact algorithm for PARTIAL VERTEX COVER (PVC for short). Towards this we first present a reduction from PVC to MAXIMUM EDGE WEIGHTED TRIANGLE (MWT) and then we use the subcubic algorithm known for MWT [22,23] to solve PVC. MWT problem is formally defined as follows.

MAXIMUM EDGE WEIGHTED TRIANGLE (MWT)
Input: A graph $G = (V, E)$, a weight function $f : E(G) \to \mathbb{Z}^+$ and a positive integer W.
Question: Does there exists a triangle in G of weight at least W?

Given an instance (G, t) of PVC, we generate an instance (G', f, W) of MWT as follows. Let $V_1, V_2,$ and V_3 be an arbitrary partition of $V(G)$ such that for each partition V_i, $|V_i|$ is at most $\lceil \frac{n}{3} \rceil$. Note that here we split the vertex set into 3 instead of 2 partitions because we are aiming at an algorithm running in time better than $\mathcal{O}(2^n)$. Now, corresponding to each set S_{ij} in 2^{V_i}, where $i \in [3]$, $j \in [2^{|V_i|}]$, add a vertex v_{ij} in $V(G')$. Now, for each $v_{ij}, v_{pq} \in V(G')$, add an edge $v_{ij}v_{pq}$ (where $i, p \in [3]$) to $E(G')$, if and only if $i \neq p$. Next, we assign a weight function, $f : E(G') \to \mathbb{Z}^+$ as follows. Let $Y, Z \subseteq V(G)$, we define $E_{YZ} = \{uv \in E(G) \mid u \in Y, v \in Z\}$. For an edge $v_{ij}v_{pq} \in E(G')$,

$$f(v_{ij}v_{pq}) = \sum_{v \in S_{ij}} d_G(v) + \sum_{v \in S_{pq}} d_G(v) - |E_{S_{ij}S_{ij}}| - |E_{S_{pq}S_{pq}}| - 2|E_{S_{ij}S_{pq}}|$$

Now, we choose an appropriate value for W. Let $W = 2t$. This completes the reduction. In the following lemma, we prove that the instance (G, t) of PVC and (G', f, W) of MWT are equivalent.

Lemma 8. (G, t) *is a* YES *instance of* PVC *if and only if* (G', f, W) *is a* YES *instance of* MWT.

Proof. In the forward direction, let (G, t) be a YES instance of PVC and X be one of its solutions. Let $S_{1j} = V_1 \cap X, S_{2p} = V_2 \cap X$, and $S_{3q} = V_3 \cap X$. We claim that the vertices corresponding to S_{1j}, S_{2p} and S_{3q}, i.e. $\{v_{1j}, v_{2p}, v_{3q}\}$ form a triangle T in G' of weight at least $2t$. It is to be noted here that $S_{1j} \cup S_{2p} \cup S_{3q}$ covers t edges, where

$$t = \sum_{v \in S_{1j}} d_G(v) + \sum_{v \in S_{2p}} d_G(v) + \sum_{v \in S_{3q}} d_G(v) -$$

$$(|E_{S_{1j}S_{1j}}| + |E_{S_{2p}S_{2p}}| + |E_{S_{3q}S_{3q}}| + |E_{S_{1j}S_{2p}}| + |E_{S_{2p}S_{3q}}| + |E_{S_{1j}3q}|).$$

Now, weight of the triangle T is

$$f(v_{1j}v_{2p}) + f(v_{2p}v_{3q}) + f(v_{3q}v_{1j}) = 2 \sum_{v \in S_{1j}} d_G(v) + 2 \sum_{v \in S_{2p}} d_G(v) + 2 \sum_{v \in S_{3q}} d_G(v)$$

$$- 2|E_{S_{1j}S_{1j}}| - 2|E_{S_{2p}S_{2p}}| - 2|E_{S_{3q}S_{3q}}|$$

$$- 2|E_{S_{1j}S_{2p}}| - 2|E_{S_{2p}S_{3q}}| - 2|E_{S_{1j}S_{3q}}|$$

$$= 2t$$

Conversely, let G' has a triangle $\{v_{1j}, v_{2p}, v_{3q}\}$ of weight at least W. Let $X = S_{1j} \cup S_{2p} \cup S_{3q}$. By the similar arguments as above, X covers at least $\frac{W}{2}$ edges in G. This completes the proof. □

Theorem 3. *There exists an exact algorithm for* PVC *with running time* $\mathcal{O}(2^{\omega n/3})$, *where ω is matrix multiplication constant.*

Proof. To construct an instance of MWT from PVC, we generate a graph G' in which for every subset of $V_i \subseteq V(G)$, where $i \in [3]$, we added a vertex in G'. Hence, G' consists of $3 \cdot 2^{n/3}$ vertices and $3 \cdot 2^{2n/3}$ edges by construction. Therefore, the construction of G' takes $\mathcal{O}(2^{2n/3})$ time. Using the algorithm presented by Williams [22] for finding maximum weighted triangle in the graph, we get $\mathcal{O}(2^{2n/3} + 2^{\omega n/3}) = \mathcal{O}(2^{\omega n/3})$ time algorithm for PVC. □

6   Conclusion

We explored kernelization of PARTIAL d-HS and PARTIAL d-EXACT SC problems in this paper, along with giving an exact algorithm for the PVC problem. For PARTIAL d-HS, we gave a kernel with $\mathcal{O}(dt)$ elements, and $\mathcal{O}(dt^2)$ sets. For PARTIAL d-EXACT SC, we gave a kernel with $\mathcal{O}((k-1)^{d+1}d^{d+2})$ elements, and $\mathcal{O}((k-1)^{d+1}d^{d+1})$ sets. The exact algorithm for PVC, ran in time $\mathcal{O}(2^{\omega n/3})$. Interesting open problems are to improve these bounds. Polynomial kernel for PVC parameterized by k (solution size) remains open for planar and bipartite graphs.

References

1. Abu-Khzam, F.N.: A kernelization algorithm for d-Hitting set. J. Comput. Syst. Sci. **76**(7), 524–531 (2010)
2. Amini, O., Fomin, F.V., Saurabh, S.: Implicit branching and parameterized partial cover problems. J. Comput. Syst. Sci. **77**(6), 1159–1171 (2011)
3. van Bevern, R.: Towards optimal and expressive kernelization for d-Hitting set. In: International Computing and Combinatorics Conference, pp. 121–132 (2012)
4. Bläser, M.: Computing small partial coverings. Inf. Process. Lett. **85**(6), 327–331 (2003)
5. Bonnet, É., Paschos, V.T., Sikora, F.: Parameterized exact and approximation algorithms for maximum k-set cover and related satisfiability problems. RAIRO - Theoret. Inf. Appl. **50**(3), 227–240 (2016)
6. Cygan, M., Fomin, F.V., Kowalik, L., Lokshtanov, D., Marx, D., Pilipczuk, M., Pilipczuk, M., Saurabh, S.: Parameterized Algorithms. Springer, Cham (2015). https://doi.org/10.1007/978-3-319-21275-3
7. Downey, R.G., Fellows, M.R.: Fundamentals of Parameterized Complexity. Texts in Computer Science. Springer, London (2013). https://doi.org/10.1007/978-1-4471-5559-1
8. Erdös, P., Rado, R.: Intersection theorems for systems of sets. J. Lond. Math. Soc. **1**(1), 85–90 (1960)
9. Flum, J., Grohe, M.: Parameterized Complexity Theory. Texts in Theoretical Computer Science. An EATCS Series. Springer, Heidelberg (2006). https://doi.org/10.1007/3-540-29953-X
10. Fomin, F.V., Lokshtanov, D., Raman, V., Saurabh, S.: Subexponential algorithms for partial cover problems. Inf. Process. Lett. **111**(16), 814–818 (2011)
11. Gandhi, R., Khuller, S., Srinivasan, A.: Approximation algorithms for partial covering problems. J. Algorithms **53**(1), 55–84 (2004)
12. Guo, J., Niedermeier, R., Wernicke, S.: Parameterized complexity of vertex cover variants. Theory Comput. Syst. **41**(3), 501–520 (2007)
13. Halldórsson, M.M.: Approximating k-set cover and complementary graph coloring. In: Cunningham, W.H., McCormick, S.T., Queyranne, M. (eds.) IPCO 1996. LNCS, vol. 1084, pp. 118–131. Springer, Heidelberg (1996). https://doi.org/10.1007/3-540-61310-2_10
14. Karp, R.M.: Reducibility among combinatorial problems. In: Miller, R.E., Thatcher, J.W., Bohlinger, J.D. (eds.) Complexity of Computer Computations. The IBM Research Symposia Series, pp. 85–103. Springer, Boston (1972). https://doi.org/10.1007/978-1-4684-2001-2_9
15. Khuller, S., Moss, A., Naor, J.: The budgeted maximum coverage problem. Inf. Process. Lett. **70**(1), 39–45 (1999)
16. Kneis, J., Langer, A., Rossmanith, P.: Improved upper bounds for partial vertex cover. In: Broersma, H., Erlebach, T., Friedetzky, T., Paulusma, D. (eds.) WG 2008. LNCS, vol. 5344, pp. 240–251. Springer, Heidelberg (2008). https://doi.org/10.1007/978-3-540-92248-3_22
17. Lawler, E.L.: Combinatorial Optimization: Networks and Matroids. Courier Corporation, Mineola (1976)
18. Niedermeier, R.: Invitation to Fixed-Parameter Algorithms. Oxford Lecture Series in Mathematics and its Applications, vol. 31. Oxford University Press, Oxford (2006)

19. Niedermeier, R., Rossmanith, P.: An efficient fixed-parameter algorithm for 3-Hitting set. J. Discret. Algorithms **1**(1), 89–102 (2003)

20. Papadimitriou, C.H.: Computational Complexity. Wiley, Hoboken (2003)

21. Vazirani, V.V.: Approximation Algorithms. Springer, Heidelberg (2013). https://doi.org/10.1007/978-3-662-04565-7

22. Williams, R.: A new algorithm for optimal 2-constraint satisfaction and its implications. Theor. Comput. Sci. **348**(2–3), 357–365 (2005)

23. Williams, V.V.: Multiplying matrices faster than coppersmith-winograd. In: Proceedings of the Forty-fourth Annual ACM Symposium on Theory of Computing, STOC 2012, pp. 887–898 (2012)

Author Index

Abedin, Paniz 615
Aghamolaei, Sepideh 675
Agrawal, Akanksha 751
Ahn, Hee-Kap 143
Aronov, Boris 567

Baharifard, Fatemeh 675
Banerjee, Niranka 650
Barbay, Jérémy 156, 180
Barequet, Gill 120, 130
Ben-Shachar, Gil 120
Bera, Debajyoti 579
Bhattacharyya, Arnab 542
Bilò, Vittorio 280
Blum, Johannes 230

Cai, Yinhui 205
Cardinal, Jean 365
Carneiro, Alan Diêgo Aurélio 84
Carosi, Raffaello 268
Cellinese, Francesco 280
Chan, Chun-Hsiang 218
Chan, T.-H. Hubert 441, 688
Chang, Jou-Ming 1
Chen, Guangting 205
Chen, Jianer 713
Chen, Yong 205
Cheng, Kun 416
Chikhi, Rayan 467
Choudhary, Pratibha 751
Conte, Alessio 328

Dantsin, Evgeny 592
Das, Gautam K. 516
de Berg, Mark 567
De, Minati 130
Demaine, Erik D. 365
Deng, Yunyun 14
Dey, Sanjana 529
Dhannya, S. M. 638
Dinesh, Krishnamoorthy 738

El-Mabrouk, Nadia 403
Eppstein, David 365
Erlebach, Thomas 242
Eulenstein, Oliver 168, 378

Feng, Qilong 726
Fu, Bin 492, 726
Fukunaga, Takuro 51

Gadekar, Ameet 542
Ganguly, Arnab 615
Ghodsi, Mohammad 675
Ghosal, Pratik 316
Goblet, Axel 554
Goebel, Randy 205
Goodrich, Michael T. 130
Górecki, Paweł 168
Grossi, Roberto 328
Guo, Longkun 14
Gurski, Frank 255

Hanaka, Tesshu 428
Hearn, Robert A. 365
Hellmuth, Marc 403
Hescott, Benjamin 626
Hon, Wing-Kai 615
Huang, Peihuang 14

Italiano, Giuseppe F. 480
Ito, Takehiro 428

Jain, Pallavi 751
Jallu, Ramesh K. 516, 529
Jayapaul, Varunkumar 650
Jena, Sangram K. 516
Jiang, Haitao 26
Jovičić, Vladan 467

Kanesh, Lawqueen 751
Kao, Shih-Shun 1
Kelk, Steven 554
Khan, Fitra 192

Kim Thang, Nguyen   480
Kita, Nanao   293
Kobayashi, Koji M.   108
Kratsch, Stefan   467
Krithika, R.   341

Lei, Hansheng   192
Li, Shi   96
Li, Shuai   39
Li, Zimao   416
Liang, Zhibin   441
Lin, Guohui   205, 353
Lin, Mugang   726
Liu, Longcheng   205
Liu, Weiwen   39
Luo, Kelin   242
Luo, Wenchang   353

Malchik, Caleb   626
Manoussakis, Yannis   480
Marino, Andrea   328
Markin, Alexey   168, 378
Markovic, Aleksandar   567
Matsui, Yasuko   76
Medvedev, Paul   467
Meesum, Syed M.   391
Melideo, Giovanna   280
Melissinos, Nikolaos   602
Merkle, Daniel   403
Mihalák, Matúš   554
Milanič, Martin   467
Mincu, Radu Stefan   662
Misra, Pranabendu   341
Mizuta, Haruka   428
Monaco, Gianpiero   268, 280
Moore, Benjamin   428

Nagamochi, Hiroshi   504
Nakano, Shin-ichi   76
Nandy, Subhas C.   529
Narayanaswamy, N. S.   638
Nekrich, Yakov   615
Nip, Kameng   63
Nishimura, Naomi   428
Nøjgaard, Nikolai   403

Ochoa, Carlos   156
Oh, Eunjin   143
Otiv, Samir   738

Pagourtzis, Aris   602
Pai, Kung-Jui   1
Paluch, Katarzyna   316
Pérez-Lantero, Pablo   180
Pham, Hong Phong   480
Popa, Alexandru   662
Protti, Fábio   84
Pu, Lianrong   26

Qingge, Letu   26
Quweider, Mahmoud   192

Rajgopal, Ninad   542
Ramya, C.   638, 701
Rao, B. V. Raghavendra   701
Raskhodnikova, Sofya   467
Rehs, Carolin   255
Rojas-Ledesma, Javiel   180

Sadakane, Kunihiko   615
Sahlot, Vibha   751
Sankoff, David   26
Sarma, Jayalal   738
Satti, Srinivasa Rao   650
Saurabh, Saket   751
Shah, Rahul   615
Souza, Uéverton S.   84
Stamoulis, Georgios   554
Storandt, Sabine   230
Su, Bing   353
Subramanya, Vijay   428
Sun, Bintao   688
Suzuki, Akira   428

Tale, Prafullkumar   341
Thankachan, Sharma V.   615

Uno, Takeaki   454

Vadali, Venkata Sai Krishna Teja   378
Vaidyanathan, Krishna   428
Varma, Nithin   467
Versari, Luca   328

Wang, Yingying   416
Wang, Zhenbo   63
Wasa, Kunihiro   454
Wieseke, Nicolas   403
Winslow, Andrew   365, 626

Woeginger, Gerhard   567
Wolpert, Alexander   592
Wu, Guangwei   713
Wu, Ro-Yu   1

Xiao, Mingyu   504
Xu, Chenyang   305
Xu, Jinhui   96
Xu, Yao   353
Xu, Yinfeng   242

Yamanaka, Katsuhisa   76
Ye, Minwei   96
Yen, Hsu-Chun   218

Zhang, An   205
Zhang, Guochuan   305
Zhang, Liyu   192
Zhang, Shengyu   39
Zhu, Binhai   26

in the United States
masters